5th World Congress on DISASTER MANAGEMENT

5th World Congress on DISASTER MANAGEMENT

Volume 2

NATURE AND HUMAN INDUCED DISASTERS

Edited by

Dr. S. Ananda Babu
President and Convenor
DMICS-WCDM

Routledge
Taylor & Francis Group
LONDON AND NEW YORK

DMICS
Disaster Management, Initiatives and Convergence Society
Envisioning a Disaster Resilient Future

World Congress on Disaster Management

First published 2023
by Routledge
4 Park Square, Milton Park, Abingdon, Oxon OX14 4RN

and by Routledge
605 Third Avenue, New York, NY 10158

Routledge is an imprint of the Taylor & Francis Group, an informa business

British Library Cataloguing-in-Publication Data
A catalogue record for this book is available from the British Library

Library of Congress Cataloging-in-Publication Data
A catalog record has been requested for this book

ISBN: 978-1-032-35546-7 (hbk)
ISBN: 978-1-003-34193-2 (ebk)

Table of Contents

List of Tables

Preface

The Fifth World Congress on Disaster Management (WCDM) was organised in Delhi from 24th to 27th November 2021 jointly by the Government of Delhi, the Indian Institute of Technology Delhi, and the Disaster Management Initiative and Convergence Society which created the platform of WCDM. Over the years, WCDM has emerged as the largest conference on disaster management in the developing world.

The theme of the Fifth WCDM was *Technology, Finance and Capacity.* 7 Plenary Sessions, 50 Technical Sessions and Special Technical Sessions were organised around this overarching theme. While eminent thought leaders and experts delivered keynote addresses and participated in the panel discussions of the Plenary Sessions. It is the Technical Sessions that received the longest traction as Call for Papers was issued for these sessions months in advance and more than 600 researchers, practitioners and policy makers responded with abstracts of their ideas. These were reviewed by experts and subsequently, 250 of these abstracts were developed as full papers. This is the second of five-volume series of compendium of these papers.

The papers have been published in the same form these were received without any peer review to provide a flavour of the raw ideas that emerged from the Technical Sessions of the conference. Some of these papers presented by the young researchers and practitioners may not have the rigours of academic disciplines, but these do reflect the cross current of thoughts that went around in these sessions of the Conference. These provide new ideas and insights that provide value to the current discourses on the subject.

These papers have been arranged under five broad themes–first: Disaster Risk Management, second: Nature and Human Induced Disasters, third: Coronavirus Pandemic, fourth: Disaster Response and fifth: Challenges and Opportunities of Disaster Management. Understandably the papers do not cover every aspect of the themes, these discuss only those aspects that the authors have chosen to highlight. The present volume is a compilation of 56 papers on the theme of Nature and Human Induced Disasters, which is further divided into seven sub-themes, first: Living with Floods: Case Studies in Flood Risk Management, second: Planning and Managing Risks of Floods and River Erosion, third: Cyclone Risk Mitigation and Management, fourth: Managing Risks of Drought, fifth: Climate Change and Disasters: New Dimensions, sixth: Technological Disasters and seventh: Other Disasters.

The Conference secretariat has brought the papers together, but his credit lies solely and exclusively on the authors.

Dr S. Ananda Babu
Convener
Fifth World Conference on Disaster Management

Acknowledgement

DMICS, the organizer of the 5th World Congress on Disaster Management (WCDM), expresses its deep appreciation to Government of Delhi (GNCTD), NDMA, NIDM, DRDO, UNICEF, ICMR, GSI, and Knowledge Partners, delegates who have supported the 2021 World Congress on Disaster Management with either earmarked or unearmarked contributions. A special acknowledgement goes to Indian Institute of Technology, Delhi for hosting the 5th WCDM for its strong support.

DMICS would like to express our deepest appreciation to Hon'ble Union Minister of Defence, Shri Rajnath Singh for inaugurating the 5th WCDM and delivering the inaugural address; Mami Mizutori, UN-DRR Chief and Special Representative of the UN Secretary General for her special message and Dr Balram Bhargava, Secretary, DHR and Director General, ICMR for his key note address. We would also like to convey our heartfelt appreciation and gratitude to Prof. V. Ramgopal Rao, Director of IIT Delhi, Dr. Parvez Hyatt, for their strategic support and guidance. Our indefatigable team members, Col. Sanjay Srivastava Dr. Chandrappa, Dr. A. Kishan, Prof. B. Gopal Rao, Mr. Bhaskar Rao Volam, Dr. B. Ram, Prof. V. Prakasam, Dr. A. Gayathri Devi, Dr. R. K. Srivastava, Mr. Amit Kumar, Mr. Mohan Kasthala, Mr. Pavan Parlapalli, Mr. Manish Vishnoi, Ms. Suparna Dutta, Ms. Parul Sharma, Ms. Megha B Bhati, M. Zoya Khan, Ms. Areeba Naaz, Mr. Rohit Kumar Azad and Ms. Shweta Zanjrukia, deserve a huge appreciation for working tirelessly for the last one year in ensuring that 5th WCDM reaches new horizons in bringing together the Government Duty bearers, Policy makers, Planners, Researchers, Academicians, Practitioners and other key stakeholders.

DMICS extends its sincere gratitude to the large number of organisations, individuals and volunteers who have contributed to the 5th WCDM, through technical support and in numerous other ways. DMICS would like to express its deepest appreciation also to the members of the Scientific and Technical Committee (S&TC) along with its Chair, technical leads and collaborative organisations of Pre-Conference Webinars, Plenaries, Special Feature Events, Special Technical Sessions and Exhibitiors.

Part 1

Living with Floods—Case Studies in Flood Risk Management

CHAPTER 1

Assessing Vulnerability, Coping Mechanism & Adaptive Capacity of Community During Urban Floods: A Case of Kerala Floods 2018

B. Sneha Singh
Consultant, Ashoka Trust for Research in Ecology and the Environment, Bangalore, Karnataka, India

Anil Kumar Roy
Associate Professor, Faculty of Planning, CEPT University, Ahmedabad, Gujarat, India

Abstract

The study assesses vulnerability, adaptive capacity and coping mechanism of community during the 2018 urban floods in Kerala. The study aims at bridging the gap by assessing vulnerability, adaptive capacity and coping mechanism of community as suggested by IPCC framework. The methodology and indicators for assessment have been conceptualized to suit socio-cultural aspect of the community in Kerala. These indicators could be measured against several variants which will help assess the current level of a community's resilience. The paper brings out recommendations based the quantitative analytical data on ground to improve the resilience of the community during urban floods.

Keywords: Urban floods, assessing vulnerability, coping mechanism, adaptive capacity, community

1. Introduction

Disasters that cities are facing today is not merely due to the act of nature. The natural disasters now being called "Man-made". Developing countries are more affected during disasters owing to their large urban agglomeration, environmental degradation and lack of planning and preparedness (EY, 2017).

India being a developing country, is more vulnerable to disasters due to its physical location, topography and exposure of its citizens to various types of risk including their livelihood. Also disasters have a huge impact on the development of the country as the economic losses are high. The trend and preliminary assessments reveal that the severity of the droughts and intensity of floods in various parts of India might increase (NATCOM, 2004). From the above gathered facts it is clear that our Indian cities need to be resistant to nature shocks. While upgradation of technology and infrastructure is the most sought after solution for transition to a more resilient India. It is essential to look at strengthening the grassroot level – "The communities" and make them more resilient.

The Community is always the first hand responder to a disaster. Traditionally disaster management strategies have considered communities as victims and rendered them helpless. While conscious planning efforts are gearing towards making the cities smart, it is time to make the urban communities and people smart (100 Resilient Cities). Disaster risk assessment happens at a large scale and seldom percolates to the resilience at community level which creates a gap (NIDM, 2014-2015).

2. Methodology

The methodology was formulated to carry out this research bearing in mind the complexities and the limitations involved in the assessment process. A framework was conceptualized to suit the community in Kerala for assessment of resilience along with the indicators for the assessment of Vulnerability, Adaptive Strategy and Coping Mechanism. Qualitative and quantitative variables were converted into a questionnaire.

Fig. 1.1 Settling the regional context

Source: (Govt. Of Kerala, 2015)

A pilot survey helped refine the questionnaire for accurate measurable output.

2.1 Study Area

The study was conducted at Aluva, a municipality which was worst affected in the Ernakulam district. The town has 26 wards out of which 10 wards were majorly affected during the floods of 2018. In total 3679 households were affected during the floods. 18 camps were organized by the government in Aluva that sheltered 12000 people from different wards. In total 2000 houses were partially damaged and 16 houses were fully damaged (*stakeholder interaction*).

2.2 Source of Data

Major data sources for this study was collected from:

Primary survey:

Primary data was collected by conducting a questionnaire survey in a total of 120 households across 13 most affected wards in Aluva, Kochi. Random sampling method was applied to select the sample of household for conducting the survey. Equal sample size of 10 households was considered in all wards except 2 wards which were least affected during the floods where a sample size of 5 households was considered. Interaction with government officials at various

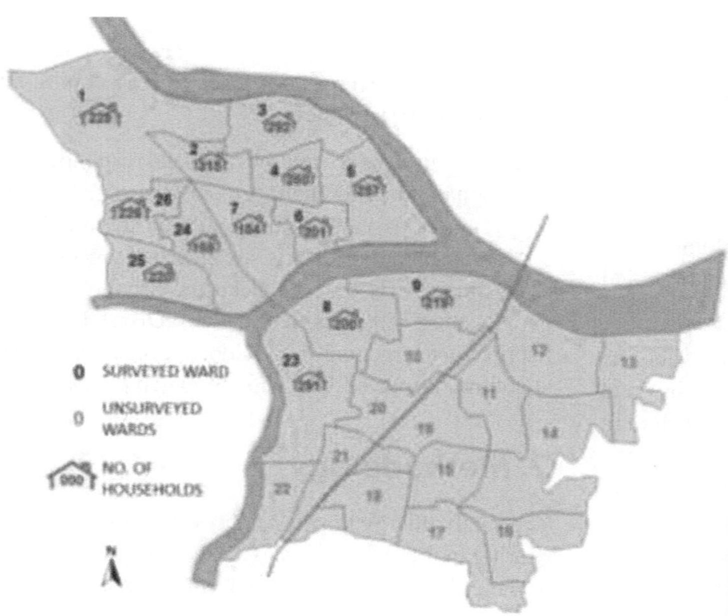

AREA OF STUDY ZONE : 2.9 Sq.Km

WARD NO	TOTAL NO. OF HOUSEHOLDS	NO. OF HOUSEHOLDS SURVEYED
1	228	10
2	315	10
3	292	10
4	260	10
5	257	10
6	201	10
7	184	10
8	200	10
9	219	10
24	168	10
25	220	10
23	291	5
26	226	5
Total	**3061**	**120**

Fig. 1.2 Map of Aluva showing household data across wards

Source: Primary survey, Jan – Feb 2019

offices like Diaster emergency operation center, Aluva municipal corporation, West village office, Aluva and the ward counselors authorities helped gather information on the actual scenario at various stages of the flood.

Secondary source:

Literature study helped understand the basic concepts of resilience, vulnerability, adaptive capacity and coping mechanism. It helped conceptualize a framework for the study. Case studies helped solve the challenges of quantifying the qualitative aspects of this study.

3. Community & settlement pattern in Aluva, Kochi:

The community of people living in each of the wards in Aluva have a homogeneous religious constitution with strong social coherence. Among all the wards that were surveyed the majority of the households, 46% were Muslims followed by 30% Hindus and 24% Christians. There is a strong sense of belonging among the community as most families have been living in the same place for many decades now. There a huge variation in topography across Aluva and the settlement pattern was such that houses were built in low lying areas around a place of worship generally located on a higher plain. Tapping on that potential these civil societies became camps during the floods. The supplies to these places could come easily as they were more accessible to roads with lesser inundation levels.

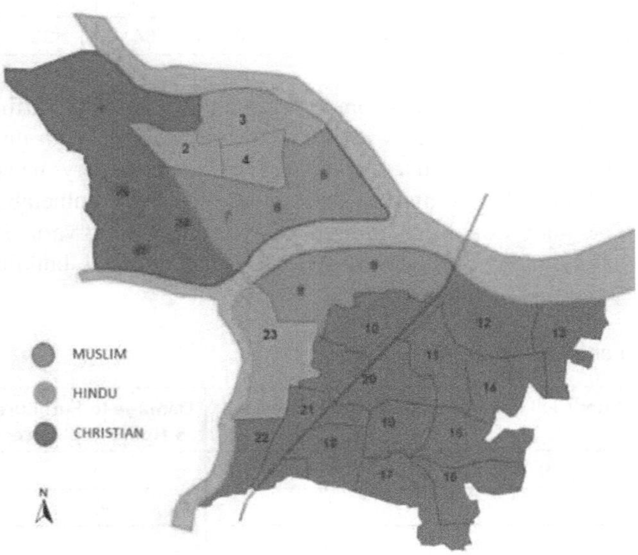

Fig. 1.3 Map of Aluva showing distribution of community across wards

Source: Primary survey, Jan – Feb 2019

4. Assessing Vulnerability of a Community During the 2018 Urban Floods

The framework of study was based on the concept of vulnerability being a function of risk and hazard ((Blaikie, 1994) has been considered as the as major indicators with variables chosen carefully bearing the community of Aluva, Kochi in mind.

4.1 Exposure

Exposure is the degree to which the community experience flooding (Marshall N.A, 2010). The exposure of community to floods varies with access to road network and infrastructure facility like condition of house (kutcha /pukka) and access to water and electricity. It was found that kutchha roads were inundated for long duration cutting accessibility while 3% of Kutchha houses did not have direct access to water and electricity. Level of inundation and ward density comparison shows that muslims and christians were more at risk due to exposure (Refer Fig. 1.4)

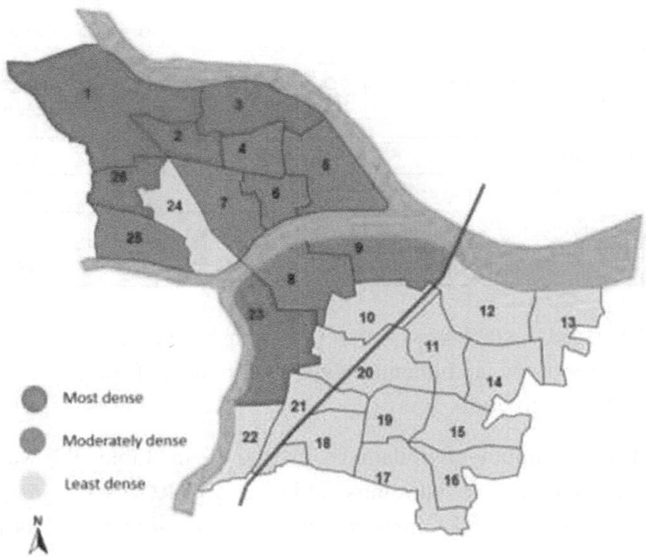

Fig. 1.4 Exposure based on density across wards in Aluva

Source : Primary survey, Jan – Feb 2019

4.2 Physical Vulnerability

The map on Figure 1.5 shows the level of inundation. Table 1.1 shows that 41% of the houses were most affected being inundated for 3-4 weeks, 29% moderately affected houses were inundated for 1-2 weeks while 30% were least affected houses which were inundated for less than a week or a couple of days. The water levels raised beyond 10'0" in low lying areas. 97% of the households live in a pukka houses. None

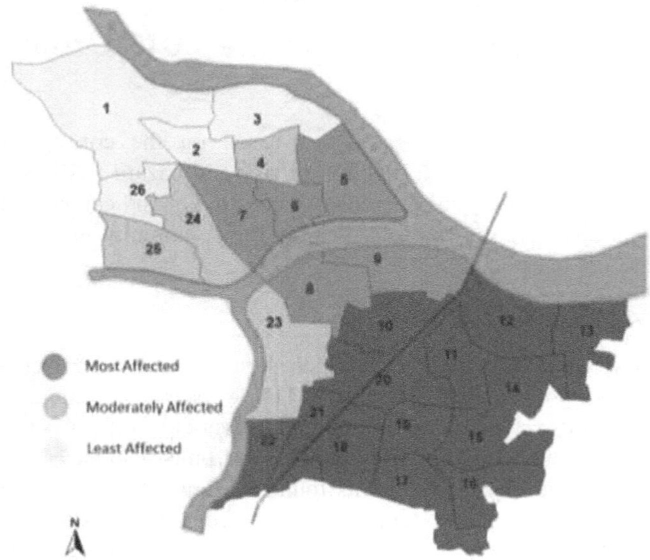

Fig. 1.5 Map of Aluva showing duration of inundation across wards

Source : Primary survey, Jan – Feb 2019

of these houses had access to water and electricity supply during the floods however supply was resumed based on severity of inundation and height of building. On an average the households did not have supply of water and electricity for almost 2 weeks.

Electricity supply was completely stopped by the department as precaution as majority of the wards were inundated. This affected the water supply into the houses. However the study establishes that supplies to taller buildings like apartments resumed within a week when compared to low height houses where supply resumed after 2 weeks. 26% of the houses had power back up / generator facility which was of no use during the floods. Most generators were inundated another reason was unavailability of fuel to run the generator.

The maximum civil damage as shown in Table 1.2 was incurred by 30% buildings falling above 20 years of age. Most houses had temporary metal sheet covering on roof which caused accidents and prevented access for air lifting Traditional buildings with wooden roofs covered with mud tiles were affected as compared to contemporary newer buildings. Hence making them more physically vulnerable.

Table 1.1 Duration of inundation and Infrastructure access

Description	Duration of Inundation	No. of Households	Type of house		No Water & Electricity		
			Pukka house	Kutcha house	1 Week	2 Week	> 2 Week
Most Affected	3-4 Weeks	50	47	3	4	19	27
Moderately Affected	1-2 Weeks	35	34	1	14	20	1
Least Affected	Less than 1 week	35	35	—	16	15	4
Total		**120**	**116**	**4**	**34**	**54**	**32**

The study establishes a direct relationship among variables like typology, architecture style, age, height of the building and infrastructure.

The maximum civil damage as shown in Table 1.2 was incurred by 30% buildings falling above 20 years of age. Most houses had temporary metal sheet covering on roof which caused accidents and prevented access for air lifting Traditional buildings with wooden roofs covered with mud tiles were affected as compared to contemporary newer buildings. Hence making them more physically vulnerable. The study establishes a direct relationship among variables like typology, architecture style, age, height of the building and infrastructure.

Table 1.2 Age of the building and severity of damage

Description	No. of Households	Major Civil Damage	Minor Civil damage	Damage to vehicles	Damage to Furniture & Home appliances
1 – 10 years	25	8	9	5	4
10 - 20 years	51	14	11	16	10
20 - 30 years	27	11	9	6	1
30 – 40 years	7	3	2	1	1
> 40 years	10	4	2	1	3
Total	**120**	**40**	**33**	**29**	**19**

Source: Primary survey, Jan – Feb 2019

4.3 Economic Vulnerability

Table 1.3 Community, remittances and income range

Description	No. of households	Remittances (Gulf)	Income Range	
			Above 10 lakhs	8–10 lakhs
Muslim	55	23	14	9
Hindu	33	19	8	11
Christian	32	5	3	2
Total	**120**	**47**	**26**	**22**

Source: Primary survey, Jan – Feb 2019

Income range is higher due to remittances coming from the gulf suggests Table 1.3. The annual household income of the 39% families across communities having bread winners working in the Gulf have an income rage above 8 lakhs. 40% of these NRI's are 2nd generation working in the Gulf after the parents came back to India. Also after analyzing the income range 50% of the muslim households have an annual household income of 8lakhs and above followed by 29% hindus and Christians suggesting lesser economic vulnerability of muslims (Table 1.4).

Table 1.4 Income pattern across communities

Description	No. of households	Muslims	Hindu	Christian
Above 10 lakhs	34	14	11	9
8.1 - 10 lakhs	48	20	14	14
5.1 - 8 lakhs	23	13	5	5
< 5 Lakhs	15	11	1	3
Total	**120**	**58**	**31**	**31**

Source: Primary survey, Jan – Feb 2019

There is a direct link between income of household and their assets. 100% households across all income range owned a mobile phone. 44% of the households has a landline connection at their home. The information of red alert given out on television and mobile. As mobile signals were poor due to heavy rains even before the floods having landline facility was handy for communication.

Table 1.5 shows that 45% of the self-employed households were found to be more vulnerable than 32% of those who are private employees. 15% of the retired households are most vulnerable due to age and no income. 17% of private employees and 23% of self-employed households have had no work for at least a month after the flood. Self-employed into agro based business dealing with perishable items and farmers have still been able to get back to work therefore unable to recover from loss incurred making them most vulnerable economically. A cross linkage of variables among economic and social indicators (income and occupation) is emerging which has a bearing on the economic vulnerability.

Table 1.5 Income pattern & occupation pattern

Description	No. of Households	Private Employee	Self Employed	Govt. Employee	Retired
Above 10 lakhs	34	9	18	0	1
8.1 - 10 lakhs	48	18	18	7	12
5.1 - 8 lakhs	23	8	9	2	5
< 5 Lakhs	15	3	10	1	0
Total	**120**	**38**	**55**	**9**	**18**

Source: Primary survey, Jan – Feb 2019

4.4 Social Vulnerability

Table 1.6 Religion & caste diversity

Description	No. of Households	General	BC	OBC	SC/ST
Muslim	55	—	—	55	—
Hindu	33	21	7	4	1
Christian	32	—	13	19	—
Total	**120**	**21**	**20**	**78**	**1**

Source: Primary survey, Jan – Feb 2019

The cast categorization has been adopted from the municipal census records of the Aluva, Kochi. Divide based on caste within community or across has not been recorded. A very strong fact that emerged during conversation with the communities in every ward is that the people have a very strong sense of belonging.

Table 1.7 Family type in Aluva

Description	No. of Households
Nuclear (Young couple, young children, older parents)	26
Nuclear (Middle aged couple with grown up children)	53
Nuclear (Old couple – male family head)	30
Nuclear (Old couple – female family head)	4
Joint family	7
Total	**120**

Source: Primary survey, Jan – Feb 2019

It has emerged from the survey that joint families are less vulnerable as compared to nuclear families. However factors like gender and age are vital factors. Families with more number of babies, older members and woman are more vulnerable as compared to those families with lesser household members in the age 0-6 years and above 60 years.

5. Assessing Adaptive Capacity & Coping Mechanism of Community During the 2018 Urban Floods

The framework has been formulated considering main indicators of societal capacity, economic capacity and management and institutional capacity to adapt and cope to an impact by disaster like flood. The is an adaption from the indicators used for a case study on flooding in Indonesia (Hidajat)

5.1 Adaptive Capacity

Adaptive capacity is "the ability of a community to take action to changing population / climate". There is a difference between knowing what could be done and having the ability to take caution (Susan l. cutter, 2003).

Table 1.8 shows that only 8 % of the total number of people (537) among all the households fell ill during the floods. The number of people who fell ill were mostly old people above 60 years of age, children and toddlers below 6 years of age. There were 5 families who lost members of their family members during and after floods. Casualties were mostly due to accidents and ill health even before the floods.

Table 1.8 Age and illness

Description	No. of people	No. of people who fell sick
0-6 years	54	8 (15%)
7-14 years	98	4 (4%)
15-59 years	273	11 (4%)
Above 60 years	112	24 (21%)
Total	**537**	**47 (8%)**

Source: Primary survey, Jan – Feb 2019

28% of the families had taken refuge in the camp, 48% had gone to their relatives places and 19% stayed at both camp and then at relatives place. 5% of the households did not evacuate their homes. These families were those living in apartments where water did not enter the flats. As the income levels are going up the number of families who stayed at the camp were lesser. This indicates stronger social network among higher income level households. The reason for this could be that households with better economic capacity would have access to better facilities where their relatives could stay.

The social network among the muslim community is found to be the highest followed by hindu households. This fact is established considering 65% of muslim and 51% of hindu families stayed at relatives place during the floods. The study shows that societal capacity among the muslims and hindus is strong (Table 1.9).

Table 1.9 Community and place of relocation

Description	No. of Households	Evacuated the house	Stayed in camp	Relatives place	Both camp & relatives place
Muslims	55	51	13	36	2
Hindus	33	33	16	17	—
Christians	32	24	11	7	6
Total	**120**	**108**	**40**	**60**	**8**

Source: Primary survey, Jan – Feb 2019

Table 1.10 Income and duration of relocation

Description	No. of Households	1 week	2 weeks	3 weeks	1 month	Upto 2 months
Above 10 lakhs	34	2	2	12	10	3
8.1 - 10 lakhs	48	1	2	17	18	3
5.1 - 8 lakhs	23	4	5	4	6	1
< 5 Lakhs	15	0	0	6	6	1
Total	**120**	**7**	**9**	**39**	**40**	**8**

Source: Primary survey, Jan – Feb 2019

Table 10 shows that 69% of the households stayed out of their home for 3-4 weeks during the floods while 6% of the families had to be away for upto 2 months. Most families could not move back as their house was severely damaged and repairs to get the house to livable condition took time. However it was recorded across communities that they were able to adapt to the flood situation with assistance from the community itself and external help.

Table 1.11 Family type and level of awareness

Description	No. of House-holds	Know of nearest medical facility / emergency point of contact	Knew about red alert	Adhered to the Red alert
Nuclear Family	97	89	90	71
Old couple (Male family head)	13	13	3	10
Old People	3	1	1	3
(Woman Head)				
Joint family	7	7	7	7
Total	**120**	**110**	**101**	**91**

Source: Primary survey, Jan – Feb 2019

Most of the households were very familiar with their surrounding and the neighborhood they belonged due to. Table 1.11 shows that 92% of the households were aware of medical facility that is close to them and its proximity from their house. 85% of the households were aware of the red alert. Out of which only 75% of the households took the warning seriously and adhered to the instructions given by the officials. The nuclear household with woman head are the only one who were unaware and not very familiar about their surroundings. This point proves the fact that joint families, nuclear families having middle aged couple with grown up children have a stronger adaptive capacity. Old aged people and single woman take a longer time to adapt to the situation and recover from the impact. The factors like age and gender to have a bearing on the adaptive capacity.

The study reports that communities most of the families did not know or follow any traditional practices. Barely 3 households out of 120 said they consumed water that is boiled with cumin seeds and a native herb to keep up their immunity which would keep them healthy. The intake of seafood is reduced during monsoon season. However other case studies suggest that traditional practices are followed in Kerala involving stocking of dry food, growing herbal shrubs and clearing the housing before monsoon season. (A. T. M. Zinnatul Bassar, 2017)

Aluva has a literacy rate of 98% (Govt. Of Kerala, 2015). Digital literacy levels were also found to be high among the age group of 15-59 years followed by children between 7-14 years. Digital literacy varies with age and gender. Men and those above teenage were more digitally sound hence having better adaptive capacity (Table 1.12).

Table 1.12 Age and digital literacy

Description	No. of people	Operate computer	Operate mobile	Operate television
0-6 years	54	2 (3%)	15 (28%)	18 (33%)
7-14 years	98	56 (57%)	68 (69%)	98 (100%)
15-59 years	273	204 (75%)	254 (93%)	269 (98%)
Above 60 years	112	31 (28%)	84 (75%)	108 (96%)
Total	**537**	**293 (55%)**	**421 (78%)**	**493 (92%)**

Source: Primary survey, Jan – Feb 2019

Adequate economic resources among all the communities due to general habit of saving (96% households). Hindu families preferred converting assets into jewellery as a saving means. This is attributed to the cultural wedding practices. Muslims had an edge due to higher number of families getting remittances hence having higher economic capacity. Families with lesser economic resources got loans at subsided interest rate of 8.45% p.a. from nationalized banks & private financial institutions like Muthoot finance (refer Table 1.13).

Table 1.14 shows that 96% of households lived in their own house while 98% of the households have their own piece of land in Aluva and other parts of Kerala. 30% of the inherited properties belong to higher income groups. The survey has established that the number of male and female property owners is almost equal the reason could be Kerala's matriarchal society. It also establishes an interdependence between social networks and economic capacity. These facts suggest strong economic capacity of the households across all the communities and are the reason why the communities were able to cope relative well despite the severe impact of the floods.

Most of the wards in Aluva, Kochi have housing societies comprising of a few houses situated in close proximity to each other. These housing societies are registered with the

Table 1.13 Income and means of monetary support

Description	No. of Households	Own Saving	Borrow from Relatives & Friends	Govt. / NGO	Remittances
Above 10 lakhs	34	34	0	0	14
8.1 - 10 lakhs	48	45	2	0	12
5.1 - 8 lakhs	23	21	2	3	4
< 5 Lakhs	15	12	2	5	2
Total	**120**	**112**	**6**	**8**	**32**

Source: Primary survey, Jan – Feb 2019

Table 1.14 Income and ownership pattern

Description	No. of Households	Own House	Own Land	Inherited Property
Above 10 lakhs	34	34	34	12
8.1 - 10 lakhs	48	44	46	15
5.1 - 8 lakhs	23	23	23	5
<5 Lakhs	15	15	15	5
Total	**120**	**116**	**119**	**37**

Source: Primary survey, Jan – Feb 2019

municipality of Aluva as they play a vital role in connecting the neighborhood with the municipality function. Through a set of committee members who are elected by all the houses which are part of the society. The society looks into aspects like door to door garbage collection, setting up of facilities like lighting on private access lanes, security and organizing festival celebration during Onam and other occasions like New year. Some societies get funds from the ward counsellor for setting up better infrastructure facility. 51% of the households mostly Christian and Hindu were part of the housing society. However being a part of a housing society did not help during the floods.

The municipality of Aluva, Kochi had put in their best efforts to provide everybody with medical facility. Every camp was stocked adequately with medicines, first aid and basic medical equipment with a doctor and a nurse on duty round the clock.

Almost all the household got medical assistance during floods irrespective of their income as shown in Table 1.15. Study has recorded that those at camps had better access to medical facility as compared to those that went to the relatives place.

Table 1.15 Medical facility available to households during floods in Aluva, Kochi

Description	No. of Households	Had Access to medical facility	Access to Doctor	Tetanus shot	Distribution of medicines
Above 10 lakhs	34	30	11	20	15
8.1 - 10 lakhs	48	46	23	26	17
5.1 - 8 lakhs	23	20	10	11	9
< 5 Lakhs	15	13	7	10	6
Total	**120**	**109**	**51**	**67**	**47**

Source : Primary survey, Jan – Feb 2019

Table 1.16 Communication of red alert

Description	No. of Households	Announcement by government officials	Electronic Media - TV/Radio	Social Media - Watsapp/Facebook/ Instagram	Neighbors/ Friends
Muslim	55	24	27	10	33
Hindu	33	6	24	9	12
Christian	32	13	17	4	15
Total	**120**	**43**	**68**	**23**	**60**

Source : Primary survey, Jan – Feb 2019

Table 1.16 shows that 48% of the households found out from neighbors and friends again establishing the fact that the communities are closely knit and have a good social network among them. The government had taken timely action and systematic approach reach out to the public about the flood and provide medical assistance. That establishes good level of management and institutional capacity of the government.

5.2 Coping Mechanism

"Coping Mechanism refers to the manner in which people and organizations use existing resources to achieve various beneficial ends during unusual, abnormal and adverse conditions of a disaster event or process. The strengthening of coping mechanisms usually builds resilience to withstand the effects of natural and other hazards" this is as defined by European Spatial Planning Observation network (2003). Table 1.17 shows that 80% of the households had 4 meals regularly while only 30% ate 3 square meals a day. Most households consumed non- vegetarian food across communities. Some even reared live stock in their homes for eggs, meat and milk. It was reported during the survey that most camps provided regular food while a few camps provided only "Kanni with kappa" which is a traditional rice porridge with boiled tapioca. Nutrition levels of higher income families better than lower income families as the lesser percentage of higher income families had to skip food.

Table 1.17 Income & food intake during floods

Description	No. of Households	Skipped food	Skipped 1 meal	Skipped 2 meals
Above 10 lakhs	34	12	12	—
8.1 - 10 lakhs	48	24	22	2
5.1 - 8 lakhs	23	16	13	3
< 5 Lakhs	15	12	8	4
Total	**120**	**64**	**55**	**9**

Source: Primary survey, Jan – Feb 2019

Majority of the people had to skip 1 meal and belonged to the age 15-59 where as children below 6 years of age did not skip meals. Both men and woman had to skip meals in the camp.

Table 1.18 Age & food intake during floods

Description	No. of people	Skipped food	Skipped 1 meal	Skipped 2 meals
0-6 years	54	—	—	—
7-14 years	98	27	24	3
15-59 years	273	51	46	5
Above 60 years	112	37	33	4
Total	**537**	**115**	**103**	**12**

Source: Primary survey, Jan – Feb 2019

The on ground study revealed that it took most of the households almost a month to get back to living normally post the flood. Most joint families have been able to resume living normally the nuclear families are still struggling to get back to normal. The regular nuclear families had better adaptive capacity when compared to the other types of nuclear family. The only reason for this could be weaker economic capacity. As the comparison with income levels show that the number of households with lesser annual income still recovering are more. 7 households out of 120 households have been able to recover within less than a month. These are those households that were located in areas which were less affected by floods. The above mentioned facts show that the coping capacity of joint families is better than nuclear families. Age also plays a major role in coping. Older people have lower coping capacity as they take a longer time to cope to the situation post the disaster (refer Table 1.19).

Economic capacity plays a crucial role in bringing life back to normal post a disaster as it involves money in repairing the house. However due to the strong social network and good societal capacity many families have been helped by NRI relatives and friends which has helped in coping to the situation post flood.

Table 1.19 Family type and duration to get back to normal post flood

Description	No. of Households	1 week	2 weeks	3 weeks	1 month	1.1 - 2 months	Not back to normal as on Feb'2019
Nuclear Family (Regular)	97	1	1	4	38	19	18
Old couple (Male head)	13	0	0	0	8	2	3
Old couple (Woman head)	3	0	0	0	0	2	1
Joint family	7	0	0	1	2	3	1
Total	**120**	**1**	**1**	**5**	**48**	**26**	**23**

Source: Primary survey, Jan – Feb 2019

Table 1.20 Income of family and during to get back to normal

Description	No. of Households	1 week	2 weeks	3 weeks	1 month	1.1 - 2 months	Not back to normal as on Feb' 2019
Above 10 lakhs	34	1	1	3	17	8	4
8.1 - 10 lakhs	48	0	1	3	23	12	9
5.1 - 8 lakhs	23	1	0	2	10	24	6
< 5 Lakhs	15	0	0	0	3	5	7
Total	**120**	**2**	**2**	**8**	**53**	**49**	**26**

Source: Primary survey, Jan – Feb 2019

It is to be noted that self-employed people who are drivers, tailors and some of them who are into agriculture had not resumed work as on February 2019. Those people who are into occupation which get impacted during the flood have lesser coping capacity. On the contrary those self employed households like contractors, carpenters, painters who are into construction field have been busy with renovation and repair works. These households have been able to able to earn money just after the floods and hence cope better.

The Table 1.21 reveals that the number of families who got help from relatives is lesser as compared to the one who got help from neighbors. This further establishes the strong societal capacity and close knit factor of the community. The communities got very little help from the Government and other NGO for cleaning their house after floods. Some wards had volunteers who had come from parts of Kerala which was not affected by floods to help those who were affected by floods.

Table 1.21 Help got by households after returning home

Description	No. of Households	Relatives	Support from neighbors	Paid Labour/ volunteers	Cleaned ourselves
Muslim	55	6	13	6	30
Hindu	33	2	7	1	23
Christian	32	3	8	3	18
Total	**120**	**11**	**28**	**10**	**71**

Source: Primary survey, Jan – Feb 2019

Table 1.22 Age of the building and duration of repairs

Description	No. of Households	7 - 8 weeks	Less than 2 weeks	3 - 6 weeks	Repairs going on as on Feb' 2019
1 – 10 years	25	4	6	10	3
10 - 20 years	51	9	10	22	8
20 - 30 years	27	5	8	9	4
30 – 40 years	7	1	0	4	2
> 40 years	10	1	1	4	4
Total	**120**	**20**	**25**	**49**	**21**

Source: Primary survey, Jan – Feb 2019

The duration for repairs depends on the extent of repairs to be done due to impact based on age of the building. It was observed that 17% of the houses are still under repair (Table 1.22). The number of houses falling in the age range of 20-30 years and above is higher. The coping capacity of those living in newer contemporary houses was better as compared to those living in older, traditional houses. Lesser duration of repairs due to lesser impact.

Most families have incurred minimum loss in the range of 1-3 lakhs across all income ranges and communities (Table 1.23). The losses were more if the house was older and traditional.

Table 1.23 Monetary Damage & Income

Description	No. of Households	Less than 1 lakh	1 - 3 lakhs	3.1 - 5 lakhs	5.1 - 7 lakhs	Above 7 lakhs
Above 10 lakhs	34	7	10	6	5	5
8.1 - 10 lakhs	48	5	24	12	2	2
5.1 - 8 lakhs	23	3	11	6	1	1
< 5 Lakhs	15	0	9	4	2	0
Total	**120**	**15**	**54**	**28**	**10**	**8**

Source: Primary survey, Jan – Feb 2019

Table 1.24 Duration of inundation and compensation from government

Description	No. of Households	Received compensation amount
3-4 Weeks	50	49
1-2 Weeks	35	34
Less than 1 week	35	25
Total	**120**	**108**

Source: Primary survey, Jan – Feb 2019

The government had given all the families living in the town of Aluva, Kochi a sum of Rs.10,000 whose house was inundated for a period of over 48 hours (2 days). However the survey recorded that not all families had received the compensation amount (Table 1.24). A kit consisting of a shirt, towel, cleaning material like detergents and disinfectants and a pouch with ayurvedic medicines and herbs was also distributed to the households. Compensation was fixed only based on inundation. Factors like economic capacity & extent of damage due to age, typology of the building not considered which is a lag.

6. Institutional Support During the Urban Floods in Kerala, 2018

The efforts taken by Government institutions during all stages of the flood was gathered by interacting with officials holding positions at various levels in the governance hierarchy.

6.1 District Level

The Disaster management act 2005 says that all state governments are mandated under section 14 of the act to establish a state disaster management authority. The Disaster Emergency Operations Center (DEOC) was in charge of the operations as per the during the 2018 floods in the Ernakulam District. During the floods the main function of the DEOC was to coordinate among the all the agencies for rescue operations across all towns, after the red alert was issued. After the floods they allocated funds to municipality for distribution of compensation, mobilized volunteers to the municipality for cleaning, rebuilding houses for those whose houses were fully damaged during floods through the "Rebuild Kerala" initiative.

6.2 Municipality Level

As per the act the municipality of each town followed orders given by the Collector's office during the 2018 floods. The municipality officials worked closely with the two village offices under Aluva Municipality which is the East village office and the West village office. The village officers report to the assistant secretary at the municipality office. They were in charge of spreading the red alert across the wards and setting up camps before, during the floods. After the floods they distributed compensation kits. A 24 hour "Sahaya Kendra" was opened to address all flood related queries.

6.3 Ward Level

The ward counselors had got in touch with all the households through phone and sent out messages through social media (watsapp) informing them about the red alert before the flood. The ward counselors moved to the camp with their families and remained there for a month almost as their houses were inundated for more than a week. Hence they could not extend support during the flood time as they were also affected.

Physical damage to property and infrastructure was inevitable due to the magnitude of flood but proactive measures did help minimize causalities. There was a discrepancy in primary survey data and what was said by the officials during interaction. The overview gained has helped analyze the flood situation holistically.

6.4 NGO " Kudumbashree"

This NGO which is run by woman to cater to community needs at neighbourhood level. During the floods Kudumbashree volunteers were posted at camps in Aluva. 15 kudumbashree staff were volunteers at the Government Ayurveda hospital and 11 at Government girls higher secondary school. The other volunteers helped in cleaning activity of public offices like post office and clinics. Few others helped in giving counselling to woman and children who were traumatized due to the impact of flood. This NGO conducted community cooking across all the affected parts of Kerala to provide food to camps and helped the collectors office in packing the kits to be distributed.

The government has played a major role in informing everyone about the disaster forecast which has prevented loss of life to a great extent. Physical damage to property and infrastructure was inevitable in a flood situation however the proactive measures did make a difference. The NGO "Kudumbashree" supporting the government in its efforts has only benefitted the affected people and help provide relief.

7. Recommendations

The analysis of the study conducted in Aluva, Kochi on assessing vulnerability, adaptive strategy and coping mechanism of community during the urban floods in 2018 has helped in identifying gaps that are currently existing in the system. There is an immediate need to address the gaps in assessment of vulnerability to improve the resilience of the urban community. The following recommendations

Physical Vulnerability: The location of dwelling unit and age of the building the household resides in plays a major role in determining physical vulnerability of community.

Findings:

- Houses closer to water bodies & low lying areas more vulnerable
- Older houses, traditional more vulnerable as compared to new contemporary structures.

Recommendation:

Amendment in development control rules by the town and country planning department (GDCA) suggesting change in typology of residential buildings along water bodies to be on stilts. Municipalities must be adhere to the protocol of cross check ground reality at all stages of construction to avoid breach of rule.

Economic Vulnerability: Variables like remittances and occupation have a direct bearing on economic vulnerability.

Findings:

- The average income range of families getting remittances is high.
- Retired persons without pension more vulnerable

Recommendation:

Monetary support to senior citizens without pension/ single woman can be provided through CSR programs, private financial institutions like Muthoot finance and civil societies.

Social Vulnerability: Vulnerability changes based on family type, age and gender.

Finding:

- The nuclear families are more socially vulnerable as compared to Joint families.
- Women, old people, babies were found to be more vulnerable during disaster.

Recommendation:

Socially vulnerable families must be identified considering family type, age and gender.

A socially vulnerable map at ward level could be generated by the disaster management cell (Collector's office) with information gathered by community participation. These maps when be shared with the community increases their awareness improving vulnerability. Camp volunteers who can cater to the emotional needs of children and the aged as the impact of flood does have a mental impact which needs counselling.

Adaptive capacity & Coping Mechanism: The community was able to adapt and cope well due to strong sense of belonging. However the variation was based on health, well being, economic capacity, awareness and institutional capacity.

Finding:

- Old people, children, toddlers and woman have lesser adaptive capacity and found it difficult to cope with the impact of the flood.
- Inadequate medical facility for pregnant woman and babies reported
- Self employed families into occupation like farming and agro based business have lesser coping capacity as they are still out of work and hence unable to recover from financial losses.
- Communities close knit and aware of their surroundings. They helped each other during and floods helping them cope better
- Red alert not adhered by some families.
- Housing societies did not have a role to play during flood time
- No early warning facility
- Varies levels of governance hierarchy

Recommendation:

Preparedness measures of the community must be improved by conducting workshops and awareness programs.

Special flood preparedness programs for woman which will improve their resilience. Stocking dry food items (traditional practices), getting the house repaired and cleaned before the monsoon season, innovative methods like making a portable stove for cooking during floods.

Installation of early warning system.

Better coordination across government agencies during floods.

8. Conclusion

The assessment of vulnerability, adaptive capacity and coping mechanism of the flood affected communities and those lying in vulnerable zones must be made mandatory for urban development functions. The recommendations suggested as a part of the study would enable bridge the existing gaps in vulnerability assessment of the community thereby improving the resilience of the city.

However outcomes of this study suggest that the variable under each indicator would vary depending on the hazard and the location. At the same time it simplifies the complication of comprehending the qualitative and quantitative aspects associated with a research of this kind. A set of indicators could be fixed upon for assessing communities during floods if more such studies that are specific to urban floods are carried out for similar geographical conditions. A comparison of many similar case studies across India will help establish a standard framework with indicators specific for hazard, location and culture. A standardized assessment methodology for community would be a very useful tool to improve resilience of India at a national level.

References

1. 100 Resilient Cities. (n.d.). *Resilient Bangkok,* Bangkok.
2. A. T. M. Zinnatul Bassar, A. T. (2017). Women Participation in Flood Risk Management: A Case Study in Char South Baladoba in Kurigram District. *American Journal of Environmental Protection. Vol. 6, No. 1, 2017.*
3. Birkmann, J. (n.d.). *Measuring Vulerability to Natural Hazards: towards diaster resilient societies* . TERI Press.
4. Blaikie, P. (1994). At Risk: Natural hazards, people's vulnerability and disasters.
5. Brooks, N. (2004). *The determinants of vulnerability and adaptive capacity at the national level and the implications for adaptation.* Elsevier.
6. Chelleri, L. (2011). *A review essay on Understanding and integrating the reislience perspective for urban systems.*
7. EY. (2017). *India's Growth Paradigm.*
8. Govt. of Kerala. (2015). *Detailed Report of Draft Master Plan.* Aluva.
9. Hidajat, C. B. (2006). Community based risk index: Pilot implementation in Indonesia.
10. Hidajat, C. B. (n.d.). Community based risk index: Pilot implementation in Indonesia .
11. *http://www.onefivenine.com/india/villag/ernakulam.* (n.d.). Retrieved from www.onefivenine.com: http://www.onefivenine.com/india/villag/ernakulam
12. Kurian, V. (2018, August 12). *Hindu BusinessLine.* Retrieved from Hindu BusinessLine: https://www.thehindubusinessline.com/news/national/relief-and-rescue-operations-in- kerala-earn-praise-from-experts/article24670004.ece
13. *maps.com.* (n.d.). Retrieved from maps.com: https://d-maps.com/pays.php?num_pay=426&lang=en
14. Marshall N.A, P. M. (2010). Social vulnerability to climate change in primary producers: A typology approach.
15. NATCOM. (2004). *India's Initial National Communication to the United Nations Framework Convention on Climate Change.*
16. NDMA. (n.d.). Retrieved from www.ndma.gov.in
17. NIDM. (2014-2015). *Annual Report.*
18. Susan l. cutter, B. j. (2003). Social Vulnerability to Environmental hazrads. *Social Science quarterly.*
19. UNC Institute for the Environment. (March 2009). *Community Based Vulnerability Assessment : A Guide to Engaging Communities in Understanding Social and Physical Vulnerability to Disasters.*
20. UNISDR. (2015). *Global Assessment Report on disaster risk reduction.*

CHAPTER 2

Flooding Problems in Periyar River Basin, Kerala—The Effects of Land Use Land Cover Changes

Kashish Sadhwani
Department of Civil Engineering,
Indian Institute of Technology Bombay, Mumbai-400076, India
kashishksh@gmail.com

T. I. Eldho
Department of Civil Engineering,
Indian Institute of Technology Bombay, Mumbai-400076, India
Corresponding Author: eldho@civil.iitb.ac.in

Subhankar Karmakar
Environment Science and Engineering Department,
Indian Institute of Technology Bombay,
Mumbai-400076, India
skarmakar@iitb.ac.in

Abstract

Floods are one of the most extremely dangerous natural hazards as they cause not only damage to property but also loss of life. The study of such extreme events gives us an opportunity to analyze the causes, be it natural or anthropogenic activities, that may have led to worsening its effect. Change in land use land cover (LULC) is one of the major reasons for worsening the flood hazards. A severe flood event occurred in August 2018 in Kerala, India, where 13 out of 14 districts of the State were affected by it, and the Periyar river basin was one of the worst affected regions. The present study aims at analyzing the impact of LULC change in the Periyar river basin (PRB) and its effects on the flooding using a semi-distributed hydrological model SWAT (Soil Water Assessment Tool). The study also investigates the possible impact of further landcover change (till 2030) in the region if another flooding event of similar intensity occurs in the future. The historical landuse maps are classified from LandSAT satellite images using the maximum likelihood method and used to predict future landuse maps using the MLP-ANN technique. The SWAT model is developed to analyze streamflow at a daily scale for monsoon season and calibrated and validated with available data. The results suggested that LULC change has a significant impact on the streamflow, and if a similar landcover change pattern occurs in the future, then there are chances of further increasing floods and other related problems in the future.

Keywords: Flood hazards, Landuse land cover change, Kerala Floods, Hydrological modeling, SWAT

1. Introduction

Floods cause substantial damage to lifestock, infrastructure, agriculture, and the environment. Extreme rainfall are reported as the major cause of floods in past studies (Teegavarapu 2012; Ivancic and Shaw 2015; Wasko et al. 2019). The frequency of extreme precipitation events has increased in India in past years and is expected to increase more due to the warming climate (Goswami et al., 2006; Rajeevan et al., 2008; Mukherjee et al., 2018; Roxy et al., 2017). The recent devastating flood events occurred in Uttarakhand (2013), Gujrat (2015), Chennai (2015), Bihar (2017), Kerala (2018) and Assam (2020). The overall damage caused due to floods between 1953 to 2016 is estimated approximately as Rs. 347581 crores (http://www.cwc.gov.in). These statistics are alarming as the flood-related damages are showing an increasing trend. Increase in population, growth in development projects and economic activities in the vicinity of flood plains rapid urbanization, coupled with ongoing global warming can be linked to the rise in flood damages (National Disaster Management Authority, India, 2008).

The 2018 August flooding in Kerala, 13 out of 14 districts experienced severe flooding in (CWC 2018). As per Indian Meteorological Department (IMD), Kerala received 2346.6 mm of rainfall from 1 June 2018 to 19 August 2018. This rainfall was about 42% above the expected 1649.5 mm of rain. The increase in rainfall during June, July, and 1st to 19 August was 15%, 18%, and 164%, respectively, above average. The flood of August 2018 was accounted as the second most disastrous flood in Kerala after the event of 1924. Generally, flooding events are ascribed to extreme precipitation or changing climate. During the August 2018 floods in Kerala, extreme precipitation and reservoir operation were considered as a significant cause of the damage (Sudheer et al. 2019, Mishra and Shah 2018, Hunt and Menon 2020). However, past studies have shown that LULC change showed a significant impact over change in the streamflow, which results in flooding (Zhang et al. 2012, Shi et al. 2013, Petelet-Giraud et al. 2017, Marhaento et al., 2018, Sinha and Eldho 2018, Sinha et al. 2020). Thus it becomes important to study the role of LULC change in flooding as it will improve our understanding of the hydrological dynamics involved in the flooding. This can help in planning flood management and mitigation strategies.

The present study focuses on identifying the effect of LULC change in the Kerala 2018 flooding event. It includes analyzing the streamflow variation due to LULC change by varying LULC and keeping rest parameters fixed. The study is further extended by comparing the possibilities of changes in the future by using future LULC projections.

2. Study Area

Periyar river is the second-longest river of Kerala with a length of approximately 244 km, and it originates at Sivagiri peak at an elevation of 2438 m above mean sea level (MSL) and joins the Arabian Sea towards the west. The watershed area ranges from longitude 76^0 E to 77^0 30' E and latitude 9^0 16' N to 10^0 20' N (Fig. 2.1). The soil texture over the region varies slightly from clayey to loamy with significant proportions of loam as per the National Bureau of Soil Survey and Land Use Planning (NBSS & LUP) (Fig. 74.2b). The elevation details and slope map of the region are shown in Fig. 2.2a and Fig. 2.2c, respectively. The elevation varies from the highest point of 2695m above mean sea level at Anamudi Peak (https://www.keralatourism.org/munnar/anamudi-peak-kerala.php) to the lowest point near the sea.

The slope of the watershed has been categorized into four classes 0-2%, 2-8%, 8-16%, and above 16%. The average annual rainfall of PRB is approximately 3200 mm (CWC, 2018). There are two hydrological observation stations managed by CWC at Neeleshwaran (10^0 12' N 76^0 5' E) and Vandiperiyar (9^0 32' N 77^0 4' E) in the watershed. There are three major dams with significant storage in PRB namely, Mullaperiyar (443.23 million cubic meters (MCM)), Idukki (1460 MCM), and Idalmalayar (1018 MCM) (Mohankrishnan & Verma, 1997). Details of the reservoirs are mentioned in Table 2.1.

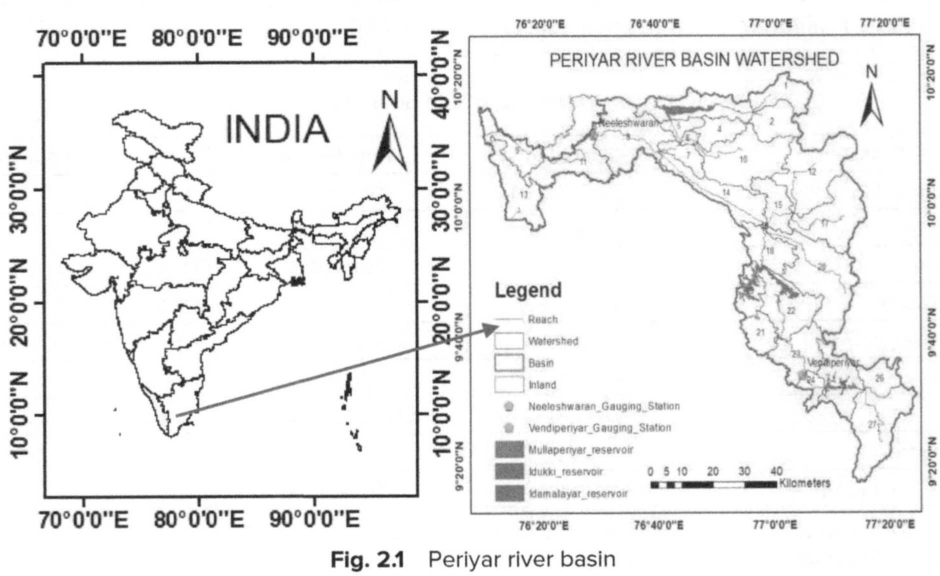

Fig. 2.1 Periyar river basin

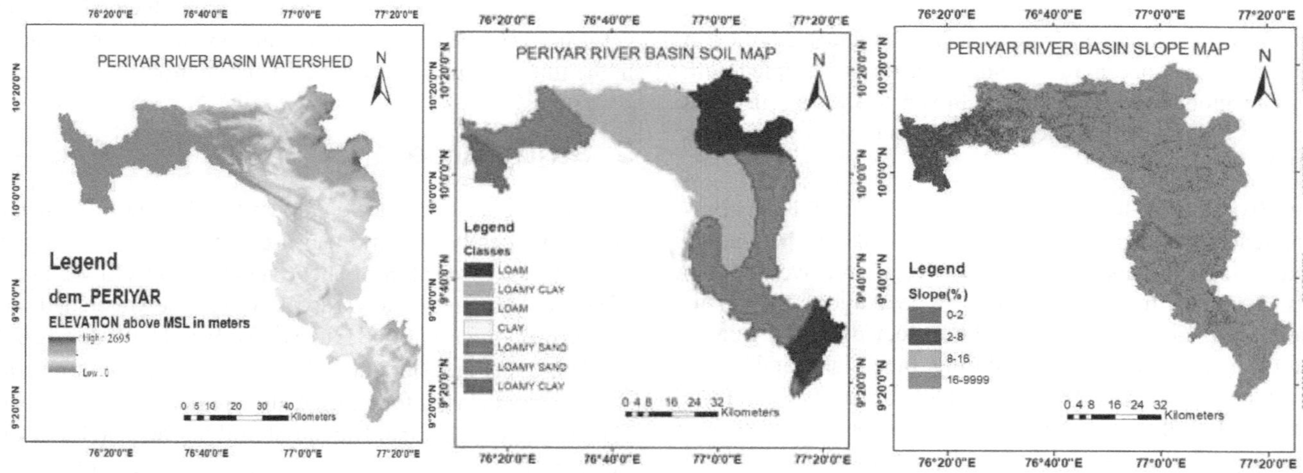

Fig. 2.2 (a) DEM (b) Soil Map (c) Slope map of PRB

Table 2.1 Details of Reservoirs

Name of Reservoir	Built in year	Full Reservoir Level (m)	Full Reservoir Capacity (MCM)	Effective Volume (MCM)	Surface Area at FRL (sq. km)
Idamalayar	1985	169	1089	1017.80	28.3
Idduki	1973	168.91	1996	1460	60.03
Mullaperiyar	1895	53.64	443.23	324	20.55

Source: Kerala State Electricity Board; Mohankrishnan and Verma, 1997

3. Material and Methods

Soil and Water Assessment Tool (SWAT) is used for hydrological modeling of PRB. The SWAT model is useful for simulating ungauged basins and particularly with limited data in the river basin scale (Arnold et al. 1998). For calibration of the SWAT model, the sequential uncertainty fitting principle (SUFI-2) algorithm (Abbaspour et al., 2007) is used. The methodology adopted for the complete study was divided into three steps: a) LULC classification and preparation of input data for SWAT model; b) Model calibration and validation at daily scale; c) Analysis of LULC impact on streamflow of Periyar river basin for Kerala flood 2018. The SWAT input data include collection and preparation of Digital Elevation Model (DEM), soil data and LULC data, and Meteorological data (Table 2.2). The

Table 2.2 Data Sources

S. No.	Data Type	Resolution	Time Period	Data Source
1	Meteorological Data Precipitation	0.250 x 0.250	2014-2018	IMD
	Temperature	1.00 x 1.00	2014-2018	IMD
	Wind Speed	0.50 x 0.50	2014-2018	CFSR
	Solar radiation	0.50 x 0.50	2014-2018	CFSR
2.	Digital Elevation Model	30m x 30m	2005	CartoDEM (bhuvan.nrsc.gov.in)
3.	Soil Texture	30 arc second	2012	NBSS & LUP
4.	LULC	30m x 30m	1988, 2002 and 2016	LandSAT (Earthexplorer.usgs)
5.	Gauge Discharge	Daily	2014-2018	Central Water Commission
6.	Reservoir water level	Daily	2014-2018	(CWC)

CFSR: Climate Forecast System Reanalysis (CFSR) data https://rda.ucar.edu
IMD: Indian Meteorological Department
NBSS & LUP: National Bureau of Soil Survey & Land Use Planning

LULC for the years 1988, 2002, and 2016 are prepared by using maximum likelihood classification of LandSAT images on the GIS platform, and the soil data was obtained from NBSS & LUP. Future LULC map for 2030 is prepared from these maps using TerrSet Land Change Modeler (LCM) that works on Markov Chain transition technique and Multi-layer Perceptron Artificial Neural Network (MLP-ANN) (Sinha and Eldho 2018, Anand et al. 2018, Sinha et al. 2020).

Meteorological data, including precipitation, temperature (minimum and maximum), solar radiation, and wind speed are obtained from IMD and CFSR correspondingly (Table 2.2). The hydrological data such as observed runoff is collected from a gauging location at Neeleshwaram for the years 2014 to 2018 at a daily time scale from CWC. All these data were extracted for the Periyar watershed, meteorological data was re-gridded to 0.25^0 x 0.25^0 resolution using linear interpolation, and spatial data was projected to WGS 1984 UTM Zone 43 N projected coordinate system. Then the watershed is delineated using the stream burn function to get the specific stream route. The watershed area obtained after delineation is 4792.83 sq. km. The HRUs are generated with LULC 2016, and then the model is calibrated and validated after adding the weather input files and the details of reservoirs (Figs 2.1 and 2.2). It is assumed that there is no water release from Idukki and Mullaperiyar dam into the Periyar river as water from the Idukki reservoir is directed towards the Thodupuzha river at Thriveni sangamam at Moolamattom Power Station for power generation (source: http://www.kseb.in), which is outside the basin watershed area. Similarly, for the Mullaperiyar reservoir, the water stored is mainly used for irrigation purposes, and thus it is assumed that there is no water release from this reservoir into the Periyar river. For Idamalayar dam, the maximum flow rate from the spillway is 3012.8 cumecs (Mohankrishnan and Verma 1997), and the same was assumed to be the monthly average annual outflow rate from the reservoir directed towards the Idamalayar Power Station. The model is calibrated and validated for daily streamflow for the years 2014 and 2017, respectively. The calibrated model is then used for analyzing the impacts of LULC change on streamflow. The effect of change in streamflow due to variation of LULC for Kerala floods 2018 is calculated by changing LULC with 1988, 2002, and 2016 for historical; and 2030 for future change analysis, respectively.

4. Results

4.1 LULC Change

The classification for land use map is done for the years 1988, 2002, and 2016 (Fig. 2.3), and six landuse classes were observed: Water, Built-up, Plantation, Cropland, Barren, and Forest. The overall accuracy and kappa coefficient (κ) are in the ranges of above 86% and 0.82, respectively. The projected LULC for 2030 was prepared using the Land Change Modeller (LCM) tool. The change between land use classes from 1988 to 2030 is compared and tabulated in Table 3. Changes from 1988 to 2016 showed that the area covered by water bodies is slowly decreasing from 3.992% to 3.673%. The same is the case with cropland and barren land; it reduced from 4.41% to 3.06% and 3.95% to 2.62%, respectively. Urban area is increased significantly from 1.18% to 5.31% in 2016, accounting for 4.13% change. It can be observed that significant transitions are plantation to urban; forest to urban; and agriculture to urban. Other than this, forest to plantation and barren to plantation change has also been observed. A similar transition change has been observed for future LULC 2030. The landcover for urban is expected to increase up to 7.79% by 2030. Plantation and cropland are expected to increase by 0.84% and 0.11% after 2016. Similarly, water; barrenland; and forests are expected to decrease after 2016 by 0.023%, 0.21% and 3.22% respectively (Table 2.3).

4.2 SWAT Model Calibration and Validation

The SWAT model for PRB is calibrated and validated at a daily scale for streamflow at the Neeleshwaram gauging station. The goodness of fit is analyzed by considering coefficient of determination (R^2) and Nash-Sutcliffe efficiency (NSE) (Legates & McCabe, 1999). During calibration, the most sensitive parameters identified were alpha factor for bank storage (alpha_bnk), baseflow alpha- factor (alpha_bf), SCS curve number (CN2), threshold depth of water in the shallow aquifer required for return flow to occur (gwqmn), groundwater delay (gw_delay), USLE soil erodibility factor (USLE_K), water use in reach (wurch), groundwater revap coefficient (gw_revap), soil evaporation compensation factor (ESCO), surface runoff lag time (surlag) and available water capacity of the soil layer (sol_awc). During the calibration period (2014), R^2 and NSE values are observed as 0.81 and 0.67, and during the validation period (2017), the values are 0.72 and 0.60, respectively. The scattered plot of observed and simulated streamflow is shown in Fig. 2.4. The calibration and validation results show a good representation of PRB for daily flow and can be accepted for further analysis (Moriasi et al. 2015).

4.3 Impact of LULC Change on Surface Runoff

For the impact assessment of LULC change effects on floods, the rainfall of severe floods in Kerala from 31 July 2018 to 31 August 2018 is used (Figure 5b). To understand the effect of LULC change during the August 2018 rainfall, the LULC of 1988, 2002, 2016, and predicted 2030 are used in the analysis. The assessment of change in the surface runoff

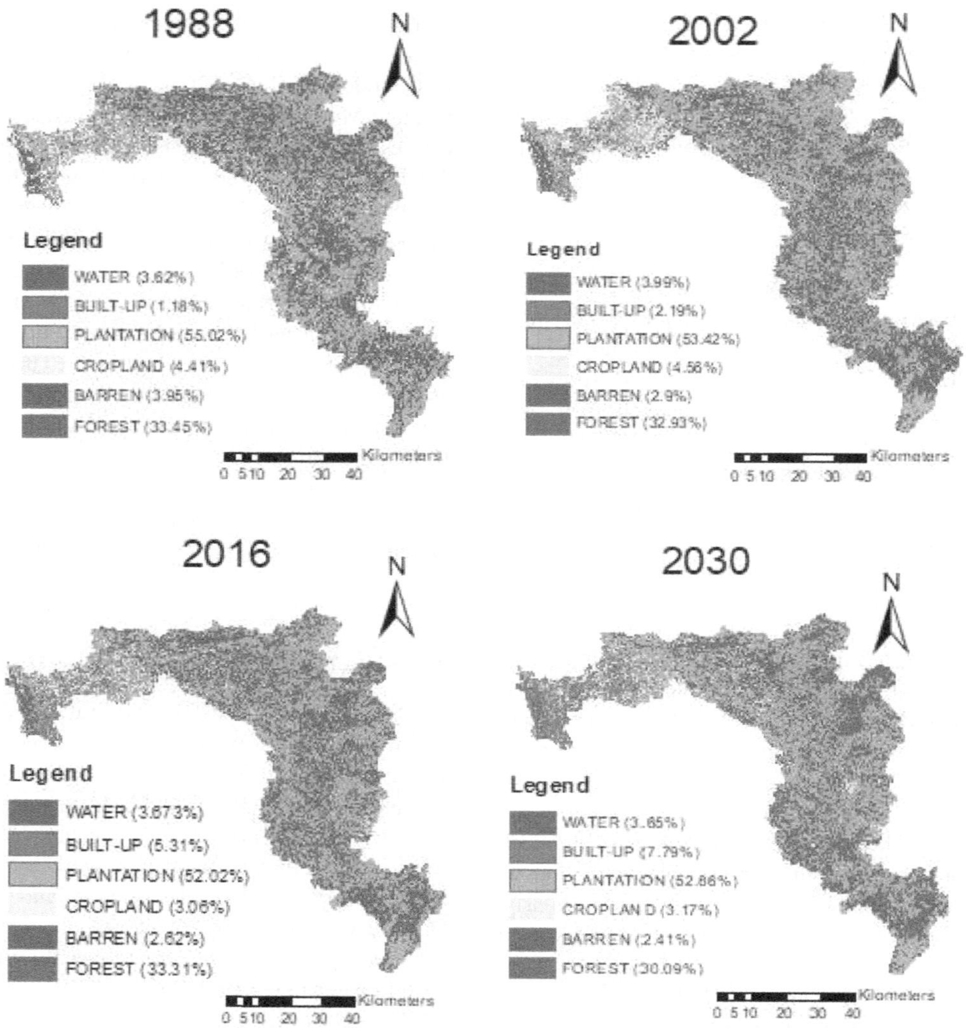

Fig. 2.3 LULC a) 1988; (b) 2002; c) 2016; d) 2030 for PRB

Table 2.3 LULC change for PRB 2002 to 2030

Class (%) (area in km²)	1988	2002	2016	2030	% Change		
					2002-1988	2016-1988	2030-1988
Water	3.992 (191.23)	3.99 (191.23)	3.673 (176.04)	3.65 (174.93)	0.37	0.053	0.03
Built-up	1.18 (104.96)	2.19 (104.96)	5.31 (254.49)	7.79 (373.36)	3.12	4.13	6.61
Plantation	55.02 (2560.32)	53.42 (2560.32)	52.02 (2493.23)	52.86 (2533.48)	-1.4	-3	-2.16
Cropland	4.41 (218.55)	4.56 (218.55)	3.06 (146.66)	3.17 (151.93)	-1.5	-1.35	-1.24
Barren land	3.95 (139.47)	2.91 (139.47)	2.62 (125.57)	2.41 (115.5)	-0.29	-1.33	-1.54
Forest	33.45 (1578.27)	32.93 (1578.27)	33.31 (1596.49)	30.09 (1442.16)	0.38	-0.14	-3.36

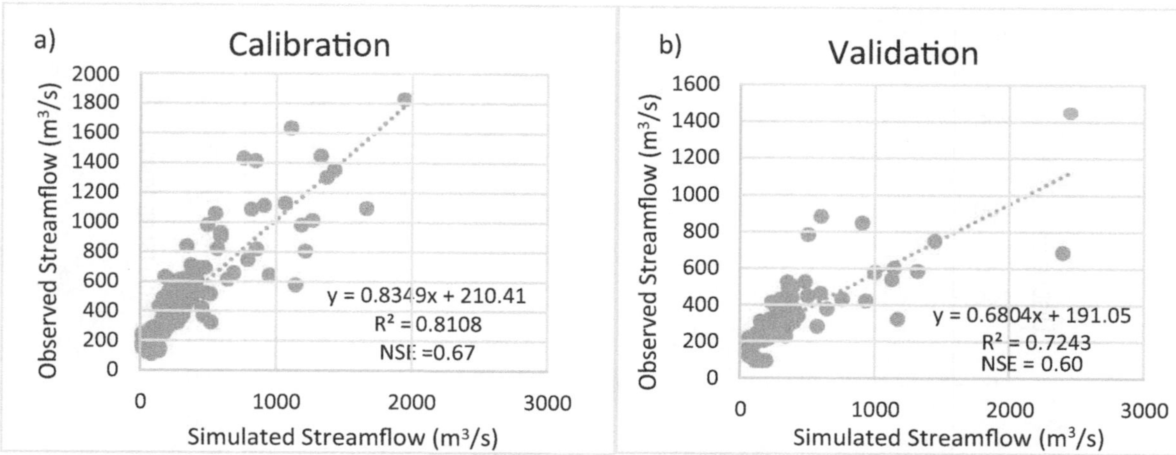

Fig. 2.4 Scattered plots between the observed and simulated monthly streamflow values for (a) calibration and (b) validation period.

for Kerala floods 2018 is done by comparing the runoff at the outlet. As shown in Fig. 2.5, due to LULC change, the surface runoff has increased considerably from 1988 to 2016 (Fig. 2.5a), and a significant change is observed in peak discharge (Fig. 2.5c). The possible reason is a rapid increase in the built-up area which led to the rise in rapid flow. Other contributing factors are a decrease in plantation and forest cover. On comparing the change with respect to future LULC 2030, the surface runoff is expected to increase during the floods. The major reason for this change is the increase in the built-up area. Also, the combined effect of change in forest cover, plantation, barren land, and cropland are responsible for such change in runoff (Fig. 2.5 and Table 2.3). A significant change in runoff is observed during the peak flow of 16 August 2018 flooding event. The change in runoff due to LULC change is calculated for flooding event (16 August 2018), and post-flooding event, and the results are tabulated in Table 2.4. The increase in peak discharge due to LULC change varied by 2.31%, 6.72%, and 9.03% for the LULC of 2002, 2016, and 2030 respectively, when compared with LULC of 1988 (Table 2.4). The post-flood effect also shows an increment in the runoff. The spatial variation of LULC change impact on surface runoff during peak flood is represented in Figure 6. Results show an increment up to 12% by the year 2030. Such a change is alarming as it may result in amplifying the damaging effects of floods. These results suggest the importance of landuse practice in watershed management and its impact on basin hydrology. Understanding these factors plays a vital role in watershed management and planning.

5. Discussions

Changes in landcover classes impact runoff in different ways. An increase in urbanization increases flood peaks (Zope et al. 2017), whereas an increase in vegetation cover reduces runoff (Yan et al. 2013, Sinha and Eldho 2018, Sinha et al. 2020). With the rapid growth in population, there is a rapid growth in urbanization. The usual trend of transition among landcover classes is agricultural land or barren land is directly converted to urban areas. At the same time, forests or plantations are converted to grasslands and then used for urban development. This transition is slow but alters the hydrology of a watershed. Such changes lead to watershed mismanagement that results in flooding in the case of extreme rainfall. Considering the change in the runoff for the LULC change from 1988 to 2016, the major change identified is increased urbanization which rise significantly by 4.13%. Thus, the change in urbanization is identified as the major cause of change in streamflow. Similar results are observed in past studies (Sinha and Eldho 2018, Anand et al. 2018, Sinha et al. 2020). The other supporting factors for increased runoff are reductions in agriculture and plantation by 1.35% and 3%, respectively (Table 2.3). For 2030, there is a decrease in agricultural land and plantation cover by 1.24% and 2.16%, respectively, and an increase in built-up area by 6.62% (Table 2.3). Such change will give to a further rise in flood intensity, and damage will be more serious. Comparing the change at the sub-basin scale provides more information regarding regions more prone to severe floods (Fig. 2.6). Mostly the areas near the river channel and high urban density are showing the maximum increase in runoff. Such regions are at severe risk of flood hazards, and control over land management practices in these areas is of high importance. Spatial analysis shows that the overall change in runoff due to LULC is because of variation in transition among different land cover types. Each landcover class shows some impact on hydrology of a watershed and any change in it will lead to various further changes. Thus landcover change practices must be considered while deciding watershed planning and management policies.

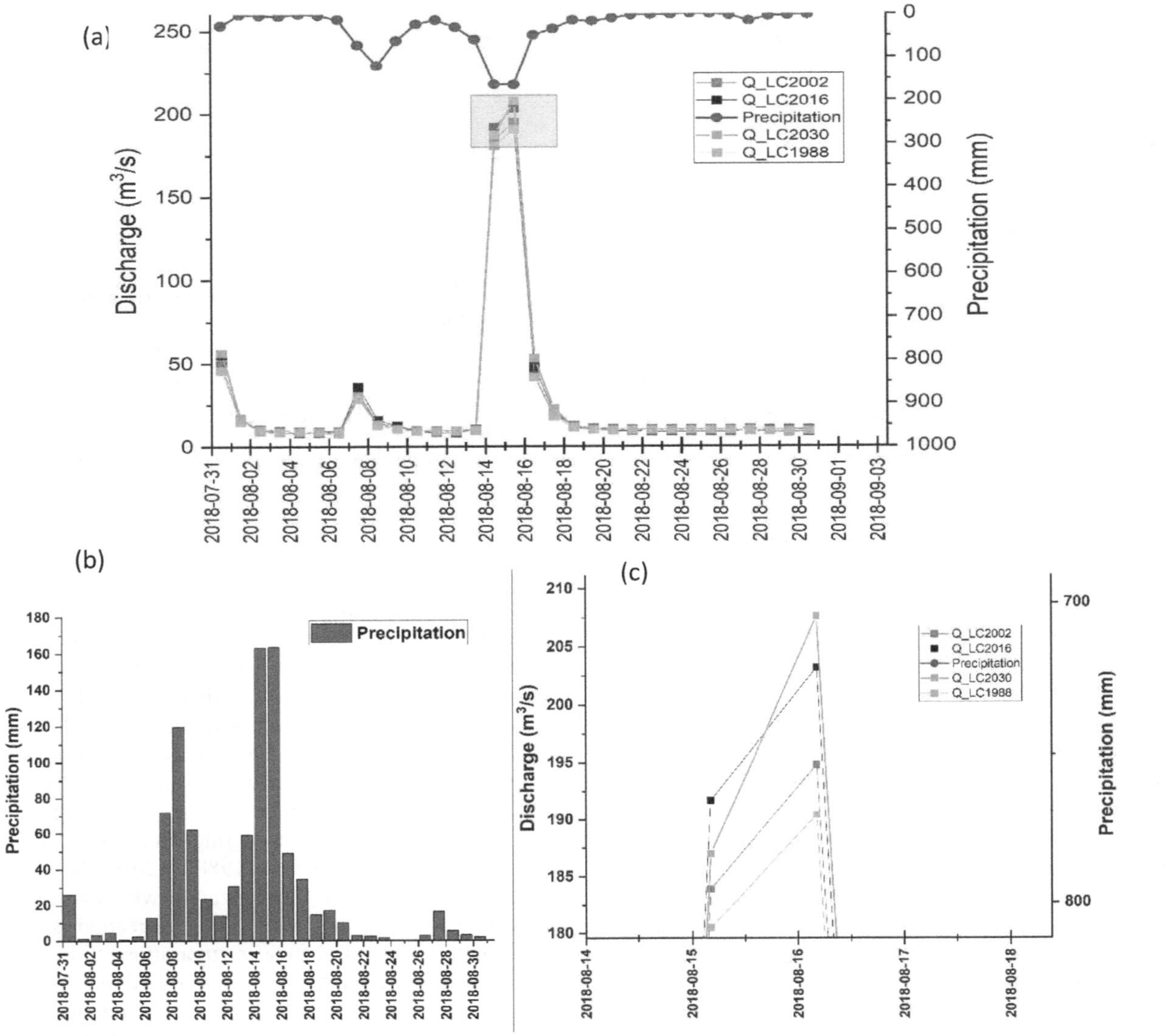

Fig. 2.5 LULC change impact on (a) Outflow (b) Precipitation and (c) Maximum runoff ; with LULC 1988, 2002, 2016 and 2030 during August 2018 in PRB

Table 2.4 Variation in Outflow during August 2018 in PRB by changing LULC from 2002 to 2030

Date	Precipitation (mm)	Outflow (cumecs)				% change in runoff		
		LULC 1988	LULC 2002	LULC 2016	LULC 2030	2002-1988	2016-1988	2030-1988
16-Aug-18	163.21	190.4	194.8	203.2	207.6	2.31%	6.72%	9.03%
17-Aug-18	48.94	41.59	42.58	47.85	52.97	2.38%	15.05%	27.36%
18-Aug-18	34.36	17.84	19.01	20.4	22.55	6.55%	14.34%	26.40%

6. Conclusions

In this study, the impact of LULC change on streamflow considering a case study of floods occurred in 2018 in Periyar river basin, Kerala is investigated using SWAT model. The model is well calibrated and results of goodness of fit are in acceptable range. The future LULC map is prepared using MLP- ANN and showed similar trend as past. To analyse the impact of LULC change on surface runoff, SWAT model was run for the rainfall of the month of August 2018 on daily sacle

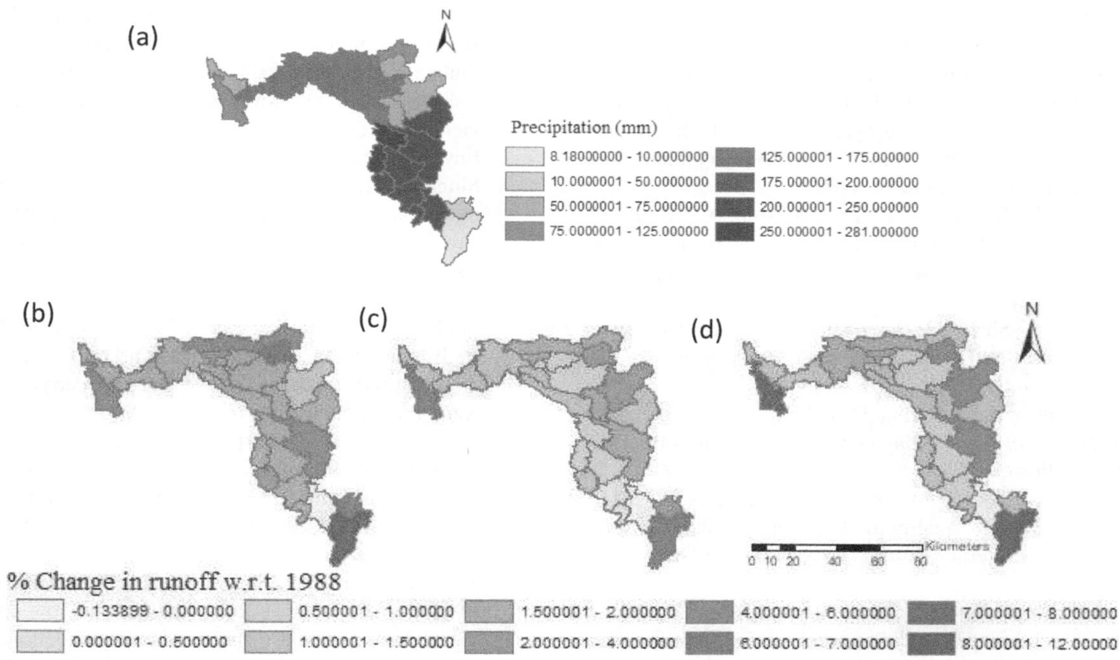

Fig. 2.6 Spatial variation of (a) Precipitation; %Change in Runoff (b) LULC 2002-1988 (c) LULC 2016-1988 and (d) LULC 2030-1988 for PRB on 16 August 2018

and by changing LULC for 1988, 2002, 2016 and 2030. The results indicate that surface runoff is higher in 2016 and 2002 as compared to a possible situation in 1988. Also, in case of 2030, the surface runoff is expected to increase even more during the peak flow due to variation in land cover classes. Built-up is identified the most influencing landcover classes that alter runoff. The combined change in plantation, forests and cropland also alter surface runoff. The study suggests if another extreme rainfall event similar to August 2018 occurs in future and LULC change follows a similar pattern then the damage caused due to landcover change component will be more during peak flow. Results also suggest that LULC change shows a significant effect on runoff and must be considered during watershed management and planning. Urban floods can be reduced using this information while deciding the flood mitigation strategies. It is recommended that LULC change shall be actively used during designing flood early warning systems and land planning and management.

7. Acknowledgements

The authors wish to acknowledge the sponsors of the project entitled "Impact of Climate Change on Water Resources in River basins from Tadri to Kanyakumari" by INCCC, Ministry of Water Resources, Government of India. We wish to express our sincere gratitude to Central Water Commission and Indian Meteorological Department, India, for providing us hydrological and meteorological data. The authors also acknowledge the National Bureau of Soil Survey and Land Use Planning, India, for giving soil data.

References

1. Abbaspour, K. C., Yang, J., Maximov, I., Siber, R., Bogner, K., Mieleitner, J., ... & Srinivasan, R. (2007). Modelling hydrology and water quality in the pre-alpine/alpine Thur watershed using SWAT. Journal of hydrology, 333(2-4), 413–430.
2. Anand, J., Gosain, A. K., & Khosa, R. (2018). Prediction of land use changes based on Land Change Modeler and attribution of changes in the water balance of Ganga basin to land use change using the SWAT model. *Science of the total environment*, *644*, 503–519.
3. Arnold, J. G., Srinivasan, R., Muttiah, R. S., & Williams, J. R. (1998). Large area hydrologic modeling and assessment part I: model development 1. *JAWRA Journal of the American Water Resources Association*, *34*(1), 73–89.
3. CWC (2018), Kerala Floods of August 2018, Central Water Commission, New Delhi, 2018.
4. Hunt, K. M., & Menon, A. (2020). The 2018 Kerala floods: a climate change perspective. *Climate Dynamics*, *54*(3), 2433–2446.
5. Goswami, B. N., Venugopal, V., Sengupta, D., Madhusoodanan, M. S., & Xavier, P. K. (2006). Increasing trend of extreme rain events over India in a warming environment. *Science*, *314*(5804), 1442–1445.

6. Ivancic, T. J., & Shaw, S. B. (2015). Examining why trends in very heavy precipitation should not be mistaken for trends in very high river discharge. *Climatic Change*, *133*(4), 681–693.
7. Legates, D. R., & McCabe, G. J. (1999). Evaluating the use of "goodness-of-fit" measures in hydrologic and hydroclimatic model validation. *Water Resources Research*, *35*(1), 233–241. https://doi.org/10.1029/1998WR900018
8. Marhaento, H., Booij, M. J., & Hoekstra, A. Y. (2018). Hydrological response to future land-use change and climate change in a tropical catchment. *Hydrological sciences journal*, *63*(9), 1368–1385.
9. Mishra, V., & Shah, H. L. (2018). Hydroclimatological perspective of the Kerala flood of 2018. *Journal of the Geological Society of India*, *92*(5), 645–650.
10. Mohanakrishnan, A., & Verma, C. V. J. (1997). *History of the Periyar Dam with century long performance*. Central Board of Irrigation & Power.
11. Moriasi, D. N., Gitau, M. W., Pai, N., & Daggupati, P. (2015). Hydrologic and water quality models: Performance measures and evaluation criteria. *Transactions of the ASABE*, *58*(6), 1763–1785.
12. Mukherjee, S., Aadhar, S., Stone, D., & Mishra, V. (2018). Increase in extreme precipitation events under anthropogenic warming in India. *Weather and climate extremes*, *20*, 45–53.
13. Petelet-Giraud, E., Cary, L., Cary, P., Bertrand, G., Giglio-Jacquemot, A., Hirata, R., ... & Aurouet, A. (2018). Multi-layered water resources, management, and uses under the impacts of global changes in a southern coastal metropolis: When will it be already too late? Crossed analysis in Recife, NE Brazil. Science of The Total Environment, 618, 645–657.
14. Rajeevan, M., Bhate, J., & Jaswal, A. K. (2008). Analysis of variability and trends of extreme rainfall events over India using 104 years of gridded daily rainfall data. *Geophysical research letters*, *35*(18).
15. Roxy, M. K., Ghosh, S., Pathak, A., Athulya, R., Mujumdar, M., Murtugudde, R., ... & Rajeevan, M. (2017). A threefold rise in widespread extreme rain events over central India. Nature communications, 8(1), 1–11.
16. Soil Conservation Service (SCS) (1972) National Engineering Handbook, Section 4: Hydrology. Department of Agriculture, Washington DC, 762 p
17. Shi, P., Ma, X., Hou, Y., Li, Q., Zhang, Z., Qu, S., ... & Fang, X. (2013). Effects of land-use and climate change on hydrological processes in the upstream of Huai River, China. *Water resources management*, *27*(5), 1263–1278.
18. Sinha, R. K., & Eldho, T. I. (2018). Effects of historical and projected land use/cover change on runoff and sediment yield in the Netravati river basin, Western Ghats, India. Environmental earth sciences, 77(3), 1–19.
19. Sinha, R. K., Eldho, T. I., & Subimal, G. (2020). Assessing the impacts of land use/land cover and climate change on streamflow of a humid tropical river basin in Western Ghats, India. International Journal of River Basin Management, 1–12.
20. Sudheer, K. P., Bhallamudi, S. M., Narasimhan, B., Thomas, J., Bindhu, V. M., Vema, V., & Kurian, C. (2019). Role of dams on the floods of August 2018 in Periyar River Basin, Kerala. *Current Science (00113891)*, *116*(5).
21. Teegavarapu, R. S. (2012). *Floods in a changing climate: extreme precipitation*. Cambridge University Press.
22. Wasko, C., Sharma, A., & Lettenmaier, D. P. (2019). Increases in temperature do not translate to increased flooding. *Nature communications*, *10*(1), 1–3.
23. Yan, B., Fang, N. F., Zhang, P. C., & Shi, Z. H. (2013). Impacts of land use change on watershed streamflow and sediment yield: an assessment using hydrologic modelling and partial least squares regression. *Journal of Hydrology*, *484*, 26–37.
24. Zhang X, Liu Y, Fang Y, Liu B, Xia D. Modeling and assessing hydrologic processes for historical and potential land-cover change in the Duoyingping watershed, southwest China. Phys Chem Earth A/B/C 2012;53–54:19–29
25. Zhang, X., Liu, Y., Fang, Y., Liu, B., & Xia, D. (2012). Modeling and assessing hydrologic processes for historical and potential land-cover change in the Duoyingping watershed, southwest China. *Physics and Chemistry of the Earth, Parts A/B/C*, *53*, 19–29.
26. Zope, P. E., Eldho, T. I., & Jothiprakash, V. (2017). Hydrological impacts of land use–land cover change and detention basins on urban flood hazard: a case study of Poisar River basin, Mumbai, India. *Natural Hazards*, *87*(3), 1267–1283.
27. http://www.cwc.gov.in/sites/default/files/statewisflooddatadamagestatistics.pdf
28. https://ndma.gov.in/Natural-Hazards/Floods

CHAPTER 3

Attitudinal Capacity and Flood Risk Management of the People of Kerala after the Flood Disaster

M. V. Bindu
NCSCM, MoeF & Climate Change
Anna University, Chennai-25
bindumv@yahoo.com

Abstract

In many disaster susceptible areas, preparedness is an important factor for disaster risk reduction. There are various management practices were adopting for disaster risk reduction or mitigation. It is the perception and preparation of reducing disaster risks through logical efforts to analyse and manage the causal factors of disasters, through reduced exposure, decreased vulnerability of people and property, wise management of land and the environment, and also through improved preparedness for adverse events. A combination of all the strengths and resources available within a community, society, or organization can reduce the level of risk or the effects of a disaster. The preventive approach is one among them. This is through creating awareness, education, capacity development, etc. It will automatically increase the capacity of the people to handle/ cope up with the disasters. This approach will support the vulnerable and elderly people. Because due to the disaster risks these people are more affecting through several behavioural changes, they could not able to adapt easily. Capacity is the capability, it may include physical, institutional, social, economic and attitudinal. Mostly the peoples do not cope with the plans and strategies that are not in streak with their culture, ideology, or religion and this can limit their capacity and increase their vulnerability to disasters. If there is a positive attitude towards the involvement of women in the capacity making, or awareness on these types of issues will motivate and get a mutual benefit for the whole of the community and can be observed as an attitudinal capacity of a community. The perception also changes this brings we feeling and attitude of collectivism as opposed to individualism in a society.

In this context, the goal of the study is to determine the levels of disaster awareness and attitude capacity of the vulnerable people's priorities of the flood-affected areas at Pathanamthitta and Alappuzha district of Kerala. The level of awareness and attitude capacity and challenges of the people of two districts of Kerala, after the flood have been studied. The findings showed differences in attitudes in both districts in terms of awareness, capacity among the vulnerable people. In case of the capacity and awareness level, peoples in the Alappuzha district exhibits a higher performance than that of Pathanamthitta district. But literacy and cultural performance were more advanced in the Pathanamthitta district than that of Alappuzha District. Most of the people from Alappuzha District are prone to the flood disaster yearly.

Keywords: Capacity, Attitude, Risk management and flood

1. Introduction

Disasters are unpredictable events that kill and affect people, demolish properties and disrupt the environment. Disaster is defined as a sudden ecological phenomenon with sufficient magnitude to require external assistance. Alternatively, disaster is defined as a sudden extraordinary event that brings great damage, loss, destruction and injury to people and their environment (Stanhope & Lancaster, 2008). During such events, the residents play a vital role in dealing with these situations. Attitude and awareness of the peoples is important and capacity and we feeling make them more adaptive to each disaster. A combination of all the strengths and resources available within a community, society, or organization can reduce the level of risk or the effects of a disaster. The preventive approach is one of them. This is through creating awareness, education, capacity development, etc. It will automatically increase the capacity of the people to handle/ cope up with the disasters. This approach will support vulnerable and elderly people. Capacity building is not limited to individuals only it extends to the capacity development of organisations and institutors entrusted with the responsibilities of various phases and aspects of the risk management cycle. This involves setting up institutional goals, objectives strategies and plans of action with strong mechanisms for monitoring and evaluations.

In kerala disaster is a sudden event and very complex in nature causing, mortality harm of people and properties or environment and causing morbidity to the local society. Kerala experienced various magnitudes of disasters mostly natural. i.e., flood, landslides, cyclones biological incidents etc. Floods and landslides are the most common of all natural disasters due to heavy rain and generally causing greater mortality. Floods are defined as the condition that occurs when water overflows the natural or artificial confines of a stream, river, or other body of water, or accumulates by drainage over low-lying areas (Hogan & Burstein, 2007). Floods account for 50% of disasters and deaths related to disasters. Landslides and droughts are also significant though their effects are limited to much smaller areas in the state. Preparedness is defined as an umbrella strategy that involves forecasting and taking precautionary measures prior to an imminent threat (Seyedin, et al., 2011). A common perception is that preparedness is only for response; however, preparing for recovery after a disaster or emergency is no less important (Singh & Subramaniam, 2009). Pathanamthitta and Alappuzha were one of Kerala's worst-hit districts in the flood disaster during 2018. Thousands of houses in the districts were flooded by the rising water levels from river Pamba, Achankovil and Manimala due to the continuous heavy rain and opening of dams. Ranni and Thiruvalla Thaluk in Pathanamthitta districts and Kuttanadu thaluk in Alappuzha districts were severe flood affected areas. The flood waters have receded in the eastern parts of the district, but the areas in the western areas of Pathanamthitta are in the grip of severe floods. But in Kuttanadu thaluk the land fully inundated with flooded water. Life of people in these areas has turned hell, with the water levels steadily rising in upper kuttanadu and the authorities shifted several people to the relief camps. Most of the roads and transportation comes to halt. Almost all the rivers in the districts have breached the banks.

In this concern, a research study was conducted to know the perception of preparedness for disaster reduction through attitudinal capacity and awareness level of the residents of the disaster-prone areas of Pathanamthitta and Alappuzha districts of Kerala.

2. Materials and Methods

Flood-affected Thaluks of Pathanamthitta and Alappuzha districts of Kerala were selected for the study (Fig. 3.1).

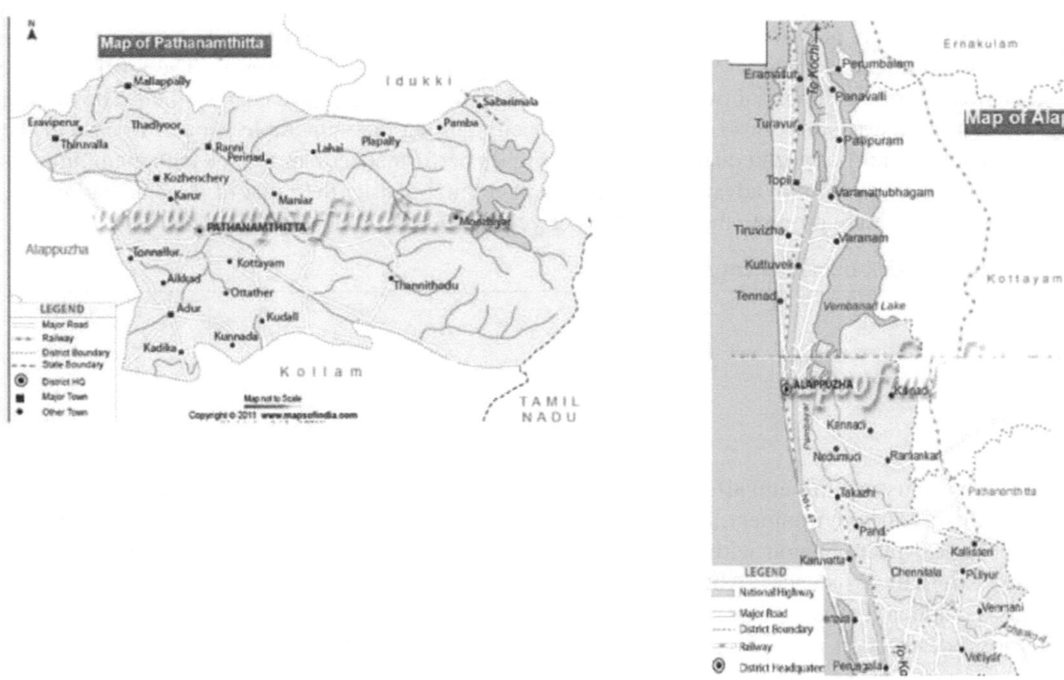

Fig. 3.1 Geographical area of the study area

The research study was formally conducted by collected the primary data with questionnaire survey, interview and group discussion from the residents of the flood affected thaluk and the village areas. The survey was performed of local households across different regions. local households within the regions were randomly selected for one-on-one, face-to-face questionnaire surveys. If some respondents in the vulnerable group declined to participate then moved on to the next person in the household. 95% of the respondents have participated in the survey. Secondary data collected from news and previous reports published by other agencies and the database profiles of the disaster management departments, thaluks and concerned villages/ wards of the selected Districts. The qualitative and quantitative data were analysed for further study.

3. Result and Discussion

After collecting the basic information and attitudinal capacity of respondents, the variables were analysed to explore the effects of the respondents' flood experience and flood attitude on their disaster preparedness behaviour. Effects of attitudes on disaster preparedness and the influence of residents' flood disaster attitude on their daily disaster preparedness behaviour are complex. The difference between attitude levels of Pathanamthitta and Alappuzha districts personnel is associated with the positive influence of education level, awareness and responsibilities. The level of awareness is an important factor in reducing disaster damages. Evaluation of age groups shows high results for kuttanadu personnel and student groups. This result is expected because older groups are supposed to be more sensitive and responsible to the problems in their area communities in regard to their experiences.

In Kuttanadu area, usually, the people will have an attachment to the place where they have lived for a long time; thus, residents will have a tendency to prepare when they encounter flood disasters. And also indicating that residents with low personal income and education are more likely to take on disaster preparation. The attitude behaviour of vulnerable people in Kuttanadu is more (Fig. 3.2), in addition, the smaller the number of family members with income in a household also showed a high capacity for taking flood preparedness measures and people themselves ready to move to a relief canter. On exploring this finding with interviewees, it turns out that they assessed themselves as living in a frequently flood-prone area, where, despite taking precautionary measures, flooding related losses were estimated unavoidable. Table 3.1.

Thus, even where residents anticipate that floods will occur, they remain uninclined to take disaster preparedness measures.

Table 3.1 Number of affected households

Districts	Total number of households affected	% Damages
1. Pathanamthitta		
a. Ranni Thaluk	1526	15-76
b. Thiruvalla Thaluk	12653	15-78
2. Alappuzha District Kuttanadu Thaluk		
a. Lower kuttanad	11321	15-80
b. Upper Kuttanadu	2611	15-85
Total	**28111**	

This result is consistent with the effect of respondents' previous experience on flood preparedness and capacity of villages has a significant positive effect on their daily disaster preparedness behaviour (Ning et al,.2013) claim That is, the higher their trust in the reliability of flood reporting and the capacity of villages, the more likely they are to adopt disaster preparedness measures.

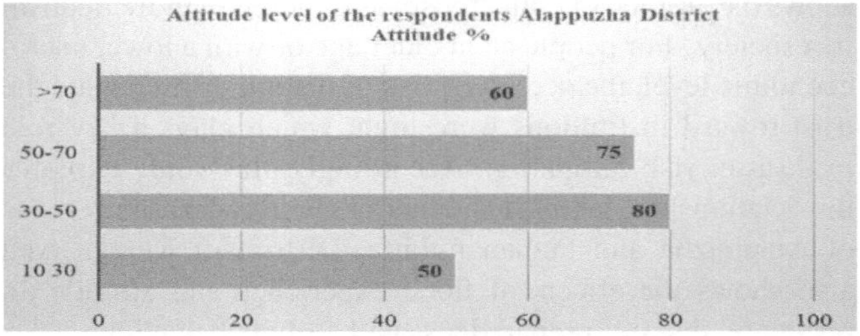

Fig. 3.2 Attitude level of respondents Alappuzha districts

In Pathanamthitta district even though the people are educated but the disaster preparedness and attitude behaviour of the residents remains at a relatively low level, and the vast majority of residents in Ranni thaluk of Pathanamthitta district has no disaster preparation (Fig. 3.3). They stayed in their home itself. The main reason is that the flood disaster that happened in that area was infrequent and sudden. The study is similar to the research of Danso & Addo, 2017. This finding fully shows that as long as the residents think that the flood remains controllable, they will actively take disaster preparedness measures to reduce flood losses. The distance between home and river in this area shows an insignificant effect, which is inconsistent with the conclusion of Baker (2011) and Xu et al. (2018). The residence has high trust in government institutions and media and has a high acceptance rate of early warning (Samaddare et al. 2012) and preventive measures. The attitude behaviour and capacity of the people of Thiruvalla thaluk were somewhat different because these areas are more or less adjacent to the upper Kuttanadu thaluk of Alappuzha district.

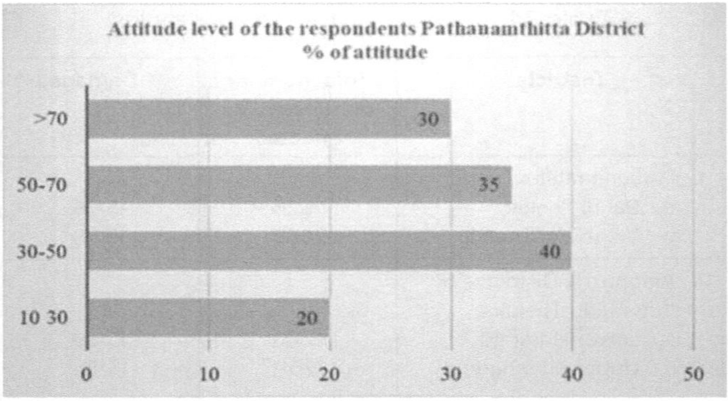

Fig. 3.3 Attitude level of the respondents Pathanamthitta districts

The vulnerable people especially 60% women and children were interested in these areas and shows willingness to move to the rescue centre. Ejeta et al. (2018) showed that emotion can affect disaster preparedness behavior; residents' preparation intention is lower, resulting in lower disaster preparedness measures. However, when the residents are united, more disaster prevention measures will be taken. This shows the attitude of collectivism as opposed to individualism in a society. For people at an older age or with a lower socio-economic level, the perceived risk of natural disasters and the trust toward institutions were high, which plays a key role in disaster risk mitigation. Accordingly, this study explores the relationship, taking residents of the flood-prone regions of Alappuzha and Pathanamthitta, District of Kerala. And also shows the effects of flood experience and attitude on residents' disaster preparedness and attitude behaviour.

4. Conclusion

The conclusion of this study exhibited that in both areas, residents' disaster preparedness knowledge, behaviour, attitude and capacity are different. In Pathanamthitta, the attitude and capacity of the people are it is unacceptably poor even though they are high income and educated. On the other hand, low education and income level, along with years of residence within the local area, have a significant positive effect on Alappuzha residents' preparedness behaviour. Disaster preparedness measures accepted by the family are simple, handy, storing valuable articles at height, increasing the floor level of ground-floor residences, and improving drainage measures. Another notable thing is that the flood experience and attitude of these residents have a significant effect on their disaster preparedness behaviour. However, residents who have experienced flood disasters resulting not just in property damage but also in human losses and have confidence in the ability of their village to proactively control floods are more likely to exhibit better flood preparedness. Based on the above research conclusions, the following recommendations are proposed for the formulation of disaster prevention and mitigation policies. Structural flood control measures in Alappuzha District, the importance of family peoples attitude development towards disaster preparedness should also be emphasized to local residents. And to enhance the capacity of flood mitigation, appropriate government departments should provide counseling and psychological guidance to comfort residents' fear and feelings of helplessness regarding recurring the disaster., improve local residents' trust in the media and the government's flood control capacity, and guide residents' disaster preparedness behaviours to improve residents' confidence level and attitude level with leadership capacity and thus improve we feeling of the residents to overcome these type of situations. The relevant departments of local governments should deliver flood disaster mitigation education to local residents, enhance the residents' objective cognition of flood knowledge, improve the willingness of rural residents to be active in flood control and disaster reduction, and then take effective disaster preparedness measures. These measures can be expected to improve living conditions, people's trust in the media, and the government's flood control capacity.

5. Acknowledgment

The author is accredited to the services of the officials of the district disaster management Authority, thaluks, villages, and local residents of the Pathanamthitta and Alappuzha districts for carrying out the study.

References

1. Ejeta, L. T., Ardalan, A., Paton, D., & Yaseri, M. (2018). Emotional and cognitive factors influencing flood preparedness in Dire Dawa town, Ethiopia. *Natural Hazards*, 93(2), 715–737.
2. Samaddar, S., Misra, B. A., & Tatano, H. (2012). Flood risk awareness and preparedness: The role of trust in information sources. Paper presented at the IEEE International Conference on Systems.
3. Baker, E. J. (2011). Household preparedness for the Aftermath of Hurricanes in Florida. *Applied Geography*, 31(1), 0–5.
4. Xu, D., Peng, L., Liu, S., & Wang, X. (2018). Influences of risk perception and sense of place on landslide disaster preparedness in southwestern China. *International Journal of Disaster Risk Science*, 9(2), 167–1.
5. Danso, S. Y., & Addo, I. Y. (2017). Coping strategies of households affected by flooding: A case study of Sekondi-Takoradi Metropolis in Ghana. *Urban Water Journal*, 14(5), 539–545.
6. Ning Yan, Tao Maoxuan, Hu Junfeng, Li Yubo, Cheng Yulan, Zhang Gang, walnut, Li Li, Chai Yan, Nan Haitao, Zhao Hongqi, Wang Junjie. (2013). Emergency preparedness status

and influencing factors of urban and rural residents in four districts (counties) of Shaanxi Province. *Chinese Journal of Preventive Medicine*, 4, 347–3.

7. Singh& Subramaniam (2009). Health emergency and disaster preparedness in Malaysia. *Southeast Asian J trop med public health*, 40 (1).
8. Seyedin H, Ryan J, Sedghi S. (2011). Lessons learnt from the past and preparedness for the future: How a developing country copes with major incidents. *Emerg Med J.* 28, 887–91
9. Hogan, D. E., & Burstein, J. L. (2007). Disaster medicine (Second Edition ed.). Philadelphia, PA: Lippincott Williams & Wilkins.
10. Stanhope M., and Lancaster J. (2008). Publish health nursing, population cantered health care in the community. 7th ed., Canada: Mosby Elsevier, pp. 545–455.

CHAPTER 4

Kosi Floods in Bihar: A Study from Anthropological Perspective

Rahul Kumar Yaduka
School of Development Studies,
Ambedkar University, Delhi
rahulk.18@stu.aud.ac.in
88263-24382

Abstract

The response of the state's disaster management apparatus to the floods in north Bihar is largely limited to strengthening of embankments and temporary rehabilitation of people, distribution of food material, construction material etc. (Mishra, 2008a). This paper attempts to study Kosi Floods from an anthropological perspective that provides a nuanced way for the policy makers to understand Kosi Floods as a humanly accentuated, socio-culturally negotiated and techno-politically attempted problem. Extant literature suggests that momentary event of the disaster is just the tip of the iceberg whereas the major extent of the problem lies with the way our social, economic, and political life is organized (Barrios, 2017). The distribution of impact of the disaster in the society is also dictated by its socio-cultural matrix further underlining the human element in the construction of disaster (Pritchard & Thielemans, 2014). This paper discusses the implications of human interventions and limitations of modern scientific and technological solutionism. It will also discuss reflections on democratization of knowledge in developing a better flood policy. Using snowball sampling, in-depth interviews were conducted with a diverse range of people including relevant officials and scholar-activists. Focused group discussions were also conducted with the flood-affected victims of 4 locations comprising both rural and urban localities in 2 districts of Bihar i.e., Supaul and Madhepura during December month of 2020 during initial field visits for doctoral research.

Keywords: Floods, Kosi, Policy, Disaster

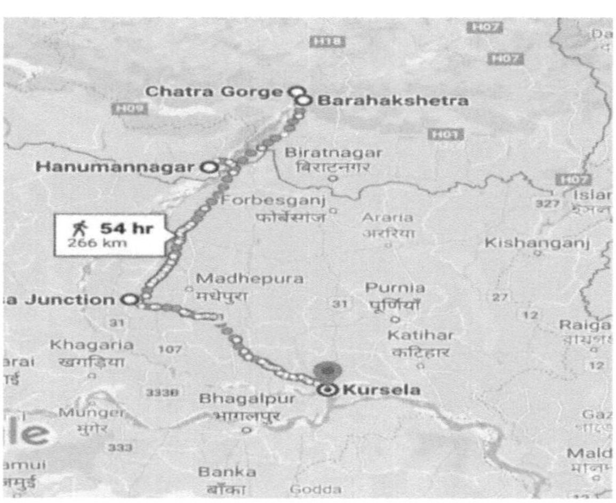

Fig. 4.1 Map showing the course of the river Kosi and major locations on its way

Source: Google Maps

1. Physical Characteristics of the River Kosi

As can be followed from the google image added above, the river Kosi originates at a height of around 7000m in Himalayas and drains Tibet, Nepal and India. The river emerges from the mountains at Barahkshetra in Nepal near Chatra gorge and enters India near Hanuman Nagar-Bhimnagar border in Supaul district of Bihar. It travels in south-west direction till Saharsa district and then takes a south-east turn and flows to join Ganga as a left-hand tributary in Katihar near Kursela. The drainage area of Kosi is around 95156 sq. km out of which only 20% is in India. The river is ill-known for the silt load that it carries.

The peak discharge volume (913000 cusec) of the river Kosi can be manyfold higher than the average discharge volume (55000 cusec) of the river (Kumar, 2015). From this deviation,

Table 4.1 Table showing historical minimum, maximum and average discharge values in some river of Bihar

DISCHARGE OF IMPORTANT RIVERS OF BIHAR, PATNA **DATE –16.06.2021, TIME:-8:00 AM**

S. No.	Name of River	Time	Today Discharge (Cusec)	Trend	Max^m Discharge (Cusec) during Flood 2021 Time, Date	Maxm. Discharge (Cusec) during Flood 2020 Time, Date	Max^m Discharge (Cusec) /Year
1	Kosi River (a) Barah	6.00AM	63775	R	71500 8.00 AM 16.06.2021	287800 12.00 MIDNIGHT 20.07.2020	913000 05.10.1968
		8.00AM	71500	R			
		10.00AM					
		12.NOON					
		2.00PM					
	Kosi River Birpur Barrage, D/S Discharge [Design Discharge-9.50 Lacs Cusec]	6.00AM	NA	-	52960 8.00 AM 15.06.2021	342970 8.00 AM 21.07.2020	788200 05.10.1968
		8.00AM	52100	R			
		10.00AM					
		12.NOON					
		2.00PM					
2	Gandak River Valmikinager Barrage, D/S Discharge [Design Discharge-8.50 Lacs Cusec]	6.00AM	350000	S	372800 8.00 AM 16.06.2021	436500 10.00 AM 21.07.2020	639750 31.07.2003
		8.00AM	372800	R			
		10.00AM					
		12.NOON					
		2.00PM					
3	Sone River Indrapuri Barrage, D/S Discharge [Design Discharge-14.58 Lacs Cusec]	6.00AM	16604	F	16604 6.00AM 16.06.2021	226406 4.00 AM 21.08.2020	1448000 23.08.1975
		8.00AM	16604	S			
		10.00AM					
		12.NOON					
		2.00PM					
4	Falgu River Udera Asthan Barrage, D/S Discharge [Design Discharge---1.59 Lacs Cusec]	6.00AM	NIL	-	NA	26219 8.00AM 27.08.2020	95092 5 AM 06.08.2017
		8.00AM	NIL	-			
		10.00AM					
		12.NOON					
		2.00PM					
5	North Koel River, Mohammad Ganj Barrage, Design (Discharge 8.47 Lacs cusec)	6.00 AM	13802	R	13802 6.00 AM, 16.06.2021	88280 12.00 MIDNIGHT 26.08.2020	792000 12 NOON 12.08.2016
		8.00 AM	13802	S			

Source: Water Resource Department, Government of Bihar

one can get a sense of devastation brought about by the river in the flooding season. In the past 200 years or so, it has shifted more than 160 km westwards (Mishra, 2008a). The figure below demonstrates how Kosi has spanned a very substantial part of Bihar and this sets Kosi apart from any other river. Also, it should be note that Kosi is a transboundary river. It will be useful when we try and understand the difficulty in developing a policy to manage Kosi and even today how it affects the institutional arrangement put in place to administer the policy that exist between India and Nepal.

2. Historicization of the Kosi Plan 1953

Two points in terms of approach of the research need to be said at the beginning. The author understands that various dimensions like political economy, ecological setting, colonial administration etc. cannot be studied in isolation. The theoretical approach is to transcend the scholarly binary of constructivism and realism and make an attempt to study the Kosi flood management policy by foregrounding the topic rather than method at the center (Forsyth, 2001). Secondly, it is important here to clarify that region of north Bihar is also called as Kosi region because of the prominent role that the river Kosi has been playing in this region since ages.

As we have established in the initial section of the paper itself, flooding is a "natural" phenomenon. It is embedded in the river's properties. We have also explained the difficult terrain of the river Kosi. Here, it needs to be clearly stated that people did cope with the floods using local and traditionally learnt mechanisms. Embankment were one of them and not the only way. The classification of floods was much more elaborate and the corresponding response pattern was also more contextual and nuanced. There was no blanket definition of "floods" and response to it (Mishra, 2001). What we see from folklores also substantiates the fact that people did have an understanding of unconventional nature of the river Kosi (Mishra, 2008). The river has been called both as "mother" and as a "witch".

Fig. 4.2 Image showing different channels through which the river Kosi has flown at different points of time starting from Purnea in 1731 to Nirmali in 1963

Source: Burele et al., 2015

As has been argued elsewhere that knowledge and social order are co-produced, based on this understanding of the river, what one had at the maximum were a few small-scale local interventions that local people made in the river course and that too only during the peak flood season. Dinesh Kumar Mishra bring examples of small technological interventions by people to manage the months in which river overflowed. Small embankments were built to stop only minor floods. The topography of the villages, location of tanks, construction of houses everything was in accordance with the needs of river to find its path (Mishra, 2001). The idea of taming the river was beyond comprehension for a people who prayed the rivers as mother. Initially, state was not actively participating in the construction and management of embankment, neither was there a vision or policy. The argument made by Mishra and Hill is also corroborated by Singh in his work (Singh, 2008). The scale of interventions made by the local Zamindars was always limited as compared to later intervention with a systematic policy.

2.1 Ecological Setting of the Region and Colonial Intervention on the River Kosi

As the colonial rule started expanding geographically and in terms of its entrenchment, some large scale and systematic changes were brought about in the ecology of this region. These changes have much to do with the logic of colonial political economy. Nationalist historian Bipan Chandra explains how through various stages, colonialism has morphed its form but the essential content of economic exploitation prevailed (Chandra, 1980). Hetukar Jha explains how the processes of deforestation for the export of timber, clearing of land to expand area under agriculture and migration of Santhals from relatively drier areas led to massive changes in the forested region of Kosi. The famines in different parts of the province and the country led the colonial state to expand the area under cultivation. This large-scale human intervention in the ecology of Kosi region was one of the reasons of disturbing the fine balance between forests, rivers, land building, soil erosion and floods. These escalation in human processes led to increase in the intensity and unpredictability of floods (H. Jha, 2017). Praveen Singh also underlines the significance of ecological setting of the region that plays a critical role in shaping the decisions made by the colonial state in dealing with the rivers of the region (Singh, 2008). Christopher Hill in his work *"River of Sorrow: Environment and Social Control in Riparian North India, 1770–1994"* also puts into perspective the role of the larger ecology of the region in determining and in fact limiting the governability of the region (MacKenzie, 1999).

Rohan D'Souza focuses more on the political economic drivers of the steps taken by the colonial state and hence argues for the case of a special kind of hydrological approach emerging in the times of colonial state called as colonial hydrology (D'Souza, 2006). If we recall what we discussed about geography of the river Kosi, we see that to define a land title will be difficult if the river changes the channels so much. Delineation of the river and the adjunct land can go only hand in hand.

Permanent settlement introduced by Cornwallis necessitated delineation of land and river course and hence embankments came prominently in the scene. It was at this historical moment that colonial state through its officers like Lord Cornwallis envisaged to impose embankments on the river Kosi on a large scale. Here, the Eurocentric mind of Cornwallis and his experiences with relatively moderate rivers in Europe and elsewhere shaped his imagination of taming the rivers. Also, for a euro-centric mind like that of Cornwallis, the complex and organically-developed land tenure system of Kosi region that was in tune with changing course of rivers like Kosi and which could factor-in the disastrous floods in particular years was not comprehensible as it was flexible, not written and was based on discretion of society (Scott, 1998). Since the Permanent Settlement Act brought in the construction of embankment as a corollary, the onus of constructing and managing it came on the newly-created landowners called Zamindars. Hence, they enthusiastically engaged into securing their lands from floods by embanking rivers in the nearby channels. (Singh, 2008).

Embankments increase the waterlogging on the outer side and affect the agricultural productivity poorly and standing waters also give rise to diseases like Malaria. Also, constraining the river on one side by the construction of the

Fig. 4.3 Image waterlogging due to poor drainage on the countryside

Source: Photographed by the author during field study near Bhaptiyahi in Supaul district on the eastern embankment (2020))

embankment is bound to increase its flow and hence flooding is other "unprotected" areas. Gradually, it was witnessed that large Zamindars started "competitive construction of embankments" and also indulged into disputes among themselves. These disputes came to the notice of the colonial state at the level of provincial government which gradually brought in the idea of "scientific approach" to construction of embankments (Singh, 2008). Apart from this motivation to extract land revenue, temptation to develop roads to curb nationalist activities after 1857, development of water transportation etc., were also factors that created support for taming the river Kosi (Mishra 2008).

Colonial state did not act as a single coherent, monolithic entity. There were differences in the way different actors thought this goal can be pursued. Zamindars, state revenue officials and colonial water engineers engaged into conflicts and negotiations because of goals that were in tension with one another. Zamindars and revenue officials were largely concerned with revenue collection. Their effort of embanking river was in line with their primary goal whereas irrigation engineers were concerned with mindless and rampant construction of embankments at local levels by Zamindars in total disregard for larger implications of their local actions. Yet another important observation and inference the author draws from the above discussion is that the usual juxtaposition between colonial state's scientific approach and native people's recourse to traditional water management system is not tenable. The colonial state did negotiate with the compulsions of the local ecology as and when required. In fact, counter-intuitively, native princely state rulers engaged in rampant and mindless construction of large-scale embankments disregarding their traditional knowledge system. With time, the experience with embankment and its pitfalls slowly changed the outlook of at least the colonial state. With the end of 19th century, a broader understanding developed that river should be left unto themselves. The colonial attempt at settling the land and revenue system under permanent settlement system also suffered a setback. (Mishra, 2008)

Table 4.2 Tabulation of maximum discharge in the river Kosi in different years immediately after independence of the country in 1947

Year	Maximum Discharge (cusecs)
1947	3.12,350
1948	4,70.442
1949	3,95,640
1950	3,40,661
1951	2,56,947
1952	3,06,443
1953	2,30,000

Source: Ray, 1953

This table borrowed from the work of Kumud Bhushan Ray before the Kosi Project kicked off empirically demonstrates how the prediction of colonial engineers actually came to be true and the capacity of the river Kosi to carry discharge gradually reduced leading to more devastating floods (Ray, 1953). The Calcutta Flood Conference, 1897, Orrisa Flood Committee, 1927 and Patna Flood Conference, 1937 all broadly came out with the conclusion that the actual problem is the drainage congestion and not the flood. This tilted the balance against the embankments.

2.2 The Revival of Embankments and the Post-colonial Indian State

The post-colonial Indian state put into the dustbin the 3 decades of debate that created a consensus against the embankment. Here, Mishra in his work *Trapped! Between the Devil and Deep Waters* argues that the main tool to gain legitimacy for post-colonial state was welfarism. It wanted to appear to do something to help people from floods. Embankment were designed to serve this purpose of state. Baghel argues that post-colonial Indian state had the job of fostering nationalist feelings in an otherwise divided society and large-scale development projects could be one way out (Baghel, 2014).

However, this loss of faith in embankment was revived when success stories of TVA in the US reached different parts of the globe. There was also an election in 1937 where after initial flipflops, Indian Nation Congress under the leadership of Srikrishna Sinha formed the provincial government. In 1936, Orrisa and Bihar were also separated. This can be deemed to be the point of inflection on the curve of Kosi Flood Policy. After this, what is seen is a lot of activity like preparation of reports, studies in laboratories, visits of technical experts to China, conference of victims of Kosi embankments etc. The point of discussion undergoes a change from 'whether embankments' to 'how to best construct embankments and rehabilitate people'(Sebastin, 2009). With occasional flirting with ideas like Belka Reservoirs and Barahkshetra Dam, finally it was concluded that keeping in mind constraints of resources and need of immediate relief, at least some temporary relief must be given to the people. This made the embankments the final choice. Global agencies led by the US helped in globalizing and normalizing the discourse on river control (Sneddon, 2015). This was an important part of the cold war geopolitics and the key role was played by US Bureau of Reclamation. Kosi Project in particular and Mega hydraulic structures in general became the temples of modernization and nation-building in the post-colonial India. This structural reliance on embankments mixed with poor institutional administration. Since then, destruction on all accounts be it in terms of area under floods, area under siltation and water logging, deaths etc. has only increased but still the Kosi flood management approach continues unabated without any significant rethinking (Mishra, 2008a).

3. Critical Appraisal of State of Research on Kosi Floods

We have seen above that before Kosi Plan 1953 gained its form, the Kosi issue was there in public discourse for at least a hundred years. There exists comments of colonial bureaucrats and engineers accepting futility of embankments as a robust flood control mechanism (Mishra, 2008b). Works by (Ray, 1953) and (Sinha, 2008) focus on the physical properties of the river Kosi to demonstrate the difficult nature of this river, the high amount of silt load that it carries and almost impossibility to straightjacket it within the walls of embankments. Bureaucrat-turned-academic researcher Ramaswamy Iyer also argues against the viability of structural solutions to deal with river Kosi (Iyer, 2008). Most extensive field study spanning decades in the Kosi region has been done by engineer-turned-public scientist Dinesh Kumar Mishra under the aegis of '*Barh Mukti Abhiyaan*'. Using archival resources from colonial times, legislative assembly debates of Bihar coupled with ethnographic studies, surveys, interviews with both laypersons and involved engineers, he has presented the most elaborate history of evolution of Kosi flood control policy (Mishra, 2008b)(Mishra, 1997)(Mishra, 2001)(Mishra, 2008a). Mishra echoes the voice of his predecessors that embankments have actually exacerbated the flood intensity and made it more unpredictable. Ajaya Dixit underlines similar problems with the emphasis on structural

Table 4.3 Table showing relation between different actors involved in the formulation and implementation of Kosi flood management policy

Actor Matrx Linkage map of the Kosi Project

Actors	GoN	GoI/CWPC/CWC	Flood control/disaste mgt	WRD GoB	Irrigation Dept. GoB	Kosi High Level Committee	Kosi Project Authority	Research institutions	Kosi Pint Vikas Pradhikar	Bureaucrats	Engi.Pro-emba	Engi. Anti-emba	Contractors	Politicians	Barh Mukti Abhiyan/BMA	Kosi Mukti sangharsh Samiti	People entrapped	people outside
GoN		M	W	W	W	M	M			W	M		M	M			S	S
GoI/CWPC/CWC	M		S			M	S	M		S	S	W		S			W	W
FlodControl/GoB	W	S		S	S	W	S	W	W	S	S	W	M	S		W	W	W
WRD GoB	W	W	S		W	W	M	M	W	S	S		S	S			W	W
I. Dept GoB	W		S	W		W	M	M	W	S	S	W	S	S			W	W
KHLC	M	M	W	W	W		M	M		S	S		M	M				
Kosi Pro. Autho	W	M	M	W	W	M		M	W	S	S		S	S			W	W
Research Institution		M	M	M	M	S	S			M	S			M				
KMSS			W	W	W		W							W			W	W
Bureaucrats	W	S	S	S	S	S	S	M			S		S	S		W	W	W
Eng.Pro-Embank	M	S	S	S	S	S	S	S		S		W	S	S	W		W	W
Engi. Anti-emba														W	S	M	M	M
Contractors	M		M	S	S	M	S			S	S			S				
Politicians	M	S	S	M	M	W	M		M	S	S		S				M	M
BMA								W	W	W	W	S				S	S	S
KMSS			W	W	W					W		M			S		S	S
People entrapped							W		W	W	W	M	W	M	S	S		M
People outside							W		W	W	W	M	M	M	M	M	M	

Legends	
(light shading)	The link which got initiated with the projec
(dark shading)	Link already existed, became crucial
(medium shading)	Link existed independent of the project

Strength of linkages	
W	Weak
M	Medium
S	Strong

Source: Sebastin, 2009

measures to deal with Kosi floods (Dixit, 2009). Kusha breach of 2008 attracted the researchers to further evaluate the efficacy of both structural measures as well as critically analyse the administrative mechanism to enforce Kosi treaty signed between India and Nepal (Shrestha et al., 2010).

The above actor linkage matrix has been produced that shows the various actors involved in the policy process of Kosi river flood and how they relate and operate with each other. As can be deciphered from the matrix, the strong linkages exist within the arms of government and technical experts and people's movements have strong relations within themselves. All cells that represent an interface between the state officials and public are weak (Sebastin, 2009). This partially explains why there has been no substantive change in imagining the Kosi flood situation over the years. Pankaj Kumar Jha also falls within the existing school that decries structural measures of flood control and argues for a more organically developed solution of 'living with floods' (P. K. Jha, 2017). Ravi Baghel pushes the envelope of research considerably by focusing on why despite criticism mounting against the mega hydraulic interventions in the rivers, they still continue unabated. He analyses the issue from the perspective of governmentality and knowledge/power connection. He traces the genealogy of hydraulic mission as it evolved over centuries and concludes that the discourse set by expert community is overpowering the local and traditional knowledge and experts act as "reasons of state" (Baghel, 2014).

4. Anthropological Study of Disasters

Gaining prominence since 1970s, anthropological approach to study disasters rejects the "naturalness" of the disaster and does not look at the disaster as a momentary crisis only. It visualizes disasters as a creation by conjunction of human population and a potentially destructive agent mediated by the vulnerability of the population which is socially and economically produced. Rather than being a revelatory event, it is a processual phenomenon and not a temporal and spatially demarcated event. In its constitution, disasters reveal complex interaction of biological, physical and socio-cultural systems. It also understands that hazards are systemic features of some natural environments whereas vulnerability is socio-culturally constructed. Disasters both manifest and test the process of adaptation of a society with the natural environment. Institutionalization of maladaptive responses may increase a society's vulnerability to the hazards. Disasters also provide a moment of development and political power consolidation. Role of enunciatory communities comes into picture who contest the idea of disaster itself (Barrios, 2017).

4.1 Locating Disaster in the Body of the Embankment

It is important here to discuss briefly how the design of embankment itself contains the seeds of its own destruction. As has been discussed and concluded by various researchers in this area (Dixit, 2009; Iyer, 2008; Mishra, 2008a; Shrestha et al., 2010; Sinha, 2008) the embankment basically tries to guide the path of the river to disallow it to flood the neighboring areas. The assumption is that the straightjacketing will increase the velocity of the river which will increase its potential to do the undercutting. This will lead to increase in depth of river bed leading to discharge capacity of the river and hence areas nearby will not flood. But the evidence has proven the contrary. Embankments were breached on major scales 8 times since its construction. The reasons were not unpredictable though. The embankments to start with promise only temporary and limited flood protection. The plan document itself declared that embankments are only a temporary solutions and real solution lies in construction of high dam at Barahkshetra in Nepal. Even that dam was going to help at best for 100 years. Since embankment don't alter any factor that leads to flooding in Kosi, as explained above, it won't provide any significant and long-term relief to the people of Kosi. Ray also suggests alternatives to embankment like creation of parallel river banks to stabilise the channel using permeable screens. This is much cheaper than the project being developed to tame Kosi. Yet another alternative discussed was to build guide banks. Still, we see that the government moved ahead with the idea of embanking rivers shunning a century old debate.

4.2 The Political Thickness of Embankments

When the post-colonial state embarked upon the construction of embankments, the major task before it was to demolish the general consensus against them due to decades of intense debate and empirical verifications. At this point, what became completely clear was the politics of scientific knowledge and technological intervention. What was sold to public was that the decision and design of embankments was based on 'expert advice' and calculations of Poona Laboratory. Engineers of repute like K. L. Rao and Kanwar Sain were sent to China to study the performance of embankments on the yellow river. On return, they reported to the government completely against the facts of what was happening in China. Similarly, people supposed to be trapped within the embankments were told wrongly by local leaders like Lalit Narayan Mishra that Poona Laboratory has studied that water won't rise inside the embankments. Later researchers in the interviews have found that all these reports were prepared under extreme political pressure (Mishra, 2008b). Bruno Latour has famously called laboratories as 'theatres of proof' (Shah et al., 2019). To this extent, the entire process of embankment construction was highly anti-democratic. What is interesting, however, is the part that was decided politically and popularly. The alignment of embankments became a question of life and death for the people who were going to be trapped within the embankments. They were decided based on bargaining power of the villagers, electoral concerns of the local politician, empathy of senior engineers, pity of chief minister, muscle power of collectivities etc. Here, societal context played a significant role in guiding science to work. Hence, as the domain of Science and Technology Studies argues, science is socially constructed and it is not an impartial, neutral observation of facts out there. It is constituted during the process of observation by socially situated human agents. It renders special kind of relationship between human and human and human and nature (Fischer, 2000).

4.3 Problematization of "Naturalness" of the Disasters

Since phenomenon like earthquakes, floods, hurricanes, tsunami etc. originate in the nature, these have been called in common parlance as 'natural disaster'. The terminology is deceiving as such because human and natural have been counterpoised to each other. If something is natural, it cannot be anthropogenic. But the problems start when not only the origin but the entire process is termed as natural. A series of events are triggered after the onset of disaster in which society deals with its impacts. It exposes the vulnerable, tests the resilience, causes displacement, encourages adaptation. All these are human activities. So, disasters study from the analytical framework of "political of disasters" give us deeper insight about how the cause and impact of so-called natural disasters are not so natural. There is a huge anthropogenic component. The disasters reveal decades or probably centuries of decisions taken in domain of society, economy and polity.

If we locate this understanding given by this analytical frame in the context of Kosi floods and try to re-historicize the evolution of policy making around it, we see that what was historically seen as anger of goddess Kosi is now related with the design flaw of the embankments that attempt to tame the river. An attempt is being made in this paper to trace the series of human decisions at various levels and in various domains at various historical moments that come together to convert river Kosi into "sorrow of Bihar".

The moments of disasters though reveal the contradictions within social, economic and political functioning of a society, surprisingly it also becomes a moment to further entrench those institutions and perpetuate the development patterns that created the disaster in the first place (Barrios, 2017). The governments normally look at the disaster in terms of simple reconstruction to restore normalcy but other actors like NGOs, trade unions, farmers' organization etc. contest this developmentality of state and demand long term transformation of things as they are (Oliver-smith, 2009). Since the state wields the ultimate authority, its version ordinarily prevails. At the center of this power contestation between local elite, dominant social groups, central authority, media, experts and NGOs, are people who have been rendered victim, passive, voiceless recipient of the natural calamity.

As seen in Kusha breach in 2008, in the initial aftermath of the disaster, government officials were not reachable. People testify that they had been informing officials about the impending disaster for some days but no empathy was shown. When it does take cognizance, it responds by sterile rhetoric and knee-jerk response. Over a period of time, government of Bihar has created a new conceptual category to shift the blame of breaches on. This is called 'anti-social elements. These are actually those people who deliberately make the cuts in the embankments to allow it to breach. They are the inhabitants within the walls of embankments. Now, if we recall the issues with the design of embankments, it creates instantaneous and definite contradiction between the interests of those on opposite sides. So, when the water flows above its normal level, people, to save their life, cut the embankment. But it becomes disastrous for those on the other side. Rather than reflecting upon this inbuilt issue with the structure, government conflate the interest of one particular section with that of national interest and simply declare the other as acting against the interest of larger society and nation. On the level of technical experts, the resolution to construct more and strengthen the existing embankments is taken. Past government is blamed for faulty choices in public but of ministry of water resources show that all governments have increased the length of embankments (Mishra, 2008a).

The process of relief, resettlement and rehabilitation is of interest to the anthropological approach to study disasters. The post-disaster relief and rehabilitation devalues the social life of the victims possibly because of its singular concern for the biological life. Interviewees confess to the researchers that the land provided to the displaced people makes complete rupture in their way of life. Often, the agricultural land that is source of livelihood remains trapped within the embankments and housing plots are allocated outside. The common grazing land is gone and hence animal husbandry suffers. Change of flow of river also deeply affects the fishing industry.

Oliver Smith focusses on political economy of the disasters. He argues that vulnerability is socially constructed and perceived (Oliver-smith, 2009). The same flood waters of Kosi don't impact everyone equally. The nature interacts with the society through its social, economic, political and cultural structures. In his work on forced settlements also, Oliver Smith explains that migration as an adaptation measure is possible for relatively resourceful sections. The lower sections are condemned to suffer more materially as well as psychologically. Pritchard and Thielemans (2014) present an analysis of how most vulnerable sections of the population in Kosi region gets disproportionately marginalized during post-disaster reconstruction. As we have discussed above, the locally powerful caste groups could get the alignment of embankment changed and hence remain outside it. This bargaining power was not available to lower castes. Their houses were often kuccha or semi-kuccha that made it more vulnerable to flood waters. During post-disaster relief and rehabilitation also, their negotiation power with official machinery was also less. So, the politically powerful groups could divert resources towards them (Pritchard & Thielemans, 2014).

5. Epistemic Foundations of the Mega Hydraulic Interventions

We have studied above how the science, an allegedly apolitical enterprise, was put to use by modern state to give verdict in favor of irrational alignment of embankments. The idea of embankment may be familiar to indigenous population but the idea of taming was rather foreign. Myths about Kosi reveal that the river has been visualized as a youthful energy that cannot be 'tamed'. It can only be allowed to take its course. Myths also have their own explanatory framework. People, hence, learned to design their way of life in line with the demands of Kosi. But the modernist science didn't accept this agency of nature. It looks at feminine nature waiting to be intervened and appropriated by masculine humans using technology. Modern science created this dichotomy between human and nature and looked at nature as something given or 'out there'. This gave confidence to the human beings and allowed them to conceptualize such terms like 'taming'. Moreover, this approach to look at nature was also related to the idea of 'utility'. The colonial administrators looked at water flowing into sea as 'waste' (Mehta, 2003). This approach verifies the instrumental role of anything other than human beings. The idea to create dams, embankments, roads, railway tracks ignoring the course of river somewhere deep down came from this arrogance of acting upon a passive nature by human agency.

Amita Baviskar in her work '*In the belly of the river*' explains how the approach of people of Anjanvara towards nature was articulated in the Gayana. It naturalizes everything and, in this process, gets naturalized. This epistemological difference in approach towards nature subsequently determines how nature is operated upon (Baviskar, 1995). In the traditional knowledge systems, human being has been seen as just one component of the ecosystem but in the modern scientific knowledge system, human being has been bestowed with agency and all other components are for the sake of human use. Overtaken by this ontological approach, colonial state determined the role of river in terms of its use value for the human beings. River water flowing to sea as such is a waste and hence this gives the reason and in fact legitimacy to the state that personified modern scientific outlook to intervene and tame the river.

However, it is not to suggest that all colonial subjects shared this belief. There were people who realized with the regular failing of embankment that embankments are a rebuke to the natural flow of river and this insult will be avenged by the nature may be in coming generations (Sinha 2008). The 1928 flood report reaffirmed that embankments are at the core of flood disaster in Orissa and the challenge is not to control the flood but to let the water reach the sea as soon as possible. If not, it will lead to distress and calamity in the future (Dixit 2009). This was the idea of nature that manifested into the logic of political economy of colonial state and manifested materially into the structures of development that it created.

6. Observations in the Field

The understanding of the Kosi flood situation that developed from study of the primary documents and the secondary literature has been corroborated from the field which the author has been visiting for last 4 years for his doctoral thesis. The location of the field is in Supaul district of Bihar.

One of the major reasons responsible for poor socio-economic indicators of this region is the lack of connectivity of the hinterland with the urban area of Supaul district. In the field visits, the author found that there was almost non-existent road connectivity and the author was taken to interior regions

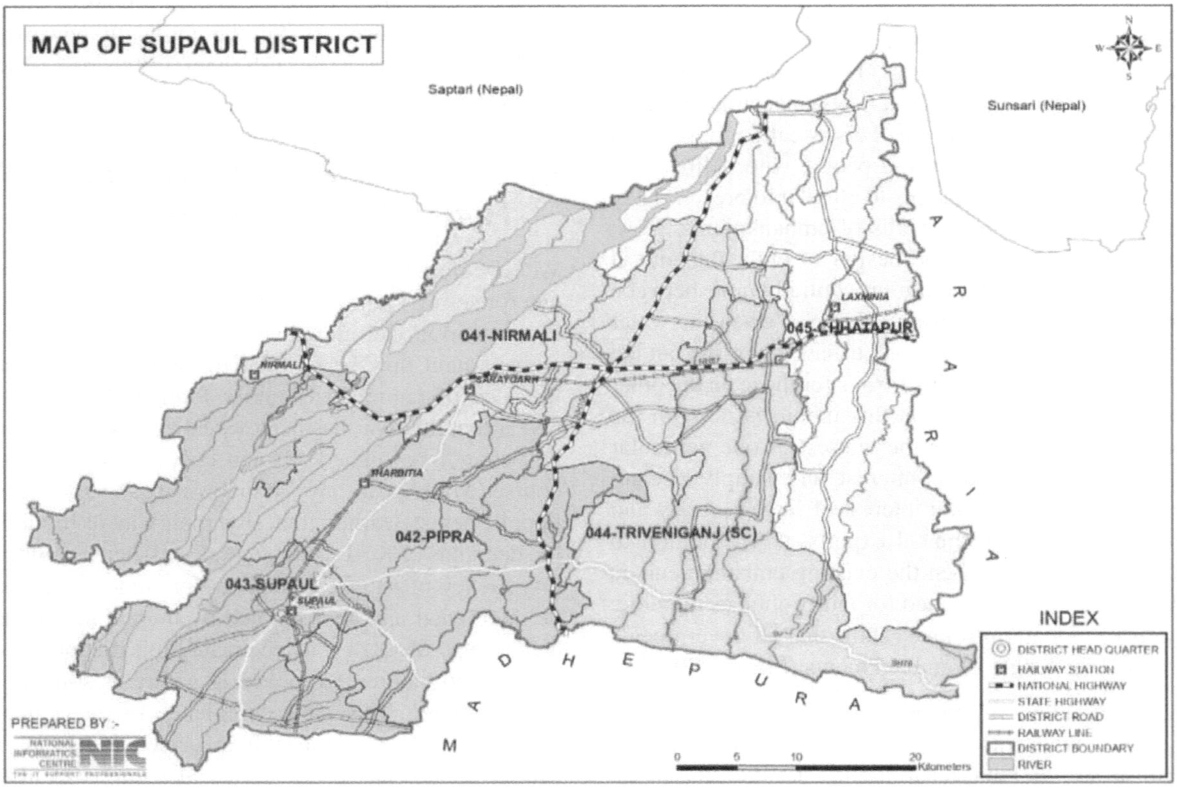

Fig. 4.4 Map displaying administrative divisions of Supaul district in Bihar

Source: CEO Bihar, Official Website

by volunteers of *Kosi Navnirman Manch* using motorcycle and small boats. This connectivity or lack of it does not augur well for women of the region specially during medical emergencies. People had to use multi-modal transportation combining bicycle, walking, boats and finally buses to reach the town area of Supaul. The condition of the schools was dilapidated. There were no children in the schools as the aberrant flow of the river Kosi did cut the villages in haphazard manner and disconnected the hamlets and schools. During focused group discussion conducted with flood-affected people of Jhameli village in Hulas Block of Supaul District in December, 2017, it was understood that there was complete lack of trust in people towards the government and administration. The headman Jhameli Sada recalled, in an in-depth interview, how no one from the village was ready to go with boats of NDRF team in last floods.

The author further attended a two-day conference of people affected by the Kosi floods at *Baarh Ashray Sthal* in Supaul District in the last week of December, 2019. There were multiple session discussing the issues of Kosi floods, the apathy of the government-administration etc. Most of the attendants were ordinary village people with very low level of formal education. What was rather surprising as well as illuminating for the author was the deep insight lay people have about the flood situation in the region. There were presenting critical arguments against the approach of the government and also offered alternatives that reflected their deep understanding of how the river Kosi flows.

In an in-depth interview with the founder of people's movement *Kosi Navnirman Manch* working under the larger banner of *National Alliance of People's Movement*, the author learnt that a signature campaign was conducted with hard work for months together to collect around 6000 signatures of people affected by the Kosi floods. The idea behind this signature campaign was to put pressure on the MLAs just before Legislative Assembly elections. This petition was not taken cognizance of by the Chief Minister's Office of Bihar (CMO).

This signature campaign was followed up by a *Kosi Mahapanchayat* in Gandhi Maidan of Supaul district of Bihar. Among the invited dignitaries were water experts, activists as well as political leaders. The extent of apathy can be sensed from the fact that no MLA candidate from the district bothered to attend the event where at least 5000 poor flood affected people from hinterlands has flocked to discuss their plight. When probed about the reason of such an apathy just before elections, the author was told that the Kosi flood victims are fragmented to become critical voice in the elections. There was a sense in people that it is probably because of their inability to be consolidated as a vote bank due to territorial delimitations and their inability to affect the fate of any candidate that they could be side-lined and their plights ignored.

Fig. 4.5 Image showing motorcycle being loaded on a small boat to cross a stream of the river Kosi near Khokhnaha village lying between the embankments and a government-run school in the village

Source: Photographed by the author during field study in Supaul district (2020)

Fig. 4.6 Image showing widespread silt brought by the river Kosi that makes agriculture and transportation difficult

Source: Photographed by the author during field study in Supaul district (2020)).

Figure 4.7 was taken by the author and they show the condition of islands created by the deposition of silt brought by the river. The shocking reality is that the displaced farmers have to pay *lagaan* on this piece of land if they wish to cultivate it. There was a separate fight that people affected by it were waging against this unjust law. People were technically given a Hobson's choice that if they want to cultivate the land, they had to pay lagaan or else land title will be transferred to the state. Hence, people in their anticipation of being able to cultivate the land sometime in future pay lagaan on a piece of land that does not accrue any benefit to them.

Fig. 4.7 Image showing the condition of silt-laden land inside the embankments

Source: Photographed by the author during field study in Supaul district (2020)

In the last visit to the field in December, 2020, the author travelled alongside the embankment and got a sense of how more interventions are being made in the river channel which defies the original argument that was made in the paper. These all interventions like guide dams, ring dams etc. further constrain the free flow of the river.

7. Findings of the Study

One point can be argued that the use of embankment to tame rivers was primarily and originally guided by the political-economic needs of the colonial state. The eurocentrism was only running as an undercurrent. This is important to understand because by making a ghost out of all that is scientific and technological also side by side bails out the political decision-making elites. This paper is not advocating the idea of technological solutionism but discovers that larger part of the blame should be on the politics around the process of policy making.

Second, and very importantly, one must accept the limitations of framework of social constructivism and political ecology and give due space to classical realist approach towards nature wherever required. All that is wrong with the Kosi river flood management policy is not a failure of the human interventions alone. This paper has illustrated at great length that the ecological setting of the region and the nature of river Kosi irrespective of large-scale human interventions have always set a limit on governability of land and water in this region of Bihar.

Third, the policy discourse around Kosi appears to suffer from a kind of fatigue where in state has accepted its limitations and is just involved in some kind of rituals annually before, during and after the monsoon floods. A glance at the website of Water Resource Department, Government of Bihar gives a sense that the steps taken by the state are largely focused on strengthening, elongation or heightening of the embankments. The floods are assumed to be a reality with a fatal acceptance and not an outcome of human interventions over the years. Year after year, same response is being given when the floods come. Relief is the second most important aspect of how the state deals with the aftermath of the floods. In the popular parlance, this relief distribution has always been seen as a huge source of corruption. Reason of the above state of affairs where the river Kosi and people affected by its floods finds themselves is an assemblage of politicians, engineers, bureaucracy and contractors who interests are entrenched in continuation of this approach of Kosi flood management (Jha, 2017).

Fourth, the notion of participation of the people at any level be it policy making or policy execution is almost non-existent. People are deemed to be passive recipient of the policy made by the state. What is striking in this case is people's passive acceptance of their fate and focus more on getting "relief" from the government. They are visualized by the policy discourse as victims of the flood who need to be given some relief. This relief in terms of food packets, construction material etc. is also euphemized as "fourth crop" apart from three crops cultivated in a year.

People's voice doesn't come into dominant policy discourse largely i.e., taming the river using embankments. As has been corroborated by the field studies in last 4 years, it has become clear that the administration is apathetic towards pain of people. There prevails an air of fatal acceptance of Kosi flood situation. The flood victims are largely rural and all administrative departments are located in urban area of Supaul. The experience of *Kosi Navnirman Manch*'s signature campaign and *Kosi Mahapanchayat* demonstrates the level of indifference of administration and politicians.

8. Conclusion

Kosi river floods are natural but Kosi floods disaster is a construct. It is not natural in the way dominant narrative suggests. There are components of political decisions, social power play, cultural dominance, logic of political economy, motivations of biopolitics, urge of governmentality and so on, all of which are part of our social world. Right from the idea of taming the river to current politics of relief distribution, we see that the decisions in the domain of politics, economics, science and technology, society and culture together create a rubric in which floods convert themselves into a disaster and distribute its impacts unequally. The decision about who is relatively more vulnerable and who will benefit from embankments are man-made. Nature doesn't dictate these terms. The re-historicization of the policy evolution around the river Kosi and the deployment of an anthropological gaze helps in problematizing the depiction of disasters as natural. Rendering it "social" again by the above study would pave a path for locating human agency in accentuating this disaster and hence, for a solution.

References

1. Baghel, R. (2014). *River control in India.*
2. Barrios, R. E. (2017). What does catastrophe reveal for whom? the anthropology of crises and disasters at the onset of the anthropocene. *Annual Review of Anthropology*, *46*, 151–166. https://doi.org/10.1146/annurev-anthro-102116-041635
3. Baviskar, A. (1995). In the belly of the river: tribal conflicts over development in the Narmada Valley. *Studies in Social Ecology and Environmental History.*
4. Burele, S. A., Sharma, N., Ahmad, Z., & Gupta, I. D. (2015). Mathematical model to study channelization by lowering the river bed. *International Conference on Hydraulics, Water Resources and River Engineering, December*, 1–11.
5. Chandra, B. (1980). Colonialism, stages of colonialism and the colonial state. *Journal of Contemporary Asia*, *10*(3), 272–285. https://doi.org/10.1080/00472338085390151
6. D'Souza, R. (2006). Water in British India: The Making of a "Colonial Hydrology." *History Compass*, *4*(4), 621–628. https://doi.org/10.1111/j.1478-0542.2006.00336.x
7. Dixit, A. (2009). Kosi embankment breach in Nepal: Need for a paradigm shift in responding to floods. *Economic and Political Weekly.*
8. Fischer, F. (2000). Citizens, Experts, and the Environment. *Citizens, Experts, and the Environment.* https://doi.org/10.2307/j.ctv11smwd9
9. Forsyth, T. (2001). Critical realism and political ecology Book section Critical realism and political ecology. *After Postmodernism: An Introduction to Critical Realism*, 146–154. http://eprints.lse.ac.uk/4760/1/Critical_realism_and_political_ecology_(LSERO).pdf
10. Iyer, R. R. (2008). Floods, Himalayan rivers, Nepal: Some heresies. *Economic and Political Weekly.*
11. Jha, H. (2017). Public Arguments – 5. *Tata Institute of Social Sciences, Patna Center, April 2017.*
12. Jha, P. K. (2017). *3rd International Conference on Public Policy (ICPP3) Panel T16 P11 Session II P11 Sustainable Development, Public Policy and the Local Searching for an Alternative Decentralized Flood Policy in India : Hydrological Flood Policy and Local People.*
13. Kumar, D. (2015). Flood And Sediment Management In Kosi River. *XVth World Water Congress, 25 - 29 May, Edinburgh, Scotland. Sub-Theme 10: Management of Water Resources,* 1–26.
14. MacKenzie, J. M. (1999). Christopher V. Hill. River of Sorrow: Environment and Social Control in Riparian North India, 1770–1994. (Monograph and Occasional Paper Series, number 55.) Ann Arbor, Mich.: Association for Asian Studies. 1997. Pp. xii, 200. $33.00. *The American Historical Review*, *104*(3), 888. https://doi.org/10.1086/ahr/104.3.888
15. Mehta, L. (2003). Contexts and Constructions of Water Scarcity. *Economic and Political Weekly.*
16. Mishra, D. K. (1997). The Bihar Flood. *Economic&Political Weekly.* https://doi.org/10.2307/4405794
17. Mishra, D. K. (2001). Living with Floods: People's Perspective. *Economic and Political Weekly.* https://doi.org/10.2307/4410885
18. Mishra, D. K. (2008a). Bihar floods: The inevitable has happened. In *Economic and Political Weekly.*
19. Mishra, D. K. (2008b). *Trapped ! Between the Devil and the Deep Waters.* 221.
20. Oliver-smith, A. (2009). Anthropology and the Political Economy of Disasters. In *The Political Economy of Hazards and Disasters.*
21. Pritchard, B., & Thielemans, R. (2014). "Rising Waters Don't Lift All Boats": a sustainable livelihood analysis of recursive cycles of vulnerability and maladaptation to flood risk in rural Bihar, India. *Australian Geographer.* https://doi.org/10.1080/00049182.2014.930001
22. Ray, K. B. (1953). Beginning Kosi Project This Year: Will It Benefit Bihar? *The Economic Weekly*, 1265–1267.
23. Scott, J. C. (1998). Seeing Like a State. James C. Scott. In *Yale University Press.* https://doi.org/10.2307/2647576
24. Sebastin, J. (2009). *Understanding Public System Management In Bihar : The Case of Kosi Floods.* 48. Shah, E., Boelens, R., & Bruins, B. (2019). Contested Knowledges Water Conflicts on Large Dams and Mega-Hydraulic Development. In *MDPI.* https://doi.org/10.3390/books978-3-03897-811-4
25. Shrestha, R. K., Ahlers, R., Bakker, M., & Gupta, J. (2010). Institutional dysfunction and challenges in flood control: A case study of the Kosi flood 2008. *Economic and Political Weekly.*
26. Singh, P. (2008). The colonial state, zamindars and the politics of flood control in north Bihar (1850- 1945). *Indian Economic and Social History Review*, *45*(2), 239–259. https://doi.org/10.1177/001946460804500203
27. Sinha, R. (2008). Kosi: Rising waters, dynamic channels and human disasters. *Economic and Political Weekly.*
28. Sneddon, C. (2015). Concrete Revolution: Large Dams, Cold War Geopolitics, and the U.S. Bureau of Reclamation. In *The University of Chicago Press.* https://doi.org/10.1353/tech.2017.0128

CHAPTER 5

Operation Research on Flash Flood in River Dhauli Ganga On 7th February, 2021 at Tapovan—Raini Village, Joshimath, Chamoli

Aditya Pratap Singh
Deputy Commandant, 8th Battalion, NDRF
Ghaziabad (UP)

Abstract

On 7th of February 2021, due to some glacial activity at Ronti peak, a rock-slide slipped into the Ronti Gad River which caused heavy discharge in Rishiganga. A massive flash flood ravaged through the Valley of Rishiganga River and completely swept away the Rishiganga Hydro-power Project near village Raini and inflicted substantial damage to the Tapovan Vishnugad Hydro-power Project. At Rishiganga power project 56 personnel were expected to be working when the flash flood hit the incident site. Apart from this 12 person from Raini village were also reported missing. At NTPC Tapovan hydro-power project 136 personnel were expected to be working inside the tunnel and at the barrage area.

Immediately after the incident few videos went viral. After confirming from district administration, NDRF immediately operationalized its two teams by road. In the next two hours, three more teams were airlifted from Hindon airbase, Ghaziabad to Jollygrant, Dehradun by C-130 aircraft and further to Joshimath by ALH & MI-17 Choppers. NDRF team carried out day and night search & rescue operations at Rishiganga hydro-power project, Raini village, NTPC Tapovan barrage and inside NTPC tunnel Tapovan. Apart from this several kilometres manual search was also conducted on the both banks of River Dhauliganga. Out of 204 missing persons, 84 dead bodies have been retrieved so far. Search & rescue operation is still continued.

An attempt has been made to study the preparedness & response mechanism. The study finds the various gaps in the existing response mechanism. An endeavour had been made to identify the measures to fill up those gaps to minimize the loss of life and property in case of similar disaster in future. The study also provides a closer insight to the events which happened after the disaster. Study is based on factual records of various agencies, interviews of eye witnesses and the rescuers who worked there without caring their own lives. The study strongly recommends adopting IRS system throughout the country in letter and spirit to achieve the goal of disaster resilience community envisaged in Sendai Framework.

Keywords: Rishiganga, Dhauliganga, NTPC, IRS, Sendai Framework, Nanda Devi and Flash flood

Fig. 5.1 River Dhauli Ganga ravaged the Tapovan Hydroelectric power project

Source:-Screen shot from a viral video shot by locals

"Himalayas are full of surprise"

1. Introduction

Chamoli district lies in high mountain ranges with steep topography, including the second highest peak in India, Nanda Devi (7,816 mtrs). District falls in Alaknanda basin which is the largest of the UKD river basins. Catchment area receives heavy amount of discharge due to presence of various glaciers. River Rishiganga is a tributary of the Dhauliganga River. It originates from the Raikhana Glacier (5375 mtr). In 1935, a British geologist Tillman discovered the Rishi Ganga gorge which is considered one of the most difficult gorges of the world.

Usually most of the disasters in mountains occur in the monsoon season except earthquakes and Forest fires. Every year NDRF pre-positioned its teams before onset of the monsoon on the request of state governments to deal with any emergency. This pre-positioning reduces the response time and enables NDRF teams to reach at incident site in golden hours. The official data of number of operations conducted by NDRF during pre-positioning shows that maximum numbers of operations have been conducted at Chamoli district.

Table 5.1 Number of operations in Uttarakhand by NDRF in last five years

Location	Number of operation					Total
	2016	2017	2018	2019	2020	
Uttarkashi	08	—	—	04	04	**16**
Chamoli	03	03	06	10	06	**28**
Almora	07	01	01	01	—	**10**
Pithoragarh	02	04	09	04	02	**21**
Rudraprayag	—	—	09	—	—	**09**
U S Nagar	01	—	—	—	—	**01**
Total	**21**	**08**	**25**	**19**	**12**	**85**

Source: 8th Battalion NDRF Ghaziabad

In 2013, Kedarnath flash flood devastated the District Rudra Prayag. Badrinath and Hemkunt Sahib areas of Chamoli district were also severely affected by the flash flood.

To understand the vulnerability profile of any area we must study various factors affecting the magnitude of disaster. These factors are directly related to loss of human life and property e.g. geology, geography, land use & topography of that area. Rainfall pattern, landslide susceptibility zonation, seismic zonation are few other important factors which must be taken care off while studying vulnerability profile of the area. Apart from these factors narrow gorges, high altitude and settlement pattern also affect the magnitude of disaster.

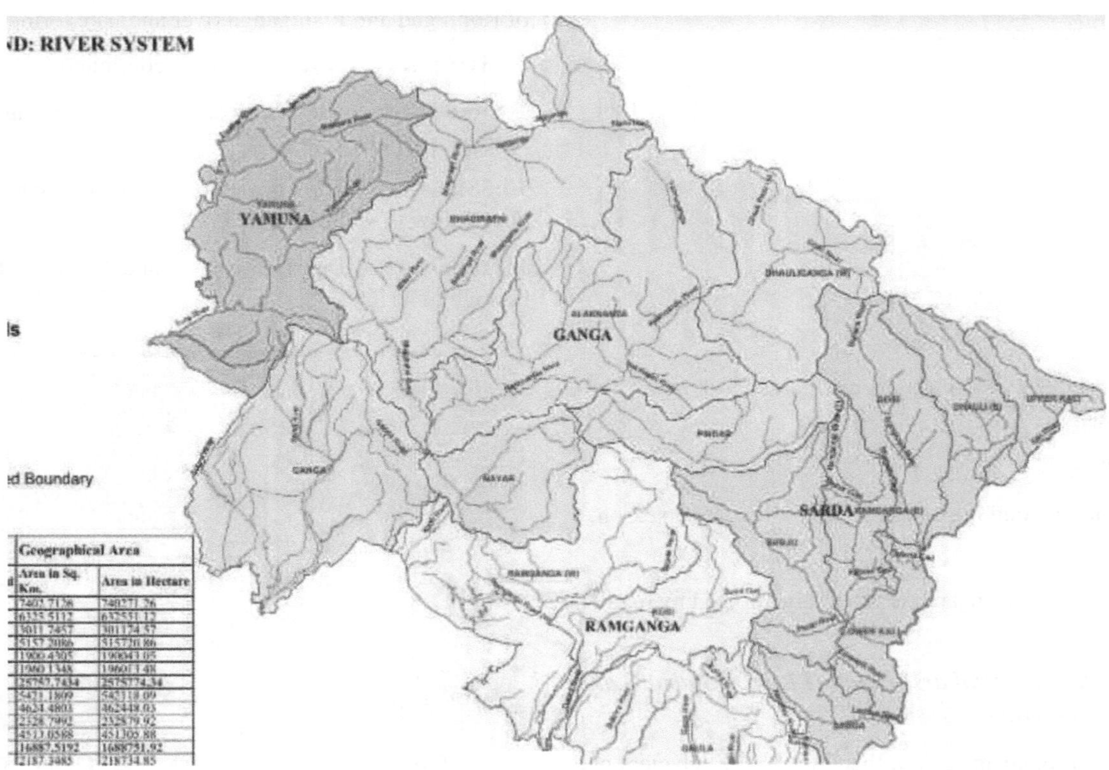

Fig. 5.2 River system of Uttarakhand

Source: www.pdfnotes.co/wp-content/uploads/2021/02/rivers-compressed.jpg

Vulnerability Profile of the Affected Area (Tapovan, Chamoli)

(a) Seismic Zone

Tapovan lies in seismic zone V of India and is situated at the junction of the middle and higher Himalayan region.

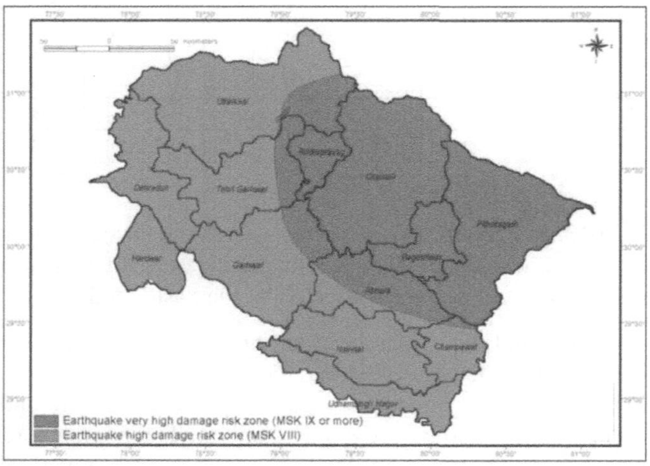

Fig. 5.3 Seismic Zone of Uttarakhand

Source: SDMP Uttarakhand

(b) Rainfall

Average annual rainfall is much lesser then the national rainfall average.

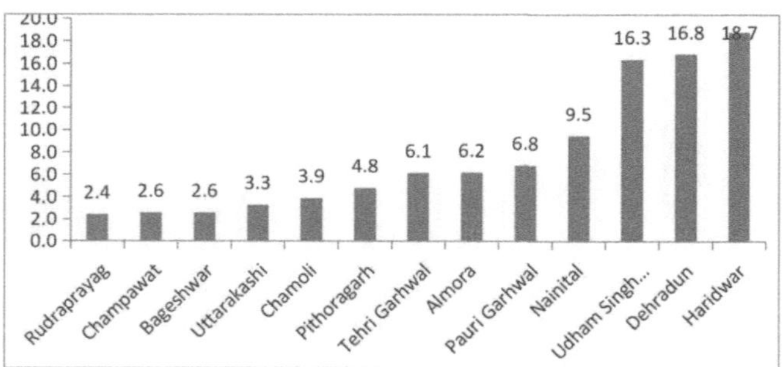

Fig. 5.4 Rainfall data of Uttarakhand

Source: Google

(c) Cloud Burst

Area is prone to cloud bursts especially in monsoon season.

(d) Land Slide

During monsoon season, area is adversely affected by regular landslides.

(e) Land slide Lake Outburst Flood (LLOF), Glacial Lake Outburst Flood (GLOF)

In the entire Himalayan region, there are 16,000 glaciers out of which 1,000 are in Uttarakhand only.

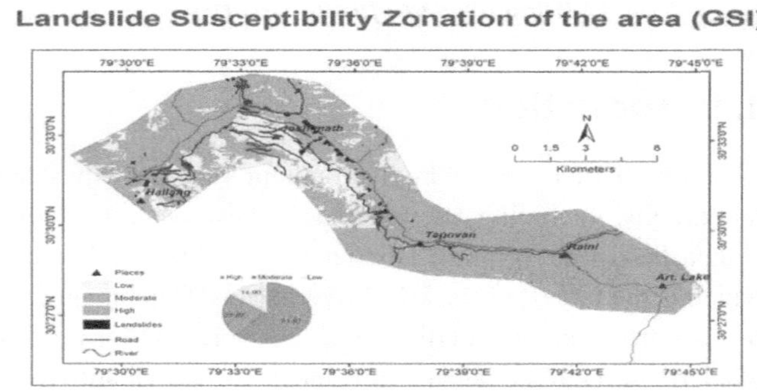

Fig. 5.5 Landslide susceptibility Zonation from Helang to Artificial lake at Peng near Raini village

Source: GSI

2. Event

On 7th of February 2021, a massive flash flood ravaged through the valleys of the Rishiganga & Dhauliganga rivers. The flood completely swept away the Rishiganga Hydropower Project near village Raini and inflicted substantial damage to the Tapovan- Vishnugad Hydropower Project downstream.

3. Time Line

10:18:43 hrs – Debris blocked Rishiganga at the confluence of Ronti gad and Rishiganga river and lake formed thereafter.

10:31:51 hrs - Active debris flow continued

10:36:00 hrs – Debris flow reached Tapovan-Vishnugad barrage of NTPC

10:58:35 hrs – Debris flow subsided. Duration of sliding and debris flow is **~50mins**

11:19:00 hrs – Debris flow reached Vishnugad- Pipalkoti barrage of THDC

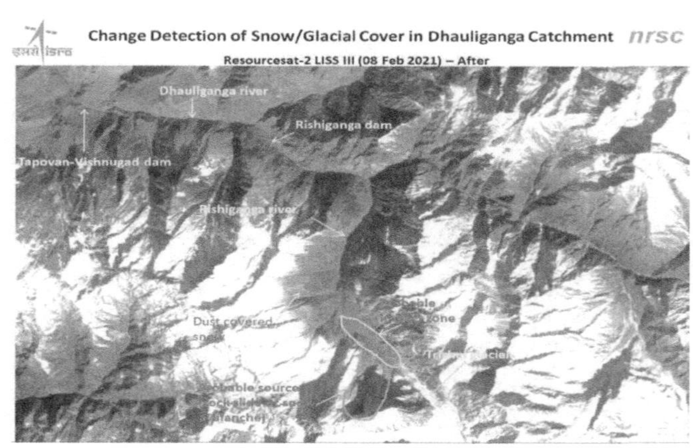

Fig. 5.6 Probable impact zone at Rongti Gad

Source: NRSC ISRO

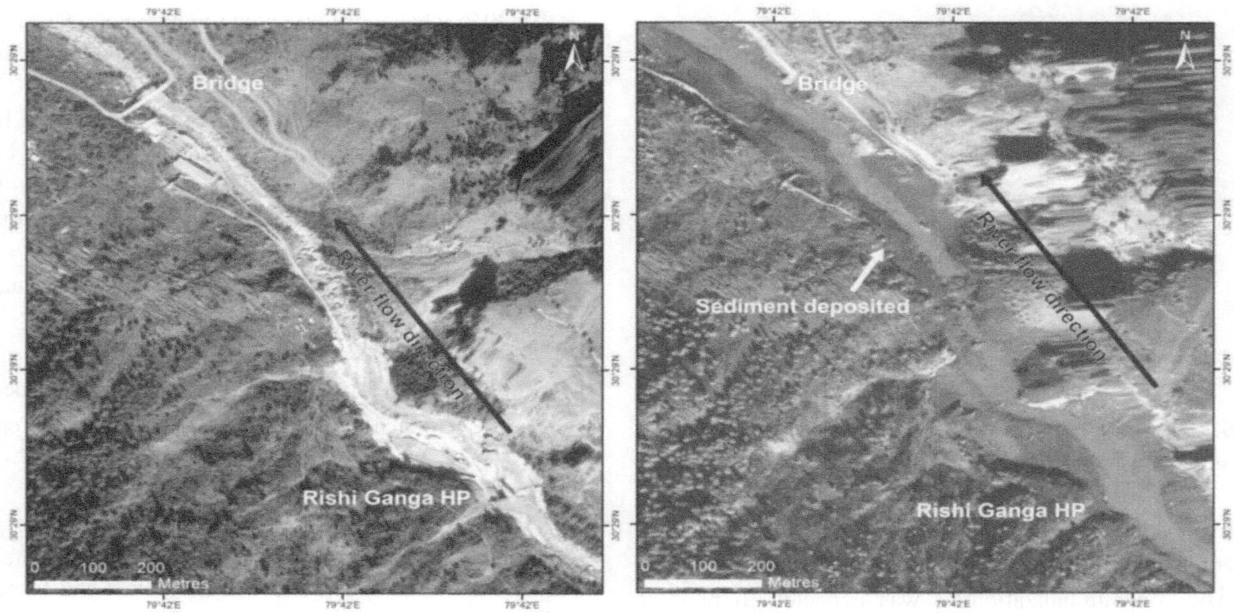

Fig. 5.7 Sediment deposition in Rishiganga River after the incident (Before & after)

Source: NRSC ISRO

4. Damage Assessment

- The 13.2 MW **Rishiganga Hydropower project** was situated on the left bank of river Rishiganga, 14 Kms downstream from the impact site. Project has faced difficulties in commissioning as it had history of substantial damages in earlier flash floods. Rishi Ganga power project washed away in 2016 flash flood. In June 2020 project get commissioned and operationalized. But on 7th February, 2020 the rolling debris with high discharge hit the power project and flattened it completely.
- At Raini village, Rishiganga River joins Dhauliganga. The under construction **Tapovan-Vishnugad power**

Fig. 5.8 Tapovan hydroelectric power plant (Before & after the sediment deposition by Dhauliganga river)

Source: NRSC ISRO

Fig. 5.9 Barrage gates of Tapovan hydroelectric power plant
Source: 8th Battalion NDRF Ghaziabad

project is approximately 8 km downstream from this confluence and on the left bank of the Dhauliganga river. After operationalization, it was expected to produce 520Mw electricity. Its barrage is 200 meters wide and 22 meters high with each of the four gates measuring 14 meters in height and breadth. Project was near completion. The work of Tapovan Vishnugad HEP started in 2006. It was expected to be completed by 2012 but delayed due to damage in 2013 Kedarnath disaster. In 2016 when work was about to complete, Rishi Ganga River wreaked havoc and there was delay due to damage in the project.

- Excessive deposition of sediments in river Channel resulting in the rise of river bed level up to 11 to 13 mtr.
- Five suspension bridges washed away hence Raini, Jugju, Juagwad, Peng & Lata village cut off from rest of the district.
- Approximately 13 villages affected due to damage of utility supply lines.
- Loss of livelihood.
- Psycho-social impacts: Due to loss of life & property.
- As per official figures, initially 204 persons were reported missing.

4.1 Cause of Event

According to the preliminary survey, huge rock/rocks fell into a glacial rivulet (Raugnti stream). It blocked and formed an artificial dam (Debris dam). This dam breached and caused the flash flood. But nobody has come up with evidences that glacial lake burst has caused the disaster. What triggered to the fall of rock is also not clear. The duration of the process was also not known, whether it took years or it was a sudden outbreak.

Fig. 5.10 Source of rockslide and its impact site at Rongti Gad

Source: NRSC ISRO

Initial reports suggested that the Ronti-Gad is formed at the confluence of Ronti, the bigger one and Nanda- Koti. A rockslide of about 550 mtrs width near Ronti peak, into the Ronti Gad River caused heavy discharge in Rishiganga.

5. Rescue Operation

On 07th February, initial information received from social media about the incident. Information was confirmed from district administration and 8th Battalion, NDRF immediately operationalized its two teams; one each from Dehradun and Ghaziabad.

In the next two hours, three more teams were airlifted from Hindon airport by C-130 aircraft to Jollygrant, Dehradun and further to Joshimath by choppers. The Air force must be credited for conducting 06 back to back sorties from Jollygrant to Joshimath to airlift three NDRF teams. These teams were carrying search, cutting, lifting equipment, deep diving sets with highly skilled deep divers and search dogs.

NDRF team from Dehradun arrived at the incident site at 2 AM next morning after an overnight journey of 14 hours. Without wasting a moment, team joined the rescue work along with other agencies already present there.

After due risk assessment by district administration, five NDRF teams were tasked at various work-sites. Initial situational reports received from district is as follows:

Table 5.2 Details of probable missing person

Location	NTPC Tapovan	Rishiganga power project	Raini village	Live victim already rescued
No of missing person	136	56	12	12

Source: Based on initial assessment of situation by District administration Chamoli

5.1 Important Jargons & Abbreviations

To understand the rescue operation inside the Tapovan adit tunnel, it is very important to know the jargons which are used in the HEP and layout plan of the tunnel.

Fig. 5.11 NDRF Team movement

Source: 8th Battalion NDRF Ghaziabad

Fig. 5.12 NDRF Team movement at Hindon airbase and Joshimath helipad respectively

Source: 8th Battalion NDRF Ghaziabad

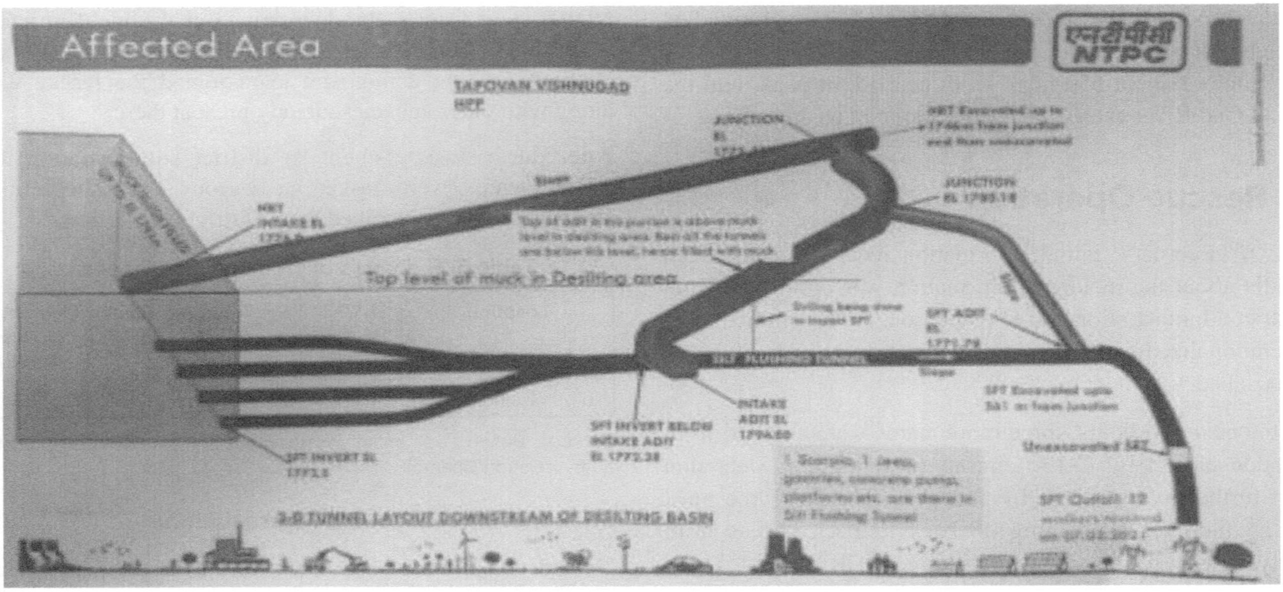

Fig. 5.13 Tapovan tunnel layout

Source: Provided by NTPC, Tapovan to all concern stakeholders at one of the planning meeting at EOC

Head race tunnel: - A tunnel which carries water under pressure from a reservoir down to the turbine hall of a HEP.

Barrage: - The barrage which block the water body to make a reservoir. It has gates to regulate the water level.

De-silting Chambers: - DSCs are provided in order to separate suspended solid particles from water, which cause abrasion at the hydraulic machinery. (Turbine of HEP)

Silt flushing tunnel: - SFT lies below the DSC to flow back water again into the river by passing the barrage.

Outfall: - An outfall is the discharge point of a waste stream into river.

Intake Adit: - An entrance or a passage for going inside for construction or repairing work, a horizontal excavation with gradient towards HRT and SFT.

Gentry: Travelling formwork (Gantry) is a formed structure to give final shape to the tunnel for any shape and size of tunnel.

Tunnel Boring Machine :- TBM also known as a "mole", is a machine used to excavate tunnels with a circular cross section through a variety of soil and rock strata.

Crown of a tunnel: - Top surface or roof.

5.2 Predictions

(a) **Rishiganga power project** – Approx. 56 personnel were expected to be working in and around Rishiganga power project.

(b) **NTPC Tapovan**: - Approximately 136 personnel were expected to be working inside the tunnel and barrage area. 05 personnel were working inside Tapovan Adit, 30-35 personnel were expected to be working at the end of SFT where 80mtr stretch was yet to be excavated. 03 personnel were expected to be working at the dead end of HRT. All were missing.

(c) The splash of sludge was several meters in height which washed away several people from the residential areas along the river. 12 people from Raini village were also reported missing. Those dead bodies were expected to be found littering around the riverbed of Dhauliganga River.

Note: Information based on State emergency operation centre Dehradun (SEOC) report.

5.3 Objectives

(a) To carryout search and rescue operation at Rishiganga power project, Raini village.

(b) To carryout search and rescue operation inside Tapovan Adit, Barrage area and SFT area.

(c) To carryout search and rescue operation along both banks river Dauliganga river.

Table 5.3 NDRF Team deployment

S. No.	Work Site	Team ID
1	Tapovan Tunnel	8G
2	Tapovan Barrage	8J
3	Rishiganga Project, Raini Village	8A
4	Manual Search along Dhauliganga River	8M

Source: 8th Battalion NDRF Ghaziabad

Fig. 5.14 Outside & inside view of Tapovan Tunnel

Source: 8th Battalion NDRF Ghaziabad

Fig. 5.15 Tapovan barrage

Source: 8th Battalion NDRF Ghaziabad

Fig. 5.16 Rishiganga power project, Raini village

Source: 8th Battalion NDRF Ghaziabad

Fig. 5.17 Manual search along the river Alakhnanda

Source: 8th Battalion NDRF Ghaziabad

At Tapovan Barrage, search was started to find the missing personnel at the barrage and in the DSCs. They were filled with thick muck and forming big bowls of sludge. This swamp was not thick enough to be approached by any machines. High powered suction pumps were brought in but it could not suck the muck as it was too thick to be sucked by suction pumps. Moreover, cobbles and pebbles were hitting the blades of suction pumps while sucking the muck. A welded frame was applied to protect the fan of suction pump but it didn't work.

Fig. 5.18 Night operation at Tapovan Barrage

Source: 8th Battalion NDRF Ghaziabad

Level of sludge subsided with the time & on 14th Feb rescuers located few dead bodies stuck in cage of rebars. NDRF rescuers used a rope throw gun to fix the rope in the rebars at the far end. Wooden ply boards were used on the surface as a firm base to walk on. In an overnight operation, NDRF rescuers were able to extricate all five dead bodies from this muddy bowl.

At Rishiganga power project, NDRF team carried out technical and canine search. After confirming that no live victim is present there, Poclain machine and JCB were brought in to take out the sludge from incident site. Total 16 dead bodies were retrieved from this work site. And along Dauliganga River, one dead body was retrieved at THDC Dam, Helang.

Tapovan Tunnel Rescue Operation

HRT is approx 1700 mtrs long tunnel with gentle gradient towards Vishnugad. Last portion of HRT was yet to be excavated; it means there was a dead end towards Vishnugad.

Fig. 5.19 Operation at Raini Village

Source: 8th Battalion NDRF Ghaziabad

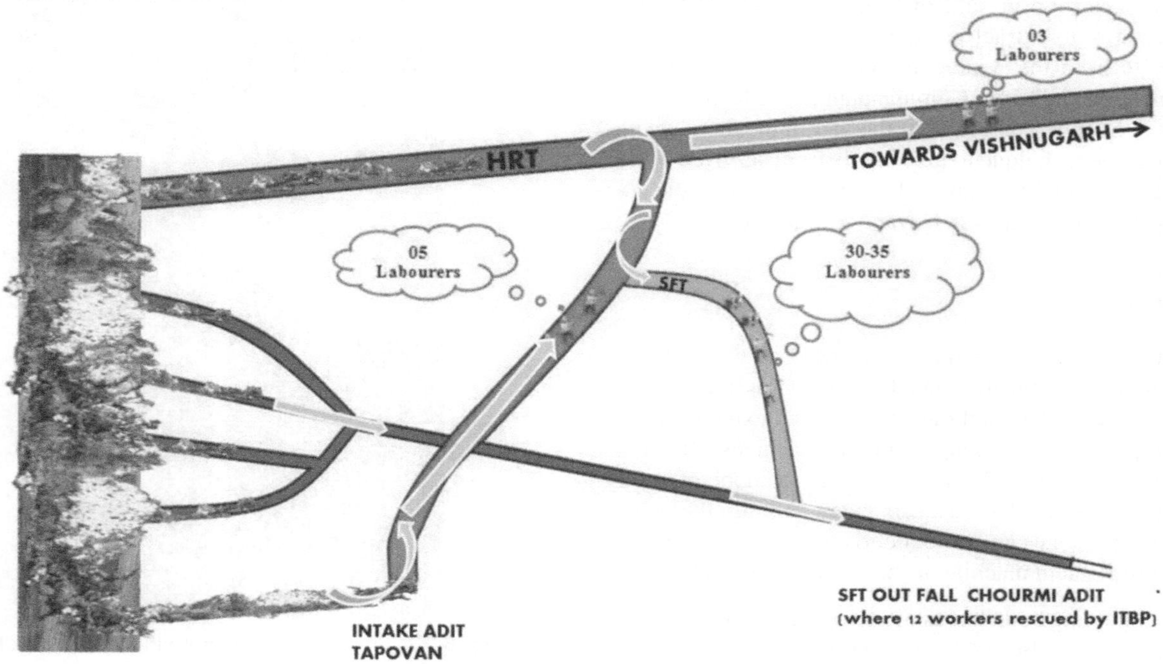

Fig. 5.20 Flow of sludge inside Tapovan tunnel

Source: Sketch prepared by author of this research paper based on layout plan & information provided by NTPC, Tapovan

SFT also have a gradient from DSC towards SFT outfall through which waste water again fall into the river bypassing the barrage. Almost 80 mtrs stretch of SFT was yet to be excavated which means that there was a dead end in SFT also. Excavation work was continuing before the occurrence of the incident from both sides i.e. from DSC and SFT outfall at Chaurmi.

Tapovan Adit provides an entry to HRT. Tapovan Adit joins HRT at a junction, 240mtrs away from the entrance. It has gradient of 1:13 towards HRT which means that after every 13 meter there is a depression of 1mtr. A small size **branch tunnel** joins Tapovan Adit to SFT. SFT runs 12 meters below the Tapovan Adit. Technically Adit & branch tunnel have no role in electricity generation. They are erected to connect SFT and HRT to carry out construction, evacuation & maintenance work. Branch tunnel is 99mtrs long and descends from Tapovan Adit towards SFT. Generally if there is an emergency inside the tunnel, all workers working inside would run towards exit of Tapovan Adit.

6. Planning of Operation

First aim was to excavate the sludge to reach inside the tunnel up to 72 mtr mark where Intake Adit tunnel was passing over the SFT. We planned to drill a borehole at the cross action of Intake Adit and SFT. This was the only option left with us to know the probability of finding of live victims inside SFT. In case of positive result, plan was to breach the crown of SFT from the surface of Adit by drilling a bigger size bore and sending in deep divers to conduct search and rescue operation. In case of negative result, resume excavation work to reach to branch tunnel which connects Adit to SFT (180mtr) and further excavation to reach HRT (240mtr). Project workers informed that one Bolero, JCB, Poclain machine were also parked inside. Branch tunnel is having lesser diameter. One alleged gentry was reported inside. Hence there may be chances that any obstacle (Bolaro, JCB, Poclain or Gentry) might have blocked the mouth of branch tunnel to avoid entering of sludge.

Project workers told that there could have been another way to enter into the tunnel. Tunnel is passing beneath the mountains and at one place it is only 40 feet from the surface. That area was earmarked on the company's map since it is the minimum distance from the surface to SFT. In SFT maximum workers were suspected to be trapped. It was a viable option to breach the tunnel vertically down at this place. But this option was straight away rejected. They argued that it would take many hours, day or probably week to breach the tunnel from this place. At that time we were expecting worker alive inside and there was no harm to use these options also, but due to some technical issue these idea was rejected.

Making a sketch on ground in front of an actual structure is the best way to extract out the information from the eye witnesses to brief the senior officials. I learnt this effective communication skill from Commandant Sh. P. K. Srivastava. He did it in Shahberi and located the victim location exactly. I used the same technique to take inputs from survived NTPC workers and brief DM Chamoli about planning to conduct the rescue operation on Tapovan barrage and in SFC.

7. Execution

To reach some vantage points it was necessary to extricate the sludge/muck from the tunnel but this was not possible manually. One poclain and one JCB worked exhaustively as available space was not enough for more machines. Operators were working day & night without rest in the hope of saving their colleagues. NDRF rescuers were cutting the wires, angle irons, pipelines and other debris which was hampering the speed of Poclain machine in excavating the muck.

With every passing hour, hopes of saving lives were withering away. On 13th Feb, Borehole machine was called in to make a borehole of 300 mm diameter. NDRF prepared a Go pro camera to peep inside the tunnel. One rescuer's torch was broken and its lamp was attached to the Go pro camera, it was attached to the strings so that it can rotate to 360 degree (As used in puppet show. A specially designed wired camera was also brought in from Ghaziabad. Drill machine was quite heavy and its mast was almost touching the crown of tunnel but somehow the drill machine operator managed to operate it.

Fig. 5.21 Operation inside the Tapovan tunnel
Source: 8th Battalion NDRF Ghaziabad

Fig. 5.22 Making a borehole inside the Tapovan tunnel
Source: 8th Battalion NDRF Ghaziabad

Fig. 5.23 Inserting the camera inside the borehole
Source: 8th Battalion NDRF Ghaziabad

Drilling a hole of 300 mm diameter in a confined space required courage. Since six days people have been witnessing shocking videos of a big disaster. The splash of sludge didn't give time to evacuate and now there were rumours that a big glacial debris lake has formed in higher regions which could burst any time. IMD warned about heavy rains which can trigger a GLOF or LLOF. If flood occurs, sludge was expected to reach faster than before. Inside the tunnel, lack of oxygen was reducing the efficiency of workers; Dust and smoke generated by drill machines was suffocating lungs and burning the eyes. Visibility was decreased to few meters only (1-2 mtr).

Usually in tunnels continuous water seepage from rock strata create a water logging which is pumped out using submersible pumps. This process is known as "mine dewatering". Every day approximately 3000ltrs water used to be pumped out in HRT only. After the incident of 7th February water was accumulating inside even with greater rate since river Dhauliganga was flowing 12mtr up from its normal course and no dewatering process took place for last 6 days inside the tunnel. This was a big hazard if waterlogging exceeds the

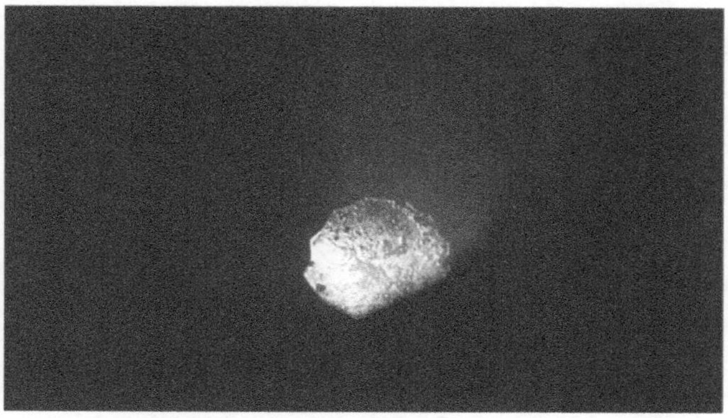

Fig. 5.24 Inside view of the borehole
Source: 8th Battalion NDRF Ghaziabad

Fig. 5.25 Dead body retrieval by NDRF
Source: 8th Battalion NDRF Ghaziabad

capacity of tunnel it would come out as a splash to wash away everybody working inside.

Working with all above mentioned hazard following probabilities were expected after making a hole:

(a) There may be a sudden blow out of trapped air due to positive pressure inside blocked SFT.
(b) A water Jet may spur out with force though it was learnt later that according to design of tunnel, it was a remote possibility but it should have been predicted and explained by NTPC engineers. But nobody had any idea about the outcome of making of borehole.
(c) Empty tunnel in this case making a rat-hole and sending deep divers inside the tunnel was the option to save lives.
(d) SFT is also fully blocked with muck/sludge or water.

A rescuer never loses hope. We prayed that any TBM, Gentry or vehicle may block the mouth of branch tunnel and in such cases there may be probabilities of survival of victims at least inside the SFT. Many of us used to hear screams in our dreams. Rescuers used to imagine that victims might be hearing the drilling sound inside the tunnel. Our efforts from outside would give them strength for survival, but all hopes shattered. When the drill bit made the final blow and the crown of SFT breached, water came out into the borehole approximately up to river bed level. Everything was clear now; it was no more a rescue operation, only a salvage operation as there was sludge all around. No more sounds of screaming from inside disturbed anybody in their dreams.

NDRF and SDRF are the premier agencies as far as disaster response is concern hence DM Chamoli thinned out other agencies which were not adequately equipped with specialised equipment. We started excavating the tunnel again to reach the mouth of the branch tunnel at 180 mtrs. After excavating 120mtrs, we started finding the dead bodies, amputated body organs and remnants of cloths. 14 dead bodies were found within the 120 to 150mtrs stretch of tunnel. Water seepage inside the tunnel hampered the excavation work. Waterlogging formed a 2-3mtrs deep pool and despite continuous dewatering it was still maintaining the same water level. Since tunnel was gradually descending, Poclain machine half submerged into this pool hence excavation used to stop till dewatering lower the water level.

Fig. 5.26 Water logging inside the Tapovan tunnel
Source: 8th Battalion NDRF Ghaziabad

At 180mtrs the mouth of branch tunnel was found completely filled by sludge. Till now 20 dead bodies have been extricated from Adit & branch tunnel. Total length of 99mtrs has been excavated and work is still continuing to excavate HRT and SFT. Due to heavy rain the link road to Tapovan Adit tunnel was washed away by River Dhauliganga & hence excavation work has been stopped by concern agencies. One NDRF team is still there at the disposal of district administration.

8. Relief

- Zip line or Jhula Bridge for Raini, Jugju, Juagwad, Peng & Lata village. Ropes throw Gun & Drone available with NDRF were used for this purpose.

THE ECONOMIC TIMES | Politics

English Edition • | 27 March, 2021, 11:50 AM IST | E-Paper

136 missing people whose body not found to be 'presumed dead': Uttarakhand Official

Synopsis
Sixty-eight bodies have so far been recovered after the February 7 disaster. There is a tunnel in which around 35 people are suspected to have been trapped, but rescuers have not been able to reach it till now due to the sludge. Authorities say clearing the debris from the tunnel could take months.

The notification was issued on the basis of an SOP issued by the Centre in this regard on February 21, Negi said.

The Uttarakhand government has issued a **notification** authorising 136 people still missing after the flash flood in **Chamoli district** over two weeks ago to be declared "presumed dead". Sixty-eight bodies have so far been recovered after the February 7 disaster. There is a tunnel in which around 35 people are suspected to have been trapped, but rescuers have not been able to reach it till now due to the sludge. Authorities say clearing the debris from the tunnel could take months.

A glacial lake was formed upstream of Tapovan over Rishiganga following the devastating flash flood which had demolished a 13.2 MW hydel project along the river besides causing extensive damage to the NTPC's Tapovan-Vishnugad hydel project along Dhauliganga.

Uttarakhand Health Secretary Amit Singh Negi said on Tuesday that the presumption of death can be made by officials in the affected district only in case of missing people who in all likelihood have died in the disaster but their bodies have not been found.

- BRO built a belly bridge in the Raini village.
- ITBP helped in ration distribution at Lata village, which was initially done by choppers.
- Food stalls at worksite were provided by district administration.
- Enquiry centre at Tapovan site was installed by district administration.
- Medical facilities & ambulance services were provided by district health service in a very smooth manner.
- A team consisting of ADM, SP & Tehsildar level officials were sent by UP Govt for identification and speedy transportation of dead bodies to their respective districts of UP. Similar action was taken by few other states also e.g. Jharkhand.
- After few days Uttarakhand Govt announced that all missing persons would be presumed dead for the matter of compensation. This came as a big relief for the family members of victims.

Fig. 5.27 Aerial view of Glacial Lake at Peng, near Raini Village

Source: NRSC ISRO

Fig. 5.28 Outside view of Tapovan tunnel

Source: 8th Battalion NDRF Ghaziabad

9. Challenges

(i) **Glacial Debris Lake**: Formation of Debris Lake is a natural process and such incidents can't be prevented by human interventions. This is common signature action of Himalayan fragility. However close monitoring by establishing early warning systems & mitigation measures like imposing scientific settlement pattern can minimize the loss of human life and property downstream. Lake Issue was hyped so badly that number of agencies had to recce the area and various agencies tried to play with nature with their own innovative & non-scientific techniques.

(ii) **Coordination**:-For initial 1-2 days there was chaos all round. Disaster linked various agencies with different resources & work procedures. It is quite difficult to avoid ad-hocism in such circumstances. There are always conflicts of jurisdiction, command & responsibility. Further, there is a little knowledge about capabilities of each other hence agencies underestimate and sometimes overestimate each other.

India has adopted incident command system (ICS) from USA. We customised ICS with slight modification according to the Indian response mechanism and named it as Incident Response System (IRS). In IRS all response agencies work under a unified command. Designated Responsible Officer (RO) deputes an Incident Commander (IC) for a particular incident. IC is responsible for overall coordination, conduct and safety of that operation site. All nodal officers of various rescue agencies must report to him, irrespective of ranks in their respective organizations.

To avoid non-essential elements at incident site, IRS recommends the concept of staging area. Stakeholders are expected to gather at staging area with all available resources. Staging area is headed by staging area manager, who on the demand of Operation Section Chief (OSC) or IC sends manpower or equipment at the incident site. All these measures must be adopted strictly.

(iii) **Lack of initial information**:-Except from few initial videos of incident, not much was known to rescuers about the incident and damage. Number of persons trapped, whereabouts of trapped and what are the resources available especially with the construction companies NTPC, Ritwik and HCC was not known.

(iv) **Decision dilemma**: - Three teams were airlifted from Hindon by C-130 and further movement from Hindon to Joshimath was through choppers. Manpower and equipment had to be split to make it fit for one MI-17 loads. Total load was to be airlifted by six sorties. We were asked to pick only most essential items and left the non-essential items. It was very difficult situation to sort out the most suitable equipment when nature of operation was not very clear.

(v) **Early warning in tunnel**: - There was no early warning system was available for the safety of workers /rescuers working inside and outside of the tunnel. Warning of another GLOF was circulated by the district administration. Learning from the past NDRF advised to install a siren which can be heard at all the worksites. A siren was placed outside the tunnel which was not audible to workers who are working inside the tunnel.

(vi) **PTSD**: - Life of rescuer is always very challenging. In each operation a rescuer learn new rescue techniques. Apart from this he also experiences different mental traumas caused by closely observing sufferings and pain. After each operation NDRF organizes debriefing session for professional betterment of rescuers.

(vii) **Psycho-social aspect**: - Usually bystanders are not allowed to come close to the rescue site as it may lead to some secondary disaster. One fine day pain of restrictions exploded and wife of victims (an AGM rank official of Ritwik) became hysteric. She posed herself as a goddess and came in front of Tapovan Adit tunnel. She cursed the company who erected the tunnel & blessed the rescuers. This incident was in front of public hence caught the attention but there might be numerous incidents which were not reported or recorded. The survivors of the victims are in hope that they will receive dead bodies of their near and dear one to perform the last rites and dispose-off the mortal remains as per their respective religious believes. More than half affected families are still waiting for the dead bodies.

(viii) Those who survived that day are also having traumatic nights. I met few people in Dhak, Peng & Tapovan village. They are experiencing the nightmares after the event. It is very necessary for district administration to provide psycho-social care by providing due assurance, insurance and by informing them about the mitigation measures which have been adopted aftermath of the incident. Distribution of relief material, shelter and food are the direct measure of psycho social care but we must use indirect methods (which are long term in nature) like preparedness & mitigation measures.

(ix) Theoretically an operational briefing is required before any operation which includes initial information about the incident. Based on this initial information, we decided the objectives. Finally the objectives determine the immediate action plan (IAP). IAP is propagated in the form of orders to the rescuers. This briefing is very important for junior leadership to make their subordinate rescuers clear about their respective roles at given work sites. But in case of sudden disaster like this, there was a little time or we can say no time to follow the procedures.

10. Lesson Learnt

(i) IRS system must be adopted by district administration in letter and spirit. NDMA and SDMA's of all state must issue strict necessary guidelines to the state & district authorities in this regards.

(ii) **Communication**: Compatibility of communication equipment with respect to other organization will always remain a challenge. Nobody is authorized to enter into allocated radio frequencies of other organization. Using cellular Networking not full proof arrangement. Hence alternative means like satellite communication must also be encouraged without

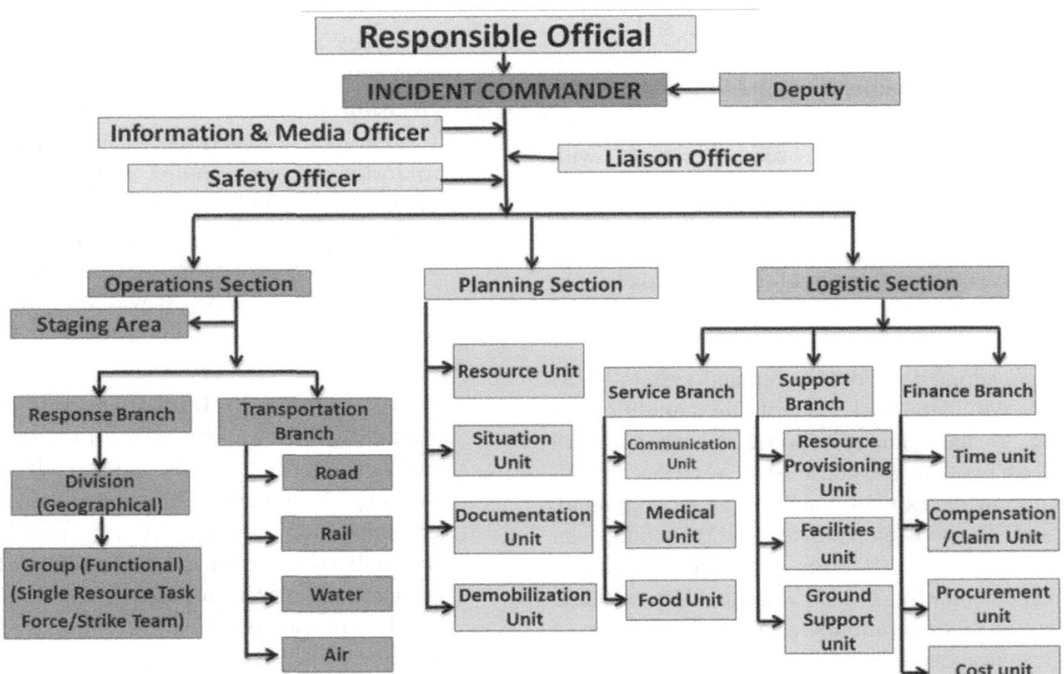

Fig. 5.29 IRS chart

Source: NDMA

ignoring the radio communication. Redundancy must be achieved in communication.

(iii) **Flow of information**: Information about number of missing, trapped and suspected to be dead was fluctuating with the time since unauthorised information was floating on social media.

(iv) Local community must be encouraged and empowered to act as a watch dog regarding environmental degradation activities. Bhutan is a classic example. Few primitive societies are also engaged in environmental protection around themselves. We must learn lesson from them.

(v) As a hunter and gatherer our ancestors were familiar with various natural indicators for prediction of disasters. Few techniques may sound non-scientific and primitive in nature but there are lot of reasons to believe that these predictions are very effective and accurate sometimes. For example earlier settlement pattern in the hilly region across the rivers was based on natural indicators. Contrary, in 2021 having all modern communication facilities, very few were able to communicate the danger to the needy one. Most of the people were engaged in video recording of event.

(vi) Exceptions are always there. Smt Mangshri Devi of Dhaak village saw the dust storm before the approaching storm sludge. She also heard the roar. Without wasting a single second she called to his son Vipul Kairuni who was working in Ritwik Company as a heavy machinery driver. He was working in an under construction SFC with many other workers. After many attempts by mother he picked call. After hearing the roars he ran towards the shortest exists, the only fixed ladders available in the death trap. Around 20-22 more workers also followed him. (Based on the memory shared by Sh Vipul Kairuni of Dhaak village).

(vii) **Prevention, mitigation & Preparedness**:

- There is head for Disaster risk assessment (HVCRA) on regular basis of the hydroelectric power plant. This also must be updated on the regular basis in the DM plan of the district.
- Monitoring and every washing system need to be installed and it must be integrated with Hooter, Sirens or multi-SMS system.
- More community involvement is required in this early warning system
- There is need to study the impact of seismic activities over the river system and especially over glacial bodies.
- Effective preventive and mitigation measures are required to be implemented strictly e.g. Afforestation by community, settlement pattern and Land use.

(vii) **Response & Recovery:-**

- Delay in sharing the information. It took 5 min for debris flow to reach Tapovan from Raini Village had it been conveyed promptly by villagers downstream, many lives would have been saved.
- Multi hazard zonation maps, vulnerability and risk assessment maps must be included and updated in DDMP.
- DDMP must be integrated with SDG's, SFDRR and climate change action plans.
- Most of the DM plan for construction sites covers the evacuation plan after the commissioning or operationalization of facilities. There are no DMP or EAP (Emergency action plan) during construction stage.
- Workers who are working in such Hazardous work places must be trained in first aid and survival skills.

(ix) A survey was organised for those rescuers of NDRF who responded in the Chamoli flash flood disasters. Analysing the survey following conclusion may be drawn: (https://docs.google.com/forms/d/e/1FAIpQLSceliexaIi79ljSwXO74dygEoVUDYDItuVL3hVesbP3OiVtDg/viewform?usp=sf_link)

- Rescuer encountered initial chaos and found that civil-military coordination was not adequate and need improvement.
- Rescuer interacted continuously at various sites with civilians and workers of construction agencies. They came to know that there was lack of very basic knowledge of evacuation drill was available with them.

(x) Another survey was conducted for general people. Total 272 responses where received from various part of country. Based on their response following conclusion are drawn: (https://docs.google.com/forms/d/e/1FAIpQLSftFciP4T3eTQpoEuyZUZHXCZA68DSmQeQH5SW9TmTVOM1xTg/viewform?usp=sf_link)

- Most of the people know about Kedarnath and Chamoli disaster but many of them were confused about the causes of these disaster.
- More than 72 % people know about GLOF and LLOF but more than more than 84% people have never attended any workshop seminar awareness session before.
- Almost 50% agreed that they are not aware about the availability of emergency logistics or early warning system in this regard.

- There was good response about do's & don'ts, before, during and after disasters because question of this segment were more or less based on common sense.

11. Recommendations

(i) Damage assessment is a process which prioritises the task to be assigned to rescue agencies. It is done by district administration for triage of worksite and it is called as *Sectorization* (As per INSARAG terminology). This process helps the district administration to distribute worksites to various agencies according to their capabilities. After the arrival of specialised agencies, other inadequately equipped agencies must be withdrawn. They must be engaged for some other task like relief and rehabilitation.

(ii) All development sectors must imbibe the principle of disaster risk management. Therefore all developmental activities must be closely monitored by district administration. All mitigation measures must be strictly followed by sectors and if found guilty restrictions must be imposed on them.

(iii) Recommendation of Dr Ravi Chopra reports must be implemented strictly.

(iv) Stability of debris dam near the crash zone needs to be monitored closely but we must also learn not to interfere in natural process as well.

(v) Narrow gorge and steep gradient in the characteristics of Rishiganga gorge. Reoccurrence of such event may lead to even more tragic incident since the riverbed downstream is laden with debris. Hence close monitoring early warning and community participation in such process are very important factor to minimize the losses.

(vi) Follow IRS in letter and spirit before, during and after the disaster.

(vii) Solar, Wind energy, Gharat (water mills) and local small hydroelectric power project (e.g. Kinnaur- 1Mw) must be encouraged.

(viii) Early approached roads to higher reaches must be earmarked and continuous effort must be done to keep it approachable by cutting bushes regularly.

(ix) Blast activities must be banned in the entire region as recommended in Dr. Ravi Chopra report.

(x) Update all the existing DM plans & Emergency Action Plan of Hydroelectric Plants and respective DDMPs must also be amended accordingly.

(xi) Prevention, Mitigation response and relief related to GLOF and LLOF must be included in school syllabus.

(xii) Training and Mock drills related to GLOF and LLOF must be conducted frequently.

(xiii) Early warning sensors which can raise an alarm after the certain water level may be installed and it should be manned 24x7. It must be integrated with siren. Traditional alarm systems like beating of drums, temple's bell and blowing of crouching shell must be encouraged in inaccessible areas

(xiv) **Media management**:-After any disaster, media person are under immense pressure to report the latest updates. Lack of information from reliable sources and lack of authentic footage often results in rumours. It also demoralizes the rescuers. It is recommended that video feed and photograph especially portraying progress of rescue and relief measures must be released to maintain transparency.

(xv) Based on hazard, risk, vulnerability & capacity (HRVC) analysis state administration must include the chapters related to do's & don'ts and response mechanism pertaining to various disasters in syllabus of school children.

12. Conclusion

Tejpal, a local driver told that "Nanda-devi ecological biosphere" is a fortress of Goddess Nanda surrounded by eight peaks and it cannot be breached. As and when there is human intervention whether in form of construction or expedition, we face fury of the Goddess. She is not very happy with approaching climbers and doesn't allow anybody to climb beyond its shoulders.

An unknown event triggered a massive destruction in Chamoli district. NDRF showed its worth again by prompt action and assisted the district administration in all manner possible be it response and recovery phase or rehabilitation. NDRF team is still present at Tapovan and providing assistance to district administration including psycho-social confidence. NDRF is the single largest force of the world dedicated to disaster response. Equipped with more than 310 state of the art equipment, NDRF is not only provides prompt disaster response but also works in tandem with district administration in preparedness and rehabilitation. Though the disaster management is a state subject but NDRF having a vast disaster response may be optimally utilized in preparedness also. This paper analyses the preparedness before Tapovan disaster thoroughly. These outcomes have been formulated as lesson learnt and challenges. Based on these facts few recommendations have also been suggested. In nut shell it is suggested that natural disaster can't be averted but preparedness and use of early warning systems may minimize the loss of lives and properties downstream.

Fig. 5.30 Picturesque view of Nanda Devi peak

Source: Google

Role of administration, response of various stakeholders and response by specialised agencies must be coordinated by a unified command. NDMA guidelines on IRS must adherence to in principle.

Apart from this community involvement in the key to success which has been already endorsed by previous reports like Ravi Chopra. SFDRR- 2015 also lays down the 10 point agenda to achieve the CBDRR community based disaster risk reduction. CAP, SSP and conduct of mock exercise is the only way we can involve the community to the maximum level. Unpredictability is the most common feature of Himalaya. So community must be made more predictable as far as preparedness and training of community is concerned. Disaster preparedness is critical because all of Uttarakhand falls in seismic zone IV and V. He mentioned in his report that the lessons from Kedarnath disaster Uttarakhand are relevant for all Himalayan states in India. Risk assessment, Risk communication and risk informed decision making is a command subject which is dealt with district administration hence they must be trained for that.

13. Abbreviation

CSSR-	Collapse Structure Search & Rescue
GLOFF-	Glacier Lake Outburst Flash Flood
ITBP-	Indo-Tibetan Border Police
IMD-	Indian Meteorological Department
CWC-	Central Water Commission
SDRF-	State Disaster Response Force
NTPC-	National Thermal Power Corporation
DDMP-	District Disaster Management Plan
DDMA-	District Disaster Management Authority
IRS-	Incident Response System
SFDRR-	Sendai Framework on Disaster Risk Reduction
INSARAG-	International Search and Rescue Advisory Group

References

1. Himalayan Tragedy: Purnima S. Tripathi
2. A Wake up call : T. K. Rajlakshmi
3. An interview of Sh. Kakachandra Sen, Director, Wadia Institute of Himalayan Geology
4. Dr Ravi Chopra report on Kedarnath Disaster
5. LBSNAA report on Kedarnath Disaster
6. NDMA GLOF guidelines
7. SEOC, UKD
8. NTPC
9. Survey/Initial information gathered from local
10. IMD/CWC/GSI
11. NTPC website:- Tapovan & Raini power generation capacity
12. 8th BN NDRF, Ghaziabad

CHAPTER 6

Multi-Stakeholder Planning and Coordination for Flood Risk Reduction in Bihar

Vivek Kumar Singh
Consultant – Planning and Coordination
Centre of Excellence in Disaster Management (CoEDM), Patna, Bihar
Corresponding Author: v.singh@dmi.ac.in

Geetanjali Kumari
Consultant – Health
Centre of Excellence in Disaster Management (CoEDM), Patna, Bihar

Amrita Dhiman
Coordinator
Centre of Excellence in Disaster Management (CoEDM), Patna, Bihar

Amritanjali Kumari
Assistant Professor
Birla Institute of Technology, Meshra, Ranchi, Jharkhand

Abstract

Floods are one of the most devastating and recurring natural disasters in Bihar. The Bihar Disaster Disaster Management Plan (BDMP) indicates that 28 of the total 38 districts in the state are prone to floods. Furthermore, the areas in South Bihar that were historically not flood-prone have also experiencing flood in recent years due to various climatic and human factors. This paper aims to share the experiences of various departments to manage extreme flood events and their adverse impacts on people, properties and livelihood. To capture stakeholders' rich and diverse experience, and experience sharing online workshop/ webinar was organised in Patna on 29 June 2021, which has been designed exclusively for the mixed group of participants viz policy planners, engaged in preparedness, capacity enhancement and execution of emergency response at Emergency Operation Centre and the site. The paper examines the multi-stakeholder innovative planning and coordination approach to bring together all concerned departments who have a tryst with the flood. Finally, the paper concludes that tackling the flood risk in Bihar, a multi-stakeholder collaborative approach is required to bring together tacit and explicit knowledge. An innovation platform would provide a convergence point to bring together all stakeholders.

Keywords: Flood Risk Reduction, Multi-stakeholder, Planning and Coordination, Sendai Framework, Innovation Platform

1. Introduction

Flood is one of the most devastating and recurring natural disasters, irrespective of the state of human development and natural conditions [1] in Bihar [2] [3]. It impacts the physical, economic, social, geographical, psychological, and cultural spheres of human endeavour in society [4]. The Bihar Disaster Management Plan [2] indicates that 28 of the total 38 districts in the state are prone to floods, and areas in South Bihar that were historically not flood-prone, have also started experiencing flooding in recent years various natural and human factors. The catastrophic floods endanger lives and cause human tragedy and heavy economic losses in north Bihar [2]. In addition to economic and social damage, floods can have severe environmental consequences as well. Due to the impact of climate change, the coming years are likely to see higher flood risks and greater economic damages in Bihar. Furthermore, the emerging trend and course of the SARS COV-2 (Covid-19) pandemic has aggravated the community level vulnerabilities and associated challenges. The changing scenario – both familiar and unfamiliar risks – requires integrated and innovative ways to reduce disaster risk, build resilience and achieve sustainable and inclusive development.

Though floods are natural phenomena, their severity and impacts could be reduced and limited through adequate preparedness and risk management approaches. The frequent and devastating nature of floods in Bihar requires innovations, need-based local efforts and inclusion of explicit knowledge along with tacit knowledge in the flood-prone areas. Furthermore, the emergence of COVID-19 requires a paradigm shift in tackling and coping with flood challenges under the complex situation of adherence to spread prevention protocols and vaccination schedules of the affected population [5].

Multi-stakeholder planning and coordination for community-level disaster risk resilience have become a significant policy objective for many governments [6] [7] [8] for several reasons. First, there is a paradigm shift towards community-based disaster risk management, which follows from recognizing the need to empower members of local communities in the disaster risk assessment process to spread support and increase local ownership of risk [9]. Second, the international agreements to prevent or reduce disaster risks at various level has increasingly shaped national and regional guidance and principles. The Sendai Framework for Disaster Risk Reduction 2015–2030, adopted in March 2015 by United Nations. The Sendai Framework aims to reduce the consequences of disaster risks, including their impact on lives, livelihoods and infrastructure.

Further, it shifts attention from managing "disaster" to managing "risk" and emphasizes learning from the effects of past hazardous events. According to the UNDRR, this represents a fundamental step in disaster risk prevention. Furthermore, climate change, increased urbanization, poverty, conflict, and globalization expose many people to disaster-prone areas [10]. The costs of disasters to governments and citizens have the potential to multiply and grow.

The paper examines the multi-stakeholder innovative planning and coordination approach and provides a platform to bring together all concerned departments who have a tryst with the flood. Finally, the paper offers a distinct opportunity and practical approach to share and transfer the knowledge to enhance the capability to deal with the grassroots level of tyranny in rural and urban flood-prone areas, particularly in the face of new COVID – 19 challenges.

2. Methodology

The current paper is based on two strategies: desk study and primary data gathered from workshop presentations and group works. The desk study helped review theories, views of different authors, information on the subject matter and key concepts related to the flood. Meanwhile, the paper aims to share the experiences of various departments to manage extreme flood events and their adverse impacts on people, properties and livelihood. To capture stakeholders' rich and diverse expertise, and experience sharing workshop was planned on 15 April 2021 in Patna, Bihar. However, due to the COVID outbreak and under the preventive protocols issued by the Government of Bihar, we were unable to organize the workshop in physical mode. Hence, On June 29, 2021, the Centre of Excellence in Disaster Management (CoEDM) at the Development Management Institute (DMI) organized a workshop on "Experience Sharing for Flood Risk Reduction in Bihar" in the online mode.

The workshop aimed to provide an experience sharing, reflection and co-creation platform to all concerned departments to deal with the tyranny of flood at the grassroots level with particular reference to the Covid-19 pandemic in the state. The Centre also aimed to document the discussions and insights received in the online workshop as a multi-stakeholder action-oriented report while capturing the various department's innovations and good practices in tackling the flood risk in Bihar.

The workshop has been designed exclusively for the mixed group of participants viz policy planners, engaged in preparedness, capacity enhancement and execution of emergency response at Emergency Operation Centre and the site. Nodal officers of the Government of Bihar at the state level from 15 departments and representatives of development partners are invited who are concerned with flood hazard and vulnerability reduction. In addition to the state nodal officers, five district-level government officials (ADM/District Nodal Officer - Disaster) are invited who are responsible and directly involve in flood crisis management.

Following the presentation by the keynote speaker and various department senior officials, the participants were divided into two groups (Table 6.1) led by the moderators from CoEDM, ADPC and UNICEF for a further detailed discussion about a riverine and urban flood in Bihar.

3. Result and Discussion

3.1 Flood Situation in Bihar

Bihar is a land-locked state demarcated by the river Ganges into north and south and further divided into regional

Table 6.1 Participant's Group Formation

Riverine Flood	Urban Flood
• Disaster Management	• Urban Development
• Rural Development	• Human Resources
• Agriculture	• Water Resources
• Animal Husbandry and Fisheries	• Health and Family Welfare
• Public Health Engineering	• ADPC Representative
• Panchayati Raj	• Bihar State Power Holding Company
• Rural Works	• UNICEF Representative
• Jeevika	• Representatives from Darbhanga, Muzaffarpur and Bhagalpur Districts
• Representatives from Sitamarhi, Supaul and Purnea districts	

Source: Based on the grouping of departments during the webinar.

blocks by rivers flowing down from the Himalayas and the Chhotanagpur Plateau to find repose in the Gange [2]. The Northern Gangetic plain of Bihar is the courtyard of rivers flowing from the Himalayas, like the Ghaghara, the Gandak, the Burhi Gandak, the Bagmati, the Kamla, the Kosi and the Mahananda. And the Southern Gangetic plain is the backyard to the rivers that descend the Chotangapur Plateau like the Karmansa, the Sone, the Punpun, the Phalgu, the Kiul and the Sakri. More than a dozen major rivers flow through the state, dividing it into seven "geo-cultural zones" (Table 6.2).

The rivers like Ghaghra, Gandak, Burhi Gandak, Bagmati, Kamla, Adhwara group of rivers, Kosi and Mahanada, originates from the Himalayas, and a considerable portion of their catchment falls in the glacial regions in Nepal and Tibet.

Table 6.2 Seven river zones in Bihar

SN	River Zones	Districts
1.	Ghaghara-Gandak	East and West Champarans, Gopalganj, Siwan and Saran
2.	Gandak-Bagmati	Sheohar, Sitamarhi, Muzaffarpur, Vaishali, Samastipur & Begusarai
3.	Bagmati-Kosi	Darbhanga, Madhubani Supaul, Saharsa & Khagaria
4.	Kosi-Mahananda	Madhepura, Araria, Purnia, Kishanganj and Katihar
5.	Karmnasa—Sone	Buxar, Kaimur, Bhojpur and Rohtas
6.	Sone—Punpun	Patna, Jehanabad, Arwal, Gaya, Nalanda, Aurangabad and Nawada
7.	Punpun-Sakri	Sheikpura, Lakhisarai, Jamui, Banka, Munger and Bhagalpur

Source: Bihar Disaster Management Plan 2014.

Among the flood-affected districts - Sitamarhi, Supaul and Kishanganj are 90% affected, five districts- Bhagalpur, Darbhanga, Khagaria, Madhepura, Saharsa, get around 70% affected and the rest of the districts vary from 55% to 25%. These floods destroy thousands of human lives apart from assets and livestock worth millions on an annual basis.

For nearly five months, the state suffers through the ravages of flood. Along with Ganga and Kosi, other major rivers, like

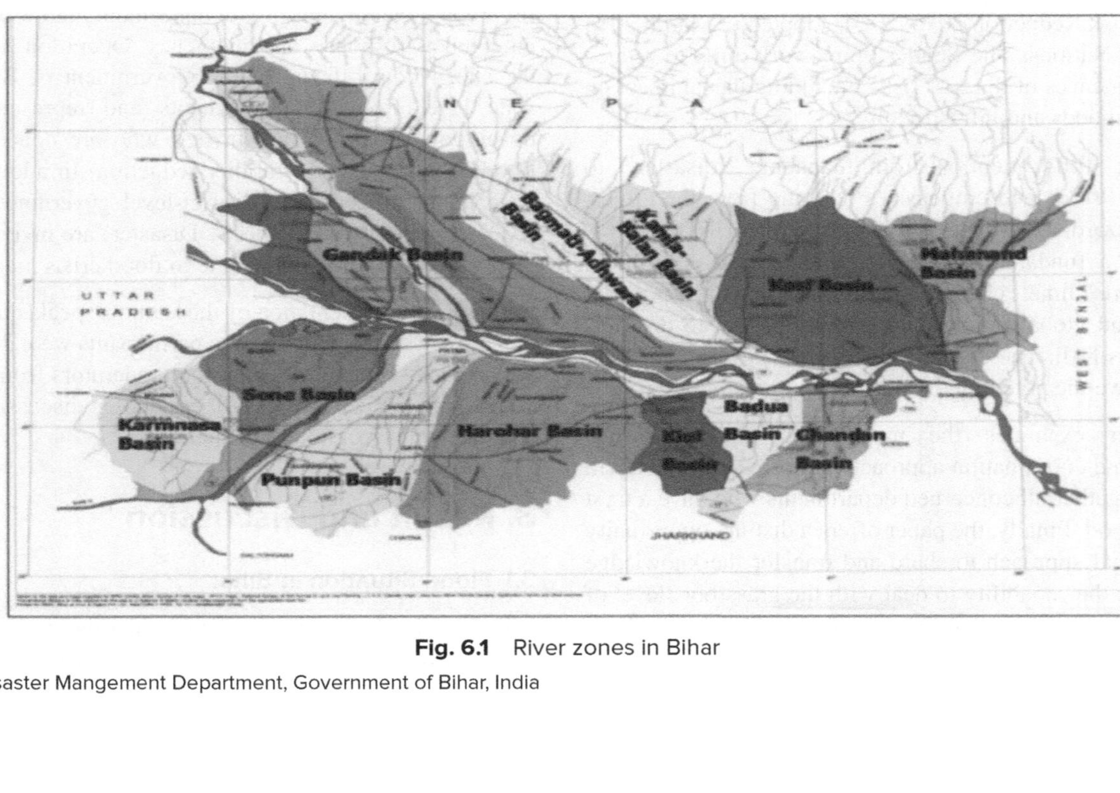

Fig. 6.1 River zones in Bihar

Source: Disaster Mangement Department, Government of Bihar, India

Son, Saryu, Gandak, Budhi Gandak, Bagmati, etc., get flooded during the monsoon months and affect the surrounding areas.

Most of the rivers mentioned above are snow-fed and perennial inflow and positioned to receive very copious rainfall during monsoon when the discharge of these rivers is 50 to 90 times larger than fair-weather flow in the region [11]. This causes frequent and large scale flooding in North Bihar. As such, 73.63 per cent of North Bihar's geographical area is considered prone to floods.

4. Phases in Flood Preparedness Planning

Flood preparedness planning involves putting in place a set of appropriate arrangements in advance for an effective response to floods [12]. Some of the commonly identified flood preparedness activities are; Public awareness-raising on flood preparedness, response and mitigation measures in flood-prone areas; Stockpiling of emergency relief materials, i.e., ready made food items, fodder for livestock, emergency medicines, materials for temporary shelter etc.; Installation of community-based early warning system for issuance of timely and effective flood warnings; Management of safe areas for temporary and quick removal of people and property from a threatened location; Transportation to secure sites/evacuation centre; Ensuring access to health and sanitation facilities; Conducting drills and rehearsals [12] [13]. The key for flood preparedness planning is to have clarity and agreement on the roles and responsibilities of relevant stakeholders - such as government agencies, disaster management organizations, Red Cross, voluntary groups, and community members. In the workshop, participants discussed the pre, during (emergency) and post planning steps and requirements in Bihar to mitigate/ reduce the flood losses.

4.1 Pre-Disaster Planning

Pre-disaster planning is one of the core areas where the concerned departments need to collaborate individually. There are certain areas where their role during the flood is already cut out, but the discussion in the workshop focused on emphasizing the need for identifying the areas where they should work in association. There are overlapping areas, which, if thought before the disaster, would reduce the burden on a particular department and make the task smoother and faster.

Broadly the pre-disaster planning can be achieved through the following action steps, which reflected the presence of the experts:

- Preparation of State and District Profile - Information about the state/district, past and floods and the community are essential building blocks for an effective flood preparedness programme. It should include location, population, high-risk areas, Geography, Livelihood details, demographics, and basic infrastructure.
- Hazard, Vulnerability and Capacity Assessment - Hazard, Vulnerability and Capacity Assessments (HVCA) are undertaken to formulate a Risk Assessment. It helps create a pool of flood data and analysis that will give a clear picture of the potential damage and loss in a particular location and the existing capacities to prepare for floods. This information can be collected through using a participatory tool with the involvement of the concerned community.
- Institutional mechanism for flood preparedness planning and flood risk management enables designation of clear roles to various agencies, with limited duplication of tasks, and be an effective arrangement for collaboration and coordination between:
  - Agencies working at the local level,
  - Local and national systems/plans,
  - Local systems with linkages to others in adjoining areas.

4.2 Emergency Response Planning

During the flood, a well-organized flood emergency response plan can make a formal advanced plan, such as a flood emergency response plan, that those involved are aware of the most important steps to take and have adequate resources on hand. Flood emergency management contributes to flooding risk reduction during emergency. Flood risks can be reduced by decreasing the magnitude of hazards, preventing exposure of individuals and their activities to flooding and diminishing the vulnerability of flood-prone communities. Further, the flood emergency response measures can reduce exposure to flooding to reduc/mitigate the loss of people, property and livelihood.

4.3 Post Disaster Planning

Disaster Management consists of five main phases - prevention, preparedness, response, mitigation and recovery to mitigate/reduce the loss of life, livelihood and property. While significant knowledge exists on the first four phases, recovery remains less developed and poorly understood.[1] Recovering from disasters is a challenging process. Disasters undermine hard-won development gains and create more significant pressures than are found in 'normal' development settings: heightened urgency; the need for rapid reaction and

[1]https://www.recoveryplatform.org/assets/tools_guidelines/National%20Post%20Disaster%20Recovery%20Pl ans%20and%20Coordination.pdf

speedy results; media scrutiny; lack of experience, capacity and expertise; a multiplicity of national, local and international actors whose efforts all require coordination; and resource scarcity and few reliable funding sources to meet recovery needs. Responding to these demands requires a clear road map, an efficient, well-organized process, knowledge gained from past failures and successes, rapid decision-making and implementation capabilities and skilled coordination.

It is hence evident that Flood Prepared Planning plays a pivotal role in flood mitigation. Further, the collaborative efforts of all the concerned departments enable and empowers all the affected people.

5. Stakeholder for FRR in Bihar

The transition from a response-based paradigm to an anticipative, prevention-based approach remains a stubborn challenge in Disaster Risk Reduction [14]; [15] and flood management is not an exception. The Sendai Framework (2015-2030) recognizes that the state has the primary role in reducing the disaster risk but that responsibility should be shared with other stakeholders, including local government and private sectors. There is a need to create a common platform to bring all the lead stakeholders under one umbrella. By definition, in simple terms, stakeholder means - "individual or group that has an interest in any decision or activity of an organization."[2] In Bihar, various national, state, and community-level government departments and national and international organizations (Table 6.3) are involved in flood risk reduction and management. The disaster risk reduction has to be made a shared responsibility, and the role of any particular department must be known to all. The primary responsibility of the concerned department varies between the phases of the flood. Some department may have a key role in preparedness while other may have a significant role during recovery. Hence it is essential to have a strategic multi-stakeholder approach for flood mitigation. The need and role of the multi-stakeholder process for flood risk reduction was an intrinsic part of the presentation and discussion and the pivotal point of organizing the workshop, named Multi-Stakeholder Action-Oriented Workshop on Disaster Risk Reduction.

Identification of appropriate stakeholder and their role and responsibilities plays a vital role in flood risk management. A review conducted by Sir Michael Pitt into the 2007 summer floods in the UK concluded that a lack of clarity in the responsibilities of government agencies and non-governmental stakeholders was one of the key factors that contributed towards the extent of the flooding [16]. Therefore, appropriately informed stakeholder decision-making is an important aspect of reducing and managing disastrous events.

Previous research has demonstrated which stakeholders should be involved in DRM activities and at what stages of the Design-Construction-Operation Process (DCOP) should be involved in flood risk reduction. The pre-construction phase was identified as the critical phase in the DCOP when DRM activities, and specifically flood mitigation, can be a integrated approach to incorporate tacit and explicit knowledge. During this phase in particular, various stakeholders and emergency/risk managers should make hazard mitigation considerations.

5.1 Multi-Stakeholder Approach for Flood Risk Reduction

Enhancing community level community resilience has increasingly involved national and regional governments adopting a multi-stakeholder approach because of the potential interagency benefits [6]. In this article, we are interested in describing and explaining the developments of stakeholder initiatives in flood risk management in Bihar and the responses these initiatives evoke from governmental actors. Based on literature about stakeholder involvement in water management and flood risk management, we expect the goals, resources and strategies stakeholders employ and how governmental actors respond to explain the role and impact of stakeholder initiatives and plans on decision-making regarding flood risk management. If the goal or strategy of stakeholders is to fight governmental plans for flood risk management, the response by governmental actors is supposed to be negative to stakeholder initiatives and results in lower impacts. If stakeholders have the resources (time, knowledge, expertise, etc.) to develop their flood management plans and initiatives, governmental actors are supposed to respond to stakeholder initiatives positively.

In Bihar, several stakeholders have a responsibility for managing flood risks and improving community resilience. They include government authorities, private and non-governmental organizations, and communities at risk. Such multi-stakeholder approaches, enshrined in international frameworks, recognize that no single agency has all the necessary power, expertise, resources, and structure to address the challenge of reducing new or existing disaster risks [17].

Since time immemorial, residents of north Bihar have developed the synergy to live with flood-affected areas. They have developed tacit knowledge to deal with the adverse situation during the flood. Tacit knowledge is generally inherent in society and transferred from one generation to

[2]ttps://asq.org/quality-resources/stakeholders

Table 6.3 Stakeholders and their activities in FRR

SN	Stakeholders	Activities
1	National government: Line departments	• Formulate preparedness related policies and issuance of government decrees • Flood forecasting and dissemination to local governments and humanitarian organizations • Arrange financial support during emergency • Deploy NDRF team during an emergency
2	State Government and District Administration	• Prepare Flood-prone district profile • Conduct Hazard, Vulnerability, and Capacity Assessment (HVCA) in Flood-prone regions • Dissemination of warning, based on forecast • Preparing for search-and-rescue • Arrange logistics and transport to safe areas • Deploy SDRF team during an emergency • Identify, organize and maintain safe flood areas • Ensure access to clean water sources at safe flood areas and appropriate sanitation facilities • Prepare to provide temporary shelter, relief and basic medical services • Arrange emergency kits • Understand and use formats for rapid needs-assessment so that they can inform the national government and request additional relief if needed • Identify key sources/humanitarian organizations that will assist in relief and coordinate with them throughout the year and more frequently before the flood season • Clarify and document roles and responsibilities within the government and those delegated to humanitarian organizations • Prepare an FPP in collaboration with humanitarian organizations • Review flood preparedness plans following floods and revise for forthcoming floods as and when appropriate
3	Community (volunteer groups such as food and shelter group, evacuation, information, etc.)	• Evacuation to safe areas upon receiving the warning • Protecting household goods by storing on higher levels • Use personal boats for helping those in need • Contribute to stockpiles of food and fodder • Recruit and register young volunteers so that they are readily available and trained for emergencies • Recruit and register swimmers • Recruit and register-trained community members who can • disseminate flood forecasting information to other members of the community
4.	International, National and Local NGOs	• Provide technical assistance in forecasting • Assist with transport to safe areas • Facilitate organization to provide water/sanitation, temporary shelter, medical needs or food, as required or requested by government • Provide relief as required or requested by the government • Facilitate preparedness planning

Source: Adopted and modified from ADPC Manual on Flood Preparedness

another. In the meantime, due to technological advancement, various explicit knowledge is also being utilized to deal with the tyranny of the flood. In multi-stakeholder approach provides a platform to bring together both tacit and explicit knowledge. The people who face the problem must have their voice during the planning.

Flood has a multi-pronged impact on the lives of affected people, animals and the surrounding peripheral areas in Bihar. It has enormous social, economic, physical, emotional repercussions, and the damage is multi-dimensional. The disaster also impedes the developmental aspiration of any nation or state, and since being natural, it cannot be stopped but can be met with better preparedness. No proof is required to demonstrate that a multi-department determination with action is required to address any disaster's direct and indirect impact. The lead role of departments also varies during preparedness, response and recovery phases. Hence it is important to have a multi-stakeholder strategy, to respond better and manage the disaster and its outcomes.

With clarity on their role, the collective responsibilities of departments would improve the service delivery during and immediately after the flood, reach the disaster-struck areas, and provide relief in terms of food, medicine, and rescue. It would also minimize the time and cost of services reaching the people, thereby improving the overall welfare of people.

5.2 Identification of Actionable Points through Multi-Stakeholder Approach

- Review of existing communication system between the department to expedite the actions during floods.
- We are creating a common platform of stakeholders to reduce response time, thereby reducing the losses at all levels.
- Emergency Response Plan, i.e. organizing a meeting as part of preparatory planning, should have multi-stakeholder participation. The effort should not encompass only government departments but also other key agencies like Municipality, Resident Welfare Association, Private School etc. for the preparedness phase
- Capacity enhancement Plan provides the skill to combat the blow of disaster in all three phases. This includes the multi-stakeholder training and orientation programs and also organizing mock drills with concerned stakeholders. Organizing a Mock drill is an effective strategy as it realizes the participants about the time and motion required during the floods.
- Response activity is highly effective with a multi-stakeholder approach. During the disaster times, the coordinated and symmetrical efforts of the State Emergency Operation Centre, District Emergency Centres, Emergency Support Functions, and Communication & Information Response would immensely help expedite the response and relief activities.
- Induce COVID appropriate measures while planning, response and recovery.
- A thorough damage assessment, listing of challenges, and lessons learnt should be documented and shared in the post-disaster phase. It must be included in the coming year's multi- stakeholders preparedness plan. Further, it would result in developing an improvised plan; innovations would be shared and replicated. This becomes more pertinent in the availability of services in the health care system, educational institutions and others.
- Search and rescue operations by the departments during flood times should be done in coordination. This becomes more significant in resources limited and time intrinsic situations during floods.
- Recovery work by nature is multi-sectoral dependent, but mainly it's been observed that during the disaster, multi-stakeholders efforts are observed, which diminishes during the disaster recovery phase. In the absence of such coordination, the revival of service delivery takes time, quality is compromised and becomes more effortful.
- Coordinated efforts are highly requited for search, rescue and evacuation activities. The NDRF and SDRF are in places they have less information about the whereabouts of these areas or people. Here roles of DEC, SEOC, Resident Welfare Agency and others play a critical role.
- Covid has also raised some additional challenges where the mandates set for ideal Covid behaviour would not be compromised in case of urban flooding. Hence, such a situation must be an important part of response planning with pre-thought action plans.
- Ensuring the reach of relief and services post-disaster is one of the intrinsic parts of disaster planning. This is more important for urban areas due to less availability of space for relief and evacuation purposes.
- The recovery phase may be taken as an opportunity to follow the "Build Back Better" principle. Recovery phase planning should put considerable effort in assessing and planning in terms of (if) reoccurrence of similar disaster but where the damage can be minimized.
- To improve the preparedness plan and to make it more meaningful, post-disaster documentation should be done. This will help clarify what worked and what did not work, identify gaps in the planning, and find the areas of response efficiency. The documents created should be kept as institutional depository so that irrespective of the turnaround of a team, every person involved over a period of time gets the benefit of such records.
- Each department has its intrinsic capacity building program for its team. Here the role of DMD gets significant by developing strategic capacity building programs for all the concerned departments.
- At the district level, the bottom-up approach should be applied to address disaster-related issues

6. Concluding Remark

The transition from a response-based paradigm to an anticipative, prevention-based Flood Risk Reduction (FRR) requires an innovative multi-dimensional approach that combines tacit and explicit knowledge. The discussions confirmed and highlighted the willingness of all the concerned departments that disaster mitigation is an exhaustive and multi-faceted endeavour. Disaster does not have only developmental implications. Instead, it has a humanitarian angle also to it; it is indeed in the interest of all to collaborate and get the comparative advantage of each other. It also envisaged bringing a paradigm shift in the outlook of the concerned departments to follow the COVID norms by prioritizing to prevention, control and prediction of flood risk in Bihar.

The presentations and deliverance by the participants were candid and dealt with self-introspection, which realized the existing gaps and the core expertise in mitigating disasters, especially floods. The disaster was looked at from the perspective of 'Build Back Better", which lays the importance of strategic planning within the departments and amongst the departments. Furthermore, tackling the flood risk requires tacit and explicit knowledge in a multi-stakeholder collaborative approach. To tackle the tyranny of flood, an innovation platform would provide a convergence point to bring together all stakeholders and tacit and explicit knowledge.

REFERENCES

1. A. Novita Sari, A. Susilo, and E. Susilo, "The role of stakeholders in flood management: study at Ponorogo, Indonesia," *Int. J. Eng. Sci.*, vol. 2, no. 10, pp. 1–10, 2013, [Online]. Available: www.theijes.com.
2. DMD, "Government of Bihar Disaster Management Plan," 2014.
3. CoEDM, "Multi-Stakeholder, Action-Oriented Report on Flood Risk Reduction (FRR) in Bihar," Patna, 2021.
4. B. S. Agbola, O. Ajayi, O. J. Taiwo, and B. W. Wahab, "The August 2011 Flood in Ibadan , Nigeria : Anthropogenic Causes and Consequences," no. August 2011, 2013, doi: 10.1007/s13753-012-0021-3.
5. E. Paul, G. W. Brown, and V. Ridde, "COVID-19: Time for paradigm shift in the nexus between local, national and global health," *BMJ Glob. Heal.*, vol. 5, no. 4, pp. 1–5, 2020, doi: 10.1136/bmjgh-2020-002622.
6. J. Adekola, D. Fischabcher-Smith, and M. Fischabcher-Smith, "Inherent Complexities of a Multi-stakeholder Approach to Building Community Resilience," *Int. J. Disaster Risk Sci.*, vol. 11, no. 1, pp. 32–45, 2020, doi: 10.1007/s13753-020-00246-1.
7. D. Chandler, "Beyond neoliberalism: Resilience, the new art of governing complexity," *Resilience*, vol. 2, no. 1, pp. 47–63, 2014.
8. E. Ntontis, J. Drury, R. Amlôt, G. J. Rubin, and R. Williams, "Community resilience and flooding in UK guidance: A critical review of concepts, definitions, and their implications," *J. Contingencies Cris. Manag.*, vol. 27, no. 1, pp. 2–13, 2019, doi: 10.1111/1468-5973.12223.
9. A. Maskrey, "Revisiting community-based disaster risk management," *Environ. Hazards*, vol. 10, no. 1, pp. 42–52, 2011, doi: 10.3763/ehaz.2011.0005.
10. I. Kelman, "Axioms and actions for preventing disasters," *Prog. Disaster Sci.*, vol. 2, pp. 4–6, 2019, doi: 10.1016/j.pdisas.2019.100008.
11. R. Sinha, M. Burton, and G. Tiwari, "Strengthening the institutional framework for flood and water resources management in Bihar: Developing a strategy for reform (Phase I)," no. September, p. 51, 2012, [Online]. Available: http://www.waterwoes.iitk.ac.in/Pdf/Sinha et al. 2012_PhaseI.pdf.
12. J. Alam and M. Muzzammil, "Flood Disaster Preparedness in Indian Scenario," *Int. J. Recent Trends Eng. Technol.*, no. April 2011, pp. 33–38, 2011, [Online]. Available: https://www.researchgate.net/publication/280082664_Flood_Disaster_Preparedness_in_Indian_Scenario.
13. Asian Disaster Preparedness Center, "Manual on Flood Preparedness Program for Provincial and District Level Authorities in the Lower Mekong Basin Countries," p. 104, 2009.
14. UNISRR, "Sendai Framework for Disaster Risk Reduction 2015 - 2030," 2015.
15. A. Ogra, A. Donovan, G. Adamson, K. R. Viswanathan, and M. Budimir, "Exploring the gap between policy and action in Disaster Risk Reduction: A case study from India," *Int. J. Disaster Risk Reduct.*, vol. 63, no. June, p. 102428, 2021, doi: 10.1016/j.ijdrr.2021.102428.
16. L. Bosher, A. Dainty, P. Carrillo, J. Glass, and A. Price, "Attaining improved resilience to floods: A proactive multi-stakeholder approach," *Disaster Prev. Manag. An Int. J.*, vol. 18, no. 1, pp. 9–22, 2009, doi: 10.1108/09653560910938501.
17. P. Hickman, "A Flawed Construct? Understanding and Unpicking the Concept of Resilience in the Context of Economic Hardship," *Soc. Policy Soc.*, vol. 17, no. 3, pp. 409–24, 2018, doi: https://doi.org/10.1017/S1474746417000227.

CHAPTER 7

Urban Floods in the Time of Pandemic—Hyderabad 2020

Vijaya Kumari Nunna
Mahatma Gandhi National Council of Rural Education Hyderabad – 500004

Abstract

The city of Hyderabad witnessed severe floods during the month of October, 2020 due to deep depression in the Bay of Bengal. Simultaneously heavy rainfall was reported from the states of Puducherry, Karnataka, Maharashtra and Andhra Pradesh. In Hyderabad the continuous torrential downpour resulted in loss of life and tremendous damage to property. Water levels in some neighborhoods reached 12 feet, with over 80 people having lost their lives and about 40,000 families being displaced. When the waters receded from areas, they left behind a trail of huge mounds of trash and mangled vehicles, and houses covered in mud and post-rain gathering up to 20,000 tons of waste. Standing crops in several parts of Telangana were also damaged. In the present study, the extensive damage caused due to the floods have been discussed along with the mitigation and management strategies to prevent as well as reduce the impact of floods.

Keywords: Floods, Impact, Displacement, Relief and Rehabilitation, Mitigation, Management

1. Introduction

Hyderabad has a long history of existence. It is more than 400 years old. Hyderabad is located on the river Musi a tributary of Krishna River (NIUA, 2016). The Musi, along which the modern city of Hyderabad was established in 1591, was once feared for its torrential waters, especially during the monsoons. The Thugyani Sitambar or the Great Musi Flood of September 1908, is considered an epochal moment in the city's history. The river's catchment areas received 98.57 cm rainfall causing the waters to rise over 60 feet. An estimated 15,000 were killed and homages are still offered every year to a tamarind tree which saved the lives of over 150 people who climbed on it to escape the waters. The Nizam of Hyderabad invited Sir Mokashagundam Visweswaraiah to develop a flood mitigation system for the city. A modern drainage system was developed and two major reservoirs of Osman Sagar and Himayat Sagar were built that provided a source of drinking water and irrigation purpose reservoir (GHMC Report, 2018). A City Improvement Board was established in 1914 to oversee a planned and phased development of the city.

Urban flooding is a man-made disaster due to overburden of drainage systems and unregulated development of land with least concern towards natural topography and hydrological factors leading to reduced natural seepage and terrain alteration. Urban flooding is of common occurrence these days in India. A number of metro cities in the country including Mumbai, Chennai, Delhi and Bengaluru have suffered from it. The latest victim of urban flooding is Hyderabad. The city as well Telangana received unusually excessive rainfall on October 13-14, 2020, due to a deep depression that developed in the Bay of Bengal. In the present study the impact assessment as well as the strategies adopted to mitigate and manage the floods have been report.

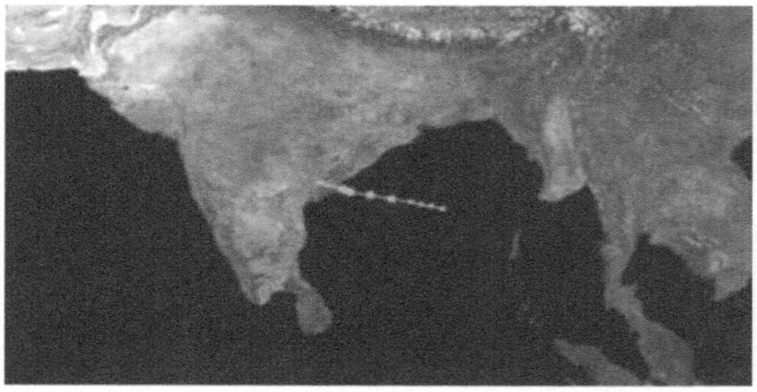

Fig. 7.1 Imagery showing the storm direction
Source: NRSC Data Centre, 2020

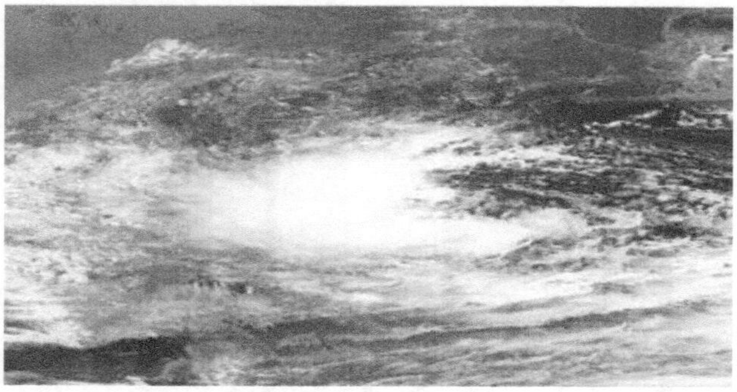

Fig. 7.2 Depression in Bay of Bengal

Source: NRSC Data Centre, 2020

2. Study Area

Hyderabad has spread in all directions in recent years. The population of the city has grown exponentially. It is 10 million today, up from 1 million in 1950 and 0.1 million at the beginning of the 19th century. New roads, buildings and infrastructure have appeared. New housing colonies have been established. Hyderabad gradually transformed in to a metropolis, a centre for IT Hub and a Pharmaceutical Capital of India.

Fig. 7.3 Location Map

However, with time the Covid pandemic struck India in the month of March and a two month Nationwide lockdown for the months of April and May was declared late March, 2020. During this period and later on also many people lost their livelihoods and even suffered and struggled to survive the pandemic.

In Hyderabad, the month of October was worse with the disastrous impact of continuous torrential downpour caused unprecedented flooding and related incidents in several places of the city. The 2020 Hyderabad floods were a series of floods associated with Deep Depression in the Bay of Bengal that caused extensive damage and loss of life as a result of flash flooding in Hyderabad, India in October 2020.

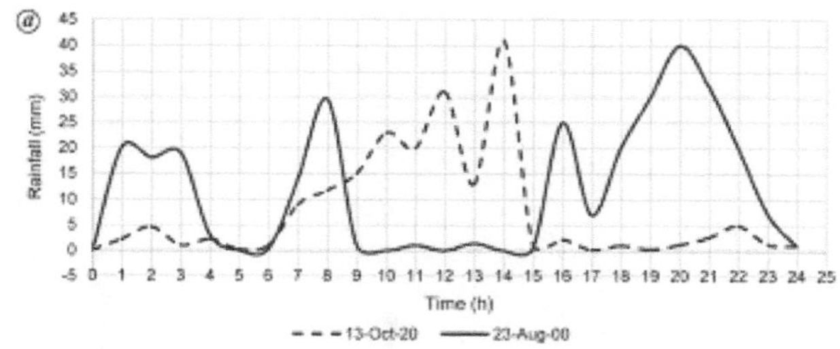

Fig. 7.4 Rainfall in Hyderabad

Source: Swathi, 2020

3. Impact Assessment

Heavy damage to property, roads and human lives has been reported. The city which received an average rainfall of 69.1 cm in a year, had seen parts of it receive 28 cm on this particular day alone. However, as heavy rains continued to beat down upon the city for over a week, lakes and drains which have been neglected and encroached upon for decades, could not keep up with the inflows. Overflowing waters from lakes flooded surrounding colonies with a force that brought down walls, overturned parked vehicles of all sizes and carried them afar, and mangled steel shutters of shops. Residents spent a sleepless night in low-lying areas as water entered their houses and damaged household articles. As heavy rains inundated and further damaged the roads, there were traffic snarls in many key industrial, commercial and residential areas in almost all parts of the city.

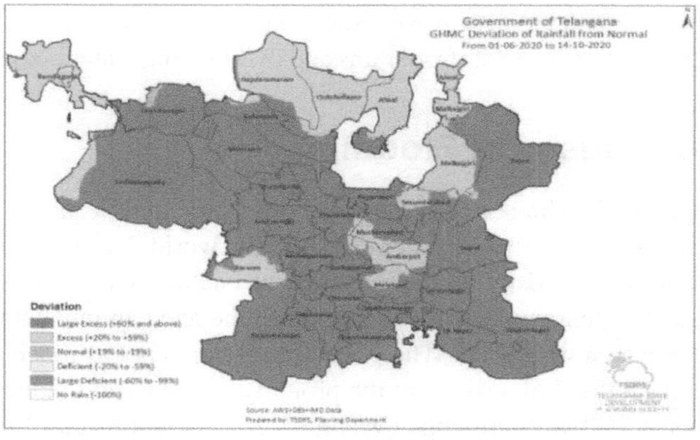

Fig. 7.5 Record Rainfall and Flooding

Source: NRSC Data Centre

The force of the water caused structural damage to the Nayapul bridge and washed away metallic barricades erected on the Chaderghat and Moosarambagh bridges. With huge inflows, the water level in Hussain Sagar lake in the heart of the city almost reached the full level. The level stood at 513.4 feet as against the full tank level of 513.7 feet. The river inundated more parts of the city when the gates of the

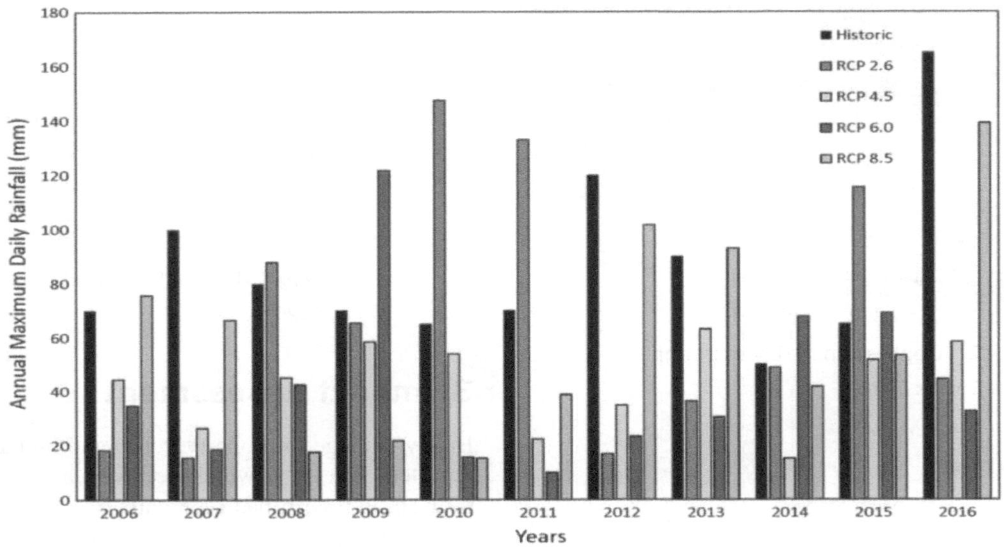

Fig. 7.6 Urban floods from 2006 to 2016 in Hyderabad

Source: NIUA, 2016

Himayat Sagar lake were lifted after more than a decade. The Musi's fury knew no bounds as it crashed through every encroachment that stood in its natural path. Loss of life was reported by drowning, electrocution, and by being buried under a wall collapse. Rainfall reached over 110 millimetres (4.3 in) in parts of Hyderabad, with heavier rainfall amounts outside of the city. With over 80 people having lost their lives and about 40,000 families being displaced, post rain gathering up-to 20,000 tons of waste. Several low lying localities were deluged in the water from rains and overflowing water bodies.

4. Causes of Flooding

Urban flooding is a global phenomenon and it poses a challenge to urban planners all over the world. Urban floods can be caused by direct and indirect Factors. Urban floods can happen in a very short period of time and can inundate a large area with water. The volume of water is not as much as a flash flood of a river but the property damages and indirect financial losses are huge (Vazhuthi, 2020). As a result of the higher flow rates, flooding happens very rapidly, even in minutes (Jainer, 2020). Urban drainage systems, or "storm water management," for redirecting water flow has been to achieve maximum convenience at an individual site by removing excess surface water as quickly as possible after a rainfall and containing and disposing of that water as quickly as possible through a closed/open conveyance system, according to Intergovernmental Panel on Climate Change (IPCC, 2007).

Human habitations have sprung up in low-lying areas throughout India, frequently encroaching on drainage channels, not only in Hyderabad but in all Indian cities and towns (Ahmed, 2013). Others point out that it is an example of bad planning, encroachment of lakes and poor urban infrastructure and preparedness. Hyderabad lost 33 lives, nearly 40,000 families got badly impacted and suffered a property loss of Rs. 6.7 billion (Rs 670 crores) as per official estimates. It is this lackadaisical attitude that has led to the number of encroachments increasing, to 28,000 in the last 13 years. Some experts claim that it is a clear case of the impact of climate change causing extreme weather events.

5. Relief and Rehabilitation Measures

For the safety of people immediate Relief and rehabilitation measures were implemented. 360 National Disaster Response Force personnel, as well as Indian Army forces were deployed. The Telangana government requested the Central government to provide relief to Hyderabad and surrounding areas. The state Government sought immediate release of ₹1,350 crore, in which ₹600 crore for farmers and ₹750 crore on relief and rehabilitation works in the Greater Hyderabad Municipal Corporation area. On 14 October, the Telangana Government declared a two-day holiday for all nonessential workers due to flooding, and urged everyone to stay home. Amid the possibility of further flooding, more than 2,100 families were evacuated near Gurram Cheruvu. More than 150,000 meal packets were distributed to flood-affected areas.

Furthermore, 60 teams were tasked with spreading bleach in cellars and open areas to prevent the spread of waterborne and vector-borne diseases.

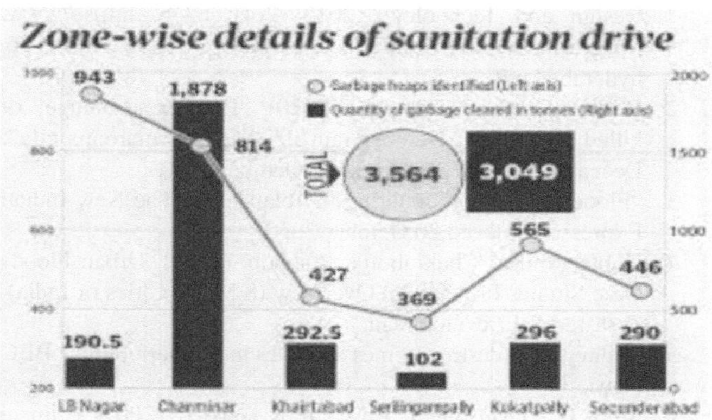

Fig. 7.7 Waste management

Source: Gupta et al, 2009

6. Flood Mitigation and Management Strategies

Extreme rainfall events occurred in Hyderabad in 2000, 2010 and 2016 and now in 2020. Sudden spells of heavy downpour, cloud bursts and flash floods over a week and especially on the intervening night of October 13-14, exposed the fragile and inefficient urban infrastructure of the capital of Telangana. Extreme weather events are wake up calls for urban planners to firm up strategies and implement measures to minimize the damages of such events, whose occurrence could only increase in future, say experts.

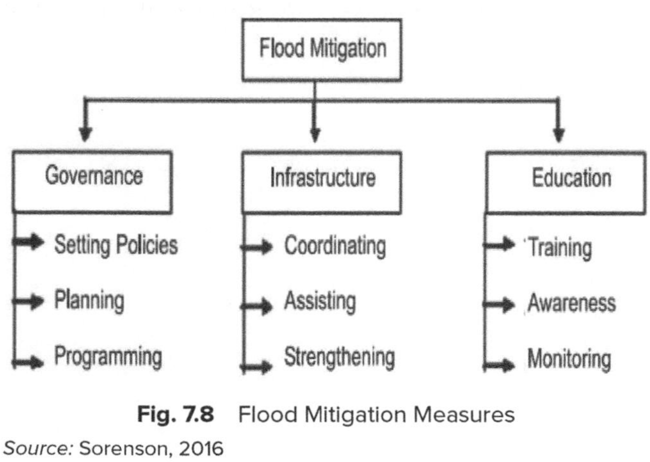

Fig. 7.8 Flood Mitigation Measures

Source: Sorenson, 2016

Preventing construction and diverting floods away from residential areas and channeling floods is an excellent flood prevention technique, but incorporating this into an action plan is challenging for any developing community (Sorensen, 2016). With the Krishna and Godavari river basins too coming into the picture with water supply being brought to Hyderabad from long distances through pipelines, there is an urgent need to have an expert commission to devise land use plans, drainage and sewage systems and disaster mitigation.

Urban storm water management may not be efficient unless it was handled in a conjunctive manner. A multidimensional approach is essential to solve this problem. Some of the management options that can be implemented are as follows:

1. Risk mapping of the areas of the city should be done to assess the vulnerability, related to urban floods using GIS technology.
2. Limit, reduce and/or mitigate for impervious surfaces throughout the watershed by Use of new engineering technique like pervious pathways, pervious parking lots should be considered and implemented wherever possible to minimize the surface run-off.
3. Town Planning department of GHMC should regularly monitor the the prohibited areas to prevent encroachments.
4. Conserve and protect areas for groundwater recharge. Utilize conjunctive management to enhance groundwater storage: in some instances, diversions of surface water for groundwater infiltration could enhance flood management by reducing peak flows.
5. GHMC/HMWSSB should rethink its policy of charging amount for rainwater harvesting structures while according permission to plan. They should keep this as a check point for giving occupancy certificate.

A well-established and comprehensive drainage management and sewage infrastructure is essential for any city that is frequently disrupted by floods, and a well-established and robust drainage management and sewage infrastructure is important for any city that is frequently disrupted by floods (Sorensen, 2016). There is a need for exploring the concept of climate proofing cities, as per the report of the Centre for Urban Governance and Infrastructure Development, Environment & Energy at the Administrative Staff College of India (ASCI). In addition to disaster management, cities too require a certain amount of resilience built in. Hyderabad is a fit case to identify 'hot spots' for flooding, design materials and systems that will withstand such events. Hyderabad and even Telangana state do not have a disaster management plans. They are not fully utilizing the doppler radar, which can give advance and precise warnings. The focus is on disaster response, even while allowing unplanned growth as reported by National Disaster Management Authority (NDMA).

There are varying solutions for the urban flooding and the options have to choose according to the need and why the viable solution is best should be evaluated and facilitated with the practical applicability of the solution. The planning process should not be unilateral phenomenon but shall involve the stake holders responsible for implementing the provision of flood risk reduction plan and allied responsibilities (Gupta, 2009).

7. Conclusions

The city of Hyderabad experienced severe flooding due to torrential rainfall in the month of October, 2020. Heavy overnight rains wreaked havoc in parts of Hyderabad and its suburbs as water entered some residential areas and flooded roads, and caused significant damage to property. With storm water drains overflowing, water entered residential colonies and the parking lots in several commercial complexes had completely submerged. Hyderabad lost 50 lives, nearly 40,000 families got badly impacted and suffered a property loss of Rs. 6.7 billion estimates. The unusually heavy rainfall was due to two reasons viz. the deep depression in the Bay of Bengal that moved onto the land in Hyderabad and the second was the rain bearing clouds of the withdrawing Southwest monsoon also pouring out.

This unusual heavy rains that caused flooding in Hyderabad in mid-October, exposed the faults in the urban infrastructure. Some experts claim it is a clear case of the impact of climate change causing extreme weather events. Others pointed out that it is an example of bad planning, encroachment of lakes and poor urban infrastructure and preparedness. The extreme weather events are wake up calls for urban planners to firm up strategies and implement measures to minimize the damages of such events, whose occurrence could only increase in future.

References

1. Ahmed Z, Rao DRM, Reddy KRM, Raj YE. Urban Flooding–Case Study of Hyderabad. Global Journal of Engineering, Design and Technology. 2013; 2(4): 63–6. https://www.longdom.org/articles/urban-flooding--case-study-of-hyderabad.pdf
2. Desk, DC Web (15 October 2020). "Hyderabad deluge: 19 killed as heaviest October rain in a century maroons city". Deccan Chronicle. Retrieved 15 October 2020.
3. "Flood toll 32 and counting in Telangana". The New Indian Express. Retrieved 20 October 2020.
4. Gupta, Anil & Chakraborty, Prabodh. (2009). Urban Floods Case Studies Project: An Overview (8 Major Cities of India). Disaster and Development.
5. "Heavy rains destroy homes and lives in southern India". BBC News.
6. Metz, B., Davidson, O., Bosch, P., & Dave, R. (2007). Climate change 2007 Mitigation. Intergovernmental Panel on Climate Change, 1-863. Doi: 10.1017/cbo9780511546013.
7. National institute of urban affairs. (2016, September). Urban Flooding, India - Urban Climate Change Factsheets. Smartnet. https://smartnet.niua.org/content/dea2681d-4921-4057-9b7a-3ba3d6b6ac26.
8. Rajeev, M. (16 October 2020). "Hyderabad Floods Telangana Seeks Immediate Aid of ₹1,350 Crore". The Hindu. Retrieved 17 October 2020.
9. Sorensen, J., Persson, A., Sternudd, C., Aspegren, H., Nilsson, J., Nordström, J. … Mobini, S.(2016, August 04). Re-Thinking urban Flood Management-Time for a regime shift. RetrievedMay 2, 2021, from https://www.mdpi.com/2073-4441/8/8/332/htm
10. Vadlamudi, Swathi (15 October 2020). "Rain fury leaves capital battered, bruised". The Hindu. ISSN 0971-751X. Retrieved 15 October 2020.
11. "With parts of Hyderabad still flooded, more rain forecast". Indian Express. 18 October 2020. Retrieved 20 October 2020.

CHAPTER 8

Himalayan Flash Floods, the Complexities and Challenges in Mitigation: Case Study of Chamoli Disaster

Agraj Upadhyay
Defence Geoinformatics Research Establishment (DGRE), DRDO, Chandigarh

S. K. Joshi
Program Office, DRDO HQ,
Defence Research & Development Organization, New Delhi

P. K. Satyawali
Defence Geoinformatics Research Establishment (DGRE), DRDO, Chandigarh

Abstract

Indian Himalayas often faces natural calamities like earth quakes, landslides, flash floods, ice and snow avalanches and other extreme weather events. Himalayan flash floods have surprised and tested the natives of Uttarakahnd time and again. On 7th February 2021, on a clear day, one such massive and unexpected flash flood event occurred near Tapovan in Chamoli district of Uttarakhand. Tapovan lies in the environs of the Nanda Devi national park in Garhwal Himalayas. This event damaged Rishiganga hydropower station and a BRO bridge (Near Raini village), about five other small bridges and another under construction hydro power project of NTPC at Tapovan. This event also led to significant loss of life, mostly people working in hydro power projects.

This paper presents an investigation of possible causes and flow dynamics of the subsequent debris flow event of Chamoli disaster using satellite images, field observations and flow simulations. Possible mitigation measures along with their limitations and challenges in their implementation are also presented. In addition, the paper also highlights the requirement of multiple mitigation measures for management of geo-hazards in the region and discusses the complexities and challenges in their implementation.

Keywords: Rock ice avalanche, debris flow, field observations, dynamic simulations, mitigation measures

1. Introduction

Indian Himalayan region often faces natural calamities like earth quakes, landslides, flash floods, ice and snow avalanches and extreme weather circumstances. Some of these events are unprecedented in nature. One such flash flood event occurred in June 2013 near Kedarnath shrine in Uttarakhand. As per NIDM report, this event occurred due to very heavy rainfall followed by Chorabari Glacial Lake outburst and possible landslides that devastated the Kedarnath region completely with a significant damage to life and property.

Another such unprecedented and unexpected event occurred on Sunday morning, 7th February 2021 in the environs of the Nanda Devi National Park in the outer Garhwal Himalayas of Uttarakhand but unlike Kedarnath event it occurred on a clear day. At first instance this appeared as a huge debris flow event when, around 1000 hrs residents of Pang and Raini villages heard loud sounds and witnessed a huge amount of mass consisting of rocks, mud, water, ice blocks etc. moving in Rishiganga River. The event was captured on mobile camera by some of the villagers and immediately transmitted further on social media for information which helped administration to alert people downstream well in time. This event damaged Rishiganga hydropower station (Near Raini), a BRO bridge (near Raini), about five other small bridges and another under construction hydro power project at Tapovan. This event also led to significant loss of life mostly people working in hydro power projects. On receiving the first information, the response teams of the local administration came into action and also simultaneously, various scientific agencies started to analyse the event and causes behind it.

By the time teams of scientists visited the location of the event, the event was already over. On the basis of some visual evidences (aerial photographs) and interview of local people, preliminary characterisation of the event was carried out. On 8th February-2021 and subsequent days various teams of scientists flew to the Nandadevi biosphere reserve and observed a large scar over a rocky slope near Raunthi peak. The scar was surrounded by the glaciated ice. Local people had witnessed flowing ice blocks in moving debris mass. On the basis of the above information, a team of scientists from DGRE and other institutes identified it as a rock-ice fall event.

Later on, detailed analysis of the causes of the event has been carried out by various scientific groups around the world (Shrestha et. al. 2021, Shugar et. al. 2021, Pandey et. al. 2021, Martha et. al. 2021). Shershtha et. al. March 2021 presented first analysis on Chamoli event and categorised event as rockslide with initial release volume of 25 million m^3 having 85% rock mass and 15% ice mass. In addition to energy calculations of the event they also discussed the possible role of precedent weather conditions, climate change and various other factors as causes of the event. In April 2021, Pandey et. al. presented another study describing the event as rockslide caused by wedge failure with a release volume of 23 million m^3. They also discussed the potential cause of the rockslide and the process of the event. In April 2021, Rao et. al. studied the seismic signals captured by their sensor network in Uttarakhand region during this event and suggested that seismic sensor network along with satellite data interpretation, numerical modelling and geomorphic analysis may be used for early warning of rock slide and flood events in Himalayan region. In May 2021, Martha et. al. 2021 studied this event through photo geological reconstruction. By cut and fill method using DEM they suggested release volume of about ~29.3 million m^3. They also showed that the event was initiated at ~10:08:45 h (IST), travelled with a velocity of 10.6 m/s and lasted for ~50 minutes.

In continuation to above studies, this paper presents an analysis of causes of the Chamoli event and its flow dynamics on the basis of evidences collected during aerial reconnaissance (by DGRE), field visits, available satellite data and simulation studies. It also discusses about the complexities and challenges in mitigation and early warning systems of such type of events. In addition, the paper also highlights the requirements of multiple mitigation measures for all Himalayan geo-hazards and flash flood events and discusses the complexities and challenges in their implementation. Various mitigation measures like, identification of glacial lakes and hanging glacier, regular monitoring of discharge of all the rivers, installation of automatic weather stations and Doppler radars, mountain weather forecasting, avalanche and landslide prediction, installation of acoustic sensors for monitoring of glacial cracks etc. are identified. Measures like use of advance engineering methods for construction of roads in mountainous regions, strengthening of communication facilities etc. are also suggested to manage with these types of Himalayan disasters in more effective manner.

2. Methodology

Analysis of the Chamoli Rock-ice fall generated debris flow event has been carried out on the basis of videos of the event available on internet, satellite images, snow-met data of DGRE, field visit up to the confluence of Rishiganga and Raunti rivers after the event (during last week of March-2021) and dynamic simulation study. During the field visit approximate height of debris flow marks was also measured using laser range finder (Vectra-IV). In addition, dynamics of the flow was assessed on the basis of terrain variations, leftover deposits on the side slopes and riverbed and debris flow marks all along the flow path between confluence of Rishiganga and Raunthi rivers and Tapovan. Interaction with the local people was also carried out to understand the various aspects of event. Dynamic simulation of event was carried out using DGRE avalanche simulation software tool based on 2D shallow water equations, which is commonly used for avalanche dynamics simulations. Complexities and challenges in mitigation of such events were analysed on the basis of capabilities and limitations of existing technologies and limitations in their implementation in remote locations.

3. Event Analysis

3.1 Causes of the Event

Rock slope failure mechanism

In present analysis we tried to establish the similarity of the current event with possible rock slope failure mechanisms available in the literature (Raghuvanshi 2019) and analysed various evidences to understand the rock failure in present event.

Photographic evidences: Visual analysis of the available photographs (Figure-1) shows that:

1. The rock failure surface has undulating surface with multiples failure planes from where rock mass had detached.
2. The line partitioning the rock failure surface and rock-ice interface failure surface is irregular in shape.
3. Fragmented rock mass is accumulated in the lower part of the rock failure surface.
4. Ice failure surface (rock-glacier ice interface) above and on sides of rock failure surface is also visible.
5. Crown crack (Bergschrund) above failed Rock-ice mass is also visible in areal (Feb-2021) as well as satellite images of 2017 (Fig. 8.1(a) and (b))
6. Shape of the failure surface is like an inverted triangle.

Satellite image evidences (Fig. 8.1(b))

1. Glacier mass present in the area where rock slide had occurred.
2. Crown crack (Bergschrund) is visible above the rock slide area.

Snow met data (DGRE, Auli station, Elevation: 2600 m):

1. Snow fall of about 25 cm was recorded during 4-5th February-2021.
2. On 6th February-2021 Minimum and maximum temperatures were recorded about -5 and 6° C respectively
3. On 7th February-2021 Minimum and maximum temperatures rose to about 1 and 13 degrees respectively.
4. Maximum temperature at Auli as recorded by DGRE is 25°C.

Analysis of above evidences/ data

1. **Failure surface:** From photographic evidence it appears that the failure has occurred from the rock surface and also from rock-ice interface (figure-1(a)).
2. The **irregular shape of the partition line of rock failure surface and rock ice interface** surface suggests 1. Either presence of a crown crack in the failed rock mass through which glacier melt water might have seeped in and froze as ice, 2. Or the top surface of the failed rock mass was irregular with lower slope angle over which glacier mass was lying.
3. **Possibility of weakening of the failure plane:** In past, the glacier mass lying over the failed rock mass must be of much higher extent and thickness. No failure of the rock mass with higher loads (previously) suggests weakening of the rock failure surface over the years due to various natural processes.
4. **Type of failure:** On the basis of visual analysis and literature available on rock slope failure (Raghuvanshi et. al. 2019), failure appears to be a plane mode rock failure with convergent lateral release surfaces.

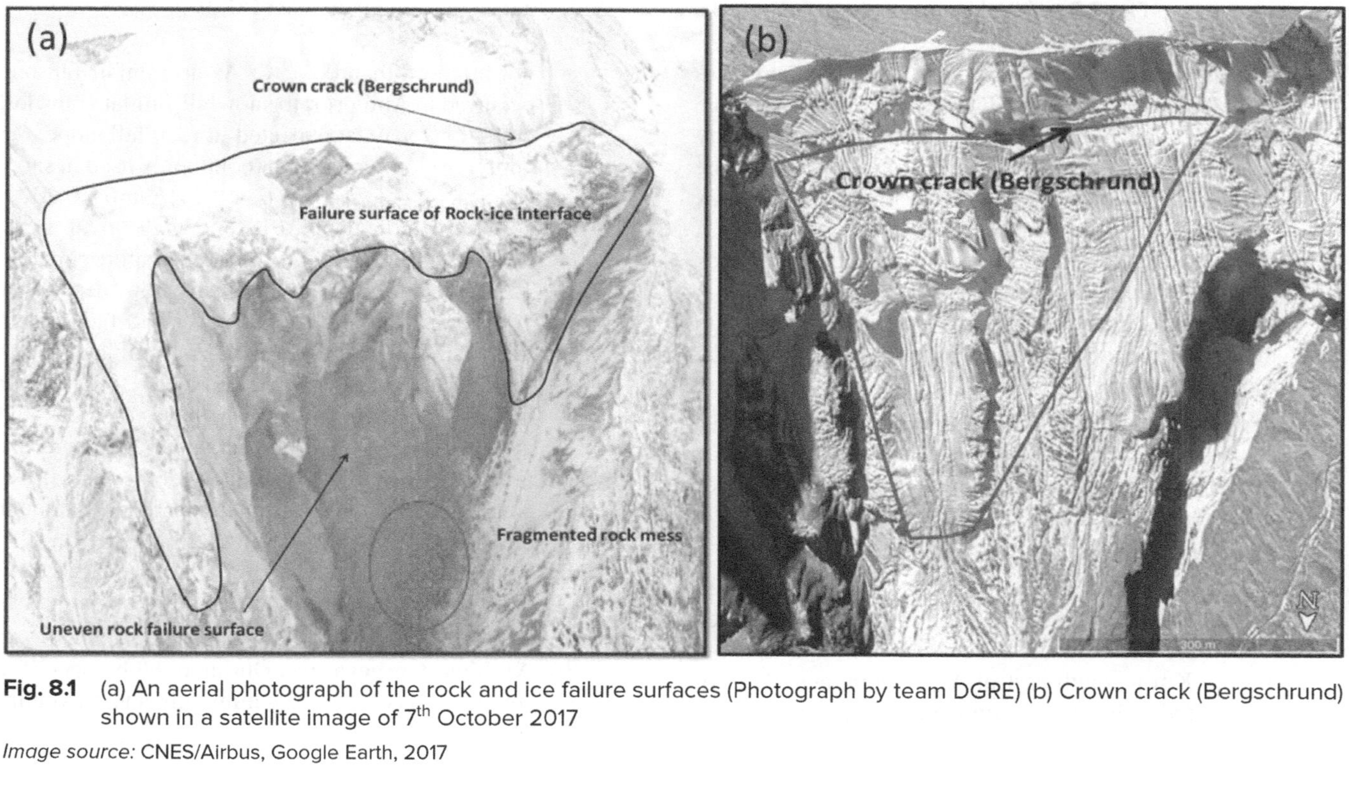

Fig. 8.1 (a) An aerial photograph of the rock and ice failure surfaces (Photograph by team DGRE) (b) Crown crack (Bergschrund) shown in a satellite image of 7th October 2017

Image source: CNES/Airbus, Google Earth, 2017

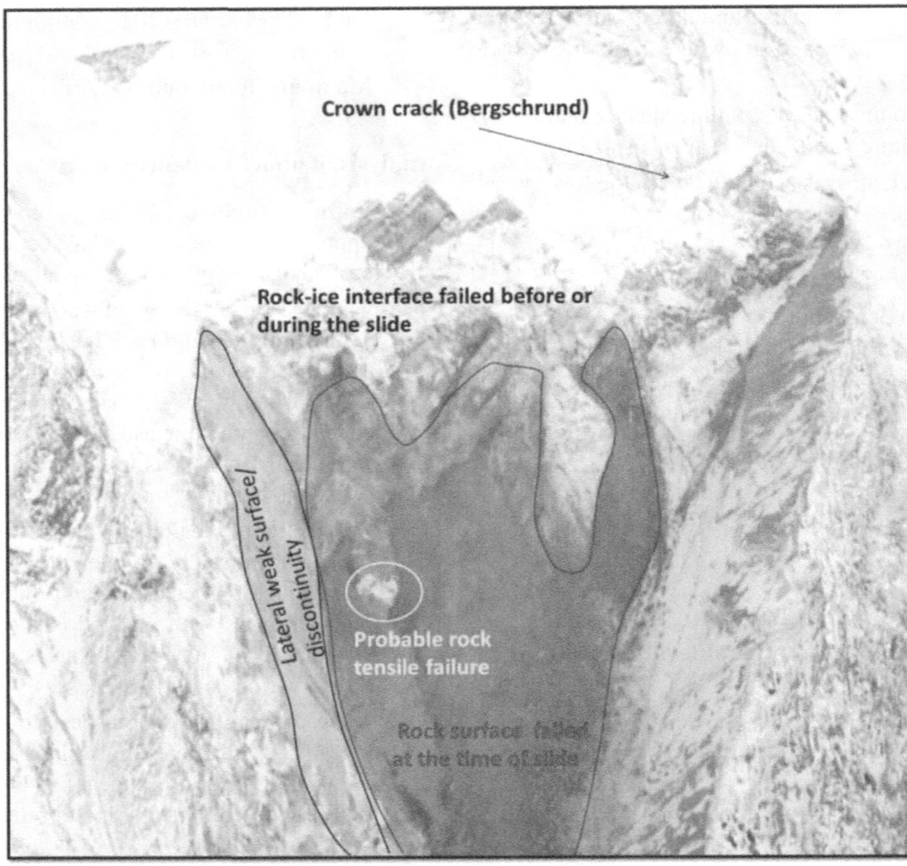

Fig. 8.2 Visual interpretation of various failure surfaces (Aerial photograph by team DGRE)

5. **Visual characterisation of failure surfaces (Fig. 8.2):** Visual inspection of the failure surface suggests existence of a weak surface or discontinuity on the left flank. In the middle surface appears to be failed during the rock slide. Accumulation of rock fragments over the surface suggests recent failure. A white scar in the centre seems as if a patch of the rock mass has failed under tension. Traces of ice with rock surface in background suggest presence of a crack with interstitial ice or glacier ice hanging over an irregular rock fracture line.
6. **Presence of Bergschrund:** Formation of a crown crack (Bergschrund) suggests that the glacier ice lying over the rock mass moving faster in comparison to ice mass above the crown crack.
7. **Analysis of Snow-met data:**
   (a) A snow fall of about 25 cm occurred at Auli during 4-5th February-2021 may be extrapolated to the rock fall area. The rock fall area is about 24 Km in south east of Auli station and effect of latitude change on ambient temperature should not be more than 0.25^0 C. As no rainfall fall had occurred in Auli prior to snowfall similar snow fall conditions may be expected in rock fall slope. On applying slope correction (commonly used in snow science, $d_{corrected} = d \cos\theta \times (0.291/(\sin\theta - 0.202 \times \cos\theta))$, d: snow depth, θ: slope angle) fresh snow thickness over the rock mass may be approximately 15 cm. Hence, assuming fresh snow density of 100 Kg/m^3, additional load over the failed rock mass (surface area 0.29 Km^2 approximately) due to fresh snowfall may be expected to be about 4350 Ton, which would have helped the acceleration of already active rock fracturing processes working to dislodge/ fail the rock slope. This snow load may be even higher if snow amount due to previous snow fall spells (5 cm on 13th Decemebr-2020 and 12 cm on 6-7 January-2021) is also taken into account, assuming no snow loss due to melting, wind drift etc.
   (b) **Ambient temperature:** During 6-7 february-2021 (on the basis of Auli data minimum and maximum

temperatures in the rock slide area are expected to be sub-zero and close to zero respectively. With an assumed average lapse rate of about 6^0C per Km, average maximum temperature in the Rockall region should also be close to zero. The above expected ambient temperature conditions suggest that the glacier mass lying over the failed rock mass may be in temperate thermal regime.

On the basis of available photographic evidences, satellite data and Snow-met data it appears that

- The failed rock mass had existing lateral weak planes/ discontinuities and a crown crack.
- A glacier ice mass was attached to the rock mass above and below the crown crack (refer Fig. 8.3).
- The glacier had a temperate thermal regime and moving faster than the ice mass on higher elevation and led to the formation of crown crack.
- Due to formation of the crown crack, the weight of the glacier supported at the crown was transferred to the rock-ice interface below Bergschrund.
- Opening up of the crown crack led to generation of more melt water especially during summer season. The melt water percolated along rock-ice interface and seeped into the crown crack of the rock leading to formation of interstitial ice patches during winter. Freezing of the interstitial water would also have helped in opening up of the crown crack. During summer, melt water would have also seeped into discontinuities of the attached rock mass and weakened its structure over the years in a cyclic manner. Since, melt water had a way through crown crack of the rock it would not have seeped through rock-ice interface below the crown crack. This would have made rock-ice interface above crown crack weaker in comparison to this interface below crown crack.
- Prior to failure, driving force for failure of the rock was its own weight and weight of the ice whereas intact rock mass along failure surface was resisting it.
- The fresh snow fall deposits on 4-5th February-2021 would have generated additional driving force on the rock mass and led to a critical condition (driving force exceeds the resisting force) for rock failure/ slide. The rock slide also carried glacier ice (sitting over it) and fresh snow along with it. Hence, initial event may be characterised as rock-ice-snow avalanche (rock-slope failure accompanied with glacier ice mass and snow). A schematic diagram of this process is shown in Fig. 8.3.

3.2 Dynamics of the Detached Mass and Resultant Debris Flow

A trajectory of debris flow event from rock-ice-snow mass release point to under construction Tapovan hydro-power plant is shown in Fig. 8.4(a). On the basis of the terrain information and field observations, following deductions on the dynamics of the released mass have been made:

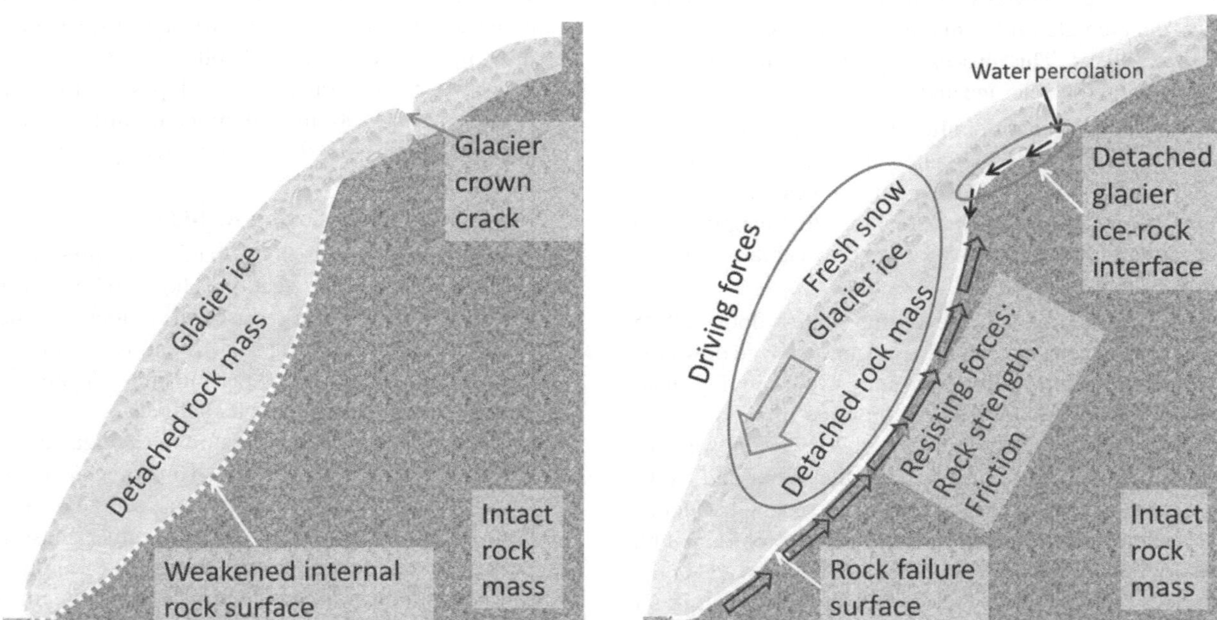

Fig. 8.3 (a) Probable status of detachd rock mass (a) prior to snowfall on 4th February- 2021 (b) After 5th February on cessation of snowfall

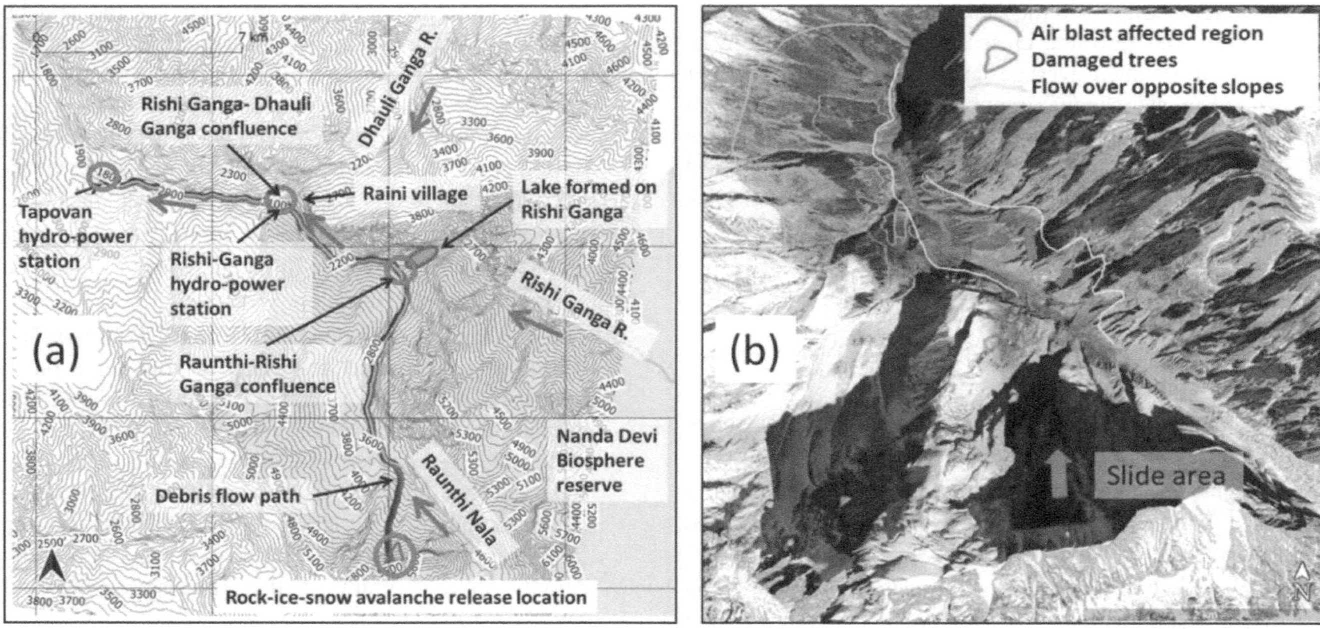

Fig. 8.4 (a) Trajectory of debris flow event from release location to under construction Tapovan Hydro-power plant (b) Initial trajectory of the flowing mass causing damage to trees (Satellite image: Maxtar technologies, Google Earth 2021)

- The dislodged mass of rock, glacier and snow travelled about 2.5 km on an average slope of approximately 40 degrees with a vertical fall of approximately 1740 m.
- Before reaching the valley, the released mass would have attained significantly high velocity.
- On hitting the valley at very high velocities, the released mass disintegrated into smaller and irregular sized rock and ice blocks. This process would have led to partial melting of the glacier ice and snow.
- Due to high momentum, after hitting the valley bottom, a part of the released mass also climbed to the opposite slope (about 400 m). During this process, the moving mass created an air blast leading to snapping and uprooting of the trees up to 3.2 km downstream from the impact zone (Fig. 8.4(b)).
- Due to considerable slope of the valley floor (20-25 degrees), the partially wet released mass moved down the valley. During this valley movement, the moving mass entrained the mass (soil mass, moraine material, snow, ice, left over ice avalanche debris of 2017 event, water of Raunthi stream etc.) lying on the valley floor and banks of the flow channel. Narrow pre-event channel widened significantly due to channel scouring and entrainment. The frictional heating within the moving mass and at the interface with the ground additional snow/ ice melt water would have generated. Assuming average water channel (Rainthi, Rishiganga and Dhauliganga) flow width 12 m and depth 0.5 m (on the basis of approximate estimation using satellite image) along the complete flow path of about 22 Km (between initial impact point and under construction tapovan hydropower station), contribution of river water into debris flow volume (~29 million m^3) appears to be insignificant (<1%).
- Presence of convergent-divergent sections and curves along the flow path (Fig. 8.5(a)) led to splashing of the moving mass (consisting of soil mass, rocks, ice blocks, snow, water etc.) on the side slopes up to significant heights (up to 90 m). Significant unstable sediment deposits along the debris flow reach were observed during the fieldwork.
- **Debris dam formation on Rishiganga**

  A sharp change in flow path and presence of a convergent-divergent section ahead of the Raunthi-Rishiganga confluence reduced the velocity of the front of the moving mass (Fig. 8.5(b) and (c)). This resulted in accumulation of incoming tail mass (from Raunthi) near the Raunthi-Rishiganga confluence. As a result, the accumulated mass climbed into the Rishiganga river for ~ **450 m** over a slope of about **10** degrees (Fig. 8.6(a)). During this process, the heavier mass settled on the valley floor of Rishiganga river and the lighter mass flowed back into main channel on decrease of inflow from Raunthi and increase of flow in the downstream. Marks of material back flow over the dam formed on Rishiganga observed during the field visit confirm this process (Fig. 8.6(b)). This process led to the formation of a debris

Fig. 8.5 Characteristics of flow channel (a) A sharp narrow bend just before Raunthi-Rishiganga confluence (b) A sharp narrow bend just after Raunthi- Rishiganga confluence (c) A narrow channel downstream of bend in Fig. 8.5(b) with rocky side slopes (d) A nearly straight channel downstream of narrow channel shown in Fig. 8.5(c)

Images source: DGRE

flow dammed lake in the Rishiganga River (Fig. 8.6(a)). The dam material consists of angular rock debris of varying sizes, large amount of fine silt material produced due to grinding and large amount of fragmented glacier ice. Ṭill Raunthi-Rishiganga confluence, the flowing mass seems to have dry mass in it. At present, dam length is about **375-380 m** and width at the front of the lake is about **112-115 m**. Length of the lake up to last visible point is about **430 m**.

- After the Rishiganga-Raunthi confluence, the mass travelled with average flow height of 30-40 m and maximum splash height of about 120 m. The splash process also threw large boulders and ice blocks on side slopes. During the flow along the channel, the process of melt water generation and mixing of the various phases of the mass continued leading to increase in water content in the debris flow.
- Due to the presence of a narrow flow section just upstream of BRO bridge location near Raini village, a large splash of moving mass blew off the Rishiganga hydro power project (HEP-1) and BRO bridge.
- **Debris back flow in Dhauliganga river:**

  Immediately after hitting the BRO bridge at Raini, the moving mass hit the T-junction where the Rishigana meets the Dhauliganga (figure-6(c)). Due to the normal impact at the junction the velocity of the flow front is reduced but continuous inflow led to increase/maintain the debris height near the confluence. This would have led to formation of a gravity wave both upstream and downstream of the Dhauliganga river. The moving mass travelled about **1.2 km** into the Dhauliganga river, upstream of the confluence, destroying a suspension bridge over Dhauliganga **about 140 m** from the

Fig. 8.6 (a) Lake formed on Rishiganga river due to debris deposition (b) Highest flow mark showing back flow of material over the earthern dam over Rishiganga river (c) T- junction near Raini village (b) Back flow in Dhauliganga and lake formation due to deposits

Images source: DGRE

confluence. Most of the debris mass, with up to **24 m** flow height from the present bed level at mid-point, flowed back to the downstream of the confluence leaving heavier mass at the river bed (Fig. 8.6(d)). The deposition of mass over the river bed led to a small lake formation.

- After Rishiganga-Dhauliganga confluence the moving mass seemed to have increased water content and debris flow moved to the Tapovan dam site (HEP-2) with average flow height of **20-30 m**. Increase in ambient temperature (due to significant decrease in elevation), frictional melting of ice/ snow and water of Dhauliganag river seems to be the source of increase in water content in the moving mass.
- Along its flow path the moving mass deposited significant mass over river bed with deposition depth varying from **5-18 m** (as suggested by local people) approximately. The exact quantity of the deposits in the river may be estimated using DEM differencing technique if good quality data is available.
- The geometry of the flow channel also led significant amount of energy dissipation and reduced impact downstream of the Raini village.
- Due to presence of the wall like obstacle at Tapovan (Tapovan dam) significant amount of solid phase of debris flow mass was entrapped and deposited near the dam.
- During the settlement process of sediments near the Tapovan dam site (in desilting tanks) fine sediments surfaced over coarse sediments as observed during the field visit.

3.3 Simulation Results

A simulation study was carried out to understand the dynamics of the above described rock- ice-snow avalanche generated debris flow event. For this study SASE avalanche flow simulation software was used. This software solves depth averaged mass and momentum conservation equations for the estimation of height and velocity of the flow. In this approach frictional dissipation is considered at the flow and terrain interface. Vollmey's two parameter model is used to model the flow resistance. For avalanche flow simulation these two parameters are commonly known as Coulomb (μ) and turbulent friction (ξ) parameters. The turbulent friction parameter corresponds to Chezy's constant ($C = \sqrt{\xi}$), which is commonly used for flow resistance modeling in river flows. Coulomb forces are considered to model the frictional dissipation due to relative movement of the granular mass with respect to the terrain whereas turbulent friction force is used to model the granular frictional dissipation within the moving mass. To model the rock-ice-snow avalanche generated debris flow (as in present study) friction parameters corresponding to a large avalanche with a return period of 300 years ($\mu = 0.155$, $\xi = 3000$ m/s^2) were considered. The higher value of ξ indicate lower flow resistance. The equivalent value of Chezy's constant value is about 54.77 m$^{1/2}$/s. To stop the flow, stopping % percentage momentum of 0.1% was considered in place of 5% used for snow avalanches. For simulations, a 5 m resolution elevation data (Cartosat DEM of DGRE) was

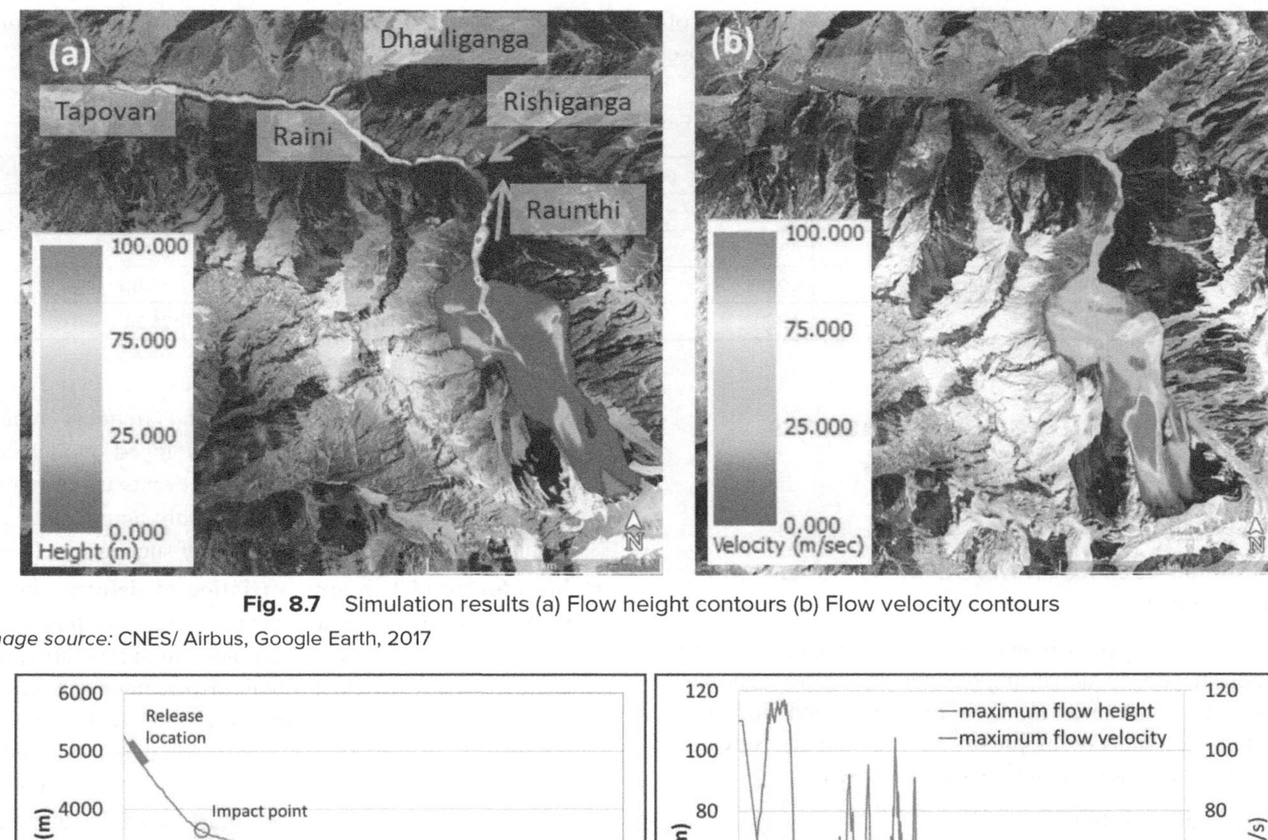

Fig. 8.7 Simulation results (a) Flow height contours (b) Flow velocity contours

Image source: CNES/ Airbus, Google Earth, 2017

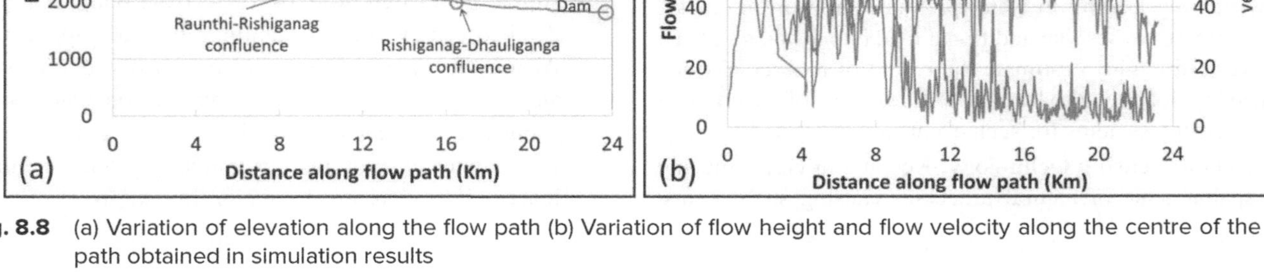

Fig. 8.8 (a) Variation of elevation along the flow path (b) Variation of flow height and flow velocity along the centre of the flow path obtained in simulation results

used but for faster computation a coarse resolution mesh of 30 m was used. Simulations were carried for a release area of about 0.29 Km^2 with an average depth of 110 m. Release area was demarcated on the basis of satellite image showing the detached area and release depth was approximated on the basis of values reported in literature (Shrestha et. al. 2021). This resulted into a release volume of about 32 million m^3, slightly higher than the values suggested in the literature.

With these parameters, simulation of the event was carried and results in the form flow height and flow velocity contours are shown in Fig. 8.10. The variation of the terrain elevation, maximum flow height and maximum velocity along the center of the flow path are shown in Fig. 8.8. The average and maximum values of flow height and velocity at three different locations of the flow path are given in Table 8.1. Values of flow height obtained in simulation seems reasonably close to the rough estimates made during field visit on the basis of maximum flow height marks and deposition on the river, especially in first two locations. Average flow height in section-3 (Raini village to Tapovan) seems to be on higher side. No direct method for validation of flow velocity is available; hence it is difficult to validate the same. From simulation results it appears that the model approximately captures the process of mass climbing on opposite slopes during the initial phase but quantitatively values of climb are on higher side. The model also captures very well the flow of mass into Rishiganga and Dhauliganga rivers with flow lengths of approximately 1.4 Km and 0.55 Km respectively. These values are very close to the values measured during the field visit. This is important to highlight that flow simulation model used here is highly simplified, but provides a preliminary insight into the otherwise a very complex natural mass movement phenomena.

Table 8.1 Average and maximum values of Flow height and velocity at the center of the flow path

Location	Flow height (m)		Flow velocity (m/s)	
	Average	Maximum	Average	Maximum
Slide point-Rishiganga-Raunthi confluence	43.62	104.23	47.49	116.97
Rishiganga-Raunthi confluence to Raini village	49.03	65.13	12.78	43.89
Raini village to Tapovan Dam	43.31	57.31	6.44	21.24

4. Possible Mitigation Measures and Challenges for their Implementation

4.1 Chamoli Rock-ice-snow Avalanche Generated Debris Flow

From above analysis it is evident that the magnitude of the Chamoli disaster was of unprecedented nature and need extraordinary measures for their prediction and mitigation measures. An analysis of possible mitigation measures for such events and their limitations are discussed.

Early warning systems

Rock-ice-snow avalanche occurrence process is very complex which involve long term fracturing of rock , ice/snow melting leading to water pressure build up in cracks, short term processes like additional load due to snow fall etc. It is very difficult to estimate the extent of rock cracks and critical overload require for rock failure. Some of the possible early warning systems for such events are discussed.

- **Remote sensing technology:** Keeping in view time and spatial scale of natural processes causing such events, very high resolution, accurate and frequent remote sensing data along with well-developed models are required for early warning of such events using remote sensing technology. As on today this approach is not matured enough for early warning of such events.
- **Monitoring of temporal variation of deformation of rock faces** may be one approach to predict the possibility of occurrence of the event but need high resolution and frequent data to detect small displacements expected in the process. Application of LidAR technology to monitor small rock face deformations is one possibility but due to large geographic and inaccessible area along with frequent data gathering requirements this approach is also not suitable for early warning.
- Another approach may be **detection of acoustic signal generated due to rock fracturing** but need detailed study before putting it into application. Again, this approach may not be feasible due to large and inaccessible geographic area. As on date, power supply, communication facilities and quick access for maintenance of instrumentation systems in these inaccessible areas are also major limitations.

Event warning systems:

Event warning systems are the systems deployed to detect occurrence of the event to provide a lead time for evacuation of the people or to take immediate safety measures for infrastructure present in the hazard zone of the event. Possible event warning systems for a Chamoli like disaster may include following:

- **Initial mass movement detection using radars, by change detection using a programmable camera or using geophones**. This approach may not be suitable keeping in view large and inaccessible geographic area and multiple potential sites where monitoring of mountain slopes may be needed. Early detection of rock fall and landslide event using a good network of seismic sensors along with satellite data interpretation, numerical modelling and geomorphic analysis is a promising approach and efforts are underway to develop such systems (Purnachandra Rao et. al. 2021).
- **Detection of increase in mass flow at intermediate locations using various systems like trip wire sensor, ultrasonic flow depth sensor, change detection using programmable cameras etc.** Such systems should be located far away from the infrastructure to be protected to get a maximum lead time. This approach is more suitable as monitoring systems may be deployed in accessible areas and in common flow path which is fed by multiple catchment of a large geographic area.

Application of above systems require suitable and accessible location, power supply for sensors and data acquisition system, communication system for transmission of data and an effective model for generation of alarm on the basis of multi-sensor data. In addition, necessary training is also required for understanding the alarm and evacuating the people as quickly as possible.

Structural Measures

Sometimes it is not possible to place event warning systems far away from the area to be protected. In such cases it is not possible to get sufficient lead time for necessary procedures for evacuation and infrastructure protection. In such cases some structural measures in the form of rigid or flexible structure may be placed in the flow path to retard the mass movement and increase the lead time. Also early warning systems provide only time for evacuation process but safety of the infrastructure may not be guaranteed. Above structural measures will not only retard the moving mass but also stop a reasonable solid mass flowing with the debris flow which will reduce the possibility of deposition of huge debris near the object of protection (as it happened near Tapovan). Structural approach has its own limitations as debris deposited in one event may leave structure ineffective for the next event and cleaning of deposits may be needed to keep the structure effective for the future events. This approach has been used on different scale for mitigation of debris flow hazards.

In addition to the above, mechanical systems or methods to speed up the evacuation process may be explored. Such systems may help in compensating for the limited lead time available for the event. Since, Hydropower plants and personnel working in them were the worst effected in Chamoli disaster, new design and construction guidelines for hydro power plants in such natural disaster prone locations should be revisited to minimise the damage in future.

4.2 For Himalayan Flash Floods in General

In addition to Chamoli like disaster, Uttarakhand is also prone to various other geo-hazards. Requirements for monitoring, early warning and mitigation of these events are briefly discussed below.

Rain induced flash floods

Rain induced flash floods generally occur due to high intensity rain fall in a particular region. Early prediction of such possibility may help local administration in taking further necessary steps to minimise the loss of life and property. A good network of weather monitoring systems, Doppler radars and application of state of the art mountain weather forecasting models may help in early prediction of such events.

Glacial lake outburst floods (GLOF)

With rising global temperatures, glaciers in Indian Himalaya are receding leading to formation of glacial lakes within their moraine deposits. Continuous melting of glacier ice results in increase of water level in these lakes leading to increase in water pressure on moraine dams. When water pressure exceeds the dam strength, dam breach occurs leading to flash flood like situation in the downstream and create considerable damage.

With known moraine dam strength (through field measurements) critical water level (through hydrostatic pressure estimates) for dam breach may be estimated. Continuous monitoring of water level in the lake, through remote sensing technology or ground based sensors, may provide an early warning about the possibility of the lake burst. In addition, provision of water flow depth sensors in the probable flow path of the flood water, may help in providing the event warning. Also, hydrological modelling of potential GLOF scenarios may help in generation of GLOF plain mapping for land use planning.

Landslide lake outburst flood (LLOF)

Landslide lakes form when flow of a water channel is blocked due to a landslide mass. In some cases water slowly erodes the mass and makes a channel. One such event occurred in

Chenab river near Junda village of Lahual and Spiti district of H.P. in August 2021when water was trapped only for few hours. However, in some case huge water lakes are formed behind the landslide deposits and may form flash flood like events if this dam is breached.

Such events are easy to detect by monitoring the water level in the downstream of the landslide location as water discharge reduces significantly. However, the most important step to prevent such events is prevention of formation of large lakes behind the deposits by releasing the accumulated water. These earthen landslide dams are generally very strong and difficult to work for creating a channel to release the water (Ex. Earthen dam formed on Rishiganga river during Chamoli disaster). There is a requirement of development of a methodology to assess the stability of such dams to predict their breach possibility and easily deployable technology to create a water channels to prevent build-up of water pressure in the dam.

Hanging glaciers

Rise in global temperature rise and climate change have been frequently cited as reasons for receding of Himalayan glaciers. Some of the glaciers are receded up to an extent where glacier mass is left lying over smaller high valleys and generally classified as hanging glaciers. These hanging glaciers moves down slowly but sometimes break and cause ice fall or ice avalanches. The mass in these events may not travel very far like flash flood events but may cause damage in the downstream. Hanging glaciers may be mapped using high resolution satellite images but prediction of their failure and fall is not straight forward. Acoustic sensors based systems may be tried at few places for detection of glacial crack propagation and ice fall/ avalanche prediction.

Snow avalanches

Snow avalanches form when a snow mass detaches over a mountain slope and travels down to valley damaging infrastructure and life. In Uttarakhand, these hazards are generally limited to snow bound regions during winter season only. Avalanche forecasting, artificial avalanche release, avalanche hazard mapping based land use planning and structural control are some of the common techniques used to mitigate these events.

Landslides

Landslides are very common in mountainous states of Indian Himalaya. Geology of the mountains along with climatic conditions causes these events. Mapping of landslide prone areas and monitoring using sensor systems are commonly used techniques to mitigate the landslide hazards. Prediction of landslides is not yet established to be used for application. Though Geo-hazards are difficult to predict and mitigate, reasonable expertise and effective methodologies are developed for mitigation of some of them. Individually mitigation measures for these events are worked on in detail but cascading of multiple events is leading to complex events is a new field of study. A combination of above discussed methods and techniques may be needed to early warn, predict or mitigate complex events. In addition, use of advance engineering methods for construction of roads in hilly regions is also required to minimise the disturbance to fragile geology of the Uttarakhand Himalayas. Strengthening of accessibility and communication facilities is also required to effectively manage these types of Himalayan disasters.

5. Conclusions

A detailed analysis of possible causes of the Chamoli event and its flow dynamics on the basis of evidences collected during aerial reconnaissance, field visits, available satellite data and simulation studies was presented. The analysis suggests that the Chamoli disaster was the result of complex and long term natural processes leading to rock slope failure which may have accelerated due to the formation of crown crack in overlying glacier, melt water percolation, recent snow fall etc. These natural processes led to detachment of large rock mass with overlying glacier mass which may be categorised as rock-ice-snow avalanche. On the basis of field evidences, a good description of event flow dynamics is presented. Presence of glacier ice, ice/ snow lying on the valley bottom and their melting due to frictional heating and ambient temperature increase during mass movement are suggested as a potential cause for conversion of rock-ice-snow avalanche event into debris flow event. Considerable variations in flow section along the path led to variation of flow height, deposition of debris and formation of lakes during the dynamics of the event. Avalanche dynamics simulations give reasonably good insight into the flow height and velocity variation along the flow path. Debris flow process into Rishiganga and Dhualiganga rivers, which led to lake formation, could also be simulated reasonably well.

On the basis of studies on Chamoli disaster and field observation it can be concluded that Rock-ice-snow avalanche generated debris flow event is a highly complex natural event for which deployment of an effective early warning system is very difficult keeping in view large and inaccessible geographical area. Deployment of event detection systems in combination with structural measures at intermediate locations may provide a promising solution for Chamoli like complex natural disasters. Such solutions have not been attempted in India at a large scale but this is an opportunity to deploy such system to test their efficacy and enhance the confidence for extended applications. In addition need of mechanical systems and other methods to speed up the evacuation process to compensate for the limited lead time available for the evacuatio is also highlighted. Since, Hydropower plants and personnel working in them were the worst effected, a review and improvement in the existing design and construction guidelines for hydro power plants in such natural disaster prone locations is suggested.

In addition to Chamoli like disaster, a general discussion on monitoring, early warning and mitigation of other geo-hazards like rain induced flash floods, GLOF, LLOF, hanging glaciers, snow avalanches and landslides has also been presented. Requirement of early warning and mitigation of complex events, generated as cascading of multiple events, has also been highlighted. Need of eco-friendly methods of construction in fragile Himalayan mountains and improvement in accessibility and communication facilities are also emphasised for effective disaster management in Uttarakhand Himalayas.

6. Acknowledgements

Authors are thankful to Defence Research and Development Organisation for all the support provided for this work. Authors are also thankful to NDMA, all other agencies and people for providing direct or indirect support during field visit to the Chamoli disaster affected areas near Tapovan.

References

1. Shrestha A.B., Steiner, J. and other (2021) Understanding the Chamoli flood: Cause, process, impacts, and context of rapid infrastructure development. ICIMOD.
2. Shugar, D. H., Jacquemart, M. and others (2021) A massive rock and ice avalanche caused the 2021 disaster at Chamoli, Indian Himalaya. Science, 373(6552), 300–306.
3. Purnachandra Rao N., Rekapalli R. and others (2021) Seismological rockslide warnings in the Himalaya. Science, 372 (6539), 247.
4. Pandey P., Chauhan, P. and others (2021), Cause and Process Mechanism of Rockslide Triggered Flood Event in Rishiganga and Dhauliganga River Valleys, Chamoli, Uttarakhand, India Using Satellite Remote Sensing and in situ Observations, Journal of indian society of remote sensing, 49(5), 1011–1024.
5. Martha, T. R., Roy, P., Jain, N. *et al.* Rock avalanche induced flash flood on 07 February 2021 in Uttarakhand, India—a photogeological reconstruction of the event. Landslides 18, 2881–2893 (2021).
6. Raghuvanshi, T.K. , Plane failure in rock slopes – A review on stability analysis techniques, Jouranl of King Saud University, Science 31 (2019), 101-109.

Part 2

PLANNING AND MANAGING RISKS OF FLOODS AND RIVER EROSION

CHAPTER 9

Cloudburst Induced Flood Assessment in the North-Western Himalayan Region—A Case Study of Upper Beas Basin

Sachchidanand Singh
Indian Institute of Technology Roorkee
Mitthan Lal Kansal
Indian Institute of Technology Roorkee
Corresponding Author: mlk@wr.iitr.ac.in

Abstract

Cloudbursts are a major natural hazard in the Indian Himalayan Region (IHR). During the monsoon season in India, clouds develop in the Bay of Bengal and move up the Gangetic plains, eventually reaching the Himalayas and falling in the form of heavy rain. The torrential precipitation leads to landslides and debris flash floods in mountainous places. These floods are so mighty that they uproot any infrastructure in their path, resulting in the loss of lives and properties. The mechanisms behind cloudburst driving processes such as orographic lifting, rainfall dispersion, precipitation thresholds, and source are still a subject of current research. The frequency of such events has grown significantly in recent years, possibly due to increased human activity and climate change occurrences. There are very limited research to explain the cloud burst phenomena. This study examines the cloudburst occurrences in Himachal Pradesh between 2001 and 2020. The impact and spatial distribution of hydrometeorology and topography to these occurrences are examined using remote sensing information. Furthermore, the resulting flash-floods due to such occurrences are estimated using empirical, rational, and unit hydrograph methods for the upper Beas River basin and compared. Effective database management and decision support systems can aid in reducing the negative effects of cloudburst in the future.

Keywords: Cloudbursts; GIS; Remote sensing; Hydrometeorology.

1. Introduction

In the Indian Himalayan states, natural disasters such as cloudbursts, flash floods and landslides are recurrent events. These are predominant in the monsoon season and cause severe harm to life and properties, thus, requiring frequent monitoring. The cloudburst events are characterised by very high-intensity rainfall (>100 mm/hour), occurring within a shorter duration of time, over a smaller area (20-80 sq. km.), during the monsoon season and may lie up to 15 km height in cloudburst-like systems (Dimri et al., 2017; Kumar et al., 2018; Sati, 2018). They generally occur in inaccessible hilly areas and are not reported until a significant socioeconomic disturbance is caused by the subsequent flash flooding and landslides (Kansal & Thakur, 2017; Sachchidanand Singh et al., 2020). It is estimated that more than 6,000 people have died in the Himalayan states over the past two decades due to cloudbursts induced floods and landslides.

According to the All India Meteorological Department (IMD), there has been a sharp increase in cloudbursts in recent years. Because of its unique geographical location, the Himalayan mountain range greatly impacts the weather and climate. It functions as a powerful barrier to the circulation pattern, serving as a heat source in the summer and a sink of heat in the winter. Its varied snow cover and vegetation influence monsoon precipitation. The Himalayas' surface boundary conditions dictate the monsoon's performance, which profoundly affects water resources and agricultural productivity (Das, 2005). The circumstances and physical mechanisms leading to floods in the Himalayas are highly complicated with various terrestrial and atmospheric surface processes. The region is responsible for the extensive incidence of landslides and significant atmospheric precipitation. High unpredictability and climate change enhanced the rainfall in the Himalayas and caused flash floods and landslides (Kuksina et al., 2017; Mishra, 2020; Sati, 2018; So & Shin, 2018).

The Himachal Pradesh hills, particularly Chamba, Kinnaur, Kullu, Mandi, Shimla, Sirmaur and Una, are prone to natural disasters such as flash floods and torrential rains and landslides. These have become a regular occurrence in recent years. The loss of human life due to natural disasters can be attributed to increasing human interference in eco-sensitive Himalayan areas. Landslides caused by heavy rains in recent days have already blocked the roads in several places in the Himachal. Landslides also blocked many roads in several other parts of the state.

This paper examines the characteristics and spatiotemporal variation of cloudbursts causing factors over the Indian Himalayan region, especially Himachal Pradesh, using satellite-derived hydro-meteorological information. Also, the resultant flash-flood discharge estimation using the empirical formula and unit hydrograph method is carried out. This would aid in the development of a more complete knowledge of the process behind these cloudburst occurrences and give some insight into their detection in the future.

2. Study Area

Himachal Pradesh is a mountainous state located between latitudes 30° 22' and 33° 12' North and longitudes 75° 47' and 79° 4' east and covers around 55673 sq. km. It is bounded on the North by Jammu and Kashmir, on the west and southwest by Punjab, on the south by Haryana and Uttar Pradesh, and on the east by Tibet. The elevation of the state varies between 350 and 6700 metres above mean sea level. Significant variations in geophysical characteristics account for substantial diversity in the climate and precipitation of the state's many sub-regions. The state is physically a part of the Himalayan system. It may be split into three topographic zones from south to north: 1) the Shivaliks, or the outer Himalayas, made up of consolidated sediments, experiences a huge amount of soil erosions and lies in the elevation range of 350 to 1500 m, 2) the Inner Himalayas, or mid-mountains, with an elevation range of 1500-4500 m and having loamy soil characteristics, and 3) the Alpine zone, or larger Himalayas, with an elevation of more than 4500 m, and experiences scanty rainfall. Himachal Pradesh's climate varies per altitude, ranging from semi-tropical to semi-arctic (Census, 2011). It has three seasons: the rainy season, which runs from July to September, the winter season, which runs from October to March, and the summer season, which runs from April to June. The annual rainfall varies from 2900 mm to 3800 mm. The state is traversed by five perennial rivers: the Sutlej, Chenab, Ravi, Beas, and Yamuna. The state is traversed by five perennial rivers: the Sutlej, Chenab, Ravi, Beas, and Yamuna. Unscientific deforestation of sensitive mountains, widening of roads, construction of tunnels for hydropower projects and changing rain and earthquake patterns are cited as reasons for increased landslides in the district of Kinnaur. This year, Himachal Pradesh recorded an increase of 116% in landslides, and the number of cloudbursts increased to 121% during the monsoon season compared to 2020 (Bisht, 2021).

For the analysis of peak flood, the upper Beas basin was delineated by taking the outlet near Kullu town, as shown in Fig. 9.1. The upper Beas basin has an elevation range of 1060 m to 6000 m and a maximum drainage density of up to 0.95 km/ sq. km. From the land use land cover classification, it was found that forests cover around 46.3% of the region, followed by 34.5% by plantations and grasslands, 12% by snow and Ice, Waterbody by 1.2%, Cropland covers 5.7%, and the built-up area covers 0.3%.

3. Datasets and Methodology

3.1 Cloudbursts Incidents

Several agencies, authors, and news reports, including the National Institute of Disaster Management (NIDM), the Himachal Pradesh State Disaster Management Authority (HPSDMA), and the India Meteorological Department, documented the occurrences of recent cloudbursts in the region. Some of the major incidents in the state were as follows:

(i) July 16, 2003- around 40 were reported killed in the cloudbursts and resulting flash-floods event at Shilagarh Gursa, Kullu, Himachal Pradesh.

(ii) On August 14, 2007-A, cloudbursts and subsequent flash-flood events killed around 52 people near Ganvi village, Bhavi, Shimla district.

(iii) August 7, 2009- Cloudbursts caused damage to several government buildings and bridges in the Nirmand, Kullu and 2 people were reported dead in Dharampur, Mandi.

(iv) July 20, 2011- due to cloudbursts in the upper Manali region, 2 people were found dead, and 22 were reported missing.

(v) In the monsoon, 2013 – around 20 incidents of cloudbursts and induced flash-floods were reported in numerous locations of Himachal Pradesh. Majorly affected regions were Bagi village, Parbati valley,

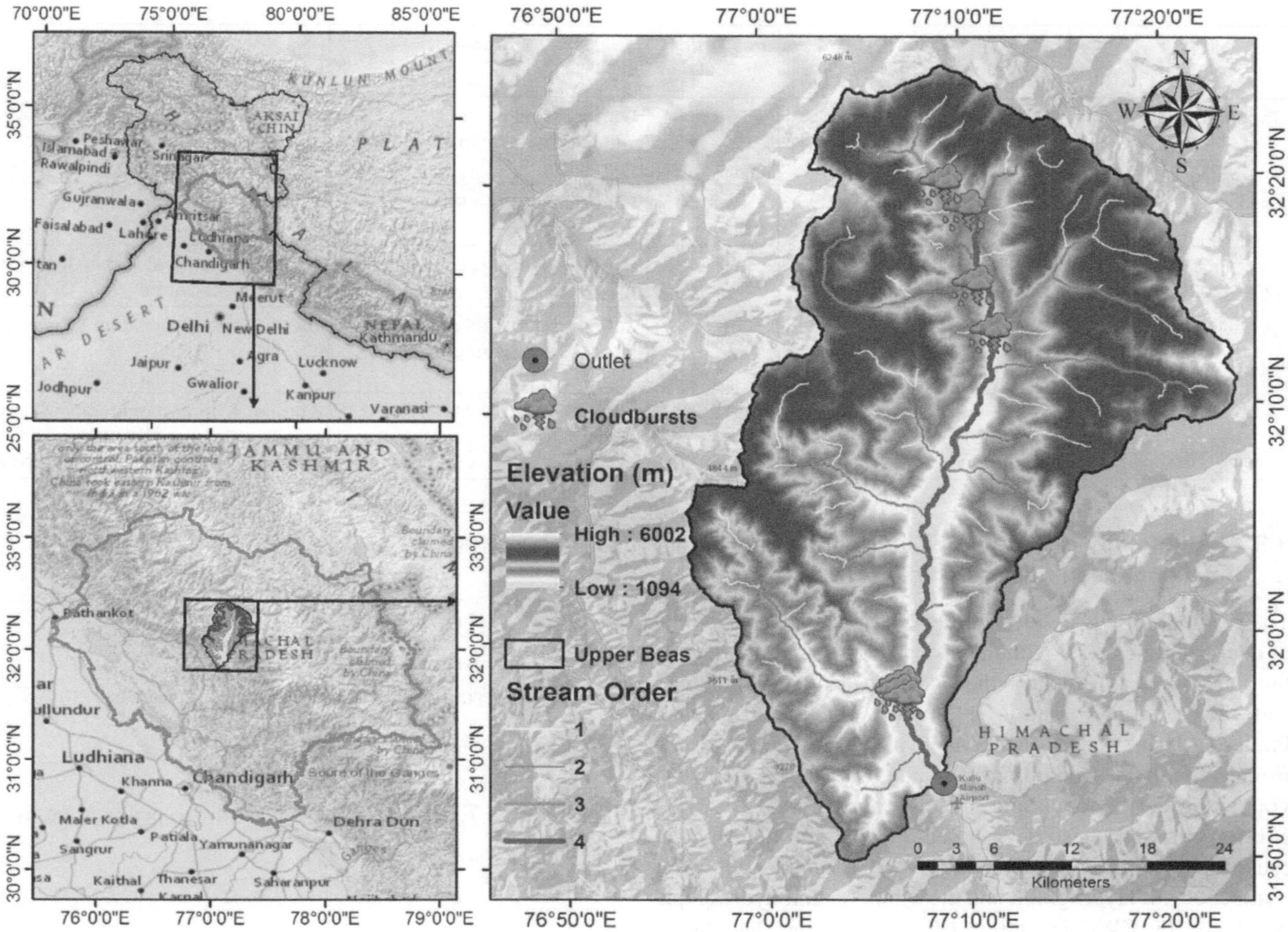

Fig. 9.1 Major cloudburst locations in the Upper Beas basin

Kinnaur, Sirmaur, Kangra, Mandi, Chamba, Hamirpur, and Shimla district. Nearly 43 people and around 24145 livestock were reported dead. Over 200 crores loss was estimated (TNN, 2013).

(vi) On September 23-24, 2018, the Beas basin in the Kullu district suffered from heavy rainfall incidents, leading to the blockage of roads and the destruction of bridges. As per the report, the Manali region had the most rainfall in the state with 127 mm, Dharamsala received 125 mm, Una received 124 mm, and Dalhousie received 90 mm. Shimla, the state capital, received 47 mm of rain. Snow fell in the Kullu, Kinnaur, and Lahaul-Spiti regions' higher elevations. This paper has explained this incident in detail using a flood hydrograph.

(vii) In the monsoon, 2019- A series of cloudbursts incidents caused a loss of 175 lives and 470 crores. The recorded rainfall was around 102.5 mm in the month of august, which alone was 1065% more than 24hr normal rainfall. Kinnaur, Bilaspur, Kangra, Kullu, and Chamba were the majorly affected districts.

(viii) July 12 & 29, 2021- Cloudbursts accompanied by the flash floods hit the Kullu district, taking away 14 lives at Manikaran. The bridge at Jhalma washed away in Lahaul Spiti. On August 25, 2021, in Kaurik village, one labour was killed due to cloudbursts induced landslides.

3.2 Datasets Used

The spatial datasets used in this paper were obtained from various data portals and information from the published papers (Gorelick et al., 2017; S. Singh et al., 2018). To study the impact of cloudburst, hydro-meteorological and topographical information is required, as provided in Table 9.1.

Table 9.1 The list of datasets utilised in this study

	Data	Resolution/ Scale	Purpose	Sources
Topographical	Digital Elevation Model (DEM)	30 m	For representing the land surface elevation in the model domain	Shuttle Radar Topographic Mission (SRTM) (http://earthexplorer.usgs.gov/)
	Land Use Land Cover	1:50000	For estimating the Runoff coefficient	Indian Space Research Organization (https://bhuvan-app1.nrsc.gov.in/
Hydro-meteorological (2001-2020)	GPM IMERG Late Precipitation L3 Half Hourly	0.1°	For estimating Accumulated precipitation	NASA GES DISC at NASA Goddard Space Flight Centre
	GSMaP Operational: Global Satellite Mapping of Precipitation	0.1°	For assessing the Precipitation intensity and hyetograph	JAXA Earth Observation Research Centre
	ECMWF / Copernicus Climate Change Service – Air Temperature	0.25°	For assessing the Mean Air Temperature	European Centre for Medium-Range Weather Forecasts- ECMWF / Copernicus Climate Change Service

3.3 Methodology Adopted for Cloudbursts and Resulting Flash-floods Estimation

The methodology is divided into two parts; at first, the study of characteristics of cloudbursts is carried out by identifying the factors responsible for the occurrence of a cloudburst, and their spatial variation is studied for the Himachal Pradesh. Then, the flash-floods peak discharge resulting from a cloudburst scenario is estimated using seven methods, viz., empirical, rational, and unit hydrograph methods. The envelop curve was also determined. For the unit hydrograph estimation, SCS Dimensionless, SCS Triangular, Snyder Synthetic Unit Hydrograph and the synthetic unit hydrograph developed by CWC are utilised to estimate the flash-flood peak discharge Beas river basin for the incident in September 2018., apart from Empirical by Dickens and Rational method. India has been divided by the Central Water Commission (CWC) into seven hydro-meteorologically homogeneous zones, with which the western Himalayan area is in zone 7. The resulting discharge value is further compared with the SCS triangular unit hydrograph for validation. The formula used is as follows:

1. Empirical: Dicken's Formula, given by

$$Q_p = C_d \times A^{3/4} \tag{1}$$

where, Q_p = peak discharge(m^3/s), A = Basin Area (sq. km.) C_d = Dicken's coefficient varies from 6-30, and 11 to 14 for northern India.

2. The Rational formula is given by,

$$Q_p = 0.0028 \times C \times i \times A \tag{2}$$

where C is the runoff coefficient(rainfall/runoff) estimated from the land-use pattern, i = mean intensity of rainfall (mm/hr), and A = is the Basin area in ha.

3. Envelope Curve, given by:

$$Q_p = \frac{3025A}{(278 + A)^{0.78}} \tag{3}$$

A = Area (sq. km.)

4. SCS Triangular Unit Hydrograph, given by,

$$Q_p = 2.08 \times \frac{A}{T_p} \tag{4}$$

where T_p(hr) = time of rise = time to peak, given by,

$$T_p = \frac{t_r}{2} + 0.6 \times t_c \tag{5}$$

$$T_b = 2.67 \times T_p \tag{6}$$

where, T_b = base length (hr), t_r = duration of effective rainfall, and t_c = time of concentration (min), given by Kirpich eq. as,

$$t_c = 0.01947 \times L^{0.77} \times S^{-0.385} \tag{7}$$

where L = maximum length of water travel (m), S = slope of the catchment

5. Dimensionless Unit Hydrograph, as per Table 9.2, given by

Table 9.2 The ordinates of SCS dimensionless unit hydrograph

t/T_p	q/Q_p	t/T_p	q/Q_p
0	0	1.5	0.66
0.1	0.02	1.6	0.56
0.2	0.08	1.8	0.42
0.3	0.16	2	0.32
0.4	0.28	2.2	0.24

t/T_p	q/Q_p	t/T_p	q/Q_p
0.5	0.43	2.4	0.18
0.6	0.6	2.6	0.13
0.7	0.77	2.8	0.1
0.8	0.89	3	0.07
0.9	0.97	3.5	0.04
1	1	4	0.02
1.1	0.98	4.5	0.01
1.2	0.92	5	0
1.3	0.84		

The value of Tp and Qp is as per the SCS method.

6. Snyder - Synthetic Unit hydrograph, given by:

$$t_p = C_t \times (LL_{ca})^{0.33} \tag{8}$$

$$t_r = \frac{t_p}{5.5} \tag{9}$$

$$Q_{ps} = 2.78 \times \frac{C_p A}{t_p} \tag{10}$$

when non-standard rainfall duration (t_R) is adopted,

$$t'_p = \frac{21}{22} t_p + \frac{t_R}{4} \tag{11}$$

$$Q_{ps} = 2.78 \times \frac{C_p A}{t'_p} \tag{12}$$

$$T_b = 5 \times \left(\frac{t_r}{2} + t'_p \right) \tag{13}$$

$$W_{50} = \frac{5.87}{q^{1.08}} \tag{14}$$

$$W_{75} = \frac{W_{50}}{1.75} \tag{15}$$

where, C_t = a regional constant varying from 0.3-6, C_p = a regional constant, vary from 0.3-0.9, q = Peak discharge per unit catchment area in $m^3/s/km^2$

7. CWC Synthetic Unit Hydrograph-Zone 7, given by,

$$t_p = 2.498 \times \left(\frac{LLc}{\sqrt{S}} \right)^{0.156} \tag{16}$$

where L = Basin length in km., L_c = distance from measuring station to point opp. To centroid of the basin along the main watercourse

$$q_p = 1.048 \times (t_p)^{-0.178} \tag{17}$$

$$W_{50} = 1.954 \times \left(\frac{LLc}{\sqrt{S}} \right)^{0.099} \tag{18}$$

$$W_{75} = 0.972 \times \left(\frac{LLc}{\sqrt{S}} \right)^{0.124} \tag{19}$$

$$T_B = 7.845 \times (t_p)^{0.453} \tag{20}$$

$$Q_p = q_p \times A \tag{21}$$

where W_{50} and W_{75} are the widths of the unit hydrograph at 50% and 75% peak discharge.

$q_p = Q_p/A$ = peak discharge per unit basin area in cumec/sq.km.

4. Results

4.1 Cloudbursts Influencing Factors and Their Spatial Variation

Active monsoons are the most prevalent extreme rainfall event that occurs each year in the Himalayas. Summers are influenced by south-westerly winds, while winters are influenced by the Mediterranean region's rainfall and snowfall. As per the analysis of the satellite datasets from 2001 to 2020, the average annual rainfall of the state ranged from 190 mm to 2190 mm, with a mean annual rainfall of 900 mm. It was discovered that the mean monthly precipitation ranged up to 200 mm during the monsoon season, which may result in the occurrence of frequent flash floods. It was estimated that most of the rainfall (38%) falls in the range of 800-1200 mm, falling in the southern part of the state, followed by 30% in the range of 1200-1600 mm, and only 4% of rainfall falls in the range of more than 1600 mm. The higher amount of rainfall falls in the north-western region of the state, while the least amount of rainfall is observed in the eastern part of the state (Fig. 9.2a).

Increased temperature results in a rise in atmospheric humidity, which results in severe rainfall. As per the analysis of the satellite datasets from 2001 to 2020, the mean air temperature of the state ranged from -12ºC to 25ºC, with a mean annual rainfall of 4.5ºC. Temperatures might fall during a rainfall event due to the passage of cold air connected with the rainfall event and evaporative cooling (Funk et al., 2015). It was estimated that most of the air temperature (39%) falls in the range of more than 10ºC, falling in the western and southern part of the state at a lower elevation, followed by 33% in the range of less than -5ºC, falling in the eastern part of the state at higher elevation (Fig. 9.2b).

Mountains function as a barrier to moisture-carrying winds, forcing them to rise. Precipitation is most on the windward side of the mountains. The amplitude of rainfall is influenced by the interaction of mountain ranges with the large-scale atmospheric flow and localised incidents. The Indian Summer Monsoon (ISM) produces abundant rainfall on the southern

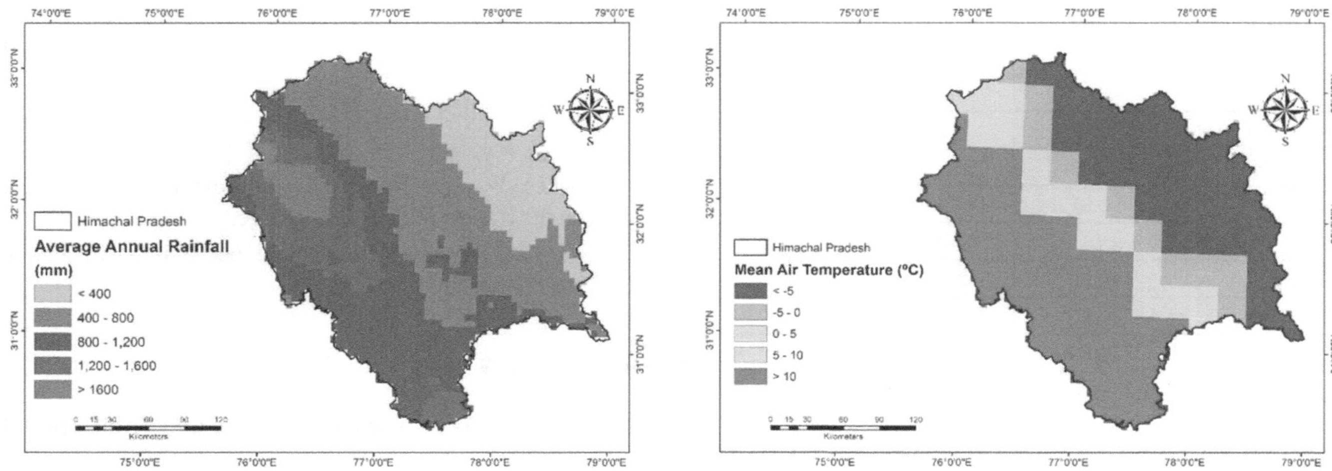

(a) (b)

Fig. 9.2 The spatial distribution of average annual rainfall (a) and mean air temperature (b) over the state

Himalayan slopes but is deficient in the northern Himalayas. The rainfall distribution gets complicated as a result of the altitudinal variance. According to some research, rainfall increases with altitude up to about 2500 m.a.s.l. and then declines. According to studies conducted in the Himalayas on monsoon rainfall characteristics, rainfall decreases with elevations between 2800 and 4500 m for Nepal (Bookhagen & Burbank, 2010; Nandargi & Dhar, 2011). Cloudbursts, extreme precipitation, and inclement weather are almost exclusively linked with low-pressure systems. The development of high- and low-pressure systems is influenced by temperature variations in the environment and temperature differences between the atmosphere and the water within oceans and lakes. The pressure mechanism triggers local weather events. Air rises in regions with low surface pressure. The mean monthly surface pressure variation from 2001 to 2020 indicates that pressure varies over time. The lowest pressure is most frequently encountered in June and July, which results in cloudbursts in these months (Sachchidanand Singh & Kansal, 2022).

Numerous anthropogenic activities (Fig. 9.3), such as road construction, building construction, hydropower development, booming tourism, and forest invasion, all

Fig. 9.3 The list of anthropogenic activities in the Himalayan states that degrade the river ecosystem

represent a severe danger to the climate. Due to the rapid response of landscapes, ecosystems containing vegetation, water, snow and glaciers, and soils to environmental change, mountain areas are especially vulnerable to climate change (Focus, 2013; Haeberli & Beniston, 1998; Immerzeel et al., 2014). Over the last decade, rapid population expansion has harmed the already vulnerable hill state environment by rising demand for the development and compelling people to participate in additional agricultural and other developmental projects in sensitive spots. According to the census, the population climbed to 68.65 lakhs in 2011 from 60.78 lakhs in 2001, while the average population density in Himachal Pradesh increased to 123 in 2011 from 109 in 2001. As the tourist inflow expanded, several multi-storey hotels and other facilities were constructed around the city, even in environmentally sensitive regions.

4.2 Peak Discharge Estimation Using eq. 1 to 21—A Case Study of the Upper Beas Basin

Description of the event:

On September 24 2018, the event of a massive cloudburst was reported at Manali, Kullu district, HP, that triggered flash-floods and landslides. The water in the Beas river rose abnormally and washed away vehicles in Kullu and Manali. As per the GSMap datasets, the rainfall intensity for September 24. 2018, was obtained as per Table 9.3 below:

Table 9.3 The hourly rainfall intensity obtained for the upper Beas basin

Time	mm/hr	Time	mm/hr
01:00	0.0	13:00	0.00
02:00	0.0	14:00	0.89
03:00	2.0	15:00	1.34
04:00	160.0	16:00	2.71
05:00	0.0	17:00	0.30
06:00	0.0	18:00	0.00
07:00	0.0	19:00	0.00
08:00	1.6	20:00	0.00
09:00	3.7	21:00	0.00
10:00	3.6	22:00	0.00
11:00	2.1	23:00	0.00
12:00	0.6	24:00:00	0.00
Mean Intensity (mm/hr)		**7.45**	

The mean intensity was obtained to be 7.45 mm/hr. To estimate the Runoff depth, the rainfall-runoff coefficient is calculated from the land use land cover distribution of the Kullu district obtained from the ISRO-Bhuvan portal, as shown in Table 9.4 below:

Table 9.4 The land use land cover distribution with the runoff coefficient

S No.	Class	Area Ratio (Ai)	C	Ci * Ai
1	Deciduous Broadleaf Forest	0.002	0.10	0.000
2	Cropland	0.057	0.50	0.029
3	Built up Land	0.003	0.90	0.003
4	Mixed Forest	0.011	0.10	0.001
5	Water Channel	0.012	0.90	0.011
6	Plantations	0.124	0.50	0.062
7	Grassland	0.221	0.50	0.111
8	Evergreen Broadleaf Forest	0.451	0.10	0.045
9	Snow & Ice	0.120	0.80	0.096
			Total	**0.356**

The runoff coefficient (C_{av}) was obtained = 0.36. Using Table 4 and C_{av}, the total rainfall was estimated to be 179mm and the phi index (ø) was obtained as 4.8 mm/hr. Thus, the Effective rainfall (ER) for the cloudburst event was 15.5 cm from 3 a.m. to 4 a.m. The upper Beas basin occupies an area (A) of 1417 sq. km., having basin length (L) = 80 km., centroidal length (Lca) = 33 km and Basin Slope (S) = 0.051.

Estimation of Peak Discharge:

1. Using Eq. 1, the peak flood Qp was obtained by adopting Cd= 11 to 14 for northern India,

   Thus Qp ranged from 2541 m^3/s to 3234 m^3/s.
2. Using Rational formula, Eq. 2, taking C = 0.36, i= 7.45 mm/hr, and A = 141700 hectares,

   Qp = 1065 m^3/s.
3. Using envelope curve, Eq. 3, Qp is estimated to be 12984 m^3/s.
4. Using Eq. 4 to Eq. 7, the values obtained for the SCS Triangular unit hydrograph are:

   tc = 6.1 hr, tr = 1 hr, Tp = 4.16 hr, tp = 3.66 hr, Tb = 11.11 hr and Peak discharge Qp = 708 m^3/s. The resulting unit hydrograph is shown in Fig. 9.4.

   Thus, for the effective rainfall of 15.5 cm, the resulting peak discharge would be 10974 cumecs.
5. Using Table 9.2, the ordinates of the SCS unit hydrograph are obtained as per Table 9.5, as shown below:

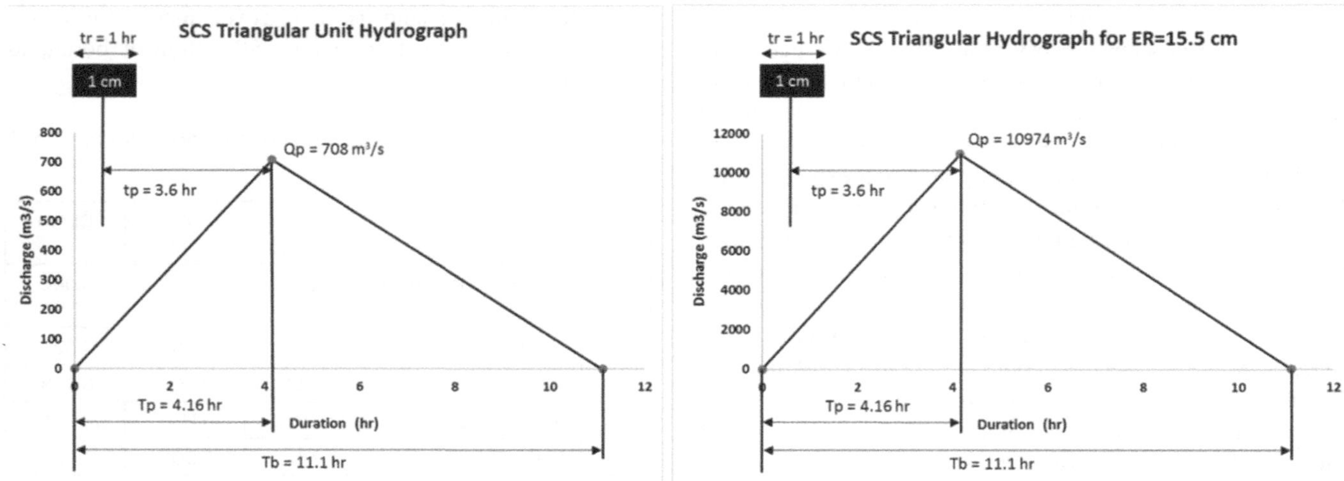

Fig. 9.4 The SCS unit hydrograph and DRH for the upper Beas basin

Table 9.5 The actual ordinates of SCS Unit hydrograph

t (hr)	*q* (m^3/s)	*t* (hr)	*q* (m^3/s)
0.00	0.00	5.83	531.35
0.42	14.17	6.24	467.59
0.83	56.68	6.66	396.74
1.25	113.36	7.49	297.56
1.66	198.37	8.32	226.71
2.08	304.64	9.16	170.03
2.50	425.08	9.99	127.52
2.91	545.52	10.82	92.10
3.33	630.54	11.65	70.85
3.75	687.22	12.49	49.59
4.16	708.47	14.57	28.34
4.58	694.30	16.65	14.17
4.99	651.79	18.73	7.08
5.41	595.12	20.81	0.00

So, the resulting unit hydrograph is as per Fig. 9.5.

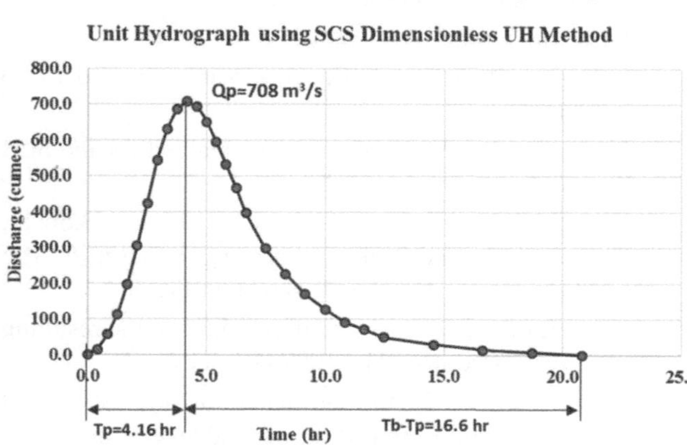

Fig. 9.5 The ordinates of unit hydrograph and cloudburst hydrograph

Using this unit hydrograph, the hydrograph for the Cloudbursts event was estimated using ER = 15.5 cm, as shown in Fig. 9.6 below:

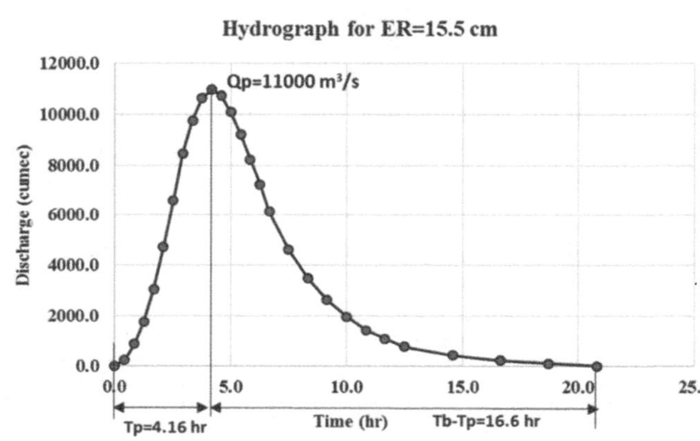

Fig. 9.6 The direct runoff hydrograph for the effective rainfall of 15.5 cm in 3-4 hours

The peak flood value was obtained to be 11000 cumecs within 5 hr from the rainfall event at around 8 a.m., resulting in flash floods and landslides.

6. Using Eq. 8 to Eq. 15, the values obtained for the Snyder Synthetic unit hydrograph are shown in Table 9.6.

   On varying the coefficient C_t and C_p, the peak discharged ranged from 719 m^3/s to 1718 m^3/s. The Tb ranged from 11.8 to 22.4 hrs, W_{50} ranged from 4.8 to 12.2 hrs, and W_{75} ranged from 2.7 to 6.9 hrs. Thus, the unit hydrograph peak would range from 719 to 1718 m^3/s. However, the peak value of 966 m^3/s was taken for the estimation, as it was close to the values obtained from other methods, having C_p = 0.25,

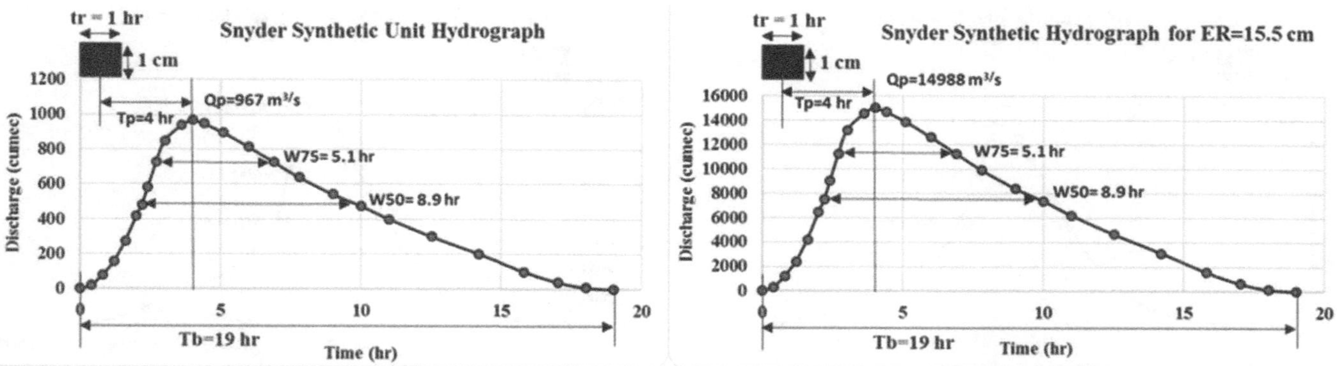

Fig. 9.7 The unit hydrograph and DRH, as obtained from the SUH method.

Ct=0.85. The resulting cloudburst hydrograph had a peak of 14988 m^3/s, as shown in Fig 9.7.

Table 9.6 The value of C_t and C_p and obtained Peak discharge using the Snyder method

Ct	Cp	tp (hr)	tp' (hr)	Qps' (m^3/s)
0.15	0.75	2.02	2.18	1356.7
0.2	0.75	2.69	2.82	1047.6
0.25	0.75	3.37	3.46	853.2
0.3	0.75	4.04	4.11	719.6
0.15	0.8	2.02	2.18	1447.1
0.2	0.8	2.69	2.82	1117.4
0.25	0.8	3.37	3.46	910.1
0.3	0.8	4.04	4.11	767.6
0.15	0.85	2.02	2.18	1537.6
0.2	0.85	2.69	2.82	1187.3
0.25	0.85	3.37	3.46	966.9
0.3	0.85	4.04	4.11	815.6
0.15	0.9	2.02	2.18	1628
0.2	0.9	2.69	2.82	1257.1
0.25	0.9	3.37	3.46	1023.8
0.3	0.9	4.04	4.11	863.6
0.15	0.95	2.02	2.18	1718.5
0.2	0.95	2.69	2.82	1326.9
0.25	0.95	3.37	3.46	1080.7
0.3	0.95	4.04	4.11	911.6

7. Using Eq. 16 -21, the Synthetic unit hydrograph was obtained from the CWC method as given below;

   t_p = 10.77 hr, q_p = 0.69 (cumec/sq. km.), T_b = 20 hr, W_{50} and W_{75} were obtained as 4.94 hr and 3.11 hr, respectively. The peak discharge for the unit hydrograph was obtained to be 973 cumecs. The resultant graph is shown in Fig. 9.8. Thus, for ER=15.5 cm, the peak obtained would be 15077 cumecs.

Summary:

It is observed that the empirical and rational methods underpredicted the result, while the unit hydrograph method estimated the results close to each other. However, the base length and time to peak obtained were different in all the cases. The overall peak discharge value is shown in Table 9.7.

5. Conclusion

Cloudburst and related disasters are common in Himachal's Himalayan area, affecting thousands of people each year and wreaking havoc on people's lives, property, livelihoods, infrastructure, and environment. Most cloudburst occurrences occur during the monsoon season, and most of these occur at night. Southwest and Mediterranean winds and regional moisture dominate during this period. When combined with localised humidity, they drop over the low-pressure zone and are further orographically elevated, resulting in a cloudburst and subsequent flash floods. This flash-flooding was measured using the empirical and unit hydrograph methods; however, empirical and rational methods only give the value of peak and underestimate the result. Thus, a modification factor is required to assess the resulting flow due to a cloudburst event. Furthermore, using unit hydrograph by CWC, SCS, Snyder and Envelope curve method could prove useful in this regard. While cloudburst events are difficult to forecast, a dense network of rain gauges is required to precisely understand this phenomenon, particularly in locations prone to cloudburst. Also, the ground station information combined with satellite imagery can be a highly effective technique for assessing cloudbursts.

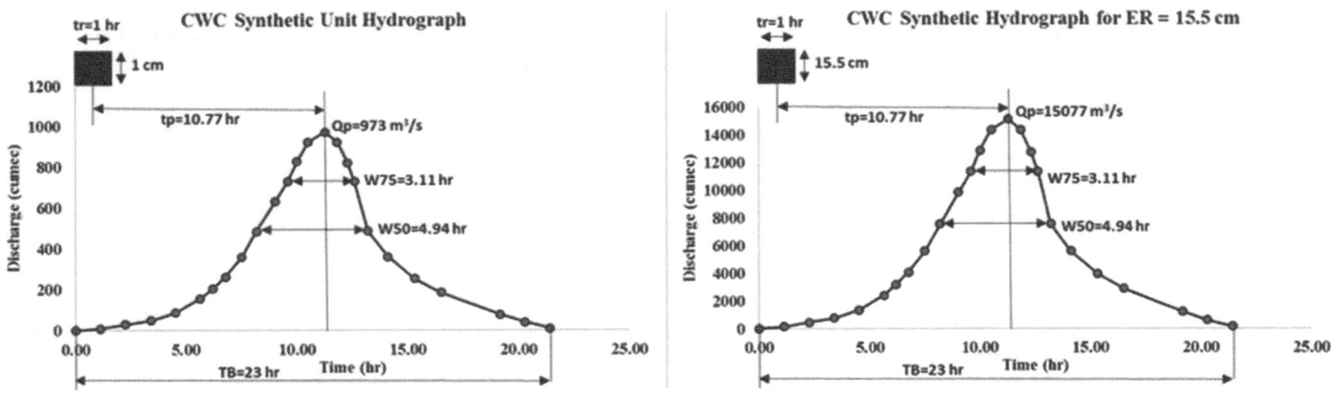

Fig. 9.8 The unit hydrograph and DRH as obtained from the CWC method for Zone 7.

Table 9.7 The peak value obtained using each method

S No.	Analysis	Method	Unit Hydrograph Peak (m^3/s)	Peak (m^3/s)	Modified Peak (m^3/s)
				(ER=15.5 cm)	
1	Flood Peak	Dicken's Formula	—	2540 to 3234 (Cd = 11 to 14)	m = 4: 10162 to 12936
2		Rational Formula	—	1065 (i_{mean} = 7.45mm/h)	m = 10: 10650
3		Envelope Curve	—	12984	-
4	Flood Hydrograph	SCS Triangular UH	708	10974	-
5		SCS Dimensionless UH	708	11000	-
6		Snyder Synthetic UH	967	14998	-
7		CWC Synthetic UH	973	15077	-

References

1. Bisht, G. (2021). *Scientists raise concern over frequent landslides, flash-floods in Himachal.* The Hindustan Times. https://www.hindustantimes.com/cities/others/scientists-raise-concern-over-frequent-landslides-flashfloods-in-himachal-101627767198387.html
2. Bookhagen, B., & Burbank, D. W. (2010). Toward a complete Himalayan hydrological budget: Spatiotemporal distribution of snowmelt and rainfall and their impact on river discharge. *Journal of Geophysical Research: Earth Surface*, *115*(3), 1–25. https://doi.org/10.1029/2009JF001426
3. Census. (2011). *Himachal Pradesh Profile* (Vol. 2011).
4. Das, S. (2005). Mountain weather forecasting using MM5 modelling system. *Current Science*, *88*(6), 899–905.
5. Dimri, A. P., Chevuturi, A., Niyogi, D., Thayyen, R. J., Ray, K., Tripathi, S. N., Pandey, A. K., & Mohanty, U. C. (2017). Cloudbursts in Indian Himalayas: A review. *Earth-Science Reviews*, *168*, 1–23. https://doi.org/10.1016/j.earscirev.2017.03.006
6. Focus, N. I. N. (2013). *Floods spur mountain study Forest management plans in a tangle.*
7. Funk, C., Peterson, P., Landsfeld, M., Pedreros, D., Verdin, J., Shukla, S., Husak, G., Rowland, J., Harrison, L., Hoell, A., & Michaelsen, J. (2015). The climate hazards infrared precipitation with stations - A new environmental record for monitoring extremes. *Scientific Data*, *2*, 1–21. https://doi.org/10.1038/sdata.2015.66
8. Gorelick, N., Hancher, M., Dixon, M., Ilyushchenko, S., Thau, D., & Moore, R. (2017). Google Earth Engine: Planetary-scale geospatial analysis for everyone. *Remote Sensing of Environment*, *202*, 18–27. https://doi.org/10.1016/j.rse.2017.06.031
9. Haeberli, W., & Beniston, M. (1998). Climate change and its impacts on glaciers and permafrost in the Alps. *Ambio*, *27*(4), 258–265. https://doi.org/10.2307/4314732
10. Immerzeel, W. W., Kraaijenbrink, P. D. A., Shea, J. M., Shrestha, A. B., Pellicciotti, F., Bierkens, M. F. P., & De Jong, S. M. (2014). High-resolution monitoring of Himalayan glacier dynamics using unmanned aerial vehicles. *Remote Sensing of Environment*, *150*, 93–103. https://doi.org/10.1016/j.rse.2014.04.025
11. Kansal, M. L., & Thakur, A. (2017). Flood Estimation for a Cloudburst Event in an Ungauged Western Himalayan Catchment. *International Journal of Hydrology*, *1*(6), 163–168. https://doi.org/10.15406/ijh.2017.01.00030
12. Kuksina, L. V., Golosov, V. N., & Kuznetsova, Y. S. (2017). Cloudburst floods in mountains: State of knowledge, occurrence, factors of formation. *Geography and Natural Resources*, *38*(1), 20–29. https://doi.org/10.1134/S1875372817010036

13. Kumar, A., Gupta, A. K., Bhambri, R., Verma, A., Tiwari, S. K., & Asthana, A. K. L. (2018). Assessment and review of hydro-meteorological aspects for cloudburst and flash flood events in the third pole region (Indian Himalaya). In *Polar Science* (Vol. 18, pp. 5–20). https://doi.org/10.1016/j.polar.2018.08.004
14. Mishra, A. K. (2020). Remote-sensing monitoring of a cloudburst event in north India. *Remote Sensing Letters*, *11*(4), 383–387. https://doi.org/10.1080/2150704X.2020.1723170
15. Nandargi, S., & Dhar, O. N. (2011). Extreme rainfall events over the Himalayas between 1871 and 2007. *Hydrological Sciences Journal*, *56*(6), 930–945. https://doi.org/10.1080/02626667.2011.595373
16. Sati, V. P. (2018). *Cloudburst-Triggered Natural Hazards in Uttarakhand Himalaya : Mechanism , Prevention , and Mitigation. January.*
17. Singh, S., Dhasmana, M. K., Shrivastava, V., Sharma, V., Pokhriyal, N., Thakur, P. K., Aggarwal, S. P., Nikam, B. R., Garg, V., Chouksey, A., & Dhote, P. R. (2018). Estimation of Revised Capacity In Gobind Sagar Reservoir Using Google Earth Engine And GIS. *The International Archives of the Photogrammetry, Remote Sensing and Spatial Information Sciences*, *XLII–5*(November), 589–595. https://doi.org/10.5194/isprs-archives-XLII-5-589-2018
18. Singh, Sachchidanand, Dhote, P. R., Thakur, P. K., Chouksey, A., & Aggarwal, S. P. (2020). Identification of flash-floods-prone river reaches in Beas river basin using GIS-based multi-criteria technique: validation using field and satellite observations. *Natural Hazards*, *105*(3), 2431–2453. https://doi.org/10.1007/s11069-020-04406-w
19. Singh, Sachchidanand, & Kansal, M. L. (2022). Cloudburst—A Major Disaster in The Indian Himalayan States. In S. Kolathayar, I. Pal, S. C. Chian, & A. Mondal (Eds.), *Civil Engineering for Disaster Risk Reduction* (pp. 115–126). Springer Singapore. https://doi.org/10.1007/978-981-16-5312-4_9
20. So, D., & Shin, D. Bin. (2018). Classification of precipitating clouds using satellite infrared observations and its implications for rainfall estimation. *Quarterly Journal of the Royal Meteorological Society*, *144*(June 2017), 133–144. https://doi.org/10.1002/qj.3288
21. TNN. (2013). *20 incidents of cloudbursts in Himachal Pradesh, monsoon going strong*. The Times of India. https://reliefweb.int/report/india/20-incidents-cloudbursts-himachal-pradesh-monsoon-going-strong

CHAPTER 10

PARA-FM: Graphical User Interface for Emergency Flood Planning and Management

Utsav Rai
Embedded Systems and Robotics Research Group of TCS Research, Bangalore, 560066, India
rai.utsav@tcs.com

Durgakant Pushp
Embedded Systems and Robotics Research Group of TCS Research, Bangalore, 560066, India
durgakant.pushp@tcs.com

Aditya Choudhary
Embedded Systems and Robotics Research Group of TCS Research, Bangalore, 560066, India
choudhary.aditya@tcs.com

Amit Kumar
Embedded Systems and Robotics Research Group of TCS Research, Bangalore, 560066, India
amitkumar.parmar1@tcs.com

Chaitanya B.
Embedded Systems and Robotics Research Group of TCS Research, Bangalore, 560066, India
c.bodduluri@tcs.com

Rolif Lima
Embedded Systems and Robotics Research Group of TCS Research, Bangalore, 560066, India
rolif.lima@tcs.com

Abhinay N. S.
Embedded Systems and Robotics Research Group of TCS Research, Bangalore, 560066, India
ns.abhinay@tcs.com

Rudrashis Majumder
Department of Aerospace Engineering, Indian Institute of Sciences, Bangalore, 560012, India.
rudrashism@iisc.ac.in

Shuvrangshu Jana
Department of Aerospace Engineering, Indian Institute of Sciences, Bangalore, 560012, India.
shuvrangshuj@iisc.ac.in

Kaushik Das
Embedded Systems and Robotics Research Group of TCS Research, Bangalore, 560066, India
kaushik.da@tcs.com

Debasish Ghose
Department of Aerospace Engineering, Indian Institute of Sciences,Bangalore, 560012, India.
dghose@iisc.ac.in

Abstract

The world has been seeing an increase in the occurrence of floods due to changing climatic conditions and also due to manual actions like rapid and arbitrary urbanisation. Though it is hard to prevent floods, the losses, ecological as well as economical, it could be minimised if timely and appropriate measures are taken. Preventive measures like evacuation of people, livestock and other wildlife will help in handling floods, whose occurrence could be forecasted but to handle emergency flood situations it is very important to build the required infrastructure like emergency response and health care teams, medical and other resources needed to be distributed in flood affected areas. In order to allocate the resources, efficient algorithms have to be designed so that the resulting allocation is optimal and can have maximum impact in handling the situation. These algorithms form the backbone of the resource allocation task and to provide a smooth and easy interface with these algorithms, a simple and user-friendly Graphical User Interface (GUI) is needed. The developed GUI, (PARA-FM, Planning and Resource Allocation for Flood Management) will be able to handle all the inputs from the users that are needed by the algorithms and displays the output from these algorithms in a compact way. The user can modify the entries as per his/her need and the GUI generates the final output for the algorithm by converting user inputs into comma-separated-values (CSV) format. The tabular structure of GUI allows the user to transition between tabs for accessibility to different input tables which make the process perspicuous for the user. This GUI also features google maps API which is used for representation of flood affected regions with interactive options on the map such as, interactive markers, highlighted affected regions and details generated by the algorithm. The final results of resource allocation for both state and district level are then represented in the form of a stacked bar graph.

Keywords: Graphical user interface, Flood management system, Decision support system, Resource allocation

1. Introduction

The deleterious effects of flood impact the economy and society of any country with the damage to both lives of inhabitants as well as infrastructures, vegetation and properties. This disruption caused by flood eventually leads to decrease in labour force, food availability and plummeting the economy. This will cause mass migration of people for which both local and government assistance becomes necessary and to ensure the safety and good health of the public in such difficult times, proper resource management becomes the utmost requirement. With such a chaotic situation, a systematic and well-defined procedure can be effective in saving time and eventually saving a lot of lives. The celerity in the process of resource management could meet the required needs without any unnecessary delay and confusion.

As multiple authorities will be engaged in relief allocation scenarios with different relief items, an integrated software-based framework will help track and optimally allocate the different resources.

With people's lives at stake any process of disaster management should be quick and easy to operate considering the gravity of the situation. An easy-to-use graphical user interface (GUI) provides the available data and options to the user in simple interpretations that enables him/her to be more effective by simply having to spend less time understanding the instructions and nature of the system. The GUI acts as an intermediator for decision support system (DSS), that saves the user from understanding the byzantine nature of internal working of the software and providing a simplified means to the user to choose options and set values that are crucial for DSS to work. This simplification saves a tremendous amount of time and effort to directly interact with the DSS, since not all operators might be trained or aware of the working of DSS. The GUI also acts as a tool that takes input from users in low level format and converts it into more complex and desired format required by the DSS which is thus a time saving feature and saves users from engaging in more redundant processes in order to make DSS work.

The objective of the paper is to provide an understanding on how to decouple a complex set of instructions to more simplified and user-friendly steps by introducing a GUI based intermediate method that helps saving time and effort to directly engage with complex steps required to make the software-based decision support system.

This paper highlights the key features of the GUI used for providing input to the DSS. It first illuminates the software architecture of the decision support system used for resource allocation in section 3. Section 4 elucidates the development process of the GUI used followed by section 5 that explicates the working of the GUI using illustrations with real scenario data.

2. Literature Survey

The majority of existing work dealing with disaster management and planning are an application of geographic information science (GIS) (Cutter, 2003) as all disasters are common geographic events. To plan according to the severity of the disaster, GIS helps by providing spatial and geographical data for analysing which helps in managing disaster. One such application of GIS is presented in (Amoako & Lin, 2018) which uses remote sensing and mapping of flood prone areas to manage flood in four stages: prevention and mitigation, preparation, damage assessment, and prediction. The information generated with the help of GIS helps policy makers and disaster management agents to act accordingly for saving damages to lives and infrastructure. More application of GIS can be seen in work presented in (Kwang & Osei, 2017), (Assilzadeh, Levy, & Wang, 2010), (Tanavud, Yongchalermchai, Bennui, & Navanugraha, 2000) and (Pradhan, Suliman, & Awang, 2007), dealing with disaster management of flood (Kwang & Osei, 2017), landslide ((Assilzadeh et al., 2010) and (Tanavud et al., 2000)), and forest fire (Pradhan et al., 2007) respectively. All such applications are involved with geographical data directly and to enable such application in real world scenario an interface is required to bridge the gap of understanding complexities involved in such systems and user's ability to work with such data efficiently. To enable the user to be fully capable of driving such applications to its fullest potential, a graphical user interface helps in removing the barrier of learning complex data manipulation and arrangements required for decision support systems and present users with self-explanatory options and interface which eventually drives the underlying algorithm.

In (da Silva, Humberto, Alencar, Ferreira, & de Almeida, 2020), a multidimensional decision model is pre- sented based on utility theory. The paper (Mollah, Sadhukhan, Das, & Anis, 2018) proposes a cost optimization model based on mixed-integer programming and genetic algorithms for shelter allocation and resource distribution. In (Jana, Majumder, Menon, & Ghose, 2021), a hierarchical framework is presented for the distribution of resources and allocation of tasks in disaster management.

Paper (Newman et al., 2017) presents a detailed survey on decision support systems employed for natural hazard risk reduction. There are very few works where GUI is developed for the end-users to get the automated decisions from the DSS. Papers using GUI can be found in (Newman et al., 2017). In (Mandiak, Shah, Kim, & Kesavadas, 2005), a GUI is developed for post-disaster data fusion. In (Balis et al., 2017), a GUI model is discussed for flood DSS. In (Esmaili, Dahlan, & Desa, 2016), a GUI-based DSS is discussed to get the requirement of ambulances after an earthquake takes place. Other papers like (Ahmad & Simonovic, 2006), (Hinkel & Klein, 2009) are concerned about building GUI.

3. Generic Software Framework for DSS

A typical software framework for disaster management at a disaster management station should be able to process the information related to relief requirements and disaster status from the ground station/higher authority, and it will be able to allocate the relief among the various relief nodes in an optimal manner. A typical relief allocation framework is shown in Fig. 10.1. Generally, a disaster management station will have a number of local relief stations. GUI of the resource allocation software takes the input of relief requirements from the different relief stations and the resource allocation algorithm to allocate the resources to the relief stations. The input of GUI consists of disaster- related information about the ground stations, available resources, and the demand from the individual ground stations. The importance of the resource allocation algorithms is selected based on the previous observations; however, the parameter of the resource allocation algorithm can be modified by the disaster authority. Finally, the allocated resources to different relief stations are presented in GUI output for resource distribution by the disaster management authority.

4. GUI Development

The GUI presented in this paper is a database powered website which features tabular data representation for both input to the web-based GUI and output to the DSS. The GUI is also capable of taking input in comma separated values (CSV) format which is a universally accepted data format. Since handling large numbers of entries could be redundant and time taking, directly importing CSV files to populate tables in the

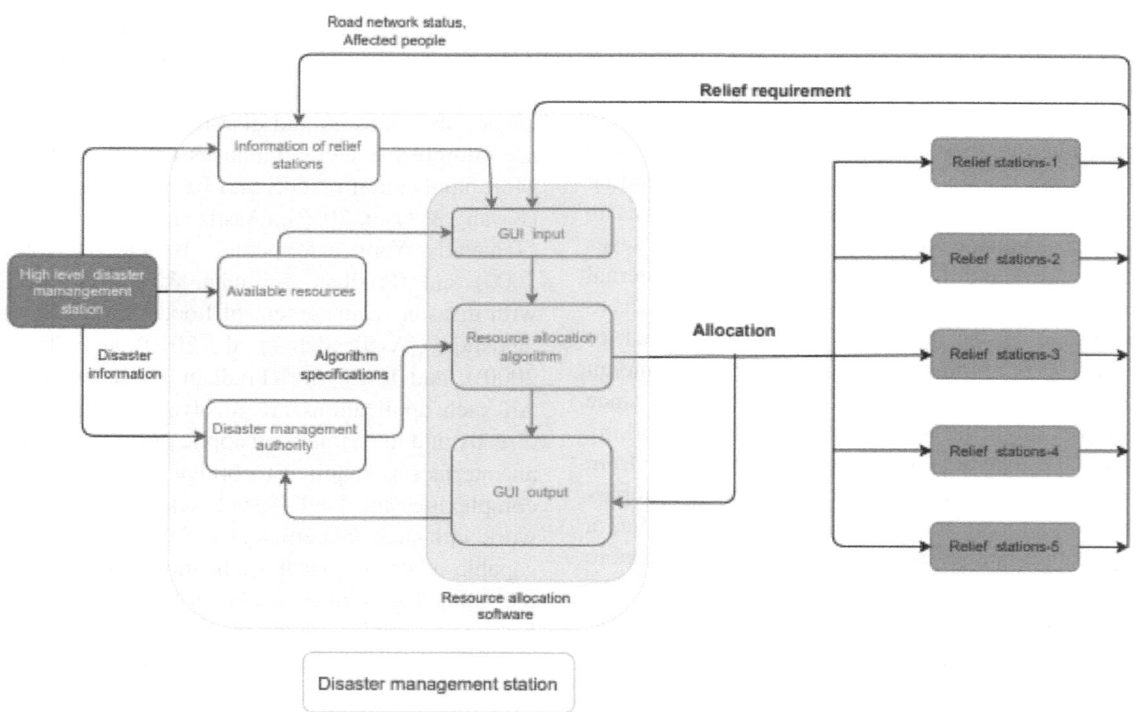

Fig. 10.1 Software framework for disaster management

GUI saves from a lot of repetitive and time taking tasks. The imported entries from the CSV file to the tables in GUI then can be further manipulated or used to add more information based on the current situation, demand and requirement. This interface allows one to manage a complicated set of entries which can be overwhelming at first if one decides to manage it without any tool. The nested table data structure allows to manage individual entries for instance each city affected or sub-region could have its own resource allocation defined by its separate table. These tables are dynamic in nature which allows users to modify the entries or nested tables within each row. As for any GUI, a good user accessibility not only makes the process easier but faster as well, our GUI features options like quick access drop downs where multiple options could be selected for a single field. For modifying any field, an editable text box with options to either add a completely new row or delete a particular row without disturbing the entire table makes data management easier. To make the data more presentable and sensible, graphs are used to represent the overall gist of the tabular data provided by the user. Since demographic data is involved, user is provided with a map of the affected region/area currently in focus or authority. Once all necessary fields are filled, the user can simply export all the tables as CSV files which can then be used as input to the DSS.

5. Illustration

The GUI begins with the dashboard page that allows users to login with the credentials to avoid any unwanted logins as shown in Fig. 10.2. On successful login, the user lands on his/her dashboard page that presents information related to the region currently affected and the current progress done in order to support disaster management activity as seen in Fig. 10.3. The dashboard also features some geographical data of the flood affected region relevant for planning and management. On map, information like severity of the disaster is presented over the region where flood has affected and its current population along with some demographic information which facilitates the overall process of disaster management.

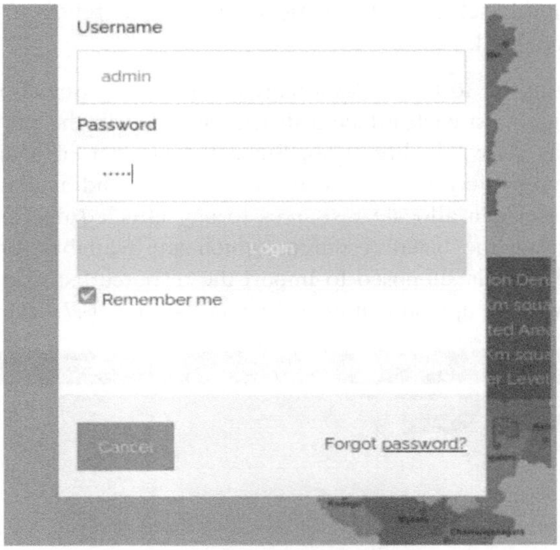

Fig. 10.2 Login window

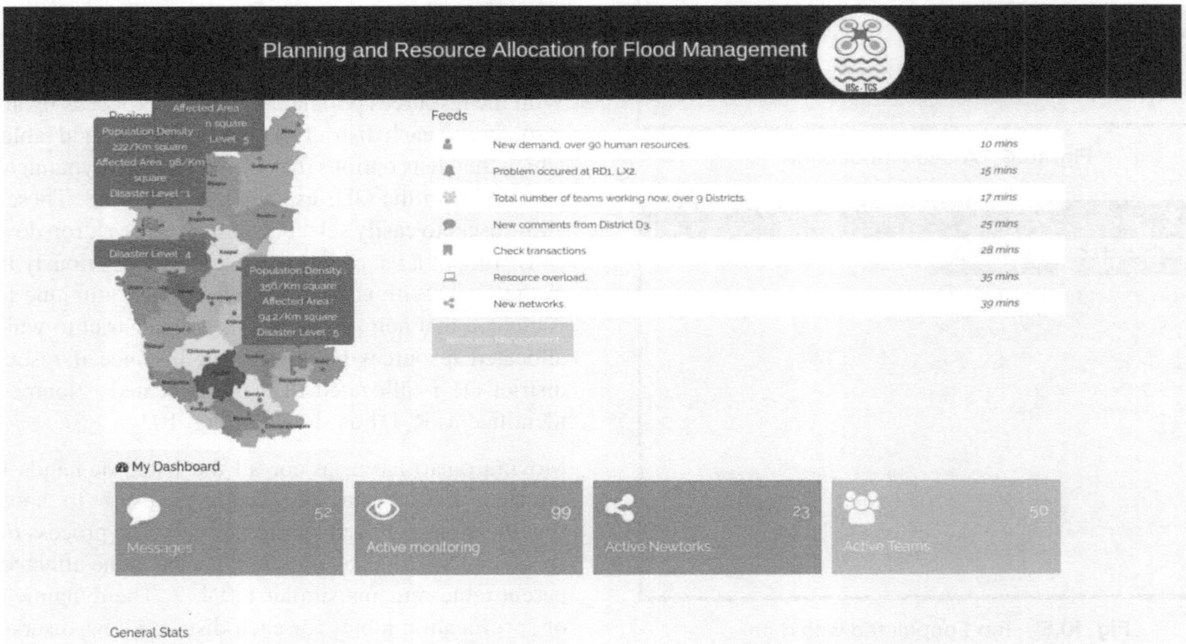

Fig. 10.3 Dashboard page presenting topical data

The dashboard leads to a table page which consists of five tabs (Fig. 10.4) each with the purpose to get user input in sequential order. The first tab named 'State' allows users to upload the CSV file of a particular state consisting of all its districts. The browse option opens the file manager in which the user can locate the desired state file and by clicking on the upload button the district table is populated as shown in Fig. 10.5. Each table in the GUI is normalized to at least first normal form or 1NF with most of the major or parent tables normalized to at least third normal form or 3NF. As seen in Fig. 10.5 the parent table with a list of districts is paired with primary keys for each row, using which a new table is created by simply selecting the districts which are currently affected by the flood.

Moving on to tab 2 *Resources*, is where the user allocates resources to each affected district selected in the previous tab as shown in Fig. 10.6. Since the number of available resources may vary based on the situation and availability, the user can allocate resources to any affected districts by selecting the listed resources which are available for that state, user is supposed to import the resource list using the browse and upload feature as seen in the Fig. 10.7.

Fig. 10.4 A 5-tab input page

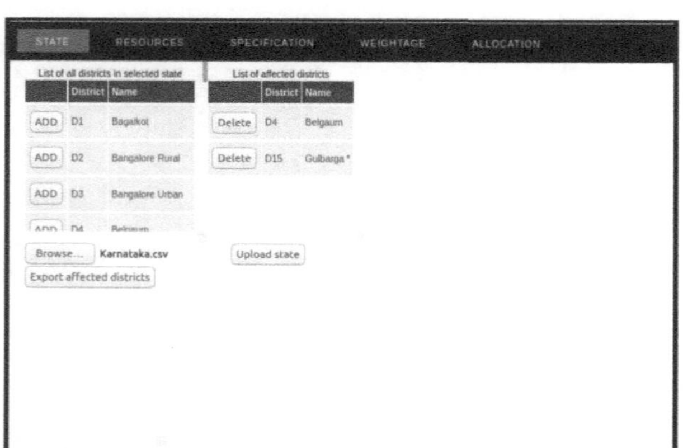

Fig. 10.5 Tab 1 populated with data

Fig. 10.6 Resource tab

Fig. 10.7 Adding available resource table

With the resources populated, the user proceeds by allocating resources to each district by clicking on the 'add table' button which then lists options that allows users to dynamically form new tables in the GUI as shown in Fig. 10.8. These options allow users to easily select resources using a drop down menu consisting of a list of resources that user previously imported in the GUI. With each new table formed a unique identifier is created that helps in identifying the district to which these allocated resources belong to, for instance if resources for district D1 is allocated then the allocated resource table is identified as R_D1 as shown in Fig. 10.9.

Moving on to tab 3, as for all the listed demands for each affected district the DSS requires users to specify the specifications for each district as well. The process of adding specifications for each individual entry in the affected district parent table remains similar to tab 2. The dynamic creation of specification tables for each district allows users to input

Fig. 10.8 Dynamic table insertion for allocating resources to each affected district

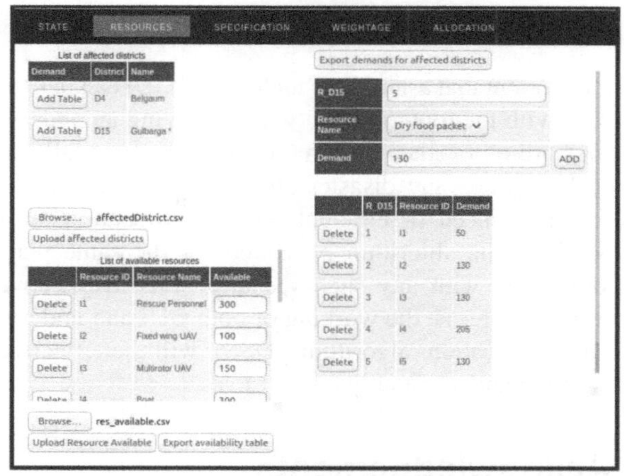

Fig. 10.9 Listing demand for resources needed by individual districts

Fig. 10.10 Adding specifications for each affected district dynamically

Fig. 10.11 Adding weights to each affected district and its specifications as required by DSS

the data in the desired format easily as it can be seen in Fig. 10.10.

Tab 4, 'weightage' table allows users to dynamically edit the previously populated fields in tables of affected districts and specifications. This tab is required in order to input the weightage for each affected district and to each of its respected specifications as shown in Fig. 10.11.

The final tab shows the result or output of the DSS which takes in the data user entered in tab 1 to 4. As seen in Fig. 10.12, the algorithm efficiently allocates the resources to each affected district. The resource demanded is represented by a red bar and the respective allocated resource computed by the algorithm is represented by a blue bar in the bar graph. The concluded table shows the total resource allocated per iteration for the current state user has provided the input.

Fig. 10.12 Result of resource allocation for each affected district from the input provided by the user

6. Conclusion

As it is evident that a simple to understand GUI can help users to work with the input in hand without having an underlying understanding of the working of the resource allocation algorithm. Since any disaster management situation shares the common need of efficient resource allocation as soon as possible, the algorithm presented for the same purpose is compatible with any such scenario, thus irrespective of what the disaster is, the working of both GUI and underlying resource allocation algorithm (DSS) will be effective in saving time and eventually a lot of lives.

7. Acknowledgements

The IISc team wishes to acknowledge partial funding support from the EPSRC-GCRF project (EP/P02839X/1).

References

1. Ahmad, S., & Simonovic, S. P. (2006). An intelligent decision support system for management of floods. *Water resources management*, *20*(3), 391–410.
2. Amoako, E., & Lin, S. (2018). Application of remote sensing technique and geographic information science for flood risk mapping-a case study of the offinso district, kumasi-ghana. *J. Remote Sens. GIS, 7.*
3. Assilzadeh, H., Levy, J. K., & Wang, X. (2010). Landslide catastrophes and disaster risk reduction: A gis framework for landslide prevention and management. *Remote Sensing*, *2*(9), 2259–2273.
4. Balis, B., Bartynski, T., Bubak, M., Harezlak, D., Kasztelnik, M., Malawski, M., ... Wilk, B. (2017). Smart levee monitoring and flood decision support system: reference architecture and urgent computing management. *Procedia computer science*, *108*, 2220–2229.
5. Cutter, S. L. (2003). Gi science, disasters, and emergency management. *Transactions in GIS*, *7*(4), 439–446.
6. da Silva, L. B. L., Humberto, J. S., Alencar, M. H., Ferreira, R. J. P., & de Almeida, A. T. (2020). Gis-based multidimensional decision model for enhancing flood risk prioritization in urban areas. *International Journal of Disaster Risk Reduction*, *48*, 101582.
7. Esmaili, E., Dahlan, H. M., & Desa, M. I. (2016). Model-driven decision support system for estimating number of ambulances required during earthquake disaster relief operation. *Journal of Theoretical & Applied Information Technology*, *90*(2).
8. Hinkel, J., & Klein, R. J. (2009). Integrating knowledge to assess coastal vulnerability to sea-level rise: The development of the diva tool. *Global Environmental Change*, *19*(3), 384–395.
9. Jana, S., Majumder, R., Menon, P. P., & Ghose, D. (2021). A decision support system (dss) for hierarchical allocation of resources and tasks for disaster management. *International Conference on Dynamics of Disasters.*
10. Kwang, C., & Osei, E. (2017). Accra flood modelling through application of geographic information systems (gis). *Remote Sensing Techniques and Analytical Hierarchy Process. J Remote Sensing & GIS*, *6*(191), 2.
11. Mandiak, M., Shah, P., Kim, Y., & Kesavadas, T. (2005). Development of an integrated gui framework for post-disaster data fusion visualization. In *2005 7th international conference on information fusion* (Vol. 2, pp. 7–pp).
12. Mollah, A. K., Sadhukhan, S., Das, P., & Anis, M. Z. (2018). A cost optimization model and solutions for shelter allocation and relief distribution in flood scenario. *International Journal of Disaster Risk Reduction*, *31*, 1187–1198.
13. Newman, J. P., Maier, H. R., Riddell, G. A., Zecchin, A. C., Daniell, J. E., Schaefer, A. M., ... Newland, C. P. (2017). Review of literature on decision support systems for natural hazard risk reduction: Current status and future research directions. *Environmental Modelling & Software*, *96*, 378–409.
14. Pradhan, B., Suliman, M. D. H. B., & Awang, M. A. B. (2007). Forest fire susceptibility and risk mapping using remote sensing and geographical information systems (gis). *Disaster Prevention and Management: An International Journal.*
15. Tanavud, C., Yongchalermchai, C., Bennui, A., & Navanugraha, C. (2000). Application of gis and remote sensing for landslide disaster management in southern thailand. *Journal of Natural Disaster Science*, *22*(2), 67–74.

Chapter 11

Flood Response to Geomorphic Setup and Coastal Land Use Patterns: A Case Study in Krishna River Delta, Andhra Pradesh, India

M. V. Ramana Murty
Scientist, AP Space Applications Centre (APSAC), Labbipet, Vijayawada - 520010
Corresponding Author: moidavrm64@gmail.com

K. Mruthyunjaya Reddy
Scientist, National Remote Sensing Centre, Balanagar, Hyderabad- 500037

K. V. Swamy
Asst. Professor, Adikavi Nannaya University, Rajamahendravaram - 533296

Abstract

Recurring floods in the low lying coastal areas is often threatening human life and their activities. As strong economic activities are operating in the coastal zone, there is a need to minimize the threats from natural hazards. Coastal geomorphology and the processes acting on them play a vital role in determining the severity of the coastal hazards. Satellite Remote Sensing with its intrinsic advantages has become a decisive tool to monitor natural disasters at local, regional and global scales. The synoptic and temporal coverage of satellite images facilitate valuable information useful in the assessment, monitoring, and management of coastal resources and natural hazards management. The present work deals with Krishna river deltaic region in parts of Krishna, Guntur, Prakasam, and West Godavari districts of Andhra Pradesh, India. This delta represents one of the most fertile regions of South India, and is highly prone to floods & cyclones. The extents of inundation during three different events are extracted, and geomorphic setup of the delta is analysed. Finally, the multi seasonal land use/ land cover (LULC) is mapped. The flood response to geomorphic setup and coastal land use patterns is discussed.

Keywords: Coastal flooding, geomorphology, land use, hazard, Remote Sensing, GIS

1. Introduction

Geomorphic processes provide major insight into the formation of various features that landscape worldwide. The coastal landforms are amongst the most dynamic, and are continuously worked by the action of waves, tides and currents. Further, the coasts are geomorphologically a complex system, which responds in various ways to extreme weather events (**Balica, et. al., 2012**). Coastal flooding is one of the most damaging natural hazards, hindering the coastal development. Storms modify coastal landforms within a short span of time (**Nordstrom and Jackson, 1995; Ramana Murty, et.al., 2012; Nayak, 2017**). Understanding the flood geomorphological is possible by studying the real processes that operated in the past (**Baker, 1994**).

The sediment laden river discharges, give rise to advancing of coastline significantly in the deltaic environment. Progressive shifting of river courses and coastline are important features in the delta evolution. A host of marine and fluvial landforms are conspicuous present now much inland from the present shoreline in the deltaic environment. In view of unabated environmental degradation and strong economic activities operating in the coastal zone, there is a need to minimise the threats to human life and activities due to the natural hazards. The common extreme events in the coastal zone are due to tsunami and storm surge, which gets amplified in shallow waters and cause flooding. Tsunamis occur far less often than storm surge, but are capable of causing huge damage to life and property. Tsunamis can reach far inland unlike storm surge. The tsunami can erode and transport significant volume of sediments including heavy minerals from the shelf to the beach region **(Narayana, et. al., 2007).** The withdrawal carries huge quantities of sediments back into the sea. In the process, the beaches too undergo significant changes. Not all the locations show a beach recovery after the tsunami stresses, as the inner shelf processes and longshore currents redistribute the sediments eroded at the coastline **(Fagherazzi and Du, 2007)**. Further, there is no single way of reducing the effects of tsunami and other coastal hazards, as the coastal region constitutes varied morphology and socio-economic setup which might demand a region specific strategy. Coastal engineering structures, in some instances, increase the tsunami vulnerability **(Kurian et. al., 2006)**. The sustainable use of the coast demands both spatial and temporal flexibility. **Pethick and Crooks, (2000)** provided a first order approximation of the temporal variability that may be expected in landform components of the shoreline system. Thus, allowing the management to provide more realistic objectives for long-term sustainability from both natural and artificial processes.

Land use mapping based on multi-season data can be an important spatial database for natural resources inventory and disaster management. Common land use/ land cover classes include: Built-up land, agriculture, forest, wasteland, wetland, water bodies, etc. The changes in land use practices can alter the flood severity. The spatiotemporal changes in the land use pattern for over four decades in Oshiwara River catchment indicated an increase in the built-up area by about 74.84%; and reduction in open space by 42.79% and water body by 62% **(Zope, et.al., 2016).** The predominant crop growing seasons along eastern coast of India are also associated with landfall of tropical storms. The crop damages from cyclone storms can be minimized by adopting suitable management practices. A shift towards growing short duration crops or storm resistant crops in recurring flood prone areas **(Kumar, et. al., 2014)** is needed.

Satellite Remote Sensing, due to its intrinsic advantages has become a decisive tool for resources inventory at local, regional and global scales. These observations have become pivotal in securing global environment, lessening disaster losses, and accomplishing sustainable development. With the availability of high-resolution data from different platforms, the scope and applications of remotely sensed data in combination with Geographic Information System (GIS) have been proved to be vastly advantageous. By virtue of its synoptic overview, remote sensing technology in combination with conventional approach facilitates in tackling numerous forms of natural assets for societal benefits. With the emergence of space-based data products, contemporary and critical understanding of coastal zone is emerging. The synoptic and temporal coverage of satellite images facilitate valuable information useful in the assessment, monitoring, and management of coastal resources and natural hazards management. Satellite images captured by sensors were widely used **(Ramana Murty, et. al., 1993, Srinivasa Rao, et. al., 2013; Klemas 2015)** for studying historical flooding. The extent of coastal flooding either due to the storm surges or tsunamis can be effectively mapped based on the satellite images. Since much of the information is likely to have a spatial component, GIS has obvious relevance to accomplish this task.

The coastal geomorphology and the processes acting on them play a vital role in determining the severity of the coastal hazards. A study on the inundation patterns of the 26th December 2004 tsunami along the Tamil Nadu coast has revealed that different coastal geomorphic units respond differently to tsunami hazard **(Ramasamy, et. al., 2006).** The response mechanism of coastal geomorphic assemblages to tsunami hazards for Kachchh coast is elaborated by **Shukla, et. al., (2010)**. These studies indicate that the severity of coastal flooding and the associated damages are predominantly landform dependent. The geomorphic units like estuary, creek, etc., act as carriers/ conveyors of storm run-up. The tidal flats and mudflats act as facilitators and are the zones of regular inundation. The beach ridges and wide sandy beaches, on the other hand, absorb surge energy and act as a barrier. **Ramasamy, et. al., (2006)** and **Shukla, et. al., (2010)** discussed on the response mechanism of coastal geomorphic assemblage for guiding disaster risk reduction activities. **Ramana Murty, et. al., (1993); Rao, et. al. (1995); and Mruthyunjaya Reddy, et. al., 2006)** discussed on various coastal landforms from flood hazard perspective for different regions of Southern India.

Ancient shoreline or strandline which is out of reach of present wave action is identified by several researchers within the Krishna-Godavari delta region **(Nageswara Rao and Vaidyanadhan, 1978)**. **Mahender Reddy and Shah (1991)** identified five stages of Krishna delta evolution represented by four to five strandlines. The strandline-I represents the oldest one, while the strandline-IV represents the youngest. The region within two consecutive younger strandlines is more likely to be flooded, compared to the region near oldest strandlines; provided all other factors remain same. The land use patterns are also closely associated with different strand plane. **Naylor, et. al. (2017)** highlighted the complex interactions between geomorphic processes, landform and landscape characteristics on the one hand and flood and erosion risks from moderate to extreme weather events on the other hand. Recent studies indicate that there is a strong potential for geomorphological processes and current landscape topographies to amplify or diminish the risks of inundation, erosion and resultant effects.

2. Study Area

The present work deals with Krishna river deltaic region (Fig. 11.1) in Andhra Pradesh, India. This delta represents one of the most fertile regions of South India, and is highly prone to floods & cyclones. The area of study is distributed in parts of Krishna, Guntur, Prakasam, and West Godavari districts of Andhra Pradesh, India. It lies between Longitude 79° 56' 59.8" E & 81° 39' 32" E; and Latitude 15° 33' 09.7" N & 16° 54' 30.9" N covering an area of about 13,477 km^2. There are 87 mandals distributed across four districts of Krishna river delta region.

3. Methodology

The methodology adopted in the present study for Krishna river delta region includes, mapping of various themes like land use/ land cover, geomorphic setup, inundation etc., to address flood response. The extent of inundation from multiple flood events of May 1990, October 2005 and September 2016 are considered for analysis. Each of the three inundation events are classified into three different categories. The Land Use/ Land Cover (LULC) mapping addresses the geographical spread of built-up land, croplands, wastelands, water bodies, wetlands, and forest vegetation using satellite imagery. These six major categories of the delta are further divided into 22 sub-classes. The LULC classes are interpreted from IRS LISS III images for three different seasons (kharif, rabi, and zaid) based on 2015-16 years' data. These thematic layers are extracted through on-screen delineation using image interpretation keys, ancillary and legacy data. The interpreted themes are verified on the ground before map finalization.

The geomorphic units/ landforms are also interpreted based on the IRS LISS-III data in GIS environment. The landforms of the Krishna river delta region are classified based on their origin (marine, fluvial, fluvio-marine, aeolian, denudational, and structural landforms). The geomorphic setup is considered to have a strong bearing on storm flow and runoff. There are eight distinct geomorphic characteristic features i.e., facilitator, carrier/ conveyor, accommodator, absorber, barrier, outflow/ hold, impact/ influence, and run-off zone. Different landforms respond differently to storm flow/ inundation. In other words, the low tidal flats, tidal marsh, and mudflats get inundated easily compared to the adjacent elevated beach ridges.

3. Results and Discussions

The deltaic plains are bestowed with lush coconut gardens, extensive croplands, evergreen mangroves, etc. Coastal hazards, particularly tropical storms claim several lives, damage croplands, destroy infrastructural facilities, disrupt communication networks, etc., depending upon their intensity. These storms are the most destructive recurring natural hazards along the coast. Some of the earliest and recent cyclonic storms that hit the Krishna river delta front in east coast of India were on 12th October 1779 close to Machilipatnam, on 1st November 1864, 23rd October 1949, 19th November 1977, 9th May 1990, etc., causing enormous damage to the life and property of the region. However,

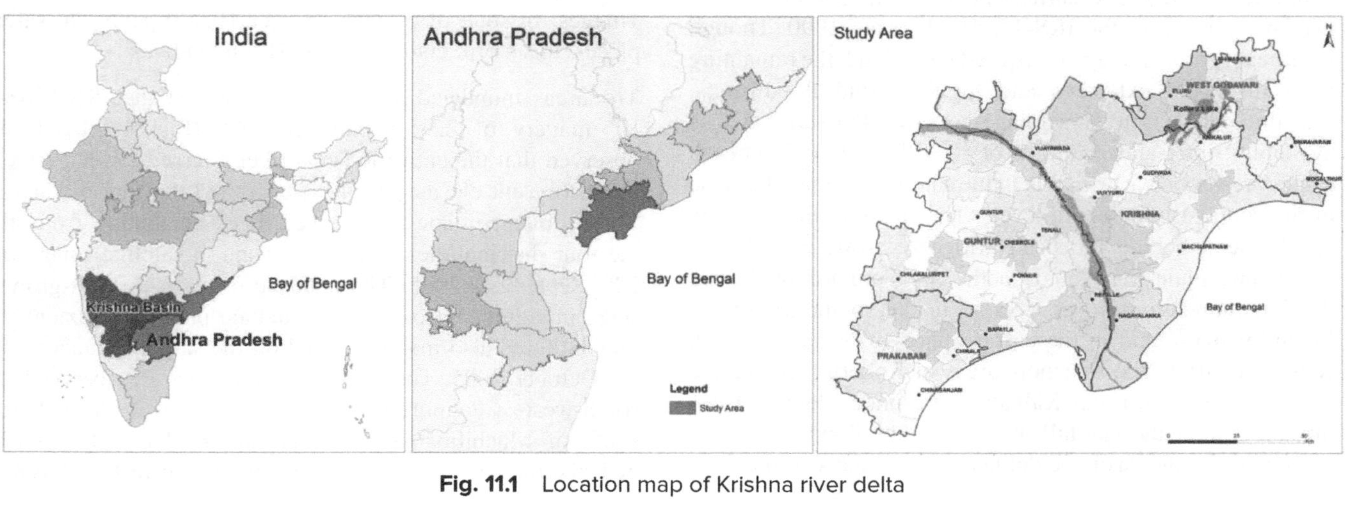

Fig. 11.1 Location map of Krishna river delta

the loss of human lives during the recent event is relatively lesser, when compared with the past cyclones. This is mainly due to the improvement in the dissemination of advance warnings and timely evacuation of population to safer areas. Flood response to geomorphic setup and coastal land use patterns is deliberated on inundation during different events; geomorphology based flood hazard zonation; land use; and impact of flood hazard on land use patterns.

3.1 Inundation during Different Events

The availability of space borne sensors from earth observation satellites on a regular basis has enabled to address various disaster management issues since 1972. Floods are among the most destructive natural disasters affecting the human population and infrastructure. River floods and coastal storm surges affect the lives of more people than most other weather related disasters **(Klemas, 2015)**. Satellite-derived flood inundation maps produced in near-real time provides valuable information to identify the affected areas. This facilitates the local Government in taking up rescue and relief operations. The extent of flood inundation in a given area during multiple past events is an important indicator for vulnerability assessment **(Ramana Murty, 2020)**. The flood/cyclone events of May 1990, September 2005, and September 2016 are analyzed.

(a) May 1990 cyclone event: The system that originated over the South-West Bay of Bengal intensified into a cyclonic storm on 5th May 1990. This severe cyclonic storm crossed the coast near the mouths of river Krishna (south of Machilipatnam) on 9th May 1990 and took a toll of 967 lives. The loss of human lives is relatively smallest because of timely evacuation of people from the flood affected areas. Else, the loss could have been colossal, as the storm surge associated with the cyclone raised the sea level by about four meters above the normal tide level (**Subramanian and Mohanan, 1992**). The earliest post-cyclone satellite image was available from the IRS-1A of 11th May 1990. Though significant parts of image is covered with cloud, the remaining parts displayed extensive and highly turbid floodwaters. The immediate following image covering the same area is available from Landsat TM as of 18th May 1990. The beach ridges are very clearly visible in the satellite imagery because of the contrast in moisture levels. The standing flood waters can be seen behind the beach ridge & swale complex in the deltaic plain between Bapatla and Ponnuru in Guntur district. Inundation is also present in the southern parts of Machilipatnam, and in the eastern parts of Nagayalanka in Krishna district. The foreshore areas of Kolleru lake and the area between Eluru and Kaikaluru are under deluge. Even nine days after the landfall of cyclone, the flood waters are present and could easily be delineated from the satellite data.

Flood mapping not only involves in demarcation of farthest extent of flooding during an event, but also the duration of its extent. Thus, the temporal satellite datasets within a flood event can be used for updating in terms of progression, recession, and persistence. In the present study, a flood map is prepared based on satellite images of 11th May and 18th May 1990 showing three distinct inundation classes (Fig. 11.2a). They include: Very Highly Flood Affected Areas (VHFAA), Highly Flood Affected Areas (HFAA), and Moderately Flood Affected Areas (MFAA). The VHFAA and HFAA are observed near Chirala, Bapatla, Ponnur, Nagayalanka, and Machilipatnam. The VHFAA and HFAA are also observed along Budameru river near Vijayawada and around Kolleru lake near Kaikalur. The total extents of VHFAA, HFAA and MFAA in Krishna delta are about 340 km^2, 364 km^2, and 2108 km^2 respectively. The significant extent of VHFAA is observed in Guntur district, which constitutes about 174 km^2. Similarly, a considerable extent of HFAA and MFAA are observed in Krishna district representing an area of about 201 km^2 and 1464 km^2 respectively.

(b) September 2005 flood event: A low-pressure area formed over the Andaman Sea and adjoining east-central Bay of Bengal on 16th September 2005, became a depression on 17th September, and deep depression on 18th September. This system was concentrated into a cyclonic storm on the same evening and crossed the coast near Kallingapattanam (north-east Andhra Pradesh) in the morning of 19th September 2005. It further moved in a west northwesterly direction and weakened into a deep depression over south Chhattisgarh and adjoining Andhra Pradesh on 20th September 2005. Under the influence of this storm, widespread rainfall occurred over Andhra Pradesh. **Singh, et. al., (2012)** studied the September 2005 event along with other tropical storms during 1998-2008, and concluded that the maximum rainfall occurred within a radius of 150 km from the storm centre. The torrential rains triggered by the cyclonic storm have inundated several villages. In spite of evacuating over 10,000 people to safer places, this storm claimed about 31 human lives.

The areas inundated are mapped based on the IRS LISS-III imagery of 22nd September 2005 (Fig. 11.2b). It is observed that the entire Krishna river course between either banks (except elevated islands) exhibited the turbid waters. The sediment plume is observed to be emanating from all the four distributaries of river Krishna in satellite image of 22nd September 2005. The paddy crop is in the full-grown stage in significant parts of the deltaic plain. The extent of inundation is also mapped based on the satellite imagery of 2nd October 2005. The flood affected areas are observed in south-west of Ponnuru village, around Nagayalanka village, south of Machilipatnam, and around Kolleru lake (north of Kaikalur). The total extent of inundation in the Krishna

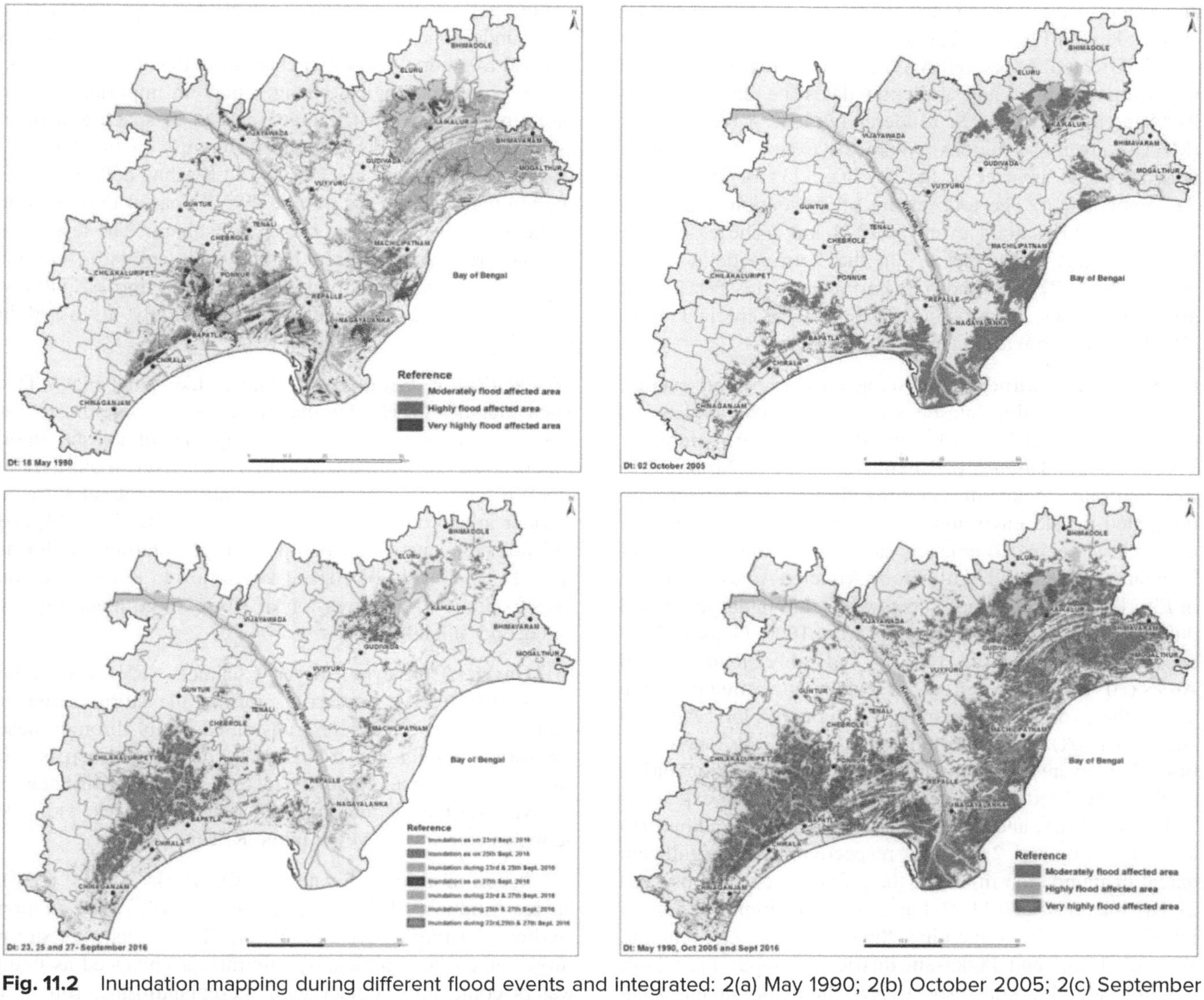

Fig. 11.2 Inundation mapping during different flood events and integrated: 2(a) May 1990; 2(b) October 2005; 2(c) September 2016; 2(d) Integration all three flood events

river delta region is about 1,408 km^2, of which the maximum extent of inundation is observed in Krishna district (777 km^2), followed by Guntur (367 km^2), West Godavari (143 km^2) and Prakasam (121 km^2) districts.

(c) September 2016 Cyclone event: The optical satellite imagery is being extensively used in mapping the inaccessible flood affected areas. One of the challenges in using optical data is the presence cloud cover during cyclone event. In such events, the space borne Synthetic Aperture Radar (SAR) data is the most preferred option for monitoring the flood condition. The SAR sensor has shown great potential for flood mapping as it can penetrate through the cloud. However, the image characteristic of SAR data differs significantly from optical data for inundation mapping. **Babu Madhavan, et. al., (1999)** delineated geomorphic features such as meander scars, palaeo channel scars and other landforms with greater ease by SAR images compared to IRS-1A LISS II data.

The National Remote Sensing Centre (NRSC) has been extensively using the satellite data for operational flood mapping and monitoring in near real-time. The inundation maps are being generated within 5-6 hours on receipt of raw satellite data products, and disseminated to the concerned Departments **(Srinivasa Rao, et. al., 2013)**. Further, the flood inundation layer in GIS compatible format is also provided to the end user for value addition. The individual geo-referenced inundation layers of 23rd, 25th, and 27th September 2016 are further analyzed. The geographical extent of inundation in all three data sets (i.e. 23rd, 25th and 27th September 2016) at a same location is delineated (Fig. 11.2c) and identified as Very Highly Flood Affected Areas (VHFAA). The Highly Flood Affected Areas (HFAA) are those regions that are under inundation on any of the two dates (i.e. 23rd, 25th, and 27th September 2016). The Moderately Flood Affected Areas (MFAA) are those regions, where the extent of inundation is restricted to any one of the days i.e., 23rd or 25th or 27th September 2016. The total extent of inundation during the September 2016 flood event is 1274 km^2. The extents of VHFAA, HFAA, and MFAA in the Krishna delta region are 620 km^2, 77 km^2 and 576 km^2 respectively. The worst affected regions during September 2016 event is Guntur (304 km^2) and Prakasam (259 km^2) districts.

(d) Combined inundation during all three flood events: The temporal satellite datasets acquired for three flood events during 1990 and 2016 over Krishna delta region are combined. The flood inundation layers extracted (Fig. 11.2a, b and c) from the satellite data during three different flood events are integrated in GIS environment. The integrated flood extents are re-classified into three categories based on the frequency of inundation. The Very Highly Flood Affected Areas (VHFAA) in Fig. 11.2d, include those areas which are under inundated during all the three flood events of May 1990, October 2005 and September 2016 (Fig. 11.2d). The Highly Flood Affected Areas (HFAA) are those regions that are under inundation in any of the two flood events, and Moderately Flood Affected Areas (MFAA) are under inundation during only one of the three flood events. The maximum extent of inundation during all the three flood events is about 4,242 km^2. The extents of VHFAA, HFAA, and MFAA in Krishna delta are about 924 km^2, 381 km^2 and 2,936 km^2 respectively. The significant extent of inundation in all the three flood events is observed in Krishna district (2,007 km^2). The maximum extent of inundation in the remaining three districts i.e., Guntur, West Godavari and Prakasam districts is about 1225 km^2, 518 km^2, and 492 km^2 respectively. The predominant extent of VHFAA is about 441 km^2 in Guntur district. The Krishna district constitutes the maximum extents of HFAA (about 231 km^2) and MFAA (1609 km^2).

3.2 Geomorphology Based Flood Hazard Zonation

The coastal region constitutes a dynamic environment with areas of rapid changes both due to natural processes and human interference. The study of coastal geomorphology is important in understanding the processes responsible for changes under different environmental conditions. Various geomorphic features formed due to the marine, fluvial, aeolian, denudational, and structural origins are delineated. The marine landforms include shoal, tidal flat, sandy beach, spit, mangrove swamp, mudflat, coastal plain, and beach ridges; point bar, channel bar and flood plain are of fluvial origin; and coastal dunes of aeolian origin are observed to have undergone changes of varying degree. Landforms respond to gradual and rapid changes in the near shore zone. These geomorphological changes play an important role in assessing the extent of damage caused in space and time, as different landforms respond differently to flood hazard **(Ramana Murty, et. al., 1993).** The Krishna delta region is studied mainly from the perspective of coastal flooding due to the occurrence of storm surge and tsunami. The responses of landforms are analyzed in terms of facilitator, carrier/ conveyor, accommodator, absorber, barrier, impact, outflow, and runoff. Each of these aspects is briefly discussed in following section from A to H.

(a) Facilitator: Geomorphic features like Tidal Flats (TF), Tidal Marshes (TiM), Mudflats (MF), etc., come under this category, as they are the first recipients of coastal flood waters. The geomorphic features like river mouth, tidal creek, estuary, etc., are capability of allowing the tidal waters to enter inland during storm surge. Hence, the TF, TiM, and MF are delineated and identified as facilitating landforms (Fig. 11.3). Devoid of coastal vegetation in seafront region may facilitate the progression of storm surge much rapidly and deeper inland.

(b) Carrier/Conveyor: River Channels (RV) and Tidal Creeks (TC) serve as carriers during the storm surge or tsunami. The surge can be transmitted easily through these landforms and can cause large scale destruction across and along up to a considerable distance inland. The distributaries of river Krishna, Upputeru river, Goguleru creek, etc., are acting as carriers/ conveyor in the Krishna delta.

(c) Accommodator: Swales (SW), Back Waters (BW), Palaeo Mudflats (PMF), and Saline Flats (SF) are common landforms acting as accommodators. The intensity of storm surge cannot be significantly minimized/ enhanced as flood waters come into contact with these landforms. But they have the potential to accommodate the flood waters for longer duration. The other fluvio-marine landforms mapped and grouped under accommodator category are channel bar (lower and upper), Palaeo Channel (PC) and Ridge & Swale Complex (RSC). The mangrove swamp can act as an accommodator to storm surge on a very limited scale, but they are good absorbers. The landforms categorized under accommodator class are observed within the Krishna river as channel bars; SF near Machilipatnam; PC near Vuyyuru; moderately elevated RSC near Bapatla; and BW and SW near Mogalthur.

(d) Absorber: Sandy Beaches (SB) and Mangrove Swamps (MS) have the capacity to absorb the storm surge intensity. A wide sandy beach is a better absorber than the narrow ones. Mangroves are well known for effectively reducing the surge

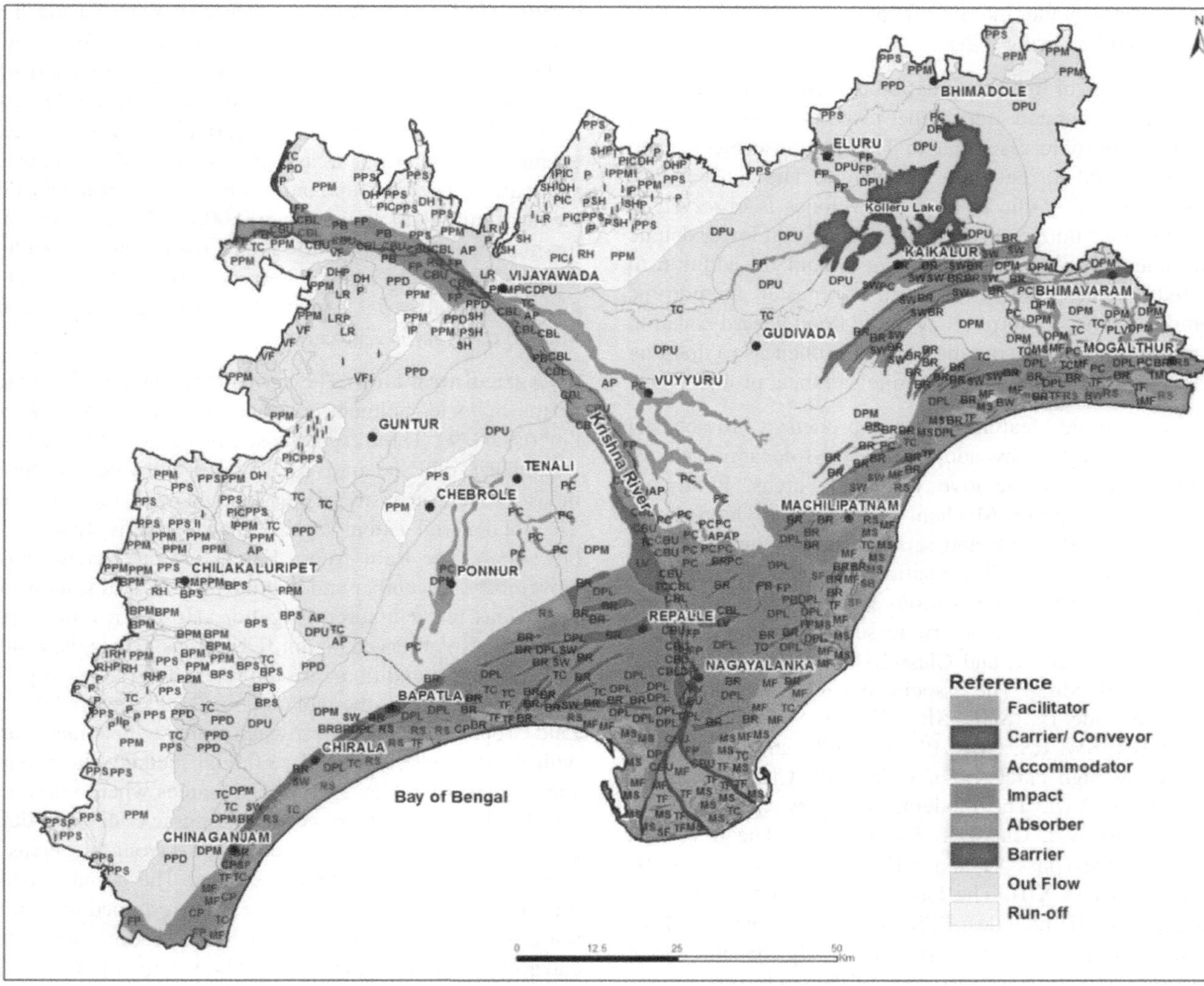

Fig. 11.3 Geomorphology and its proneness to coastal flooding

velocity and associated energy. The presence of mangroves can be seen in the Krishna river mouth regions.

(e) Barrier: Beach Ridges (BR) and Coastal Dunes (CD) act as natural barriers during the storm surge. The amount of resistance these landforms can offer depends upon the height and extent of these landforms in relation to the surge height. Landforms like Spits (SP), Natural Levee (NLV), and Paleao-levee (PLV) are also grouped under barrier category.

(f) Impact: The impact of storm surge is extremely significant on the landforms formed due to fluvial, marine and fluvio-marine processes. These landforms include Coastal Plain (CP), Flood Plain (FP), Point Bar (PB) deposits and Deltaic Plain - Lower (DPL). The impact regions can be observed between Bapatla in southwest up to Mogalthur in northeast.

(g) Outflow: The outflow region is generally not affected by the storm surge. The landforms comprising the outflow region includes- Alluvial Plain (AP), Deltaic Plain Middle (DPM), Deltaic Plain Upper (DPU), Pediplain Deep (PPD), and Valley Fill (VF). The extent of the outflow region is present between Chinaganjam and Bhimadole, covering Tenali, Guntur, Gudivada, and Eluru.

(h) Run-off: The structural, denudational, and certain fluvio-marine landforms represent run-off category. The landforms like Structural Hills (SH), Residual Hills (RH), Denudation Hills (DH), Pediment (P), Pediment Inselberg Complex (PIC), Inselberg (I), Linear Ridge (LR), Pediplain Shallow (PPS), Pediplain Moderate (PPM), Buried Pediplain Shallow (BPPS), and Buried Pediplain Moderate (BPPM) are least influenced by the storm surge and even by tsunami. Most of these landforms under the run-off category are far away from the coast and bordering DPU.

Thus, the zones of likely inundation caused due to the storm surge are deliberated under the response mechanism of coastal geomorphic assemblage. This response mechanism also holds good for tsunamis inundation. The consequence of coastal flooding due to tropical cyclones is observed to have enormous impact on the landforms within lower deltaic environment than on the undulating upland. In addition to the study on response mechanism of coastal geomorphic assemblage, the geomorphology based flood hazard zonation is attempted by addressing the issue of proneness to flooding due to storm surge, rainfall, riveraine and local problems.

The geomorphic features based proneness to coastal flooding is analyzed by adopting a two-fold approach for the identification of relative vulnerable areas. Firstly, in terms of the Response Mechanism of Coastal Geomorphic Assemblage (RMCGA); and secondly the Geomorphology based Flood Hazard (GFH) zonation. Depending on the proneness to flooding, various flood hazard classes are derived based on the geomorphic setup (Table 11.1). The GFH Class-I, Class-II, and Class-III are under the extreme category of flooding. The associated landforms of extreme category include TC, SHL, SB, TF, TM, SP, MS, BW, SF, MF, CP, BR, SW, RS, CBL, PB, FP, DPL, and CD. This is followed by high category of GFH with Class-IV, Class-VI and Class-VIII. The moderate category GFH includes Class-V, Class-VII, Class-IX, and Class-X. The low GFH classes include Class-XI and XII. Finally, the least GFH category is Class-XIII. The landforms associated with this hazard Class-XIII includes: PPS, BPM, BPS, DH, P, I, PIC, RH, SH, and LR. These landforms are mostly concentrated in the western and north-western parts of the study area.

3.3 Land Use Patterns

In the present study three seasons (kharif, rabi, and zaid) data is used for Land Use/ Land Cover (LULC) mapping of Krishna river delta based on IRS-P6 LISS-III satellite data of years 2015-16. The entire Krishna delta region is broadly divided and mapped into seven main categories namely built-up land, agricultural lands, forest land, wastelands, wetlands, water bodies, and other LULC classes for 2015-16 (Fig. 11.4). These seven broad categories (Level-1) are further classified into various sub-units (Level-2 & Level-3) and described. The percentage distribution of LULC classes is presented in the form of pie diagram. Each of the LULC class/ sub-class is discussed (A to G).

(a) Built-up Land: The built-up land is outlined as the landscape that has been transformed by human actions. The built-up land type includes buildings, transport network and utilities in alliance with water, vegetation and vacant lands. The total extent of built-up land is 778.6 km^2 (5.76%) during 2015-16. The built-up land is sub-divided into rural and urban (compact/ sparse). The principal activity of over 80% people in Built-Up area (Rural) class is agriculture. The total aerial extent of built-up rural areas is 544.6 km^2 (4.03% to the total study area). The built-up land (urban) consist of towns, cities having a minimum population of 5000, at least 75 percent of the male working population is involved in non-agricultural activity and having a population density of at least 400 per km^2. The total aerial extent of built-up Urban (Compact/ Sparse) category is 234 km^2.

(b) Agricultural land: The agricultural category is found to be the major constituent covering an area of about 9,384.3 km^2 (69.43 %). Three cropping seasons exist in the study area viz., kharif, rabi and zaid. Kharif Crop cultivated during SW monsoon season is mostly associated with rain-fed crops. The total aerial extent of the kharif crop is 1,640.4 km^2 (12.14 %). The Rabi cropped areas are cultivated between November / December and February / March. It is associated with areas under assured irrigation. However, rabi cropped areas are also observed in rain-fed regions, where the residual soil moisture conditions are sufficient for rising crop. The total aerial extent of the rabi crop is 421.9 km^2 (3.12 %). The Zaid Cropped areas are mostly associated with irrigated areas with fertile soils, restricted to fluvial/ deltaic plains during summer (April – May). Further, the areas where kharif crop couldn't be raised due to the water logging conditions during monsoon season, are observed to be favourable areas for growing crop during summer season. The aerial extent of the zaid crop is 18.1 km^2 (0.13 %). The cropped area in two seasons (double cropped areas) is the leading agricultural category with an extent of 6,434.3 km^2 (47.60%). The presence of cropland in all the three seasons constitutes about 113.4 km^2 (0.84 %). The fallow lands are the areas of cultivation, but are temporarily allowed to rest, un-cropped for one or more seasons. The aerial extent of the fallow land is 271 km^2 (2.01 %). The agricultural plantation areas are trees planted adopting agricultural management techniques. These plantations include cultivation of coconut, mango, citrus fruits, casuarinas, orchards and other horticultural nurseries mostly under irrigated conditions. The agricultural plantation occupies an area of about 485.4 km^2 (3.59%). The casuarina plantations are widely grown along the coast in the sandy areas. These plantations act as windbreakers, resist coastal erosion and defend the abutting croplands from cyclonic storms.

(c) Forest lands: These are the areas which bear an association predominantly with trees and other vegetation types (within the notified forest boundaries) suited for producing timber and other forest produce. The forest cover occupied an area

Table 11.1 Landforms and their distribution in different flood hazard classes

Geomorphic Unit/Landforms	Extent km²	Flood Hazard Classes												
		1	2	3	4	5	6	7	8	9	10	11	12	13
Tidal Creek (TC)	18.25	√												
Shoal (SHL)	0.19	√												
Sandy Beach (SB)	2.48	√												
Tidal Flat (TF)	101.04	√												
Tidal Marsh (TM)	3.32	√												
Spit (SP)	6.15	√												
Mangrove Swamp (MS)	204.04	√												
Back Water (BW)	8.32	√												
Saline Flat (SF)	68.87	√	√											
Mud Flat (MF)	278.39		√	√	√									
Coastal Plain (CP)	114.08		√	√		√								
Beach Ridge (BR)	534.19	√		√		√				√	√	√	√	
Swale (SW)	156.91		√	√			√	√			√			
Ridge Swale complex (RS)	300.93			√								√	√	
Channel Bar Lower (CBL)	138.84			√					√					
Channel Bar (CBU)	117.40								√	√				
Point Bar (PB)	442.59			√					√	√				
Flood Plain (FP)	187			√					√	√				
Natural Levee (LV)	98.53									√		√		
Alluvial Plain (AP)	217.72					√						√		
Palaeo Channel (PC)	292.43				√	√	√	√						
Palaeo levee (PLV)	1.59											√		
Deltaic Plain Lower (DPL)	1267.75			√	√	√	√	√		√	√			
Deltaic Plain Middle (DPM)	2553.04				√	√	√	√		√	√			
Deltaic Plain Upper (DPU)	2268.68				√	√					√			
Coastal Dune (CD)	7.11			√										
Pediplain – Deep (PPD)	1221.37					√		√				√	√	
Pediplain – Moderate (PPM)	1381.58												√	
Pediplain – Shallow (PPS)	734.71												√	√
Buried Pediplain (BPM)	69.45												√	√
Buried Pediplain – Shallow (BPS)	298.87												√	√
Valley Fill (VF)	91.12											√	√	
Denudational Hill (DH)	111.18													√
Pediment (P)	86.67													√
Inselberg (I)	18.74													√
Pediment Inselberg Complex (PIC)	22.67													√
Residual Hill (RH)	12.78													√
Structural Hill (SH)	58.98													√
Linear Ridge (LR)	12.52													√

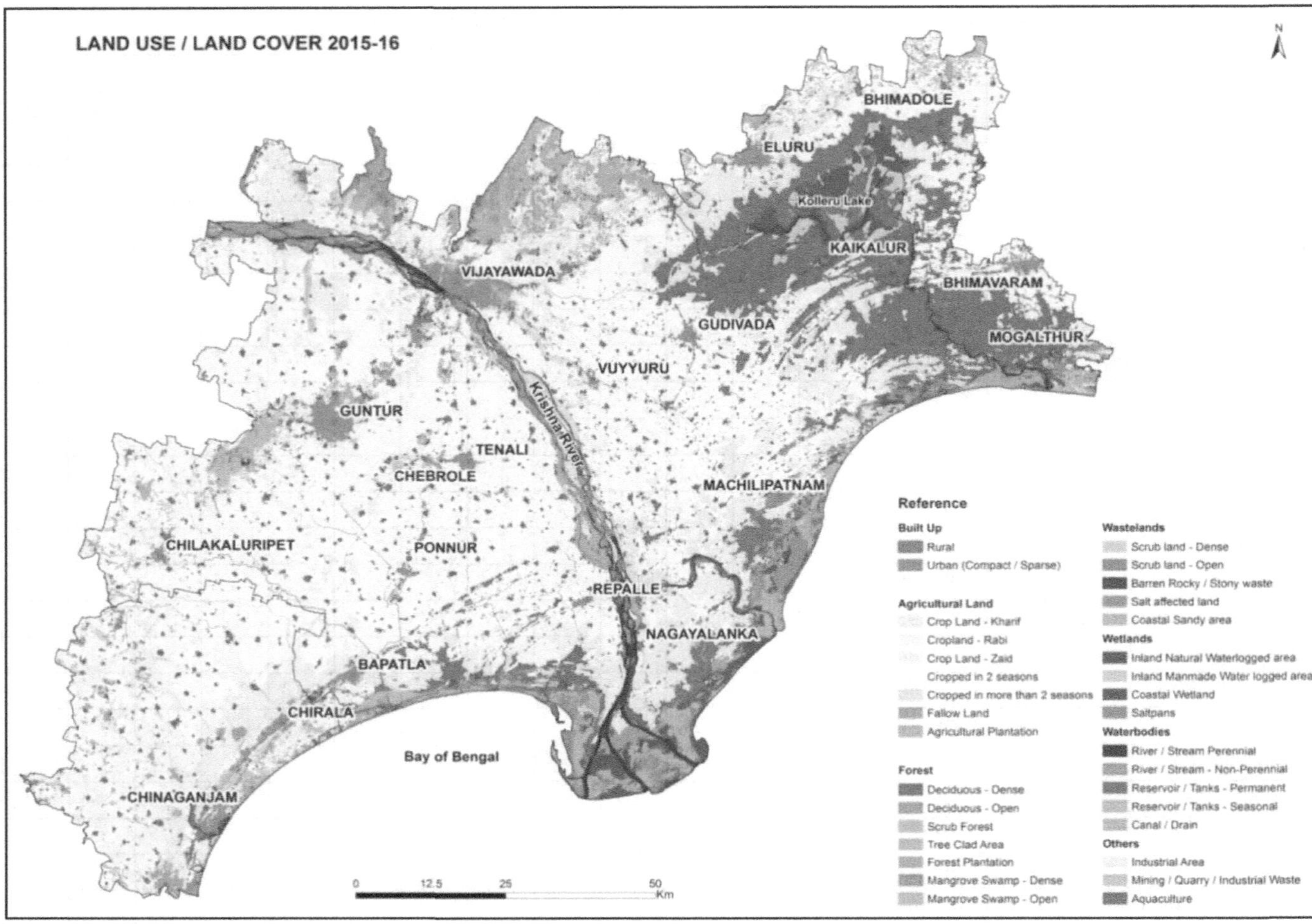

Fig. 11.4 Land use/ land cover map (2015-16)

of about 579.9 km^2 (4.29%). The forest class is sub-divided into seven groups, namely deciduous-dense (20.3 km^2), deciduous-open (94.9 km^2), scrub forest (86.6 km^2), tree-clad area (88.8 km^2), forest plantation (45.7 km^2), mangrove swamp-dense (109.8 km^2), and mangrove swamp-open (133.8 km^2). Mangroves are known for acting as a protective shield against storm surge and fierce winds during landfall. In view of this, large scale afforestation with suitable species is being taken up within the potential areas of salt affected lands and coastal sandy areas.

(d) Wastelands: These are presently underutilized and which are deteriorating for lack of appropriate water and soil management or on account of natural causes. The total extent of all categories of wastelands is 132.3 km^2 (0.99%). Five categories of wastelands are mapped, which encompass dense scrub land (50.9 km^2), open scrub land (26.6 km^2), barren rocky (9.8 km^2), salt affected land (9.8 km^2), and coastal sandy area (38.8 km^2).

(e) Wetlands: These are areas where the water table is at or near the surface of the land, or where the land is covered by water. Wetlands are under phenomenal stress due to rapid urbanization, industrialization and agricultural escalation; exhibited by the shrinkage in their areal extent, and drop in the hydrological, and ecological functions they perform. The wetlands category make up an area of about 468.5 km^2 (3.46%). There are four subcategories of wetlands i.e., inland – natural (143.7 km^2) and manmade (103.7 km^2) wetlands, coastal wetlands (173.4 km^2), and saltpans (47.7 km^2).

(f) Water Bodies: This category comprises areas with surface water, either impounded in the form of ponds, lakes, and reservoirs or flowing as streams, rivers, canals, etc. The major river Krishna separates the Krishna and Guntur districts in the study area. This river bifurcates into two distributaries near Etimoga, and further bifurcated into two more distributaries, and finally all the three distributaries empty into the Bay of Bengal. The water body category occupies an area of about

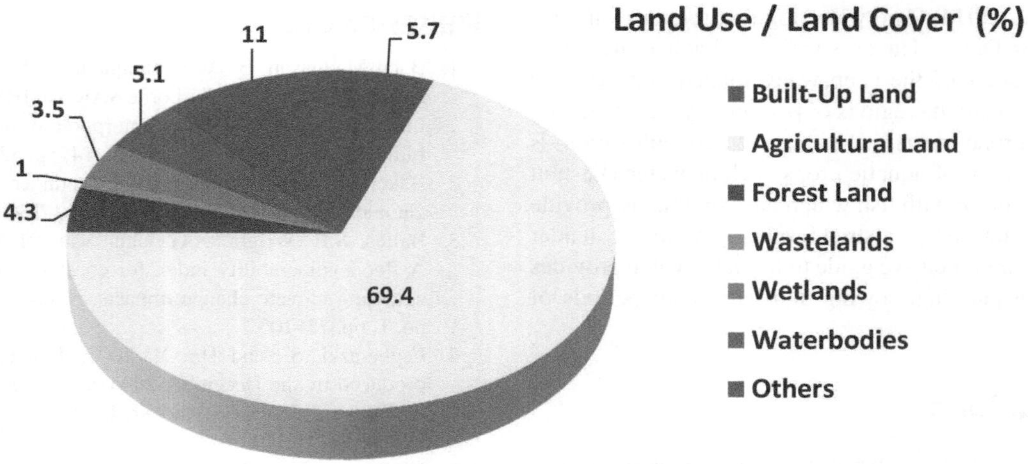

Fig. 11.5 Percentage distribution of Land use/ land cover (2015-16)

691.2 km^2 (5.12%). Perennial river/ stream (190 km^2), non-perennial river/stream (217.3 km^2), reservoir/tanks (53.5 km^2) and canal/drain (124.9 km^2) are sub-categories of water bodies.

(g) Other land use classes: They occupies an area of about 1,480.9 km^2 (10.95%) during 2015-16. They mainly include industrial area with an aerial extent of about 44.9 km^2; mining/ industrial waste (25.8 km^2); aquaculture (1,410.2 km^2); and rail & road.

3.4 Impact of Flood Hazard on Land Use Patterns

It is evident that the satellite derived observations on land use/ land cover pattern form important databases for various emerging areas of environmental monitoring. The extreme natural event has large impact on human settlements and their activities. Major floods during November 1977, May 1990 and subsequent events give rise to a new focus on these problems. The effects in terms of flood characteristics are important for land use management. The flood prone areas are being occupied/ encroached with inadvisable land use practices. Any adverse changes in land use pattern can increase the flood severity. Unplanned urbanization, mangrove deforestation, encroaching drainage networks, etc., can enhance flood severity. There are many non-structural flood management strategies from land use perspective. Apart from the various cyclone mitigation measures, there is a need for taking up shelterbelt plantation along the coastal area with suitable tree species to mitigate the impact of strong cyclonic winds, and thus checking soil erosion and inward sand drift. The role of mangroves in protecting the coastal zone from natural hazards such as storms, tsunamis, and coastal erosion is well known. The conservation and spread of these ecologically sensitive areas help in lessening the loss of life, and damage to property in areas behind mangroves. Despite their important social and ecological roles, the degradation of these forests continues. Conversions of mangroves into brackish water aquaculture, industrialization and urban development have destroyed many such areas.

The agricultural sector plays a very important role in the growth of Indian economy. The land put to agricultural activity is found to be the major land use category covering over 9,000 km^2, which constitutes about 70% of the Krishna delta region. Cropped areas are widely distributed in the irrigated areas of deltaic plain. The predominant crop growing seasons include kharif and rabi, which are associated with Southwest and Northeast monsoons respectively. The cyclone and its associated flooding cause severe damage to agricultural land. The stagnation of water and salinisation are the prime causes of crop damage. The pulses and vegetable crops are highly susceptible to flood. In the case of paddy, the stage of crop, and duration of deluge decides the extent of damage to crop lands. The fierce winds during a storm can cause extensive damage to banana and sugarcane crops. Plantation with suitable variety along the palaeo channels, beach ridges, ridge & swale complex, etc. can protect the adjacent cropland. The extent of crop damage during November 1977 cyclone is to the tune of about 13,500 km^2, while during May 1990 the crop loss is limited to about 5,000 km^2.

The frequency distribution of cyclonic storms in the Bay of Bengal indicates that there are two peaks: One in the month of May and the other in November. Keeping in view of the frequency of tropical storm crossing, the cropping seasons needs to be adjusted accordingly. Further, there is a need to shift towards growing short duration crops or storm resistant crops in recurring flood prone areas. The coastal flood tolerant rice varieties such as Swarna Sub-1 may be grown by the farmers to reduce the yield loss under such conditions

(Kumar, et. al., 2014). The water logging tolerant varieties of rice such as Durga, Hangseswari, Varshadhan are being suggested for areas of the deep water logged and cyclone/ flood prone areas. In the regions of perennially water logging and cyclone prone areas, where even rice cultivation is difficult, cultivation of aquatic crops such as water chestnut either as sole crop or with fish which has potential to provide great income to farmers. A district level crop weather calendar can also be a comprehensive guide to farmers, which provides information on planting, sowing and harvesting periods of major crops.

4. Conclusions

Flooding is a major challenge in low-lying areas like deltas and coastal plains. Suitable flood proofing activities for protection of coastal region against recurring events is needed. Land use optimization can be an important aspect in coastal disaster mitigation. The land use patterns and flood response to geomorphic setup is analysed for the Krishna river delta region, Andhra Pradesh, India. Satellite Remote Sensing provides valuable information useful for resource inventory and disaster management. Since much of the information needed is having a spatial component, mapping is GIS environment is most appropriate. The extent of flood inundation during the past events can form an important aspect of baseline data. In the present study, the flood/ cyclone events of May 1990, September 2005, and September 2016 are analyzed. The maximum extent of inundation during all the three flood events is about 4,242 km^2. The Krishna delta region is also studied from the perspective of landforms responses in terms of facilitator, carrier/ conveyor, accommodator, absorber, barrier, impact, outflow, and runoff. Depending on the proneness to flooding, various flood hazard classes are also derived based on the geomorphic setup. The satellite derived observations on land use/ land cover pattern form important databases for various emerging areas of environmental monitoring. Unplanned urbanization, mangrove deforestation, encroaching drainage networks, etc., in the coastal zone can enhance flood severity. The landfall frequency of storms (Bay of Bengal) has two peaks one in month of May, and the other in November. Hence, there is a need to avoid these months and shift towards growing short duration crops or storm resistant crops in recurring flood prone areas.

5. Acknowledgements

The authors are thankful to National Remote Sensing Centre for providing flood inundation layers for analysis. The first author expresses his sincere gratitude to Dr. B.Sundar, IFS, Vice Chairman, APSAC for constant encouragement.

References

1. Babu Madhavan, B., Venkataraman, G., Krishna Mohan, B., and Shah, S.D., 1999. Airborne SAR and IRS 1A LISS II data Interpretation of coastal Geomorphology in Godavari Delta, India. Geocarto International, vol. 14, pp. 52–61.
2. Baker, V.R. 1994, Geomorphological understanding of floods. Geomorphology, vol. 10 (1-4), pp. 139–156
3. Balica, S.F., Wright, N.G., and van der Meulen, F. 2012. A flood vulnerability index for coastal cities and its use in assessing climate change impacts, Natural Hazards, vol. 64, no. 1, pp. 73–105.
4. Fagherazzi, S., and Du, X. 2007. Tsunamigenic incisions produced by the December 2004 earthquake along the coasts of Thailand, Indonesia and Sri Lanka. Geomorphology, vol. 99, pp. 120–129.
5. Klemas, V. 2015. Remote sensing of floods and flood-prone areas: An overview predicting and determining coastal storm impacts. Journal Coastal Research, vol. 31, no. 4, pp. 1005–1013.
6. Kumar, A., Brahmanand, P. S., and Nayak, A.K. 2014. Management of Cyclone Disaster in Agriculture Sector in Coastal Areas. Directorate of Water Management. Bhubaneswar. 108 p.
7. Kurian, N. P., Pillai, A. P., Rajith, K., Murali Krishnan, B.T., and Kalaiarasan, P. 2006. Inundation characteristics and geomorphological impacts of December 2004 tsunami on Kerala coast. Current Science, vol. 90, pp. 240–249.
8. Mahender Reddy, K. and Shah, B.M. 1991. Evolution of the Krishna delta, Andhra Pradesh, Memoirs Geological Society of India, no. 22, pp. 57–64
9. Mruthyunjaya Reddy, K., Ramana Murty, M.V., and Ramana Murty, P.V. 2006b. Storm vulnerability and coastal landforms in Nellore district, Andhra Pradesh, India. ISPRS TC-IV International Symposium on Geo Spatial Databases for Sustainable Development, Goa, 27-30 Sept., 2006, pp. 212–214
10. Nageswara Rao, K., Vaidyanadhan, R., 1978. Geomorphic Features of Krishna Delta and its Evolution. University of Delhi, Proceedings of National Symposium on Morphology and Evolution of Landforms, pp. 120–130.
11. Nageswara Rao, K., Saito, Y., Naga Kumar, K.Ch.V., Kubo, S., Pandey, S., Li, Z., Demudu, G., and Rajawat, A.S. 2020. Holocene evolution and Anthropocene destruction of the Krishna Delta on the east coast of India: Delta lobe shifts, human impacts, and sea-level history. Marine Geology, 427
12. Narayana, A.C., Tatavarti, R., Shinu, N., and Subeer, A. 2007. Tsunami of December 26, 2004 on the Southwest coast of India: Post-tsunami geomorphic and sediment characteristics. Marine Geology, vol. 242, pp. 155–168.
13. Nayak, S. 2017. Coastal zone management in India - present status and future needs. Geo-spatial information science, vol. 20, no. 2, pp. 174–183
14. Naylor, L. A., Spencer, T., Lane, S. N., Darby, S. E., Magilligan, F. J., Macklin, M. G., and Möller, I. 2017. Stormy geomorphology: geomorphic contributions in an age of climate extremes. Earth Surf. Process. Landforms 42, 166–190

15. Nordstrom, K. F., and Jackson, N. L. 1995. Temporal scales of landscape change following storms on a human-altered coast, New Jersey, USA. Journal of Coastal Conservation, vol. 1, no. 1, pp. 51–62
16. Pethick, J. S. and Crooks, S., 2000. Development of a coastal vulnerability index: a geomorphological perspective. Environmental Conservation, vol. 27, pp. 359–367.
17. Ramana Murty, M.V. 2020. Flood vulnerability analysis in Krishna river delta, Andhra Pradesh, India, using geospatial tools. Unpublished Ph.D. thesis submitted to Adikavi Nannaya University, 269 p.
18. Ramana Murty, M. V., Rao, R. S., and Bhagavan, S. V. B. K. 1993. Geomorphlogical Studies for Disaster Mitigation - A Case Study of the Krishna Delta Andhra Pradesh, India. International Journal of Remote Sensing, vol.14, no.17, pp. 3269–3274.
19. Ramana Murty, M. V., Ravi Kumar, Ch., and Bhagavan, S. V. B. K. 2012. Morphological changes along Godavari delta front - a case study on impact of coastal hazards on Kakinada sand spit, Andhra Pradesh, India. National conference in recent advances in Geo-sciences, Engineering & Technology. 20–21 December 2012, JNTU, Kakinada, pp. 129–138.
20. Ramasamy, S. M., Kumanan, C. J., Saravanavel, J., and Selvakumar, R. 2006. Geosystem responses to December 26, 2004, Tsunami and mitigation strategies for Cuddalore-Nagapattanam coast, Tamil Nadu. Jour. Geol. Soc. of India, vol. 68, pp. 967–983.
21. Rao, R. S., Ramana Murty, M. V., and Bhagavan, S. V. B. K. 1995. Remote Sensing aided flood hazard zoning in Pennar river delta, Nellore district, India, Hydrocoast 95' International workshop on Water related problems in low-lying coastal areas. A contribution to the UNESCO-IHP project H-2-2, 13–17 Nov., AIT, Bangkok, pp.236–244.
22. Shukla, S. B., Prizomwala, S. P., Ukey, V., Bhatt, N., and Chamyal, L. S. 2010. Coastal Geomorphology and tsunami hazard scenario along the Kachahh coast, Western India. Indian Jour. of Geo-Marine Sciences, vol. 39, no. 4, pp. 549–556.
23. Srinivasa Rao, G, Bhatt, C. M., Manjusree, P., Sharma, S.V.S.P., Begum, A. 2013. Flood monitoring and management using remote sensing. Bulletin of the NNRMS (B)-38, Satellite Remote Sensing & GIS Applications in Water Resources, ISRO, Bangalore, pp 81–88.
24. Subramanian, S. K., and Mohanan, G. 1992. Some features of the Bay cyclone of May 1990. Vayu Mandal, vol. 22, pp. 97-102 Singh, C., Das, S., Verma, R.B., Verma, B. L., and Bandyopadhyay, B.K. 2012. Rainfall estimation of landfalling tropical cyclones over Indian coasts through satellite imagery. Mausam, vol. 63, no. 2, pp. 193–202.
25. Smith D. E. 1968. Strandline. In: Geomorphology. Encyclopedia of Earth Science. Springer, Berlin, Heidelberg. https://doi.org/10.1007/3-540-31060-6_350
26. Zope, P. E., Eldho, T. I., and Jothiprakash, V. 2016. Impacts of land use–land cover change and urbanization on flooding: A case study of Oshiwara River Basin in Mumbai, India. Catena, vol. 145, pp. 142–154.

CHAPTER 12

Flood Hazard Mapping for Disaster Management: Multisectoral Approach for a Riverine Region

Deeksha Biswas
Department of Environmental Planning,
School of Planning and Architecture, Bhopal, India
deekshabiswas@gmail.com

Anurag Bagade
Department of Environmental Planning,
School of Planning and Architecture, Bhopal, India
anuragbagade@spabhopal.ac.in

Abstract

According to IPCC, South Asia is one of the most vulnerable and highly impacted regions of the world with India, being the fifth most vulnerable to climate change. Districts of Northeast India are on the forefront of being affected by recurring disasters of various nature and intensity. Rapid urbanisation and large-scale migration due to unequal distribution of resources and infrastructure has aggravated the causes for equality and inaccessibility of varying degrees. Far flung and border cities, districts of India have been focal points for unforeseen rapid development. Assam is one of the most flood-affected state in India. Topographically challenged, this region has been affected by climate change and frequent disasters in the past decades. Thus, making the local communities vulnerable and backward. Flooding in the areas of Upper Assam has caused massive loss to both life and property. Disaster Management through mapping would help in identification of critical areas as well as better planning of developmental activities. This paper uses a multisectoral approach in flood assessment and mapping at the sub-district level in the district of Dibrugarh, Assam through Flood Hazard Mapping. This paper focuses on the usage of geo-spatial tools through geomorphological, hydrological and embankment assessment for rapid assessment of disaster risk.

Keywords: Disaster Management, Flood Hazard Mapping, Multisectoral Approach, Riverine Region

1. Introduction

According to the Sixth Assessment Report by IPCC, the large scale indicators of climate change such as changes in regional precipitation patterns, global surface temperature changes are indicators of climate sensitivity. With extreme weather events such as droughts, floods, storms and rise in sea level will attribute to the future disasters. The region specific changes in intensification and increase of frequency will make settlements across the world on the forefront of climate change induced disasters. Many regions including the Asian region has been categorized under medium to high confidence in terms of heavy precipitation where the South Asian region is expected to experience increase in summer and annual fluctuations with enhanced interannual variability (IPCC, 2021).

In previous special reports of IPCC on mitigation of risks of extreme events and disasters have denoted hazards to be potential occurrence of a threat or any type of adverse impacts which can cause damage to infrastructure, livelihood, environmental resources as well as loss of life. Extreme events such as riverine flooding poses risk to communities (IPCC, 2012).

According to special reports of IPCC, both frequency and magnitude of flooding as well as vulnerability to human and natural ecosystem will increase over time (Reisinger, et al., 2020). It has been suggested that the magnitude or frequency of flood events should not be characterised as changes in flood risk but be covered under climate-related hazard. The uncertainty in terms of magnitude and likelihood of occurrence is susceptible to changes over time due to socio-economic conditions and earlier measures taken with respect to adaptation, mitigation, management.

The natural hazards with the highest frequency and widest geographical distribution across the world is floods. Flooding is the most common occurring phenomenon which can be result of both climatic and non-climatic processes. It is often assumed that flooding mainly occurs due to excessive rainfall and the lack of landform's capacity to convey excess water. There are different types of floods: riverine floods, flash floods, urban floods, coastal floods which have impacted ecosystems. Most of the floods, which were previously occurring as small events, have turned into major flood hazards. Recently, monster floods occurrence has become more frequent (UNISDR, 2017). Not only are floods the most recurring and frequent disaster but they are also the most prevalent and economically damaging natural disaster.

India is severely affected by floods during the months of monsoon which are primarily between the months of June to September (NRSC, 2013). Among the riverine regions, most flood prone areas in the India are part of the Brahmaputra, Ganga and Meghna river basins for North and North-east India. State of Assam, located in the north-eastern region of India, is prone to multiple natural disasters such as earthquakes, landslides, draught, cyclones, and severe floods. The annual flood in Assam has causes heavy damage to both life and property to thousands of residents living in the river basin/floodplains (Sharma, Johnson, Hutton, & Clark, 2010).

The risk of physical harm and the disparity in socioeconomic status enhances the level of vulnerability (Sharma, Johnson, Hutton, & Clark, 2010). The proposed approach of the study tried to spatially aggregate factors to create a risk surface that policymakers and decision makers at different administrative levels can use. Hazard, such as flooding and the resulting bank erosion exacerbated by climate change, is known to be well anchored in a large scientific effort. This is due in part to the complicated growth of the underlying concept of hazard, which reflects the evolution of a number of strands, each reflecting a different discipline tradition. The concept of vulnerability and risk as a description of a society's condition in relation to an imposed danger such as flooding, or the associated damages caused. Land loss due to erosion is on the rise, which has clearly contributed to the impoverishment of the riverine people. The associated uncertainties of climate change in terms of hydrological and hydraulic river behaviour may worsen the Brahmaputra's channel instability.

District of Dibrugarh, Assam as shown in Fig. 12.1, is selected for this study. It is bounded by Tinsukia district in

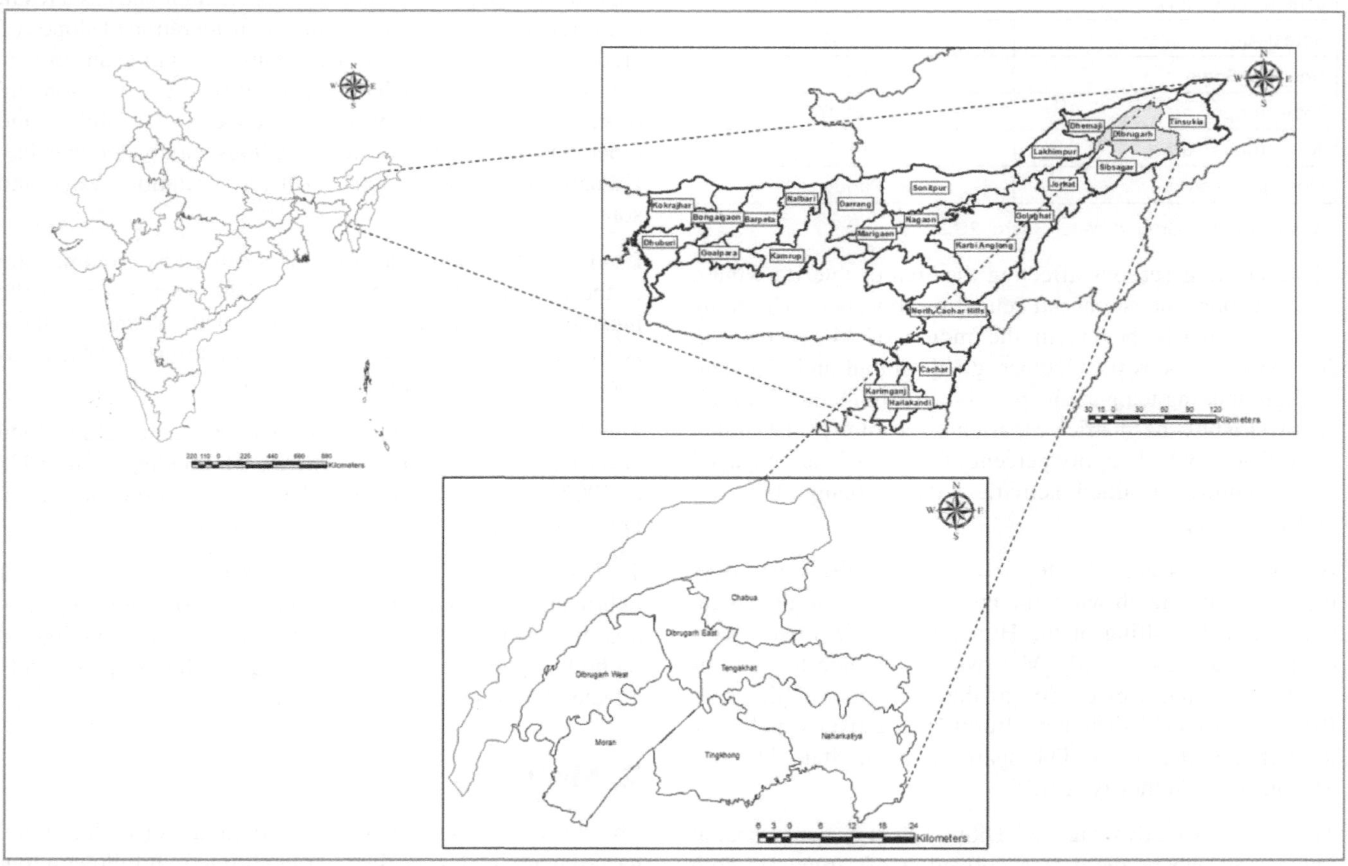

Fig. 12.1 Key map showing the location of the study area-Revenue circles within Dibrugarh district, Assam, India

the east, Sibsagar district in the west, Dhemaji district is separated by the river-Brahmaputra in the north. It is partially surrounded by Sibsagar district and state of Arunachal Pradesh in the south. The district is part of the Brahmaputra River basin with Burhi Dihing, Disang, Dibru, Sessa and Lekhijan being the important tributaries as shown in Table 12.1. All of these tributaries are perennial and meandering. Lower order streams have a dendritic structure, and higher order older streams have a subparallel pattern.

Table 12.1 District information at a glance

Information	Statistics
Geographical area	3,381 sq.km.
Population (2011 Census)	13,27,748
Rural population	10,83,984
Urban population	2,43,764
Density	393 (per sq. km.)
Average Annual Rainfall (mm)	2,819
Major physiographic units	Flood plain, Terrace deposit
Major Drainages	Brahmaputra, Burhi Dihing, Sessa, Disang, Dibru
Land use (sq. km.)	
Forest Area	233.41
Net Area Sown	1,394.98
Total cropped area	1,589.17
Total cropped area	194.19
Most grown crop	Paddy

Source: (Central Ground Water Board, 2013)

There are four seasons affecting the area of interest-winter, pre-monsoon, monsoon and retreating monsoon. The rainy season primarily begins in the month of May, peaks in July and continues till October, good rainfall and abundant sunlight making tea gardens possible on high lands. Rainfall characteristics-the district experiences a total of 193 rainy days. Due to which eighty percent of the people are engaged in agriculture or allied activities. Paddy being the most cultivated crop.

Brahmaputra is glacially fed river wherein the water rises by the end of March when the pre-monsoon rainfall sets in, coupled with melting of the Himalayan snow. First effects of floods are felt in early May which continue to most of the flooding takes place due to the back flow of the river Brahmaputra and its tributary-Burhi Dihing river which flow through the district of Dibrugarh. (Assam State Disaster Management Authority, 2013)

The locational advantage of Dibrugarh has made spatial imbalances where there is a collection of structures both temporary and ill-planned permanent dwelling units with little or minimum or no amenities and services present for the residents especially in unplanned areas. As most of the district is still rural, high rate of migration coupled with the drive to be the commercial center of trading, manufacturing and commercial activities center, Dibrugarh city has grown at an unprecedented rate into its rural surrounding. The sub-district of Borboruah is the only geographical area where the city grows as it is bounded by the river, Brahmaputra in the north, the tea estates in the south and defence area in the east, leaving only the south-west direction for growth. Furthermore, the yearly flood and the consequent erosion have been so severe that district's very existence is in jeopardy, and as a result, neither the government nor any private investors are ready to take on any substantial projects or erect any significant structures. Several issues, such as erosion, unplanned settlement, transportation, housing, and infrastructure affect and in turn get affected by flood events.

Dibrugarh is undergoing rapid population growth, which has led in higher population density. The district's growing urban population and people's lifestyles are having an impact on the district's natural and urban environments. The main cause of land degradation is population pressure, which leads to intensive land usage without effective management. People are drawn to sensitive places such as the highlands as a result of overcrowding. Land use without consideration of slope and erodibility in such regions will result in significant erosion and related difficulties. The impact of road construction and other landscape disturbances on erosion, landslides, and other mass movements in steep terrain is widely documented. Human intervention has an effect on vegetation, water, and soil.

Each year, the annual devastation is caused by Brahmaputra in the region of Upper Assam and Dibrugarh is one of the most affected districts (Chakraborty, 2020). Therefore, there is a need to study the existing flood condition of these areas and discuss the need for hazard mapping.

The district has been previously studied for creating flood hazard maps based on the severity of flooding from 1998 to 2007 as shown in Fig. 12.2 based on data procured from Dibrugarh Water Resource Division, Assam.

The objective of the study is to create a hazard map for Dibrugarh district by using a multi-criteria, multi-sectoral approach which takes into account the criteria highlighted by both experts as well as stakeholders who are perennially affected by these devastating floods.

2. Methods

The study uses weighted overlay technique for multisectoral hazard mapping for the district of Dibrugarh. It follows a four stage methodology. Initial stages identifies different sectors

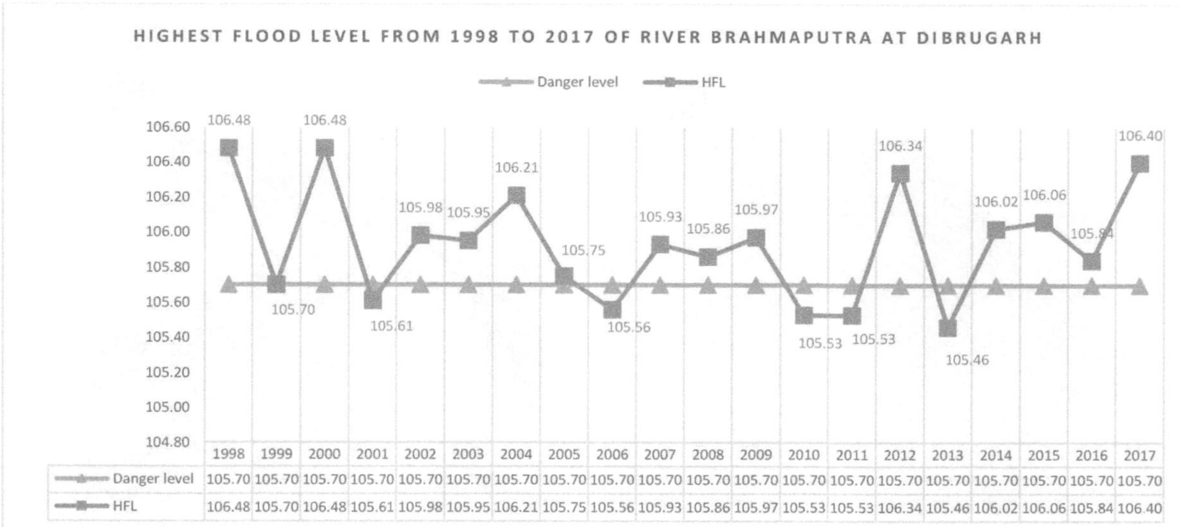

Fig. 12.2 Highest Flood level from 1998 to 2017 of Brahmaputra River in Dibrugarh

Data source: Dibrugarh Water Resource Division, Assam

relevant for the case through secondary and primary data. It also identifies various open-sourced data for mapping. The second stage includes processing of dataset such as ASTER DEM and Landsat-8 images, courtesy of the U.S. Geological Survey. The imagery for the months of January 2015 and July-2015 were used for assessing vegetation health and land cover changes during the flood event as shown in Fig. 12.3 and Fig. 12.4. The assessment was developed based

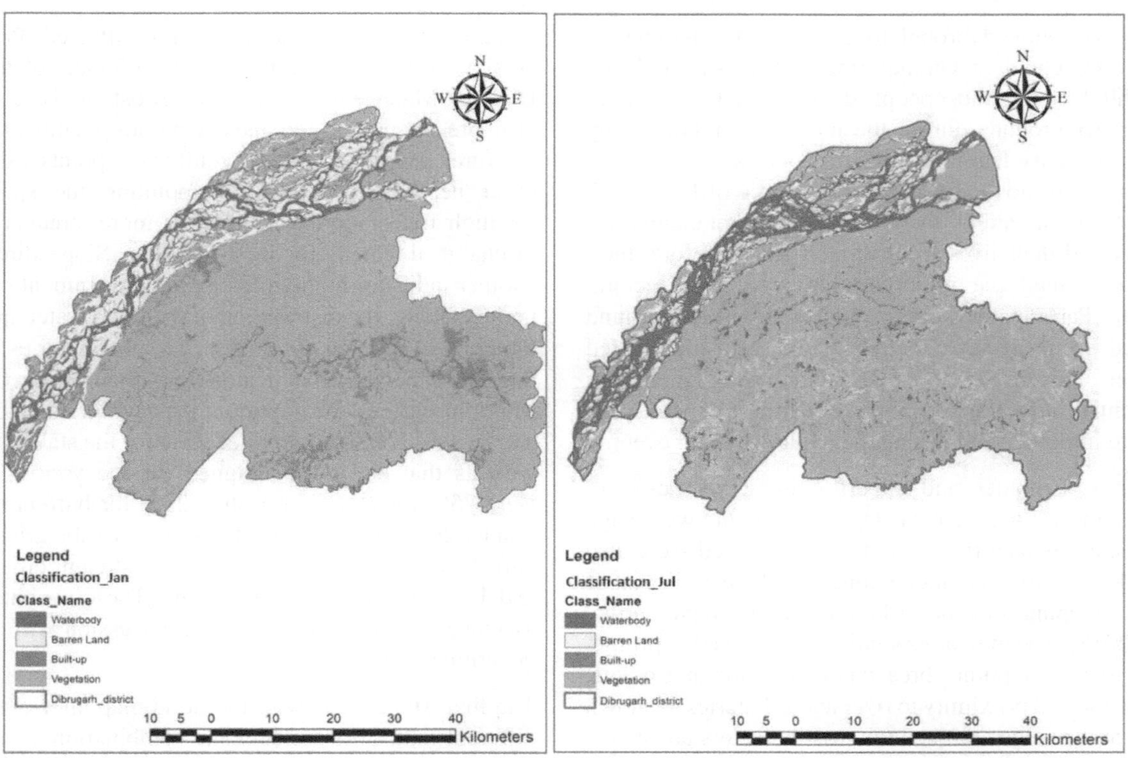

Fig. 12.3 Land Cover Supervised Classification of Dibrugarh district, Assam, India (Image date-January 2015, July 2015)

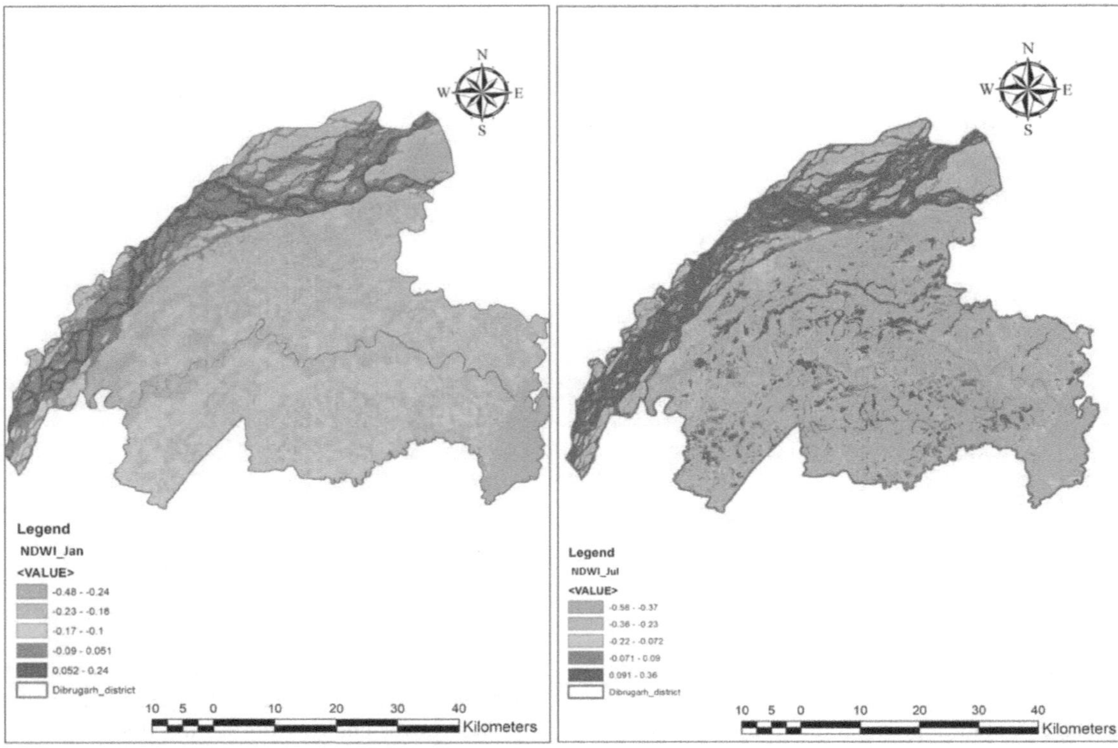

Fig. 12.4 NDWI analysis of Dibrugarh district, Assam, India (Image date-January 2015, July 2015)

on indicators developed through focus groups discussion with experts and stakeholders. On inquiring with the stakeholders, it was highlighted that most people do not leave their villages to seek higher grounds during the monsoon months. They stated that they have learnt to live with floods as there is very little incentive to move and seek other source of livelihood. The compensation paid by the government is not enough for them to rebuild their lives from scratch and therefore, their perspective formed the important basis for multisectoral assessment. Participatory approach has tested on ground with mixed group of stakeholders. It was also highlighted that women and children are left behind while men seek employment in urban areas. With very little to spare, poor families are impoverished during the annual disaster event.

After both stakeholder and expert opinion was used to identify the indicators, the next stage in assigning weightage to the selected indicators. The indicators selected were also evaluated by a similar technique done for Dhemaji district of Assam for mapping flood hazard (Hazarika, Barman, Sarma, & Borah, 2018). The indicators studied were elevation profile, proximity to erosion points, breach points, confluence points, slope direction and proximity to river and tributaries as shown in Fig. 12.5; the value assigned to these indicators are shown in Table 12.2. The elevation profile was taken as maximum inundation takes place in low-lying region and Dibrugarh being a low-lying district is severely affected. Proximity to erosion points were highlighted by officials of the District Disaster Management Authority. Most of these points are also breach points and have temporary embankments for the rural people. Buffer of confluence points using stream order networks helped in pinpointing the spatial zones for high indentation when two or more streams combined discharge threatens to flood the area. Slope direction was another indicator highlighted by the government officials for understanding the distribution of the flood water and possible direction of spillage. Proximity of river/tributaries is possibly the most common indicator for flood hazard zonation. Multiple buffers are important especially from the damage and loss of livelihood point of view for the stakeholders. The weights that had been assigned for the various indicators were-25% for elevation profile, 25% for buffers for erosion points, 20% for proximity to river and tributaries, 20% for buffer of confluence points of the stream order network and 10% for direction of the slope. These weights has been assigned after the focused group discussion with locals and government officials.

The final stage deals with the flood map after overlying the different indicators includes the combination of past trends, existing data base and previous hazard maps of the study area. It helps in identifying the strategies which are critical

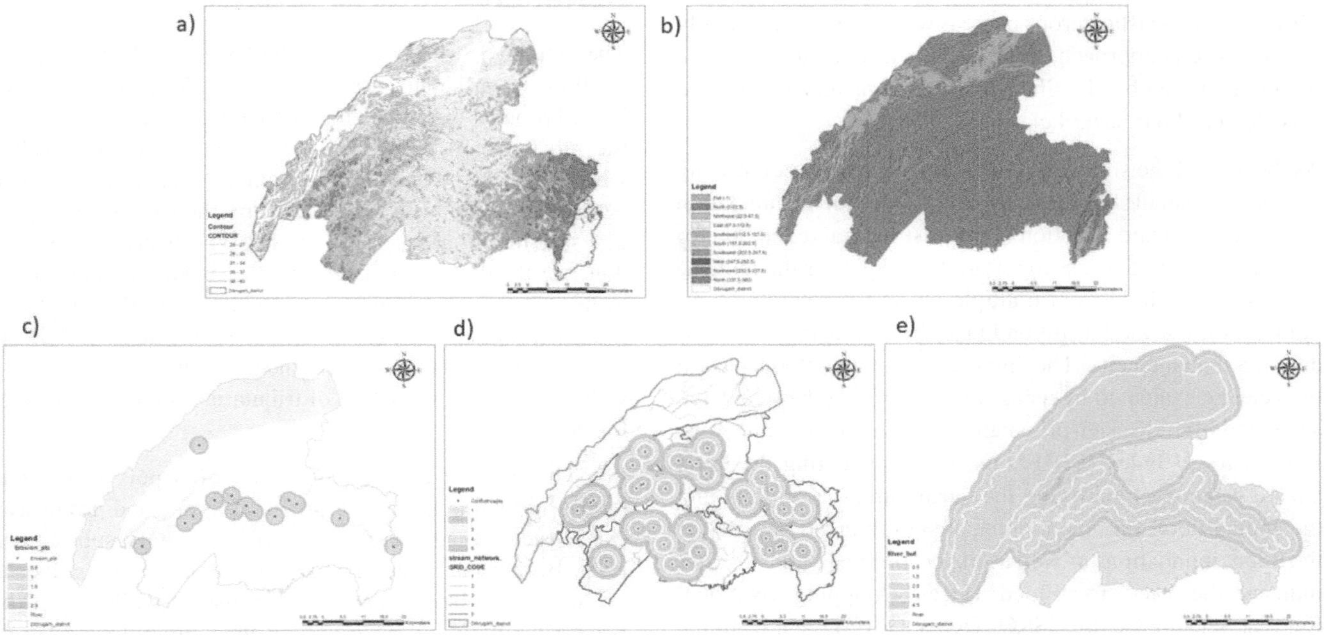

Fig. 12.5 (a) Elevation Profile from Aster Dem for the district of Dibrugarh, (b) Direction of the slope in the riverine region of Dibrugarh district, (c) radial buffers of the erosion points within the district based on expert opinion on water bodies, (d) radial buffer of confluence points of tributaries Burhi Dihing, Sessa, Dihang, Dibru, e) proximity to rivers of the Upper Brahmaputra basin within the district of Dibrugarh-Brahmputra, Burhi Dihing, Sesaa, Disang, Dibru rivers.

Source: Authors

Table 12.2 Ranking of Indicators for Flood Hazard Zonation

Indicators	Very high	High	Moderate	Low	Very low
Elevation	Less than 27 m	28 m-30 m	31-34	35-37	38-40
Slope Direction	Flat, N, NE	E. NW	W	SW, SE	S
Proximity to erosion points	Less than 0.5 km	1 km	1.5 km	2 km	2.5 km
Proximity to confluence points	1 km	2 km	3 km	4 km	5 km
Proximity to river	Less than 0.5 km	1.5 km	2.5 km	3.5 km	4.5 km

N-North, NW-North west, NE-North east, S-South, SW-South west, SE-South east, E-East, W-west

Source: Authors

for resource mobilization especially during the monsoon season in these poorly connected areas. The methodology takes into account the 'local' perspective while understanding the geographical, socio-economic constraints. The study was aimed at creating flood hazard zonation maps and highlighting the need for creating the maps at the regional, sub-regional and at the local level. The rank based hazard map that has been generated for this research, makes an attempt to bridge the gap between practice and research. It also highlights the need for data and informative related to all hazards readily available and easily comprehensible by local people.

One of the steps of the method used was using the Land Cover Supervised classification and NDWI classification for flood detection in the district, for which the Landsat-8 image has been use for the months of January and July-2015 as shown in Fig. 12.3 and Fig. 12.4. For finding the flood zones, from Fig. 12.2, it can be seen that there are other years where the water level of the river has risen above the danger level. Landsat 8 data has been used for the aforementioned analysis.

Further, in the study, it was noted that this approach was extremely critical in identifying areas for maximum flooding as well as the pattern that emerged while taking into account the latest flood data from the government records. The utilization of the flood hazard maps is critical for helping in the further understanding and planning of development related activities, support disaster relief, rescue efforts and plan for flood resistant crops in areas where floods perennially effect. For this approach, there are five classes under which hazard areas can be classified, starting from very low hazard class to very high hazard class.

As the area of the study lies in the riverine region comprising of multiple small and large tributaries, frequent inundation makes flood hazard zonation is a non-structural technique for disaster mitigation as an important technique for mapping as it enables stakeholders at multiple levels for assessing their vulnerability to the hazard and take the adequate actions for disaster management. The importance of the flood hazard map can be evaluated by using it for observing the inundated areas and high disaster-prone areas at regular time intervals. Flood Hazard Index (FHI) is used for calculating the flood hazard category, identifying the hazard areas as well as find the annual flood variations in the area of interest. Severity of the flood, inundations as well as the occurrence of the disaster indicate the need for Flood Hazard Mapping (National Remote Sensing Centre, ISRO (DOS), Government of India, 2020).

For the study undertaken, the overall methodology as shown in Fig. 12.6 has been adopted for preparation of flood zonation map.

3. Result

The weighted overlay analysis has been used for preparing the flood hazard map as shown in Fig. 12.7. It shows the most hazard prone areas within the district, the red colour indicates the most hazard prone areas, followed by orange, yellow while green is the least hazard prone. It was noted that there is the magnitude of being affected by the river Brahmaputra is lesser then that of its tributaries. The hazard assessment done for the district indicates that the maximum damage due to flooding can be seen in areas which were demarcated as erosion points and confluence points within the district. Based on the analysis there is a presence of confluence regions as well as periodic inundation of tributaries is the main reasons for flooding.

In the hazard map there are majorly six hotspots locate along the Burhi Dihing river, The distribution of flood prone areas can be seen spread over the entire district. At the sub-regional level, Borboruah is the most affected revenue circle, located in the north-western corner of the district while Tengakhat is the least flood hazard prone revenue circle, located in the south-eastern part of the district.

This assessment also highlights that the people living in villages in the nearby areas are more likely to be vulnerable to the risk posed by flood hazard. As most of the region lies

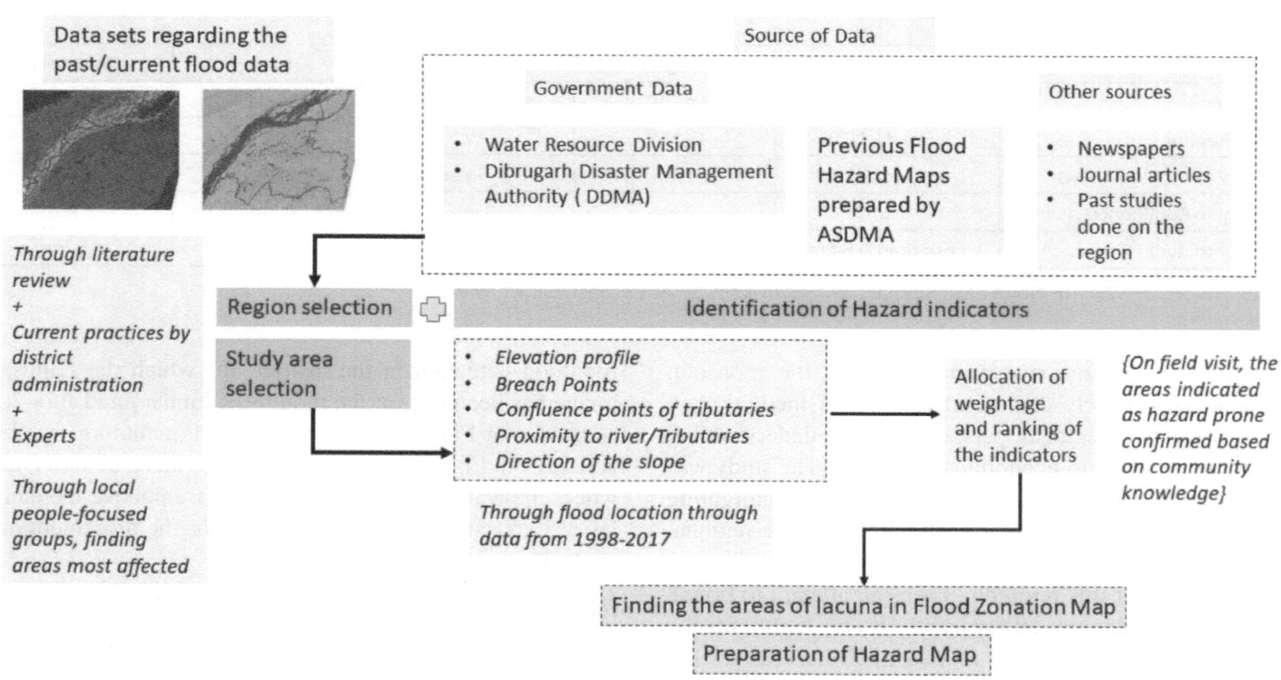

Fig. 12.6 Methodology used in the study for creating flood hazard map

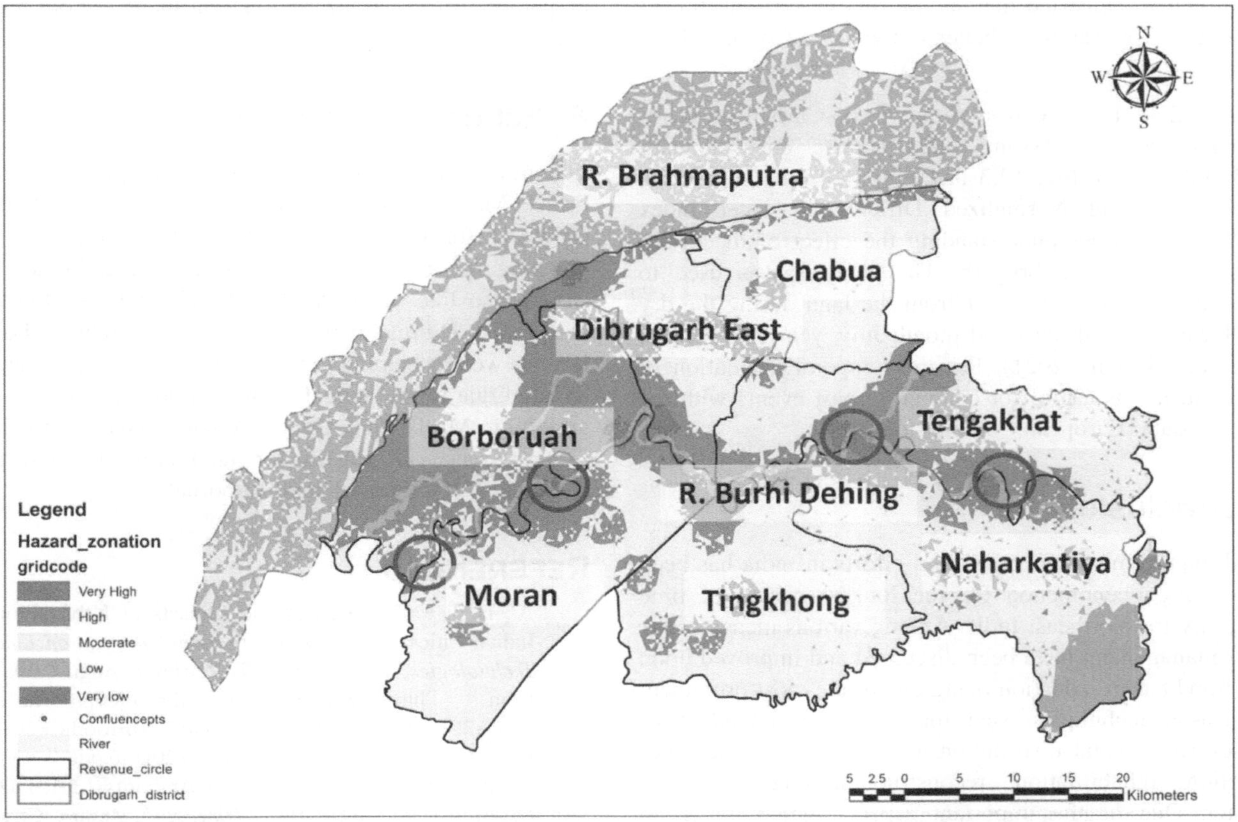

Fig. 12.7 The flood hazard map showing the most flood prone areas in the district of Dibrugarh

Source: Authors

in the low lying region, people's source of livelihood and their settlements are highly influenced by the meandering nature of the tributaries. The study also creates further need for investigation in the accessibility of all-weather roads, availability of flood and clean drinking water among many other services which are critical for the survival of people in these high hazard prone areas. The annual occurrence of floods also indicate that this region has to re-think ways for better flood management and enhance the mitigation strategies.

4. Discussion

For assessing the potential threat by flooding in the study area, there is a need for creating flood hazard maps using the indicator based methodology which takes into account parameters and indicators not only from traditional approach but also looking into stakeholder perspective. The multi-sectoral criteria which has been adopted for this study has taken into consideration the past studies done on the district, the flood data for the past decade as well as used stakeholder and community based approach for identification of possible hotspots within the area. Here, officials for the District Disaster Management Authority highlighted that the stakeholders are usually only brought into discussion during the rehabilitation and compensation aspect of flooding and not while preparing for hazard mapping or pre-disaster evaluation. The uniqueness of this approach is that it gives the researcher the flexibility of utilizing qualitative data and relies on more on knowledge based of the locals than the standard approach of flood models.

For creating a flood hazard map, several steps were considered for finding the possible techniques for flood mapping, different approaches have been done previously at the state level as well as at the district level using NDWI approach where waterbodies were recognised especially during flooding. Another approach was using Digital Elevation Model (DEM) using which the regional scale of flooding was found out. The waterbodies and flood affected areas were identified using these tools and, in the analysis done by Jain, Saraf, Goswami, Ahmad, Dibrugarh district was one of the most affected districts. (Jain, Saraf, Goswami, & Ahmad,

2006). It was suggested that such techniques are useful for planning and research on better mitigation of flood related events.

Figure 12.3: Land Cover Supervised Classification of Dibrugarh district, Assam, India (Image date- January 2015, July 2015). Fig. 12.3 and Fig. 12.4 the supervised classification and Normalized Differential Water Index (NDWI) for further understanding the effect of the floods on the district of Dibrugarh. The analysis was used to differentiate areas with water from the land during the dry month-January and the flood month-July, year 2015 (Earth Observing System, 2021). For the temporal evaluation of the event, it was useful for comparing past events with the recorded data (European Commission, 2011).

5. Conclusion

Flood management and mitigation efforts in India has been an ongoing phenomenon for the longest period of time especially for Northeast India. Among various measures for better management have been discussed and improved upon with flood hazard zonation being one of the most prominent pre-disaster techniques used for better understanding of the severity, spatial distribution as well as from the relief operation, rehabilitation, reconstruction phase of post disaster. One of the important aspects which has been highlighted in this study is the indicator selection done from both the expert's and stakeholder's perspective.

The approach of using a multisectoral, multi-criteria approach for flood hazard mapping. Although this techniques of flood management has been used in similar studies of other districts of Assam but not for Dibrugarh district. It has been advised that local knowledge and tradional methods of flood management should be backed up for creating a need for mapping of resources (Hazarika, Barman, Sarma, & Borah, 2018). These areas face weak connectivity issues through both roads and communication-warnings. With disasters becoming more frequent, pre-disaster efforts are also foiled with previously unknown geographies becoming disaster prone.

Moreover, this strengthens that idea of mapping for hazards using community knowledge coupled with GIS based remote sensing techniques which can provide a platform for both the stakeholders and the researchers to have flexibility in choosing indicators and factors for the study. The accuracy of the map can only be measured when flood data is calibrated into the previous maps at regular intervals. Capacity of both the stakeholders and the institutions working in this field of disaster management need to enhance and more collaborative effort needs to be on the forefront when it comes to preparation of flood hazard zonation maps at both regional and local level.

6. Acknowledgement

The authors are extremely thankful to the Deepjyoti Hatikakoti, Project Manager, Dibrugarh Disaster Management Authority (DDMA) for his support and insight into flood hazard mapping and the lacunas which need to be addressed. They would also like to thank officials of the Water Resources Division at Dibrugarh for providing an insight at the local level as well as helping in mapping areas which are severely affected due to the annual event of flooding. They are also grateful to Mr. Diganta Gogoi, Personal Assistant to Deputy Commissioner, Dibrugarh for facilitating focused group discussions at Dibrugarh and Borboruah.

References

1. A., H. I. (2020). Causes and Impacts of Urban Floods in Indian Cities: A Review. *International Journal on Emerging Technologies, 11*(4), 140–147. Retrieved August 20, 2021, from https://www.researchtrend.net/ijet/pdf/Causes%20and%20Impacts%20of%20Urban%20Floods%20in%20Indian%20Cities%20A%20Review%20Drj6.pdf
2. Assam State Disaster Management Authority. (2013). *District Disaster Management Plan- Dibrugarh, Assam.* Guwahati: Assam State Disaster Management Authority. Retrieved April 2021, from http://www.asdma.gov.in/dmp.html
3. Central Ground Water Board. (2013). *Ground Water Information Booklet, Dibrugarh District, Assam.* Ministry of Water Resources, Central Ground Water Board. Guwahati: Ministry of Water Resources. Retrieved July 31, 2021, from http://cgwb.gov.in/District_Profile/Assam/DIBRUGARH.pdf
4. Chakraborty, A. (2020, May 26). Assam: First wave of flood and erosion hit Dibrugarh. Dibrugarh, Assam, India. Retrieved August 4, 2021, from https://nenow.in/north-east-news/assam/assam-first-wave-of-flood-and-erosion-hit-dibrugarh.html
5. Dhar, O., & Nandargi, S. (2003). Hydrometeorological Aspects of Floods in India. *Natural Hazards, 23*, 1-33. doi:https://doi.org/10.1023/A:1021199714487
6. Discussion on proposed flood management & erosion protection works held in Dibrugarh. (2021, February 4). Dibrugarh, Assam, India. Retrieved August 4, 2021, from https://www.sentinelassam.com/north-east-india-news/assam-news/discussion-on-proposed-flood-management-erosion-protection-works-held-in-dibrugarh-523372
7. Earth Observing System. (2021). *Index Stack (NDVI, NDWI, NDSI).* Retrieved August 10, 2021, from Earth Observing System: https://eos.com/make-an-analysis/index-stack/
8. European Commission. (2011). *Product fact sheet: NDVI-Europe.* DESERT Action – LMNH Unit. European Commission. Retrieved August 12, 2021, from https://edo.jrc.ec.europa.eu/documents/factsheets/factsheet_ndwi.pdf

9. Hazarika, N., Barman, D., Sarma, A. K., & Borah, S. B. (2018). Assessing and mapping flood hazard, vulnerability and risk in the Upper Brahmaputra River valley using stakeholders' knowledge and multicriteria evaluation (MCE). *Journal of Flood Risk Management, 11*, S700–S716. doi:DOI: 10.1111/jfr3.12237
10. IPCC, 2021. [. (2021). *IPCC, 2021: Summary for Policymakers.* IPCC, Working Group I. Cambridge University Press. Retrieved August 15, 2021, from https://www.ipcc.ch/report/ar6/wg1/downloads/report/IPCC_AR6_WGI_Full_Report.pdf
11. IPCC, 2012. [.-K. (2012). *Managing the Risks of Extreme Events and Disasters to Advance Climate Change Adaptation:A Special Report of Working Groups I and II of the Intergovernmental Panel on Climate Change.* Intergovernmental Panel on Climate Change (IPCC). Cambridge UK and New York, NY, USA: Cambridge University Press. doi:doi:10.1017/CBO9781139177245.005
12. Jain, S. K., Saraf, A. K., Goswami, A., & Ahmad, T. (2006, September 22). Flood inundation mapping using NOAA AVHRR data. *Water Resources Management* , 949–959. doi: 10.1007/s11269-006-9016-4
13. National Remote Sensing Centre, ISRO (DOS), Government of India. (2020). *Flood Hazard Atlas- Bihar-A Geospatial Approach.* Deaprtment of Space . Hyderabad: National Disaster Management Authority, Govt. of India; Disaster Management Department, Govt. of Bihar; Bihar State Disaster Management Authority. Retrieved April 2021
14. NRSC. (2013). *Flood Hazard Atlas for Assam State.* Guwahati: Assam State Disaster Management Authority. Retrieved July 21, 2021
15. Reisinger, A., Howden, M., Vera, C., Garschagen, M., Hurlbert, M., Kreibiehl, S., . . . Ranasinghe, R. (2020). *The concept of risk in the IPCC Sixth Assessment Report: a summary of cross Working Group discussions.* Guidance for IPCC authors. Retrieved August 12, 2021, from https://www.ipcc.ch/site/assets/uploads/2021/01/The-concept-of-risk-in-the-IPCC-Sixth-Assessment-Report.pdf
16. Sharma, N., Johnson, F. A., Hutton, C. W., & Clark, M. (2010). Hazard, Vulnerability and Risk on the Brahmaputra Basin: A Case Study of River Bank Erosion. *The Open Hydrology Journal,, 4*, 211-226. doi:1874-3781/10
17. Tripathi, P. (2015, September). Flood Disaster in India: An Analysis of trend and Preparedness. *Interdisciplinary Journal of Contemporary Research, 2*(4). Retrieved August 12, 2021, from https://www.researchgate.net/publication/292980782_Flood_Disaster_in_India_An_Analysis_of_trend_and_Preparedness
18. UNISDR. (2017). *Words into Action Guidelines: National Disaster Risk Assessment-Hazard Specific Risk Assessment.* UNISDR. Retrieved August 3, 2021, from https://www.unisdr.org/files/52828_04floodhazardandriskassessment.pdf

CHAPTER 13

Benefit-Cost Analysis of Flood Management, A Case Study of Jammu and Kashmir

Kowser Ali Jan
Ph.D Research Scholar, Department of Economics,
Annamalai University, Chidambaram-608002

R. Balaji
Assistant Professor, Department of Economics,
Government Arts College, Ariyalur-621 713

Abstract

A disaster hurts those affected. It also spares many in the affected areas, yet those spared may be indirectly affected. The analytical framework of prevention and coping has proved useful in many circumstances. Historically and currently, there has been limited quantitative information available on flood management in Jammu and Kashmir. This study focuses on the cost-benefit Analysis (CBA) of flood management by District Disaster Management Kulgam, and the assessment is based on secondary pooled data collected from government offices, NGOs, published Journals, and local and national newspapers. It also described the scenario, the approach adopted, and the sources of flood damage cost information. The estimated total benefits account for 78686.18 lakh of rupees, and that of total costs account for 2218.75lakh of rupees. The Benefit-Cost ratio greater than one (>1) shows that Flood Management in District Kulgam was economically feasible and successfully managed. The State of Jammu and Kashmir takes essential prevention and management measures to bring down the damages due to floods to significant status.

Keywords: Cost-benefit analysis, nature, flood management, disaster

1. Introduction

Change is the law of nature. It is a continuous process that uninterruptedly involves phenomena, big and small, material and non- material, making our physical and socio- cultural environment. It is a process existing worldwide with daily variation in terms of amount, intensity, and scale. Change can be a gradual or slow action like the evolution of landforms, and it can be as fast and swift as volcanic eruptions, tsunamis, earthquakes, flash floods, etc. Except for inevitable floods triggered by dam collapse or landslides, floods are climate phenomena determined by the circumstances of geology, geomorphology, relief, soil, and vegetation[1]. The World becomes highly vulnerable to natural hazards. While floods are natural events, the river basins' condition has changed dramatically about human activities and human interference in biological processes, such as changes in drainage patterns resulting from urbanization, agricultural practices, or deforestation[2]. Floods are relatively sudden in occurrences and often occur in well-identified regions and within the expected time in a year. Floods occur commonly when water in the form of surface run-off exceeds the river channels' carrying capacity and streams and flows into the adjacent low-lying floodplains. Floods in the South, Southeast, and East Asian countries, particularly in China, India, Pakistan and Bangladesh, are frequent and equally disastrous. Various states of India face heavy loss of lives and property due to recurrent floods in the past decade [3].

On September 6, 2014, Jammu and Kashmir saw gigantic floods that left thousands of people abandoned inside their houses. Two hundred fifty towns across Kashmir had been affected, one hundred sixty individuals had passed on, and many domiciles were bore away[4]. One can scarcely observe the lofty Chinars, the fragrant pine trees, and the rich sobbing willows that gave harbour to those slammed by the destinies. The charming magnificence of Mughal gardens, reviving springs, and impressive cascades lament the condition of the destroyed land, the contaminated streams, and the enduring individuals. River Jhelum, flowing five feet above the danger mark in 2014, had inundated over hundred villages[5]. The most affected districts were Anantnag, Kulgam, Shopian, Pulwama, Bandipora, Ganderbal, Srinagar, and Badgam. The condition heightened as virtually all the significant emergency medical institutions were unfavourably influenced and delivered dead. Almost ten days after the floods, three significant healthcare facilities in the city stayed shut for patients, while two partially continued their administrations to specific patients (outpatient care just)[6]. The floods straightforwardly influenced more than two thousand six hundred villages in the State and turned down thirty per cent of the town zones. The water levels ascended to twenty feet in the city while numerous towns were sliced up off because of the decimation of the main extensions and streets driving into these towns[7]. The state government declared a calculated loss of Rs1 trillion because of the floods, with housing sphere deprivations of rupees thirty thousand crore and business sector losses of seventy thousand crores [8].

2. Flood management in Jammu and Kashmir

John Stuart Mill "What has so often excited wonder [is] the great rapidity with which countries recover from a state of devastation; the disappearance, in a short time, of all traces of the mischiefs done by earthquakes, floods, hurricanes, and the ravages of war"[9]. Despite the gains ground in science and technology, humanity cannot still cope with nature's anger. The natural order of life is interrupted by destruction. It has challenges and adverse health effects. We can not avoid disaster entirely, but disaster management is necessary. The flood's aftermaths can be prevented by taking practical action strategies for disaster prevention. The measures to reduce the flood catastrophe will involve. 1. Design strategies to resilience these risks and enforce them. 2. A solid plan for post-flood recovery and reduction[10].

As mentioned above, heavy rains in September 2014 caused devastating floods and landslides in UT of Jammu and Kashmir, affecting some parts of Jammu and most of Kashmir valley. More than 700 villages remained submerged. Srinagar, the capital city, got submerged. 6.48 lakh hectares of Agriculture/horticulture land got affected. About 3 lakh houses got fully or partially damaged[8]. The case of the Kulgam district was not different. The situation warrants a close probe to identify the factors at work from a policy perspective—the various research issues delineated below would provide a rationale for the proposed study in the Kulgam district. The flood devastation is severely affected the Kulgam district. In light of the above reflections, the present researcher intends to analyze flood management in-depth. Accounting for the damage assessment and documenting administration records in management and rehabilitation works gives specific policy directions.

2.1 Cost-Benefit Analysis

The United States Flood Control Act of 1936, which specified the Federal Government of the United States' participation, was the first used CBA in the World in flood management to control flooding on major rivers of the country. According to that, flood management is economically feasible "if the benefits to whomsoever they accrue are more than the estimated costs"[11].

2.2 The Essence of CBA Lies in

(i) To identify the items of benefit and cost in the flood management from an economic point of view, i.e., allowing all the benefits accruing to and all the costs incurred by the private or public sector;

(ii) Adopting appropriate prices for measuring the benefits and costs in monetary terms; and

(iii) To adjusting the future prices of costs and benefits to present values to make them comparable.

Because benefits and costs are stemming from many different effects, a systematic procedure is required to ensure that each is considered and evaluated correctly. The practise so far has been to include only the direct effects, even though some other effects are also being taken into account.

This study has three goals. The first is to analyze the direct and indirect benefits Loss due to flood] incurred due to the 2014 flood in Kulgam district. In the case of a flood management project, direct benefits include the reduction in physical damage. Thus, flood damage to crops, cattle, houses, commercial and industrial buildings, infrastructure etc., avoided because of construction of embankments, flood retention reservoirs or flood warning systems, are direct benefits. Indirect benefits may occur by avoiding disruption to business, transport networks, and public services and avoiding the accident and emergency response and recovery costs.[11].

The 2nd goal of the study is to compute the direct and indirect costs of flood management in the Kulgam district after the 2014 flood. The State Government adopted many methods to tackle future disasters in the district. As benefits, the costs of flood management projects can be defined as primary and secondary. However, cost estimation is comparatively more exact and easy. Construction measures and their prices are at the core of cost analysis. One method for estimating costs is to use a comprehensive checklist of cost items readily

Table 13.1 Classification of Direct and Indirect Benefits to the society through Cost of Damage Due to Flood Approach

S. No.	Type of Benefits	Components
1	Direct benefits: (i) Tangible Costs (ii) Intangible costs	Physical damage to property: Housing, Agriculture, Commercial Buildings and Livestock
		(a) Primary: Loss of human life (b) Secondary: Illness of flood victims
2	Indirect benefits:	(a) Primary: Disruption of traffic and trade (b) Secondary: Reduced purchasing power

available in the professional literature. However, any such list should be updated and modified according to the latest thinking and local requirements. Thus, the cost estimates are based on actual market prices even if these are distorted and not suitable for CBA[11].

Table 13.2 Types of Costs

Structural: Mitigation costs	Non-structural: Adaptation costs
Reshaping of land surface	Flood defence
Protection from erosion	Forecasting
Levees, dykes, floodwalls	Warning
Dams and reservoirs	Floodproofing
Flood ways and diversion works	Evacuation
Polders and fills	Relocation
Drainage works	

The study's third and last goal is to compare the cost and benefit of flood management in the Kulgam district. Cost-benefit analysis compares the reduced risk due to measure implementation with the costs of implementing the measures. Implementing measures does not change the probability of a weather event occurring; however, the propagation and related damage cost do.

3. Data Construction

The study is based on the secondary pooled data on the cost and benefits of the flood from the publishing record of District Disaster Management plan Kulgam 2017-18, research journals, Government offices, and newspapers. The data was treated for a cost- benefit analysis to derive the necessary results—the analysis was based on past damage data. In the case of flood management in the Kulgam district, important information was available as the number of damaged houses during the 2014 flood, relief amount, construction on river banks, damaged roads, etc. However, some data, particularly on the social impact, were lacking, and the analysis followed the pooled data approach based on estimates of past impacts. Impacts assessed comprise direct and indirect economic impacts. The method of cost-benefit analysis follows these steps: 1. An economic valuation of damage costs. 2. Identification of measures. 3. Cost-estimation of measure implementation. 4. Reduced risk costs as a result of measure implementation. 5. Cost-benefit analysis.[12].

3.1 Limitations

This study focused on the estimation of the effects of measures for flooding and their related benefits. There is no intention to list all available measures to reduce the risk of flooding, as this is highly dependent on local conditions. This analysis is the first of its form in the Kulgam district while using the CBA model for flood management. Since relevant in-data has been provided using secondary data, the quality of the case study results fluctuates accordingly.

4. Analysis of Benefits of Flood Management in the Kulgam District

The Vishaw tributary of River Jhelum flowing through Kulgam District caved embankments and eroded habitants, healthcare system, schools, Govt. offices, and many other buildings and infrastructure. The most significant tangible cost category which was hit badly was considered to be housings. Most of the houses were made with bricks and cement. On average, two thousand seven hundred five households got damaged during the 2014 floods in twenty-two villages, according to the rapid assessment report 2014 and District Disaster Management Plan Kulgam (2017-18).

4.1 Direct Benefits

Damage of Houses and Assets

The Centre for Science and Environment of India stated that Kashmir floods were caused by mismanagement, heavy rains, unplanned urbanization, and lack of preparedness. [13]. There is an African proverb, "where water is the boss, there must the land obey," which is perfectly suited to our valley. The illegal construction on river banks will pave more risk when one disobey the water bodies and their nature. The modern architecture of the housing sector in the valley does not suit the climatic conditions. The housing damage in district Kulgam in the 2014 floods was devastating. The housing building's total damage cost has been estimated with rupees nine thousand nine hundred fifty lakh calculated on the current market price of infrastructure material. In contrast, the total cost of damages of assets (durable and nondurable assets) has been calculated as five hundred fifty-two lakh of rupees. Flood damage has the potential to step up into a grievous emergency if not treated immediately.

Agriculture

Flood 2014 was one of the devastating floods in the history of Kashmir, as mentioned above. The study revealed that human interference and non-eco-friendly development in Jammu and Kashmir were the main reason for such a disaster[14]. Agriculture is the mainstay of the Kulgam district. Almost eighty per cent of people directly depend on agriculture for their livelihood. It had been seen that mainly the villages along the Nallai Veshaw River were suffering from river erosion, whereas other villages suffer from flooding, inundation, and sedimentation. The findings of the total cost of agricultural land damage, loss of economic trees, crops, and apple cultivation revealed: Out of two thousand seven hundred and five affected households, thirty-seven per cent are large farmers, and sixty-two per cent of small or marginal farmers were affected by the flood. The total estimated cost of land damage is fifty-five thousand seven hundred and eighty-five lakh of rupees. Since the August- September season for apples in Kashmir has been considered the ripe growth season, the Floods of 2014 hit this sector very hard. The cost of apple loss is figured by adding the cost of apple loss of big cultivators with that of small cultivators, and it is figured as two thousand one hundred twenty-three lakh of rupees. Regarding loss of crops, rice, wheat, maize, and vegetables, the estimated total cost of loss of crops accounts for rupees one thousand and eighty-nine lakhs.

Animal husbandry is the second-largest source of income after agriculture in the district. A vast number of livestock deaths occurred because of the flood, and unidentified diseases after the flood-impacted the living standard of people. There was also infrastructure damage to cow yards in the vicinity. The estimated total cost of loss of livestock and livestock infrastructure accounts for rupees lakhs.

Loss of Commercial Buildings

Since the district has a large rural population, there is not much commercial property in the villages. But 2014 flood-hit small business shops as well. Commercial businesses contain many kinds of property, inventory, machinery, equipment, products, and goods. The total number of commercial buildings damaged was one thousand one hundred ninety lakh. Fifty-six per cent of the households, as per data available, were not attached to commercial activities. So the commercial business affected by the 2014 flood was less[21].

4.2 Indirect Benefits

Indirect benefits come from the effects of technological linkages of a Management project. These could be described as "Management externalities." Flood management projects may benefit users of transportation by reducing interruptions to traffic.

Reduction in Purchasing power.

The devastation caused by the 2014 flood was colossal. It destroyed everything that came to its way-residential houses, schools, colleges, hospitals, paddy fields, orchards, government establishments, businesses, etc. It has depicted thousands of people roofless and jobless. The destruction around has left the entire society traumatized[15]. The socioeconomic dimension analysis leads to the distribution of income among victims. The flood has destroyed the sources of income of the families. The income of households was reduced to half when comparing pre-flood income. The total cost reduction in purchasing power is accounted for one hundred forty-seven lakhs of rupees in the district. The cost reduction in purchasing power in income group (A) is estimated as rupees forty-two lakhs (20.6), and it is rupees one hundred five lakhs of rupees in case of income Group B. Before the flood, the monthly income ranges from 10000 > (A) to 20000 > (B)

Transport

Flooding of roads can cause physical damages and interruptions in both passenger and freight traffic. In district Kulgam, main roads were blocked by landslides. Roads connecting the Kulgam town with rural areas of the district were washed away by floodwater. It is not easy to attain reliable values for flooded roads. Current values for road damages have been maintained using the restoring cost data from Jammu and Kashmir Home Department. The total cost of damaged roads and bridges accounted for wight hundred twenty lakh of rupees. It indicates that the main roads of Kulgam district were washed away by floodwater in 2014, resulting in interrupted traffic for two to three years in the district[23]

5. Analysis of the Direct and Indirect Costs Incurred due to Flood in Kulgam District

As benefits, the costs of a project can also be defined as primary and secondary. However, the cost estimate is relatively more accurate and accessible. Construction measures and their prices are at the core of cost analysis. One method for estimating costs is to use a comprehensive checklist of cost items readily available in the professional literature. However, any such list should be updated and modified according to the latest thinking and local requirements. Thus, the cost estimates are based on actual market prices.

5.1 Primary Costs

Prime Minister's National Relief Fund (PMNRF)

The government and key stakeholders were engaged to restore the livelihood of the population. Alternative livelihood strategies took place in different areas of District Kulgam. Prime Minister's relief fund was one of the successful rehabilitation plans for constructing damaged houses.

Reconstruction of damaged houses

The Sanction rate of reconstruction of fully damaged pucca and kutcha houses was Rs.1 lakh and rupees fifty thousand per household, respectively. The reconstruction rate of partially damaged pucca houses and kutcha houses was twenty-five thousand and rupees five thousand each. Altogether, Rupees five hundred sixty-five crore was released for Jammu and Kashmir to reconstruct twenty- one thousand eighty-eight fully damaged Pucca & Kutch houses and one lakh ninety-six thousand six hundred fifty-two severely and partially damaged pucca and kutcha houses[16]. The total amount released for Kulgam district is four crores twenty-three lakh. However, there is a huge gap between damaged cost and the relief incentive by the government[22].

Infrastructures costs for Disaster Management

Because of flood mitigation, many govt agencies took steps to construct approach walls, reshape land surfaces, protect from erosion, levees, dykes, and floodwalls in district Kulgam. The data we used were collected from Irrigation and Flood Control Department Jammu and Kashmir. Infrastructure cost is estimated as seven hundred lakh of rupees[21].

Cost of Floodways and Diversion works

In the Kulgam district, the general requirement was drainage systems to cope with rainfall. Flooding in Nallah Vishow often results in overloading in the small canals leading to overflows. Irrigation and Flood Control Department, Jammu and Kashmir, constructed and restored canals to cope with further damage in the future. The data were gathered from Irrigation and Flood Control department Jammu and Kashmir. The total restoration cost accounts for rupees eight hundred sixty-two lakh, and there is a need for further diversions in the NallahVishow and restoration of these canals from time to time.

Flood Warning system

The Flood Control Department has initiated to development of a flood monitoring station in the vicinity of Nihama. In this station, a rain gauge and other tools will help deal with anticipating floods. Four number Gauges to install on the various Nallahs to know flood discharges. Since there was no cost-related information available, as per the District Disaster Management Plan (2017-18), the vicinity covered five Kanals of land, which costs twenty-five lakh as per the present market price[21].

5.2 Secondary Costs

Allocation of Costs in Multiuser disaster Management.

The Kulgam District has been divided into eight beats of the flood. Therefore Flood Control Division Kulgam keeps the availability of flowing items as per the need. The allocation of costs in Multiuser disaster Management among different items and total cost for Multipurpose disaster Management is estimated as one hundred ninety-three lakh of rupees[20].

Cost of Availability of Emergency drugs.

The most significant diseases experienced during a flood are diarrhoea, fever, and skin diseases. To cope with these types of diseases in the future, the Local Flood Zonal Committee keeps the Medicines available. The cost of the availability of emergency drugs allotted is eleven lakh of rupees[20].

6. Cost-Benefit Analysis

The findings of Cost-benefit Analysis reveal: There are eleven components in the estimation of benefits of CBA. The estimated total benefits account for seventy-eight thousand six hundred eighty-six lakh of rupees. Likely, there are six components in the estimation of costs of CBA. The estimated total costs account for two thousand two hundred nineteen lakh of rupees. The estimated Benefit-Cost Ratio is 35.5:1. The CBA result shows that Flood Management in District Kulgam was economically feasible and successfully managed. The State takes essential prevention and management measures to bring down the damages due to floods to significant status. To protect the district from another disaster, govt should pump more money into its disaster security.

7. Conclusion

As discussed under various perspectives, it is clear from the study that the 2014 floods hit the economic status of people's livelihoods in the district Kulgam badly. Among the categories of property damaged, Land value damage is the highest. All the households own apple cultivation land; there was significant damage to the farmers. The study region's average financial assistance was one lakh per household, which was less than the damages caused. The study has further demonstrated that floods in one sector could also affect other social sectors. In the Agriculture sector, the loss of crops led to the reduction of purchasing power of households. Furthermore, the business sector suffered through interrupted traffic for years due to road damage connecting with the district headquarter.

REFERENCES

1. *Effects of ocean sprawl on ecological connectivity: Impacts and solutions—ScienceDirect*. (n.d.). Retrieved June 7, 2021, from https://www.sciencedirect.com/science/article/pii/S0022098117300618
2. *Evaluation of impacts of climate change and human activities on streamflow in the Poyang Lake basin, China—Zhang—2016—Hydrological Processes—Wiley Online Library*. (n.d.). Retrieved March 25, 2021, from https://onlinelibrary.wiley.com/doi/abs/10. 1002/hyp.10814
3. Dewan, A. M. (2013). Hazards, Risk, and Vulnerability. In A. Dewan (Ed.), *Floods in a Megacity: Geospatial Techniques in Assessing Hazards, Risk and Vulnerability* (pp. 35–74). Springer Netherlands. https://doi.org/10.1007/978-94-007-5875- 9_2
4. Bansal, D. N. (2017). *Causes and Effect of Kashmir Flood*. 7. *Jammu and Kashmir was unprepared for flood fury*. (n.d.). Retrieved March 3, 2021, from https://www.downtoearth.org.in/news/jammu-and-kashmir-was-unprepared-for- flood-fury-46233
5. Sandrp. (2015, July 6). Dams, Rivers & People News Bulletin, July 06, 2015. *SANDRP*. https://sandrp.in/2015/07/06/dams-rivers- people-news-bulletin-july-06-2015/
6. *The 2014 Kashmir Flood: The Extreme of the Extremes*. (2015, March 14). Greater Kashmir. https://www.greaterkashmir.com/news/gk-magazine/the-2014-kashmir-flood-the-extreme-of-the-extremes/
7. Yatoo, I. A., Chourey, Prof. B., & Aalum, K. (2018). A Brief Survey on Impact of Flood on Children, Water Sanitation and Hygiene in Kashmir Valley. *International Journal of Engineering and Management Research, 8*(5). https://doi.org/10.31033/ijemr.8.5.1
8. (*Economics Survey Kashmir 2014-15). Planning Development & Monitoring Department*. (n.d.). Retrieved June 7, 2021, from http://jkplanning.gov.in/jk-economic- survey.html
9. Stein, S., & Stein, J. (2014). *Playing against Nature: Integrating Science and Economics to Mitigate Natural Hazards in an Uncertain World*. John Wiley & Sons.
10. Lee, T. (2014). *Global Cities and Climate Change: The Translocal Relations of Environmental Governance*. Routledge.
11. Associated Programme on Flood Management, World Meteorological Organization, & Global Water Partnership (Eds.). (2007). *Economic aspects of integrated flood management*. Associated Programme on Flood Management.
12. *World Meteorological Organization*. (n.d.). World Meteorological Organization. Retrieved February 10, 2021, from https://public.wmo.int/en
13. The Centre for Science and Environment of India stated that Kashmir floods were *caused by mismanagement, heavy rains, unplanned urbanization, and lack of preparedness. (Malay Das, 2019)*
14. Waza, Mehraj Ud Din, and Sonam Sharma. 2018. 'Preparedness and Response to Disasters: A Case of 2014 Floods in Jammu City'. *International Journal of Research in Social Sciences* 8(3): 111–21.
15. *PM approves Rs 426.83 crore for J-K flood relief—India News*. (n.d.). Retrieved June 7, 2021, from https://www.indiatoday.in/india/story/prime-minister-jammu-kashmir-flood-relief-242036-2015-02-26
16. *Financing Disaster Management: Options for the GST Council and the Fifteenth Finance Commission | Economic and Political Weekly*. (n.d.). Retrieved April 3, 2021, from https://www.epw.in/journal/2019/25/special-articles/financing-disaster- management.html
17. *Jammu and Kashmir was unprepared for flood fury*. (n.d.). Retrieved March 3, 2021, from https://www.downtoearth.org.in/news/jam mu-and-kashmir-was-unprepared-for- flood-fury-46233
18. *Hazards Risk Assessment Methodology Guidelines 2012-2013*. (2018, December 21). https://www.publicsafety.gc.ca/cnt/rsrcs/p blctns/ll-hzrds-ssssmnt/index-en.aspx
19. *Disaster Management | District Kulgam, Government of Jammu and Kashmir, Picturesque place nestling on the bank of river Veshaw | India*. (n.d.). Retrieved March 29, 2021, from https://kulgam.nic.in/disaster- management/
20. *Official website of Kashmir Irrigation and Flood Control Department, Srinagar*. (n.d.). Retrieved March 29, 2021, from https://www.ifckashmir.com/
21. *Prime Minister's National Relief Fund*. (n.d.). Retrieved September 3, 2021, from https://pmnrf.gov.in/en/about/projects- in-progress
22. *Public Works Department—Jammu & Kashmir*. (n.d.). Retrieved July 3, 2021, from http://jkpwdrb.nic.in/

CHAPTER 14

Tackling River Erosion: A Post-Disaster Transformative Framework

Siji Chacko
Director, Conference Development Office (CDO),
Indian Social Institute (ISI), New Delhi

Anil Kumar
Public Policy Researcher and Analyst
based in New Delhi

Abstract

The paper examines how river erosion, locally known as *Katav*, in Gandak River affects the lives and livelihood of people, especially the Musahar community in the Gangetic Plain in Bihar, India. Due to excessive siltation of the river bed, its drainage capacity and channels get disturbed, which makes difficult for the river to maintain its equilibrium flow. It forces the river to find a new course, causing recurrent floods to accommodate the excess drainage. Exploring the context of *Katav*, based on the lived experiences of Musahars and their multiple displacements, the paper analyses the vicious cycle of vulnerability in the disaster context. Recognising the primacy of victims' narratives of their experience of recurrent floods, *Katav* and multiple displacements, the study uses a qualitative phenomenological paradigm in the 'life-world' of the Musahars, by entering into the field of their perceptions and experiences. It draws attention on the need for a specific context-based approach in the implementation of Disaster Risk Reduction (DRR) and Disaster Risk Management (DRM) programmes. Further, it suggests what is required for long-term resilience at community level beyond an academic exercise, and discusses the '*Post-Disaster Transformative Framework*' that has emerged from the study.

Keywords: Vulnerability, Adaptation, Resilience, Musahars, River erosion.

1. Introduction

India is blessed with many major rivers and their numerous tributaries. Among them, Ganga and Brahmaputra Rivers stand out not only due to the recurrent floods that ravages livelihood of the people across many states, but also for the phenomenon called river bank erosion. This paper examines how floods affect the lives and livelihood of the Musahar community, the poorest community in the entire Gangetic Plain, in the state of Bihar, India. Despite the availability of vast literature examining the huge resource poverty and social vulnerabilities of the Musahars, there has been hardly any attempt to look at and capture the challenges facing the community due to climate change, especially floods and river erosion, in a comprehensive way. Going beyond an analysis of the emerging literature on climate change, it brings on board the 'lived experiences' of Musahars in their context of the recurrent floods and *katav* (erosion of the river banks) and multiple displacements. Recognising the primacy of victims' narratives of their experience of recurrent floods, *katav* and multiple displacements, the paper also locates itself on the study (Chacko, 2019) undertaken using a qualitative phenomenological paradigm in the 'life-world' of the Musahars, by entering into the field of their perceptions and experiences.

Located in the eastern part of India, bordering the states of West Bengal (in the eastern side), Uttar Pradesh (west), Jharkhand (south) and Nepal (north), Bihar has a total geographical area of 93,595 sq. km. Bihar is the third most populated state of India with total population of 10,38,04,637 as per 2011 census. Nearly 90 per cent of the state's population lives in rural areas. Being a largely agrarian state, more than 80 per cent of the workforce is engaged in agriculture and allied activities. *Katav*, being 'unprecedented, unexpected and uncertain', often causes severe stress, which aggravate with longer duration and prolonged negative impacts. The paper draws attention on the need for a specific context- based approach in the implementation of Disaster Risk Reduction (DRR) and Disaster Risk Management (DRM) programmes. Further, it suggests what is required for long-term resilience at community level beyond an academic exercise, and discusses the 'Post-Disaster-Transformative Framework' that has emerged from the study.

2. Disasters and Vulnerabilities

Disasters are no more considered as purely 'natural'. They manifest as a mix of natural and 'human-made' (Smith, 1996; Perry and Quarantelli, 2005). A few scholars consider disaster as involving a combination of reasons which are, whatever their origin (natural, technological, social and political) and results, both physical and social (Kreps, 1985), and therefore, result in an inevitable social state/condition, i.e., crisis or vulnerability (Barton, 1969; Pelanda, 1982; Gilbert, 2005). Scholars and practitioners understood that disasters are not just physical, rather inherently social phenomena and that the source of disaster is embedded in the social structure or social systems (Granovetter, 1985; Quarantelli, 2005).

The poor are more vulnerable to disasters of any sort. None can deny the fact that resource poverty and high population density exacerbate the impacts of natural disasters. Further, the intersection of race, class and caste adds to the existing physical vulnerabilities. A review by Singh *et al.* (2014) finds that some people within the social systems like ethnic minorities, disempowered castes and classes are relatively disaster-prone, and that lack of access to resources and power could also be a major factor. It is not the just scope, magnitude, or intensity of the hazard event that determines the severity, but factors such as class, ethnicity, poverty and inequalities (Varley, 1994). Exposure to hazards varies in time and space, according to people's 'positionalities' by class, race, ethnicity, age, gender and nationality. Based on various empirical evidences, drawing attention to socio-political origins of disasters, geographers, anthropologists and social scientists have understood that disasters affect different class, ethnicities and caste differently (Cannon, 1994). The poor and politically powerless communities of ethnic and racial minorities are located in challenging geographies (Bullard, 1990).

Disaster can lead to displacement, which in turn could escalate vulnerability. Displacement is the *forced* or *obliged* movement, evacuation or relocation of individuals or groups of people from their homes or places of habitual residence in order to avoid the threat or impact of a disaster.[2] It is also regarded as forceful removal of a person from his/ her home or community due to conflict, natural hazards, instability, developmental projects, etc (Belcher and Bates, 1983). They are those displaced from, or feel obliged to leave their usual place of residence, because their lives, livelihoods and welfare have been placed at serious risk as a result of adverse environmental, ecological or climatic processes and events (Gorlick, 2007). It is a more reactive measure of last resort or a survival response to severe and immediate threats.

According to embeddedness theory, all actions are embedded within a socio-political, cultural context that enables or constraints action (Granovetter, 1985). Disaster is embedded in a complex mix of physical, socio-political and environmental factors that determine the potential for people to be exposed to particular type of a natural hazard. The consequences of structural inequalities are very important to understand embeddedness (Hewitt, 1983; Bolin and Stanford, 1998). Accordingly, a disaster is the outcome of the interaction between a society with its environmental features, which are primarily (though not solely) determined by the societal structures (material as well as social). The social dimensions of disaster vulnerability can be compared to structural forms that occur before, during, and after the events. They are very closely related to unequal distribution of resources and power relations operative in a society. Disasters are 'symptomatic of the conditions of the society', pursuing the notion that vulnerability to, and the consequences of, disasters are potentially unevenly distributed across social groups. Disaster is a collective of intersecting processes and events—social, environmental, cultural, political, economic, physical, technological—transpiring over varying lengths of time as totalizing events (Hewitt, 1983; Smith, 2006; Wisner, 1993; Blaikie et al., 1994; Cannon, 1994).

According to 'the political economy of disasters', vulnerability is political in nature as the material production and distribution of vulnerability is the result of political processes (Bertrand, 2005). The net effect of a disaster situation is the final outcome of impact, measured in direct and indirect financial loss. Politicians and policy makers often neglect vulnerable people's interpretations and demands, despite the fact that empirical evidence states otherwise; this is partly because vulnerable people very often lack the political power to make their demands heard and their needs met. Continued vulnerability diminishes people's ability to cope with disasters on a daily basis. Hence, it is not only the exposure to hazards that puts people at risk, but also socio-economic and political processes in society that generate vulnerability. These create the conditions that adversely affect the ability of

[2]UN Guiding Principles on Internal Displacement, 1998, available at http:// goo.gl/uVW0DW.

communities or countries to respond, to cope with or recover from the damaging effects of disaster events.

Post-disaster relief and rehabilitation have revealed the vulnerability of Dalits and ethnic minorities. They experience discrimination, segregation, and denial of access to relief and entitlements" (Zakour and Gillespie, 2013; Jha and Raghavan, 2008). Although belonging to a lower caste itself is a cause of disparities; the impact is all the greater when the lower-caste families happen to be very poor and marginalised (Srinivasan, 1962). People experience social discrimination along castes and having belonging to a particular caste, of the lower strata (class too), has profound implications for people's exposure and recovery from disasters (Ray-Bennett, 2009). Many scholars have reiterated that disasters could be said as by-products of inequalities existing in a society and also the impacts are asymmetric due to carious contextual factors. As mentioned above, there have also been a few empirical studies that reiterated the poor, marginalised, ethnic minorities and the races with poor socio-economic indicators are more vulnerable. However, there are not many studies that showcase caste-related pre-existing vulnerabilities escalating post-disaster conditions.

The development of 'vulnerability to disaster' paradigm has its roots in the articulations of Fanon, (1961), Edward Said (1978), Paulo Freire (1970). At the core of the agenda is a clear understanding on oppressive social structures and unequal power relations. Freire (1970) regards human beings as fundamentally constrained and de-humanised by structural systems of oppression. Yet emancipation and liberation are possible for the poor and the marginalised. For Freire (1970), everyday life, experiences of oppression and struggle themselves constitute as learning/pedagogical spaces and this is reflected in everyday as a valid inquiry. While disasters[3] generally create uncertainties and crisis situations to everyone, they make it extremely difficult for some communities to cope with and recover. The aftermath could even leave people with much longer duration impact on their lives. Although largely downplayed, river bank erosion is one such devastating phenomenon, which is caused by various factors.

3. Floods, River Erosion and Displacement

Flood is the one of the major natural disasters in the country. As per the estimate by the Ministry of Water Resources, Government of India, the country has 40 million hectares of floodplains, with floods causing immense damage to property and human life every year (NRSC, 2020). People who live near riverbanks are often the victims of bank erosion which threatens their personal safety and shelter as well as sources of their livelihoods (Brouwer *et al.*, 2006). The extent of impacts on human settlements in any riverbank erosion situation depends on the population density and economic conditions of the people in that region (Tuhin *et al.*, 2017). According to Das *et al.* (2009), vulnerability of a community to water-induced hazards, such as floods, is determined by factors like location (living on flood plains, proximity to rivers), physiography and hydro-geomorphology (altitude, state of rivers, etc.), structural interventions (embankments, dams, etc.), economic conditions of exposed communities (purchasing capacity, savings, local inflation and supply chain of essential commodities), social conditions (literacy, social security and ethnic diversity), general awareness of problems and solutions.

People across countries are affected by riverbank erosion to different extents, with varying degrees of environmental and socio-economic impacts. For instance, in Papua New Guinea, people have been displaced permanently due to the erosion of the Busu River (Sekac and Jana, 2014), while in Serbia, farmers and their settlements are affected by the erosion of the banks of the Kolubara River (Dragicevic et al., 2013). Ahmed & Fawzi (2009) says how reduction in agricultural production has resulted with the loss of agricultural land to erosion along the Nile in Africa. In New Zealand, rive erosion is evident across the banks of Waikato, Avon and Heathcote rivers (Mabin, 2007; Roper- Lindsay, 1994). Further, the Murray, Darling and Gordon rivers in Australia (Hughes & Prosser, 2003; Bradbury, 2013); and the River Danube in Europe (Jones *et al.*, 2007) are known to erosions. Even the United States is not free of river erosion, with the Mississippi, Missouri and Haw rivers leading in river erosion (Briaud *et al.*, 2007; Macfall, Robinette and Welch, 2014). Although erosions in all these countries have led to the loss of agricultural land and threatened physical infrastructure such as roads, houses, etc., the erosion along Gandak River is unique with the presence an extremely vulnerable community like the Musahars. With considerable section of the Musahar community living along the river banks, their vulnerability to river bank erosion are relatively high. The banks of Gandak River is one of the worst examples of riverbank erosion in India, with the most severe long-term impacts on human life.

Running across the state of Bihar is a web of rivers originating from Nepal, Uttarakhand, Uttar Pradesh and Madhya Pradesh. The prominent rivers originating in Nepal are Mahananda,

[3]A few hazards can turn out to be disasters. This study considers that any hazard could be considered a disaster when it causes significant damage and disruption to a group of people or community (especially those who are vulnerable). Disaster is an occurrence that could cause ecological disruption, loss or deterioration of life and health and the affected community requires support as it is not able to cope with the situation and recover by themselves.

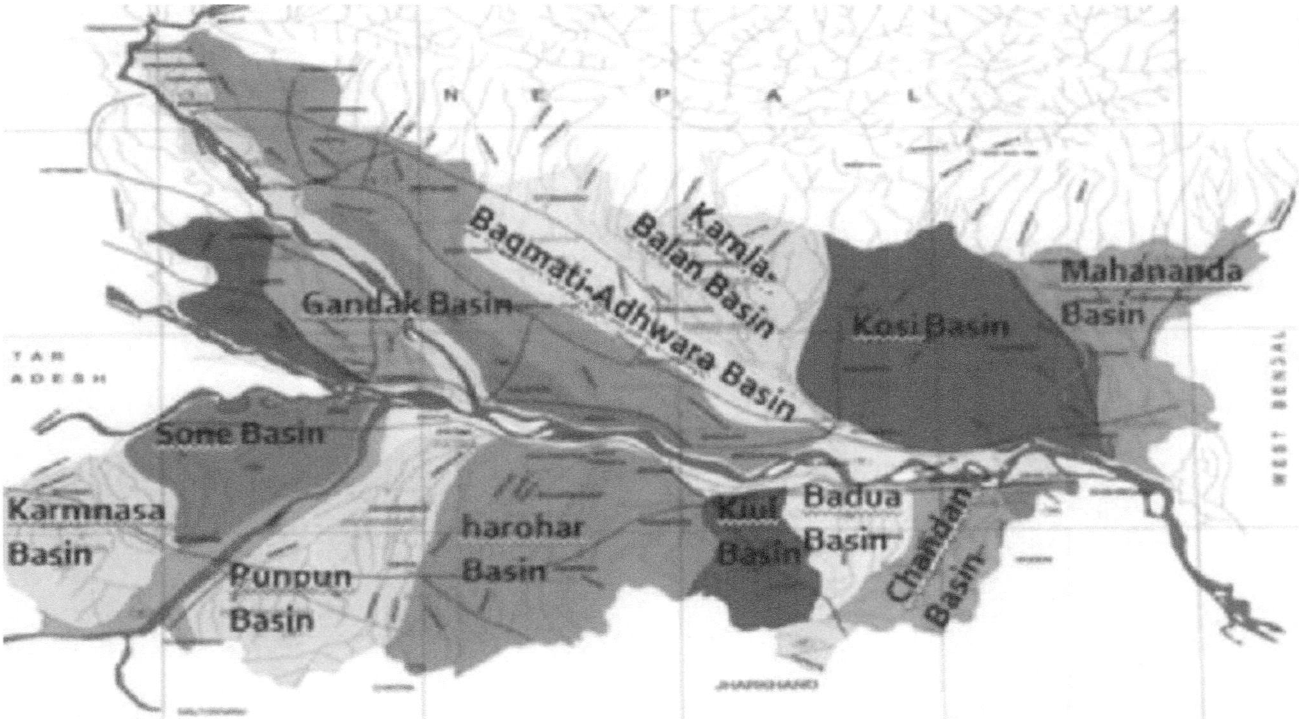

Fig. 14.1 River Basins of Bihar

Source: http://fmis.bih.nic.in/Riverbasin.html

Kosi, Kamala Balan, Bagmati, Gandak, Burdhi Gandak and rivers known as Adhwara group of rivers. All the rivers finally join the River Ganga at different locations. Bihar has 12 river basins. With 17.2 per cent of India's flood-prone areas, the state is one of the most flood-prone states in the country. Nearly 76 per cent of the population of the state, and 73.06 per cent of its area, is under the constant threat of devastation by floods.[4] According to 'A Report on Bihar Floods 2016' published by Bihar State Disaster Management Authority (BSDMA, 2017), of the 38 districts of the state, 28 are flood-prone among which 15 are extremely flood prone. The floods of 2007 which affected 22 districts of the State are very vivid in the memories of people.[5] In 2017 alone, of 38 districts in Bihar, 14 districts were affected by floods with a total 65.37 lakh population and 93,641 villages disturbed by the same.[6] The state is also one of the most disaster-prone states. It is highly vulnerable to hydro-meteorological natural disasters.

In addition to floods, the settlement in vulnerable areas and inadequate or no investment on mitigation and preparedness measures have further increased the vulnerability. The Kosi River (The Sorrow of Bihar), one of the major tributaries of the Ganga River, is known for rapid and frequent avulsions of its course and causes extensive damages to life and property almost every year. Recurrent, excessive floods and the situation of riverbed have reduced the drainage capacity of the channels of Gandak River, forcing it to change its course further causing floods to accommodate the excess drainage. It results in breaking one river bank and making another, and further eroding the banks (Chacko, 2019) Citing Census 2011, the study maintains that out of the total 1483 villages, there are 118 uninhabited villages in the West Champaran District, and the non-habitation has been mainly due to *katav*.

West Champaran (also known as Paschimi Champaran) is particularly prone to flooding, and 80 per cent of its area vulnerable to inundation. According to a report prepared by the District administration, 17 of its 18 blocks are hazard-prone of which 10 are at high risk of both severe floods and erosion[7] and 178 of its 315 Gram Panchayats are flood-prone.[8] As per

[4] *Bihar Floods: A Field Study,* 2013.
[5] *A Report on Bihar Floods 2016,* Published by BSDMA (2017).
[6] *Annual Report 2016-17-* National Disaster Management Authority (NDMA), New Delhi.
[7] Hazard, Vulnerability and Capacity Assessment Report, prepared by the District Administration, Government of Bihar, October 2013.
[8] District Paschim Champaran: Flood-2017, review meeting by Hon'ble Chief Minister, Shri Nitish Kumar, Date: 16-8-2017 (handbook)

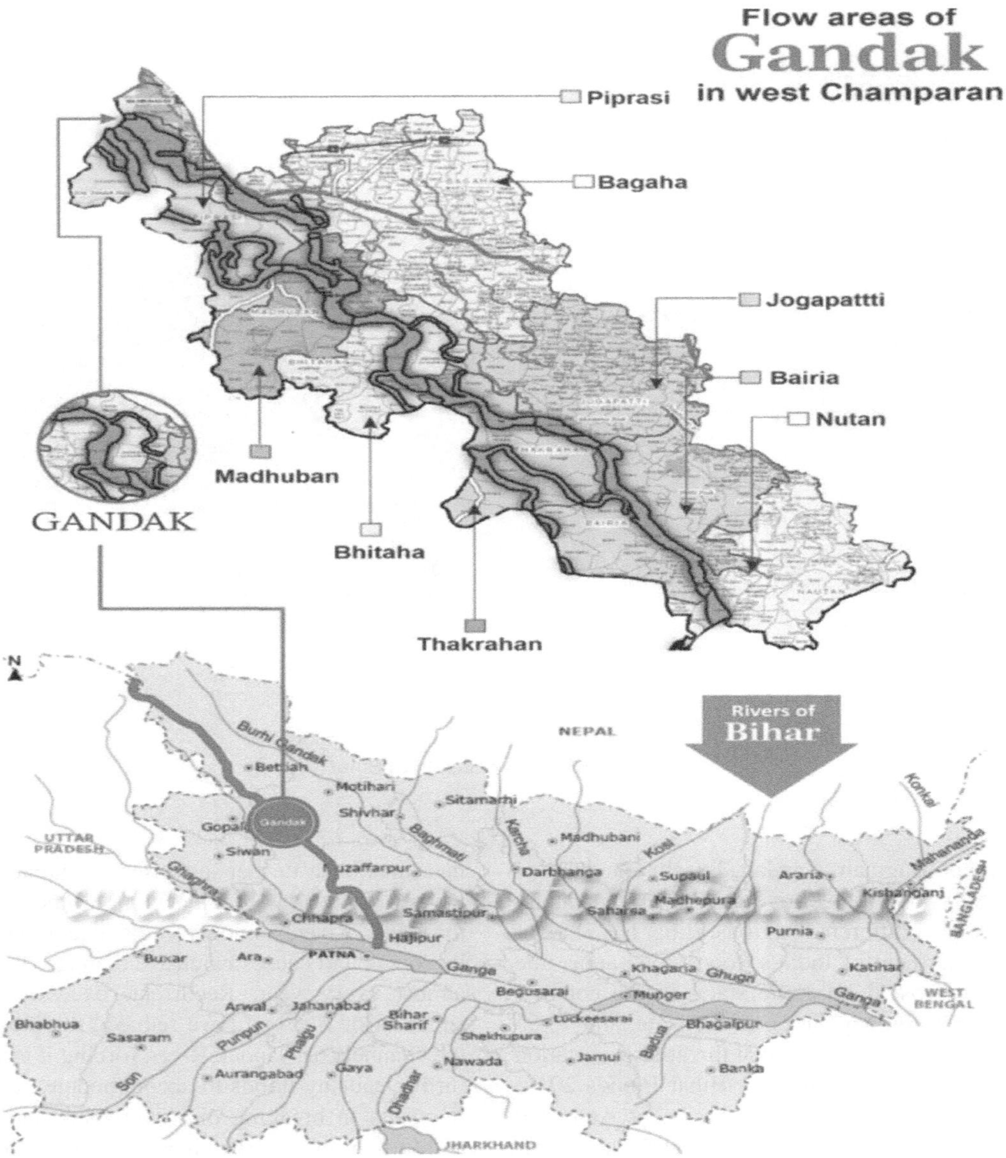

Fig. 14.2 Rivers of Bihar and River Gandak

Source: https://www.jagranjosh.com/general-knowledge/list-of-major-rivers-in-bihar-1480142371-1.
Note: Customised to highlight the area of the flow of river Gandak in West Champaran District, Bihar.

the 2011 Census, there are 118 of 1483 villages in the district which are uninhabited.[9] Non-inhabitation is mostly due to the erosion of the river banks. West Champaran District is placed in the 'vulnerable' category in the mapping of disaster-prone districts.[10] In 2017, due to heavy rains, high water discharge from Gandak affected 13 of the 18 blocks of the district, affecting 135 Panchayats[11] and a total population of 542,000 and damaging 10,502 houses.[12] The main rivers of the district are Gandak, Sikrahna and Burhi Gandak (known in different names). The Gandak River, the most significant river flowing

[9]Source: Census 2011 reflected in https://westchamparan.nic.in/about-district/demography
[10]Roadmap for Disaster Risk Reduction 2015-30, Government of Bihar, 8 March 2016
[11]Of 178 flood prone panchayats, 33 are on the bank of the river Gandak
[12]District Paschim Champaran Flood-2017, Review Meeting by Hon'ble Chief Minister of Bihar, Date: 16-8- 2017

through the district, originates near the Nepal-Tibet border at an altitude of 7620 m situated northwest of Dhaulagiri. It rises in the central mountain basin of Nepal known as Sapt-Gandaki from the seven streams which unite to form this river. The total length of the river is 630 km of which 370 km lie in Nepal and Tibet. After travelling nearly 100 km in Nepal, it enters the plains of West Champaran District near Valmikinagar at Triveni on the Indo-Nepal border. It then flows in a southern direction and forms the Uttar Pradesh--Bihar border to some distance after which it flows through West Champaran, East Champaran, Siwan, Saran, Gopalganj, Muzaffarpur and Vaishali districts of the state, eventually joining the Ganga on its left bank opposite to Patna city.[13]

From time immemorial, the Gandak River, which has a drainage area of 7620 sq.km, has been regarded as a gift of nature by the people of North Bihar. Entering India at the border of Gorakhpur district (Uttar Pradesh) and West Champaran District (Bihar), it courses downward from a north-west to south-east direction touching the two districts and then through five districts of North Bihar, sustaining much vegetation and life systems in its wake. However, what used to be a boon for the people has, in the past five decades has become a bane as the river gets frequently flooded causing much suffering to the people, especially those of Gorakhpur and West Champaran. Coming with its torrential waters, down the hills of Nepal, the Gandak gets suddenly flattened and its huge volume of waters spread into the expansive flat terrain of West Champaran district. This results in the displacement of thousands of people from their homes.

Being snow-fed, the Gandak is a torrential stream until it leaves the hills after which it becomes wider and less turbulent.[14] Due to the special nature of this river, with alluvial soil and flood, there is frequent erosion of the river's banks. Due to excessive siltation of the bed, the drainage capacity of the river and its channels are disturbed as a result of which it becomes difficult for the river to maintain its equilibrium flow. The disturbed condition forces the river to find a new course, causing the floods to accommodate the excess drainage. Breaking one bank and making another has become the nature of rivers like the Gandak. The problem is acute in the monsoon as the water puts enormous pressure on the natural banks. Flood-led riverbank erosion in West Champaran has adversely affected the life of the people. While this phenomenon results in physically displacing many families, the impacts are distressful and traumatising for the affected people. Floods and *katav* are closely related that the latter follows the former. Frequent changes in the course of the river results in the erosion of its banks, which causes the displacement of people living in proximity to the river. Most of the affected victims are the poorest of the poor who already lead a precarious existence on account of their vulnerable socio- economic situation.

While *katav* might affect all irrespective of any backgrounds and affiliations, for those who already suffer from the many pre-existing vulnerabilities of extreme poverty, assetlessness, illiteracy, caste-based structural and systemic oppression, the experience could be very challenging. People's existing vulnerabilities, such as social exclusion due to caste factors, intertwined with the spiral of poverty, cripple resilience and intensify their plight. Thus, they become more vulnerable to future *katavs* and the associated experiences. The people in the area suffer from absence of entitlements and lack of realisation of rights related to their concerns.

4. Methodology

Recognising the primacy of victims' narratives of their experience of recurrent floods, *katav* and multiple displacements, the paper also locates itself on the study (Chacko, 2019) undertaken using a qualitative phenomenological paradigm in the 'life-world' of the Musahars, by entering into the field of their perceptions and experiences. This is required because analysing the vulnerability implicit in 'everyday life' is essential to understand disaster and its impact (Wisner *et al*, 1994; Quarantelli, 2005). The study, which was conducted over a period of five years and carried out in different phases (floods and non-flood times), used multiple theoretical lenses to unearth and understand the phenomenon. It used purposive heterogenous sampling, with the application of the principle of 'maximum variation' for data gathering. In this context, 12 Musahar hamlets in which the people experienced *katav* were studied from five consecutive blocks, where the River Gandak flows.

The study was conducted in West Champaran District focusing on the recurrent floods, *katav* and multiple displacements with special reference to the Musahars. It has been an attempt to bring the narratives from the 'periphery' to the centre by understanding Musahars who, with pre-existing socio-economic vulnerabilities, live with katav. While looking at how katav itself has affected people, it goes beyond looking at aspects like the long-term impact brought upon by social exclusion-led discrimination, exploitation and injustice on the marginalised in the post-disaster setting.

As per the 2011 census of India, there are 30,57,396 persons belonging to the Musahar community, which is 1.8 per cent of India's scheduled caste population and 0.23 per cent of the total population. Musahars are found mainly in four states—

[13]Flood Management Information System (FMIS), Water Resource Department, Bihar.

[14]District Census Handbook: Paschim Champaran, Village & Town Directory- Directorate of Census Operations, Bihar- Series II, Part XII.

Bihar (89.2 per cent), Jharkhand (1.7 per cent), Uttar Pradesh (8.4 per cent) and West Bengal (0.7 per cent). They live mostly in rural areas (96.3 per cent) and work as agricultural labourers (95 per cent) and, with an illiteracy rate of 79 per cent, are among the least educated of caste groups (Singh, 2016). As per the data available with the *Mahadalit Mission*[15] of the Government of Bihar, the *Bhuinyas*[16] and Musahars together have a population of 26,80,539 in Bihar. Being socially and economically marginalised, they are one of the poorest and occupy the lowest rung of the Dalit hierarchy. Of the total population of 358,725 SCs, in the district, 75,964 are Musahars.[17]

Adopting a phenomenological framework, the study explored the long-lasting negative impacts of the phenomenon of katav on the lives, which makes katav itself a 'disaster'. It used purposive heterogenous sampling, with the application of the principle of 'maximum variation' for data gathering. It was conducted in different phases (floods and non-flood times) over a period of five years. The study was conducted in 12 Musahar hamlets: Srinagar and Rakhahi (Thakrahan Block), Siswa Mangalpur and Kalaiyastan (Jogapatti[18] Block), Ranha Bandh (also known as Godayya), Asaram-Pathkhouli, Gambhirpur and Saraiya (Bairia Block), Bankatwa (also known as Mangalpur Gudaria), Bhediharwa and Pujaha Patijirwa (Nautan Block) and Rajwatia (Bagaha-I Block) of West Champaran District of Bihar.

5. Results and Discussions

The Musahars went through the experience of being 'uprooted', living in an ambiguous, broken and dislocated state of liminality due to *katav*. They go through distressing experiences of survival and sustenance, including the struggle for an address and identity. Being 'unprecedented, unexpected and uncertain', the katav events often cause severe stress, which aggravate with longer duration and prolonged negative impacts. In spite of having to encounter the raw realities of structural violence, political--bureaucratic nexus, with no other option left, they still turn to the 'everyday state'.

While the 'embeddedness' nature of disasters has been captured by academicians and researchers, there has been much scope and need for a deeper analysis of the applications of embeddedness, especially in relation to vulnerable and marginalised communities affected by a disaster, that too in a prolonged disaster context. Although a few speak of a serial character of disasters, stating it as 'embedded' reality, no one seems to have gone deep into how a particular community (socio- politically and economically exploited) encounter a particular disaster context, especially when it is prolonged. The narratives clearly highlight that the already existed vulnerabilities of the Musahars, due to various socio-economic factors influenced their exposure to risk as well as the encounters with recurrent floods, *katav* and displacement. With the oppression that they have suffered (previously) coupled with the experiences surrounding *katav*, they seem to internalise it and consider themselves as further inferior. Both, physically and psychologically they go through difficult times and are forced to suffer these experiences in a silent manner.

Class-caste segmentation is an essential characteristic of Indian society. The caste system, with its myriad forms of exclusion, discrimination, humiliation, social power and influences of exploitation and subjugation, has played a significant role in the context of *katav*. What is obvious from the narratives of the Musahars is that this caste exclusion and inequality in society have been structurally built into whereby at times the dominance is visible and other times they are subtle. It is the caste dominance as a factor expressed in various ways that has not helped the Musahars to come out of their struggles related to *katav*.

Musahars repeatedly suffered multiple shocks that severely affect their lives and livelihoods. They live in unsafe locations, such as river banks, embankments or roadsides. With every event of displacement and relocation, their vulnerability increased further, which undermined their preparedness and capacity to withstand a future disaster. The narratives of the experiences of the Musahars reveal that caste system operative in the subtle practices of discrimination and exclusion has increased their vulnerability. Caste as a factor indirectly contributed to the slow recovery from their present situation. Based on the experiences of vulnerability suffered by the Musahars (in the disaster and post disaster setting), it is clear that Musahar's vulnerability to *katav* is directly tied to their previous and present conditions. It is obvious from their experience that pre-existing vulnerabilities, which are caused by social processes of exclusion (lack of resources), lead to the intensification of the effects of a disaster, particularly *katav*. *Katav* has created conditions in which it is virtually impossible for them to escape through normal ways, except with a intervention through the government or so. Living with

[15]With the aim of reaching out to the most marginalised among the Scheduled castes in Bihar, the Government of Bihar constituted a commission known as State Mahadalit Commission in 2007 for the welfare of the Mahadalits. Of late, the composition of the group has been changed a little to include other groups.
[16]At times also written as Bhuiyas. The etymology of the word 'bhuinya' comes from its root from Sanskrit 'bhuu', meaning land. A few scholars connect 'Bhinyas' trace their origins to tribals.
[17]Source: Bihar Mahadalit Mission: http://www.mahadalitmission.org/BMVM-about-us-list.php
[18]Also at found in some documents as Yogapatti.

the poverty and uncertainty of their day today life, mixed with various discriminations that they suffer, Musahars are more exposed to further vulnerable conditions. What Hari Manjhi (Srinagar) expressed as the following is very revealing of the present situation:

> "We feel that all the pain of the world is concentrated in this river and given to us. Our voice is not heard because they are suppressed by government officials and other caste groups. When we try to voice our pain to officials and other responsible people, they tell us, 'You people have already put up with such misery for the last 4–5 generations. Does it matter if things do not improve? After all this is your fate?'", *as translated from Hindi.*

The very comments by the dominant caste groups wanting the Musahars to believe that all their experiences are due to their 'fate' itself is oppressive and discriminative. It is generally said that 'untouchability, deprivation and humiliation do not leave the Musahars. They carry the baggage of bondage right from their birth to their death. In the deeply fractured social structure of Bihar, which is mostly along caste lines, the Musahars affected by *katav* suffer from a precarious financial status, denial of human dignity, citizenship rights, violation of their human rights, and chronic and abject poverty.

The vulnerability of Musahars is also associated with poverty, which is almost an outcome of the discrimination that they suffer. With lack of resources and capacities, when they are isolated, insecure and defenceless to floods, *katav* and multiple displacements, they face the risk, without any capacity to resist, cope with and recover from hazards. The caste structures and practices put some communities at a disadvantage in their everyday lives. There is both physical and geographical marginalisation as they live in geographically marginalised areas. They are economically marginalised due to loss of homes and livelihoods; social and cultural marginalisation because of their position in the hierarchy; and political marginalisation because their voices are disregarded by those in power (cited by Gaillard, 2010). These narratives from the 'periphery' bring to light that the Musahars are particularly vulnerable not only to disasters of an immediate nature, but also to small-scale, high-frequency hazards which easily makes them more vulnerable because of their impaired ability to withstand the outcomes and impact that a disaster brings (cf. Collins, 2009; Wisner *et al.*, 2012).

There is no doubt that geography creates and perpetuates vulnerabilities. The profiles of the 12 Musahar hamlets tell their own stories. Their locations—poor connectivity and access to roads, poor infrastructure and health services, non-functioning of essential institutions—add to their vulnerability. The locations of the hamlets is indicative of the vulnerability of the people living in them. The risk of being affected by a disaster is ever-present. The lack of resources increases the marginalisation of the people. Living in such places is not a matter of choice; rather, it is the sense of hopelessness that there are no alternatives which drives such communities to live in conditions of constant risk.

Periodic floods and *katav* makes it very difficult for the Musahars to lead secure and meaningful lives. Often, they live under the constant threat of *katav*. In a few hamlets like Rajwatia and Bankatwa, two generations have been victims of *katav*, making it a cradle-to-grave existence with the threat. With context established and the profiles located in the context and discussions on the experiences of the Musahars with floods and *katav* show how events unfolded which eventually forced them to live on roadsides and embankments.

The Musahars in some groups had the earliest experiences of being displaced by *katav.* They continue to suffer from the consequences even today, living under the constant threat of breach of the embankments they are living in. They are yet to be resettled by the state government because of which they are compelled to continue with their precarious existence. The way pain engulfed the people's life with katav is evident from what emerged in the interactions in the community. *Nadi ki avaj itna hola ki jan lekar bhagata (the noise which the river made as it engulfed our homes was unimaginable).* Many of them (especially women) had tears when they recollected such experiences. The intensity and magnitude of *katav* was such; "*Within two days, we saw nearly 1100 sq ft. land being cut [eroded by the river] ... in five days, the entire area was gone.*"

When a person belonging to Musahar community says that he/she lost everything, it means that the very little (usually the Musahars do not own much assets such as land, cattle, savings, including human capital) the person or family had was washed away by *katav*. While the material losses were articulated as; "*We abandoned everything else we had, We had 'mal, jal, bakari' (cattle) and everything but all went into the water.*" A few shared, *in two days we lost everything [to erosion].* Some of them ended up in relief camps (as shared). But beyond these material losses and the challenges of finding a place and rebuilding their houses, they were able to see *katav* as a complete loss which would make it difficult for them to rebuild their lives: "*All our houses collapsed and we had to begin our life over again.*" And they were aware of the resource crunch that would make it impossible to ensure recovery. Accepting some sort of temporary inabilities, they expressed; "*We do not have anything to lose... We are destined to die along with our children.*" In fact, on the other hand they had 'everything to lose'. *Katav* even took away their rightful departure from this world as one expressed in the context of a woman who died with a child birth: "*We had to push the body into the water as there was no place where we could cremate the body".* Not having a 'space' even for cremation continued as a reality as they always had to struggle for a 'homestead land'.

Their overall experiences of encountering the event/s of *katav* were expressed in *Khubparesanibhayi (it was so difficult and very frightening)* and also in *BadiDasaHola (It is very painful).* Perhaps along with *katav* came also the shattering of their life in general, death in particular, but over and above *katav* that engulfed them like a 'tufan' (heavy storm) eroded their hope and aspirations of in the first instance. The incapabilities of the community to fall back to normal life is expressed as '*budhinahikhe*' (we do not have intelligence' '*hum log hoshiarnahikhe*' (we are not intelligent).'*eh hamarabhagyaba*' (this is our fate), '*sirfhamara pet bachgaye'* (only our stomachs are left), etc. Recovery from the immediate impacts of the hazard which turned out to be a disaster meant that they had to restart again.

Musahars experienced utter helplessness and powerlessness as they had to encounter *katav* mostly as 'unexpected' and therefore, unprepared. The immediate powerlessness is felt mostly because of the loss of land and belongings and are pushed away from their familiar and owned places to a 'no man's land'. *Katav*---a hazard that turned out to be a disaster occurred in a particular context of embeddedness, which included factors such as ecological, social, political and economic. In the everyday context, there were structures, agencies, institutions, policy frameworks, ideologies, social capital and status with their caste identity that influenced and determined their identity and experience at the occurrence of floods, katav and multiple displacements. Pre-existed, present and post social structures and systems had a major role to play in shaping their experiences. Katav and its dehumanising constituent elements had been by-products of the many inequalities existing in a society and the impacts are asymmetric due to various contextual realities.

In the case of Musahars affected by *katav*, there was a pre-existing vulnerability and liminality arising out of many factors including their location in geographically vulnerable areas. Their houses were precariously close to the flood-prone areas, as they occupied space that was not generally preferred by dominant groups. 'Geographies of inequalities' and socio-economic compulsions pushed them to the fringes, making them more prone to being continued victims of recurrent floods, *katav* and multiple displacements. Being pushed to the geographical 'peripheries' meant on one hand having to face the recurring floods, katav and multiple displacements. There was pre-existing chronic poverty, assetlessness and marginalisation. Even before encountering the hazard events per se, there was a dominance, exploitation and voicelessness experienced by the community, especially being in that location.

It was a situation beyond their control, an experience of 'horizonlessness'. There was a liminality, vulnerability and structures and systems interface which was expressed as, "*Badi Dasa Hola (it is very painful), Khub paresani bhayil (it was so difficult).* At this stage there were interactions of the agencies and structures with the Musahars. If agencies like the government, social systems and networks were responsive and proactive, the intensity of their experience would have been lesser. But the institutions and structures ignored Musahars to a large extent whereby they continued to a liminal stage, a state of ambiguity, suffering and confusion, not knowing which direction to follow. The vulnerable and liminal situations put together, they experienced larger 'horizonless-ness'. The more vulnerable they were, the more intense the experience of liminality. As expressed, "*Kate gaye aur hatte rahe (When one place was eroded, we kept moving to another).* Their experiences of multiple displacement (some families have been displaced as many as five times) were also reflected in various phrases, such as *Nala-nuli hote hote* (wading through several gutters or natural drain channels), *kat-thei gaye aur hat-thei rahe* (the river kept eroding, and we kept moving), *bandh ki patari par zindagi* (life on the embankment as though on a log of wood), etc.

Frequent displacement and relocation had rendered the Musahars without an identity or address, both essential for claiming state welfare schemes. Suddenly they felt like, '*Sadhu ka kutha na ghar ka, nag ghat ka' (The seer's dog neither belongs to house nor to the banks*). They faced this state of ambiguity, grief, a sense of loss, depression, apathy and further confusion as in the liminality state. Their ambiguity was mostly due to their loss of an identity, the experience of which got more intensified as they moved from place to place with their dismantled huts. There was a resource asymmetry that existed as pre/present/and post-vulnerability factor that in turn increased their proneness to the hazard event and the same increasing their exposure. Such a resource asymmetry compelled the Musahars to locate in and occupy geographically risk-ridden and vulnerable areas, which always left them with the possibility of being hit by the said hazard events. This perpetuated their vulnerable conditions in the pre-present and post disaster context. This asymmetry was caused by and further perpetuated when they were interacting with the agencies, state, ideologies, institutions and caste factors. The 'embeddedness' context in its negative impacts could exacerbate the already existing risks and vulnerabilities. The structures that were rooted, perpetuated and manifested in injustice, exploitation and oppression did not allow them to have proper access to land anywhere. The poor socio-economic capital of the community was taken advantage of, leaving them in a challenging situation where they could not raise their voice in spite of what they had to undergo.

5. Conclusions

The aim of the study was primarily to explore the context of *katav* as experienced by the Musahars. It also tried understand their lived experiences and analyses the vicious cycle of vulnerability in the disaster context. The study also examined the Musahars' experience of the State, in living with *katav* and asserting their rights and entitlements. The findings of this challenge the existing broad-based approach to disasters, which are primarily focused on the physical event, disruption and immediate consequences. This approach is limited mostly to the rescue, relief and some general rehabilitation without regard to the need for a longer-term settlement, particularly of communities that are at a constant risk of being displaced multiple times. Hence, there is a need to focus more on the social dimensions and pre-existing vulnerabilities of the affected communities, especially in the assessment of impact, and planning for support. The parameters for measuring vulnerability, which are widely used by governments, are generic in nature (number of women, children, elderly, differently-abled, etc.). They do not include communities that are severely disadvantaged and for whom targeted interventions are necessary. There is a need to go beyond the 'conventional' ways of mapping hazard and vulnerability assessments with focus on intensity, extent, impacts, coping and adaptation capacities, and mechanisms. Paying attention to the specific contexts and communities beyond quantifiable terms is very significant. Adaptation effectiveness will depend on how 'effectiveness' is framed and looked at, which will significantly impact adaptation implementation and outcomes. Further, it depends on the role of people impacted by river erosion and how they exercise influence in adaptation decision-making.

The findings of this study of the Musahars' experience of *katav* challenges the existing broad- based approach to disasters which are primarily focused on the physical event, disruption and immediate consequences. This approach is limited mostly to the rescue, relief and some general rehabilitation without regard to the need for a longer-term settlement, particularly of communities that are at a constant risk of being displaced multiple times. Thus, there is a need to focus more on the social dimensions and pre-existing vulnerabilities of the affected communities, especially in the assessment of impact, and planning for support.

The parameters for measuring vulnerability, which are widely used by governments, are generic in nature (number of women, children, elderly, differently-abled, etc.). But these do not include communities that are severely disadvantaged and for whom targeted interventions are necessary. There is a need to go beyond the 'conventional' ways of mapping hazard and vulnerability assessments with focus on intensity, extent, impacts and coping capacities and mechanisms. A comprehensive post- disaster need assessment is a need of the hour. Paying attention to the specific contexts and communities beyond quantifiable terms is very significant. Assessing vulnerability to climate change is important as it helps to understand climate risks and provides information, on the location-specific measures to be taken to adapt to climate change. Hence, a vulnerability assessment is the first step in adaptation planning. The analysis also helps the in understanding the major drivers of vulnerability and target the adaptation actions accordingly (DST and SDC, 2020). A targeted approach would help in enhancing the efficiency and effectiveness of adaptation actions.

The experiences of the Musahars show that there is a need to consider the relative vulnerability of various communities so that interventions and responses are targeted for optimum effectiveness. There is also a need for developing a Disaster Risk Reduction (DRR) and resilience framework for the poor and marginalised, as well as risk-informed social protection for the marginalised and poor, especially the 'excluded' groups. There was an 'executive order' of the state government to cater to the resettlement of the *katav*-affected and displaced people, which was more in the nature of a 'resolve'. But such a 'resolve' may not be as binding as a sturdy and effective policy, legal and institutional framework. Effective legal framework would fix better responsibilities. Thus, such framework is the need of the hour. Having a proper system to plan, monitor, implement and review is very important. Instead of the half-hearted measures of a 'resolve', it is very important to bring forth such articulations under a legal framework whereby, the affected citizens could approach it further from a 'rights' and 'legal' perspective, rather than left to the mere good intentions and generosity of the implementing authorities. Bringing such a framework is a need of the hour. Interventions in a disaster must be seen as a right of the people affected and as a duty by the state rather than as the 'generosity' of the politicians and bureaucrats.

For effective documentation for disaster response and planning, it is essential to listen, document and bring to the forefront voices from the 'periphery' and 'subaltern'. This will help in the development of tools and mechanisms for assessment of the experiences of vulnerabilities of such people so as to plan purpose-specific policies, programmes and intervention models for them. There is a need to build capacities, institutional and structural mechanisms to address social vulnerability among diverse groups. Community participation must be strengthened through capacity building and empowering local leadership from within the communities. This is possible with investing in their social and human capital enhancement.

There is also an urgent need to address the vulnerability of the Musahars because of which they cannot recover from disasters. Thus, ad hoc and reactive actions must be replaced by long-term and sustainable plans and programmes for their resettlement in safe places including locations where livelihood options are available. As the district disaster plan of West Champaran recognises, Bihar could include '*katav*' as a unique disaster hazard that calls for immediate attention. The state must build accounting structures and mechanisms to monitor the progress of the resolution (2003) for settling the *katav*-affected people. Livelihood is a major concern for the *katav* affected. A comprehensive and sustainable livelihood plan, which takes into consideration the local knowledge and skills would be of greater relevance. There is a need for a proper warning and information sharing system whereby, the people could be prepared for sudden *katav* or its aftermath.

The findings clearly show that because of existing social vulnerability, which are determined by caste and other socio-economic factors, the *katav* disaster[1] is not a one-time event. Rather, it is a 'life-long' or 'life-changing' event the Musahars are condemned to live with. The study also shows that the Musahars have been among the worst affected people due to recurrent floods, *katav* and multiple dislocations. The repercussions are long-term and often last their lifetimes. There is an urgent need to do a proper mapping at various levels. First of all there needs to be complete mapping/survey of the people affected by *katav* for the past 50 years, at the Gandak River basin area. This survey should be guided by the concerned disaster management department of the district. Meanwhile the revenue and land department could identify the land belonging to the government, but under the illegal occupation of some landed gentry. Homestead land could be allotted to people who are affected. This mapping also would help identifying the disaster affected vulnerable communities and provide them with proper identity card to avail special provisions.

Any planning for the resettlement of the *katav* affected should include the aspects of socio- cultural identity. Recognition and redistribution both should find a place in any planning and execution. A comprehensive plan for resettling the *katav* affected would mean, for allotment of homestead land, integrating with the existing schemes and programmes, including housing. For a community where the community feelings are stronger and a need to stay together they could be staying together in clusters. What is very important is proper customised monitoring systems and structures that comprehensively cater to the concerns of the people. Political representation and power structures play a very important role. Therefore, concerted engagements to ensure representation of Musahars in various decision- making bodies is very significant. There is a greater need for integrating community development approach with community organisation approach. Building human resource capital and building leadership of the marginalised communities would be crucial in addressing the overall issues. Instead of approaching the disaster interventions from an immediate crisis perspective and water-tight compartment approach alone, there is a greater invitation to identify the root causes. There is a need for preventive, curative and rehabilitative approach.

What seems to stand out at the onset of such events is that the people who already suffer from marginalisation and exploitation struggles more than others to cope with and recover from the occurrence of *katav*. Government parameters for the katav affected seem to be still very broad and general, occasionally recognising them as mere 'flood victims' who deserve minimum relief occasionally. The lived experiences of the Musahars assert the need for a specific context-based approach, taking into consideration a social transformative agenda in the planning and implementation of DRR and DRM programmes.

The 'global goal on adaptation' (UN 2015, Article 7.1) aims to ensure an 'adequate adaptation response', thereby enhancing adaptive capacity, strengthening resilience, and reducing vulnerability to climate change, which ultimately contribute to sustainable development. We need to go beyond the normative views on adaptation outcomes, from one matrix to another delivering a set of benefits, and zero in on what really contributes to effective outcomes 'leaving no one behind' when confronted with vulnerabilities. The proposed transformative framework calls for Community-based adaptation (CbA), which "is a bottom-up approach that focuses on increasing the participation and agency of vulnerable communities in adaptation prioritization and implementation". It should be a process that fundamentally changes human thinking and practices in the face of climate change and overtly challenges the power structures that generate vulnerability" (Chandni *et al.*, 2021). Few *et al.* (2017) mooted two main interpretations of this: 'transformational adaptation' and 'transformative adaptation'. The former argues that 'incremental' adaptation that seeks to modify existing human-environment relations will be insufficient for society to manage the consequences of climate change. Hence, fundamental changes on how we utilise and manage resources, where and how people live, what systemic limitations are imposed on people's agency, and how we interact with nature are required in order to avoid catastrophic impacts. The second interpretation, which is 'transformative adaptation', put forward that adaptation requires, and presents an opportunity for society to fundamentally challenge and change the relations of social power and the socio-structural constraints that deepen vulnerability, while at the same time reducing the negative consequences of climate change impacts.

The UN Sendai Framework for Disaster Risk Reduction 2015-2030 (UN, 2015) is of immense relevance in this context. It recognised the role of all stakeholders, including women, persons with disabilities and their organisations, older persons, migrants, academia, scientific and research entities and networks, business, professional associations and private sector financial institutions, and media. Accordingly, the Framework stated that women and their participation are critical to effectively managing disaster risk and designing, resourcing and implementing gender-sensitive disaster risk reduction policies, plans and programmes; and hence, adequate capacity building measures need to be taken to empower women for preparedness as well as to build their capacity to secure alternate means of livelihood in post-disaster situations. Persons with disabilities and their organisations are critical in the assessment of disaster risk and in designing and implementing plans tailored to specific requirements, taking into consideration, inter alia, the principles of universal design. Recognising the years of knowledge, skills and wisdom of older persons, which are invaluable assets to reduce disaster risk, they should be included in the design of policies, plans and mechanisms, including for early warning. Migrants contribute to the resilience of communities and societies, and their knowledge, skills and capacities should be useful in the design and implementation of disaster risk reduction. Further, the Framework has proposed engagement of academia, scientific and research entities and networks, business, professional associations and private sector financial institutions, and media too in disaster risk reduction. As cited in Blackburn (2018) "Participatory development geographers have long argued that local action can be an engine for progressive social–political change, when situated within a wider commitment to a radical politics of development and an evolved, responsive governance framework. Participatory, plural, and deliberative governance is posited to have multiple benefits, including promoting societal equality and fairness through plural deliberation and representation, taking account of local and traditional forms of knowledge, and increasing the legitimacy of decision-making". Transformations in social, economic, political and social–ecological systems are required to forge positive development futures that protect local interests, by taking into account local knowledge and voices, and provide inclusive, equal and just protection for extremely vulnerable communities like Musahars in a disaster setting like *katav*.

REFERENCES

1. Ahmed, A. A. and Fawzi, A. (2009), Meandering and Bank Erosion of the River Nile and its Environmental Impact on the Area between Sohag and El-Minia, Egypt. *Arab Journal of Geosciences*, 4, 1–11. doi: 10.1007/s12517-009-0048-y.
2. Barton, A. H. (1969), "Communities in Disaster", *A Sociological Analysis of Collective Stress Situations.* NY: Doubleday.
3. Belcher, J. C., and Bates, F. L. (1983), Aftermath of Natural Disasters: Coping through Residential Mobility. *Disasters*, 7 (2), 118–128.
4. Blackburn, Sophie (2018), "What Does Transformation Look Like? Post-Disaster Politics and the Case for Progressive Rehabilitation", *Sustainability*, 10, 2317; doi:10.3390/su10072317.
5. Bolin, R., and Stanford, L. (1998), "The Northridge Earthquake: Community-based Approaches to Unmet Recovery Needs. *Disasters*, 22 (1), 21–38.
6. Bradbury, J. (2013), *Lower Gordon River Erosion Monitoring, Tasmanian Wilderness World Heritage Area. Nature Conservation Report Series 2013/08.* Retrieved from http://dpipwe.tas.gov.au/Documents/LGR%20 Monitoring-2013-14.pdf
7. Briaud, J. L., Chen, H.-C., Chang, K. A., Chung, Y. A., Park, N., Wang, W., & Yeh, P. H. (2007). *Establish guidance for soils properties-based prediction of meander migration rate. Technical report FHWA/TX-07/0-4378-1.* Retrieved from http://tti.tamu.edu/documents/0-4378-1.pdf
8. Brouwer, R., Aftab, S., & Brander, L. (2006). *Socio-economic vulnerability and adaptation to environmental risk: A case study of climate change and flooding in Bangladesh. PREM working paper 06/01.* Retrieved from http://www.prem-online.org/archive/16/doc/PREM06-01.pdf
9. Cannon, T. (1994), Vulnerability analysis and the explanation of 'natural' disasters. *Disasters, Development and Environment*, 1, 13–30.
10. Chacko, Siji (2019), "Eroding Lives of the Musahars. Lived Experiences in the Context of Floods and Rive Bank Erosion", *Social Action*, 61 (1).
11. Chandni Singh, Soundarya Iyer, Mark G. New, Roger Few, Bhavana Kuchimanchi, Alcade C. Segnon and Daniel, Morchain (2021), "Interrogating 'Effectiveness' in Climate Change Adaptation: 11 Guiding Principles for Adaptation Research and Practice", *Climate and Development*, DOI: 10.1080/17565529.2021.1964937.
12. Dragicevic, S., Tosic, R., Stepić, M., Živković, N., and Novković, I. (2013). Consequences of the river bank erosion in the southern part of the Pannonian Basin: Case study – Serbia and the Republic of Srpska. *Forum Geografic*, *XII*, 5–15. doi: 10.5775/fg.2067-4635.2013.008.i
13. DST and SDC (2020), *Climate Vulnerability Assessment for Adaptation Planning in India Using a Common Framework*, Indian Institute of Technology Mandi and Indian Institute of Technology Guwahati, Department of Science and Technology (DST), Government of India, and Swiss Agency for Development and Cooperation (SDC), New Delhi.
14. Few, R., Tebboth, M. G. L., Morchain, D., Spear, D., Mensah, A., and Bendapudi, R. (2017), Transformation, adaptation and development: Relating concepts to practice. Palgrave Communications, 3(1), 17092. https://doi.org/10.1057/palcomms.2017.92

15. Gilbert, C. (2005), "Studying Disaster: Changes in the Main Conceptual Tools", In *What is a Disaster?* Routledge, pp. 21–30.
16. Gorlick, B. (2007), Environmentally displaced persons: A UNHCR perspective. *Environmental Refugees: the Forgotten Migrants", Discussion Event, New York.*
17. Granovetter, M. (1985), "Economic Action and Social Structure: The Problem of Embeddedness. *American Journal of Sociology*, 91(3), 481–510.
18. Hewitt, K. (1983), "The Idea of Calamity in a Technocratic Age. *Interpretation of Calamity: From the Viewpoint of Human Ecology. Allen & Unwinn, Boston*, 3–32.
19. Hughes, A. O., & Prosser, I. P. (2003). *Gully and riverbank erosion mapping for the Murray-Darling Basin. Technical report 3/03, March.* Canberra, Australia: CSIRO Land and Water.
20. Jones, W., Eldridge, J., Pedro Silva, J., and Schiessler, N. (2007), *LIFE and Europe's rivers: Protecting and improving our water resources.* Luxembourg: European Communities.
21. Kreps, G. A. (1985), "Disaster and the Social Order", *Sociological Theory*, 3(1), 49–64.
22. Mabin, M. (2007), *Waikato River Bank Erosion Study Volume 1.* Retrieved from www.waikatoregion.govt.nz/PageFiles/11266/Waikato%20River%20erosion%20 study.pdf
23. Macfall, J., Robinette, P., and Welch, D. (2014), "Factors Influencing Bank Geomorphology and Erosion of the Haw River, A High Order River in North Carolina, since European Settlement. *PLoS One*, *9*. doi: 10.1371/journal. pone.0110170.
24. NRSC (2020), *Flood Hazard Atlas - Bihar - A Geospatial Approach,* National Remote Sensing Centre (NRSC), Indian Space Research Organisation, Dept. of Space, Govt. of India in Association with National Disaster Management Authority Ministry of Home Affairs, Govt. of India, Disaster Management Department, Govt. of Bihar, Patna & Bihar State Disaster Management Authority, Government of Bihar, Patna.
25. Pelanda, C. (1982), *Disaster and order: Theoretical problems in disaster research.* Sociology of Disasters Department, Institute of International Sociology.
26. Perry, R. W. and Quarantelli, E. (eds.). (2005), *What is a Disaster: New Answers to Old Questions.* New York, International Research Centre on Disasters.
27. Quarantelli, E. L. (ed.). (2005), *What is a Disaster?: A Dozen Perspectives on the Question.* Routledge.
28. Roper-Lindsay, J. (1994), "Tales of the Riverbank – Examples of Bank Restoration on Urban Rivers in Restoration of Aquatic Habitats", In K. J. Collier (ed.), *Restoration of Aquatic Habitats* (pp. 125–143). Wellington, New Zealand: Department of Conservation.
29. Sekac, T. and Jana, S. K. (August 2014), Change detection of Busu river course in Papua New Guinea- impact on local settlements using remote sensing and GIS technology. *International Journal of Scientific & Engineering Research, 5,* 891--899. https://www.researchgate.net/publication/304216714
30. Singh, S. R., Eghdami, M. R., and Singh, S. (2014), *The Concept of Social Vulnerability: A Review from Disasters Perspectives.*
31. Smith, A Oliver (1996), "Anthropological Research on Hazards and Disasters. *Annual Review of Anthropology*, 25(1), 303--328.
32. Smith (1996), Anthropological Research on Hazards and Disasters. *Annual Review of Anthropology*, *25*(1), 303--328.
33. Smith, N. (2006), "There's No Such Thing as a Natural Disaster", *Understanding Katrina: Perspectives from the Social Sciences*, 11.
34. Tuhin K Das, Sushil K Haldar, Debaprasad Sarkar, Marion Borderon, Stefan Kienberger, Ivy Das Gupta, Sangeeta Kundu and Debarati Guha-Sapir (2017), Impact of Riverbank Erosion: A Case Study, *Australasian Journal of Disaster and Trauma Studies*, Vol. 21, No. 2.
35. UN (2015), "Sendai Framework for Disaster Risk Reduction 2015-2030", Third UN World Conference in Sendai, Japan, 18 March 2015.

CHAPTER 15

Flood Inundation Mapping Using C-band Synthetic-aperture Radar and Random Forest Algorithm: A Methodological Basis

Swapnil Singh Parihar
Defence Geoinformatics Research Establishment,
Defence Research and Development Organisation,
Chandigarh 160036, India
swapnilsparihar@gmail.com

Shafique Matin
Madhya Pradesh Agency for Promotion of Information Technology,
Bhopal 462011, Madhya Pradesh, India
shafiquematin@gmail.com

Abstract

The synthetic aperture radar (SAR) sensors are sensitive to parameters such as scattering mechanism, surface roughness, the dielectric of the target feature, and incident angles. The usability of C-band Sentinel 1 satellite with co-polarized (VV) and Cross-polarized (VH) data proved to be sufficient for accurate flood monitoring. The focus of this paper is to discuss the utility of texture analysis as a powerful input for random forest classifiers to demonstrate the application of SAR data in dynamic flood mapping.

Sentinel 1 interferometric Wide (IW) mode Level-1 ground range detected (GRD) product is pre- processed for this research. A refined lee filter was used for speckle filtering to preserves the texture information and edges. The geocoding was carried out by the Range Doppler orthorectification method using the Shuttle Radar Topographic Mission (SRTM) 3 Sec digital elevation model (DEM) and bilinear interpolation method for resampling. For this study, 11 GRD products consisting of a pre- event, flood, and post-flood scenes for the year 2019 were processed to identify the dynamic flood extent of the Kosi River in Bihar, India. In the present study, Grey Level Co-occurrence Matrix (GLCM) stack proved to be a suitable input dataset for supervised random forest classification.

The proposed approach demonstrates that a random forest classifier is a robust ensemble technique and gives satisfactory results for flood inundation mapping using microwave remote sensing data. The accuracy assessment, performed for all the classified images resulted in an overall accuracy was found to be greater than 88.96 %. The average cohen's kappa coefficient is 0.82, which indicates substantial agreement. The study also iterates that the adopted methodology can be useful for disaster management applications. However, since the microwave satellites are sensitive to the dielectric of soil, the overestimation of flood extent due to surrounding saturated/waterlogged soil has to be identified rigorously for accurate results. The study concludes that the variation of GLCM descriptors as input feature bands could affect the classification differently, which is primarily associated with the variation of the backscatter throughout the year. In addition, the availability of a high spatial and temporal resolution SAR sensor would better map the dynamics of disaster events like floods, forest fire.

Keywords: Flood inundation, Random Forest, Sentinel 1, Kosi River, GLCM texture analysis.

1. Introduction

In India, monsoon floods are annual events in various parts of the country having proximities to the major river channels such as Ganga, Brahmaputra, Yamuna, and Narmada. Floods in the Kosi River have become annual events causing the loss of lives and resources each year. Kosi River is also called the Sorrow of Bihar due to the recurrent destruction caused during the monsoon season. These floods have altered the fluvial system over the years causing heavy siltation. The Kosi mega fan is prone to floods during monsoons resulting in temporary and permanent waterlogged patches spreading across the course of the river. Mitigation of flood hazards in Bihar is primarily response-driven with no preparation to avoid the loss of life and property (Pandey et al., 2010). Kosi River has shown significant variability in flood dynamics both spatially and temporally. Hence, the synoptic mapping of the flood can help in determining the extent of the disaster and quantifying the impact.

Before the launch of the microwave satellites RISAT 1 and Sentinel 1, the temporal frequency of available satellites was not satisfactory to map the dynamic extent of the flooding. The research majorly relied on integrated use of different datasets such as field surveys, gauge station observations, past data, high-resolution optical imagery, and flood risk maps for hydrological analysis of the inundation.

The inability of optical imaging satellites to penetrate the cloud cover limits the application of optical remote sensing in flood monitoring. The SAR sensors have quickly gained popularity amongst researchers due to their all-weather, day-night capabilities. The freely available Sentinel 1 data at a medium spatial resolution of 10 m is useful to map the flood extent. The revisit of sentinel 1A is 12 days while the combined revisit of Sentinel 1A and Sentinel 1B constellation is 6 days. Microwave satellites have been used extensively in the past few years and have found applications in domains such as wetland monitoring, soil moisture studies, DEM generation, land subsidence, and disaster management applications such as earthquake mapping, volcanic eruptions, and flood mapping, etc.

The backscatter from the SAR sensors is affected by incidence angle, polarisation, surface roughness and di-electric (Singh et al., 2020). The advantage of the SAR sensors in capturing the backscatter based on scattering mechanism from a target can be employed to delineate water from non-water pixels. The specular scattering from the water surface separates the open water from the other targets in a scene. The water appears dark in the SAR imagery since there is less or no return pulse received at the satellite sensor from the open water surfaces.

Sinha et al. (2008) computed an index for flood risk and developed a flood map using the analytical hierarchy process (AHP) by using multiple datasets including DEM, LISS III imagery, and topographical maps. For short-lived floods, the use of data with a high temporal resolution is well suited. Hence, the use of the Soil Moisture Active Passive (SMAP) mission gives the ability to monitor the soil moisture with revisit ability of 2-3 days operating in L Band (Rahman et al., 2017). The high temporal resolution SMAP soil moisture product comes at the cost of coarse spatial resolution. Using SMAP soil moisture data for flood analysis can be done by comparative analysis of moisture in flooded and non-flooded conditions. Tsyganskaya et al. (2016) used TerraSAR-X and ancillary data to apply the fuzzy logic-based algorithm in the change detection method for flood mapping. While Brivio et al. (2002) proposed an approach to detect floods using ERS-1 SAR images in conjunction with the digital topographic data. Tripathi et al. (2020) used Sentinel 1 SAR data with MODIS NRT (Near Real-Time) flood data to apply a threshold/ binarization technique for creating flood inundation maps. Thresholding [Bhatt et al., 2013; Cheng et al., 2019] is a quick and extensively used mapping technique but in some cases, it may overestimate the flood by classifying the saturated soil as flood pixels or underestimate the flood by leaving the inundated vegetation areas as unclassified. The thresholding technique is based on the principle that the water bodies have low radar backscatter compared to the surrounding targets. Agnihotri et al. (2019) stated that the co-polarised band of Sentinel 1 has greater accuracy than the cross-polarized (VH) band while using a threshold method. Martinis et al. (2017) proposed an improvement of the automatic Sentinel 1 flood Service (S- 1 FS) in arid areas by adding an exclusion layer built to remove the areas with low backscatter, which may result in overestimation of flood extent. Vanama et al. (2019) mapped the Chennai flood area using Normalized Change Index (NCI) by using co-registered pre-flood and post-flood images. The more recent studies are making use of machine learning algorithms and automation to detect floods without training data.

The texture analysis may be helpful to map flooded areas since spatial characteristics change due to differences in backscattering (Wakabayashi et al., 2019). Kupidura (2019) compared several texture analysis methods for assessing classification accuracy. He concluded that while some GLCM indicators improve the classification accuracy, some indicators could reduce the accuracy of the classifiers.

Due to the soil properties of the Kosi alluvial fan, the fine sediments have high moisture content during monsoons, which makes it difficult to identify the flood from the waterlogged soil. In SAR data, the waterlogged areas produce the backscatter similar to the low-intensity backscatter from the flooded areas. Hence, compared to the thresholding method, the random forest algorithm serves as a robust machine learning technique that learns from the training datasets and the feature bands generated from the GLCM image analysis to map the flooded area.

The pre-processed scenes were used to perform the texture analysis, which gives multiple bands based on the randomness, statistical properties, and general contrast of the backscatter. It helps to increase the feature bands, which were used in the random forest algorithm. The supervised binary classification was done based on the feature bands, training datasets, and the number of trees. The primary objective of the study is to demonstrate the utility of the methodology in the delineation of the river. The use of texture analysis in conjunction with the random forest algorithm can be helpful in the classification of flood boundaries.

2. Study Area

River Ganga flows across the state of Bihar and one of the major tributaries (in terms of discharge) of Ganga is Kosi. The Kosi River forms a megafan in the eastern Gangetic plains, also known as "saptKosi" due to its seven tributaries. It is a transboundary river flowing through the southern plains in Nepal before entering India in Bihar. It branches out and finally meets Ganga near Kursela. The alluvial fan built by the river is one of the largest in the world. The fan is approximately 180 km long and 150 km wide having shown a lateral shift of about 140 km in the past 250 years along the faults (Agarwal et al., 1992). The shift is attributed to the geological faults in the region causing the migration of the river from east to west. The shift of the river has caused numerous floods along its changing course over the years.

Several measures were taken to contain the flood in past years. The construction of the hydraulic structure in Nepal called "Kosi Barrage" was done with the intent of flood control, power generation, and disaster risk reduction. In 1955, engineers have constructed an embankment along the course of the river to contain the rapid westward shifting and flood risk management. A major embankment breach in the year 2008 in Nepal caused the shift of the river 180 km to the east resulting in a devastating flood in Bihar. The flood resulted in a high amount of silt deposition in the flood plain. Silt deposition leads to the rise of the riverbed and hence making the river more prone to flood events.

The Kosi River provides the water necessary for the irrigation in the surrounding areas but the yearly heavy rainfall causes high runoff in monsoons and permanent water logging (Pandey et al., 2010). The poor maintenance of the embankment and breach of the barrage has also been the reason for the flash floods in past years.

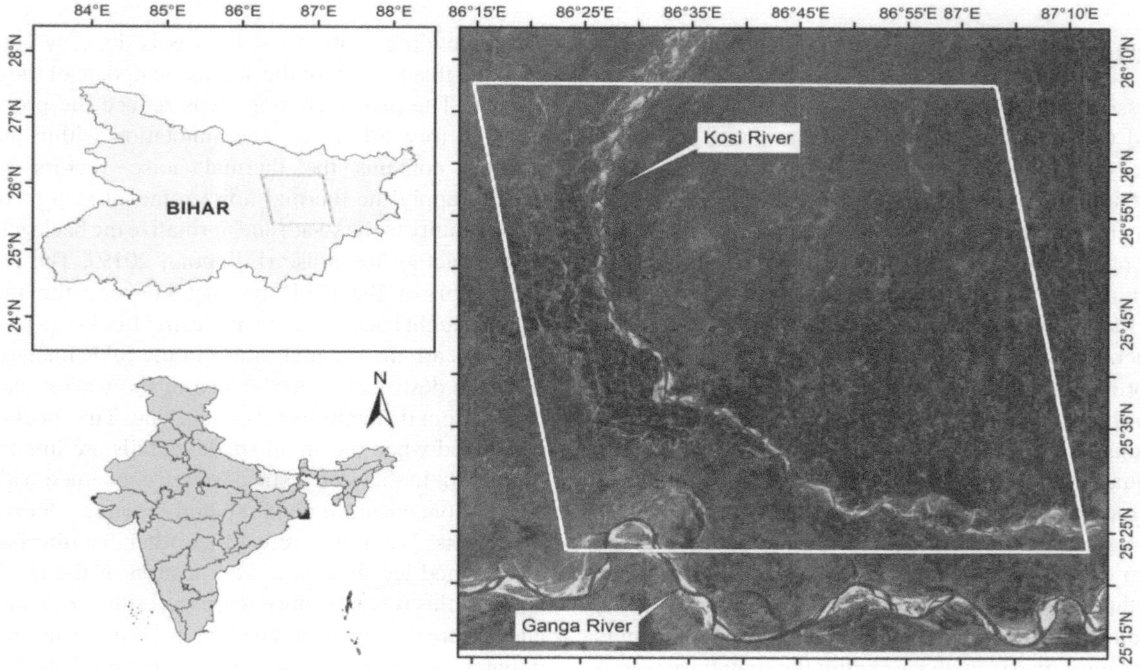

Fig. 15.1 The study area is in Bihar state (India). The red bounding box indicates the study area overlaid on Sentinel 2 colour composite.

3. Dataset

The launch of microwave satellites has improved spatial and temporal imaging for various applications. The freely available C-Band Sentinel 1 SAR satellite operates at a frequency of 5.35 GHz. The acquisition mode of the data is interferometric wide (IW) swath. The Level 1 Ground Range Detected (GRD) multi-looked products were used for the study. The Sentinel 1 satellite mission consists of a constellation of two satellites; Sentinel 1A and 1B in the same orbital plane having a combined revisit of 6 days. However, for our study area, only Sentinel 1A data was available hence the temporal resolution was 12 days.

The temporal dataset was used to analyze the pre-monsoon, monsoon, and post-monsoon dynamics of the Kosi River and to map the extent of flood over time. The pre-processing necessary for the SAR imagery is carried out in SNAP 8.0 software. 11 images were used for the analysis of the flood dynamics for the year 2019.

Table 15.1 Details of Sentinel 1 dataset used for this study

Acquisition Date	Polarization	Resolution	Orbit	Condition
12-04-2019	VH+VV	10m	Ascending	Pre-Flood
11-06-2019	VH+VV	10m	Ascending	Pre-Flood
05-07-2019	VH+VV	10m	Ascending	Pre-Flood
17-07-2019	VH+VV	10m	Ascending	Flood
29-07-2019	VH+VV	10m	Ascending	Flood
10-08-2019	VH+VV	10m	Ascending	Post-flood
22-08-2019	VH+VV	10m	Ascending	Post-flood
03-09-2019	VH+VV	10m	Ascending	Post-flood
15-09-2019	VH+VV	10m	Ascending	Post-flood
27-09-2019	VH+VV	10m	Ascending	Post-flood
09-10-2019	VH+VV	10m	Ascending	Post-flood

4. Methodology

The processing was carried out for both Single Look Complex (SLC) and GRD products to assess if the phase increased the accuracy of the classification. The trial resulted in no significant improvement in the use of SLC products. SLC product increased the computational complexity while the preserved phase information in the SLC product did not add significant improvement to the results of the study. Therefore, GRD level 1 sentinel 1A product was used for the delineation of the river. Sentinel 1 GRD product contains the ground range coordinates that are obtained by projecting the slant range coordinates onto the earth ellipsoid model. The GRD product consists of square pixels representing the amplitude and the square of amplitude i.e.; intensity thereby eliminating the need for multi-look processing of the GRD scene. The proposed method has a resemblance to past researches for the delineation of a feature. The adopted methodology is based on using the outputs of texture analysis as an input in the supervised random forest algorithm. The different indicators in co-occurrence matrix analysis exploit the spatial variations in the backscatter and texture of the SAR imagery.

The geocoding of the SAR images is done by applying the orbit file that allows for the automatic update of the orbit state vectors. The pre-processing steps reduce the probability of carrying forward errors. The annotation within the product datasets contains the thermal noise vectors which are used to apply the thermal noise removal step to reduce the discontinuity in the swath and normalize the backscatter signal by subtracting the noise (Filipponi, 2019). The radiometric calibration of the GRD product converts the intensity to sigma naught backscatter values using Look-Up Table (LUT). Speckles are the granular noise in the SAR images resulting from the destructive interference of the waves when signals reflect from different targets in a scene. The speckle filtering is avoided when the small spatial details are important to be preserved. In this study, since we are concerned with the river delineation, losing a small spatial structure does not affect the results. We used a refined lee filter for filtering speckle. The refined lee filter is a modification of the lee filter. It is used for this research since it preserves linear features, edges, and texture information. The final preprocessing step is range doppler terrain correction. The objective of this final step is to compensate for the geometric errors such as topographic variations, the tilt of sensor, and distorted distances in the SAR images. This orthorectification is done using SRTM 3sec DEM with bilinear interpolation resampling method to obtain the terrain corrected image.

GLCM is a two-dimensional matrix of the second-order texture statistics which are widely used for feature extraction by analysis of the spatial distribution of grey level values (Zayed et al., 2015). The texture bands derived from the SAR imagery improve the random forest algorithm. Texture is a descriptive statistic hence it is difficult to standardise the ranges of the GLCM descriptors for different scenes.

The studies in the past have demonstrated that the use of GLCM derived texture features can help in classification algorithms (Caballero et al., 2020; Hu et al, 2018; Nkeumoe Numbisi et al., 2018; Gkanatsios et al., 2018). Random forest is an ensemble method, it requires numerous decision trees for accurate analysis. This is made possible by using GLCM texture analysis. The texture analysis is applied to the images to make use of the variation in the backscatter that records the spatial information in the scene. The GLCM feature bands exploit the information from the SAR backscatter signal based on the orderliness, contrast, and statistical relation of a pixel relative to its surrounding pixels. The GLCM analysis calculates the co-occurrence of the grey levels concerning spatial distance and in an orientation such as 0°, 45°, 90°,135° (Wurm et al., 2017). The textural descriptors derived from the GLCM image analysis such as homogeneity, entropy,

Table 15.2 The various Grey Level Co-occurrence Matrix (GLCM) descriptors with their formulae

GLCM Band	Mathematical Expression	Description
Homogeneity	$\sum_{i,j=0}^{n-1} \frac{P_{i,j}}{1+(i-j)^2}$	Measures uniformity in the image; Closeness of element from the GLCM diagonal
Entropy	$-\sum_{i,j=0}^{n-1} P_{i,j} \ln(P_{i,j})$	Measure of disorder/randomness of the backscatter intensity
Angular Second Moment (ASM)	$\sum_{i,j=0}^{n-1} [P_{i,j}]^2$	Measures localized homogeneity of distribution of grey levels
Contrast	$\sum_{i,j=0}^{n-1} P_{i,j}(i-j)^2$	Measures localized change in the grey level; directly proportional to grey level variability
Mean	$\mu_i = \sum_{i,j=0}^{n-1} i(P_{i,j})$ $\mu_j = \sum_{i,j=0}^{n-1} j(P_{i,j})$	Measure of average of grey level distribution of image
Variance	$\sigma_i^2 = \sum_{i,j=0}^{n-1} P_{i,j}(i-\mu_i)^2$ $\sigma_j^2 = \sum_{i,j=0}^{n-1} P_{i,j}(j-\mu_j)^2$	Measures the deviation of grey level from the mean; Sum of the square of the difference between intensity of a pixel and its surrounding pixels.
Maximum probability	largest $P_{i,j}$ entry in matrix	Largest entry in the GLCM matrix corresponding to the highest backscatter
Energy	$\sqrt{ASM}$	Square root of ASM; high value occurs when matrix is well ordered
Correlation	$\sum_{i,j=0}^{n-1} P_{i,j}\left[\frac{(i-\mu_i)(j-\mu_j)}{\sqrt{(\sigma_i^2)(\sigma_j^2)}}\right]$	Measure of linear dependency of grey levels on surrounding pixels

$P_{i,j}$ is the probability from the GLCM matrix, $P_{i,j} = \frac{T_{i,j}}{\sum_{i,j=0}^{n-1} T_{i,j}}$; where $T_{i,j}$ is the pixel value of cell i,j

Adapted from: Caballero et al., 2020; PS et al., 2016

dissimilarity, angular second moment, contrast, mean, variance for VV and VH polarization aided in increasing the number of decisions trees without classification bias. The spatial relation of the pixels was computed for a 5 X 5 moving window.

The focus of this study is to make an efficient random forest model for the classification by providing it with sufficient input to eliminate the redundancy of decision trees. The input descriptors were further increased in number by using non-linear filters and filters to enhance the edges and discontinuities of the features.

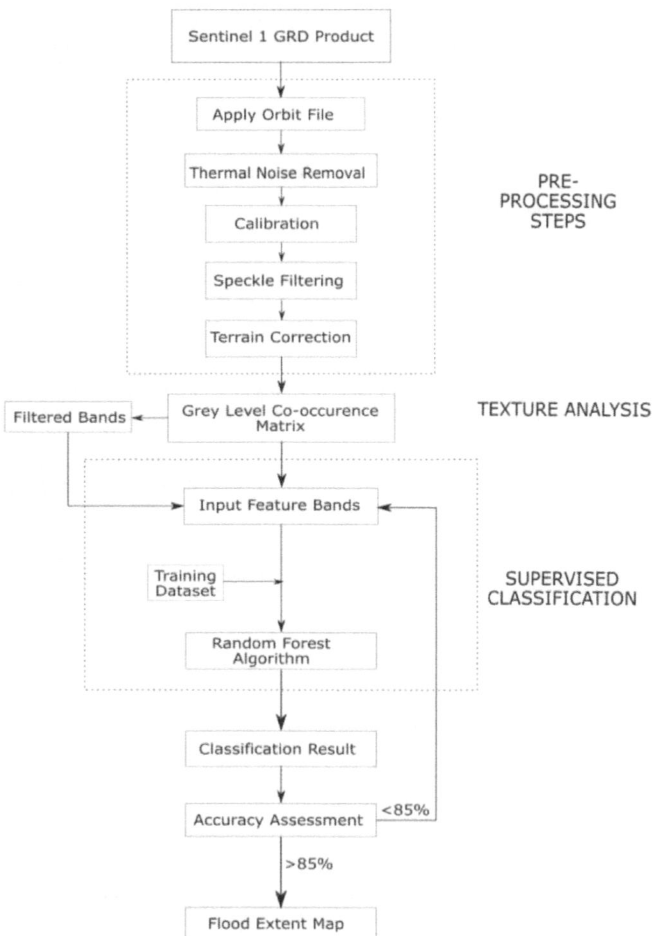

Fig. 15.2 Schematic of methodology adopted in this study for flood extent delineation

Random forest (Breiman, 2001) is an efficient tool for regression and classification. It is a non- parametric algorithm that works on an ensemble of decision trees where the class is assigned based on the majority voting technique. This method uses a large number of decision trees to determine or predict the class of unknown pixels (Fletcher, 2016). The method works on the bootstrap technique to create subsets of data from the training data selected randomly with replacement. Each decision trees vote for the class of the unknown pixel. The algorithm is generally treated as a black-box model where each subset tree is learning from the bagged data using randomly selected features. The random forest algorithm gives better results compared to a simple thresholding method and is efficient for quick and accurate post-disaster analysis. Relative to other algorithms, the random forest algorithm is less likely to overfit since it randomly selects the training data for decision trees. The overfitting of the random forest algorithm can be avoided by creating a greater number of trees. Although, it is found from this study that the accuracy of the model is not directly proportional to the number of trees. It is not possible to obtain the optimal number of trees for the high performance of the classifier (Mishina et al., 2015). The number and choice of input feature bands affect the learning of the model. The choice of input feature bands for the temporal analysis cannot be generalized as some texture bands either generate redundancy or deteriorate the classification by creating learning bias. The trial of combinations of different feature bands shows that the spatial similarity in the input feature bands may result in biased outputs. It was observed that not all the texture analysis bands are useful as input feature bands and can result in misclassification of the flood pixels. The training samples were taken for binary classification. The classification was carried out for pre-flood, flood, and post-flood conditions using a supervised random forest algorithm. The number of input feature bands for classification of all the scenes ranged from 18-26 bands by removing the bands based on visual discretion.

The accuracy assessment of the classification was done by creating a confusion matrix for the binary classification to calculate the overall accuracy and Cohen's kappa value. Some of the classification results were inaccurate as indicated by the overall accuracy and hence the iteration of the classification was done using a different set of input feature bands.

5. Results

The actual river extent can be seen from the pre-flood classified results. The 2 classified scenes from July 2019 indicate the increase in the discharge of the river due to the flood. The medium-resolution SAR dataset gives satisfactory results for the flood delineation and the method used in this study proves to be a robust technique for the classification.

It is observed that some areas along the river were misclassified into flood class for the monsoon scenes. The probable cause can be the GLCM feature bands that accentuate the saturated soil pixels to resemble the river water pixel or in some cases, the roughness of the water surface having similarity to the surrounding roughness.

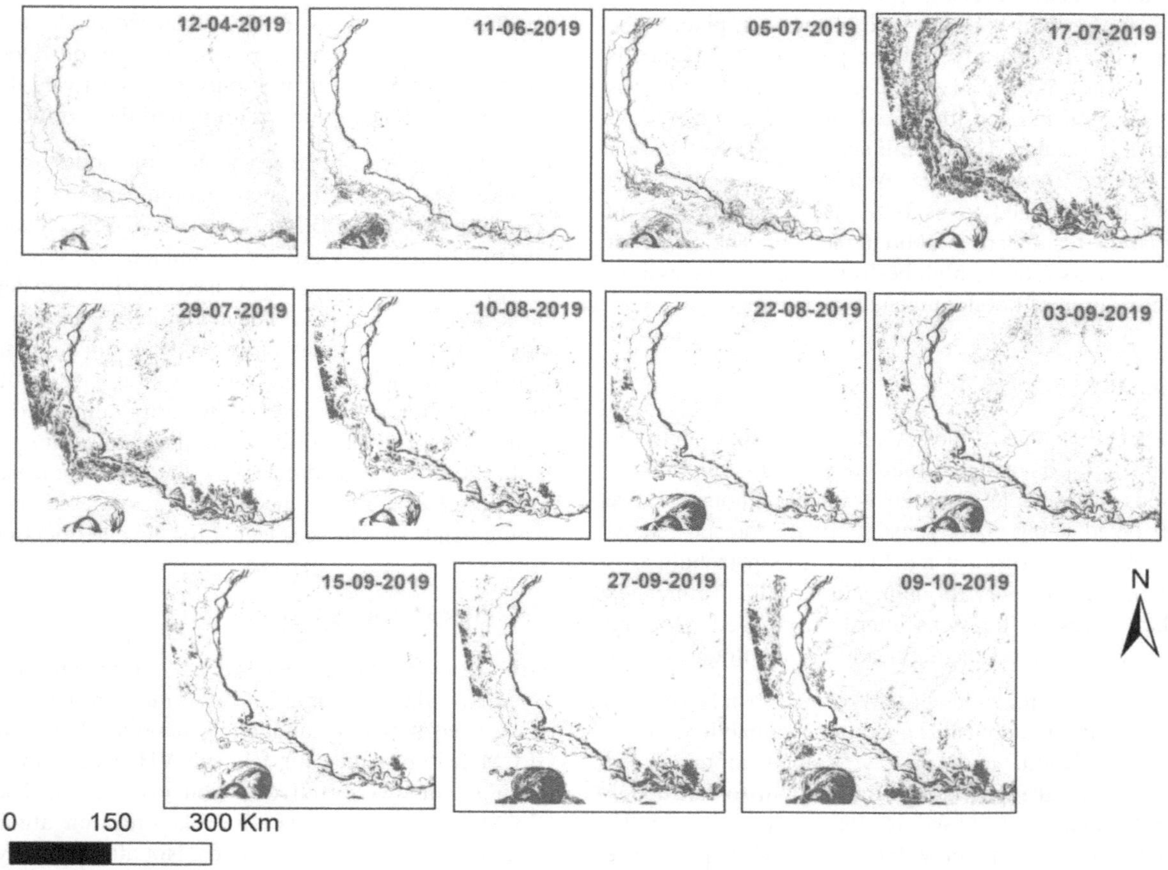

Fig. 15.3 The classification results of the Random Forest algorithm shows the temporal change in the flood extent of the Kosi river in the year 2019

The Kosi river is lined with an embankment on either side which controls the flood inundation and the westward shifting of the river. The embankment serves as a hotspot for the accumulation of the floodwater and causes permanent and temporary water logging near the embankment.

Through the 11 classified SAR scenes, the dynamic change of the river course is evident. The nearby- inundated areas spread approximately 6-7 km perpendicular to the course of the river due to the monsoons and the increased discharge of the river due to the rain. The river discharge starts to normalize from august 2019 leaving some permanently waterlogged patches along the river course.

The accuracy of the classification was assessed using the random stratified points on the classified scenes. The 2 X 2 confusion matrix for the binary classification was computed for all the scenes. If the accuracy was found to be less than 85%, the combination of the input feature bands was altered keeping the training data the same to feed into the random forest algorithm. Along with the random stratified point taken for accuracy assessment, set of points were manually marked along the the river to gauge the accuracy of classification along the boundary of the river.

Table 15.3 The accuracy assessment results for classified images

Image Date	Overall Accuracy	Kappa Coefficient
12-04-2019	88.96	0.77
11-06-2019	94.28	0.61
05-07-2019	88.31	0.76
17-07-2019	95.27	0.90
29-07-2019	92.72	0.85
10-08-2019	96.36	0.95
22-08-2019	96.66	0.93
03-09-2019	97.17	0.83
15-09-2019	92.72	0.85
27-09-2019	89.81	0.79
09-10-2019	93.45	0.87

It was found that the classification had to be repeated on the scenes with a higher percentage of the dark pixels. The scenes having a comparatively large percent of dark pixels resulted in some ineffective GLCM output bands. It was compensated by using non-linear and linear filters on the VV-VH polarisation bands. The overall accuracy of the classifier is found greater than 88.96 %. While the average kappa coefficient is calculated to be 0.82. The kappa coefficient depicts that the results are in good agreement. The results of this study demonstrate the high performance of the classifier for the delineation of flood inundated areas.

6. Discussion

The primary advantage of the SAR sensors is their ability to collect data irrespective of the cloud cover or sun illumination. The periods of probable cloud cover such as monsoons are better analyzed using SAR sensors. In the present study, the applicability of a robust machine learning technique is explored on SAR imagery for delineation of flood-inundated areas. The contrast of the backscatter from flooded areas and other features is used for the delineation of the flood extent.

It is demonstrated that the GLCM feature bands such as entropy, dissimilarity, contrast, maximum probability, mean, variance, correlation, and angular second moment can serve as a powerful input tool in the random forest model. The GLCM analysis is based on the computation of the directional relation of a pixel with the surrounding pixel. The computation is done based on the distance calculations and statistics from the designated GLCM diagonal.

The processing in this study demonstrates the impact from the number of decision trees in a random forest, choice of feature bands and number and size of training samples control the accuracy of the classification. The conclusions that can be drawn from this study are: (1) There is no generalized standard optimal number of decision trees for more accurate classification; (2) Random forest method can provide better results compared to a simple threshold method; (3) The accuracy depends on the number of training samples and the distribution of training samples; (4) Some of the descriptors from the texture analysis decrease the accuracy of the classification.

The chances of misclassification may increase due to the higher backscatter in VV polarization, which causes less contrast between the target and the surrounding pixels (Gulácsi et al., 2020).

The uncertainties in the backscatter cannot be completely removed (Stephens et al., 2012) but can only be interpolated using ancillary data as the ambiguities are not systematic. Some models are difficult to generalize if the domain knowledge is a constraint. Past researches have demonstrated that there is a need for a more advanced model to estimate the probability of accurate classification of flood pixels. The flood in this study was only mapped twice due to the constraint of the temporal resolution of the satellite.

The improvement in the adopted methodology may be possible by using the boosted random forest algorithm. The limitation of the algorithm is the user-based decision of the selection of feature bands for the random forest model input. Also, the accuracy of the supervised classification depends on the accurate training data. The methodology used in the study is difficult to automate as supervised classification makes use of analyst-defined training data. The simplicity of the method and correlation of the pixel to its surrounding pixels can be used as the principle to delineate the themes in a scene. Du et al., (2015) suggested that since the textural measurements have correlation, not all the descriptors are used and suitable feature bands should be selected for the classification.

7. Conclusion

In this research study, the ability of GLCM bands being used to increase the performance of the random forest algorithm is analyzed for flood inundation mapping. We obtained the GLCM feature bands for VV and VH band using a 5 X 5 moving window for all direction orientation. The use of GLCM certainly improves the classification algorithm by increasing the input feature bands. Not all the GLCM feature bands serve as useful input for the classifier as experimentally observed in this study.

The approach for the classification of Sentinel 1 C Band GRD product is useful during the monsoon season as the SAR sensors have all-weather capabilities. The fact that SAR images contain information as the texture is exploited for classification. The random forest classifier proved to be a robust machine learning method for flood mapping and can be used for other disaster management applications for accurate and quick analysis.

SAR image processing for delineation of floods are usually binary outputs and hence may not discern the uncertainties associated with the classification models (Dasgupta et al., 2018).

The presently available methods range from using simple histogram-based approaches to the use of machine learning. The current challenge is to parametrize the uncertainties that affect the classification arising from the acquisition time, incidence angle, sensor characteristics. Also, future flood inundation studies can focus on developing models that take into consideration the probability of each pixel classified into flood class (Dasgupta et al., 2018). The probabilistic approach to classification and the boosted random forest algorithms may give better results by eliminating the misclassification of the unknown pixel as flood class and by providing accurate results even with scares decision trees, as it is capable of computing classifiers by boosting respectively (Mishina et al., 2015).

Future research can also aim to decrease the dependency to choose classification algorithms for flood mapping. The automation of the process of flood mapping and assessment for an accurate dynamic analysis may further be explored. Comparative studies can be conducted to test the classification accuracy of L-Band flood mapping in contrast to C Band.

8. Acknowledgement

We would like to thank Dr. Jyoti Sarup (National Institute of Technology, Bhopal) for her valuable suggestions. We acknowledge and express our gratitude to Abhilash Singh (IISER, Bhopal) and Zafar Beg (IISER, Bhopal) for their insightful comments and timely recommendations.

References

1. Agarwal, R. P., & Bhoj, R. (1992). Evolution of Kosi river fan, India: structural implications and geomorphic significance. *International Journal of Remote Sensing*, *13*(10), 1891–1901.
2. Agnihotri, A. K., Ohri, A., Gaur, S., Das, N., & Mishra, S. (2019). Flood inundation mapping and monitoring using SAR data and its impact on Ramganga River in Ganga basin. *Environmental monitoring and assessment*, *191*(12), 1–16.
3. Bhatt, C. M., Rao, G. S., Begum, A., Manjusree, P., Sharma, S. V. S. P., Prasanna, L., &Bhanumurthy, V. (2013). Satellite images for extraction of flood disaster footprints and assessing the disaster impact: Brahmaputra floods of June–July 2012, Assam, India. *Current Science*, 1692–1700.
4. Breiman, L. (2001). Random forests. *Machine learning*, *45*(1), 5–32.
5. Brivio, P. A., Colombo, R., Maggi, M., &Tomasoni, R. (2002). Integration of remote sensing data and GIS for accurate mapping of flooded areas. *International Journal of Remote Sensing*, *23*(3), 429–441.
6. Caballero, G. R., Platzeck, G., Pezzola, A., Casella, A., Winschel, C., Silva, S. S., ... & Delegido, J. (2020). Assessment of multi-date sentinel-1 polarizations and GLCM texture features capacity for onion and sunflower classification in an irrigated valley: An object level approach. *Agronomy*, *10*(6), 845.
7. Cheng, S., Zhao, W., & Yin, Z. (2019, December).PS-InSAR Analysis of Collapsed Dam and Extraction of Flood Inundation Areas in Laos Using Sentinel-1 SAR Images. In *2019 IEEE 4th Advanced Information Technology, Electronic and Automation Control Conference (IAEAC)* (Vol. 1, pp. 2605–2608). IEEE.
8. Dasgupta, A., Grimaldi, S., Ramsankaran, R. A. A. J., Pauwels, V. R., Walker, J. P., Chini, M., ... & Matgen, P. (2018). Flood mapping using synthetic aperture radar sensors from local to global scales. *Global Flood Hazard: Applications in Modeling, Mapping, and Forecasting*, 55–77.
9. Du, P., Samat, A., Waske, B., Liu, S., & Li, Z. (2015). Random forest and rotation forest for fully polarized SAR image classification using polarimetric and spatial features. *ISPRS Journal of Photogrammetry and Remote Sensing*, *105*, 38–53.
10. Filipponi, F. (2019). Sentinel-1 GRD preprocessing workflow. In *Multidisciplinary Digital Publishing Institute Proceedings* (Vol. 18, No. 1, p. 11).
11. Fletcher, R. S. (2016). Using vegetation indices as input into random forest for soybean and weed classification. *American Journal of Plant Sciences*, *7*(15), 2186–2198.
12. Gkanatsios, I. (2018). Classification of sea ice types for the East part of Greenland waters using SENTINEL 1 data.
13. Gulácsi, A., & Kovács, F. (2020). Sentinel-1-imagery-based high-resolution water cover detection on wetlands, Aided by Google Earth Engine. *Remote Sensing*, *12*(10), 1614.
14. Hu, J., Ghamisi, P., & Zhu, X. X. (2018). Feature extraction and selection of sentinel-1 dual-pol data for global-scale local climate zone classification. *ISPRS International Journal of Geo- Information*, *7*(9), 379.
15. Kupidura, P. (2019). The comparison of different methods of texture analysis for their efficacy for land use classification in satellite imagery. *Remote Sensing*, *11*(10), 1233.
16. Martinis, S. (2017, July).Improving flood mapping in arid areas using Sentinel-1 time series data.In *2017 IEEE international geoscience and remote sensing symposium (IGARSS)* (pp. 193–196).IEEE.
17. Mishina, Y., Murata, R., Yamauchi, Y., Yamashita, T., & Fujiyoshi, H. (2015). Boosted random forest. *IEICE TRANSACTIONS on Information and Systems*, *98*(9), 1630–1636.
18. Nkeumoe Numbisi, F., Vancoillie, F., & De Wulf, R. (2018). Multi-date sentinel1 SAR image textures discriminate perennial agroforests in a tropical forest-savannah transition landscape. In *Mid- term symposium'Innovative sensing: from sensors to methods and applications'* (Vol. 42, pp. 339–346). Copernicus Gesellschaft.
19. Pandey, A. C., Singh, S. K., &Nathawat, M. S. (2010). Waterlogging and flood hazards vulnerability and risk assessment in Indo Gangetic plain. *Natural hazards*, *55*(2), 273–289.
20. Rahman, M. S., Di, L., Shrestha, R., Eugene, G. Y., Lin, L., Zhang, C., ...& Yang, Z. (2017, August). Agriculture flood mapping with Soil Moisture Active Passive (SMAP) data: A case of 2016 Louisiana flood. In *2017 6th International Conference on Agro-Geoinformatics* (pp. 1–6).IEEE.
21. Sharma, E. K., Priyanka, E., Kalsh, E. A., & Saini, E. K. (2015). GLCM and its Features. *Int. J. Adv. Res. Electron. Commun. Eng*, *4*(8), 2180–2182.
22. Singh, A., Gaurav, K., Meena, G. K., & Kumar, S. (2020). Estimation of soil moisture applying modified dubois model to Sentinel-1; a regional study from central India. *Remote Sensing*, *12*(14), 2266.

23. Sinha, R., Bapalu, G. V., Singh, L. K., &Rath, B. (2008). Flood risk analysis in the Kosi river basin, north Bihar using multi-parametric approach of analytical hierarchy process (AHP). *Journal of the Indian Society of Remote Sensing*, *36*(4), 335–349.
24. Tripathi, G., Pandey, A. C., Parida, B. R., & Kumar, A. (2020). Flood inundation mapping and impact assessment using multi-temporal optical and SAR satellite data: a case study of 2017 Flood in Darbhanga district, Bihar, India. *Water Resources Management*, *34*(6), 1871–1892.
25. Tsyganskaya, V., Martinis, S., Twele, A., Cao, W., Schmitt, A., Marzahn, P., & Ludwig, R. (2016).A fuzzy logic-based approach for the detection of flooded vegetation by means of synthetic aperture radar data. *The international archives of photogrammetry, remote sensing and spatial information sciences*, (B7), 371–378.
26. Vanama, V. S. K., &Rao, Y. S. (2019, July). Change detection based flood mapping of 2015 flood event of Chennai city using sentinel-1 SAR images. In *IGARSS 2019-2019 IEEE International Geoscience and Remote Sensing Symposium* (pp. 9729-9732). IEEE.
27. Wakabayashi, H., Motohashi, K., Kitagami, T., Tjahjono, B., Dewayani, S., Hidayat, D., & Hongo, C. (2019). Flooded Area Extraction of Rice Paddy Field in Indonesia Using Sentinel-1 SAR Data. *International Archives of the Photogrammetry, Remote Sensing & Spatial Information Sciences.*
28. Wurm, M., Taubenböck, H., Weigand, M., & Schmitt, A. (2017). Slum mapping in polarimetric SAR data using spatial features. *Remote sensing of environment*, *194*, 190–204.
29. Zayed, N., & Elnemr, H. A. (2015). Statistical analysis of haralick texture features to discriminate lung abnormalities. *International journal of biomedical imaging*, *2015*.

CHAPTER 16

Public-Private-People Partnership (4P) Framework for Managing Floods in Raigad [MH]

A. N. Chavan
Department of Civil Engineering,
Dr. Babasaheb Ambedkar Technological University, Lonere

S. M. Pore
Department of Civil Engineering,
Dr. Babasaheb Ambedkar Technological University, Lonere

S. R. Bhagat
Department of Civil Engineering,
Dr. Babasaheb Ambedkar Technological University, Lonere

Abstract

The major community getting affected due to disasters are the people living in those affected areas, the ecosystem they are living in and the socio-economic activities they depends on. And none other than the solicitude and familiarity with the surroundings of these people to tackle the disasters they have been constantly dealing since ages would prove it the most lethal component among all the units assigned. It would be an in-engineering way of dealing with disasters if it does not involve retrospection and resource mobilisation in a pro-active manner aiding-in of the support from locale. The theories of 4P framework has backed the implementation of Public-Private-People Partnership (4P) in disaster management practices and this paper is also an attempt to study the how this framework can be implemented in Raigad district of Maharashtra which is often been a victim to the disasters in the form of floods, cyclones and bio-chemical outspreads due to geographic, operational and economic characteristics of the region. Among these disasters, this paper particularly emphasize on enabling the use of 4P in flood management and related practices towards the same. Initial research was conducted to get insights over the current practices and agencies involved in carrying them out during floods originated in west flowing rivers such as Savitri, Amba and Kundalika to identify the useful aspects that could become a part of new procurement method on implementation. An applicability and Viability structure has been formed to verify the use of proper procurement method and does the 4P model fits-in within the proposed system has been addressed. Roles of each party in the collaboration during various phases of disaster management are acknowledged and a probable channel for delivering the given duties is proposed in order to devise a framework adhering to the characteristics of each component and their scope of involvement. Significance of 4P implementation during disaster planning, preparedness, post disaster response mechanism and disaster risk reduction is discussed and highlighted the importance of involving all the managerial elements during entire disaster management cycle. Based on the disaster characteristics and scope of availability of all parties, a framework has been computed and analysed for ensuring the applicability, suitability, and efficiency from economical, social, cultural, technological and operational point of view. At the later part, the policies, protocols and guidelines over specific requirements of implementation and operations, payment mechanism, contract duration and critical issues has been discussed to shape the research idea and aftermaths of the same.

Keywords: Disaster management, 4P, People, applicability and viability structure, risk reduction

1. Introduction

India is prone to a large number of natural as well as man-made disasters. 58.6 per cent of the landmass is prone to earthquakes of moderate to very high intensity; over 40 million hectares (12 per cent of land) is prone to floods and river erosion; of the 7,516 km long coastline, close to 5,700 km is prone to cyclones and tsunamis; 68 per cent of the cultivable area is vulnerable to drought and hilly areas are at risk from landslides and avalanches. Maharashtra, being the most prospering state in the country, has major contribution towards the GDP and also characterises a huge scale of socio-economic activities. Disasters disrupt progress and destroy the hard-earned fruits of painstaking developmental efforts in quest for progress.

The Maharashtra state has ample opportunities towards international trade, rich in thousands of years of history and culture, the vast bio-diversity and political impacts on national level. Still, a major region of the state is prone to flooding on yearly basis disrupting all the developmental, socio-economical, and ecological activities. The characteristics of Western Ghats and off shore – on shore oceanic activities at the Arabic sea impose complex climatic conditions and give rise to cyclones, floods, and tsunamis; this gets worse with the addition of man-made disasters and calamities.

In the 21st century itself, we have witnessed few of the major disasters in the form of 2005 floods, 2002 Tsunami, 2019 Floods, Nisarg Cyclone 2020, Tauktae 2021, etc. The frequency and nature of these disasters are alarming not only for the stakeholders involved in those particular disasters but also for all the mankind irrespective of the geography and their vulnerability towards the potential risk.

The 2005 Maharashtra floods impacted many parts of the Indian state of Maharashtra including large areas of the metropolis Mumbai, a city located on the coast of the Arabian Sea, on the Western coast of India, in which approximately 1,094 people died. It occurred just one month after the June 2005 Gujarat floods. The term 26 July, is used to refer to the day when the city of Mumbai came to a standstill due to flooding.

The Maharashtra government has pegged the total damages caused by Cyclone Nisarga which had hit Raigad, Ratnagiri and Sindhudurg districts in June at Rs 6,048 crore. According to officials, in Raigad district, around 2,285 houses have been damaged completely while over 2 lakh houses have been partially damaged. The crops and orchards on around 12,000 hectare agriculture land have been damaged in Raigad.

2. Floods of Raigad District

Apart from above disasters, Floods are the prominent disasters to which the West coast of Maharashtra has been a regular victim. Raigad, one of the closest district to the financial capital of the country and consisting Mumbai Goa Highway, one of the most important National highway passing through most of its urban local bodies, has gone through various floods that affected the area socio-economically, culturally and at various stages of development as it halted the progress and inhibited setbacks for various economic sectors such as tourism, trade, telecommunication, manufacturing, etc.

Raigad consists of various rivers flowing across the districts covering the major Tehsils and ULBs that includes Ulhas River, Savitri River, Kundalika River, Amba River, etc. These rivers are prone to flooding almost every year and causes immense damage to structures, buildings, public places and industries. It also causes loss of livestock and threatens human lives too. Officials at the government departments delivers the best services to curb the situation but the nature and geography of the area does not permit them to carry out disaster management in a full-fledged manner. The lack of human resources, equipment and complex characteristics does not allow various response teams to do their duties at full flow and these restrictions gives indications towards implementing something newer, something comprehensive and equally effective approach in disaster management. This is where the need of 4P implementation arises and that is the main theme of this article.

Following images shows the map of Raigad district having the structure of major rivers and the important local bodies situated across the channel of respective rivers. The major towns situated across the banks of these rivers are Mahad, Mangaon, Roha, Karjat, Nagothane, Khalapur, etc. These towns often get drown in the floods and the day-to-day activities get halted for several weeks.

2.1 Reasons to Flooding

Flooding in Raigad is strongly related to the occurrence of torrential rainfall caused by the monsoon winds coming from Arabian Sea. Floods in and around various ULBs are caused by the small rivers coming down from the adjacent Western Ghats Ranges to the west of the town. Due to meteorological reasons, July and August were the most hazardous months. These were the month with highest total monthly rainfall, with maximum peak discharge.

As the population has grown exponentially, so has the demand for housing, food, and cloth. As more cities and towns are developed across the banks of river, they have resulted in narrowing the river channel and clogging it at the same time which causes the erosion and change in HFL level every now and then. The analysis of long term rainfall data indicates that normal annual rainfall over the Raigad district ranges from 2200 mm to more than 3000 mm in the plains and it is above 5000 mm in the hills.

Flooding always occurs in coastal areas. Due to the High tides or storms water level start to rise. flooding occur when water level is became higher than the level of coastal lowland The most common cause of floods in cities is the lack of pervious surfaces and trees.

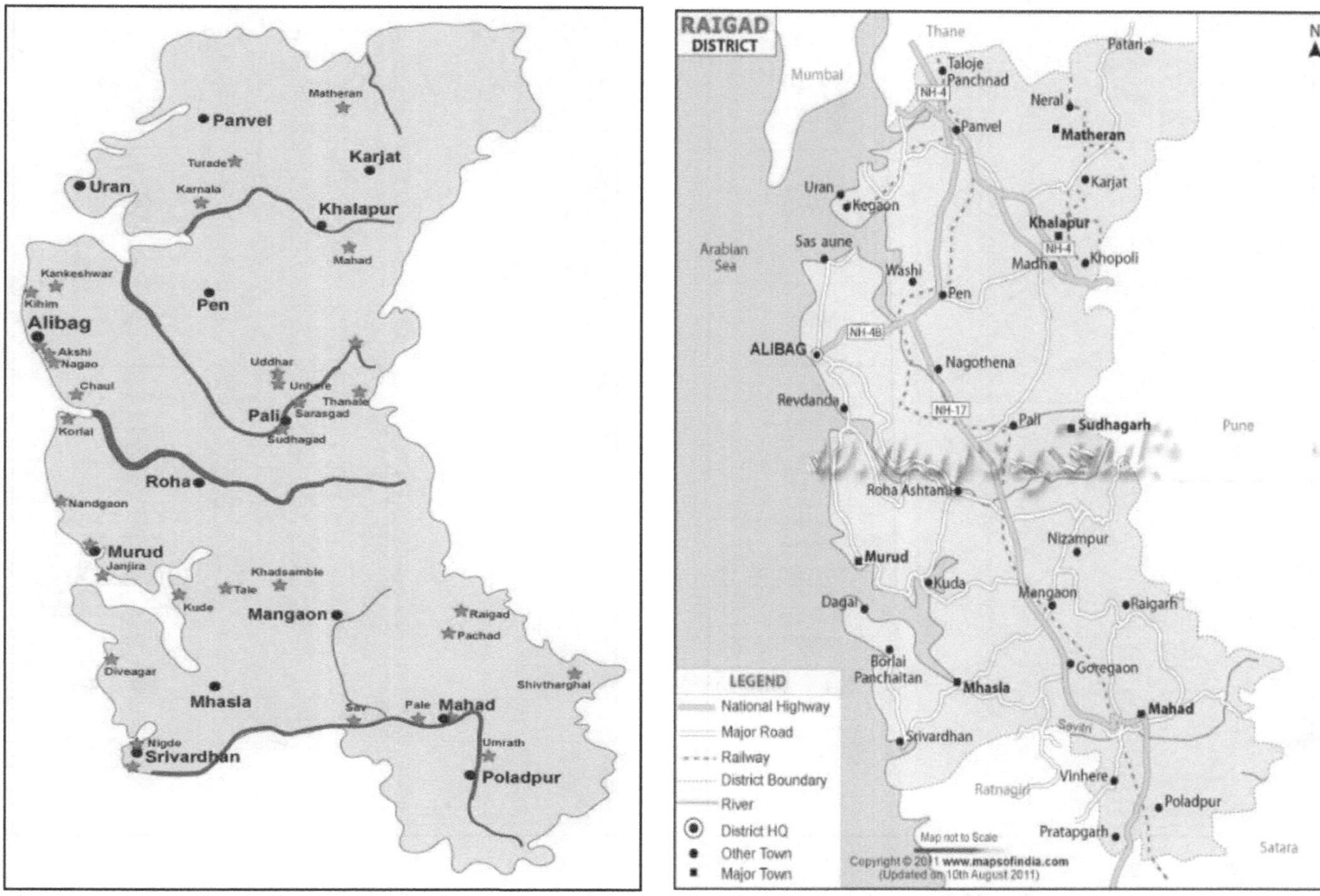

(a) (b)

Fig. 16.1 (a) River Map of Raigad (b) Major ULBs in Raigad

Source: (a) Konkanyatra.in (b) Maps of India

In a natural setting, about 80% of the rainwater is absorbed by the soil and the tree canopies. And, the rest of the water is flown to the natural waterways. However, in cities, we have disturbed this storm water management system by building concrete structures everywhere. As a result, when rainwater falls on the surface, it does not find a way to be absorbed naturally, resulting in water logging. When deforestation occurs in a particular area, there are no more trees to help soak up precipitation and reduce water flow over the landscape. Without these natural protections, there is an increased risk of flooding and erosion whenever it rains.

2.2 Impacts of Floods on Raigad

Floods are known to impact almost every sector of the society irrespective of the characteristics it possesses. These sectors are socio-economic activities, developmental activities, educational activities, cultural activities, agriculture sector and various modes of communication, and most importantly threat livestock and to human life. Particularly, Raigad gets impacted in following ways;

These floods destroy the road network, bridges, telecommunication towers, electricity towers and other modes of communication which result in loss of contact from place to place and create obstacles during carrying out of response phase. This loss of contact disturbs all the trade, tourism and socio-economic activities for several weeks.

Water contamination gives rise to different diseases and infections among the community and threatens the human life beyond the control of stakeholders with reference to the socio-economic and geo-political characteristics of the area. It also creates shortage of water for domestic and industrial purposes.

The district is having a vast bio-diversity spread across most of its parts; it gets disturbed and destroyed resulting in loss of wildlife habitat and unbalancing of the ecosystem.

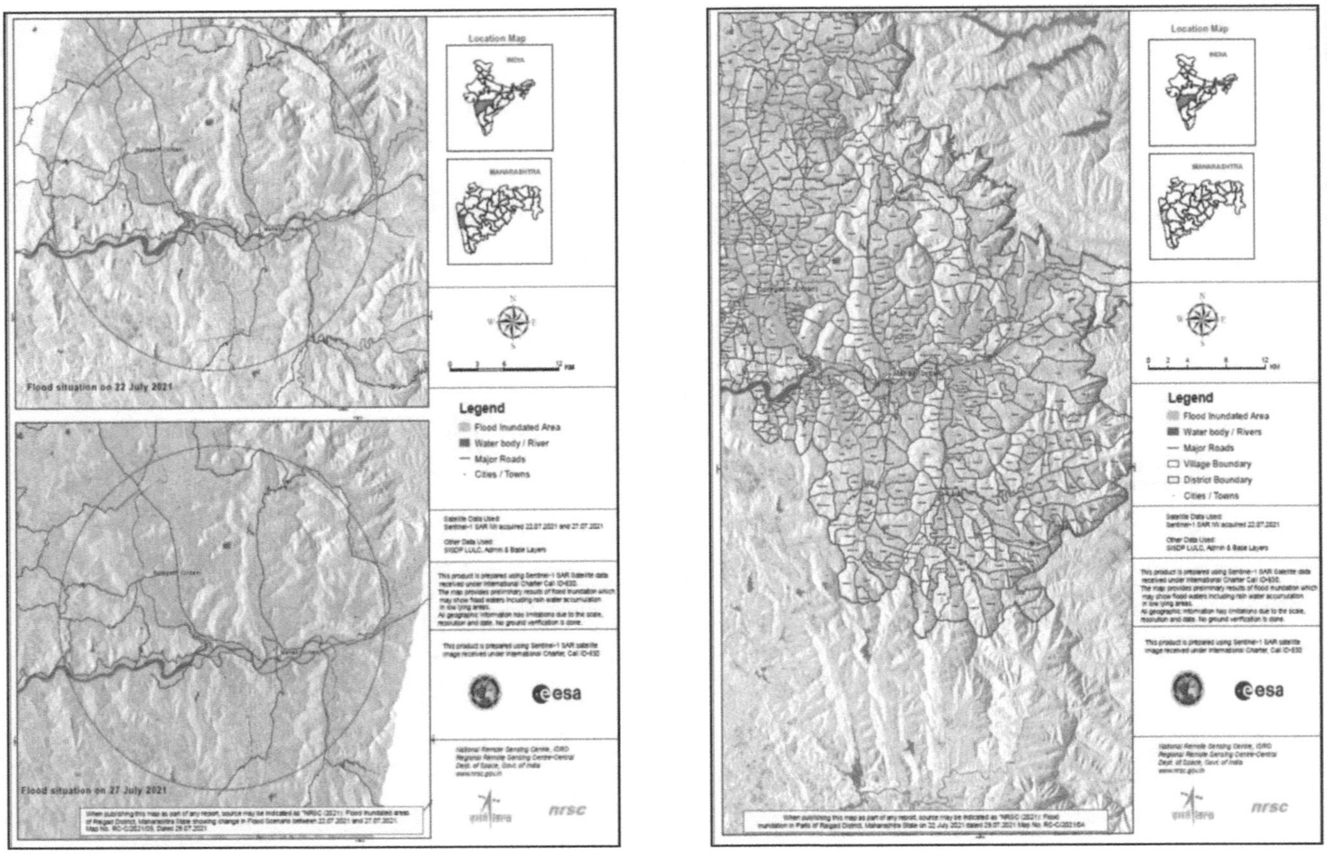

(a) (b)

Fig. 16.2 (a) Flood Inundation 2021 (b) Flood Inundation of Savitri River

Source: www.nrsc.gov.in

Floods damage property significantly. Flood waters damage land by eroding shore lines and stripping soils, as well as taking out whatever natural vegetation in the path of the flowing water. Floods also damage personal property, such as vehicles and homes, creating hazardous living conditions if water is not cleaned up immediately.

3. Management Plans and Their Implementation Status

The implementation of flood management guidelines provided in the NDMP 2019 involves a collective effort from central government, state government, NDMA, SDMA, NDRF, SDRF, District authorities and officials. The early warning systems triggers the deployment of NDRF teams and early response groups to mitigate and prepare for the impacts of disasters and subsequent response work would be carried out to prevent threat to human life, livestock and assets. There are various NGOs actively working in the region to counter the disaster situations and help the administration and officials from all possible ends. Apart from this the timely guidelines would be conveyed to all the stakeholders and involved agencies to curb the disaster. There are private agencies established in the area which also possess the expertise required for search and rescue operations.

4. 4P as a Tool for Disaster Management

Background of the 4P belongs to a very popular method of procurement called as Public-Private- Partnership (3P) which facilitates the government/organisations to complete a particular project with the contributions from the private entities/organisations within the legal framework of a contract. These contracts vary from project to project depending upon numerous aspects involved in the project during its development and service period. India has systematically rolled out a PPP program for the delivery of

(a) (b)

Fig. 16.3 (a) Response work by NDRF (b) Rescue of locals from SAR teams

Source: (a) www.thewire.in, (b) www.indianexpress.com

high- priority public utilities and infrastructure and, over the last decade or so, developed what is perhaps one of the largest PPP Programs in the world. With close to 2000 PPP projects in various stages of implementation, according to the World Bank. India is one of the leading countries in terms of readiness for PPPs.

In the context of Disaster Management, It is not that India hasn't been keen on implementing the PPP guidelines in Disaster Management sector; the use of PPP framework for construction of disaster infrastructure is already been witnessed in various projects but it is still short of achieving the said targets and capacity building provisions. Considering India's expertise and adaptability in the field of PPP, it's conglomeration with societal components such as local community, academicians, media, NGOs, welfare groups, and researchers has to be encouraged and drafting needs to be carried out within the legal framework of disaster management. This component of the society is referred as 'People' in various journals and research works carried out recently in the field of financial structuring of the disaster management. This forms the basis of implementing 4P in the disaster management. With this background the article further discusses the essential elements of 4P and its applicability in the disaster management cycle.

4.1 Elements for Creating Applicability and Viability Structure

The applicability and viability structure is devised to investigate the suitability of 4P in carrying out the project called disaster management effectively. This structure is based on the work done by Antonio Vives and et al over financial structuring of Infrastructure projects in PPP. They have identified the various forms of local conditions and project modalities for selecting the most feasible procurement method for a project. This particular structure has included 8 local conditions and kept it flexible for others to modify it as per the needs of the projects under consideration of PPP. This article has identified new local conditions to fit the 4P and its attributes. The structure of 4P for managing floods is computed as shown in Table 16.1:

5. 4P Frameworks and their Suitability

Depending upon the availability of assets, resources, outsourcing and involvement of stakeholders in different stages of DM Cycle, the 4P is classified in four different frameworks. The suitability of these frameworks is as shown in Table 16.2.

5.1 BLCT: Build, Lease, Coordinate, Transfer

This framework allows public sector organisation or government to transfer the rights of property/asset which have the potential to fulfil the functional requirements for disaster management to the competent private sector entity/ SPV. The organisation would develop and build the required facility and again lease the built facility to the government so that that government would make use of it to deliver required services and perform disaster management deliverables. Meanwhile, government should collaborate with the community and make them prepare for the disasters by awareness, trainings, workshops and skill development sessions that would aid in to deal the disasters from two fronts. This framework is suitable where government wants to use the rapid development technology available with the private sector and lack of time in arranging the funds for the same so as to raise the facility

Table 16.1 Applicability and Viability Structure

	Local Conditions	4P essentials for Flood Management in Raigad
Existing local conditions	Legal Framework	Established under the PPP act India, Department of Economic affairs, NDMP guidelines, NDMA, Ministry of Home Affairs
	Political Risk	Apart from the involved institutions such as NDMA, NDRF, District collectorate, and local community the political risk remains under uncertainty irrespective of the method of procurement
	Fiscal capacity	Availability of public funds, Financing from government funded subsidy stream
	Macroeconomic factors	Economic volatility that includes currency devaluation which seems unlike considering the geo-political nature of the district
	Institutional capacity	Institutions established by central government and state government to look after the regulation and strict adherence with the contract terms.
	Willingness to pay	The beneficiaries of the project are already under the impact at various economic levels due to the disaster. This seems likely that they would contribute apart from the regular taxes and charges. However they are likely to pay during preparedness stage of disaster management.
	Size and location	The entire district suffers same disasters at same time with same nature of impacts. Therefore the implementation should be carried out considering the needs of masses.
Newly added local conditions	Involvement of Stakeholders	Earlier procurement methods do not have the involvement of all the stakeholders. 4P seems to involve all the stakeholders whether they are at receiving end of disasters or at the dealing with disasters. It basically refers to the local community and affected people.
	Aid of empirical technology and delivery of services	This refers to the application of research carried out by researchers, academicians and welfare groups with prime focus on flood mitigation, preparedness, response and recovery for the Raigad district.
	Information transfer	The information regarding the area specific early warnings, policies, guidelines, training sessions, workshops, research, and advancements in disaster technology and structural resilience has to be transferred effectively with the use of print media, digital media and social media platforms.

Source: Antonio Vives et al.

Table 16.2

Local Conditions	4P frameworks			
	BLCT	MLCT	MOCT	LOCT
Assets provisions	Public property	Public property	Public property	Private Property
Facility Establishing	Private	Private	Private	Public
Operative human Resources	Public People	Public People	Private People	Public People
Preparedness Stage	Public Private People	Public Private People	Public Private People	Public Private People
Response stage	Public People	Public People	Public Private People	Public Private People
Recovery stage	Public People	Public People	Public People	Public People

Source: Author

as earliest and which would divert the additional load on the available workforce which is already engaged in revamping other facilities.

5.2 MLCT: Modify, Lease, Coordinate, Transfer

This framework would allow private sectors to modify one of the assets owned by the government/ public sector and make it functional so that government authorities would coordinate with the local community and people to deal with the disasters and implement the disaster management cycle effectively.

5.3 MOCT: Modify, Operate, Coordinate, Transfer

This framework allows the private sector to modify and operate within the boundaries of asset owned by the public sector or

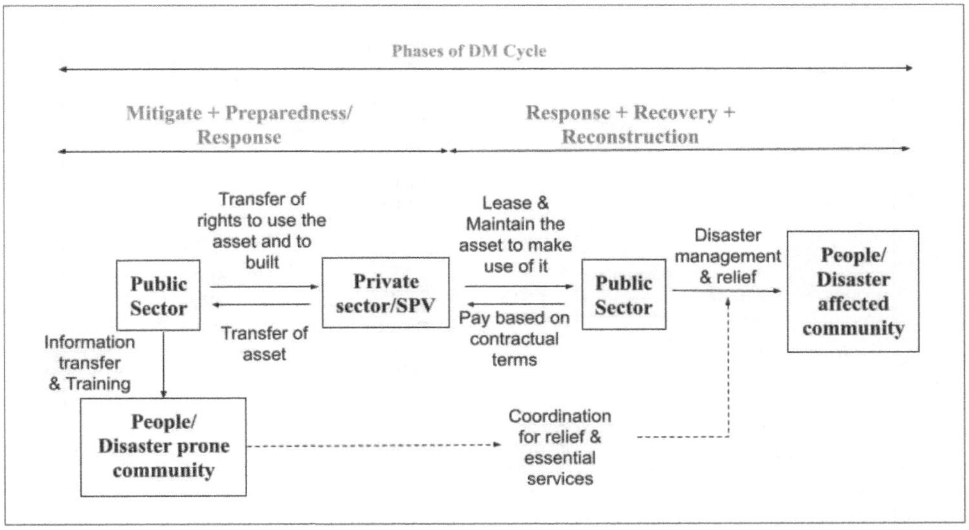

Fig. 16.4 BLCT: Build, Lease, Coordinate, Transfer

Source: Author

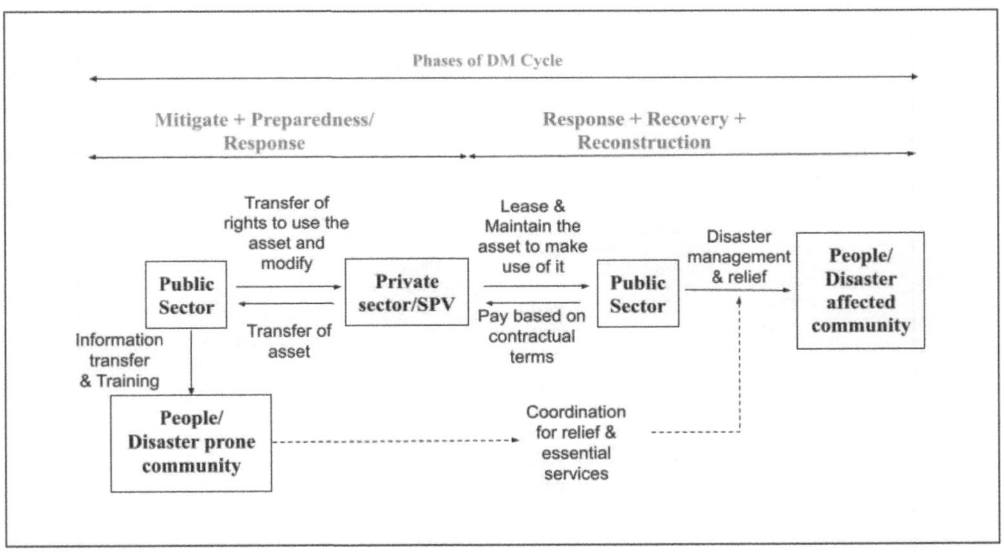

Fig. 16.5 MLCT: Modify, Lease, Coordinate, Transfer

Source: Author

government. A competent private sector organization that has the potential resources to perform the disaster management work right from the preparedness phase and then able to transfer the rights back to the public sector once the DM cycle completes. The private sector entity can make use of third party special purpose vehicle to perform the tasks on its behalf along with coordination from the community and people. This framework is suitable in cases where the private sector is competent and having sufficient resources with them to carry out all the essential tasks.

5.4 LOCT: Lease, Operate, Coordinate, Transfer.

By virtue of this framework, a private sector entity would lease their assets/ property to the government and subsequently government would make use of that asset to perform the activities related to disaster management. This framework is cost effective and suitable in the view of response phase of the DM cycle.

This type of framework is applicable to scenarios where private parties have well established premises that would

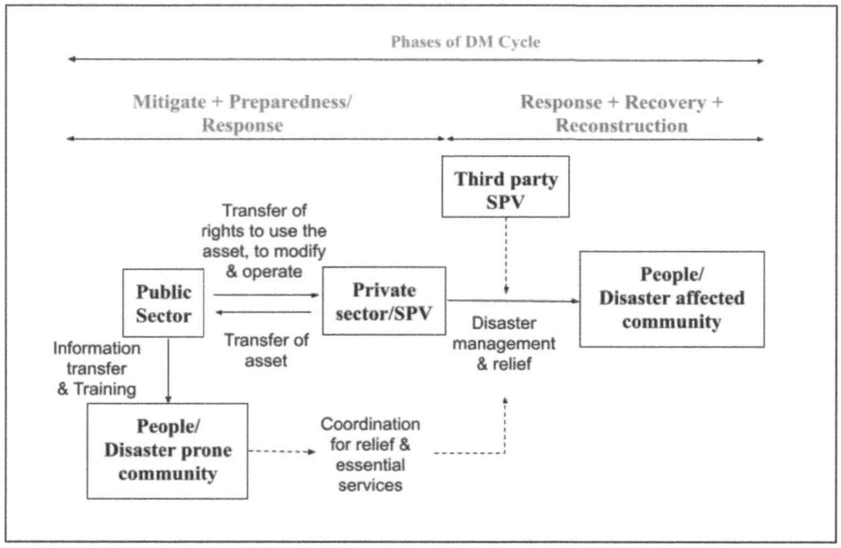

Fig. 16.6 MOCT: Modify, Operate, Coordinate, Transfer

Source: Author

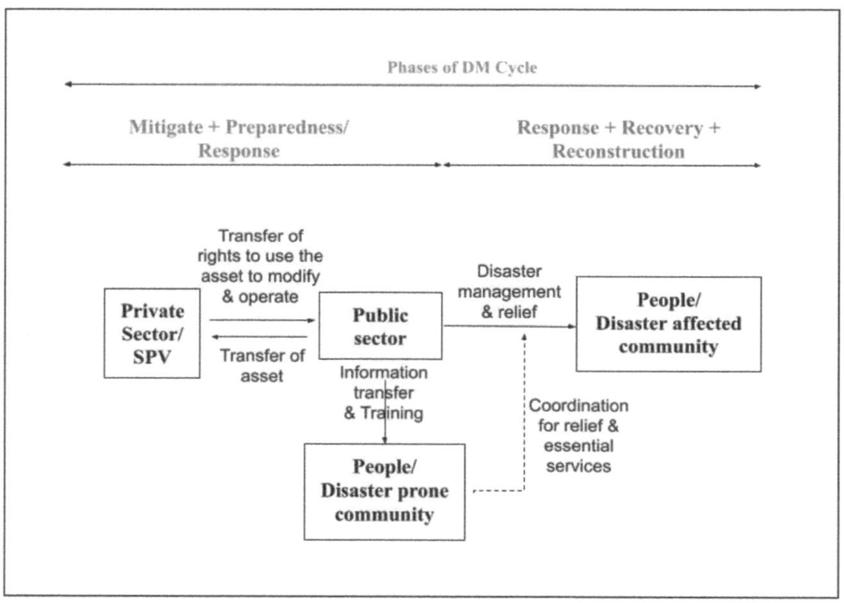

Fig. 16.7 LOCT: Lease, Operate, Coordinate, Transfer

Source: Author

be effective from economy, geography and functional characteristics.

6. Significance of Implementation

4P has a great potential to counter the obstacles in managing the disasters provided that the framework is backed up by the adequate relevant policies, guidelines, and legal systems. It could be very effective in reducing down the corruption, discrepancy in DM activities, centralisation at various stages, and also it can improve the communication among various stakeholders right from the beginning of a particular DM cycle. One of the most important advantage lies with the presence of active practitioners and researchers in the field which would ensure the smart use of recent advancements in economic and effective manners. It will also help towards better transfer of risks and responsibility sharing among various stakeholders. 4P allows proceeding swiftly in carrying out response and reconstruction tasks. As 4P involves proactive planning and constitutes relevant expertise it can help in meeting the sustainable long term development for rapid and visible solutions.

7. Policies and Protocols

National Disaster Management Plan 2019 has laid useful guidelines that would provide directives towards managing the disasters and envisage this dynamic document to be a comprehensive guide for future disaster management practices. The main constraint in disaster management practices is the interest of private entities in undertaking low profit works. The attention of private sector has to be drawn actively by adding some policies that would make private parties invest more into disaster management projects and make it a part and parcel of their company profile. There should be some relaxation in taxation, awarding subsidies, signing of MoUs, flexibility in payment modes and outsourcing parameters given that would make them participate actively in DM cycle.

References

1. Magar P (2020) " Flood Inundation Mapping of Savitri River Floods in and around Mahad Town (2005 & 2007 Storm Events) by Using GIS Tools, North Konkan, Maharashtra, India" International Journal of Innovative Science and Research Technology, pp 57-63, Volume 5, Issue 1, January – 2020
2. National Disaster Management Authority (2019) "National Disaster Management Plan 2019" A publication of National Disaster Management Authority, Government of India, New Delhi, November 2019
3. Rufat S, Tate E, Burton C and Maroof A S (2015) "Social vulnerability to floods: Review of case studies and implications for measurement", International Journal of Disaster Risk Reduction, 2015
4. United Nations (2015) "Sendai Framework for Disaster Risk Reduction 2015-2030" United Nations Office for Disaster Reduction, Geneva - Switzerland
5. Zhang J and Kumaraswamy M. (2012) "Public-Private-People Partnerships (4P) for Disaster Preparedness, Mitigation and Post-disaster Reconstruction" The University of Hong Kong, 2012

CHAPTER 17

Role of Kerala Fire & Rescue Services Department in Handling Kerala Floods 2018

B. Sandhya IPS
Director General of Police
Director General, Kerala Fire and Rescue Services
Phone: 9497920001
dg.frs@kerala.gov.in

Nousad M.
Director (Technical)
Kerala Fire and Rescue Services Thiruvananthapuram
Phone: 9446540101
dt.frs@kerala.gov.in

Abdul Rasheed K.
Regional Fire Officer (Academy Director)
Kerala Fire and Rescue Services Academy Thrissur
Phone : 9447519656
adoac.frs@kerala.gov.in

Abstract

Kerala state experienced a devastating flood event during the month of August 2018. The flood resulted in extensive flooding and landslides across most districts of the state, causing severe damage to both the built and natural ecosystems. Rescue of the people affected by the Kerala floods 2018 and evacuation of those who were vulnerable to the flood and related incidents was the biggest challenge. KF&RS is the trained Governmental Agency to respond first, to any such disaster, in Kerala. In June 2018, when there was heavy rain which was much more than what is expected, the Department started precautionary measures in the different flood prone districts like Ernakulam. The Department Opened State control room at Headquarters, Thiruvananthapuram and District Control rooms at all District Headquarters, especially for Monsoon risk management.

During August 2018, heavy rain affected almost all districts of the State. Precautionary measures were proper in those areas which were predicted to be very much prone to flood. When the shutters of Edamalayar Dam and Idukki Dam were opened during the first ERE from 8th to 10th of Aug 2018, red alert was declared in Idukki and Ernakulam Districts. The Department anticipated severe flood in the banks of river Periyar and various low lying areas of Ernakulam District and evacuated the people from such places to different relief camps. As the equipment available with the Department was not sufficient to work in case of heavy floods, improvised water rescue facilities were prepared for rescue operations using the available resources and the situation was effectively handled by the Department.

Another ERE occurred from 14th to 19th of August 2018, which was above all predictions and anticipations. The high-altitude areas like Ranny in Pathanamthitta District were totally flooded on 14th August 2018. It was totally unexpected and quite difficult to effect rescue operations since most of the equipment was placed in the flood anticipated places and all of them were in use. The Fire and Rescue teams started rescue operations throughout the State mobilizing the maximum resources and during the night of 14th August alone, rescued more than 1500 trapped persons, from Ranny area only. Since the flood affected almost the total area of more than 10 out of the 14 districts, providing assistance from the unaffected area, as usual, was not practicable.

When the coastal area suffered tidal attacks and wind, the hilly areas were suffered by landslides and heavy current of water. Nearly 341 landslides were reported from 10 districts, in which Idukki district alone registered 143 landslides. Fire and Rescue personnel alone saved the life of 39866 persons who were trapped due to flood and evacuated 87571 persons from the flood affected areas to relief camps. Total 113 personnel from Fire and rescue Department suffered injuries during rescue operations and three suffered snake bites.

The devastating incident resulted in a total of 450 casualties and more than 1.4 million people were moved to relief camps. The flood affected almost 5.4 million people, that is, one-sixth of the State's total population and the Department personnel assisted the public in cleaning their homes and roads,and to prevent epidemic which is normally expected after a flood.

The proper planning and timely action of Kerala Fire and Rescue Services Department helped to limit the death toll to much less than that of what is expected of such a devastating flood. The community preparedness activities of the Department were also found to be successful during the floods.

Keywords: ERE – Extreme Rainfall Event; GOK – Government of Kerala; Golden Hours – The first few hours after any incident; KF&RS – Kerala Fire and Rescue Services

1. Introduction

The Asian continent has witnessed, in the past, many calamities in the form of landslides, floods, tsunamis, famines, cyclones, earthquake etc. Gujarat Earthquake (2001), Indian Ocean Tsunami (2004), flood in Tamil Nadu, and Bihar (2017) are few disasters that have created havoc in India in the recent past. The most recent being Kerala Flood (2018), (Disaster Management Laws in SAARC countries edited by S. Sivakumar, Manohar Tairani, and Lisa P Lukose – Page 1).

The onslaught of flood and course of recovery in Kerala as described by Muralee Thummaruludy states thus:

People from all walk of life – fisherman to IT professionals, house wives and college students – came together to lend a helping hand. As the state administration was overwhelmed by a large number of rescue calls, many of them without our complete information and often in duplicates, young professionals sitting around the World quickly established a 24×7 help desk that received the calls, eliminated duplicates, found exact coordinates from Google Earth and relayed precise information to the rescue teams. Relief camp came up all around and there was no shortage of resources. While a few camps did have a shortage of food or dress in the initial days, the bigger problem was that an excess of supplies came from near and far. Cast, religion, political ideology did not matter in the rescue and relief operation. It was Kerala at its best. While Kerala did exceptionally well in the post-disaster rescue and relief, its performance in the pre-disaster situation was not as exemplary.

It is noteworthy to mention that the Kerala flood has witnessed a standout contribution of fisher folk community who braved the precarious flood situation and lent a helping hand in saving people who were marooned due to the deluge. As per the report, 'the fishing community of the state rendered phenomenal voluntary assistance towards search and rescue in the flood affected areas. Nearly 669 boats that went out with 4537 fishermen estimated to have saved at least 65000 lives'. The newspapers reported that the fisherman spent their own money to transport their mechanized boats and fuel. In areas where airlifting was not feasible, the people volunteered to rescue stranded and elderly people in small boat and even in vessels. The selfless service rendered by the fisherman and the local people was appreciated across the world. However, the fishermen response was that 'we didn't do it for monetary benefits. Instead, we could empathize with them. We did it for our own people'. The primary reason for the success of the disaster management operations in Kerala was the people's response and their collective spirit. All rescue operations witnessed mass participation of people.

The State of Kerala could efficiently manage the flood calamity mainly with the framework of decentralized administration. Senior IAS officers were deputed to each District to help in coordinating rescue and relief measures. Along with them, an elected representative, or minister of the State Government was deputed to each district to provide the necessary support. While the District Collectors were the overall in charge for their districts, the Superintendents of Police handled the responsibility of rescue operations. All the staff of local bodies such as Panchayats and Municipalities were directed to work in relief operations; they were given freedom to use their funds in the manner they deemed fit.

The chief minister, Pinarayi Vijayan, handled the situation with great leadership, strong determination, immense patience, clear vision and perseverance. He faced media every day with data, facts and information to tackle the situation. In his own words 'history will speak of Kerala not as a State that got defeated but as one that overcame the floods'.

The preparedness of a society to deal with emergencies will be reflected in the aftermath of every unexpected event. In a State like Kerala which is very much prone to multiple hazards due to the geographical features, such preparedness is of much importance. The State of Kerala, one of the thickly populated States of India which maintains high standard of living, has an emergency Department, the Kerala Fire and Rescue Services, which is responding to any emergencies as the first responder. Geographically, Kerala is having long coastal area, Valleys/mid land and Hilly area which extent from North to South.

Floods are very often in few coastal Districts of Kerala during the monsoon season. But it was totally unexpected that the entire state is affected by the ERE and resultant floods. There were two consecutive EREs during August 2018, from 8th to 10th and from 14th to 19th. During the second ERE, almost all the areas were totally flooded, and the condition lasted for many days. Nearly 341 landslides were reported from 10 Districts in the hilly areas. Despite all these, the death toll was limited to 450. This shows the extent to which the Sate Fire and Rescue Services Department was prepared to face emergencies.

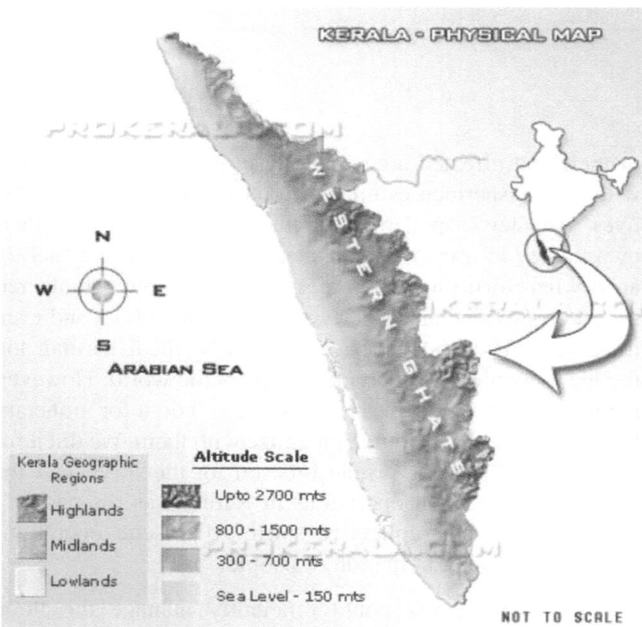

Map 17.1 Kerala – Physical

2. Details of the Event

In Kerala, the rain fall was very much above the normal during 2018, which resulted in extensive floods. Rate rainfall was higher from the month of June itself. During August 2018, Kerala state experienced some of the most severe EREs on record. During the first 19 days of August 2018, Kerala received 164% above normal rainfall, of which the major share was from two EREs during 8– 10 and 14–19 in August 2018. This resulted in extensive flooding and landslides across most districts of the state, causing severe damage to both the built and natural ecosystems. Rescue of the people affected by the Kerala floods 2018 and evacuation of those who were vulnerable to the flood and related incidents was the biggest challenge that Kerala Fire and Rescue Services ever faced.

During June 2018, when there was heavy rain which was much more than what is expected, KF&RS Department, being the trained Governmental Agency to respond first to any such disaster, started precautionary measures in the different flood prone districts like Ernakulam. The SCUBA vans, Rubber Dinghies with OB engine, Lifebuoy, Life Jackets, necessary

Fig. 17.1 The extent of flood

Fig. 17.2 Rescue in Land slide area

Fig. 17.3 Dam opened Picture 4: Heavy current of rainwater through inhabited areas

Ropes and such available equipment were kept ready to face any emergency. During June 2008 the flood affected various places like Aluva, Angamali, Muvattupuzha, Perumbavoor, Kothmangalametc and the Department successfully carried out rescue operations and evacuation of the affected. The Department Opened State control room at Headquarters, Thiruvananthapuram and District Control rooms at all the 14 District Headquarters, especially for Monsoon risk management.

Situation was different during August 2018. Heavy rain affected almost all districts of the State. Precautionary measures were proper in those areas which were predicted to be very much prone to flood. The Department was aware of the fact that the equipment available with the Department will not be sufficient to work in case of heavy floods. Hence improvised water rescue facilities were prepared with inflated vehicle tubes, big barrels, bamboo and such available things. Due to heavy rain, it was decided to open the shutters of Edamalayar Dam in the river Periyar on 9th August 2018, and orange alert was declared in different places of Ernakulam District. Coordination meeting was conducted by the Regional Fire Officer, Ernakulam and evacuation started anticipating flood in the banks of river Periyar. SCUBA Teams of all the 14 districts were put in to action. Due to the abnormal increase

Fig. 17.4 Aerial view of flood – entire building except the weather protection roofing is under water

in the water level, shutters of Idukki Dam, which is the largest reservoir in Kerala, was opened on 10thAugust 2018 and red alert was declared in Idukki and Ernakulam Districts. All the five overflow gates of the Idukki Dam were opened, for the first time in 26 years. The Department anticipated severe flood in the banks of river Periyar and various low lying areas of Ernakulam District and evacuated the people from such places to different relief camps. Similar arrangements were made in all the other districts, where flood was anticipated. The heavy rain (ERE) lasted till 10th of August 2018, but the situation was effectively handled by the Department. The Director General along with the Director Technical visited Ernakulam during this time and evaluated the situation. Other Districts in the coastal area were also affected by the ERE from 8th to 10th of August, but the Department carried out effective rescue operations as all such happenings were anticipated as part of the heavy rains.

The second spell of ERE started on 14th of August 2018 which caused flood in all the areas including Hilly areas. When the high-altitude areas like Ranny in Pathanamthitta District were totally flooded on 14th August 2018, it was totally unexpected and quite difficult to effect rescue operations since most of the equipment were placed in the flood anticipated places and all of them were in use. Being the first responders to effect rescue operations during any emergency, the Fire and Rescue personnel started rescue operations throughout the State. With the available facilities, during the night of 14th August alone, from Ranny area only, the Departmental personnel rescued more than 1500 trapped persons. The second spell of the ERE from 14th to 19th of August was above all predictions and anticipations. Since the flood affected almost the total area of more than 10 out of the 14 districts, providing assistance from the unaffected area, as usual, was not practicable.

3. Disaster Response

For any emergency, rescue operation during the golden hours is important in saving lives. During the initial stages of the flood, no other rescue agencies were there, and KF&RS carried out all the Rescue and evacuation activities with the help of Police and Disaster Management Department. Evacuating the isolated people to safer places was quite difficult owing to the lack of proper facilities to cross the heavy water currents which isolated such areas. KF&RS Department put their maximum effort, but the Rubber Dinghies available with Fire and Rescue Services Department was not sufficient to evacuate such a large number of people. The improvised water rescue facilities prepared by the Department with inflated vehicle tubes, big barrels, bamboo and such available things, were of great use for the rescue and evacuation of the affected. Besides this the Departmental personnel used all the available things, including catering utensils, to rescue and extricate people. The Community Rescue Volunteers, a group of volunteers trained by KF&RS, helped a lot in such activities. Later, the Central government and other States also provided support to GOK's disaster response and relief efforts. 57 teams of National Disaster Response Force and 435 boats were deployed for search and rescue in addition to five companies of paramilitary forces, armed forces and coast guards with 40 Helicopters, 20 aircrafts, 2 ships, 10 columns and 10 teams of Engineering Task Force. 300 Fire and Rescue Personnel from Odisha along with Rubber Dinghie and OB Engine, 50 from Andhra Pradesh and 10 from Tamilnadu Joined KF&RS for Rescue Operations. In addition, the fishing community of the State rendered voluntary assistance towards search and rescue. Nearly 669 fishing boats were there in the rescue operation with 4357 fishermen who saved many lives in the critical situation.

The second ERE ended by 19th August and Rescue and evacuation continued for two more days. From there onwards the water level gradually decreased. The devastating incident resulted in a total of 450 casualties and more than 1.4 million people were moved to relief camps. The flood affected almost 5.4 million people, that is, one-sixth of the State's total population. The Department personnel assisted the public in cleaning their homes which were totally affected with mud and dirt. Assistance was also given by the Department for the efforts of the State to prevent epidemic which is normally expected after a flood.

4. Discussion

The first thing to say is the preparedness of the State to deal with such Disasters. KF&RS Department anticipated flood during monsoon and preparations were in place for all Fire and Rescue Stations, irrespective of whether the area is prone to flood or not. Such preparedness helped in the initial stages for early warning and evacuation. But in the later stages, the absence of a specialized team to deal with such situation was evident. Since all the Districts were affected, the usual practice of utilizing the resources available in the neighboring unaffected Districts was not at all possible. Every State shall form some kind of special task forces under the Emergency/ Rescue department to deal with such situations.

Effective Prevention Mechanism of KF&RS helped a lot to reduce loss of life. Awareness creation programs and community-oriented disaster response plan of KF&RS were found to be very effective. The Community Rescue Volunteers, an initiative by KF&RS, worked effectively in evacuation and Rescue operations. Resources may be limited in such situations but ensuring maximum

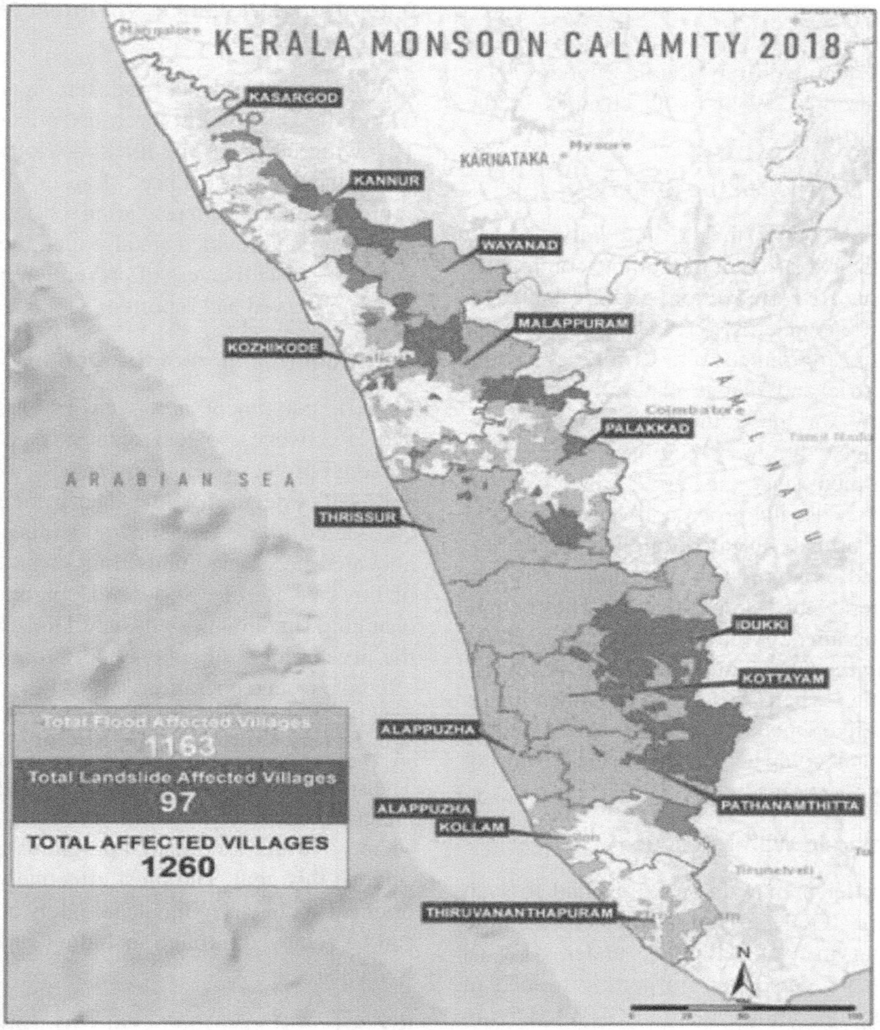

Map 17.2 Flood and Landslide affected areas

and effective utilization of the available resources is the way to overcome such limitations. KF&RS concentrated on the improvised use of all the available resources for rescue and evacuation. Inflated vehicle tubes, Barrels, Bamboo and even the big cooking utensils were used for evacuation. The Community Rescue Volunteers helped in identifying and collecting all such locally available things.

The fisherman community helped a lot for evacuation from the waterlogged areas. There are different such groups in the community who may be identified for special tasks like cutting of wood, trees etc. Similar groups may be identified and listed for future. Inter Departmental Coordination is another important aspect to be discussed. Vehicles with high ground clearance were provided by Motor Vehicles Department and Kerala State Road Transport Corporation helped in the transportation of the rescued and evacuated people. For the coordination of different such Departments during disasters, there shall be some established coordination system.

Because of the proper planning and timely action of KF&RS Department, the loss of life was much less than that of what is expected of such a devastating flood. The arrangements made as part of monsoon preparation for fast evacuation from the flood anticipated areas helped a lot to reduce loss of life. When the situation worsened much beyond what was expected, the Departmental personnel rose to the occasion and made improvised use of all the available things around for effective rescue and evacuation. The community preparedness activities of the Department are also found to be successful during the floods.

5. New Initiatives

Based on the lessons learned from Kerala Floods 2018, KF&RS Department brought about many changes in the functioning of the Department.

5.1 Established Civil Defense in the State

Accepting the effectiveness of the 'Community Rescue Volunteer Service' of the Department to promote community participation in Disaster Risk Reduction, GOK established 'Civil Defence' in the State under KF&RS Department. 6200 Civil Defense Volunteers are trained so far. Training is being provided at Station, District and State level. GOK Established Civil defence Academy for the training of Civil Defence volunteers. CD Volunteers are very helpful in prevention activities of the Department since they are familiar with the local circumstances. They usually provide inputs on the local resources which can be utilised during different emergencies. CD Volunteers worked continuously for many days in incidents like the Pettimudy land slide (Munnar). They help in creating awareness programs & videos for the public. HAM Radio operators among the CD Volunteers provide necessary communication facilities where normal communication is difficult. Professionals like Doctors, Engineers etc. are there in CDV whose personal competency can be used during emergencies and for prevention activities.

5.2 Advanced Training in Water Rescue

As part of ensuring better preparedness to respond to such emergencies in future, GOK have established Institute for Advanced Training in Water Rescue, under KF&RS Department. The KF&RS personnel are getting trained in advance water rescue operations there. SCUBA Diving teams are given special training here and special allowances are sanctioned to those complete advanced training. Every event is critically analysed and new prevention strategies are formulated here.

5.3 Strengthening of Prevention Mechanism

As part of implementing more effective prevention mechanism, the Department started **Safety Beat system** in all Fire and Rescue Stations to find out the possible hazards in the Station area, identify the risk and to plan accordingly. Every Station area is divided into a number of beats and an officer is given charge of each beat area. Every potential hazard in the area is listed and situation analyzed at regular intervals. Awareness classes conducted as and when required and Community Participation is ensured by the beat officers. Buildings are inspected from fire protection and safety point of view and irregularities are reported to the concerned. Multi Hazard Maps are prepared for every District and for the entire State.

5.4 Internal Vigilance and Intelligence Wing for the Department

An 'Internal Vigilance and Intelligence wing' is formed within the Department to ensure effectiveness the activities. This wing intimates the likelihood of disasters to concerned Department in advance. Lapses like lack of proper coordination, delay in providing service etc. and shortcomings in the service provided are also identified and reported by this wing. The effectiveness of prevention mechanism of KF&RS is also observed and reported.

5.5 Modernization of the Department

Modern equipment and vehicles will increase the speed and effectiveness of emergency rescue operations. Modern Personal protective Equipment will help to build confidence and self esteem for the Departmental personnel. GOK considered the need for modernisation of KF&RS and allocated necessary funds. Initiatives for the modernization of Fire and Rescue Services Department, to adapt with the changing situation is on its way. Discussions are on regarding the procurement of robots and Drones for fire fighting and effective rescue operations.

5.6 Safety Culture in the Society

The society need to be changed a lot for effective prevention mechanism. Safety Culture needs to be inculcated in the society. Awareness creation programs are being conducted to achieve this goal. The most effective way to make a safety-oriented society is to inculcate safety culture in students. This can be achieved through including such topics in the School Syllabus.

5.7 Separate Planning for Different Seasons

Every season will bring some different tasks for the rescue team and planning for every season is very important. Monsoon preparation plan, summer preparation plan and drowning prevention plan are prepared separately and updated every year. Plans are prepared station wise and integrated to District and State plans.

5.8 Customer Satisfaction Survey

It is quite logical that only the person who requested the service can give proper feedback of the service delivery of the Department on account of whether the requested service was delivered as expected. As part of the efforts to further improve the service delivery of the Department, a survey was conducted among those who have recently received service from this Department with an aim to assess whether the services rendered by the Department reach the people on time and whether the public is satisfied with the services of this Department. Opinion of the public regarding the service received is recorded and suggestions are collected to improve the quality of service. In the completed survey, the percentage of overall satisfaction was 92.

5.9 Formation of a Special Task Force

A proposal is submitted to GOK for the formation of a Special Task Force to deal with emergencies occurring in the State. It is proposed to station the Special Task Force at 3 strategic locations, Trivandrum, Kochi and Calicut. They can assist the Fire Stations during any major emergency and can be used for VVIP security.

6. Disaster Response Improvement during the Floods 2019, 2020, and 2021

In the light of the experience during 2018, Kerala Fire and Rescue Services Department changed the mode of preparation for Monsoon. Heavy rains and flood continued during the monsoon season in the year 2019, 2020 and 2021, which caused heavy damage to the State. Huge landslides were there during 2019 in Puthumala inWayanad District and Kavalappara in Malappuram District. Similar incident happened in2020 also at Pettimudy, Munnar in Idukki District. Heavy rain, flood and landslide were there in 2021 also. But in all such instances, KF&RS Department effectively managed the rescue operations in a time bound manner. With the help of Civil Defense volunteers and Safety Beat Officers, the Departmental Officers could reach the incident spot immediately after the incident and start rescue Operations. It is only because of the proper planning and preparedness of the KF&RS Department, based on experience, the monsoon incidents during 2019, 2020 and 2021 were not that much destructive in terms of life and livelihood.

7. Conclusion

Kerala Floods 2018 is an indicative of the possible hazards in future. It reveals that we need to be prepared to face any unexpected challenges, like the floods 2018. Every time it need not be a flood. It can be fire, cyclone, earthquake or even a new type of disaster that the world never ever heard of. The only thing that matters will be our planning and preparedness to deal with such Disasters, especially in the golden hours.

References

1. Central Water Commission, 2018. Report on Kerala Flood & Solutions
2. Kerala Post Disaster Need Assessment – Floods and Landslides – August 2018.(UN Organisations, Asian Development Bank, Government of Kerala, The World Bank, European Union Civil Protection and Humanitarian Aid, 2018).
3. Centre for Social Studies and Culture (CSSC), 2018. Rainfall, Landslides and Deluge in Kerala, 2018.
4. ‘Pralayam 2018’– Report of Kerala floods 2018, Department of Information and public relation, Government of Kerala.
5. ‘Athijeevanam 2018’ – Report of the Activities of Kerala Fire and rescue Services Department during Kerala Floods 2018
6. Disaster Management Laws in SAARC countries edited by S. Sivakumar, Manohar Tairani, and Lisa P Lukose

Part 3

Cyclone Risk Mitigation and Management

CHAPTER 18

Impact and Management of Super Cyclone Amphan in Odisha

D. Panda
Associate Professor, P. G. Department of Geography,
Utkal University, Bhubaneswar
damodar.geography@utkaluniversity.ac.in

M. Devi
Assistant Professor, Department of Physics,
School of Appied Sciences, KIIT University, Bhubanesar.
mdevifpy@kiit.ac.in

Abstract

Odisha located on the east coast of India is highly vulnerable to tropical cyclones of various intensities. The frequency, intensity and magnitudes of these cyclonic storms have increased in the recent past leading to heavy loss of life and property. It has been observed that very severe cyclonic storms have changed their timing of occurrence from post monsoon to pre monsoon season. In this paper an attempt has been made to study the impact of the most recently occurred super cyclonic storm amphan in Odisha. The research work is based on the information of the Special Relief Commissioner, Odisha, IMD and, NRSC. The composite index has been used to assess the vulnerability of the districts to super cyclone Amphan. Amphan, after its origin in the Bay of Bengal made a land fall across Sunderban in West Bengal on the evening of 20th May 2020. The cyclone passed the Odisha coast with heavy rainfall and high wind speed. Due to its impact northern coastal and adjoining districts were affected with damages of infrastructure, power and agriculture. The state government took a lot of preventive measures to reduce the impact on the loss of life and hardship of likely affected people after getting the information from the IMD. The major steps taken were communication of warnings to the people through mass media, electronic media and local announcements, evacuation of vulnerable people to safer places, mobilization of manpower and machine and relief materials. Due to effective preventive measures loss of life could have been minimized and standing crops were saved.

Keywords: Amphan, preparedness, impact, response, vulnerability and mitigation

1. Introduction

1.1 Natural Disasters

With the effects of natural hazards rising in terms of loss of life and injuries in poorer nations (ISDR, 2002; World Bank, 2005; CRED, 2007), institutional disaster reduction approaches (ISDR, 2004; UNDP, 2004; DFID, 2005) and approaches adaptable to individual social and livelihood experiences are required. Vulnerability factors have been central to the development of disaster research (Khan, 1974; Burton, Kates and White, 1993; Hewitt, 1997; Twigg and Bhatt, 1998; Weichselgartner, 2001, Wisner et al., 2004; IFRC, 2006). However, despite good progress in cyclone preparedness, exemplified by the existing comprehensive disaster management policies of the Government of India, localised vulnerability factors in cyclone hazards arguably remain only partly considered. Indigenous cyclone response experiences could be more central to cyclone interventions in the country. Enhanced local-level adaptation to cyclone hazards through established coping will be crucialfor resilience to any suspected increase in cyclone hazards.

Burton, Kates and White (1993, p. 48) presented the view that hazard response 'encompasses all the ways in which a society may act to reduce the effects or increase the benefits of a hazard'. These include actions immediately before hazards (such as warning signals, emergency evacuation or other precautionary measures), or long-term actions (such as planting crops less susceptible to drought), or adaptation to extreme environments (such as building villages on levees). Hence, human responses for living with natural hazards have been differentiated as both short- and long-run actions involving the use of 'adjustment' and 'adaptation' respectively. For example, 'designing a house to resist a storm surge would be an adjustment; locating and organizing a community over a long period of time so that its houses are beyond the storm surge would be an adaptation' (Burton, Kates and White, 1993, p. 49). Adaptation also refers to the manifestation of adaptive capacity (Smit and Wandel, 2006), meaning the capacity of a system to deal with, accommodate, adapt to and recover (Jones, 2001; Smit et al., 2000)

Meanwhile, 'coping is the manner in which people act within the limits of existing resources and range of expectations to achieve various ends. In general this involves no more than "managing resources", but usually it means how it is done in unusual, abnormal and adverse situations' (Wisner et al., 2004, p. 113). 'Thus coping can include defence mechanisms, active ways of solving problems and methods for handling stress' (Murphy and Moriarty, 1976, quoted in Wisner et al., 2004, p. 113). Hewitt (1997, p. 36) defined coping as 'the actions of ordinary people or disrupted remains of institutions, in contrast to official and planned response'.

The geographical location and physical environment make Odisha Coastal Zone vulnerable to frequent cyclonic storms. The high wind speed together with torrential rain and storm surges associated with the cyclones bring damage to the coastal settlements. The state generally experiences cyclones during the pre-monsoon and post-monsoon season. During their course the storms turn towards north and northeast direction .Re-curving storms mostly strike Odisha. The post-monsoon cyclonic storms are very common and destructive. Out of 30 districts 14 coastal districts are highly vulnerable to cyclonic storms. The East Coast of India is one of the six most cyclone-prone areas in the world. The Odisha coast is highly vulnerable to cyclonic storms in comparison to other eastern coastal States. It has been observed that the magnitude, intensity and frequency of the cyclones are increasing in the recent past. Due to the impact of the climate change, the cyclones are changing their path, life cycle, magnitude and frequently, thus posing serious challenge to predict using modern forecasting techniques. Some of the most severe cyclonic storms recently affected the state are, supercyclone in 1999, Phailin in 2013, Hudhud in 2014, Titil in 2018, Fani and Bulbul in 2019.

Odisha is one of the state of India which first developed a state disaster management authority (OSDMA) after the super cyclone of 1999. It coordinates the state disaster management plan from the village level to the state level with trained manpower and machines. Odisha Disaster Rapid Action Force (ODRAF) in line with National Disaster Response Force (NDRF) of the government of India. With different funding agencies multipurpose cyclone and flood shelters are constructed in vulnerable areas for shelters of the evacuated people.

2. The Study Area-Odisha

Odisha is located in the east coast of India with coast line of 480 km along the Bay of Bengal. The state has an area of 155707sqkm and consists of 30 districts. The state has been divided in to four physiographic regions. (i) the northern plateau, (ii) the central river basin, (iii) the eastern hills and (iv) the coastal plain. It is drained by major rivers in to the Bay of Bengal. The state experiences monsoon climate with warm summer and cold winter. The coastal Odisha is economically developed in comparison to other parts of the state due to fertile soil which supports high population density (Fig. 18.1).

2.1 Data Base and Methodology

The study is based on the information of the Special Relief Commissioner, Govt. of Odisha, Indian Meteorological Department, Govt. of India and satellite images with flooding areas of the state due to the impact of Amphan. The data collected are analysed using various stattstical methods, GIS and composite index for understanding the impact and management. The visuals of the damages are used from print and electronic media to assess the intensity of the impact in terms of damages.

2.2 The Tropical Super Cyclone Amphan

The Super Cyclonic Storm (SuCS) "AMPHAN" was the first SuCS over the BoB, after the Odisha SuCS of 1999. It originated as a Low Pressure Area which occurred near Equatorial Easterly wave over south Andaman Sea adjoining southeast Bay of Bengal (BoB) on 13th May. It concentrated into a depression (D) over southeast BoBand further intensified into a deep depression (DD), Severe Cyclonic Storm (SCS), Extremely Severe Cyclonic Storm (ESCS) , Super Cyclone (ScCS) and moved north- northwestwards .It crossed Sundarbans, during 1530-1730 hrs IST of 20th May, with maximum sustained wind speed of 155 – 165 kmph gusting to 185 kmph. Moving further north-northeastwards, it weakened into an SCS over Bangladesh.

Analysis

Impact of Amphan

Due to the impact of Amphan 10 northern coastal and adjoining districts (Jagatsinghpur, Balasore, Kendrapara, Jajpur, Mayurabhanj, Puri, Bhadrak, cuttack, Khordha, and Keonjhar), 97 blocks, 9838 villages, 22 ULB and 45.26 lakh

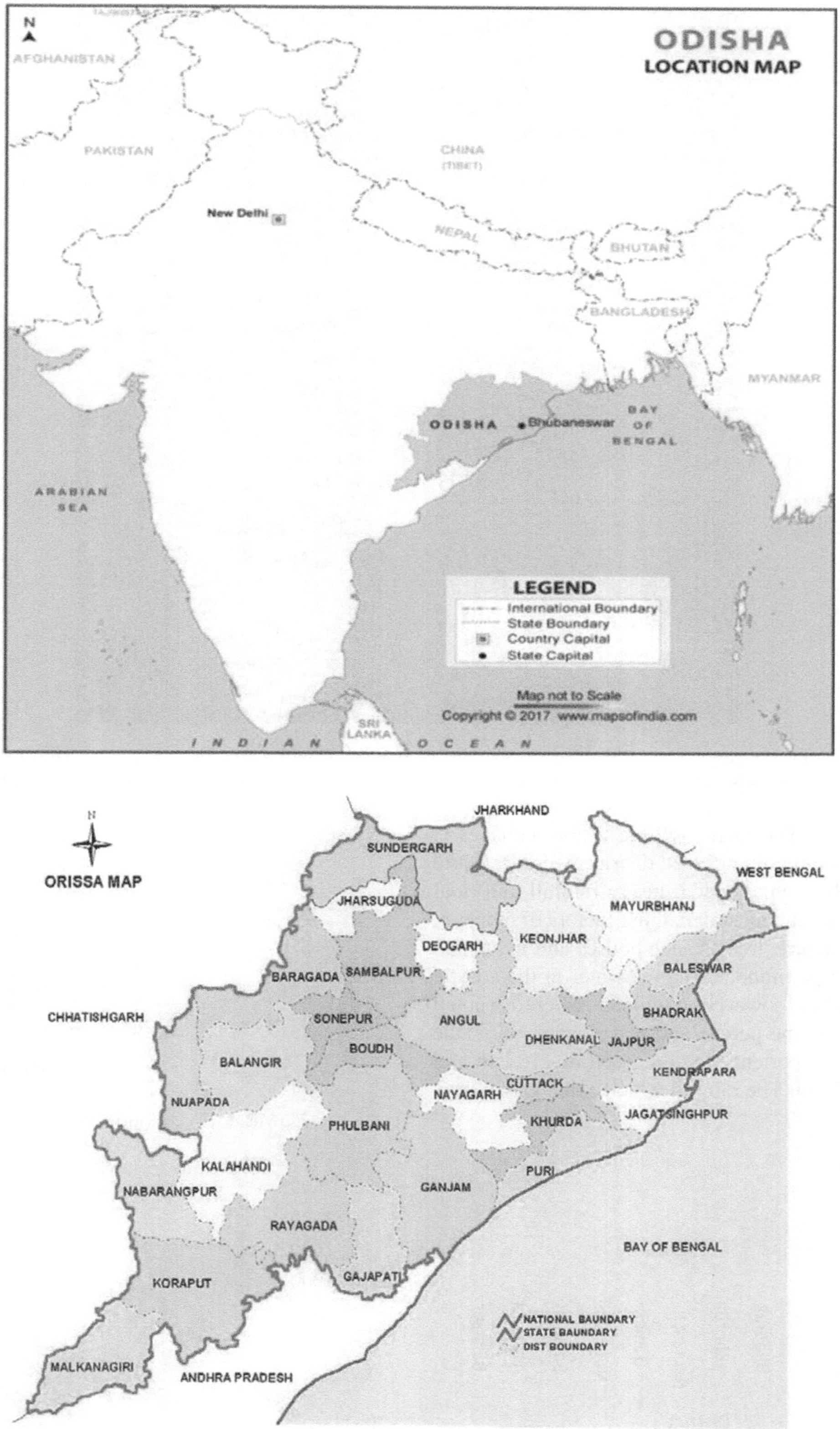

Fig. 18.1 The study area- Odisha

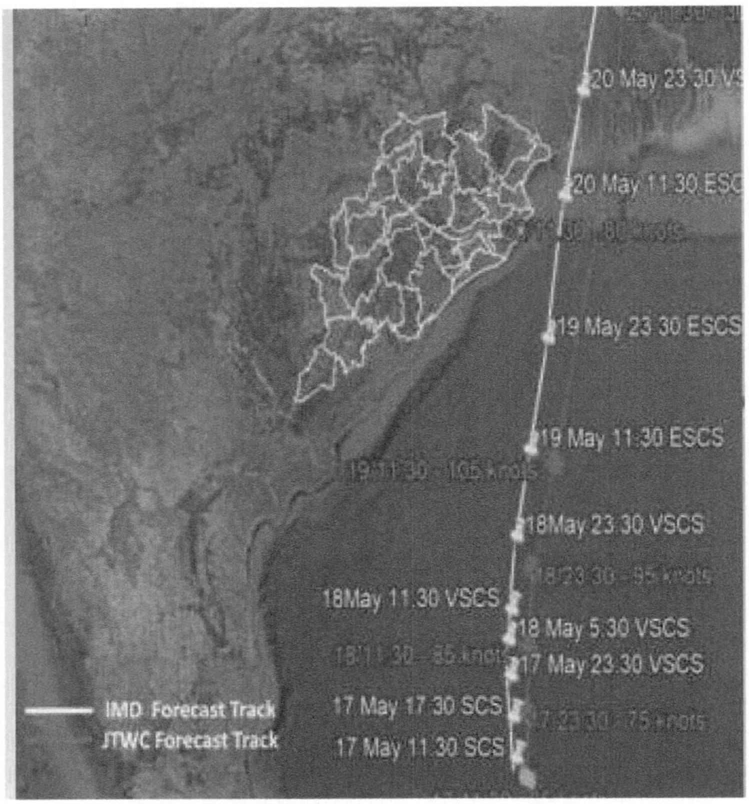

Fig. 18.2 Track of the Super Cyclone Amphan

Source: Regional Specialised Meteorological Centre-tropical Cyclones, New Delhi, Indian meteorological Department (IMD), Govt. of India

people were affected. The total loss due to the damages is Rs. 254.25 crore. The highly affected districts were Balasore and Kendrapara. High wind speed, heavy rainfall and tidal surges at the coast are the most damaging factors of amphan. It damaged infrastructure, trees, kacha houses and inundated crop lands. Tidal surge inundated saline water in the coastal cultivable land. Road and power communication were disrupted leading to hardship to the people and rescue operation. Due to The coastal parts experienced wind speed of 110 kph and the interior parts 75 kph.The rainfall which extends two days

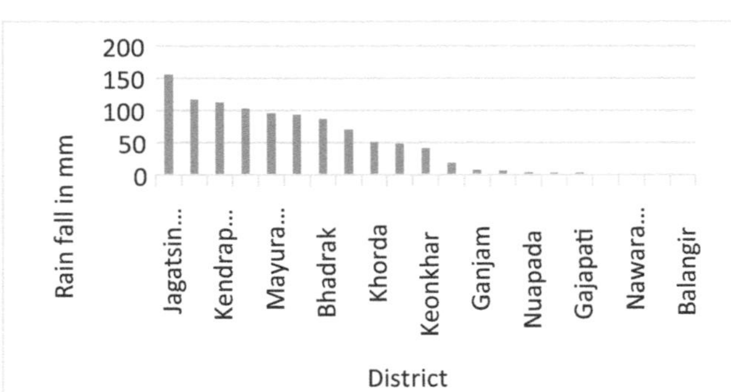

Fig. 18.3 Rainfall in mm from 20-21 May 2020 in the districts of Odisha

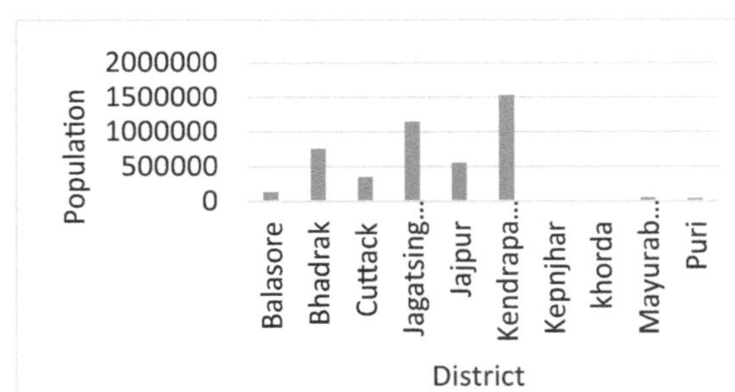

Fig. 18.4 Population affected in the Districts

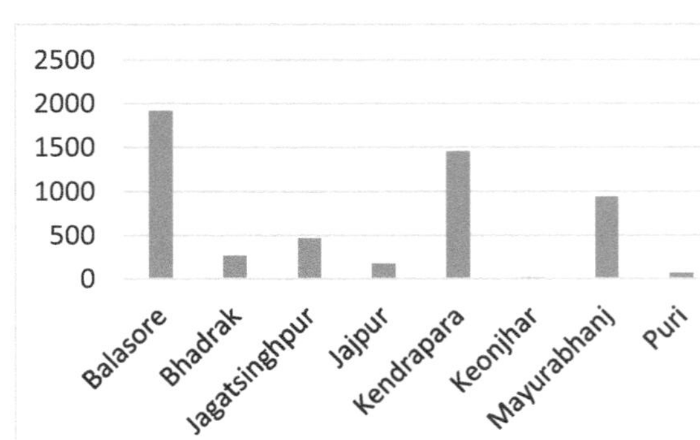

Fig. 18.5 Number of houses damaged in the districts

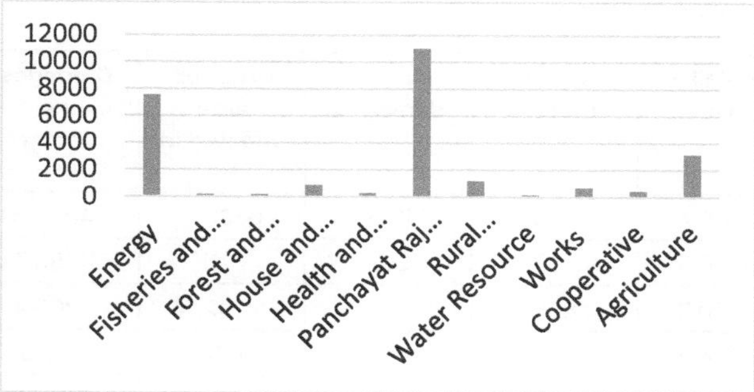

Fig. 18.6 Extent of Damage lakh

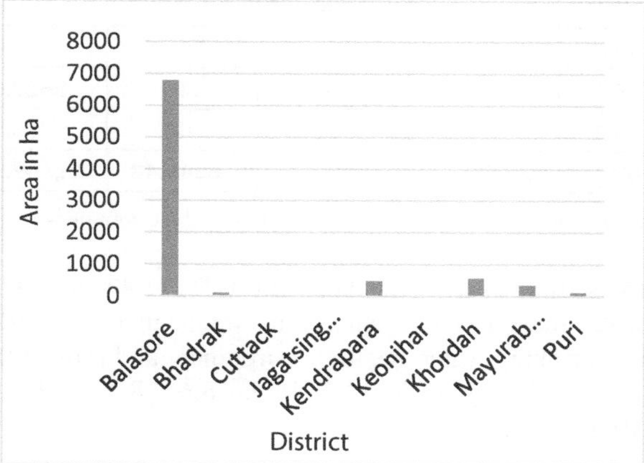

Fig. 18.7 Croped area affected in Ha

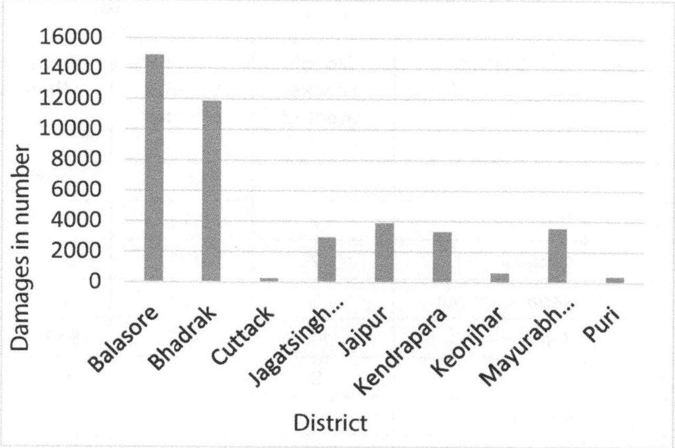

Fig. 18.8 Damage in Energy sector in the district

20 and 21 May with more than 100 mm at the coastal part and less in the interior parts. Jagatsinghpur received 155.4 mm rainfall which is 199% of the monthly rain fall of the district. Maximum villages affected in in Balasore district followed by, Kendrapara, Bhadrak and Jagatsinghpur districts. Highest number of villages is affected in Balasoredistrict. Out of total population of 45.26 lakh. Livestocks are the livelihood of the poor villagers, any loss of lifestock adversely affect their livelihood and economic condition. About 20,778 number of livestocks were killed with maximum in Jajpur district followed by Cuttack, Balasore and kendrapara districts. Total coast of the livestock loss is Rs, 8.20 lakh. Power sector always suffers heavily due to high wind speed by damaging the power transmission sysiem. Most of the electric poles were damaged due to uprooting trees and broken branches. Its reinstallation and reparation needs huge money and time. Total loss of the power sector is 75.14 crore. Balasore district suffers heavily in power sector followed by Bhadrak, Jajpur, Mayurabhanj and Kendrapara district. The coastal region of the state is agriculturally developed, due to heavy rainfall and

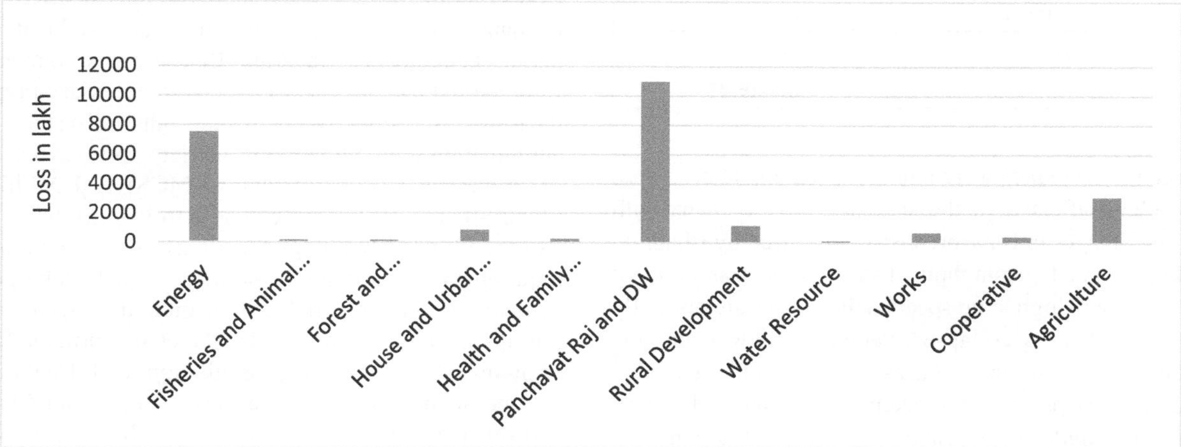

Fig. 18.9 Department wise Damage Rs. In lakh

Table 18.1 Impact of Amphan in Odisha

Sl.No	District	No. of Blocks affected	No. of Villages affected	Population affected	Rainfall in mm	Livestock affected	Houses damaged	Cropped area affected in Ha	Damages in energy sector
1	Balasore	12	2898	130000	108.7	1010	1915	6789	14853.34
2	Bhadrak	7	1373	750000	86.5	7	263	111	11856.74
3	Cuttack	7	474	344077	70.3	1865	—	12.74	266.4
4	Jagatsinghpur	8	1308	1136971	144.4	878	464	10/74	2914.1
5	Jajpur	10	666	547526	102.4	17001	173	—	3860.27
6	Kendrapara	9	1592	1522901	112.4	17	1451	478	3281.26
7	Keonjhar	7	18	258	41.4	—	14	11	638.68
8	Khordhar	5	62	3172	49.7	—	—	570	—
9	Mayurabhanj	21	983	49800	95.6	—	937	349	3522.65
10	Puri	11	464	40857	93.2	—	65	118	376.5
Total		**97**	**9838**	**4525562**	**9157**	**20778**	**5282**	**6349.48**	**41,569.94**

Source: Final memorandum on Super Cyclonic Storm Amphan,May ,2020,Special Relief Commissioner (SRC),Govt. of Odisha,Bhubaneswar.

Table 18.2 Department wise loss due to Amphan

Sl. No	Departments	Loss in lakh
1	Energy	751.30
2	Fisheries and animal resources	160.40
3	Environment and forest	157.87
4	Housing and urban development	836.91
5	Health	256.15
6	Panchayat Raj department	10977.80
7	Rural development	1165.24
8	Works department	131.77
9	Cooperative department	420.30
10	Water resources	131.77
11	Agriculture	3115
Total		**25425.89**

tidal surge the croplands were inundated leading to crop loss. Total cropland affected in the state was 10726.18 ha with maximum damages in Balasore district followed by Bhadrak, Mayurabhanj and Jagatsinghpur district. Total loss is about 4.56 crore. Due to high wind speedkacha houses are normally weak structured easily collapsed. People with low economic condition built traditional houses using locally available raw materials. Steps are being taken by the state and central government to built under Prime Minister abash yojana and state govt. Biju paccaghar yojana. Total number of houses damaged were 5282 with maximum damaged in Balasore followed by Kendrapara, Mayurabhanj and Jagatsinghpur districts. Total coast of the damage was Rs.21.47 crore. Other damages include roads, canals, embankments both river and saline, drinking water, health, environment and forest (Table 18.1 & 18.2, Fig. 18.3, 18.4, 18.5, 18.6, 18.7, 18.8, 18.9, 18.10 & 18.11).

Management of the impact super cyclone Amphan

Preparedness

The state government and the central government after getting the information from IMD on 16th May 2021 about the likely impact of the super cyclone in the northern coast of the state state, regular monitoring with necessary preventive measures were taken. Concerned departments of the government and Collectors of the 12 coastal and adjoining district Gajapati, Ganjam, Nayagarh, Puri, Khordha, Cuttack, Jagatsinghpur, jajpur, kendrapada, Bhadrak, Balasore and mayurabhanj) were alerted and advised to undertake necessary preparatory measures to effectively handle the disaster. District administration of vulnerable districts were asked to keep Multipurpose Cyclone Shelters (MCS) and Multipurpose Flood Shelters (MFS) in readiness to identify other suitable RCC roof public buildings for use as safe shelters in case of requirement. To accommodate more than 12 lakh people in 12 districts were prepared with maintaining social distancing norms for Covid-19. It was also directed to arrange for Food, Drinking Water, Lighting, Sanitation and Health facility at each shelter and keep the Tower Light, and Chain Saw and other emergency equipment available in MCS/MFS in readiness to meet any eventuality. Collectors of the districts likely to be affected to assess the situation and identified the people living near the coast/low lying areas, people

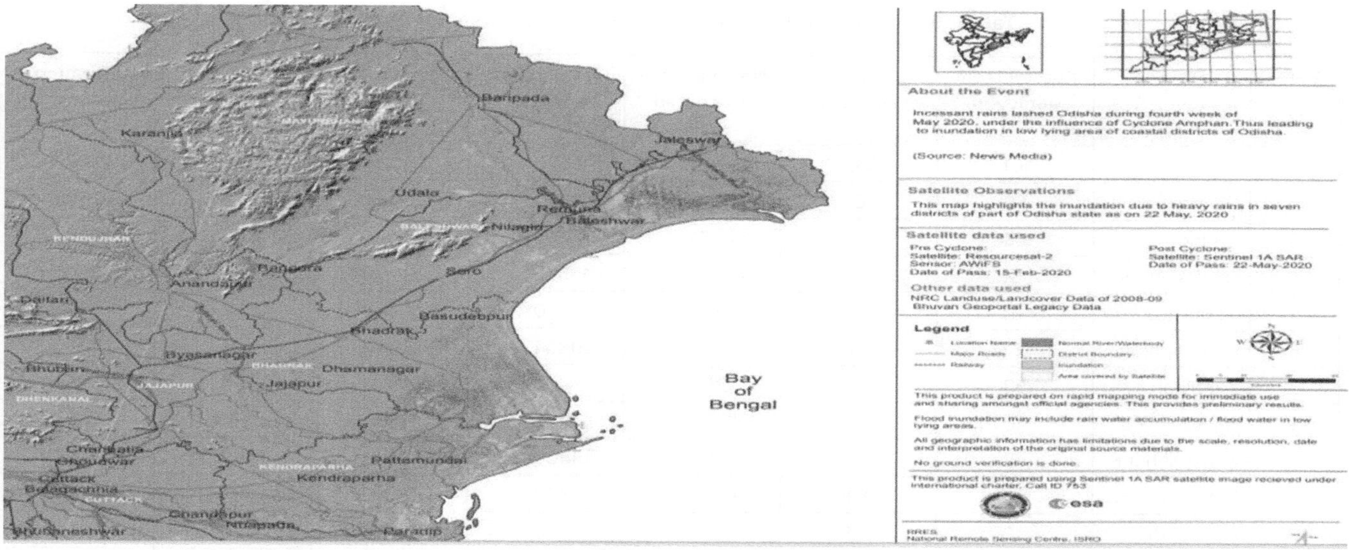

Fig. 18.10 Satellite image showing flood inundated areas due to Amphan

Source: National Remote Sensing Centre(NRSC)),Hyderabad,Department of Space,Govt. of India.

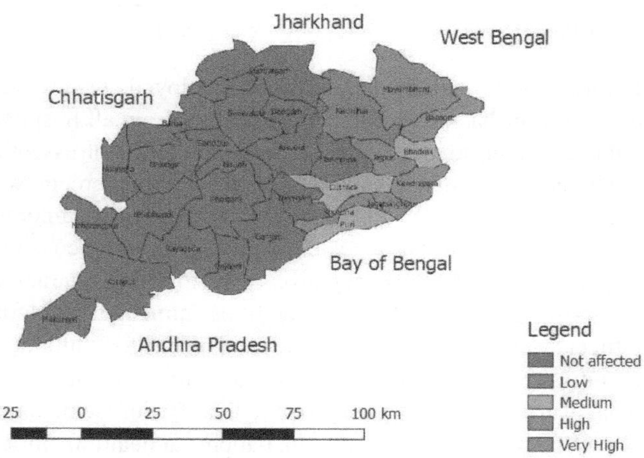

Fig. 18.11 Vulnerability map of Odisha due to Amphan

Source: Prepared by the author

living in vulnerable condition in hut or kutcha house & take steps to evacuate them to MCS/MFS or other identified safe buildings and organize all basic needs like food, drinking water, lightning, sanitation and health facilities for them. ollectors of Balasore, hadrak, Kendrapada and Jagatsinghpur were requested to take immediate steps to remove all hoardings in ULB areas and roadside hoardings put in rural areas. Senior IAS Officers and IPS officers were deputed to provide guidance and support to the District Administrations in management of the Cyclone. 2,00,346 people were evacuated to the safe shelters including1885 pregnant women were shifted to hospitals/Maa Gruhas.20 NDRF teams, 16 ODRAF ,OFDC teams and 231 Fire Service team were deployed in 6 coastal and 6 nearby districts. Balance 4 ODRAF teams, 17 teams of Fire Service and 20 OFDC teams were kept standby to be pressed in to service .Power-Advance action was taken by Odisha Power Transmission Corporation Limited vulnerable condition in buildings and organize all basic needs like food, drinking water, lightning, anitation and health facilities for them. ollectors of Balasore, Bhadrak, Kendrapada and Jagatsinghpur were requested to take immediate steps to remove all hoardings in ULB areas and roadside hoardings put in rural areas. Senior IAS Officers and IPS officers were deputed to provide guidance and support to the District Administrations in management of the Cyclone. 2,00,346 people were evacuated to the safe shelters including1885 pregnant women were shifted to hospitals/ Maa Gruhas.20 NDRF teams, 16 ODRAF ,OFDC teams and 231 Fire Service team were deployed in 6 coastal and 6 nearby districts. Balance 4 ODRAF teams, 17 teams of Fire Service and 20 OFDC teams were kept standby to be pressed in to service .Power-Advance action was taken by Odisha Power Transmission Corporation Limited (OPTCL) and Distribution Companies (DISCOMs) for mobilizing technical gangs and materials to take up restoration immediately after the cyclone. Gangs with Gen Sets, gas cutters and vehicles were kept in readiness for immediate deployment. As power supply was likely to be shut down before the cyclone, the life line institutions including Government offices were asked to make power back up arrangement (Genset with sufficient fuel). 24X7 Control room was set up in OPTCL/ GRIDCO for sharing of information on 24X7 basis. District wise nodal officers were allotted for monitoring the situation.

Table 18.3 Preparedness and response to the impact of Amphan

Sl. No	District	Rescue team deployed	No. of shelters	No. of people evacuated	No of relief beneficiary	Relief in Qtl	No. of medical team deployed	Restoration of water supply
1	Balasore	101	720	49564	2526	408	22	2877
2	Bhadrak	63	561	31017	31017	14.86	8	5280
3	Cuttack	45	428	10961	10961	137.07	0	1226
4	Jagatsinghpur	51	299	16840	13575	116.27	16	354
5	Jajpur	41	171	7140	0		36	450
6	Kendrapara	61	622	39864	39864	1284.7	37	2530
7	Keonjhar	53	2	258	0		0	0
8	Khordha	40	103	3172	1189	5.53	21	549
9	Mayurabhanj	79	1029	27216	3500	401	89	1819
10	Puri	32	364	12712	2893	31.97	32	1250
11	Nayagarh	0	1	55	0		0	0
12	Ganjam	66	0	0	0		0	0
13	Gajapati	18	14	1547	0		0	0
Total		**650**	**4314**	**200346**	**105525**	**2399.34**	**261**	**16305**

Source: Final memorandum on Super Cyclonic Storm Amphan,May ,2020,Special Relief Commissioner (SRC),Govt. of Odisha,Bhubaneswar.

Telecom: Meetings were held with all Telecom Service Providers (TSPs) for taking necessary preparedness measures to keep the Telecom/Internet Service functional during cyclone situation and to restore the communication in case of any damage due to cyclone. The TSPs made necessary arrangements for functioning of all the DG sets with fuel reserve of 24 hours. Movable DG sets were kept ready at central locations for movement. 23 All the TSPs kept Cell-on-Wheels in readiness for deployment in case there is damage to mobile towers in district due to impact of AMPHAN. All the telecom operators were advised to be ready with adequate numbers of technical manpower for restoration of any affected telecom towers and Optical Fiber Cables (OFC),

Drinking water-276 DG sets were kept ready for Rural Pipe Water Supply (PWS) projects in 12 districts for uninterrupted water supply in absence of electricity. 805 vehicles loaded with PVC tanks and 75 water tankers were in readiness for use. 199 mobile repair units (each comprising of two SEMs) and 106 local electricians were kept ready for taking up repair work on emergency basis. 2724 no. of overhead tanks were kept loaded with water. 1828 bags of bleaching powder were supplied for disinfection.

Urban drinking water-113 DG sets, 95 tankers and 246 PVC tanks were arranged for deployment in the ULB areas of Balasore, Bhadrak, Kendrapada, Jagatsinghpur and Mayurbhanj districts for ensuring uninterrupted water supply. 20 storage reservoirs were kept for continuous pumping operation to ensure availability of adequate stock for 24 hours post cyclone.

Health-Mobile health teams were deployed in affected areas. Power back up arrangement was made in all hospitals with sufficient quantity of fuel. Sufficient stock of essential medicines, anti snake venom was kept. Arrangement was made for screening all migrant workers in Temporary Medical Centres (TMC) located within 3 km of the coastline and to send the COVID negative cases to home quarantine and positive cases to COVID facilities immediately. Nodal Officers were identified for districts and blocks for continuous monitoring of disaster response activities. Contingency plans were made for opening of Medical Relief Centers and providing treatment to the trauma patients at health facilities. Contingency plan for sanitation was in place and stocking of bleaching power/ ORS/ Halogen tablets at PHC, CHC, DHHs and up to ASHA level was ensured.

Fishermen-Warnings were issued to the fishermen and were advised not to venture into sea. District Administration along with officials of Fisheries Department was in close coordination with the marine fishermen in the matter. All 4115 number of boats that had gone to sea for fishing returned to shore by 17th May and no boat was allowed enter into sea from 18th. 22,050 number of boats in 6 coastal districts and Chilika area of Khurdha district berthed in safe places like fishing harbor and fish landing centres. Awareness campaign on protection of fanimal resources from possible cyclone situation was launched in electronic as well as print media. Supply of required medicines and vaccines to the CDVOs for distribution to vulnerable areas was expedited. People were advised to shelter the cattle and districts and

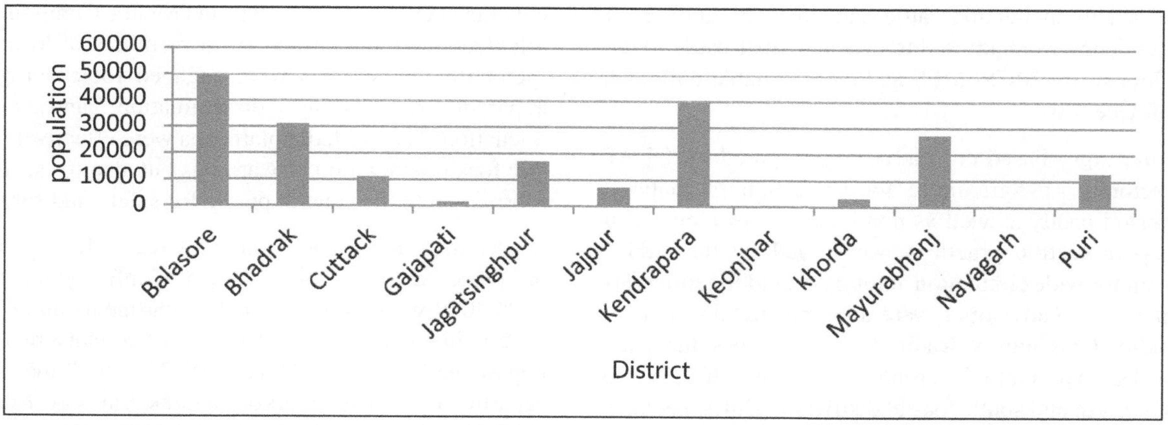

Fig. 18.12 People evacuated due to the impact of Amphan

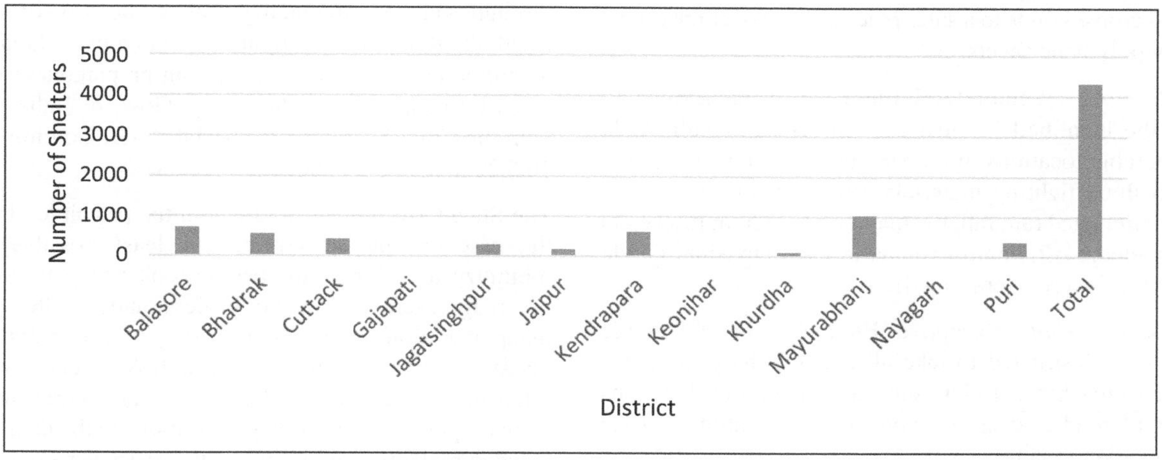

Fig. 18.13 Number of Shelters for evacuated people

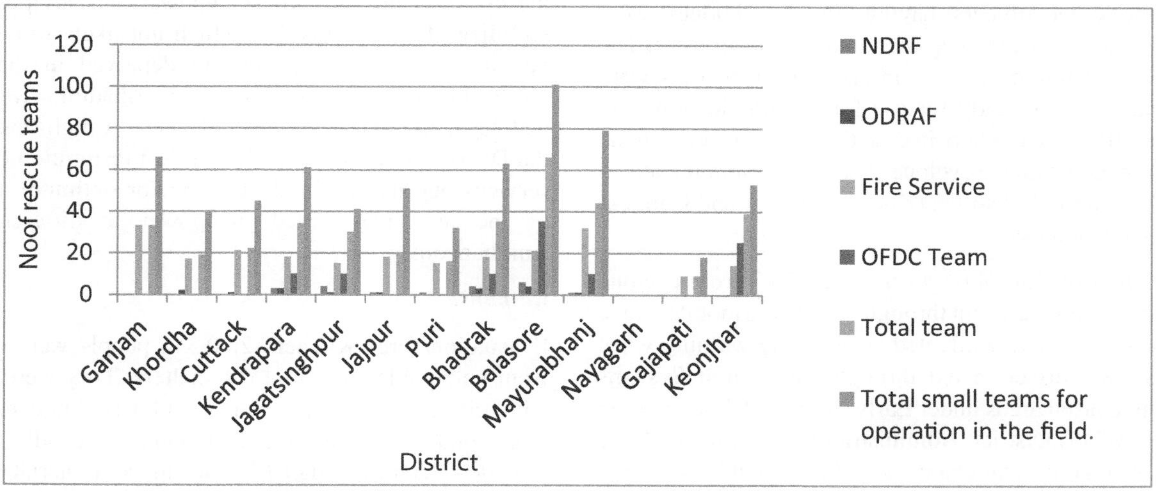

Fig. 18.14 Rescue teams deploid in the vulnerable areas

Chilika area of Khurdha district berthed in safe places like fishing harbor and fish landing centres. Awareness campaign on protection of animal resources from possible cyclone situation was launched in electronic as well as print media. Supply of required medicines and vaccines to the CDVOs for distribution to vulnerable areas was expedited. People were advised to shelter the cattle and domestic animals at safe places. They were advised to evacuate their cattle to the ground floor of the MCS/ MFS and other suitable buildings in case of necessity.

Agriculture: Agro based crop advisory, prepared by OUAT and Directorate of Agriculture for protection of matured and harvested paddy as well as non paddy crops along with other crop protection practices was issued by the A&FE Department for wide circulation among the field functionaries and farmers. Crop advisories were disseminated through All India Radio, Community Radio Stations across the state and Door Darshan. Out of a crop coverage area of 2,65,800 hectares in north and south coastal districts 1,37,400 hectares were harvested in a week time. Awareness camps and mike announcements were done in different villages to take the harvested crops/ grains to a safer place and to cover the crops/ grains by polythene sheets.

Water Resources: Senior level Engineers were deputed to each of the identified 12 districts. All critically vulnerable and vulnerable locations of embankments were restored and sufficient flood fighting materials were kept in position. In view of anticipated rain fall, the major and medium reservoirs were depleted.24x7 Control rooms in each Division, Circle and CE Office were operationalized.

Rural Development: Executive Engineers of vulnerable districts were instructed to take all precautionary measures with men, machine, roller, etc. in vulnerable locations of different roads, so as to avoid any dislocation in road communication system.

Communication: The Collectors, State Disaster Response Forces and other officers having Satellite Phones were directed to recharge and keep the phones ready for use. The Digital Mobile Radio installed in six coastal districts were also tested and kept ready for use for communication with Collectors, BDOs and others in case the telephone/cell phone networks are down due to cyclone. The V-sat communications at State HQ and in 3 districts (Khordha, Jajpur and Cuttack) were also in readiness.

Dissemination of Warnings: As many as 2.17 crore cyclone warning messages were sent through SMS to the mobile phone subscribers of the likely affected areas. Early warning voice messages were disseminated through Alert Siren Systems installed in coastal areas under Early Warning Dissemination System (EWDS).District Administration disseminated the warning through official channels and PRI members. Public address systems were used for dissemination at local level. Fisheries and ARD Department communicated the warning messages to marine/Chilika fishermen. All cyclone bulletins and warning messages of IMD were disseminated with key State Level Officers, Collectors and Media. Community Level Volunteers and Task Force Team members of Multipurpose Cyclone/Flood Shelters were engaged for ensuring family preparedness, warning dissemination and expediting evacuation. Social Media platforms were effectively used not only for dissemination of warnings but also for sending vital information to the general public for safety and survival.

Review Meetings: The Hon'ble Chief Minister reviewed the preparedness through Video Conferencing (VC) on 16.05.2020 with the concerned Departments and Collectors of 12 vulnerable districts. The Chief Secretary reviewed the preparedness through VC on 19.05.2020 with the concerned Departments and Collectors of Bhadrak, Balasore, Kendrapada and Jagatsinghpur districts. Special Relief Commissioner & Additional Chief Secretary, Disaster Management reviewed through video conferencing with Collectors of Balasore, Bhadrak, Kendrapara, Jajpur, Jagatsinghpur, Cuttack, Puri, Khordha and Mayurbhanj regarding preparedness measures taken by them. Cabinet Secretary, Govt. of India discussed with the Chief Secretary & Special Relief Commissioner on 16.5.20 and 18.5.20 about preparedness.

The State Government in line with the principle of restoring the essential public services in least possible time to minimize hardship to the people took necessary steps well before the cyclone. A host of professionally skilled groups of manpower along with required equipment pertaining to key disaster response sectors like search & rescue/ evacuation, relief line/tree cutting/road clearance and energy restoration were prepositioned in strategic locations in the district jointly decided by Collector and SP of the districts to start working for restoration of the respective services immediately after the cyclone subsides. The manpower and equipment were mobilized from the districts which not likely to be affected due to the current cyclone and deployed in the 4 most vulnerable districts i.e., Balasore, Bhadrak, Kendrapada and Jagatsinghpur. This not only gave a high lead time to the District Administration for earliest restoration of the key services but also allowed planning for optimum utilization of the available skilled manpower at different placed simultaneously.

Response

Evacuation/ Free Kitchen: 2,00,346 people were evacuated from vulnerable areas to 4314 shelters. They were provided with dry food, drinking water and free kitchen. Besides this, Chuda, Gud, Biscuits, matchbox and candle were also provided. Hon'ble Chief Minister made an aerial survey of the affected areas on 21.5.2020. Hon'ble Prime Minister along with Hon'ble Chief Minister visited the cyclone affected districts on 22.05.2020 by helicopter and took a review meeting with the Hon'ble Governor, Hon'ble Chief

Minister and Senior Officers of the State Govt. about the damages and relief measures.

Repair/Restoration Activities

Clearance of Rural Roads: All the roads blocked due to falling of trees and debris in Balasore, Bhadrak, Kendrapada, Jagatsinghpur and other affected districts were cleared within 24 hours with the help of NDRF, ODRAF, Fire Service and OFDC personnel. Restoration of Power: 44,57,182 no. of consumers were affected in Balasore, Bhadrak, Jagatsinghpur , Jajapur, Mayurbhanj, Ganjam Districts. Power was restored in five days in all affected areas (Table. 18.3, Fig. 18.12, 18.13 & 18.14)

3. Conclusion

The whole coastal districts are highly vulnerable to cyclonic storm. As the Amphan did not hit the Odisha coast, the impact was comparatively very low in comparison to the previous cyclonic storms. The impact of the cyclone is due to heavy rainfall and high wind speed. Maximum damages in the rural infrastructure, energy and agriculture. The composite index of the damages show the Balasore and Kendrapada districts are severely affected followed by other coastal district and least affected by Keonjhar district. Due to awareness and adequate preparations no death was reported. To minimize the loss of life and property, the cyclone and flood resistant infrastructure are proposed by the state government with financial aid from the government of India.

References

1. Alam. E and Collins. A. E (2010) Cyclone Disaster Vulnerability and Response Experiences in Coastal Bangladesh, Disaster, 34(4), pp. 331–354. USA.
2. Bahinipati. C. S. (2014) Assessment of Vulnerability to Cyclones and Floods in coastal odisha, India: A District level Analysis, Current Science, vol. 107, No. 12, pp. 1997–2007.
3. Indian Meteorological Department (2020).The Super cyclone Amphan, New Delhi
4. Kalsi, S. R. (2003). Monitoring and Forecasting of Tropical Cyclones and Their Associated Effects, Disaster management, pp. 48-84, Universities press, Hyderabad.
5. Mohapatra, M. (2015) Cyclone hazard proneness of Districts of India, Journal of Earth System Sciences 124(3), pp. 515–526.
6. Mohapatra, M., Mandal, G. S., Bandopadhyaya, B. K., Tyagi, A., & Mohanty, U. C. (2012) Classification of Cyclone Hazard Prone districts of India, Natural Hazard, 63, pp. 1601–1620.
7. National Institute of Disaster Management (2001) Manual on Natural Disaster Management in India. New Delhi.
8. Special Relief Commissioner (2020) Memorandum of the Supercyclone Amphan, Bhubaneswar.

CHAPTER 19

At the Cross-roads in Cyclone-prone Odisha: *Titli*—The Butterfly's Duel with Development

Geeta Vaidyanathan
Professor, Sri Sri University,
Corresponding Author: g vaidyanathan o yahoo.com,
geeta.v@srisriuniversity.edu.in

Ramani Sankaranarayanan
CTx GREEN, Kitchener, Ontario, Canada;
ctxgreen@gmail.com

Abstract

Cyclone Titli (translated as butterfly) devastated the homes and livelihoods of marginalized indigenous communities living in remote hilly villages of Ganjam and Gajapati districts of Odisha, India. Strong winds of 150 kmph accompanied by very heavy rainfall caused multiple disasters in the form of cyclone, floods and landslides on 11-12 Oct-2018. While the Odisha state machinery is adequately equipped to handle single hazards, the cascading effect of the unprecedented cyclone in these remote areas, made disaster management extremely difficult. The ongoing rehabilitation and rebuilding process in the hills of Ganjam and Gajapati districts is analysed in this paper using a conceptual framework based on the Disaster Management Cycle, with an emphasis on the preparedness phase. Mitigation measures that are put in place, as a step towards disaster-risk-reduction and enhancing the community's capacity to adapt to future disasters, are discussed in the context of the Sendai Framework's mandate of 'Building Back Better.'

Keywords: Resilient habitat, Disaster risk reduction, Mitigation, Odisha

1. Introduction

In the past four decades, storms and floods have accounted for more than two-thirds of all natural disasters globally. The frequency of disastrous events is also on the rise. Fig. 19.1 shows nearly a two- fold increase in global disaster events in the last twenty years (2000-2019) compared to the previous twenty years. Climate-related disasters are often freak events and although human mortality may be decreasing due to early warning systems and better preparedness, the losses to habitat, livelihoods and the economy continue to be tremendous. This is especially exaggerated in the case of the cascading effect of disasters, further exacerbated by their systemic nature that is intrinsically linked to loss of forest, soil cover and biodiversity, loss to agriculture and breakdown of critical infrastructure. Such direct and indirect damage results in heavy losses to the livelihood base in underserved rural areas and economic migration to far away cities, especially of men. A remittance economy and associated distress is a consequence, including more load on the women of the household entrusted with the responsibility of taking care of children, elders, the usual subsistence agriculture, in addition to keeping the household running. This paper is set in this context, where there is a clarion call at the international, national and regional levels of governments for "strengthening disaster risk reduction governance." (Centre for Research on Epidemeology of Disasters, CRED, 2020).

The coast of Odisha, in south-east India, is among the most cyclone-prone areas in the country (Table 19.1). The super cyclone of 1999 was followed by Phaillin (2013), Hud-Hud (2014), Titli (2018) and the most recent Fani (2019), each causing widespread damage to property and the natural resource base. Titli made landfall as a very severe cyclonic storm on 11-Oct-18 around 4:30 to 5:30 am, between North

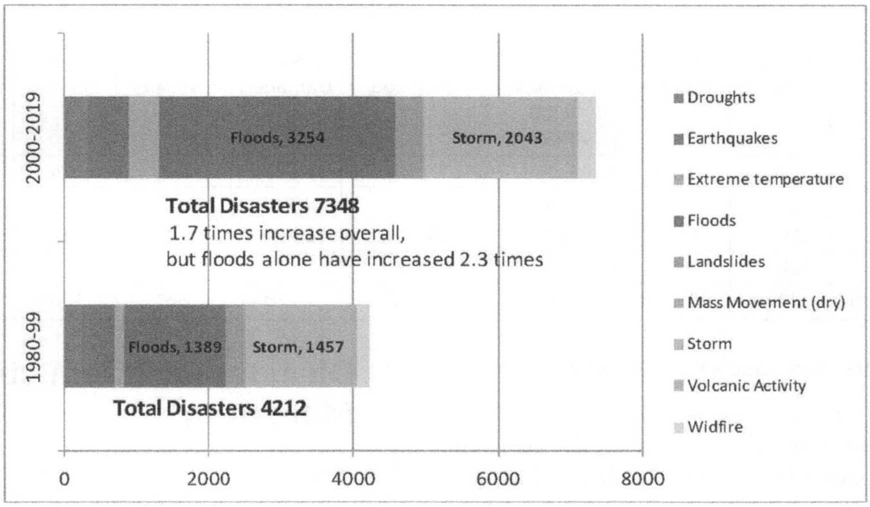

Fig. 19.1 Number of disasters globally by type, between 1980-99 and 2000-2019 (CRED, 2020)

Table 19.1 State-wise distribution of Cyclonic storms: 1891-2000 (IMD as cited in (Special Relief Commissioner, RDM, GoO, 2018))

State	No	Percentage	
		of total	per coast
East Coast	**308**	86%	
West Bengal	69	19%	22%
Odisha	98	27%	32%
TamilNadu	62	17%	20%
Andhra Pradesh	79	22%	26%
West Coast	**51**	14%	
Gujarat	28	8%	55%
Maharastra and Goa	18	5%	35%
Karnataka	2	1%	4%
Kerala	3	1%	6%
Total Cyclones	**359**		

Andhra Pradesh and South Odisha coast (18.8°N and 84.5°E), with a sustained wind speed of 140-150 kmph, winds gusting to 165 kmph, wreaking havoc all along its path. Titli continued its path in Odisha, in a north-westerly direction and lay over the hill villages of Ganjam and Gajapati districts for five hours, maintaining wind-speeds of 130-140 kmph, accompanied by heavy rainfall. Between 11th and 12th October, rainfall recorded was about 300 mm in Gajapati and 230 mm in Ganjam (Special Relief Commissioner, RDM, GoO, 2018). A cloudburst resulted in landslides causing heavy damage to the natural resource base and destroying entire settlements. Villagers in the affected region have no living memory of anything even remotely similar, neither a cloud burst nor cyclonic winds of this speed. Titli destroyed entire habitations, and most villages did not have a single community building that could act as a cyclone shelter. In one village, over 100 people took refuge in the only *"Pucca"* or permanent house with a Reinforced Cement Concrete (RCC) roof, that had a floor area of about 25 m°, allowing for standing-room only (about 0.25m° per person, for over half a day at least if not more).

Table 19.2, excerpted from an official document, sheds light on the magnitude of the disaster. The intensity and the trauma of the people affected was more than the numbers could capture. This is especially true for three *panchayats* in the Patrapur block of Ganjam, namely Ankuli, Burataal and Tumba, and some of the adjoining *panchayats* of Gajapati district, home to Saura Tribals, one among 13 Particularly Vulnerable Tribal Groups (PVTG) of Odisha. Disaster preparedness of the Odisha state authorities in India has received accolades in the first two decades of the 21st Century as loss to life has decreased compared to the 1999 cyclone and rapid emergency measures followed by recovery mechanisms have become the norm. With Titli in Odisha, the government rose to the occasion and worked bottom-up not only in the prevention and recovery phase but also in the rebuilding effort (Special Relief Commissioner, RDM, GoO, 2018). This paper focuses on the Post-Titli rebuilding/rehabilitation efforts undertaken in a few selected villages in Ganjam and Gajapati districts, with the objective of assessing these efforts against the call and guidelines for *"Building Back Better"* (BBB), in the Sendai Framework for Disaster Risk Reduction (United Nations office of Disaster Risk Reduction, UNDRR, 2015).

Table 19.2 Area affected by Very Severe Cyclone, Titli (Special Relief Commissioner, RDM, GoO, 2018)

SN	District	Blocks Affected	Villages Affected	Urban Local Bodies ULB Affected	Population Affected	Houses Damaged	Human Casualty	People Missing	Agriculture Land Affected in ha	Agriculture area where Crop Loss>33% ha
1	Gajapati	7	1,636	2	577,817	34,278	43	4	52,310	44,528
2	Ganjam	22	3,250	18	3,950,000	14,314	8	8	193,768	193,768
	Ganjam+Gaj a pati	**29**	**4,886**	**20**	**4,527,817**	**48,592**	**51**	**12**	**246,078**	**238,297**
	Total in Odisha	128	8,691	36	6,031,154	62,372	59	12	396,770	304,009
	%Manjam +Gaja pati	**23%**	**56%**	**56%**	**75%**	**78%**	**86%**	**100%**	**62%**	**78%**

1.1 Hazard multipliers and cascading effect of Titli

Strong winds uprooted many trees, and the torrential rains caused severe landslides in many of the hilly areas of Ganjam and Gajapati districts. Tumba, Ankuli and Burataal *panchayats* of Patrapur block in Ganjam district (Table 19.3), and Koinpur and Gangabada *panchayats* in Rayagada Block of Gajapati district were severely affected.

The direct damage was in the form of loss to natural habitat and soil erosion, with rainfall causing landslides in the hills and flooding in the lower basins of the river Mahendra-Tanya. Spring sources and intake wells were damaged and pipelines ruptured leading to disruption in drinking water supply. Most of the villages were connected through stand-alone solar Photovoltaic (PV) systems and a micro grid. Community-based young caretakers switched off the power supply in anticipation, a wise decision, as power lines and wires were blown away. Roads became pathways for the water, as cross-drainage culverts became non-functional due to blockages caused by trees. In general there was seemingly an infrastructure failure, but the community and the State Government responded and quickly cleared away the fallen logs and trees and re-established connectivity of roads at least. Disease outbreak was stemmed and dry food and plastic sheets for immediate roof cover were handed out to the community.

2. Disaster Risk Reduction and Mitigation

The context for this paper is the Sendai Framework for Disaster Risk Reduction 2015-2030 and its call for substantially reducing risk from disasters and losses accruing to the livelihood base that includes social- economic-natural and cultural assets thus encompassing both the natural and cultural habitat (UNDRR, 2015). One important factor in the analysis in this paper, with respect to the Titli cyclone, is the understanding of the Disaster Management Cycle (Fig. 19.2) as an inclusive planning process that allows for anticipating, reducing disaster risk, and finally enhancing resilience (National Disaster Management Authority, NDMA, November 2019).

Building resilient economies and communities are mandates of disaster management agencies at the National and State levels (https://ndma.gov.in, https://osdma.org, https://nidm.gov.in//) and of institutions working on risk-reduction, while trying to achieve SDG goals (https://CDRI.world). Resilient infrastructure becomes the backbone for risk reduction and includes (1) Physical services like water supply and sanitation, power, grid and decentralised, telecommunication including radio and cellular networks, road and other transport linkages (2) Basic Social services including housing, schools, health

Table 19.3 Incomplete land slide vulnerability info from the Ganjam Disaster Management Plan (Odisha State Disaster Management Authority, OSDMA, 2019)

3.9 Land *fide Vulnerability:

Vulnerable Areas in the Dist.

Sl. No.	Land Slide Zone/ Area/Location	Area in Sq. Km	No. of susceptible Villages/Wards	Vulnerable Population In Nos.	Houses	Vulnerable Infrastructure				
						School/AWC Buildings	Hospitals	Roads (in Km)	Other Critical Infrastructure

In the year 2018, in the mean time of Cyclone TITLI and subsequent very heavy rainfall Ankuli, Buratal & Tumba Gram Panchayats of Patrapur Block have severely affected.

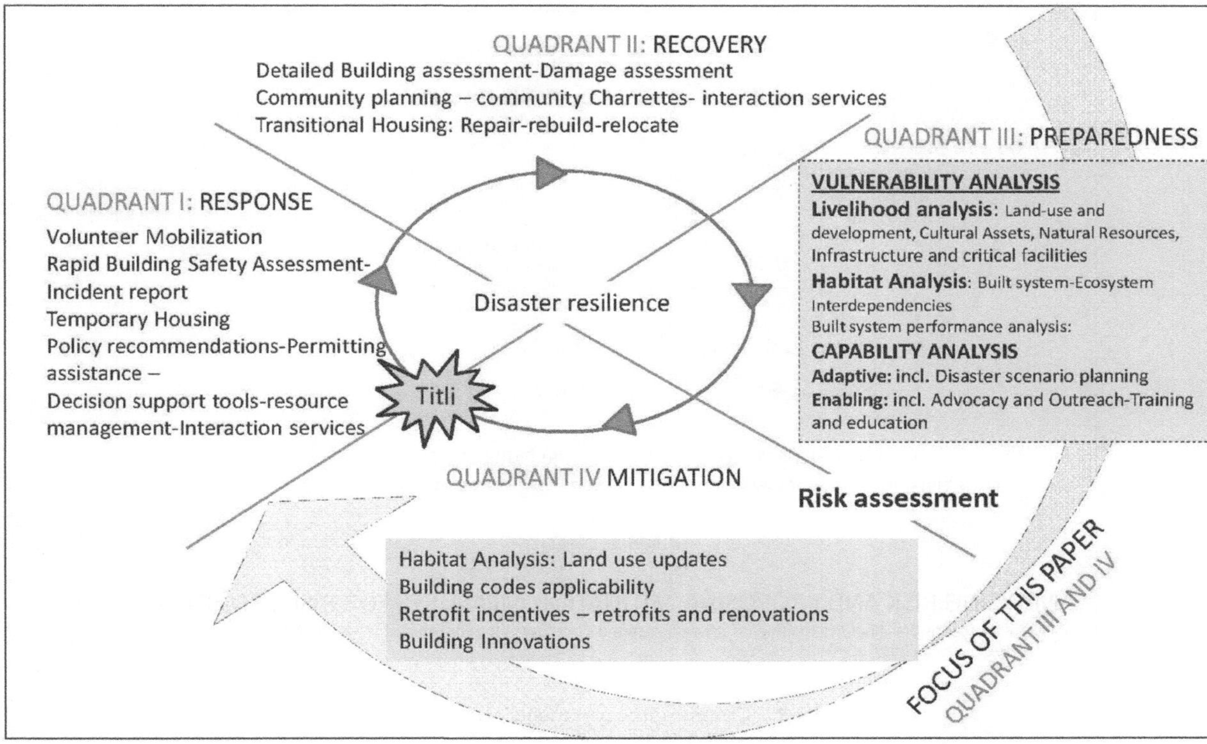

Fig. 19.2 Role that can be played by an architect (adapted from AIA 2017), in disaster management of a cyclone like Titli

and wellness centres, community infrastructure including government services at the block level (3) Ecosystem-based services in the form of watershed and spring sheds, catchments, reservoirs including ponds, lakes, other water-bodies, estuaries, mangroves and wetlands, which regulate water but are being overwhelmed by rapid construction activities.

Four priority areas identified by the Sendai Framework are (1) Understanding disaster risk (2) Strengthening disaster risk governance to manage disaster risk 3) Investing in disaster risk reduction for resilience and (4) Enhancing disaster preparedness for effectively responding and *Building Back Better* (BBB) in all aspects of recovery, rehabilitation and reconstruction (UNDRR, 2015). The guiding principles of the Sendai Framework are based on, among other values, enabling local decision-making for reducing risk, empowerment of local authorities and communities, and importantly addressing risk cost- effectively rather than relying primarily on post disaster response and recovery. In other words, the stress is on hazard and vulnerability assessment to gain a better understanding of the disaster risk and accordingly proposing mitigation measures to enhance preparedness by enabling local institutional capacity. The role of the enabler, in building the adaptive capacity of the communities becomes very important for risk reduction (Wickham, Kinch & Lal, 2009).

While looking at disaster risk reduction and climate resilience, it is also important to build on social equity, economic prosperity, health, safety and wellness and protection of natural resources. In this context, it is important to emphasise the role that architects could play in bringing an integrated and holistic approach to building resilience within the community, while looking at both the built system and the ecosystem, the cultural and natural habitat (American Institute of Architects, AIA, March 2017).

Figure 19.2 (AIA, March 2017), lays out the activities under each of the four phases of a disaster management cycle, namely Response, Recovery, Preparedness and Mitigation. This is shown as a four quadrant approach. The focus of this paper is on Quadrant III, "Preparedness," which leads to mitigation measures outlined in Quadrant IV, in preparation of future disasters (like Titli in this case). The list under "Preparedness" defines the vulnerability assessment for selected villages of Ganjam and Gajapati in their Post-Titli phase.

A conceptual framework shown in Fig. 19.2 and detailed in Fig. 19.3 was used to analyse the disaster risk and the resilience of the community, drawing upon definitions as per UNISDR (2009). Figure 19.3, identifies components of hazard risk and a vulnerable system. The overlap between hazard risk and a vulnerable system defines the disaster

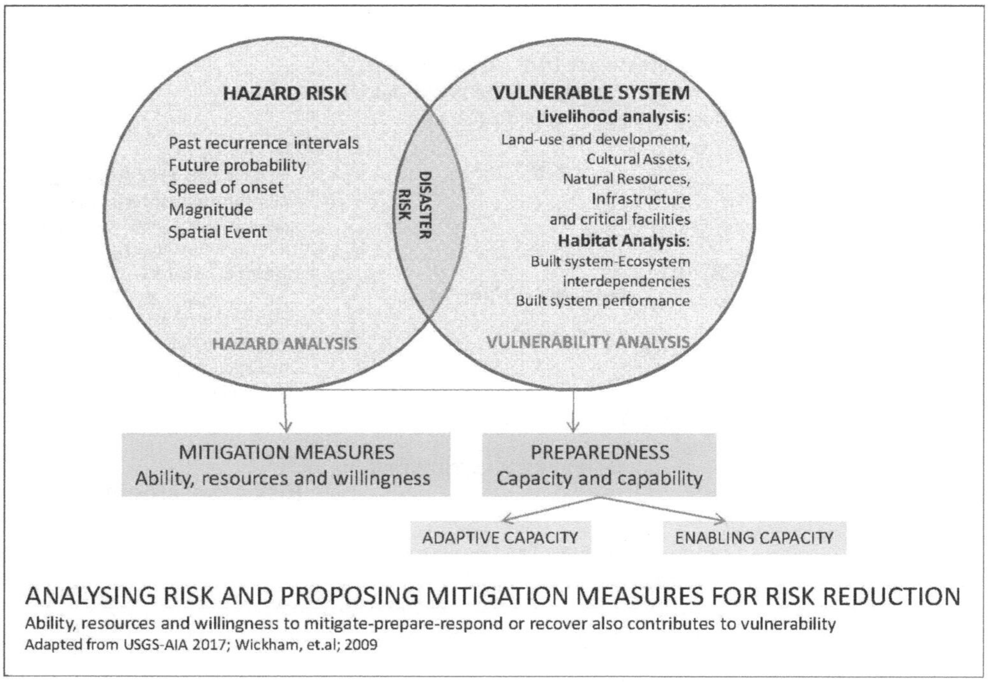

Fig. 19.3 Components for analysing disaster risk

risk. Assessment of this risk is a first step to develop a risk-reduction strategy to enable preparedness by implementing mitigation measures to reduce the vulnerability of a system. In other words, ability, resources and willingness to mitigate, prepare, respond and recover, strengthened by enabling local adaptive capacity contributes to the reduction of vulnerability (AIA, March 2017).

3. Methods

Primary survey was carried out through transect walks, informal focus group discussions, interviews with *Sarpanch (Panchayat* elected representative) homeowners and masons and observation-cum-rapid measurements of the housing in-progress in several villages (Fig. 19.5). These were done in three stages: (i)immediately after the cyclone in November-December 2018, (ii) in February 2020 and (iii) in July-August 2021, to get the latest update.

Secondary data from internet sources as well as data collected by the Non-Governmental Organization, NGO, Gram Vikas, as a part of their Titli Rehabilitation work, was used in the analysis. Other secondary sources analysed include (a) Google Earth maps of the study villages to get a wider perspective of the land form, pre and post-Titli (b) Vulnerability of the housing stock from the Vulnerability Atlas (Building Material and Technology Promotion Council, BMTPC, 2019) (c) Compendium of rural housing typologies proposed for Odisha (United Nations Development Program, UNDP, 2016) (d) Census data for the *Panchayats* (Census of India 2011) cross verified using the Rural Water Supply database (https://eialshakti.gov.in/IMISReports/), (e) Ganjam Disaster management Plan as well as the report on Titli (Special Relief Commissioner, RDM, GoO, 2018) (fJ Reports by the Rural Works Division on damage to rural roads, and (g) Pre-Titli information for these *Panchayats,* from a detailed livelihood study conducted by CTx GREEN during 2004-2010, in partnership with Gram Vikas. (Vaidyanathan, 2009, Vaidyanathan, G; Sankaranarayanan, R; 2019).

4. Results and Discussion

Titli, a high magnitude *very severe cyclonic storm,* with no past history of occurrence, future probability likely low at the same location and a speed of onset that was sudden, rates high as a hazard risk. With this knowledge, in order to build better, we need to analyze the livelihood base and habitat, both components of the Vulnerable System (Fig. 19.3) and determine the *Disaster Risk* as ability of the vulnerable system to cope with the hazard (overlap of the two, as shown in Fig. 19.3). Enabling the adaptive capacity of the local community can further contribute to disaster risk reduction.

4.1 Vulnerability Analysis

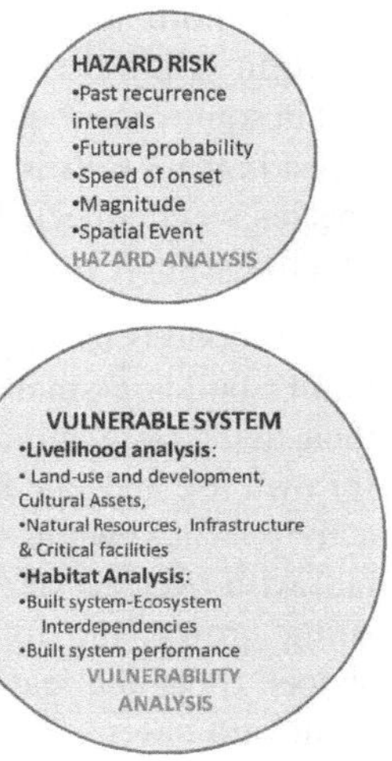

The study region includes Burataal and Ankuli *panchayats,* both in Patrapur block of Ganjam district, and the neighbouring panchayats of Laxmipur and Koinpur in Rayagada block of Gajapati district, Odisha. This region is home to Saura tribals, an indigenous community identified as a PVTG and accorded priority by the Government of India in terms of protection and improvement of socio-economic status to reduce their vulnerability (Ministry of Tribal Affairs, Government of India, GoI, ND). To this effect, currently there are government programs such as the Odisha PVTG Empowerment and Livelihoods Improvement Program (OPELIP) that are being implemented.

The topographical map, Fig. 19.4**,** shows the extent of the Burataal-Ankuli area, and their relationship to the villages (Marmaria, Billosahi and Madaba) in the neighbouring Gajapati district. All the villages being discussed are part of the Mahendra-Tanya watershed with a similar natural habitat. They are also culturally similar being inhabited by the Saura tribals. Hence the system conditions for vulnerability assessment can be considered the same, despite differing administrative boundaries. For the habitat analysis, information from randomly selected villages of Ankuli and Burataal *panchayat* is presented. To gain better insights, a resettled village, Maramaria, in Ganjam district's Burataal *panchayat* was compared with two resettled villages in Gajapati district, Billosahi and Madaba. Billosahi is part of the Alarasingi Revenue village, where Gram Vikas has initiated some spring-shed work as a step towards developing

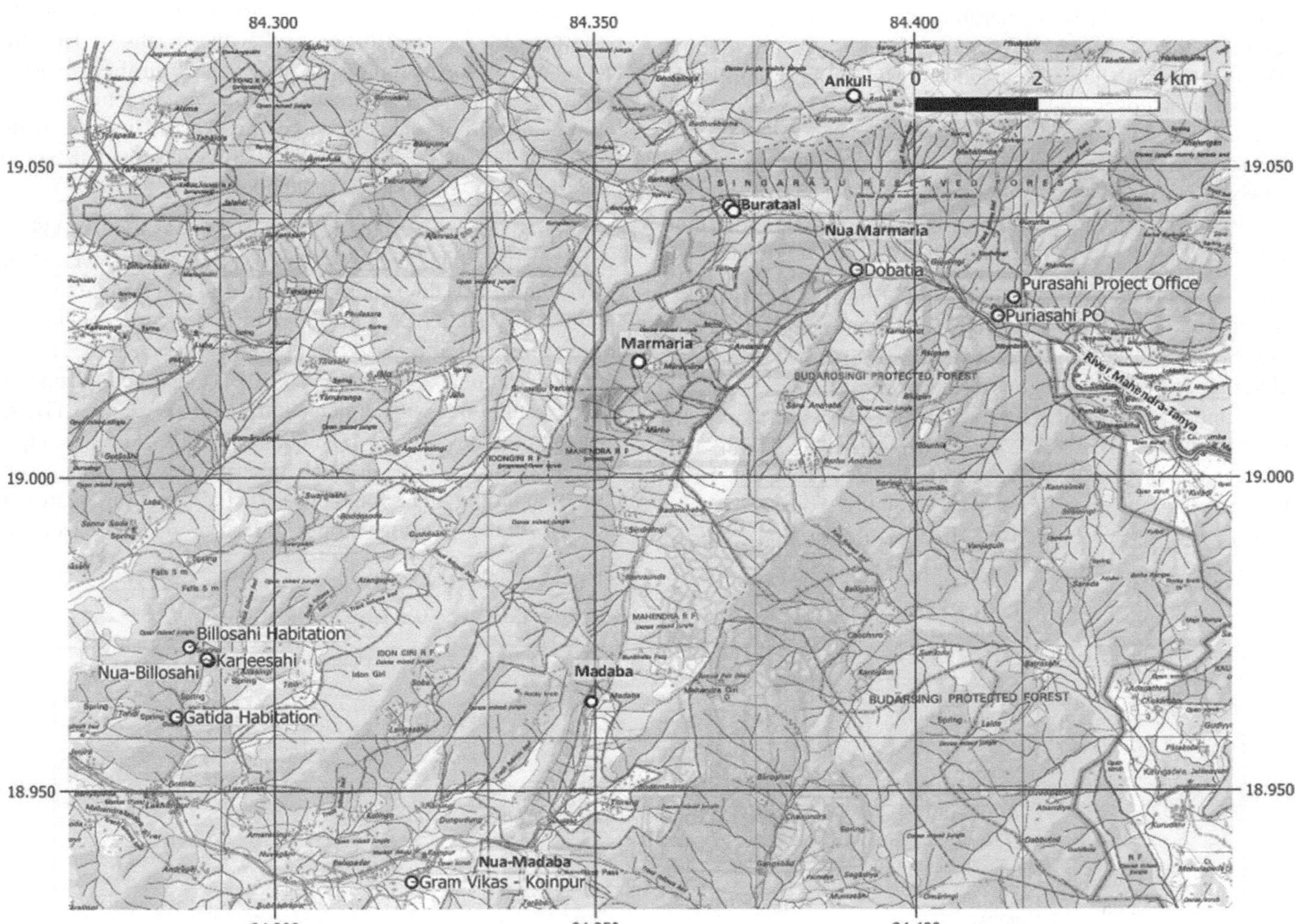

Fig. 19.4 Topographical map showing the resettled villages studied (Topo Sheet no: 74A-8c)

Water-Secure gram panchayats in the long term (Gram Vikas, 2019). Madaba in Koinpur *panchayat,* is currently a large resettlement of 66 houses.

4.2 Livelihood Analysis

Gram Vikas, a field based NGO working with tribals since 1979, has done extensive work in about 22 of the 50 villages over a 4500 ha area (roughly about 50% of the actual spread of the Burataal and Ankuli *panchayat* as is indicated in Table 19.4 and Fig. 19.5)

The land-use map of this region in Figure 5(b) was developed using a LISS IV image from the Indian Remote Sensing satellite IRS P6 with a resolution of 5.8 m (date of pass April 2004) and ground trothed data in 2004-2005. The land use plan clearly shows that the community lives in forested areas, most often reserve forests, and practices a form of slash-and-burn agriculture called *bogodo.* The 4500 ha falls under eight categories (1) Agricultural land 17% (2) Bogodo land 5% (3) Old bogodo/degraded forest land 3%, (4) Dense forest 21% (5) Sparse forest 41 % (6) Scrub land 13%, (7) Barren rocky area less than 1% and (8) Water bodies 1% (Mishra, 2005). The area has patches of degraded forest land in the midst of a good forest cover, and the settlements (red dots) are usually located close to springs, and agricultural land is developed in the vicinity. The land use now would be different due to the newly constructed road completed in 2020 that has reduced the forest cover, further aggravated by cyclone Titli that destroyed many of the old stands in the Reserve Forest.

The livelihood system of the Saura tribals is intrinsically linked to their natural habitat. They continue to practise *bogodo,* which keeps them occupied from April, when the clearing and burning begins in the peak of summer, almost until mid-January when the last of the harvests is brought to the village. There are some villages like Ankuli and Dhanabada, in the upper reaches of the hills, which have terraced lowland like farms with irrigation sources, where paddy cultivation is practised. In the middle land, *Sauras* practise horticulture, growing cashew, lemon, papaya and jackfruit. They also own livestock, mainly poultry and goats and in some cases bullocks, used for ploughing lowlands. The forest in these hills is rich in biodiversity, and up to 105 plant species were collected and identified in 2004-2005 as a part of a forest survey (CTx GREEN, 2004).

The Bogodo is usually a clearing in the forest, and while the family builds a small homestead there to watch and guard their fields from animals, collection of minor forest produce for own use and sale is also an important income-generating activity. Thus, while their fields are a veritable biodiversity lab, with over 22 varieties of cereals, pulses and vegetables being grown, the forest offers tree-borne oilseeds, fruits, shoots, roots and leaves that are used and also sold directly or in rare cases as value-added products. They spend long hours in their farm homes, going back to the village in some cases only at night, to sleep. A point being emphasised here is that almost all the people in these communities, have one house in the village which usually has a *patta* or legal land-deed, and another of a more temporary nature in their *bogodo* or farmstead. It is worth noting that the deaths in Titli are attributed to those people who stayed back in their *bogodo* homes, which being isolated stand-alone structures were badly damaged by the severe cyclone. All the *bogodo* fields were completely destroyed when Titli stuck, and in some cases, came into the path of landslides.

Livelihood mapping undertaken on 03-04 March 2005 is reproduced in Fig. 19.6 as a baseline reference. Watershed development was then considered integral and critical to their livelihood system. Other critical infrastructure considered important that emerged from Gram Vikas' Rural Health and Environment Program (RHEP), included drinking water, sanitation with toilets and bathing rooms, TBRs, (www.gramvikas.org; Vaidyanathan, 2002) and energy services. Access to health and education were considered part of building the social infrastructure and safeguarding the forest and the non-timber forest produce as well as

Table 19.4 Overview of the study region (Office of the Registrar General, 2011)

Villages in Panchayat Patrpaur Block, Ganjam district	Gram Panchayat	Total Geographical Area (in Ha)	Households hh	Population of Villages	Focus Village
43	**Ankuli**	4733	801	3831	Burusahi and surround
24	**Burataal**	4027	620	2970	Marmaria
Resettled village comparison in Ganjam and Gajapati District					
Revenue village	**Gram Panchayat - District**	**Geog Area (ha)**	**Households hh**	**Population**	**Remarks**
Marmaria	Burataal - Ganjam	128	23	107	New Marmaria: 23 new hh
Alarasingi	Lakshmipur-Gajapati		48	100	New Billosahi: 23 hh; Data in table includes old village, 18hh Karjeesahi 19 hh + Bhogabandhu 11hh
Madaba	Koinpur - Gajapati	331	64	272	New-Madaba total hh 88

Location	Latitude N ° ′	Longitude E ° ′	Elevation m
Tadakasahi Project office	N 19° 07.672′	E 84° 25,485′	96.9 m
Taparada	N 19° 06.887′	E 84° 25.108′	500 m
Raikhal	N 19° 06.619′	E 84° 24.894′	578.7 m
Jaliora	N 19° 05.741′	E 84° 24.929′	718 m
Ankuli	N 19° 03.804′	E 84° 23.931′	925 m
Kuruguda	N 19° 03.501′	E 84° 23.004′	912 m
Buratal	N 19° 02.615′	E 84° 22.265′	554 m
Dobatia	N 19° 01.726′	E 84° 24.924′	615.8 m
Purasahi Project office	N 19° 01.726′	E 84° 24.928′	616 m

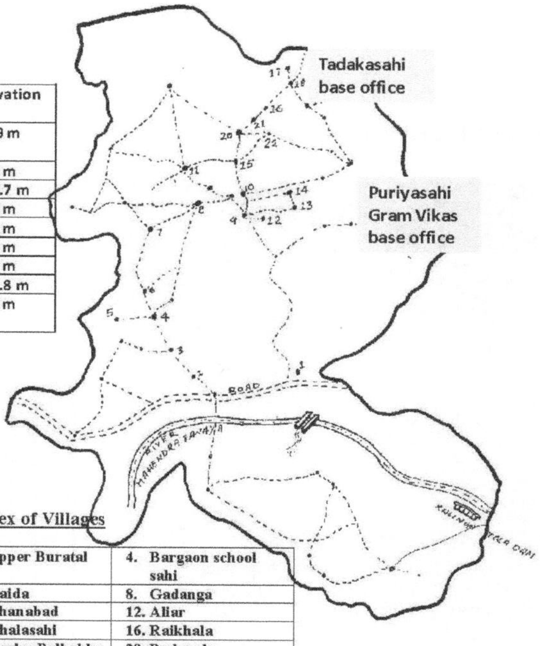

TUMBA PROJECT –Index of Villages

1. Guruda Proj Office	2. Lower Buratal	3. Upper Buratal	4. Bargaon school sahi
5. Bargam Upar Sahi	6. Badua	7. Gaida	8. Gadanga
9. Ankuli	10. Titrising	11. Dhanabad	12. Aliar
13. Gokhnasahi	14. Phulasahi	15. Khalasahi	16. Raikhala
17. Masanibada	18. Tadekasahi	19. Raghu Balhabha	20. Badanala
21. Taparada	22. Jaliar		

a) Villages covered as a part of the survey

b) Land-use map of Ankuli and Burataal Panchayat under Gram Vikas field area (2005)

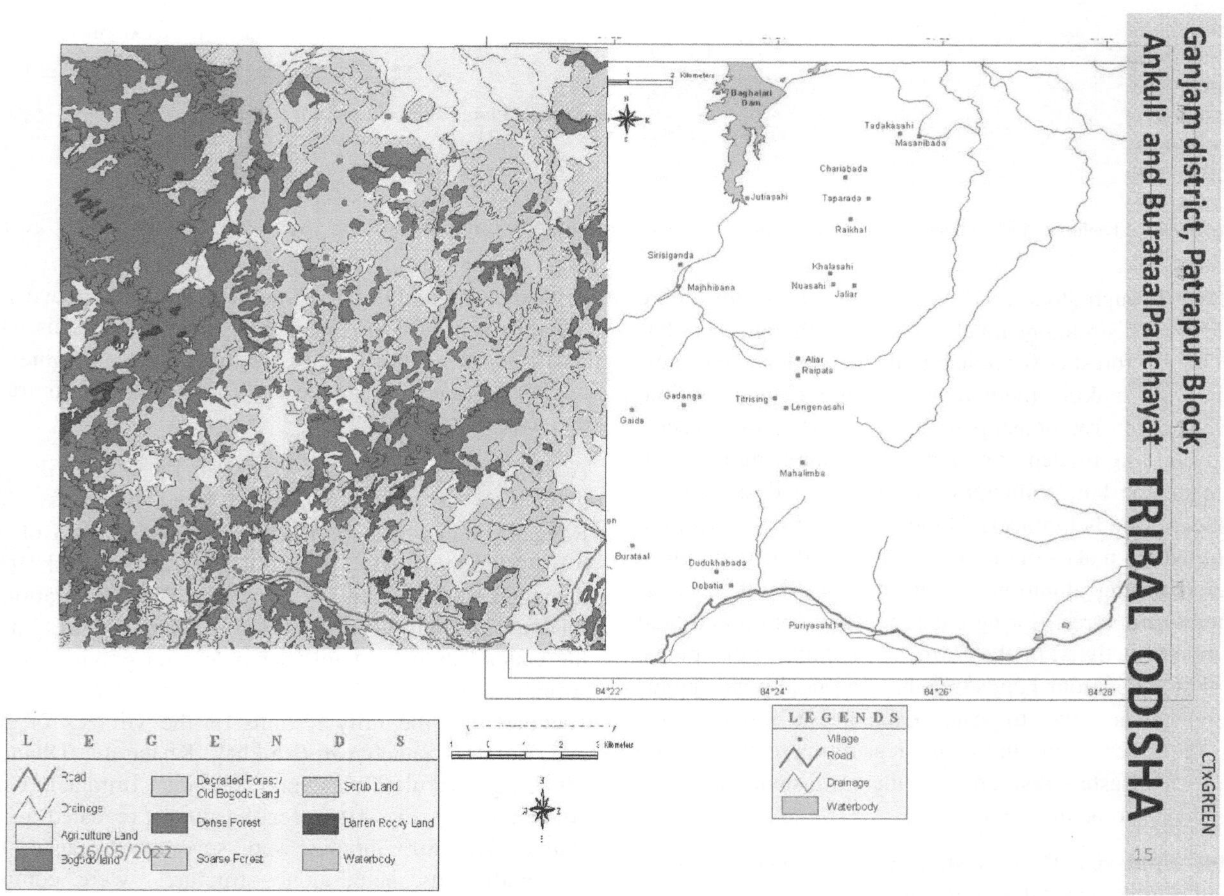

Fig. 19.5 Map of the study region

Source: (a) (CTx GREEN, 2018) (b) (Mishra, 2005)

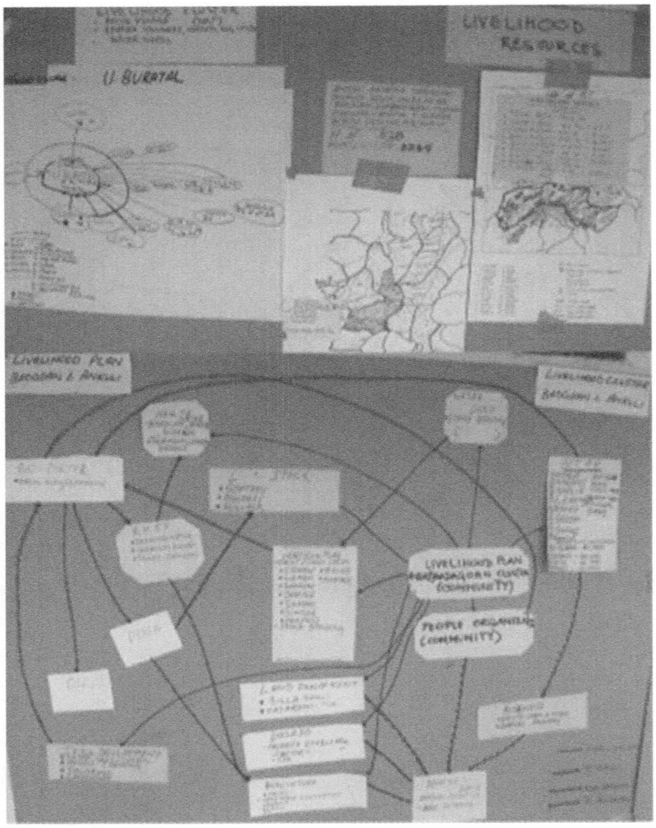

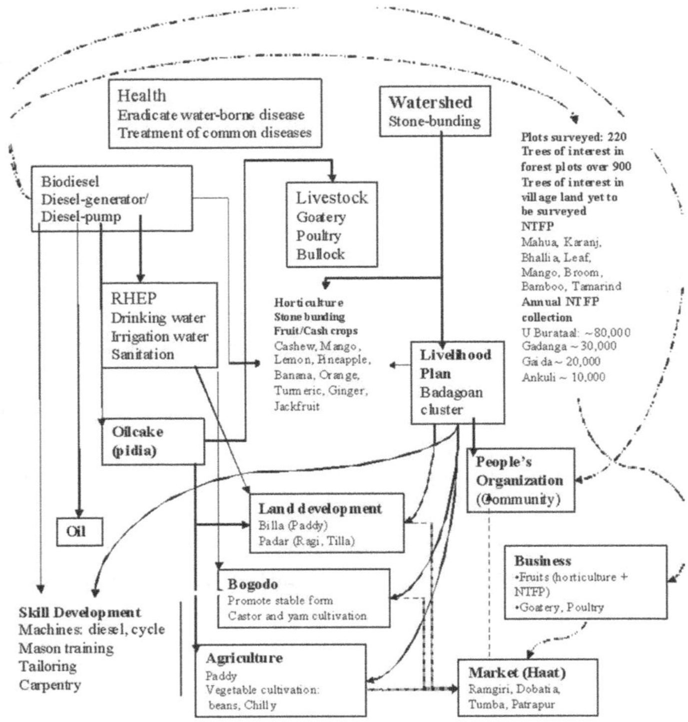

Livelihood Plan: Burataal cluster Incl. Badagoan, Ankuli

Fig. 19.6 Livelihood Plan based on local resources developed in the community, in the year 2005 (Vaidyanathan, 2009)

their land through stone bunds was a way to ensure secure livelihoods. This remains much the same, although there has been a loss of forest cover reducing the dependence on forest produce, and making them more vulnerable to migration. This weakening relationship with the forest is of concern, as road building related continued earth quarrying and Titli have made the land vulnerable to erosion and subsequent land-slides. Another concern is the continued bond with the *bogodo,* which makes the farmstead homes their main place of stay requiring attention to how they are built, even as temporary structures. Spring catchment protection and land development activity in the form of pitching, gully plugs, earth or stone contour-bunds/trenches and field ponds/bunds are important activities to ensure resilience of water supply, roads, and other social and ecological infrastructure. Some thought on disaster- resilience of their transient farmstead homes may also be important.

Habitat Analysis: Built system performance analysis: internal and external dependencies

The term habitat in the context of this paper reflects the built system and the ecosystem, along with their internal inherent dependencies and external interdependencies between each other (Yeang, 1995). The built form being discussed here is the house in the village, as the farmstead home in the bogodo, although occupied for a longer period, is transient and temporary in nature.

A rapid survey conducted Post-Titli by Gram Vikas in 57 habitations of four *gram-panchayats* in the worst affected Raygada block of Gajapati district revealed the 36% of houses were fully damaged, 23% had damaged roofs and 19% had damaged walls (Gram Vikas, 2019). These habitations had toilets, of which 80% were fully destroyed, and almost all the water supply systems, dependent on gravity-flow from springs, were seriously damaged. A transect-walking survey conducted in the hilly terrains of the villages of Ankuli and Burataal panchayat (Raikhal, Khalasahi, Dhanabada, Titirising, Ankuli, Tala Burataal, Upar Burataal, Guruda, Andanada, see Fig. 19.5a for the route), revealed a similar status. Over 80% homes were severely damaged, toilets and bathrooms, stand-alone structures, were completely destroyed. Water supply systems were disrupted, although hand-pumps were functional in most cases. Electricity supply through the solar grid was discontinued in time in most

Table 19.5 Vulnerability assessment of housing, Ganjam district (Building Material and Technology Promotion Council, 2019)

Distribution of houses by predominant material of Roof and wall and level of damage risk

Table No OR 19	State: Odisha										District: Ganjam	
Wall/Roof		No. of Houses	%	EQ Zone				Wind Velocity m/s				Flood Prone
				V	IV	III	II	55&50	47	44&39	33	
				Area in %				Area in %				Area in %
							100	56.8	43.2			4.7
A1-Mud and Unburnt brick wall	Rural	127,879	14%									
	Urban	16,800	2%									
	Total	144,679	16%				L	VH	H			VH
A2: Stone wall not packed with mortar	Rural	6,886	0.8%									
	Urban	1,526	0.2%									
	Total	8,412	0.9%				L	H	M			VH
Total Category A	Rural	153,091	16.9%									
B. Burnt Brick wall and stone wall	Rural	469,247	51.8%									
	Urban	158,725	17.5%									
	Total	627,972	69.3%				VL	H	M			H/M
Total Category B	Total	627,972	69.3%									
C1-Concrete Wall	Rural	8,964	1.0%									
	Urban	6,110	0.7%									
	Total	15,074	1.7%				VL	L	VL			L/VL
C2 Wood Wall	Rural	28,356	3.1%									
	Urban	2,549	0.3%									
	Total	30,905	3.4%				VL	VH	H			H
Total - Cateogry C	Rural	45,979	5.1%									
x- Other Materials	Rural	71,891	7.9%									
	Urban	7,024	0.8%									
	Total	78,915	8.7%				VL	VH	H			VH
Total Houses		905,957										
Roof												
R1 - Light wieght sloping roof	Rural	341,073	37.6%									
	Urban	59,732	6.6%									
	Total	400,805	44.2%				VL	VH	VH			VH
R2 - Heavy weight Sloping Roof	Rural	26,137	2.9%									
	Urban	8,630	1.0%									
	Total	34,767	3.8%				VL	H	H			H
R3- Flat Roof	Rural	346,013	38.2%									
	Urban	124,372	13.7%									
	Total	470,385	51.9%	*Damage Risk as per that for the wall supporting it*								
Total Houses		905,957										

Probable Maximum Precipitation at a station of the district in one day for arial extent of 1000 sqm = 726 mm

Housing Category: Wall Type	Housing Category: RoofType
Category A: Buildings in field stone, rural structures	Categry R1-Light Weight (Grass, Thatch
Unburnt brick houses, clay houses	Bamboo, wood, mud, plastic, Polythene
Category B: Ordinary brick buildings of the large block and Prefabricated	GI metal, Asbestos sheets, other materals
Half timbered structures, building in natural hewn stone	Category R2: Heavy weight (toles, stone/slate)
Category C: Reinforced building, wall built wooden structures	Category R3: Flat roof (brick, concrete)
Category X: Other materials not covered in A, B, C. Theses are generrally lig	EQ Zone V: Very High Damage Risk Zone MSK >IX
Notes	EQ Zone IV: High Damage Risk Zone MSK >VII
Flood prone area includes that protected area which may have more severe	EQ Zone III: Moderate Damage Risk Zone MSK >VII
damage may be severe under heavy rains and checked drainage	EQ Zone II: Low Damage Risk Zone MSK >VI
Damage risk for wall types is indicated assuming heavy flat roof	Level of Risk: VH: Very High, H= High
in categories A, B and C (Reinforced Concrete) building	M= Medium; L= Low, VL: Very low
Source of Housing Date: Census of Housing GOI, 2011	Total No of houses including Vacnat/Locked houses
BMTPC: Building Material and Technology Promotion Council	Peer Group, MoHUA, GOI

P.291

villages avoiding fatalities due to shocks, but distribution lines, electric poles, home-meters were blown away. Solar panels in some of the micro-grids in the villages were damaged, but in most villages, the panels were well tied in place, and the control room with the batteries was also intact. Communities took refuge in the only brick-and- concrete houses or community buildings, as the wind blew away Corrugated Galvanized Iron (CGI) roofs and the lashing rain and flooding water eroded walls.

The 2011 housing census data for Ramagiri tehsil, under which the habitations of the study are listed shows that only 9.1% of the homes were in dilapidated condition, but among these houses over 90% used either thatch or CGI or Asbestos sheet (30%) and only 9.5% had a concrete roof. Similarly with respect to the walls, 89% used non-permanent material (67% used mud) and only 11% used burnt bricks. Referring to Table 19.5: "*Distribution of houses by predominant material of Roof and wall and level of damage risk,*" for Ganjam developed by the Building Material and Technology Promotion Council (BMTPC, 2013), one can assess the vulnerability of housing for Ramgiri Tehsil, as described above - walls all fall under Category A or B and the roofs mainly under R1. This sets their vulnerability as very high for wind speeds of 50-55m/s (180-200 kmph) and also with respect to floods. In Table 5 we see a direct correlation between the percentage of brick-and-concrete houses and their low vulnerability to cyclone and flood damage, which is also very intuitive. However, beyond this, the house has to additionally be looked in relation to its local habitat, especially how it is sited, as the damage was also due to the cascading effect of the cloudburst that resulted in landslides.

If we study the Google earth images of the Burataal area, Figure 7, over a period starting a year before Titli in 2017, a few days before the cyclone (11-Oct-18 was when it hit the region), a month or so after and then two years after, the pictures are very revealing. The forest cover has reduced, bare patches have been exposed due to erosion, and there are clearings for pathways and housing settlements (02-July-20 and 02-Feb-21). The settlement demarcated by an ellipse in the top-left corner of Fig. 19.7 is New/Nua- Marmaria, consisting of 27 houses resettled in a Burataal patch of forest-land. The river below, Mahendra-Tanya used to be a narrow stream, but has become much wider, having swallowed some agricultural land along its banks. Base rocks are exposed along the bed by erosion, some brought down by landslides.

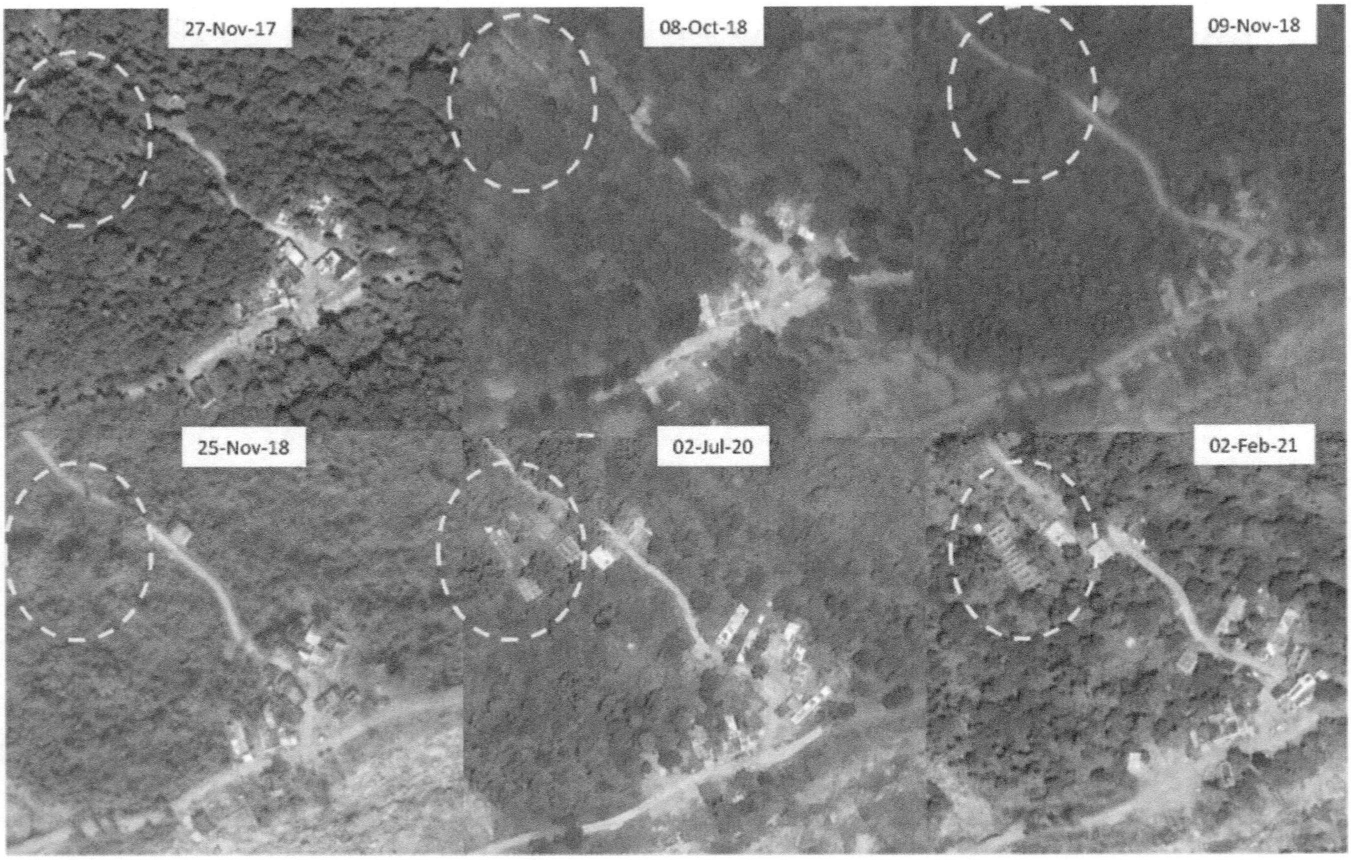

Google Earth images of a sample village (Marmaria) studied in the Burataal Panchayat, Images 1 and 2 are pre-Titli, Image 3 is four weeks after Titli, Image 4 is six weeks post-Titli, Image 5 is about 20 months after when the resettlement of village Marmaria had started, and the last figure shows further development

Fig. 19.7 Time lapse images of villages around Burataal, Ganjam district showing resettled Marmaria village

Money to the tune of Rs. 1,05,000 was granted for repair of each damaged house as an immediate relief. In addition, almost entire villages were sanctioned houses under the Biju Pucca Ghar Yojana (BPGY), where a core self-built housing of 25m° is allocated Rs 1,30,000. Money from Mahatma Gandhi National Rural Employment Guarantee Scheme's (MNREGS) 100 days employment to the tune of Rs 17,000/- was directed towards the self-building process. Additionally, there was an incentive of Rs. 20,000-10,000 for people completing the work within 4-6 months. Besides the housing, Rs. 12,000 is allocated for a toilet, Rs 12,000 for a compost pit. Every village has also been allocated money for retaining/guard walls. Details of the scheme were communicated to the community by the elected village leader, who has played a key role all along, with the Government's *Gram Sewak* in achieving the target of completing the rebuilding process.

At least 800 houses have been sanctioned in these two *panchayats* post-Titli as a part of the rebuilding activity. The idea of a core house to be completed has been ignored in most cases and houses have doubled and trebled in size, with the homeowners stating in unison, that one cannot build a house for the amount sanctioned. While finishing in terms of flooring and plaster maybe forgiven, many houses do not have doors and windows (critical in the case of another cyclone), and some have not even completed their roofs. Masons during this time were in short supply it seems, while the demand was huge, resulting in work dragging on. The entire building process, which moves rapidly in the first three months, often grinds to a halt after, with work extending sometimes to a year or two or more. Most villagers have spent over and above the allocated amount, anything between Rs 30,000 and Rs 1,00,000 or more, either by mortgaging land, selling assets or taking loans. The last option is the most treacherous as the interest rates are much higher than any credit card company in the urban context. While there was a lull in migration due to the pandemic, when most of the building activity was happening, now, many young males have again migrated to cities, some to repay their loans. Many houses have been rebuilt in the village, close to their old homes, and some villagers have retained their old homes, have repaired and continue to live in them, while they build their *"Pucca Ghar."* Where villages were vulnerable to landslides, the community has been advised to relocate. Land has either been allocated by the government or has been organized by the villagers themselves. The allocation for the new house is usually a plot approximately 12 feet x 60 feet, and the villagers have built-up to the extent of 60-70%. The plans are 12feet/4m wide with at least 2-3 continuous rooms about 10 feet deep, a 6-feet front veranda and scope for a lean-to kitchen at the back. There is very little space left for a backyard, something that was integral to their old homes. The new homes are currently used by children who need internet connectivity for online school lessons during the COVID-19 pandemic, as these houses are better connected to the network than their old home, and by those people in the village who do not have land and are in some form of service-oriented profession. In the current scenario, in these resettled villages, each family has a homestead in their *bogodo* fields where they spend most of the time, their old homes with their backyard gardens, repaired and used in conjunction with their *bogodo* is their second home, and a third incomplete house (as of Aug-21) built as a part of the Titli rehabilitation. These houses, as we will see in the case studies, do not have toilets or bathing spaces, just a single piped water connection in front (or back) of their houses, and in most cases the retaining/guard walls are not yet in place. These new homes are seen as a place for refuge in case of extreme weather events. Luckily such an emergency has not happened in the last 2-3 years.

Built system performance analysis

A very detailed guideline is available for the Biju Pucca Ghar Yojana (BPGY-2017) issued by the Panchayati Raj and Drinking Water (PR&DW) Department, Government of Odisha, in "suppression of all former guidelines." It defines *Pucca* as a house with foundation, wall and roofs, materials like flyash bricks, burnt bricks, stone in cement mortar for walls and CGI or asbestos roofs, RBC or RCC for roofs, with a minimum life span of 30 years. There is a clarification that all buildings approved must be of RCC or of equivalent strength, approved by the PR & DW departments. There is also clause that innovations will be encouraged. A compendium of rural housing typologies is available, developed by UNDP for by the PMAY(G) and the Ministry of Rural Development, MoRD. There are building typologies with cost and construction details for 130 zones in 18 states, making each design context specific. The zone specific building typology for Ganjam is extracted below (Figure 8), with recommendations for the built form and construction technologies. An ideal tool-kit inclusion, it ***could*** have been adapted further by adding a recommendation for the site, when the construction process began Post-Titli.

There is provision for this made by the PR & DW department under the BPGY scheme. In the guidelines, under Section 4 (4.14-4.17), *Components of the Scheme,* there is a clear mandate for training officials, preparing IEC material, imparting housing and habitat literacy to the beneficiaries and provision for incentives to SHGs/CRPs and CBOs "for facilitating construction. Such an adaptation was especially

'Personal communication, Nayak,K. Feb.'20; Aug.'21; Patra,S. Aug.'21

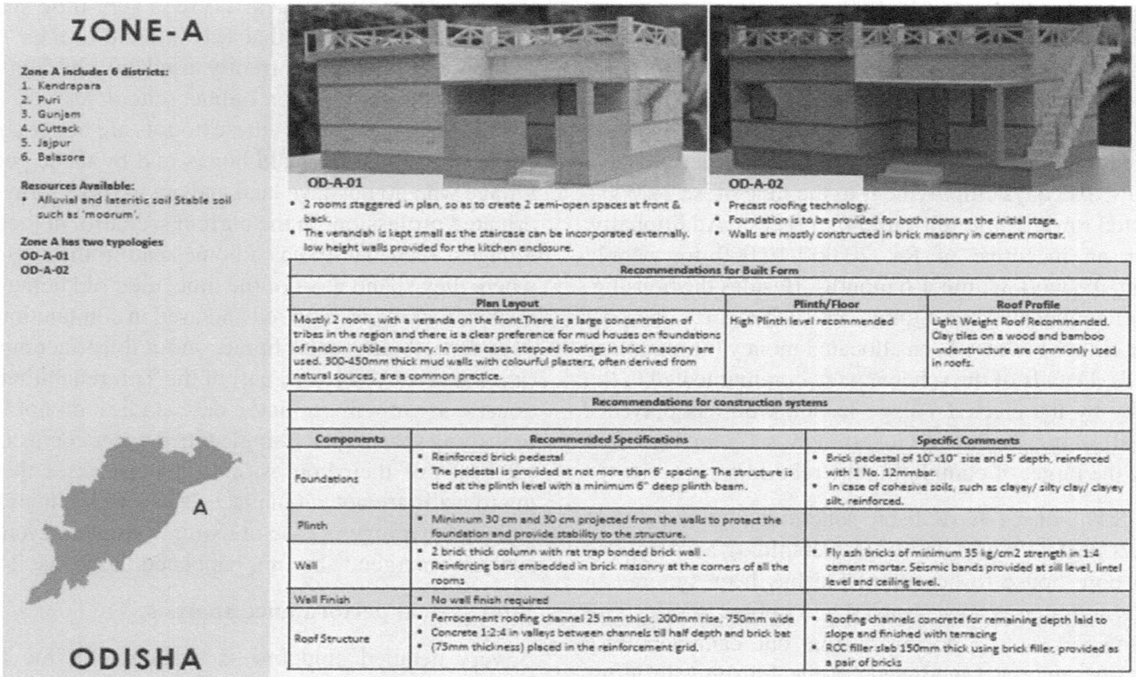

ZONE-A

Zone A includes 6 districts:

1. Kendrapara
2. Puri
3. Gunjam
4. Cuttack
5. Jajpur
6. Balasore

Resources Available:

- Alluvial and lateritic soil Stable soil such as 'moorum'.

Zone A has two typologies
OD-A-01
OD-A-02

A

ODISHA

OD-A-01

- 2 rooms staggered in plan, so as to create 2 semi-open spaces at front & back.
- The verandah is kept small as the staircase can be incorporated externally. low height walls provided for the kitchen enclosure.

OD-A-02

- Precast roofing technology.
- Foundation is to be provided for both rooms at the initial stage.
- Walls are mostly constructed in brick masonry in cement mortar.

Recommendations for Built Form		
Plan Layout	**Plinth/Floor**	**Roof Profile**
Mostly 2 rooms with a veranda on the front.There is a large concentration of tribes in the region and there is a clear preference for mud houses on foundations of random rubble masonry. In some cases, stepped footings in brick masonry are used. 300-450mm thick mud walls with colourful plasters, often derived from natural sources, are a common practice.	High Plinth level recommended	Light Weight Roof Recommended. Clay tiles on a wood and bamboo understructure are commonly used in roofs.

Recommendations for construction systems		
Components	**Recommended Specifications**	**Specific Comments**
Foundations	• Reinforced brick pedestal • The pedestal is provided at not more than 6' spacing. The structure is tied at the plinth level with a minimum 6" deep plinth beam.	• Brick pedestal of 10"x10" size and 5' depth, reinforced with 1 No. 12mmbar. • In case of cohesive soils, such as clayey/ silty clay/ clayey silt, reinforced.
Plinth	• Minimum 30 cm and 30 cm projected from the walls to protect the foundation and provide stability to the structure.	
Wall	• 2 brick thick column with rat trap bonded brick wall . • Reinforcing bars embedded in brick masonry at the corners of all the rooms	• Fly ash bricks of minimum 35 kg/cm2 strength in 1:4 cement mortar. Seismic bands provided at sill level, lintel level and ceiling level.
Wall Finish	• No wall finish required	
Roof Structure	• Ferrocement roofing channel 25 mm thick, 200mm rise, 750mm wide • Concrete 1:2:4 in valleys between channels till half depth and brick bat • (75mm thickness) placed in the reinforcement grid.	• Roofing channels concrete for remaining depth laid to slope and finished with terracing • RCC filler slab 150mm thick using brick filler, provided as a pair of bricks

UNDPs PMAY (G) Rural Housing typology for Ganjam district, Odisha
excerpt from Pahal-Prakriti hunar lokvidya, a compendium of rural housing typologies (Pp. 160-168)
to aid decision makers and homeowners in the construction of PMAY-G houses that are appropriate, affordable, disaster-resilient and aesthetic in the context of specific regional attributes (p.9)
Developed by UNDP for Prime Minister Awas Yojana (Gramin) and Minstry of Rural Development

Fig. 19.8 Extract for Ganjam, UNDP's compendium of Rural housing typologies (United Nations Development Program, UNDP, 2016)

feasible as UNDP also had funded post-Titli mason training programs in the region. However, the case studies that follow will indicate how the well-meaning self-build informal building process was driven by masons who had learnt their trade in urban centres of Kerala, Tamilnadu and Andhra Pradesh/Telengana and had returned home to rebuild their home (staying back during the COVID-19 lockdown). They built as they would in the cities, oblivious of the context, investing huge funds in frame structures and foundation, instead of building with the lay of the land, as they may have done intuitively with familiar materials (Fig. 19.9).

Disaster Scenario planning for three high-risk villages

The current rebuilding process in the region was analysed to gain an understanding of the self-built housing process and a common pattern emerged, namely

(a) Almost everyone was exceeding the stipulated 25m° core housing

(b) Row housing, sharing common walls, seemed to be the preferred way (upto 8 houses in one cluster), with brothers building next to each other, and in some cases sharing a staircase

(c) All houses were envisaged to be more than one floor, with the upper level also being used as an extension of the house, not as a mere terrace

(d) Whether building as row housing or not, most houses avoided windows on the side wall, and had windows only in the front (opening into the veranda). Some houses provided ventilators in the outer wall, but most provided them on the inner walls. Light was sometimes brought into the middle rooms, by inserting a glass brick in the flat RCC slab.

(e) All houses had columns, but were building the walls together, and the slab appears to be resting on the wall too, which makes for a kind of hybrid load-bearing-frame construction

(f) Almost none of the home-owners were building toilets or bathing rooms, even those in villages that were earlier Open Defecation Free, ODF, due to the RHEP intervention by Gram Vikas (www.gramvikas.org; Vaidyanathan, 2002).

RBC: Reinforced Brick Concrete; SHGs: Self Help Groups; CRPs: Community Resource Persons; CBOs: Community Based Organizations

Old CGI roofs and mud wall shave been replaced with flat RCC roofs and concrete blocks, Aerated Concrete blocks or burnt bricks for walling with RCC frame structures. A typical work-order received by each homeowner is also included.

Fig. 19.9 Collage showing current building trend.

(g) Kitchen, a woman's domain, is visualised in almost all homes as being outside the main house (or in the case of houses where the old house is close by, in them), and water connections are either in front or behind the house.

(h) Solar PV is the usual way for home-lighting these days, but many of these villagers are likely to get grid connectivity soon, since they are now connected by a road.

Three resettled villages are described briefly with respect to their site configurations and internal layouts and the house-habitat interdependencies (see Fig. 19.10): (1) Marmaria village, initiated by the local *Sarpanch* (village elected representative) (2) Madaba Village, initiated by the government but catalyzed by aforeSt *watcher* in the village (3) Billosahi village, where there are signs of possible integration with spring-shed development promoted by Gram Vikas.

Marmaria: The old village is located in the upper reaches of the hills and relocated as road connectivity to the village is poor making it difficult to get construction materials up to the village. There was resistance to begin the new housing, as villagers wanted a site closer to their village or get sanctions for road construction. The *Sarpanch* tried to catalyze the construction process by initiating the first nine (9) out of 23 houses in collaboration with the homeowners. The sarpanch was confident of completing the construction within the four month time-frame stipulated, to claim the early completion incentive of Rs. 20.000. He estimated the available outlay of Rs, 1,88,000 as follows: *Rs. 1,30,000 for the house incentive of Rs. 20,000 for completion within 4 months Toilet Rs. 12,000sCompost pit Rs. 12,000s, Rs. 14,000 MNREGA funds for the house owner helping in house construction* The villagers say they have also paid Rs. 15,000 — Rs. 17,500 for their plots, sometime in the form of sweat equity.

The houses are in two distinct clusters, 20 houses are arranged on either side of an open court 40 feet deep, while four houses are on slightly higher ground (see Figure 7). All houses are 2 rooms with veranda, one room leading into the other and opening into the courtyard, where each house is provided with a tap (house carpet area is 32m°). The houses have concrete block or brick in cement mortar walls, RCC slab and columns (23x23cm) in corners. About 10-15% of the homes are partially occupied. All 26 houses are at the superstructure level and at least one room of two has a roof slab. Differing plinth and slab levels are implemented without consideration of how the water will be drained from the roof or from the site. Home owners feel there is no space for a backyard garden, and see these houses only as utilitarian. Most house owners continue to occupy their old homes, saying they would have preferred a land close to their village

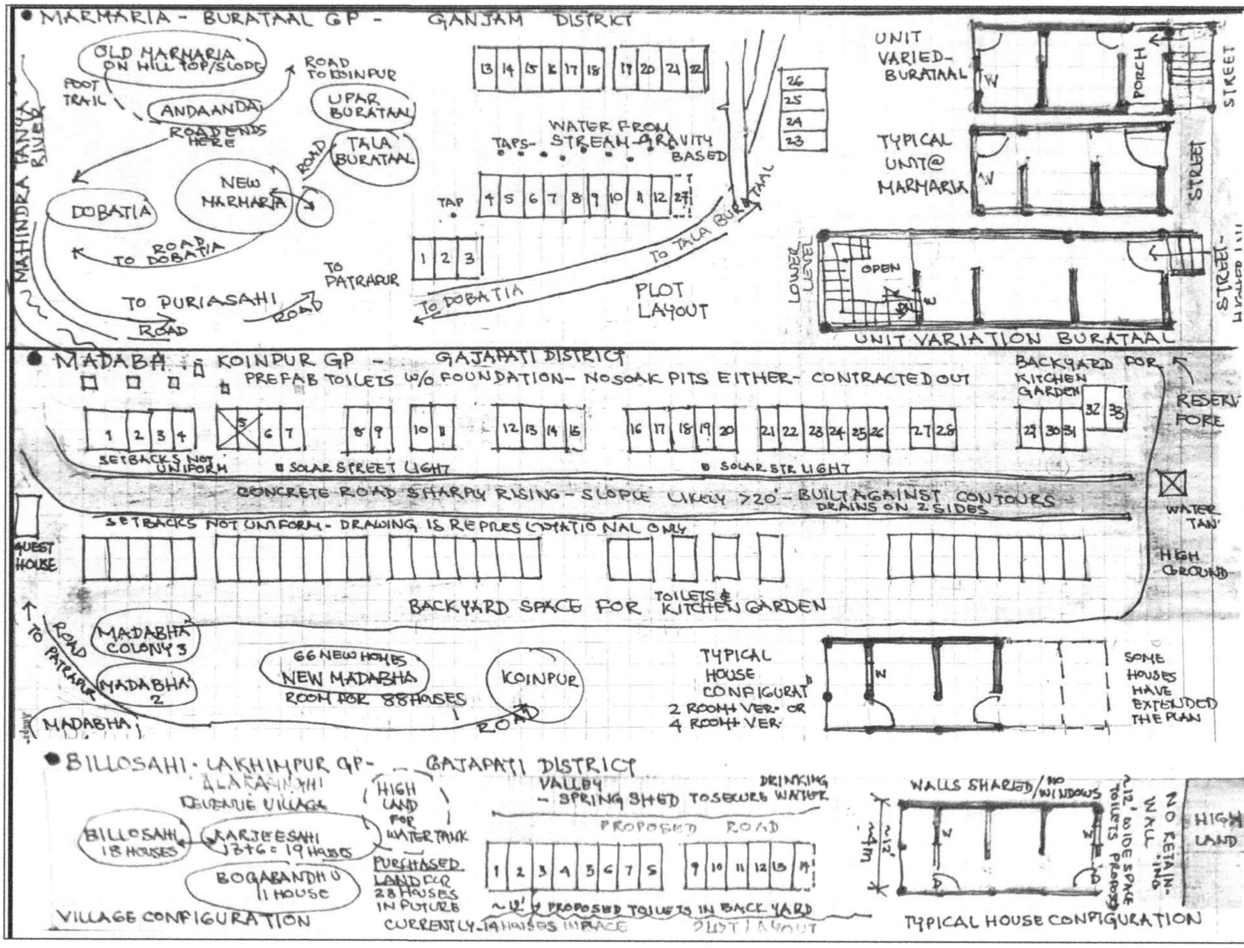

Fig. 19.10 Comparing the three resettled villages: pages from author's field notebook.

where they have their farmlands. They say they will retain their old homes, and come down to Nua (New) Marmaria when there is another storm.

Madaba: The existing village with 66 houses is along the main road connecting Koinpur and Jarada, and is divided into three settlements. Their hunt for a land for resettlement took them to several potential sites. One stipulation for the villagers was they wanted a location similar to their existing one, on the roadside, the other was that it should be in close proximity to their old village. After several failed attempts, aforeSt *watcher* "in the village requested 4 ha land in the Reserve forest that he felt could be granted as a part of the Forest Rights Act. After several rounds of the Collectorate"*, they were finally able to get a piece of land on reserve forest land, measured exactly to fit their houses (88 instead of 66 in number, forecasting future growth). Each one was allotted a plot, 12feet x 60feet, and the land was demarcated. The villagers themselves cleared the forest and levelled the space, marking out 88 plots, 44 on either side of a 40 feet wide road. The layout of the houses is against the contours and the road rises sharply from the entrance. The cacophony of the slabs, sharply changing heights and the lack of a retaining wall at the top of the hill, where the site was excavated, makes it look like an unplanned siting. However, the home owners justify their layout saying this was the only way they could have got space for a homestead garden at the back. Prefab-precast toilets have been built at the back of the houses, as a part of a tender given to a company from outside the State, who have constructed only the toilets, no bathing rooms or soak pits.

Personal Communication, K.Nayak, 10-Feb-20

Forest watcher is usually recruited from the community, village and works closely with the Forest Guards appointed by the Forest Department

** Personal Communication, Krupasindhu Pata Bhuyan, 24-Aug-21

* Office of the Administrative authorities in the district headquarters

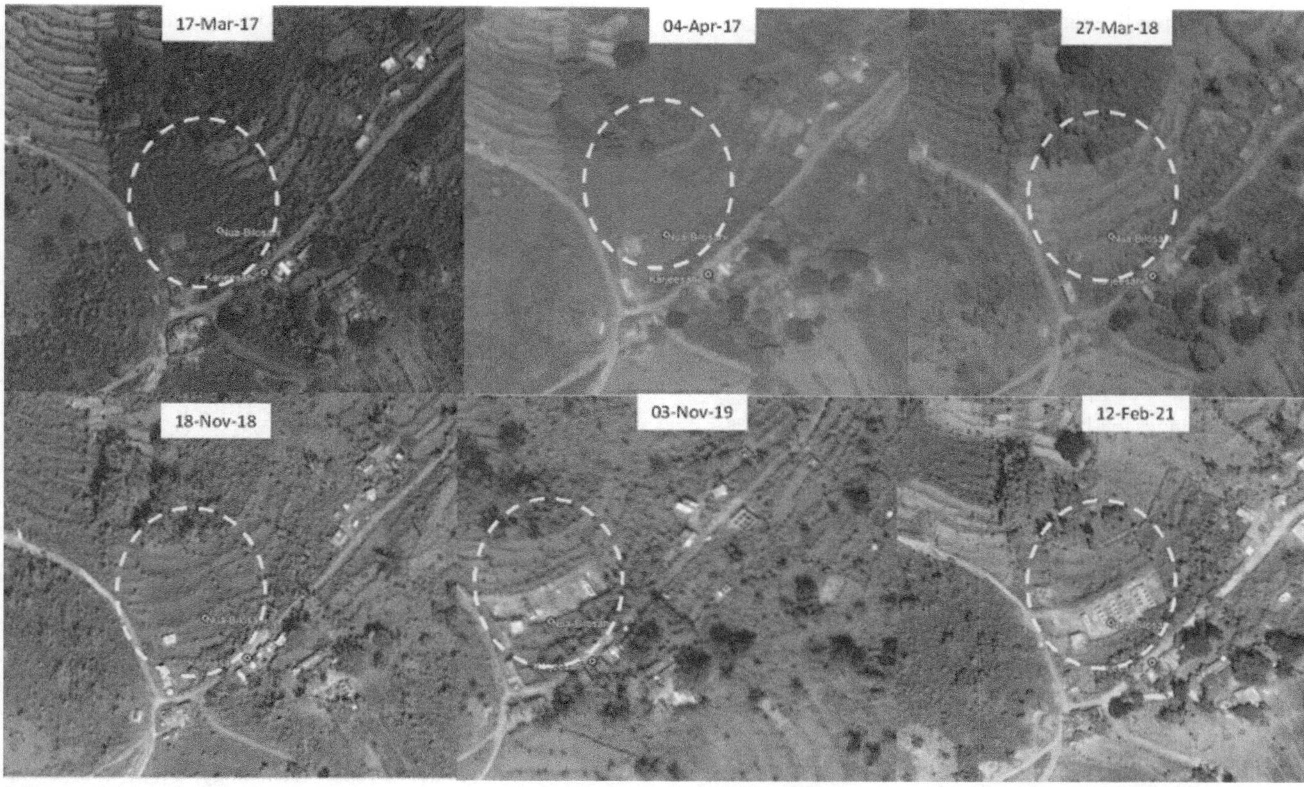

Google Earth images of Billosahi village cluster studied in the Laxmipur Panchayat, of Gajapati district. 1, 2 and 3 are pre-Titli, Images. Image 4 is 5 weeks after, Image 6 is about 2 years after when the resettlement of village Marmaria had started, and the last figure shows further development

Fig. 19.11 Time lapse images of Billosahi village, Gajapati district

Taps are provided in front of the houses, not near the toilet or proposed kitchen. Houses are set back from the central road that has a drain on either side. The houses are mainly 2 rooms and a veranda, although a couple of unfinished houses appear to be larger. Some of the houses are occupied, mostly by people without farmland. Other families have split-up, with some family members living in the old village and others, especially young children attending online classes during the COVID19 pandemic, living here. Most of the villagers still live in their farmstead homes returning to the old village at night, since it is closer. The old village is about 5 km by road from the newly settled one.

Billosahi: Residents of this small habitation of 18 houses were asked to relocate as their old village was considered vulnerable to landslides by government officials. The villagers have purchased 12x60 feet plots less than a km from their old village, from one of their own community members. Fourteen out of 18 houses are laid out along a proposed road all in single file, three rooms and veranda, leaving a 12-feet corridor-like backyard-space that needs a retaining wall. Other plots are scattered around the site. However, what is different about this habitation compared to Madaba or Marmaria, is that it is in close proximity to the old habitation, and more importantly, the village community has been working on securing their water sources through a spring-shed work initiated by Gram Vikas. This work offers potential for ensuring a sustainable source of water for domestic purposes that could also supplement livelihood activities as needed. This spring-shed development is in the valley between the old and new village and could become the starting point for planning of secure habitations in conjunction with their natural habitat (Figure 11). There is scope for integrating the household services with land development and catchment protection activity. Such site development activity, when taken up in conjunction with the house layout, could contribute to building safely in harmony with the natural habitat and reduction in disaster risk.

Advocacy and Outreach, Training and education

Resilient habitat planning and disaster-resistant building systems are needed that combine technology advances with local training. Policy instruments as well as elements for a good toolkit already exist. However, there is still a long way to go in making what the government calls in the guidelines to BPGY (2017), *"habitat and housing literacy."* The house is certainly a shelter from the elements, but also much more. Working on the house design with homeowners, community, block officials, village leaders, is important to ensure secure and healthy homes, while also considering the environment as a client. This will make the homes functional, avoid waste and ensure site development as a first step to building safe houses (unlike as seen in the study, illustrated on the right). Training of masons would therefore include these elements for planning safely and holistically, while including disaster-resilient construction materials and techniques. The requirement for Building Codes cannot be stressed enough, for eg., *IS14243 Part 2: 1995, Selection & Development of sites for Building in Hill areas Guidelines,* is essential for the study region to ensure locating buildings safely. Masons, returning from the city and building in rural areas, need to be reminded that the built habitat has to be integrated into the watershed development of the area safeguarding their aquifer and avoiding excessive soil erosion, and planning roads and toilets judiciously. Billosahi in Gajapati has made a small start, with the assistance of Gram Vikas that needs to be reinforced and replicated.

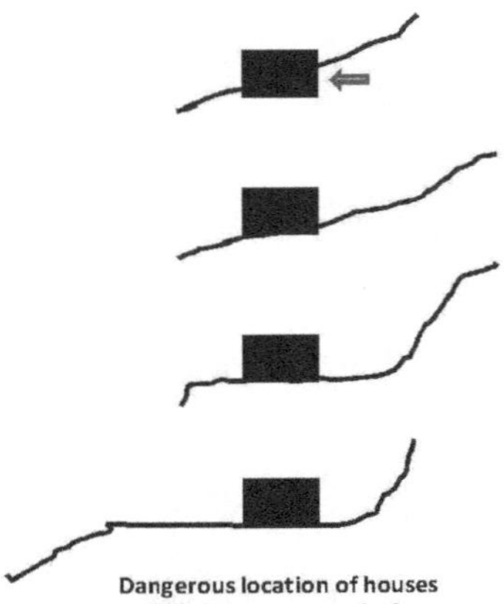

Dangerous location of houses on hill slopes, as seen during the study

4.3 Conclusion

With enhanced technology and early-warning-systems, prediction of hazards is becoming possible. This coupled with the reach of cellular phones, makes it possible to communicate to the last mile, or almost. However, the unplanned growth in the rural areas is making the risk of disaster high and resilience low, especially to extreme and unprecedented climate events like Titli that also emerge with very short notice. Increasingly, there is an attempt to address the priorities of the Sendai framework by understanding disaster risk and strengthening disaster risk governance through some level of decentralization aided by Smartphone applications that attempt building resilient physical infrastructure. However, these efforts can be effective only if the community is enabled with sensitive and holistic decision-making processes that integrate social equity, health, safety and wellness, while building on their innate sense of living-with- nature. Mitigation measures must be put in place, respecting the need for development but without compromising the ecosystem. The focus on physical hardware of relief and rehabilitation currently ignores the integration of the house with the habitat and the critical linkages to social and ecosystem- based services. Also missing is the need for an expanded client base that includes the *entire* family.

4.4 Acknowledgements

We are grateful, to CTx GREEN, Canada, the Executive and Board members who considered it important to go back to study the Tumba villages Post-Titli; to ICT Private Ltd., Delhi, for expertise assistance; to Gram Vikas, for logistical support and knowledge-exchanges and to Sri Sri University colleagues, especially Dean, Prof. Srilakshmi Swamy for her empathy and Amrita Singh, with whom I taught the course on "Disaster-Risk-Reduction & Mitigation." The usual disclaimer applies.

References

1. American Institute of Architects, AIA. (March 2017). *Disaster Assistance Handbook, 3rd Edition.* Washington DC: AIA.
2. Building Material and Technology Promotion Council, BMTPC. (2019). *Vulnerability Atlas ofIndia, 3rd Edition.* New Delhi: Ministry of Housing and Urban Affairs.
3. Centre for Research on Epidemeology of Disasters, CRED. (2020). *Human cost of Disasters. An overview of the last 20 years. 2000-2019.* Switzerland: UN Office of Disaster Risk Reduction, UNDRR. Retrieved Aug 10, 2020
4. CTx GREEN. (2004). *Understanding traditional agricultural systems.' Bogodo in Tumba. Workshop proceedings, August 25-August 27, 2004.* Odisha, India.
5. CTx GREEN. (2018). *Tumba-map-Titli-report - Internal Trip Report.* Mohuda Village, Odisha: CTx GREEN.
6. Gram Vikas. (2018). *Damage and Need Assessment Survey, Cyclone Titli affected villages.* Bhubaneswar: Gram Vikas.
7. Gram Vikas. (2019). *Cyclone Titli Early Recovery Interventions.* Bhubaneswar.
8. Ministry of Tribal Affairs, Government of India, GoI. (ND). Retrieved 09 07, 2021, from https://tribal.nic.in/ST/StatewisePvTGsList.pdf.
9. Mishra, B. (2005). *Land use land cover assessment of Tumba region, a part of Ganjam district by using remote sensing and GIS techniques. (Internal report).* Mohuda, Orissa, India: CTxGREEN.
10. National Disaster Management Authority, NDMA. (November 2019). *NAtional Disaster Management Plan, NDMP.* New Delhi: National Disaster Management Authority, Government of India.
11. Odisha State Disaster Management Authority, OSDMA. (2019). *Disaster Management Plan, Ganjam, Odisha. Volume II.* Ganjam, Odisha: District Disaster Management Authority.

12. Office of the Registrar General, G. o. (2011). *Home/2011 Census Data.* Retrieved July 21, 2021, from Census of India: https://censusindia.gov.in/2011-common/censusdata2011.html
13. Panchayati Raj and Drinking Water Department (PR & DW), GoO. (2017, June 05). Guidelines of Biju Pucca Ghar Yojana. *PR-RH-BPGY-004-2017.* Bhubaneswar, Odisha: Government of Odisha, GoO.
14. Special Relief Commissioner, RDM, GoO. (2018). *Final Memorandum. Very Severe Cyclonic Storm, Titli and subsequent fioods.* Bhubaneswar: Revenue and Disaster Management Department, Government of Odisha, GoO.
15. United Nations. (2009). *UNISDR Terminology on Disaster Risk Reduction.* Geneva: United Nations International Strategy for Disaster Reduction. Retrieved from https://www.unisdr.org/files/7817 UNISDRTerminologyEnglish.pdf
16. United Nations Development Program, UNDP. (2016). *Pahal-Prakriti Hunar Lokvidya. A compendium of Rural Housing Typologies. Pradhan Mantri Awas Yojana, Gramin.* Ministry of Rural Development, Government of India.
17. United Nations office of Disaster Risk Reduction, UNDRR. (2015, June 03). *https.'//www.undrr.org/implementing-sendai-framework/what-sendai-framework.* Retrieved from UNDRR Web site: https://www.preventionweb.net/files/resolutions/N1516716.pdf
18. Vaidyanathan, G. (2002). In Gandhi's footsteps: Two unusual development organizations foster sustainable livelihoods in the villages of India. *Alternatives,* 28(2), 32-37.
19. Vaidyanathan, G. (2009, December 07). *Energizing Sustainable Livelihoods. A Study of Village Level Biodiesel Development in Orissa, India.* University of Waterloo, Geography. Waterloo, Canada: University of Waterloo. Retrieved from University of Waterloo: http://hdl.handle.net/10012/4857
20. Vaidyanathan, G., & Sankaranarayanan, R. (2019). Bridging the chasm — Diffusion of energy innovations in poor infrastructure starved communities. *Renewable and Sustainable Energy Reviews, 99,* Page: 243-255.
21. Wickham, F., Kinch, J., & Lal, P. (2009). *Institutional capacity within Melanesian countries to effectively respond to climatechange impacts, with a focus on Vanuatu and Solomon Islands.* Samoa: The Secretariat of the Pacific Regional Environment Programme (SPREP).
22. Yeang, K. (1995). *Designing with Nature. The ecological basis for Architectural Design.* Mc Graw Hill, Inc.

CHAPTER 20

Resilient Settlements Amid Cyclonic Storm—Empirical Evidence on Factors at Play from Southwest Coastal Bangladesh

Sumaiyah Binte Mamun
Department of Architecture,
Bangladesh University of Engineering and Technology,
Dhaka, Bangladesh

Khandaker Shabbir Ahmed
Department of Architecture,
Bangladesh University of Engineering and Technology,
Dhaka, Bangladesh

Abstract

Bangladesh is vulnerable for tropical cyclone due to geographical position, wide and shallow continental shelf and the funnelling shape of the coasts. With the advancement of cyclone warning systems, shelters, and awareness in people regarding tropical storm reduced loss of life, but the amount of property has risen quite significantly. But recorded evidence shows that *Ghorkumarpur,* a small coastal village at Shamnagar Upazila has surprisingly survived the strong thrust of cyclonic wind not only during Sidr, Aila and Amphan but also at previous storms with its thatched houses, whereas neighbouring villages were washed away. The dynamics of cyclonic wind gust and obstacle aerodynamics at roughness layer has a correlation with wind pressure within settlement form. The gust of wind speed, threshold of comfort, discomfort and danger at settlement level depends on settlement porosity factor at interfacial layer. Settlement form and features plays a vital role to decrease or increase frictional force within boundary layer which determines destruction or minimization of the damage and discomfort caused by cyclonic wind. Observation from filed survey and *in-situ* measurements at *Ghorkumarpur* village indicates several physical parameters of the village land use morphology has helped the survived settlement to act like a bluff body during cyclonic storm and minimized the damage from cyclonic wind by increasing frictional force within the settlement boundary layer. Literature review supports different settlement morphology parameters at interfacial has a possibility to form defensive landscape against extreme wind events. Existing ranges of these parameters has been measured during field survey at *Ghorkumarpur,* village. For a better understanding of cyclonic wind movement through settlement porosity, selected sample among survived settlements form of the village is modelled and simulated within cyclonic wind turbulence using CFD techniques. Findings from CFD simulation shows that wind speed around the survived homesteads, during category 5 cyclone remains below danger level (20 m/s). Results indicates a defensive landscape morphology has been created at *Ghorkumarpur* village which has an impact in reduction of vulnerability due to cyclonic wind. The present case study aims to investigate the parametric ranges behind forming defensive landscape morphology against cyclonic wind at rural settlement level.

Key words: cyclone, coastal settlement form, defensive landscape, resilient settlement

1. Introduction

Records from 1990 show that Bangladesh has faced cyclones at every three years on an average [1, 2]. Bangladesh has 3,739 cyclone shelters for 10,979,846 people living at 13 vulnerable coastal districts, which serves only 27.5% of the total endangered population [3]. The value of property or fixed asset damage was estimated for previous major cyclones based on CRPD report and local newspaper (Fig. 20.1.):

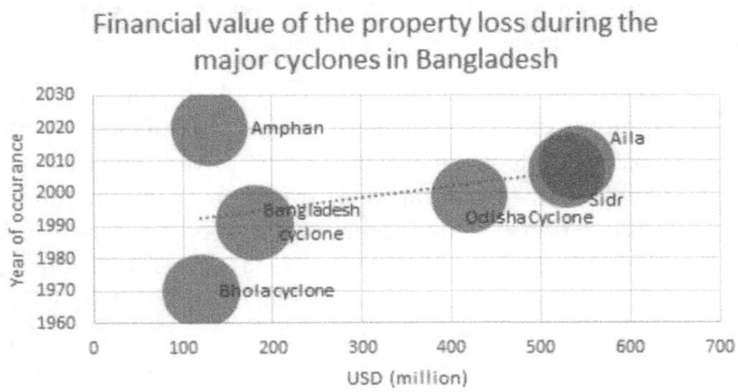

Name of Cyclone	Damages (USD M)	Year
Bangladesh cyclone	180	1991
Odisha cyclone	420	1999
Sidr	530	2007[4]
Aila	540	2009[5]
Amphan	130	2020[6]

Fig. 20.1 Financial value of the property or fixed asset damage (in USD)

It is observed that damage and losses to private assets, such as housing sector (50% of total loss) and livelihoods always outweighed the losses and damage to public infrastructure [7]. The most affected sectors as reported by World Bank [7] are housing, agriculture, transport, water control structures, education and industry.

According to the World Bank's estimation, by 2050 cyclone exposed areas in Bangladesh will increase by 26% and the affected population will grow as high as 122% [8]. In pandemic situation like Covid 19, overcrowding at cyclone has create another dimension of danger instead of safety. Also, it was always difficult for olds, people with physical limitation, pregnant women, lactating mother to adjust at cyclone shelter due to lack of facilities and overcrowding. Household survey at Mongla municipality after Alia, revealed that despite having options of availing shelter at safer places, only 19 % of the respondents had left their houses for taking shelter in safer places like cyclone centres [9]. According to Emergency 2007 Cyclone Recovery and Restoration Project, by the year 2025, 7,124 new shelters are required in 14 coastal districts under Cyclone Preparedness Program [7] which depends on huge foreign funding. One type 3 (capacity: 750 person per floor, 200sqm per floor) [10] cyclone shelter costs USD 67,500. But behaviour analysis shows even if the estimated cyclone shelters are built in due time, that will not be properly used as a larger portion of endangered population will not be ready to leave their homes for a safer shelter [11]. It was also found through the survey that 73.8% of respondents over 50 years old did not leave or move anywhere outside their house, despite having poor housing conditions [9]. An alternative approach to cyclone shelter is in high demand in the country to reduce impact of cyclonic wind and storm surge on different coastal settlement form along with reducing damage on ecology and livelihood. Recently, few initiatives have been taken from NGOs along with government support to create a strong house hold among the community, which will be able to house 5 or 6 families or more during cyclones within 600sqft to 800sqft area. But there is also, a growing concern and knowledge gap regarding how ecology and livelihood can be protected at coastal area with a growing population of 57.9 million in 2050 [12]. So, it is important to look into the knowledge gap for creating the conditions of a cyclone resilient settlement, where inhabitants of costal belt don't lose their belongings and property amid a cyclonic storm, and therefore reduce rising the financial and ecological damage after each cyclone.

2. Problem Statement

Whenever a cyclone landfall, the lower part of the cyclone, among the 3 vertical parts hits the ground [13]. Also, Frictional force gets active at this part near surface of earth and only at higher altitude geostrophic wind is found, vertically that is the middle part of a cyclone. In this part air moves from all direction towards the centre and moves at upward direction [13]. Turbulence at middle part causes major devastation after landfall. This circular wind flow is known as gradient wind and when a cyclone landfall the most effective force working at surface layer is frictional force (Fig. 20.2). Frictional force acts opposite to wind direction and slows down the wind speed and even it creates changes at wind direction [13].

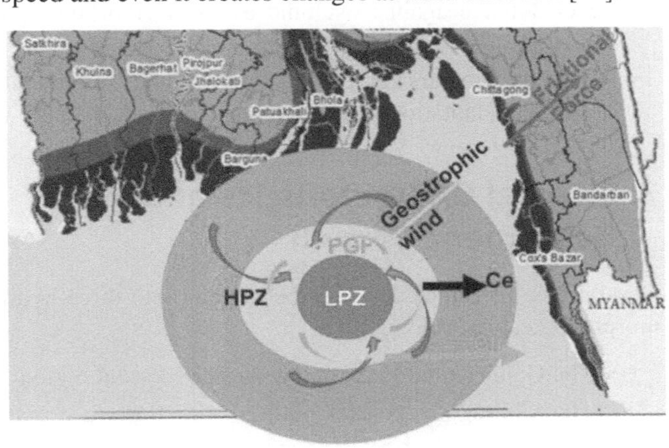

Fig. 20.2 Frictional force (Modified after: Ali & Choudhury, 2014)

From perspective of atmospheric boundary layer (ABL) over land in mesoscale, land types are considered into two categories [14],

Land Types	Land covered with mixed vegetation at length scale	ABL (atmospheric boundary layer)
Type A	<10km	Wind turbulence dominated by local advection movements [15].
Type B	>10km	ABL adjusts to new surface condition and creates inland breeze, sea breeze, river breeze etc. [15].

All the coastal settlement can falls under type A from ABL perspective and therefore is governed by frictional effect, which works opposite direction to local advection movement [16]. Now, wind will create discomfort at pedestrian level or interfacial layer or within 10m(33ft) of height, according to Beaufort scale [17], depending on wind speed, frequency and sustaining time [18]. Wind is considered comfortable for pedestrian if wind speed remains within 5m/s at least 20% of time and it can be acceptable up to 6m/s mean wind speed [19]. At interfacial layer average wind speed decreases with the increase of surrounding obstacle density [20].

At settlement level, interfacial layer context can be determined from built environment and surrounding contextual parameters [21]. Hence, this paper looks into the ratio of the different parameters found from literature review along with settlement typology at Ghorkumarpur village from southwest coastal region of Bangladesh, where the coastal inhabitants has faced minimal damage and tries to understand how settlements from this village has survived extreme cases. Research questions of this paper are:

- What are the parametric ranges behind forming defence against cyclonic event at *Ghorkumarpur* village?
- How are these parametric ranges forming settlement level defence against cyclonic event at Ghorkumarpur village?
- Which parameters can be replicated at other settlements to design adaptation strategy against cyclonic events?

3. Coastal Characteristics and Settlement Growth

Coastal area of Bangladesh can be divided into three hydro morphological sections [22]

(i) The Ganges tidal Plain or the Western coastal region
(ii) The Chittagong coastal plain or the Eastern coastal region
(iii) The Meghna estuarine plain or the Central coastal region

Yearly, 40,000 million cubic feet of silt settle at this plain forming chars and islands and the new land formation at coast is protected naturally by mangroves, which also helped the system of natural land formation at the shoreline [24]. Civilization at Ganga plain could not occur before Iron Age as the land was full of dense forest [25]. Large- scale inhabitation occurred at most of the areas in east Ganga plain dated from 2500BP, with massive deforestation [25]. Other than a part of Ganga tidal flood plain and Chittagong coastal plain most of our significant coastal settlement were habitable after massive deforestation of Sundarbans both at Ganga tidal flood plain and Meghna estuarine plain during British period (1793 to 1879)[24]. Among the three coastal zone the Ganges tidal plain or south western coastal zone covers the largest part of coastal area containing the Sundarbans, Satkhira, Khulna, Bagerhat, Pirojpur, Barguna and part of Patuakhali district which is in total 17,066 sqkm[26]. This zone is almost at sea level with numerous tidal rivers and creeks crisscrossing the region [23]. Except protected Sundarbans part most of this region have been cultivated for more than 150 years [24] and west zone of this area is serving as a production line for shrimp industry for last 20 years [23].

4. Methods and Analysis

4.1 Methodology

Shamnagar Upazila of Satkhira district has both, the most affected village at every cyclone, Gabura at nearby area. This gives a unique opportunity to study filed level strength and weakness of settlement form at the same kind of cyclonic exposer. Here both sample settlement has a protection from wide(≥10m) mangrove cover and falls within low elevation (≤50m), and hence it will get some protection from cyclonic storm [27]. At any cyclonic event, both wind force and storm surge create the major damage to property. This paper will mostly focus on effect of wind force on property during cyclonic event and will try to investigate both micro and macro scale of design parameters for possible cyclonic wind resilient coastal settlement. Household design parameters for shock absorption is the topic of discussion in micro scale of and parameters for defensive landscape and protection of local ecology is the topic argued at macro scale.

For analytic analysis, the study is proposed to be divided into two basic stages.

(i) In literature review, factors that have a co-relation with wind speed within settlement form are summarized along with parameters.

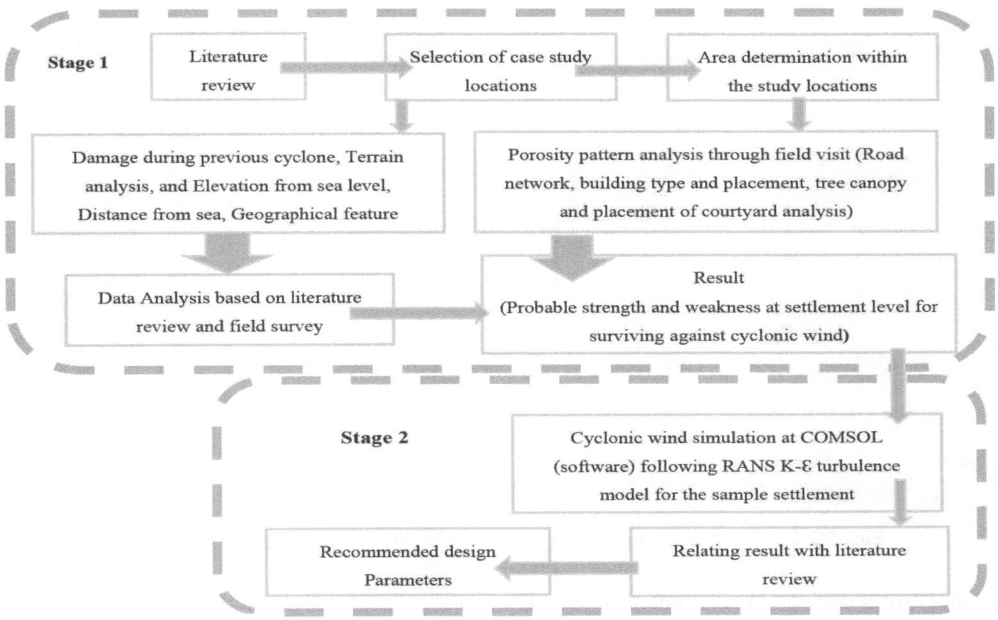

Fig. 20.3 Methodology diagram

(ii) At data analysis part, findings from field studies are discussed and probable strength and weakness at settlement level for surviving against cyclonic wind are assumed for stage 2.

(iii) Stage 2 is the simulation part of this study. For this research RANS k-ε model, COMSOL 5 was chosen due to its competency to available resources and for validation purpose as Bangladesh University of Engineering and Technology has institutional license for this software.

4.2 Analysis

"D- Form Data" for Shyamnagar upazila after Aila, which was submitted to ministry of disaster management, with all the damage information, has been collected. In literature review references [28, 29] about *Ghorkumarpur* village at Padmapukur union reports minimal damage at this village after Sird and many previous cyclones that passed over this land. To observe field scenario and to find parameters of factors found from literature review, detail field study on *Ghorkumarpur* village has been conducted. Padmapukur union is under jurisdiction of Shamnagar Upazila at Satkhira district. Fig. 20.4 is representing analysis from "D- Form Data" after Aila of Shamnagar upaziala at Satkhira district.

Figure 20.4 confirms, during Aila Gabura and Padmapukur union had the most devastating experience in Shyamnagar upazila. Village Gabura of Gabura union is selected to study weakness within settlement level and village Gharkumarpur of Padmapukur union is selected to study resilience at settlement level. At Ghorkoomorpur only 52.4% household [28] faced very little damage during cyclone Sidr. When Aila struck that area, there were a few households left standing after enduring the effect of storm surge whereas, other villages, for example Gabura, were completely washed away. During Amphan and Yaas, nothing happened to Ghorkoomorpur but, Gabura suffered from water logging due to failure of embankment at few points.

Typology of settlements at these two villages has been analysed by using Debouverie's statistical method (1943) for studying rural settlement type. The formula used for Debouverie's method is

$$K = \frac{X \times L}{H} \quad (8)$$

Here, K = Index of concentration, X = Total no of dwellings per settlement, L = Total no of settlement in the region, H = total no of dwellings in the region. Results found from the above equation has been classified into 4 types which are-Compact type ($K > 1.5$), Semi compact type (K = 1.00 to 1.50), Hamletted (K = 0.50 to 1.00), Dispersed ($K < 0.050$). This method is used for settlement analysis as it will indicate which neighbourhoods will create more frictional force at interfacial layer due to higher value of concentration index. Field survey (year: 2015) was conducted by walking through roads and alleyways of *Ghorkumarpur* and *Gabura* village tracking with GARMIN GPSMAP 62s navigator. Received data from GPS navigator helped to get exact location and terrain data of *Ghorkumarpur* and *Gabura* village.

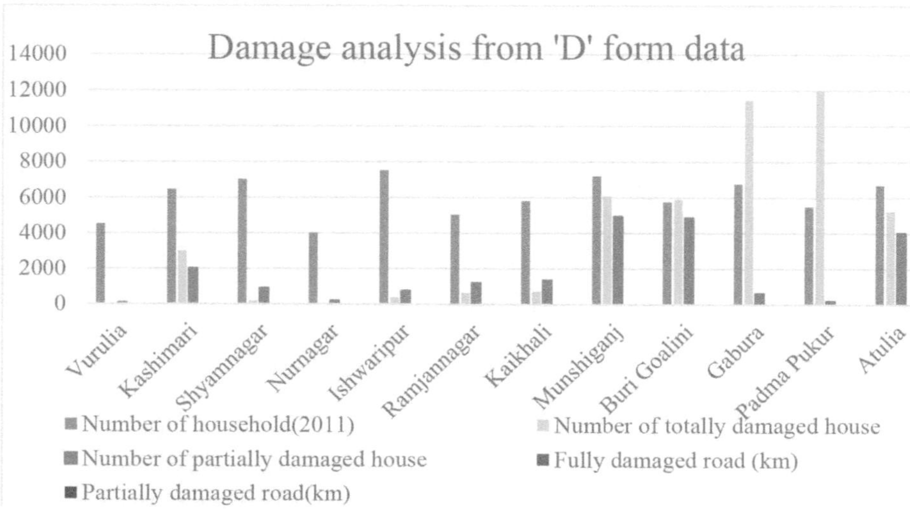

Damage Summary	Gabura	Padmapukur
Fully Damaged Houses	11,450	12,000
Partially Damaged House	630	200
Fully Damaged Road	15 km	11 km
Partially Damaged Road	19.75 km	18.5km

Fig. 20.4 D- Form Data after Aila of Shamnagar Upazila , Satkhira District

Every household at Ghorkumarpur was counted from all the neighbourhood. Specific information on terrain level, surrounding trees, courtyard size and placement, adjacent street and alleyway length and width, location of saved homestead at study area are collected and measured from physical survey and questionnaires survey. To prepare the base map for analysis of settlement form containing all the features is prepared using both high resolution satellite image and field survey. Both vulnerability and resilience are studied on basis of geographical feature analysis and damage data analysis from previous cyclonic storm.

4.3 Summary of Literature Review

At settlement level, interfacial layer context can be determined from built environment and surrounding contextual parameters [21]. Factors having correlation with wind speed within a settlement form, are summarized along with parameters in reference to literature review at Table 20.1.

Table 20.1 Correlation of wind pressure with parameters of features affecting wind speed within settlement from

Parameters of sublime factors affecting wind speed	Correlation with wind speed
Street segment length up to 4 times of building height [32]	Negative, minimize wind speed
Street segment length over 8 times of building height [32]	Positive, increase wind speed
Street width less than 2 times of building height with perpendicularly placed buildings to street [32]	Positive, effective street length becomes 10 times of building height
Street width less than 2 times of building height with 'T crossing' road network [32]	Negative, reduces wind speed
Street width more than 6 times of building height [32]	Positive, increases wind speed
Placement of trees along the street [32,33]	Negative, decreases wind speed and increase the potential for longer street
Terrain elevation [32,21]	Positive, Higher elevation increases wind speed whereas lower elevation decreases wind speed
Placement of dense orchard homestead and coastline [28]	Negative, reduces wind speed
Longer street with effective street with of 0.4 times of building height [32]	Negative, increases wind shelter
Placement of trees and shrubs for long quay [32,21]	Negative, decrease wind scream
Irregular building face [32,21]	
Plain building face [32,21]	Positive, increases wind speed

Parameters of sublime factors affecting wind speed	Correlation with wind speed
Length of building is less than 3 times of building width [32,34,21]	Negative, increases wind shelter
Plinth level determination according to terrain and highest surge level[33]	No correlation, reduces damage from storm surge
Use of building wings [32]	Negative, decrease wind scream
Length of courtyard is less than 2 times of courtyard width and courtyard width is less than building height [32,28,34]	Negative, reduces wind speed

4.4 Findings from Field Survey

Figure 20.5 is showing satellite image of *Padmapukur* and *Gabura* union after Aila devastation. In fig:3, a triangular patch of green is clearly observed at location of *Ghorkumarpur* village which indicates survival of greenery and households in a situation where nearby villages faced a complete wipe out due to Aila. Figs 20.5 and 20.6 shows a closer look at *Ghorkumarpur* village.

The red line at Fig. 20.6 is illustrating path way of Aila. The balck and white image is the survived houses after Sidr at *Ghorkumarpur* Village [28]

Fig. 20.5 Satellite Image of Padmapukur Union and Gabura Union after Aila -3.1.2010

Source: Google earth

Fig. 20.6 Left: Satellite Image (Source: Google earth) of Ghorkumarpur village after Aila -3.1.2010 and Right: household image at Ghorkumarpur village After Sidr, [Credit: Nahiduzzaman & Haas, 2008]

Geographical Features of *Ghorkumarpur* and *Gabura*

Geographical Features	Ghorkumarpur	Gabura
Position Longitude: Latitude:	 22°21'49.52"N 89°13'39.22"E	 22°16'49.57"N 89°16'32.23"E
Highest terrain elevation	5.486m	3.962m
Lowest terrain elevation	0m	0m
Average terrain Elevation	2.438m	1.067m
Soil type	Saline (35 ppm) (August, 2015)	Saline (40 ppm) (August, 2015

Settlement Pattern of *Ghorkumarpur* and *Gabura*

At *Ghorkumarpur* village there are total 1,110 dwelling units, distributed among 19 zones. Number of dwellings at 19 settlement zone and their index of concentration(K) according to Debouverie's statistical method (1943) are:

Zone	Concentration (K)	Dwelling units (no)
A	0.90	53
B	1.01	59
C	1.63	95
D	1.16	68
E	0.22	13
F	1.44	84
G	1.23	72
H	0.91	53
I	1.42	83
J	1.14	67
L	0.51	30
M	0.55	32
N	0.80	47
O	1.40	82
P	0.72	42
Q	1.39	81
R	1.40	82
S	0.31	18
T	0.84	49

Among 19 zones Zone S and E is dispersed type, zone A, H, T, L, M, P, N are hamletted type and zone B, D, F, G, I, J, R, Q, O are semi compact type settlement and only zone C is the compact type settlement (Fig. 7(a)).

Debouverie's statistical method (1943) could not be applied at Gabura village as all the dwelling unit that was present before Alia was not yet build at the time of field survey (2015). According to nearest neighbourhood analysis method all 3 settlement zone at Gabura village is nearly regular pattern(Fig. 7(b)).

Among 95 dwelling units of zone C, 31 units faced no damage and these units are located at inner core of zone C, others got partially damaged like loss of fence, broken trees, debris falling over roof (Fig. 20.8).

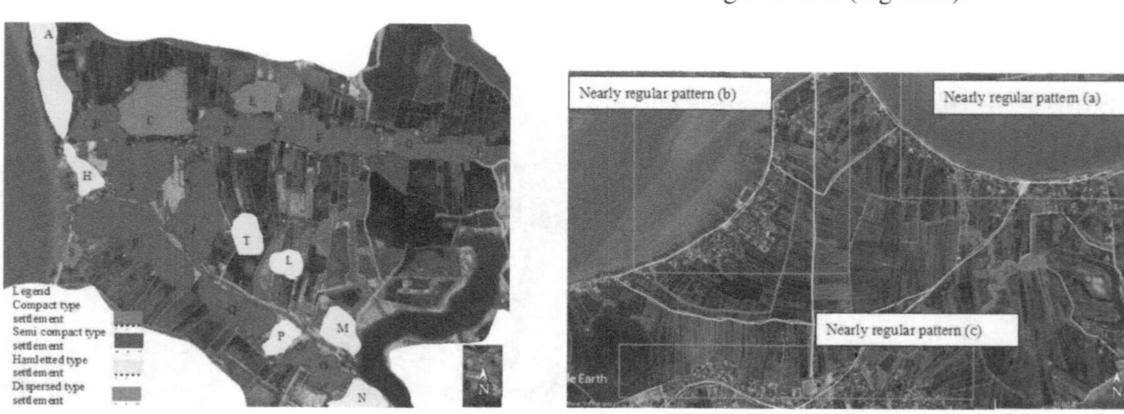

(a) (b)

Fig. 20.7 Settlement type of Ghorkumarpur village(a) and Gabura village (b)

Source: Google earth, 23. 6. 2019

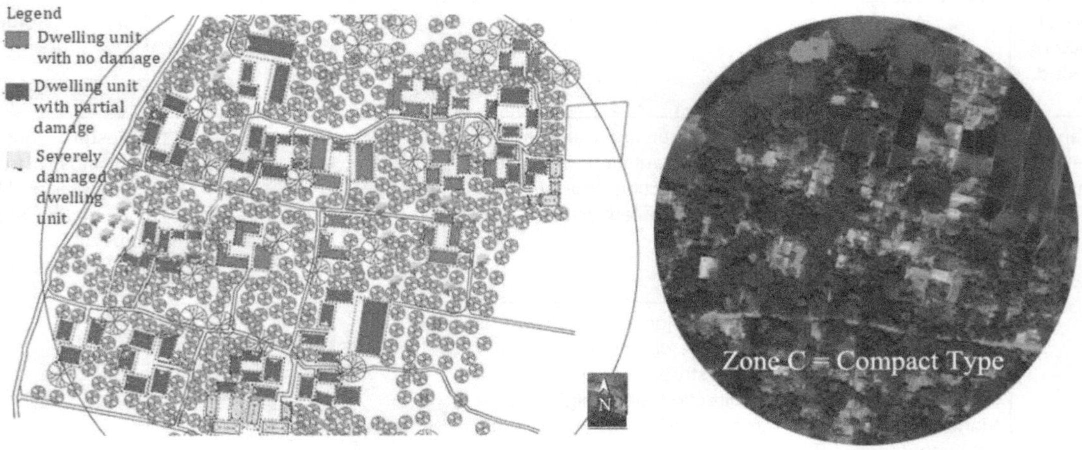

Fig. 20.8 Survived home stead at clustered pattern neighbourhood during Aila at Ghorkumarpur village [field survey, 2015 and Google earth image]

Interfacial layer features of *Ghorkumarpur*

Features of a settlement form that was found from literature review having correlation with wind speed within interfacial layer was measured during. Findings from onsite parametric measurement are summarised in Table 20.2.

Table 20.2 Findings from field survey

Parameters of sublime factors affecting wind speed	Field observation	Summary
i. Alleyway network	Alleyway pattern at zone C along with courtyard connection and zone D portraits grid tree (GT). which is similar to T crossing network and zone T is portraying radial tree (RT) type network. [Marshall et al., 2014; Marshall, 2005]. 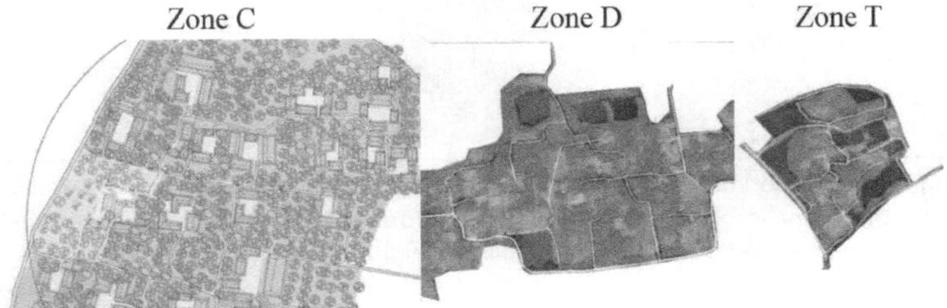 *Gabura* has linear tree (LT) pattern street network [Marshall et al., 2014; Marshall, 2005; Image: Google Earth] 	Position of survived and partially damaged homesteads after Aila[field survey] and Sird at zone C, zone D, heavily damaged homesteads at zone T and devastating experience of Gabura indicates T crossing type road network has a negative correlation with wind speed.

Parameters of sublime factors affecting wind speed	Field observation	Summary
ii. Street and alleyway orientation	Major street network of Ghorkumarpur faces SE-NW orientation and east west orientation, which are perpendicular to cyclonic wind direction and maximise friction within settlement form for turbulent wind.  [Image: Google Earth]	Western coastal zone gets hit mostly from southwest direction. Major street orientation at perpendicular direction with cyclonic wind will create more frictional force.
	For Gabura village number of major street network is 4 and faces EW, NS, SE-N, SW-N direction. Except EW direction street other 3 major street network is almost parallel to cyclonic wind direction and might create tunnel effect which increases wind scouring within settlement form. 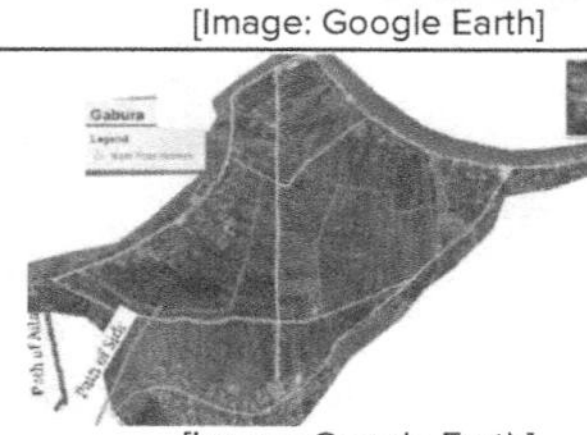 [Image: Google Earth]	
iii. Length and width ratio of dwelling units	Survived homes after Aila were measured at Ghorkumarpur village. It is observed that width and length ratio of sample survived dwelling units is from 1:1.5 to 1:2.25. ratio 1: 2.25 ratio 1: 1.15 ratio 1: 1.65 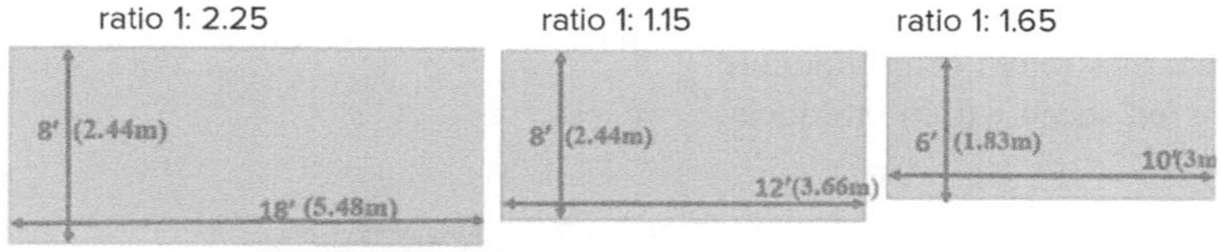  [Credit: Author]	For low rise steep pitch gable-roofed buildings where L : H > 3, a huge negative pressure coefficients increase is found on leeward side roof and wall [37]. Maintaining width and length ratio within 1:2.25 for dwelling units acted positively for reducing wind pressure on dwelling units. No such example was studied from Gabura village as no survived homes after Aila or Sidr was found there.
iv. Plinth level of dwellings units	Plinth level of dwelling units at Ghorkumarpur ranges within 4.8m(16ft) to 6m (20ft) from sea level. 1.4m(4.5ft) plinth height is observed for 3.6m(12ft) terrain elevation, 1.5m (5') plinth height for 3.4m(11ft) terrain elevation and 0.6m(2ft) plinth height for 5.4m(18ft) terrain elevation.  [Image Credit: Author] 	Dwelling units at lower terrain have higher plinth height and units at higher elevation has lower plinth height in relations to ground level. According to locals during Aila Ghorkumarpur faced storm surge height up to 5m(16.5ft), though it was recorded 3m surge at western coastal belt[38].

Parameters of sublime factors affecting wind speed	Field observation	Summary
v. Size and shape and placement of courtyard	At Ghorkumarpur village courtyards are centrally placed, connected with pathways at alternate ends and surrounded by three or four dwelling units. Courtyards are either southern facing or eastern facing. Some homestead had more than 4 dwelling units around courtyard. In proportion, zone C courtyards are within ratio of 1:1.5 to 1:2. Among 22 courtyards of zone C, 17 courtyards have approximate 1:1.5 ratio for width and length and only5 have approximate 1:2 ratio for width and length. For larger courtyards, trees are also planted within courtyard area. 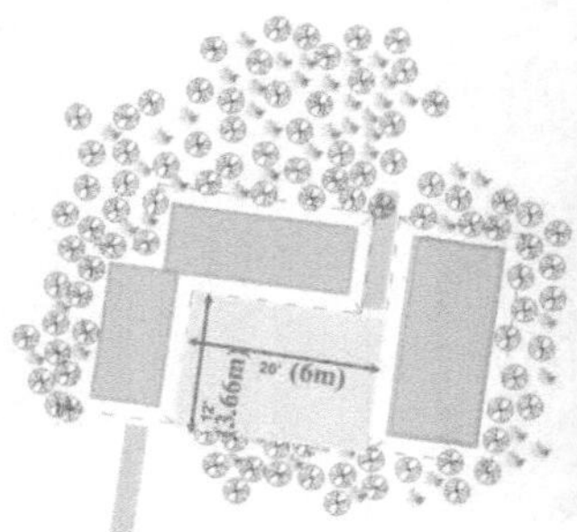  Plan of courtyard Image of courtyard [Credit: Author]	It is observed that courtyards are either rectilinear or "T shaped" not only at zone C but all though the village. At any observed case width of courtyard did not cross one dwelling unit height. Every homestead is surrounded by orchards and for courtyards with 3 dwelling units at 3 sides are enclosed by orchards at other side where there is no dwelling unit.
vi. Typology and placement of trees along the street and coastline	Except mangroves some identified trees were Debdaru, Lour, Menda, Kanthal, Shimul, Koroi, Kewra, Pyera, Kalo jam, Boroi,.Aam, Nim, Mehagoni, Dhundal, Bel, Jambura, Supari and Tal.  [Image Credit: Author (top and bottom), Google Earth (middle)]	Along coastline mangrove trees are observed. Except for streets beside agricultural field, lines of tropical trees and shrubs are seen across and along streets and alleyways. Shrubs were seen at first layer beside alleyways or secondary path networks and trees were at second layer beside alleyways.

Parameters of sublime factors affecting wind speed	Field observation		Summary
vii. Typology and placement of trees around a home stead	 [Image Credit: Author]	Maze at front yard of homestead	Trees are placed all around every home stead and ponds. Most of the homestead has one pond at backyard. Dense orchards are mainly found at south, east and west side of home stead Position of orchards created a defensive fortification of trees around every homestead at zone C. Also, dense orchard at southern, eastern and western sides means these lines of trees are at perpendicular position with cyclonic wind direction. Position of tree lines create sturdy obstacle layers for cyclonic wind to reach dwelling units.
		Maze and shrubs as fencing of homestead. Cactus was observed to be used as fence at many households. [Image Credit: Author]	
		Mahogany, Indian lilac, Date, Coconut, Palm type trees, Mango, Guava, black berry at backyard of homestead. [Image Credit: Author]	

<table>
<tr><th>Parameters of sublime factors affecting wind speed</th><th colspan="3">Field observation</th><th>Summary</th></tr>
<tr><td rowspan="6">viii. Layout of homestead</td><td colspan="3">Homesteads were found to have east- west direction at Ghorkumarpur village. At zone C, 4 types of homestead layout were found.</td><td rowspan="6">In general, all homestead has a central rectilinear courtyard with varying number of dwelling units, ranges from 3 to 7. All the homestead had orchard around them. Without a few most homestead had a pond at backyard. Number of dwelling units increases with increase in family member. Dwelling units include 2 or 3 sleeping unit, kitchen, granary or store and washroom. It was found at October, 2015 wind was flowing from southeast direction at 3.4 m/s at major street level, but interestingly courtyard showed no wind movement at anemometer.</td></tr>
<tr><td>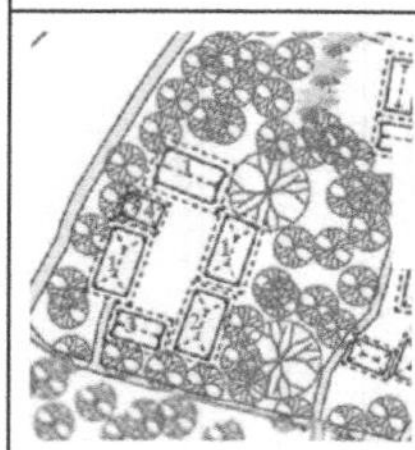</td><td>sleeping unit (1)
kitchen (2)
granary or store (3)
washroom (4).</td><td>
Sleeping unit [Credit: Author]</td></tr>
<tr><td>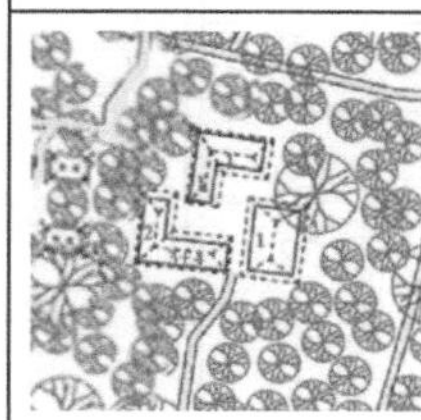</td><td>sleeping unit (1)
kitchen(2)
granary or store (3)
washroom(4).</td><td>
Kitchen and courtyard [Credit: Author]</td></tr>
<tr><td>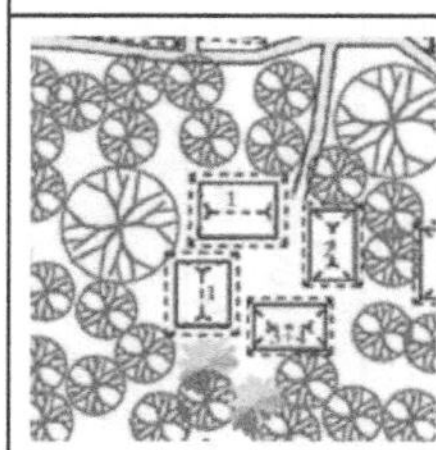</td><td>sleeping unit (1)
kitchen(2)
granary or store (3)
washroom(4)</td><td>
Outside of dwelling unit [Credit: Author]</td></tr>
<tr><td></td><td>sleeping unit (1)
kitchen(2)
granary or store(3)
washroom(4)</td><td></td></tr>
<tr><td colspan="3">During field visit at Ghorkumarpur, wind flow and temperature were measured within dwelling unit, courtyard and outside street.</td></tr>
</table>

Parameters of sublime factors affecting wind speed	Field observation	Summary

Time	Location points	Wind speed at street (m/s)	Temp. at street (°C)	Wind speed at alleyway (m/s)	Temp. at alleyway (°C)	Wind speed at courtyard (m/s)	Temp. at courtyard (°C)	Wind speed at inner house (m/s)	Temp. at inner house (°C)
9:24 am	734	2.9	32.7						
9:27 am	740	2.4	32.7						
9:33 am	745			0.2	32.3				
9:37 am	746						34.1	0	33.4
9:45 am	749					0	33.1		32.8
9:50 am	752			2.1	33.5				
9:57 am	755			2.1	34.1				
10:03 am	756			2.6	34.7				
10:10 am	759			0.3	34.4				
10:20 am	770			1.4	34.2				
10:22 am	782	3.4	33.8						
10:33 am	793					0	35.3		
11:00 am	794							0	35.1
11:15 am	818					0	36.4		

Image Source: Google Earth

Parameters of sublime factors affecting wind speed	Field observation	Summary
ix. Alleyway and roadside geometry of homestead	Alleyways in *Ghorkumarpur* creates parallel line to cyclonic wind. Homesteads are placed along 2 sides of alleyway. Also a line of shrubs and one or two lines of trees are planted along alleyways. At *Ghorkoomorpur* mostly fruit trees are planted along both sides of alleyway. 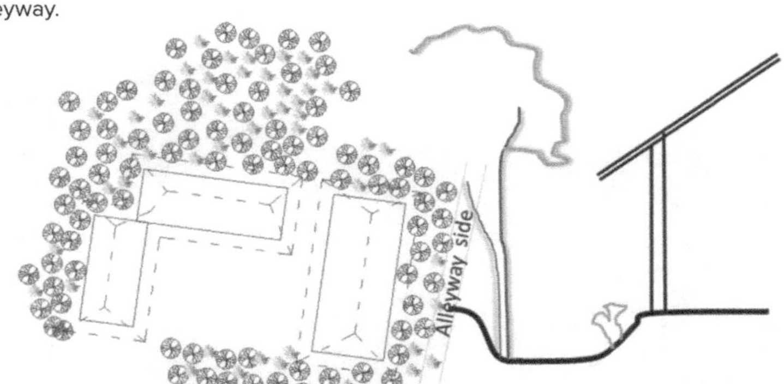 [Image Credit: Author]  	Trees can reduce 25±5% wind pressure if planted along streets and for trees planted across the street this reduction is 20±20% [39]. Placement of trees along alleyways might also reduce wind speed during cyclone time and may work as a factor in saving homestead.
x. Ratio between alleyway's width and building height	All alleyway width (S_y) at *Ghorkumarpur* ranges from 0.6 m (2') to 0.9 m (3'). Dwelling unit height depends on its location and without plinth level ranges from 2.6m to 3.3m. Depending on terrain elevation, total dwelling unit height ranges from 4.6m (15') to 5.33m (17.5') with varying plinth level. Ratio of alleyway width and building height ranges from 1:5 to 1: 8.75. 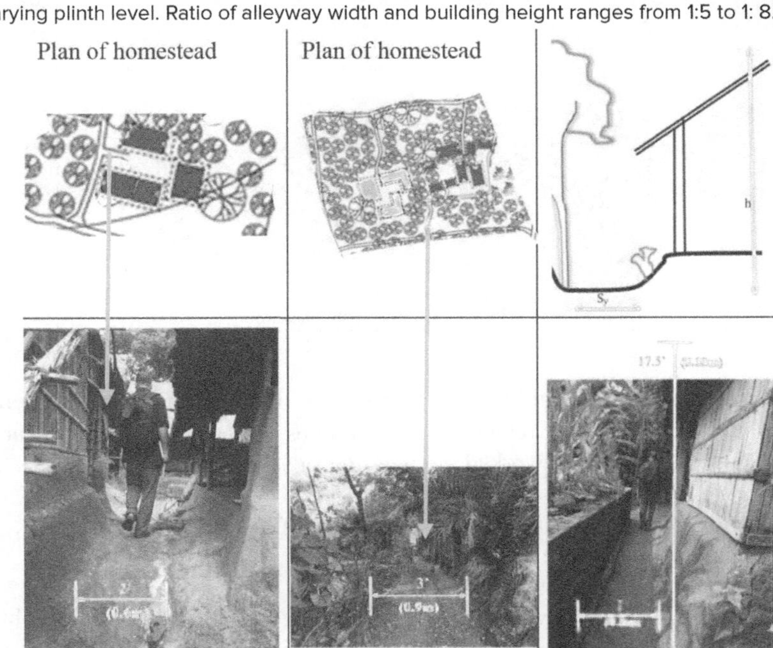 [Credit: Author] [Credit: Author] [Credit: Author]	At *Ghorkumarpur* dwelling units at lower terrain has higher plinth level and total height of dwelling unit becomes higher than units at high terrain level. Wind gust reduces, if street width is less than 2 building height [32]. In case of *Ghorkumarpur* village, alleyway width is less than one building height. Narrow alleyway width contributes to reduce wind gust during cyclone time as well as normal time.

Parameters of sublime factors affecting wind speed	Field observation	Summary
xi. Ratio between alleyway segment length and dwelling unit height	Alleyway segment length at Ghorkumarpur typically ranges from 4.6 m(15') to 6.71m (22'). Uninterrupted straight alleyway segment length at zone C, D, F, G, I, J, R, Q, O ranges from 24.4m (80') to 6m (19.66'). Dwelling unit height rages from 4.6m to 5.33m. Ratio between dwelling unit height and alleyway segment length at Ghorkumarpur ranges from 1:1.3 to 1:5.3.  [Image Credit: Author (Foreground) & Google Earth (Background)]	According to tree canopy ratio (street segment length = 100%, where space between tree crowns ≤0.4h, h = dwelling unit's height [32]), for up to 1.8 m (6') alleyway or street width, wind gust reducing alleyway and street segment length could be highest 36.58 m (120'), if dwelling unit height remains 4.6m.
xii. Alleyway's length to alleyway's width ratio	*Ghorkumarpur* village has narrow streets and alleyways. Most of the long street segment at *Ghorkumarpur* is 36.58m (120'). Longer street means streets with segment length of 8 building heights [5]. Inner alleyway with 0.6m width (2') and 0.9m (3') is most common in this village.  [Image Credit: Author (Foreground) & Google Earth (Background)]	For longer streets, reduction of street width than 2 building heights ensures better wind shelter at high wind event [32]. 1:40(for alleyway width 0.9m) and 1:60 (for alleyway width 0.6m) width, length ratio of alleyway at *Ghorkumarpur* will reduce wind gust at settlement level.

Relatively, all the positive parameters worked contributed to the uniqueness of *Ghorkumarpur*. Literature review supports, settlement features of *Ghorkumarpur* have worked behind lowering wind gust during extreme event. The case of *Gabura* remain untouched for several parameters during field visit and as till now, the village Gabura could not be rebuilt and many of the villagers either migrate or became internally displaced. A major difference between these two villages found from field survey, is *Gabura* has straight long roads, light vegetation, nearly regular pattern linear settlement connecting long linear shrimp gher, but *Ghorkumarpur* village does not cultivate shrimp near settlement arena and have dense vegetation.

4.5 Simulation Model

Turbulence modelling:

The module used turbulence models based on the Reynolds-averaged Navier-Stokes (RANS) model, where Reynolds number is $7.2X10^8$. The following assumes that the fluid is incompressible and Newtonian in which case the Navier-Stokes equations take the form, 9:

$$\rho\frac{\partial u}{\partial t}+\rho(u\cdot\nabla)u=\nabla\cdot[-pI+\mu(\nabla u+(\nabla u)^T)]+F \quad (2)$$

Turbulence effects for cyclonic wind are modeled using the standard two-equation *k*-ε model. The flow near walls is modeled using wall functions.

Boundary conditions:

The inlet boundary conditions for the velocity is a no-penetration condition u . n = 0 and a shear stress condition

$$n\cdot\sigma-(n\cdot\sigma\cdot n)=-\rho u_\tau\frac{u}{u^+}$$

where $\sigma=\mu(\nabla u+(\nabla u)^T$ is the viscous stress tensor.

The normal inflow velocity is specified as $\mathbf{u}=-\mathbf{n}U_0$, where **n** is the boundary normal pointing out of the domain and U_0 is the normal inflow speed. The value of $U_{0\,=\,75\text{ m/s (270 km/h)}}$. The tangential stress component is set to zero. As the reference pressure p_{ref}, defined at the physics interface level, is equal to 0, the value of the Pressure p_0, at the boundary, is the absolute pressure. The default boundary condition to model solid wall is 'No Slip'. A no slip wall is a wall where the fluid velocity relative to the wall velocity is zero. For a stationary wall that means that $\mathbf{u}=0$.

Computational grid (mesh):

Mesh calibrated for Fluid dynamics has been chosen for the simulation. Maximum and minimum element size are fixed at 2.68m and 0.671m respectively. Maximum growth rate had been fixed at 1.3, Curvature factor 0.9 and resolution of narrow regions at 0.4.

Location and Morphology of the modeling area:

An area of Ghorkumarpur village equivalent to the size of (150X115) from zone C was selected for the simulation study. The height of the control domain for simulation study was 10 m roughly three times height of the houses in the village. Fig 20.9 is showing the modeling area of *Ghorkumarpur* village.

Terrain level is modelled at 0 level. Plinth of every home stead has been considered at 1m. Height of homestead without roof is 2.8 m and roof height is considered 1m for modelling. 2 types of trees were considered for mesh model, regular and large type of trees with globular canopy. Shrubs are not modelled for reducing complexity due to lower computational power capacity. For large trees, height and diameter of tree truck is considered as 6.3m and 0.9m, with canopy of 5.6m diameter. Regular trees height and diameter of tree truck is considered as 5.5m and 0.44m with canopy of 2.8 m diameter.

Simulation result:

Table 20.3 demonstrations wind speed curves at different level along Z axis (XY plane). Level 1m to 3m is within pedestrian level range. Dense obstruction is effective up to 7m level, and only some large tree canopy reaches over 10m and densification for those tree canopy works till 9m level. Red contours are showing areas with maximum wind speed and blue contour lines are displaying areas with minimum wind speed. From images of wind speed along XY plane it is observed main raise of wind speed happens after 6m level.

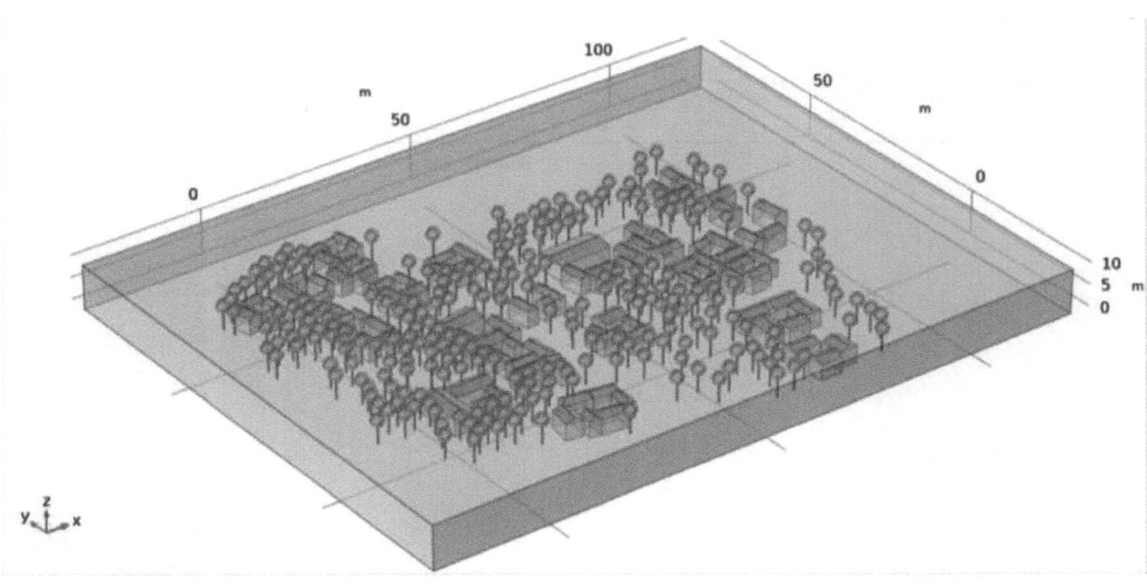

Fig. 20.9 Modelling area of Ghorkumarpur village (Created on COMSOL)

Table 20.3 Findings from CFD simulation (at different wind speed within inner area of sample settlement with turbulent flow (XY plane)

(a) Wind speed: 1 m level

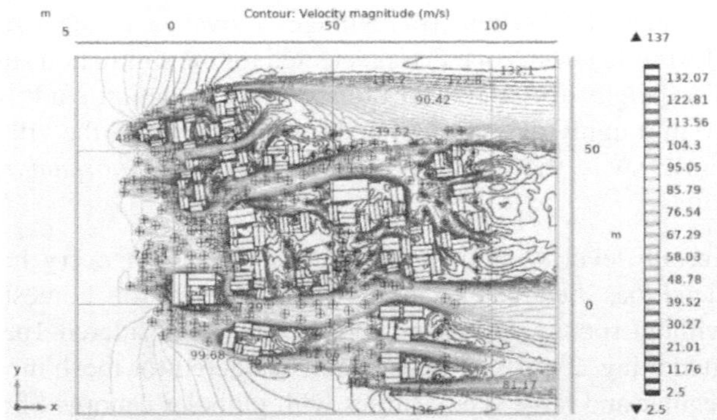

Wind Speed	m/s
Maximum	113.6 to 139.01
Minimum	0.0589 to 21.01
Average	61.53

(b) Wind speed: 2 m level

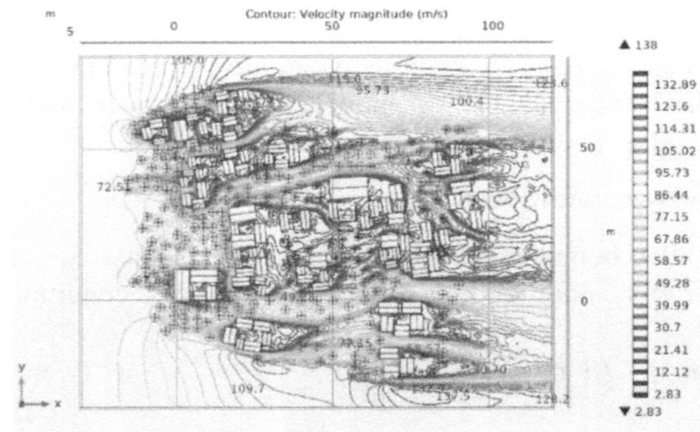

Wind Speed	m/s
Maximum	139.86
Minimum	0.087 to 21.41
Average	62.951

(c) Wind speed: 3 m level

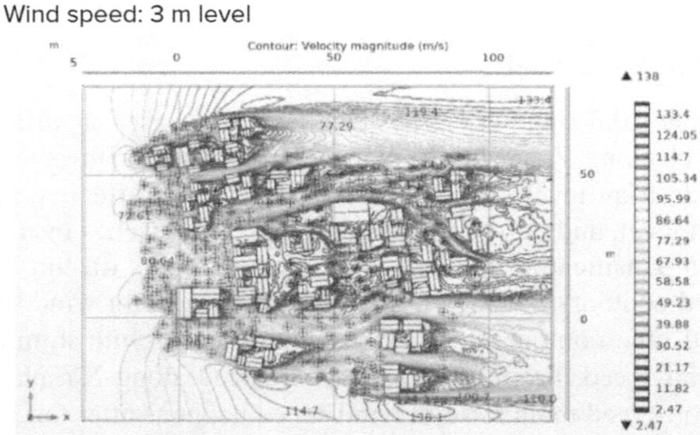

Wind Speed	m/s
Maximum	124.05 to 140.42
Minimum	0.078 to 21.17
Average	64.85

(d) Wind speed: 5 m level

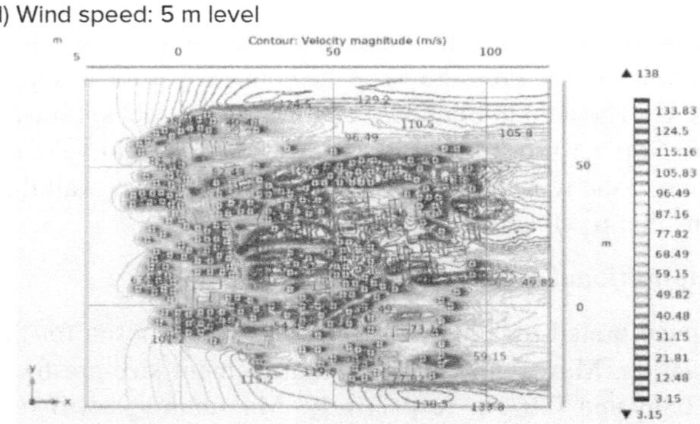

Wind Speed	m/s
Maximum	115.16 to 140.83
Minimum	0.236 to 21.81
Average	69.213

(e) Wind speed: 6 m level

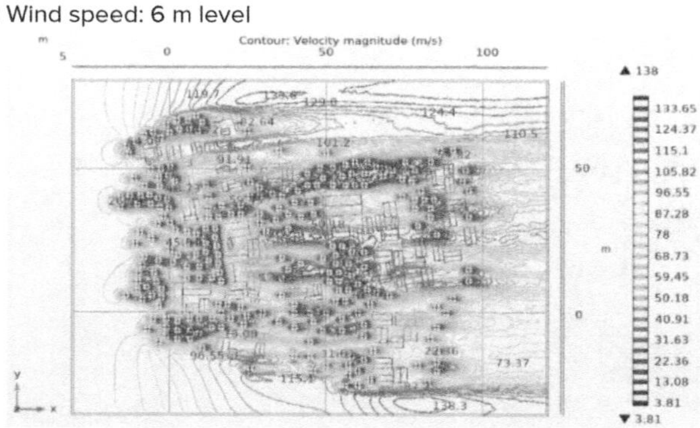

Wind Speed	m/s
Maximum	115.16 to 140.83
Minimum	0.236 to 21.81
Average	77.29

(f) Wind speed: 8 m level

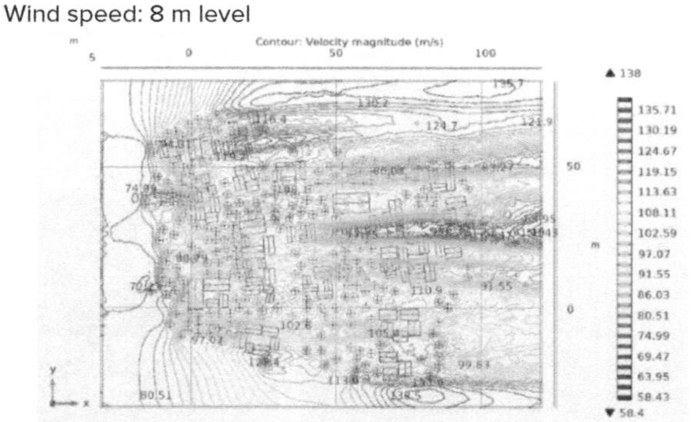

Wind Speed	m/s
Maximum	139.85
Minimum	57.037
Average	102.39

(g) Wind speed: 9 m level

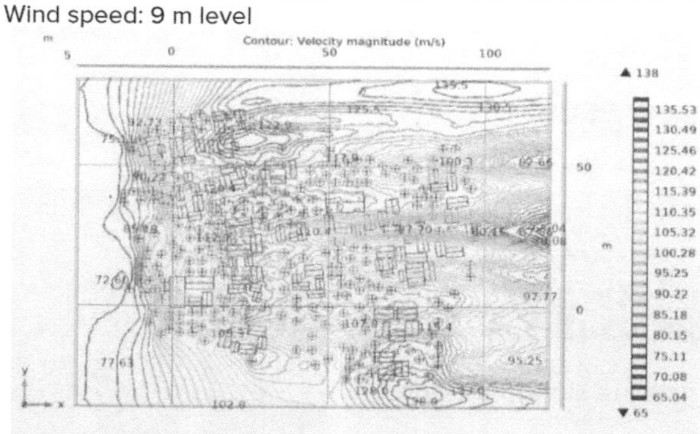

Wind Speed	m/s
Maximum	139.30 m
Minimum	63.77 m
Average	107.86 m

(f) Wind speed: 10 m level

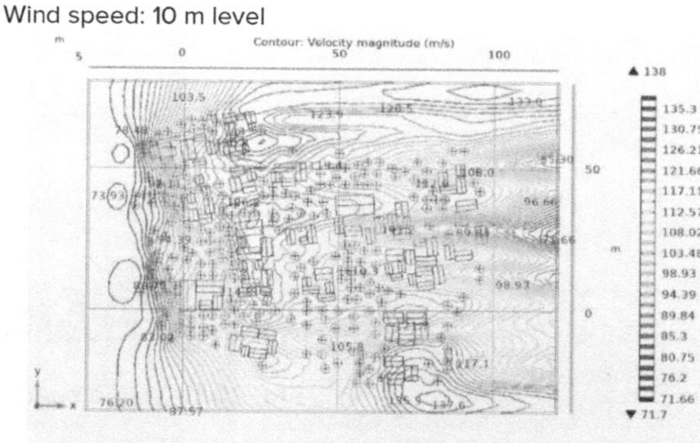

Wind Speed	m/s
Maximum	138.71
Minimum	70.522
Average	109.43

a, b and c are showing wind speed up to pedestrian level at XY plane. Due to absence of shrub layer in the model, simulated wind speed at pedestrian level will be higher than actual wind speed during cyclonic event. Simulation result indicates at pedestrian level, within interfacial layer courtyards and inner core always have wind speed near danger probability, even during category 5 cyclone. At 5m level (d) from ground canopy layer starts. The gaps between canopies have the lower speed of wind at this level. Canopy layer gradually ends within 8m level (f) to 10m level (h).

Scenario within inner area of selected settlement zone gets more clear from sections (Fig. 20.10 and Fig. 20.11) along Y (XZ plane) and X(YZ plane) axis. At *XZ plane* and *YZ plane* it is observed that wind speed remains within 40m/s up to canopy layer which is like category one cyclone.

Analysing contour lines from Fig. 20.10, Fig. 20.11, Fig. 20.12 it can be specified that, simulation result indicates inner core courtyards are the safest open spaces having wind speed within 6m/s and at canopy layer gaps between canopies have the lowest wind speed within interfacial layer.

Interestingly, simulation shows every courtyard of zone C of *Ghorkumarpur* village has comfortable wind speed up to roof level of dwelling units and safe for regular activity, during category 5 cyclone and daily activity will not get hamper inside boundary of homestead.

5. Results

Impact of cyclonic wind thrust at settlement level depends not only on the forecasted wind speed but also on surrounding terrain and settlement geometry. From analysis of study villages, reference sample settlement and CFD simulation 10 design parameters for cyclonic wind resiliency has been identified which is supported by field survey and literature review. From CFD analysis, it is observed that, due to defensive landscape feature of *Ghorkumarpur* the whole settlement at neighbourhood scale work as a bluff body against wind during cyclonic event. Findings and design parameters supported by this research is discussed under section 5.1 and 5.2.

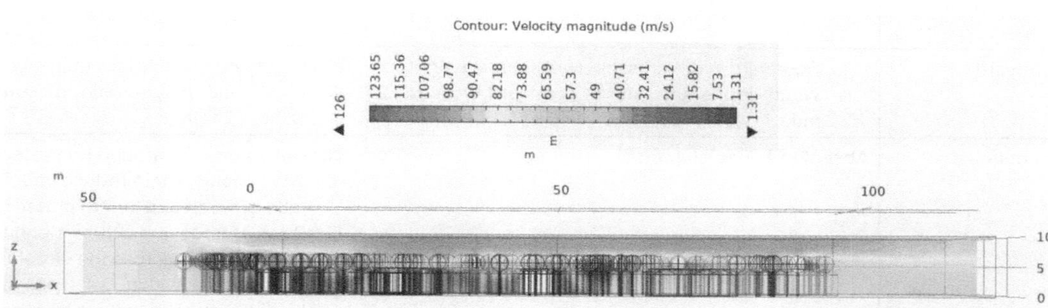

Fig. 20.10 Wind speed at 5m grid line along XZ plane

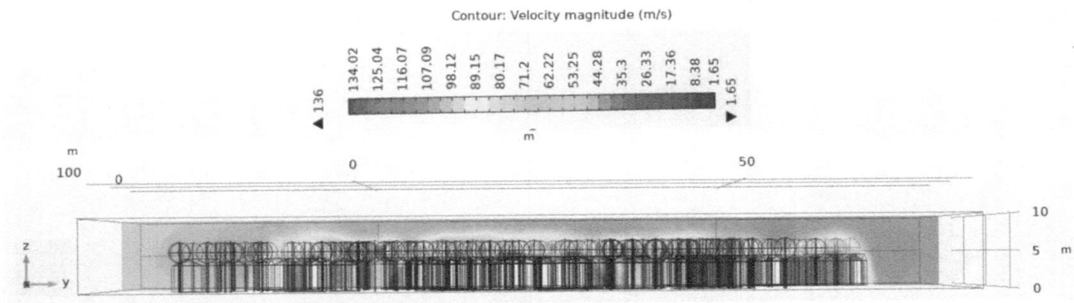

Fig. 20.11 Wind speed at 20m grid line along YZ plane

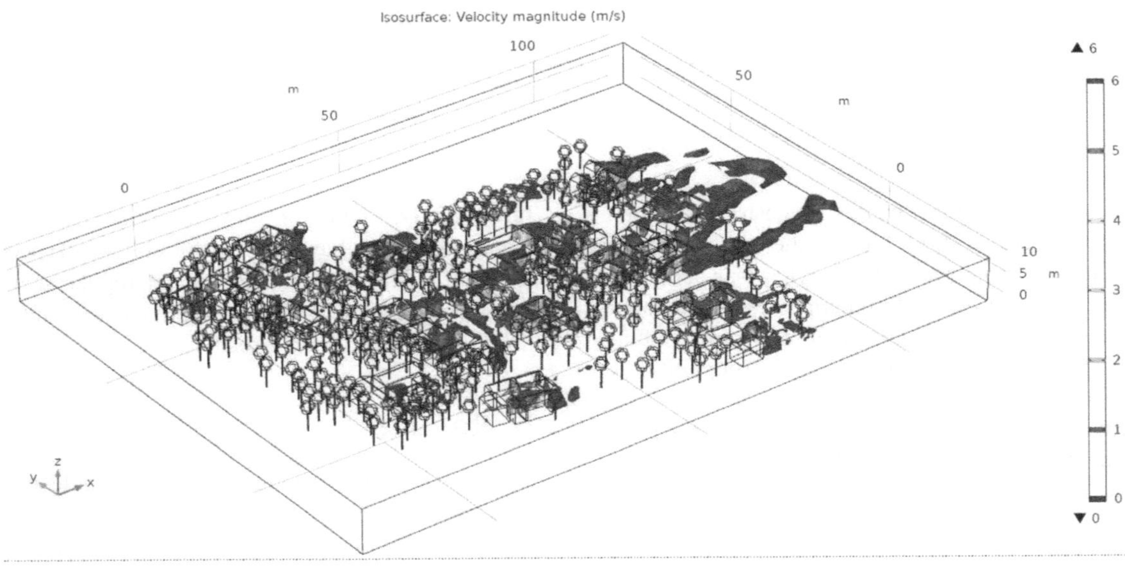

Fig. 20.12 Zones having wind speed up to 6m/s when U_0 =75 m/s (270 km/h).

5.1 Findings: Field Survey Analysis, Simulation and Literature Review

Settlement feature	Positive correlation with cyclonic wind (had devastation during cyclone)	Negative correlation with cyclonic wind (experience minimal damage during cyclone)
i. Settlement type	Depressed and hamlatted type settlement	Nonlinear, compact settlement type
ii. Alleyway network	Radial Tree network and Linear Tree	Grid Tree type
iii. Street orientation	Parallel to cyclonic wind direction	Perpendicular to cyclonic wind direction
iv. Width and length ratio of dwelling unit	Ratio crossing 1:25	Ratio between 1:1.5 to 1:2.25
v. Size of courtyard	i. Courtyard width crosses dwelling unit height ii. Width and length ratio of courtyard crosses 1:2 ratio	i. Courtyard width is less than dwelling unit height ii. Width and length ratio of courtyard remains within 1:2 ratio
vi. Placement of trees	Absent of dense orchard	i. Dense orchard around homestead ii. Within courtyard to reduce effective length if the ratio of width and length crosses 1:2. iii. Rows of trees and shrubs along the street iv. Rows of tress across the street to limit effective street width within 0.4 times of dwelling unit height, if street segment length crosses 4 times of dwelling unit height. v. Rows of trees along coast line

Settlement feature	Positive correlation with cyclonic wind (had devastation during cyclone)	Negative correlation with cyclonic wind (experience minimal damage during cyclone)
vii. Layout of homestead	Linearly placed dwelling units along alleyway	Centrally placed courtyard with dwelling units and dense orchard around it.
viii. Ratio between dwelling unit height and alleyway segment length	Ratio crosses 1:8 with reduced effective street width within 0.4 times of dwelling unit height.	Ratios from1:1.3 to 1:5.3, with reduced effective street width within 0.4 times of dwelling unit height.
ix. Ratio between alleyway width and alleyway length	Ratio crosses 1:100 with effective street width within 0.4 times of dwelling unit height.	Ratio 1:40 when alleyway width is 0.2times of dwelling unit height ratio 1:60 when alleyway width is 0.13 times of dwelling unit height

Though the study is focused on parameters of settlement form that influences the wind speed and flow pattern within settlement form, but height of plinth level of dwelling unit from sea level also plays major role for surviving during cyclone as it is related with survival from storm surge. Settlement form of coastal area in Bangladesh can be considered as mostly linear type, along the coastal belt [40]. In linear type settlement homestead follows a single row along roadside, riverbanks or canals [41]. As there is not much difference between our rural settlement patterns throughout the coastal belt, it is projected that findings from this field study will be applicable throughout our coastal belt.

5.2 Design Decisions for Implementation at Coastal Settlement for Cyclonic Wind Resiliency

(i) For coastal area plantation of mangrove trees and other deep-rooted trees are a must for gaining resilience.

(ii) All major coastal road network should be NW-SE direction to gain maximum resistance against cyclonic wind.

(iii) Every coastal settlement comprises of dwelling units, orchards, road network and ponds should be designed as to create compact settlement.

(iv) Every road network should follow tributary tree crossing keeping the road length within 4 to 6 building height as none of the street width crosses 2building height at coastal area.

(v) In case of narrow alleyway with alleyway width up to 0.4building height street length can be allowed 100 times of street width.

(vi) Trees with deep root needs to be placed along both side of streets to minimize the scouring of cyclonic wind and effective street width.

(vii) Roads that connect the settlement with surrounding terrain will be considered as roads of infinite length and needs to have roadside orchard at terrain side.

(viii) No building at coastal area should have length 3 times longer than building width. Guidelines on Building elements and components for separate building code for coastal areas applicable for both rural and urban areas.

(ix) Width of courtyard should not cross building height and courtyard length should not cross 2 times of courtyard width.

(x) Concept of defensive landscape can be implemented by local authority including community and it can also be designed as a sustainable system of livelihood for locals.

6. Conclusion

The outcome of the study is useful for practical implementation. A significant reduction of housing loss has been found through analysis which will help us to think about adaption using nature-based solution at existing coastal settlement areas of Bangladesh. The study also states the importance of green infrastructure along the coast and within settlement for reducing porosity by creating a dense fragmented surface for turbulent wind. Present condition of case study zone is also an important finding of this study. The consequence of this research opens up the scope for future studies on the issue.

Reference

1. DDM (Department of Disaster Management), Summary of Cyclonic Storm "AILA", Government of Bangladesh, Dhaka, 2009.
2. K. M. Tasnim, T. Shibayama, M. Esteban, H. Takagi, K. Ohira, R. Nakamura, Field observation and numerical simulation of past and future storm surges in the Bay of Bengal: case study of cyclone Nargis, Nat. Hazards 75 (2) (2015) 1619–1647
3. Emergency Cyclone Preparedness Program updated on 2014, under the ministry of Disaster Management and Relief
4. M. Z. Hossain, M. T. Islam, T. Sakai and M. Ishida "Impact of Tropical Cyclones on Rural Infrastructures in Bangladesh". Agricultural Engineering International: the CIGR Ejournal. Invited Overview No. 2, Vol. 10, . April, 2008
5. Md. Shamsuddoha, Dr. Mahmudul Islam, Mohammed Atikul Haque, M. Forruq Rahman,Erin Roberts, Anna Hasemann and

Stephen Roddick, Local Perspective on Loss and Damage in the Context of Extreme Events:Insights from Cyclone-affected Communities in Coastal Bangladesh, Center for Participatory Research and Development (CRPD), June 2013

6. "Cyclone Amphan: Bangladesh estimates damage worth $130 mn, 20 dead", Business Standard May 22, 2020 for Amphan data
7. Document of The World Bank, Report No: PAD807, International Development Association project appraisal document on a proposed credit in the amount of SDR253.7 million (us$375 million equivalent) to the People's Republic of Bangladesh for a Multipurpose Disaster Shelter Project,November 19, 2014
8. Ahsan Uddin Ahmed, Saleemul Haq, Mahbuba Nasreen, Abu Wali Raghib Hassan, Sectoral inputs towards the formulation of Seventh Five Year Plan (2016 – 2021) Climate Change and Disaster Management, Final report, Janary 2015
9. Bishawjit Mallick, Necessity of acceptance? Searching for a sustainable community-based disaster mitigation approach – The example of a coastal city in Bangladesh, Solutions to Coastal Disasters Conference 2011
10. Cyclone Shelter Construction, Maintenance and ManagementPolicy2011, Ministry of Disaster Management and Relief Bangladesh Secretariat, Dhaka
11. Christopher Mayhorn, Cognitive Aging and the Processing of Hazard Information and Disaster Warnings, Natural Hazards Review,vol.6, issue: 4, November 2005
12. M. H. Minar, M. Belal Hossain, M. D. Shamsuddin,Climate Change and Coastal Zone of Bangladesh: Vulnerability, Resilience and Adaptability, Middle-East Journal of Scientific Research, January, 2013
13. Ali A., Choudhury G.A. Strom surge in Bangladesh, 2014, University Press Ltd.
14. W. James Shuttleworth, Insight from Large-Scale Observational Studies of Land/Atmosphere Interactions, Surveys in Geophysics 12(1): 3–30 · March 1991
15. J. R. Garratt, Review: the atmospheric boundary layer, Earth-Science Reviews 37 (1994) 89–134
16. John D. Holmes, Wind Loading of Structures, Spon Press, London,2001
17. W. D. Janssen, B. Blocken, T. van Hooff, Pedestrian wind comfort around buildings: comparison of wind comfort criteria based on whole-flow field data for a complex case study, Building and Environment, October 25, 2012
18. Hemant Mittala, Ashutosh Sharmaa, Ajay Gairolab, A Review on the study of urban wind at the pedestrian level around buildings, Journal of Building Engineering, 18(2018), 154–163
19. Stathopoulos Ted, Wind and Comfort, EACWE 5 Florence, Italy, 19th–23rd, July 2009
20. NAHB Research Center, Inc. Upper Marlboro, MD, Monitoring of Near Ground Wind in a Built- up Suburban Environment, September 1998
21. Agarwal Ankush, Cyclone Resilient Building Architecture, UNDP, Disaster Risk Management Program, March 2007
22. MCSP- Multipurpose Cyclone Shelter Programme (1993), Final Report, Planning Commission, Government of Bangladesh, United Nations Development Programme,World Bank.
23. Hugh Brammer, Climate change, sea-level rise and development in Bangladesh, The University Press Limited, 2014
24. Iftekhar Iqbal, The Bengal Delta Ecology, State and Social Change, 1840–1943 2010
25. Gupta Anil K., Anderson David M., Pandey Deep N., Singhvi Ashok K., Adaptation and human migration, and evidence of agriculture coincident with changes in the Indian summer monsoon during the Holocene, CURRENT SCIENCE, VOL. 90, NO. 8, 25 APRIL 2006, 1082–1090
26. Bangladesh bureau of statistics, Statistical pocketbook Bangladesh 2015
27. Jacob P. Hochard, Edward B. Barbier & Stuart E. Hamilton , Mangroves and coastal topography create economic "safe havens" from tropical storms, 2021 www.nature.com/scientificreports
28. Nahiduzzaman K. M., Haas T., Micro Climate House Design: A way to adapt to climate Change? The case of GharKumarpur village in Bangladesh, Theoretical and Emperical research in Urban Management, Year 3, Number 9, 2008
29. Forest department ,2000, Sundarbans Bio diversity conservation project: Inception Report, vol.2, Khulna
30. Hemant Mittal, Ashutosh Sharma, Ajay Gairola, A review on the study of urban wind at the pedestrian level around buildings, Journal of Building Engineering,Volume 18, July 2018, Pages 154–16323
31. Walter Frei, Which Turbulence Model Should I Choose for My CFD Application?, COMSOL Blog, July 6, 2017
32. Bottema M. A, Wind climate and urban geometry , PHD thesis, 1993
33. Kabir Saimum , A study of geo climatic vulnerability and its response in the settlement pattern in coastal areas of Bangladesh , Masters of Architecture thesis submitted to BUET, July 2012
34. Holmes Jhon D. Wind Loading of structures, Spon Press, 2003
35. Wesley E.Marshall, Daniel P.Piatkowski, Norman W.Garrick, Community design, street networks, and public health, Journal of Transport & Health, Volume 1, Issue 4, December 2014, Pages 326–340
36. S. Marshall, Streets and Patterns, 2005, Spoon press
37. M. Kanda, E. Maruta, Characteristics of fluctuating wind pressure on long low-rise buildings with gable roofs, Journal of Wind Engineering and Industrial Aerodynamics, Volume 50, December 1993, Pages 173–182
38. Staff Writer (May 26, 2009). "18 killed as Cyclone Aila lashes coast". The Daily Star. Retrieved May 26, 2009.
39. G. Th. Visser, 1987, research on the usability of the ''W'UlBtek ' report for the prediction of BET wind climate in public space (in Dutch), Report 87-180, IMET-TNO, Apel Doom, NL
40. Manik Kumar Saha, Niaz Ahmed Khan, Changing Profile of Cyclones in the Context of Climate Change and Adaption Strategies in Bangladesh, Journal of Bangladesh Institute of Planners, Vol.7, December 2014, pp 1–12
41. Kerry Emanuel, Increasing destructiveness of tropical cyclones over the past 30 years, Nature, Vol.436(Auguest,2005), pp 686–688

CHAPTER 21

Climate Change Induced Cyclones in Arabian Sea and Mitigating the Emerging Risks for Small Scale Fishermen Community

Vaishnavi
Student, Jamshetji TATA School of Disaster Studies,
TISS Mumbai, India

Katyayini Sood
Student, Jamshetji TATA School of Disaster Studies,
TISS Mumbai, India

Abstract

Tropical cyclones are the most frequent natural disaster in India which brings with it storm surges, strong winds, floods due to gushing waves, erosion etc. along with loss in lives, infrastructure, property and further economic loss. With the changing climate and rising temperature, it becomes essential for the communities to build resilience towards the catastrophic events occurring frequently. 88% lesser lives were lost due to cyclones in the last decade (2010-19) as compared to the 2000-2009 decade. Although mortality rates have decreased because of the better early warning system and immediate response, the increased severity of these cyclones has led to greater damage to infrastructure and property of the coastal communities. The strong winds associated with the cyclones cause damage to boats, engines, gears etc used for fishing, electric poles, which eventually hamper the supply chain of fishing communities. These economic losses sometimes have a prolonged impact on the marginalised fishing community due to the widespread destruction of resources. Thus the paper attempts to explore the vulnerabilities of the fishermen community using the Pressure and Release (PAR) Model. The model covers the root causes and the dynamic issues within the community along with suggesting ways of building resilience for future events.

Keywords: Arabian Sea, Small scale fisheries, Mitigation strategies, Sea safety, Pressure and release model

1. Introduction

"Much has been learnt from the creative disaster prevention efforts of poor communities in developing countries. Prevention policy is too important to be left to governments and international agencies alone. To succeed, it must also engage civil society, the private sector and the media."

Kofi Annan UN Secretary-General
International Decade for Natural Disaster Reduction (IDNDR) Programme Forum
Geneva, July 1999

Climate change has been a challenge for the entire globe and forces the international community to rethink its development discourse. India, a developing nation in the tropical region, faces severe consequences because of increasing climate change. It has a 7,516.6 km long coastline making it vulnerable to various hydro-meteorological disasters. Out of the total Indian landmass, 56% is prone to cyclones (SP et al., 2014). Highly destructive, they cause immense damage to mankind, the mortality is although decreasing in the eastern coast, the western In the recent decade, the number of cyclones hitting the western coast has been on the rise; researchers like Evan (2011), Rajeevan (2013) Wang (2012) have attributed this to the evident climate change. This has posed a problem for the states like Kerala, Goa, Maharashtra, and Gujarat.

Nandi (2021), in an article, states that the frequency of cyclones in the Arabian sea is rising by nearly 52% in the last two decades. Another article published in Mint (2021) elaborated that the Arabian sea would merely witness two or three cyclones that were not powerful enough to disrupt the local life and create an emergency in earlier times, but with every passing year, the uncertainty regarding the event is increasing. The states have immense dependence on the sea for their livelihood. The Arabian Sea coast is vital to India's economy since it supports roughly 21.7 lakh tonnes of fish production, higher than the eastern coast. Being less prone in the previous years, the infrastructure and livelihoods are not resilient enough to bring life on track immediately. The impact is not uniform because of the wide gap between the well-to-do fishermen community and the marginalized community. The strong winds associated with the cyclones cause damage to boats, engines, gears, etc., used for fishing, electric poles, which eventually hamper the supply chain of fishing communities. These economic losses sometimes have a prolonged impact on the fishing community due to the widespread destruction of resources.

Small-scale fisheries can be broadly characterized as a dynamic and evolving sector employing labour intensive harvesting, processing and distribution technologies to exploit marine and inland water fishery resources. Export-oriented production, however, has increased in many small-scale fisheries during the last one to two decades because of greater market integration and globalization. While typically men are engaged in fishing and women in fish processing and marketing, women are also known to engage in near shore harvesting activities and men are known to engage in fish marketing and distribution. Other ancillary activities such as net-making, boatbuilding, engine repair and maintenance, etc. can provide additional fishery-related employment and income opportunities in marine and inland fishing communities. Small-scale fisheries operate at widely differing organizational levels ranging from self- employed single operators through informal micro-enterprises to formal sector businesses. This sub-sector, therefore, is not homogenous within and across countries and regions, and attention to this fact is warranted when formulating strategies and policies for enhancing its contribution to food security and poverty alleviation.

(FAO, 2004)

The dependency of communities on local fisheries is assessed in the paper to determine the impact on food security and livelihood. There are various reasons for their vulnerability. The paper attempts to identify the pattern, reason, and impact of the increasing cyclones in the Arabian sea, later focuses on the marginalized fisherman community and identifies the root causes, the dynamic causes and the unsafe conditions which lead to the progression of vulnerability among them. The last section highlights the mitigation strategies for landfall and sea safety. In the last decade (2010-19), 88 percent fewer lives were lost due to cyclones than in the previous decade (2000-2009), which can be ascribed to increased IMD forecasting ability. While mortality rates have dropped as a result of improved early warning systems and rapid reaction, the ferocity of these cyclones has resulted in more damage to coastal populations' infrastructure and property. And gives concluding remarks on how the vulnerability among them can be reduced, and they can be bought in a bounce-back state because there is no way to escape the growing disaster.

2. Reasons for the Upsurge in the Number of Cyclones

As the cyclone Tauktae approached the Indian mainland, the increase in the intensity drew attention. Rising Temperatures: The favourable conditions for the development of cyclones are warm waters, that is, temperature more than 26 degrees Celsius in depth that is more than 60m, and air circulation, which is unstable. The convection and condensation lead to the formation of cumulonimbus clouds (NDMA). It is estimated that 90% of the heat which is caused by the greenhouse gases is absorbed by the ocean, which has ultimately led to an increase in the sea surface temperature of the Arabian sea.

This has also increased the Tropical Cyclone Potential Heat, (TCHP) Leipper and Volgenau 1972; Shay et al. 2000; Lin et al. 2008) defines it as an integrated heat content per unit area relative to 26°C . Deshpande (2021) has developed a flowchart to establish the relation between the increase in temperature and its relation with cyclones of greater intensity. IIMS in a report stated that in the earlier times it was just the Bay of Bengal which was a warm pool, but due to global warming the temperature of the Arabian sea has also reached the 28 degree celsius which was nearly 1.2- 1.4 degree celsius less. El Nino, another climate induced phenomenon causes unfavourable conditions for cyclone formation in the Bay of Bengal Region but favours the formation in the Arabian sea. The studies further anticipate that the frequency, duration and the intensity of the tropical cyclones in the Arabian sea would further increase and lead to immense destruction in the region. As the SST rises, the saturation of vapour pressure rises, causing water vapours and latent heat to form, intensifying cyclones.

SST is important in enhancing cyclonic severity, as well as floods and other natural calamities, according to Evan and Camargo (2011) and Emanuel (1987). Rather than a constant positive association between maximum storm intensity and SST, observations suggest the existence of a temperature threshold required for the generation of significant hurricanes (Michaels et al., 2006).

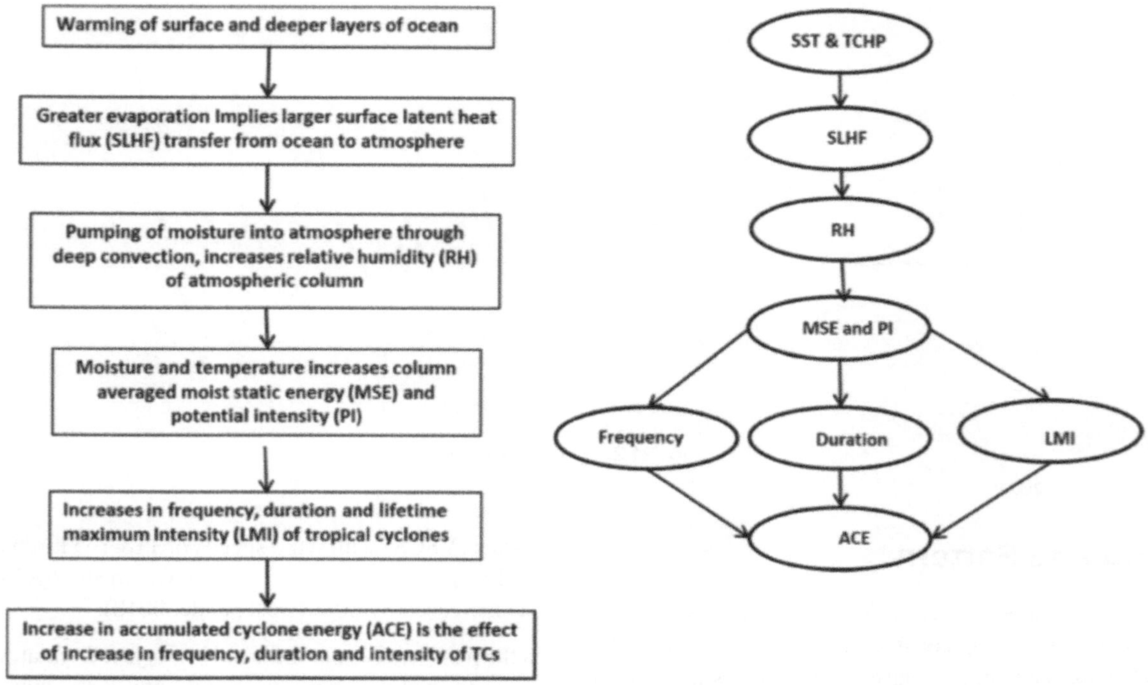

Fig. 21.1 Possible reasons of increase in frequency

Source: Deshpande et al. (2021)

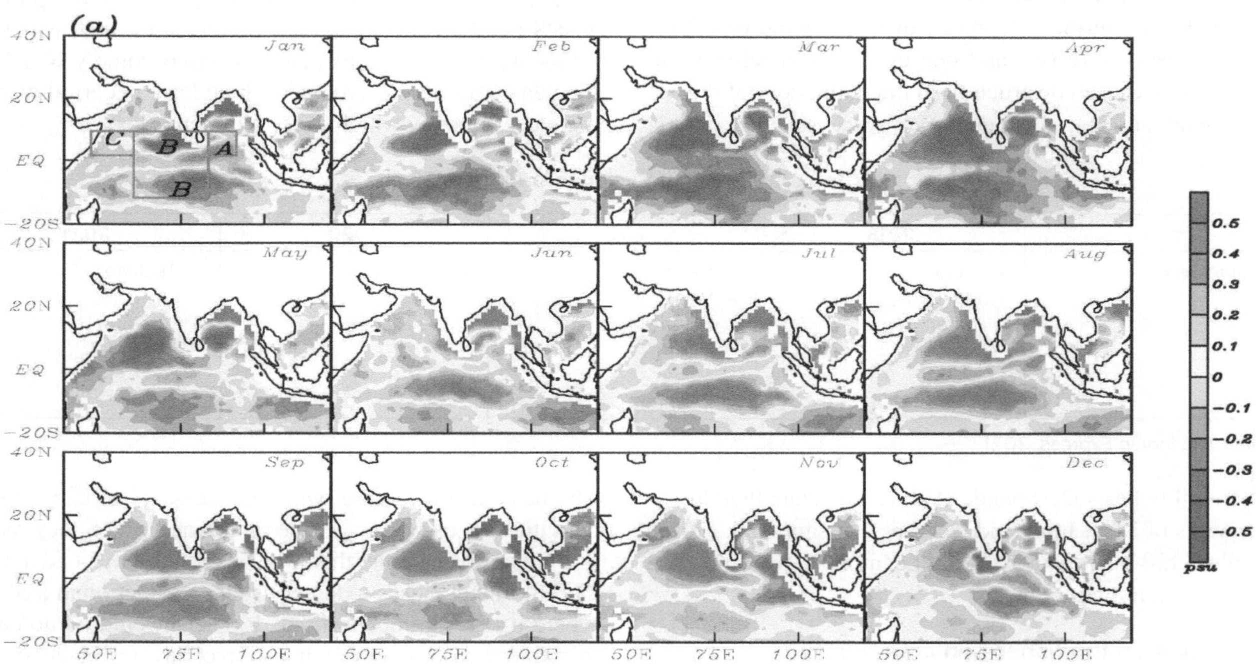

Fig. 21.2 SST in Arabian Sea

Source: Yuan et al., 2018

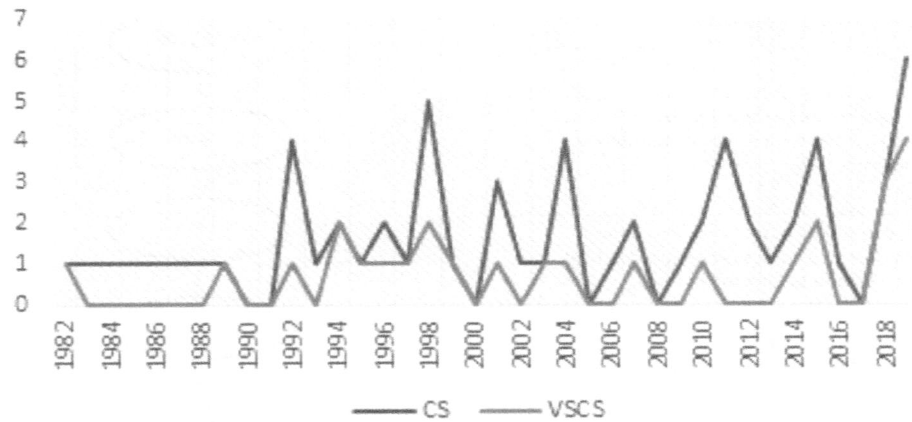

Fig. 21.3 Cyclone Trend

Source: Deshpande (2021)

2. Trend and Pattern

The number of cyclones under the Very severe cyclonic storm and cyclonic storm have considerably increased since the year 1982. This is evident because of the reasons mentioned above. The cyclones under both the categories have increased. It was the year 2019 in which a maximum number of cyclones were witnessed. As the coastal population in the country is high, the hazard further becomes a disaster. In the year 2020 and 2021 there were two and one cyclone respectively but unfortunately caused destruction in the entire coastal region. In the Arabian Sea, the number of very severe cyclonic storms (VSCS) with wind speeds of 118-165 km/h has grown by 150 percent. The duration for which the cyclone occurs has increased by 80% (Deshpande, 2018).

In the past years as per the trend it was just Gujarat, which was prone to cyclones but now Kerala and Karnataka have also been added to the list. The time which was taken to become a VSCS cyclone has now reduced, for example the Mekunu cyclone and Nisarga took nearly 4-5 days to develop into an VSCS but Tauktae which was the recent cyclone which had hit the western coast took just 2 days to become VSCS. Since the year 2018 all the cyclones which have occurred were of a greater intensity which caused immense loss.

Table 21.1 List of Cyclones in Arabian Sea

	2018	2019	2020	2021
Arabian Sea	• Sagar • Mekunu • Luban	• Pabuk • Vayu • Hikaa • Kyarr • Maha • Pawan	• Nisarga • Gati	• Tauktae

Source: The Indian Express, 2021

On a monthly basis Deshpande et al. (2021) state that during the months of May, June, and October. Hussain et al. (2011) compiled 120-year data to find the months in which the cyclonic activity was higher.

2.1 Impacts on the fishermen community

Economic Impacts: The cost of assets damaged by the recent Taukae was 1500 crore rupees in total, this will further rise to thousands of crores if the intensity does not fall. The boats which are essential to carry out their livelihoods were broken; this broke the entire supply chain for some commodities. The basic infrastructure which is necessary to carry out the activities are also damage, the remittance which they receive is comparatively lesser than what is required to mend the loss. This ultimately leads to lesser fishery production in a country where it is a significant industry. There would be no capital to restart their business. Exhaustion of savings in meeting the everyday needs during the closure period is another major impact of the cyclone Difficulty in taking loans because what was to be kept as collateral was no longer under their asset.

Psychological Impacts: Some fishermen because of losses face mental health issues, which include acute stress

Table 21.2 Monthly Analysis for 120 years

Cyclone Data (120 years)	Frequency
Total annual	194
January	1
February	0
March	0
April	8
May	33
June	45
July	9
August	2
September	11
October	43
November	37
December	9

Source: Hussain et al. (2011)

disorder, sleep disorder, post-traumatic stress disorders (PTSDs), generalized anxiety disorders, suicidal ideation and depression. Due to lack of funds and intuition of further losses the survivors will not be able to receive treatment. The counselling services are not provided. Few people might fear returning to the sea.

Environmental Impact: Cyclones can and do cause havoc on marine habitats along the shore. They devastate coral reefs, churn up the water column, rearrange bottom sediments, and pollute the environment through storm-water runoff. Hurricanes can also drive fish to flee to deeper water from nearshore estuaries and coastal ocean settings. Although no one has investigated how storms affect fish in deeper water, most people believe they are largely impervious to the effects of storms. The marginal farmers, without assistance, cannot restore the ecosystems.

Socio-Cultural impact: The marginal fishermen are mainly the indigenous people who will try to relocate themselves in various industries, which will reduce the number of people who have indigenous knowledge of the subject, which is essential in building resilience along with modern techniques which help the community to build back better.

3. Importance of Disaster Management Plan for Cyclones in India

Climate change and the increased Sea Surface Temperature (SST) of the Indian Ocean has led to exaggerated frequency and intensity of cyclones originating in Arabian Sea and Bay of Bengal. The coastline of India covers 7500 km, which consists of 96 coastal districts (that either touch the coast or are in close proximity), with approximately 262 million people having an exposure to tsunamis and cyclones. According to *Parida, Roy Chowdhury and Kumar Sahoo, 2021*, 313 cyclones have crossed India's western and eastern coasts between 1891 and 2020, of which 130 were categorised as severe cyclonic storms. The eastern coast experienced 282 cyclones whereas the western coast had 31 cyclones. The

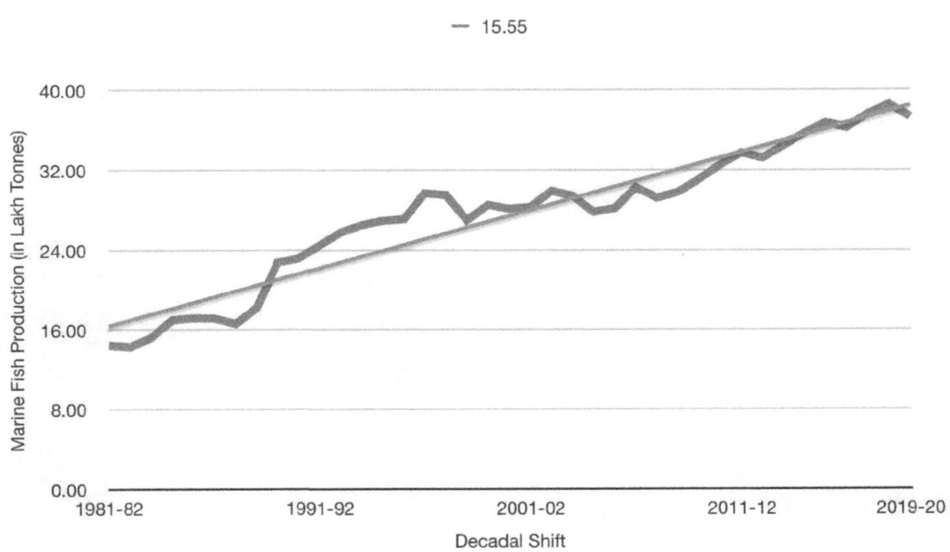

Fig. 21.4 Decadal growth of Marine Fishing in India (1980 - 2020)

Source: Handbook on Fisheries Statistics, 2020

coast of Odisha had seen 97 cyclones, followed by Andhra Pradesh (79), Tamil Nadu (58), West Bengal (48), Gujarat (22), Maharashtra/Goa (7), and Kerala (2).

For the coastal communities, fishing is one of the most important livelihoods for economic stability. Fishing can probably be considered as the most dangerous occupation in the world. And the people vulnerable to the sea accidents are mostly amongst the poorest section of the society. Yet there has been rise in population depending on marine fishing. The figure above (Fig. 21.4.) shows the production of marine fish in lakh tonnes and how the production grew between 1980 to 2020. The upright trend or growth of marine fishing in India denotes the dependency of coastal population on fishing for their livelihood.

The figure above shows the economic benefits of the western coastal states due to marine fish production. The fishermen communities have been relying on export market for the last few decades. Therefore it symbolises the importance of mitigating the risks of small scale fishermen communities. With the increased occurrence of intensified cyclones in the western coast of the country, and prevalence of small scale communities, preparing mitigatory strategies becomes the call of the hour.

Safety of fisher community is not only be the responsibility of the government, through guidelines and regulations, but also requires activities at the grass-root level. The governments and regional bodies shall coordinate their programmes and produce a holistic approach in order to conduct activities among schools, villages and fishermen etc. The projects on fisheries livelihood, climate change and coastal zone management, coastal vulnerability, etc shall consist of safety for fishermen as an integral objective. Safety of the fisher men community is a cross-cutting issue which needs to be approached with a holistic perspective (*Danielsson, P., et al. 2010*). The continuum of response, relief and development concept needs to guide the civil societies, coastal communities and the disaster managers whereas the long-term developmental objectives are to be taken care of throughout the process of recovery and rehabilitation *(FAO, 2014)*. The primary principle of rehabilitation must be to "build a culture of safety and resilience at all levels" *(Brown and Poulain, 2013)*. A free flow of information between the administration and the fishers can produce an effective and better management within the fisheries sector.

India had also adopted Sendai Framework for Disaster Risk Reduction 2015- 2030 (SFDRR). SFDRR aims at reducing loss of lives significantly because of disasters by providing safety on land and at sea through collaborative and coordinated steps at all levels. Although governments disburse compensatory packages to the family of missing and dead fishermen but the vulnerabilities still remain *(Lakshmi, A., 2018)*. The psychological trauma suffered by the survivors and the livelihood troubles faced by the widow women are yet of concern. The harsh impacts of these hazards cannot be completely prevented, but their intensity or severity can

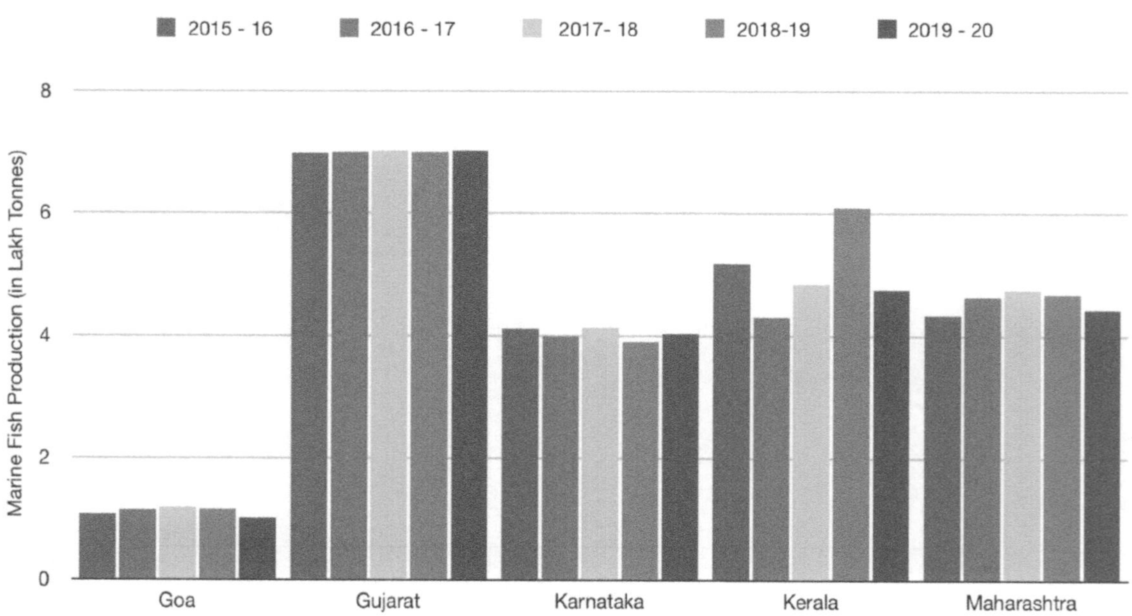

Fig. 21.5 State-wise (West Coast) Marine Fish Production in India (2015 - 2020)

Source: Handbook on Fisheries Statistics, 2020

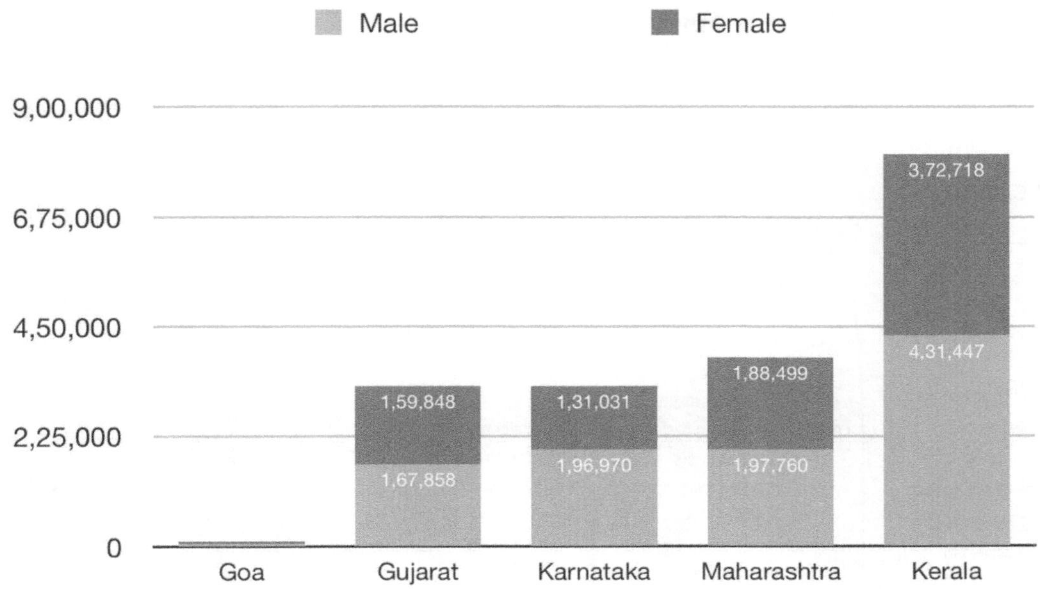

Fig. 21.6 Gender wise Fishermen Population

Source: Handbook on Fisheries Statistics, 2020

be significantly reduced through actions and strategies. Hazard resistant infrastructure, engineering techniques, even improved ecological policies and public awareness and preparedness can be included under mitigation measures.

Besides it's a vicious cycle of climate change and disasters. Because of the rising sea surface temperature there is a disturbance in fish catch which eventually decreases the net primary production. The same trend of rising sea surface temperature intensifies the cyclones. These cyclones affect the livelihood, food security, supply chain, the regular fish catchment area etc, putting them under a vicious cycle if climate change is not controlled *(Kumari, V., et al. (2020)*.

According to the Department of fisheries, Ministry of Fisheries, Animal Husbandry and Dairying, Government of India, 2020, there are approximately equal number of females and males with fishing as livelihood in the Western coastal states of Goa, Gujarat, Karnataka, Maharashtra and Kerala as shown in *Fig 3.3*. Thus, it is crucial to identify different vulnerabilities faced by different groups which depend on gender, culture, etc. These different characteristics influence their capacities of mitigation, preparedness, prevention and response to the disasters (*UNISDR*, 2006). Women and men play different roles in society and therefore during a disaster situation, often have varied information. Women of the fishing communities are generally devoid of decision making authorities in the families and communities. It makes them vulnerable, directly as well as indirectly, to disasters (as was in the case of cyclone Ockhi) *(FAO, 2017)*. The wives who lost fishers in their homes are unable to cope with the financial pressures they. There are a number of cases in the marginalised communities where women have never worked outside their homes, despite them being better educated than their husbands.

With the increasing frequency of the cyclones, it becomes important that dissemination of information, arrangements of institutions and early warning communication system shall be designed to reach the last mile. It should be tailored to meet every vulnerable section of the community. Communication and coordination shall be established vertically as well as horizontally between the stakeholders.

4. Explanation through Pressure and Release (PAR) Model

The Pressure and Release (PAR) Model is utilised to describe about the vulnerabilities faced by the communities. The model is inclusive of the root causes, dynamic pressures experienced as well as the presence of unsafe conditions. Followed by the release model which explains the methods that can be utilised to reduce the pressure and provide safer conditions. The model also attempts to suggest ways to address the root causes of vulnerabilities.

4.1 Landfall

According to the data of India meteorological Department (IMD), 2013, cyclones originating near the Indian coastal

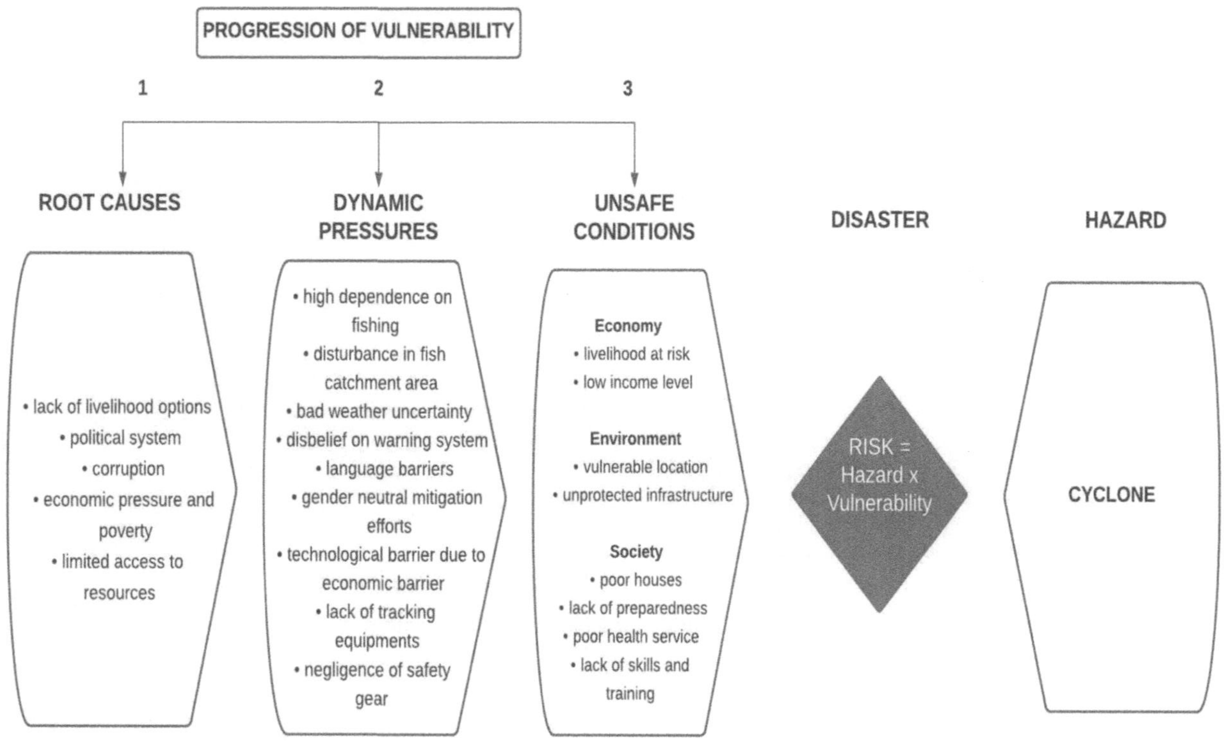

Fig. 21.7 Faces of Vulnerability; Pressure and Release Model (Part I)

Source: Brink, Ebba. (2018)

states account for 7% of the global tropical cyclones. Approximately 5 to 6 tropical cyclones are formed in the Arabian Sea and Bay of Bengal i.e. the northern Indian Ocean, of which 2-3 turn into severe cyclones. Recently Indian coast had experienced Cyclone Tauktae on the western coast and Yaas on the eastern coast, earlier in May, 2021. Massive damage to infrastructure, houses, agricultural sector was caused. Economic loss count to 4.3 billion U.S. dollars with over 200 human deaths. Moreover, the high speed winds also cause uprooting of trees in urban areas after the landfall, further depleting the green cover.

The factor of rapid intensification brings in the importance of preparedness among the local community. Therefore it calls for attention on building knowledge and preparedness among the locals themselves. With the changing dynamics of the disasters lately, it becomes crucial to develop a safety culture through community engagements and efforts.

4.2 Sea Safety

While focusing on the mitigation efforts and preparedness at the site of landfall of cyclone, we tend to miss out on the threats at sea, further risking the safety and security of fishermen at sea. Safety at sea needs to be considered as an integral component of fisheries management. An effective management of fishery system will benefit in different ways, which is already demonstrated by several countries (*Danielsson, P., et al. 2010*). Although it has been observed in the South Asian countries that the policies on small scale fisheries are often negligent of safety of the fishermen. But sea safety becomes a critical issue for the long haul fishers (30-40 day trips) during cyclones. When it comes to the cyclones originating in Arabian Sea, the recent years have seen an increase in the frequency of cyclones. Not only has the number of cyclones but also their intensities have increased. In case of cyclones in Bay of Bengal, their origin is far away from the coast which gives a good lead time to the local community and institutions to respond. Whereas Cyclone Ockhi, in the Arabian Sea saw rapid intensification close to the coast. Similarly Cyclone Tauktae underwent rapid intensification from a 'depression' to 'very severe cyclonic storm' in a span of two days, which generally takes four to five days. For a better and improved safety of fishermen and safety at sea, collection of reports on accidents post cyclone, becomes of prime importance. After a thorough analysis, it was observed that challenges were faced both by the communities and the officials.

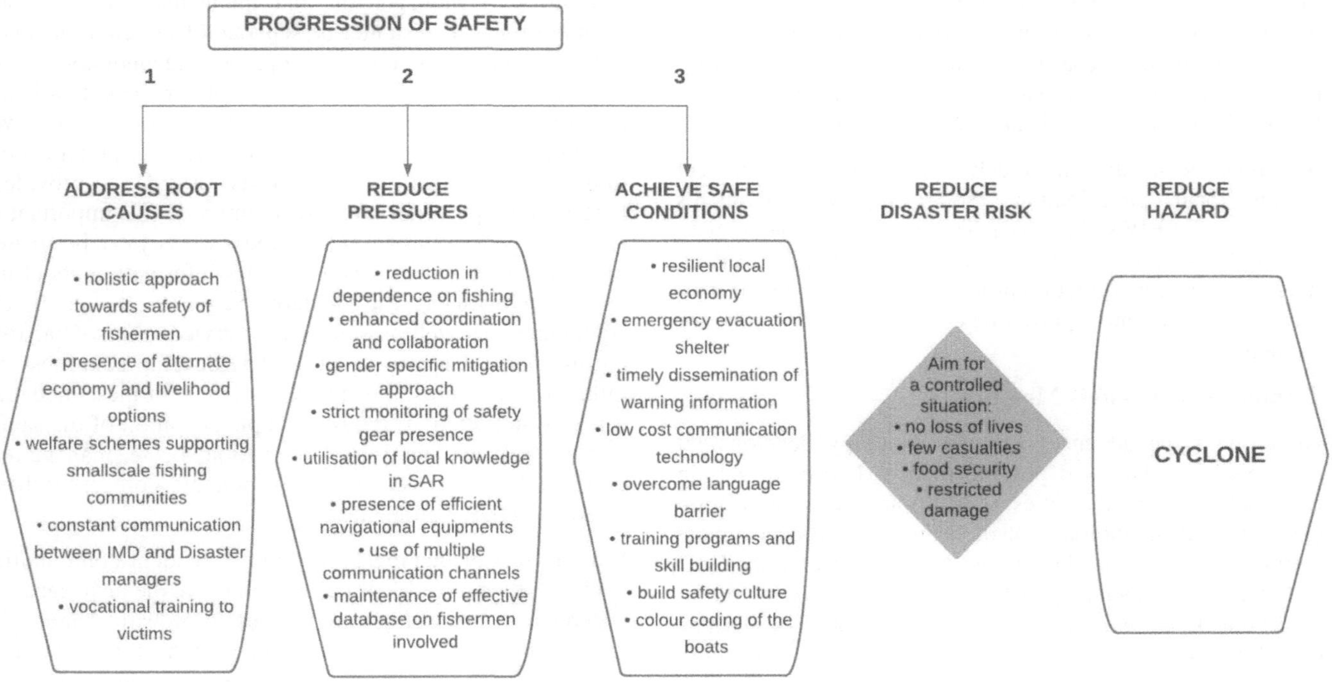

Fig. 21.8 Faces of Safety; Pressure and Release Model (Part II)

Source: Brink, Ebba. (2018)

The paper also tries to broadly categorise the challenges into four sub categories which are described below.

Communication Channel and Early Warning System

The most prominent issue in case of mitigation and risk management for the safety at sea was found to be the communication channels and transmission of early warnings to the fishermen already at sea. In case of cyclone Ockhi, IMD had issued warnings but the warnings were not disseminated father on by the transmitters. The delay in transmission of early warning about the cyclone led to loss of lives of the fishermen already at sea. Therefore, It requires a better collaboration and coordination among the different stakeholders responsible during cyclone risk management i.e. disaster managers and IMD, between states and between departments.

The language of communication is also a concern. A centralised method of communication throughout the coast can prove to be ineffective. Thus it is important to disburse the responsibility of communication, further to the local levels. Besides loudspeakers and port signals can also be put into use. At the same point of time, besides providing early warnings, the fishermen need to be informed of the risks involved and the details of actions to be followed.

Participative Local Coordination

An active participation of the local communities and a bottom-up approach in warning system allows a multidimensional response to the problems during disasters. With the inclusion of the civic communities, vulnerability can be reduced even through a traditional approach, which eventually also strengthens local capacities *(International Strategy for Disaster Reduction UNISDR, 2006).*

Disaster management needs to be strengthened among the local community organisations *(such as religious organisations, gram panchayats, self help groups, cooperative societies etc).* But it can never be the sole responsibility of the institutions and the officers in building responsibility of safety. Thus building capacity within local organisations with the help of community consultation shall be prioritised in order to mitigate risks of the marginalised fishermen. Initiatives for the safety of fishermen shall be voluntarily taken by the fishing village communities and the other fishermen. In order to have a wider reach with the safety message to the fishermen, a local grassroots level or a participatory approach is needed. Local fishermen's knowledge can be utilised by the coastal guards and navy for location identification. In the search and rescue (SAR) operations, the local fishermen can be interviewed for the specific locations to avoid loss of time and for effective rescue.

Strengthening of management techniques at the local levels of district and village level through frequent training workshops. Also the Fisheries Departments can be given the regulatory role under fisheries' risk management. Such an approach will eventually build an overall safety culture among the locals.

A gender perspective approach also needs to be present during local participation. Self Employed Women's Association (SEWA) like organisations can help widows to get employment even if the community retaliates *(Lakshmi, A., 2018)*. The rehabilitation process by the government shall include employment opportunities and vocational training for women.

Technology and Safety Measures

Along with the advances brought about by development in technology, it is also proving to be a barrier to certain marginalised communities. Multiple communication channels and technology shall be utilised to disseminate warnings and establish better coordination between the IMD and disaster managers. According to *Monirul Islam, M., Sallu, S., Hubacek, K. and Paavola, J., 2014*, because of absence of radio signals and preferred language of communication, the delivery and received information gets delayed, which thereby increases their probability of cyclone exposure. The technological barrier is also associated with economic barrier. Since the communities do not have effective navigational instruments and powerful engines, it becomes difficult to return to the shore in a span of few hours. The boats are easily damaged by the heavy winds of the cyclone and also puts the lives of fishermen under threat. Two significant factors poses a challenge and complicate timely delivery of relief and rescue of fishers at sea: first, lack of tracking equipments for the fishing vessels at sea and second, not sufficiently strict about enforcement of sea safety gears. The fisheries management plan need to be inclusive of a mechanisms of Monitoring, Control and Surveillance (MCS), that involves notification of landing points, colour coding of fishing vessels, maintenance of database of all yards etc. There is already a government run fishing harbour at Cochin which asks for different colour coding of boats from different states and territories, for example, Kerala (blue and orange), Lakshadweep islands (white) and Tamil Nadu (green) *(Islam, M., et al., 2014)*.

Innovations in technology has made it possible to predict cyclones much in advance along with accuracy. Therefore the onus lies on the local village authorities to disseminate the information in advance. Certain villages of Tamil Nadu issue warnings on their own. Similar steps need to be encouraged in all the fishing villages of the Indian Coast.

Long Term Strategies to be adopted for Disaster Risk Reduction

The Sendai Framework (SFDRR) not only focuses on reducing risks but also avoiding developing new risks. Therefore Coastal Area Management and intensification of coastal economic activities cannot be separated from environmental security and sustainable development. Technology cannot always protect the lives of people, which makes it essential to allow communities to respond and take control whenever required. *Danielsson, P., et al. 2010* suggests preparedness through data collection and analysis. In order to provide a long-term preparedness for the community, it is important to identify where the problem lies. And inventory can be created of the fisheries sector which includes information about the fishermen involved, the available resources, technological equipments, as well as supporting services. This data from the inventory can also be utilised to develop solutions and mitigation measures. Besides development of these methods, it is also important to focus on implementation of the same. Therefore, it is equally important to advocate, manage and evaluate the progress of the implementation process to have an effective DRR strategies.

With a view of making the fishing dependent communities resilient, there needs to be a decline in the fish capacity dependence in developing countries in order to handle the diminishing fish stocks. There are a number of fishermen competing for too few fish count. Because of such pressurised competition fishermen are forced to go deep into the seas, for longer periods and even in bad weather. This leads to increased risks, making fishing at sea even more dangerous. Therefore one of the solutions is to look for alternative economy. An alternative method of livelihood for the fishermen and the community can reduce the pressure on the fish resources *(Roshan, M., 2021)*. Co- management in marine capture fisheries can also be introduced. Another local initiative is participatory MCS which can enable the government to manage and control fisheries in an efficient manner. This can also positively impact accidents at sea by reducing the collisions between vessels.

5. Conclusion

With the present situation of climate change and increasing average sea surface temperature of the Indian Ocean, occurrence of frequent and intensified cyclones on both the Eastern and Western Indian shoreline has become inevitable. Thus it is not only mitigation and preparedness we need to focus on, but also the safety culture which needs to be imbibed among the communities vulnerable (physically and socially) to the cyclonic disasters. While formulating mitigatory strategies, holistic approach needs to be taken into account, which is inclusive of gender differences and importance of community participation to reduce accidents at sea as well as inculcating safety culture within the community in the long run. Such an approach shall be included under curriculum in the schools of the coastal regions, which can enable building an attitude of safety and preparedness.

Mobilisation of the community in disaster preparedness is essential, which demands for establish meant of village level committees, raising of awareness through training programmes, setting up of safer places and supply of food and medicine during emergencies. Fishing communities and specifically the small-scale fishing communities are the most vulnerable community when it comes to suffering from flooding, cyclones and tsunamis, etc. Therefore while implementation of disaster preparedness and formulation of mitigating strategies, safety of fishermen and their families should be of prime importance.

REFERENCE

1. 2019. Cyclone Ockhi – Disaster risk management and sea safety in the Indian marine fisheries sector. Rome, FAO. 72 pp. Licence: CC BY-NC-SA 3.0 IGO. Available at: <https://sdma.kerala.gov.in/wp-content/uploads/2019/12/085-FAO_ICSF-OCkhi-Study-Manas- Roshan.pdf >
2. Birkmann, J., 2007. Risk and vulnerability indicators at different scales:Applicability, usefulness and policy implications. *Environmental Hazards*, 7(1), pp.20-31.
3. Brink, Ebba. (2018). Adapting Cities: Ecosystem-based approaches and citizen engagement in municipal climate adaptation in Scania, Sweden.
4. Chaston Radway K, Manley M, Mangubhai S, Sokowaqanilotu E, Lalavanua W, Bogiva A, Caginitoba A, Delai T, Draniatu M, Dulunaqio S, Fox M, Koroiwaqa I, Naisilisili W, Rabukawaqa A , Ravonoloa K, Veibi T (2016) Impact of Tropical Cyclone Winston on Fisheries-Dependent Communities in Fiji. Report No. 03/16. Wildlife Conservation Society, Suva, Fiji. 79 pp
5. Danielsson, P., Kuyateh, M. and Ravikumar, R., 2010. SAFETY AT SEA FOR SMALL- SCALE FISHERIES IN DEVELOPING COUNTRIES. Available at: <http://www.fao.org/3/ al960e/al960e.pdf >
6. Deshpande, M., Singh, V., Ganadhi, M., Roxy, M., Emmanuel, R., & Kumar, U. (2021). Changing status of tropical cyclones over the north Indian Ocean. *Climate Dynamics*. doi: 10.1007/s00382-021-05880-z
7. Dof.gov.in. 2019. *Blue Revolution | Department of Fisheries, GoI*. [online] Available at: <http://dof.gov.in/blue-revolution>.
8. Evan, A., & Camargo, S. (2011). A Climatology of Arabian Sea Cyclonic Storms. *Journal Of Climate*, *24*(1), 140-158. doi: 10.1175/2010jcli3611.1
9. Hussain, Ansari, & Abbas. (2011). Persistency Analysis of Cyclone History in Arabian Sea. Retrievedfromhttps://www.researchgate.net/publication/237845012_Persistency_Analysis_of_Cyclone_History_in_Arabian_Sea
10. *International Strategy for Disaster Reduction (UNISDR)*, 2006. Developing Early Warning Systems: A Checklist. EWC III.
11. Lakshmi, A., 2018. *Report of the National Workshop on Small-scale Fisheries, Cyclone Ockhi and Disaster Risk Management*. [online] Fao.org. Available at: <http://www.fao.org/ 3/ca3205en/CA3205EN.pdf >.
12. MASTERS, J., 2021. *Tropical Cyclone Tauktae is fifth-strongest cyclone on record in the Arabian Sea » Yale Climate Connections*. [online] Yale Climate Connections. Available at: <https://yaleclimateconnections.org/2021/05/tropical-cyclone-tauktae-is-fifth-strongest- cyclone-on-record-in-the-arabian-sea/ >.
13. Muhammad, S., Memon, A., Muneeb, M., & Ghauri, B. (2016). Seasonal and spatial patterns of SST in the northern Arabian Sea during 2001–2012. *The Egyptian Journal Of Remote Sensing And Space Science*, *19*(1), 17-22. doi: 10.1016/j.ejrs.2015.12.007
14. mint. 2021. *Tropical cyclones in the Arabian Sea: Why are they increasing?*. [online] Available at: <https://www.livemint.com/news/india/tropical-cyclones-in-the-arabian-sea- why-are-they-increasing-11621260501162.html >.
15. Monirul Islam, M., Sallu, S., Hubacek, K. and Paavola, J., 2014. Limits and barriers to adaptation to climate variability and change in Bangladeshi coastal fishing communities. *Marine Policy*, 43, pp.208-216.
16. NDMA. (2021). Cyclone I NDMA, GoI. Retrieved 27 August 2021, from https:// ndma.gov.in/Natural-Hazards/Cyclone
17. Parida, Y., Roy Chowdhury, J. and Kumar Sahoo, P., 2021. *Tauktae, Yaas and planning for the next*. [online] The Hindu. Available at: <https://www.thehindu.com/opinion/op-ed/tauktae-yaas-and-planning-for-the-next/article36067714.ece>
18. Roshan, M., 2021. *Cyclone Ockhi: Disaster Risk Management and Sea Safety in the Indian Marine Fisheries Sector*. [online] Icsf.net. Available at: <https://www.icsf.net/images/monographs/pdf/english/issue_165/165_Ockhi_Study_Manas_2018.pdf >.
19. Tasdik Hasan, M., Adhikary, G., Mahmood, S., Papri, N., Shihab, H., & Kasujja, R. et al. (2020). Exploring mental health needs and services among affected population in a cyclone affected area in costal Bangladesh: a qualitative case study. *International Journal Of Mental Health Systems*, *14*(1). doi: 10.1186/s13033-020-00351-0
20. Tezzo, X., Belton, B., Johnstone, G., & Callow, M. (2018). Myanmar's fisheries in transition: Current status and opportunities for policy reform. Marine Policy, 97, 91– 100. doi:10.1016/j.marpol.2018.08.031
21. The Indian Express. 2021. *Explained: How does Tauktae compare with other cyclones in severity and damage?*. [online] Available at: <https://indianexpress.com/article/explained/ how-tauktae-compares-with-other-cyclones-in-severity-damage-7324998 >.
22. Vinaya Kumari, P., Jayappa, K. S., Thomas, S., & Gupta, A. (2020). Decadal variations of sea surface temperature in the eastern Arabian Sea and its impacts on the net primary productivity. Spatial Information Research. doi:10.1007/s41324-020-00340-y
23. Yuan, X., Salama, M., & Su, Z. (2018). An Observational Perspective of Sea Surface Salinity in the Southwestern Indian Ocean and Its Role in the South Asia Summer Monsoon. *Remote Sensing*, *10*(12), 1930. doi: 10.3390/rs10121930

CHAPTER 22

Vulnerability Assessment of a Cyclone Affected Community Kapateswar, Puri Sadar

Chestha Khurana
Faculty of Architecture and Design,
Dehradun Institute of technology, Dehradun, India
cheshtha9@gmail.com, Ar.cheshkhurana@gmail.com

Meghna Chatterji
Faculty of Architecture and Design,
Dehradun Institute of technology, Dehradun, India
meghna.apj@gmail.com, meghna.chatterji@dituniversity.edu.in

Abstract

The Indian East coast is four times more susceptible to tropical cyclones than the western coast in terms of the intensity of the cyclonic happenings. This makes the eastern coast ever more vulnerable to huge loss of lives, homes, infrastructure, flora & fauna, etc. Furthermore, the Vulnerability Assessment reports extensively deal and simplify these as annual statistics represented as percentages of regional, state-based, and likewise.

The paper intends to focus on the existing condition of a heritage village, Kapateswar in Puri Sadar situated adjacent to river Bhargavi, ravaged by the cyclone in depth. The study documented the unguarded conditions of the people struggling to survive in the region through manual data collection, visual analysis, and interviews.

Considering the differences in the type of damages to the houses, the paper developed a vulnerability indicator to categorize it along with multiple parameters including architectural techniques & joineries, materials used- comparison of now and then, type of vegetation, and others.

The paper leads to the conclusion for the need to focus on the requirements for explicit solutions in terms of built structures, community-level management, and planning techniques, and the provision of disaster shelter in the area with the comparative study with an unscathed heritage Shiva temple, existing within the village premises.

Keywords: Vulnerability assessment, Vernacular techniques, Cyclone resilience, Cyclone in Odisha

1. Cyclones in India

India is a country of immense diversity and it hosts a range of variegated climates along its peninsular landmass. The variations in latitude, altitude, reliefs, distance from the sea and the presence of the Himalayas in the northern region among other factors determine the regional climate in India. And the seasonal shift brings in the Monsoon rains.

The human settlements along these different regions have had to adapt to their regional variations. The architecture too has evolved regionally to respond to them through the local craftsmanship and materials to fulfil the climate-resilient needs in the best way possible, referred to as vernacular architecture. And still, we struggle to fulfil the basic need for safe shelters.

India with its coastal belt of 483 Km, has an extreme hydrometeorological and topographical characteristic, making it prone to many naturally occurring water-based disasters in the coastal belt. These disasters include cyclones, storm surges, floods, varying in their intensities according to the places they occur. The Bay of Bengal has ideal conditions for tropical cyclones to brew and thus, the number of cyclones on the eastern coast is in the ratio of 4:1 as compared to that of the western coast of India, of which 2 to 3 cyclones can be severe every year leaving a trail of catastrophe with loss of life & property, infrastructure, flora & fauna and much more for people as it makes landfall.

Nevertheless, the tremendous growth of India in terms of disaster management, the occurrence of severe cyclones from the year 1891 to 2019 are recorded to be 110 in number, making it extremely vulnerable for cyclonic happenings

With the changing weather conditions, a disaster such as cyclones has become the major halt for the country. Every disaster that happens, brings the livelihood back by 20 years. The amount of destruction caused by a cyclone in the terms of Infrastructure, livelihood, economy, flora & fauna occurs in unmeasurable. Almost 12% of lives lost every year is due to the reason that these disasters are not well looked after and pre managed. And global climate change is only making it ever-more- frequent and extreme.

Along, the cyclone that battered the east coast, is the state of Odisha, severely impacted by the most recent cyclone FANI(2019). To date, the most disastrous super cyclone that occurred in Odisha was in the year 1999, brought a tremendous loss in the terms of human lives, Infrastructure, livelihood, economy, flora & fauna. Along with more than 10,000 people were found dead under the stack of debris and the fallen trees in the state. Whereas, if we compare it to the recent severe cyclones such as Fani 2019, it took away about 45 lives and great damage to Infrastructure, livelihood, economy, flora & fauna.

The government has succeeded marvellously in saving the human loss through the capacity building concept being followed by the 2000 cyclone. There are about 840 Multipurpose disaster shelters in the state of Odisha at present spread majorly along the coastal belt to shelter the vulnerable population of its local region. The cyclone shelter provides temporary shelter along with food to the people during extreme weather conditions. After the cyclone settles, the locals reportedly either are lucky enough to get compensation amount for the government or are preparing them for many more cyclones to survive.

1.1 About Cyclone Fani, 2019

Cyclone FANI, in 2019 was a major cyclone that occurred in the region that made landfall in the Puri region of Odisha. The wind speed was 297km/hr. It caused a large amount of destruction in Puri and Khurda districts and left marks of extensive damage to the life, property, infrastructure and livelihood of more than 1.65 crore people in 12 districts it passed through. The following is the damage report of various blocks and villages affected in the vulnerable state.

Although the pucca buildings, made of concrete usually, can withstand the high wind load, it was found that 15,118

Fig. 22.1 Destruction caused by cyclone fani, 2019 – Puri, Odisha

Source: The indian express Publication: 07th August 2019

Table 22.1 Destruction caused by cyclone fani, 2019 – Puri, Odisha

District	Population Affected (no.) by flood and cyclone	Blocks affected (no.)	Villagers affected (no.)	ULB affected (no.)
Angul	5,709	5	67	0
Balasore	11,33,374	12	2,535	4
Bhadrak	10,15,742	7	910	4
Cuttack	30,96,874	14	2,095	4
Dhenkanal	93,699	8	652	4
Ganjam	20,00,000	22	2,673	12
Jagatsinghpur	5,00,000	8	256	2
Jaipur	21,92,630	10	1,865	2
Kendrapara	15,22,902	9	1,592	2
Keonjhar	7,160	9	163	1
Khordhar	25,02,008	10	1,669	5
Mayurbhanj	1,73,095	26	1,126	2
Nayagarh	3,44,806	8	1,031	5
Puri	**19,68,228**	**11**	**1,772**	**4**
Total	1,65,55,509	159	18,388	51

Source: Disaster report issued By State Emergency Operation Centre, Odisha State Disaster Management Authority

Table 22.2 Destruction caused by cyclone fani, 2019 in Puri region

	Structure	Type of Damage	No. of huses affected
Rural	Kuccha	Completely	937
		Substantially	14,181
		Partially	67,174
	Pucca	Completely	21,015
		Substantially	75,556
		Partially	1,22,940
Urban	Kuccha	Completely	379
		Substantially	5,495
		Partially	27,352
	Pucca	Completely	3,125
		Substantially	12,571
		Partially	17,178
		Total	

rural and 5,814 urban pucca buildings were severely damaged/destroyed due to fallen vegetation (the high planted commercial trees and other structures falling on them. Puri has the highest number of destroyed pucca buildings.

In the affected districts, 61,074 rural and 27,352 urban pucca buildings need repair and retrofitting. The rural region of Puri has about 69% of the total number of affected houses among the 14 districts. The wind speed was so high that many kutcha buildings (i.e. with kutcha walls) have been levelled to the ground, rendering many families temporarily homeless, compelling them to stay with relatives/neighbours or in makeshift tarpaulin tents. During the field visits, it was observed that the slums are severely affected in the urban areas. More than half the slums affected (829 slums of 150811) are situated in the three most severely affected districts of Puri, Khurda, and Cuttack.

Approximately, 93%12 of the houses in these urban slums were either kutcha or semi pucca.

Fig. 22.2 Malilsahi village, Puri sadar

Source: www.earth.google.com

Further, in cities such as Puri and Cuttack, there are several large-scale slums are situated on the river/coastal side. These factors increased the vulnerability of the slums and therefore the damages were largely concentrated in these areas.

2. Aim, Scope & Limitations

This paper cites the vulnerability assessment of a cyclone-affected community Kapateswar, Puri and identifies the local resilient strategies with the tendency to withstand the cyclones in the region, through a comparative study in the area.

It includes a comparative study of the collected data with the cyclone resilient settlement in order to compare the traditional and the present ways of local survival constructions. For the last couple of decades, people of these areas have been trying to cope up with climate change and its tremendous effects. This research aims to understand the various factors in detail so as to come up with a local solution to withstand cyclones.

This research is based on both qualitative and quantitative approaches, carried out through primary and secondary data collection. The study area of Kapateswar village was selected on the basis of the neighbourhood characteristics comparison. The people of the village are extremely prone to various natural hazards and rural settlement of the regions present noticeable resilience strategies to live with disasters. Comparative analysis of both the cases gives the combined strategies as discussed in a further chapter.

2.1 Scope

- To analyse the types of damages caused by the cyclones through a detailed data collection and its assessment.
- To analyse the local traditional strategies of construction and compare through quantitative analysis of the traditional and present types of construction techniques.
- To understand local techniques and their feasibility to withstand the cyclones.
- To understand the need of the changes to withstand cyclones in the region through need assessment, derived by the detailed data analysis.

2.2 Limitations

- This data is limited to the detailed study of vulnerability Kapteswar Village, Puri Sardar, Puri, Odisha.
- The study only highlights the condition infrastructural and built challenges to withstand the cyclones in the region.
- The facts mentioned about the existence of the temple are as per the finding after a formal conversation with the people in and around the village.
- The Kapateswar village is divided into two, Kapateswar temple village and Maliksahai as the fisherman's village. Kapateswar temple village has been taken into consideration for this assessment.

3. Vulnerability Assessment

The vulnerability here is the diminished ability of an individual or a group to be able to prepare, withstand, or recover from any natural or manmade occurring disaster. This paper highlights the various ways to assess the vulnerability in the cyclone-affected community. Vulnerability assessment differs as per the present situation and stage of the disaster. The following are broadly categorized vulnerability assessments, namely,

- Pre-disaster vulnerability assessment
- Post-disaster vulnerability assessment

This assessment can be further divided into two sections to understand the present situation of the community through a quantitative survey and analysis of the data collected to come up with a solution to increase its resilience from the uncontrollable cyclones in the region.

3.1 Situational Assessment

This section of the research sites the present situation of the community through visual analysis, topographical mapping, livelihood, types of construction, and the damage of the built, through data collection and its analysis through comparative study. It provides the extent of a disaster, priority needs of the affected community and the vulnerable people. It further identifies secondary threats, resource availability, and the capacity of the local response to create resilience to these cyclones. The following are the various factors discussed in detail to analyse the vulnerability level of the community to withstand cyclones.

The various parameters are, Social-economic, Topographical conditions, Livelihood of the community, Built resilience, Community building and infrastructure and Vegetation as a barrier to adding resilience.

3.2 Need Based Assessment

This part of the research focuses on the articulation of the level of damage and the need for immediate assistance required to safeguard the community. It talks about the conclusive understanding of the reasons for damage and immediate action that need to be taken for it. It provides the extent of damage by identifying the threats to the community through data assessment. The need assessment allows being prepared for any extremity by creating resilience from any kind of threats identified from this data collection and analysis, hence making the community less vulnerable.

In the end, need assessment further gives solution to the vulnerability assessment falling in three categories, namely,

Low cost resilience

This section focuses on the low-cost solutions of construction in the terms of material choices, sustainable construction techniques, long-term solutions, planning of flora and fauna, et cetera for the vulnerable people to which stand the tremendous climatic conditions in the area through gaining resilience.

Awareness

This section focuses on creating awareness among the people for the need of change for a better good of the community. The solutions of this particular section would focus on indoctrinating the people of the village about various techniques and and preparing them to help oneself for the situation

Sustained livelihood

This section would give a solution to create opportunities for the locals to sustain a lifestyle in the area which is highly reliable and approachable in the situations of danger.

4. The Study Area

The criteria for the selection of the area of study is based on the frequently occurring disasters in India and the destruction caused. The study is limited to understand the various parameters that play a role in increasing the effectiveness of these intractable natural calamities and quantitate the need for the change through thorough data analysis. Kapateswar is a heritage village is located in (Fig. 22.4(b) & Fig. 22.5) Puri Sadar, Puri. It is about 19 km from the eastern coast of India. Its revenue village is Jankadeipur located 500 m away. The village is surrounded by many other small villages namely Raghurajpur, the traditional Pattachitra Artist village- 500m, Pipli, Artist village- 5km away, Birabalabhadrapur, traditional brahmin village- 4.5 km away, and many more.

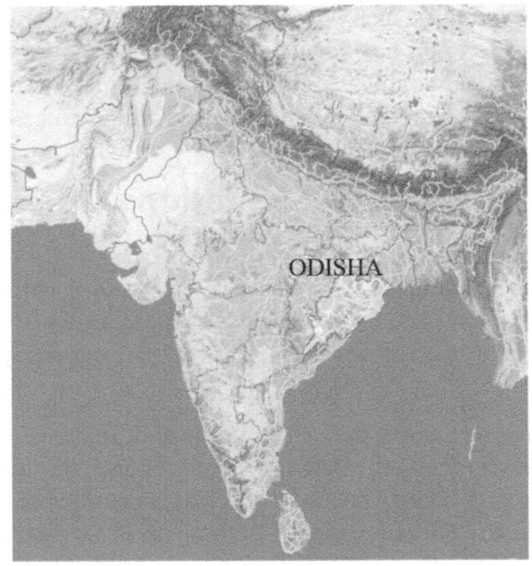

Fig. 22.3 Terrain Map of India

Source: www.googlemaps.com

The village is divided into two parts as per their sets of beliefs and traditions namely, Kapateswar village for the Hindus and Maliksahai for the fishermen. There exists a heritage building known as Kapateswar shiv temple. The village is about 15 km from the Puri Bus stand. The village is accessible from the diversion towards Raghurajpur from the Puri- Cuttack road. The road diverts to the village near the Chandanpur bus stand. The local area had a self-sustaining character with all the

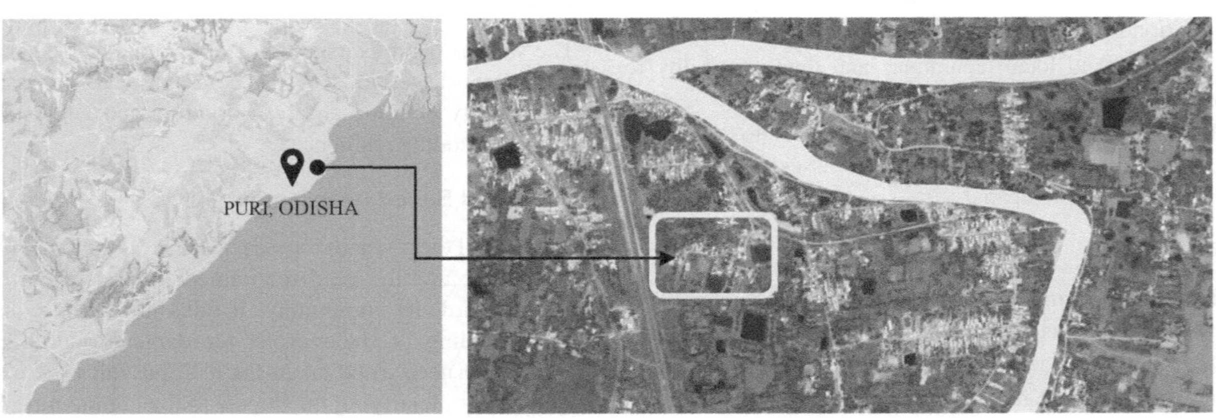

(a) (b)

Fig. 22.4 (a) Map of odisha, India, (b) Kapateswar village, Puri sadar

Source: (a) www.googlemaps.com (b) www.earth.google.com

Fig. 22.5 Dhobi ghat & fishing – river Bhargavi, Kapateswar

Source: Onsite survey

Fig. 22.6 Brick opposite to the village

Source: Onsite survey

Fig. 22.7 Backside of Kapateswar village showing traces of cyclonic distruction, 2020

Source: Onsite survey

Fig. 22.8 Affected dwelling units, Kapateswar

Source: Onsite survey

Fig. 22.9 View of Kapateswar shiv temple, Kapateswar

Source: Onsite survey

accessible availabilities daily basic needs. The entrance was through the chandanpur market lane which led to kapateaswar

At the entrance, the diverting road to Kapateswar, there exists s man-made pond which is actively being used by the nearby communities for their day-to-day occupation. Beside the man-made pond, runs the river Bhargavi, a stream of Mahanadi. River Bhargavi is being used for various purposes like fishing and the as the dhobi Ghats (Fig. 22.5). The condition in and around the village seems to be severely damaged at the first instance. Multiple trees were uprooted. On the opposite side of the village entrance, exist a brick kiln run by the locals of the village.

4.1 Kapateswar Shiv Temple

Kapateswar shiv temple is believed to be one of the oldest temples of Odisha that exists. As per the locals, the temple predates even before the famed Jagannath temple, Puri.

The temple shows the characteristics of vernacular architecture through the use of its materials and is in a habitable condition used by the community on daily basis. The temple acts as the heart of the community, hosting attraction for its important festivals that take place at the village every year. Dwadashi Mela (fete), celebrating the foundation of the temple attracting almost 5000-6000 people from in and around the village. The open space beside the temple is set with 200-250 stalls of local handicrafts and food during the festival. Other similar festivals are Jagar Mela (Maha Shivratri), Dussehra-rath rally, Rath yatra with foot fall in the range of 800-2000 people every year. The temple located about 1300 mm below the ground level of the rest of the village. (Fig. 6.12)

5. Methodology

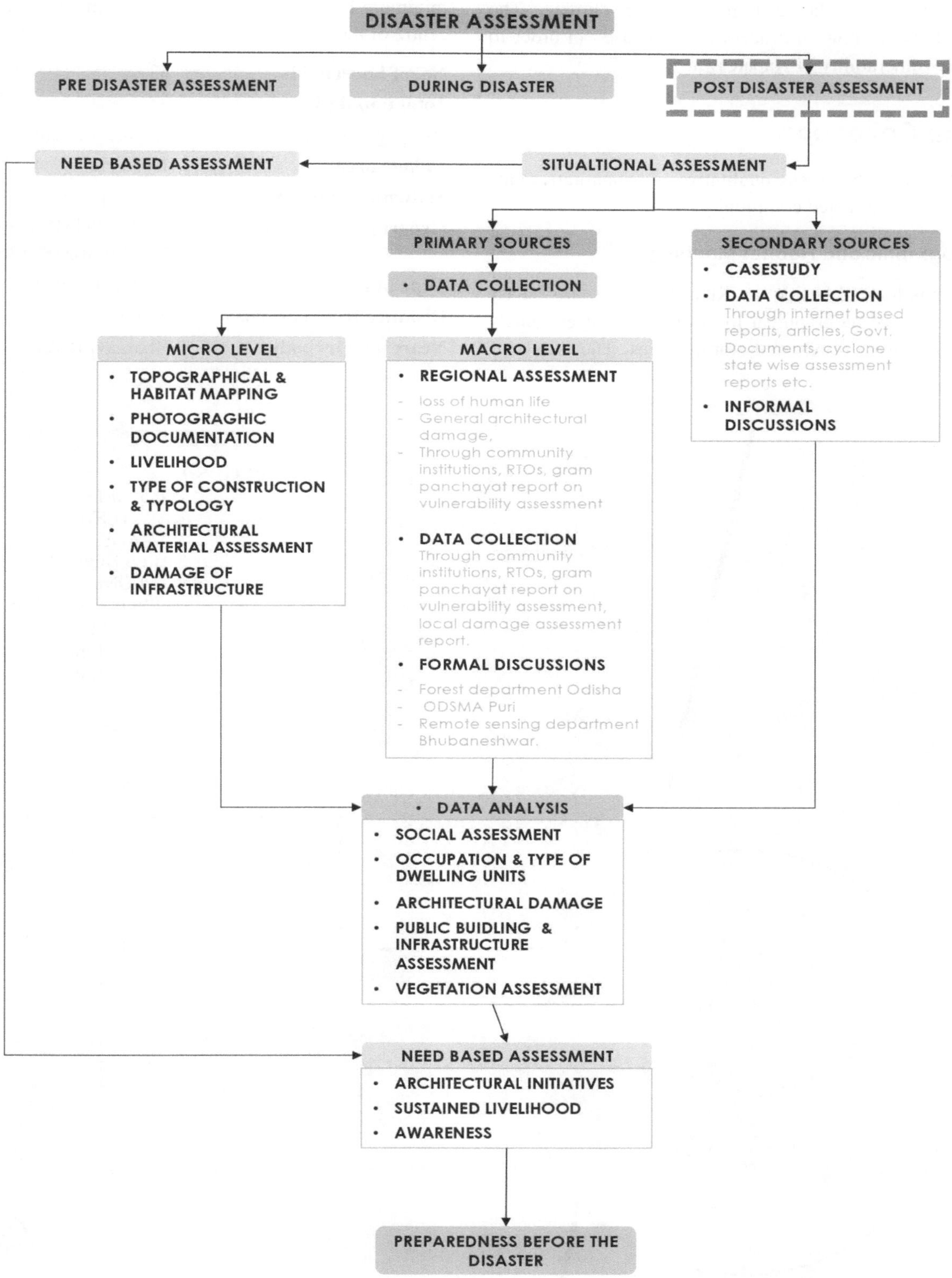

Fig. 22.10 Methodology for the vulnerability assessment

The methodology intends to showcase the process of gathering information and analysing it to come to a solution that can withstand the extremity of the climate. The Fig. 22.10 above shows a process that was used in order to access the vulnerability of Kapateswar.

6. Data Collection

This section sites the on-site qualitative and quantitative data collection of the effected community.

6.1 Topographic and Habitat Mapping

Kapteswar is located beside the River Bhargavi, a tributary of Mahanadi. Its geographical location makes it extremely prone to cyclones, floods and storm surges. The village is located about 15 km from the coastal line with the range of about 8m to 19 m above the sea level as its maximum and minimum heights. Kapteswar Shiva temple is located at the centre of the village.

No. of households	52
Total population	356
Heritage building	Kapateswar shiv temple
Village area	15.4 acres
Settlement pattern	Linear pattern
Geographic coordinates	19.887054518805073, 85.82194625871266
Type of climate	Tropical climate
Distance from east coast	20 km
Nearest waterbody	Bhargavi River

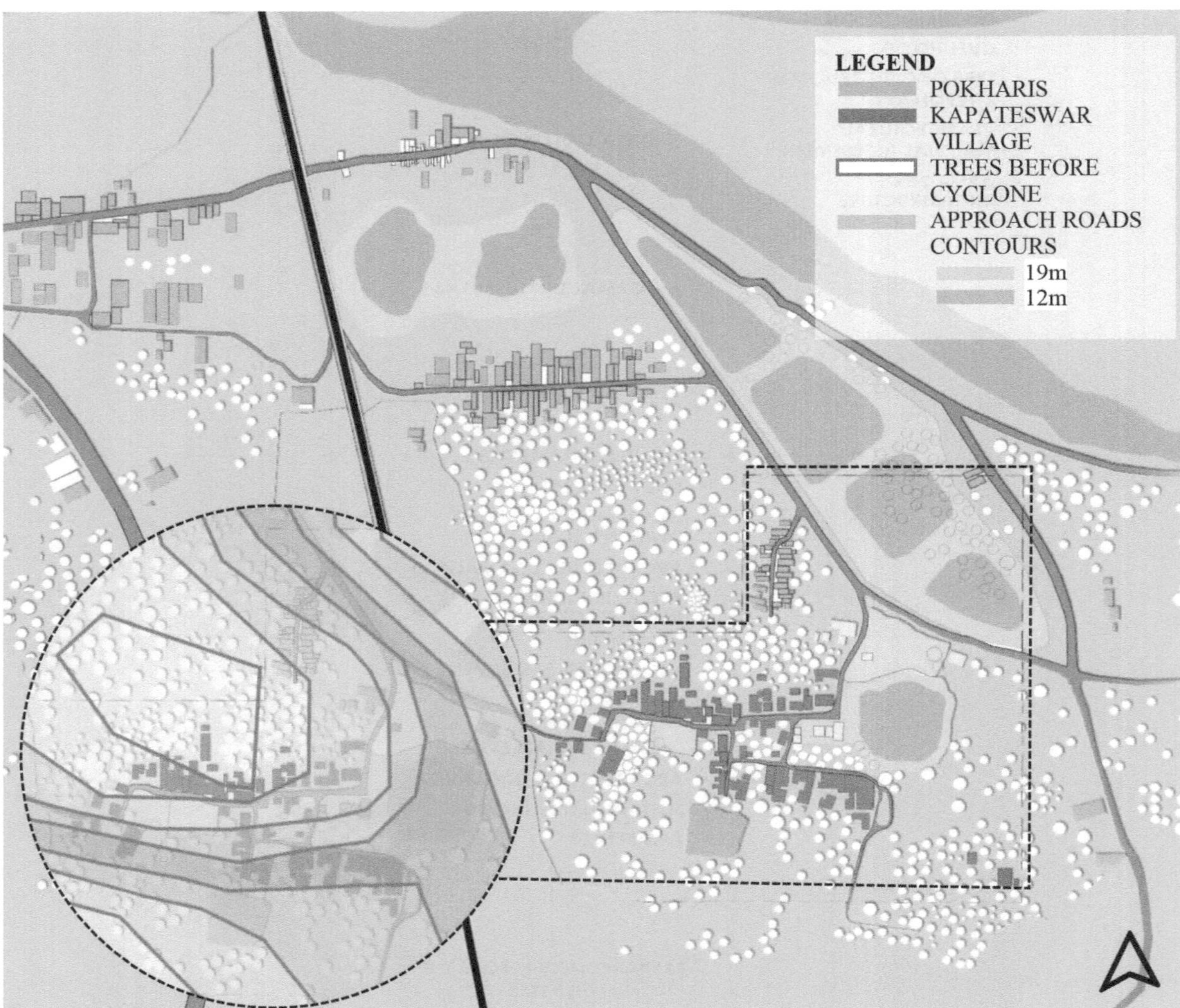

Fig. 22.11 Existing site plan- Kapateswar village, Puri

6.2 Photographic Documentation (On-site)

Fig. 22.12 Temple lane elevation, Kapateswar, Puri

Fig. 22.13 South side elevation Kapateswar, Puri

Fig. 22.14 North side elevation, Kapateswar, Puri

The following are the images of the dwelling units at the village, damaged by the cyclone Fani, 2019

Fig. 22.15 Present condition of the dwelling units

6.3 Livelihood

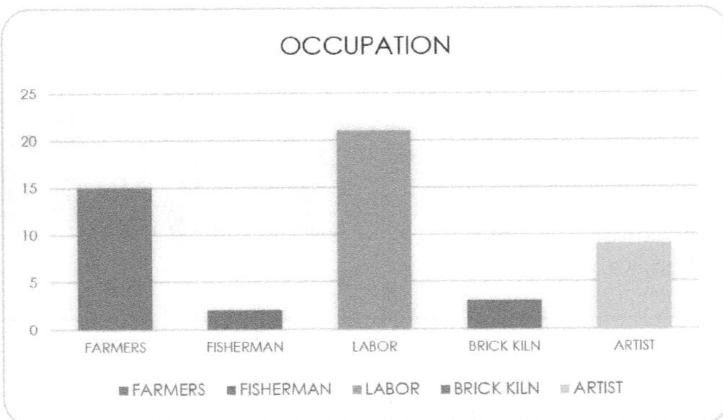

Fig. 22.16 Occupation of the village

Source: Onsite documentation analysis

6.4 Type of Construction and Typology

The community has a self-sustaining character. The major occupation is farming, labour work local brick kilns and traditional Pattachitra artists (Fig. 22.16).

The farmers were involved in the cultivation of crops like Pan, Supari coconut farming, mango trees, etc. some of the labourers were working locally, while others would travel to Puri or nearby cities for everyday work. Many women of the house would usually be involved in the traditional art and craft of Kapteswar.

As per the onsite findings, the character of extensive use of local materials for the construction with about 67% of the dwellings constructed in local materials, and their condition was worse than a Kaccha house with multiple damages (Fig. 22.17). Out of which, 85 % of the buildings can be categorised as residential. Mixed-use dwelling units, where the front of the house is workshop space for the Pattachitra artist and storage spaces for the farmers (Fig. 22.18). The backyard of most of the houses is used as an open kitchen or a cowshed, usually a temporary structure. The village consisted of various types of private dwelling units and a set of public buildings. The built structures of the village can further be divided into 4 categories namely,

- Residential dwellings
- Mixed-use
- Commercial
- Religious & Community buildings

6.5 Materials Used Before Cyclone and after Cyclone

This section showcases the List of materials that were being used by the community for construction of the houses before the before and after cyclone. The material used are both local and transported. (Fig. 22.19). The materials

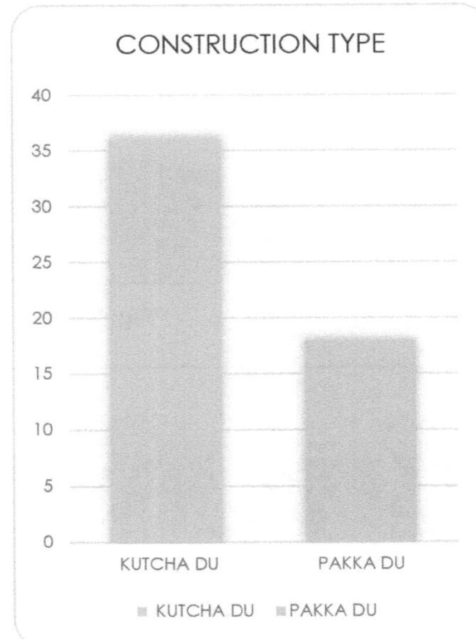

Fig. 22.17 Type of Constructions

Source: Onsite documentation analysis

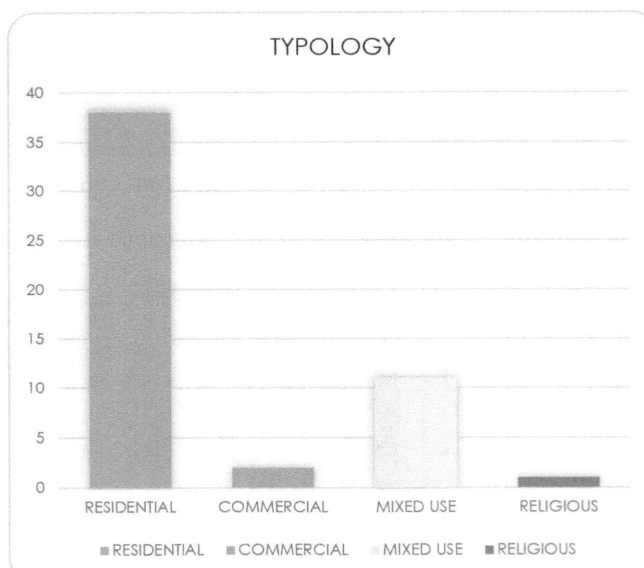

Fig. 22.18 Type of constructions

Source: Onsite documentation analysis

assessment is done by sub categorising the materials as per its use- materials used for roof building, walls, flooring and the foundation of the DUs. Immaturity of the materials being used for construction can be categorised into two, the local materials and the transport ones.

The local materials such as the Balia mud (the local River mud) was being used extensively for construction as a mortar & plaster for the vernacular typologies. Another important

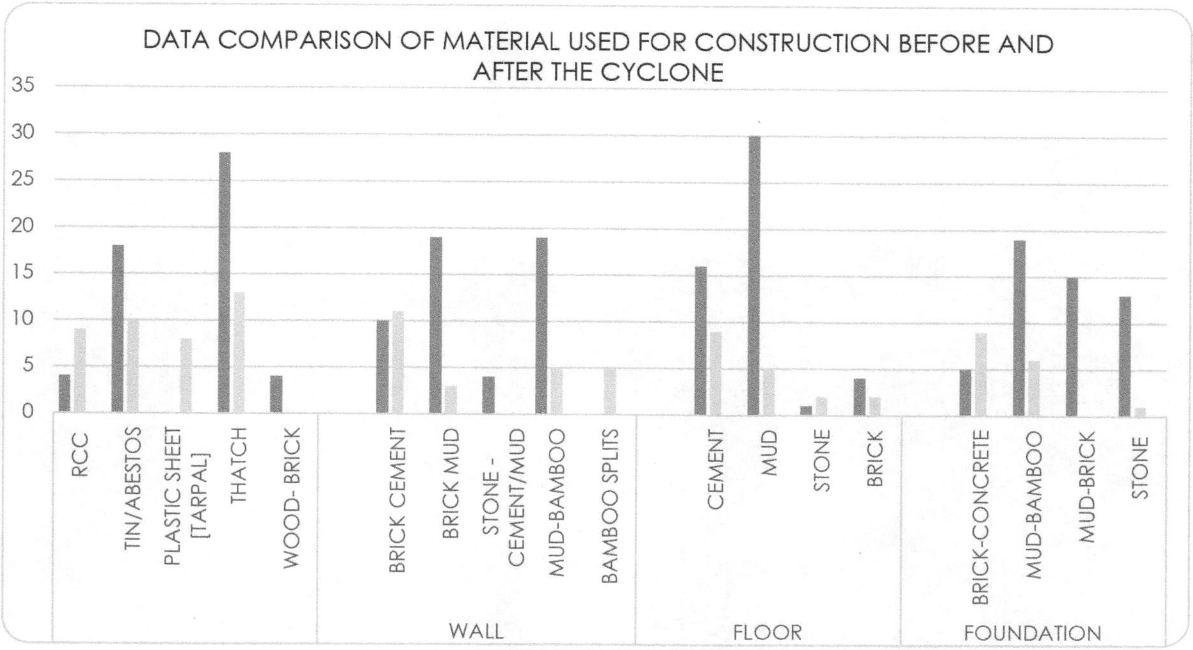

Fig. 22.19 Materials used for construction before and after cyclone

Source: Onsite documentation analysis

local material was the laterite stone, being used extensively for both kutcha and pakka typologies of dwelling units. The stone was transported from Cuttack, Which is about 30 km from the site. The modern construction work using the modern techniques of construction with RCC roofing and Bricks cement walls. The materials for these modern constructions were transported from the nearest cities. The after cyclone materials showed temporary structures with extensive use of the tarpal, bamboo sticks, mud, etc as the main materials for immediate relief.

6.6 Damage to Infrastructure and Public Buildings

As per the findings, Kapteswar village has only suffered the electrical damage during the cyclone due to damaging of the tower about 1 km away from the site. The temple Mandapam, at the entrance of the village was severely affected by the cyclone winds. This Mandapam, was being used as a platform for the religious pujas and celebrations during the time of festivals and rallies. There is no other infrastructure provision that was found to be damaged.

The following were the public buildings that were found to be the part of Kapteswar village, the Kapteswar Shiva temple, medical room for the pregnant women,

Kapteswar Shiva Temple

Kapateswar Shiv Temple is located at the centre of the village. As per the findings the temple has existed for many decades. It show the characteristics of a traditional temple architecture. The temple has been constructed with the local materials and is maintained by the locals of the village. Every year, on the occasion of Dwadashi mela, the temple is repainted and decorated for the festival. The materials that have been used to construct are local materials such as the local laterite stone, along with lime mortar, and a damaged wooden roof, which now is replaced by thatch. The foundation to the plinth is done in the local stone.(Fig. 22.20). During the cyclone, the thatch roof was found to be damaged causing no other damage to the temple.

7. Data Analysis/Comparative Analysis

7.2 Social Assessment

This section narrates the post-disaster impact on the qualitative areas of the community. The social assessment was conducted through manual interviews and surveys focusing on the various factors,

Socio economic impacts

The post disaster situation of the village has shown traces of social economic impacts through the damaged dwelling units and vegetation. The vegetation is the direct source of income to these people and it the sector which is the most affected. As per the findings, the total damage of vegetation is 60% of the pre disaster vegetation.

Fig. 22.20 Kapateswar shiv temple, Puri Sadar

Quality of life

The people of the village are lead a simple life, struggling to fulfil their daily needs. There are 3 handpumps for clean water. The people of the village depend on the rainwater harvesting through Pokharis for all house hold purposes. The quality of santitation is quite low due to lack of maintance and their livelihood.

Work life issues

After the cyclone, the laborers lost their local jobs due to lack of opportunities and now, have to travel far off to work. The farmers the farmers have faced lot of destruction in terms of commercial crops such as coconut trees , Supari, pan farming lands and tamarind trees.

During disaster safety & health care

As per the findings, the nearest hospital from the site is about 17 km away. The nearest village, Raghurajpur consists of a community building which is sometimes used as a multipurpose cyclone shelter for the vulnerable people of the area. At present, the Community building is not being maintained as such showing traces of damage due to the cyclone.

7.2 Occupation and Types of Dwelling

The prominent occupations of the people in Kapteswar labourers, farmers, artist. As per the findings, the labourers were found to be 30% more vulnerable. The kutcha houses are highly vulnerable to the cyclones as per their condition in terms of construction and usage of materials. Most of the villagers have constructed their homes out of need to survive after the cyclone without even thinking about the consequences of any further extremity. (Fig. 22.21)

7.3 Architectural Damage Assessment

The Fig. 22.22 shows the amount of damage to both kutcha and pukka houses through the damage indicator. The total number of fully damaged houses are 15 and severely damaged are 14.

The Fig. 22.23 shows the assessment of the damage in detail through the materials study.

7.4 Vegetation Assessment

The vegetation is one sector which majorly gets affected by the cyclone. After the two most devastating cyclone in the eastern coast, super cyclone 1999 and cyclone FANI 2019,

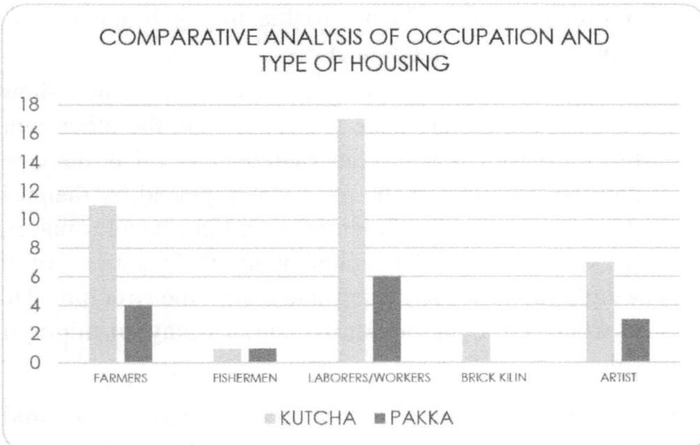

Fig. 22.21 Occupation & type of housing

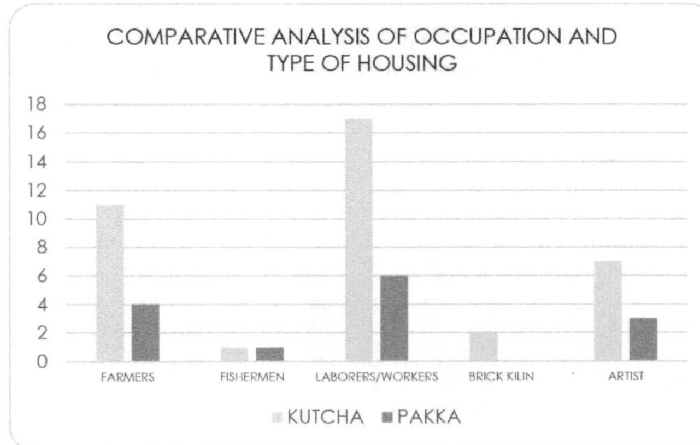

Fig. 22.22 Amount of damage and construction

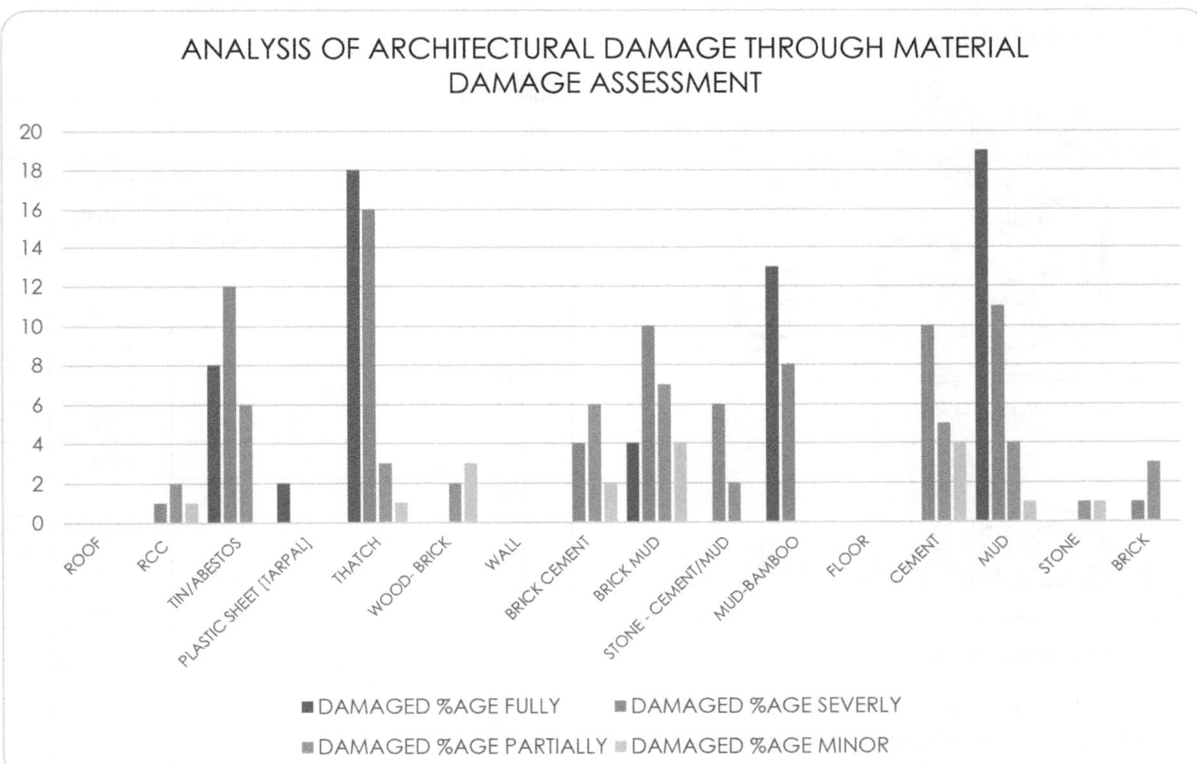

Fig. 22.23 Per material damage assessment

the amount of vegetation loss is up to 70%. The following are the various plantations in and around Kapateswar, Puri sadar majorly focusing on the economic benefits:

Mango Plantations, banyan tree, Tamarind, Supari trees, Paan Plantation, coconut trees, Cotton plantation, Rice fields, Jowar

7.5 Public Building to Dwelling Unit Assessment

The temple

While comparing the damage analysis of Kutcha or pukka dwelling units to the existing traditional architecture example that has been which standing cyclones for over a century, it can be stated that,

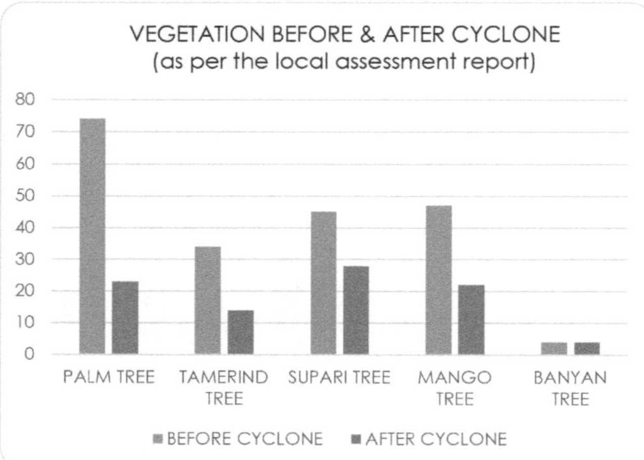

Fig. 22.24 Vegetation recorded before and after the cyclone
Source: Onsite documentation analysis

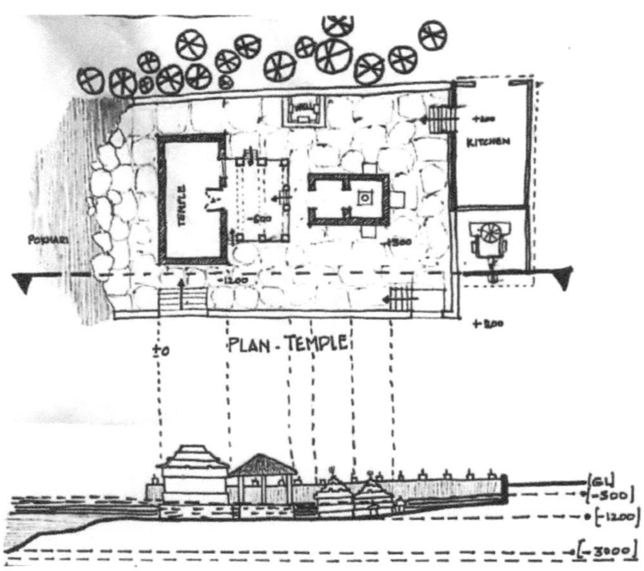

Source: Onsite sketch of Kapateswar temple

- Stone, being the heavy material and unaffected by stagnant water or the cyclonic winds, keeps the foundation intact hence, showing no traces of destruction.
- As per the findings, the temple complex is flooded with water during the rainy season up till the plinth level.
- Plinth of Kapteswar temple is made of stone along with lime mortar which is highly permeable and does not get damaged with wetness.
- The columns of the temple are of the local stone along with lime mortar. The traces of small amount of cement is seen on the exterior of the stone structure to compensate the washed of limewater. The middle structure is made of stone and the roof is carved out of the wood shows traces of minor damage. The due to this, the top of the roof was replaced by the thatch later.
- The two roofs of the temple are in stone which shows no traces of damage due to cyclone. On the other hand, the third roof that is constructed in wood is damaged and the covering of the rood was replaced by thatch in the later years. The thatch being a light weight material doesn't withstand the cyclone without any support of the weight of strings to keep it intact with the structure. This is where the construction techniques play an important role.

The following (Table 22.3) shows a comparative study of the materials being used by the people of Kapateswar for construction and their damage due to the cyclone to the materials used in the traditional Kapteswar temple, showcasing the vernacular techniques of construction in the area.

Table 22.3 Percentage damage per material used for construction

	Materials	Damaged %Age		
		Fully or Severely	Partially	Minor
1	Roof			
	RCC	1.92	3.84	1.92
	Tin/Abestos	38.46	11.53	0
	Plastic Sheet [Tarpal]	3.84	0	0
	Thatch	65.38	5.76	1.92
	Wood- Brick	0	3.84	5.76
2	Wall			
	Brick Cement	7.69	11.53	3.84
	Brick Mud	26.92	13.46	7.69
	Stone - Cement/Mud	11.53	3.84	0
	Mud-Bamboo	40.38	0	0
3	Plinth And Floor			
	Cement	19.23	9.61	7.69
	Mud	57.69	7.69	1.92
	Stone	0	1.92	1.92
	Brick	1.92	5.76	0

8. Need Based Assessment

This section of the paper concludes the basic needs of the community to withstand the cyclone in the future and gain resilience. The following derived parameters that can play an important role in decreasing the vulnerability of Kapateswar to the extreme climate

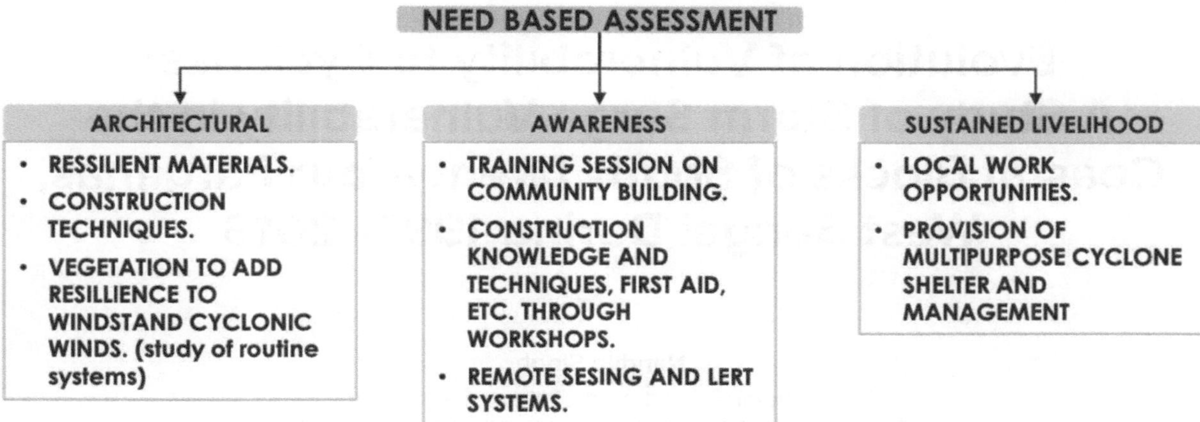

REFERENCES

1. Business insider, June, 2020, businessinsider.in , *Article on India's western coast is hit by fewer cyclones than the eastern coast: Scientific reasons explained.*
2. Indian express, 07 August, 2019, Indianexpress.com, *The indian express Publication about the Cyclone Fani- Beyond shelter in a storm, Odisha eyes sturdier homes, power and trees.*
3. India today, May 3, 2019, indiatoday.in, *20 years before Cyclone Fani, Super Cyclone killed 10,000 in Odisha. As Cyclone Fani hammers Odisha, we take a look at the devastating 1999 Super Cyclone that left Orissa in tatters*
4. *Rawal V., Prajapati D., 2018,* Unanati.org, t*oolkit for assessing damage after disasters.*
5. Wikipedia, September, 2018, *vulnerability, an article on the Vulnerability and its types.*
6. National Institute of Disaster Management (NIDM), June, 2020, *Report 2020, Disaster Risk, Vulnerability Assessment and Management*
7. Inter-American development bank, September, 2015, *Report 2015, Indicators of disaster risk and risk management*
8. Google earth, 2021, *Kapateswar village location*

CHAPTER 23

Evolution of Vulnerability to Cyclones: A Study of Storm Surge Vulnerability in the Coastal Blocks of South Twenty-four Parganas, West Bengal During 1997–2019

Nandita Singh
PhD scholar, Department of Natural Resources,
TERI School of Advanced Studies, New Delhi- 110070
nanditasingh.2911@gmail.com

Neeti Neeti
Assistant Professor, Department of Natural Resources,
TERI School of Advanced Studies,
New Delhi- 110070

Abstract

The coastal zones are a fragile, dynamic and productive environment, with a plethora of economic activities and a growing population. They are constantly influenced by natural and anthropogenic processes and prone to coastal hazards like sea level rise, erosion, cyclones and accompanying storm surges. The coastal zone of West Bengal is no exception, and among the most vulnerable to storm surges in India. The present study attempts to analyse the changing patterns of storm surge vulnerability during the period 1997-2019 in the coastal blocks of South Twenty-four Parganas employing the Coastal Vulnerability Index approach, employing five bio-physical and two socio- economic variables. The study finds that several blocks in the area are highly vulnerable in the study period, while others have become increasingly vulnerable to the hazard.

Keywords: coastal vulnerability index, storm surge, remote sensing, GIS.

1. Introduction

Coastal zone is defined as "that part of the land affected by its proximity to the sea and that part of the ocean affected by its proximity to the land, that is an area in which processes depending on the interaction between the land and sea are most intense" (Sorenson and McCreary, 1990, as cited by Mahapatra et al. 2013). It's a dynamic zone, continually affected by daily tides, tsunamis and storm surges, erosion and accretion, sea level rise etc. Coastal zone are the most fragile, dynamic and productive ecosystem and it is under pressure from both anthropogenic activities and natural processes (Mahapatra et al. 2013). Coastal areas are important regions in terms of their industrial friendliness, in terms of them being transportation hubs, having high population, tourism prospects etc. The downside of urbanisation and development presents itself as the challenges that coastal zones now face like pollution, coastal erosion, landfill, dredging, loss of habitat, global warming induced sea level rise etc. The mangroves forest, coral reefs, sea beaches, sea weeds, tidal flats make coastal zones a unique ecosystem supporting a variety of flora and fauna and protecting the shoreline. This calls for proper management of coastal zones and vulnerability assessment is instrumental in this regard as it helps to identify vulnerable regions or populations in coastal areas.

The Intergovernmental Panel on Climate Change (IPCC CZMS, 1992) defines vulnerability of coastal zones by their degree of incapability to cope with the impacts of climate change and accelerated sea-level rise. The Coastal

Vulnerability Index (CVI) is a rapid and effective way to rank coastal regions according to their levels of vulnerability using physical as well as socio-economic variables. The CVI allows these variables to be related in a quantifiable manner yielding numerical data that cannot be directly equated with particular physical effects. It is however useful in highlighting the regions where the various effects of sea level rise may be the greatest (USGS 1999). The coastal vulnerability index is a simple and reliable method to identify the relative vulnerability of coastal zones to change or perturbations. The earliest application of CVI can be traced back to 1989 when Gornitz and Kanciruk applied it to US coast considering the physical variables of relief, rock type, land form, relative sea level change, shoreline displacement, tidal range and maximum wave height (Aboudha and Woodroffe, 2006). Thereafter, Theiler and Hammer-Klose in 2000 developed a CVI for USA using geomorphology, shoreline erosion and accretion, coastal slope, relative sea level change, mean wave height and mean tidal range (Aboudha and Woodroffe, 2006). These classical works have been the basis of many future works employing diverse variables to develop Vulnerability indices for the coastal ecosystems. The work of Boruff et al. 2005 marked a shift towards the socio- economic aspect of vulnerability when a social vulnerability index was developed, again for the US coast using census-derived social data (Aboudha and Woodroffe, 2006).

Remotely sensed data has contributed to a large extent in gathering information regarding the variables in order to compute Coastal Vulnerability Indices. While remote sensing technology aides in data collection, the aspects of data analysis is enabled by Geographical Information systems which come with various software packages capable of analysing the data so collected to provide useful information. One such package is the Digital Shoreline Analysis System which comes with ArcGIS and is useful for calculating shoreline change rates by providing shape files shorelines for different years. Contributions of moderate resolution data like IRS-P6, AWiFS and Landsat are important from a geomorphology perspective. SRTM DEMs have been used to calculate coastal slope. Landsat MSS, TM, ETM+ images have been used by Kumar et al, 2010, Duriyapong and Nakhapakorn, 2011, Kumar and Kunte, 2012, Joevivek et al, 2013, mostly to extract data on shoreline change rates.

The present study aims to analyse changing patterns of vulnerability to storm surges in the coastal blocks of South Twenty-four Parganas. Towards this, CVI has been calculated for the years 2004, 2009, 2014 and 2017 for thirteen coastal blocks of the district using five bio-physical (elevation, geomorphology, sea level change rate, shoreline change rate, storm surge height) and two socio-economic (land use land cover, population density) variables.

2. Study Area

The study area comprises of the coastal blocks of South Twenty-four Parganas, one of the three coastal districts of West Bengal. Administratively, the district is divided into 29 blocks, out of which thirteen coastal blocks for the focus of this study.

Geographically, the area extends from 21°32'56"N to 22° 23' 46" N and 88° 2' 27" E to 89° 6" 18" E. The average elevation of the region is approximately 5.2m above sea level. The Hoogly and Ichamati rivers for the western and eastern boundaries of the area, which is interspersed by several others like Matla, Bidyadhari, Datta etc. Geomorphically, the area broadly comprises of coastal plains, mudflats, mangroves and swamps. Aquaculture ponds are present towards the northern reaches. There is a dense network of creeks, particularly so, in the Gosaba and Kultali blocks.

The area owing to its location, bio-physical and socio-economic characteristics is vulnerable to coastal hazards like erosion, sea level rise, cyclones and accompanying storm surges.

3. Materials and Methods

The present study applies the Coastal Vulnerability Index approach to analyse the evolution of vulnerability to tropical cyclone storm surge in the coastal district of South Twenty-four Parganas, West Bengal. Five bio-physical variables (namely elevation, geomorphology, shoreline change rate, sea level change rate and storm surge height) and two socio-economic variables (namely population density and land use) were employed to calculate the overall vulnerability of the thirteen community development blocks. The details of the data employed for deriving each of these variables is provided in Table 23.1. The importance of the risk variables and the methods used to derive them is presented in the subsequent section.

3.1 Elevation

Coastal regions having higher elevation offer greater resistance from inundation due to storm surges, tsunamis or sea level rise. Areas having a low coastal slope, but high elevation are less vulnerable to coastal hazards as compare to low elevation areas. Digital elevation models of approximate 30m resolution from the Shuttle Radar topography Mission were used to construct elevation map of the study area and block-wise average elevations were extracted in a GIS environment.

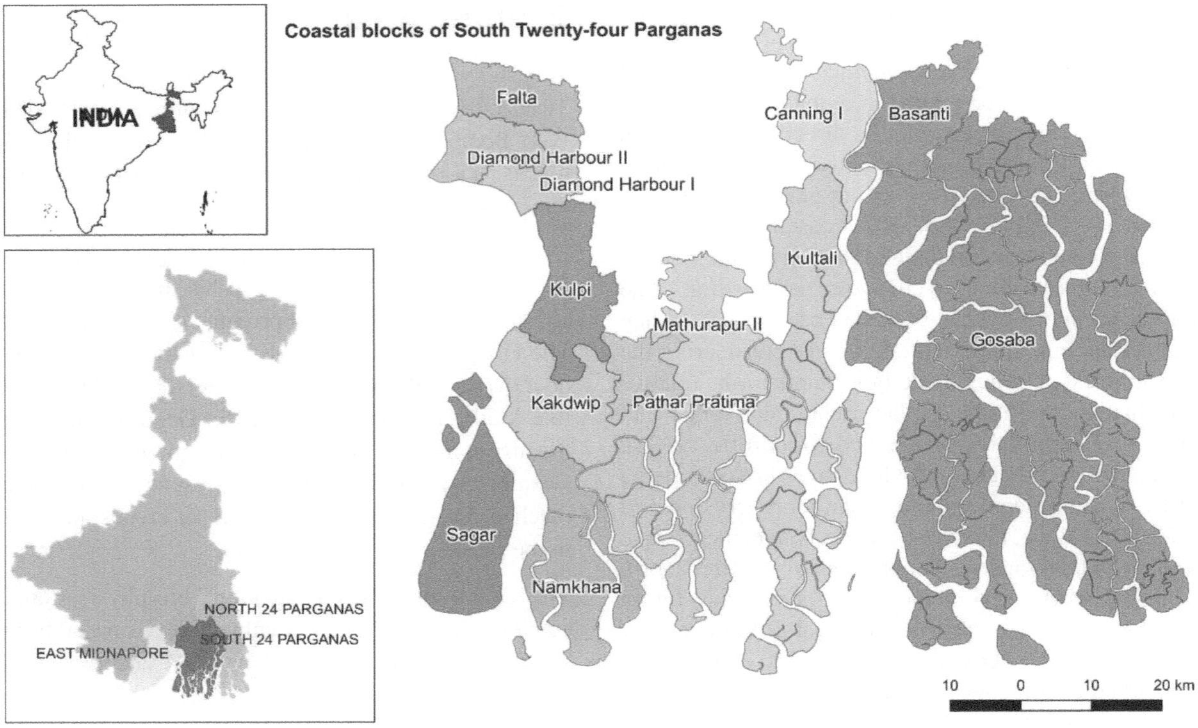

Fig. 23.1 The study area map depicting the coastal blocks of South Twenty-four Parganas.

Table 23.1 Details of data employed for deriving bio-physical and socio-economic variables for CVI calculation

Component	Variable	Time period	Data
	Elevation (m)	2000	SRTM DEM
	Geomorphology	2007, 2017	Landsat 7 imagery
Bio-physical	Shoreline change rate (m/yr)	1997-2002, 2002-2007, 2007-2012, 2012-2017	Landsat 5 and 7 imageries
	Sea level change rate (mm/yr)	1916-2016	PSMSL database
	Storm surge height (m)	1998, 2002, 2019	Literature
	Land use land cover	1997, 2002, 2007, 2012, 2017	Landsat 5 and 7 imageries
Socio-economic	Population density (people/km^2)	1999, 2004, 2009, 2014, 2019	Population estimate, Census 1991, 2001, 2011

3.2 Geomorphology

The processes that act on a coastal area are reflected in its geomorphology. Different landform categories are affected by coastal hazards to different degrees. While beaches and mangroves are considered highly vulnerable, estuaries represent areas of medium vulnerability. Low vulnerability coastal landforms include cliffs, aquaculture ponds, inundated coasts, etc. Geomorphology maps for 2017 were created by onscreen digitization technique using visual interpretation of satellite imagery, with the help of published web maps and literature.

3.3 Shoreline Change Rate

The rates at which shorelines are retreating or advancing has a bearing on the vulnerability of a region as erosion means diminishing distance between coastal population and the sea, thereby increasing its vulnerability to coastal hazards. Coastal erosion may result in loss of life and property, and coastal habitats. Automatically extracted shorelines for the years 1997, 2002, 2007, 2012 and 2017 were used for calculation of shoreline change rates using the End Point Rate (EPR) method in Digital Shoreline Analysis System (DSAS), following which their block wise average values were extracted for CVI development.

3.4 Sea Level Change Rate

Coasts with higher rates of sea level rise are considered more vulnerable to coastal hazards as higher sea levels would mean increased wave action during coastal storms. Rapidly rising sea levels would mean a higher shift to the zone of wave action, more so on milder slopes. Sea level change rate for 16 tide gauge stations along the east coast of India and adjoining Bangladesh were calculated from long term data obtained from Permanent Service on Mean Sea Level (PSMSL) portal. Inverse Distance Weighted (IDW) interpolation was used to create a surface from this point data. Block wise values of average rate of sea level change was extracted from the IDW surface using a vector shoreline.

3.5 Storm Surge Height

Wave action is responsible for shoreline modification as it can cause rapid widespread erosion. Higher surges also mean greater inundation, thereby leading to greater threat to life and loss of public/private property. Storm surge heights used in this study were obtained from literature pertaining to cyclones occurring during 1999-2019 at specific locations in South Twenty-four Parganas.

3.6 Land Use Land Cover

An understanding of land use land cover is essential to observe the role it plays in elevation/diminishing the vulnerability to coastal hazards. In this study, percentage of total block area that is under agricultural land use was used as a risk variable. Areas having a higher percentage of area under agricultural land use are considered more vulnerable as damage from coastal hazards would cause subsequent loss of livelihood for the dependent population. For this purpose, land use land cover maps for the years 2002, 2007, 2012 and 2017 were created using Random Forest Classifier, followed by extraction of area under agriculture class for index development.

3.7 Population Density

Population may exert a negative pressure on vulnerability, with densely populated areas being more vulnerable to coastal hazards. Population for the years 1999, 2004, 2009, 2014 and 2019 were estimated from Census data for the years 1991, 2001 and 2011. Thereby, population densities for the coastal blocks were calculated for CVI development.

3.8 Coastal Vulnerability Index

The variables thus derived were ranked into five vulnerability classes (very low, low, moderate, high and very high), according to characteristics mentioned in Table 23.2. A risk ranking of 1 would mean very low vulnerability, while that of 5 would mean very high vulnerability. Equal weightage were given to all risk variables. The CVI was then calculated for the years 2004, 2009, 2014 and 2019 for each of the thirteen blocks of South Twenty-four Parganas using the square root of product mean (Eq. 1).

$$CVI_{ss} = \sqrt{\frac{x_1 \times x_2 \times x_3 \times x_4 \times x_5 \times x_6 \times x_7}{7}} \quad (1)$$

Where x_1 = elevation

x_2 = geomorphology

x_3 = sea level change rate

x_4 = shoreline change rate

x_5 = storm surge height

x_6 = percentage area under agriculture, and

x_7 = population density

4. Results

4.1 Elevation

The average elevation of the study area ranged from 4.29 m to 6.3 m. Falta and Diamond Harbour I were categorised under low risk, while all the remaining blocks belonged to moderate risk categories.

Table 23.2 Risk ranking of variables of storm surge vulnerability

Variable	Very low	Low	Moderate	High	Very high
Elevation (m)	—	>6.0	3- 6	0- 3	—
Geomorphology	—	Coastal plains	—	Estuaries, creeks	Beaches, mangroves, mangrove swamp, mudflats
Shoreline change rate (m/yr)	>4	1–4	-1–1	-4- -1	<-4
Sea level change rate (m/yr)	<1.8	1.8–2.5	2.5–3.0	3.0–3.4	>3.4
Storm surge height (m)	—	<1	1–2	>2	—
Area under agriculture (% of block area)	<40	40–60	60–80	>80	—
Population density	<100	100–250	250–500	500–1000	>1000

4.2 Geomorphology

Falta, Diamond Harbour I and II were categorised as low risk based on presence of extensive coastal plains in this region. Mathurapur II and Canning I were classified as high risk due to presence (though to a lesser extent than very high-risk blocks) of mangrove, swamps and mudflats. Kakdwip, Namkhana, Pathar Pratima, Kultali, Basanti, Gosaba and Sagar blocks were categorised as high risk owing to the presence of mudflats, mangrove, mangrove swamps, beaches, creeks and estuaries.

4.3 Shoreline Change Rate

Several fluctuations were observed in average shoreline change conditions across the coastal blocks of South Twenty-four Parganas during the study period. In 2004, widespread erosion in the district, thus all the blocks were categorised as high risk according to Table 23.2 classification scheme. In 2009, widespread accretion was observed, the lowest rate of shoreline change being 6.9m/year, thus all the blocks were categorised a very low risk in terms of shoreline change rates. Canning I, Gosaba, Kultali, Pathar Pratima, Diamond Harbour I and II, Kulpi and Kakdwip were categorised as high or very risk blocks.

4.4 Sea Level Change Rate

The rate of sea level rise averaged over each block ranged from 0.16 to 4.34mm/year. Diamond Harbour I and II, Falta, Basanti, Canning I and Gosaba were classified a very high risk owing to sea level rise rates of more than 3.4 mm/year.

4.5 Storm Surge Height

Storm surge height information from literature was used to categorise the coastal blocks into various risk classes. The storms considered for this were the Kultali cyclone (2002) for vulnerability analysis of 2004, Aila (2009) for 2009 and 2014 and Fani (2019) for 2019. For the years 2004, 2009 and 2014, all the blocks were categorised as high risk. Storm surge height information was available for the Extremely Severe Cyclonic Storm Fani, according to which Diamond Harbour I and II, Kulpi, Sagar, Falta, Namkhana and Kakdwip blocks were categorised as high-risk blocks.

4.6 Land Use Land Cover

The information on percentage area under agriculture was derived from land use land cover maps. In 2004, 2009 and 2019 all the blocks were categorised as very low or low risk. In 2014 however, Diamond Harbour I and Gosaba were classified as high risk due to increase in percentage area under agriculture.

4.7 Population Density

The coastal blocks of highly dense South Twenty-four Parganas district were categorised as high and very high risk during all the assessment years, with Namkhana classified as moderate risk only in 2004. Canning I, Diamond Harbour I and II, Kulpi and Falta have constantly been classified as very high risk owing a high population density.

4.8 Coastal Vulnerability Index

The coastal vulnerability index was calculated according to Equation 1 for the years 2004, 2009, 2014 and 2019. The 25th, 50th and 75th percentiles of CVI values were calculated for each of the above years and the blocks were classified into low (1), moderate (2), high (3) and very high (4) vulnerability classes (Fig. 23.2). It is evident that Basanti and Gosaba blocks consistently figured in very high vulnerability blocks. Probable explanations may lie in the low elevation of these blocks, together with geomorphic characteristics like extensive presence of mangroves, estuarine environments, widespread creek networks that act as conduits for storm surges to affect widespread areas. Kakdwip, deteriorated from high vulnerability in 2004, 2009 and 2014 to very high vulnerability in 2019. Similar observations were made for Sagar and Namkhana blocks which were categorised as highly vulnerable in 2019. It is thus observed that the vulnerability of the coastal blocks comprising the Sundarban region of South Twenty-four Parganas to storm surges has either remained consistently high, or has increased over the years.

5. Conclusions

The present study aimed at analysing the evolution of vulnerability in the coastal blocks of South Twenty-four Parganas district of West Bengal. Remotely sensed data and Geographic Information Systems based analysis aides in creation of multi-temporal databases. This enables the observation of changing vulnerability to coastal hazards like storm surges across space and time. The observations regarding blocks like Basanti and Gosaba which remained very highly vulnerable all along the study period, or those like Sagar and Namkhana which recorded deterioration in termed of increased vulnerability highlight the need for better protection and mitigation measures in these areas.

References

1. Abuodha, P., & Woodroffe, C. (2006). Assessing vulnerability of coasts to climate change: A review of approaches and their application to the Australian coast.

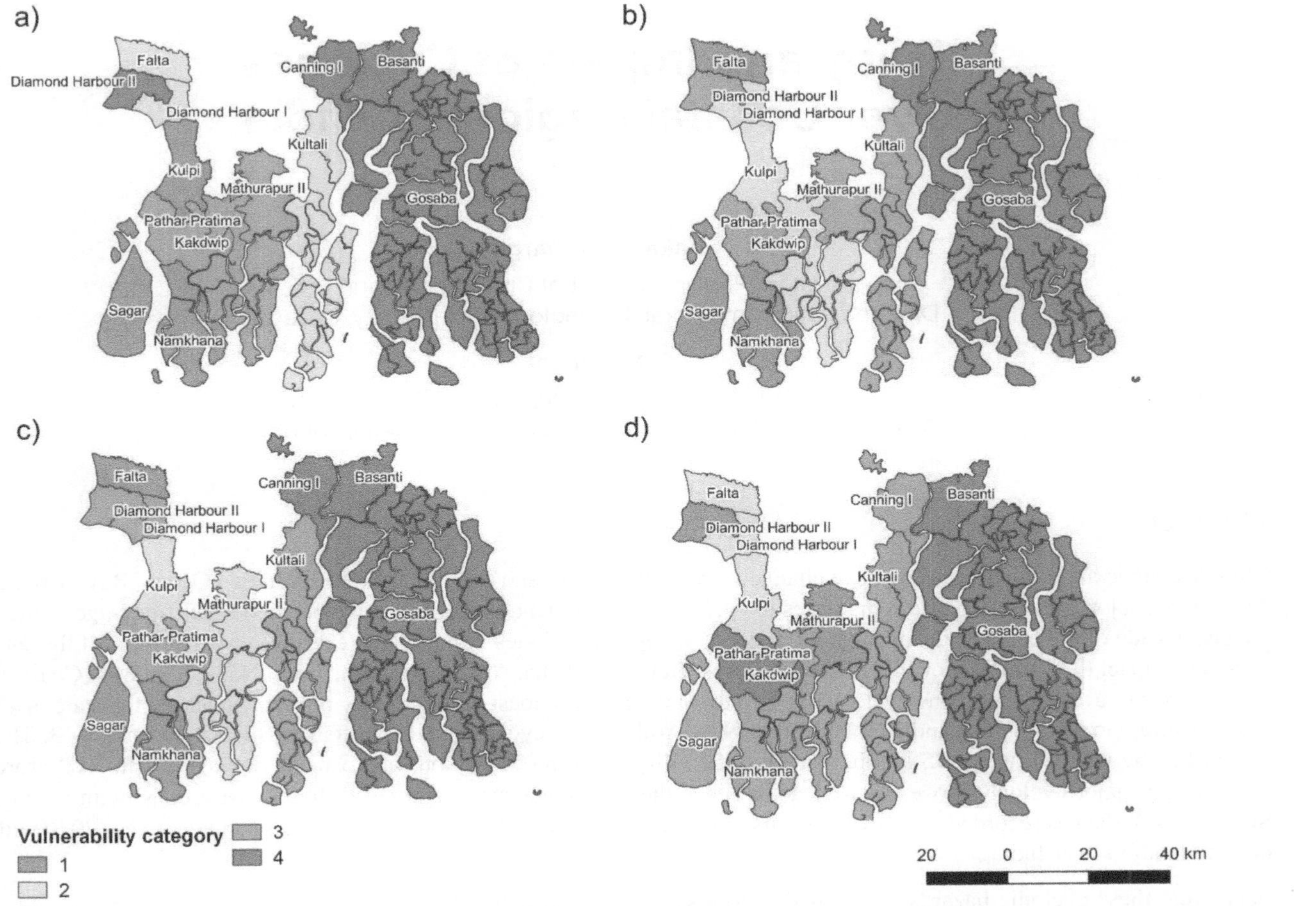

Fig. 23.2 Changing patterns of vulnerability to storm surges in coastal blocks of South Twenty-four Parganas; a: 2004, b: 2009, c: 2014 and d: 2019

2. Boruff, Bryan & Emrich, Chris & Cutter, Susan. (2005). Erosion Hazard Vulnerability of US Coastal Counties. Journal of Coastal Research - J COASTAL RES. 21. 932-942. 10.2112/04-0172.1.
3. Duriyapong, F., & Nakhapakorn, K. (2011). Coastal vulnerability assessment: a case study of Samut Sakhon coastal zone. Songklanakarin Journal of Science and Technology, 33, 469-476.
4. IPCC (Intergovernmental Panel on Climate Change), 2001. Technical summary: climate change 2001: impacts, adaptation, and vulnerability. A Report of Working Group II of the Intergovernmental Panel on Climate Change. Retrieved from http://www.grida.no/climate/ipcc tar/wg2/pdf/wg2TARtechsum.pdf.
5. Joevivek, V., Saravanan, S., & Chandrasekar, N. (2013). Coastal Vulnerability and Shoreline Changes for Southern Tip of India-Remote Sensing and GIS Approach. Journal of Earth Science & Climatic Change, 2013, 1-10.
6. Kumar, A.A., & Kunte, P. (2012). Coastal vulnerability assessment for Chennai, east coast of India using geospatial techniques. Natural Hazards, 64, 853-872.
7. Mahapatra, M., Ratheesh, R., & Rajawat, A. S. (2013). Sea level rise and coastal vulnerability assessment: a review. International Journal of Geology, Earth & Environmental Sciences, 3(3), 67-80.
8. Thieler, E., & Hammar-Klose, E. (2000). National assessment of coastal vulnerability to sea-level rise; preliminary results for the U.S. Gulf of Mexico Coast.

CHAPTER 24

Study and Impact of Cyclones Over Coastal Regions of India

Aakanksha Darge
Assistant Professor, Dept of Civil Engineering
Dr. Babasaheb Ambedkar Technological University, Lonere

S. R. Bhagat
Head of Department, Dept of Civil Engineering
Dr. Babasaheb Ambedkar Technological University, Lonere

Abstract

Climate change and global warming have changed almost the nature and behaviour of North Indian Ocean, Bay of Bengal and Arabian Sea a lot, as witnessed from the series of cyclones with tremendous intensity, havoc wind speed and huge rainfall since the last decade of the century. On average around 2 to 4 tropical cyclones impact India every year, while most of these tropical cyclones impact the east coast of Indian states of West Bengal, Odisha, Andhra Pradesh, Tamil Nadu. The West Coast of India is less prone to cyclones with one cyclone out of 2 to 4 hits the west coast with majority of them attacking the state of Gujarat, Maharashtra, Goa, Karnataka and Kerala. In the 1976, Arabian Sea cyclone made landfall in Saurashtra on June 3. The coast was lashed by wind of up to 175 kmph, damaging 51 villages killing 70 persons and causing damage worth Rs 3 crore. That was the last major cyclone in over four decades. Since then, cyclones have returned to India's west coast with a vengeance. Since 2019, India has recorded eight cyclones, five in the Arabian sea alone, while five cyclones formed in year 2020 with three making landfall over India.

Very often these cyclonic hazards had turned in to disasters affecting the life, livelihood and property of the people. Strong wind, torrential rain, flooding and unusual storm surges accompanied with the cyclones cause severe devastations with the destruction of dwellings, damage to infrastructure and standing crops besides loss of life along the track of its movement and adjacent areas. This study shows Coastal region's exposure to these extreme events, people's perception and human response, adaptations, its risk mitigation, management and preparedness has undergone a sea change in the twenty-first century keeping at pace with the scientific innovations and international guidelines.

Keywords: Climate Change, Coastal Regions, Management, Preparedness

1. History of Cyclones in India

Tropical cyclones are one of the deadliest natural disasters worldwide. The word '*Cyclone*' was coined in 1848 by Henry Piddington, British meteorologist, and is derived from the Greek word '*kuklos*' i.e. the coil of a snake as the air flow of the storm resembles it. The nomenclature of tropical cyclones is different in different parts of the world. In the Atlantic and eastern Pacific, they are known as '*hurricanes*' and in western Pacific as 't*yphoons*. In the Indian region, they are simply known as 'tropical cyclones. Almost all these storms form within 25° latitude on both sides of the equator except over the 5°N to 5°S equatorial region.

The genesis of tropical cyclones over the Indian seas is highly seasonal, with primary maximum in the post-monsoon season (mid- September to December) and secondary maximum during the pre-monsoon season (April and May). A few cyclones form in transitional monsoon months June and September also. On an average about 5–6 tropical cyclones (maximum sustained wind of 34 knots or more) form in the Bay of Bengal every year, of which 2–3 reach severe stage (maximum sustained wind of 48 knots or more). The post-monsoon storms are more devastating in nature. The Bay of Bengal is a potentially active region for the formation of tropical cyclones. There are eight maritime states along the eastern and western coast of India that have deeply suffered from tropical cyclones (Fig. 24.1), out of which the eastern coastal states- Orissa, Andhra Pradesh, Tamil Nadu and West Bengal, have suffered the most (Fig. 24.2).

Fig. 24.1 Cyclone occurrence during 1891–2013

Source: IMD

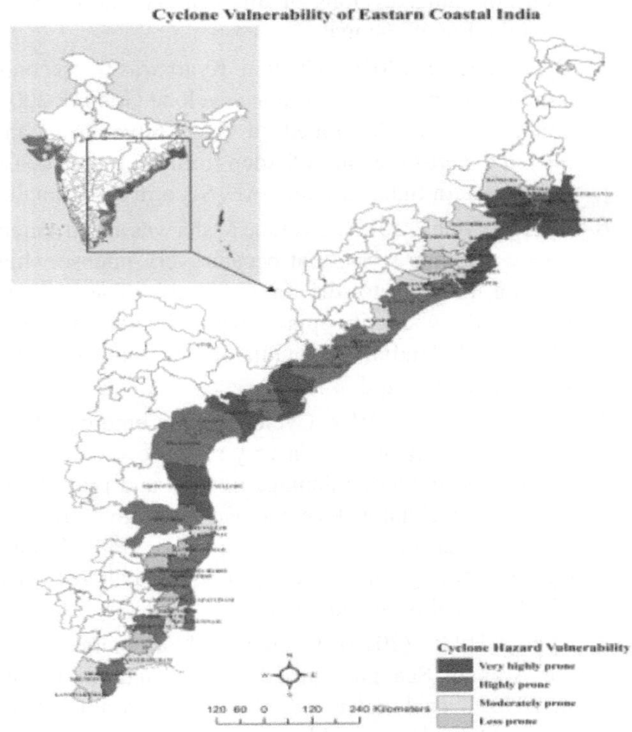

Fig. 24.2 Cyclone vulnerability of Eastern Coastal India

Some of disastrous tropical storms of the 20th century include the Orissa Super Cyclone (October 1999), Kandla cyclone (1998), Porbandar cyclone (1976), Bombay cyclone (1948). The False Point Cyclone (Orissa) which occurred in 1885 was reported to be associated with the highest storm surge of 22 ft. Between 1981 and 2002, the states located in the eastern coast i.e. West Bengal, Odisha, Andhra Pradesh, Tamil Nadu have faced 69, 98, 79 and 54 cyclones respectively (NDMA 2008). Highly prone Orissa faced 98 cyclones which were distributed across the districts of Balasore (32), Cuttak (32), Puri (19) and Ganjam (15). On the other hand, West coastal states have faced much lesser number of cyclones than their Eastern counterparts - Maharashtra (13), Gujarat (28), Kerala (3), Karnataka and Goa 2 each. Therefore, the eastern coast is more prone to Cyclone than the western coast.

2. List of Cyclones Hitted to Indian Coastal Regions

1. Cyclone Tauktae (2021): It is the first cyclonic storm of 2021 that emerged from the Arabian Sea. It hit southern Gujarat on 17 May 2021 and is classified as a Very Severe Cyclonic Storm (VSCS). Warnings were issued for Porbandar, Amreli, Junagarh, Gir Somnath, and coastal Ahmedabad districts while Junagarh, Gir Somnath and Amreli are on alert. Mumbai airport was closed till 2 p.m. on 17 May 2021 after the city experienced rainfall and gusty winds. Mumbai's s Bandra-Worli sea link was closed. Also, five temporary shelters were put up in Mumbai's 24 wards and three NDRF teams are on alert.
2. Cyclone Yaas (2021): The cyclonic storm, Cyclone Yaas, formed in the Bay of Bengal and hit West Bengal and adjoining Odisha coasts in May 2021. The name of the cyclone is given by Oman. Deep Depression over East-central Bay of Bengal intensified into Cyclonic Storm 'Yaas' and about 600 km of Port Blair.
3. Cyclone Nisarga (2020): Cyclone Nisarga was the second pre-monsoon cyclone that emerged from the Arabian Sea. It hit Alibag in Mumbai and weakened in 6 hours. It was the first cyclone to impact Mumbai since Phyan of 2009. The cyclone caused 6 deaths and 16 injuries in Maharashtra.
4. Cyclone Amphan (2020): Cyclone Amphan was a powerful tropical cyclone that led to the destruction of lives and property in the Indian states of Odisha and West Bengal. Cyclone Amphan was the first pre-

monsoon super cyclone of this century that emerged from the Bay of Bengal.

5. Cyclone Kyarr (2019): Cyclone Kyarr was the second-strongest tropical cyclone since cyclone Gonu in 2007. Cyclone Kyarr developed in the Arabian Sea and moved towards the Gulf of Aden from the Indian coast. It hit Western India, Oman, UAE, Socotra and Somalia.
6. Cyclone Maha (2019): Cyclone Maha was an extremely severe cyclonic storm that became very intense while moving parallel to the Indian coast. The cyclone weakened when it approached Gujarat. Cyclone Maha made landfall near Gujarat as a depression that weakened afterward.
7. Cyclone Vayu (2019): Cyclone Vayu emerged from the Arabian Sea and was a very severe cyclonic storm that caused moderate damage to lives and property in the state of Gujarat. Cyclone Vayu was the strongest cyclone that hit the state since the 1998 Gujarat Cyclone. Along with India, cyclone Vayu also affected Maldives, Pakistan and Oman.
8. Cyclon Hikka (2019): Cyclone Hikka emerged from the Arabian Sea and turned intense and hit Oman. In 2019, 4 cyclones emerged from the Arabian Sea– Kyarr, Maha, Vayu, and Hikka.
9. Cyclone Bulbul (2019): Cyclone Bulbul was a very severe cyclonic storm that hit West Bengal in India. It caused huge rainfall, floods, etc. that resulted in the destruction of lives and property. Outside India, it hit Bangladesh
10. Cyclone Fani (2019): Cyclone Fani was the strongest tropical storm that hit Odisha since the 1998 Odisha Cyclone. Cyclone Fani emerged from the Indian Ocean and caused huge destruction of lives and property in Odisha, West Bengal, Andhra Pradesh, and East India. Outside India, it hit Bangladesh, Bhutan and Sri Lanka.

One of the reasons why tropical cyclones are more prone to the bay of bengal is that its surface temperature is more than that of the arabian sea. Tropical cyclones generally need a temperature of around 28 degrees celsius. In the last century, out of the 1019 cyclonic disturbances in the indian subcontinent, 890 were along the eastern coast, and of these, 260 cyclonic disturbances had their landfall along the odisha coast. The cyclones which hit the state in the last two decades are the 1999 super cyclone, phailin 2013 and titli 2018.

Another peculiarity with the Bay of Bengal is that it is known for its potential in generating dangerous high storm tides. When these tides and cyclonic storms coincide, it becomes a major killer. According to the India Meteorological Department, "Out of 10 recorded cases of very heavy loss of life (ranging from about 40,000 to well over 2,00,000) in the world due to tropical cyclones, nine cases were in the Bay of Bengal and the Arabian Sea."

When a cyclone approaches to coast, a risk of serious loss or damage arises from severe winds, heavy rainfall, storm surges and river floods. The effect of a storm surge is most pronounced in wide and shallow bays exposed to cyclones such as in the northern part of Bay of Bengal. On an average, five or six tropical cyclones occur every year, of which two or three could be severe. Most cyclones occur in the Bay of Bengal followed by those in the Arabian Sea and the ratio is approximately 4:1. The incidence of cyclonic storms, with wind speeds between 65 Km/h and 117 Km/h and severe cyclonic storm with wind speeds between 119 Km/h and 164 Km/h, reaching Tamil Nadu and Andhra Pradesh is high during the north east monsoon season ie. October – December, where as the highest annual number of storms, severe storms occur in the Orissa - West Bengal coast.

The yearly distribution of tropical cyclones in the north Indian Ocean indicates large year to-year variations in the frequency of cyclonic disturbances and tropical cyclones, but no distinct periodicity. However, the trend indicates a slight decrease with time. The annual average of cyclonic disturbances in the North Indian Ocean is about 15.7 with a standard deviation of 3.1. The annual number of cyclonic disturbances range from seven in 1984 to twenty three in 1927. The annual average of tropical cyclones has varied from one in 1949 to ten in 1893, 1926, 1930 and 1976.

A severe super cyclonic storm with winds of upto 250 km/hour crossed the coast in Orissa on October 29, 1999. This may have been the worst cyclone of the country in the Orissa region and was responsible for as many as 10,000 deaths, rendering millions homeless and extensive damage to property and environment.

3. Case Study For Cyclone Fani

Extremely Severe Cyclonic Storm Fani was the strongest Tropical Cyclone to stike Odisha, an eastern state of India since 1999 Odisha Super Cyclone. It commenced from west of Sumatra on 26th April from a Tropical depression. Since its conditions were favorable, it rapidly intensified into an Extremely Severe Cyclonic Storm and reached its peak on 3 rd May 2019. The Fani, a rare summer cyclone, hit the Odisha coast close to Puri on 3 rd May 2019 between 0800 and 1000 hours. As reported by Indian Meteorological Department (IMD), the maximum sustained surface wind speed of 175–180 kilometres per hour (kmph) gusting to 205 kmph was observed during landfall at Satpada.

Generally, extremely severe cyclones hit India's east coast in the post-monsoon season (OctoberDecember). IMD data on cyclones that hit India between 1965 and 2017 show that the country had come across 39 extremely severe cyclones in these 52 years. Of these, nearly 60 per cent (23) were between October and December.

Arrival of Fani in the month of May is an unusual time. What makes Cyclone Fani special is its trajectory. Fani started developing around the Equator and moved upwards. The long journey allowed it to gather a lot of moisture and momentum, resulting in strong winds.

Fig. 24.3 Cyclone Fani

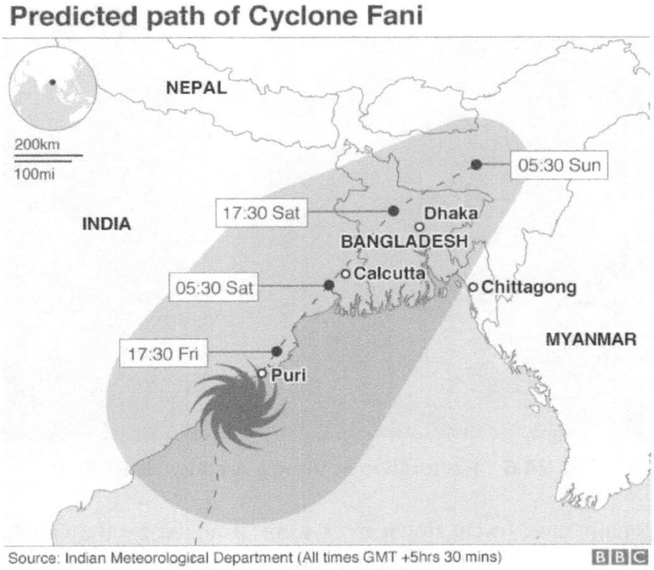

Fig. 24.4 Predicted Path of Cyclone Fani

4. Preparedness

Naval Ships & Aircrafts kept ready by Indian Navy at two air bases to face the aftermath of the storm & help the rescue & relief operation. 300 power boats, 02 Helicopters & many

Fig. 24.5 Effect of Cyclone Fani

Fig. 24.6 Restoration in Village Area by NDRF Team

chain saws to cut down trees were arranged by the Odisha Government. 1.55 million people evacuated towards 9,177 shelters in 24 hrs. 25,000 tourists were evacuated by 23 special trains and 18 buses. All fishing activities were suspended two days prior to the landfall. Twenty Odisha Disaster Rapid Action Force (ODRAF) units, 335 Fire Service units and 25 units of the National Disaster Response Force (NDRF) were deployed.

District and Block Emergency Operation Centres were activated 24 × 7. Nearly 4.5 lakh polythene sheets (temporary shelter materials) were pre-positioned at district/ sub-district levels and 1.5 lakh kept ready for airdropping after the cyclone. Nearly 600 pregnant women were shifted to Maa Gruhas/ delivery points before the landfall. More than 45,000 volunteers, 2000 emergency workers, 100,000 officials, youth clubs & other civil society organisations such as National Disaser Response Force(NDRF) , Odisha Disaster Rapid Action Force (ODRAF), Panchayat Raj Institute (PRI) agencies worked together round the clock to evacuate million people in 24 hrs.

5. Disaster Effect and Impacts

Approximately 72 people have been killed by Fani in India; 64 in Odisha. The cyclone adversely affected electricity supply and telecommunication in several coastal areas of Odisha, and to a lesser extent, of West Bengal. Puri and Khordha district in Odisha were the worst hit. The Jagannath Temple in Puri suffered minor damage, the repairing cost were estimated to be ₹51 million. The KIIT University also suffered a damage of about ₹300 million. Total damage in Odisha were estimated at ₹120 billion, mostly in property damage and the relief. After the cyclone, Odisha required ₹170 billion for rebuilding the infrastructure. There were massive environmental devastation, as well.

6. Case Study for Cyclone Nisarga

Nisarga a Category-1 tropical cyclone made land- fall in Alibag (Western coast, Maharashtra) on 3rd June, 2020 with a speed of 102 km/h. It was categorized as a severe cyclonic storm on IMD scale and under SSHWS as. It caused 6 fatalities and an estimated economic loss of 665$ USD. It was the first cyclone to hit Mumbai after Cyclone Phyan, 2009. In India where 60% of population depends on agriculture, such natural calamities cause great psychological stress along with economic loss.

6.1 Preparedness

On 1 June, Union Home Minister Amit Shah, held a preliminary review meeting with officials of National Disaster Management Authority, National Disaster Response Force (NDRF), India Meteorological Department and the Indian Coast Guard. On the same day, 33 team of NDRF were deployed in the coastal region of both the state. Fishermen from Maharashtra were alerted to return from the sea. As a precaution, 100,000 people were evacuated ahead of the storm.

6.2 Disaster Effects and Impacts

Nisarga pounds state of which Raigad district of Maharashtra is the worst affected district. Four people were killed and severe others injured as Cyclone Nisarga slammed the coastal districts of the state with wind speeds of up to 120 km/h. Raigad and Palghar bore the maximum brunt of the storm experiencing strong winds, heavy rainfall and raging sea surge. Tidal waves measuring up to 6-8 feet lashed parts of the coastal areas of Maharashtra.

Tin roofs of houses blew away in some places and several trees and electric poles uprooted. In some districts, officials reported extensive damages of crops. Power supply was disrupted to more than 25 lakh consumers in four districts of Raigad, Palghar, Thane and Pune. Besides some casualties, lot of destructions of public and private properties, there are reports of an extensive damage to the crops in districts like Raigad, Palghar, Ratnagiri, and also many parts of the district plunged into darkness after disruption of power supply.

6.3 Early Warning System

As tropical cyclone cannot be tamed to reduce their adverse effects, one has to learn to live with them. Effective Cyclone Disaster Prevention and Mitigation Plan require:

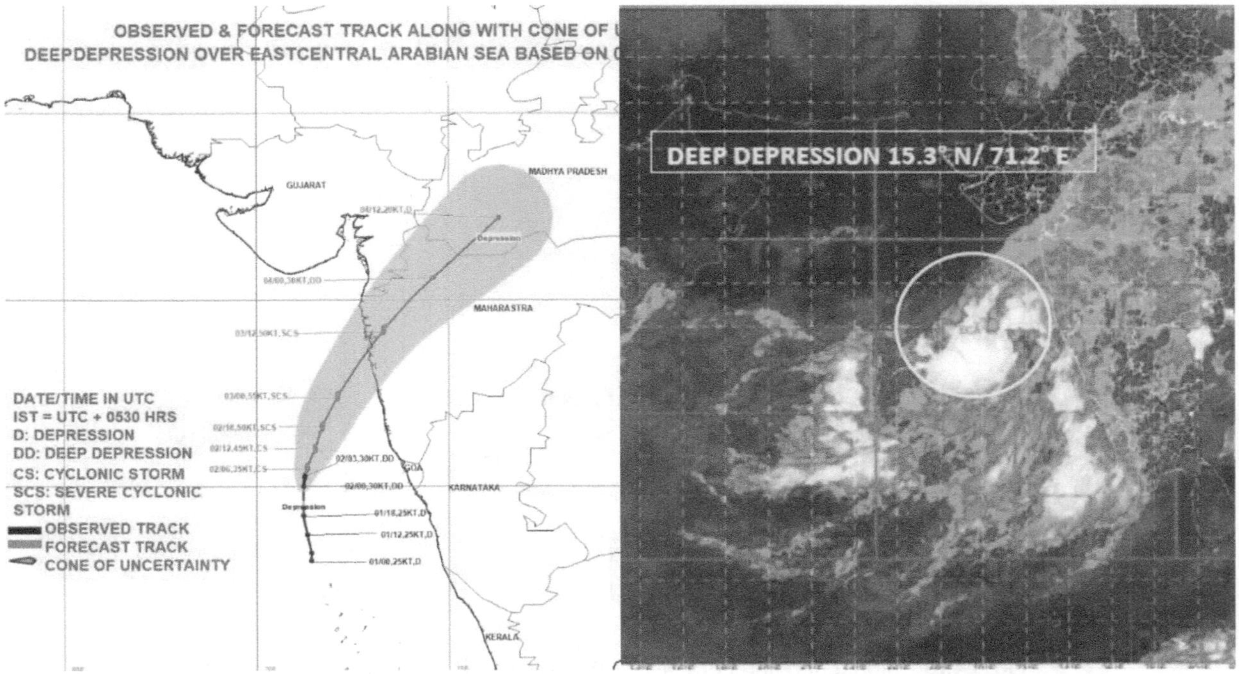

Fig. 24.7 Predicted path of Cyclone Nisarga

Fig. 24.8 Tree fallen on Kaccha House Due to Nisarg Cyclone

- Hazard analysis
- Vulnerability analysis.
- Early Warning and Mitigation
- Community preparedness and planning at all levels to meet the exigencies.

India Meteorological Department (IMD) monitors and predicts CDs over NIO and provides early warning services for management of the cyclone. Various components of early

Fig. 24.9 NDRF Team Working After Cyclone

warning system for a cyclone include (i) monitoring and prediction, (ii) warning organization, (iii) warning generation, presentation & dissemination, (iv) coordination with disaster management agencies,(vii) public education & reaching out and (viii) post-event review. The entire early warning system of cyclone is depicted in Fig. 24.10.

6.4 Forecast System

Global Forecast System (GFS T-382): 7 days

Regional Forecast System (WRF): 3 days

Mesoscale Forecasting System (WRF, ARPS & HWRF-high resolution): 48 hours

Nowcasting: Processing of DWR observations

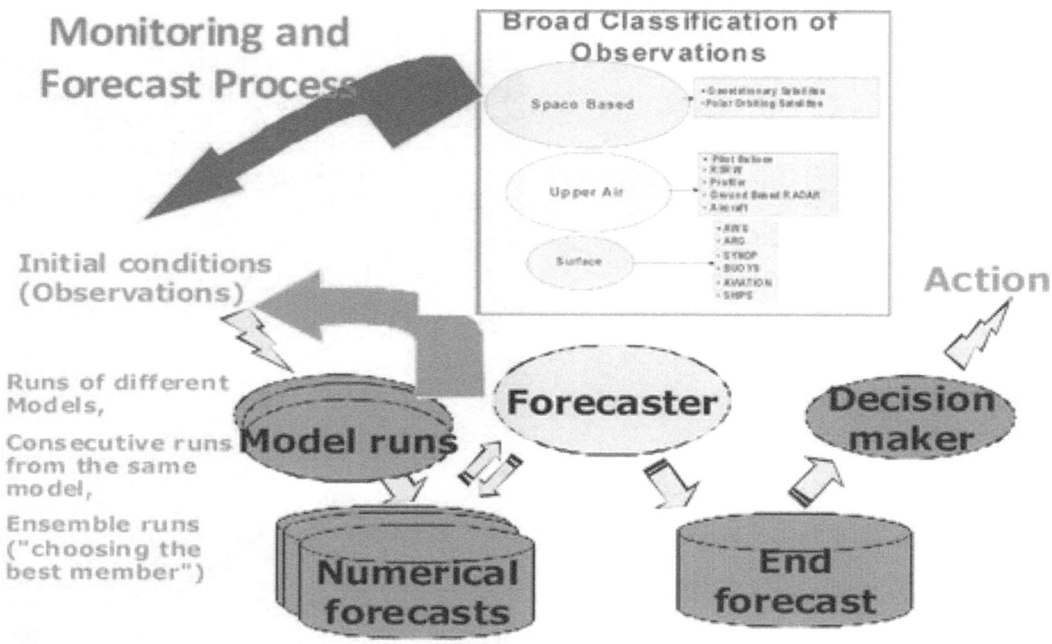

Fig. 24.10 Monitoring and Forecasting Process of Tropical Cyclones over NIO

India Meteorological Department (IMD) has developed state of art tools for cyclone warning services, and has demonstrated its capability to provide early warning for Cyclones with high precision and has earned accolades globally and nationally for very effective state of art early warning system for monitoring and prediction of cyclones. The cyclone forecast accuracy has significantly improved in recent years as has been demonstrated during cyclones Phailin (2013), Hudhud (2014), Vardah (2016), Titli (2018), Fani& Bulbul (2019) and Amphan, Nisarga&Nivar (2020). During recent years, the loss of life due to cyclone has been drastically reduced being limited to double digit figures.

Further, enhancing the accuracy of weather forecasts and their more effective & timely dissemination in the ensuing years is a continuing process. For this purpose, under the Umbrella Scheme entitled 'Atmospheric & Climate Research – Modeling Observing Systems & Services (ACROSS), a sub-scheme has been planned viz., 'Upgradation of Forecast Services'. The various components of this sub-scheme include;

1. Development of an advanced operational forecast system
2. Upgradation and sustenance of communication systems for weather services
3. Capacity building & outreach etc.
4. In order to cater to the needs of Cyclone Warning Services and Marine weather services, there are seven established Cyclone Warning Centers covering the east & west coasts of our country. Among these, three are Area Cyclone Warning Centres (ACWCs) located at Chennai, Mumbai and Kolkata and remaining four are Cyclone Warning Centres (CWCs) located at Ahmadabad, Thiruvananthapuram, Visakhapatnam and Bhubaneswar. Area of responsibility of ACWCs and CWCs is shown in the Table 24.1.

This along with further improvements in the observational network and numerical modeling capability planned under ACROSS are also expected to increase the accuracy of weather forecasts. Also, Indian National Centre for Ocean Information Services (INCOIS) has set up Storm Surge Early Warning System (SSEWS) for the Indian coasts in active collaboration with IMD. Prime objective of this service is to save the lives of coastal community by forecasting cyclone induced storm surges and inundation extent.

Further, the Government of India has initiated the National Cyclone Risk Mitigation Project (NCRMP) with a view to address cyclone risks in the country. The overall objective of the Project is to undertake suitable structural and non-structural measures to mitigate the effects of cyclones in the coastal states and UTs of India. National Disaster Management Authority (NDMA) under the aegis of Ministry of Home Affairs (MHA) will implement this Project in coordination with participating State Governments. The Project has identified 13 cyclone prone States and Union Territories (UTs), with varying levels of vulnerability.

Table 24.1 Area of responsibility of ACWCs and CWCs

Centre	Coastal area*	Maritime State/UT
ACWC Kolkata	State: West Bengal UT: Andaman & Nicobar Islands	State: West Bengal UT: Andaman & Nicobar Islands
ACWC Chennai	State: Tamil Nadu UT: Puducherry	State: Tamil Nadu UT: Puducherry
ACWC Mumbai	State: Maharashtra & Goa	State: Maharashtra & Goa
CWC Thiruvananthapuram	State: Kerala & Karnataka UT: Lakshadweep	State: Kerala & Karnataka UT: Lakshadweep
CWC Ahmadabad	State: Gujarat UT: Dadra-Nagar Haveli- Daman-Diu	State: Gujarat UT: Dadra-Nagar Haveli- Daman-Diu
CWC Visakhapatnam	State: Andhra Pradesh	State: Andhra Pradesh
CWC Bhubaneshwar	State: Odisha	State: Odisha

7. Conclusion

The present study has focused on the historical background of number of cyclones hitted to Coastal regions with meteorological variability, India's preparedness and awful aftermath. The discussion helps to understand that the probabilities of intensification of a cyclonic storm into a severe or more intense cyclone over the northern Bay of Bengal and Arabian Sea is maximum during the monsoon season (June to September), differs from the intensification period of the cyclones that originate over the southern BoB and AS mainly during pre-monsoon (March to May) and post-monsoon (October to December) seasons. Generally, there are two possible ways of obtaining the complete resilience against such intense Cyclones, namely, the pre-storm preparedness plans understanding the risk based on early warning and also resilient infrastructure. India has already abled to achieve its first goal. IMD has made the work easier for the Indian government. But it is quite impossible to manage the second goal, resilient infrastructure for every citizen in a highly populous country like India. For that reason, the government has to build sufficient cyclone and flood shelters so that evacuated people get protection during the storm event and the death toll can be reduced.

REFFERENCES

1. "Cyclone Warning in India Standard Operation Procedure", By India Meteorological Department, March 2021.
2. "Disaster Management in India", By Ministry of Home Affairs, Govt. of India, May 2011.
3. Gautam Kumar Das "Nisarga - A Debutant Cyclone Ravaged Mumbai and Maharashtra" Article From Frontier, June 6, 2020.
4. Parna Sarkhel (2019),"A Review of Cyclone and its Impact on the Coastal Belts of Odisha", International Journal of Engineering Research & Technology (IJERT), PP 759-762, Vol. 8 Issue 05, May-2019.
5. Poorva Jage (2020), " A Study of Psychological Impact of Recent Natural Disaster 'Nisarg' and Socio-Economic Factors Associated with It on People in Coastal Maharashtra", National Journal of Community Medicine, Volume 1, Issue 11, November 2020.
6. Soumen Chatterjee (2020) "Analytical Study of North Indian Oceanic Cyclonic Disturbances with Special Reference to Extremely Severe Cyclonic Storm Fani: Meteorological Variability, India's Preparedness with Terrible Aftermath", Natural Hazards and Earth System Sciences, Sept 23, 2020.

Part 4

MANAGING RISKS OF DROUGHT

CHAPTER 25

Machine Learning Tools and Techniques for Prediction of Droughts

Rashmi Nitwane
Research student, Dr. B. A. M. University, CS & IT
Vaishali D. Bhagile
Research Guide, Dr. B. A. M. University, CS & IT Department
R. R. Deshmukh
Research Guide, Dr. B. A. M. University, CS & IT Department

Abstract

Due to climate change and global warming there is increase in occurrences of natural disasters in recent times. Droughts are one of the worst disasters that has long lasting effects on the food security and water resources of the country. The onset of drought is slow and recovering from a drought situation takes considerable amount of time and resources. Development of drought prediction tools is still in early stages. Since drought prediction is dependent on various climatic factors therefore there is need for integration of various technologies. With the advent of technologies like data science, machine learning and artificial intelligence prediction models are developed for creating early warning systems so as to plan ahead and reduce the loss incurred by this natural disaster. The existing prediction tools have not achieved required accuracy for proper mitigation planning. In this article we discuss the advantages and challenges of integrating machine learning tools with remote sensing techniques using the climatic data observed over the area for particular time span.

Keywords: convolutional neural network, generative adversarial network, markov chain models, recursive multistep neural network, Deep belief network, Empirical Mode Decomposition

1. Introduction

Droughts have now become a Global concern as different regions experience droughts at different timescales and with different intensity. There is a need of prediction tools since drought intensity and the losses increases with the duration of time and therefore the success of mitigation plans depends upon timely information on its onset and propagation in terms of temporal and spatial extent [3][4]. This information is obtained with effective continuous drought monitoring using drought indices. In various parts of the world Drought monitors are used [2]. These drought monitors are not forecasters but they analyze drought conditions like spatial extent , intensity etc.[38] In India Agricultural drought monitoring is done by Aridity Anomaly Index .This index is not applicable to long term forecasting as well as for multi seasonal events also for Meteorological drought events Standard Precipitation Index is used in India [39]. The main challenge in the development of Integrated drought monitor using remote sensing and machine learning algorithms in India is the satellite monitoring of the earth atmosphere which gives either high temporal resolution (geostationary) or high spatial resolution (low-orbit) but never both [1] moreover state of art machine learning algorithms should be developed for the accurate drought mapping as well as analyzing remote sensing datasets. Several studies have proposed drought monitoring tools and forecasting methods to predict and quantify the risk associated with the recurrent occurrences of drought.

2. Machine Learning for Remote Sensing

Remote Sensing typically gives us different temporal, spatial and spectral resolution data [5]. Apart from multispectral data hyperspectral imaging systems with very narrow bands are employed for identifying specific spectral signature for agriculture and surveillance application for analysis[6][7]. We can define the flow of machine learning process by given diagram.

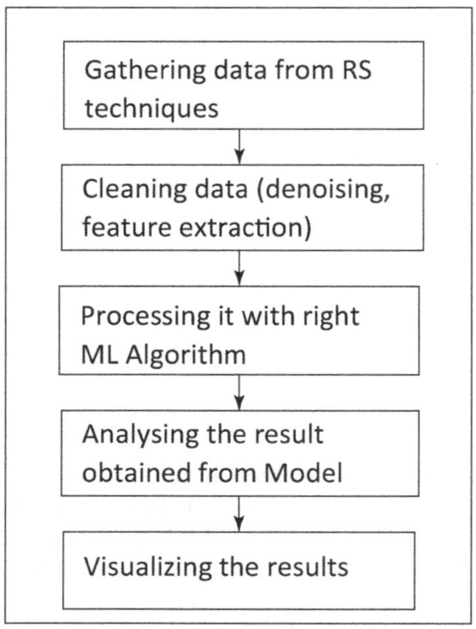

Fig. 25.1 Machine learning Process

An integrated drought monitoring system can be divided in to five essential components

(i) Determination of applicable climate indicators and resultant trigger level.
(ii) Identification of data requirement and its sources.
(iii) Acquisition and analysis of reliable data with appropriate tools.
(iv) Synthesis of data and generating useful practical products
(v) Information dissemination

A useful drought indicator needs to look at not only climate factors but the soil, water impact information also, here it is important to note that the major gap in our drought tracking lies in the area of impact estimation and documentation. [8] [9] [10]

2.1 Challenges with Applying Machine Learning Techniques in Remote Sensing

Despite its great potential Machine learning is not used for many remote sensing task directly. Some of the challenges are as listed below.

(i) [8][9] Some Remote sensing images especially hyperspectral contains hundreds of bands that can cause a small patch to be a really large data cube also there is spectral mixing in hyperspectral datasets.
(ii) [10]The high dimensionality of Remote sensing data can provide rich information and complex data model but it is expensive and difficult to create sufficient amount of labeled data for reliable supervised learning.
(iii) [11][21] Atmospheric noise and interference introduces error and variance in the acquired training data.
(iv) [12][13] Without proper feature extraction tool overfitting can occur in the learned model which eventually affects the performance and reliability of the model.

In the next section we have reviewed various researchers who have used artificial intelligence, machine learning strategies for drought prediction in their respective regions using remote sensing datasets or historical meteorological data.

3. Literature Survey

[Soumyabrata et.al] [14] provided an overview of recent developments in machine learning and remote sensing using examples of ground-based image analysis. They focused attention on the linear Dimensionality Reduction (DR) techniques because of their lower computational complexity and simple geometric interpretation. They concluded that because of high dimensionality of data it is expensive to use raw data for analysis. They also provided the experimental results in segmentation, classification and denoising of sky/cloud images. [15][16][17][18][19].

[Junfei chen et.al][23] in his study proposed a short term drought prediction model based on deep belief networks (DBN) and evaluated the performance of deep belief network by comparing it with the back propagation neural network model. They concluded that DBN model outperforms the neural network model and is a useful tool for short term drought prediction.

[Lohani and Loganathan][20][22]used a non homogeneous Markov chain model to characterize the stochastic behavior of drought and an early warning system in the form of decision tree enumerating is proposed for drought management.

[Yang et.al][24][25][26] proposed chaotic Bayesian method based on multiple criteria for decision making to forecast non linear hydrological time series.

[Mishra and Desai][36][37][40] compared linear stochastic model (ARIMA/SARIMA), recursive multistep neural network (RMSNN) and Direct multistep neural network (DMSNN) for drought forecasting by using SPI series as drought index in the Kansabati River basin in India.

[Hinton][27][28] proposed recently a Deep Belief network which is a probabilistic generative model in which bottom layer is observable and the multiple hidden layers are created by sacking multiple restricted boltzmann machine (RBM) on top of each other .he derived a way to perform a fast greedy algorithm of deep belief network one layer at a time, with the top layer forming undirected bipartite graph.

[John Nay][29][30] created a user friendly predictive software tool that increases the capacity of data scarce agricultural systems to prepare for and respond to drought in future his tool predicts future vegetation health vales and a high spatial resolution using open source tools .this tool applies gradient boosted machine model to moderate resolution imaging spectro- radiometer (MODIS) datasets openly available on NASA land processes distributed active archive center server . His model learn potentially complex relationships between past remotely sensed variables and future vegetation health as measured by EUI. He applied his tool on two locations Srilanka and California and concluded that gradient boosted model effectively handled the missing data and was able to obtain sufficient output of sample predictive power at both location.

[Agana & Homaifar][31][32][33] observed that different architectures of artificial neural networks have shown great promise in modeling non-linear time series however the issue of nonconvex optimization ensues when two or more hidden layers are required. They proposed hybrid method composed of Empirical Mode Decomposition (EMD) and Deep Learning approach .They applied the model to predict different time scale drought indices across the Gunnison River Basin in Upper Colarado. They concluded that hybrid EMD-DBN provided better forecasting results.

[Hao,Z et.al][34] provided a synthesis of drought prediction based on statistical, dynamical and hybrid methods. The statistical prediction methods uses empirical relationships of historical records taking different influencing factors as predictors, for statistical prediction techniques like Time-series modeling. Linear Regression Model, Markov chain Model and conditional probability model is used. The success of statistical prediction depends upon the predictor values. In the dynamical methods of drought prediction seasonal climate forecast, multi model ensemble forecast and hydro climatic forecast techniques are used and in hybrid methods of drought prediction there is comparison and combination of statistical and dynamical methods. Their focus was on the drought severity at the seasonal time scale. They reviewed drought prediction methods and challenges and concluded that still more work is to be done in this area of forecasting droughts with high precision.

[X.X.Zhu et.al] [35] focused his attention on the deep learning algorithms for Remote Sensing (RS) big data analysis. The Deep Learning techniques were originally rooted in the machine learning fields for classification and recognition tasks and they have recently appeared in Geoscience and RS community. The data warehouse of RS images are increasing daily, including images with different spectral and spatial resolutions. They observed that if sufficient training sample set is available, a deep network turns out to be Supervised approach also a Deep learning algorithm adopts a deep architecture to deal with the complicated relationship between original data and specific class label. They concluded that from the four perspectives i.e. i) image preprocessing, ii) pixel based classification, iii) target recognition and iv) scene understanding Deep learning techniques have significant success in the areas of target recognition and scene understanding.

4. Conclusion

We observed that researchers are having varying opinions on the use of AI & ML with Remote sensing. The researchers tried to implement early warning systems by understanding the stochastic behavior of drought while others compared between the two existing models success ratio on the river basins some other researchers proposed short term drought prediction tool like seasonal drought prediction model based on Bayesian Framework to characterize hydrological droughts whereas for long term drought prediction Wavelet Linear Genetic Programming Model (WLGP) was proposed.

It is important to note that researchers demonstrated that classical Linear Genetic Programming Model is unable to learn the non-linear structure of drought phenomena which is more than the duration of three months. Moreover, the remotely sensed data should be properly denoised and tools for much better feature extraction should be implemented as without proper feature extraction the accuracy of the prediction suffers.

Lately the focus has shifted to the hybrid methods of the predicting drought but its efficiency and accuracy needs to be tested on various grounds. However, the Deep learning model which is relatively new but promising technology proved to be effective in certain cases. but still the development of the prediction tools using various drought indices is in initial stages and researchers are continually experimenting to find more and more accurate methods

5. Acknowledgement

This work is supported by Department of Science & Technology under the funds for Infrastructure under Science & Technology (DST-FIST) with sanction no. SR/FST/ETI- 340/2013 to Department of Computer Science

& Technology., Dr. Babasaheb. Ambedkar. Marathwada University. The authors would like to thank Department and University authorities for providing the infrastructure and necessary support for carrying out the research.

6. References

1. Soumayabrata Dev, Bihan Wen, Student member IEEE, Yee hui Cee,Stefan Winkler, Senior member, IEEE. "Machine learning techniques and applications for ground based image analysis", ,IEEE Geoscience and Remote sensing magazine, June 2016.
2. Mark. D. Svoboda, Michael Hayes, Donald. A. Wilhite, Tsegaye Tadesse, National Drought Mitigation center, University of Nebraska, Lincoln "Recent advances in Drought monitoring"
3. Schmidhuber J. Deep learning in neural networks: An overview.Neural Networks, 2015; 61: 85–117.
4. Kamilaris A, Prenafeta-Boldú F X. Deep learning in agriculture: A survey. Computers and Electronics in Agriculture, 2018; 147: 70–90.
5. Simonyan K, Zisserman A. Very deep convolutional networks for large-scale image recognition. Cornell University Library, Available at: https://arxiv.org/abs/1409.1556.
6. Simonyan K, Zisserman A. Two-stream convolutional networks for action recognition in videos. Advances in Neural Information Processing Systems, 2014; 1–4: 568–576.
7. Fan Y, Qian Y, Xie F, Soong F K. TTS synthesis with bidirectional LSTM based recurrent neural networks. Proc. Interspeech, 2014; pp. 1964–1968.
8. Alex G, Schmidhuber J. Framewise phoneme classification with bidirectional LSTM and other neural network architectures. Neural Networks, 2005; 18(5-6): 602–610.
9. Alec R, Metz L, Chintala S. Unsupervised representation learning with deep convolutional generative adversarial networks. Cornell University Library, Available at: https://arxiv.org/abs/1511.06434.
10. Suárez P L, Sappa A D, Vintimilla B X. Infrared image colorization based on a triplet dcgan architecture. IEEE Conference on Computer Vision and Pattern Recognition Workshops (CVPRW), 2017.
11. Jin J Q, Fu K, Zhang C H. Traffic sign recognition with hinge loss trained convolutional neural networks. IEEE Transactions on Intelligent Transportation Systems, 2014; 15(5): 1991–2000.
12. Haykin S S. Neural Networks and Learning Machines. Pearson Schweiz Ag, 2008.
13. M. Galun, E. Sharon, R. Basri, and A. Brandt, "Texture segmentation by multiscale aggregation of filter responses and shape elements," in Proc. International Conference on Computer Vision (ICCV), 2003, pp. 716–723.
14. S. Dev, Y. H. Lee, and S. Winkler, "Categorization of cloud image patches using an improved texton-based approach," in Proc. IEEE International Conference on Image Processing (ICIP), 2015.
15. J. Arenas-Garcia, K. Petersen, G. Camps- Valls, and L.K. Hansen, "Ker-nel multivariate analysis framework for supervised subspace learning: A tutorial on linear and kernel multivariate methods," IEEE Signal Processing Magazine, vol. 30, no. 4, pp. 16–29, July 2013.
16. C. Munyati, "Use of principal component analysis (PCA) of remote sensing images in wetland change detection on the Kafue Flats, Zambia," Geocarto International, vol. 19, no. 3, pp. 11–22, 2004.
17. R. Doerffer and D. Murphy, "Factor analysis and classification of remotely sensed data for monitoring tidal flats," Helgoländer Meere-suntersuchungen, vol. 43, no. 3-4, pp. 275–293, Sept. 1989.
18. T. V. Bandos, L. Bruzzone, and G. Camps- Valls, "Classification of hyperspectral images with regularized linear discriminant analysis," IEEE Transactions on Geoscience and Remote Sensing, vol. 47, no. 3, pp. 862–873, March 2009.
19. U. Schumann, R. Hempel, H. Flentje, M. Garhammer, K. Graf, S. Kox, H. Losslein, and B. Mayer, "Contrail study with ground-based cameras," Atmospheric Measurement Techniques, vol. 6, no. 12, pp. 3597–3612, Dec. 2013.
20. A. Chatterjee, A. M. Michalak, R. A. Kahn, S. R. Paradise, A. J. Braverman, and C. E. Miller, "A geostatistical data fusion technique for merging remote sensing and ground-based observations of aerosol optical thickness," Journal of Geophysical Research, vol. 115, no. D20, Oct. 2010.
21. F. Yuan, Y. H. Lee, and Y. S. Meng, "Comparison of cloud models for propagation studies in Ka-band satellite applications," in Proc. IEEE International Symposium on Antennas and Propagation, 2014, pp. 383–384.
22. Tao Liu. A novel text classification approach based on deep belief network. In International Conference on Neural Information Processing, pages 314 {321. Springer, 2010.
23. Ren Zhang, Furao Shen, and Jinxi Zhao. A model with fuzzy granulation and deep belief networks for exchange rate forecasting. In 2014 International Joint Conference on Neural Networks (IJCNN), pages 366-373. IEEE, 2014.
24. Junfei Chen, Qiongji Jin, and Jing Chao. Design of deep belief networks for short- term prediction of drought index using data in the huaihe river basin. Mathematical Problems in Engineering, 2012,
25. Takashi Kuremoto, Shinsuke Kimura, Kunikazu Kobayashi, and Masanao Obayashi. Time series fore-casting using a deep belief network with restricted boltzmann machines. Neurocomputing, 137: 47–56, 2014.
26. Chongli Di, Xiaohua Yang, and Xiaochao Wang. A four-stage hybrid model for hydrological time series forecasting. PloS one, 9(8): e104663, 2014.
27. Maarten Jansen and Adhemar Bultheel. Asymp-totic behavior of the minimum mean squared error threshold for noisy wavelet coe cients of piecewise smooth signals. IEEE transactions on signal process-ing, 49(6): 1113–1118, 2001.
28. Norden E Huang, Zheng Shen, Steven R Long, Manli C Wu, Hsing H Shih, Quanan Zheng, Nai-Chyuan Yen, Chi Chao Tung, and Henry H Liu. The empirical mode decomposition and the hilbert spectrum for nonlinear and non-stationary time series analysis. In Proceed-ings of the Royal Society of London A: Mathematical, Physical and Engineering Sciences, volume 454, pages 903 {995. The Royal Society, 1998.

29. John Nay, Emily Burchfield & Jonathan Gilligan (2018) A machine-learning approach to forecasting remotely sensed vegetation health, International Journal of Remote Sensing, 39:6, 1800- 1816, DOI: 10.1080/01431161.2017.1410 296
30. Burchfield, E., Nay, J. J., and Gilligan, J.: APPLICATION OF MACHINE LEARNING TO THE PREDICTION OF VEGETATION HEALTH, Int. Arch. Photogramm. Remote Sens. Spatial Inf. Sci., XLI-B2, 465–469, https://doi.org/10.5194/isprs-archives- XLI-B2-465-2016, 2016.
31. Domenico Labate, Fabio La Foresta, Gianluigi Occhiuto, Francesco Carlo Morabito, Aime Lay-Ekuakille, and Patrizia Vergallo. Empirical mode decomposition vs. wavelet decomposition for the extraction of respiratory signal from single-channel ecg: A comparison. IEEE Sensors Journal, 13(7): 2666{2674, 2013.
32. T. van Zyl, Big Data: Techniques and Technolo-gies in Geoinformatics, chapter 7 Machine Learning on Geospatial Big Data, pp. 133–148, CRC Press, 2014.
33. Agana, N.A.; Homaifar, A. EMD-Based Predictive Deep Belief Network for Time Series Prediction: An Application to Drought Forecasting. *Hydrology* **2018**, *5*, 18. https://doi.org/10.3390/hydrology5010018
34. Hao, Z., AghaKouchak, A., Nakhjiri, N. *et al.* Global integrated drought monitoring and prediction system. *Sci Data* **1,** 140001 (2014). https://doi.org/10.1038/sdata.2014.1
35. X. X. Zhu et al., "Deep Learning in Remote Sensing: A Comprehensive Review and List of Resources," in IEEE Geoscience and Remote Sensing Magazine, vol. 5, no. 4, pp. 8–36, Dec. 2017, doi: 10.1109/MGRS.2017.2762307.
36. Mishra, Ashok & Desai, V.. (2005). Drought forecasting using stochastic models. Stochastic Environmental Research and Risk Assessment. 19. 326- 339. 10.1007/s00477-005-0238-4.
37. Mishra, Ashok & Desai, V.. (2006). Drought forecasting using feed-forward recursive neural network. Ecological Modelling. 198. 127–138. 10.1016/j.ecolmodel.2006.04.017.
38. Rhee, Jinyoung & Im, Jungho & Park, Sumin. (2016). DROUGHT FORECASTING BASED ON MACHINE LEARNING OF REMOTE SENSING AND LONG-RANGE FORECAST DATA. ISPRS - International Archives of the Photogrammetry, Remote Sensing and Spatial Information Sciences. XLI-B8. 157-158. 10.5194/isprsarchives-XLI-B8-157-2016.
39. https://imdpune.gov.in
40. M. Pesaresi, V. Syrris, and A. Julea, "Benchmarking of the Symbolic Machine Learning classifier with state of the art image classification methods - application to remote sensing imagery," JRC Technical Report EUR 27518 EN, EC, JRC, IPSC, 2015.

CHAPTER 26

Assessment of Rainfall Variability and Drought Features in Bundelkhand Region of Madhya Pradesh, India

R. V. Galkate
National Institute of Hydrology, Central India Hydrology Regional Centre,
Bhopal, Madhya Pradesh
Corresponding author: rgalkate@yahoo.co.in
9977000340

R. P. Pandey
National Institute of Hydrology,
Roorkee, Uttarakhand.

Shalini Yadava
Rabindranath Tagore University,
Bhopal, Madhya Pradesh

Abstract

Drought is a creeping disaster that can be referred to as a regional occurrence of relatively below normal water availability, either in the form of precipitation, river flow, surface and groundwater storage for a certain period. The Bundelkhand region of Madhya Pradesh in India has always been in attention due to many regional issues such as water scarcity, frequent droughts, lack of economic and development activities which are posing threat to agricultural productivity, livelihood and social aspects in the region. The present study has been carried out to analyze the variability of rainfall in the Bundelkhand region using suitable indicators and its relationship with the drought frequency, duration, and magnitude using long-term rainfall data from 1980 to 2017 of five stations Sagar, Damoh, Panna, Tikamgarh and Chhatarpur. The Durbin Watson test revealed good autocorrelation in rainfall time series with slight negative bias at Damoh and Chhatarpur.

The standard deviation, coefficient of variation and interquartile range (IQR) analysis shows very high rainfall variability at Sagar, Damoh and Chhatarpur and high spread in rainfall data as compared to Tikamgarh and Panna. The Bundelkhand region has a very high rainfall concentration in four monsoon months. The average Precipitation Concentration Index (PCI) of the region is 31.9% which indicates a very high monthly concentration in the Bundelkhand region. The proportion of negative Standard Rainfall Anomalies (SRA) indicating dry years have been seen very high in the region and found ranged from -1.41 at Damoh (year 2017) to -2.41 at Tikamgarh (Year 1989) indicating high interannual fluctuations in rainfall. The stations Sagar, Damoh and Chhatarpur where the rainfall variability has been observed very high, are also appeared as severe drought-prone stations with a very high drought frequency causing 1 drought year after every 3 to 5 years. Thus high variability, negative rainfall anomalies and high PCI values of rainfall are observed indicative of severe and frequent drought possibilities. This indicates the seriousness of droughts and the need of developing and managing water resources and adopting suitable conservation practices for drought mitigation.

The Standardised Precipitation Index (SPI) analysis was carried out for assessment of drought frequency, severity, and magnitude of dry events on 1, 3, 6, 9, 12-month time scales. The drought severity was found higher in the eastern part covering the Chhatarpur and Panna district of the Bundelkhand region. The analysis has shown more moderate and severe dry events on a 6-month time scale as compared to remaining other time scales whereas more extreme dry events were evident on a 12-month time scale. It was seen that evidence of drought also increased with the increase of the SPI time scale. However extreme drought events were seldom in the region on a 1-month time scale. The drought frequency has been seen as very high at all five stations of Bundelkhand. The probability distribution analysis indicated these all stations as drought-prone. Similar results are seen for SPI analysis which has shown that these stations were experienced a significant drought on a time scale of 6, 9, 12-month. However, SPI outcomes seem to be is more sensible to severe drought events than the probability distribution analysis and gives the best result using only rainfall data without other meteorological parameters. The assessment of rainfall variability and its relation with drought frequency and magnitude is very important to comprehend and monitoring water scarcity situations for decision making and adopting appropriate strategies for mitigation.

Keywords: Drought, Rainfall variability, Standard Rainfall Anomalies, Precipitation Concentration Index, Standardised Precipitation Index.

1. Introduction

Drought is the most common natural hazard, defined as a regional occurrence of below normal water availability for certain periods, whether in the form of precipitation, river flow, surface, and groundwater storage. It is a common recurring climatic phenomenon that can occur everywhere on the earth in all climatic zones. Drought-related water shortages can have serious implications on water resources, agriculture, irrigation, livelihood, the economy, and society (Solh and Ginkel, 2014). Droughts are complex, less understood, poorly documented and their impacts are often silent but prolonged than other natural disasters. Drought can affect each section differently, resulting in varying degrees of severity and influence on each component. Drought is the most severe natural threat to food security and the life of billions of people resulting in increased migration of population worldwide (Hao et al., 2014; FAO, 2017). Climatologically, drought can be classified into five types like meteorological, agricultural, hydrological, socio-economic, ecological drought. Droughts are generally associated with the climatic variability of the region over space and time (Jain et al., 2015). Climatic variability can be referred to as the year-to-year fluctuation of climate above or below long-term average values of climatic variables which is highly unpredictable at multiple scales both in space and time (Cristiano et al., 2017; Gouda et al., 2017).

In recent years, frequent droughts and associated variability in hydro-meteorological variables due to climate change has aggravated the rise in water demand (Mishra and Singh, 2010; Mishra and Singh, 2011). To find the appropriate solutions to meet water scarcity challenges it is important to examine the temporal and spatial variability of rainfall in the context of drought frequency, magnitude, and severity in drought-prone regions. The coefficient of variation which is dependent on inter-annual rainfall variability can specify a complex picture of rainfall, both in terms of the spatial and temporal timescale from daily to decade (Conway, 2000; Seleshi and Zanke, 2004; Woldeamlak et al., 2007; Bewket and Conway, 2007). In recent times more emphasis is given to establishing the relation between rainfall variability, rainfall anomaly, wet and dry spell during monsoon and drought characteristics using the suitable indicators to understand the insight of the drought process in different climatic regions worldwide (Azam et al., 2018; Mallik et al., 2018; Malik et al., 2019). Many researchers have analyzed the variability of climatic parameters and drought aspects over the various agroclimatic zone for drought monitoring and formulation of strategies to minimize adverse impacts of droughts on water resources, agriculture and society (Martins et al., 2012; Shah et al., 2015).

In India, more than 70% population is engaged in agricultural activities and around 65% of agriculture is rainfed in the country. The Indian farming system is mostly dependent on the southwest monsoon and variation associated with it hence agricultural production, economy, and rural livelihood are more susceptible to droughts. (Sathyan et al, 2018; Vittal et al., 2016). Under these circumstances assessment of droughts, their characteristics and their relation with climatic variability are very important from water resources management point of view in the country for improving agricultural production. The Bundelkhand region of Madhya Pradesh in India has always been in attention due to many regional issues such as water scarcity, frequent droughts, lack of economic and development activities which are posing threat to agricultural productivity, livelihood, and social aspects in the region. (Pandey et al., 2010). The present study has been carried out to understand the temporal and spatial variability of rainfall in the Bundelkhand region using suitable indicators and its relationship with the drought frequency, duration, and magnitude.

2. Materials and Methodology

The present study has been carried out using long-term daily rainfall data of 38 years from 1980 to 2017 of five stations of the Bundelkhand region of Madhya Pradesh. The daily rainfall data was collected from the Indian Meteorological

Department, Pune. The study involves the identification of autocorrelation and randomness in climatic time series using standard statistical tests, examination of rainfall variability using statistical tests for the coefficient of variation, standard deviation, and application of indices like Standardize Rainfall Anomaly (SRA) index and Precipitation Concentration Index (PCI). Drought severity characteristics were studied using Departure Analysis, Probability distribution analysis and Standardize Precipitation Index (SPI).

2.1 Study Area

Bundelkhand region of Madhya Pradesh state is located in the central part of India and covers five northern districts, Sagar, Damoh, Chhatarpur, Tikamgarh, and Panna. This region lies between 23.14 and 25.55 latitude and 78.05 and 80.67 longitude covering an area of 37,479 km^2 as shown in Fig. 26.1.

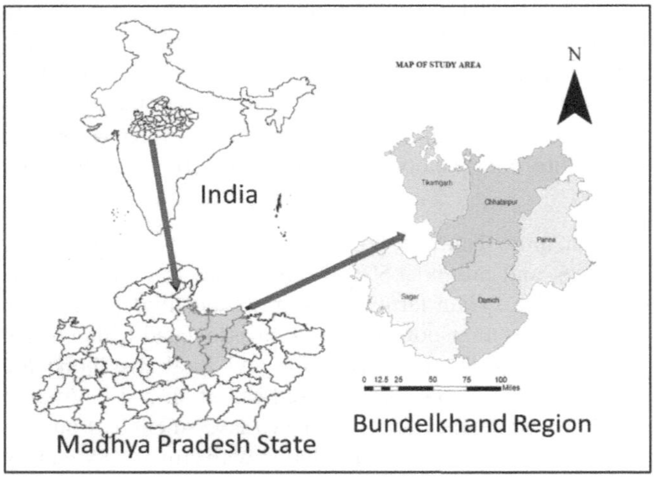

Fig. 26.1 Map showing Study area: Bundelkhand region of Madhya Pradesh

This region is characterized by its poverty, low agricultural productivity, frequent droughts, and lack of economic activities. It is endowed with the Vindhyan mountain range and tributaries to the Ganga river system like Betwa, Dhasan, Ken, Son, tones, etc. This region is dominated by hot as well as semi-humid types of climate and receives 930 to 1230 mm average annual rainfall. Despite having good average annual rainfall, it is always under the threat of water scarcity. The Bundelkhand region has long been regarded as a drought-prone region of the country, but drought frequency and severity have increased in recent decades (Gupta et al., 2014). The Bundelkhand region of Madhya Pradesh state faces the problems of recurrent droughts that are unpredictable both in their occurrence and duration; hence predictions and preparedness against droughts would be key elements for minimizing their impacts (Galkate et al., 2015).

2.2 Rainfall Variation Analysis

First of all the rainfall data of 38 years from 1980 to 2017 of five stations Sagar, Damoh, Chhatarpur, Tikamgarh, and Panna of Bundelkhand region was tested for autocorrelation and randomness in data set using Durbin Watson (DW) statistical test (Makridakis et al., 1998). In the DW test, the Null Hypothesis is that there is no autocorrelation in the data set and rejected if the DW tests statistic value is between 1.5 to 2.5. The DW statistic is defined as equation 1.

$$DW = \frac{\sum_{t-2}^{T}(e_t - e_{t-1})^2}{\sum_{t-1}^{T} e_t^2} \tag{1}$$

where *et* is the residual at the time *t* and *et−1* that at time *t − 1*. After the DW test and its analysis, the Statistical analysis of annual rainfall data was carried out to compute mean, median, standard deviation, coefficient of variation to assess spatial and temporal variation in rainfall. The statistical mean denotes the central tendency of the data, the standard deviation is a measure of variation or dispersion and the coefficient of variation is the standardized measure of the dispersion of a probability or frequency distribution.

2.3 Precipitation Concentration Index (PCI)

PCI is a foremost indicator used for the assessment of the temporal distribution and seasonal changes in precipitation. In this study, the coefficient of variation (CV) and the Precipitation Concentration Index (PCI) has been used as statistical descriptors for rainfall variability assessment (Bewket and Conway, 2007). The *PCI* values are calculated using equation 2 as given by Oliver (1980).

$$PCI = \frac{\Sigma pi^2}{(\Sigma pi)^2} \times 100 \tag{2}$$

Where, Pi = the rainfall amount of the ith month ($i = 1$ to 12 Months). According to Oliver (1980), *PCI* can be classified into three groups i.e. uniform monthly distribution of rainfall where PCI is less than 10, high concentration where *PCI* values lie between 11 to 20, and very high concentration where *PCI* value is greater than 21.

2.4 Standardized Rainfall Anomalies (SRA)

SRA is denoted as normalized anomalies, calculated by dividing the annual rainfall anomalies by the standard deviation, and exhibits inter-annual fluctuations in the time series (Bewket and Conway, 2007; Ayanlade, 2018). Rajeevan et al. (2006) have used high-resolution daily gridded data for the Indian region for analysis of breaks and active monsoon spells using the SRA index. In the present study, the SRA has been computed for the annual rainfall time series at all five using equation 3.

$$SRA = \left(\frac{Pt - Pm}{\sigma} \right) \tag{3}$$

Where *SRA* is standardized rainfall anomaly, *Pt* is annual rainfall in year *t*, *Pm* is long-term mean annual rainfall throughout observation and σ is the standard deviation of annual rainfall throughout the observation. Based on SRA analysis driest and wettest years at all stations of the Bundelkhand region were identified and compared with the drought frequency analysis outcome.

2.5 Rainfall Departure Analysis

The departure analysis of annual rainfall has been carried out to evaluate drought frequency, return period, and severity at five stations. A year is considered a drought year if the annual rainfall deficit is greater than or equal to 25 % of its long-term mean (Appa Rao, 1986; India Meteorological Department, 2002). The droughts are further classified according to their severity levels like moderate drought when annual rainfall deficit is between 25% to 50%, severe drought when rainfall deficit is between 50% to 75%, and extreme drought when rainfall deficit is more than 75%. Yaseen et al. (2013), Ramaraj et al. (2015), Galkate et al. (2015) and Jaiswal et al. (2019) have carried out an analysis to identify droughts, their frequency and magnitudes in selected areas of Madhya Pradesh and India. In the study drought years were identified using percentage annual rainfall departure (*D*) which have been calculated using equation 4. The frequency and return period of droughts were calculated for further analysis.

$$D(\%) = \frac{Xi - Xm}{Xm} \times 100 \tag{4}$$

Where *Xi* and *Xm* are annual rainfall and mean annual rainfall

2.6 Probability Distribution Analysis

The probability analysis of rainfall is important for predicting the relative frequency of occurrence of the rainfall with reasonable accuracy. An area can be considered drought-prone if the probability of exceedance of 75% of normal rainfall is less than 80% (Central Water Commission, 1982). The percentage probability of exceedance of 75% of mean annual rainfall has been worked out to delineate the drought proneness of five stations of the Bundelkhand region. Jaiswal et al., 2019 and Galkate et al., 2015 have carried out an analysis to identify drought-prone areas in Madhya Pradesh. Weibull's distribution was fitted to the ranked annual rainfall data for computing the probability of exceedance (*P*) using Equation 5.

$$P = \frac{m}{N+1} \times 100 \tag{5}$$

Where *m* is the rank and *N* is the number of years of data used.

2.7 Standardized Precipitation Index (SPI)

Several drought indices have been developed by researchers to detect the onset and assess the severity of drought. A drought index integrates any one or more meteorological and hydrological parameters like rainfall, temperature, evapotranspiration, runoff, groundwater and other water supply indicators into a single number and gives a comprehensive picture for decision-making (Zargar et al, 2011). Some of the important rainfall-based indicators are Palmer Drought Severity Index (PDSI), Rainfall Anomaly Index (RAI), Bhalme and Mooley Drought Index (BDMI) and Standardized Precipitation Index (SPI). Among the indices, the percentage departure and Standardized Precipitation Index are more commonly used in India (Alley, 1984; Alley, 1985; Jain et al. 2015). These indices are normally continuous functions of any one or more hydro-meteorological variables, such as precipitation, temperature, soil water, streamflow, groundwater, potential evapotranspiration, etc. (World Meteorological Organization, 1975). The Standardized Precipitation Index (SPI) was designed by McKee et al. (1993) to quantify the precipitation deficit for multiple time scales (Kazem et al., 2011; Khan and Gadiwala, 2013). The SPI can be calculated on 3, 6, 9, 12, and 24-month time scales, which correspond to soil moisture deficit, meteorological and hydrological droughts (Agnew, 2000; Livada and Assimakopoulos, 2007; Almedeij, 2014). When the SPI is normalized to represent the varied behavior of precipitation for time steps shorter than 12 months, the algorithm becomes more complex (Patel et al., 2007; Sebenik et al., 2017). To address this issue, long-term precipitation records from stations are fitted to a gamma distribution, which has been shown to adjust well to the precipitation distribution. Rainfall deficits are represented by negative SPI values, whereas rainfall surpluses are represented by positive SPI values. Many drought severity studies have been conducted globally and in India, yielding several valuable and pragmatic conclusions in understanding the drought characteristics (Pandey et al., 2008; Nandintsetseg and Shinoda, 2013; Jain et al., 2015; Tommaso 2017). Among the various drought indices, the SPI (McKee et al., 1993, 1995) is considered to be one of the most robust and effective drought indexes (Capra and Scicolone, 2012). It has been effectively used for widespread application worldwide as it can be evaluated for different timescales and allows the analysis of different drought categories (Baran et al., 2017; Jalpa et al., 2014; Kumar, 2016). The drought classification based on SPI is given in Table 26.1.

Table 26.1 Drought classification based on SPI values

SPI Values	Class
SPI > = 2	Extremely Wet
1.5 to 1.99	Severely Wet
1.0 to 1.49	Moderately Wet
–0.99 to 0.99	Near Normal
–1.0 to –1.49	Moderately Dry
–1.5 to –1.99	Severely Dry
SPI <= –2	Extremely Dry

In the resultant outcome, a drought event occurs any time the SPI is continuously negative and reaches an intensity of -1.0 or less. The event ends when the SPI becomes positive. Each drought event, therefore, has a duration defined by its beginning and end, and intensity for each month that the event continues. The positive total of the SPI for all the months within a drought event can be termed as the drought magnitude. The monthly rainfall data of 38 years from 1980 to 2017 of five stations Sagar, Damoh, Chhatarpur, Tikamgarh and Panna of Bundelkhand region have been used to obtain SPI values on 1, 3, 6, 9 and 12-month timescales to analyze the frequency and probability of occurrence of dry events on different all time scales.

3. Results and Discussion

The present study has been carried out to assess rainfall variability and drought features in the Bundelkhand region of Madhya Pradesh using statistical tests and suitable indices like Rainfall Departure Analysis, Probability Distribution, SRA, PCI and SPI. The long-term rainfall data of 38 years for the period from 1980 to 2017 of five stations namely Sagar, Damoh, Chhatarpur, Tikamgarh and Panna have been used for analysis.

3.1 Rainfall Variation Analysis

The annual rainfall data of all stations have been tested for the existence of autocorrelation and randomness in data sets using the Durbin-Watson (DW) test. If the DW test value ranges between 1.5 to 2.5 then the data set is considered to be good, normal, autocorrelated and sufficiently non-random. The descriptive statistics of annual rainfall and rainy days with their mean, median, standard deviation (SD), coefficient of variation, interquartile range (IQR) and Durbin-Watson (DW) tests statistics have been worked out and shown in Table 26.2. The spatial variation of annual rainfall and rainy days over the Bundelkhand region is shown in Fig. 26.2.

Table 26.2 Descriptive statistics of rainfall in Bundelkhand region (period of 1980-2017)

Stations	Annual Rainfall						Annual Rainy days		
	Mean (mm)	Median (mm)	SD	CV (%)	IQR	DW	Mean	SD	CV (%)
Sagar	1230	1128	380	36.6	529	2.40	52	13	24.8
Damoh	1189	1112	355	29.8	487	2.63	52	10	19.8
Chhatarpur	963	909	351	36.4	423	2.59	42	10	24.3
Tikamgarh	935	913	321	34.3	351	2.21	41	10	23.7
Panna	1180	1129	339	28.7	388	2.18	49	8	16.2
Average	1099	1038	349	33.2	430	2.40	47	10	21.8

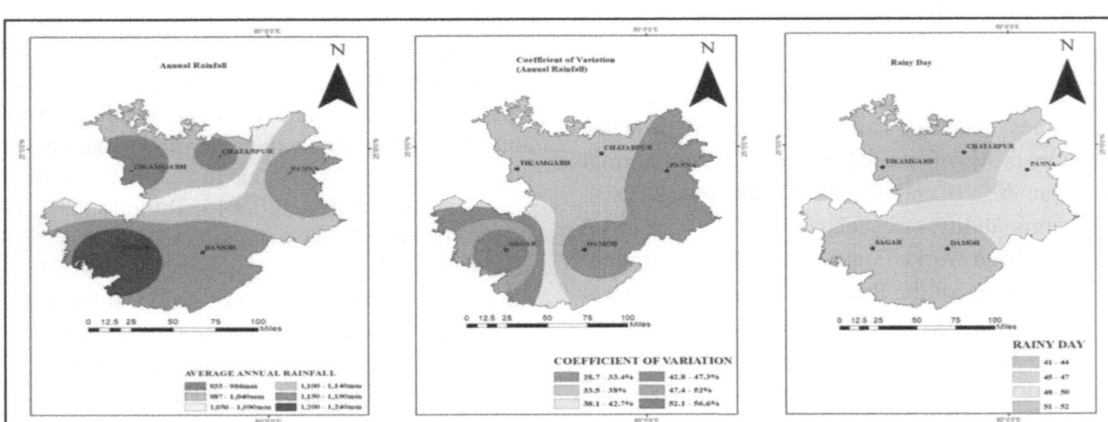

Fig. 26.2 Spatial variation in annual rainfall and rainy day in Bundelkhand

The spatial and temporal variability of rainfall can be analyzed from the statistics shown in Table 26.2 and represented in Fig. 26.2. The DW test indicated autocorrelation in data sets at Sagar, Tikamgarh and Panna stations of the Bundelkhand region, however slight negative randomness has been observed at Damoh and Chhatarpur as DW statistic values have been found more than 2.5. The average annual rainfall in the Bundelkhand region has been observed as 1099 mm and varying from 935 mm at Tikamgarh to 1230 mm at Sagar. The median rainfall values are also seen in a similar pattern of mean rainfall, with the lowest value of 935 mm at Tikamgarh to the highest value of 1128 mm at Sagar. The interquartile range (IQR) which is the difference between rainfall at 75th and 25th percentile has been seen lowest at Tikamgarh (351mm) and highest at Sagar (529 mm) station. IQR analysis indicates a high spread in rainfall data set at Sagar, Damoh and Chhatarpur stations as compared to Tikamgarh and Panna. The average standard deviation of annual rainfall in the region is observed 349 mm and the coefficient of variation is observed 33.2% indicating very high temporal variation in rainfall in the region as a whole. Despite having good rainfall, Sagar (380), Damoh (355) and Chhatarpur (351) stations have very high values of standard deviation as compared to other stations. Coefficient of variation values in annual rainfall time series have been seen very high at Sagar (36.6%) and Chhatarpur (36.4%) indicating very high interannual variation. Similarly, the average number of rainy days have been observed 47 in the region, ranging from 41 at Tikamgarh to 52 at Sagar and Damoh. The highest standard deviation for rainy days has been observed at Sagar. However, the coefficient of variation in rainy days is observed highest at Sagar (24.8%) and Chhatarpur (24.3) and lowest at Tikamgarh (16.2) which is very similar to the rainfall pattern at all five stations. From Fig. 26.2 it is seen that, despite having higher average annual rainfall and rainy days in Sagar and Damoh districts in the southern part of the Bundelkhand region, these two districts have very high deviation and variation in rainfall as compared to the northern part.

3.2 Precipitation Concentration Index (PCI)

The PCI is considered as one of the main indicators for the assessment of the temporal distribution and seasonal changes in precipitation. The average rainfall contribution of monsoon and non-monsoon season, the highest monthly rainfall contribution to the annual total annual rainfall, the Precipitation Concentration Index (PCI), and its trend (Linear) in the Bundelkhand region are shown in Table 26.3.

From the analysis of Table 26.3, it is observed that the average contribution of monsoon rainfall is 91.7% and non-monsoon rainfall is 8.4% indicating the concentration of rainfall in four monsoon months from June to September in the Bundelkhand region. The ultimate concentration of rainfall can also be seen from the contribution of the single largest monthly total to annual total rainfall at each of the stations which have been seen ranging between 42 to 45.9%. The average PCI observed in the region is 31.9% which is also indicative of a very high monthly concentration at all five stations in the Bundelkhand region. However, a non-significant decreasing trend has been observed in PCI at all stations of the region and the Panna station has shown a significant decreasing trend (at 95% confidence level).

3.3 Standard Rainfall Anomalies (SRA)

The interannual fluctuation in rainfall time series at all five stations of the Bundelkhand region has been studied and analyzed using the SRA index. The highest SRA value indicates the driest year and the lowest value indicates the wettest year. Based on the SRA analysis of annual, monsoon and non- monsoon rainfall time series, driest and wettest years were identified at each of the stations for all three categories over the period from the year 1980 to 2017 and are given in Table 26.4.

The identified driest and wettest years were then compared with the rainfall departure analysis. The proportion of negative SRA indicating dry years has been seen very high

Table 26.3 Average contribution of seasons and highest monthly rainfall and PCI

Stations	Monsoon Rainfall (%)	Non-Monsoon Rainfall (%)	Highest Monthly Rainfall (%)	PCI (%)	Trend of PCI
Sagar	90.2	9.8	44.9	32.1	-0.67
Damoh	91.8	8.2	42.0	31.8	-0.40
Chhatarpur	92.4	7.6	43.7	34.6	-0.93
Tikamgarh	92.5	7.5	44.8	32.5	-0.75
Panna	91.6	8.7	42.2	28.5	-1.84
Average	91.7	8.4	43.5	31.9	-0.92

Table 26.4 Distrctwise driest and wettest year in Bundelkhand region

Stations	Driest year				Wettest year			
	Annual rainfall	SRA of the year	Monsoon rainfall	Non-monsoon rainfall	Annual rainfall	SRA of the year	Monsoon rainfall	Non-monsoon rainfall
Sagar	1981	–1.8	1981	1992	2006	2.5	2006	1997
Damoh	2017	–1.4	2017	1998	2013	2.4	2013	1987
Chhatarpur	2017	–1.6	2017	2004	1985	2.6	1985	1987
Tikamgarh	1989	–2.4	1989	1988	1982	2.4	1982	1982
Panna	2007	–1.9	2007	1988	1980	3.4	1980	1987

for all stations and found ranging from -1.41 at Damoh (year 2017) to -2.41 at Tikamgarh (Year 1989) as shown in Table 26.4.

3.4 Rainfall Departure Analysis

The graphs showing annual rainfall departure at five stations, Sagar, Damoh, Chhatarpur, Tikamgarh and Panna of Bundelkhand region are shown in Figure 26.3. The identified drought years (Deficit between 25 to 50%), severe drought years (Deficit greater than 50%) and drought frequency evaluated from departure analysis have been compiled and given in Table 26.5 and its spatial representation has been shown in Figure 26.3.

From the analysis of Figure 26.3 and Table 26.5, it is observed that the drought frequency in the Bundelkhand region is varying from 18.4% at Tikamgarh and Panna to 28.9% at

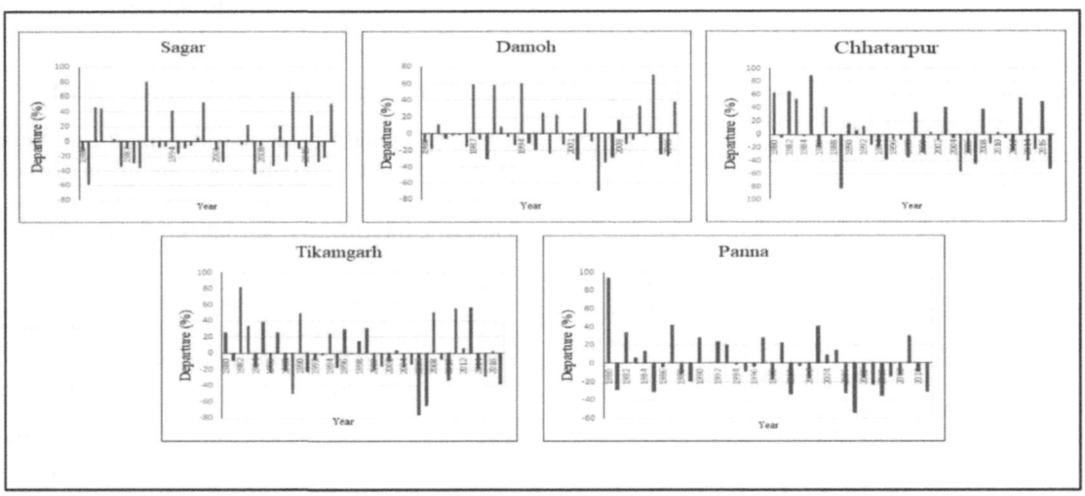

Fig. 26.3 Annual rainfall departure (%) at five stations of Bundelkhand region Table

Table 26.5 Drought frequency analysis in Bundelkhand region

Stations	Drought Return Period	Drought Frequency (%)	Drought Years	Severe Drought years
Sagar	1 in 3-4 years	28.9	1981, 1984, 1986, 1988, 1989, 2002, 2004, 2007, 2010, 2012, 2015, 2017	1981
Damoh	1 in 4-5 years	20.5	1989, 1998, 2002, 2006, 2007, 2015, 2016, 2017	2017
Chhatarpur	1 in 4-5 years	23.2	1989, 1995, 1998, 2000, 2005, 2006, 2007, 2012, 2014, 2017	2017
Tikamgarh	1 in 5-6 years	18.4	1989,2000, 2006, 2007, 2010, 2015, 2017	1989
Panna	1 in 5-6 years	18.4	1981, 1985, 2000, 2006, 2007, 2010, 2015	2007

Sagar. From the analysis of return periods of droughts, it is observed that the Sagar station has a chance of occurring a drought year after every 3 to 4 years, Damoh and Chhatarpur stations have a chance of occurring a drought year after every 4 to 5 years; however, stations Tikamgarh and Panna have the chance of occurring a drought year after every 5-6 years. The drought frequency has been observed very high at Sagar but other stations also have the threat of frequent droughts.

3.5 Probability Distribution Analysis

In this study, the approach of probability analysis of rainfall has been applied to understand whether the station is drought-prone or not depending upon the probability of exceedance of 75% of average rainfall.

A typical rainfall probability distribution graph for Sagar stations is shown in Fig. 26.4. The probability of exceedance of 75% of normal rainfall and drought status of all five stations is presented in Table 26.6.

From Table 26.6 it is observed that the probability of occurrence of rainfall equivalent to 75% of the normal annual rainfall at all stations of the Bundelkhand region ranges from 69% at Sagar to 79% at Tikamgarh and Panna. Thus all stations are under the category of drought-prone stations as the probability of occurrence of rainfall equivalent to 75% of the normal annual rainfall is less than 80% at these stations. Thus whole Bundelkhand region of Madhya Pradesh is the drought-prone region where Sagar and Chhatarpur seem to be more prone to droughts.

3.6 Standardized Precipitation Index (SPI)

The analysis of meteorological drought has been carried out to evaluate using SPI on 1, 3, 6 9, and 12- month time scale for long-term monthly rainfall time series of all stations of Bundelkhand. The typical SPI graphs for Sagar on different timescales are shown in Fig. 26.5. Further, the SPI values have been categorized based on their severity and their frequency and probability have been worked out and given in Table 26.7.

The 1, 3 and 6-month SPI correspond to soil moisture deficit and meteorological droughts whereas 9 and 12 months SPI represents hydrological drought. From the analysis of Table 7, it is observed that the probability of occurrence of near-normal events at all stations ranges from 65.6 to 84.9 % which is higher than other categories of drought events. The

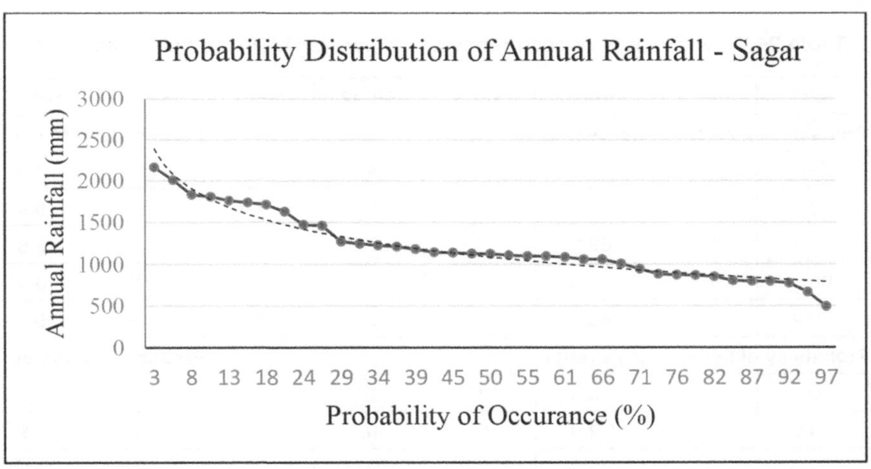

Fig. 26.4 Annual Rainfall Probability distribution analysis at Sagar

Table 26.6 Probability distribution analysis of rainfall

Name of stations	Mean annual rainfall (mm)	Rainfall at 75% probability level	Probability of occurrence of 75% of normal (%)	Status
Sagar	1230	923	69	Drought Prone
Damoh	1189	891	77	Drought Prone
Chhatarpur	963	713	75	Drought Prone
Tikamgarh	935	701	79	Drought Prone
Panna	1180	885	79	Drought Prone

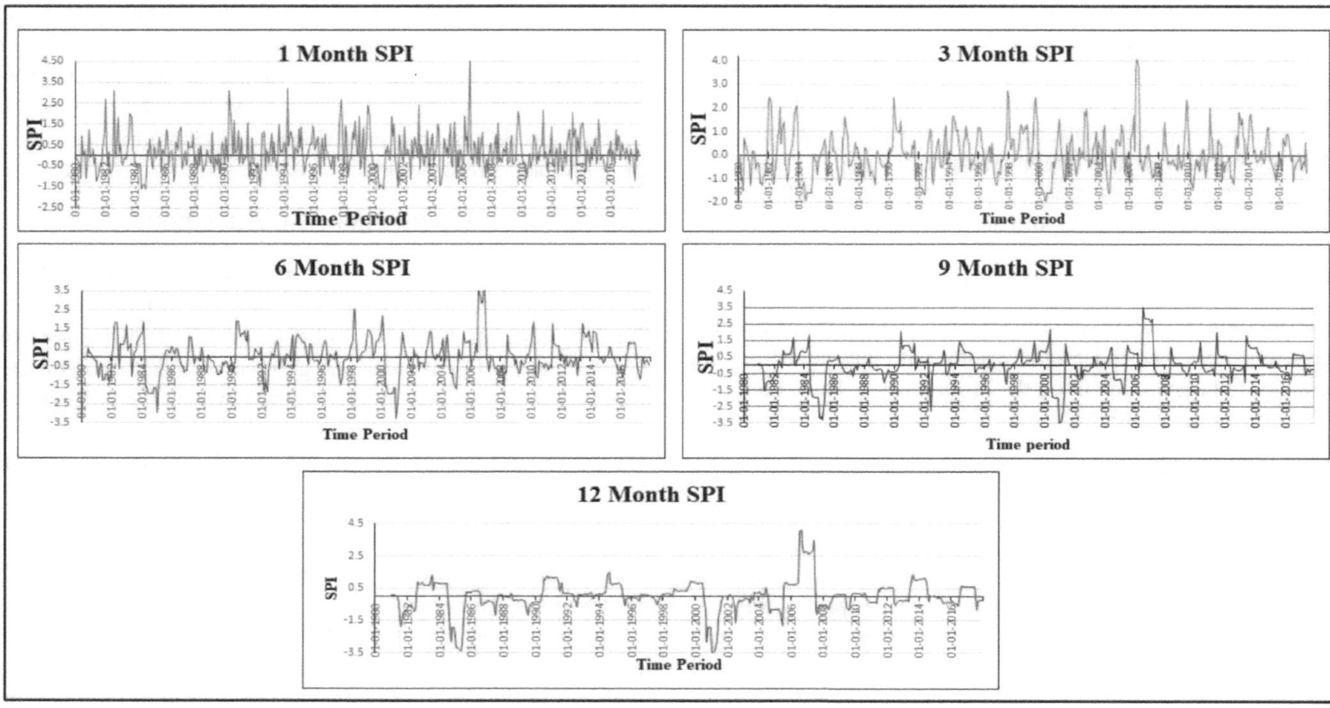

Fig. 26.5 SPI graphs of Sagar station on a different time scale

Table 26.7 Probability of occurrence of dry events (%) on a different time scale

Station	SPI 1	SPI 3	SPI 6	SPI 9	SPI 12	SPI 1	SPI 3	SPI 6	SPI 9	SPI 12
	Probability of Near normal events					Probability of Moderate dry events				
Sagar	79.0	73.0	75.9	82.5	84.9	4.4	8.1	4.0	1.3	2.0
Damoh	79.2	71.3	66.0	70.8	71.9	2.9	5.0	9.9	8.1	7.2
Chhatarpur	79.4	76.8	69.5	71.5	68.9	2.4	3.1	8.8	5.9	6.1
Tikamgarh	80.3	77.4	71.3	73.9	73.0	2.2	2.9	6.4	4.6	3.5
Panna	79.2	70.6	66.7	65.6	65.4	2.9	8.3	9.4	10.8	10.1
	Probability of Severely dry events					Probability of Extremely dry events				
Sagar	0.9	4.0	5.3	3.5	2.2	0.0	0.0	0.7	2.2	2.9
Damoh	1.5	1.5	2.9	0.9	0.4	0.0	0.7	1.5	2.6	2.9
Chhatarpur	2.0	2.6	2.6	3.1	3.1	0.0	0.7	1.8	2.4	2.6
Tikamgarh	1.8	2.6	1.8	1.8	1.3	0.9	1.3	3.3	4.2	5.3
Panna	0.9	2.6	3.3	3.1	2.2	0.4	0.9	1.8	2.4	2.6

probability of moderate dry events at all stations for all SPIs ranges from 1.3 to 10.8%. Panna station experiences more moderate dry events as compared to others on almost all time scales. The probabilities of moderate dry events were observed 8.3 and 9.4% for SPI 3 and 6 while 10.8 and 10.1% for SPI 9 and 12 respectively at Panna which is indicative of more chances of conversion of moderate meteorological drought events into moderate hydrological droughts. Similarly, Damoh and Chhatarpur have shown a higher probability of moderate drought events for SPI 6, 9 and 12 indicating more chances of meteorologic and hydrological droughts. The Sagar also has a high probability of moderate dry events for SPI 1 and 3 indicating more chances of occurrence of moderate meteorological droughts. The probability of occurrence of severe and extremely dry events indicates the intense drought events and are important from the water resources management point of view. The probabilities of occurrence of severely dry events were seen very high at Damoh and Chhatarpur for SPI-1 (1.5 and 2.0%) and SPI-3 (1.5 and 2.6%) indicating more chances of meteorological drought events. The Sagar has a probability of 4% to encounter severe meteorological droughts. The SPI-12 analysis indicated a high probability of hydrological droughts at Sagar, Chhatarpur and Panna. The probability of occurrence of extremely dry events was seen very high at Tikamgarh i.e. 4.2 and 5.3% for SPI-9 and SPI-12 respectively indicating a high frequency of meteorological and hydrological droughts. Contrarily, the Sagar, Damoh and Chhatarpur have rare chances of coming across extreme dry events as the probability of occurrence has been observed as zero. On comparison of SPI results at different time scale, it was seen that years 1989, 2006 and 2017 have experienced severe to extreme droughts at almost of the stations in Bundelkhand region.

4. Conclusions

The present study has been carried to analyze rainfall variability and its relation with drought characteristics with the help of suitable statistical and drought indices using long-term rainfall data of five stations i.e. Sagar, Damoh, Chhatarpur, Tikamgarh and Panna of Bundelkhand region of Madhya Pradesh. Auto-correlation analysis using Durbin-Watson has shown slight negative randomness in rainfall time series at Damoh and Chhatarpur. Very high variability has been seen in rainfall at all five stations of the Bundelkhand region with a wide range in mean annual rainfall from 935 mm at Tikamgarh to 1230 mm at Sagar. The standard deviation, coefficient of variation and interquartile range (IQR) statistics were seen very high at Sagar, Damoh and Chhatarpur indicating very high variability and high spread in rainfall data set as compared to Tikamgarh and Panna. The standard deviation and coefficient of variation in annual rainy days were also seen higher at Sagar and Chhatarpur.

The Precipitation Concentration Index (PCI) has been used for the assessment of the temporal distribution and seasonal changes in precipitation. The rainfall concentration seems to be uneven as monsoon rainfall contribution is almost 91.7% of the total annual rainfall in the Bundelkhand region. The average contribution of the single largest monthly rainfall is observed around 43.5% of the total annual rainfall. The average PCI values are observed as 31.9% which is indicative of very high monthly concentration at all stations in the Bundelkhand region. However, the non-significant decreasing trend has been seen in PCI at most of the stations. The interannual fluctuation in rainfall time series has been analyzed using the Standard Rainfall Anomalies (SRA) index and driest and wettest years are identified. No single year could be seen as the common driest year in the region except the year 2017 which has been seen as a common driest year at Damoh and Chhatarpur.

Departure analysis of rainfall is useful in identifying drought years, their frequency and return period. Drought frequency has been seen vary high at Sagar station followed by Damoh and Chhatarpur, where the chances of occurring a drought year are 1 in every 4 to 5 years. Stations Tikamgarh and Panna have seen chances of occurring a drought year after every 5 to 6 years. The very high drought frequency in the Bundelkhand region has urged a need for strategic planning for drought mitigation to abate their impacts on water resources, agriculture and rural livelihood. Probability Distribution analysis of rainfall indicated all stations of the region under the drought-prone category and the whole Bundelkhand region can be considered as a drought-prone area where Sagar and Chhatarpur seem to be more prone to droughts. The high variability of rainfall at stations Sagar, Damoh and Chhatarpur has been seen as one of the main reasons for frequent and severe droughts.

Standardized Precipitation Index (SPI) was used in the study to characterize droughts in terms of their severity and probability of occurrence. The Sagar has shown a high probability of moderate dry events for SPI 1 and 3 indicating more chances of occurrence of moderate meteorological droughts. The Panna station has been found to experience more moderate dry events as compared to others on almost all time scales. The probability of occurrence of severe and extremely dry events indicates the intense drought events and are important from the water resources management point of view. The probabilities of occurrence of severely dry events were seen very high at Damoh and Chhatarpur. The probability of occurrence of extremely dry events was seen very high at Tikamgarh indicating a high frequency of meteorological and hydrological droughts. In a comparison of SPI results at different time scales it was seen that years 1989, 2006 and 2017 have experienced severe to extreme droughts at all stations in the Bundelkhand region. Evidence of drought events has been seen to increase with the increase of the SPI time scale. However extreme drought events were seldom in the region on a 1-month time scale. The SPI analysis shown that all stations of the region have experienced significant droughts on a long-term time scale i.e. 6, 9, 12-month and these results are seen in agreement with the frequency and probability analysis of rainfall. However, SPI outcomes seem to be is more sensible to severe drought than the probability distribution analysis and gives the best result using only rainfall data without other meteorological parameters. The assessment of rainfall variability and its relation with drought frequency and magnitude is very important to comprehend and monitoring water scarcity situations for decision making and adopting appropriate strategies for mitigation.

5. Acknowledgement

The authors are thankful to the National Institute of Hydrology and Rabindranath Tagore University Bhopal for providing facilities to undertake this research work. The authors are also thankful to India Meteorological Department, Pune and Madhya Pradesh Water Resources Department, Bhopal for providing data.

REFERENCE

1. Agnew, C.T. (2000). Using the SPI to identify drought. Drought Network News, 12(1): 6–12.
2. Alley, W. M. (1984). The Palmer drought severity index. Limitations and Assumptions. Journal Climate Applied Meteorology, 23: 1100–1109.
3. Alley, W. M. (1985). The Palmer drought severity index as a measure of hydrological drought, Water Resources. Bull., 21, 105–114.
4. Almedeij, J. (2014). Drought analysis for Kuwait using Standardized Precipitation Index. The Scientific World Journal. Article ID 451841. DOI: https://doi.org/10.1155/2014/451841.
5. Appa Rao, G. (1986). Drought climatology, Jal Vigyan Samiksha, Publication of High Level Technical Committee on Hydrology, National Institute of Hydrology, Roorkee, India.
6. Ayanlade, A., Radeny, M., Mortonc, J. F. & Muchaba, T. (2018). Rainfall variability and drought characteristics in two Agro-climatic zones: An assessment of climate change challenges in Africa, Elsevier, Science of the Total Environment 630 (2018) 728–737. https://doi.org/10.1016/j.scitotenv.2018.02.196.
7. Azam, M., et al., (2018). Spatial and temporal trend analysis of precipitation and drought in South Korea. MDPI, 10, 765.
8. Baran, T., Guner, U. & Dikbas, F. (2017). Drought analysis with SPI index and entropy. European water, 60: 263–270.
9. Bewket, W. & Conway, D. (2007). A note on the temporal and spatial variability of rainfall in the drought-prone Amhara region of Ethiopia. International Journal of Climatology, 27 (11). pp. 1467-1477. ISSN 1097-0088.
10. Bewket, W. & Conway, D. (2007). A note on the temporal and spatial variability of rainfall in the drought-prone Amhara region of Ethiopia. International Journal of Climatology, 27: 1467–1477.
11. Capra, A. & Scicolone, B. (2012). Spatiotemporal variability of drought on a short-medium time scale in the Calabria region (Southern Italy). Theoretical and Applied Climatology 3:471–488.
12. Central Water Commission. (1982). Report on identification of drought prone areas of 99-Districts, New Delhi.
13. Conway, D. (2000). Some aspects of climate variability in the northeast Ethiopian highlands-Wollo and Tigray. SINET: Ethiopian Journal of Science 23: 139–161.
14. Cristiano, E., Veldhuis, M. C. & Giesen, N. V. (2017). Spatial and temporal variability of rainfall and their effects on hydrological response in urban areas. Hydrology and Earth System Sciences, 21, 3859- 3878. DOI: https://doi.org/10.5194/hess-21-3859-2017.
15. FAO, (2017). Drought. Food and Agriculture Organization of the United Nations, Disponivel em: <http://www.fao.org/emergencies/emergency-types/drought/en>.
16. Galkate, R. V., Pandey, R. P., Thomas, T., Jaiswal, R. K. & Nayak, T. R. (2015). Meteorological and hydrological aspects of drought in central India preparedness and management of water related disasters', 3rd India Water Week 2015, Water Management for Sustainable Development, 13–17 January 2015, New Delhi, India, 2015.
17. Gouda, K. C., Sahoo, S. K., Samantray, P. & Shivappa, H. (2017). Comparative study of monsoon rainfall variability over India and the Odisha State. *Climate;* 5(4):79. https://doi.org/10.3390/cli5040079.
18. Gupta, A. K., Nair, S. S., Ghosh, O., Singh, A. & Dey, S. (2014). Bundelkhand drought: retrospective analysis and way ahead. National Institute of Disaster Management, New Delhi, 148.
19. Hao, Z., Agha K. A., Nakhjiri, N. & Farahmand, A. (2014). Global integrated drought monitoring and prediction system, Scit. Dat. 1, 1–10. <http://dx.doi.org /10.1038/sdata.2014.1.
20. India Meteorological Department. Southwest monsoon (2002) End of season report. [Online]. Available: <http://www.imd.gov.in/section/nhac/dynamic/mid1.htm>; 4 October 2002 [accessed February 2013].
21. Jain, V. K., Pandey, R. P., Jain, M. K. & Byun, H.R. (2015). Comparison of drought indices for appraisal of drought characteristics in the Ken river basin, Weather and Climate Extremes: Elsevier, Vol. 8, pp 1–11.
22. Jaiswal, S., Galkate, R., Chandola V. K., Singh, P. & Thakur, A. (2019). Assessment of drought in Shipra river basin, Madhya Pradesh. Journal of Soil and Water Conservation 18(2): 188-195, ISSN: 022-457X (Print); 2455-7145 (Online); DOI: 10.5958/2455-7145.2019.00026.2.
23. Jalpa, K. S. & Parekh, F. (2014). Drought assessment using Standardized Precipitation Index. International Journal Science Research, 3(7), 1073–1076.
24. Kazem, N. & Zareiee, A. R. (2011). Assessment of meteorological drought using SPI in West Azarbaijan Province. Applied Science Environment Management,15(4): 563–569.
25. Khan, M. A. & Gadiwala M. S. (2013). A Study of drought over Sindh (Pakistan) Using Standardized Precipitation Index. Pakistan Journal of Meteorology, 9(18): 15–22.
26. Kumar, A. (2016). Characterizing meteorological drought using Standardized Precipitation Index for Dehradun, Uttarakhand. Indian Journal of Soil Conservation, 44(2): 157-162.
27. Livada, I. & Assimakopoulos, V. D. (2007). Spatial and temporal analysis of drought in Greece using the Standardized Precipitation Index (SPI). Theoretical and Applied Climatology, 89,143-153. DOI: 10.1007/s00704-005-0227-z.
28. Makridakis S., Wheelwright, S. C. & Hyndman, R.B. (1998). Forecasting: methods and applications. John Wiley and Sons: New York; 241–310.
29. Malik, A., Kumar, A. & Guhathakurta, P. (2019). Spatial-temporal trend analysis of seasonal and annual rainfall (1966-2015) using innovative trend analysis method with

significance test. Arabian journal of geoscience,12: 328. DOI: 10.1007/s12517-019-4454-5.

30. Malik, A., Kumar, A., & Guhathakurta, P. (2018). Temporal and spatial trends in rainfall pattern in Uttrakhand State, India.
31. Martins, D. S., Raziei, T., Paulo, A. A. & Pereira, L. S. (2012). Spatial and temporal variability of precipitation and drought in Portugal. Natural Hazards and Earth System Sciences,12, 1493-1501. doi:10.5194/12-1493-2012.
32. McKee, T.B., Doesken N.J. & Kleist, J. (1993). The relationship of drought frequency and duration to time scale. In: preprints eighth conference on applied climatology, Anaheim, The American Meteor Society, pp 179–184.
33. McKee, T.B., Doesken, N.J. & Kleist. J. (1995). Drought monitoring with multiple time scales. In: Preprints ninth conference on applied climatology, Dallas, The American Meteor Society, pp 233–236.
34. Mishra, A.K. & Singh, V.P. (2010). A review of drought concepts. Journal of Hydrology, 391: 202–216
35. Mishra, A.K. & Singh, V.P. (2011). Drought modeling – A review. Journal of Hydrology, 403:157–175.
36. Nandintsetseg, B. & Shinoda, M. (2013). Assessment of drought frequency, duration, and severity and its impact on pasture production in Mongolia. Natural Hazards (2013) 66:995–1008, DOI 10.1007/s11069-012-0527-4.
37. Oliver, J. E. (1980). Monthly precipitation distribution: a comparative index, The Professional Geographer, 32:3, 300-309, DOI: 10.1111/j.0033-0124.1980.00300.x.
38. Pandey R. P., Dash, B. B., Mishra, S. K. & Singh, R. (2008). Study of indices for drought characterization in KBK districts in Orissa (India). Hydrological Processes. 22, 1895–1907, DOI: 10.1002/hyp.6774.
39. Pandey R. P., Pandey A., Galkate R. V., Byun H. & Mal, B. C. (2010). Integrating hydro-meteorological and physiographic factors for assessment of vulnerability to drought. Water Resource Manage 24: 4199–4217.
40. Patel, N. R., Chopra, P. & Dadhwal, V. K. (2007). Analysing spatial patterns of meteorological drought using Standardized Precipitation Index. Meteorological Applications, 14:329-336. doi:10.1002/met.33.
41. Rajeevan, M., Bhate, J., Kale, J. D. & Lal, B. (2006). High resolution daily gridded data for the Indian region: Analysis of break and active monsoon spell. Current Science. Vol 91, No. 3, 296–306.
42. Ramaraj, A. P., et al., (2015). Rainfall stability and drought valuation using SPI over Southern Zone of Tamil Nadu. Journal of Current World Environment,10(3): 928–933.
43. Sathyan, A. R., Christoph, F., Thomas, A. & Lutz, B. (2018). Climate vulnerability in rainfed farming: analysis from Indian Watersheds, Sustainability 10, no. 9: 3357. https://doi.org/10.3390/su10093357
44. Sebenik, U., Brilly, M. & Sraj, M. (2017). Drought analysis using the Standardized Precipitation Index (SPI). Acta Geographica Slovenica, 57-1. DOI: https://dx.doi.org/10.3986/AGS.729.
45. Seleshi, Y. & Zanke, U. (2004). Recent changes in rainfall and rainy days in Ethiopia. International Journal of Climatology 24: 973–983.
46. Shah, R., Bharadiya, N. & Manekar, V. (2015). Drought index computation using Standardized Precipitation Index (SPI) method for Surat district, Gujarat. International conference on water resources, coastal and ocean engineering, *4* (2015),1243–1249. DOI: 10.1016/j.aqpro.2015.02.162.
47. Solh, M. & Ginkel, M. (2014). Drought preparedness and drought mitigation in the developing world's drylands, Weather Climate Extreme 3, 62–66. http://dx.doi.org/10.1016/j.wace.2014.03.003.
48. Tommaso, C. (2017). Drought analysis in New Zealand using the Standardized Precipitation Index. Environ Earth Sci (2017) 76:569, DOI 10.1007/s12665-017-6909-x.
49. Vittal, H., Ghosh, S., Kamakar, S., Pathk, A. & Murtugudde, R. (2016). Lack of dependence of Indian summer monsoon rainfall extremes on temperature: An Observational Evidence. Sci Rep 6, 31039 https://doi.org/10.1038/srep31039.
50. World Meteorological Organization. (1975). Drought and agriculture Technical Note No. 138, World Meteorological Organization No. 392, Geneva, Switzerland.
51. Yaseen, K., AL-Timimi & AL-Jiboori. (2013). Assessment of spatial and temporal drought in Iraq during the period 1980-2010. International Journal of Energy and Environment, *4*(2): 291–302.
52. Zargar, A., Sadiq, R., Naser, B. & Khan, F. I. (2011). A review of drought indices. Environmental Reviews, *19*(NA), 333–349.

CHAPTER 27

Drought Crisis In Rajasthan and Their Management

Harish Samaria
Post Doctoral Fellow Department of Geography
University of Rajasthan

1. Introduction

Muhammad Tughlakh was the first Sultan to take systematic steps to alleviate efforts of droughts by distributing grains to drought affected people in Delhi in 1343 AD .The first Scarcity Manual was prepared by the British Government in 1883, which was followed by other manuals by some provincial governments (Hirway 2001), The Royal Commission on Agriculture in 1928 recommended promotion of dry land farming to promote agriculture in famine affected regions. However, the efforts were scanty and there was an alarming increase in the frequency of during the British period (Bhatia 1967).

The term drought is associated with scarcity of water, it means different things to different people. To the **agriculturist**, it means the deficit of rainfall and soil moisture to support healthy crop growth: to the **meteorologist,** it indicates the deficiency of rainfall compared to normal' rainfall of the region and to the **hydrologist** it is the scarcity of water in surface and groundwater resources. According to the National Commission of Agriculture (1976). if the drought occurs in more than 40% of the years in an area, it is classified as chronically drought prone area. As per this classification, Indian arid zone is a chronically drought prone area.

Drought is one of the most frequently occurring national disasters in India. With its increased frequency and expanded coverage in the recent years, about one third of the country is either drought prone or under desert areas. . These areas are lagging behind in agriculture and also in overall economic growth.

Rajasthan is the largest state in India covering an area of 34.22 million hectare i.e. 10.5 % of the country's geographical area but sharing only 1.15 % of its water resources. The estimated per capita water availability in the state during 2001 was 840 m3 and is expected to be 439 m3 by the year 2050 against the national average of 1140 m3 by 2050. More than 70% of its people depend upon agricultural activities. Rajasthan experiences acute weather and consists of four distinctive seasons- Pre-monsoon, Monsoon, Post-monsoon and winter. The average temperature in winter ranges from 2° to 26° C and in peak summer the average temperature range from 28° to 48° C making the region arid and draught-prone.

Keyword: Agricultural Drought, Meteorological Drought, Hydrological Drought, Standardized Precipition index (SPI), Desest rose, Sheep Model

1. Study Area

Rajasthan is situated in the North-Western part of India and located between 23°30′ and 30°11′ North latitudes and 69°29′ and 78°17′ East longitudes . This is the largest state of India, it comprises 11% of the total geographical area of the country and consists of 33 districts, 44,795 villages and . One of oldest folded mountain chains of the world, the Aravalli is stretching from southwest to northeast of the state with a distance of more than 550 km. This mountain is playing a major role in the climatic as well as physiographic pattern of Rajasthan. The Arabian branch of Monsoon hits the Eastern slope of Aravalli and the Eastern part gets sufficient rainfall whereas the Western part remains dry. The South-Eastern part of Aravalli is characterized by humid climate, green vegetation and fertile soil whereas North-Western part has desert or semi desert condition.

2. Methodology

Daily rainfall data from Indian Meteorological Department for a period of 2002-03 and 2016 of 33 dictrict were collected and processed on Excel sheets according to the requirements

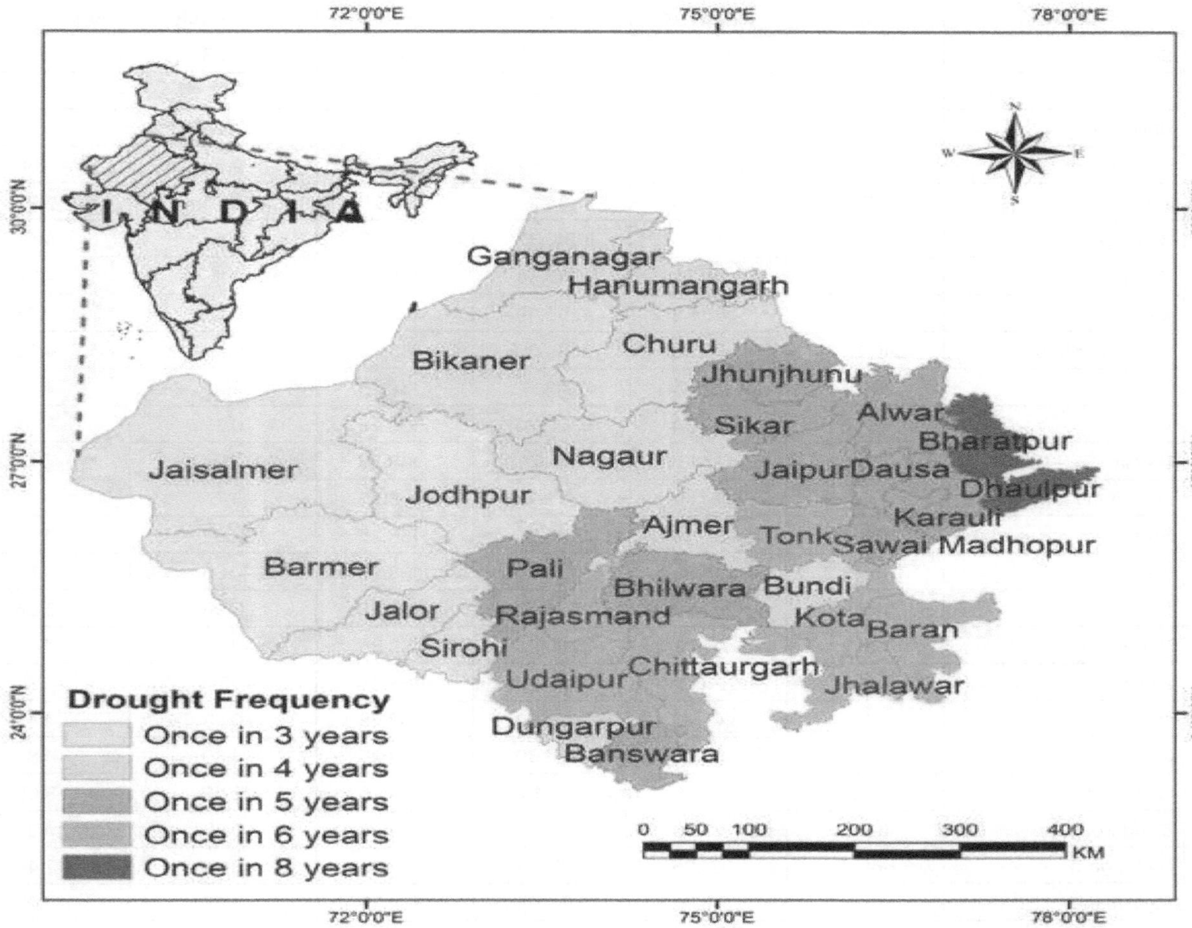

Fig. 27.1 Study area

Standardized Precipition index (SPI) : The SPI was defined by Komuscu (1999) and Me Kee et al. (1995) with Following classes

$$\text{SPI} = \frac{X - \bar{X}}{\sigma}$$

Where X = Precipitation for the station.

$\bar{X}$ = Mean precipition -334.44

σ = Standardized deviation.

Sl. No	District	X	$dx = x - \bar{X}$	dx^2	σ	SPI-16 (July)	SPI-16(August)
1	Bikaner	149.9	–184.54	34055.01	168.25	–1.09	–1.01
2	Churu	133.8	–200.64	40256.40	168.25	–1.19	–2.21
3	Sri Ganganager	93.4	–241.04	58100.88	168.25	–6.23	–2.53
4	Hanumangarh	92.7	–241.74	58438.22	168.25	–6.27	–1.04
5	Jodhpur	253.4	–81.04	6567.48	168.25	–1.09	1.98
6	Barner	205.4	–129.04	16651.32	168.25	–2.75	1.21
7	Jaisalmer	84.7	–249.74	62370.06	168.25	–6.51	–2.24
8	Jalore	334.2	–0.24	0.05	168.25	–0.24	2.68
9	Pali	564.4	229.96	52881.60	168.25	–5.54	–3.56

Sl. No	District	X	$dx = x - \bar{X}$	dx^2	σ	SPI-16 (July)	SPI-16(August)
10	Sirohi	469.2	134.76	18160.25	168.25	−2.88	o.38
11	Ajmer	248.9	−85.54	7317.09	168.25	−12.6	−2.96
12	Bhilwara	420.2	85.76	7354.77	168.25	−1.20	0.98
13	Nagaur	199.6	−134.84	18181.82	168.25	−2.94	1.86
14	Tonk	327.5	−6.94	48.16	168.25	+0.64	2.97
15	Bharatpur	232.9	−101.54	10310.37	168.25	1.84	4.46
16	Dholpur	223.7	−110.74	12263.34	168.25	2.16	4.31
17	Karauli	268.9	−65.54	4295.49	168.25	0.141	3.87
18	Sawai madhopur	409.5	75.06	5634.00	168.25	0.73	4.53
19	Jaipur	200.6	−133.84	17913.14	168.25	2.91	4.32
20	Alwar	261.5	−72.94	5320.24	168.25	0.71	2.76
21	Dausa	358.6	24.16	583.70	168.25	+0.94	2.47
22	Jhunjhunu	173.4	−161.04	259.33	−168.25	−3.77	0.67
23	Sikar	199.9	−135.54	18101.01	168.25	−2.93	1.46
24	Kota	433.3	98.86	9773.29	168.25	−1.68	3.76
25	Baran	568.8	234.36	54924.60	168.25	−5.98	−2.56
26	Bundi	428.2	93.76	8790.93	168.25	−1.50	−1.00
27	Jhalawar	559.5	225.16	50652.00	168.25	−5.70	−1.32
28	Udaipur	379.8	45.36	2057.52	168.25	+0.93	1.42
29	Bansawara	514.5	180.06	32421.60	168.25	−4.30	−1.87
30	Chittorgarh	781.1	446.66	199505.15	168.25	−12.46	−3.06
31	Dungarpur	408.3	73.86	5455.29	168.25	−0.67	1.23
32	Pratapgarh	666.4	331.96	110197.44	168.25	−8.97	−2.45
33	Rajsamand	390.6	56.16	3153.94	168.25	−1.57	−1.07
	Total	11036.8	−13.56	931994			

Sl. No.	Districts	Drought year (2002)			Wet year (2003)		
		SPI 02-Jul	SPI 02-Aug	SPI 02-Sep	SPI 03-Jul	SPI 03-Aug	SPI 03-Sep
1	Ajmer	−1.24	−1.47	−1.5	0.49	0.23	0.25
2	Alwar	−1.54	−1.54	−1.58	0.67	0.6	0.54
3	Banswara	−1.61	−1.5	−1.63	1.15	1.22	1.07
4	Baran	−1.27	−1.31	−1.33	0.86	0.77	0.51
5	Barmer	−1.14	−1.24	−1.37	0.48	0.19	0.15
6	Bharatpur	−1.3	−1.48	−1.48	0.73	0.77	0.67
7	Bhilwara	−1.24	−1.48	−1.72	0.33	0.06	0.04
8	Bikaner	−1.33	−1.65	−1.88	0.58	0.39	0.36
9	Bundi	−1.54	−1.9	−1.81	1.05	0.69	0.58
10	Chittaurgarh	−1.32	−1.52	−1.85	0.44	0.44	0.35
11	Churu	−1.78	−1.87	−2.04	1.07	0.95	0.75
12	Dausa	−1.71	−1.88	−1.58	0.54	0.34	0.06
13	Dhaulpur	−1.59	−1.61	−1.5	0.84	0.88	0.78
14	Dungarpur	−1.33	−1.29	−1.59	0.9	0.72	0.47
15	Hanumangarh	−1.28	−1.62	−2.35	0.5	0.35	0.58
16	Jaipur	−1.61	−1.81	−2.18	0.58	0.36	0.28

Sl. No.	Districts	Drought year (2002)			Wet year (2003)		
		SPI 02-Jul	SPI 02-Aug	SPI 02-Sep	SPI 03-Jul	SPI 03-Aug	SPI 03-Sep
17	Jaisalmer	–1.21	–1.33	–1.77	0.58	0.66	0.18
18	Jalor	–1.44	–1.63	–1.4	0.65	0.47	0.6
19	Jhalawar	–1.44	–1.63	–1.97	0.68	0.61	0.39
20	Jhunjhunu	–1.6	–1.82	–1.65	0.67	0.6	0.47
21	Jodhpur	–1.39	–1.79	–1.96	0.03	0.02	0.52
22	Karauli	–1.42	–1.53	–1.88	1.01	1.01	0.02
23	Kota	–1.08	–1.32	–1.48	0.39	0.3	0.98
24	Nagaur	–1.25	–1.48	–1.49	0.53	0.42	0.29
25	Pali	–1.4	–1.64	–1.71	0.35	0.3	0.41
26	Rajsamand	–1.37	–1.6	–1.95	0.7	0.57	0.35
27	Sawai Madhopur	–1.21	–1.43	–1.61	0.84	0.71	0.46
28	Sikar	–1.44	–1.74	–1.39	0.81	0.69	0.52
29	Sirohi	–1.72	–1.87	–1.85	0.48	0.36	0.54
30	Sri Ganganagar	–1.56	–1.79	–1.92	1/04	0.89	0.41
31	Tonk	–1.31	–1.54	–1.59	0.76	0.52	0.24
32	Udaipur	–1.12	–1.29	–1.38	0.66	0.69	0.52

SPI value	Category	Probability (%)
>2.0	Extremely wet	2.3
1.5–1.99	Very wet	4.4
1.0–1.49	Moderately wet	9.2
_0.99 to 0.99	Near normal	68.2
_1.0 to _1.49	Moderately dry	9.2
_1.5 to _1.99 Very dry 4.4	Very dry	4.4
<_2.0	Extremely dry	2.3

A significant difference in SPI of drought year (2002) and wet year (2003) can be observed from the fortnightly pattern of SPI . The average SPI of Rajasthan was found more than 0.75 throughout the kharif season of 2003 whereas it was much less during the year 2002. The SPI value reached below _1.50 in the semiarid district Ajmer and humid Banswara during September, 2002. The results obtained from the fortnightly pattern of SPI significantly of districts

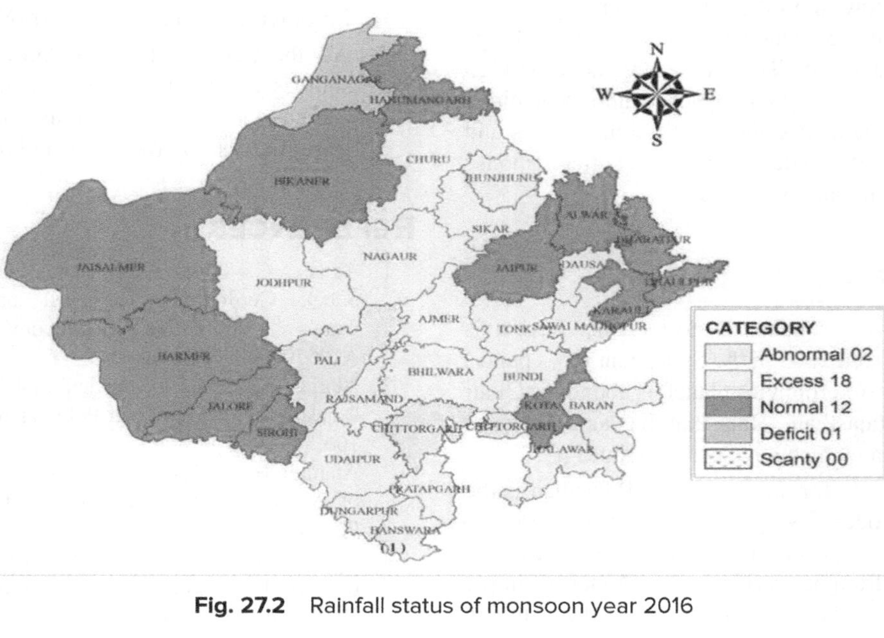

Fig. 27.2 Rainfall status of monsoon year 2016

List of districts showing highest negative change between two time periods

Rainfall indices generated	Districts showing Highest range of positive change	Districts showing Highest range of negative change
Annual rainfall	Jaisalmer, Barmer, Jalore, Udaipur, Pali	Bikaner, Swaimadhopur, Jaipur, Chittorgarh
Change in number of wet days	Pali, Nagour, Udaipur, Jalore, Chittorgarh, Dholpur, Udaipur	Ganganager, Jodhpur, Bikaner, Ajmer, Jaisalmer
Change in June month rainfall	Bikaner, Jodhpur, Alwar	Jaipur, Banswara
Change in July month rainfall	chilwara, Udaipur, Chittorgarh, Banswara	Sikar, Jaipur, Swaimadhopur
Change in August month rainfall	Jaisalmer, Barmer, Pali, Baran	Chittorgarh, Jahalawar Banswara
Change in September month rainfall	Jolore, Pali, canswara Udaipur, Baran, Alwer, Dausa, Dholpur	Bhilwara, Bikaner

3. Conclusion

The study provides a comprehensive knowledge of change in rainfall indices and climate change in Rajasthan. High negative change in rainfall indices have obtained for the eastern and southeastern parts of the state despite of having higher long time averages than other parts in the state.

Similarly SPI drought index depicts that droughts in the state varies spatially and temporally from mild to severe, for all districts. Study shows higher percentage of occurrence of mild drought in the state covering large area, whereas severe drought frequency increases on long time scale.

Short term analysis shows the loss of soil moisture and agricultural droughts and it requires short term measures like good irrigation system, and crop rotation for the critical areas.Similarly in long term analysis shows the impacts of droughts on stream flow and other surface water resources so proper conservation and management system and recharge capacities should be develop . The whole state is at the high riskl due to climate change and its impacts cannot be avoided completely but with the assessment of different indices and using advance toolsl better understanding and better response can be generated to mitigate the impacts

4. Suggestion

1. According to vijay Thakur Report (This Report is published 13 march 2003 in the Statesman news paper) says **Desest rose** (xerophytes plants) is abundane Okair thickel in the Rajasthan . Rajasthan has an abundance of flora and fauna which have medicinal usage.
2. A " **Sheep Model**" for draught prone Rajasthan is the most sutable model. Five year long intensive research on sheep reasing opportunities in rajathan and introduce the model for drought proofing. The Model is simple to understand but hard to implement. In the situation of Drought, drought effect people want to veterirany doctors for vaccination of sheep, made a common plan from for drinking weater.
3. **Famine code** needs to be reviewed . The government efforts are confined only to cows and bullocks. The sheep, goats and camels play a very important role in the desert economy and are equally vulnerable to the threat of drought. Therefore , the famine code needs to be reviewed include not only stray cattle but also all important farm animals with special emphasis on protecting the livelihood of Weaker section.
4. Water crisis worsening in the desert state of Rajasthan, the state government should focus on **community-based water management** solutions instead of predominantly engineering-based ones. The first step taken by the government is to revamp most of the ponds and Johars under the Mahatma Gandhi National Rural Employment Guarantee Act (MNREGA).
5. All the traditional **water harvesting** structures and sources should be renovated and people should be encouraged for roof-top rain water harvesting, storm-water harvesting, recycling and reuse of waste water.

References

1. Agnew, C. T., & Chappell, A. (1999). Drought in the Sahel. GeoJournal. Retrieved on August 13, 2014, from http://link.springer.com/article/10.1023%2FA%3A1007059403077#page-1
2. Ahuja, K. (1973). Agricultural Underemployment in Rajasthan. Economic and Political weekly, 8 (39), pp. 101-106.
3. Alamgir, M. (1980). Famine in South Asia: Political Economy of Mass Starvation. Cambridge, Oelgeschlager and Jain Publishers.
4. Babur Z. M., Babur Nama (Memoirs of Emperor Babur, Eng. tr. A. S. Beveridge) Orient Books Reprint Corporation Delhi, 1970, 2 vols. (Reprint).

5. Bayly, C. A. (1828).Indian Society and the Making of the British Empire (1828), 48.
6. Banerji, R. C., & Upadhyay, D. S. (1975). A survey of drought and scarcity in Rajasthan. Meteorological Centre, jaipur, Rajasthan. 42 (A1) Retrieved on October 3, 2014, from http://www.new1.dli.ernet.in/data1/upload/insa/INSA_1/20005bae_15.pdf
7. Benson, C., & Clay, E. (1998). The impact of drought on Sub-Saharan African Economies: A preliminary Examination. Wold Bank. Retrieved on October 6, 2014, from http://documents.worldbank.org/curated/en/1998/03/693685/impact-drought-sub-saharan-african- economies-preliminary-examination#
8. Bhatia, B. (1967). Famine in India: a study in some aspects of the economic history of India (1860- 1965). Delhi: Asia Publishing House.
9. Bhattacharjee, G. (2017). MGNREGA as Distribution of Doles. Economic and Political Weekly, Vol 52, Issue No. 25, Jun 24, 2017.
10. Bokil, M. (2000). Drought in Rajasthan: in search of a perspective. Economic and Political Weekly, 35(48) pp. 4171–4175.
11. Brara, R. (1989). Commons' Policy as Process: the case of Rajasthan, 1955-1985. Economic and Political Weekly,24 (40), pp. 2247–2254.
12. Drought Management Division. (2014). Crisis Management Plan. Ministry of Agriculture, Department of Agriculture and Cooperation. Government of India. Retrieved on September 23, 2014, from http://agricoop.nic.in/image default/droughtmanagement/CMP2432014.pdf
13. Narain, P., and Kar, A. 2005. Drought itt western Rajasthan; impact, coping mechanism and management strategies. Published by Arid agro-ecosysrem Director, NATp, Central Arid Zone Research Institute, Jodhpur-342 003, India 104 p.

Chapter 28

Floods as Prequels to Droughts and Lessons for Drought Resilience

I. Kavila
School of Pure and Applied Physics, Mahatma Gandhi University,
Kerala 686560 India
kindulekha@gmail.com, indulekha@mgu.ac.in

B. V. Hari
Wipro, Electronic City Phase 1, Bengaluru,
Karnataka 560100 India

Amrutha Sasidharan
Department of Physics, Sree Narayana College, Nattika,
Kerala 680566 India

Abstract

The state of Kerala is a narrow strip of land lying at the south-western tip of the Indian peninsula. Despite receiving annual rainfall of the order of 3000 mm the region is witnessing steady water table draw down and worsening dry season water shortages. The possible reasons for the anomaly of water shortage in spite of ample rainfall is explored. The impact of various distinctive features of the region like topography, high population pressure, extensive land use/land cover (LULC) changes and changes in agricultural practices etc. on hydrology is critically examined. In particular, the impact of filling up of paddy fields and marshlands and conversion of forests into plantations and the concerted effect of changes in IFDP (Intensity Frequency Duration Phase) of rainfall alongside of these, is estimated. From the insights gained, policy measures towards achieving drought resilience are drafted and presented.

The rivers of the region are self-contained, originating from the Western Ghats, and all except three flowing into the Arabian sea. The land, which slopes from the Ghats to the sea, is in its topography, similar to an island. Such a feature is common to landmasses in general; most rivers drain into the sea. The hydrological analysis and the measures suggested here for drought resilience, have relevance to more general situations, due to this.

Keywords: Hydrology – Land use land cover changes – aridization – Policy measures

I. Introduction

Most regions of the world are witnessing rapid and large anthropogenic changes in land use under increasing population pressure. A large fraction of such changes in land use have been at the cost of natural forests. Vitousek et al (2012) had pointed out that the impact of large scale changes in land use could go beyond global climate change. Here, some of the possible impacts of high population pressure on the local hydrology is analysed, by exploring the impact of some of the main land use changes that have taken place for habitation and livelihood purposes in the Indian State of Kerala. The significance of various local, traditional agricultural practices is also examined briefly from a hydrological point of view.

The small State of Kerala in South India, enjoys abundant rainfall. The Travancore region of Kerala has historically been one of the most densely populated areas of the world with one of the highest rates of population growth (Nair 1953; Namboodiri 1968). Presently, the northern regions of the state have comparatively higher population growth rate, compared to the southern portions that include the erstwhile

Travancore. The state lies between 8°17' and 12°47' N latitude and 74°51' and 77°23' E longitude, at the western tip of the Indian peninsula. Kerala is a hydrologically isolated, narrow strip of land, sandwiched between the Western Ghats on the East and the Arabian Sea on the West. The state receives abundant monsoon rainfall of order 3000 mm annually. The region provides ecosystem services to a dense populace, housed practically over its whole extent. The 2011 census reports for India, records the population density of Kerala at 860 / km2 (Census of India 2011, 2) over the 38,863 km2 of its area. This is less than the population density of other such extensively densely populated regions of the world like Bangladesh and the Indian states of Bihar and Uttar Pradesh (Fig. 28.1). However, Kerala constitutes a hydrologically isolated region. Its rivers are short and rapidly flowing; their catchment areas lie practically completely within the state boundaries. The topography of the land is similar to that of a continent or island in that the land slopes from mountains to the sea. Thus, Kerala constitutes a meaningful location for studying and drawing general conclusions regarding the hydrological impact of large scale changes in land use under population pressure.

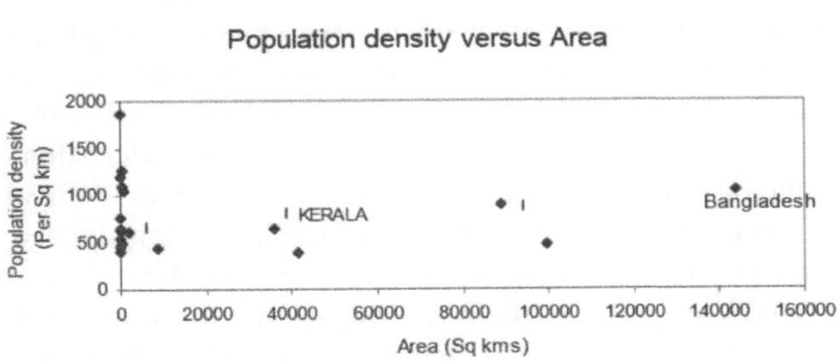

Fig. 28.1 Population density per square kilometre versus area in square kilometre of the most densely populated regions of the world (data courtesy wikipedia.org, figure by authors)

2. Methods

Over the half century from 1961 to 2011 the population of Kerala saw a *70%* increase, its area under rice cultivation a *75%* decrease, and the area under rubber a *364%* increase (Table 28.1). The per capita available cultivable area went down to less than *30* cents over this period (*Economic Review (1961 to 2011)*). Ramsar Convention on Wetlands Reports (2018) point out that the world has lost around *75%* of the wetland resources of the world just over the last century. Due to the large and world- wide demand for latex, greater than *2* Mha of land have been put under rubber cultivation in densely populated tropical South East Asia, over the last decade itself (Singh et al 2021).

Here, the loss to ground water due to the stoppage of detainment of water in wetlands and its infiltration from there is examined and a rough estimate of the loss made, using values for the relevant parameters as determined for paddy fields. Further, changes to hydrologically relevant soil characteristics under land use change from forest to rubber plantation is explored using data sourced from Anilkumar et al (2016). The significance of traditional agricultural practices for the hydrology of the region is also briefly examined. Increase in population, with concomitant increase in built up area and decrease in forest cover, is known to lead to increased runoff. From the analysis presented here, a picture of increased runoff, decrease in water storage as well as changes to the persistent flow emerging from the water stored in the soil emerges. Finally, the results are discussed and conclusions presented.

Table 28.1 Population, area under paddy cultivation and area under rubber cultivation in Kerala from years 1961–2011 (data extracted from Economic Review (1961 to 2011), table by authors)

Year	Population (in Million)	Area under paddy (in Mha)	Area under rubber (in Mha)
1961	19.604	0.752	0.115
1971	21.347	0.875	0.188
1981	25.453	0.807	0.248
1991	29.098	0.560	0.384
2001	31.836	0.347	0.474
2011	33.387	0.213	0.534

Loss of wetlands and decrease in seepage to groundwater from wetlands: The clayey soil of wetlands capture, detain and store rainfall. The water is slowly released in the summer months and is an important component of the persistent flow during the dry season. Loss of the world's wetlands thus constitute an annually compounding deficit in the hydrological services provided by ecosystems suffering the loss. The puddled soil of rice fields, with high clay content and organic matter, may be taken as a proxy for the soil of wetlands. In the case of Kerala, we examine whether the loss of paddy land which had been under Mundakan cultivation could have a significant impact on water security of the region. Large tracts of low lying land which had been under paddy cultivation has been levelled in Kerala for conversion to other use. Over less than half a century the area under paddy cultivation, which had gone up to *0.875* Mha, decreased by about *75%* (Table 28.1).

Saturated hydraulic conductivity of the soil in rice fields, after puddling, range from ~*10-7* to *10-4* cm/s (Naphade and Ghildayal 1971; Aimrun and Amin 2009). The contribution to ground water via seepage from paddy fields during the approximately *150* days of Mundakan cultivation when the fields are flooded, is thus greater than about *1.29* cm over a year. Rate of the consequent loss to ground water recharge is greater than *0.07* BCM/yr (billion cubic meter per year). It is seen that, the conservative estimate of the annual loss to ground water of the order of a mm/year made under the assumption that the seepage contributes to ground water over an area ten times larger than that of the paddy field, is non-negligible. If we take a similar estimate as a crude, conservative estimate of annual loss to the water table of a stable ecosystem which contained wetland area, this points to an annual loss of order a mm/year to the water table. This loss can accumulate and become significant over the long run, if measures for amelioration are not taken.

The impact of rubber plantations on hydrological properties of soil: The impact of land use change to rubber plantations, has been extensively studied, with forest cover / forest soil as controls. Various studies have examined evapotranspiration rates over rubber plantations and contiguous natural forests, in view of reports of downstream dry season water shortages and water table draw down in areas which have undergone such a change (see for example Tan et al 2011). A recent review by Singh et al (2021), points out that, the results of such comparative studies consistently point to the impact of such a replacement of natural forests with rubber plantations being negative with respect to ecological processes and ecosystem functions.

Depth resolved analysis of core sample data, from a hundred rubber plantations and twenty one natural forests / sacred groves contiguous with the plantations, was done to explore possible differentials in the dependence of field capacity water on soil organic matter (SOM) content by mass. Linear fits were made to the plots of field capacity water of the sample against SOM content. The slopes of the fitted lines, for forest soil and rubber growing soil, obtained for various depth ranges, are given in Table 28.2.

Table 28.2 Range in depth to the middle of the core sample, slope of the linear fit to the plot of field capacity water of the sample against SOM content of the sample for samples with centres at these depths for forest soil, and for rubber growing soil (values as calculated using data from Anil Kumar et al 2016, table by authors)

Range in depth to the middle of the core sample (cm)	Slope of the linear fit for forest soil	Slope of the linear fit for rubber growing soil
0–20	1.64	0.13
20–40	1.89	–0.59
40–60	2.40	–0.36
60–80	6.74	–2.44
80–100	–1.11	–4.29
100–140	–0.32	–3.50

Traditional agricultural practices of Kerala and water security: Natural ponds are rare in Kerala since water can move underground more easily seaward due to the high drainage properties of the soil. Since the slope from the Western Ghats to the sea is large due to the narrow width of the land (average width *70* km), overland flow can reach the sea practically in under *48* hours from the hill regions. These mandate that measures for capturing as well as detaining the water received as rainfall are essential for ensuring water security for the region. Traditional agricultural practices, as may be seen from the analysis reported below, seem to be aimed at capturing and keeping the water in the soil for as long as possible to avoid dry season water shortages.

3. Results

The population of Kerala increased by *70%* over the half century from 1961 to 2011. The area under rice cultivation underwent a *70%* decrease over the same period. The increase in number of consumers by about *14* M, constitutes an increase in direct water consumption itself of the order of *0.2* billion cubic meter per year (BCM/yr), assuming a minimum consumption rate of *40* lpcd (litre per capita per day). This ignores possible increases in per capita water consumption due to changes in life style and also indirect consumption from water usage in industries / agriculture. A crude estimate of the rate of drawdown of the water table in areas along the direction of flow from wetlands which have been drained / otherwise converted, obtained as above is of the order of a mm/yr. This points to the possibility of net drawdown of the water table by as much as *5* cm over a period of fifty years.

From Table 28.2 it may be seen that opposing trends are shown by rubber growing soils and forest soils in the top *0.2* to *0.8* m of soil. Forest soils show consistently increasing trend of residual water content with SOM, while rubber growing soils show a decreasing trend in the *0.2* to *0.8* cm depth regime. At almost every depth the range of SOM content of rubber growing soils is actually larger than for forest soil too. The larger range is due to the SOM content of samples from rubber growing soils running to higher values compared to forest soils. In the top 20 cm layer it is likely that due to better aeration, litter from rubber trees suffer degradation to similar extends as forest litter.

4. Discussion

From the results presented above it may be concluded that loss of wetlands can cause water table drawdown at rates that are not insignificant. The mean rate of change in ground water level observed for some ground water monitoring wells across Kerala is given in Table 28.3. The data are for a period of thirty odd years for each of the monitored wells / aquifers. It is noticed that the draw down rates are consistent with what may be expected, from the results presented above, for the annual loss to the water table arising from loss of wetlands.

Table 28.3 Location of ground water monitoring well/aquifer, pre-monsoon water table drawdown rate and post-monsoon water table drawdown rate in cm/yr. (data extracted from Ground Water Yearbook (2015-16), table by authors)

Location of monitoring well/aquifer	Pre-monsoon drawdown rate (cm/yr)	Post-monsoon drawdown rate (cm/yr)
Kommeri, Kannur	–0.05	–0.3
Madavur, Thiruvananthapuram	–1.3	–0.33
Kottiyur, Kannur	–1.1	–0.7
Ayur, Kollam	–1.0	–0.1
Palghat, Palakkad	–14.7	–9.3
Mannanthala, Thiruvananthapuram	–5.5	–3.7
Warkalai aquifer, Karthikappally, Alappuzha	–28.8	–21.3
Quilon aquifer, Karthikappally, Alappuzha	–26.6	–15.2
Vadakkancherry, Palakkad	3.6	0.37
Koduvayur, Palakkad	–1.9	14.8

Curve Number Method is a popular method developed by the Soil Conservation Services (SCS) of USA which gives accumulated direct runoff depth Q as a function of accumulated rainfall depth P in terms of potential maximum retention S assuming an Initial Abstraction Ia of $0.2\ S$ (Curve Number Method). A family of curves tracked by Curve Numbers CN, which encapsulate relevant characteristics of the soil via S as $CN = 25400/(254 + S)$, gives Q as a function of P. The larger the value of CN the more the losses to direct runoff. While decrease in infiltration rate can increase the value of CN, the filling of low lying areas like paddy fields, ponds and lakes can change the $Q – P$ relation by decreasing Ia which enters the $Q – P$ relation $Q = (P- Ia)2/(P+S- Ia)$. Also, deforestation can cause soil erosion which can lead to a decrease in Ia via silting of static water bodies. Losses to runoff will be increased due to the decrease in Ia. Figure 28.2 gives the change in runoff depth Q (in cm) versus accumulated rainfall depth P (in cm) for $CN = 25$ and 75 for $Ia = 0.1$ and the standard value 0.2. It is noted that the decrease of Ia from 0.2 to 0.1 can double the runoff losses for accumulated rainfall depth $P = 300$ cm.

In comparing the field water capacity of rubber growing soils and forest soils, it is seen that for rubber growing soils field

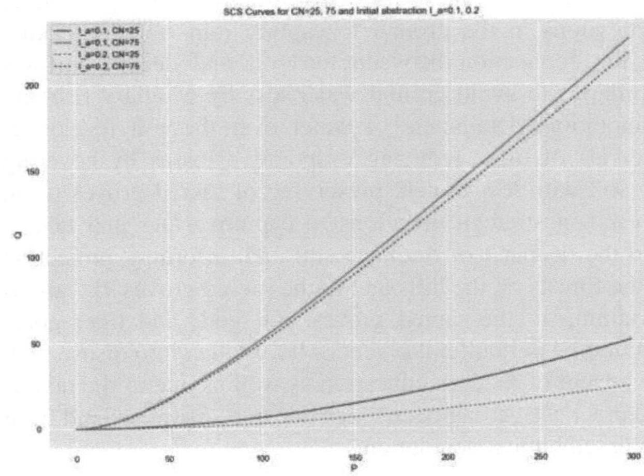

Fig. 28.2 Accumulated direct runoff Q (cm) versus accumulated rainfall depth P (cm) for Ia values 0.2 and 0.1, each for CN values 25 and 75 (figure courtesy: Kavila 2021).

water capacity is decreasing with SOM content in the top *0.2* to *0.8* m depth of soil. This is counter to expectations. The difference, between the slope of this relation, for forest and rubber growing soils, is statistically significant (Kavila and Hari 2021). It is possible that the comparatively high lignin content of the leaves of *Hevea brasiliensis*, which is both difficult to degrade as well as capable of imparting hydrophobicity to soil, might be the reason for this difference. Study of water release curves of artificially hydrophobic soils point out more rapid drainage of capillary water from hydrophobic soils (Liu et al 2012) compared to the steady drainage from non-hydrophobic soil. Rubber plantations have been likened to water pumps, due to dry season water shortage and zero flow days in rivers being experienced by downstream inhabitants, after conversion of natural forests into rubber plantations. Increased evapotranspiration from rubber plantations is often pointed out as a cause for this (Tan et al 2011; Arulbalaji and Maya 2019). The results presented here point to increased rate of desorption of capillary water from the soil of rubber plantations as an additional causative factor for the drying out of the soil. Also, soil hydrophobicity can reduce infiltration and increase runoff losses (Doerr, Shakesby and Walsh 1996)). Such impacts can be aggravated by changes in the IFDP (Intensity, Frequency, Duration, Phase) of rainfall under climate change. Increase in the variance of rainfall intensity, with increase in the incidence of heavy downpours, can compound the increase in runoff losses, even when there is no decrease in the quantum of total annual rainfall.

Kavila (2020) had explored the physics underlying some of the traditional agricultural practices of Kerala. It was concluded that the various practices were aimed at increasing the macro and meso porosity of the soil, generating depressions in the ground to capture rain water, breaking up the connection between top soil and deeper soil via capillaries to avoid ground water loss by capillary rise and evaporation, detainment of water over the soil for longer intervals of time, reducing evaporative losses by covering the soil with leaf mulch, preserving of sacred groves (mini forests) in strategic locations to capture water and take it into the deeper low drainage soil and, affording protection to the forests on the hills and to the sacred groves via taboos pertaining to the sacred groves, hill gods and tiger gods. Traditional agricultural practices have fallen into disuse with the advent of modern education as well as due to the taboos and practices getting overridden due to pressure on land from increased human population.

The results presented highlight the impact of some of the ecosystem changes happening under population pressure on water security of the region. Loss of wetlands, filling up of low lying areas, and large scale conversion of forests to rubber plantations can cause increase in runoff losses and also increase in variance in streamflow due to increase in number of zero flow days. Excess streamflow during heavy showers and consequent large runoff can also increase the variance in streamflow. Over a period of decades, more than a hundred zero flow days are being reported in Kerala for rivers which had been perennial. For example, the Vamanapuram river of Kerala (ESSO-NCESS 2017-18). In Kerala, this river is one of the few rivers which have no irrigation or hydroelectric project across it. Its watershed area however, has undergone extensive land use change from natural forests to rubber and eucalyptus plantations and homesteads (the waxiness of the leaves of Eucalypti is known to cause hydrophobicity for soils (Doerr, Shakesby and Walsh 1996)). From all the above it is possible that, in spite of receiving abundant rainfall, Kerala is moving towards land aridization, under long standing population pressure. The policy measures proposed in the light of the above are presented below.

5. Conclusions

This article examines some threats to the water security of Kerala by looking at the loss to aquifer recharge from the decrease in area under paddy cultivation and the possibility that capillary water desorption may take place more readily and faster from rubber growing soils than from forest soils. The analysis as well as the conclusions reached have wider application, since paddy fields may be taken as a proxy for wetlands in general and also, large tracts of natural forests are being put under rubber cultivation around the world, particularly in the densely populated South East Asia region. It is also pointed out that the traditional agricultural practices evince cognizance of the hydrological vulnerabilities of Kerala. The article points out effects of some anthropogenic changes, which potentially affects water security via their impact on hydrology. These have serious implications and impacts over and above those due to global climate change.

Floods and droughts are often pointed out to be a direct fallout of climate change; rising global temperatures leading to more frequent storms and heavy rains, which lead to floods, and the increased evaporation rates leading to droughts. Floods and droughts have regularly made the news in the past decade (World Bank Feature 2021). The results presented here highlight anthropogenic land use changes as a causative factor for droughts occurring as sequels to floods. The land use changes highlighted have the potential for setting up flood-drought cycles, in regions which have undergone anthropogenic interventions of a nature similar to those analysed here.

Policy measures (which are relevant *mutatis mutandi* for regions under population pressure the world over) suggested are: A) the State should seriously consider measures of population planning to reduce population pressure on the land B) measures for protecting and preserving in pristine form the natural forest cover that is still remaining should be seriously undertaken and C) measures for rewilding of the Western Ghats should be urgently undertaken

6. Acknowledgements

The data used in this study is from the project Anil Kumar K S '*Soil Fertility Assessment and Soil Health Monitoring in Traditional Rubber-growing Areas of Kerala, Tamil Nadu and Karnataka*' National Bureau of Soil Survey and Land Use Planning (NBSS&LUP (ICAR)), Rubber Research Institute of India (RRII), Rubber Board (Kottayam), Ministry of Commerce (Government of India).

IK acknowledges discussions with G. Gopinathan and K. M. Nair.

References

1. Aimrun W., Amin M. S. M. "Pedo-transfer function for saturated hydraulic conductivity of lowland paddy soils" Paddy Water Environ (2009) 7:217–225 DOI 10.1007/s10333-009-0165-y (Aimrun and Amin 2009)
2. Anil Kumar K S et al "*Soil Fertility Assessment and Soil Health Monitoring in Traditional Rubber-growing Areas of Kerala, Tamil Nadu and Karnataka*" 2016 Technical Report, (Indian Council for Agricultural Research) ICAR-National Bureau of Soil Survey and Land Use Planning, Bengaluru, India; Rubber Research Institute of India; Rubber Board, Kottayam

3. Arulbalaji P., Maya K. "Effects of Land Use Dynamics on Hydrological Response of Watershed: a Case Study of Chittar Watershed, Vamanapuram River Basin, Thiruvananthapuram District, Kerala, India" 2019 *Water Conservation Science and Engineering* 4 33–41
4. Census of India 2011 | Office of the Registrar General & Census Commissioner, India. https://censusindia.gov.in/2011census/PCA/PCA_Highlights/pca_highlights_file/kerala/Chapter-1.pdf (Census of India 2011, 2)
5. Curve Number Method engineering.purdue.edu/mapserve/LTHIA7/documentation/scs.htm (accessed 2021)
6. Doerr S. H., Shakesby R. A., Walsh R. P. D. "Soil hydrophobicity variations with depth and particle size fraction in burned and unburned Eucalyptus globulus and Pinus pinaster forest terrain in the Águeda Basin, Portugal" 1996 CATENA Volume 27, Issue 1, Pages 25-47, ISSN 0341-8162, https://doi.org/10.1016/0341-8162(96)00007-0 (https://www.sciencedirect.com/science/article/pii/0341816296000070)).
7. ESSO-NCESS Annual Report 2017-18, Earth System Sciences Organization, Ministry of Earth Sciences, Government of India p49. https://www.ncess.gov.in/images/Annual_Reports/AR_2017- 18_Final.pdf
8. Ground Water Year Book 2015–16, cgwb.gov.in, Government of India http://cgwb.gov.in/Regions/GW-year-Books/GWYB-2015-16/GWYB%20KR%202015-16.pdf
9. Kavila I. "Erstwhile social culture of Kerala pertaining to water security: lessons and the way forward" 2020 International Conference on Intangible Cultural Heritage, V Khandare (Ed.), Lokvidya Prakashan, Indian Institute of Social Sciences and Folklore Research, Parbhani, INDIA ISBN978-81-89730-35–2
10. Kavila I., Hari B. V. "A study of the impact of some land use land cover changes on watershed hydrology" 2021 Proceedings of the First International Conference on River Corridor Research and Management, IIT Jammu, India (25-27 Feb 2021)
11. Liu H, Ju z, Bachmann J, Horton R, Ren T 2012 Moisture-Dependent Wettability of Artificial Hydrophobic Soils and Its Relevance for Soil Water Desorption Curves Soil Sci. Soc. Am. J. 76:342–349 Posted online 3 Jan. 2012 doi:10.2136/sssaj11.0081
12. Nair U. Sivaraman. "Travancore-Cochin Part 1A Report" Census of India Vol XIII 1953. Manager of Publications, Government of India Branch, Civil Lines, New Delhi http://lsi.gov.in:8081/jspui/bitstream/123456789/5616/1/23765_1951_REP.pdf (Nair 1953, 6)
13. Namboodiri Krishnan N. "The changing population of Kerala." Monograph 7, Census of India Monograph Series, 1968. Office of the Registrar General, India Ministry of Home Affairs http://lsi.gov.in:8081/jspui/bitstream/123456789/5644/1/51528_1961_MON.pdf accessed on 12 June 2021 (Namboodiri 1968, 8)
14. Naphade J. D. and Ghildayal B. D. "Effect of puddling on the physical properties of rice soil' 1971 Indian J. Agric. Sc. 41. 1065–67
15. Ramsar convention on Wetlands Reports (2018) The Global Wetland Outlook - Status and Trends 2018, The Ramsar Convention Secretariat, Switzerland https://www.global-wetland-outlook.ramsar.org/
16. Singh A. K., Liu W., Zakari S., Wu J., Yang B., Jiang X. J., Zhu X., Zou X., Zhang W., Chen C., Singh R., Nath A. J. "A global review of rubber plantations: Impacts on ecosystem functions, mitigations, future directions, and policies for sustainable cultivation" 2021 Sci Total Environ. 2021 Nov 20 796 148948. doi: 10.1016/j.scitotenv.2021.148948. Epub 2021 Jul 10 PMID: 34273842
17. Tan Z-H, Zhang Y-P, Song Q-H, Liu W-J, Deng X-B, Tang J-W, Deng Y, Zhou W-J, Yang L-Y, Yu G-R, Sun X-M, Liang N-S. 2011 "Rubber plantations act as water pumps in tropical China" *Geophys. Res. Let.*, Vol. 38, L24406, doi:10.1029/2011GL050006, 2011 https://en.x- mol.com/paper/article/1247285563670597632
18. Vitousek, P. M., H. A. Mooney, J. Lubchenco, *and* J. M. Melillo *(*1997*)*, Human domination of Earth's ecosystems, *Science*, 277, 494*–*499, *doi:*10.1126/science.277.5325.494
19. World Bank Feature "Floods and Droughts: An EPIC Response to These Hazards in the Era of Climate Change" June 17 2021. https://www.worldbank.org/en/news/feature/2021/06/17/floods-and-droughts-an-epic-response-to-these-hazards-in-the-era-of-climate-change
20. wikipaedia.org: en.wikipedia.org/wiki/List_of_countries_and_dependencies_by_population_density (as accessed in 2010)

CHAPTER 29

Reliability, Resilience, and Vulnerability Risk Assessment of Extreme Hydrological Disasters Using Gravity Recovery and Climate Experiment (GRACE) of India

Sachin Bhere
Department of Civil Engineering,
Indian Institute of Technology Bombay 4000076, India
studentsachinbhere@gmail.com

Manne Janga Reddy
Department of Civil Engineering,
Indian Institute of Technology Bombay 4000076, India
Interdisciplinary Program (IDP) in Climate Studies,
Indian Institute of Technology Bombay 4000076, India
mjreddy@civil.iitb.ac.in

Abstract

The changes in the various water storages are inevitable, and it is important to understand the scale of such changes. The Gravity Recovery and climate experiment (GRACE's) terrestrial water storage is column integrated data that includes groundwater, soil moisture, surface runoff, and snowmelt, making it more convenient to analyze the total change in the water availability. This study focuses on calculating the sustainability of the terrestrial water storage using the Reliability-Resiliency-Vulnerability assessment. To calculate that, the GRACE's TWS data is used to form an index which helps to identify the unique events which otherwise would be difficult to identify due to annual cycle and seasonality. The reliability of this index is considered as the frequency of the satisfactory events, resilience is the capacity to recover from the unsatisfactory event, and vulnerability is the extreme magnitude of the events. Sustainability is calculated as the integration of Reliability, Resiliency, and Vulnerability, and then the spatial distribution is plotted. It helps to identify the changes in the groundwater storage at Punjab and Haryana along with the snowmelt at the Himalayas and surface runoff in the southern peninsular after comparing with the individual hydrological parameter of the Global Land Data Assimilation System (GLDAS) and in-situ groundwater observations.

Keywords: Reliability, Resiliency, Vulnerability, sustainability, GRACE, GLDAS, TWS

1. Introduction

The Reliability-Resiliency-Vulnerability assessment for designing the water resources system has been used for more than four decades. Hashimoto et al., 1982 proposed the use of additional criteria of resiliency and vulnerability for determining the risk, and since then, the Reliability-Resiliency- Vulnerability(RRV) assessment has been widely used for many applications in hydrology and water resources. Reliability is the oldest and widely used performance criteria, which were interpreted as the probability of the system being in good condition by (Hashimoto et al., 1982). The reliability is also interpreted as the frequency of the satisfactory events. Resiliency is measured as to how fast the system can recover from the unsatisfactory event (Moy et al., 1986). The vulnerability is interpreted in different ways by different researchers according to their applications. According to Hashimoto et al., 1982 vulnerability is the sum of the probability of extreme events. Kundzewicz & Kindler, 1995 proposed the vulnerability as the maximum magnitude of the unsatisfactory event, which become more convenient to calculate and adopted by various researchers (Maity et al., 2013; Najafi et al., 2021; Paton et al., 2014). This study focuses on the sustainability of terrestrial water storage across India using the RRV assessment. To calculate that the GRACE's TWS data is used of three data centers Jet propulsion laboratory (JPL), German Research Center for Geosciences Potsdam (GFZ), and the University of Texas at Austin Center for Space Research (CSR), which provides the monthly gravitational changes of earth leads to change in terrestrial water storage in terms of water equivalent thickness in cm (Diaper et al., 2007; Hasan, 2009; Swenson et al., 2003; Swenson & Wahr, 2002). The data centers also provide different data products for various applications, including hydrology. With the help of very precise measurements at a scale of micrometer, the distance between two satellites is measured, which are orbiting the earth at a very low orbit of about 300 to 500 km. The distance between the satellites, which is 220 km, approximately changes due to the gravitational pull of the earth, and that variation is recorded over the period of time and compared to calculate any changes in the gravity which is caused due to the changes in the mass (Bai et al., 2018; Bhere & Regulwar, 2000; Cammalleri et al., 2019; Dembélé et al., 2020; Feng et al., 2013; Khaki et al., 2017; Sharma, 2014). The K-band ranging instruments help to calculate that changes with the precision of micrometers observed by the four key instruments of the GPS receiver. The variation of the gravity field recorded by the GRACE may occur due to various influences, including the core of the earth, which can merely be observed with the longer time scale. The other influences, including the redistribution of the atmosphere, water storage, including the ocean, groundwater, snow, and surface water, can be reflected in the GRACE's changes in the gravity field and which is of the prime concern of this study. The changes in the groundwater, soil moisture, snow and surface runoff can be identified with the GRACE's TWS data. The objective of this study is to determine the spatial variation of sustainability derived from the RRV assessment. The study also extended by understanding the influence of individual hydrological parameters. The Global Land Data Assimilation System (GLDAS) data is used to identify the individual influence of snow melt and surface runoff. The groundwater changes is also understand by studying the trend of more than 5000 observation well across India. This helps to justify the low sustainability across Punjab and Haryana state.

The paper is divided into four sections. The methodology sections discuss the method, formulation of the index, RRV assessment, and detailed information of the data and its sources. The result section focuses on the finding of the trend analysis along with the spatial distribution of RRV assessment and sustainability. The discussion section is more focused on justifying the variation of sustainability across the country by comparing the individual parameters like snowmelt, surface runoff, and groundwater level changes. The conclusion section discusses the key findings and the future scope of the study.

2. Methodology

The study focuses on understanding the sustainability of terrestrial water storage (TWS) across India by calculating the reliability-resiliency-vulnerability of GRACE's TWS data. The terrestrial water storage data includes the groundwater, soil moisture, surface runoff, and snowmelt runoff. The combination of these hydrological parameters gives the overall perception of the available water. The GRACE's TWS data gives a variation or changes of these components by interpreting the changes in the earth's gravitational field. The GRACE's TWS anomaly gives the data into the water equivalent thickness in cm. With the help of water equivalent thickness, the monthly index is formed to understand the comparative monthly and sessional changes of the terrestrial water storage. The index indicates the deviation from the monthly average of 20 years. The reliability is considered as the frequency of the events which are above the zero-line of the index, indicating the desirable events considering the water availability. The resiliency is calculated by considering how many times the terrestrial water storage is recovering from deficit to the normal values. The vulnerability is the lowest value of deficit of the TWS, and the sustainability is calculated by combining the reliability, resiliency, and vulnerability. The spatial distribution indicates the sustainability over various homogeneous monsoon regions over India. The following section discusses the data used for the RRV assessment of GRACE's TWS data, along with the data used for justifying the spatial variation of the sustainability of the TWS.

2.1 Data and Sources

The Gravity Recovery and climate experiment (GRACE) is a joint satellite mission of NASA and the German Aerospace Center (DLR) with the objective to provide the monthly, spatially wide gravity field of the earth. The Spatio-temporal variation of the earth's gravity field leads to the changes of the terrestrial water storage, which include the changes of snow, surface water, and groundwater, including soil moisture. For this study, the GRACE's TWS data of 1-degree spatial resolution data of three major processing stations are used from April 2002. The Jet propulsion laboratory manages the project and data along with The German Research Center for Geosciences Potsdam (GFZ), the University of Texas at Austin Center for Space Research (CSR), which also provides the monthly data of gravitational changes. The data centers also provide different data products for various applications, including hydrology. With the help of very precise measurements at a scale of micrometer, the distance between two satellites is measured, which are orbiting the earth at a very low orbit of about 300 to 500 km. The distance between the satellites, which is 220 km, approximately changes due to the gravitational pull of the earth, and that variation is recorded over the period of time and compared to calculate any changes in the gravity which is caused due to the changes in the mass. The K-band ranging instruments help to calculate that changes with the precision of micrometers observed by the four key instruments of the GPS receiver. The variation of the gravity field recorded by the GRACE may occur due to various influences, including the core of the earth, which can merely be observed with the longer time scale. The other influences, including the redistribution of the atmosphere, water storage, including the ocean, groundwater, snow, and surface water, can be reflected in the GRACE's changes in the gravity field and which is of the prime concern of this study. The GRACE' TWS data gives such changes or redistribution of the water and snow at a monthly scale. The data suggest the changes in the available water resources in various forms. For this study, the level 3 version 4 data of the GRACE's TWS is used, which is provided by three data centers. The various studies suggested using all three data sets from data centers to minimize the error. The monthly changes of the TWS are used to form a GRACE-based index for RRV assessment. The Global Land Data Assimilation System (GLDAS) data is used to compare the sustainability calculated using the RRV assessment of the GRACE's TWS with different individual hydrological parameters. The local station-wise data of groundwater levels across India is also used to understand the influence of the changes in the groundwater on the sustainability of terrestrial water storage. The groundwater levels are collected by the various central and state government agencies and available to download on the integrated platform of the India Water Resources Information System (India WRIS). The following section describes the detailed method for calculating the GRACE-based Index, RRV assessment, and the sustainability of the terrestrial water storage.

2.2 Method

This study aims to calculate the sustainability of terrestrial water storage across India using the reliability-resilience-vulnerability assessment. This section elaborates the discussion on the method for achieving this objective. It includes a detailed discussion on the GRACE-based index formulation and the reliability-resilience-vulnerability framework for calculating sustainability. The section is further extended to understand the influence of the various hydrological parameters on TWS sustainability.

GRACE-based Index

The GRACE TWS data gives the anomaly in terms of the water equivalent thickness in cm. The index is formed by considering the monthly deficit of the TWS by the monthly mean, and then it is divided by the standard deviation. With the GRACE-based index (GI), the unique events that are different by the annual cycle or regional seasonality can be identified. The following formula is proposed by (Reager et al., 2015) for GRACE-based deficit,

$$\text{Deficit}_{TWS} = TWS_{ij} - \overline{TWS_i}$$

The deficit is calculated by subtracting the monthly average of 20 years by the monthly value of TWS.

TWSij indicates the terrestrial water storage value of ith month of jth year, which is then subtracted by the $\overline{TWS_i}$, which is indicates the mean value of TWS of ith month. The index is then formed by dividing the deficit by the standard deviation of the monthly TWS.

$$GI_{ij} = \frac{\text{Deficit}_{TWS}}{SD_i}$$

The SDi indicates the standard deviation of ith month, and by dividing it the index is formed. This index denotes the standardized monthly deviation value of the TWS data from its monthly mean.

Reliability-Resilience-Vulnerability assessment

After calculating the GRACE-based index, the unsatisfactory events (U) are identified. The events with the negative value of the GRACE-based drought index are considered the unsatisfactory events (U), and the events with the positive value are considered satisfactory events (S) of water availability.

The reliability is calculated for ath row, and bth column of the spatial distribution of the index as the frequency of the satisfactory event occurred, resiliency is calculated as the frequency of recovering from unsatisfactory events (Ur), and the vulnerability is considered as the magnitude of the extreme unsatisfactory event. The following formulas are the mathematical expression of the Reliability- Resilience-Vulnerability assessment.

$$\text{Reliability}_{(a,b)} = \sum_{i=1}^{n} S_{i(a,b)}/n$$

$$\text{Reliability}_{(a,b)} = \sum_{i=1}^{n} (Ur)_{i(a,b)}/n$$

$$\text{Vulnrability}_{(a,b)} = \min_{1=1\,\text{to}\,n} (GI_{ij})_{(a,b)}$$

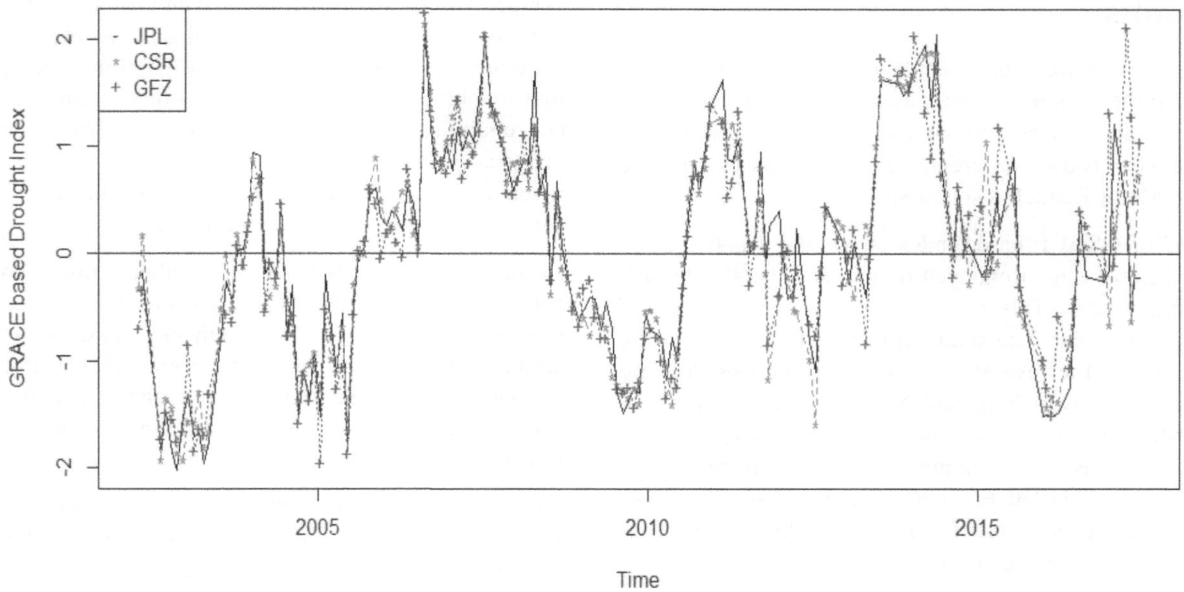

Fig. 29.1 GRACE-based index of three GRACE's TWS datasets. The events above zero line are considered as satisfactory events, and the events with negative values are unsatisfactory events.

The above formulas are for Reliability-Resilience-Vulnerability assessment of ath row and bth column of spatial distribution for all the events of ith month of jth year. After calculating the reliability resilience and vulnerability, sustainability is calculated. To calculate sustainability, it is important to understand that for higher sustainability, a high value of reliability and resilience is required, and low values of vulnerability are desired. For this, the vulnerability is normalized and rages from 0 to 1 (high vulnerability to low vulnerability). The following formula is a mathematical representation of the sustainability of terrestrial water storage.

$$\text{Sustainability} = \text{Reliability} \times \text{Resiliency} \times \text{Vulnerability}$$

The above formula will give the sustainability at each pixel using the Reliability-Resilience-Vulnerability assessment. After calculating the sustainability, the spatial distribution is plotted to understand the variation of the terrestrial water storage across India. The findings are discussed in the result and discussion section.

To understand which parameters are influencing sustainability, the trend is calculated of various hydrological parameters. The snowmelt runoff and snow depth trend is calculated to understand the changes caused due to the Himalayan snowmelt on the runoff of the Ganges basin. The trend of the root zone soil moisture, groundwater, and surface runoff is also calculated to understand the spatial variation of individual parameters to justify the changes in the sustainability of terrestrial water storage. For trend analysis, the Mann-Kendell test and Sen's slope are used. This method is widely used for calculating the trend in the field of hydrology. The following formula can calculate the statistics (St)

$$St = \sum_{k=1}^{n-1} \sum_{j=k+1}^{n} \operatorname{sgn}(x_j - x_k)$$

$$\operatorname{sgn}(x_j - x_k) = \begin{cases} +1, & \text{if } (x_j - x_k) > 0 \\ 0, & \text{if } (x_j - x_k) = 0 \\ +1, & \text{if } (x_j - x_k) < 0 \end{cases}$$

The x_k and x_j are calculated from the series of k and j. After calculating the trend, the spatial distribution is analyzed. The significance of the trend is calculated by,

$$z = \begin{cases} \dfrac{St - 1}{\sqrt{\operatorname{Var}(St)}}, & \text{when } St > 0 \\ 0, & \text{when } St = 0 \\ \dfrac{St - 1}{\sqrt{\operatorname{Var}(St)}}, & \text{when } St < 0 \end{cases}$$

The Var(St) is the variance of the statistics. Using the above formula, the trend is calculated for all the parameters for every pixel. The next section discusses the results of spatial variation of sustainability using RRV assessment.

3. Results

The trend of terrestrial water storage is calculated to understand the changes in water availability over a period of time across the country. For the analysis, the data from JPL, CSR, and GFZ is used differently. The trend analysis is done using the Mann-Kendell test and Sen's slope.

The trend of GRACE-based index shows the negative trend at the Ganges basin, Himalayan region, and parts of Punjab and Haryana state. This region shows a negative trend of more than -10 /year. The southern peninsula also shows the negative trend of terrestrial water storage. The west-flowing central river basin of Tapi and Narmada shows a positive trend, indicating the increase of surface runoff in terms of floods. Figures also show not many significant changes in the three data sets of GRACE's TWS. The trend simply shows the overall changes of the TWS and not the capability of recovering from extreme events.

The reliability, resilience, and vulnerability are calculated, and the spatial distribution is plotted across the country. Reliability is the frequency of the satisfactory events to occur, which is calculated by the sum of events that have a positive value of GRACE-based index divided by the total number of events. A high value of reliability is observed across the Ganges basin and Himalayas of around 0.6. The low reliability is observed at Punjab and Haryana, which comes under the northwest zone of rainfall of about 0.4. The reliability only signifies the number of events and not their severity; therefore, the region of Himalaya does show high reliability despite having a negative trend of terrestrial water storage. The similarity is observed in all three data of JPL, CSR, and GFZ. The western ghat also shows high reliability due to high annual rainfall. The upper Krishna basin also has high reliability as compared to the other peninsular basins. The reliability doesn't signify the trend or any changes in the terrestrial water storage as it calculates the overall number of satisfactory events and not the changes in the number of satisfactory events for a certain timespan.

The resiliency is calculated as the number of times it recovers from unsatisfactory to satisfactory events with respect to the total number of events. The resiliency shows the region's capability to recover from any extreme events. It depends on many factors like regions geography, annual rainfall, groundwater recharge rate, and climatic factors which influence soil moisture.

The Ganges basin and south peninsula zone show very low resiliency (<0.05) as compare to the other rainfall zones and basins. Mahanadi river and western ghat show a high resiliency of about 0.13.

The vulnerability is the magnitude of the most extreme event of the GRACE-based index. It simply shows the maximum deviation from the monthly mean value of the terrestrial water storage. It also signifies at what level the deficit can drop. The spatial variation ranges from -1.12 to -3.30. Mahanadi, Godavari, and the Krishna river basin show the low value of the vulnerability, which indicates there is a low deviation from the monthly mean values when compared with the other basins and rainfall zones. South peninsula, Himalaya, northwest rainfall region, and Ganges basin show the high vulnerability. The region of Punjab and Haryana shows

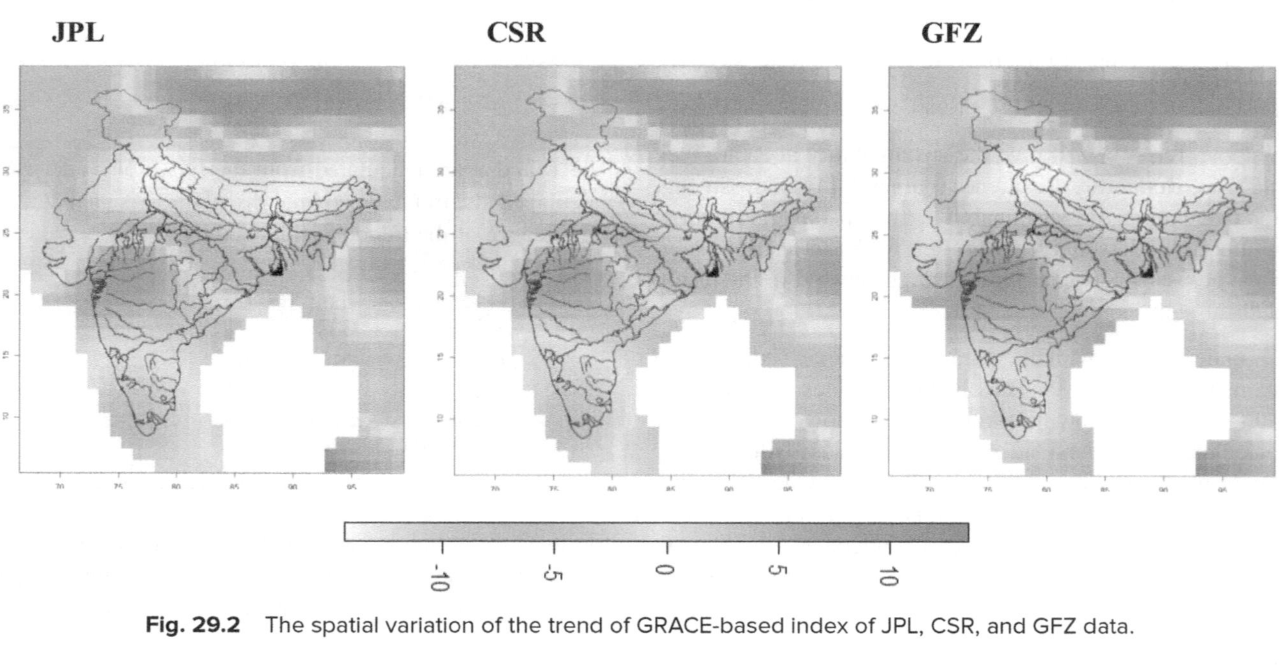

Fig. 29.2 The spatial variation of the trend of GRACE-based index of JPL, CSR, and GFZ data.

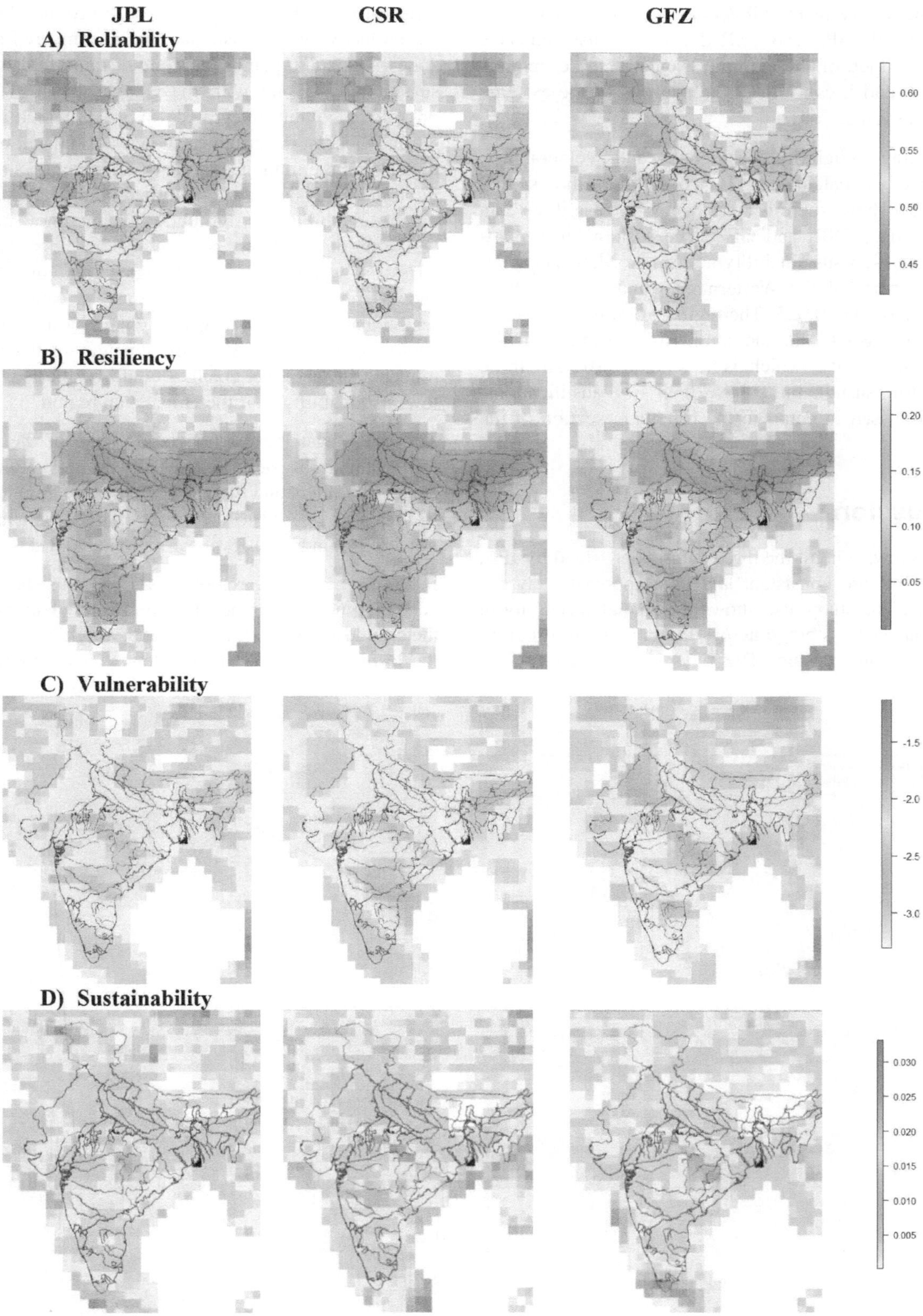

Fig. 29.3 The Reliability-Resiliency-vulnerability-Sustainability map of JPL, CSR, and GFZ datasets.

a high vulnerability in the GFZ datasets, which is slightly different from the JPL and CSR datasets. It also indicates that the magnitude of the extreme events of these datasets are different, and it can impact the analysis of the extreme hydrological events.

By integrating reliability, resiliency, and vulnerability, sustainability is calculated. By observing the spatial distribution, it shows there is consistency in these three data products of JPL, CSR, and GFZ. The sustainability maps show that the high sustainability across the Mahanadi river basin, some parts of the Western ghat and northern hilly regions of more than 0.025. The region of Punjab, Haryana, Himalayas, Ganges basin, and southern peninsula indicates the low sustainability, which is less than 0.01. All these regions of low sustainability have their own justifications, and it is discussed in detail in the following section of the discussion.

4. Discussion

This study shows that all the three datasets provided by JPL, CSR, and GFZ are consistent and give approximately the same results. The study also shows the spatial variability in the terrestrial water storage across the country in different rainfall zones and basins. The data is analyzed for six different rainfall zones, and it is observed that the northwest region has very low sustainability as compared to the other zones. The south peninsular zone of rainfall also shows low sustainability along with the Ganges basin. The GRACE provides a column integrated TWS data which includes the groundwater, surface runoff, snowmelt, and soil moisture. To understand the influence of each parameter on sustainability, the trend analysis is done and then compared spatially with the sustainability. The groundwater and snowmelt seem to be the prominent factor of influence on the sustainability of terrestrial water storage. The Fig. 29.4 shows the trend of snow depth, and it is observed that the central Himalayan region has very high negative changes in terms of snow depth. As in the GRACE, the changes in the snow depth reflected in the TWS changes could be the reason for very low sustainability near the Himalayan region.

The groundwater extraction is also one of the reasons for low sustainability across the northwest and south peninsular zones. To understand that, more than 5000 groundwater level station data is used to create the map, which indicates the trend of groundwater level across major states.

In the Fig. 29.5, the region of Punjab and Haryana have a very high negative trend of groundwater changes indicating the groundwater is extracted at a very high rate, and therefore the levels are dropped tremendously. The graph shows the

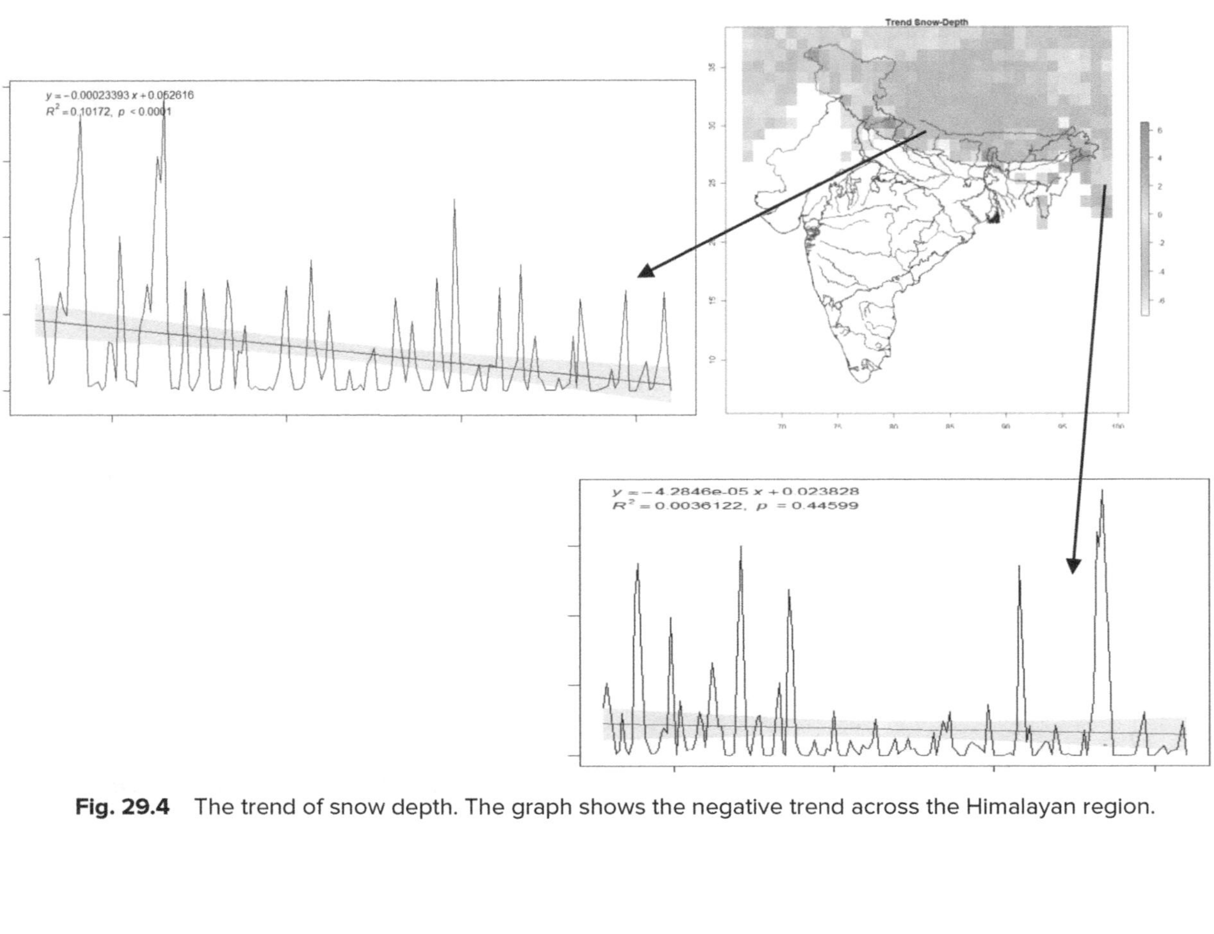

Fig. 29.4 The trend of snow depth. The graph shows the negative trend across the Himalayan region.

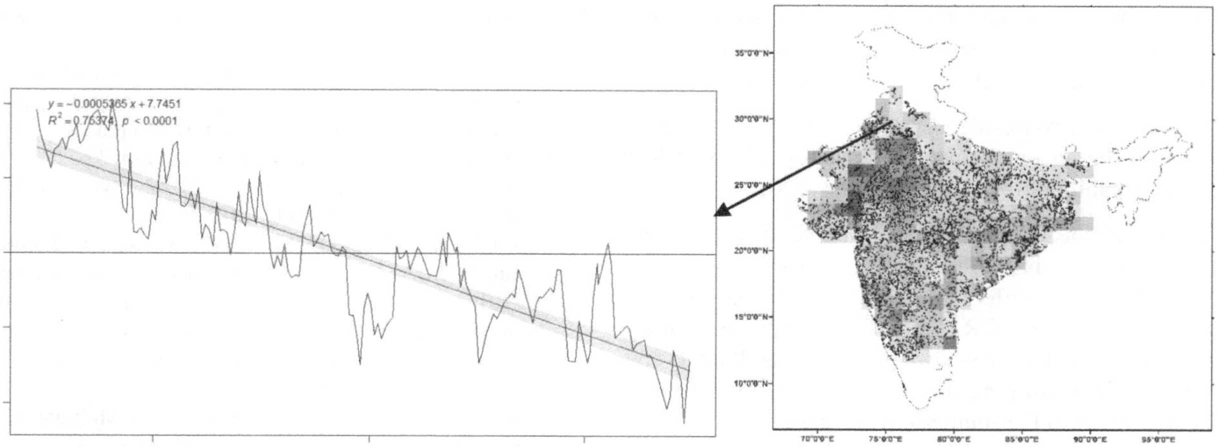

Fig. 29.5 The trend of groundwater level. The map shows the location of the groundwater observation wells, and the map is created using the IDW method to form a grid. The graph next to the map shows the negative trend in Punjab and Haryana state, which shows the drastic reduction in the groundwater levels.

groundwater level for the Punjab region, which is dropping with a very steep slope. Similarly, the region of Marathwada is also observed to be under low sustainability along with the south peninsular. There are multiple factors responsible for the low sustainability in this region; one could be the hydrological drought condition due to reduced flow in the river. It is evident that the flow is reduced in the Cauvery River due to various anthropogenic reasons. The Fig. 29.6 shows the negative trend of the runoff across the southern peninsular.

Not only the low sustainability is alarming, but the high value of sustainability should be a concern too. The high sustainability is observed across the Mahanadi river basin and parts of western ghat, which is a very high flood-prone region. The high sustainability values indicate the trend of high rainfall and stormwater runoff which is also alarming in terms of flooding.

5. Conclusion

The study proves the applicability of GRACE's TWS data for understanding the sustainability of water availability across the country. Following are some key points which are achieved in this study:

- The GRACE-based index help to identify the unique events which otherwise would have been difficult to, because of the annual cycle or seasonality. The index is used to divide the satisfactory and unsatisfactory events based on their location below the zero line.

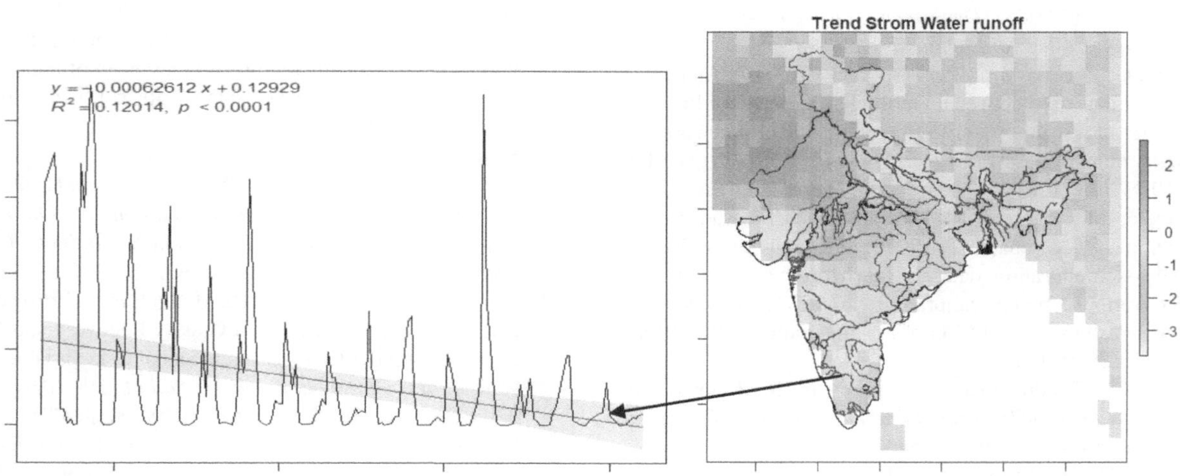

Fig. 29.6 The trend of surface runoff. The map shows the spatial distribution of the trend of the surface runoff by the GLDAS product. The graph next to it shows the negative trend of the runoff at the south peninsular zone of rainfall and the Cauvery river basin.

- The study determines the events and their spatial variability across the country, which helps to understand the changes in different rainfall zones and basins.
- The Reliability-Resilience-Vulnerability assessment proves to be an immaculate method for understanding the sustainability of available water.
- Combining the GRACE's capability to determine the changes of the last drop below the surface helps to touch down the different parameters of the different water storage. The GRACE's TWS helps to identify the changes in the groundwater storage at Punjab and Haryana along with the snowmelt at the Himalayas and surface runoff in the southern peninsular.
- The three different datasets (JPL, CSR, and GFZ) are consistent and don't show much variation spatially.

The study concluded that the water resources over India are decreasing, mainly the groundwater, which is impacting many agricultural regions by falling down the root zone soil moisture below permanent wilting point. The sustainability index can further improve by considering the actual demand of the region. Incorporating the in-situ data for downscaling the GRACE data will result in a detailed spatial analysis as the 1-degree resolution is quite coarse for analyzing the water availability across small basins or flood analysis.

References

1. Bai, P., Liu, X., & Liu, C. (2018). Improving hydrological simulations by incorporating GRACE data for model calibration. *Journal of Hydrology*, *557*, 291–304. https://doi.org/10.1016/j.jhydrol.2017.12.025
2. Bhere, S., & Regulwar, D. G. (2000). *Analysis of extreme hydrological event using Gravity Recovery and Climate Experiment (GRACE) Department of Civil engineering , Government College of Engineering ,.*
3. Cammalleri, C., Barbosa, P., & Vogt, J. V. (2019). Analysing the relationship between multiple-timescale SPI and GRACE terrestrial water storage in the framework of drought monitoring. *Water (Switzerland)*, *11*(8). https://doi.org/10.3390/w11081672
4. Dembélé, M., Ceperley, N., Zwart, S. J., Salvadore, E., Mariethoz, G., & Schaefli, B. (2020). Potential of satellite and reanalysis evaporation datasets for hydrological modelling under various model calibration strategies. *Advances in Water Resources*, *143*(March). https://doi.org/10.1016/j.advwatres.2020.103667
5. Diaper, C., Tjandraatmadja, G., Kenway, S., Kenway, S., & Kenway, S. (2007). *Sustainable Subdivisions - Review of technologies for integrated water services. June.*
6. Feng, W., Zhong, M., Lemoine, J. M., Biancale, R., Hsu, H. T., & Xia, J. (2013). Evaluation of groundwater depletion in North China using the Gravity Recovery and Climate Experiment (GRACE) data and ground-based measurements. *Water Resources Research*, *49*(4), 2110–2118. https://doi.org/10.1002/wrcr.20192
7. Hasan, S. (2009). *Terrestrial water storage change from temporal gravity variation.* http://www.hydrology.nl/images/docs/dutch/2009.04.27_Shaakeel_Hasan.pdf
8. Hashimoto, T., Stedinger, J. R., & Loucks, D. P. (1982). Reliability, Resiliency, and Vulnerability Criteria. *Water Resources Research*, *18*(1), 14–20.
9. Khaki, M., Hoteit, I., Kuhn, M., Awange, J., Forootan, E., van Dijk, A. I. J. M., Schumacher, M., & Pattiaratchi, C. (2017). Assessing sequential data assimilation techniques for integrating GRACE data into a hydrological model. *Advances in Water Resources*, *107*, 301–316. https://doi.org/10.1016/j.advwatres.2017.07.001
10. Kundzewicz, Z. W., & Kindler, J. (1995). Multiple criteria for evaluation of reliability aspects of water resource systems. *IAHS-AISH Publication*, *231*, 217–224.
11. Maity, R., Sharma, A., Nagesh Kumar, D., & Chanda, K. (2013). Characterizing drought using the reliability-resilience-vulnerability concept. *Journal of Hydrologic Engineering*, *18*(7), 859–869. https://doi.org/10.1061/(ASCE)HE.1943-5584.0000639
12. Moy, W. S., Cohon, J. L., & ReVelle, C. S. (1986). *Programming Model for Analysis of the Reliability, Resilience, and Vulnerability of a Water Supply Reservoir. 22*(4), 489-498 wreraq.
13. Najafi, M. R., Zhang, Y., & Martyn, N. (2021). A flood risk assessment framework for interdependent infrastructure systems in coastal environments. *Sustainable Cities and Society*, *64*(August 2020), 102516. https://doi.org/10.1016/j.scs.2020.102516
14. Paton, F. L., Dandy, G. C., & Maier, H. R. (2014). Integrated framework for assessing urban water supply security of systems with non-traditional sources under climate change. *Environmental Modelling and Software*, *60*, 302–319. https://doi.org/10.1016/j.envsoft.2014.06.018
15. Reager, J. T., Thomas, A. C., Sproles, E. A., Rodell, M., Beaudoing, H. K., Li, B., & Famiglietti, J. S. (2015). Assimilation of GRACE terrestrial water storage observations into a land surface model for the assessment of regional flood potential. *Remote Sensing*, *7*(11), 14663–14679. https://doi.org/10.3390/rs71114663
16. Sharma, H. (2014). *Evaluation of Groundwater Depletion Scenario and its Impacts in NW India by Geodetic Techniques and Modelling Approaches Evaluation of Groundwater Depletion Scenario and its Impacts in NW India by Geodetic Techniques & Modelling Approaches*. 76.
17. Swenson, S., & Wahr, J. (2002). Methods for inferring regional surface-mass anomalies from Gravity Recovery and Climate Experiment (GRACE) measurements of time-variable gravity. *Journal of Geophysical Research: Solid Earth*, *107*(B9), ETG 3-1-ETG 3-13. https://doi.org/10.1029/2001jb000576
18. Swenson, S., Wahr, J., & Milly, P. C. D. (2003). Estimated accuracies of regional water storage variations inferred from the Gravity Recovery and Climate Experiment (GRACE). *Water Resources Research*, *39*(8). https://doi.org/10.1029/2002WR001808

CHAPTER 30

A Survey of Drought Impact and Mitigation in the State of Telangana

Vijaya Kumari Nunna
Mahatma Gandhi National Council of Rural Education
Hyderabad – 500004

Abstract

The present paper highlights the drought situation in the districts of Telangana State. Reeling under a successive drought conditions, the country's youngest state is facing acute water crisis, with scarcity hitting not just the villages and towns but even the capital Hyderabad. A village-level survey was conducted to study and assess the drought situation in 38 mandals across 7 districts of Telangana.

Acute drinking water shortage due to drastic fall in water levels in major reservoirs, continuing crisis in agriculture and a harsh summer have caused the worst ever drought in Telangana. Dependent mostly on rain water, the state of 35 million people is witnessing migration of small farmers and farm workers to other states. An acute shortage of water and fodder is forcing people, especially the farmers, to sell their cattle at throwaway prices. According to farmers' organizations, nearly 1.4 million people have migrated from the worst affected districts of Mahabubnagar, Ranga Reddy, Medak, Nizamabad and Adilabad. The state, with the most number of farmer suicides, has been registering negative growth in agriculture. However, by planning and preparedness along with a drought contingency plan, the region can undertake measures to conserve water both before and during a drought.

Keywords: Drought, Acute water shortage, Agricultural crisis, Social migration

1. Introduction

Telangana is the 29th state of India, formed on the 2nd of June 2014. The state has an area of 1,12,077 Sq. Km. and has a population of 3,50,03,674.

Drought may come in several forms (www. 2005). The Indian government declares that a meteorological drought is taking place when less than 75% of the 'average' rainfall falls in an area over a prolonged period of time (Sainath, 1996).

1. A meteorological drought occurs when there is scanty rainfall.
2. A hydrological drought is identified when there is a significant reduction in water bodies such as rivers, ponds, tanks and groundwater.
3. Finally, an agricultural drought is when crops fail because there is insufficient moisture in the soil during crucial times in the harvest. In India this occurs when there is a meteorological drought for four consecutive weeks during the period of mid-May to mid- October, or for six consecutive weeks during the rest of the year.
4. Socio-economic drought occurs due to decrease in agricultural yield.

Of course, these types of drought are connected and its possible for a community to experience all four at the same time.

The village-level survey was conducted in 38 mandals across 7 districts of Telangana. The survey in the present study showed that the failure of rains in August and late July have created a drought situation in many parts of the state, while the villages had once been blessed with surplus production. However, planning and preparedness along with a 'drought contingency plan', a region can undertake measures to conserve water both before and during a drought.

2. Location Map

Telangana is surrounded by Maharashtra and Chhattisgarh in the North, Karnataka in the West and Andhra Pradesh in the South and East directions. Major cities of the state include Hyderabad, Warangal, Nizamabad, Nalgonda, Khammam and Karimnagar. The State comprises of 33 districts covering an area of 1,12,077 sq. km. The largest district is Bhadradri Kothagudem whereas Hyderabad is the smallest.

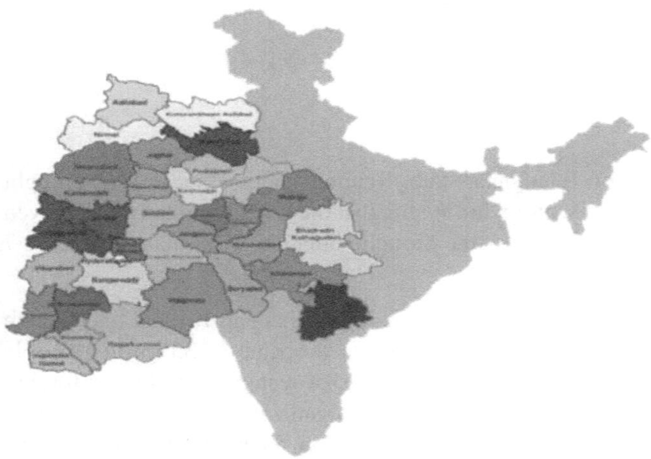

Fig. 30.1 Location map of Telangana state
Source: www, 2005

3. Drought Scenario

A report of FAO-CRIDA (2004) indicates that out of 450 mandals or revenue units comprising varying number of villages, 231 have been drought affected. The farmers' organizations claim that 368 mandals are hit. All the mandals in Mahabubnagar (64), Medak (46) and Nizamabad districts were declared drought-hit. Other mandals declared drought-hit included 33 out of 37 in Ranga Reddy, 19 of 57 in Karimnagar, 22 of 59 in Nalgonda, and 11 of 51 in Warangal. Assessing the agricultural drought situation of Telangana up to Oct 2015, the Ministry of Agriculture's Mahalanobis National Crop Forecast Centre assessed Mahbubnagar and Nizamabad districts as facing 'moderate drought'. The 'moderate drought' category is the most severe drought category under the assessment. Medak and Rangareddy districts were assessed as facing mild drought while the situation in the rest of the districts was assessed as normal.

Fig. 30.2 Drought situation in districts
Source; SANDRP, 2016

4. Impact of Drought

The severe drought in Telangana has caused acute shortage of drinking water and worsened the agriculture crisis in the state. According to farmers' organisations, nearly 1.4 million people have migrated from the worst affected districts of Mahabubnagar, Ranga Reddy, Medak, Nizamabad and Adilabad. People are mostly migrating to Pune, Mumbai, Bhiwandi, Ahmedabad and Surat looking for work.

Fig. 30.3 Agricultural drought assessment—October 2015
Source: SANDRP, 2016

The survey showed that in 25 villages, more than 25% of the cattle was sold by the farmers due to drought. Due to the distress sale, the farmers often got only 1/3rd or 1/4th of the normal price of the cattle – making a big loss. For example, in a village in Rajapet Mandal of Nalgonda district, there were 1000 cattle, out of which 350 were sold due to drought. The normal price was Rs.30,000 on an average but the average price they got on sale was only Rs.8,000. This meant that in the single village, 77 lakh rupees were lost by the farmers. The cattle used for farming and for dairy were being sold for 20-30 percent less price

Fig. 30.4 Drought scenario

Source: Agricultural Department of Telangana, 2016

A dry spell of more than 21 consecutive days without rain was experienced in 32 out of the 38 mandals surveyed, resulting in severe damage to the crops – especially maize, paddy, red gram, cotton, green gram and soya. Deficit rainfall along with severe dry spells caused withering and drying up of crops leading to reduced yield. The foodgrain production was 6.5 million tonnes in 2015-16 as against the target of 11.1 million tonnes. Rice production was 3.5 million tonnes while the state's consumption is six million tonnes. The production of pulses and oil seeds also fell drastically.

Fig. 30.5 Paddy crop withers due to dry conditions

Source: Agricultural Department of Telangana, 2016

Further, the delay in the crop loan disbursement has meant that the farmers have once again been forced to take high interest loans from money-lenders pushing them into a debt trap. Also, it has meant that the deadlines for crop insurance have passed. Less than 20% of the farmers have been covered by crop insurance, despite the big fanfare of Pradhan Mantri Fasal Bima Yojana. More than 60 farmers have already committed suicide in mainly due to the crop failure.

Projects were not started in many villages under the National Rural Employment Guarantee Programme (NREGP) due to drought conditions. And in villages where work was on, labourers were keeping away due to the intense heat. The labourers were also not paid their wages.

Water in most of the 14 major reservoirs across the Krishna and Godavari rivers has reached dead storage level. The World Health Organization has declared that drinking water with more than 1 mg of fluoride per litre is unsafe, as higher levels can cause fluorosis. This disease causes tooth decay (dental fluorosis) or the crippling of bones (skeletal fluorosis). The ground water in the state during March plummeted by 2.5 metres.

Fig. 30.6 Water scarcity

Source: FAO-CRIDA, 2004

Table 30.1 Plummeting Ground water

District	Depth to water level in mbgl in Jan 2016	Decrease compared to Jan 2015, in m.
Nizamabad	18.35	–6.22
Medak	23.82	–7.37
Warangal	10.15	+0.17
Karimnagar	12.00	–2.05
Adilabad	9.46	–1.33
Ranga Reddy	17.03	–3.73
Hyderabad	9.79	+0.12
Mahbubnagar	15.27	–3.92
Nalgonda	13.12	–2.94
Khammam	8.55	–0.53
State	13.75	–2.78

Source: http://agrisnet.tg.nic.in/weeklyreport.jsp?mode=rep

The large number of migrants escaping rural life during a drought leave a vast amount of land unutilized. This 'current fallow' land accounts for 9% of land, whereas it only accounts for 4% of land in the whole of India. The exploding population of India has placed severe pressure on the available resources, so that food production per capita has decreased. Thus, the government has initiated projects like the Integrated Land Development Programme (ILDP) to increase the cultivation of fallow lands. Under this programme the government is providing irrigation, seed subsidies, technical advice and financial support for landless farmers to develop fallow land.

Fig. 30.7 Migration of rural folk

Source: Agricultural Department of Telangana, 2016

5. Mitigation Measures of Drought

There were many ways to minimise the impact of a drought. The key words seemed to be 'planning' and 'preparedness'. Armed with a 'drought contingency plan', a region can implement measures to conserve water both before and during a drought. For instance, structures can be built to harvest rainwater during the monsoon. The filtered result can provide supplementary drinking water during dry periods. Moreover, farmers can be educated about water conservation and methods of cultivating land that optimize water efficiency. For instance, ridges and trenches can be dug in the fields so that the rain that does come slowly soaks into the soil and recharges the groundwater, rather than running off the field.

There are numerous other NGO and government programmes being implemented throughout the state to mitigate drought by providing opportunities to access safe drinking water and to earn income from alternative sources. For example, the MYTRY and UNICEF are working with the government to promote domestic de-fluoridation filter units. UNICEF have also undertaken campaigns to educate people about the causes of fluorosis (www, 2005).

Another NGO, the Rural Environment Development Society (REDS). The REDS team combats farmer suicides by highlighting the negative consequences of this act on family members. It also provides support to farmers against moneylenders by organizing meetings between both parties, REDS staff and the police.(Cheruva, 2005). Finally, it helps families get the compensation payment that the government offers those who have lost a member to a farmer suicide. Women's entrepreneurial activities are widely being supported by NGOs and the government to alleviate financial hardship in times of drought.

The government has various policies to assist communities affected by drought, at both the state and central level. An early warning system is a necessary ingredient for drought mitigation. (Bharat,2004). If a drought is identified early in the season it is possible for farmers to change crops to those that are better suited to dry conditions and produce fodder for cattle. Furthermore, if a drought situation is declared the government can intervene with programmes to alleviate problems that arise from dry conditions. In Kamareddy, Nizamabad, government officials and panchayat members have jointly prepared action plans for such events. Such a plan will include the aforementioned switch to dry land crops, if it is early enough in the season (Shivanand, 2005). As a preventative measure the government is offering 50% seed subsidies, a minimum supporting price and marketing assistance for dry crops like maize and soya bean to encourage more farmers to grow them on a regular basis.

In villages with a large number of cattle, the action plan will outline where farmers can get an extra supply of fodder – whether it is from a fodder bank, from grazing the cattle on the tank bed or through purchasing government cattle feed at a 75% subsidy. A fodder bank is when a community jointly stores fodder so that it can be used in a time of drought. Seed banks may also be built up for the same purpose, to mitigate the steep price increases during a drought.

Other drought mitigation initiatives are being promoted in the state's Neeru-Meeru Programme. As part of this programme, a 42-member panel of scientists, NGOs and people with a special knowledge and interest in this issue, known as the 'Water Conservation Mission', was created to focus on improved water management

Historically, the tanks in this region made a valuable contribution to the local ecosystem and economy. Tanks supported the local economy by providing employment for fishermen and quality clay for pot and brick-makers on the tank bed. What is more, the silt that accumulated on the tank bed over time was an effective natural fertilizer that was periodically removed and applied to fields.

The installation of water harvesting structures such as soak pits, check dams and household tanks could improve the situation. Soak pits are drainage pits constructed to allow water to slowly seep through layers of stone and sand and recharge the groundwater. A series of check dams can also replenish groundwater levels. Check dams harvest the surface run-off that comes from streams during the monsoon and spread it over the streambed.

In addition, it is particularly beneficial to have rooftop water harvesting structures in locations where safe drinking water is scarce. These structures collect rainwater from a roof, filter it and store it in a covered tank for use in time of water scarcity (Athavak, 2003).

The government also is working on ambitious projects to find a lasting solution to recurring drought. These include redesigning irrigation projects, Mission Kakatiya for restoration of irrigation tanks and Mission Bhagiratha for piped drinking water supply to every household.

In order to mitigate drought, there is a need to empower farmers to make wise decisions concerning the selection and management of crops. NGOs could provide information about dry crops for which there is a good market, such as medicinal and dye-yielding plants including Withania somnifera, Aloe barbadensis, Bixa orellana and Lawsonia inermient of their crops so that they make efficient use of the water available to them. Another option could be to cultivate aromatic grasses such as lemon grass and vetiver grass that produce essential oils that can be profitable in dryland agriculture. It would be beneficial for NGOs to actively promote such crops when there is an early-onset drought. This type of drought occurs before the season's seeds are sown, giving farmers an opportunity to switch crops to those that can withstand dry conditions.

Fig. 30.8 Dry land Farming

Source: SANDRP, 2016

Legumes are also suitable for dry land agriculture (Ramakrishna, 2005). Chickpeas and other legumes are particularly good for poor soil because they enhance biological nitrogen fixation. If farmers in water-scarce areas insist on cultivating water-intensive crops, it is recommended that they rotate them with plants that require less water, such as legumes. This can optimize both crop production and moisture conservation, as provides root systems that bring nutrients to the surface and facilitate the seepage of water into the soil. Intercropping can deliver similar benefits, as well as providing extra security in case one crop is hit by a pest, drought or disease.

6. Recommendations

The following recommendations have been suggested to mitigate the impact of drought.

- Strict implementation of Government policies.
- To replenish a depleted water supply.
- Provision of new irrigation schemes.
- Restoration of Water tanks.
- Construction of Water harvesting structures.
- The need for Community Ownership of Water Harvesting Structures.
- Smart crop management
- Re-vegetation of fallow lands
- Identification of suitable crops.
- Efficient use of the water available in Horticulture.
- Reduce water usage in Paddy cultivation
- Rehabilitation of the migrants.
- Create awareness among the farmers and community people.

Fig. 30.9 Combat Drought & Desertification

Source: www, 2004

References

1. Athavak, R. N. Water Harvesting and Sustainable Supply in India (Jaipur \ New Dehli, Ravat Publications, 2003), p.30.
2. Agricultural Report of Telangana. A survey study of Drought situation in districts of Telangana. 2017.
3. Bharat Ramaswami, Shamika Ravi, & S.D. Chopra, 'Risk Management', in State of the Indian Farmer – A Millennium Study. Vol 22 (Delhi, Ministry of Agriculture, 2004) P.52.
4. Cheruvu Bhanuja (Rural Environment Development Society), Personal Communication, April 21, 2005.
5. Combating fluorosis with household filters. Available on World Wide Web: www.whoindia.org/SDE/Sangam/07-vol01-03/Page01.pdf (Accessed May 10 2005).

6. 'DefiningDrought',AvailableonWorldWide Web: www.disasterrelief.org/Library/Dictionary/index_txt.html. (Accessed January 30 2004).
7. Dr. Y. S. Ramakrishna (Director of CRIDA), Personal Communication (Hyderabad: April 21 2005).
8. Report of the FAO-CRIDA expert group consultation on farming systems and best practices for drought-prone areas of Asia and the Pacific region, op. cit., p.11. 2004.
9. Sainath. P, Everybody Loves a Good Drought: Stories from India's Poorest Districts (New Delhi: Penguin, 1996), p. 320.
10. SANDRP, 2016. South Asia Network on Dams, Rivers and People. A report, 2016)
11. Shivanand, Personal Communication, (Kamareddy, Nizamabad: Mandal Agricultural Officer, April 24, 2005) 41. Ibid
12. 'Water Harvesting and Conservation Methods' (International Water Management Institute). Available on the World Wide Web: http://www.iwmi.cgiar.org/droughtassessment/index.asp (Accessed 27 January 2005).

Part 5

CLIMATE CHANGE AND DISASTERS—NEW DIMENSIONS

CHAPTER 31

Role of Local Sea-level Rise on Disaster Exposure in Coastal Island

Vinay S.
Energy and Urban Research Group
RCG School of Infrastructure Design and Management,
Indian Institute of Technology Kharagpur,
West Bengal, India
vinay.s@infra.iitkgp.ac.in

Aishwarya N.
Energy and Urban Research Group
RCG School of Infrastructure Design and Management,
Indian Institute of Technology Kharagpur,
West Bengal, India

Bharath H. A.
Energy and Urban Research Group
RCG School of Infrastructure Design and Management,
Indian Institute of Technology Kharagpur, West Bengal, India
bharath@infra.iitkgp.ac.in

Abstract

Coastal islands across the globe are under threat due to recurring water logging conditions caused by climatic extremes (cyclones), local topographic conditions, human interventions, an unprecedented wave actions enhancing the regional exposure and hampering livelihood of the habitants. IPCC 2014 emphasises about the coastal inundations due to sea-level rise, anthropogenic activities, and extreme climatic conditions. Sagar a coastal island located in South 24 Pargana district and part of Sundarbans mangrove forests in the mid1800s, this island has seen a largescale transformation in the landscape with now supporting human habitations, agriculture, socio- economic activities, goods, and services, etc. The island has witnessed super cyclones, unprecedented tidal variations, continually rising sea levels, etc. causing socio-economic complications. The current research aims to visualize and model (connected N8 model) the exposure conditions by using recent high-resolution remote sensing data of spatial resolution of 2.5m, 5m, and 10m along with secondary data using advanced geospatial techniques. Increasing paved surfaces in the island show a strong relationship with increasing human habitation in this island due to the submergence of the neighbouring islands. Long-term assessment of temperature and rainfall events depicts an increasing trend in annual average temperature and increasing high rainfall events in the pre and post- monsoon seasons indicating rising instances of cyclonic hazards. Assessment of long-term observation of Sea level rise demonstrates an increasing sea level of 0.46m by 2050. The results indicate that the southern and eastern portion of the island is exposed to inundation due to the sea level rise, as well as surge. Analysis of SAR data about cyclonic events Yaas and Amphan demonstrate most parts of the Island on North, South, and East are more affected by Cyclonic hazards. The assessment provides both awareness and knowledge helping to focus on investigations and preparedness levels to different water inundation levels relative to climatic circumstances.

Keywords: Physical factors, Hazard assessment, Spatio-Temporal data, Sea Level

1. Introduction

Coastal Island, landscapes are relatively new geological formations that are formed due to fragmentation of landmass, sedimentation, and deposition of soil due to the tectonic activities, waves, and ocean currents (Earle, 2019). Enriched with nutrients, the coastal areas are highly productive with fertile lands, rich aquatic richness, and diversity (Walker et al., 2005). These landscapes such as estuaries, delta, islands have evolved providing habitat, resource, feeding grounds, and nurturing biodiversity (Mahima et al., 2016; Ramachandra et al., 2020). All these natural systems have attracted human habitations resulting in over-exploitation of the resource, changing landscape, and creating an imbalance (hazards) in the natural system at the regional and global context.

Natural Hazards such as cyclones, storm surges, sea-level rise, and so forth are triggered by changing climatic patterns, flooding the low-lying coastal areas affecting livelihood, damaging assets, natural habitats, native ecosystems to name a few. Globally accelerating temperatures have led to an increase in frequency and intensity of natural disasters, sea-level rise. Eighteen percent of these global natural disasters occur in India, predominantly in the Bay of Bengal. In addition to this, sea levels on the Indian coast are rising in the range of 0.9 mm/year (southern oceans) to 6.5 mm/year (northern oceans).

The increased rate of temperatures and natural disasters are evidence of human-induced imbalances in the natural system (Fig. 31.1). The number of natural disasters has drastically increased post-1960's proportional to the increasing global temperatures. In the last century, the global temperatures have increased at a rate of 0.12°C per decade and the sea levels are rising at a rate of 3.3 mm/year. In the last two decades accelerated rising sea levels have increased the frequency and duration of coastal flooding (CDC, 2017), threatening the coastal areas and communities even after the global mean temperature stabilises (Nauels et al., 2019).

Climatic extremes such as floods due to cyclones, surges are discrete, while inundation due to sea level rise (SLR) is a continuous cumulative phenomenon. This necessitates an improved understanding of hazard exposures due to discrete flooding events and locally varying sea level rises for effectively improve management strategies and reducing risks for achieving sustainable development goals (UNDRR, 2015).In the current communication, we try to understand the role of varying levels of flood inundation exposure due to multiple natural hazards (Tidal surge, Sea level rise) in Sagar Island using hydrologically connected models.

2. Study Area

Sagar is a habited Island located in South 24 Parganas district of West Bengal state. Being located in the Bay of Bengal, this Island is prone to a large number of natural hazards such as sea-level rise, torrential rains, tropical cyclones, storm surges causing inundations, physical damage. Currently, the Island has an area of 238.9 km^2, with 44 villages, supporting a population of over 0.22 Million.

Being located in the cyclogenesis area, the island has witnessed over 36 severe to super cyclones, and numerous low-intensity cyclones/depressions. These cyclonic events cause tidal surges causing short-term regional inundations. The intensity of extreme events is increasing in the Bay of

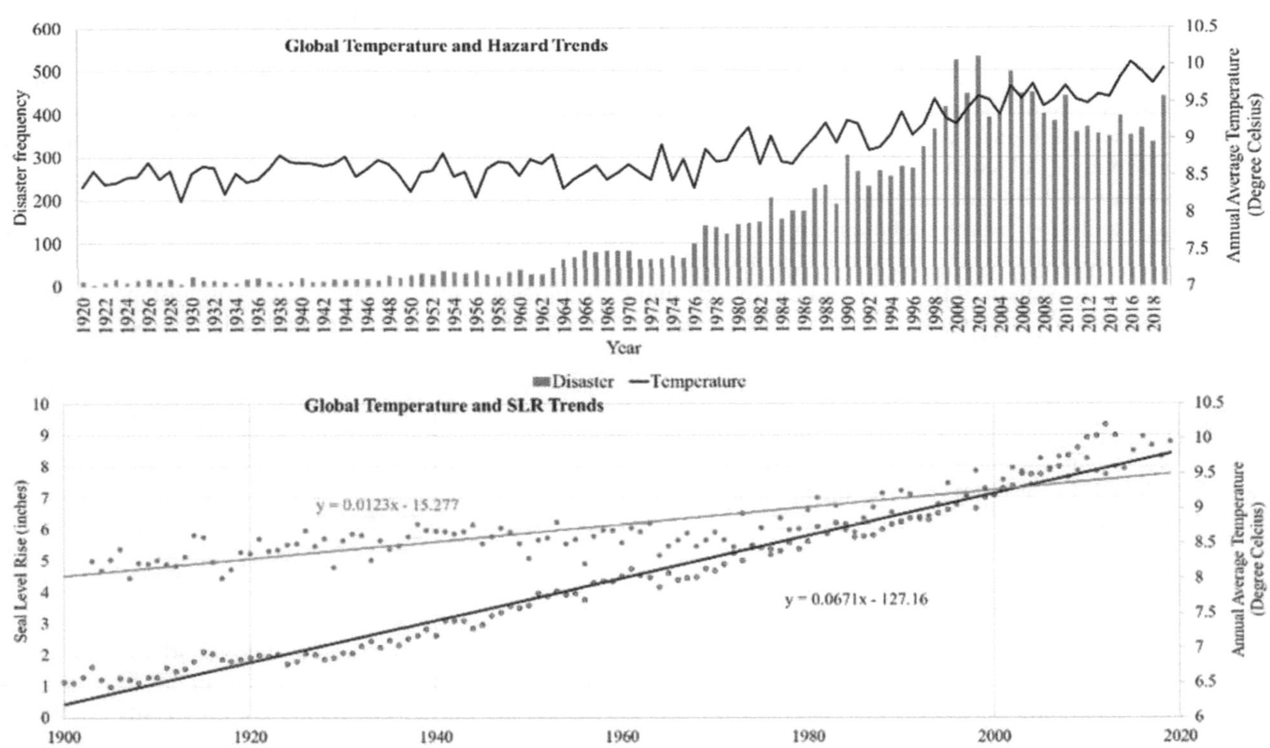

Fig. 31.1 Trends in temperature, natural hazards and SLR (CDC, 2017; EM-DAT, 2020)

Bengal (Krishnan et al., 2020) at one per decade (Aila 2009, Bulbul 2019, Amphan 2020, Yaas 2021) increasing the social, economic, physical vulnerabilities in the region resulting in damages over billions of rupees, mortalities, disrupted services loss of connectivity, damaged infrastructure, disruption of the supply chain, damage to power sources, disordered and chaotic environs (political, economic, social), increase in poverty levels and societal inequalities. Studies indicate that between 1969 and 2001, 12.16 km^2 has been eroded due to the rising sea levels and ingression of seawater (Gopinath and Seralathan, 2005), and this is a continuous process. Accelerated rise in SLR, temperature, natural hazards would further deteriorate the landscape and the livelihood of the habitants.

This makes it necessary to understand flood inundation risk due to the natural hazards and rising sea levels by evaluating potential threats that would help in planning proactive measures.

3. Data and Method

In order evaluate the variable hazard exposure in Sagar Island, the Multi-Hazard Multi-Criteria Evaluation framework is adapted (Fig. 31.3) based on principles of the SENDAI disaster risk evaluation framework (UNDRR, 2015) and Disaster Management Act of India (The Disaster Management Act, 2005, 2005).

Data Collection: Data were collated from field observations, government records, online portals such as NRSC, BHUVAN, Survey of India Topographic sheets. Table 14.1 describes the data used in the analysis.

Land use: Land use analysis for the year 1922 was extracted from the toposheets and for the year2012, 2014 2020, the standard protocol of land use analysis using Supervised Gaussian Maximum likelihood classifier (Lillesand et al., 2004; Ramachandra et al., 2017; Bharath et al., 2018) was followed. Land use was predicted for the year 2050 considering sea-level rise, and other dependent parameters using AHP (Ramachandra et al., 2020, 2021).

Digital Terrain Model: Stereo pair of satellite data was used to derive a detailed Digital Surface Model using standard protocols (Giribabu et al., 2013), and DTM was created based on morphological filtering (Ismail and Jaafar, 2013; Prakash and Bharath, 2021) of DSM considering local minima, local slope parameters. Slope obtained from DSM was used to describe exposure levels due to the terrain characteristics.

Virtual Globe data along with DSM and Resource Sat data were used to delineate the high-water levels and river network in 2020 and Toposheets were used to delineate the water

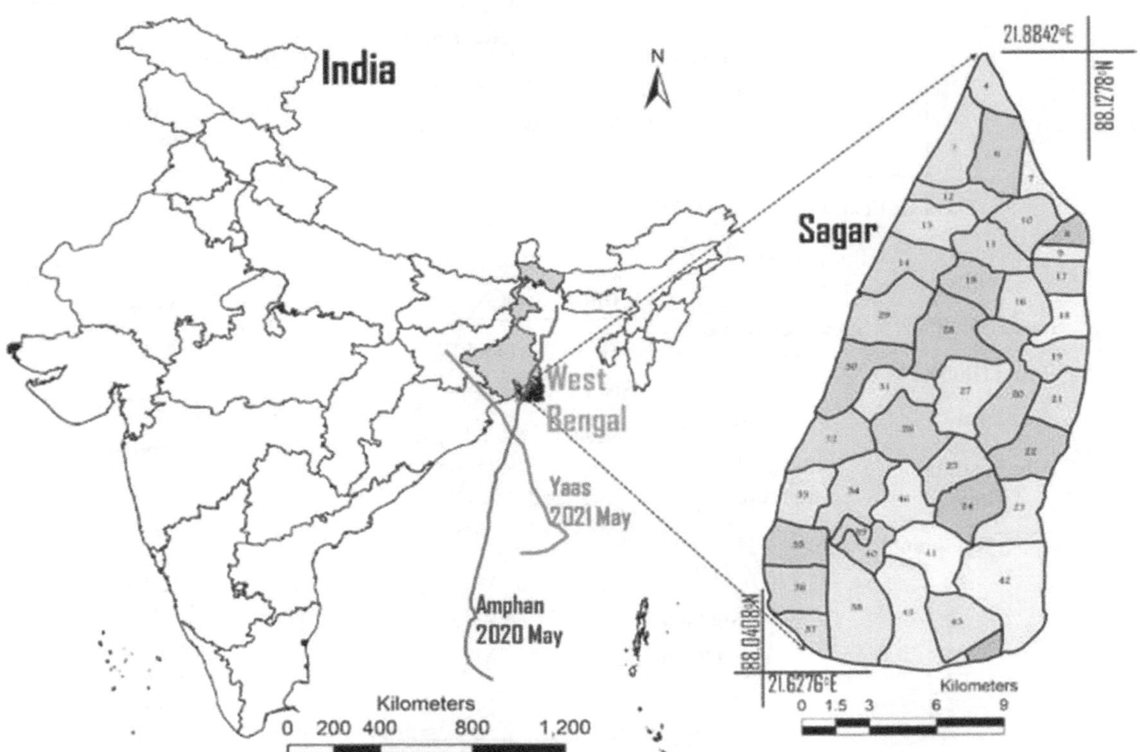

Fig. 31.2 Study Area

Table 31.1 Data used in the analysis

Data	Period	Output	Source
Resourcesat 1, 2 (5.8 m)	2020, 2012	Land use, Landscapes exposed to flood, land use prediction	(National Remote Sensing Centre, 2020)
Cartosat 1 (2.5 m) Stereo	2010	Topography, Slope, River Network, Exposure	
Sentinel 1	2020, 2012	Exposure due to Cyclonic events	(ESA, 2021)
Sea level data	60 years	Sea level Anomalies, Exposure	(PSMSL, 2020)
Tide data	40 years	Spatial rates of SLR, Exposure	
Ancillary Map	1922	Land use, Geometric rectification, Inundation,	(Ryder, 1924)
Virtual Earth data	2020	Land use, Geometric rectification Building Extraction	(Google, 2020)
Climate data	1950 – 2020	Rainfall patterns, Temperature patterns, relation with anomalies	(Terrestrial Hydrology Research Group, 2014; Indian Meterological Department, 2020; NASA, 2021)

levels for the year 1922. The difference in the shorelines was digitised to quantify the inundation and accretion levels in the Island.

Long-term observations of SLR at Sagar and neighbouring docks were collated from 1950 to 2019, covering at least 50 years. This enables understanding of the local level changes in sea levels. Similarly, long-term tidal data were collected to understand the tidal amplitude variations due to surges and their frequencies. DSM and Tidal amplitude are integrated to define the exposure hazards to variable tidal heights.

Long-term changes in rainfall and temperature patterns and trends were analysed using IMD and open source databases. This enables us to understand the anomalies and relate the same with tidal patterns.

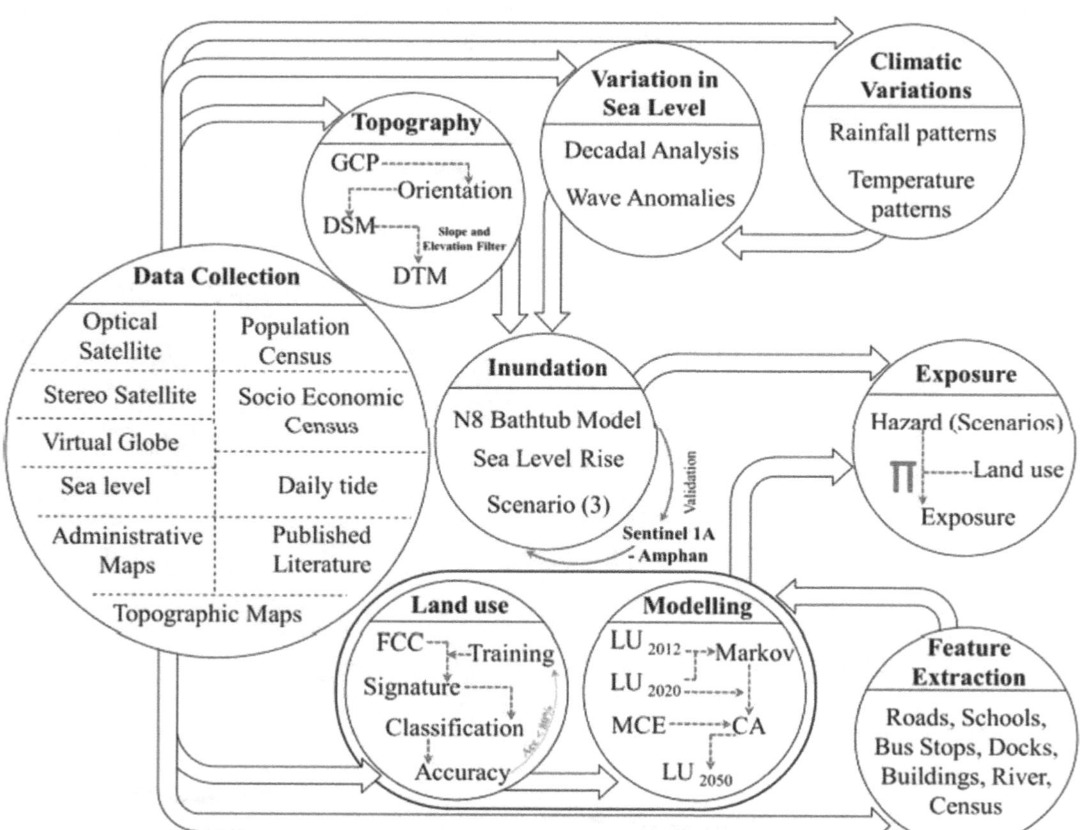

Fig. 31.3 Method

In order to understand the role of locally varying rates of sea levels, the exposure due to tidal surge hazards was simulated using N8 connectivity models under 3 scenarios (Table 31.2).

Table 31.2 Scenario development for different time scales

Year	2020	2050
Scenario 1	Surge height of 2 m	Surge height of 2 m + Sea level rise
Scenario 2	Surge height of 4 m	Surge height of 4 m + Sea level rise
Scenario 3	Surge height of 6 m	Surge height of 6 m + Sea level rise

4. Results and Discussion

Land use and Changing shorelines are as depicted in Fig. 31.4. The mangrove vegetation has reduced from 103 sq. km to 7.64 sq. km in the last century. Agriculture and Horticulture landscapes have increased to 50.2% and 27.2% respectively by 2020, built up are has increased from less than 1% to 8.2% indicating larger communities and economically viable landscapes are exposed to the disasters. Shoreline changes showed the Island is lost 28.24 sq. km are due to inundation and 9.9 sq. km land reclaimed by accretion of sand and mangroves in the last century as a consequence of changing sea levels and surges caused by climatic and tectonic actions.

Decadal assessment of long-term tidal data at Sagar indicates the rate of sea-level rise has increased from 0.28 mm/year (1940 – 1950) to 4.77 mm/year (2010 – 2020). Based on the current trend, sea levels are forecasted (RMSE 0.99) to rise at a rate of 8.04 mm/year (2050) and 11.74 mm/year (2090-2100) (Fig. 31.4). Accordingly, the sea levels are projected to rise to 460 mm by 2050 and 870 mm by 2100. The forecasted sea levels fall in line with the IPCC RCP 8.5 scenario, where the sea levels are projected to rise between 0.65 m to 0.93 m (Church and Clark, 2013; Horton *et al.*, 2020). In

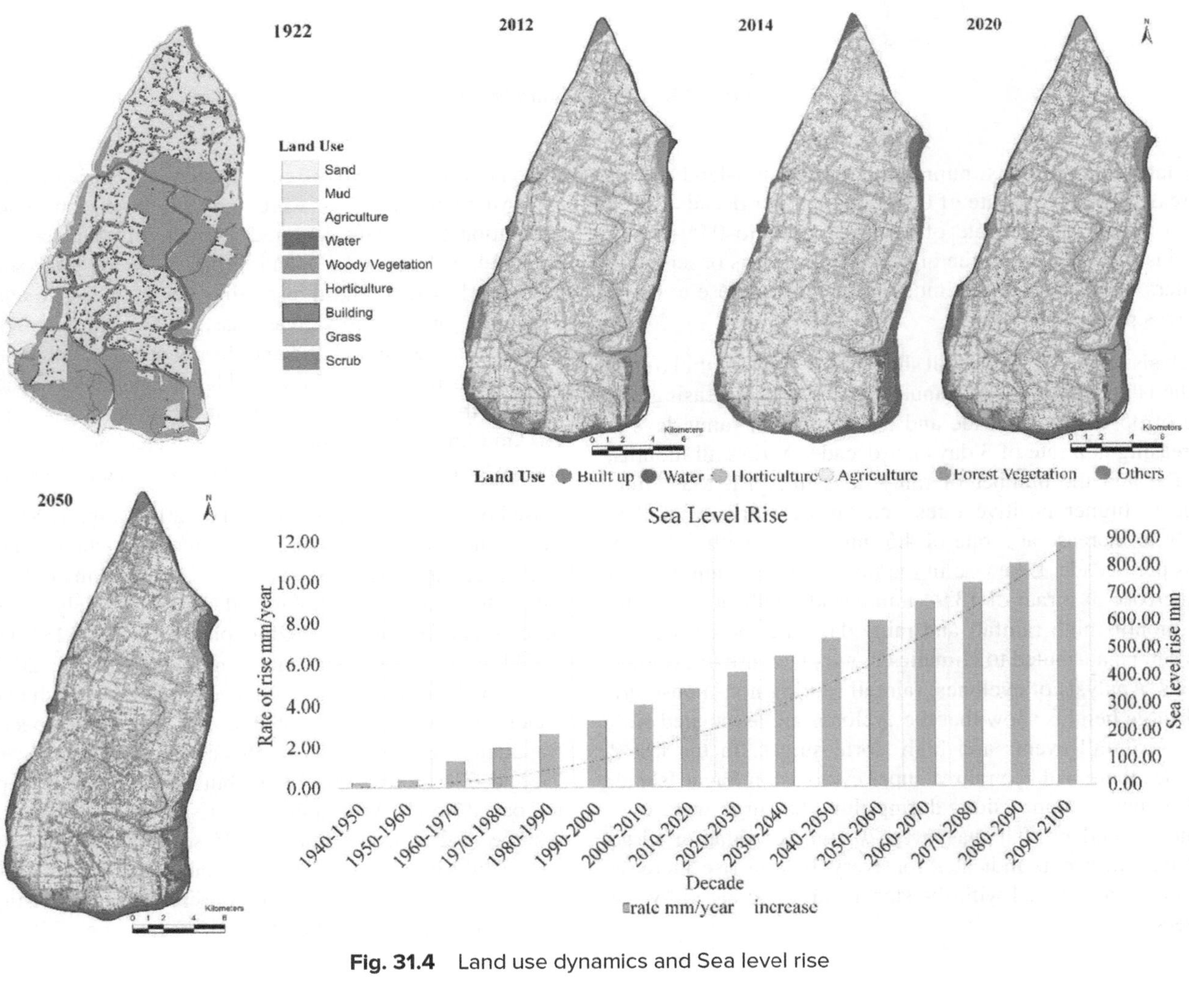

Fig. 31.4 Land use dynamics and Sea level rise

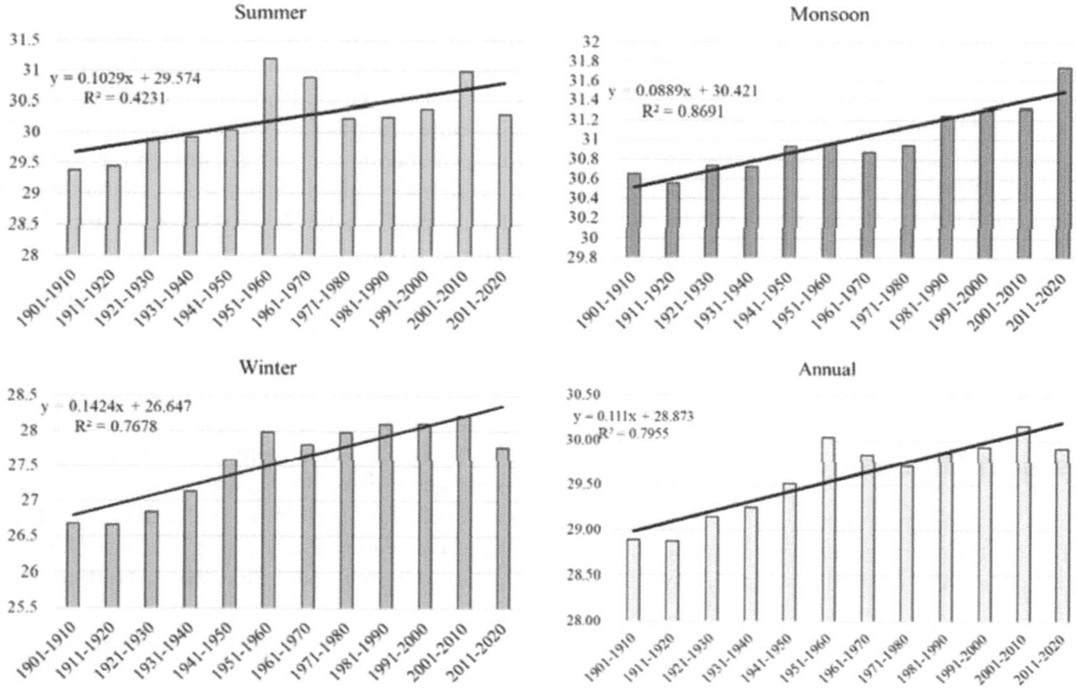

Fig. 31.5 Temperature Dynamics

the last century, maximum temperature in island shows increasing trend at a rate of 0.104 degrees per decade, while in the monsoons the rate of rise was close to 0.09-degree Celsius per decade and the highest increase was observed in winters with a rise in maximum temperatures close to 0.142 degrees per decade.

Analysis of long-term rainfall shows increasing rainfall trends in the island (Fig. 31.6). Annually, rainfall is increasing at a rate of 35 mm per decade and the number of rainy days is increasing at a rate of 3 days per decade. Across all months, rainfall and the number of rainy days in April, May, June indicate higher positive rates i.e., during April, rainfall is likely to increase at a rate of 4.5 mm/decade with 0.4 rainy days per decade. Likewise in the month of may rainfall tend to increase at a rate of 13.94 mm/decade, indicating, in the dry months both rainfall and rainy days are increasing and this can be attributed to climatic change, increase in cyclonic events. Analysis of cyclones, rainfall (during non-monsoon), and surge heights show that the cyclones are associated with heavy rainfall events and high storm surges. In the last 2 decades there had been three super cyclones in Sagar Island, and numerous depressions during this, the surge were over 3 meters and rainfall was over 70 mm. correlating surge and rainfall events indicates for every 10 mm rise increase in rainfall associated with the storm, surge increased by 1.4 meters.

Flood hazards for two different cyclones of varied intensities viz Amphan and Yaas were mapped based on moderate resolution SAR data acquired from Sentinel 1 series data. Sentinel 1 SAR GRD data products were analysed for flooding based on the thresholding technique. The analysis indicates for severe cyclone Yaas, 48 sq. km of low-lying areas would get inundated and during super Amphan, 144.6 sq. km was inundated. Cultivated landscapes encompass over 80% of the inundated areas, about 39 sq. km (Yaas) and 93 sq. km (Amphan). Engulfing of floodwaters are mostly from the eastern parts of the island and through the creeks (Yaas).

Based on the current conditions, i.e., 2020 (Fig. 31.8), 30 sq. km of the coastal landscapes get inundated when surges are of 2 m, and the area of 88 sq. km and 167 sq. km are likely to inundate due to surges of 4 m and 6 m respectively. Exposure assessment for the year 2020 depicts 1371 ha, 5233 ha, and 10211 ha of agricultural landscapes and 536 ha, 2007 ha, and 5366 ha of horticulture lands get inundated with oceanic waters due to the surges of 2m, 4m, and 6m, respectively. Population residing in 56 ha (Scenario 1), 286 ha (Scenario 2), and 674 ha (Scenario 3) of built-up spaces are exposed to coastal floods. With a 460 mm rise in sea levels by 2050, a surge of 2 m would inundate 35 sq. km, while a surge of 4 m would inundate 108 sq. km, and a surge of 6 m would inundate 203 sq. km. It can be observed that the rising sea level would aggravate the effect of hazard, i.e., 32%, 20%,

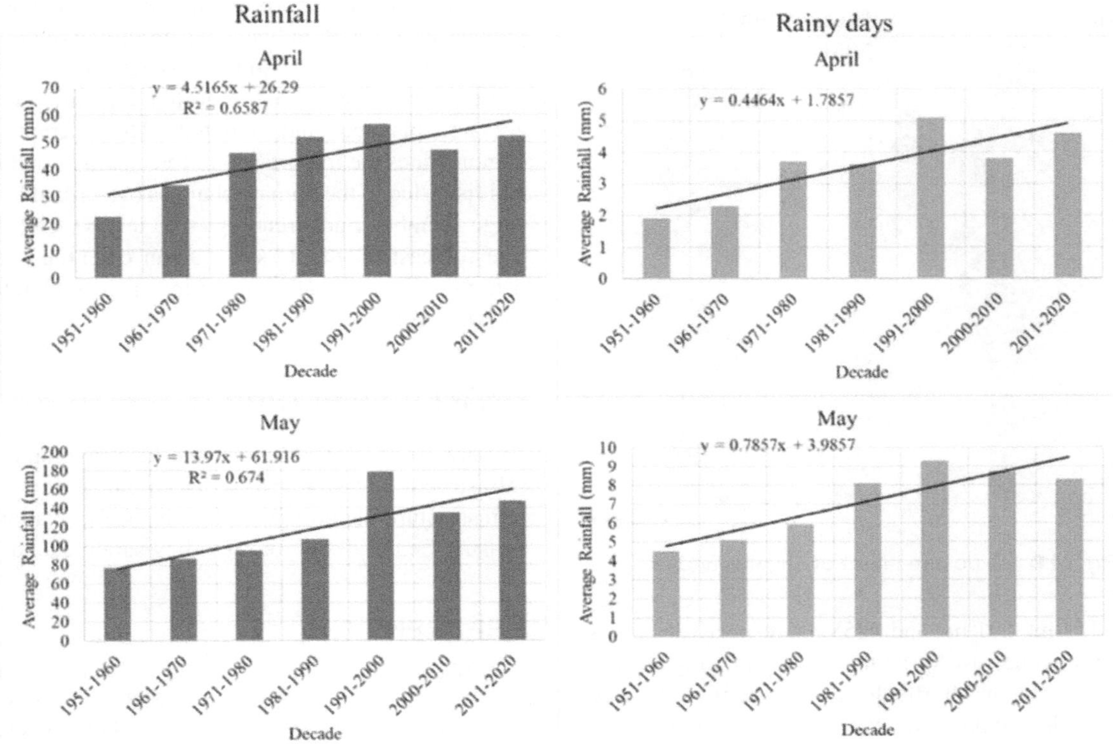

Fig. 31.6 Rainfall Patterns

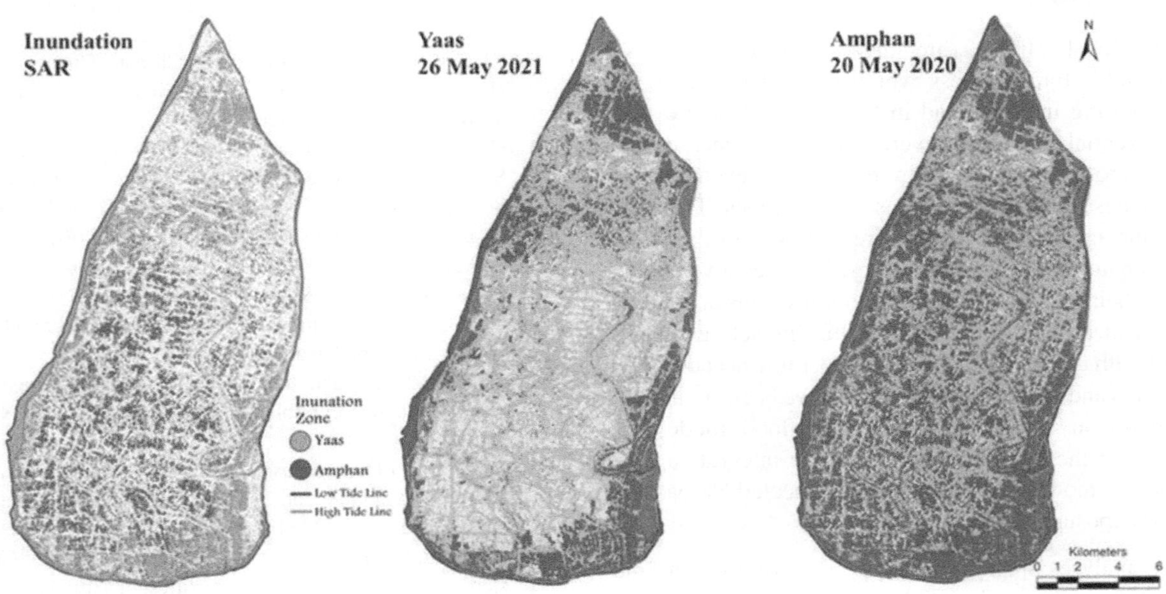

Fig. 31.7 Inundation during Amphan and Yaas cyclones

and 9% increase in area under flood for scenarios 1, 2, and 3, respectively. Spatially the eastern coast is more exposed to flooding than the western coast due to the local topography.

With the increase in sea levels and changes in land use, by 2050 population exposed would increase with increasing exposures of built-up spaces. The exposed built-up areas

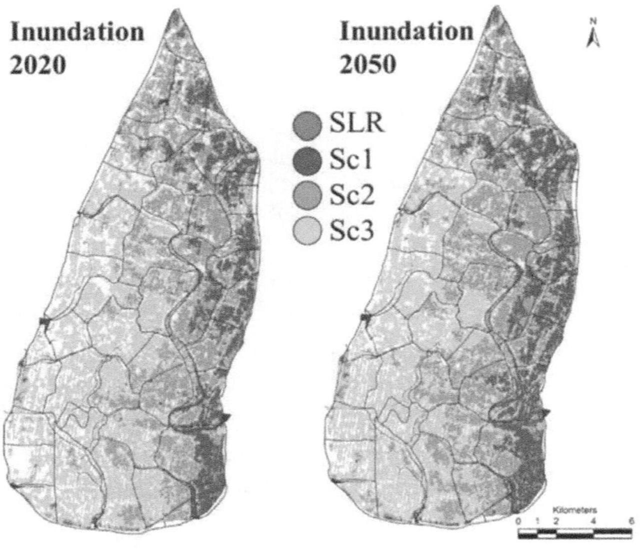

Fig. 31.8 Exposure levels of Sagar Island

increase to 151 ha, 891 ha, and 2258 ha, respectively, across the three surge scenarios. Agriculture assets ranging between 1809 ha to 9736 ha and horticulture assets ranging between 602 ha to 5718 ha are predicted to be exposed to flooding for different surge scenarios.

4. Conclusion

Changing global climate are driven by anthropogenic pressures and natural process which has led to rising sea levels, increasing intensity and instances of extreme events such as torrential rains and severe cyclones, breaching the coastlines across the globe. Extreme climatic events cause tidal anomalies inundating the coastal landscapes. The zone of inundation aggravates with an increase in sea levels. These unforeseen natural hazards have a largescale negative impact on society, their livelihood, and the economy. This at a global scale has necessitated understanding the impacts and risks associated with coastal flooding. In the last few decades with advent of GIS and remote sensing, there have been significant improvements in spatial landscape and flood modelling. The novelty of the current study is that we integrate agent-based land use models, hydrologically connected N8 bathtub models for exposure assessment.

The study analyses spatial and temporal variations in land use, temporal variations in sea level and pre-defined surge conditions to understand and simulate flooding, its impacts and risks in Sagar Island. Tidal amplitudes were pre-defined based on the severity of cyclones and surges observed in those events. With the current trends in rate of sea levels rise, 460 mm of sea level rise by 2050 would inundate additional 275 hectares, particularly along the eastern coast. Land use predictions by 2050 indicate mushrooming of built-up spaces by 2.98 times (w.r.t 2020) that would also lead to increasing coastal risks in the island if sustainable mitigation strategies are not planned and adopted. The current assessment demonstrates the importance of forecasting flood inundations and risks due to the climate changes (described by predefined surge heights and trending sea levels) and landscapes. The forecasted outputs serve as an essential component to determine sustainable adaptive management strategies at the Mouza level.

5. Acknowledgment

We are grateful to Ranbir and Chitra Gupta School of Infrastructure Design and Management for the financial and infrastructure support. We also thank the National Remote Sensing Centre, India for providing satellite data.

References

1. Bharath, H. A. et al. (2018) 'Urbanisation in India: Patterns, Visualisation of Cities and Greenhouse Gas Inventory for Developing an Urban Observatory.', in Weng, Q., Quattrochi, D., and Gamba, P. E. (eds) Urban Remote Sensing. 2nd edn. Boca Raton: CRC Press: Taylor and Francis.
2. Burrough, P. A. (1986) Principles of geographical information systems for land resources assessment, Oxford University Press.
3. CDC (2017) Coastal Flooding, Climate Change, and Your Health. Centre for Disease Control and Prevention. Available at: www.cdc.gov/climateandhealth.
4. Church, J. A. and Clark, P. U. (2013) Sea Level Change Supplementary Material, Climate Change 2013: The Physical Science Basis. Contribution of Working Group I. Available at: https://www.ipcc.ch/site/assets/uploads/2018/07/ (Accessed: 28 May 2021).
5. Earle, S. (2019) 'Shorelines', in Physical Geology. 2nd edn. Creative Commons Attribution. Available at: https://opentextbc.ca/physicalgeology2ed/.
6. EM-DAT (2020) The International Disaster Database, CRED. Available at: https://public.emdat.be/ (Accessed: 17 November 2020).
7. ESA (2021) Copernicus Open Access Hub, European Space Agency. Available at: https://scihub.copernicus.eu/dhus/#/home (Accessed: 20 February 2021).
8. Giribabu, D. et al. (2013) 'DEM generation using Cartosat-1 stereo data: Issues and complexities in Himalayan terrain', European Journal of Remote Sensing. DoI: 10.5721/EuJRS20134625.
9. Google (2020) Google Earth. Available at: https://www.google.com/intl/en_in/earth/ (Accessed: 3 March 2019).
10. Gopinath, G. and Seralathan, P. (2005) 'Rapid erosion of the coast of Sagar island, West Bengal - India', Environmental

Geology, 48(8), pp. 1058–1067. DoI: 10.1007/s00254-005-0044-9.

11. Horton, B. P. et al. (2020) 'Estimating global mean sea-level rise and its uncertainties by 2100 and 2300 from an expert survey', npj Climate and Atmospheric Science, 3(1), p. 18. DoI: 10.1038/s41612-020-0121-5.
12. Indian Meterological Department (2020) Hydrometerological Services, Ministry of Earth Sciences Government of India. Available at: http://www.imd.gov.in/pages/services_hydromet.php (Accessed: 7 August 2020).
13. Ismail, Z. and Jaafar, J. (2013) 'DEM derived from photogrammetric generated DSM using morphological filter', in 2013 IEEE 4th Control and System Graduate Research Colloquium. IEEE, pp. 103–106. DoI: 10.1109/ICSGRC.2013.6653284.
14. Krishnan, R. et al. (eds) (2020) Assessment of Climate Change over the Indian Region. Singapore: Springer Singapore. DoI: 10.1007/978-981-15-4327-2.
15. Lillesand, T. M., Kiefer, R. W. and Chipman, J. W. (2004) Remote sensing and image interpretation. 7th edn, Lloydia Cincinnati. 7th edn. New York: Wiley. Available at: http://www.osti.gov/energycitations.
16. Mahima, B. et al. (2014) 'Fish distribution dynamics in the Aghanashini estuary of Uttara Kannada, west coast of India', Current Science, 106(12), pp. 1739–1744.
17. Mahima, B. et al. (2016) 'Inventory of fishes of Gangavali estuary in Uttara Kannada, Karnataka state', Journal of the Marine Biological Association of India, 58(1), pp. 69–74. DoI: 10.6024/jmbai.2016.58.1.1872-08.
18. NASA (2021) NASA Prediction of World Energy Resources, The Power Project. Available at: https://power.larc.nasa.gov/data-access-viewer/ (Accessed: 26 May 2021).
19. National Remote Sensing Centre (2020) IRS data products, Indian Space Research Organisation, Government of India. Available at: https://www.nrsc.gov.in/ (Accessed: 22 July 2020).
20. Nauels, A. et al. (2019) 'Attributing long-term sea-level rise to Paris Agreement emission pledges', Proceedings of the National Academy of Sciences, 116(47), pp. 23487–23492. DoI: 10.1073/pnas.1907461116.
21. Prakash, P. S. and Bharath, H. A. (2021) 'Digital building height preparation from satellite stereo images', Photogrammetric Engineering and Remote Sensing.
22. PSMSL (2020) Permanent Service for Mean Sea Level - Data Explorer. Available at: https://www.psmsl.org/ (Accessed: 10 October 2020).
23. Ramachandra, T. V. et al. (2017) 'Green Spaces in Bengaluru: Quantification through Geospatial Techniques', Indian Forester, 143(4), pp. 307–320.
24. Ramachandra, T. V. et al. (2020) 'Insights into riverscape dynamics with the hydrological, ecological and social dimensions for water sustenance', Current Science, 118(9), pp. 1379–1393. DoI: 10.18520/cs/v118/i9/1379-1393.
25. Ramachandra, T. V., Vinay, S. and Bharath, H. A. (2020) 'Spatial Decision Support System (SDSS) for Urban Planning', in Bharath, H. A. and Ramachandra, T. V (eds) Urban Growth Patterns in India-Spatial Analysis for Sustainable Development. 1st edn. CRC Press: Taylor and Francis.
26. Ramachandra, T. V., Vinay, S. and Bharath, S. (2021) 'Visualisation of landscape alterations with the proposed linear projects and their impacts on the ecology', Modeling Earth Systems and Environment. DoI: 10.1007/s40808-021-01135-2.
27. Ryder, C. H. D. (1924) Topographi sheet - 24 Paragana District, Survey of India.
28. Terrestrial Hydrology Research Group (2014) Global Meteorological Forcing Dataset for land surface modeling, Princeton University. Available at: http://hydrology.princeton.edu (Accessed: 20 January 2017).
29. The Disaster Management Act, 2005 (2005). New Delhi: Government of India. Available at: https://www.ndmindia.nic.in/images/The Disaster Management Act, 2005.pdf.
30. UNDRR (2015) Sendai Framework for Disaster Risk Reduction 2015-2030, UNDRR. Available at: https://www.undrr.org/publication/sendai-framework-disaster-risk-reduction-2015-2030.
31. Walker, H. J. et al. (2005) 'Indian Ocean Coasts, Coastal Geomorphology', in Encyclopedia of Coastal Science. Dordrecht: Springer Netherlands, pp. 554–557. DoI: 10.1007/1-4020-3880-1_179.

CHAPTER 32

Spatial Variability of Climate Extreme Indices Over Mahanadi Basin (India)

R. K. Jaiswal
Scientist, CIHRC, National Institute of Hydrology,
WALMI Campus, Bhopal (India)
rkjaiswal_sagar@yahoo.co.in

A. K. Lohani
Scientist, National Institute of Hydrology,
Jal Vigyan Bhavan, Roorkee (India)

R. V. Galkate
Scientist, CIHRC, National Institute of Hydrology,
WALMI Campus, Bhopal (India)

S. Jain
Scientist, CIHRC, National Institute of Hydrology,
WALMI Campus, Bhopal (India)

Abstract

The management of water resources is an utmost important issue in developing countries like India because of considerable spatial and temporal variability of precipitation, fast-growing population, infrastructural problems, lower efficiencies of projects, distribution issues, declining groundwater, change in climatic pattern, etc. In recent times, the frequencies and intensity of extreme events and resulting natural disasters are becoming more frequent in many nations and affecting economic and environmental situation significantly. The Mahanadi is one of the major rivers of Chhattisgarh having important tributaries like Sondur, Pairy, Tandula, Sheonath, Jonk, Mand, Kelo, Ib, Kodar, Hasdeo, Arpa, Kharun, Kharang, Borai, etc. The Mahanadi is one of the major river basins in eastern India with a catchment area of 1,41, 600 sq km. well known for paddy cultivation and climate extreme can impact paddy production and thus wellbeing of farmers in the region. In the present study, an attempt has been made to compute seven extreme indices of temperature and nine extreme indices of rainfall including prescribed by the World Meteorological Organization and Expert Team on Climate Change Detection and Indices (ETCCDI) based on suitability under Indian condition were computed for gridded data of Mahanadi river basin covering by 206 grids from 1981 to 2015. ClimPACT2 software has been used for computation of these indices and results were analyzed in ARC GIS for spatial distribution. The analysis suggested that seasonal rainfall may vary in the range of 800 to 1200 mm from west to east and maximum temperature varies from 38.5 to 50°C where the higher maximum temperature in the central part of the basin. The number of consecutive rainy days (CWD) is higher in the Orrisa coast (28 days) and decreases towards the Chhattisgarh side (22 days) where water resource projects may have an adverse impact. The distribution of different extremes obtained from spatial interpolation may be useful to get important insight regarding spatial variation and may be useful for planning water resources in the region.

Keywords: Climate change, extreme indices, rainy days, dry days

1. Introduction

The economic development of society depends on the spatial and temporal distribution of water which is affected significantly by climate extremes. Climate change can alter the prevailing average, variance, frequency, spatial distribution, and extremes related to temperature and precipitation (Seneviratne et al 2012). Worldwide mean annual precipitation is obliged by the vitality spending plan of the troposphere (Allen & Ingram 2002), and extreme precipitation is affected by the atmospheric moisture. The changes in extreme precipitation are more prominent than those in mean precipitation. Singh et al. (2014) classically interpreted the impact of climate extreme based on literature reviews and concluded that tropical regions may get higher precipitation because of expanded water vapor, while mid-region might be influenced by the appropriation of expanded moisture. The warm areas may get the most extraordinary precipitation due to higher water vapour (Meehl et al 2005). The rising temperature may prompt higher outrageous occasions of heavy to extremely heavy precipitation albeit complete precipitation may continue as before. This can be additionally expounded as because of outrageous precipitation; the number of stormy days might be decreased, and dry spells become more prominent. The expected climate extremes have direct links with drought, extreme heat, flood, tornadoes, hurricanes, wildfires, etc. which are becoming common phenomena nowadays. The large scale social and economic distress due to climate change is largely seen worldwide due to the shift in precipitation regime and global warming. The year 2010 is considered the hottest year, second-highest number of disastrous events like extreme heatwaves and heavy precipitation and fifth-greatest economic losses (Huber & Gulledge 2011).

Menang (2017) computed climate extreme for observed data of precipitation and temperature in Cameroon as a part of the World Climate Research Program Grand Challenge on Weather and Climate Extremes using Climdex software. The analysis suggested that the analysis is important to understand the large-scale climatic variability in Cameroon and western Africa. Yin and Sun (2017) computed twenty-six climate indices of 2419 stations in China from 1961 to 2017 and found that daily maximum and the minimum temperature reached the record highest level in recent years, while the precipitation based indices were located within the range during 2017. The extreme indices provide an insight into the variation of climate and need to be evaluated for the development of adaptation policies and management plans.

India among the worst affected countries in the world due to climate warming which prompted to rise of temperature by about 0.7°C from 1901 to 2018. Krishnan et al (2020) has reviewed the climate change impact in India and stated that during the period of 29 years from 1986 to 2015, an expansion in warm days and nights and an ascent in incredibly serious cyclonic tempests over the Arabian Sea have been recorded due to consequence of human-caused environmental change. Further, flood risk has increased in West Bengal, eastern Uttar Pradesh, Gujarat, the Konkan region along with the urban centers of Metropolitan cities like Mumbai, Kolkata, and Chennai. The rainfall pattern has confirmed skewed rain with decreasing rainfall by 6% and increased frequency of extreme rainfall greater than 150 mm by 75% during the period of 1951 to 2015. The present study has been carried out to compute and demonstrate the spatial distribution of climate-related extreme indices of temperature and rainfall in the Mahanadi basin which is an important river of east-central India during the present period which can be used to manage water resource and compare with the future climatic condition. In the study gridded data of Mahanadi basin covering nearly 1.41 lakh sq km passes through Chhattisgarh and Odisha state of India before merging to the Bay of Bengal. The area of Mahanadi basin is rich in agriculture and known as the rice bowl of India having several multipurpose water resource projects. Any significant change in climate and its extremity will affect the wellbeing of the large-scale rural population in the states of Chhattisgarh & Odisha which primarily depend on irrigation water from river Mahanadi.

2.Material and Method

2.1 Study Area and Data Used

River Mahanadi originates from Sihawa village of Dhamtari district of Chhattisgarh state (India). Seonath, Mand, Ib, Hasdeo, Ong, Parry, Jonk, and Tel are some of the important tributaries of River Mahanadi. The total length of the river is nearly 858 km having several multipurpose water resource projects mainly for irrigation of paddy crops. The location map and grids of the Mahanadi basin are presented in Fig. 32.1 and 32.2. The Central Water Commission (India) has divided the Mahanadi river into three different parts i.e. upper, middle, and lower basins. The study has been conceived with the primary aim to understand spatial distribution and range of different climate extremes in the basin under present climatic conditions. The minimum, maximum temperature, and rainfall data set of 206 grids in the Mahanadi basin from 1981 to 2015 obtained from Mishra et al (2020) has been used in the analysis.

2.2 Methodology

The climate extreme indices are important for water resource planning and disaster management point of view as ranges and its spatial distribution provide an insight into the mechanism and help in preparedness to combat any disastrous event. Most of the studies on climate change profoundly concluded that even if total rainfall changes are not significant, the occurrences of extreme events are more likely to change in almost all the part of the earth which is the most significant consequences of global warming and poses a continuing threat to agriculture production (Subash *et al.*, 2011) and human life (Jones *et al.*, 2004). In the study, seven different extreme indices from temperature and nine from rainfall were selected from the list of extreme indices of the World Meteorological Organization (WMO, 2009) and Expert Team on Climate Change Detection and Indices (ETCCDI) (Kelin Tank et al. 2009; Zhang et al. 2011; Alexander 2016) based on suitability under Indian condition (Joshi & Rajivan, 2006). After computing these indices for all grid stations, the spatial distribution was determined to assess the vulnerable areas.

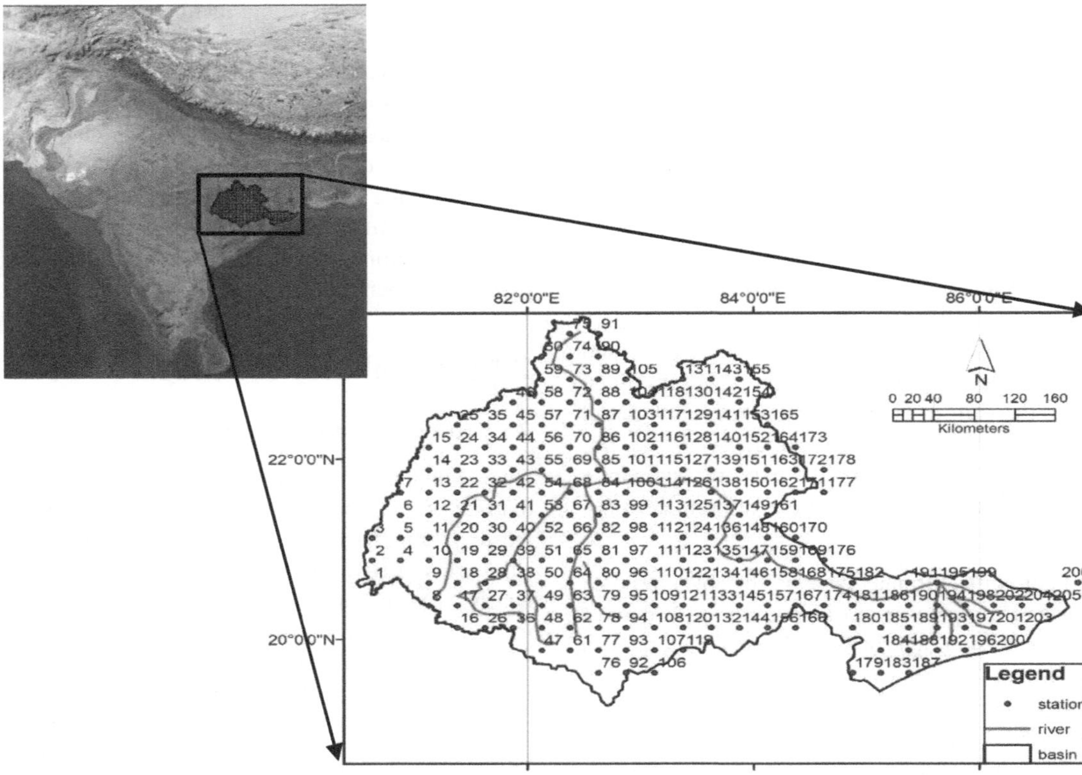

Fig. 32.1 Location map of Mahanadi basin in India

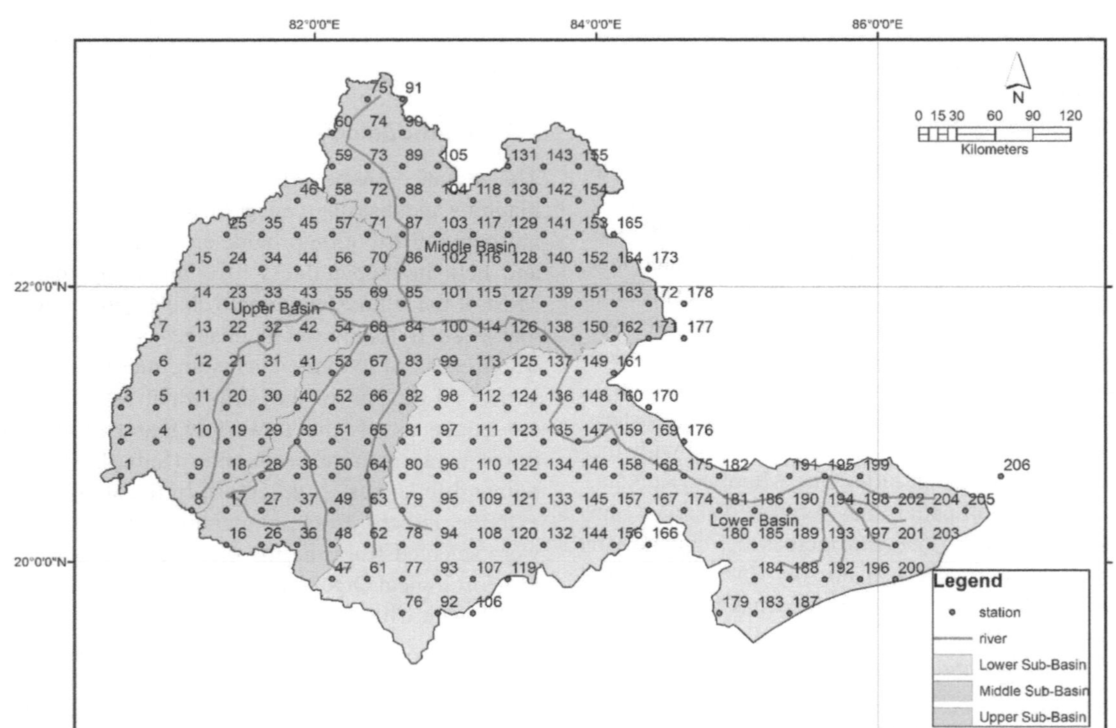

Fig. 32.2 Grids in Mahanadi basin

Description of different indices of rainfall and temperature is described in the next section.

Climate Extreme Indices

Annual total wet-day precipitation (PRCPTOT)

The annual total wet-day precipitation (PRCPTOT) may be defined as the total wet days rainfall and can be expressed by the following equation

$$PRCPTOT_j = \Sigma_{i=1}^{n} R_{ij} \quad (1)$$

where $PRCPTOT_j$ is the total precipitation for the jth year and R_{ij} is the daily precipitation greater than or equal to 1mm for the ith day in the jth year.

Number of rainy days (RD)

The number of rainy days is an important index for water resource planning can be defined as the days where rainfall is more than a threshold limit which is considered here as 2 mm. Let R_{ij} be the number daily rainfall on the ith day of jth period/year, the number of rainy days (RD_j) for i^{th} season/year may be described as:

$$RD_j = \text{Count } [R_{ij} > 2 \text{ mm}] \quad (2)$$

Maximum one-day rainfall (RX1D)

The maximum one-day rainfall (*RX1D*) is the highest in a day. The one-day maximum rainfall is important for designing drainage structures and flood studies. Let R_{ij} be the daily rainfall on the ith day of jth period/year, the maximum one-day rainfall for the jth period ($RX1D_j$) will be:

$$RX1D_j = \text{Max}[R_{ij}] \quad (3)$$

Maximum 5-days rainfall (RX5D)

The maximum 5-day rainfall may be defined as the highest rainfall amount in a consecutive five-day period. Let, $R5_{ij}$ be the total rainfall of the ith interval for the 5-days in the jth period/year. Here, i is the last day of the 5-day interval. The $RX5D_j$ can be defined as:

$$RX5D_j = \text{Max}[R5_{ij}] \quad (4)$$

Very wet days (R95P)

Very wet days (*R95P*) are the number of days having rainfall more than 95 percentile of the long period. Let R_{ij} be the precipitation series of the wet days (R_{ij} > 2 mm) in j = 1, 2, 3….. n period/year. The probability analysis of R_{wj} is used to compute 95 percentile (R_{95}). Now, the moderate wet days will be the count of the days where rainfall is greater than R_{95} in the R_{ij} series.

$$R95P_j = \Sigma_{i=1}^{n} \text{ Count}[R_{ij} > R_{95}] \quad (5)$$

Contribution from very wet days (R95PTOT)

It may be defined as the ratio of total rainfall in period comes from very wet days and the total rainfall during the period generally expressed in percentage or friction.

$$R95PTOT_j = \frac{\sum_{i=}^{n} Rij > R95P_j}{w} \quad (6)$$

Simple daily intensity index (SDII)

The simple daily intensity index *(SDII)* is the mean precipitation amount of wet days. Let, R_{wj} be the daily precipitation on a wet day (precipitation ≥ 2 mm) in the jth period and W be the number of rainfall days. The following equation can be used to compute simple daily intensity index ($SDII_j$) for the jth period/year:

$$SDII_j = \frac{\Sigma R_{Wj}}{W} \quad (7)$$

Consecutive dry days (CDD)

Consecutive dry days (CDD_j) for any j^{th} period is the longest dry period may be defined as the maximum number of days where rainfall is less than 1mm.

Consecutive wet days (CWD)

Consecutive wet days (CWD_j) for any j^{th} period is the longest wet spell that may be defined as the maximum number of days where rainfall is equal to or more than 1mm.

Mean daily maximum temperature (TxX)

The mean daily maximum temperature (*TxX*) may be defined as the average of maximum temperature for any *jth* period and can be defined as:

$$TxX_j = \frac{\sum_{i=1}^{n} TX_{ij}}{n} \quad (8)$$

where TxX_j is the mean daily maximum temperature at the jth period, TX_{ij} is the maximum daily temperature on the ith day of the jth period and n is the total number of days in a period.

Mean daily minimum temperature (TxM)

The mean daily minimum temperature may be defined as the average minimum temperature for any *jth* period and can be defined as:

$$TxM_j = \frac{\sum_{i=1}^{n} TM_{ij}}{n} \quad (9)$$

where TxM_j is the mean daily maximum temperature at the jth period, TM_{ij} is the minimum daily temperature on the ith day of the jth period and n is the total number of days in a period.

Hot days (TX90P)

Hot days ($TX90P_j$) may be defined as the fraction or percentage when the maximum temperature is higher than the 90th percentile value of maximum temperature.

$$TX90P_j = \frac{\sum_{i=1}^{n} \text{Count}[TX_{ij} > TX_{90}]}{n} * 100 \qquad (10)$$

where TX_{ij} is the maximum temperature on the ith day of the jth period and TX90 is the 90th percentile maximum temperature of the long-term period.

Warm nights (TN90P)

Warm nights ($TN90P_j$) may be defined as the fraction or percentage when minimum temperature is higher than the 90th percentile value of minimum temperature.

$$TN90P_j = \frac{\sum_{i=1}^{n} \text{Count}[TM_{ij} > TM_{90}]}{n} * 100 \qquad (11)$$

Where, TM_{ij} is the minimum temperature on the ith day of the jth period and TM_{90} is the 90th percentile maximum temperature of the long-term period.

Daily temperature range (DTR)

The daily temperature range (DTR_j) may be defined as the mean of difference of maximum and minimum temperature.

$$DTR_j = \frac{\sum_{i=1}^{n} [TX_{ij} - TM_{ij}]}{n} \qquad (12)$$

Warm spell duration indicator (WSDI)

The warm spell duration indicator is indicative of the long duration warm period and can be defined as the number of days where six or more consecutive days experience maximum temperature more than the 90th percentile value.

Cold spell duration indicator (CSDI)

The cold spell duration indicator is indicative of the long duration cold period and can be defined as the number of days where six or more consecutive days experience minimum temperature less than the 10th percentile value. For the computation of all these extreme indices, ClimPACT2 software has been used and described below.

2.3 ClimPACT2

The ClimPACT2 is an R package developed in collaboration with the World Meteorological Organization (WMO) and the Expert Team on Sector-specific Climate Indices (ET-SCI). The software having a user-friendly GUI where all input files and location of output files can be given. Batch processing possible to carry out analysis for multiple stations online. The GUI of ClimPACT2 is given in Fig. 32.3, where the whole process of computing indices for a station can be carried out in two steps i.e. Load and checking data in the first step and then calculating indices in the last step. After computing yearly values, the average has been computed for all the stations.

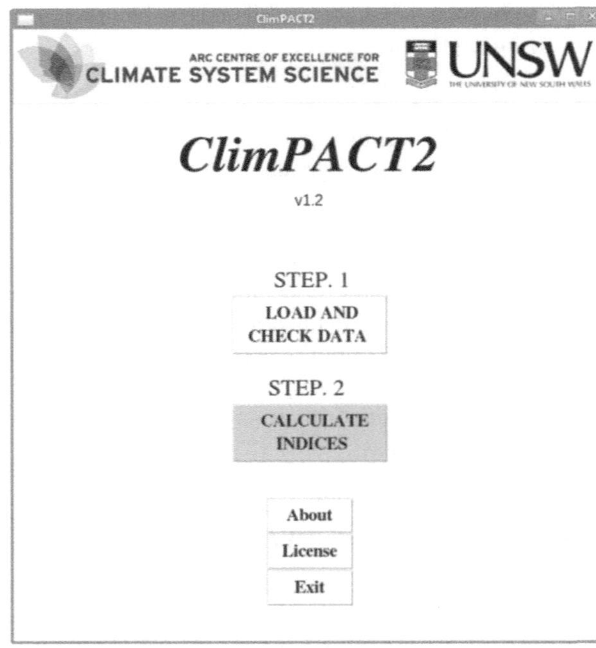

Fig. 32.3 Graphical user interface of ClimPACT2

Source: https://climpact-sci.org/assets/climpact2-user-guide.pdf

After computing the average values of different climate extreme indices, the ARC GIS has been used to plot these values as point maps. The inverse distance weighting (IDW) technique of multivariate interpolation for scattered points has been used for interpolation and adjudged the spatial distribution of different extreme indices.

3. Results and Discussion

The present study has been framed to examine the spatial distribution of different climate and temperature-based extreme indices in the Mahanadi basin of India using 206 gridded data at an interval of 0.25°x0.25°. The average yearly rainfall in the Mahanadi basin is about 1400 mm and ranges from 1100 to 1740 mm. The rainfall is increasing from east to west, where the highest rainfall is near the coast. The maximum one-day rainfall during the investigation period may vary from 60 to 136 mm which is highest in the central and coastal regions of the basin. The mean daily maximum temperature (*TxX*) is one of the important parameters from the management of water resource, crop water requirement, and heatwave point of view varies in the range of 38.5 to 50°C and higher in the central part of the basin. Similarly, the mean daily minimum temperature (*TxM*) and warm spell duration indicator (*WSDI*) has shown the higher values in the upper and central parts of the basin. The average values of different extreme indices and their range are presented in Table 32.1. The spatial distribution of some important indices has been presented in Fig. 32.4, where it can be seen that the average annual rainfall (PRCPTOT), the number of rainy days, simple daily intensity index, daily temperature range were increasing towards the lower part (coastal region), while the number of rainy days is maximum in central and hot days (*TX90P*) in the upper part of the basin.

Table 32.1 Average and variation of different climate extreme indices in the Mahanadi basin

S.N.	Extreme index	Unit	Variation in the basin		
			Average	Minimum	Maximum
1.	Annual total wet-day Precipitation (*PRCPTOT*)	mm	1407.1	1112.7	1743.3
2.	No. of Rainy Days (*RD*)	number	55	44	63
3.	Maximum one-day rainfall (*RX1D* in mm)	mm	96.2	60.2	136.1
4.	Maximum five-day rainfall (*RX5D*)	mm	260.9	186.1	342.2
5.	Very wet days (*R95P*)	number	23	6	27
6.	Contribution from very wet days (*R95PTOT*)	mm	373.4	259.2	491.1
7.	Simple daily intensity index (*SDII*)	mm/day	13.5	11.2	21.4
8.	Consecutive dry days (*CDD*)	number	49	39	83
9.	Consecutive wet days (*CWD*)	number	26	12	33
10.	Mean daily maximum temperature (*TxX*)	°C	44.6	38.5	50.0
11.	Mean daily minimum temperature (*TxM*)	°C	33.0	28.9	37.5
12.	Hot days (*TX90P*)	number	1	0	17
13.	Warm nights (*TN90P*)	number	11	11	11
14.	Daily temperature range (*DTR*)	°C	11.8	7.9	12.8
15.	Warm spell duration indicator (*WSDI*)	number	16.3	10.6	26.4
16.	Cold spell duration indicator (*CSDI*)	number	10.1	8.3	12.1

4. Conclusion

The study has been carried out to ascertain the range and variation of average climate extreme indices in the Mahanadi basin of India. The results of the analysis confirmed that temperature-related indices have higher values in the central portion of the basin while lower values in the coastal region. The rainfall-based indices are showing an increasing trend towards coastal zone and there is a significant rising trend in total precipitation, one-day and five-days rainfall, R95P, and the number of rainy days in coastal zones of the Mahanadi basin. The higher values of maximum and minimum temperature in the central part of the basin may demand a higher amount of water for different designated purposes where population density is higher than the upper and lower zones of the basin.

References

1. Alexander LV (2016) Global observed long-term changes in temperature and precipitation extremes: a review of progress and limitations in IPCC assessments and beyond. Weather Clim Extremes. 11: 4–16
2. Allen MR, Ingram WJ (2002) Constraints on future changes in climate and the hydrological cycle. Nature. 419: 2224–2232
3. Huber DG, Gulledge J (2011) Extreme weather and climate change: understanding the link and managing the risk: Science and impacts program. Center for Climate and Energy Solutions, Arlington, Available at: http://www.c2es.org/publications/extreme-weather-and-climate-change.
4. Jones C, Waliser DE, Lau KM, Stern W (2004) Global occurrences of extreme precipitation and the Madden—Julian Oscillation: Observations and predictability. J Clim. 17: 4575-4589.
5. Joshi UR, Rajivan M (2006) Trends in precipitation extremes over India, National Climate Centre, India Meteorological

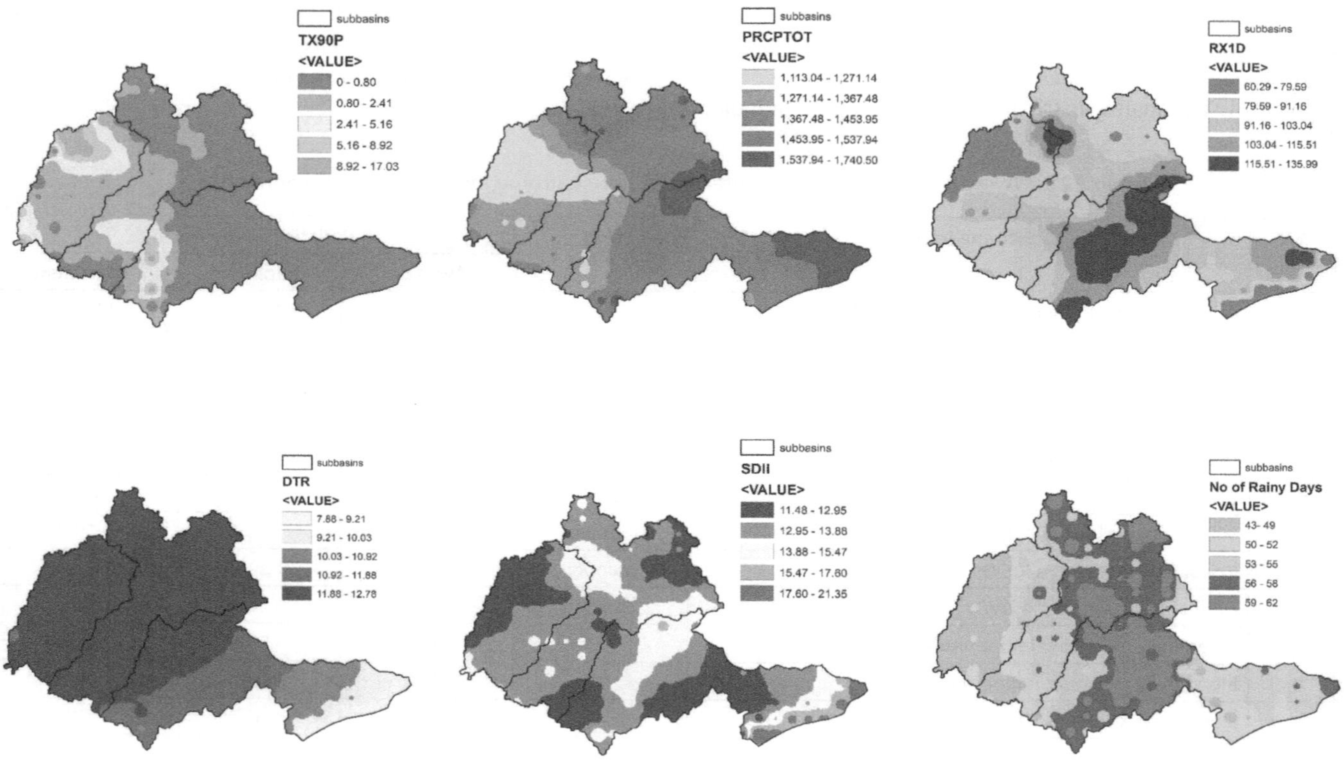

Fig. 32.4 Spatial distribution of few extreme indices in the Mahanadi basin

Department, Pune.

6. Klein Tank AMG, Zwiers FW, Zhang X (2009) Guidelines on analysis of extremes in a changing climate in support of informed decisions for adaptation. WCDMP-72, WMO-TD/ No.1500, 56 pp.
7. Krishnan R, Sanjay J, Gnanaseelan C, Mujumdar M, Kulkarni A, Chakraborty S (2020) Assessment of climate change over the Indian region, A report of the Ministry of Earth Sciences (MoES), Government of India, Springer, https://doi.org/10.1007/978-981-15-4327-2
8. Meehl GA, Arblaster JM, Tebaldi C (2005) Understanding future patterns of increased precipitation intensity in climate model simulations. Geophys Res Lett. 32: L18719. doi:10.1029/2005GL02368028.
9. Menang K P (2017) Climate extreme indices derived from observed daily precipitation and temperature data over Cameroon: the need for further assessments. Meteorol Appl. 24: 167–171.
10. Mishra V, Bhatia U, Tiwari AD (2020) (2020) Bias corrected climate projections from CMIP6 models for South Asia [Data set]. Zenodo. http://doi.org/10.5281/zenodo.3987736
11. Seneviratne SI, Nicholls N, Easterling D, Goodess CM, Kanae S, Kossin J, Luo Y, Marengo J, McInnes K, Rahimi M, Reichstein M, Sorteberg A, Vera C, Zhang and X (2012) Changes in climate extremes and their impacts on the natural physical environment. In: Managing the risks of extreme events and disasters to advance climate change adaptation (Field CB, Barros V, Stocker TF, Qin D, Dokken DJ, Ebi KL, Mastrandrea MD, Mach KJ, Plattner GK, Allen SK, Tignor M, Midgley PM (eds.)). A special report of working groups I and II of the Intergovernmental Panel on Climate Change (IPCC), Cambridge University Press, Cambridge, UK, pp. 109-230
12. Singh VP, Mishra AK, Chowdhary H, Khedun C.P. (2014) Climate Change and Its Impact on Water Resources. In: Wang L., Yang C. (eds) Modern Water Resources Engineering. Handbook of Environmental Engineering, vol 15. Humana Press, Totowa, NJ. https://doi.org/10.1007/978-1-62703-595-8_11
13. Subash N, Singh SS, Priya N (2011) Extreme rainfall indices and its impact on rice productivity—A case study over sub-humid climatic environment. Agril Water Manage. 98(9): 1373-1387. http://doi.org/ 10.1016/j.agwat.2011.04.003

14. World Meteorological Association (WMO) (2009). Guidelines on analysis of extremes in a changing climate in support of informed decisions for adaptation. Geneva
15. Yin H, Sun Y (2018) Characteristics of extreme temperature and precipitation in China in 2017 based on ETCCDI indices, Adv Clim Res. 9(4): 218-226
16. Zhang X, Alexander L, Hegerl GC, Jones P, Klein Tank A, Peterson TC, Trewin B, Zweirs FW (2011) Indices for monitoring changes based on daily temperature and precipitation data. Clim Change. 2: 851–870, DOI: 10.1002/wcc.147.

CHAPTER 33

Climate Change—Induced Natural Disaster: A Case Study of 2013 Kedarnath Disaster, Uttrakhand

Neha Gupta
Research Scholar, Department of Earthquake Engineering,
Indian Institute of Technology Roorkee

Josodhir Das
Professor, Department of Earthquake Engineering,
Indian Institute of Technology Roorkee

James Xavier Paul
Research Scholar, Department of Earthquake Engineering,
Indian Institute of Technology Roorkee

Abstract

Climate change in higher Himalaya are very uncertain, that led to a various natural disaster, such as cloudburst triggered flash flood, landslide, debris flow snow avalanches is due to extreme weather condition and are very common in Himalaya region, In the last 50 years, the intensity and frequency of heavy precipitation increased that results in increased flood and larger scale of the landslide. The Intergovernmental Panel for Climate Change (IPCC) shows that the increase in temperature, the intensity of rainfall, storm intensity, and variability are likely to result in global climate change. During the recent studies it is observed that, extreme climate change was mainly due to population growth and deforestation. The global climate change led to an increase in rainfall that may influence the chain of multiple hazards called cascading hazard as flood, landslide, and glacier outburst floods. The aim of this work is to discuss a case study of the 15-17 June 2013, Kedarnath disaster and assess how climate change pattern and rainfall acts as a cascading event for multiple hazards using remote sensing. During 15-17 June 2013 heavy rainfall occurred in Kedarnath led to cause flood in and around the adjoining region and experience massive landslides and debris flow along Mandakini and Bhagirathi basin. We used MODIS Land Surface temperature (LST) to study the climate change impact during the devastating Kedarnath disaster. All the meteorological factors collected from the satellite data, for rainfall analysis GPM satellite data and IMD gridded data are used.

Keyword: Climate change, Cascading hazard, rainfall, IPCC, 2013 Kedarnath disaster

1. Introduction

During 13-17 June 2013, a flood devastated the Himalayan village of Kedarnath, India, the destination of half a million Hindu pilgrims annually. Heavy rainfall together with rapid snowmelt lead to cloudbursts, extreme flooding, and a massive amount of landslides and mudslides, which devasted the village of Kedarnath, Rambara, and Gurikund and its surrounding areas (Dubey et al. 2013; Allen et al., 2016; Das et al., 2015; Martha et al., 2014). The town sits 11,500 feet up in a tight valley. Sharp, snowy peaks tower on three sides, and a stone temple sits at one end. The flood - which occurred on June 17, 2013 - was India's worst disaster in a decade. Several thousand people drowned. The deluge tore apart dozens of bridges, swept away miles of paved roads, and carried off herds of livestock killed more than 6000 people (sati and Gahalaut 2013). Due to rising temperatures and Black Carbon (BC) concentrations, snowfall in the Himalayan region has decreased. The melting of glaciers is increased as snowfall decreases, and temperatures rise (Rafiq et al., 2017). The decline in snowfall and warming results in the accelerated melting of glaciers (Rafiq et al., 2018). During 14 –17 June 2013, Kedarnath and neighbor regions received heavy rainfall due to heavy precipitation. Heavy rainfall, combined with excess melting of ice of the Chorabari glacier, increased the flow of glacier-fed Mandakini River, leading to heavy flooding and landslides downstream of the river (Dobhal et al., 2013; Srinivasan, 2013; IMD, 2013; Joseph et al. 2013). This devastating flood episode was recorded as one of India's worst natural disasters, which causes heavy loss in life and property. In the region, the accumulation of rainfall from multiple episodes of heavy rainfall spells during 14–17 June 2013 which triggered Chorabari lake to get breached, resulting in heavy flash floods.

The consequence of seasonal climate change results in heavy snowfall due to Westerly Disturbances in 2012-13. In the summer of June 2013, there was also heavy rainfall due to the early arrival of the monsoon. These activities resulted in heavy rainfall and consequently heavy snow melting. Therefore, cloudbursts and melting of snow caused outburst floods in Chorabari Glacier Lake and flash floods in surrounding areas (Arora et al. 2014; Dobhal et al. 2013). According to the scientists of Wadia Institute of Himalayan Geology, Dehradun, a heavy rainfall recorded as 315 mm near Chorabari Glacier during 16- 17 June 2013. As per the meteorologists, the rainfall ≥100 mm. (3.97 Inches) per hour is a Cloud Burst. The actual phenomenon of Cloudburst occurs in a small area and exists for a few minutes. But in the Kedarnath disaster, there was heavy rainfall that continued for more than two days leading to the Cloudburst. Therefore, the unexpectation of heavy rainfall in the havoc created conditions like Cloudburst and not the actual one. However, this climatic condition played a significant role in the tragedy.

Global warming and climate changes engendered highly adverse climatic conditions on our planet. As per the IPCC (Intergovernmental Panel on Climate Change) noted in its special reports on extremes, it is increasingly clear that climate change has detectably influenced several of the water-related variables that contribute to floods, such as rainfall and snowmelt (Stocker et al., 213; Hartmann et al., 2013). With higher temperatures, we have more energy in the Earth's system. At higher temperatures, the air can hold more moisture content. This can lead to an increase in precipitation intensity, duration, and frequency. Thus, an increase in temperature stimulated ice melting, especially in polar areas creating extreme adverse winters (Polar Fear) and heavy rainfall that gave rise to flooded areas (Vishwambhar Prasad sati 2013). Therefore, global warming affected western disturbances as well as the Indian monsoon. Significantly, the frontal area of the central Himalayan glaciers has shrunk by 1% (Hartmann et al., 2013). However, Himalayan glaciers have not responded uniformly to recent climate warnings. Both of the climatic conditions influence the central Himalayan glaciers: Western disturbance, as well as Indian summer Monsoon (ISM), and hence, some glaciers of east and central Himalayas are receding Mehta et al. (2014). While the western Himalayan glaciers are influenced mainly by the Westerlies only and hence, the glaciers of western Karakoram Himalaya are advancing. These conditions show that central Himalayan areas receive heavy snowfall (in winters) and monsoon rain, causing rapid snow melting (in summers). Therefore, glaciers in these areas are receding. These phenomena support the formation and expansion of glacial lakes and rapid flooding in streams and landslides, avalanches (Clague and Evans 2000; O'Connor et al. 2001; Worni et al. 2012,2013).

In mid-June 2013, the monsoon had reached Uttarakhand in advance by about two weeks (Rafiq et al., 2019, Dobhal et al. 2013; Asthana and Asthana 2014). Precipitation in the month was also more than average. Also, the convergence of the upper air trough and the low-pressure system pushed the monsoon line across northwest India well ahead of the due date (Cho et al., 2016). According to Sarkar (2013), the monsoon covered the entire nation about a month before the normal date (June 15). Increase in climate change in the Himalayas region resulted in hydro metrological hazards, especially heavy rainfall and consequently heavy snow melting that's results in Cloudburst, flood, avalanche, landslides, and mudflow (Kulkarni et al., 2007; Bolch et al., 2012; Rafiq and Mishra 2016, Mir et al., 2018; Mishra and Rafiq 2017). Kedarnath disaster 2013 and Dhuali Ganga disaster 2020 witnessed these cascading of hazards due to glacier melting and heavy rainfall subject to an impact as a result of climate change. Scientific evidence shows that the glacier is retreating due to rising temperature due to climate change (Kulkarni and Karyakarte 2014, ICIMOD 2007, 2011; IPCC SROC 2019). Increased concentration of greenhouse gases in the atmosphere due to anthropogenic activities leads to warming of the Earth's atmospheric system, while aerosols reduce surface warming. Global warming is manifested by decreases in snow cover, the lesser extent of sea ice in the northern hemisphere, thinner sea ice, shorter freezing (Rafiq et al., 2019, 2018). Warming in the Himalayas is demonstrated by the increased rate of glacial recession and the resultant formation of new glacial lakes in the high Himalayas (Mool et al., 2001). Melting of glaciers and snow caused various types of risks one of the dramatic risks is that of glacier lake outburst flood (GLOFs). The outburst of the lake has the potential for generating destruction in the valley. A report published by CRED (Fig. 33.1) showed the increasing trend of hydrometrical disasters between 1974 and 2003 in the Eastern Himalaya region.

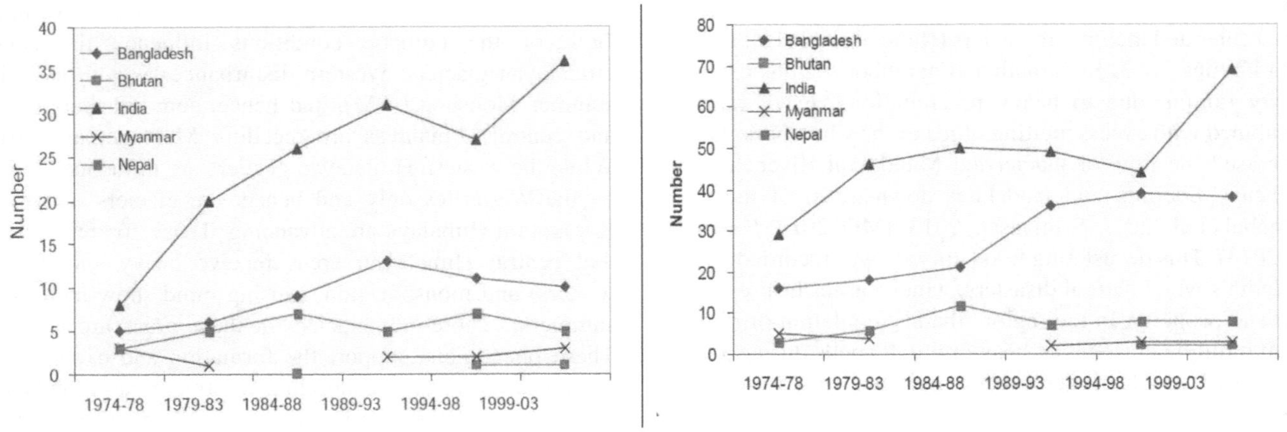

Fig. 33.1 Total number of hydrometeorological and related disasters in Bangladesh, Bhutan, India, Myanmar, and Nepal (1974-2003)

source: ICIMOD

Previously various studies have been done on the Kedarnath disaster considering different assumptions. Allen et al.,2016 discussed considering two factors hydrometeorological triggering factor and topographic factors concerning Glacier lake burst. Cho et al.2016 use the metrological Data and WRF model for simulation of June 2013. Pratap ae al., 2020 studied the variations in hydrometeorological variables during the devastating Kedarnath cloudburst in Uttarakhand. Xavier et al.,2017 explain the instability and dynamics of the abnormal circulation pattern and the intensification. No studies have been done to show the relation of rainfall and snow covers with land surface temperature. This study analyzes the impact of climate change on natural hazards in the Himalayas region through land surface temperature (LST) and rainfall accompanying the June 2013 Kedarnath event disaster. The study focuses on the following objectives 1) assessment of satellite rainfall data (GPM) with IMD gridded data of June 2013 Kedarnath event, (2) analysis of land surface temperature concerning topographic factors.

2. Study Area

The study area is Kedarnath, situated at 3850 m above sea level in Mandakini valley of Rudrapragya District, Uttrakhand. The area is observed to be located on glacial outwash deposits and is occupied by outcrops and overburden. Along the valley flanks, moraine deposits are observed. Rocks of Central Crystalline Group represented by Augen gneiss and migmatites are exposed around the Kedarnath area. Exposures of Augen gneiss are observed along the footpath section to the downstream of the Kedarnath Township on the left bank of Mandakini River. These rocks are medium-grained, greyish in color, moderately weathered, moderately jointed, and medium to thickly foliated—these dip towards the northeast to the northwest at moderate angles. The area represents highly rugged and immature topography characterized by moderate to steep slopes that are intervened by narrow valleys. The topography of the region appears to be controlled by the structural and lithological factors.

3. Dataset and Methodology

To understand the climate impact of the Kedarnath disaster, we used IMD Daily gridded data and GPM daily rainfall data for the period from 2000-2020. To prepare change in land surface temperature (LST), preceding to Kedarnath disaster, MODIS derived LST of 8 –day product which provides pixels per temperature and emissivity, was extracted. MODIS data have 1200 x 1200 pixels value and each value equal to 1 km^2 extending from 2010 to 2017. We used Terra daytime LST the daytime proves less cloud cover and easy for analyzing the higher temperature. For snow cover, we used Terra MODIS snow cover 8-day product (MOD10 A2).

4. Results & Discussion

4.1 Observation from Land Surface Temperature

Land surface temperature is one of the most important factors for estimating the effect of climate change on higher Himalayas. LST usually shows high in the river valley due to lower elevation. The LST calculated from the MODIS data of each year from 2010-2017, the increase or decrease in monthly LST calculated and used for further analysis and show correlation with snow cover and rainfall. Time series and Mann Kendell test trends have been carried out from

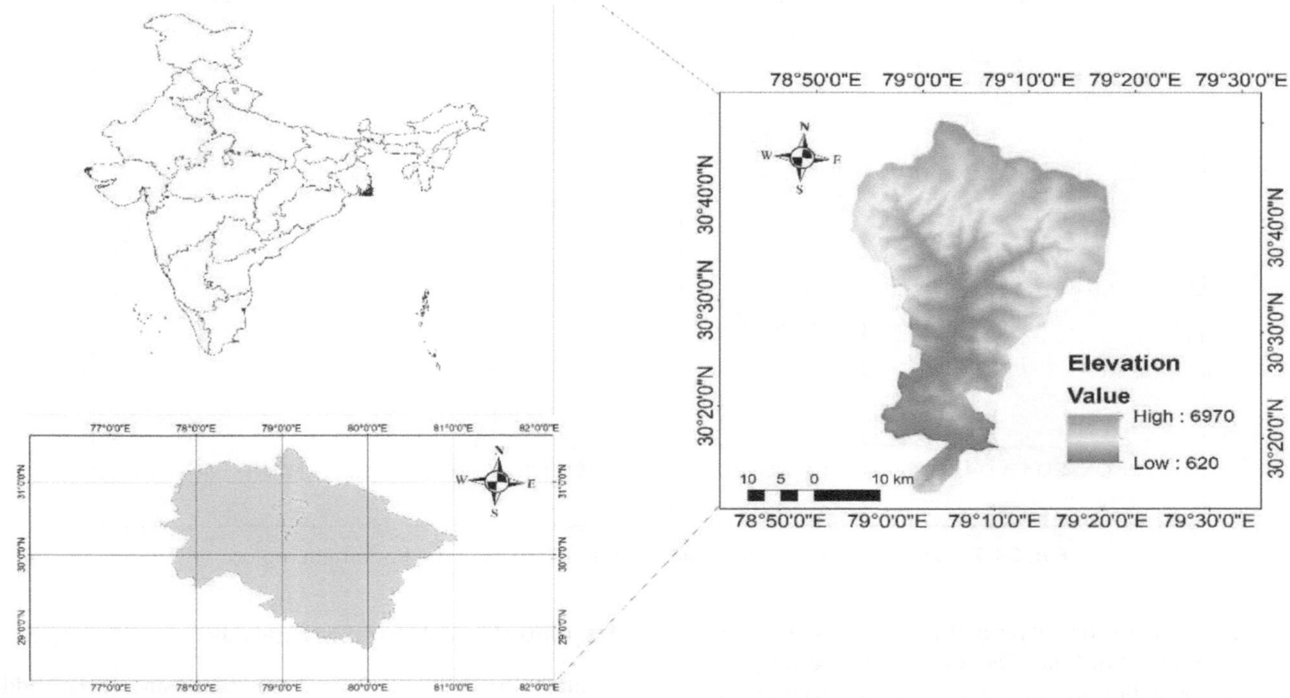

Fig. 33.2 Study area

2010–2017. It should be noted that the LST is not equal to air temperature. In June 2013, the temperature gradually increased and resulted in the maximum rainfall during the 15–17 June 2017 (figure.4). increased rainfall is due to the heating of the Earth's surface by solar radiation generate thermodynamic instability that forms rain and thunderstorm. The surface temperature increases the intensity of convection which affects rainfall. Comparing the precipitation rate with surface temperature, it is evident that the convection process is operating during 15–17 June 2013. This magnitude of rainfall had not experienced in history. As a result of heavy magnitude rainfall and snow melting chorabari lake burst caused flooding of sarswasti and Mandakini River in Kedarnath valley and surrounding area (Dobhal et al., 2013). A time-series analysis was carried out from 2010–2017, and it is observed that temperature is increasing over the time due to climate change (Fig. 33.3). From the graph, it is cleared that the maximum land surface temperature observed in June 2013 with an increasing trend. Temperature also plays an important role in melting snow; we analyze the relationship

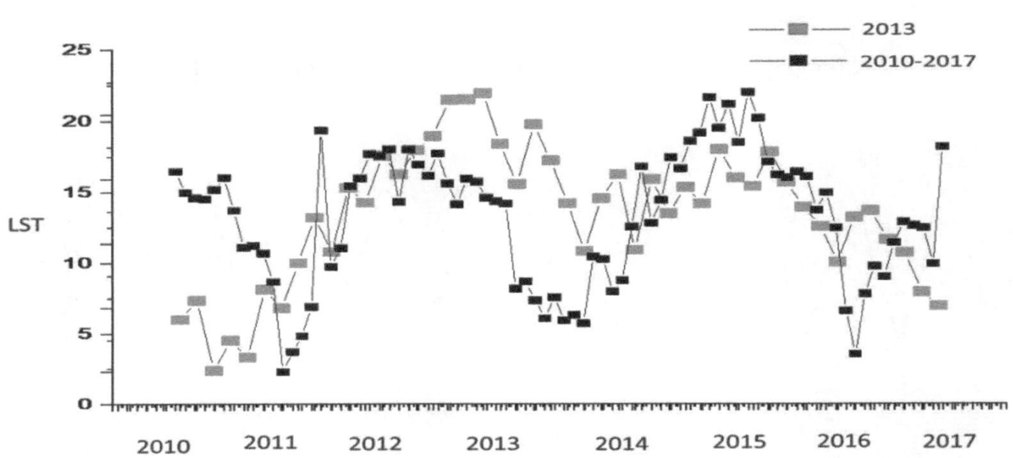

Fig. 33.3 Time series analysis of land surface temperature from 2010-2017 over Kedarnath region

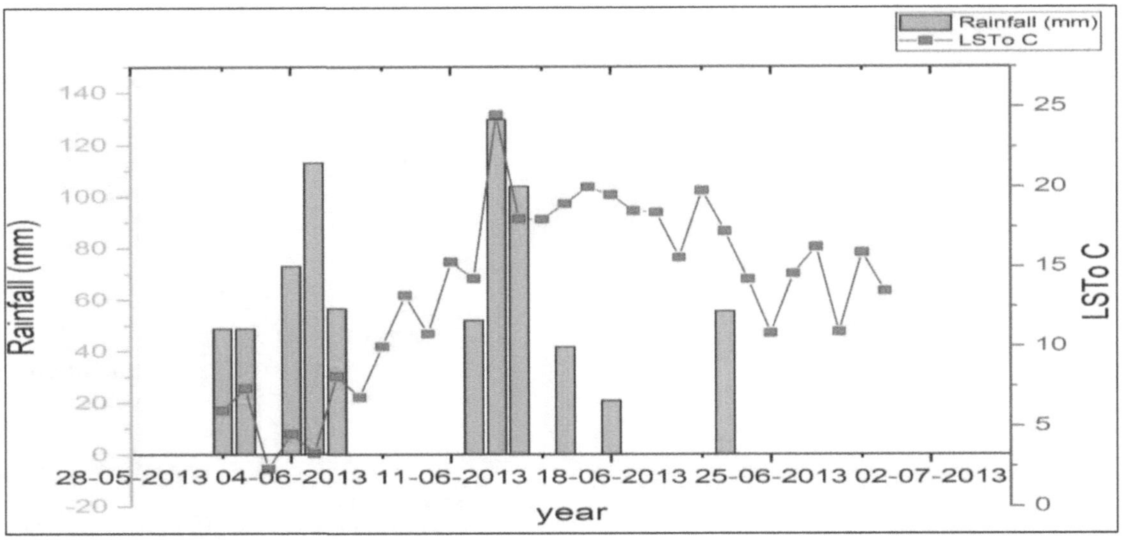

Fig. 33.4 Correlation of LST and rainfall during 2013 over the Kedarnath region

between LST and snow cover and found that the maximum LST was observed in June. The onset of this period of rapid snow cover depletion appears with usually warm temperatures in May to June. The finding suggests that global warming is increasing and potentially affects the snow-covered area in mountain terrain. This increase in surface temperature shows the melting of a glacier in decreasing trend during 2001–2018 (Karwaiya et al., 2021). LST decreases at higher elevation because air temperature decreases with an increase in elevation, which affects the heat exchange between surface and air and acts as a surface heating process. The Mann Kendall test has been carried out to determine the trend, the Mann-Kendall value statistic Z value is 0.057, the positive value shows the increase in temperature for the studied period.

4.2 Metrological Variable Analysis

Rainfall was collected from 2000-2020 using IMD gridded and GPM satellite data. The metrological dataset of rainfall, atmospheric temperature, and mean and maximum temperature. The average, maximum, and minimum atmospheric temperature was estimated at 12° C, 15° C, and lower down by 5° C on June 17, 2013 (figure 5). Fig. 33.5 shows that atmospheric temperature is increasing from 2010 - 2020. During this period, the upper atmospheric wind anomaly at 200 hPa observed from June 2013 over the Indian region found a larger positive anomaly and indicates that the wind at 200 hPa was more substantial than the long term average (Xavier et al., 2017)

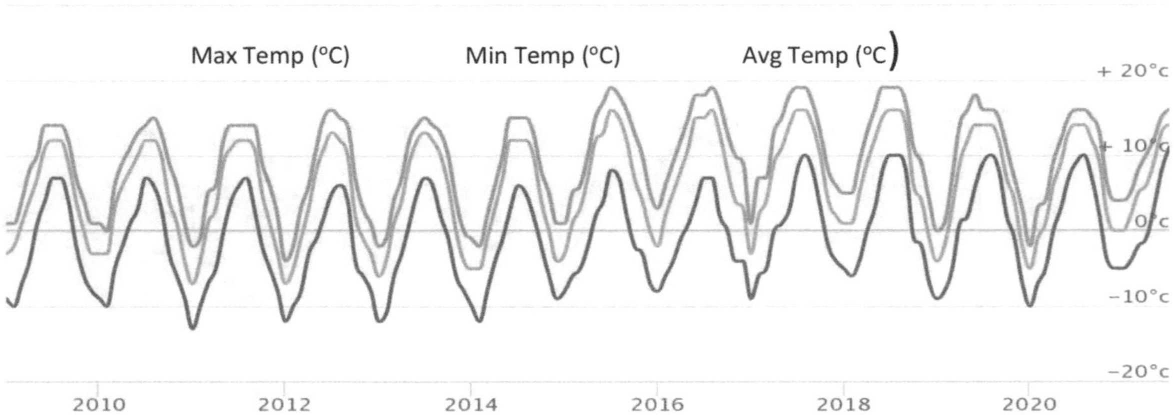

Fig. 33.5 Increasing trend of Min, Max and average temperature from 2010-2020 over Kedarnath region

Source: IMD

The continuous precipitation in the area during 15–17 June 2013, in addition to snowmelt, filled the lake at a much faster rate than usual normal rainfall (Allen et al., 2016) results in floods and landslides. GPM and IMD gridded data was used to derive precipitation over the Kedarnath catchment from 2000–2020 Fig. 33.6. The increasing trend of rainfall was observed over the time. Total rainfall of 486 mm occurred in June 2013. In 24 hr on 15-16 June 2013, 325 mm of rainfall was recorded at the metrological station (dobhal et al., 2013a). Most studies reported heavy rainfall beginning on June 15 and peaking on June 16 and 17 (Martha et al. 2014). From Figure 33.6, it is reported that no such event of heavy rainfall magnitude of 325 mm during two-day rainfall measured within the last 20 years' GPM time interval record. GPM data shows on 10-11 June, and rainfall was uniformly distributed across Uttrakhand. It is observed that the cumulative rainfall due to multiple rainy events combined with snow cover melting was associated with a sudden increase in surface temperature resulted in a heavy flash flood and landslide. The heavy monsoon normally arrives in early July but during the peak monsoon period in the northern Himalayas. Still, such magnitude of rainfall appears in early June month which was not unprecedented. The heavy magnitude of precipitation on June 15-17 was due

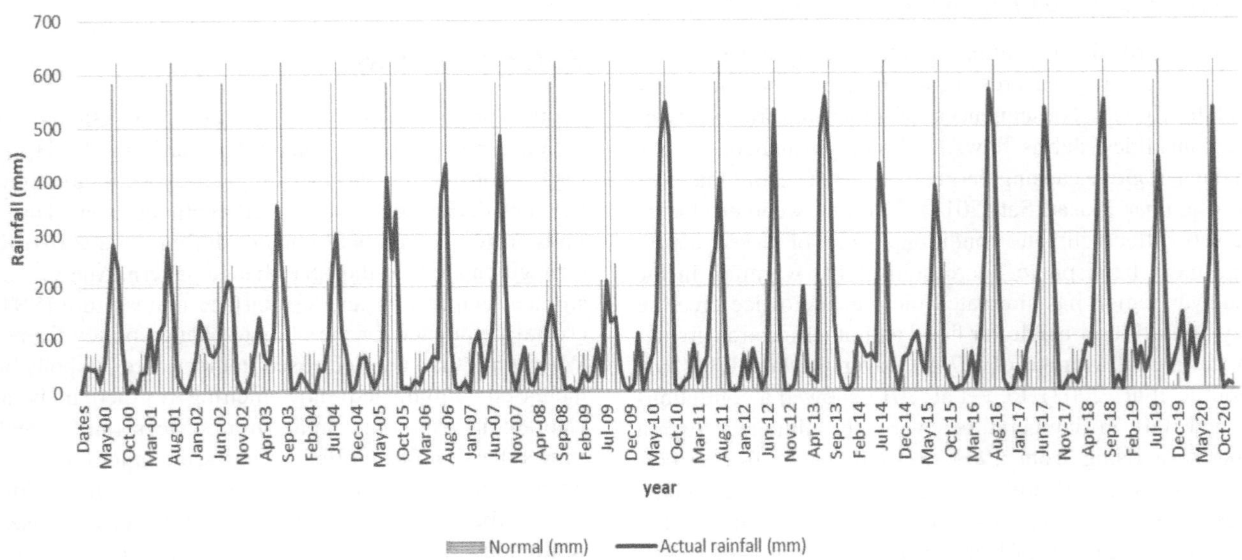

Fig. 33.6 Distribution of GPM satellite rainfall over Kedarnath region from the year 2000–2020

Source: WIRS INDIA

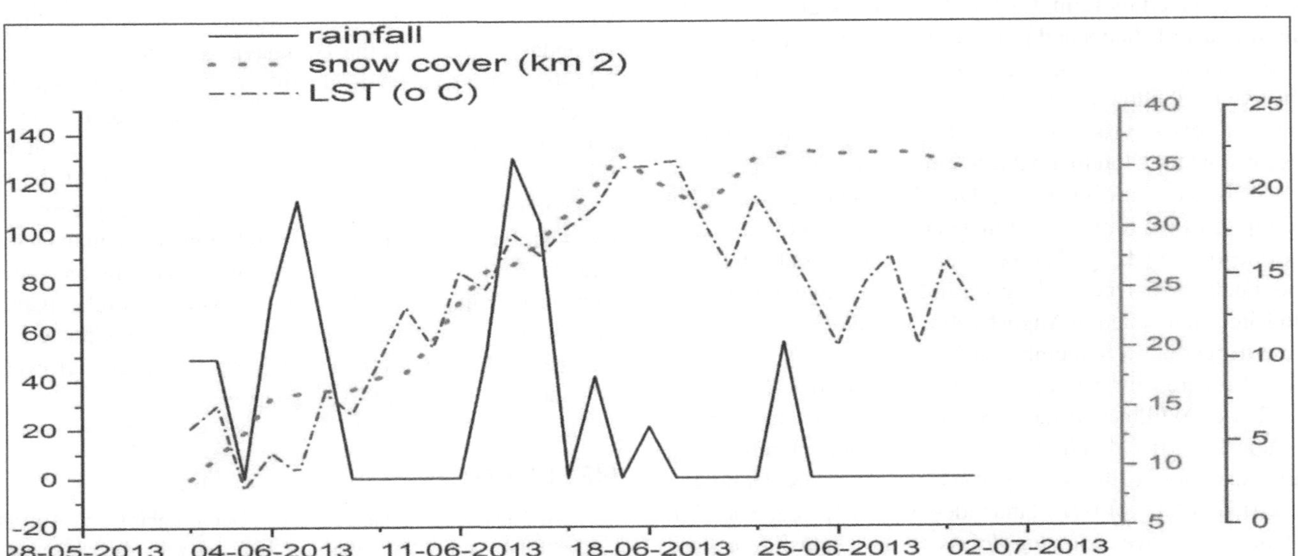

Fig. 33.7 Graph representing daily rainfall, snow cover area, and land surface temperature over Kedarnath during June 2013

to the rapid increase in land surface temperature in May and June 2013, as shown in Fig. 33.4, leading to Cloudburst and GLOF. Mean temperatures dropped at Gangotri Glacier snout to 1.6 ° C by June 17 and adiabatic lapse rate of 5.5 ° C Km^{-1}. with freezing level from 5650m a.s.l at beginning down to 5000 m a.s.l on June 16 and 4000m a.s.l on June 17 (Allen et al., 2016). GPM data from 2000 to 2020 date confirm that the precipitation is increasing. Due to the increase in global warming, snow cover decreased by 50 % from the average in the mid of May to below average in mid June such snow cover retreating is not reported previously in MODIS dataset (Allen et al., 2016)

4.3 Role of Climate Change for Cascading of Disaster

Natural hazards due to extreme weather changes may have a series of secondary hazards, heavy precipitation and heavy snowfall triggers consequences of other hazards such as flood, landslides, debris flows, and snow avalanches in the Himalaya region, causing more than 70 % economic loss (Vishwambhar Prasad Sati 2013). Extreme weather change adversely affect climate conditions were observed across the northern hemisphere. Increasing global warming in the Himalayas region has simulated the melting of ice creating heavy rainfall responsible for flood and other hazards greater than the global average of 0.74° C over the last 100 years (Solomon et al., 2007). Ren et al. 2017 showed a continuous rise and fall in temperature over the Himalaya region temperature rising from 1901 to 1940, while from 1949-1970, it showed a decreasing trend. Thus, climate change causes a severe change in the high magnitude of weather events as a conclusion of heavy rainfall causing Cloudburst, flash floods, landslides, and avalanches and increase in temperature stimulate the intensity of rainfall and expansion of area affected by rainfall (Ashutosh Pathak 2018). Various natural hazards happened in the Indian Himalaya region due to an increase in temperature in august 2010, Leh Ladakh observed Cloudburst triggered flood and landslides study revealed that the rise in temperature, rainfall, and snowfall responsible for Cloudburst due to climate change (Pepin et al., 2015). Kedarnath disaster and Dhauli Ganga are two recent examples of the impact of climate change. Both the disasters are due to the melting of snow/glaciers resulted in flash floods and landslides. Due to heavy rainfall, a large number of landslides happened in August 1998, a total of 466 landslides were triggered, in September 2012, a total of 470 landslides were triggered by rainfall in the Okimath area (Martha et al., 2014). Within two days of rainfall in Kedarnath on 15 to June 17, 2013, this had caused severe floods, landslides, a large-scale loss of lives, properties, and damages (Dobhal et al. 2013), a total of 6013 landslides were detected. The high gradient of a river carrying debris from a landslide caused bank erosion at several places resulting in new landslides. (Martha et al., 2014) all these disasters witness climate change. Aside from the LST increase caused by changes in the surface cover condition, the overall climatic warming environment provides a useful backdrop for assessing temperature differences. The near-surface air temperature is rising at a rapid rate, with an average annual temperature increase of 0.06 C between 1977 and 1994 (Shrestha and Aryal 2011) with a rise of temperature at higher altitudes (0.03-0.7° C) Shrestha and Aryal (2011). The high altitude of the Tibetan plateau north of the Himalayas is heating due to global warming is rerouting western disturbance storms (Madhura et al., 2014)

5. Conclusion

In the present paper, we analyze how climate change increase responsible for natural hazards in the Himalayas region considering the metrological and climatic condition by considering June 2013 Kedarnath disaster. This study finds impact of climate change during June 2013 disaster through understand the LST change pattern. Analysis of land surface temperature and sea surface temperature (SST) is a crucial parameter for the climate change policy framework. We found that the increased temperature responsible for increased rainfall and snow melting resulted in breaching Chorabari Lake. The relationship between rainfall and LST shows the significate effect of surface temperature for high-intensity rainfall. LST also shows the correlation with snow cover, when the LST increases, snow cover decreases. The continuous warming over the glaciered covered mountain profound impact like snow cover or glacier mass loss cause increased in flooding of the Himalaya river basin.

Climate warming plays a significant role as cascading of secondary hazards in the Himalayas region, heavy rainfall, GLOF, a massive landslide in and around the Kedarnath valley. This study concludes that climate change, responsible for rainstorm occurrences. Furthermore, the occurrence of this June 2013 event in northern Himalaya during the pre-monsoon season, as well as circulation and precipitation trends in June, necessitates the adaptation of risk management strategies in the face of increasingly frequent and strong rainstorms outside of the core monsoon months (Cho et al., 2014). More study should be focused on the climate change model for future rainfall or temperature projection considering different IPCC scenarios.

References

1. Asthana A. K., Asthana H. (2014) Geomorphic Control of cloud burusts and flash floods in Himalaya with special reference

to Kedarnath area of Uttrakhand, India. International Journal of Advancement in Earth and Environmental Sciences 2(1): 16– 24.

2. Allen, S. K., Rastner, P., Arora, M., Huggel, C., & Stoffel, M. (2016). Lake outburst and debris flow disaster at Kedarnath, June 2013: hydrometeorological triggering and topographic predisposition. *Landslides*, *13*(6), 1479–1491.
3. Bolch T, Kulkarni A, Kääb A, et al (2012) The state and fate of Himalayan glaciers. Science 336(6079): 310–314.
4. Bajracharya, S. R., & Shrestha, B. R. (2011). *The status of glaciers in the Hindu Kush- Himalayan region*. International Centre for Integrated Mountain Development (ICIMOD).
5. Cho, C., Li, R., Wang, S. Y., Yoon, J. H., & Gillies, R. R. (2016). Anthropogenic footprint of climate change in the June 2013 northern India flood. *Climate Dynamics*, *46*(3-4), 797–805.
6. Clague J. J., Evans S. G. (2000) A review of catastrophic drainage of moraine-dammed lakes in British Columbia. Quat Sci Rev 19:1763–1783
7. Dobhal, D. P., Gupta, A. K., Manish, M., & Khandelwal, D. D. (2013). Kedarnath disaster: Facts and plausible causes. *Current Science*, *105*(2), 171–174.
8. Dubey C. S., Shukla D. P., Ningreichon A. S., Usham A. L. (2013) Orographic control of the Kedarnath disaster. Curr Sci 105:1474–1476
9. Das S., Kar N. S., Bandyopadhyay S. (2015) Glacial lake outburst flood at Kedarnath, Indian Himalaya: a study using digital elevation models and satellite images. Nat Hazards.
10. IMD: India Meteorological Department: southwest monsoon end of-season report http://imd.gov.in/section/nhac/dynamic/endofseasonreport2013.pdf (last access: November 15, 2014), 2013.
11. Hartmann D. L., Klein Tank A. M. G., Rusticucci M., Alexander L. V., Brönnimann S., Charabi Y., Dentener F. J., Dlugokencky E. J., Easterling D. R., Kaplan A., Soden B. J., Thorne P. W., Wild M., Zhai P. M. (2013) Observations: atmosphere and surface. In: Stocker T. F., Qin D., Plattner G. K., Tignor M., Allen S. K., Boschung J., Nauels A., Xia Y., Bex V., Midgley P. M. (eds) Climate change 2013: the physical science basis. Contribution of working group I to the fifth assessment report of the intergovernmental panel on climate change. Cambridge University Press, Cambridge
12. Hock, R., Rasul, G., Adler, C., Cáceres, B., Gruber, S., Hirabayashi, Y., ... & Zhang, Y. (2019). High mountain areas.
13. Joseph, S., Sahai, A. K., Sharmila, S., Abhilash, S., Borah, N., Pillai, P. A., Chattopadhyay, R., and Kumar, A.: Extended Range Prediction of Uttarakhand Heavy Rainfall Event by an Ensemble Prediction System based on CFSv2, Technical Report, Indian Institute of Tropical Meteorology, Pune, India, 2013.
14. Kulkarni A. V., Bahuguna I. M., Rathore B. P. et al. (2007) Glacial retreat in Himalaya using Indian remote sensing satellite data. Current Science 1: 69–74.
15. Kulkarni, A. V., & Karyakarte, Y. (2014). Observed changes in Himalayan glaciers. *Current Science*, 237–244.
16. Martha T. R., Roy P., Govindharaj K. B., Kumar K. V., Diwakar P. G., Dadhwal V. K. (2014) Landslides triggered by the June 2013 extreme rainfall event in parts of Uttarakhand state. India Landslides. doi:10.1007/s10346-014-0540-7
17. Mir R. A., Jain S. K., Lohani A. K. et al. (2018) Glacier recession and glacial lake outburst flood studies in Zanskar basin, western Himalaya. Journal of Hydrology 564: 376–396.
18. Madhura, R. K., Krishnan, R., Revadekar, J. V., Mujumdar, M., & Goswami, B. N. (2015). Changes in western disturbances over the Western Himalayas in a warming environment. *Climate Dynamics*, *44*(3-4), 1157–1168.
19. Mool P. K., Wangda D., Bajracharya S. R. et al. (2001) Inventory of glaciers, glacial lakes and glacial lake outburst floods. Monitoring and early warning systems in the Hindu Kush- Himalayan Region: Bhutan. International Centre for Integrated Mountain Development, Kathmandu, Nepal.
20. Pathak, A. (2018). Climate Change Impact on Hazards in Garhwal Himalaya of Uttarakhand: A Case Study of Kedarnath Catastrophe.
21. Pepin, N., Bradley, R. S., Diaz, H. F., Baraër, M., Caceres, E. B., Forsythe, N., ... & Mountain Research Initiative EDW Working Group. (2015). Elevation-dependent warming in mountain regions of the world. *Nature climate change*, *5*(5), 424–430.
22. Pratap, S., Srivastava, P. K., Routray, A., Islam, T., & Mall, R. K. (2020). Appraisal of hydrometeorological factors during extreme precipitation event: case study of Kedarnath cloudburst, Uttarakhand, India. *Natural Hazards*, *100*(2), 635–654.
23. Rawat, M., Karwariya, S., Raushan, R., Kanga, S., Taloor, A. K., & Thapliyal, A. (2021). Snow cover and land surface temperature assessment of Mana basin Uttarakhand India using MODIS satellite data. In *Water, Cryosphere, and Climate Change in the Himalayas* (pp. 159–174). Springer, Cham.
24. Rafiq M., Mishra A. (2016) Investigating changes in Himalayan glacier in warming environment: a case study of Kolahoi glacier. Environmental Earth Sciences *75*(23): 1469. https://doi.org/10.1007/s12665-016-6282-1
25. Rafiq M., Mishra A. (2018) A study of heavy snowfall in Kashmir, India in January 2017. Weather 73(1): 15–17.
26. Rafiq M., Meer M. S., Mishra A. (2018) On land use and land cover changes over Lidder valley in changing environment. Annals of GIS,
27. Rafiq, M., Romshoo, S. A., Mishra, A. K., & Jalal, F. (2019). Modelling Chorabari lake outburst flood, kedarnath, India. *Journal of Mountain Science*, *16*(1), 64–76.
28. Ren, Y. Y., Ren, G. Y., Sun, X. B., Shrestha, A. B., You, Q. L., Zhan, Y. J., ... & Wen, K. M. (2017). Observed changes in surface air temperature and precipitation in the Hindu Kush Himalayan region over the last 100-plus years. *Advances in Climate Change Research*, *8*(3), 148–156.
29. O'Connor J. E., Hardison J. H., Costa J. E. (2001) Debris flows from failures of Neoglacial-Age moraine dams in the Three Sisters and Mount Jefferson wilderness areas, Oregon. US Geological Survey Professional Paper 1606
30. Sati, V. P. (2013). Extreme weather related disasters: A case study of two flashfloods hit areas of Badrinath and Kedarnath

valleys, Uttarakhand Himalaya, India. *Journal of Earth Science and Engineering*, *3*(8), 562.

31. S. Solomon, D. Qin, M. Manning, Z. Chen, M. Marquis, K. B. Averyt, et al., (Eds.), Contribution of Working Group I to the Fourth Assessment Report of the Intergovernmental Panel on Climate Change, Cambridge and New York, Cambridge University Press, 2007. Stocker T. F., Qin D., Plattner G. K., Alexander L. V., Allen S. K., Bindoff N. L., Bréon F. M., Church J. A., Cubasch U., Emori S., Forster P., Friedlingstein P., Gillett N., Gregory J. M., Hartmann D. L., Jansen E., Kirtman B., Knutti R., Krishna Kumar K., Lemke P., Marotzke J., Masson-Delmotte V., Meehl G.A., Mokhov II, Piao S., Ramaswamy V., Randall D., Rhein M., Rojas M., Sabine C., Shindell D., Talley L. D., Vaughan D. G., Xie S. P. (2013) Technical summary. In: Stocker T. F., Qin D., Plattner G. K., Tignor M., Allen S. K., Boschung J., Nauels A., Xia Y., Bex V., Midgley P. M. (eds) Climate change 2013: the physical science basis. Contribution of working group I to the fifth assessment report of the intergovernmental panel on climate. Cambridge University Press, Cambridge
32. Shrestha A., Aryal R. (2011) Climate change in Nepal and its impact on Himalayan glaciers. Reg Environ Change 11(Suppl 1): 65–77.
33. Srinivasan, J.: Predicting and managing extreme rainfall, Curr. Sci. India, 105, 7–8, 2013
34. Worni R., Stoffel M., Huggel C., Volz C., Casteller A., Luckman B. (2012) Analysis and dynamic modeling of a moraine failure and glacier lake outburst flood at Ventisquero Negro, Patagonian Andes (Argentina). J Hydrol 444–445: 134–145
35. Worni R., Huggel C., Stoffel M. (2013) Glacier lakes in the Indian Himalayas—from an area- wide glacial lake inventory to on-site and modeling based risk assessment of critical glacial lakes. Sci Total Environ 468–469:s71–s84
36. Xavier, A., Manoj, M. G., & Mohankumar, K. (2018). On the dynamics of an extreme rainfall event in northern India in 2013. *Journal of Earth System Science*, *127*(2), 1–13.
37. Xu, J., Shrestha, A., Vaidya, R., Eriksson, M., & Hewitt, K. (2007). The melting Himalayas: regional challenges and local impacts of climate change on mountain ecosystems and livelihoods. *The melting Himalayas: regional challenges and local impacts of climate change on mountain ecosystems and livelihoods.*

CHAPTER 34

Framework for Evaluating the Physical Infrastructure Vulnerability Due to Relative Sea-Level Rise in Coastal Districts of Kerala—Case of Alappuzha

R. S. Vishnu
Department of Architecture, College of Engineering Trivandrum, Thiruvananthapuram, India

Anurup K.
Department of Architecture, College of Engineering Trivandrum, Thiruvananthapuram, India
anurup@cet.ac.in

Abstract

Scientific assumptions of climate change-induced sea-level rise seems so convincing and recent projection for sea-level rise by 2050 (Climate Central, October 2019) has estimated 3 times that of the past projections (IPCC). These projections alone are insufficient for planning purposes or to spatially address its impacts. The best methodology would be through a framework that incorporates risk assessment through the estimated probabilities of various sea-level rise impacts. Here the study intends to assess the physical infrastructure vulnerability to RSLR using a composite index 'CHVI' (which is a blend of CVI & HVI) and with accessible secondary data which were tried in Alappuzha District to know the most vulnerable ULB and to estimate the efficiency of framework. An approach to evaluate the decadal RSLR was fabricated using shoreline delineation, highest mean elevation, and relative sea-level trend to comprehend the decadal risk areas of RSLR concerning the 2050 projection. Projected decadal RSLR risk to population and land area shows, 19646, 36909, 73193 population and 7%, 12%, and 23% land area is at risk in 2030, 2040, and 2050 respectively. Damage loss was calculated for the physical infrastructure to decadal RLSR with the current estimation of Indian rupee to reveal the extent/severity of the damages that could be inflicted due to RSLR. The damage loss from RSLR in the year 2050 is estimated to be 2.3 times that of the 2018,2019 Kerala Flood CMDRF. Overall, we conclude that impending RSLR impacts could enormously influence the physical infrastructure and if we combine the other sectors like environment, social, economic, health, etc. the impacts and damage could be a lot bigger. The need to plan and prepare for these impacts is important to reduce the loss and give a head start to the preparation/mitigation process for the risk areas before its past the point of no return. The framework developed assists with evaluating the vulnerability of physical infrastructure to RSLR.

Keywords: Coastal Habitat Vulnerability Index, Relative sea-level rise, Coastal Vulnerability Index, Habitat Vulnerability index, Coastal Resilience, RSLR Vulnerability assessment.

1. Introduction

1.1 General Background and Need for the Study

Future sea-level rise inundation is expected to turn out to be increasingly serious and is additionally expected to alter by environmental change with frequent flooding occasions. These environmental change driven varieties lead to significant effects along the coast.

Coastal Vulnerability Index (CVI) and Habitat Vulnerability index (HVI) assists with evaluating the coastal weakness of environmental change-induced relative sea-level rise.

The evaluation of the vulnerability of the coastal zones to environmental change is, in this manner, a subject of developing interest around the world.

Foreseeing effects of RSLR can improve the life of the forthcoming infrastructure projects and empower the decision-making procedure for existing ones.

1.2 Why Alappuzha?

Kerala Context:

- Kerala coast is about 580 km long (Alappuzha District constitutes 14% of it 83m).
- Population density in coastal area: 2500 persons/sq.km against 859 persons/sq.km in State.

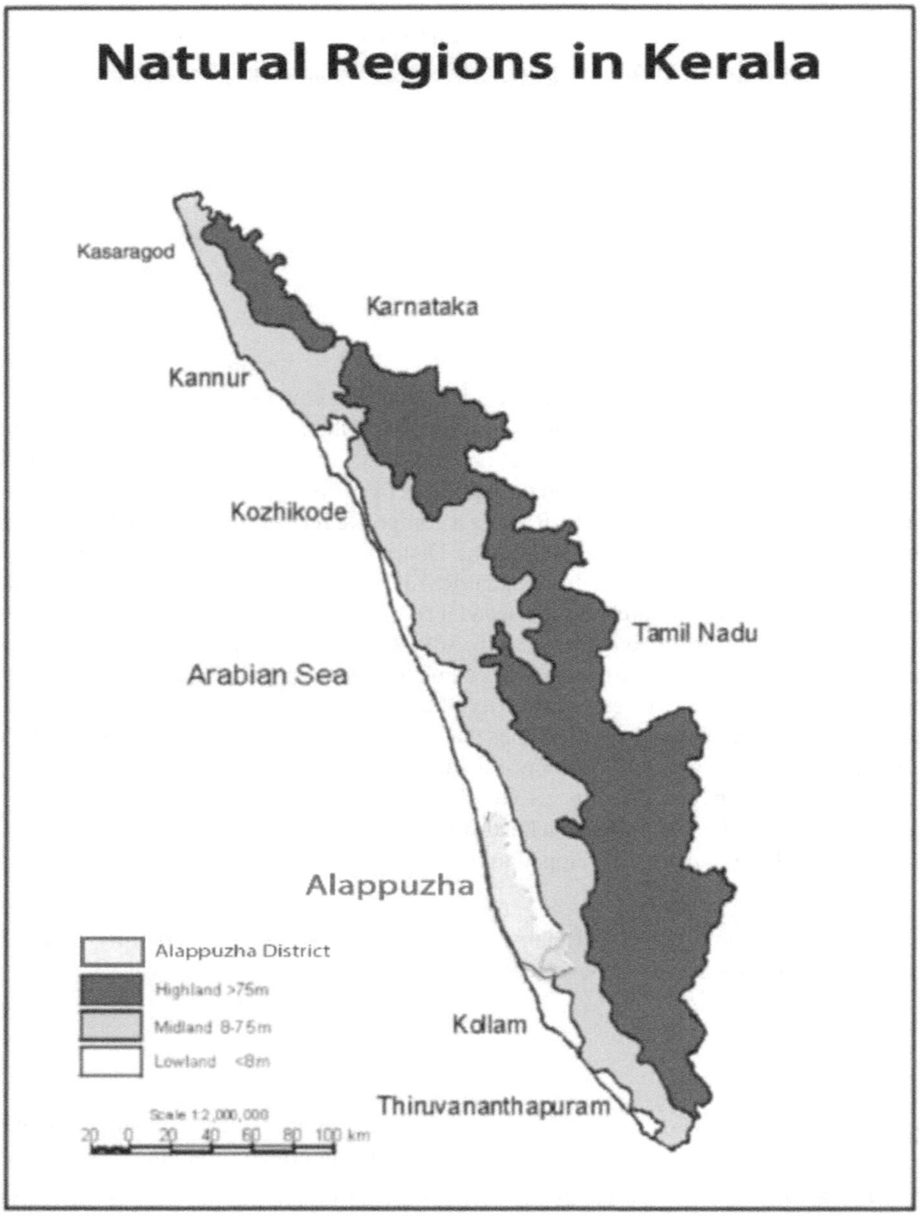

Fig. 34.1 Natural Regions of Kerala

Source: Author edited, Resource Atlas, Centre for Earth Science Studies, Thiruvananthapuram, National Academies of Sciences, Engineering, and Medicine. 2001. Growing Populations, Changing Landscapes: Studies from India, China, and the United States. Washington, DC: The National Academies Press. https://doi.org/10.17226/10144.

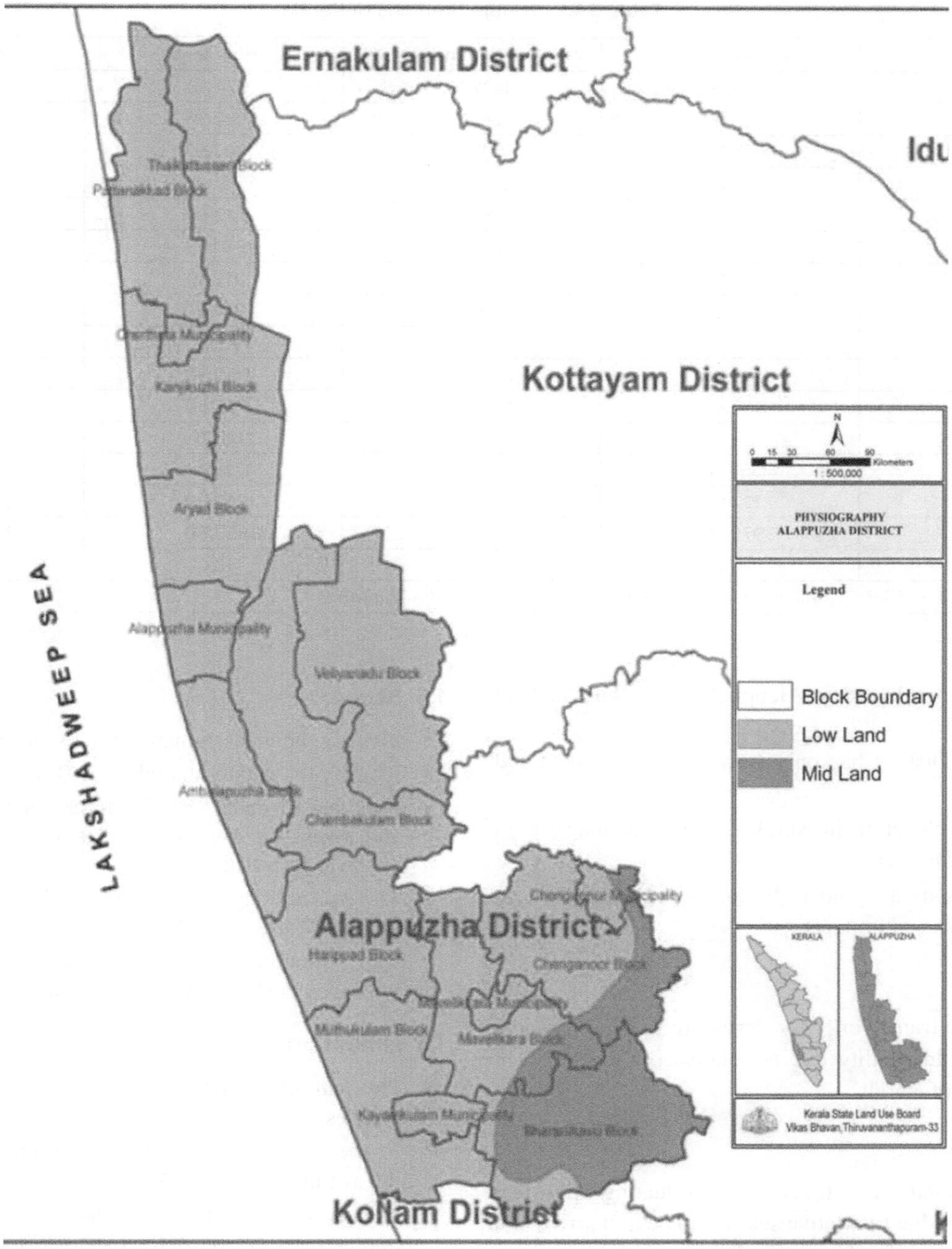

Fig. 34.2 Physiography Map of Alappuzha District

Source: KSLUB – Kerala State Land use Board

- The Kerala state is divided into 3 considering rise zones. Which is Low land, Midland, and Highland.
- The low land area, which is below 7.5m elevation, possesses roughly 10 % of the all-out region. (3,886.3 sq.km)

Alappuzha District Context:

- Out of all the districts in Kerala, Alappuzha District, in particular, requires more attention as it has the largest amount of Low Land (approx.:29 %, according to rise zones of Kerala)

Table 34.1 Natural Regions of Kerala

District	Low land	Mid land	High land	Total in (sq.km.)
Thiruvananthapuram	114.00	1497.70	574.30	2186.00
Kollam	366.00	1041.00	1087.60	2495.80
Pathanamthitta	48.90	717.10	1887.60	2653.60
Alappuzha	1131.60	285.30		1416.90
Kottayam	398.40	1287.70	508.80	2195.50
Idukki		188.80	4937.98	5126.50
Ernakulam	577.50	1587.80	212.40	2377.70
Thrissur	465.50	1546.90	981.50	2993.90
Palakkad		1562.60	2913.20	4475.80
Malappuram	104.10	2841.90	600.20	3546.20
Kozhikode	362.80	1344.70	625.80	2333.30
Wayanad			2125.70	2125.70
Kannur	157.20	1594.00	1216.8	2968.00
Kasargode	192.10	739.90	1849.50	2781.50
Total	3979.30	16231.20	18653.50	38864.00
Percentage	10.24	41.76	48.00	100.00
(Low land – less than 7.5+MSL, Mid land – 7.5 to 75+MSL, High land – above 75mMSL)				

Source: ENVIS Kerala

- 80% of the total land in Alappuzha is in the low land category.
- Alappuzha District has only Low Land and Mid Land category.
- The lowest elevation in Kerala (-2m, Kuttanad), is in Alappuzha District.
- The Low lands are said to be more vulnerable to sea-level rise induced impacts

1.3 Aim

To develop a framework for evaluating the physical infrastructure vulnerability due to relative sea-level rise in coastal districts of Kerala.

1.4 Objectives

- To understand the tools for evaluating physical vulnerability due to relative sea-level rise and arrive at a composite index for Alappuzha District.
- To analyze Alappuzha District based on secondary studies, apply the composite index in spatial tools, to identify the most vulnerable Taluk and delineate a ULB as the area of study.
- Understand the concept of decadal RSLR scenario simulation, and arrive at a method to incorporate spatial tools for delineation of decade wise land inundation.
- Evaluation of decadal physical infrastructure vulnerability of the ULB, and arriving at a damage assessment and formulate strategies and recommendations.

1.5 Scope

- Identify the most vulnerable taluk in Alappuzha District
- Identify the most vulnerable ULB in Alappuzha District.
- Simulate decadal RSLR inundation potential areas and assess the asset damage loss for decadal RSLR to physical infrastructure.
- The research tries to attain guidelines and strategies specific RSLR for vulnerable ULB.
- Gap identification of present framework and improvement/framework development.

1.6 Limitations

- Physical infrastructure in the study only considers roads, bridges, building footprint, and land area.
- Road, Bridge, and Building footprint data is limited to the availability in the OSM (Open Street map).
- Non-availability of on-field data sets due to the closing of concerned offices to the Covid-19 lockdown situation.
- The limited accuracy of the open-source dataset (DSM) for topography assessment.
- All the elevation data projected in the study is based on WGS 1984 43N datum, not MSL.
- Climate Centrals SLR 2050 projection is digitized manually using images from Climate Central's map viewer, so there could be changes in the exact area measurement.
- The administrative boundary of the Alappuzha Municipality area is manually digitized using an image from Alappuzha Municipality Masterplan.

1.7 Methodology

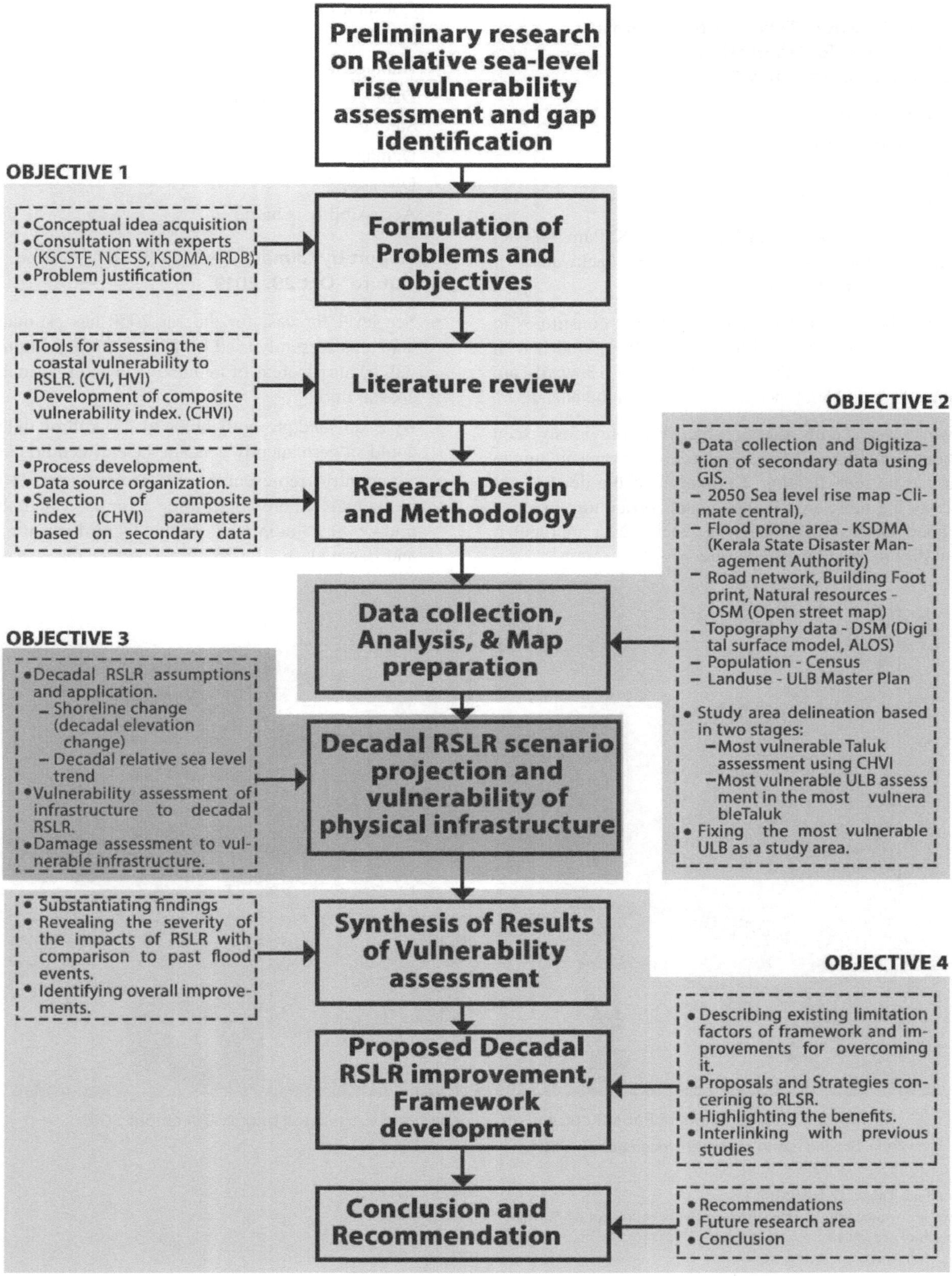

2. Literature Review

2.1 Munroe Thuruth - Tiny Islands In Kerala 'Sinking' Due To Sea Erosion, (Economic Times – July 2017)[1]

- Area: 13.4sq.km
- Population: 15,000
- Wards: 13
- District: Kollam

Munroe Thuruth Grama Panchayat is in Kollam District in Southern Kerala. An island formed by backwaters of Ashtamudi Lake and Kallada River.

It is said to be shrinking periodically. Island constitutes to about 1hectare area inhabited by humans. It is flooded with saline water in most of the areas. 10 out of 13 wards are inundated. Around 15000 people are living on the island.

Tidal waves are a daily threat, which was previously seen only about 2 months each year. Toilets are broken and human excreta can be seen floating. Cremation of the dead is not possible as the most area is inundated. Agriculture practice has stopped. Diseases like diarrhea, cancer, fever are regular. Accessibility to hospitals is a hurdle.[2]

Impacts and Threats in Munroe Thuruth:

- Constant Flooding
- Soil Erosion
- Damage to Mangroves
- Damage to infrastructure
- Saltwater intrusion
- Damage to Agriculture
- Diseases
- Accessibility is hard.

2.2 Report by Climate Central Titled "Flooded Future" Oct 29, 2019[3]

- Sea level forecast for the ear 2050 has estimated that land area presently used for living by 300 million people will be at risk areas of inundation level of normal annual coastal flood.
- By 2100, land presently home to 200 million individuals could sit permanently beneath the elevated tide line.
- Six countries represent about 75 percent of the 300 million individuals ashore confronting a similar vulnerability at mid-century. India ranks in the 3rd position after China and Bangladesh.

Fig. 34.3 Inhabitants of Pattamthuruthu at a crushed and submerged bridge, December 2018

Source: The Hindu, Sinking Island of Kerala, December 22, 2018

[1]The Economic Times, 17 July 2016.

[2]https://www.downtoearth.org.in/news/water/that-sinking-feeling-58443

[3]Climatecentral.org / 2019

CoastalDem's Improvements:

- Includes LiDAR.
- Projection estimations have revealed far more land area and people vulnerable to sea-level rise in this century than which were recently believed.

Methods Used:

- Sea Level Projections
- CoastalDEM
- Vertical datum conversion
- Exposure Analysis

CoastalDEM uncovers a significant increment in territories underneath anticipated avg. yearly flood statures in 2050:

Coastal DEM uncovers a significant increment in territories underneath anticipated avg. yearly flood statures in 2050. Rising sea levels pose more threats to India's major economy

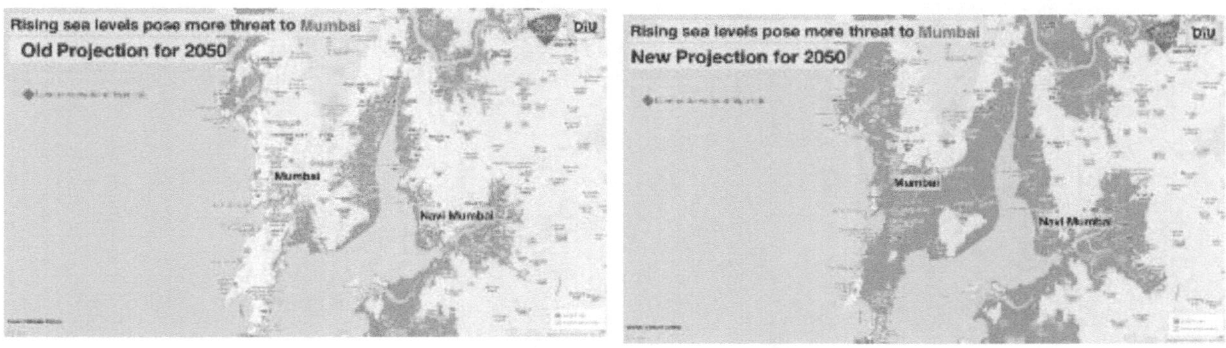

Fig. 34.4 Projected avg. annual flood heights in 2050 of Mumbai

Source: Climate Central Report – Flooded Future, 2019

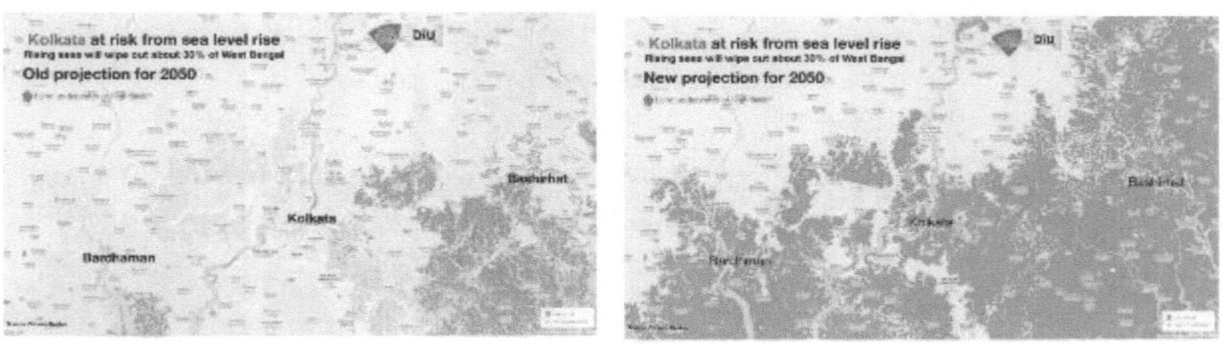

Fig. 34.5 Projected avg. annual flood heights in 2050 of Kolkata

Source: Climate Central Report – Flooded Future, 2019

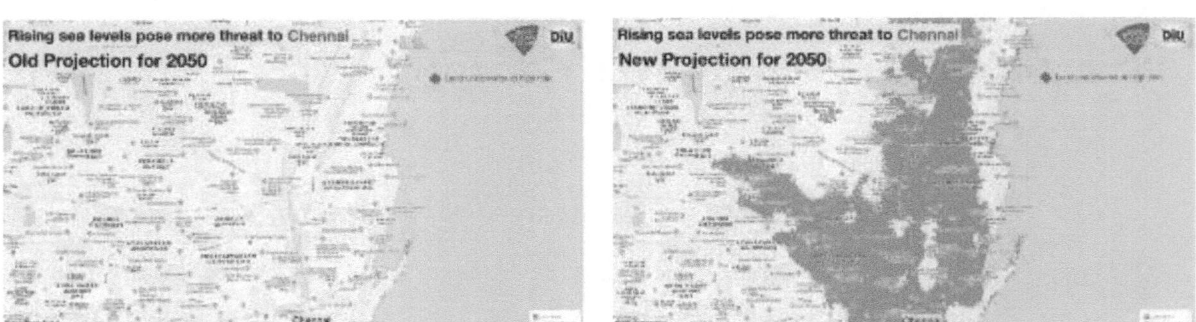

Fig. 34.6 Projected avg. annual flood heights in 2050 of Chennai

Source: Climate Central Report – Flooded Future, 2019

States like Mumbai, Kolkata, and Chennai. Kolkata is at risk from sea-level rise (up to 30% of West Bengal).

Visualizing CoastalDEM's improvements of Kerala

These estimates reveal that a large amount of land area and population will be vulnerable to sea-level rise by 2050 than which was predicted in the past. Alappuzha seems to be the most affected region in terms of land inundation possibility from the projected map of sea-level rise. The land inundated lies towards the interior part of Alappuzha along the backwaters as a result of the rise in sea level.

Fig. 34.7 Old Sea level projection for Kerala.2050

Source: Climate Central Report – Flooded Future, 2019

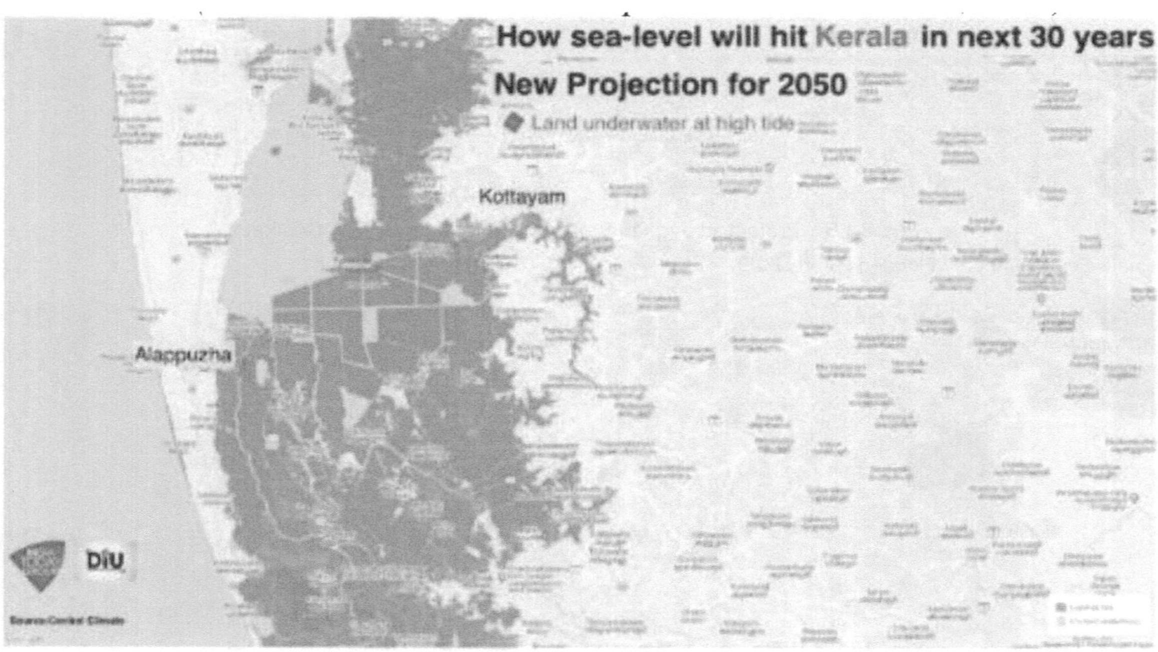

Fig. 34.8 New Sea Level projection for Kerala, 2050

Source: Climate Central Report – Flooded Future, 2019

2.3 CVI (Coastal Vulnerability Index)

CVI is a list that helps to communicate the overall vulnerability of the coastal areas to the impending physical changes because of future relative sea-level rise. It features regions where the different impacts of sea-level rise might be the best.

At the point when every region of coastline is assigned a hazard value subject to each specific information variable, the ocean side vulnerability index is resolved as the square root of the geometric mean, or the square root of the result of the ranked elements partitioned by the total number of components as:

$$CVI = \sqrt{((a*b*c*d*e*f)/6)}$$

Where,

a = 1st variable (geomorphology)

b = 2nd variable (coastal slope)

c = 3rd variable (relative sea-level rise rate)

d = 4th variable (shoreline erosion/accretion rate)

e = 5th variable (mean tide range)

f = 6th variable (mean wave height)

	Ranking of coastal vulnerability index				
	Very low	Low	Moderate	High	Very high
VARIABLE	1	2	3	4	5
Geomorphology	Rocky, cliffed coasts Fiords Fiards	Medium cliffs Indented coasts	Low cliffs Glacial drift Alluvial plains	Cobble beaches Estuary Lagoon	Barrier beaches Sand Beaches Salt marsh Mud flats Deltas Mangrove Coral reefs
Coastal Slope (%)	>0.115	0.115 – 0.055	0.055 – 0.035	0.035 –0.022	< 0.022
Relative sea-level change (mm/yr)	< 1.8	1.8 – 2.5	2.5 – 3.0	3.0 – 3.4	> 3.4
Shoreline erosion/ accretion (m/yr)	>2.0 Accretion	1.0 –2.0	-1.0 – +1.0 Stable	-1.1– -2.0	< - 2.0 Erosion
Mean tide range (m)	> 6.0	4.1 – 6.0	2.0 – 4.0	1.0 –1.9	< 1.0
Mean wave height (m)	<0.55	0.55 – 0.85	0.85 – 1.05	1.05 –1.25	>1.25

Fig. 34.9 Ranking of Variables in the Coastal Vulnerability Index

Source: USGS

3. Case Studies and Their Comparison

NAME	VARIABLES	CALCULATION	REMARK
Case Study 1 GEOGRAFIA FISICA E DINAMICA QUATERNARIA (Issn, 2019)	1. Coastal elevation (relief range) potentiality (m) 2. Coastal flooding (or) seawater inundation potentiality (km) 3. Coastal erosion (shoreline change) potentiality (m) 4. Impacts of settlements and shoreline extend (km) 5. Coastal habitats over washing potentiality (m^3/km) 6. Saltmarsh (or) swamp change potentiality (m^3/km) 7. Mudflat (or) tidal flat change potentiality (m^3/km) 8. Impact of coastal groundwater vulnerability 9. Impacts of population growth and density (persons/km^2)	Weighted vulnerability score of feature classes of each variable and a combination of variables to facilitate the computation of thresholds. (Issn, 2019)	Inclusion of Social variables into the CVI. It is an index-based model that is used to estimate the degree of magnitude to which coastal habitats cannot cope with adverse effects

NAME	VARIABLES	CALCULATION	REMARK
Case Study 2 SEA-LEVEL RISE AND COASTAL VULNERABILITY: AN ASSESSMENT OF ANDHRA PRADESH COAST, INDIA THROUGH REMOTE SENSING AND GIS (Rao et al., 2009)	**Conventional data** 1. Toposheet 2. Naval hydrographic charts 3. Wave level data, mean tidal range data 4. GPS survey data 5. IPCC (Intergovernmental Panel on Climate Change) technical guidelines, 2001 **Remote sensing data** 1. IRS-1C/P6 LISS III þ PAN merged data (February 2000, February 2006) 2. SRTM data	Conventional and Remotely sensed data Modeling technique.	The CVI provides an insight into the relative sea-level rise risk areas and the possible coastal damages due to future RSLR. The maps presented here can be seen at any rate two unique ways: • To recognize ones where physical changes are to happen as ocean level ascents; and • As a planning gadget for regulating and making sure about resources in the study region.
Case Study 3 COASTAL VULNERABILITY ASSESSMENT OF THE FUTURE SEA LEVEL RISE IN UDUPI COASTAL ZONE OF KARNATAKA STATE, WEST COAST OF INDIA (Dwarakish et al., 2009)	1. Geomorphology 2. Coastal slope (%) 3. Shoreline change (m/year) 4. Mean spring tide range (m) 5. Significant wave height (m)	Incorporation of differentially weighted rank values of the five variables.	Unlike the other two studies where grids were used, doing so may generalize the conditions to different degrees (depending on the size of the grid used). Instead, the study has analyzed the coastline in a line feature using GIS. It also has shown more accuracy to picture the vulnerability of the coast at any point.

3.1 Inference and Data Collection

From the comparison of the case studies, parameters from CVI and HVI, which could be used for assessing the coastal vulnerability of the Alappuzha District were identified and were integrated to form CHVI (Coastal Habitat Vulnerability Index) which is a combination of CVI and HVI. These parameters could be used for the Kerala context and for Alappuzha which could be calculated using Coastal Vulnerability Index formula for coastal vulnerability assessment.

Required data and Data Source:

Data	Source
Geomorphology	Kerala State Land Use Board Vikas Bhavan, Thiruvananthapuram
Shoreline change erosion/ accretion	IRDB Vikas Bhavan, Thiruvananthapuram
Near Shore Slope	DEM, USGS (open source) DSM, ALOS (open source)
Mean Wave Height	IRDB Vikas Bhavan, Thiruvananthapuram
Mean Tide range	IRDB Vikas Bhavan, Thiruvananthapuram
Relative Sea level change	IRDB Vikas Bhavan, Thiruvananthapuram
Population Density	2001, 2011 Census Data
Predicted sea-level rise inundation area	Climate Central Data (climatecentral.org), 2050
Infrastructure road, railway, buildings	TCPO Alappuzha

Selected parameters based on secondary data availability

Due to the Covid19 situation and lockdown throughout the Nation, many offices where the above-mentioned data sets were to be collected was interrupted. Parameters that can be generated using secondary data from available open sources were selected for the assessment. They are listed in the Tables 34.2 and 34.3.

Table 34.2 Parameters and rankings for Coastal Habitat Vulnerability Index

CHVI INDEX					
Parameters	Very low	Low	Moderate	High	Very high
	1	2	3	4	5
Coastal elevation potentiality (m)	>40	30-40	20-30	10.0-20.0	<10
Coastal flooding or Sea water inundation potentiality (m)	<0.5	0.5-1.0	1.0-1.5	1.5-2.0	>2.0
Impacts of settlements and shoreline extends (km)	>2.0	1.5-2.0	1.0-1.5	0.5-1.0	<0.5
Impacts of population growth and density (persons per sq.km.)	<615	616-1063	1064-1745	1746-5033	>5033
Geomorphology	Rocky cliff coarse	Medium cliffs indented coarse	Low cliffs, lateritic plain	River deposit, alluvial plain	Coastal plain, beach, mudflat

Table 34.3 Digital Datasets used in the research

DIGITAL DATASETS USED AND THEIR SPECIFICATIONS	
Data and Source	Specifications
DSM (Digital Surface Model – ALOS)	ALPSMLC30_N009E076_DSM
Landsat image for 2020 shoreline delineation (USGS- United States Geological Survey)	LC08_L1TP_144054_20200416_20200423_01_T1
Building Footprint, Roads, Waterbodies (OSM – Open Street Map)	_planet_osm_polygons/lines
Administrative Boundaries of District, LSGD (KSDI – Kerala State Spatial Data Infrastructure)	Shapefile

4. Study Area Analysis (Stage 1: Taluk Level Assessment using CHVI)

4.1 Alappuzha District—Taluk Boundary

See Fig. 34.10 on next page.

4.2 General Details

Table 34.4 Alappuzha District - Taluk Demography

DEMOGRAPHY			
ALAPPUZHA DISTRICT			
Total population	**2121943**		
Population Density	**1721.13**		
TALUK	**Total population**	**Total land area (sq.km.)**	**Population density (persons/sq.km.)**
CHERTHALA	542657	233.66	2322.42
AMBALAPUZHA	518114	146.16	3544.84
KUTTANAD	196475	279.36	703.30
KARTHIKAPALLY	421624	211.04	1997.84
MAVELIKARA	333318	221.89	1502.18
CHENGANOOR	197309	140.74	1401.94

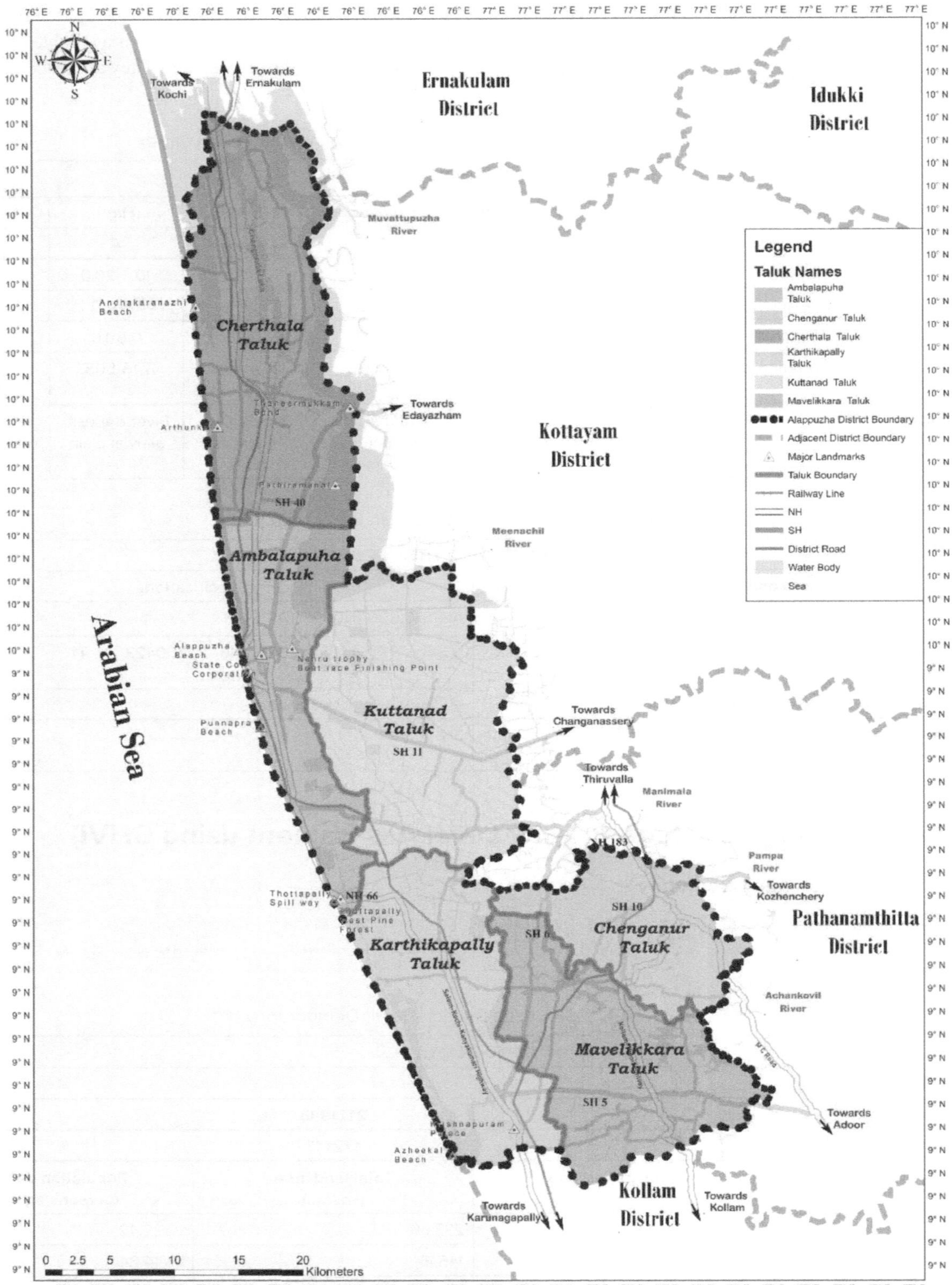

Fig. 34.10 Alappuzha District - Taluk Boundary Map

Source: Alappuzha TCPO, generated using ArcGIS

Table 34.5 Alappuzha District - Taluk Land Division

LAND DIVISION				
ALAPPUZHA DISTRICT				
Total land area (sq.km.)	**Administrative boundary area**	**Land area**	**SLR area 2050**	**In %**
	1417.37	1232.88	367.35	30%
TALUK				
	Administrative boundary area	**Land area**	**SLR area 2050**	**In %**
CHERTHALA	325.08	233.66	12.53	5%
AMBALAPUZHA	181.27	146.16	39.5	27%
KUTTANAD	306.08	279.36	237.84	85%
KARTHIKAPALLY	231.98	211.04	61.69	29%
MAVELIKARA	230.34	221.89	9.15	4%
CHENGANOOR	142.59	140.74	6.59	5%

Table 34.6 Alappuzha District - Taluk Flood Prone area

FLOOD PRONE AREA			
	Total area (sq.km.)	**Flood prone area (sq.km.)**	**%**
ALAPPUZHA DISTRICT	1232.88	718.67	58%
TALUK			
CHERTHALA	233.66	97.69	42%
AMBALAPUZHA	146.16	64.71	44%
KUTTANAD	279.36	275.12	98%
KARTHIKAPALLY	211.04	120.32	57%
MAVELIKARA	221.89	74.67	34%
CHENGANOOR	140.74	86.11	61%

Table 34.7 Alappuzha District - Taluk Building Footprint

BUILDING FOOTPRINT			
ALAPPUZHA DISTRICT	**Building footprint (sq.km.)**	**Buildings vulnerable to SLR 2050 (sq.km.)**	**Percentage of Buildings vulnerable to SLR 2050**
	35.37	3.39	10%
TALUK			
CHERTHALA	13.85	0.38	3%
AMBALAPUZHA	4.27	0.35	8%
KUTTANAD	2.25	2.25	100%
KARTHIKAPALLY	0.27	0.27	100%
MAVELIKARA	5.35	0.02	0.4%
CHENGANOOR	0.31	0.03	10%

4.3 Impacts on Population Density

Population Density map shows that the population varies from 641persons/km^2 to 2858 persons/km^2 in the 6 Taluks. The concentration of population density is high in the Ambalapuzha Taluk.

This could be because of the well-connected roadways and railway network. And maybe also because of Alappuzha Municipality, which is in the Ambalapuzha Taluk, which is a 1st order settlement and 1st order node.

Table 34.8 CHVI rank – Taluk (Impacts on population growth and density)

Impacts on population growth and density (persons/sq.km)	
Taluk	**CHVI ranking score**
Cherthala	4
Ambalapuzha	4
Kuttanad	2
Karthikapally	4
Mavelikara	3
Chenganoor	3

4.4 Impacts on Coastline Settlement

Settlements that are near to the shoreline are assessed based on the pre-defined distance in CHVI (<0.5 km, 0.5 to 1 km, 1 to 1.5 km, 1.5 to 2 km). Ambalapuzha Taluk is having major settlement area (data procured from the open street map), could be because of the well-developed and well-connected road and rail networks.

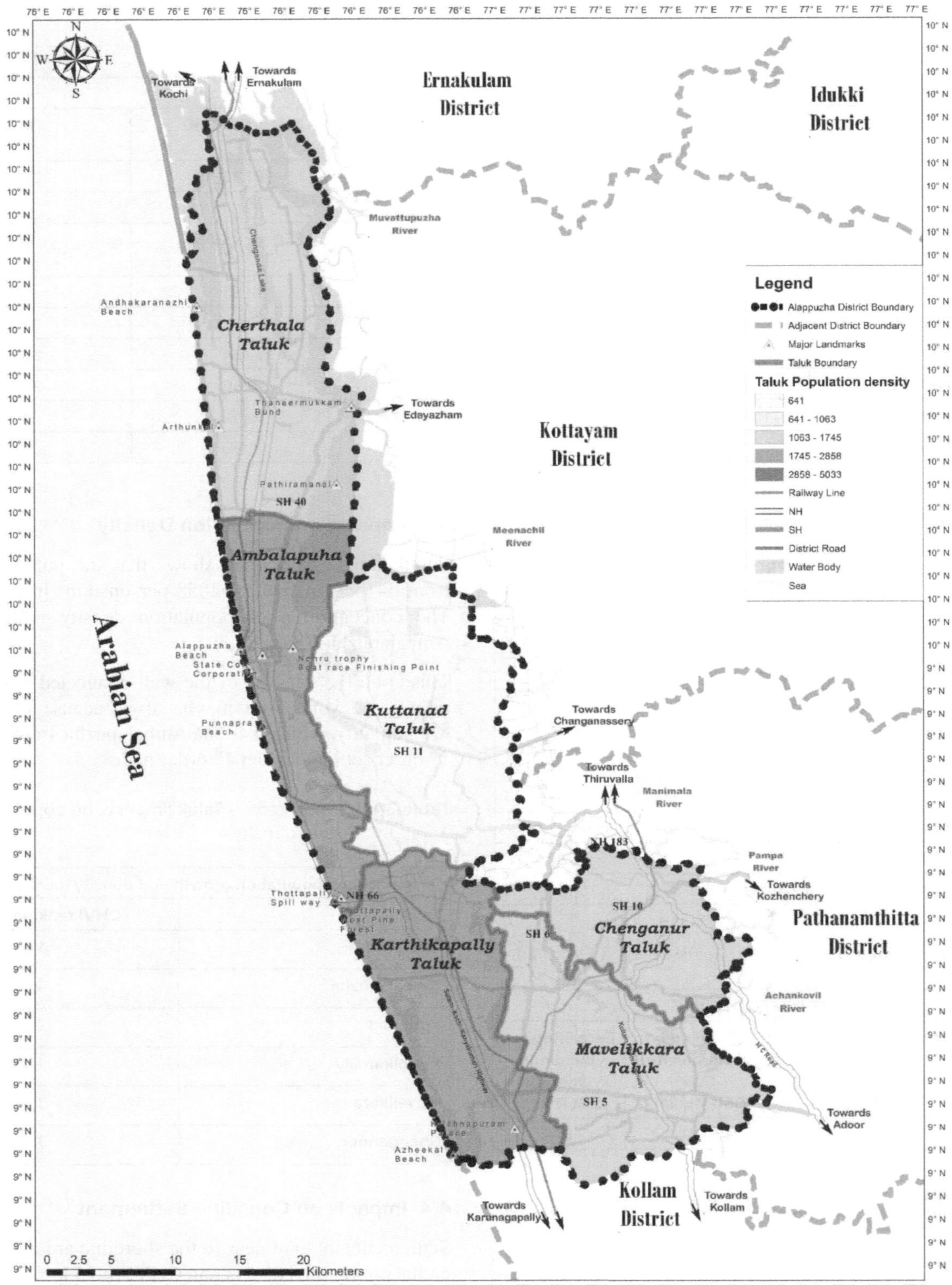

Fig. 34.11 Alappuzha District - Taluk Population Density Map

Source: Alappuzha TCPO, generated using ArcGIS

Table 34.9 CHVI rank – Taluk (Impacts of settlements and shoreline extends)

Impacts of settlements and shoreline extend (km)	
Taluk	**CHVI ranking score**
Cherthala	3
Ambalapuzha	5
Kuttanad	1
Karthikapally	1
Mavelikara	1
Chenganoor	1

4.5 Coastal Elevation Potentiality

Elevation ranges from -71 m low to 120 m high. Coastal Taluk areas are a bit higher than 1m elevation when compared to the Mid-East lying Kuttanad Taluk. -2 m elevation is identified in the Kuttanad Taluk area.

These features could make the Kuttanad Taluk area more vulnerable to flooding and relative sea-level inundation potential.

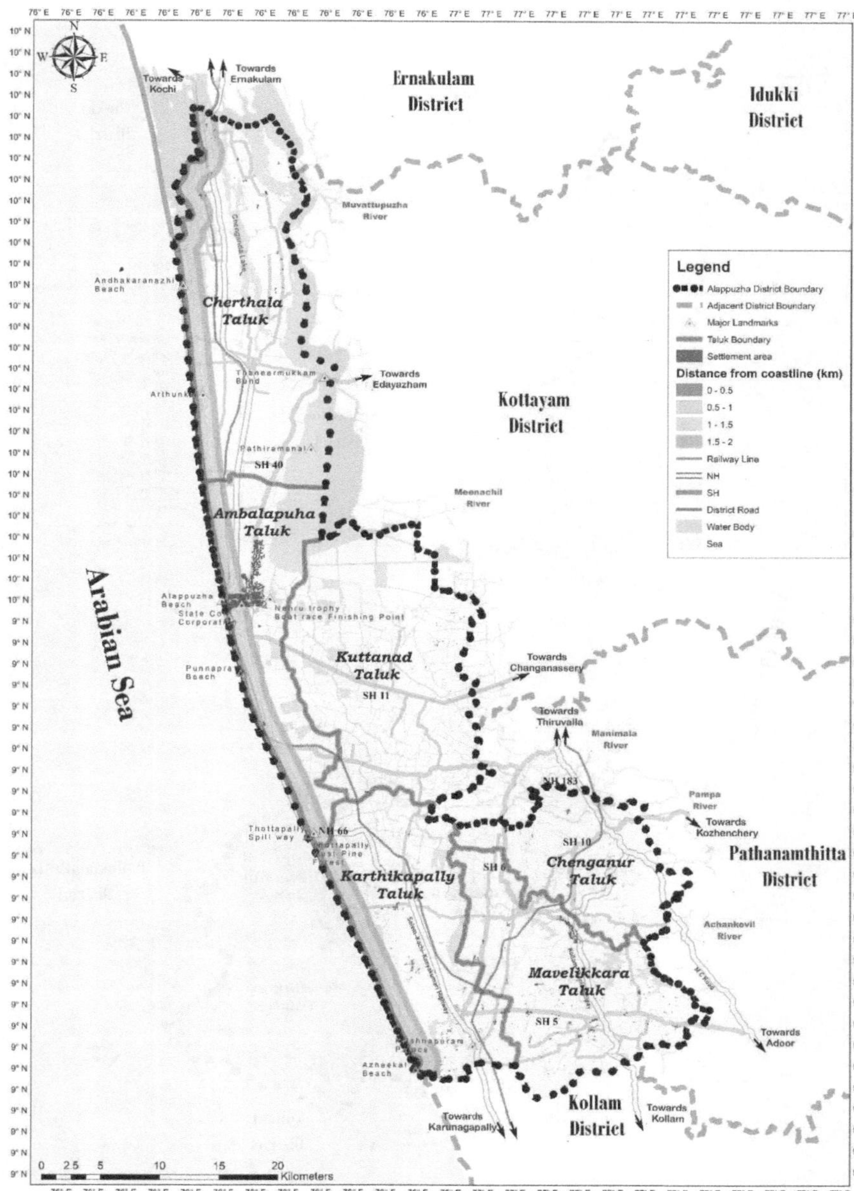

Fig. 34.12 Alappuzha District -Taluk Impacts on coastline settlement map

Source: Open Street Map, generated using ArcGIS

Table 34.10 CHVI rank – Taluk (Coastal elevation potentiality)

Coastal elevation potentiality (m)	
Taluk	**CHVI ranking score**
Cherthala	3
Ambalapuzha	3
Kuttanad	5
Karthikapally	4
Mavelikara	2
Chenganoor	1

4.6 Coastal Flooding or Sea Water Inundation Potentiality

Kuttanad Taluk is found to be 98% vulnerable to flooding, followed by Karthikapally Taluk. Previous events of floods support this data. They are to be categorized as most vulnerable to pluvial/fluvial flooding due to their low relief topography nature.

Compared to the Taluks which are connected to the coastal area, Kuttanad Taluk is found to be most affected by SLR in 2050 according to climate central's map. It could be mainly

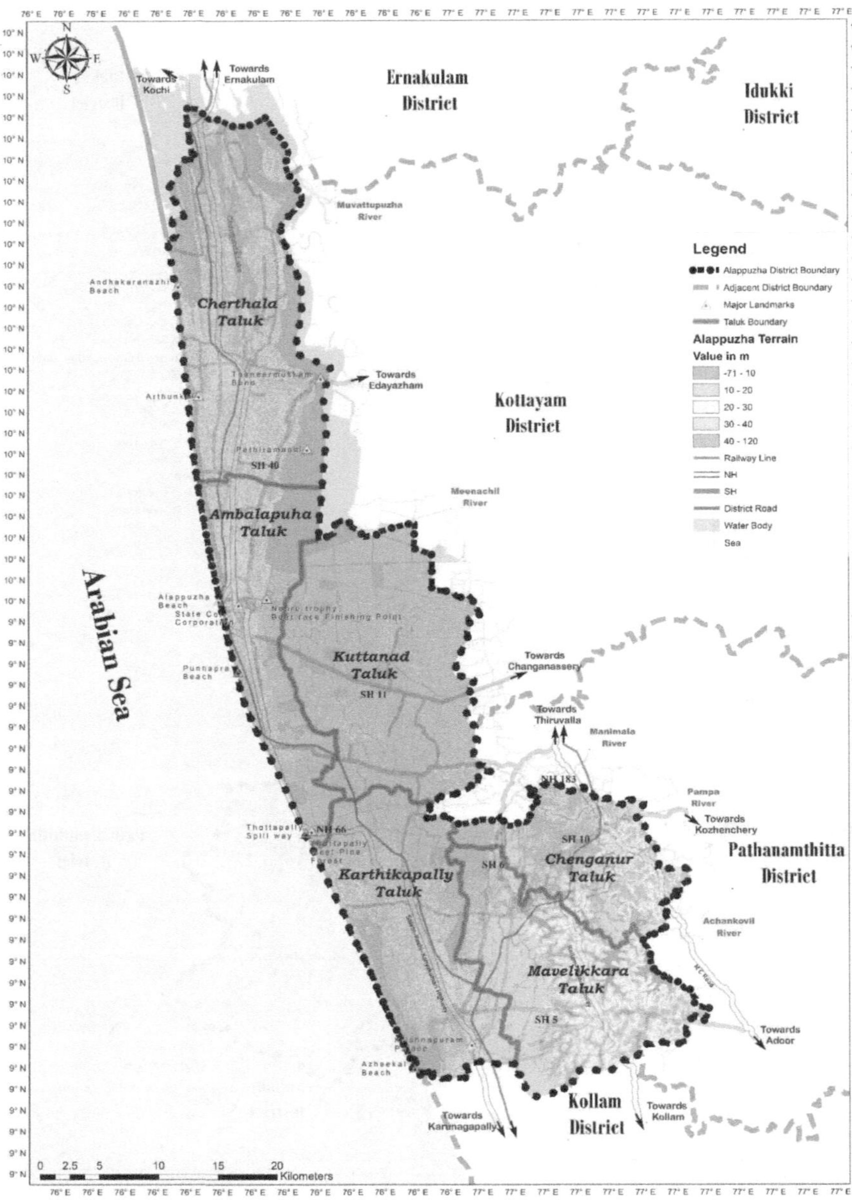

Fig. 34.13 Alappuzha District -Taluk Elevation Map

Source: DSM - ALOS, generated using ArcGIS

Table 34.11 CHVI rank—Taluk (Coastal flooding/Seawater inundation potentiality)

Coastal flooding or Seawater inundation potentiality	
Taluk	**CHVI ranking score**
Cherthala	5
Ambalapuzha	5
Kuttanad	1
Karthikapally	5
Mavelikara	1
Chenganoor	1

because of its low terrain feature and its proximity to the Vembanad Lake, which could be overflown and flooding the Kuttanad Taluk area. Around 85% of the Kuttanad Taluk area is found to be vulnerable to SLR by 2050.

4.7 Geomorphology

Alappuzha District is mainly composed of Coastal Plain and these areas are safe from flooding under normal circumstances. These areas are well-drained indicating that the area possesses highly permeable soil and low relief.

The Kuttanad Taluk is mainly composed of Alluvial plain. It could mean that the Kuttanad area was formed as a result of

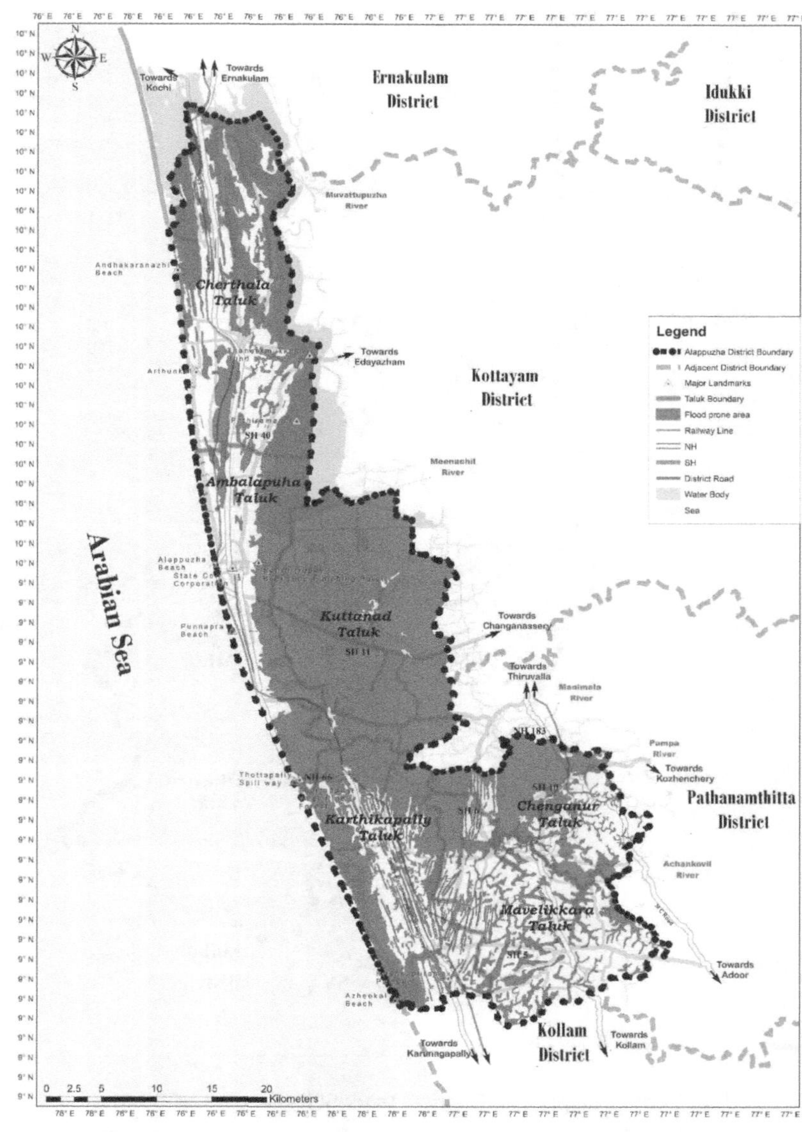

Fig. 34.14 Alappuzha District - Taluk Flood Prone Area Map

Source: KSDMA, generated using ArcGIS

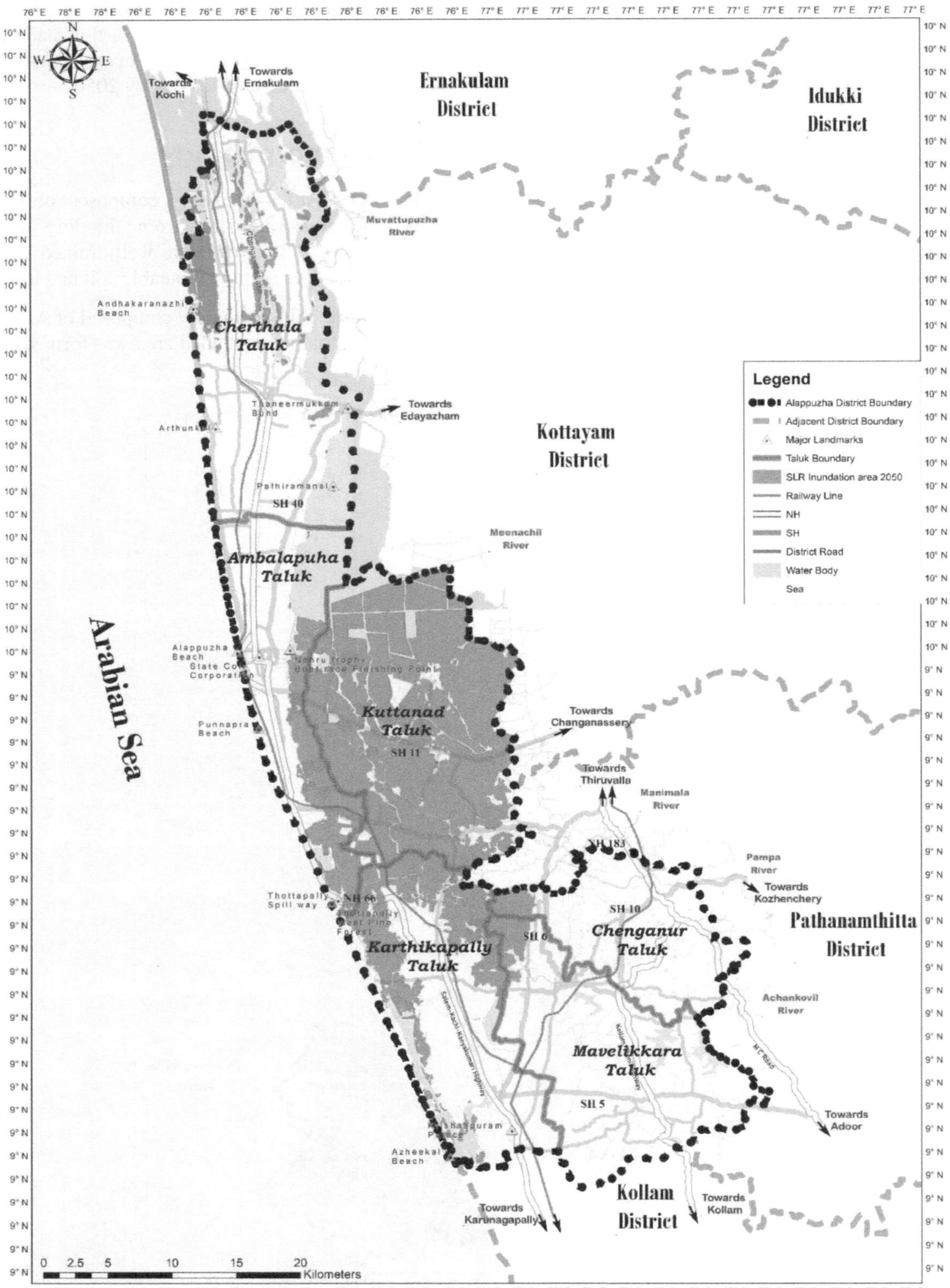

Fig. 34.15 Alappuzha District -Taluk SLR risk area 2050 Map

Source: Climate Central, delineated using ArcGIS

the deposition of sediments by rivers coming from the high land areas, these areas are vulnerable to river flooding.[4]

Table 34.12 CHVI Rank -Taluk Geomorphology

Geomorphology	
Taluk	**CHVI ranking score**
Cherthala	5
Ambalapuzha	5
Kuttanad	4
Karthikapally	5
Mavelikara	3
Chenganoor	3

4.8 Results of Stage 1—Taluk Level Assessment

Based on the assessment of different parameters form the CHVI on Alappuzha District Taluks, Coastal Vulnerability Habitat Index was calculated using the CVI equation, from the values in each parameter based on the specific data variables.

The calculation has values ranging from 1.32 to 17.32. The mean value is 7.62 and the standard deviation is 6.11. The 25th, 50^{th,} and 75th percentiles are found to be 2.13, 5.89, and 12.30.

The CHVI is segmented into low, medium, high, and very high, based on the quartile ranges of the data. CHVI values below 2.13 are considered to be low, values in between 2.13 to 5.89 as a medium, 5.89 to 12.30 as high, and above 12.30 as very high.

Based on the CHVI value, Ambalapuzha Taluk and Cherthala Taluk are under Very High category, in which Ambalapuzha Taluk is considered to be more vulnerable in terms of score and will be taken as the study area for the stage 2 assessment, which is done on the Local Bodies of Ambalapuzha Taluk to identify the most vulnerable Local Body of Ambalapuha Taluk.

The CHVI assesses the coastal vulnerability with more consideration to the areas within the proximity of the coastline. But in the case of Alappuzha District, Kuttanad Taluk, which is an area that does not share coastline because of which it didn't score more value in the CHVI but is found to be the most vulnerable to SLR by 2050 in terms of Climate Central's map.

To ascertain the vulnerability of Kuttanad Taluk a more accurate and hydrological based study is needed to be carried out incorporating the Hydrological graph of Vembanad lake which to be done using HEC-RAS / Mike21 to delineate the flood plains of Vembanad lake when 1m relative sea-level rise is occurring.

Due to the limitation of academic time and data availability for the study purpose we are not considering Kuttanad Taluk's condition and are going ahead with the most vulnerable Taluk form the CHVI, that is Ambalapuzha Taluk and will do the stage 2 of assessment on the Local Body level assessment of CHVI and identify the most vulnerable Local Body of Ambalapuzha Taluk.

Table 34.13 Alappuzha District – Taluk's Coastal Habitat Vulnerability Index

CHVI OF ALAPPUZHA DISTRICT -TALUKS						
PARAMETERS	**CHERTHALA**	**AMBALAPUZHA**	**KUTTANAD**	**KARTHIKAPALLY**	**MAVELIKARA**	**CHENGANOOR**
Coastal elevation potentiality (m)	3	3	5	4	2	1
Coastal flooding or Sea water inundation potentiality (m)	5	5	1	5	1	1
Impacts of settlements and shoreline extends (km)	3	5	1	1	1	1
Impacts of population growth and density (persons per sq.km.)	4	4	2	4	3	3
Geomorphology	**5**	**5**	**4**	**5**	**3**	**3**
Coastal Habitat Vulnerability Index (CHVI)	**13.42**	**17.32**	**2.83**	**8.94**	**1.90**	**1.34**
	VERY HIGH	**VERY HIGH**	**MEDIUM**	**HIGH**	**LOW**	**LOW**

[4]KSLUB- Kerala State Land Use Board

Hence the Most Vulnerable Taluk in Alappuzha District according to CHVI is considered to be Ambalapuzha Taluk and is represented in CHVI Map of Alappuzha District Taluks.

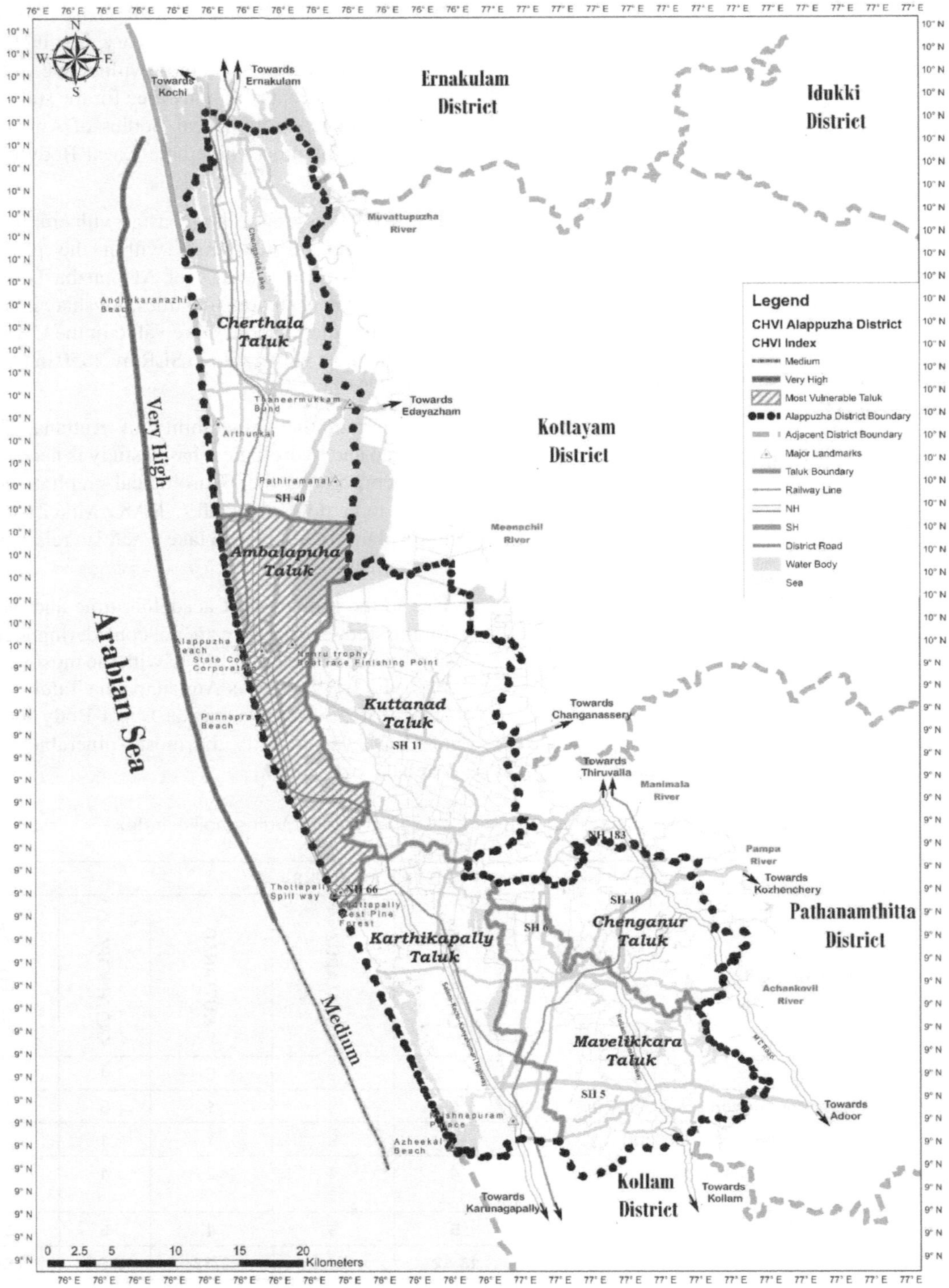

Fig. 34.16 Alappuzha District - Taluk CHVI Map

Source: Generated using ArcGIS

5. Study Area Analysis (Stage 2: Local Body Level Assessment using CHVI)

5.1 Ambalapuzha Taluk Boundary Map

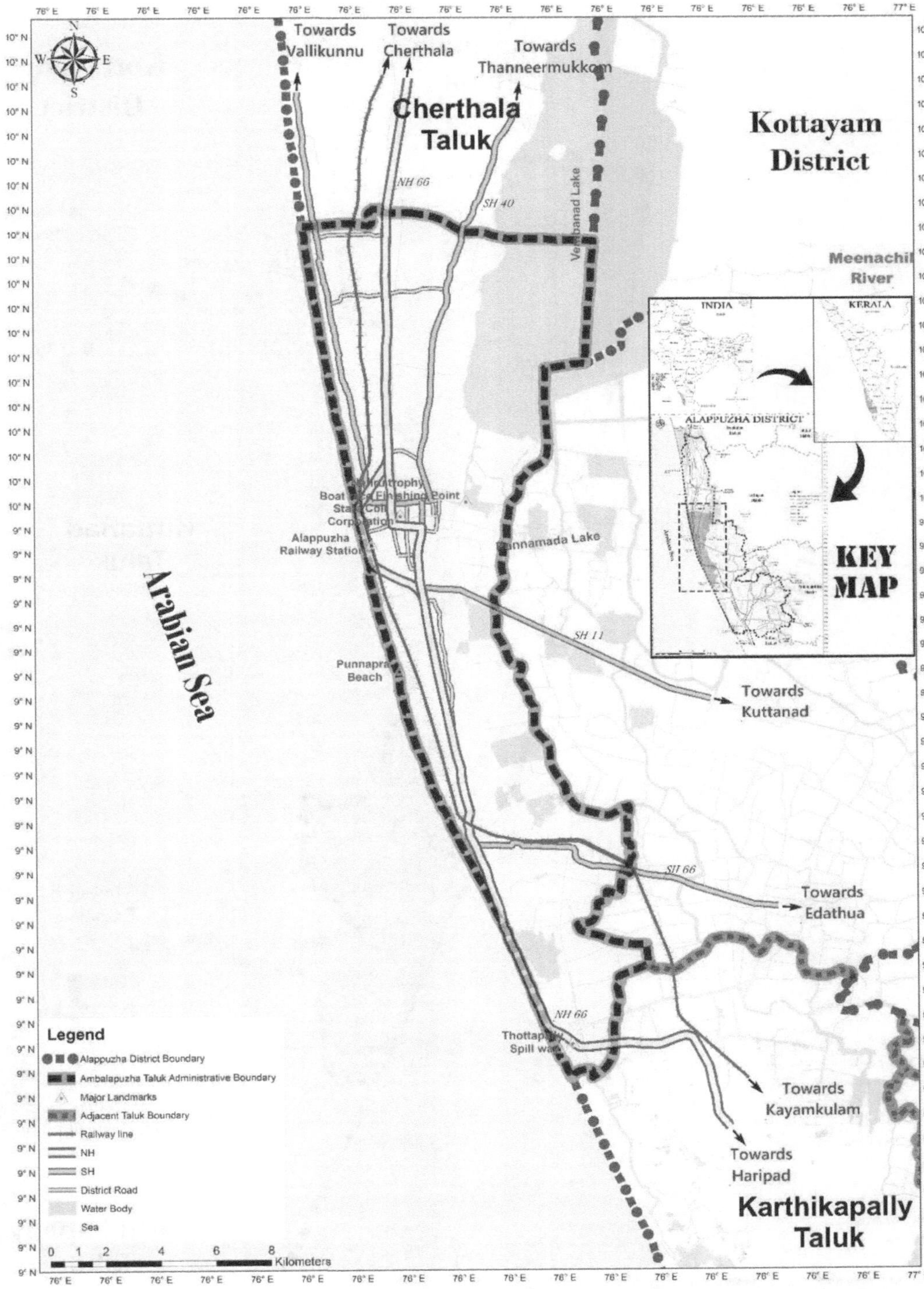

Fig. 34.17 Ambalapuzha Taluk - Local Bodies Boundary Map

Source: KSDI, generated using ArcGIS

5.2 Ambalapuhza Taluk – Local Body Boundary

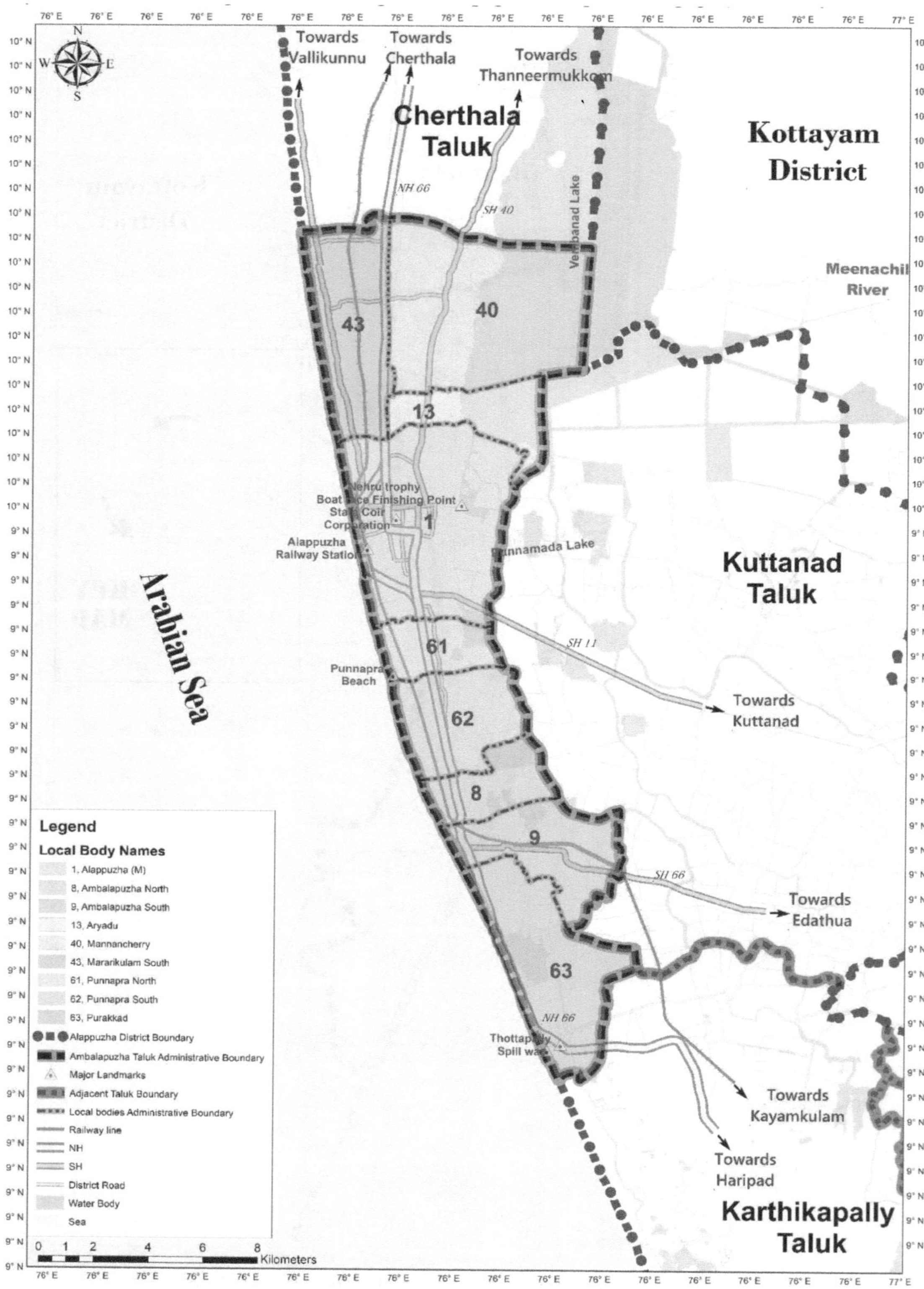

Fig. 34.18 Ambalapuzha Taluk - Local Body Boundary Map

Source: KSDI, generated using ArcGIS

5.3 General Details

Table 34.14 Ambalapuzha Taluk – Local Body Demography

DEMOGRAPHY				
Ambalapuha Taluk				
Total Population		518114		
Population Density (persons/sq.km)		3544.76		
LSGD Number	**Local Bodies**	**Demography**	**Total Land Area (sq.km)**	**Population Density (persons/sq.km)**
43	MARARIKULAM SOUTH	53263	20.42	2609.01
40	MANNANCHERRY	50990	20.37	2503.70
13	ARYAD	32265	4.25	7584.29
1	ALAPPUZHA (M)	240991	36.17	6663.04
61	PUNNAPRA NORTH	28749	7.16	4015.12
62	PUNNAPRA SOUTH	28377	15.01	1891.15
8	AMBALAPUZHA NORTH	30492	6.59	4629.45
9	AMBALAPUZHA SOUTH	23205	15.08	1538.31
63	PURAKKAD	29782	21.12	1409.94

Table 34.15 Ambalapuzha Taluk - Local Bodies Land Division

LAND DIVISION				
Total Land Area (sq.km.)	**Administrative Land Area (sq.km.)**	**Land Area (sq.km.)**	**SLR inundation Area (sq.km.)**	**in %**
AMBALAPUZHA TALUK	181.27	146.16	45.44	31%
LOCAL BODIES				
MARARIKULAM SOUTH	20.46	20.42	0	0%
MANNANCHERRY	39.75	20.37	0.06	0%
ARYAD	10.55	4.25	0.75	18%
ALAPPUZHA (M)	38.62	36.17	8.52	24%
PUNNAPRA NORTH	8.22	7.16	2.10	29%
PUNNAPRA SOUTH	15.35	15.01	5.17	34%
AMBALAPUZHA NORTH	9.16	6.59	5.04	77%
AMBALAPUZHA SOUTH	16.04	15.08	10.57	70%
PURAKKAD	23.11	21.12	13.23	63%

Table 34.16 Ambalapuzha Taluk - Local Bodies Flood-prone area

FLOOD PRONE AREA			
Administrative boundary	**Total Land Area (sq.km)**	**Flood Prone area (sq.km)**	**%**
AMBALAPUZHA TALUK	146.16	99.34	68%
LOCAL BODIES			
MARARIKULAM SOUTH	20.42	4.79	23%
MANNANCHERRY	20.37	6.29	31%
ARYAD	4.25	1.46	34%
ALAPPUZHA (M)	36.17	11.78	33%
PUNNAPRA NORTH	7.16	2.35	33%
PUNNAPRA SOUTH	15.01	6.82	45%
AMBALAPUZHA NORTH	6.59	3.17	48%
AMBALAPUZHA SOUTH	15.08	11.64	77%
PURAKKAD	21.12	16.41	78%

5.4 Impacts on Population Density

Population Density map shows that the population varies from 1282 persons/km^2 to 6240 persons/km^2 in the 9 Local Bodies. The concentration of population density is high in the Alappuzha Municipality.

This could be because of the well-connected roadways and railway network. And maybe also because of Alappuzha Municipality is a 1st order settlement and 1st order node in the Ambalapuzha Taluk.

Table 34.17 CHVI Rank - Local Bodies (Impacts of Population Growth and Density)

IMPACTS OF POPULATION GROWTH AND DENSITY	
LOCAL BODIES	CHVI RANKING SCORE
MARARIKULAM SOUTH	4
MANNANCHERRY	3
ARYAD	4
ALAPPUZHA (M)	5
PUNNAPRA NORTH	4
PUNNAPRA SOUTH	4
AMBALAPUZHA NORTH	4
AMBALAPUZHA SOUTH	3
PURAKKAD	3

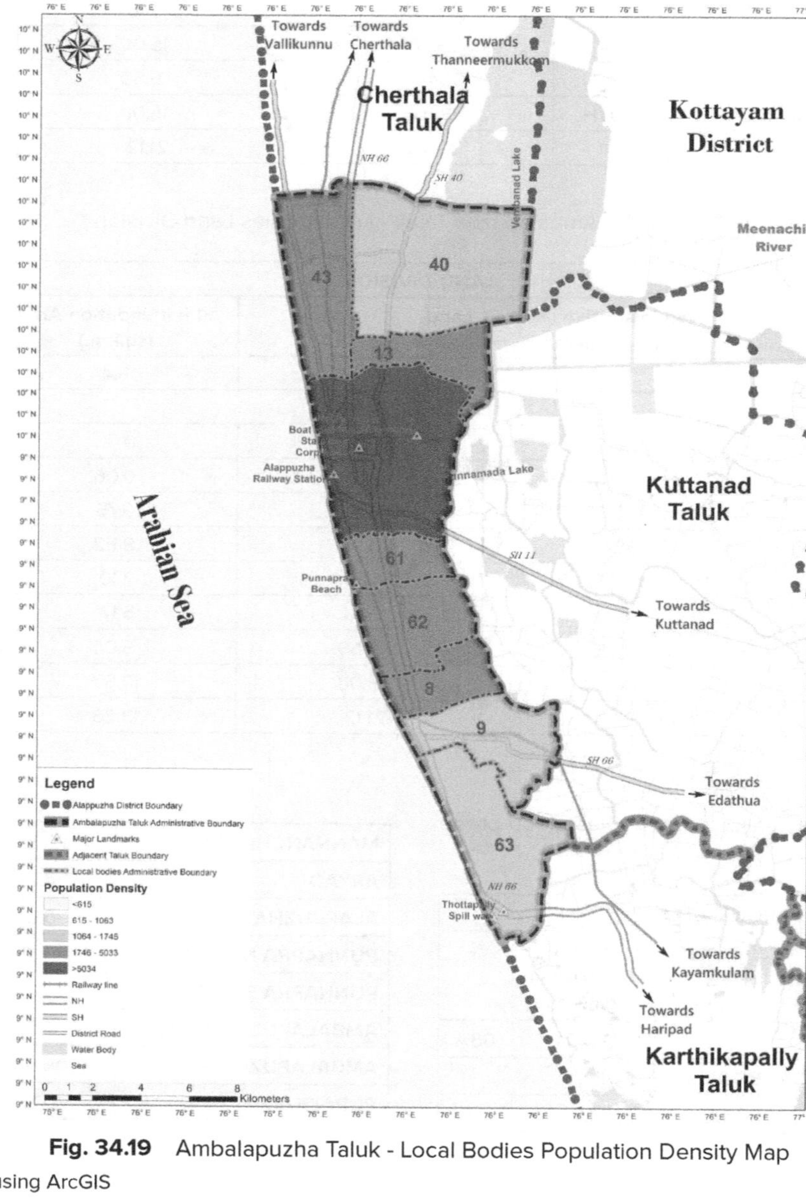

Fig. 34.19 Ambalapuzha Taluk - Local Bodies Population Density Map

Source: TCPO, generated using ArcGIS

5.5 Impacts of Coastline Settlements

Settlements that are near to the shoreline are assessed based on the pre-defined distance in CHVI (<0.5 km, 0.5 to 1 km, 1 to 1.5 km, 1.5 to 2 km). Alappuzha Municipality is having major settlement area (data procured from the open street map), could be because of the well-developed and well-connected road and rail networks.

The concentration of settlements and population living in the area makes Alappuzha Municipality area more vulnerable in terms of the population going to be displaced or affected by the relative sea-level rise.

Table 34.18 CHVI Rank - Ambalapuzha Taluk Local Bodies (impacts of settlements and shoreline extends)

IMPACTS OF SETTLEMENTS AND SHORELINE EXTENDS	
LOCAL BODIES	**CHVI RANKING SCORE**
MARARIKULAM SOUTH	4
MANNANCHERRY	1
ARYAD	1
ALAPPUZHA (M)	5
PUNNAPRA NORTH	1
PUNNAPRA SOUTH	1
AMBALAPUZHA NORTH	4
AMBALAPUZHA SOUTH	1
PURAKKAD	1

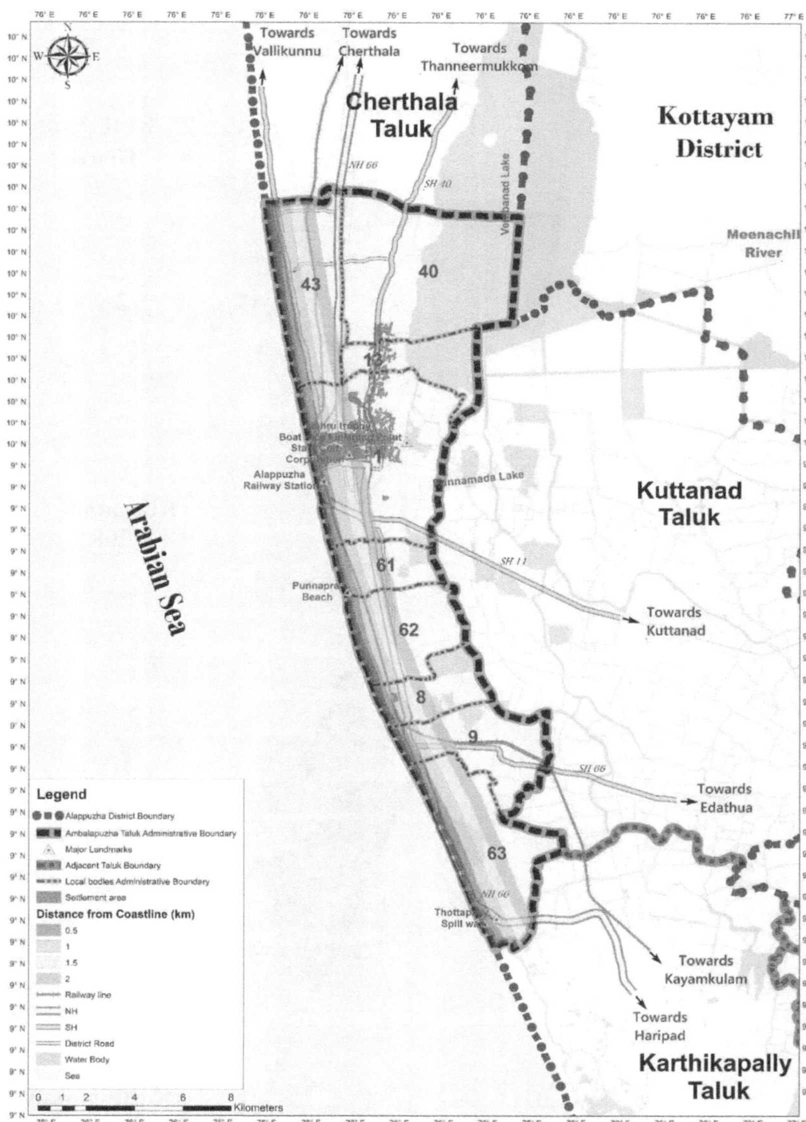

Fig. 34.20 Ambalapuzha Taluk - Local Bodies Impact on Coastline Settlement Map

Source: Open Street Map, generated using ArcGIS

5.6 Coastal Elevation Potentiality

Elevation ranges from -66 m low to 44 m high. Coastal Local Body from Northside areas is a bit higher than 10m elevation when compared to the Mid-East and South side-lying Local Bodies which are near to Vembanad Lake and has terrain features less than 10m elevation.

These features could make those areas more vulnerable to flooding and relative sea- level inundation potential.

Table 34.19 CHVI Rank - Ambalapuzha Taluk Local Bodies (Coastal elevation potentiality)

COASTAL ELEVATION POTENTIALITY	
LOCAL BODIES	**CHVI RANKING SCORE**
MARARIKULAM SOUTH	3
MANNANCHERRY	3
ARYAD	3
ALAPPUZHA (M)	3
PUNNAPRA NORTH	4
PUNNAPRA SOUTH	4
AMBALAPUZHA NORTH	4
AMBALAPUZHA SOUTH	4
PURAKKAD	4

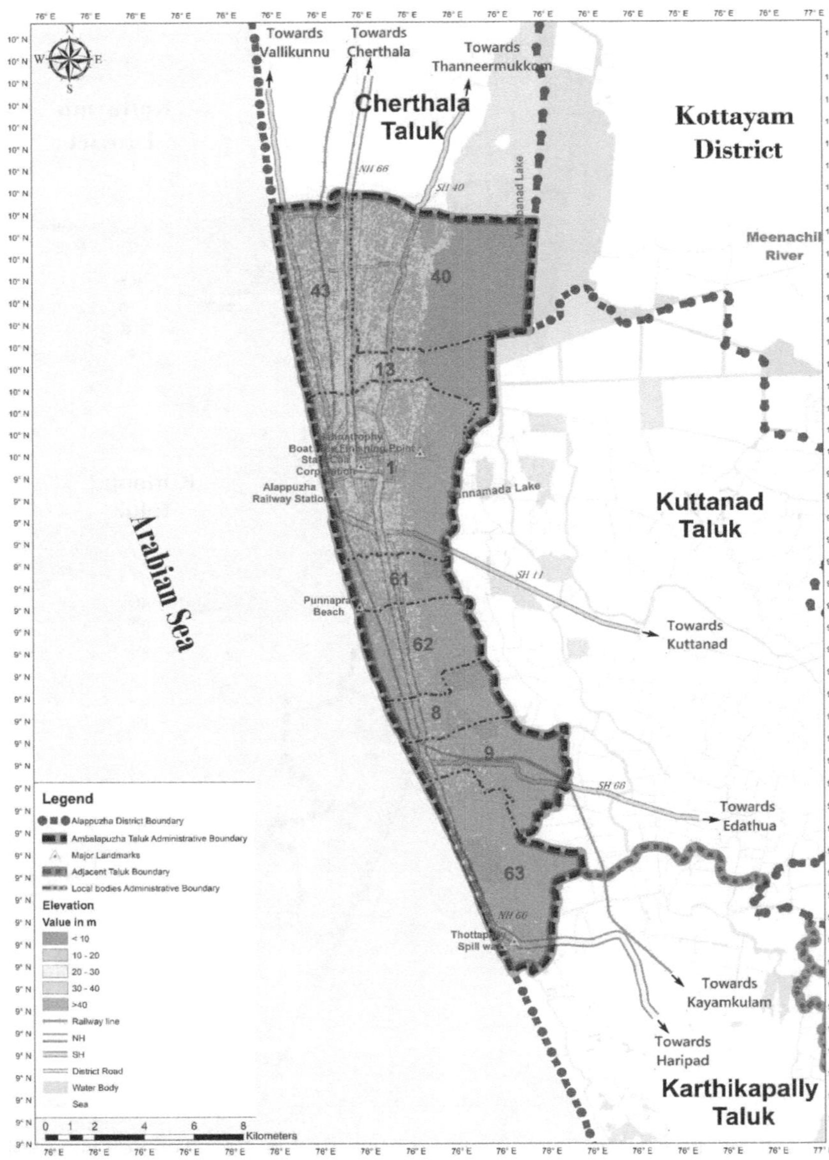

Fig. 34.21 Ambalapuzha Taluk - Local Bodies Coastal elevation potentiality Map

Source: DSM - ALOS, generated using ArcGIS

5.7 Coastal Flooding or Sea Water Inundation Potentiality

Ambalapuzha South and Purakkad Gram panchayats are highly vulnerable to flooding among the 9 local bodies. Previous Flooding events of Kerala supports this statement. They are vulnerable to flooding due to their low relief topography nature. Coastal areas of Ambalapuzha Taluk are less vulnerable to RSLR, Eastern regions of the Taluk are mostly affected by SLR in 2050 (according to area delineated using Climate Central's data). It could be mainly because of the area's low-lying terrain feature and its proximity to the Vembanad Lake, which could be overflown and flood the nearby areas. Around 31% of the Ambalapuzha Taluk area is found to be vulnerable.

Table 34.20 CHVI Rank - Ambalapuzha Taluk Local Bodies (Coastal flooding/seawater inundation potentiality)

COASTAL FLOODING/SEA WATER INUNDATION POTENTIALITY	
LOCAL BODIES	**CHVI RANKING SCORE**
MARARIKULAM SOUTH	1
MANNANCHERRY	1
ARYAD	1
ALAPPUZHA (M)	4
PUNNAPRA NORTH	4
PUNNAPRA SOUTH	4
AMBALAPUZHA NORTH	5
AMBALAPUZHA SOUTH	5
PURAKKAD	4

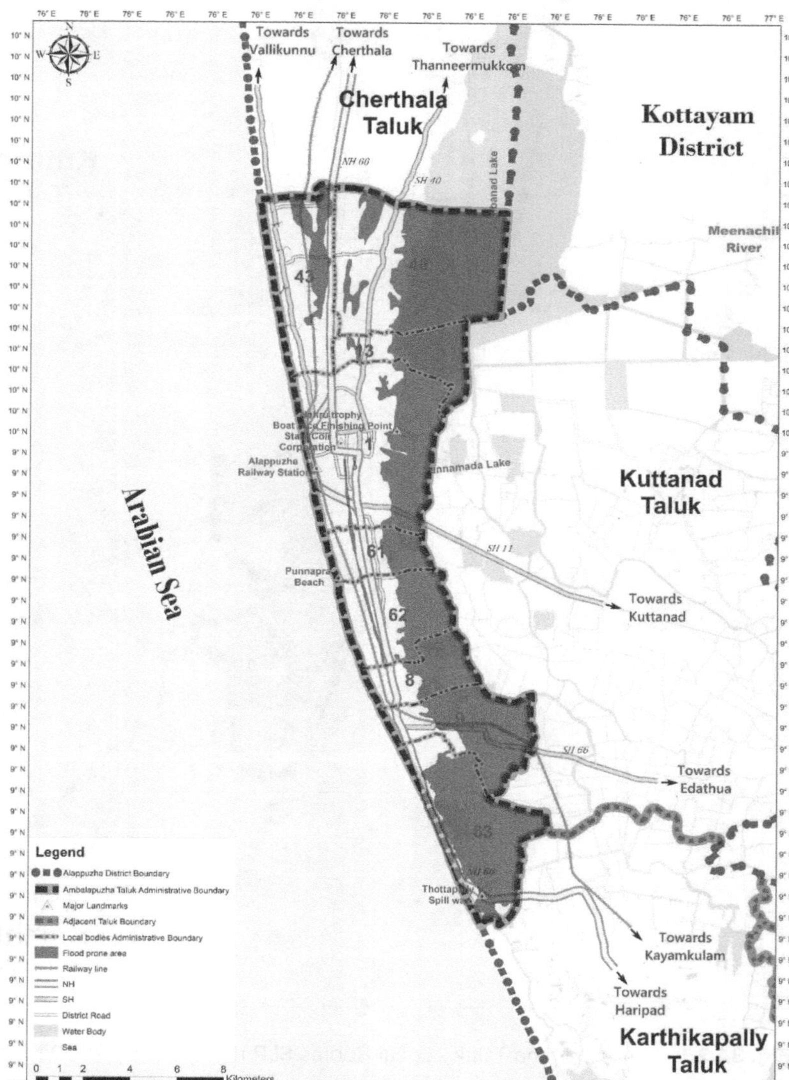

Fig. 34.22 Ambalapuha Taluk - Local Bodies Flood Prone Area Map

Source: KSDMA, generated using ArcGIS

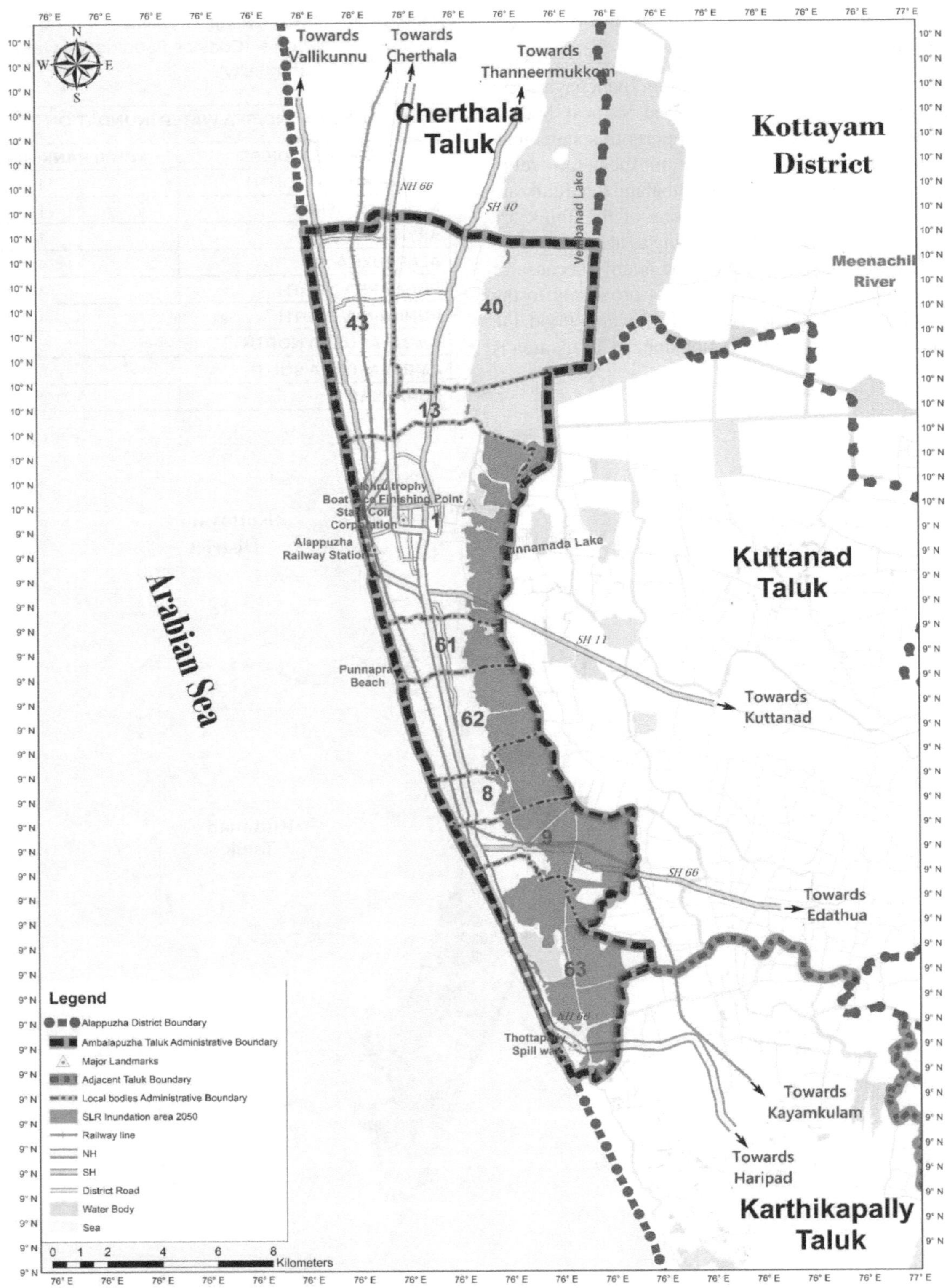

Fig. 34.23 Ambalapuzha Taluk - Local Bodies SLR risk areas by 2050 Map

Source: Climate Central, Delineated using ArcGIS

5.8 Geomorphology

Ambalapuzha Taluk is mainly composed of Coastal Plain and these areas are safe from flooding under normal circumstances. These areas are well-drained indicating that the area possesses highly permeable soil and low relief.

The Purakkad Grama panchayat is mainly composed of the Alluvial plain. These areas are vulnerable to river flooding.[5]

Table 34.21 CHVI Rank - Ambalapuzha Taluk Local Bodies (Geomorphology)

GEOMORPHOLOGY	
LOCAL BODIES	**CHVI RANKING SCORE**
MARARIKULAM SOUTH	5
MANNANCHERRY	5
ARYAD	5
ALAPPUZHA (M)	5
PUNNAPRA NORTH	5
PUNNAPRA SOUTH	5
AMBALAPUZHA NORTH	5
AMBALAPUZHA SOUTH	5
PURAKKAD	4

5.9 Results of Stage 2 – Local Body Level Assessment

Based on the assessment of different parameters form the CHVI on Local Bodies of Ambalapuzha Taluk, Coastal Vulnerability Habitat Index was calculated using the CVI equation, from the values in each parameter based on the specific data variables. The calculation has values ranging from 1.7 to 7.7. The mean value is 4.4 and the standard deviation is 2.3. The 25th, $50^{th,}$ and 75th percentiles are found to be 3.46, 4, and 4.47. The CHVI is segmented into low, medium, high, and very high, based on the quartile ranges of the data. CHVI values below 3.46 are considered to be low, values in between 3.46 to 4 as a medium, 4 to 4.47 as high and above 4.47 as very high.

Based on the CHVI value, Alappuzha Municipality and Ambalapuzha North and Ambalapuzha South are under Very High category, in which Ambalapuzha North is considered to be more vulnerable in terms of score, but considering the population getting effected and data availability, Alappuzha Municipality will be taken as the study area for further assessment. Alappuzha Municipality area will be assessed to find out decadal relative sea-level scenarios and the respective physical infrastructure vulnerable to the time series.

Hence the Most Vulnerable Local Body in Ambalapuha Taluk according to CHVI, data availability, and population

Table 34.22 Ambalapuzha Taluk - Local Bodies Coastal Habitat Vulnerability Index

CHVI OF AMBALAPUZHA TALUK – LOCAL BODIES									
PARAMETERS	**Mararikulam South**	**Mannancherry**	**Aryad**	**Alappuzha (M)**	**Punnapra North**	**Punnapra South**	**Ambalapuzha North**	**Ambalapuzha South**	**Purakkad**
Coastal elevation potentiality (m)	3	3	3	3	4	4	4	4	4
Coastal flooding or Sea water inundation potentiality (m)	1	1	1	4	4	4	5	5	4
Impacts of settlements and shoreline extends (km)	4	1	1	5	1	1	4	1	1
Impacts of population growth and density (persons per sq.km.)	4	3	4	5	4	4	4	3	3
Geomorphology	5	5	5	5	5	5	5	5	4
Coastal Habitat Vulnerability Index (CHVI)	3.5	1.7	1.7	7.7	4.0	4.0	8.9	4.5	3.6
	Very High	**Low**	**Low**	**Very High**	**High**	**High**	**Very High**	**Very High**	**Medium**

[5]KSLUB – Kerala State Land Use Board

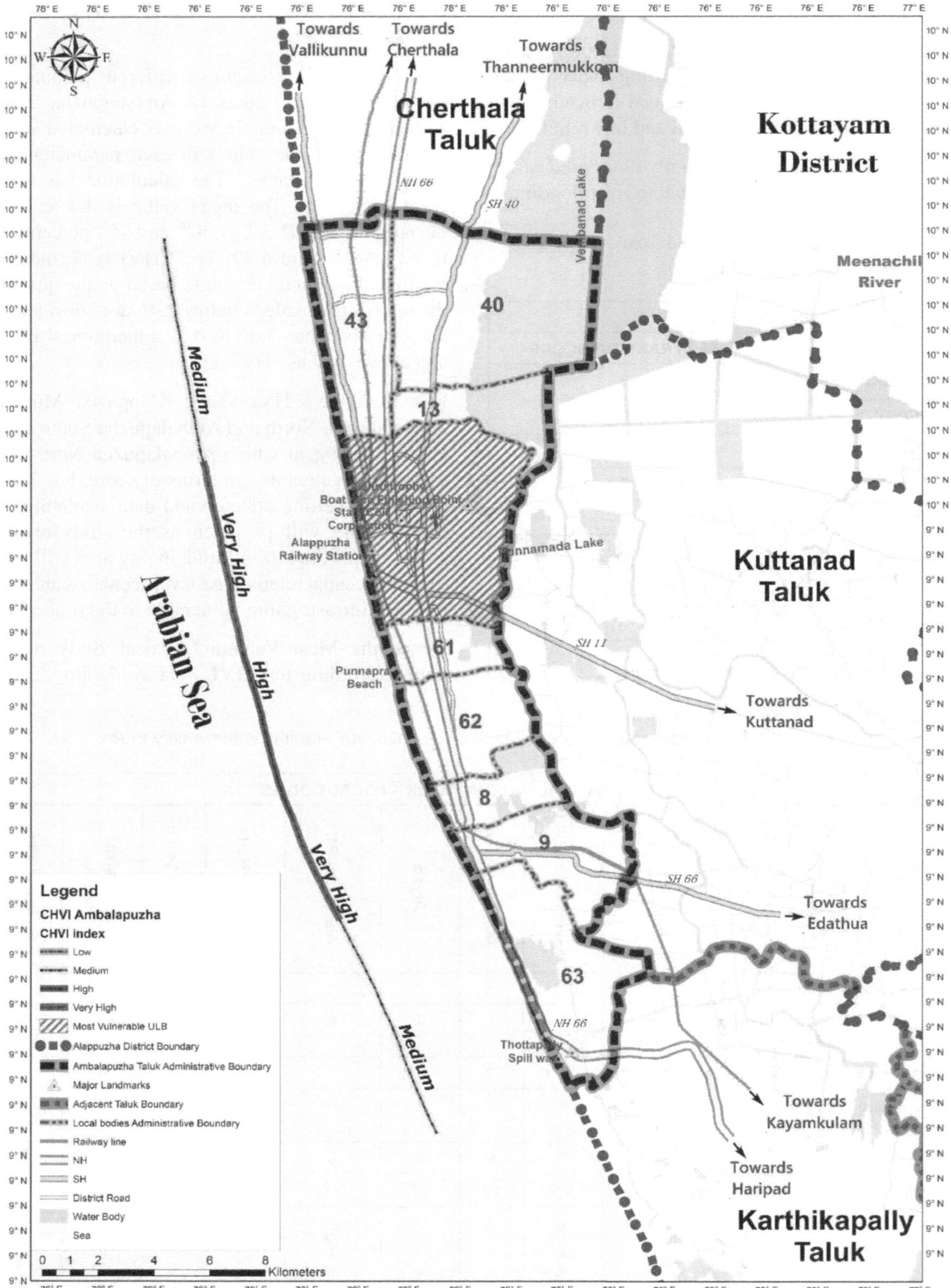

Fig. 34.24 Ambalapuzha Taluk - Local Bodies CHVI Map

Source: Author Generated, Developed using ArcGIS

displaced is considered to be Alappuzha Municipality and is represented in CHVI Map of Ambalapuzha Taluk – Local Bodies.

6. Evaluation of Decadal RSLR and Inundation Potentiality of Physical Infrastructure of Alappuzha Municipality

6.1 Approach to Decadal Relative Sea Level Rise & Inundation Potentiality of Physical Infrastructure:

Data procured:

1. SEA level rise risk areas by 2050 area of Alappuzha Municipality.
2. Road Network of Alappuzha Municipality.
3. Building footprint of Alappuzha Municipality.
4. Population Statistics of Alappuzha Municipality.
5. Decadal shoreline changes of Alappuzha Municipality.

Integrating and assessing all the above-mentioned data could result in assessing decadal RSLR and Inundation potentiality of the physical infrastructure of the Alappuzha Municipality.

Assessment is carried out in 4 steps:

1. Assessment of Decadal RSLR based on relative sea-level rise trend, decadal shoreline change, and highest inundation risk elevation form Climate Central data.
2. Assessment of Inundation potentiality of physical infrastructure by decadal RSLR.
3. Population affected by decadal RSLR.
4. Damage assessment of potentially vulnerable physical infrastructure to decadal RSLR.

Assessment of Decadal relative sea-level rise:

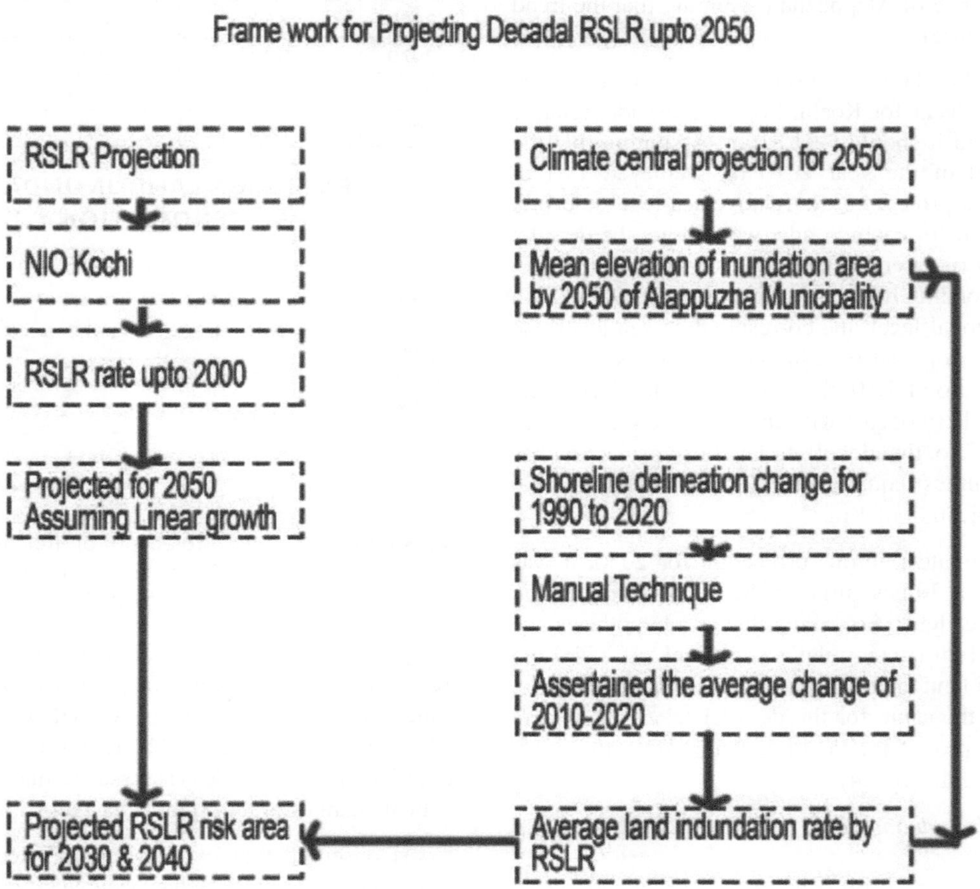

Fig. 34.25 Methodology for assessing decadal relative sea-level rise

Table 34.23 Net sea level rise trends along the Indian coast

Station	Period of analysis	Number of years of data availability	Trends in relative sea-level rise (mm yr^{-1})	GIA Correction (mm yr^{-1})	Net sea-level-rise trend (mm yr^{-1})
Mumbai	1878-1993	113	0.77±0.08	-0.31	1.08
Kochi	1939-2007	56	1.45±0.22	-0.36	1.81
Visakhapatnam	1937-2000	53	0.69±0.28	-0.24	0.93
Diamond Harbour (Kolkata)	1948-2010	61	4.61±0.37	-0.35	4.96

Source: A.S. Unnikrishnan, CSIR-National Institute of Oceanography, GOA

Estimation of decadal relative sea-level rise, avg. decadal shoreline changes and its equivalent land elevation inundation of Alappuzha Municipality:

Relative sea-level rise projection:

In the coastal length of Kerala State, only Kochi is having a tidal station (according to Permanent Service for Mean Sea Level - PSMSL). Sea level rise trend data for Kochi from the National Institute of Oceanography (NIO-GOA) - (Sea level rise trends of Indian coast) was used to assess the trend of relative sea-level rise of Alappuzha (assuming that the trend shows similar feature).

This data shows that the relative sea-level rise trend in 2000 - 2007 is 1.81mm/year for Kochi. So, for a decade, relative sea-level rise trend it should be 1.81cm. Assuming that the relative sea level of the year 2010 has achieved 1.81cm (in a decade), and projecting the relative sea-level rise in a linear growth chart for each decade, we will get the decadal relative sea-level rise trend. Therefore, it's observed that in 2050 the corresponding relative sea-level reaches 9.05 cm in Kochi. On the global level, the corresponding sea level rise is 1m for 2050 (projected by Climate Central), so there is an approximate ratio of 1:10. So at the global level 10times, the value of RSLR (9.05cm) will give us SLR value which is 90.5 cm ~ 1 m. So, the global sea-level rise projection has come out to be almost equal to the Climate Centrals global sea-level rise projection of 1 m.

Based on the Climate Centrals projection for 2050, it was said that there is a 3times surge in the areas of inundation for a 1m rise in sea level. So, taking that as a hypothesis, we shall say for 18.1cm rise in relative sea-level rise, 54.3 cm (3 times of 18.1) land area will be at risk of inundation. So, working out for the same for the decadal relative sea-level scenario, we can project the land inundation potentiality for each decade (see Table 34.24)

Decadal shoreline delineation and inundation potential elevation:

To ascertain the above calculation, another process to find the decadal elevation inundation potential using shoreline

Table 34.24 Assumed Decadal relative sea-level rise and its respective land elevation at risk of inundation projection for Alappuzha Municipality

SEA LEVEL RISE AND ITS RESPECTIVE LAND INUNDATION PROJECTION			
Year	**RSLR in Kochi (in cm)**	**SLR at Global level (in cm)**	**The land area at risk of inundation (in cm)**
2010	1.81	18.1	54.3
2020	3.62	36.2	108.6
2030	5.43	54.3	162.9
2040	7.24	7.24	217.2
2050	9.05	90.5	271.5

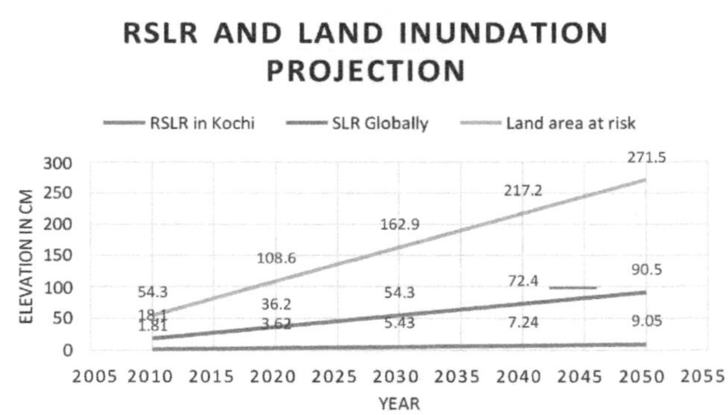

Fig. 34.26 Assumed projection of Relative sea-level rise trend and its respective land elevation at risk of inundation for Alappuzha Municipality

change was carried out. Decadal shoreline change map is created using google earth maps and Landsat images with manual delineation techniques. Google earth maps of 1990, 2000, and 2010 and for 2020 Landsat image is used. The shoreline of each decade is assessed to find the horizontal and vertical changes of the Alappuzha Municipality coastal area.

It was observed that over the past 3 decades, Alappuzha Municipality coastal area has been changing with an average length of 0.496 m (horizontal distance) and with an average elevation of 1.57 m (Vertical distance) (see Fig. 34.32). So,

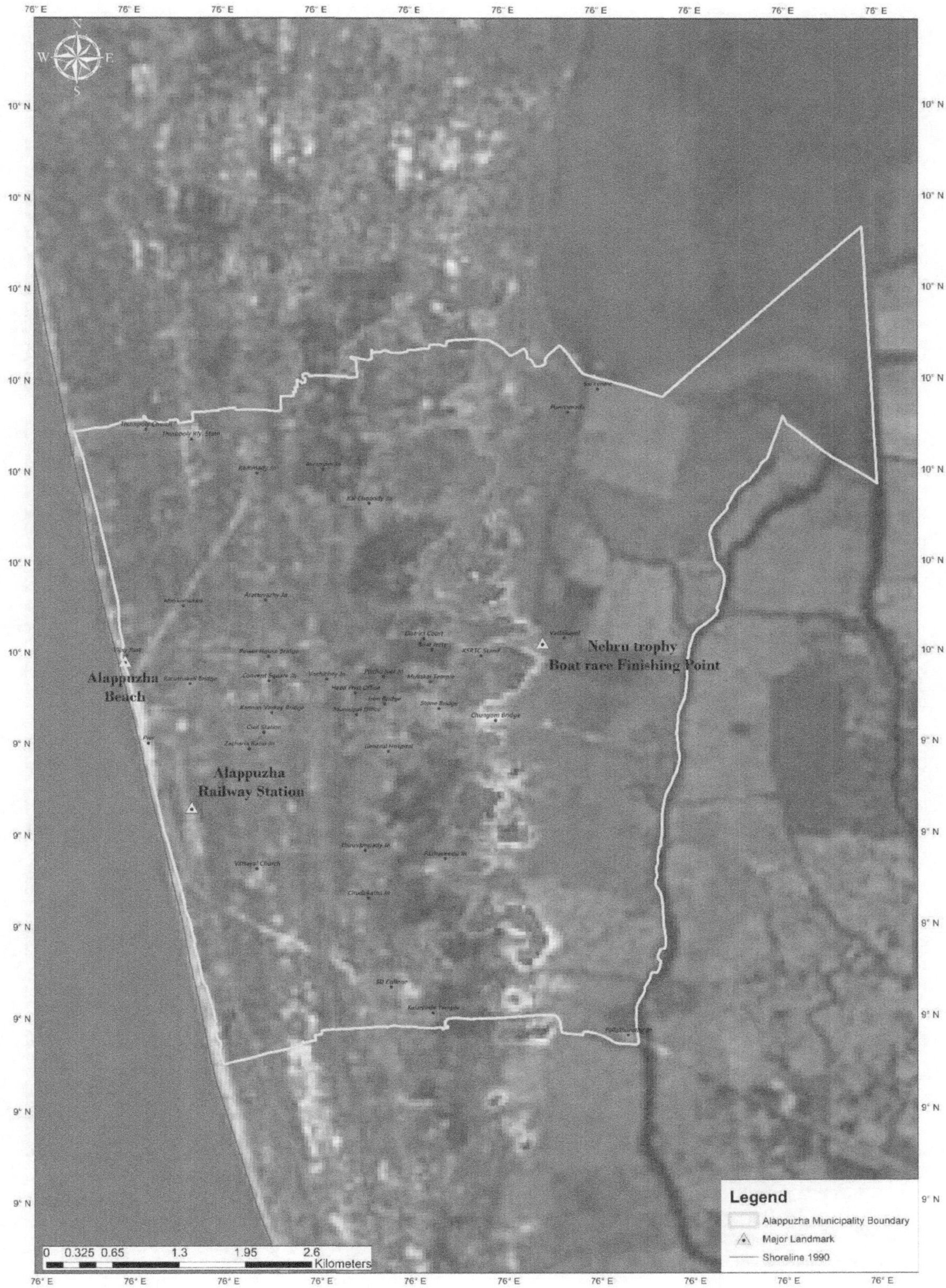

Fig. 34.27 Shoreline Map of Alappuzha Municipality 1990

Source: Google Earth Map, interpreted with manual delineation technique using ArcGIS

Fig. 34.28 Shoreline change Map of Alappuzha Municipality 2000

Source: Google Earth Map, interpreted with Manual delineation technique using ArcGIS

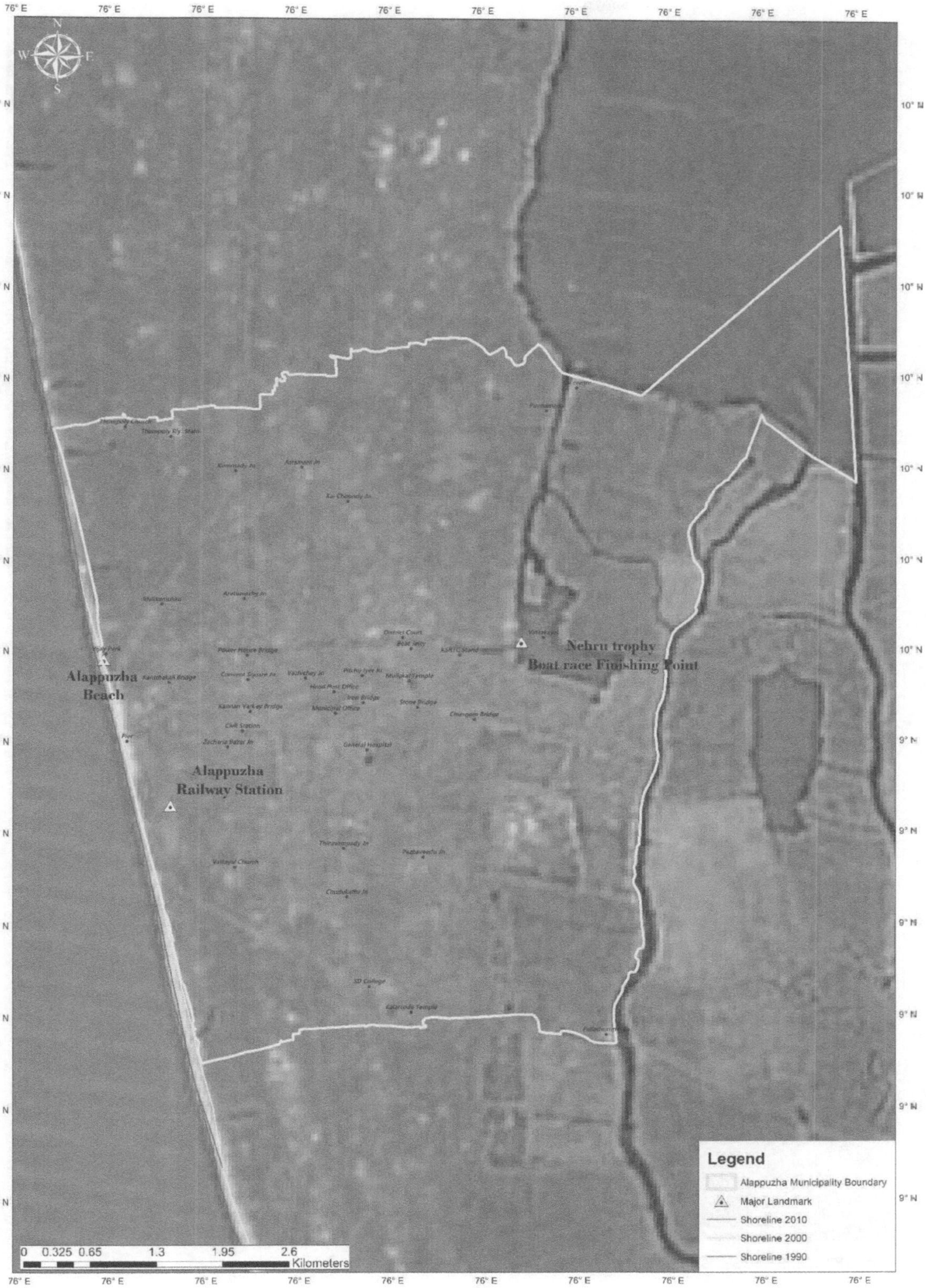

Fig. 34.29 Shoreline change Map of Alappuzha Municipality 2010

Source: Google Earth Map, Interpreted with Manual delineation technique using ArcGIS

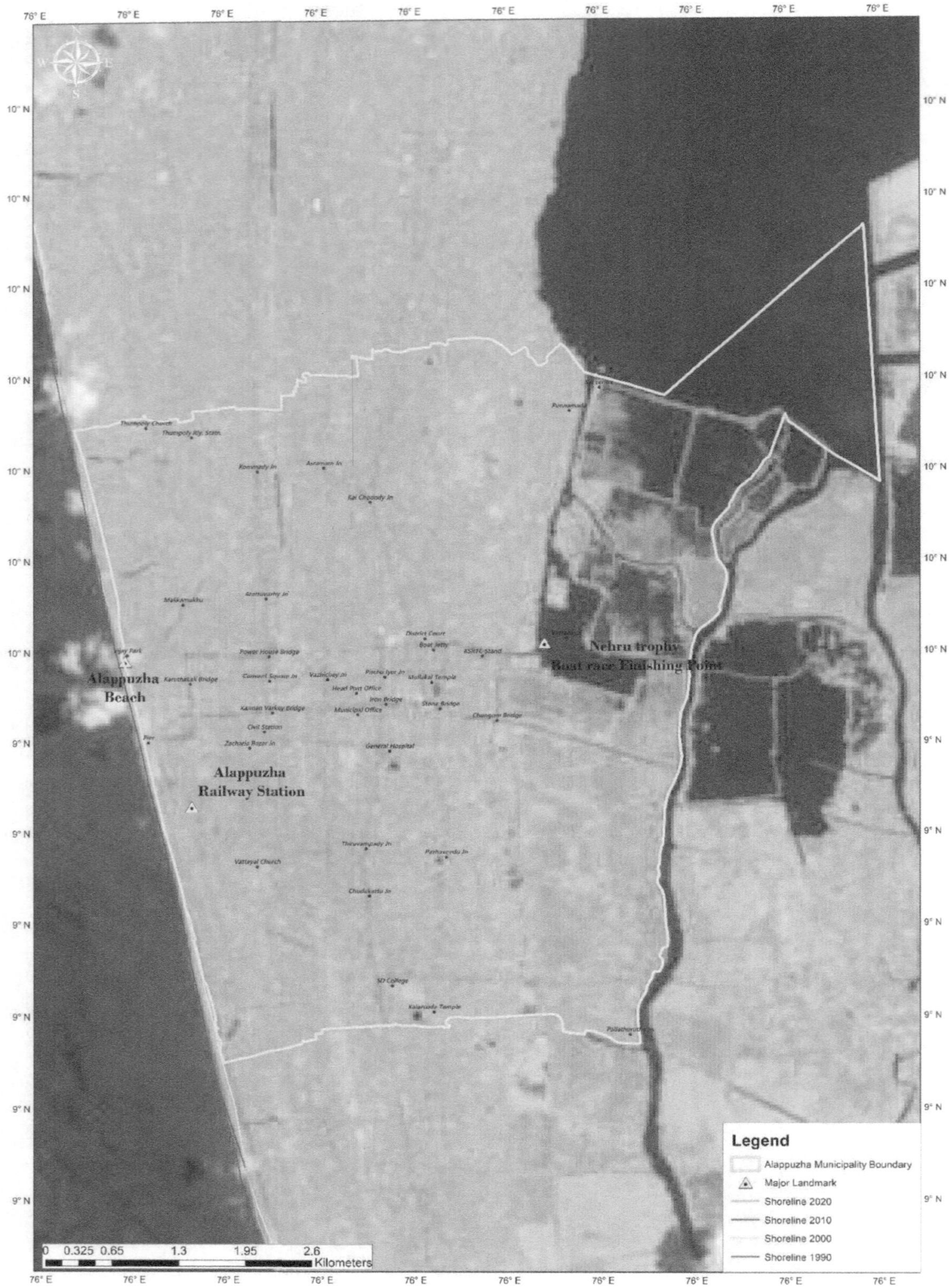

Fig. 34.30 Shoreline change Map of Alappuzha Municipality 2020

Source: Landsat Image, Interpreted with Manual delineation technique using ArcGIS

Fig. 34.31 Shoreline change Map of Alappuzha Municipality 1990 – 2020, overlaid with contour

Source: DSM - ALOS, generated using ArcGIS

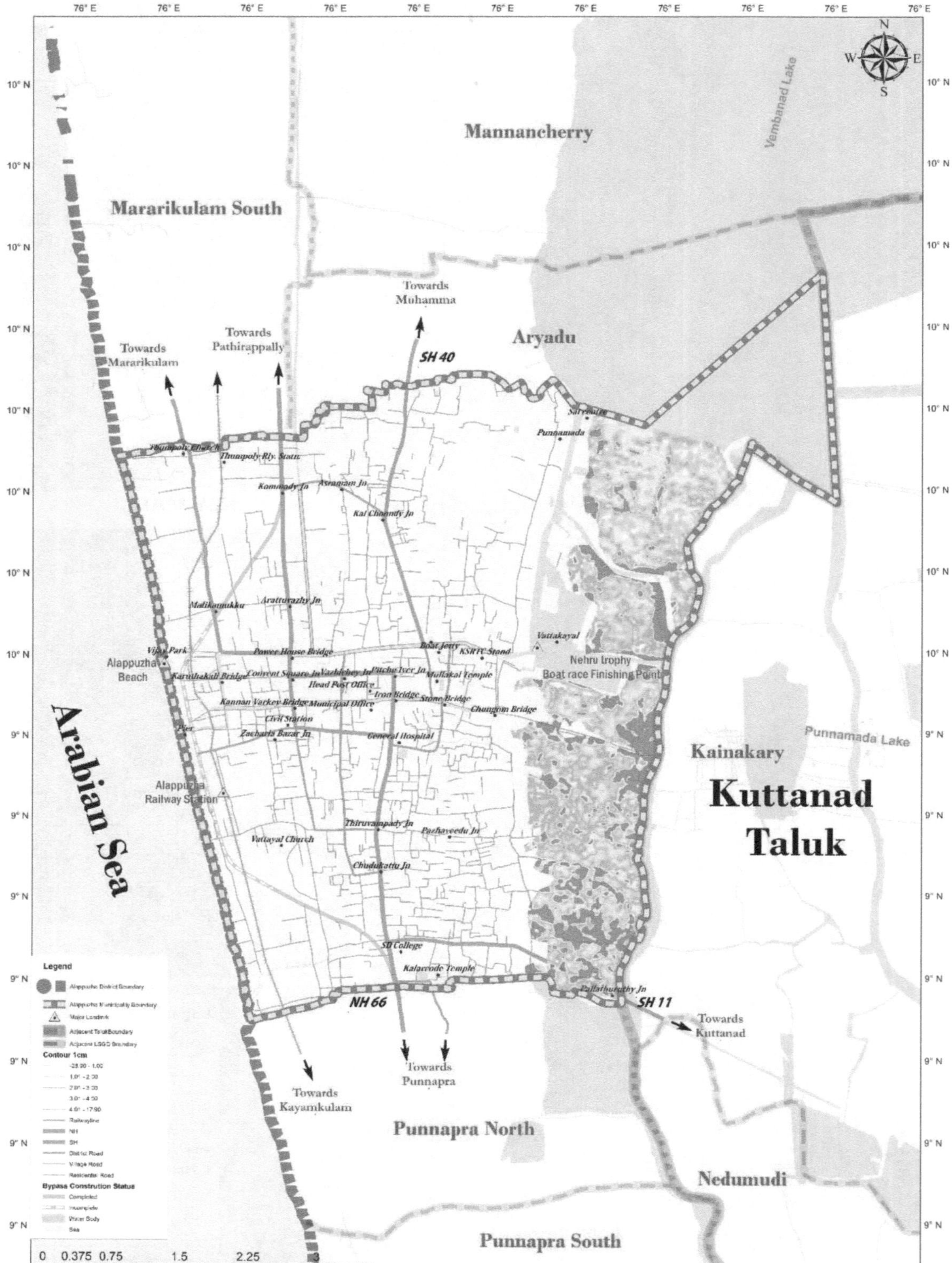

Fig. 34.32 Alappuzha Municipality RSLR 2050 Contour Map

Source: DSM - ALOS, generated using ArcGIS

dividing the elevation by 3 we get decadal relative sea-level inundation area of land that is 0.525 m. This result is almost matching with the 3 times range of Relative sea-level rise result (0.543 m).

Also, from the contour map generated for the RSLR area of 2050 in the Alappuzha Municipality, the mean elevation of the RSLR inundation area was found to be 2.87m ~ 3m (see Fig. 34.33).

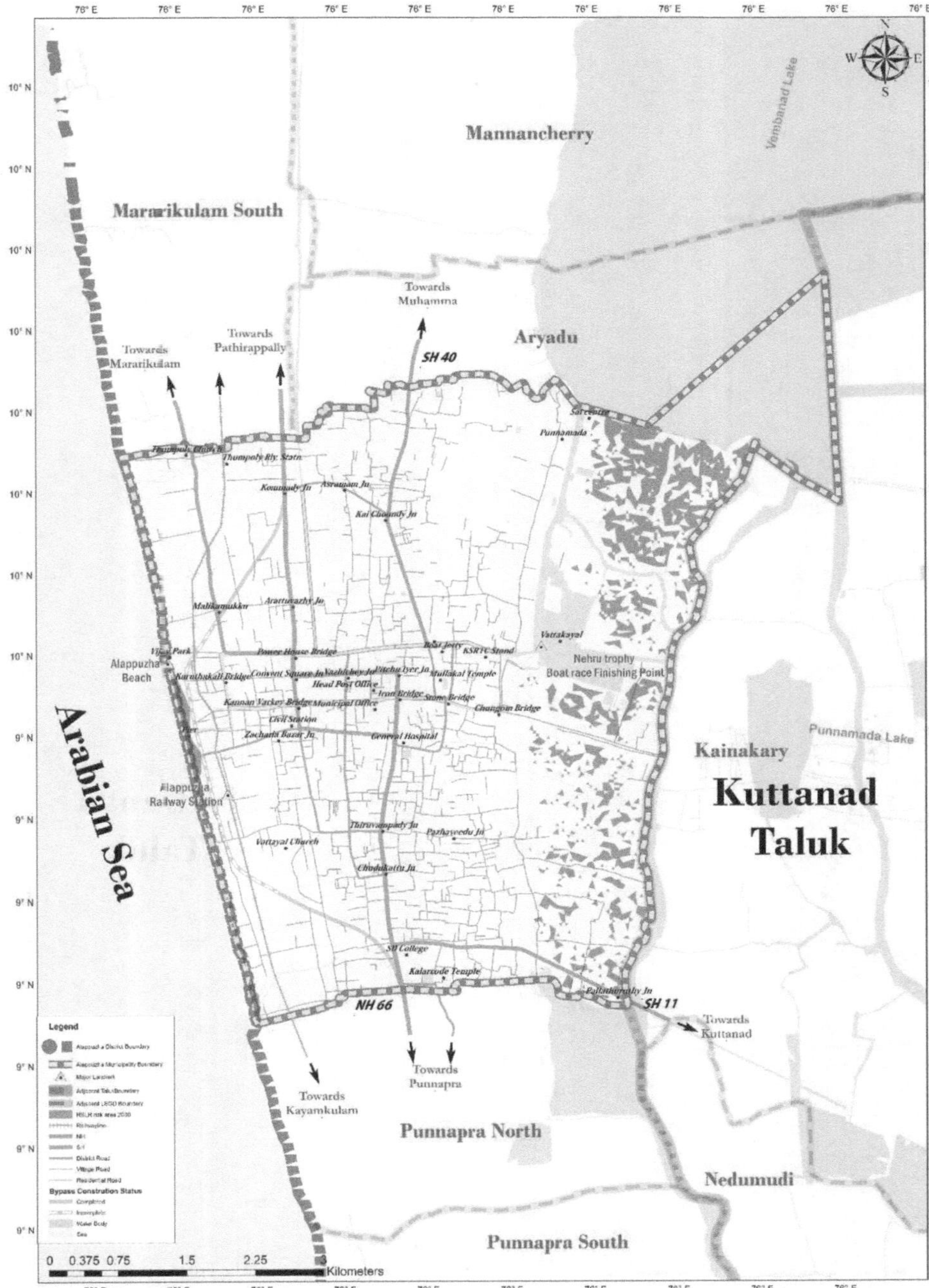

Fig. 34.33 Alappuzha Municipality 2030 RSLR risk areas Map

Source: Study analysis, generated using ArcGIS

Hence the assumption that 1m sea-level rise would inundate almost 3 times that of the elevation inland is proved. And the relative sea-level inundation areas of each decade are delineated in the Decadal RSLR Map of Alappuzha Municipality (see Fig. 34.34) which could be used to assess the inundation potentiality of physical infrastructure through the decades 2030, 2040 and 2050.

The delineated area was developed concerning two conditions:

1. The area needs to have any connection to near water bodies like canals, rivers,lakes, sea, etc.
2. The area has to be coming under Climate Centrals 2050 RLSR risk areas.

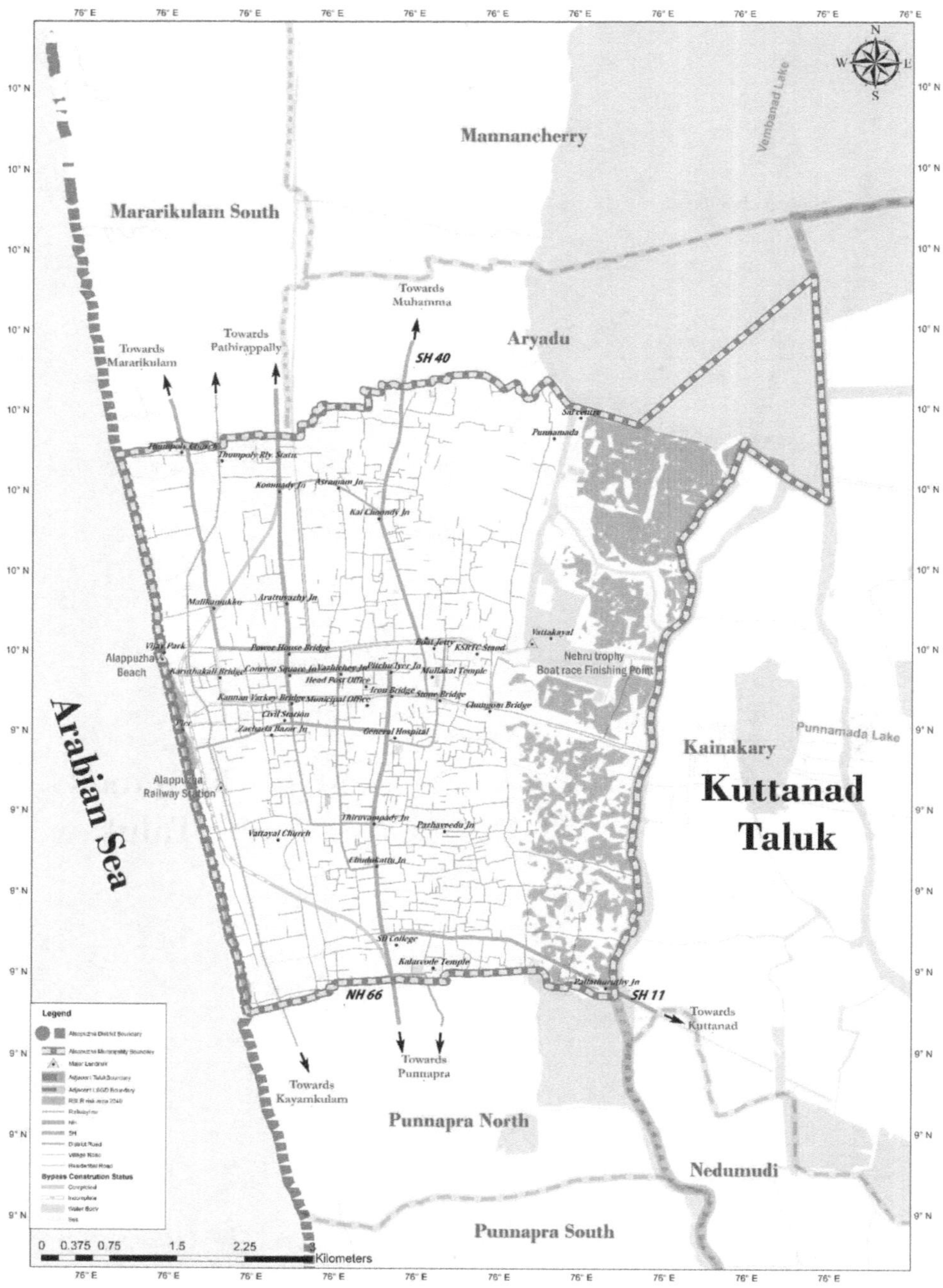

Fig. 34.34 Alappuzha Municipality 2040 RSLR risk areas Map

Source: Study analysis, generated using ArcGIS

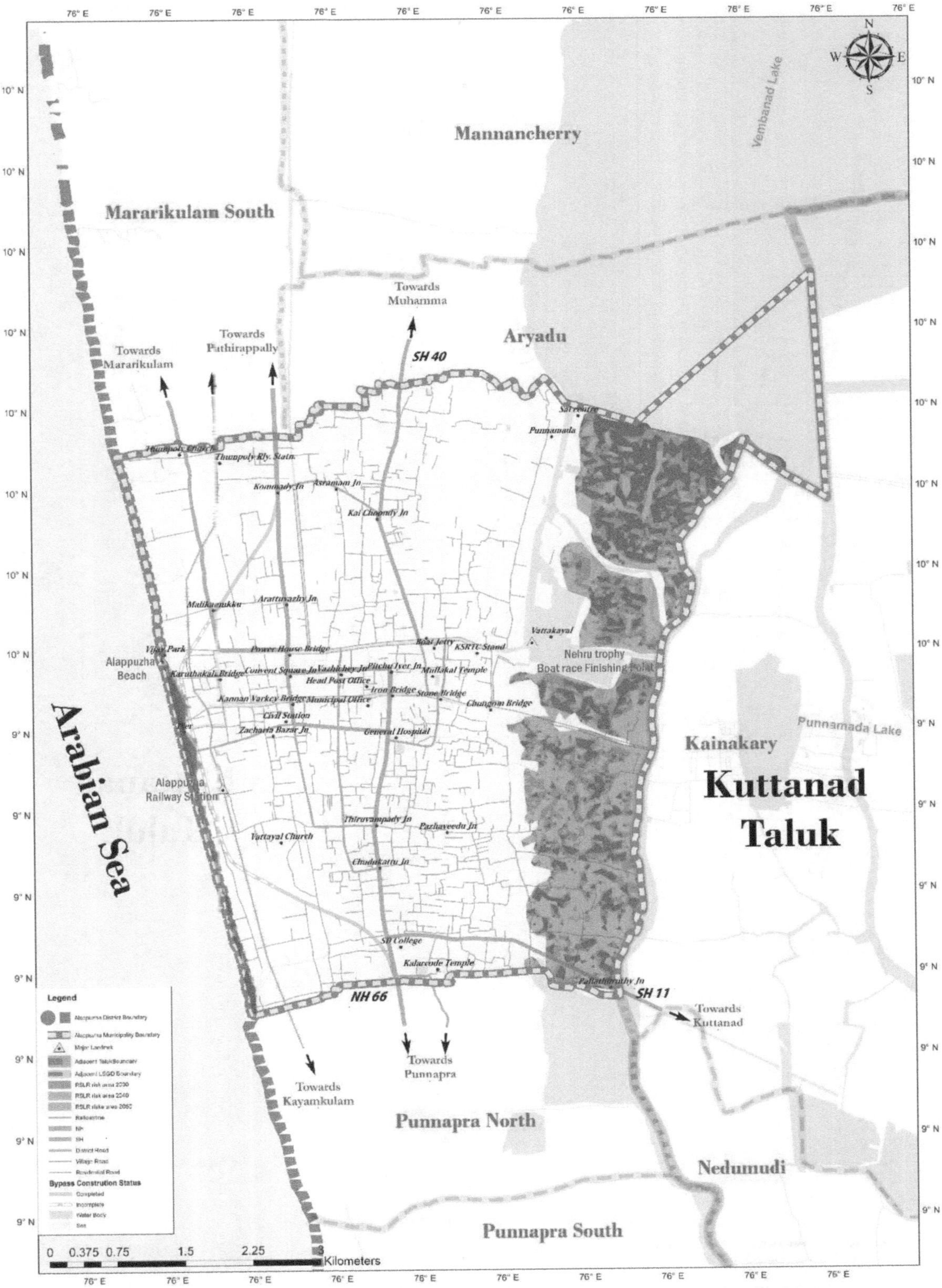

Fig. 34.35 Alappuzha Municipality decadal RSLR risk areas Map

Source: Study analysis, generated using ArcGIS

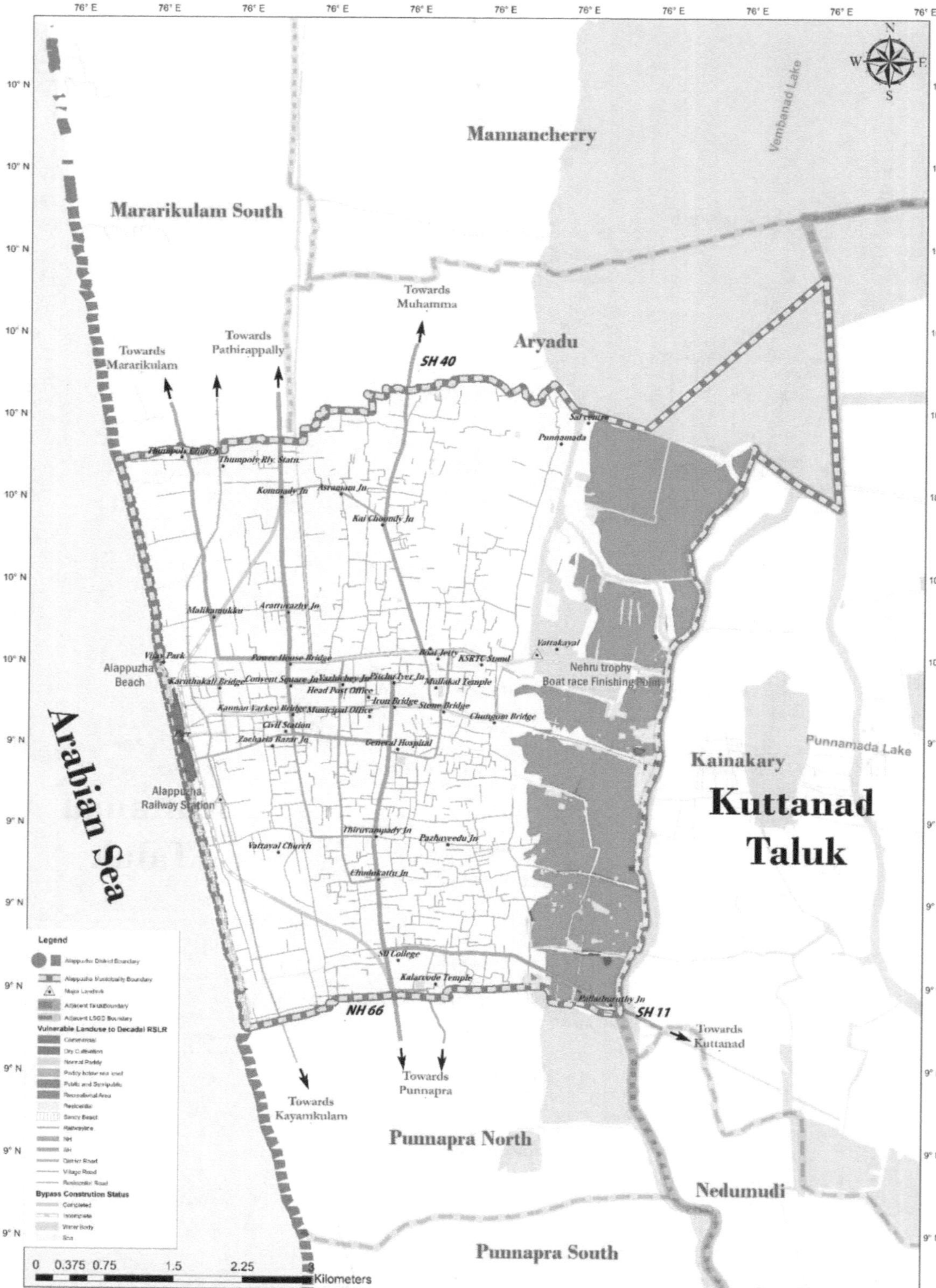

Fig. 34.36 Potentially vulnerable land use to decadal RSLR of Alappuzha Municipality

Source: TCPO-Alappuzha, generated using ArcGIS

Assessment of inundation potentiality of Physical Infrastructure to decadal RSLR:

See Figures 34.37, 34.38 and 34.39.

The population at risk of decadal RSLR of Alappuzha Municipality:

The population projection of the Alappuzha Municipality was done using an Arithmetical increase method. And with population projection, the population density was calculated. And vulnerable areas to decadal RSLR was multiplied with population density to calculate the number of populations at risk to RSLR for each decade.

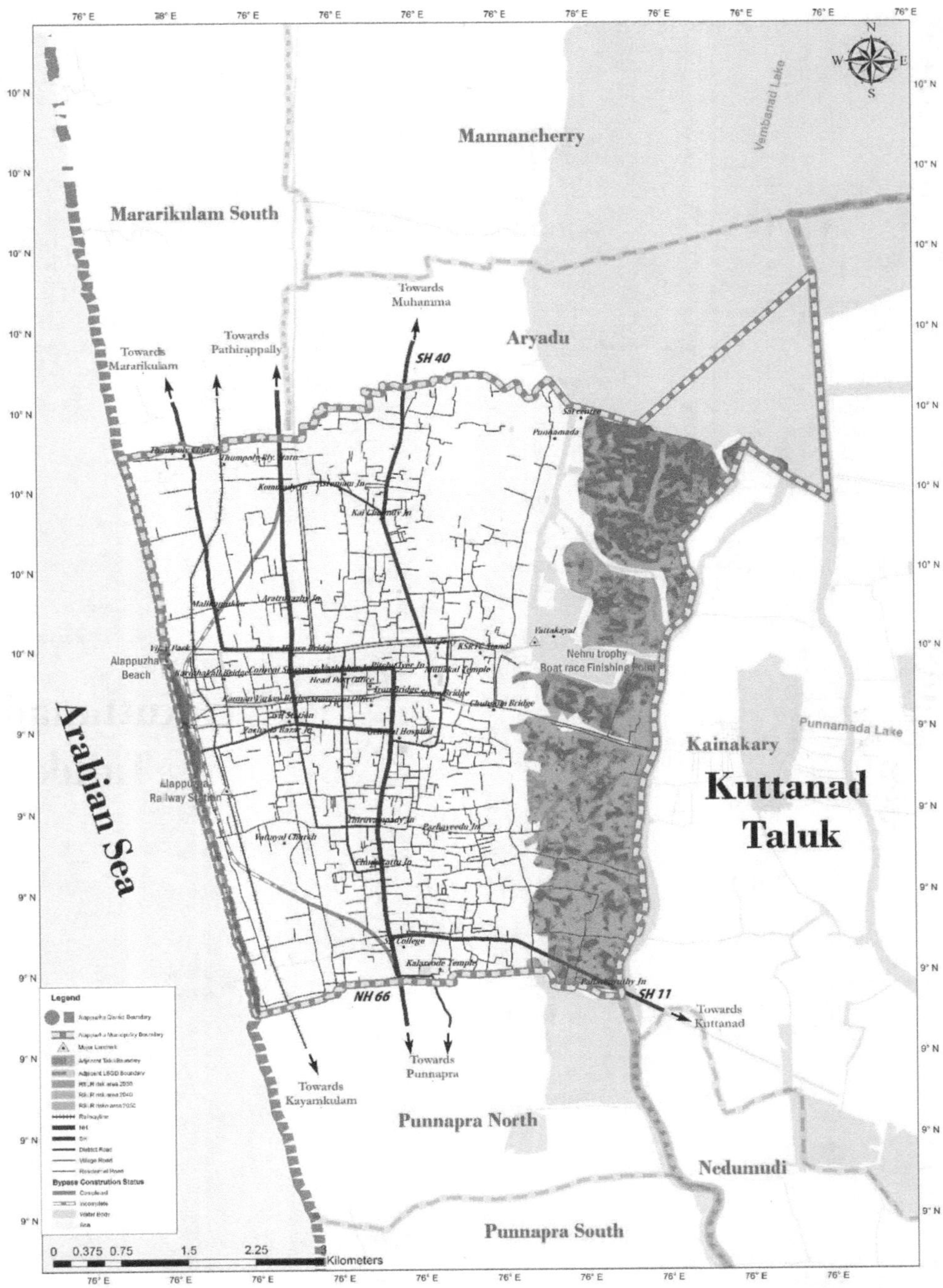

Fig. 34.37 Potentially vulnerable roads to decadal RSLR of Alappuzha Municipality

Source: Open Street Map, generated using ArcGIS

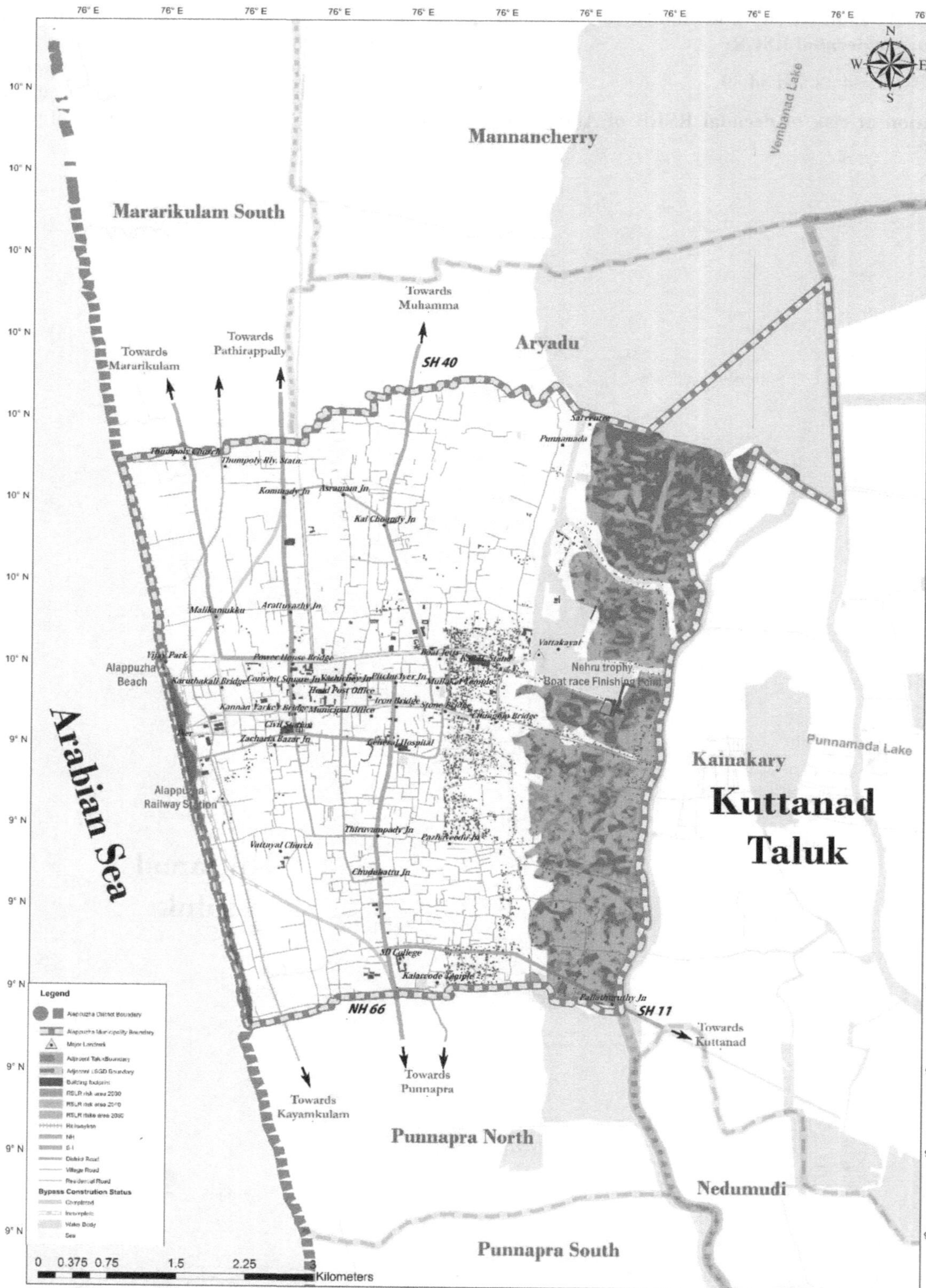

Fig. 34.38 Potentially Vulnerable Buildings to decadal RSLR of Alappuzha Municipality

Source: Open Street Map, generated using ArcGIS

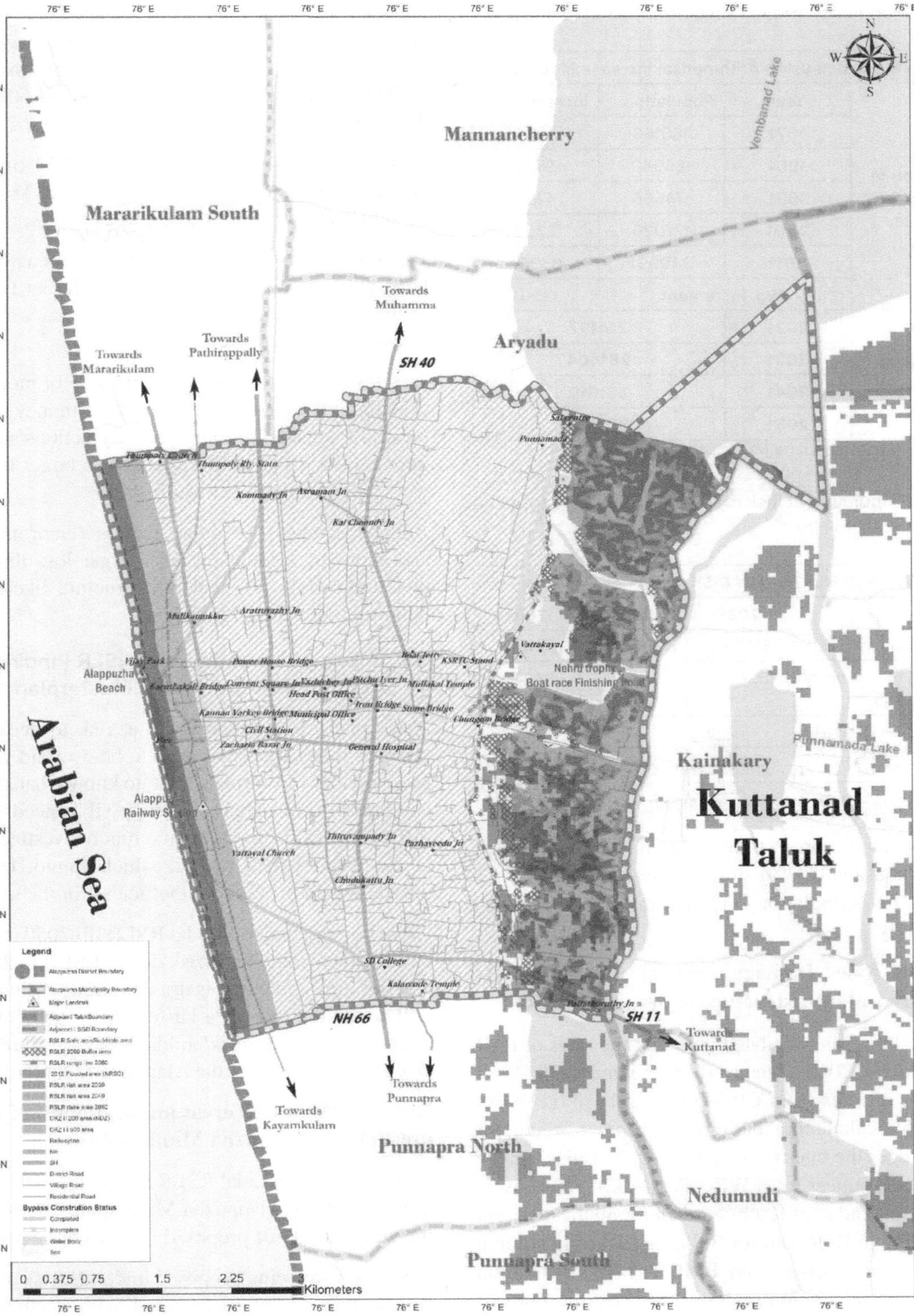

Fig. 34.39 Map showing Decadal RSLR range line, CRZ II, CRZ III, 2018 Flooded area, and Safe/buildable areas of Alappuzha Municipality

Source: KCZMA, KSDMA, Generated using ArcGIS

Table 34.25 Population projection of Alappuzha Municipality

Population Projection using Arithmetical Increase Method			
	Year	Population	Increment
The population of previous years from Census data.	1971	160166	–
	1981	169940	9774
	1991	174666	4726
	2001	177029	2363
	2011	240991	63962
	Avg. Increment		20206.25
Projected Population	2021	261197	
	2031	281404	
	2041	301610	
	2051	321816	

Source: Census data, Alappuzha Municipality Masterplan

Table 34.26 Population at Risk to decadal RSLR of Alappuzha Municipality

POPULATION IN RISK AREA			
	2031	2041	2051
Projected Population	281404	301610	321816
Population density (person per sq.km.)	8249.9	8842.3	9434.7
Affected land area (sq.km.)	2.38	4.17	7.76
Population at risk	**19646.1**	**36909.1**	**73193.2**

Source: Census data, Alappuzha Municipality Masterplan

From the population projection table, it was found that around 19.6 Thousand (6.9%), 36.9 Thousand (12.2%), 73.1 Thousand (22.7%) population is at risk in 2030, 2040, and 2050 respectively.

Assessment of Asset Damage Loss of the potentially vulnerable physical infrastructure to decadal RSLR:

Asset Damage loss was calculated based on the rates of PAR (Plinth area rates) 2019 (Sugawara & Nikaido, 2014) of Delhi schedule of rates from CPWD (Central Public Works Department) for calculating building damage loss and for Roads and Bridges the rates provided by Mr. K. Balakrishnan (Retired Asst. Engineer at PWD, Palakkad) were used. Construction Cost Index of Kochi is taken for all the different sectors according to PAR, and for Building Footprint column CCI of Alappuzha District from PRICE (PWD estimation site) is used. For calculation of Land loss costing, rates of plots with access roads in the Alappuzha Municipality area from Fair value site was used.

Estimation for different physical infrastructure vulnerable to decadal RSLR of Alappuzha Municipality (different sectors includes buildings, roads, bridges, and land area) was carried out. From the calculation of Asset Damage Loss, it was assessed that, a total of 3,418Crores loss in the year 2030, 5,999 Crores loss in the year 2040, and 11,502 Crores loss in the year 2050.

This total loss for the year 2050 is 2.34 times that of the 2018&2019 Kerala flood CMDRF (Chief Ministers Distress Relief Fund), which is 4,912 Crores.

We should also understand that this damaging loss is for a ULB alone, if we consider for the Taluk or District it would be more than that. So, imagine in the case of the whole Coastal Districts of Kerala.

This loss is based on the present value of money, it could be even more if we add the escalation of money into it. Hence if we don't consider taking necessary actions/steps to adapt or mitigate the RSLR, it could lead to a larger loss from which we won't be able to recover from.

Adding damages to other sectors (environmental, social, and economic) could reveal a larger loss than projected in this study (only physical infrastructure like roads, bridges, building footprint, and land use).

6.2 Implications of Decadal RSLR Findings in Alappuzha Municipality Masterplan

Areas that are assumed to be at risk to decadal RSLR are taken as focus areas that need to be overlaid with Alappuzha Municipality Masterplan 2031, to know what all proposals are coming in the risk areas. And if possible, need to reconsider it, if the proposal is involving too much investment as investing in an infrastructure project which cannot be used after a decade would prove to be inefficient for the money involved.

Areas coming under decadal RSLR till 2050 is the focus area. Areas which could be at risk due to RSLR in 2060 is taken as a buffer area and a line is drawn joining the endpoints which will act as the safe area buffer line which could be used to delineate the safe areas/buildable area towards the opposite side (Western side) of the Alappuzha Municipality.

Decadal RSLR risk areas implications on Proposed Land use 2031 of Alappuzha Municipality:

Overlaying the decadal RSLR risk areas on the Proposed land use 2031 of the Alappuzha Municipality helps to understand the vulnerability of proposed infrastructure proposals.

Two roads are being proposed under the risk areas of decadal RSLR, these need to assessed with accurate engineering studies to ascertain its safety to RSLR before implementing or constructing it.

Another proposal in the risk areas is tourism promoted areas, which are in a way good as recreational areas would lessen the

Table 34.27 Quantification of Asset Damage to decadal RSLR of Alappuzha Municipality

SL. NO	ANALYSIS	YEAR						
		Present	2030		2040		2050	
1	SEA LEVEL INUNDATION ON RISK AREA (sq.km.)		2.38		4.17		7.76	
		area/ km	area/ km	%	area/ km	%	area/ km	%
2	BUILDING AFFECTED BY RSLR (sq.m.)	588206	890.23	0.14%	2466.33	0.39%	64558.05	10%
3	ROADS AFFECTED BY RSLR							
	State Highway (km)	17.39	0	0%	1.08	6%	1.38	8%
	District Roads (km)	10.81	*	*	*	*	*	*
	Village roads (km)	29.47	0	0%	0	0%	2.76	9%
	Residential roads (km)	139.06	3.43	2%	5.56	4%	9.60	7%
4	BRIDGES AFFECTED BY RSLR	*	*	*	*	*	*	*
5	LAND AREA AFFECTED BY RSLR (sq.km.)	34.11	2.38	7%	4.17	12%	7.76	23%
	Residential	15.67	0.055	0%	0.110	1%	0.636	4%
	Commercial	0.61	0.001	0%	0.002	0%	0.005	1%
	Agriculture	6.61	2.11	32%	3.78	57%	6.41	97%
	• Paddy below sea level		2.11	100%	3.77	100%	6.37	99%
	• Normal paddy		0.00004	0%	0.00022	0%	0.003 129	0%
	• Dry Cultivation		0.002	0%	0.0057	0%	0.042126	1%
	Public and semi public	1.41	0.083	6%	0.087	6%	0.111	8%
	Sandy Beach	0.28	0.063	22%	0.084	30%	0.142	51%
	Recreational area	0.21	0.005	2%	0.006	3%	0.008	4%
	Industrial	0.30	0	0%	0	0%	0	0%

population at risk and infrastructure damage. But regulations like material used for construction should be implemented to make it into effect. (For example, such construction needs to be made with temporary construction materials/ water-based structures, etc.) Towards the beachside (Western side), 50% of the beach area is assumed to be at risk in 2050. NH Bypass elevated bridge is being constructed in the risk area. Since it is elevated the need to check the engineering capacity of the pillars is necessary as the chance of corrosion and tidal surges is there with the RSLR. (See Fig. 34.40)

ASSET DAMAGE LOSS CALCULATION													
NAME OF THE SECTOR	RATE	UNIT	NO. OF UNITS AFFECTED			AMOUNT in Rupees			CCI (CONSTRUCTION COST INDEX) in %	APPROX Total LOSS in Rupees		
			2030	2040	2050	2030	2040	2050		2030	2040	2050
Building Footprint	**39,724**	sq.m.	890.23	2466.33	64558.05	₹ 3,53,63,273.96	₹ 9,79,71,876.34	₹ 2,56,44,87,838.69	**148.71**	₹ 5,23,37,645.46	₹ 14,49,98,376.98	**₹ 3,79,54,42,001.26**
Residential Buildings (Up to 6 storeys, RCC framed, Every 0.3 m. higher plinth over normal plinth height of 0.45 m, Every 0.30 m. deeper foundations over normal depth of 1.20 m, R.C.C. Raft foundations)	**25,145**	sq.m.	*	*	*	*	*	*	**143.04**	*	*	*
Commercial Buildings (RCC Framed Structure, Extra for every additional story over six stories up o twelve storeys, Making stronger foundations to take load of one additional floor at a later date, Pile foundation)	**44,150**	sq.m.	*	*	*	*	*	*	**143.04**	*	*	

NH, SH Bridge over lake/river with piling (4lane, 4-10 Cr.)	**NH, SH (2 lane) bridges-** 2-4Cr.)	**Small Bridge** (7-8 Lakh/m)	**Bridges**	**Hospitals** (RCC Framed Structure, Extra for every additional storey over six storeys up to twelve storeys, Making stronger foundations to take load of one additional floor at a later date, Pile foundation)	**Schools, Colleges** (RCC Framed Structure, Extra for every additional storey over six storeys up to twelve storeys, Making stronger foundations to take load of one additional floor at a later date, Pile foundation)
7Cr	**3Cr**	**7.5L**	**1,58,895**	**45,450**	**44,150**
sq.m	**sq.m**	**sq.m**	**sq.m**	**sq.m.**	**sq.m.**
*	*	*	*	*	*
*	*	*	*	*	*
*	*	*	*	*	*
*	*	*	*	*	*
*	*	*	*	*	*
*	*	*	*	*	*
138.57					
*	*	*	*	*	*
*	*	*	*	*	*
*	*	*	*	*	*

Decadal Grand Total Loss	**Land Value** (3.6-8 Lakh)	**Village Road** (3-3.6m width, Bituminous (30Lakh/km) concrete(40Lakh/km)/interlock(45Lakh/km)	**District Road** (4-5m width - 60-65 Lakh/km)	**State Highway** (7m width)	**National Highway**	**Roads**
	5.8 Lakhs	**37.5 Lakhs**	**67.5 Lakhs**	**1.5 Cr**	**7 Cr**	
	cents	**km**	**km**	**km**	**km**	**km**
	58842.84655	**0**	*	**nil**	*	
	103142.0311	**0**	*	**1.08**	*	
	191695.3793	**2.76**	*	**1.38**	*	
	₹34,12,88,51,000.74	**0**	*	**0**	*	
	₹59,82,23,78,057.82	**0**	*	**₹1,62,00,000.00**	*	
	₹1,11,18,33,19,990.12	**₹1,03,50,000.00**	*	**₹2,07,00,000.00**	*	
	*	**144.07**				
₹34,18,11,88,646	**₹34,12,88,51,000.74**	**0**	*	**0**	*	
₹59,99,07,04,435	**₹59,82,23,78,057.82**	**0**	*	**₹2,33,28,000.00**	*	
₹1,15,02,34,73,991	**₹1,11,18,33,19,990.12**	**₹1,49,04,000.00**	*	**₹2,98,08,000.00**	*	

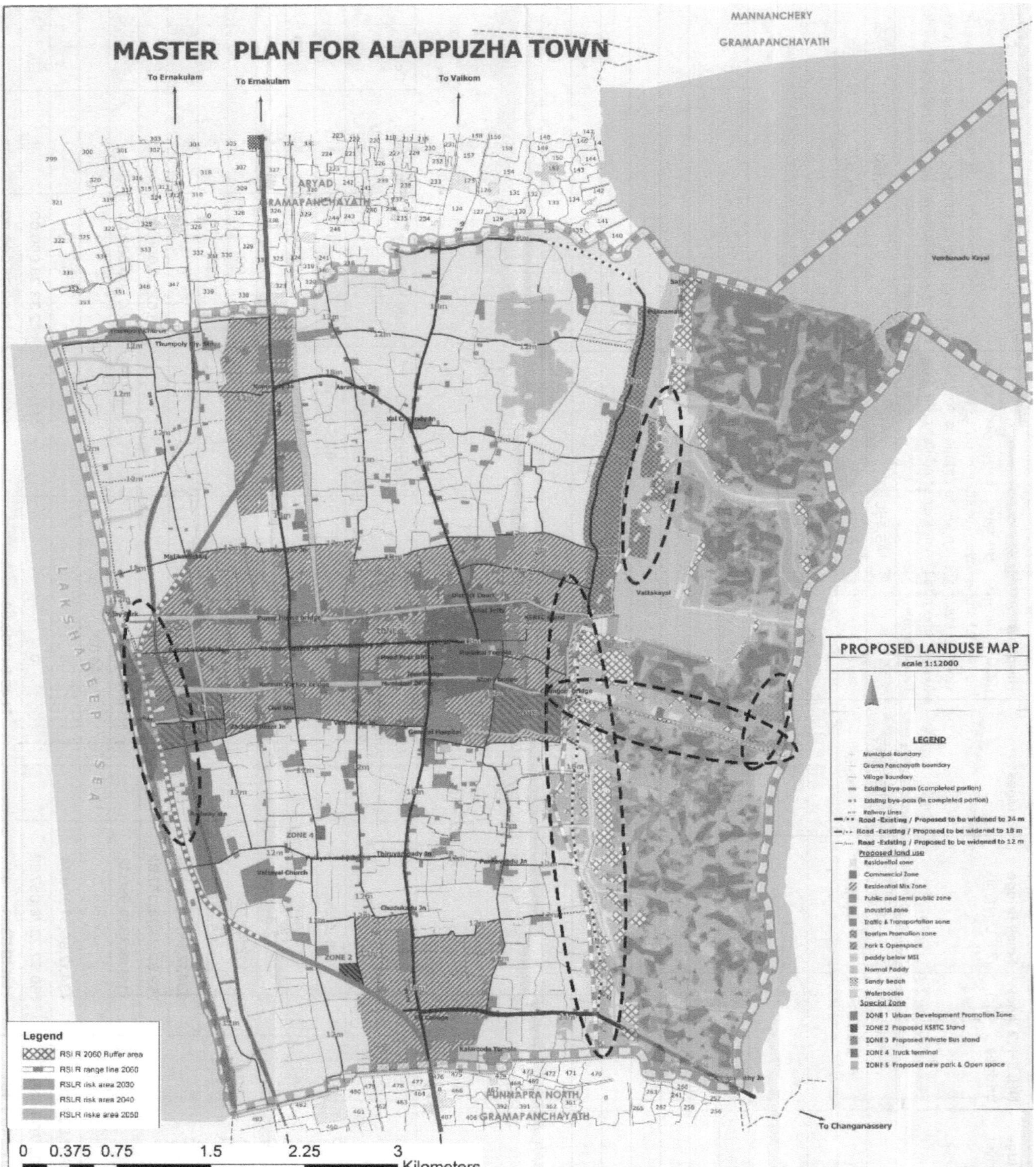

Fig. 34.40 Map showing Proposed land use and proposals 2031 overlaid with Decadal RSLR risk areas of Alappuzha Municipality

Source: Alappuzha Municipality Masterplan, Generated using ArcGIS

Encircled proposals are under the risk areas of decadal RSLR in the AlappuzhaMunicipality.

6.3 SWOT Analysis

STRENGTH

Well-connected canals:
Canals are natural drainage systems, and they help to reduce flood damage by draining excess water into the eastern sea.
(Kerala flood mitigation: Chengannur canal case study, Down to Earth, March 2020) (See Fig. 34.47)

Existing Water transportation infrastructure:
Alappuzha Municipality already has a welldeveloped water transportation system. These can be upgraded and made more efficient to promote the use of Water transport over the road, as there is conjecting in the core areas and loss of connectivity on road during a flood is observed.
(See Fig. 34.47)

Wetlands & Agriculture practices, KBSFS (Kuttanad Below Sea Level Farming System):
Paddy fields and wetlands help to capture water and regulate water underground.

Thaneermukkam barrage
(It acts as a shield to the tidal activities, and stops saltwater intrusion into the inland areas of Vembanad Lake.)

Fig. 34.41 Thaneermukkom Barrage constructed to regulate seawater into Vembanad lake in the south from its northern half

Source: TCPO Alappuzha, Alappuzha Master Plan 2031

Potential for Water-based Tourism Activities.
Promotion of canal tourism and backwater tourism (Masterplan).
Development of Nehru Trophy Area and its surroundings
Canal Beautification and Maintenance

Fig. 34.42 Kerala became the first state to provide the opportunity to enjoy seaplane rides in India

Source: TCPO Alappuzha, Alappuzha Master Plan 2031

WEAKNESS

Growth of water weeds in canals:
Since the saltwater is restricted by Thaneermukkam Barrage, the growth of water weeds took place.

Fig. 34.43 Aggressive waterweeds in canals of Alappuzha Municipality

Source: TCPO Alappuzha, Alappuzha Master Plan 2031

Low maintenance and monitoring of water resources:
Low maintenance of canals is also a reason for the growth of water weeds.

Fig. 34.44 Low maintenance of canals in Alappuzha municipality

Source: TCPO Alappuzha, Alappuzha Master Plan 2031

The land below sea level:
The low-lying region on the western side is below sea level, these areas are vulnerable to flooding and inundation by relative sea-level rise.

Low Recreational area
Recreational areas are less as per standards. Even if we take account of the proposed recreation area, it is still deficient.

Water Salinity
Water salinity and the presence of fluoride in water is an issue in the Municipality area. 80% of water need is met through the KWA pipeline (Kerala Water Authority). Salinity could hinder the growth of rice cultivation.
(See Fig. 34.48)

Pollution of Vembanad Lake
The declining area of Vembanad lake due to encroachment and pollution of Vembanad lake is an issue.
Waste from commercial and residential areas are disposed into Vembanad lake in some areas.

Fig. 34.45 Housing boat service is famous in Alappuzha, Promoting tourism

Source: TCPO Alappuzha, Alappuzha Master Plan 2031

OPPORTUNITY

Upgrading water transport
Need to develop the inland water system and increase its usage to make people used to it, keeping in mind the impending RSLR scenario.

Water-based leisure activities
Adaptation to the water-based living conditions is to be slowly introduced. Potential for waterrelated leisure activities are more in Alappuzha Municipality, this could fill the gap of recreational area.

Commercial canals
History of the Alappuzha Municipality reveals that the canals were used for commercial purposes. Rejuvenation of these activity patterns is essential in the RSLR scenario.

Fig. 34.46 Commercial canal in 1971 of Alappuzha Municipality

Source: TCPO Alappuzha, Alappuzha Master Plan 2031

Interlinkage through water transport
Alappuzha has many canals connecting to different parts of the Municipality, interlinking these canals could help in developing water transport inside the Municipality area.
(See Fig. 34.49)

Canals to reduce flood damage
Developing an inland waterway could encourage canal rejuvenation and proposals for widening and new canals. This could help regulate flood water at times needed.

Water-based Tourism
Floating and Structures on stilts are a future for RSLR risk areas of the Alappuzha Municipality. Floating markets could attract tourism too.

Fish Farming
Apart from paddy cultivation, fish farming is found to be way more reliable in the case of the RSLR scenario

THREAT

Risin sea level
Relative sea-level rise estimates are not looking well for the municipality area. Must take steps to prepare for an impending impact.

Flood prone area
Previous flood impacts imply the vulnerability of the Municipality area to flood. 38% of the total land area is prone to flood.

Vembanad lake area land encroachment
Rich biodiversity and ecological sensitivity of Vembanad lake are at risk due to land encroachment. This could also result in the lower water catchment area of the lake and during the flood, it will affect the land areas to inundate

Saltwater intrusion
Salinity in the ground would affect the water quality and impact the growth of rice cultivation (Which is a landmark of the Alappuzha region).

Loss of Connectivity during floods
During recent flood events in Alappuzha Municipality SH11 (AC road) was inundated for almost 2 months. Loss of connectivity on road was an issue

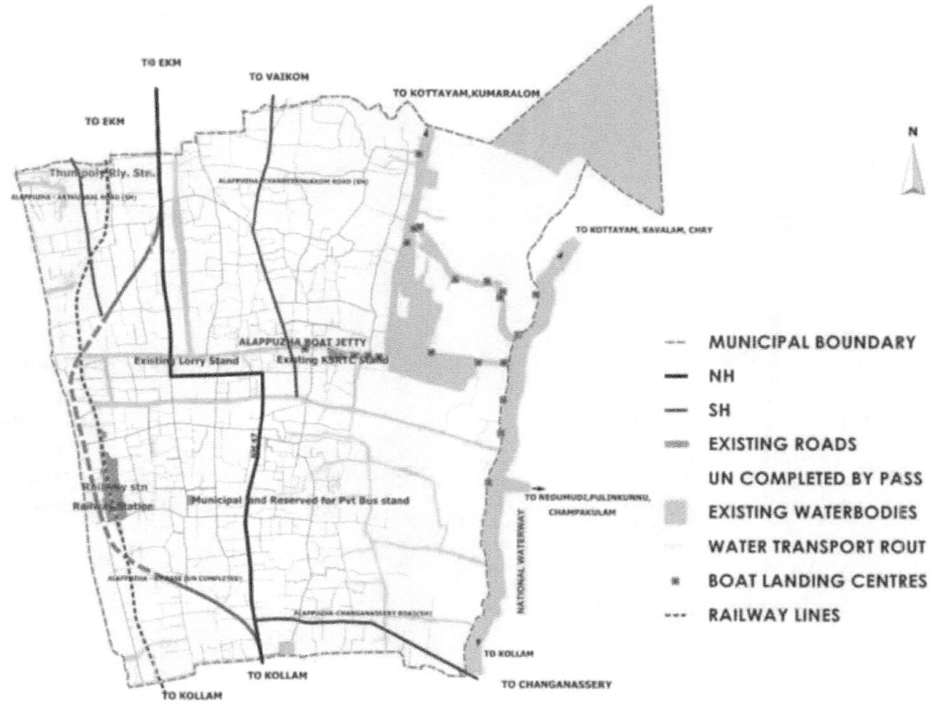

Fig. 34.47 Existing canals and Water transportation routes

Source: TCPO Alappuzha, Alappuzha Master Plan 2031

Fig. 34.48 Location of RO plants in Alappuzha Municipality

Source: TCPO Alappuzha, Alappuzha Master Plan 2031

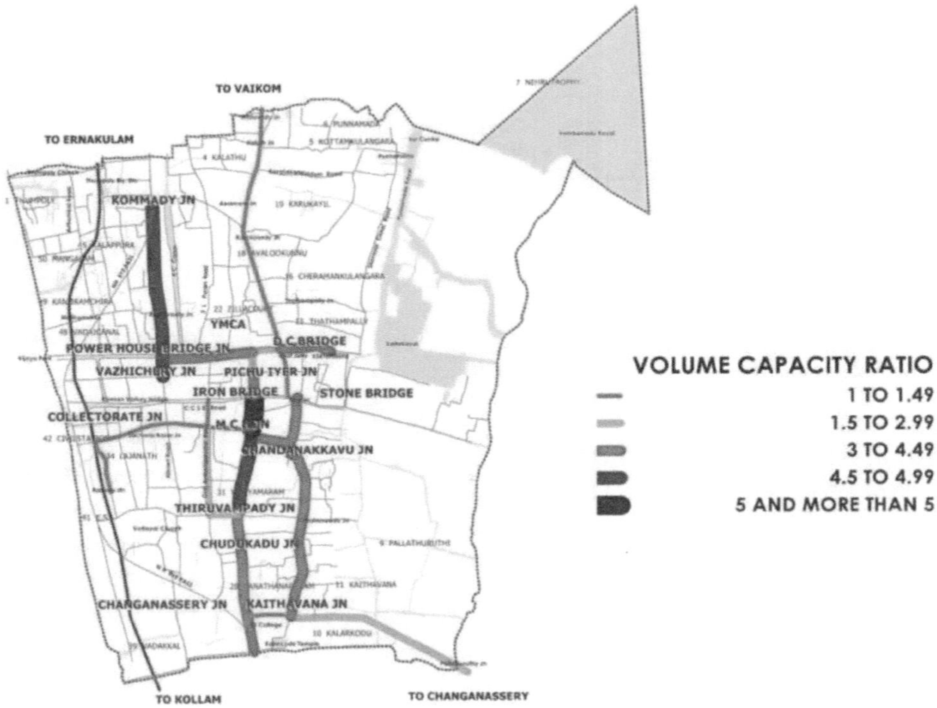

Fig. 34.49 Projected Traffic volume capacity ratio in 2031

Source: TCPO Alappuzha, Alappuzha Master Plan 2031

7. Best Practices from Secondary Study Addressing to RSLR

Name of the project	Strategies adopted	Similar Characteristics	Remarks
Case Study 1: Virginia Beach, Department of Public works, Virginia 2019 (*Sea Level Wise VBgov*, n.d.) *Source:* https://www.vbgov.com/government/departments/public-works/comp-sea-level-rise/Pages/default.aspx)	• Adaptation Framework. • Natural mitigations to preserve environmental assets and increase the use of natural and nature-based features in the city. • Prepared communities to grow educational outreach and promote economic resilience. • Engineered defenses to pursue an expansive defense network. • Watershed management.	• Low Lying region. • Adjacent river system.	Government Policy level intervention. - "Sea levelwise"

Case Study 2: New Jersey Rutgers Institute of Earth, Ocean, and Atmospheric Sciences (Willis, 2014) *Source:* https://sealevelrise.org/solutions/	• Relocating development away from the shore;
	• Accommodating natural processes by continuing occupancy and adjusting to the hazard (for example, by building houses on pilings); and
	• Protecting existing infrastructure in place:
	  ◦ Beach replenishment and dune restoration is mobile and green solution already in action at the Jersey Shore
	  ◦ Clean sediment accumulated in navigation channels has been dredged to fill up drowning wetlands
		• Low lying area
		• Effected with severe flood
			• Planned relocation from the risk areas.
			• Adaptation with the natural process

Case Study 3: **Sea level rise.org (Non-profit organization, research on sea-level rise solutions.)** (Willis, 2014) *Source:* https://sealevelrise.org/solutions/	**Elevating houses** Houses can be raised above flood levels by using six-foot tall wooden stilts or concrete blocks. **Relocating** **Building seawalls** New York City is building a $335 million flood wall in Manhattan Seawalls are built on the coast to decrease flooding from tides and storms.  **Raising roads** Raising roads above sea level can help drain water and reduce tidal flooding.  **Using beaches as barriers** Beaches and dunes can act like a natural wall that reduces the impact of storm surges 		Best Practices

<table>
<tr>
<td></td>
<td>Creating natural infrastructure
Natural structures such as barrier islands, oyster and coral reefs, mangroves, seagrass, and salt marshes can work in unison with built infrastructure, such as seawalls, to absorb storm surges
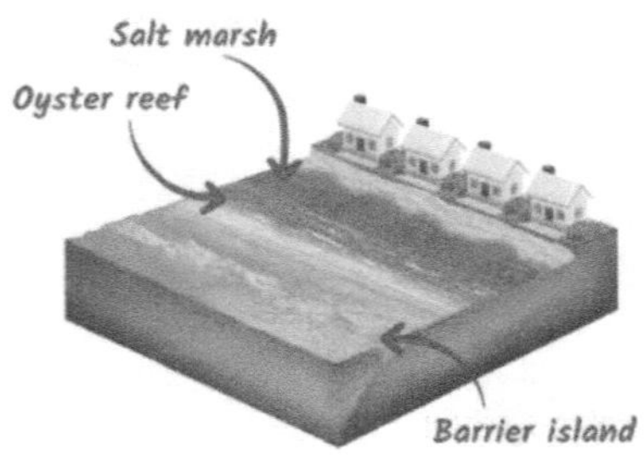
Slowing land sinkage
To inject water underground to slow land sinkage
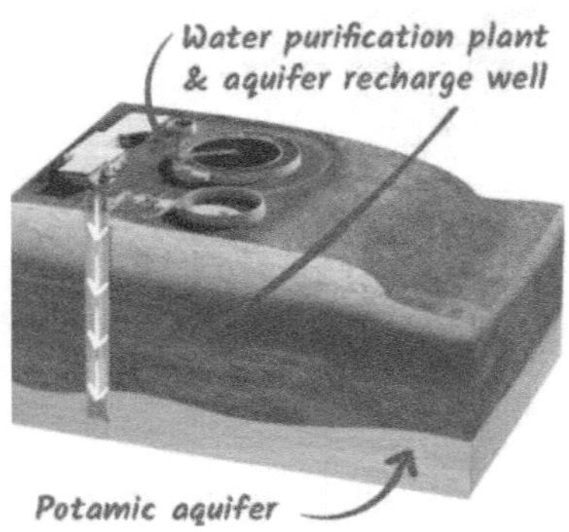</td>
<td></td>
<td></td>
</tr>
<tr>
<td>Case Study 4:
Report from the United Nation's Intergovernmental Panel on Climate Change
(Moore, 2019)</td>
<td>• Doing nothing

• Protect
(seawalls, levees, etc.),

• Accommodate
(elevating structures or insurance)

• Advance
(building a buffer out into the ocean)
</td>
<td></td>
<td>• Best practices</td>
</tr>
</table>

	• **Retreat** (moving away from the coast)  • **Ecosystem-based Adaptation** 		

8. Proposals, Strategies and Recommendations

8.1 Proposals and Strategies

The proposals are based on 3 phases with strategic actions based on the case studies and best practices for each decade from 2020. 1st Phase is 2020 – 2030, which includes Adaptation measures. 2nd Phase is 2030 – 2040, which includes Defensive measures, and the 3rd Phase is 2040 – 2050, which includes Planned retreat measures.

1st Phase: 2020 – 2030 (Adaptation)

1. **Need for a detailed understanding of the capacity and working of Thaneermukkom Barrage.**

   It acts as a shield from tidal waves. If opened, salt water may help to eradicate water weeds, improve the habitation of coastal living organisms like crabs, shrimps, etc. into Vembanad Lake.

2. **Research on the possibilities of Growing Mangroves in RSLR risk areas near the Vembanad Lakeside.**

   50 m buffer area (like that of CRZ IV-B) on the RSLR risk area of the Vembanad Lakeside area of Alappuzha Municipality could be used to try and grow Mangroves.

   Mangroves act as a natural buffer and shield from flooding and tidal waves, it also absorbs water and prevents saltwater intrusion. 1.26sq.km. area of Mangroves can be planted in the 50m buffer area.

   Mangrove plantation in Kunhuimangalam village, in Kannur as part of the Kundal Project (May 2016) supported by WTI (Wildlife Trust of India) is an example.

   Geomorphology of the buffer area is identified as an alluvial plain, and these are fit for growing Mangroves. (See Fig. 34.50)

3. **Revitalization of AC and AS canals.**

   The use of canals to lower water susceptibility during Flood events are a proven example. Flood prone areas would increase as a rise in sea level is permanent and any addition of water to it would increase the inundation area. MNREGA could be put to the task of the revitalization of canals. (See Fig. 34.50)

4. **Work on possibilities of linking canals for inner water transportation in the Alappuzha Municipality.**

   Linking canals to increase the efficiency of waterways inside the Alappuzha municipality could take the load on road transport usage and also improve the chances of maintaining connectivity through the water to other parts which otherwise is not possible through roads in cases of flooding.

5. **Start research on best practices for the desalination process for Alappuzha Municipality.**

   Water salinity and inadequate good quality water is an existing issue in Alappuzha Municipality. With the increase in RSLR, the chances of saltwater intrusion will be more severe. As a conceptual solution, a desalination plant could meet the water quality issue.

6. **Introduce the use of Sea Rice Seeds for the cultivation of rice for the below- sea-level farming system.**

   Scientists from China have developed several types of rice that can be grown in seawater intrusion land areas (Food and Agricultural Organization of the United Nations). This could help the paddy cultivation without concerning the saltwater intrusion inland. (See Fig. 34.50)

7. **Support incentives for flood resilient designs/ retrofits and minimize infrastructure vulnerability.**

   ULB can help the residents and farmers who are in risk areas by providing incentives to capacity buildings

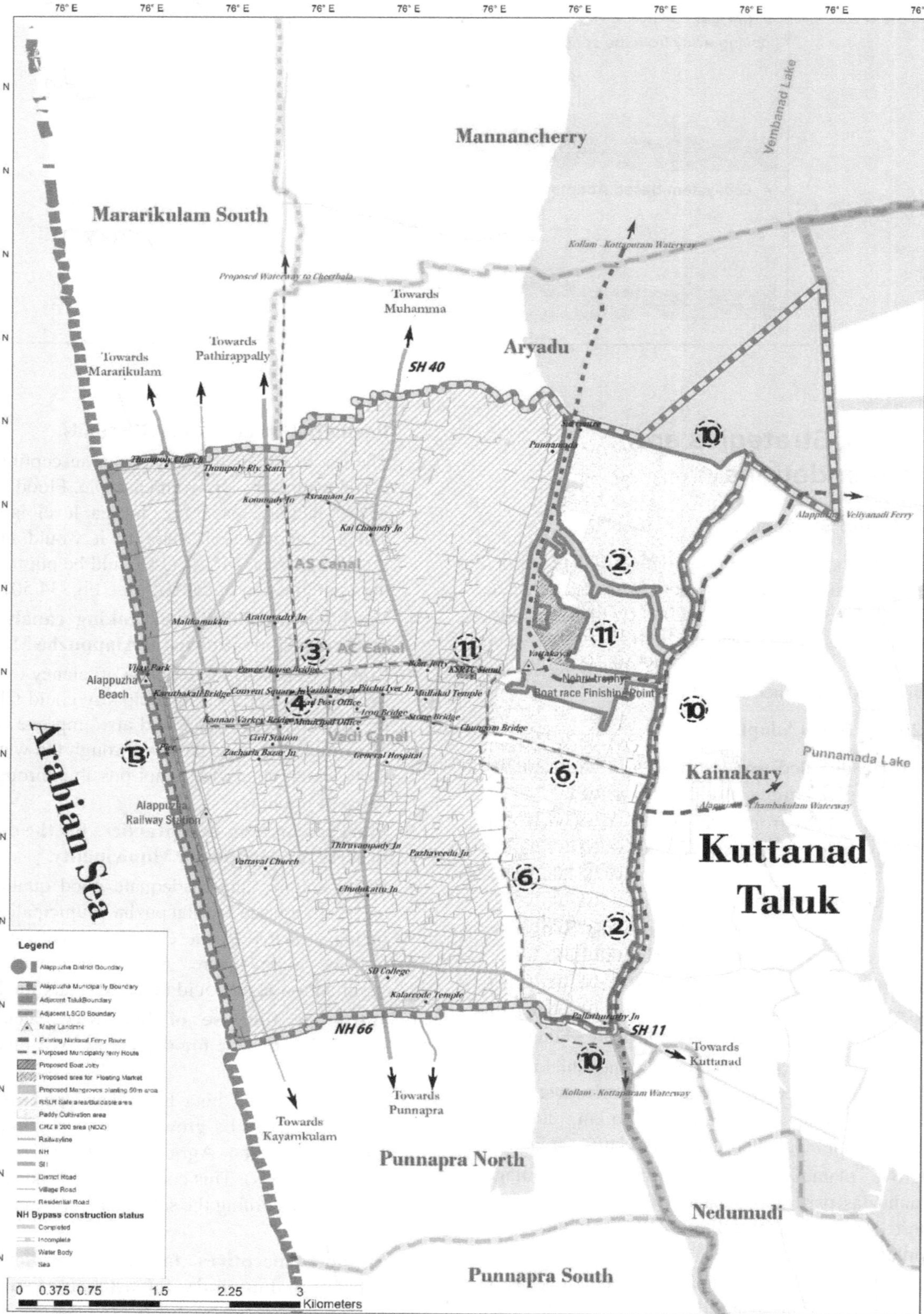

Fig. 34.50 Study Proposal Map of Alappuzha Municipality

and flood resilient constructions. In the case of farmers "Kisansree" scheme could be used to provide financial help if their land is lost to RSLR.

2nd Phase: 2030 – 2040 (Defensive)

If the decadal RSLR projections are found to be true for the year 2030, then certain steps are needed to be considered (no. 8 and no. 9) as prime importance:

8. **Adapted structures to regulate new building and development codes and standards specific for RSLR risk areas for new construction and redevelopment.**

   Elevation of land would be an option to protect critical infrastructures like airports, roads, or railways. Raising adaptation for buildings is an advantage since it does not restrict from building or rebuilding. Such structures to be made at least 2.5 m high from the then relative sea level.

9. **The strict control of the encroachments in the Vembanad lake area.**

   Need to include strict regulations on the Vembanad Lake area in coastal policies, to restrict encroachment and damage to the ecosystem. KCICM (Kerala Centre for Integrated Coastal Zone Management) can incorporate such policies in their initiatives.

10. **Levees should be made alongside the Vembanad lake areas of the Alappuzha Municipality.**

    Levees help to limit the tidal influence and impacts due to flooding. Levees near the lake edges are recommended to help at times of flood. (See Fig. 34.50)

11. **New boat jetty, Water-based activities near the Vattakayal area to promote tourism.**

    Converting old KSRTC stand to the boat jetty in the Alappuzha Municipality. Projects like Floating Market like that in Thailand could be incorporated in the tourism proposals to incorporate water-based tourism attractions. (See Fig. 34.50)

3rd Phase: 2040 -2050 (Planned Retreat)

If the RSLR projections for the year 2040 are met:

12. **Using beach barriers/Beach nourishment techniques.**

    Planning for coastal protection measures along Kerala coast (Department of ocean engineering, IIT Madras, Chennai, August 2002) has mentioned beach nourishment techniques (Dutch digging) for the coasts of Alappuzha, but these studies are outdated. These studies suggest beach protection measures specific to Alappuzha District but are outdated. Hence the need for recent studies along the beachside and lakeside of the Alappuzha District is necessary.

13. **Install geo tubes to prevent erosion near beach edges.**

    50% of beach area in Alappuzha Municipality is getting affected in 2050, these areas could also be affected by erosion. To prevent such impacts, the installation of Geotubes can be a solution. (See Fig. 34.50)

14. **Planned retreat strategies.**

    Rehabilitation to safer/higher areas. Raised structures are to be made mandatory near risk areas. These measures should also ensure that the existing pollutants are not left as it is, like sewage, garbage, septic tanks when planning to retreat to safe lands.

9. Way Forward and Conclusion

9.1 Way Forward & Framework Development

- Predictions made in the study are based on available secondary data and assumptions.

  The Developed framework for the Vulnerability assessment can be tested and calibrated for the other Coastal ULB's to assess their vulnerability to RSLR.

- According to ADRC (Asian Disaster Reduction Centre):

  Risk = (Hazard × Vulnerability)/Capacity.

  So, the need to assess the capacity of the Alappuzha Municipality could lead to more understanding of the risk of RSLR. Presence of Thaneermukkom Barrage, Sea walls are not taken into considerations in the projection of SLR by the Climate Central data. These capacities can greatly affect the outcomes of the RSLR.

- The use of Lidar in the coastal areas could improve the accuracy of elevation data of the DEM, DSM. Since the relative sea-level rise is happening in centimeters and the highest accuracy of open-source data (DSM-ALOS) is +5 meter.
- Hydrological modeling should be carried out to assess the shoreline change (erosion and accretion) and the Vembanad lake with software like Mike21. (Already being used by INCOIS)
- Apart from the vulnerability of infrastructure other impacts in different sectors like poverty, diseases, food/livelihood, access to clean water, etc. can be studied using this framework.
- Data Sharing between Departments is necessary, and the need for a DBMS (Database management system). Decadal field data is available in the IDRB - Coastal Division (The Irrigation Design and Research Board) which could be used to understand the growth of relative sea level.

- The need for a well-developed Vulnerability analysis with precise and accurate data is of the utmost necessary as the RSLR impacts are very near.

9.2 Conclusion

The study concludes that the RSLR is an impending disaster that needs to be properly addressed, ensuring improved coping capacity and better information. If not dealt with it now, it will be too late then as the RSLR scenario is seeming to be inevitable. Vulnerability assessment of the critical infrastructures could save them from the RSLR impacts. The framework has shown significant efficiency in assessing the vulnerability of physical infrastructure to the impending RSLR.

Significant damage loss is seen with the impacts of RSLR in the physical infrastructure sector (which includes only roads, bridges, land-use, and building footprint) in the study area ULB (Alappuzha Municipality). The need to ascertain the critical infrastructure damages and to prepare for the impending disaster is of utmost importance.

CHVI assesses the vulnerability of coasts to the RLSR, but in the case of Alappuzha District, the potential area at risk of inundation was observed to be distant from the coast but has a connection to sea through Vembanad Lake. So, there is a need to incorporate more parameters that could address the impacts of RSLR through estuaries and lakes which are in contact with the sea.

It is very important to know where and when the vulnerable areas are before the disaster strikes as to protect it. Because every $1 spent on disaster mitigation saves $6 in disaster relief (Natural Hazard Mitigation Saves: 2019 Report – National Institute of Building Sciences) in terms of physical infrastructure (Buildings)(MMC, 2017).

These forecasts on risk areas can help the decision-makers and authorities to plan their actions accordingly for existing infrastructure and upcoming infrastructure projects.

The framework developed is flexible enough to incorporate other sectors like economic, environmental, social, etc., which are identified as a gap. Future research can fill the gap to the existing framework.

The framework developed in the study can be tested and improved in the other coastal areas in Kerala, India to assess their vulnerability to RSLR.

References

1. Council, K. S., Technology, S., & November, K. (2014). *Coastal Zone Management Plan of Alappuzha District , Kerala. November.*
2. Dwarakish, G. S., Vinay, S. A., Natesan, U., Asano, T., Kakinuma, T., Venkataramana, K., Pai, B. J., & Babita, M. K. (2009). Coastal vulnerability assessment of the future sea level rise in Udupi coastal zone of Karnataka state, west coast of India. *Ocean and Coastal Management*, *52*(9), 467–478. https://doi.org/10.1016/j.ocecoaman.2009.07.007
3. Issn, I. T. (2019). *Bollettinodel Comitato Glaciologico Italiano-ser. 3.42*(1).
4. MMC. (2017). *Natural Hazard Mitigation Saves: 2017 Interim Report - National Institute of Building Sciences.* https://www.nibs.org/general/custom.asp?page=mitigationsaves
5. Moore, R. (2019). IPCC Report: Sea Level Rise Is a Present and Future Danger. In *Natural Resources Defense Council.* https://www.nrdc.org/experts/rob-moore/new-ipcc-report-sea-level-rise-challenges-are-growing
6. Rao, K. N., Subraelu, P., Rao, T. V., Malini, B. H., Ratheesh, R., Bhattacharya, S., Rajawat,
7. A. S., & Ajai. (2009). Sea-level rise and coastal vulnerability: An assessment of Andhra Pradesh coast, India through remote sensing and GIS. *Journal of Coastal Conservation*, *12*(4), 195–207. https://doi.org/10.1007/s11852-009-0042-2
8. *Sea Level Wise VBgov*. (n.d.).
9. Sugawara, E., & Nikaido, H. (2014). Plinth area rates. *Antimicrobial Agents and Chemotherapy*, *58*(12), 7250–7257. https://doi.org/10.1128/AAC.03728-14
10. TNN. (2019). Sea-level rise may hit central Kerala. In *The Times of India* (pp. 3–5).
11. Willis, J. (2014). Sea level rise. In *Encyclopedia of Earth Sciences Series* (pp. 743–746). https://doi.org/10.1007/978-0-387-36699-9_163

Chapter 35

Increasing Risk of Glacial Lake Outburst Floods (GLOFs) as a Consequence of Climate Change in the Himalayan Region

Ahmed Faraz Khan
National University of Science and Technology, Pakistan

Abstract

The greater Himalayan Mountains hosts the largest snow covered area outside the Polar Regions and serve as the source for some of the major systems of the world. The region acts as the life line for approximately 10% of the world's population. The terrain is geologically active, highly susceptible to climate change processes and plays a significant role in global hydro-meteorological cycles and biodiversity. With the increasing impacts of climate change the glaciers and ice caps during the past few decades, people living in the Himalayas are vulnerable to a higher risk of floods, avalanches and glacial lake outburst floods (GLOFs). A review of the research work carried out by earlier workers is drawn to understand the history and science of GLOFs and its potential risk to the communities in the Himalayan belt, particularly in Pakistan.

Keywords: GLOFs, Himalayan glaciers, climate change

1. Introduction

The Himalayan Mountain system is among the most active tectonic belts of present day on earth. The extensive Himalayan system traverses along eight countries: Afghanistan, Bangladesh, Bhutan, China, India, Myanmar, Nepal, and Pakistan and sustains ten river systems – the Amu Darya, the Brahmaputra, the Ganges, the Indus, the Irrawaddy, the Mekong, the Salween, the Tarim, the Yangtze, and the Yellow. It is an east-west trending mountain chain with exceptionally high relief. It covers an area of about 2,400 km within the 3,500 km stretch of the Hindu Kush-Karakoram-Himalayan (HKH) ranges. The mountain system is a house to many diverse cultures and an extensive diversity of flora and fauna. They act as a barrier to the northerly winds during winters and confine the monsoon system which is the major source of precipitation for South Asia (Fig. 35.1).

Due to rapid industrialization and anthropogenic activities during the past few decades there has been an increase in the global temperatures and rapid retreat of the snow cover globally. Glaciers are nature's tool for indicating variation in the air temperature over the earth's history. These ice masses are being extensively studied to understand the recent process of global warming. Following the termination of the glacier advance during the Little Ice Age (more than 100 years ago), a general trend of depletion and retreat has been observed in glaciers around the world. The Himalayan glaciers are no exception to the prevailing impacts of global warming and climate change. As a response to the warming atmospheric temperatures, lakes are formed at terminus of the ice mass or within the main body of the glaciers (Fig. 2). Due to rapid accumulation of melt water in these lakes the end moraine may burst causing a catastrophic flooding situation in the downstream riparian areas (Richardson and Reynolds, 2000).

Over the past few years the HKH region has experienced a number of glacial lake outburst floods (GLOFs) events that have resulted in considerable damage to the community, their livelihoods and infrastructure. With the increase in global temperatures such events are likely to become frequent posing a greater risk of potential losses to the Himalayan terrain. Recent studies have revealed that appearance of incipient lakes and widening of the existing ones. Pakistan heavily depends on its riverine system which has its source in the Himalayan glaciers. The northern part of the country has experienced a number of outburst floods due to melting glaciers. It is therefore important to understand the impacts of

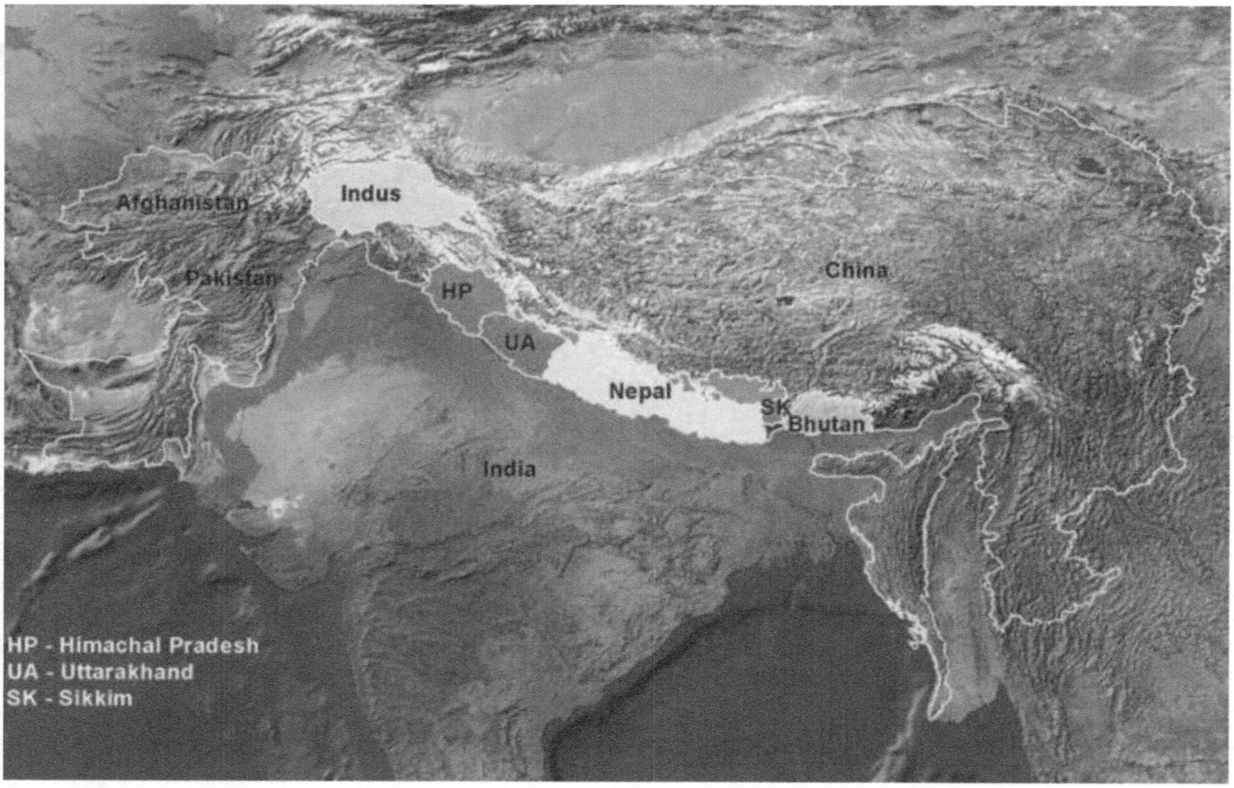

Fig. 35.1 Glaciers and glacial lakes along the entire stretch of the Himalayas (Bhutan, China, India, Nepal and Pakistan)
Source: ICIMOD and National Partners (1999-2005)

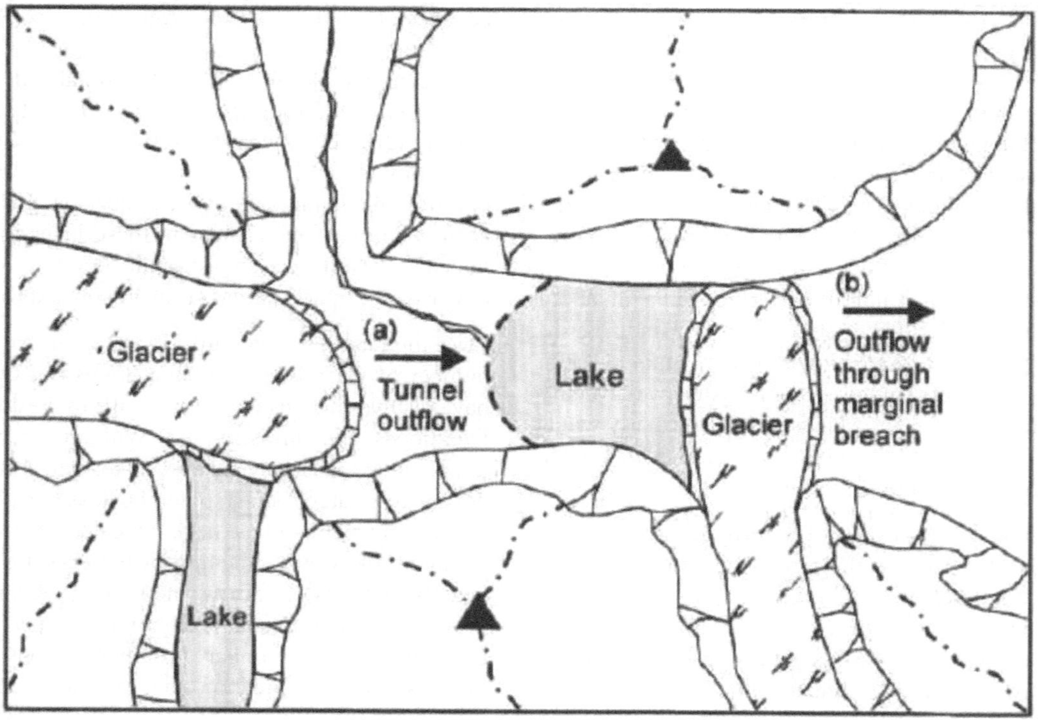

Fig. 35.2 Two patterns of ice-dammed/ glacial lakes and breach mechanisms (Walder and Costa, 1996)

climate change on these glaciers and determine the potential GLOFs sites in the country. The present study focuses on summarizing the work done on the risk of glacial lake outburst floods (GLOFs) in the Himalayas in general and the probability of such catastrophic events in Pakistan.

2. Literature Review

The global trend of rise in earth's surface temperatures and its subsequent impacts on the ice masses around the globe, researchers are now keenly focused on studying the glacial environments and their response to the changing climate. One of the most threatening processes that are taking place at an accelerated rate in the glaciers is the development of glacial lakes. As the Himalayan range serves as a homeland for quarter of the population in Asia a large population is at risk from potential catastrophic outburst flooding. There is a greater need to study the glacial lakes and the potential risk site in the area.

In the decade of 80s different scientists worked on analyzing the outburst flooding events in the Himalayan terrain. Goudie (1981) and Miller (1984) reported that a total of 339 disasters occurred in 1980 along the Karakorum Highway and the most destructive were outburst flooding events. Hewitt's work in 1982 indentified that about during the past two centuries about 35 catastrophic GLOF events have occurred in the Karakorum Range.

With the onset of new millennium research in the field of GLOFs has gained an accelerated pace keeping in view the large population that is vulnerable to the effects these hazardous events. ICIMOD and various UN organizations have been actively carrying out extensive studies on the probable hazard site of GLOFs in the Himalayas and the communities that are at a higher risk of suffering the impacts of these events. These organizations alongwith the support of governments and local communities have been organizing communities awareness and training activities to enhance their capacity for the future hazard events in their vicinity. Farukh et al., carried out meteorological data analysis for the estimation of snow cover in the Pakistani part of the Himalayas. Richardson and Reynolds (2000) worked on the carried out an overview of glacial lakes in the Himalayas. Bajracharya et al., (2007) worked on the impacts of climate change on Himalayan glaciers and glacial lakes through a study conducted by joint collaboration by ICIMOD and UN. They worked on the temperature variations and the increasing retreat of the Himalayan glaciers. This has caused an increase in the number of glacial lakes and outburst flooding events. Rasul et al., (2008) carried out a research on the melting glaciers in the HKH region of Pakistan. They analyzed the meteorological data to determine the isotherm shift in the high elevation areas and addressed the anthropogenic influences on Siachen glaciers. UNDP launched a project in 2008 to monitor the situation of the Himalayan glaciers and glacial lakes which aimed at minimizing the risk reduction of the communities in the region. Ashraf et al., (2011) carried out a research to identify the glacial lakes outburst floods in the Karakorum Range. Their prime focus was on the use of GIS and satellite data to critically analyze the effects of climate change in the region. In the same year ICIMOD carried out a study on the GLOFs and expected damages from these lakes in Nepal to extensively identify the vulnerable areas and develop an effective coping mechanism for the communities.

3. Himalayan Glaciers—The Third Pole of the World

The glaciers of the Hindukush-Karakoram-Himalayas are the third largest ice reserves outside the Polar region with an estimated area of about 33,000 sq.km. According to a study conducted by ICIMOD, UNEP and Asia-Pacific Network for Global Change Research (APN) during the time span from 1999-2003, approximately 15,000 glaciers and about 9000 glacial lakes were documented in the Himalayan region. Dyhrenfurth 1955 termed it as the "Third Pole" on the basis of high concentration of glaciated water and ice cover. These glaciers are one of nature's greatest renewable warehouses of fresh water. Asia's largest perennial rivers such as the Indus, the Ganges and the Yangtze are fed by the meltwater from the Himalayan glaciers which are source of fresh water to nearly quarter of the human race settled in the downstream areas of the river basins. Thus the HKH region is the lifeline for about one third of the world's population dwelling in the region.

4. Global Climate Change Trend

Glacial environments are highly susceptible even to minute changes in the atmospheric conditions since thermal variations are more evident at high altitudes. The earth's atmospheric temperature changes over time and is generally influenced by the amount of solar radiation (terrestrial) entering our atmosphere, volcanic activity (geothermal), and rate of fossil fuel utilized (human activity). The rise and fall in the amounts of solar energy encroaching on the earth (mainly in the far north during summer) is a foremost driving mechanism behind climate change (Milankovitch 1920). The Intergovernmental Panel on Climate Change (IPCC) reported an increase in the concentration of atmospheric CO2 from a pre-industrial value of about 280ppm to 379 ppm in 2005. Stud of the ice cores indicated that this level has exceeded by far the natural range (180 to 300 ppm) over the last 650,000 years. projection of the data for future atmospheric trend

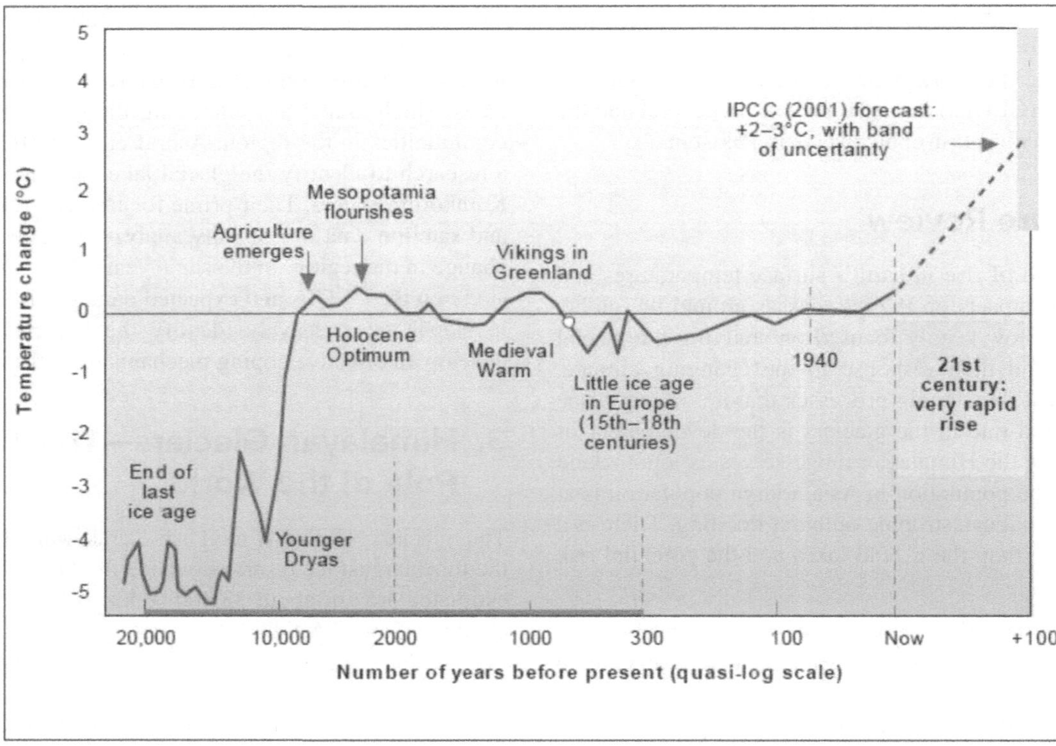

Fig. 35.3 Earth's surface temperature variation curve over the past 20,000 years

Source: Intergovernmental Panel on Climate Change

indicate that the level of atmospheric CO2 is likely to double from its pre-industrial level within the next 50-100 years (Fig. 35.3).

The rapid increase in global population and human activities has significant adverse effects on the earth's atmosphere. According to IPCC (2001a) there has been an increase in the average surface temperature of the earth between a range of 0.3ºC and 0.6ºC within a short time span of about a hundred years, with all the 12 warmest years ever documented over the globe taking place in 1990s and the recent ongoing decade. This is predicted to continue during the present century (Fig. 35.4). By 2100 the predicted increase in temperature over the Indian subcontinent is expected to be between 3.5 ºC and 5.5ºC with an even higher increase is anticipated for the Tibetan Plateau (Lal, 2002). The increasing trend of thermal variation around the globe has resulted in an accelerated rate of formation of new glacial lakes and thus an increased risk of potential glacial lake outburst floods.

Global trend is also replicated by the shifting thermal regime in different parts of Pakistan and the most evident in northern high elevation meteorological stations located in the HKH. Rasul et al., 2008 carried out the analysis of meteorological data (daily maximum temperatures) from 1981-2008 to determine the isotherm dynamics. Keeping the 30°C as the reference isoline it was observed that by the end of 1990s, there is an upward shift of about 300m the reference isoline from its position of 1981-85 in the HKH region (Fig. 35.5). According to Pakistan Meteorological Department, the northern parts of the country have experienced an increase in temperatures by 0.8 and 1.5 degrees centigrade between 1990 and 2010.

5. GLOFs and its Impact on Mountain Communities in Pakistan

Due to the rugged topography and political constraints, the Himalayan belt has long been isolated from scientific investigation. Lack of sufficient data makes it difficult to understand and indentify the potential risks from natural hazards to the community residing in the area. Researchers working on the domain of GLOFs, however, are making use of satellite imagery to monitor the response of glaciers to rising temperatures and have identified the expansion of existing lakes and the initiation of new ones. A number of glacial lakes are marked as potential risk of outburst floods in the region.

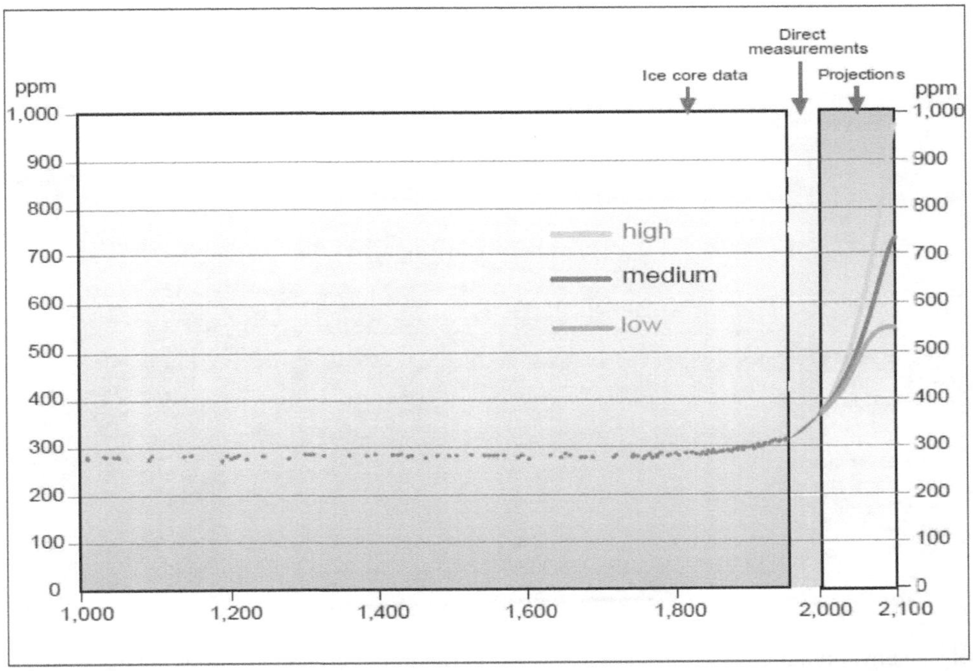

Fig. 35.4 Trend of the concentration of CO_2 in Earth's atmosphere from year 1000 to year 2000 (data obtained from polar ice cores and direct atmospheric measurements over the past few decades. 2000–2100 projections of the concentration are based on the IPCC's six illustrative SRES scenarios and IS92a

Source: Intergovernmental Panel on Climate Change

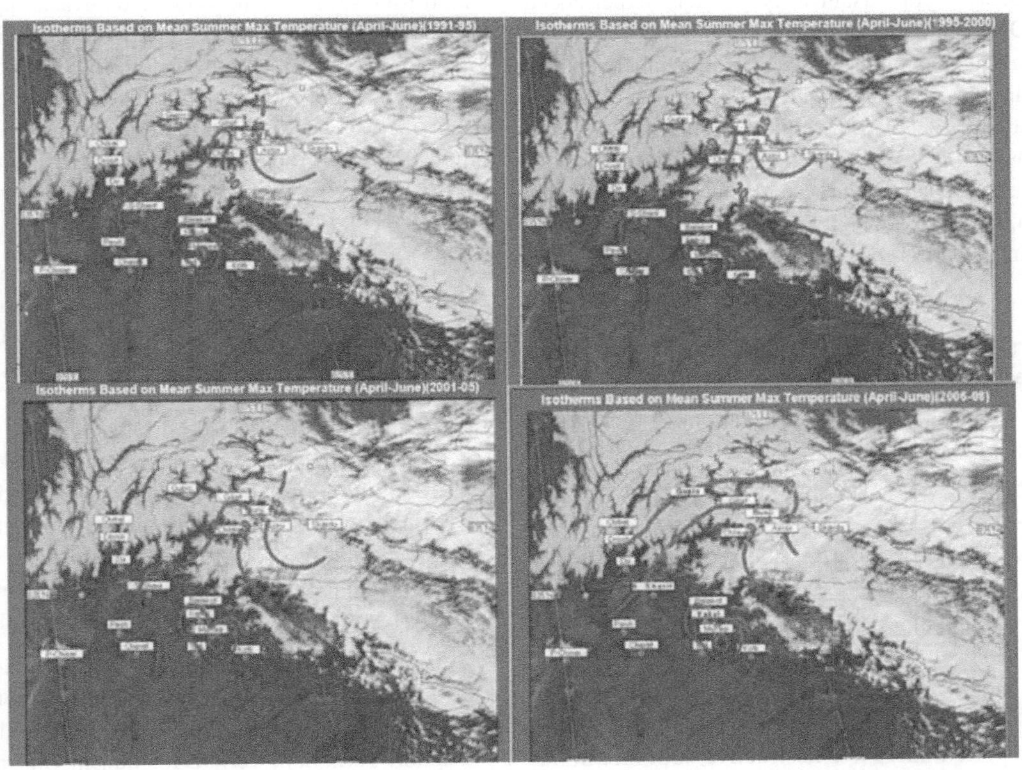

Fig. 35.5 Shifting isothermic dynamics along southern slopes of HKH ranges showing a general northward trend (Rasul et al., 2008)

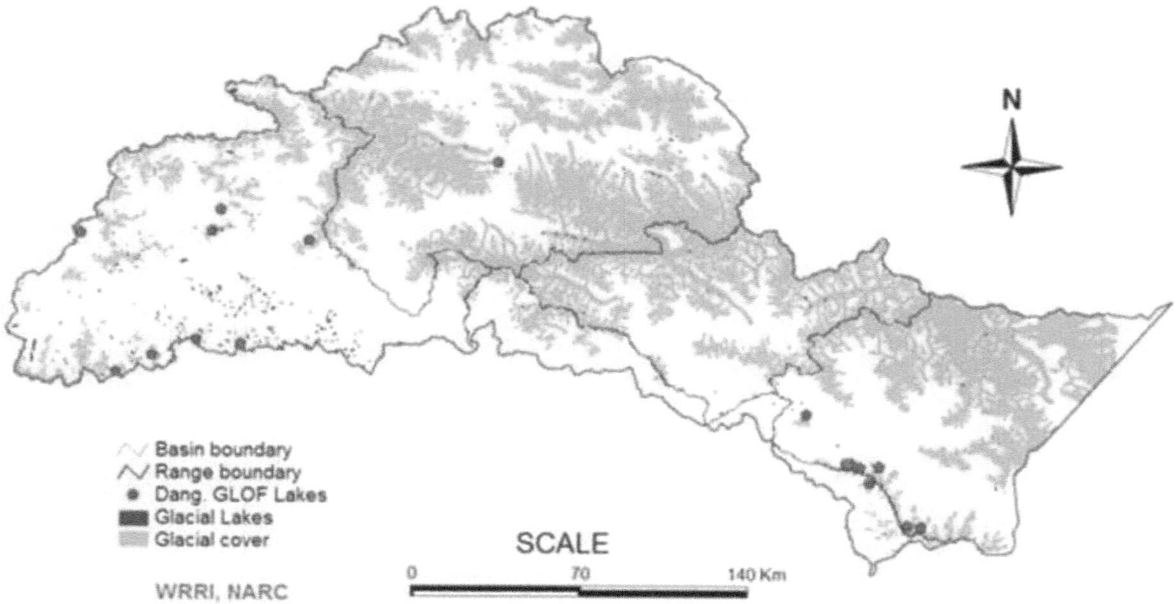

Fig. 35.6 Glacial Environment and Potential GLOF Lakes in Different Karakoram Basins, Pakistan (Arshad et al., 2011)

Glacial Lake Outburst Floods (GLOFs) is becoming a major threat to the communities settled in the Himalayan terrain. The entire belt has experienced an increased number of such devastating hazards, for instance Nepal has experienced about 24 GLOF events in the recent times. About 23 % of the Karakoram Range is under widespread glacial cover. The range accommodates about 2,398 glaciers supplied by Shyok River basin (about 37 %) followed by Hunza (34 %) and Shigar river basin (24 %). Siachen, Hispar, Biafo, Baltoro and Batura are some of the largest glaciers present in this range (Fig. 35.6). Table 35.1 gives a summary of the major glacial lakes in the different river basins of the Karakorum Range.

Table 35.1 An Overview of the Glacial Lakes in Major River Basins of Karakoram Range

Basins	Total Lakes	Lakes Area (km^2)	Major Lakes	Potential GLOF Lakes
Gilgit	614	39.2	380	8
Hunza	110	3.2	47	1
Shigar	54	1.1	11	—
Shyok	66	2.7	31	6
Indus part	43	1.6	23	1
Total	**887**	**47.7**	**492**	**16**

Hewitt, 1982 reported thirty five destructive out-burst floods to have occurred for the Karakoram Range during the past two hundred years, growth and disappearance of Virjerab Lake in Hunza due to glacial shift, and occurrence of a series of GLOF events in the upper Hunza valley within short time spans during 2008 had devastating impacts of the communities. These glacial lakes vary in their style based on their location in the glacial system and can be categorized into various risk profiles on detailed study (Table 35.2). Satellite data analysis have shown the presence of about 887 lakes in the Karakorum range out of which 492 are major lakes and 16 are spotted as probable GLOFs sites in the future (Arshad et al., 2011).

Recently research is being carried out under the Pakistan GLOF Project in different Himalayan districts of the country to probability of GLOFs events and carryout the risk profiling of the vulnerable communities in the vicinity of the glaciers. one of the findings of this study pinpointed Chitral district as a potential hazard site for GLOFs. It is the largest districts of the Khyber Pakhtunkhwa Province of Pakistan located at the foothills of the Hindukush Range. The district has been experiencing outburst flood events throughout its history. The most devastating event that is said to have occurred in the area is dated to about 350 years back. The catastrophic event initiated from Chiantar Glacier in Broghil valley.

Chitral hosts about 542 glaciers out of which 187 glaciers have developed glacial lakes. About 20 potential sites for outburst floods have been marked in the valleys of Golain, Laspoor, Yarkhoon, Terich, Reshun in Chitral. The increase in frequency of outburst floods in the region can be indicated

Table 35.2 Summary of Different Types of Glacial Lakes in Karakoram Range of Pakistan

Type	Gilgit	Hunza	Shigar	Shyok	Indus	Total
Cirque	53	—	—	2	—	55
End-moraine dammed	100	4	—	12	12	128
Lateral moraine dammed	49	3	1	3	6	62
Erosion	283	20	—	26	13	342
Supraglacial	2	55	30	11	4	102
Valley	125	24	2	8	6	165
Blocked	2	4	21	4	2	33
Total	**614**	**110**	**54**	**66**	**43**	**887**

by the fact that within a short time span of 12 years the district has witnessed more than 8 GLOF events. Some of the major ones include outburst floods of Yarkhon Lasht (2003), Brep (2005), Sonoghor valley (2007), Bindo Gol (2010), Boni (2011). The latest addition to the list of these devastating events occurred on July 31, 2013 in Reshun valley with its source from the Reshun Glacier. It resulted in devastating impacts on the community in flash of seconds. About 40 houses, 30 animal corrals, 20 shops and cultivated land including crops and orchards were wiped away. According to Dr. Rasul, Chief Meteorologist, outburst flood event generated by the Buni glacier caused severe damage to houses in its path. Gulkin glacier on the Karakorum Highway experiences an average of two to three GLOF events every year. Gulkin glacier feeds the newly formed Attabad Lake in Hunza which is another potential threat for the downstream population in case the landslide generated lake breaches. Passu glacier has also been undergoing a rapid retreat of about 20 m over the past 15 years. A similar phenomenon is being observed at the Baltoro glacier which is confirmed by the disappearance of most of the low lying tributaries. Liligo and the Silili glaciers, the low lying tributaries were once fed the Baltoro glaciers are disrupted now. Another tributary of the same glacier has vanished completely.

Due to the increasing intensity and frequency of outburst flood event, certain positives have been taken by the Government and concerned departments to closely monitor the variation in the glacial environment and device methods for community preparedness if such an event triggers in the future. For glacier monitoring in the HKH region Pakistan Meteorological Department has installed about 14 automatic weather stations at different elevation points. With the installation of these gauging stations the ground data obtained can help in assessing the real time situation of our glacial terrain and timely strategies can be devised for minimizing the increasing risk of glacial lake outburst floods in Pakistan.

6. Conclusion

Earth's climate is undergoing a rapid change due to increasing impacts of natural and anthropogenic activities during the past few decades. As a result of the shifting temperature towards a warmer trend glaciers are experiencing a much faster rate of depletion and retreat. This scenario has generated many new glacial lakes and caused the expansion of the existing ones in the glacial terrains throughout the globe. These glacial lakes are one of the major threats to mountains community as they are unstable and burst out resulting in outburst floods. The Himalayan region being the largest snow covered area outside the Polar region is also vulnerable to the increasing risk of Glacial lake outburst floods. The regions has experienced some of the devastating outburst events in the recent past. It is therefore the foremost need of the Himalayan nation to develop strategies and mechanisms for close monitoring of the glaciated terrain and cooperate in reducing the frequent devastating impacts of GLOFs on the communities vulnerable to this natural hazard.

7. Recommendations

There should be a systematic approach for monitoring the varying glacial responses along the entire stretch of the Himalayas.

Mutual cooperation and data sharing among the nations in this belt can help in effective assessment of the potential risk of glacial lake outburst floods.

Satellite imagery should be combined with the data obtained from various weather stations to develop a more realistic output of outburst floods in the region.

Community awareness and training programmes must be organized to make people familiar with the increasing risks and enhance their coping capacity.

Focus must be laid on developing an effective early warning system for GLOFs to minimize the loss of life, livelihood, infrastructure and damage to environment in the region.

References

1. Ashraf. A., Roohi. R., Naz. R., and Mustafa. N., 2011. Identification of Glacial Flood Hazards in Karakoram Range using Remote Sensing Technique and Risk Analysis. Scientific Journal of COMSATS – SCIENCE VISION. Vol.16 and Vol. 17 (January 2010 to December 2011).
2. Bajracharya. S. R., Mool. P. K., and Shrestha. B. R., 2007. Impact of Climate Change on Himalayan Glaciers and Glacial Lakes: Case Studies on GLOF and Associated Hazards in Nepal and Bhutan. Report by International Centre for Integrated Mountain Development (ICIMOD) in cooperation with United Nations Environment Programme Regional Office Asia and the Pacific (UNEP/ROAP) Kathmandu, Nepal.
3. Dawn News article, 2013. Glacial Lakes Threat to Villages. Retrieved on November 13, 2013 from http://www.dawn.com/news/788579/glacial-lakes-a-threat-to-villages-says- expert
4. Furrukh. B., Rasul. G., 2010. Estimation of Average Snow Cover over Northern Pakistan. Pakistan Journal of Meteorology, Vol. 7, Issue 13.
5. Glacial Lake Outburst Flood (GLOF) Risk Reduction in the Himalayas. Project Facts – Himalayas UNDP. March 2008 – February 2009Rasul. G., Dahe. Q., and Chaudhry. Q. Z., 2008. Global Warming and Melting Glaciers along Southern Slopes of HKH Ranges. Pakistan Journal of Meteorology, Vol. 5, Issue 9.
6. ICIMOD, 2010. Formation of Glacial Lakes in the Hindu Kush-Himalayas and GLOF Risk Assessment Report.
7. IPCC (2001a) Climate Change 2001: The Scientific Basis, Contribution of Working Group I to the Third Assessment Report of the Intergovernmental Panel on Climate Change. Cambridge: Cambridge University Press
8. Lal, M. (2002) Possible Impacts of Global Climate Change on Water Availability in India, Report to Global Environment and Energy in the 21st Century. New Delhi: Indian Institute of Technology.
9. Pakistan Meteorological department, PMD, 2012. Pakistan GLOF-Project, District Chitral, Pakistan.
10. Report on Glacial Lakes and Glacial Lake Outburst Floods in Nepal. International Centre for Integrated Mountain Development, Kathmandu, March 2011
11. Richardson. S.D., Reynolds. J. M., 2000. An overview of glacial hazards in the Himalayas. Quaternary International, Vol. 65/66, Pg. 31–47.
12. Walder, J. S., Costa, J. E., 1996. Outburst floods from glacier-dammed lakes: the effect of mode of lake drainage on flood magnitude. Earth Surface Processes and Landforms, 21, 701–723.

CHAPTER 36

Glacial Lake Outburst Flood Modelling

A. K. Lohani
National Institute of Hydrology
Roorkee-247667

Sanjay K. Jain
National Institute of Hydrology
Roorkee-247667

R. K. Jaiswal
National Institute of Hydrology
Roorkee-247667

Abstract

A glacial lake outburst flood (GLOF) is created when water dammed by a glacier or a moraine is released. Some of the glacial lakes are unstable and most of them are potentially susceptible to sudden discharge of large volumes of water and debris which causes floods downstream i.e. GLOF. Many glacial lakes are known to have formed in the Himalaya in the last half century, and a number of GLOF events have been reported in the region in the last few decades. Due to extreme hazard potential of GLOF events, it is necessary to take into account GLOF in addition to design flood while planning, designing and constructing any infrastructure, especially water resources projects, as they are located on the path of glacial lake outburst flood wave and would be the prime target in case of GLOF.

GLOF modeling may be carried out by either i) scaled physical hydraulic models, or ii) mathematical simulation using computer. A modern tool to deal with this problem is the mathematical model, which is most cost effective and reasonably solves the governing flow equations of continuity and momentum by computer simulation. Mathematical modeling of dam breach floods can be carried out by either one dimensional analysis or two dimensional analysis. In one dimensional analysis, the information about the magnitude of flood, i.e., discharge and water levels, variation of these with time and velocity of flow through breach can be had in the direction of flow. In the case of two dimensional analyses, the additional information about the inundated area, variation of surface elevation and velocities in two dimensions can also be assessed. In the present paper, methodology for simulation of glacier lake outburst floods has been discussed.

Keywords: Glacial lake outburst flood, Mathematical models, Mike11, Himalayan Region, Geographic Information System (GIS), Potentially dangerous lake, Simulation

1. Introduction

The Indian Himalayan region is home to a numerous glaciers, which is nature's renewable storehouse of fresh water. According to a report by the International Centre for Integrated Mountain Development (ICIMOD) there are 15000 glaciers and 9000 glacial lakes in the Himalaya (Mool, 2005). The Himalayan region is intrinsically linked to global atmospheric circulation, hydrological cycle, bio-diversity and water resources and play a crucial role in shaping and influencing the Indian environmental conditions. Due to global climatic change observed during the first half of the twentieth century, a tremendous impact on the high mountainous glacial environment has been observed. It has been reported by the various researchers that many of the big glaciers melted rapidly and gave birth to the origin of a large number of glacier lakes. The accelerated global warming may accelerate the retreating/melting of glaciers in the Himalayan region and thus lead to accumulation of increasing amounts of water in mountain-top lakes. As glaciers retreat, glacial lakes located at the snout of the glacier are normally formed behind moraine or ice 'dams'. Due to the inherent instability of such 'dams', the potential of sudden outbursts/ breaches is extremely high. Such lakes could be dangerous as they may hold a large quantity of water and such outbursts can lead to a very high discharge of millions of cubic meters of water and debris in a few hours. Breaching and the instantaneous discharge of water from such lakes can cause flash floods in the downstream areas and it may lead to serious damage to life, property, agriculture, livestock, forests, eco-systems and engineering and water resources structures such as Hydro power projects

Since 1986, ICIMOD has been involved in the preparation of an inventory of glacier and glacial lakes and the identification of potentially dangerous glacial lakes. Between 1999 and 2005, ICIMOD in collaboration with partners in different countries, embarked on the preparation of an inventory of glaciers and glacial lakes, and identification of potential sites for GLOF (ICIMOD, 2010). Researchers have identified a number of glacial lakes and found that they are expanding as a consequence of climate change and glacier thinning (Reynolds 2000; Ageta et al. 2000; Benn et al. 2000). A number of studies related to assessment of potential of glacial lake outburst (Watanabe et al., 1994), estimation of the rate of expansion of a glacial lake (Sakai et al., 2000), automatic detection method of the lake surface using a normalized difference water index (NDWI) (Bolch *et al.,* 2008). Some studies have asserted the expansion rate as posing a risk of a glacial lake outburst flood (GLOF) (Quincey *et al* 2007), determining the trigger and proof strength of the damming moraine is more important.

2. Methodology

In order to identify the potentially dangerous glacial lakes, a digital database of glaciers and glacial lakes is necessary. Different image enhancement techniques are useful for the identification of the individual glaciers and glacial lakes. However, complemented by the visual interpretation method (visual pattern recognition), with the knowledge and experience of the terrain conditions, glacier and glacial lake inventories and monitoring can be done. Using the combinations of different spectral band in false colour composite (FCC) and in individual spectral bands, glaciers and glacial lakes can be identified and studied using the knowledge of image interpretation keys such as: colour, tone, texture, pattern, association, shape, shadow, etc. Different colour composite images highlight different land-cover features. The lake water in colour composite images ranges in appearance from light blue to blue to black. In the case of frozen lakes, it appears white. Sizes are generally small, having circular, semi-circular, or elongated shapes with very fine texture and are generally associated with glaciers in the case of high lying areas, or rivers in the case of low lying areas.

For glacial lake mapping digital image processing have to be applied. When applying basic techniques of multispectral classification similar to those used for the normalized difference vegetation index, NDVI (Hardy and Burgan 1999), a normalized difference water index (NDWI) for lake detection can be used (Huggel 1998).

$$\mathrm{NDWI} = \frac{B_{\mathrm{NIR}} - B_{\mathrm{blue}}}{B_{\mathrm{NIR}} + B_{\mathrm{blue}}} \quad (1)$$

With the help of NDWI maps, manual delineation of lakes can be done.

3. Identification of Potentially Dangerous Glacial Lakes

The criteria for identifying potentially dangerous glacial lakes are based on field observations, processes and records of past events, geo-morphological and geo-technical characteristics of the lake and surroundings, and other physical conditions. Criteria such as area/volume of lake, breaching evidences, condition of lake and its surrounding etc. are being generally used by the various researchers (Rai et al., 2012, Jain et al., 2012, Lohani et al., 2013). For this purpose remote sensing data in conjunction with Google Earth provides a useful input. It is important to verify the location of the glacial lakes in the upstream catchment. Weather the lake is located in abalation zone of glacier and connected with mother glacier? Is there any breaching evidences and area of lake is expanded significantly in the recent past. After identifying the vulnerable lake from GLOF point of view, breaching of the biggest lake may be considered for the simulation of GLOF. Arc-GIS and ERDAS Imagine software may be used to delineate cross-sections of the stream. For this purpose the vector layer of the stream and the buffer lines along the stream on the both side of stream at the distance of 1 km can be created. This helps in creation of the cross-section layer. ERDAS Imagine Software is generally used to overlay DEM of basin and vector layer of cross-section. The Spatial Profile Viewer in ERDAS allows visualizing the reflectance spectrum of a polyline of data file values in a single band of data (one-dimensional mode) or in many bands (perspective three-dimensional mode).

4. Glacial Lake Outburst Flood (GLOF) Simulation

Glacial lake outburst floods increase to peak flow than gradually or abruptly decrease to normal levels once the water source is exhausted. Outburst flood peak flow is directly related to lake volume, dam height and width, dam material composition, failure mechanism, downstream topography, and sediment availability. In order to get the maximum GLOF peak at any location, the breaching of moraine dams of above glacial lakes have to be considered along with channel routing.

In order to get the maximum GLOF peak at the outlet, the outbursting or breaching of glacial lakes is considered along with channel routing. There is no estimate available for the estimation of volume of glacial lakes in Himalaya from their water spread areas. However, Huggel et al. (2002) has given some estimates for glacial lakes in Swiss Alps. In the absence of information on the volume of potentially dangerous glacial lakes, it is considered appropriate (Rai et al., 2012, Jain et al., 2012, Lohani et al. 2013) to use the same relationships developed for the lakes in Swiss Alps for estimating the water volume for the lakes in this area. The empirical relations used by Huggel et al. (2002) are:

$$\text{The lake volume } V = 0.104\,A^{1.42} \quad (2)$$

where V is the lake volume in m^3 A is the lake area in m^2.

$$\text{The lake depth } D = 0.104\,A^{0.42} \quad (3)$$

where D is the lake depth in m

Due to possible out bursting or breaching of lake existing in the study area may result a surge of flood at the outlet. Therefore, in order to estimate the maximum GLOF peak at the outlet, the out bursting or breaching of lake has to be considered along with channel routing. Thus, the estimation of GLOF (Sharma et al., 2009) provides the flood hydrograph of discharge from the dam breach and maximum water level at different locations of the river downstream of the dam due to propagation of flood waves along with their time of occurrence.

The essence of dam break modelling is hydrodynamic modelling, which involves finding solution of two partial differential equations originally derived by Barre De Saint Venant in 1871. These equations are:

$$(\partial Q/\partial X) + \partial(A + A_0)/\partial t - q = 0 \quad \text{(continuity equation)} \quad (4)$$

$$(\partial Q/\partial t) + \{\partial(Q^2/A)/\partial X\} + gA\,((\partial h/\partial X) + S_f + S_c) = 0 \quad \text{(Momentum equation)} \quad (5)$$

where, Q = discharge; A = active flow area; A_0 = inactive storage area; h = water surface elevation; q = lateral flow; x = distance along waterway; t = time; S_f = friction slope; S_c = expansion contraction slope and g = gravitational acceleration

The mathematical models which approximately solve the governing flow equations of continuity and momentum by computer simulation are the cost effective modern tool for the dambreak analysis. MIKE11and DAMBRK computer programs have been developed in the recent past; however these computer programs are dependent on certain inputs regarding the geometric and temporal characteristics of the dam breach. The state-of-art in estimating these breach characteristics is not as advanced as that of the computer programs, and therefore these are the limiting factors in this analysis.

5. GLOF Simulation

Potential dangerous lake upstream of the project site is generally considered for the estimation of the outburst flood at the lake outlet and project site. Lakes dammed behind moraines present a hazard of outburst floods if the moraine dams are sufficiently steep and sharp-crested to rapidly downcut if they are overtopped by a wave or running water, and if there is a source above the lake of possible icefall, avalanche, rockfall, or flood which creates a wave or a sudden rise in water level. If glacial retreat continues, it is likely that more glacial dammed lakes may appear in the future. Equations for flood flow estimates from glacier dammed are reviewed in Table 36.1:

Table 36.1 Empirical Equations for the Calculation of Peak Flows from Moraine and Ice Dam Failures

Authors	Type of Dam	Equation
Costa and Schuster (1988)	Moraine	$Q_p = 0.0013PE^{0.6}$
Costa and Schuster (1988)	Glacier	$Q_p = 0.0000055PE^{0.59}$
Clague and Mathews (1973)	Glacier	$Q_p = 75PE^{0.67}$
Costa (1988)	Glacier	$Q_p = 113PE^{0.64}$

Where PE is the potential energy of the reservoir in Joules (product of dam height (m), reservoir volume (m^3) and specific weight of water (9800 newtons/m^3), V = reservoir storage (10^6 m^3) and Q_p = peak discharge (m^3/s).

Using these equations, it is important to realize that flood waves will attenuate downstream, and that the behaviour of flood is largely dependent on the topography of the valley downstream. Floods will attenuate most quickly where floods traverse low-gradient floodplains, whereas they may travel considerable distance at high discharge in steep and narrow valleys.

Instead of above mentioned empirical equations, hydrodynamic modelling using MIKE-11 model has been successfully applied by the various researchers for simulation of the glacial lake outburst flood in the Himalayan region. River reach from glacial lake to the project site is typically represented in MIKE-11 model by a number of cross sections at an appropriate interval (generally 1 km), developed from DEM. The volume of the lake is calculated using Equation 2 taking lake area as of the lake estimated from remote sensing data. The depth of the lake is computed using equation 3. The breaching starts at top of the dam and continues up to the breach invert level. The mode of failure for glacial lake outburst generally considered as overtopping. Equation given by FERC, 1987 is used for the computation of the average breach width. Equation given by FERC, 1987 also provides the breach development time. Appropriate value of the Manning's roughness coefficient may take considering the boulder beds and hilly terrain of Himalayan.

Glacial lake outburst flow modelling process is nothing but approximation of a physical phenomenon through which the physical phenomenon and its effects can be studied for water resources structure design and flood management. In GLOF modeling, assumptions are mainly associated with the breach parameters, especially, breach width and breach depth, which has impact on flood peak and arrival times. In general, glacial lake bursting mechanism and formation of breach in glacial lakes are not fully understood. Furthermore, the high velocity associated with GLOF can cause significant scour of channels associated with bed as well as bank erosion. Change in the channel cross section due to GLOF is neglected due to limitations in modelling such a complicated physical process. Generally, GLOF creates a large amount of transported debris and this may be accumulate at constricted cross sections, where it acts as a temporary dam and partially or completely restricts the flow, resulting variation in flood peak arrival time. This aspect has also been neglected due to limitations in modelling of such a complicated physical process. These limitations have an effect mainly on the conservative side. Even with the assumptions and limitations outlined above, hydrodynamic modelling serves very useful purpose, as it provides reasonable estimate of glacial lake outburst flood, thus enabling the appropriate estimation of design flood.

6. GLOF Simulation—An Example

In the present example, GLOF modelling has been carried out to demonstrate its impact at the basin outlet. This glacial lake is at a distance of 25.0 km from the basin outlet. River from glacial lake to the outlet has been represented in the model by a number of cross sections at an interval of 1 km, developed from DEM. The locations of cross sections are shown in Fig. 36.1 and cross sections at an interval of 1 km. downstream of lake are shown in Fig. 36.2.

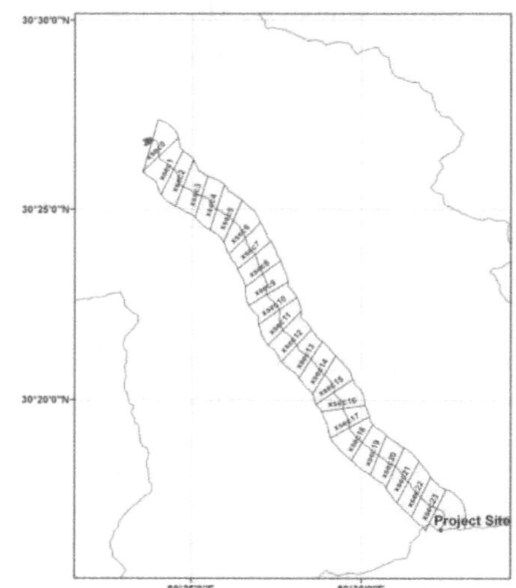

Fig. 36.1 Location of cross sections at 1 km interval downstream of lake

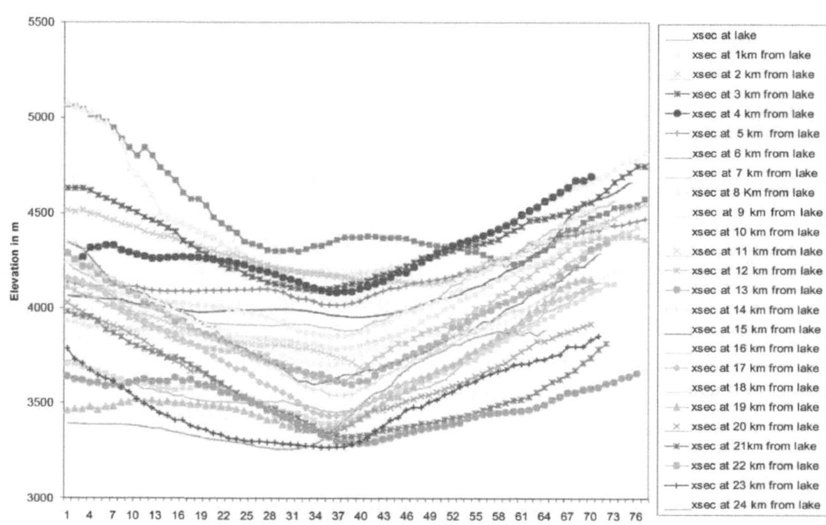

Fig. 36.2 Cross sections at 1 km interval downstream of lake

The altitude of the Lake is 4339 m. The surface area of the lake and altitude are given in Table 36.2. The volume of the lake is calculated using lake area as 125000 m^2 and it comes out to be 1.79 Mm^3. The depth of the lake is computed to be 15 m. The breaching starts at top of the dam and continues up to the breach invert level. The mode of failure for glacial lake outburst considered is overtopping. The breach invert level should be taken as two-third to three-fourth of the height of dam below its top level, therefore, breach depth has been taken as 12 m which is approximately three fourth of 15 m. For average breach width, equation given by FERC, 1987 has been used. As per this equation, breach width comes out to be in the range of 30-60 m. Using this range of width and going through the satellite image of the lake, the breach width has been taken as 50 m. The breach development time has been computed using Equation given by FERC, 1987 and it is from 0.1 to 1 hour. Mool et al. 2001 has given breach development time as 1000 to 2000 seconds. Considering the size of the lake, it is appropriate to assume the breach development time of 30 minutes to get the maximum GLOF peak. The side slope has been taken as 1H: 1V. The Manning's roughness coefficient has been taken as 0.05 considering the boulder beds and hilly terrain of Himalayan.

Table 36.2 Elevation-area relationship of Glacial Lakes

Glacial lake (volume 1.79 Mcum)	
Elevation (m)	**Surface area (m^2)**
4323.000	115000.00
4324.000	115000.00
4339.000	125000.000

The 100 year return period flood value available at outlet is distributed along the study reach on the basis of the catchment areas of the contributing tributaries. There are mainly four streams which are meeting with the stream coming from lake up to outlet. The distributed flood values have been taken as lateral inflows of magnitude corresponding to 100 yr flood with their impingement locations in the River, as given in Table 36.3.

MIKE 11 software was applied for generation of flood hydrograph for breach width of 50 m. These flood hydrographs at just downstream of the lake and near outlet are shown in Fig. 36.3. The total flood peak and its travel time from GLOF site is given in Table 36.4. It can be seen from the Table 36.3, that the GLOF peak is 808.66 cumec and the

Table 36.3 Lateral inflow considered corresponding to 100 year return period flood

Sl. No.	Location	100 year flood m^3/s	Adopted lateral inflow m^3/s	Area (km^2)
1.	Upstream stream from lake	215.11	215.1143	138.87
2.	7.85 km down stream from lake	325.37	110.2602	71.18
3.	10.25 km down stream from lake	370.31	44.93747	29.01
4.	16.75 km down stream from lake	474.62	104.3119	67.34
5.	20.21 km down stream from lake	602.63	128.0121	82.64
6.	25.21 km down stream from lake	1082.00	479.4259	309.5

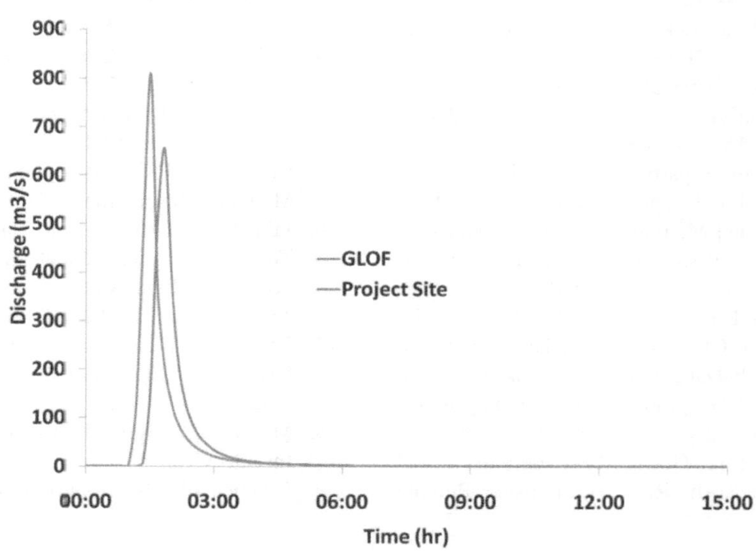

Fig. 36.3 GLOF Hydrograph at Outlet

Table 36.4 Flood peak due to glacial lake outburst for breach width of 50 m

Location	Distance (km) from Glacial Lake	Breach Width 50m	
		GLOF peak (m^3/s)	Travel time (hr-min)
Just d/s of lake	0	808.66	00-00
Outlet	25.0	652.87	0-20

same get mitigated to 652.9 cumec at the outlet for breach width of 50 m. The time of travel of flood peak from the lake site to outlet is about 20 minutes.

7. Conclusions

The integration of visual and digital image analysis with a Geographic Information System (GIS) can provide very useful tools for the study of glacial lakes. A glacial lake inventory for a basin has to be prepared. After that it is very much required to identify vulnerable lake from the prepared inventory. Once potentially dangerous lake is identified then simulation of flood using various input parameters is required. MIKE 11 software was successfully applied for generation of flood hydrograph due to GLOF by various researchers. Proper combination of the floods due to precipitation and GLOF is required to be considered for the design flood estimation of the various upcoming projects in the Himalayan region.

References

1. Ageta, Y., Iwata, S., Yabuki, H., Naito, N., Sakai, A., Narama, C. and Karma, T. (2000). Expansion of glacier lakes in recent decades in the Bhutan Himalayas. In: Debris-covered glaciers, proceedings of a workshop held in Seattle, Washington, USA, Sept 2000. International Association of Hydrological Sciences, Wallingford, 170–172.
2. Bajracharya S. R., and Mool P. K., (2006). Impact of global climate change from 1970s to 2000s on the glaciers and glacial lakes in Tamor Basin, eastern Nepal, ICIMOD.
3. Bajracharya, B., Shrestha, A.B. and Rajbhandari, L. (2007). Mountain Research and Development,Vol. 27 No. 4,Glacial Lake Outburst Floods in the Sagarmatha, 336–338.
4. Bajracharya, S. R. , Mool, P. K. and Shrestha B. R. (2008). Global Climate Change and Melting of Himalayan Glaciers. In Melting Glaciers and Rising sea levels: Impacts and implications, The Icfai's University Press, India, 28.
5. Bajracharya, S. R., Mool, P. K. and Joshi, S. P. (2009). Spatial Database Development of Glaciers and Glacial Lakes in the identification of Potentially Dangerous Glacial Lakes of Nepal using Remote Sensing and Geographic Information Systems.
6. Budhathoki, K. P., Bajracharya, O. R. and Pokharel. B.K. (2010). Assessment of Imja Glacier Lake outburst Flood (GLOF) Risk in Dudh Koshi River Basin using Remote SensingTechniques, Journal of Hydrology and Meteorology, Vol. 7, No. 179–80.
7. Campbell, J. G. (2005). Inventory of glaciers, glacial lakes and the identification of potential glacial lake outburst floods (GLOFs) affected by global warming in the mountains of India, Pakistan and China Tibet Autonomous Region, Technical Report of APN Project 2004.
8. Frey, H., Huggel, C., Paul, F.and Haeberli, W.(2010). Automated detection of glacier lakes based on remote sensing in view of assessing associated hazard potentials, 10th International Symposium on High Mountain Remote Sensing Cartography, 263–264.
9. Grabs, W. E. and Hanisch, J. (1993). Objectives and prevention method of Glacial Lake Outburst Floods (GLOFs), Snow and Glacier Hydrology (Proceedings of the Kathmandu Symposium, November 1992), IAHS Publ. no. 218,344-345.
10. Hardy, C. C., and Burgan, R. E. (1999). Evaluation of NDVI for monitoring moisture in three vegetation types of the western U.S. Photogrammetric Engineering and Remote Sensing, 65: 603–610.
11. Huggel, C. (1998). Periglaziale Seen im Luft- und Satellitenbild. Diploma Thesis, Department of Geography, University of Zurich.
12. Huggel, C., Ka¨a¨b, A., Haeberli, W., Teysseire, P. and Paul, F. (2002). Remote sensing based assessment of hazards from glacier lake outbursts: a case study in the Swiss Alps. Can Geotech J 39: 316–330.
13. ICIMOD (2001). Inventory of Glaciers, Glacial Lakes and Glacial Lake Outburst Floods, Monitoring and Early Warning Systems in the HKH Region (Nepal), 111–112.
14. ICIMOD (2005). Glacial Lakes and Glacial Lake Outburst Floods in Nepal, 57–59.
15. ICIMOD (2010).Formation of glacial lakes in the Hindu Kush-Himalayas and GLOF risk assessment. In: Ives JD, Shrestha RB, Mool PK (eds) International Centre for Integrated Mountain Development, Kathmandu, May 2010.
16. Jain, S. K., Lohani, A. K., Singh, R. D. , Chaudary, A and Thakural, A. K. (2012).Glacial lakes and glacial lake outburst flood in a Himalayan basin using remote sensing and GIS . Nat Hazards 2012) 62:887-899.
17. Lohani, A. K., Jain, S. K. and Singh, R. D. (2013) .Peak Discharge Estimates of Glacial Lake Outburst Floods ,India Water Week- 2013.
18. Mool, P. K., Bajracharya, S. R., and Joshi, S. P., (2001). Inventory of Glaciers, Glacial Lakes and Glacial Lake Outburst Floods, Monitoring and Early Warning Systems in

the Hindu Kush Himalayan Region: Nepal, ICIMOD & UNEP RRC-AP.

19. Palikhey,E.(2011). Assessment of Glacial Lake Outburst Flood using Satellite Imagery: A case study of Lumding Glacier, Nepal, 224–226.
20. Rai, N. N., Bakshi, S. S. and Garg, V. (2012). Integrated Glacial Lake Outburst Flood (GLOF) Study For Lohit Basin, Arunachal Pradesh, India Water Week 2012 – Water, Energy and Food Security : Call for Solutions, 10-14 April 2012, New Delhi.
21. Raj, K. B. G. (2010). Remote sensing based hazard assessment of glacial lakes: a case study in Zanskar basin, Jammu & Kashmir, India. Geomat Nat Hazards Risk 1(4):341-343.
22. Raj, K. B. G, Remya, S. N. and Kumar, K. V. (2013). Remote sensing-based hazard assessment of glacial lakes in Sikkim Himalaya ,Current Science, Vol. 104, No. 3, 359.
23. Yao (2010). Glacial distribution and mass balance in the Yarlung Zangbo River and its influence on lakes. Chin Sci Bull 55(20): 2075–2078.

CHAPTER 37

Comparative Analysis of Developed, Developing and Underdeveloped Countries in Regards to Climate Change and Global Policies

Aanchal Pundir
Student, Institute of Vocational Studies, GGSIPU

Abstract

The earth's average temperature is around 15 degrees Celsius. (Earth.Org, 2020) In recent years there have been natural fluctuations in the climate, but scientists say temperatures are now rising faster than many other times. This change is driven largely due to increased carbon dioxide emissions into the atmosphere and additional human activities. Climate-change consequences are projected to hinder economic development, make poverty reduction more complicated, additionally erode food security, create new poverty traps, and much more. This paper attempts at understanding climate change and its physical, psychological, and socio-economic impacts and efforts at doing a comparative analysis between developed, developing, and under-developed countries of the world, Sweden, India, and Niger, respectively. The analysis is done on the grounds of effective implementation of climate change policies in the countries and addresses the loopholes of these policies. There is also an examination and exploration of the two major subsidiaries of the United Nations Framework Convention on Climate Change- The Kyoto Protocol, and the Paris Agreement within this manuscript. The paper then strives to give solutions that are specific to the respective country followed by giving general solutions at large. After evaluating the situations of different countries, the most important conclusion that can be made is that we need a bigger push to change everything about how the world is currently running. It is down to all of us, to change our ways and make conscious efforts with our everyday activities. All the policies and solutions are required to be executed properly and effectively for climate change evolution in a coordinated way.

Keywords: climate change, renewable energy, carbon emissions, policies.

1. Introduction

"Climate risk is a systemic risk, one that can be managed with coordinated global policy action," the Swiss Re report said.

Climate change is defined by the Intergovernmental Panel on Climate Change as "any change in climate over time, whether due to natural variability or as a result of human activity." (Intergovernmental Panel on Climate Change [IPCC], 2007b). Climate change is one of the most significant developmental challenges faced by the world today and each day we come across even more distressing news about the continuing and expedited depletion of the environment, rich biodiversity, and ecological assets. Vulnerability to climate change and natural catastrophes has a strong linkage with poverty which is why the repercussions of climate change are more severe and notable for the disadvantaged. The challenge of climate change is frequently attributed to defining global concerns of our time, which acts as a key threat to sustainable advancement, and it has the potential to create multiple stresses over any country's growth trajectory.

Using natural resources wisely and judiciously is always regarded as a person's moral responsibility. Being a dynamic and developing economy, India needs to strike a balance between maintaining the essence of sustainable development of the country and environmental protection. The decisions that India, being a developing country, makes on investment in energy systems in the following years will have a significant influence on global climate change over the coming century as we enter a phase of higher economic growth. India has been an enthusiastic participant in the discussions, actions, and consultations on climate change. To mitigate climate change India has adopted many policies, acts, and projects. In 2008, India inaugurated the National Action Plan on Climate

Change, which involved eight national missions to address climate change challenges. (*India First Biennial Update Report to the United Nations Framework Convention on Climate Change*, n.d.)

Considering current regulations, they are certainly lacking to succeed at the Paris Agreement objectives of limiting warming and fortifying the resilience of vulnerable occupants. India is significantly defenseless against the effects of the worldwide environment emergencies, and especially powerless against outrageous warmth because of low per capita income, social imbalance, and a substantial dependence on agriculture. India would lose 5.8% of its working hours because of heat stress by 2050, which is comparable to 34 million full-time jobs out of a sum of 80 million around the world. (Tripathi, 2019)

The UN explains, affordable, scalable solutions are now available to enable countries to leapfrog to cleaner, more resilient economies. The pace of change is quickening as more people are turning to renewable energy and a range of other measures that will reduce emissions and increase adaptation efforts. (*Goal 13: Climate Action - SDG Tracker*, n.d.) The world's average surface temperature is projected to rise over the 21st century and is expected to transcend 3 degrees Celsius this century. (n.d.) Greenhouse gas emissions are higher than 50 percent than in 1990 and global warming is causing long-lasting modifications to our climate system. Climate change also intensifies disasters and resisting it is essential to ensure our survival and the wellbeing of future generations. The yearly average financial misfortunes from environment-related fiascos are in the hundreds of billions of dollars. This isn't to refer to the human effect of geophysical catastrophes, which are 91% environment-related, and which somewhere in the range of 1998 and 2017 killed 1.3 million individuals and left 4.4 billion harmed.

Human activities contribute to climate change by causing variations in Earth's atmosphere in the amounts of greenhouse gases, cloudiness, and aerosols. Modern human activities release too many greenhouse gases [carbon dioxide (CO2), methane (CH4), nitrous oxide (N2O), and the halocarbons] into the atmosphere at the same time, which generates sudden global warming on a scale never encountered before. These gases accumulate in the atmosphere, causing concentrations to increase with time. (*FAQ 2.1 - AR4 WGI Chapter 2: Changes in Atmospheric Constituents and in Radiative Forcing*, n.d.)

By and large, Earth is simply 7 to 9 degrees Fahrenheit hotter today than it was during the last ice age. The new expansion in normal temperature a few degrees is as of now prompting worldwide changes in precipitation, snow and ice extent, and extreme climate, like heavy downpours, heat waves, and extreme storms. Today, Earth is warming at a much quicker rate than it has warmed over 7,000 years since the last ice age. If the current pace of warming isn't eased back, Earth's temperature is on target to increase by an extra 7 degrees Fahrenheit or more by the year 2100. That implies that Earth will get hotter over a few decades rather than over a few thousand years. (*Humans Are Causing Global Warming*, 2019)

3. Impact of Climate Change Globally

The complex variations that are now affecting our planet's weather and climate arrangements, is climate change. These repercussions extend well beyond an increase in temperature, altering ecosystems and communities. Effects are probably going to contrast in both magnitude and pace of progress in various continents, nations, and regions. A few countries will probably encounter more antagonistic impacts than others. The ability to adjust to climate change can influence how climate change affects people, communities, nations, and the global population. Many global leaders reiterate the fact that by 2050 they are going to half the emissions, but science says we need to get there by 2030, which means we have little time left to act over the climate crisis before it is too late, which can be done by adjusting transitions away from fossil fuels to renewable energy.

- Ice is dissolving around the world, particularly at the Earth's poles. This incorporates mountain glaciers, ice sheets covering West Antarctica and Greenland, and Arctic Ocean ice which adds to the sea-level rise. Worldwide ocean levels are rising 0.13 inches a year (3.2 millimeters), and the ascent is happening at a quicker rate as of late.
- Rising temperatures are influencing natural life and their habitats. Numerous species are migrating, that is, a few butterflies, foxes, and snow-capped plants have relocated farther north or to higher, cooler territories.
- Precipitation (rain and snowfall) is on a rise across the globe, on average. However, a few districts are encountering more serious drought, expanding the danger of rapidly spreading wildfires, lost harvests, and deficiency of drinking water. (*Global Warming Effects*, 2021)

3.1 Physical Impact of Climate Change

Numerous global climate changes that were predicted by scientists in the past are now happening in the present and they are anticipated to continue in the future as well if immediate actions are not taken in that regard. Climate change influences and obliterates crops and food production causes sickness and death, annihilation and loss of economic livelihoods, and migration of climate refugees. Individuals in developing countries like India, China, etc might be the most vulnerable to health risks around the world, but climate change presents critical dangers to wellbeing even in affluent countries like the United States. The physical system of the planet can be seen in the dissolving of the poles, which simultaneously cause glacial regression, snow melting, warming and defrosting of permafrost, flooding in rivers and lakes, dry droughts in streams and lakes, coastal disintegration, sea-level rise, and outrageous natural phenomena. (*Impacts of Climate Change*, n.d.)

The physical impact of climate change threatens our day-to-day life experiences like warming temperatures which will lead to warmer days and more frequent and prolonged heat waves (further leading to dehydration, heat strokes, cardiovascular or respiratory diseases), changes in precipitation, increases in the frequency or intensity of some extreme weather events. The spread of climate-sensitive diseases is another major concern as it will rely upon both environmental and non-environment factors, for example, land use, financial and social conditions, pest control, admittance to medical services, and human reactions to disease hazard. Vector-Borne diseases are ailments that are communicated by disease vectors, which incorporate mosquitoes, ticks, and insects. These vectors can carry infectious pathogens, for example, viruses, bacteria, and protozoa, from animals to humans. Changes in temperature, precipitation, and extreme events increase the geographic scope of infections spread by vectors and can prompt ailments happening earlier in the year. (*Climate Impacts on Human Health | Climate Change Impacts | US EPA*, n.d.)

3.2 Psychological Impact of Climate Change

Mental health is an indispensable part of any climate change endurance strategy, requiring ethics of equity and care, and widespread awareness. The American Psychological Association says that our psychological responses to climate change, like conflict avoidance, helplessness and resignation are growing. This indicates that our conscious and unconscious mental processes are holding us back from recognizing the causes of the obstacles from what they are, operating on solutions, and nurturing our psychological resilience, but we need all those things to take on what we have created. (Wray, n.d.) Heat, severe climate disasters, and expanded struggle for limited environmental resources, compounded by past incongruities and unbalanced effects among groups and countries will influence interpersonal and intergroup conduct and may result in expanded pressure and strain. (*Psychology & Global Climate Change Addressing a Multifaceted Phenomenon and Set of Challenges*, n.d.) The World Bank predicts that by 2050 there are going to be 140 million environment displaced people in Sub-Saharan Africa, Latin America, and South Asia. Mass movement and resource shortage increase the danger for viciousness, war, and political precariousness.

Climate change is truly extraordinary at Intersectionality, it increases the stresses that marginalized communities as of now face. The unbalanced impact of Hurricane Katrina's effects on poor people, mainly the Black people group of New Orleans' Ninth Ward affirmed that race and socioeconomic factors ought to be recognized in psychological response and prevention efforts. Collectively, these occupants lacked admittance to quality education, housing, and employment opportunities available in surrounding communities. These incongruities were related to an absence of fundamental assets, shelter, transportation, and data about clearing plans during the storm. (*Psychology & Global Climate Change Addressing a Multifaceted Phenomenon and Set of Challenges*, n.d.) Climate change effects incorporate individual and social impressions of the risks, psychosocial prosperity, animosity, intergroup results, and community building. People and communities vary in their vulnerability to climate change and these varieties can raise moral issues. An early meta-analysis of studies on the connection among calamities and mental well-being impacts found that between 7% and 40% of all subjects in 36 studies showed some form of psychopathology (Clayton et al., 2017). Individual experience of extreme climate events can prompt mental and emotional health results related to misfortune, disruption, and relocation. These results incorporate intense and posttraumatic stress disorder, other pressure related issues, like complicated melancholy, depression, anxiety disorders, somatoform issues, and drug and alcohol abuse; higher rates of suicide attempts raised danger of child abuse and expanded weakness of those with past mental- health issues (*Psychology & Global Climate Change Addressing a Multifaceted Phenomenon and Set of Challenges*, n.d.). Prolonged distress results in a lowered immune system response, leaving individuals more vulnerable to pathogens in the air and water and at more danger for several physical ailments. Sleep disorders also increase in response to chronic distress. (Clayton et al., 2017)

Activists, protestors, environmentalists, and anyone genuinely concerned about climate change is shrieking for revolution all around the world in the voice of despondency. Any child born at this age will have to live in a world where the natural disasters that we know of flooding, hurricanes, and wildfires are going to be commonplace. This bothers a significant number of people in this generation by one question, which is, should they have a child in the time of climate change?

3.3 Social and Economic Impact of Climate Change

Climate change is a danger multiplier, natural catastrophes can trigger wars, disputes, and unrest. Peace and climate change are not stand-alone entities; they are linked to each other directly. Diminishing labor efficiency is a widespread issue among climate change impacts on the socio- economic framework. Heat stress brought about by blistering climate influences financial exercises, prompting monetary misfortune through loss of production. This requires an assessment of how climate-change-induced labor productivity change influences socio-economic activities (*Climate Change Impacts on Socio-Economic Activities through Labor Productivity Changes Considering Interactions between Socioeconomic and Climate Systems*, 2019b). As we know the world, the social cost of climate change can change it completely and for the most acute. Resource crisis is not a simplistic matter, species have faced extinction because of resource crunch and human beings, all said and done, are just another species on earth. We know that the worst hit by climate change are the socially, economically, institutionally, and politically marginalized. There is an increased displacement of those from more unfortunate or low-lying nations to more well-off or more secure nations looking for better conditions. The health impacts of climate change are expected to be more extreme for older people and individuals with prior ailments, not to be denied that children and women living in poor countries are also easy targets.

Extinction Rebellion protests of Greta Thunberg are not the solutions to many questions bothering the world today, they are just the symbols of a conflict on this hostile issue. Given that we are not equipped to handle the consequences of climate change, urgent actions need to be taken by all the nations across the world. Heatwaves make our efficiency and strength to work decrease. Storms, tornadoes, and hurricanes devastate a huge number of individuals, leaving them in absolute poverty. Droughts shrivel harvests, further entangling the laborious errand of taking care of the world population, which is relied upon to arrive at 10 billion by 2050 (World Population Prospects 2019, United Nations Organization). The World Bank is cautioning on the off chance that we don't accomplish something promptly, climate change could drive 100 million additional individuals into neediness by 2030 (2019). Droughts, because of expanded times of diminished precipitation, joined with other extreme weather events, are causing the loss of livelihoods and demolition of harvests, leaving nearby occupants with a predicament: leaving, looking for other occupation openings or remaining and confronting hunger. Another immediate result of these occasions is food price spikes which are obvious to have dramatic effects on low-income families and people on a worldwide scale. Costs of the absolute most fundamental harvests like rice, wheat, and maize are estimated to increase by up to 150% by 2060 (Pawlik, 2020).

4. Emphasis on SDG-13

In the year 2015, all the member states of the United Nations adopted the 2030 Agenda of Sustainable Development which gives a common outline to harmony and prosperity for individuals and the planet.

The Food and Agriculture Organization (FAO) of the United Nations is encouraging nations to both adjust to and relieve the impacts of climate change by creating national climate plans and through research-based schemes and ventures, with an emphasis on adjusting smallholder production and making the occupations of rural populaces stronger. Food production threatens to be the biggest setback of climate change; however, sustainable agriculture can be an essential solution (*SDG 13. Climate Action | Www.Fao.Org*, n.d.).

Sustainable Development Goal 13 deals with Climate Action, which gives urgent engagement to combat climate change and its repercussions. The SDG 13 objective means to activate US$100 billion yearly by 2020 to address the necessities of agricultural nations to both adjust to environmental change and put resources into low-carbon development. Supporting vulnerable areas will straightforwardly contribute not exclusively to Goal 13 yet additionally to other SDGs. These activities should likewise go inseparably with efforts to integrate disaster risk measures, sustainable natural resource management, and human security into national development plans. It is still possible, with solid political will, expanded speculation, and utilizing existing innovation, to restrict the expansion in worldwide mean temperature to two degrees Celsius above pre- mechanical levels, focusing on 1.5 degrees Celsius, yet this requires critical and aspiring aggregate action (*Sustainable Development Goals | United Nations Development Programme*, n.d.).

The world must expect, adjust and become resilient to the current and expected future effects of environmental change. Ever since the ratification of the United Nations Framework Convention on Climate Change (UNFCCC), forest-based regulation and mitigation have been consolidated in global climate agreements. To accomplish the UNFCCC objective of restricting worldwide temperature ascend to well underneath 2°C above pre-industrial levels and to seek after endeavors to restrict the temperature increment to 1.5°C above pre-industrial levels, the world should change its energy, industry, transport, food, farming, and forestry frameworks to guarantee that aggregate net discharges don't surpass one trillion tons of combined carbon, which suggests worldwide net- zero emanations by the second of the century (*SDG 13: Take Urgent Action to Combat Climate Change and Its Impacts*, n.d.).

As we know climate change is influencing all the nations in the world, sparing none, we also know that human activities are the most prominent contributors to rising greenhouse gas emissions. So, to make our economies more rebounding and streamlined the countries must make use of scalable and cheap solutions available.

5. Brief History of Climate Change

Over the most recent 130 years, the world has warmed by around 0.85oC. Every one of the most recent thirty years has been successively more heated than any former decade since 1850 (*Climate Change and Health*, 2018).

Discussing climate change is quite possibly the most pressing task confronting our planet and its occupants. The historical backdrop of climate change policy and politics alludes to the proceeding with the history of activist efforts, trends, debates, political activities, and policies as they relate to the issue of global warming and other ecological peculiarities.

UNFCCC contained central issues that have been fundamental in resulting worldwide climate change discussions and processes instead of setting out explicit national or international targets to diminish greenhouse gas (GHG) emanations. The key points are as follows:

- An intention of settling the environment to block 'dangerous anthropogenic interference with the climate system' in a time that would sanction characteristic frameworks to adjust without significant harm to food systems and economic development.
- The necessity for nations to screen and restrict their ozone-depleting substance emanations and for various national limits, assessing nations' various obligations and limits.
- Particular distress for developing nations and especially those generally incompetent against climate change impacts, for example, small island states.
- The significance of prudent measures to react to the seriousness of climate change dangers, notwithstanding genuine scientific vulnerabilities concerning climate change cycles and effects (*2.2 A Brief History of Climate Change Policy*, n.d.).

5.1 Policies at the Global Level

Adjusting to and coping with climate change is an ongoing and ever-changing process so the worldwide concern requires a strategy of global dimensions that can convert into regional and local activities as well. One of the biggest reasons why efforts to address climate change have been through international collaboration and agreement is because emissions of greenhouse gases have a global impact, in contrast to some different types of contamination. So, whether they are emitted in one part of the world, they quickly scatter equally across the globe.

The first world climate conference occurred in 1979 and the early 1980s also marked the sharp increase in temperatures worldwide. In 1988 an expert body, Intergovernmental Panel on Climate Change (IPCC) to evaluate scientific information on climate change and its possible broader impacts, and to propose a global response was set up by the World Meteorological Organization (WMO) and the United Nations Environment Programme (UNEP). The UN Framework Convention on Climate Change (UNFCCC) was adopted in May 1992 and entered into force on 21 March 1994 (*Policy Responses to Climate Change - World Nuclear Association*, n.d.). The 197 countries that have ratified the Convention are called Parties to the Convention. UNFCCC aims to prevent dangerous human intervention with the climate system (*What Is the United Nations Framework Convention on Climate Change?* n.d.).

The first Convention of Parties (COP1) to UNFCCC took place in 1955, and on 11 December 1997 after negotiations, Kyoto Protocol was adopted which obligates state parties to reduce greenhouse gas emissions by an average of 5.2% below 1990 levels by the period 2008-2012 based on the fact that human-made CO_2 emissions are encouraging global warming.

The Kyoto Protocol included the following details:

- By 2012, developed nations would diminish their aggregate emissions by 5.2% from 1990 levels, every nation is committed to a specific figure.
- The emanations covered by the Protocol are carbon dioxide, yet in addition methane, nitrous oxide, hydrofluorocarbons, perfluorocarbons, and sulfur hexafluoride.
- These responsibilities would be figured on a net basis, thinking about sinks just as sources, and every nation should measure its commitment and meet its responsibility.
- Nations may satisfy their responsibilities mutually with regional arrangements and they may improve the productivity of compliance through "flexibility mechanisms" (*Policy Responses to Climate Change - World Nuclear Association*, n.d.).

Another milestone accord on climate change is the Paris Agreement in 2015, vowed by 197 countries to set targets for their greenhouse gas cuts. The Paris Agreement within the UNFCCC is a legally binding international treaty on climate change and the goal is to restrict global warming to considerably below 2, preferably to 1.5 degree Celsius and countries strive to reach global peaking of greenhouse gas emissions as soon as possible to achieve a climate-neutral world (*The Paris Agreement*, n.d.). Today 189 countries are a part of this agreement (United Nations, n.d.).

Even though the current US President Joe Biden signed an agreement of joining back into the Paris Agreement, former President Donald Trump pulled out of the Paris deal on June 1, 2017, because he believed that signing the Paris agreement disadvantages the US to the exclusive benefit of other countries leaving American workers and taxpayers to absorb the lost jobs, lower wages, shuttered factories and vastly diminished economic production. It is very unfair at the highest level to the US, and the draconian financial and economic burdens the agreement imposes on the US include ending the implementation of the nationally determined contributions and very importantly the Green Climate Fund which is costing the US a vast fortune. (2017) Many other developing nations are even accused of not giving themselves ambitious targets, owing to the Paris Agreement and even others who are planning to increase the construction of more power plants that use coal. John Christensen, Lead Author, United Nations 2019 "Emissions Gap Report," says that CO2 emissions have been going up by two percent from 2018 which is above the average of the last 10 years, which means it started to increase again.

In August 2018, Swedish youngster and climate activist Greta Thunberg began protesting before Swedish Parliament with a sign: "School Strike for Climate." Her protest to bring concerns to light for global warming took the world by storm and by November 2018, more than 17,000 students in 24 nations were participating in climate strikes. For her exemplary contributions, she was awarded the Nobel Peace Prize in March 2019. She participated in the United Nations Climate Summit in New York City in August of 2019, broadly taking a boat across the Atlantic as opposed to flying to reduce her carbon footprint (History.com Editors, 2020). She took up the #FridaysforFuture movements to protest against inaction towards climate change and called for concrete government action. To make the global movement a local one, students and adults from all over the world began assembling and demonstrating in front of parliaments in their respective countries.

6. Comparison of Policies on Climate Change: Sweden, India, and Niger

Specific impacts of climate change on a nation depend on the climate it experiences as well as its geographical, social, cultural, economic, and political situations.

Sweden

"If we pass 2 degrees, we risk pressing a button that makes it irreversible and that we then move beyond a point where we can turn back" — *Johan Rockström, Swedish scientist (Duxbury, 2018)* With an HDI of 0.937, Sweden is the eighth-most-developed country on the planet. The normal Swede has a life expectancy of 82.7 years due to better health care the life expectancy has increased. Sweden also has a low birth rate as well as a low death rate. Sweden is known for its top-notch life, with low joblessness and destitution rates and free admittance to medical care for the entirety of its residents (*Developed Countries List*, n.d.). Sweden is the first country to pass an environment protection act in 1967, and additionally facilitated the main UN gathering on the global environment in 1972. From that point forward, Sweden is managing to develop its economy generously while decreasing fossil fuel emissions and restricting contamination. The larger part of Sweden's natural energy supply comes from renewables and thorough legislation focuses on further decreasing greenhouse gas emanations. For over 10 years, Sweden has been in the top ten of the worldwide regarded Environmental Performance Index produced by Columbia and Yale universities, with exceptionally clean air and clean water alongside its low emanations (*Saving the Climate | The Official Site of Sweden*, 2021).

In 2017, the Swedish Parliament Riksdag adopted a national climate policy framework for Sweden consisting of new national climate targets, A Climate Act, which authenticates that the Government's climate policy must be based on the climate aims and stipulates how the implementation is to be conducted and a climate policy council was also adopted, which will guarantee that the Government's policies are following the climate goals. For a long time, Sweden utilized viable approach switches that put a cost on carbon, utilizing a carbon dioxide tax joined by a political administration that upholds the development of climate-smart advancements. This has brought about decreased emissions and market-based arrangements while the economy has developed (*Sweden's Initiative Challenges World to Go Fossil-Free*, 2014).

The framework includes New Climate Goals for Sweden:

1. By 2045, Sweden is to have zero net discharges of greenhouse gases into the atmosphere and ought to from there on accomplish negative discharges, which means that Sweden overall helps to reduce the amount of greenhouse gas in the atmosphere. The reform is a key component of Sweden's endeavors to implement the Paris Agreement.
2. By 2030, discharges from domestic vehicles, barring domestic flying, will be decreased by at least 70% compared with 2010.

3. By 2030, discharges in Sweden in the areas that will be covered by the EU Effort Sharing Regulation ought to be at least 63 percent lower than in 1990.
4. By 2040, emissions in Sweden in the areas that will be covered by the EU Effort Sharing Guideline should be at least 75 percent lower than in 1990 (*The Swedish Climate Policy Framework*, n.d.).

The positive experience of combining emission reductions and enhanced economy and welfare encouraged Sweden to announce that they will become the world's first fossil fuel-free nation by the year 2045 and at the same time, they have challenged the entire world to do likewise.

Niger

Niger is a Sahelian country, three-fourths of the space of which (1,267,000 km2) is located in the desert zone, which makes it profoundly presented to environment chances, with rainfall that is exceptionally variable during the year, as far as both space and time. (*«Intended Nationally Determined Contribution (INDC) » of Niger*, 2015) Niger's strategy depends on the vision of environmentally smart farming and access to current energy administrations for everybody in 2030 (*Intended Nationally Determined Contribution (INDC) of Niger*, 2015). Niger signed the United Nations Framework Convention on Climate Change (UNFCCC) in June 1992 and ratified it on 25 July 1995 and signed the Kyoto Protocol in December 1996 and ratified it on 17 March 2004. Climate change models a massive threat to Niger's economic security and resiliency. Niger approved the Paris Agreement and presented its first Nationally Determined Contribution (NDC) on September 20, 2016. Preference adaptation steps described in its NDC focus on the agriculture, livestock, water resources, transport, and public health sectors with importance on the variation of technologies; enhanced capability in the area of research; and methodical observations and activities in the area of awareness building, education and training, and knowledge (*World Bank Climate Change Knowledge Portal*, n.d.).

Intended Nationally Determined Contribution (INDC)

The goals of Niger's INDC are to guarantee food security, battle poverty, and contribute to the decrease of world greenhouse gas (GHG) outflows so they won't increment in the abundance of 2°C in the 2050 horizon on account of green development and a low-carbon improvement procedure, the motivation behind which is to guarantee flexibility of the populace and environments. The procedures of climate-smart agriculture (CSA) are reliable with the goals of the INDC (variation, alleviation, and food security) by reinforcing grassroots advancement. Climate-smart agriculture considers climate data, early cautioning, the administration of dangers and calamities, and list-based rural climate protection. The incorporation of environmental change in local (PDC program), territorial and national planning (in the wellbeing and animal husbandry areas) are tried and affirmed great practices in Niger and are subsequently thought about in the INDC.

To effectively implement this agreement Niger also actively participated in the conference of parties on climate change and it prioritized focus on strategies and adaptation for climate change. Given the geographical location of Niger, it faces the direct impact of climate change as it is situated along the edge of dry areas. Different examinations are done regarding the National Communications on climate change in Niger notice the expanding variability of precipitation in terms of both space and time; a pattern of expanded temperatures, especially since 1996; expanded recurrence furthermore, the force of outrageous environment hazards (droughts, floods, violent breezes, and sand and residue storms, the enemies of harvests); the silting of watercourses (Niger River valley and Lake Chad) and desert gardens; dry spell misfortunes in Niger assessed at more than the US $70 million (World Bank, Climate Risk Assessment, Niger, 2012); the harm, including costs, caused to key areas of the economy by the floods in the 2000s, assessed at the US $18 million. (*Intended Nationally Determined Contribution (INDC) of Niger*, 2015).

To reinforce its people's flexibility to climate change, Niger considers adaptation to be vital, through activities with solid co-benefits that fortify the strength of ecosystems and populations and reduce GHG emissions. Niger ensures its people fight all the global climate challenges and promote growth and inclusive development. Not only this but climate change relief activities in Niger have improved access to power, fully intent on coming to 60% by 2030, on account of rural electrification and the advancement of renewable energy. Ultimately, energy proficiency will be improved by this and the Kandadji hydropower station will be created to supply 130 MW, notwithstanding a 100 MW solar project (African Development Bank Group, 2019).

India

India has introduced many policies that work towards climate change by reducing greenhouse gas (GHG) emissions. Despite being one of the largest emitters of GHGs, India's cumulative emission does not impart much harm. For a populous, developing country like India that has already been suffering from climate change and is still vulnerable to it, formulating a policy comes naturally. India is at the frontlines of facing the impacts of climate change despite its noteworthy contribution to worldwide fossil fuel byproducts. The aggregate normal since 1850 is a meager 2%, which is better compared to most developed nations. Repeating floods, shifting rainfall patterns and other disasters hamper our endeavors to destroy other menaces like hunger and poverty (Upadhyay, 2020). India is continuing to make its best endeavors to increase energy productivity and make more comprehensive utilization of renewable energy, even with limited resources in hand.

India is also taking some essential steps in regards to using renewable energy:

1. India's National Solar Mission is being increased five-fold from 20,000 megawatts to 100000 megawatts. This will mean an extra investment of 100 billion dollars and investment funds of around 165 million tons of CO2 discharges every year. Asia's one of the largest solar power plants, Rewa Solar Power Plant which boasts of generating 750 MW of electricity, is expected to decrease India's CO2 emissions by 1.5 million tonnes annually.
2. India is delivering 6 billion US dollars in intense afforestation, which will result in more carbon sinks.
3. India has allotted around 200 million US dollars for the 'National Adaptation Fund,' setting- up of Ultra Mega Solar Projects, Ultra-Modern Super Critical Coal based thermal power tech, and the improvement of Solar Parks on trenches.
4. India has set up severe standards for the cement industry. The Action Plan for cleaning river Ganga will bring various advantages. Not exclusively will it help clean the stream yet in addition help in the recovery of waterway Dolphins found in this holy river.
5. India has begun arrangements to build up a National Air Quality list and has started a National Air quality plan (Upadhyay, 2020).

The external and domestic dimensions of India's Climate Change Policy have been explained by two fundamental documents. One is the National Action Plan on Climate Change(NAPCC) adopted on June 30, 2008, and has a domestic focus, and the other is India's Intended Nationally Determined Commitments(INDC) presented to the UNFCCC on October 2, 2015. The INDC is a declaration of intention on Climate Change action proclaimed in the run-up to the Paris Climate Change summit held in December of the same year (Saran, 2019). NAPCC recognizes several measures that together elevate the country's advancement and climate change-related objectives of adaptation and mitigation, consistent with the national preference for accomplishing sustainable development. It realized the significance of a strategic transformation that would help in enhancing India's security and contribute to dealing with the threat of climate change. The shift was required from the current dependence on fossil fuels to a model of economic venture based on renewable sources of energy such as solar energy and cleaner sources such as nuclear energy.

In order to achieve key goals in the context of climate change, the national commission focuses on 8 specific areas, they are;

- Solar Energy- Strives to encourage the use of solar energy in India by making it competitive with fossil fuels.
- Enhanced Energy Efficiency—Intends to develop energy productivity of domestic, commercial, and industrial sectors in India by designing enabling policy management and promoting innovative business models for enhancing energy efficiency.
- Building a Sustainable Urban Habitat—Intends to strengthen the energy potency of domestic, commercial, and industrial sectors in India by planning to facilitate policy administration and encouraging innovative business models for enhancing energy efficiency.
- Conserving Water—Proposes to secure sustainable water supply by preserving water, reducing waste, and ensuring equal distribution of water resources throughout India.
- Sustaining the Himalayan Ecosystem—Aims to improve comprehension of climate change consequences and adjustments required in the Himalayas.
- Formulating a Green India through expanded forests—strives to preserve, improve and restore forests and react to climate change with proper adaptation and mitigation activities.
- Sustainable Agriculture—Aims at developing sustainability, productivity, remuneration, and climate flexibility of agriculture in India.
- Strategic knowledge—Intends to build a comprehensive knowledge system that notifies and promotes climate change action in India with the help of research and communication- based actions (Rattani, 2018).

National Environment Policy (NEP), 2006 was an attempt towards India's dedication to a clean environment and creating a positive enrichment to global ventures. The NEP develops on the numerous earlier policies which had discussed the difficulties of the environment and the requirement for sustainable growth before this policy. It is mandated in the Constitution in Articles 48 A and 51 A (g), backed by judicial interpretation of Article 21 (*National Environment Policy 2006*, n.d.).

India's INDC indicates a more proactive, resourceful, and forward-looking approach in the run-up to the Paris Climate summit. The objectives India has willingly pledged itself to are such that the energy intensity of India's growth will decrease by 33-35% by 2030 compared to the 2005 base year, which suggests that for every extra dollar of GDP India will be utilizing progressively and significantly lesser amounts of energy. India is one of the world's largest developing economies, which already has a large energy footprint globally, this forms a major contribution to tackling global

Climate Change. The INDC has set a target of 175 GW of renewable energy by the year 2030 on the strength of the exceptional achievement of the National Solar Mission. It is stated that this capacity may well be accomplished 10 years in advance. The government may raise India's target to 227 GW for 2030. The target of achieving 40% of power from renewable sources by 2030 is likely to be accomplished many years in advance (Saran, 2019).

Few other policies to combat the challenge of India are as follows:

- **State Action Plan on Climate Change (SAPCC):** After the formulation of the NAPCC, the requirement to gain harmony between operations at the national and sub-national levels became obvious. Therefore, the State Governments also had to prepare their State Action Plans on Climate Change (SAPCC) in line with the strategies outlined in NAPCC.
- **National Adaptation Fund on Climate Change (NAFCC):** In 2015 NAFCC was launched with an original outlay of Rs. 350 crores to reach the cost of adjustment to climate change for the State and Union Territories of India that are especially exposed and weak to the unfavorable impacts of climate change.
- **Climate Change Action Programme (CCAP):** It is a central scheme passed by the Cabinet in January 2014 at a total cost of Rs. 290 crore for five years which strives to build and strengthen the scientific and analytical capability for evaluation of climate change in the country, putting in place a relevant institutional framework for scientific and policy initiatives and implementation of climate change-related actions in the context of sustainable development (*Climate Change | The Official Website of Ministry of Environment, Forest and Climate Change, Government of India*, n.d.).

7. Loopholes in the Policies

Sweden: As per the expectations from the climate policy, with such a high aim, when the emissions reductions were assumed to accelerate, they have reduced. The highest reductions in greenhouse gas emissions since 1990 happened between 2003 and 2014, with an average annual decline rate of just over 2 percent. During the last four years (through 2018), the pace has decreased, and emissions have only dropped by less than 1 percent per year on average (*2020 Report of the Swedish Climate Policy Council*, 2020). The assessment from the Government's 2020 climate report says: The climate targets beyond 2020 won't be accomplished if current conditions and existing approaches proceed. The reports stress the significance of the requirement for additional actions as the climate report cannot be said to meet the legislative necessities in such a manner. By and large, impact assessments are lacking and in the remaining cases, some form of impact evaluation is introduced, however, there is no data on how they were led or what suspicions they depended on. The fact that the Government additionally uses various units and configurations in its impact assessments makes it challenging to compare different activities, efficiency evaluations, and the assessment of the overall impact of the decisions introduced (*2020 Report of the Swedish Climate Policy Council*, 2020).

Niger: Climate change supplements a complicating layer in a country where underdevelopment, poverty, extensive criminality, and brutality already make life so delicate. Niger suffers extremely severe difficulties– a threatening climate, profound poverty, and related social strain intensified by enormous population development, a tremendous and very sparsely populated landmass, a delicate economy, and a very young democracy that is being terrorized by the intrusion of violent extremism from adjoining nations (Ministry of Foreign Affairs of Denmark, n.d.). Niger is identified to be particularly vulnerable because of its fragile ecosystem and human actions such as gas burning that have increased the competence of climate change and its consequences in the region. Stringent steps are necessitated to address the crises of reproductive health, education, and economic dependence on agriculture in the light of climate change, to protect the country from falling into a terrible state.

India: India strives to accomplish 175GW of renewable energy by the year 2022, aiming to establish one of the biggest renewable programs in the world. But because NAPCC was formulated in a short period and under a lot of pressure, it did not address concrete strategies and merely provided broad objectives and lacked specificities. Although NAPCC has been in existence for over 10 years, research shows that for most missions it has just been three to four years since they became effective. It is likewise evident that progress has been nearly uncertain for the vast majority of the missions. Missions dealing with topics as sustainable habitat, water, and agriculture and forestry are multi-sectoral, overlaying, and multi-departmental. They are consulting and holistic and have been slow-moving. An added hurdle for the missions is their monitoring practices, which are either incompetent or missing. Overall, we can say that institutional, systemic, and method limitations— including financial restrictions, inter-ministerial coordination, lack of technical expertise, and project approval delays, stand as significant challenges in the efficient implementation of the missions (Rattani, 2018).

"To unconditionally reduce the emission intensity of all GHGs of its GDP by 30-35 percent from 2005 level by 2030"

(O'Neill, 2019). India's pledge was not considered adequate as it will not commit to the needed 50 percent decrease in global emissions by 2030.

8. Solutions and Recommendations

The index to track and observe the climate protection and performance for countries globally is The Climate Change Performance Index (CPPI), which is an autonomous source for monitoring. The CPPI was fully revised in 2015 to serve the Paris Agreement, a structure designed by the United Nations to combat the impacts of climate change. The CPPI is a trusted and authorized autonomous source for quality information, although it should be noted that the ranks demand data interpretation. The CPPI is a prevailing strategy to quantify emissions with production-based emissions. It assesses and grades 57 countries based on four key divisions:

- their Global Greenhouse Gas Emissions (GHG),
- their renewable energy,
- their total energy use,
- their climate policy.

The ranking method for nations is on a relatable scale, such as high, very high, and low, very low. The CPPI is a significant instrument for following and positioning the activities of nations worldwide in the fight against drastic climate change. Even with the simple rules and ranking framework, no country has yet had the option to place in the top three ranks (Drier, 2020).

It might be easy to talk about issues, and get oneself into arm-chair activism or sitting down with expert bodies and formulating policies for climate change, but is that enough? The major drawback that lies behind not even a single country being successful to make it to the top rank of the CPPI Index is not the lack of policies but the lack of effective implementation of these policies at regional and national levels. Today when climate change abatement is dire, the absence of plausible targets and the absence of a technique for execution is disturbing.

One of the most essential steps at this point is to align renewable energy goals. The Government should highlight the significance of collaboration throughout society to accomplish climate objectives which can be done by involving all stakeholders, from the national to the regional and local levels.

Sweden

1. Prioritize continued public investments in fossil-free, competitive industrial processes that can reduce industrial greenhouse gas emissions.
2. Create strategies to stimulate and encourage the order from houses, industries, and the public sector for zero-emission, added resource-efficient assets, and co-operations beyond all sectors (*2020 Report of the Swedish Climate Policy Council*, 2020).

Niger

1. On-ground superintendence of water and energy should be renewed on a large scale. For this, at the local level, there should be an extension in the application of solar-powered pumping plants, small dams that retain water during rainfall, permitting the water to flow into the soil and recharge aquifers.
2. Broadening its economy will be critical if Niger desires to overcome the hardship of climate change, which implies vigorously investing resources into innovation, infrastructure, and education. Added endeavors to convey innovations to farmers are additionally fundamental (Erin G, n.d.).

India

1. To prevent generating electricity production plants utilizing fossil fuel, particularly coal, and channel investments to renewable energy, clean transport, and energy efficiency to combat global warming and at the same time get financial advantages.
2. The state-level cell may include their partner Institutions and organizations for studies and emission accounting and Government organizations, Public Sector Units, and Private Companies for the execution of climate change missions (Sharma, 2010).

Common solutions to all the countries:

1. Restricting environmental degradation completely is essential and common for all the countries in the world.
2. All countries are required to decrease emissions to reach the Paris Agreement objectives, although not all countries have equal accountability because of the principle of differentiated responsibility, historical eruptions, current per-person emissions, and the necessity to progress. Few nations would even require international assistance.
3. Strengthening or restoring existing natural ecosystems can assist in combating climate change. Nature-based solutions can center on decreasing emanations from deforestation and farming methods and improving the capacity of natural ecosystems to eliminate carbon dioxide from the air. For instance, forests don't simply ingest carbon, they also safeguard us from its most wrecking impacts. Cautiously planted tree species can go about as firebreaks, keeping trees close to farmland

can shield crops from the erosive powers of intense rainfall, and forests can reduce inland floods because of the wipe-like way they retain water. Changing our land practices alone could convey 30% of the decrease in emissions that we need to accomplish the objectives of the Paris Agreement on climate activity by 2030 (*Nature 'One of Most Effective Ways' of Combating Climate Change*, 2019). It is significant to remember that expanding aspiration challenges us to engage simultaneously in energy conversion and greater investments in nature.

4. To approach the challenge of climate change in the long run it is imperative to make a substantial investment in building and maintaining low carbon infrastructure in transportation, construction, and other related sectors. This requires the course of development to be based on Clean Development Mechanism. Besides, R&D in alternative fuels and low carbon infrastructure must get due priority. To attain sustainable development, the focus should be on developing indigenous green technology (*CLIMATE CHANGE: CHALLENGES TO SUSTAINABLE DEVELOPMENT IN INDIA*, 2008).

9. Conclusion

Communities have forever relied upon the climate but are only now accepting the reality that the climate depends on their activities. Ever since the Industrial Revolution, there has been a precarious hike in greenhouse gases that have altered the relationship between individuals and the environment. In other words, not only does climate affect advancement but advancement affects the climate. The path of opportunity to pick the appropriate strategies to deal with climate change and encourage growth and expansion is closing. The more nations go with prevailing emissions trajectories, the more troublesome it will be to modify the course and adjust infrastructures, economies, and lifestyles. We need to change our vision and accomplish the aim of sustainable development. Attempts must be made to know the potential of alternative sources of energy such as wind, solar, and hydropower and ultimately progress towards transformation to clean energy. By adopting a blend of determinants which include the selection of clean technologies, equal allocation of resources, and discussing the issues of equity and justice, we can make our developmental process more peaceful with nature.

References

1. *2.2 A brief history of climate change policy*. (n.d.). 2012. Retrieved 15 May 2021, from https://www.soas.ac.uk/cedep-demos/000_P524_CCD_K3736-Demo/unit1/page_14.htm
2. *2020 Report of the Swedish Climate Policy Council*. (2020, March). Klimat Politiska Radet. https://www.klimatpolitiskaradet.se/wp-content/uploads/2020/05/2020reportofthes wedishclimatepolicycouncil.pdf
3. A. (2020, November 25). *Climate Change: What You Need To Know In 2020*. Admodito. https://www.admodito.com/climate-change-what-you-need-to-know-in-2020/
4. African Development Bank Group. (2019, January 31). *Niger: strengthening climate change resilience*. African Development Bank - Building Today, a Better Africa Tomorrow. https://www.afdb.org/en/news-and-events/niger-strengthening-climate-change-resilience-16563
5. Clayton, S., Manning, C., Krygsman, K., & Speiser, M. (2017, March). *MENTAL HEALTH AND OUR CHANGING CLIMATE: IMPACTS, IMPLICATIONS, AND GUIDANCE*. American Psychological Association. https://www.eenews.net/assets/2017/03/30/document_cw_01.pdf
6. *Climate change and health*. (2018, February 1). World Health Organization. https://www.who.int/news-room/fact-sheets/detail/climate-change-and-health
7. *CLIMATE CHANGE: CHALLENGES TO SUSTAINABLE DEVELOPMENT IN INDIA*. (2008, October). RAJYA SABHA SECRETARIAT. https://rajyasabha.nic.in/rsnew/publication_electronic/climate_change_2008.pdf
8. *Climate Impacts on Human Health | Climate Change Impacts | US EPA*. (n.d.). Environment Protection Agency. Retrieved 15 May 2021, from https://19january2017snapshot.epa.gov/climate-impacts/climate-impacts-human-health_.html
9. *Climate change impacts on socio-economic activities through labor productivity changes considering interactions between socioeconomic and climate systems*. (2019b, April 10). ScienceDirect. https://linkinghub.elsevier.com/retrieve/pii/S095965261833837X
10. *Climate Change | The Official Website of Ministry of Environment, Forest and Climate Change, Government of India*. (n.d.). Ministry of Environment, Forests and Climate Change. Retrieved 15 May 2021, from http://moef.gov.in/en/environment/climate-change/
11. *Developed Countries List*. (n.d.). World Population Review. Retrieved 15 May 2021, from https://worldpopulationreview.com/country-rankings/developed-countries
12. Drier, K. (2020, September 26). *Countries Doing The Most To Combat Climate Change*. WorldAtlas. https://www.worldatlas.com/articles/countries-doing-the-most-to-combat-climate- change.html
13. Duxbury, C. (2018, September 3). *Sweden's far right sees red on green 'populism'*. POLITICO. https://www.politico.eu/article/sweden-far-right-democrats-sees-red-on-green-populism-climate- change-heatwave/
14. Earth.Org. (2020, September 29). *What is Climate Change?* Earth.Org - Past | Present | Future. https://earth.org/what-is-climate-change/
15. Erin G, R. (n.d.). *Climate + Niger*. Purdue Climate Change Research Center. Retrieved 15 May 2021, from https://ag.purdue.edu/climate/stories-of-change/climate-niger/

16. FAQ 2.1 - AR4 WGI Chapter 2: Changes in Atmospheric Constituents and in Radiative Forcing. (n.d.). IPCC. Retrieved 21 June 2021, from https://archive.ipcc.ch/publications_and_data/ar4/wg1/en/faq-2-1.html
17. *Global Warming Effects*. (n.d.). Environment. Retrieved 14 May 2021, from https://www.nationalgeographic.com/environment/article/global-warming-effects
18. *Goal 13: Climate Action - SDG Tracker*. (n.d.). Our World in Data. Retrieved 21 June 2021, from https://sdg-tracker.org/climate-change
19. History.com Editors. (2020, November 20). *Climate Change History*. HISTORY. https://www.history.com/topics/natural-disasters-and-environment/history-of-climate-change
20. *Humans are causing global warming*. (2019, August 5). The National Academies of Sciences Engineering Medicines. https://www.nationalacademies.org/based-on-science/climate-change- humans-are-causing-global-warming
21. I. (2019, October 10). *The Impact Of Climate Change*. Iberdrola. https://www.iberdrola.com/environment/impacts-of-climate-change
22. *Impacts of climate change*. (n.d.). Sustainability for All. Retrieved 14 May 2021, from https://www.activesustainability.com/climate-change/impacts-climate-change/
23. *India First Biennial Update Report to the United Nations Framework Convention on Climate Change*. (n.d.). Ministry of Environment, Forest and Climate Change, Government of India. Retrieved 14 May 2021, from https://www.google.com/url?q=http://moef.gov.in/wp-content/uploads/2017/08/indbur1.pdf&sa=D&source=editors&ust=1620984296849000&usg=AOvVaw3yw-fcjw3Fj4Q4QMyme-f2
24. Intended Nationally Determined Contribution (INDC) of Niger. (2015, September 26). UNFCCC. https://www4.unfccc.int/sites/ndcstaging/PublishedDocuments/Niger%20First/Niger- INDC-final_Eng.pdf
25. M., D., D., D., M., M., & M. (n.d.). *Climate Action*. United Nations Sustainable Development. Retrieved 21 June 2021, from https://www.un.org/sustainabledevelopment/climate-action/
26. Ministry of Foreign Affairs of Denmark. (n.d.). *Niger*. Retrieved 15 May 2021, from https://um.dk/en/danida-en/strategies%20and%20priorities/country-policies/niger/
27. *National Environment Policy 2006*. (n.d.). Government Of India, Ministry Of Environment And Forests. Retrieved 15 May 2021, from http://www.indiaenvironmentportal.org.in/files/nep2006e.pdf
28. *Nature 'one of the most effective ways of combating climate change*. (2019, September 23). UN News. https://news.un.org/en/story/2019/09/1046752
29. O'Neill, M. (2019, November 5). *The Truth Behind the Paris Agreement Climate Pledges – "Insufficient to Address Climate Change"*. SciTechDaily. https://scitechdaily.com/the-truth-behind-the-paris-agreement-climate-pledges-insufficient-to-address-climate-change/
30. P. (2017, June 2). *Donald Trump pulls out of Paris climate deal, says it favours India, China more*. The Economic Times. https://economictimes.indiatimes.com/news/international/world- news/donald-trump-pulls-out-of-paris-climate-deal-says-it-favours-india-china- more/articleshow/58952959.cms
31. Pawlik, A. (2020, October 14). *Social Impact of Climate Change: How It Is Accentuating Social Injustices*. Impact Hub Vienna. https://vienna.impacthub.net/2020/10/14/social-impact-of- climate-change-social-injustices/
32. *Policy Responses to Climate Change - World Nuclear Association*. (n.d.). World Nuclear Association. Retrieved 15 May 2021, from https://www.world-nuclear.org/information-library/energy-and-the-environment/policy-responses-to-climate-change.aspx
33. *Psychology & Global Climate Change addressing a multifaceted phenomenon and set of challenges*. (n.d.). American Psychological Association. Retrieved 15 May 2021, from https://www.apa.org/science/about/publications/climate-change-booklet.pdf
34. Rattani, V. (2018, February 16). *COPING WITH CLIMATE CHANGE (NAPCC)*. Centre for Science and Environment. http://www.indiaenvironmentportal.org.in/files/file/coping-climate- change-NAPCC.pdf
35. Saran, S. (2019, November 8). *India's Climate Change Policy: Towards a Better Future*. Ministry of External Affairs, Government of India. https://mea.gov.in/articles-in-indian-media.htm?dtl%2F32018%2FIndias_Climate_Change_Policy_Towards_a_Better_Future
36. *Saving the climate | The official site of Sweden*. (2021, April 23). Sweden.Se. https://sweden.se/climate/
37. *SDG 13. Climate action | www.fao.org*. (n.d.). Food and Agriculture Organization UN. Retrieved 15 May 2021, from http://www.fao.org/sustainable-development-goals/goals/goal-13/en/
38. *SDG 13: Take urgent action to combat climate change and its impacts*. (n.d.). SDG Compass. Retrieved 15 May 2021, from https://sdgcompass.org/sdgs/sdg-13/
39. Sharma, S. (2010, September 1). *India's National Action Plan for Climate Change and Its Implementation*. Research Gate. https://www.researchgate.net/publication/279513871_India%27s_National_Action_Plan_for_Cli mate_Change_and_Its_Implementation
40. Sustainable Development Goals | United Nations Development Programme. (n.d.). UNDP. Retrieved 15 May 2021, from https://www.undp.org/sustainable-development-goals#climate- action
41. *Sweden's initiative challenges the world to go fossil-free*. (2014, October 3). Regeringskansliet. https://www.government.se/articles/2015/11/swedens-initiative-challenges-world-to-go-fossil- free/
42. *The Paris Agreement*. (n.d.). United Nations Climate Change. Retrieved 15 May 2021, from https://unfccc.int/process-and-meetings/the-paris-agreement/the-paris-agreement
43. *The Swedish climate policy framework*. (n.d.). Ministry of the Environment and Energy. Retrieved 15 May 2021, from https://www.government.se/495f60/contentassets/883ae8e123bc4e42aa8d59296ebe0478/the-swedish-climate-policy-framework.pdf
44. Tripathi, B. (2019, December 5). *India 5th Most Vulnerable To Climate Change Fallouts, Its Poor The Worst Hit*.

IndiaSpend. https://www.indiaspend.com/india-5th-most-vulnerable-to-climate-change-fallouts-its-poor-the-worst-hit/#:~:text=New%20Delhi%3A%20India%20is%20the%20fifth%20most%20vulnerable,vulnerable%2C%20followed%20by%20the%20Philippines%2C%20Germany%20and%20Madagascar
45. United Nations. (n.d.). *The Paris Agreement*. Retrieved 15 May 2021, from https://www.un.org/en/climatechange/paris-agreement
46. Upadhyay, B. (2020, August 26). *How Is India Planning To Tackle The Growing Menace Of Climate Change?* Youth Ki Awaaz. https://www.youthkiawaaz.com/2020/07/climate-change-indias-stance-in-tackling-this-menace/
47. *What is the United Nations Framework Convention on Climate Change?* (n.d.). United Nations Climate Change. Retrieved 15 May 2021, from https://unfccc.int/process-and-meetings/the-convention/what-is-the-united-nations-framework-convention-on-climate-change
48. *World Bank Climate Change Knowledge Portal*. (n.d.). World Bank Group. Retrieved 15 May 2021, from https://climateknowledgeportal.worldbank.org/country/niger
49. Wray, B. (n.d.). *How climate change affects your mental health*. TED Talks. Retrieved 21 June 2021, from https://www.ted.com/talks/britt_wray_how_climate_change_affects_your_mental_health/transcript

CHAPTER 38

The Most Credible 05 Elements to Tackle Displacement in Nearest 10 Years of Asia

Mahendra Jagath Kuragodage
Assistant Director Disaster Management center
Sri Lanka
jagathm@dmc.gov.lk
+94 773 957 905

Abstract

"Internal displacement is the great tragedy in our time. Internally displaced people are among the most vulnerable of the human family"

—Kofi Annan.

There are many elements to tackle displacement in Asia. This paper attempted to discuss the most credible 05 elements of them. In the South Asian region, displaced people recorded as 23.8% of the world end of 2020. With the Covid pandemic, displaced people face doubled problems and additional risks. The year 2020 was a record for the displacement of the world the highest figure was painted in ten years as 40.5m new displacement. Asian Countries fight to search for a solution to overcome this issue. This paper looks to address the problem.

Data collection was carried out by literature review, key informant interviews, and field surveys. In methodology, qualitatively analyzed done in a comparative mode for receiving temporal basis data. In analysis, the paper illustrated propose credible 05 elements, which brings the solution for displacement.

1. Recurring - after disaster returning home has to transfer in local integration or planned relocation.

   To support community-led livelihood initiatives requires empowered local governance and decentralized interventions that include the perspectives of those at risk.
2. Climate change is not the one & only factor that drives displacement risk. There are economic and social drivers forced indispensable for human migration.

   Displacement associated with disasters and the effects of climate change is one. The convergence of conflict and disasters led to the displacement of people for a second or even third time prolonging.
3. Keeping eyes on small-scale events and learning to tackle slow on-set events.

   Drought, sea-level rise, and glacial retreat are all directly affected by slow-onset events is a gradual one resulting in displacement. 54% of all events in 2009 displaced fewer than 100 people, highlighting the necessity of the most attention & prevention but took the least.
4. Embrace uncertainty challenges focusing on risk reduction.

   Must Disaster risk reduction be mainstream in all-inclusive Changing is uncertainty forcing by climate and political, social, and economic environment. The answer is embracing this uncertainty, seeing it as a challenge and opportunity alike.
5. Need to put the community in the center, with joint force has to forced mobilize political will to identify the link between conflict and climate change.

Since disaster and climate displacement are mostly happening within borders, the responsibility of tackling those first with national governments. The international community, civil society help relay the urgency of the issue. The first reflex is still often to turn to (inter)national experts for advice. While expert counsel is, of course, valuable, actively integrating local knowledge is equally important. Scarcity of natural resources because of slow-onset events such as drought may lead to aggravating already existing tensions. On the other hand, violence may increase risk. Disasters and conflict concur, triggering new displacement of already uprooted populations.

In conclusion, the author proposes to adopt these 05 elements to overcome displacement issues in South Asian countries that will happen in the nearest decade of the own setup.

Keywords: Displacement, Elements, Disaster Risk Reduction, Disasters

1. Introduction

1.1 Background

Displacement is a critical factor that is limiting the development and drawback sustainable human prevailing. Human has been struggled to settle this implication, ensuring resettlement procedures everywhere of the globe addressing long-lasting solutions. This paper illustrates that resettlement is one option and several others are having even before happening displacement which can eradicate driving factors for the displacement in Asia upcoming decade. The findings are presenting based on conception conte xt for filling the gap of policy framework in terms of restitution in sustainable and durable.

The year 2020 reported 23.8% displacement in Asia of the world. This astonishing age of year recorded the highest figure in ten years, 40.5m around the world. Policymakers, practitioners, and Researchers stand in front of Asia, giving their attention to human mobility with Internal Displacement (IDPs) no matter the driving force as development, conflict & violation, or climate change.

The largest displacement camp was erected in the Asian region (South Asia) in Cox's Bazar Rohingya camp in Bangladesh at the proceeding time. And one group of massive displacement is also reporting from the Asian region (Western Asia) in Syria,960000 populations were displaced due to the Syrian offensive to Northerners of Idlib. There was 2/3 of the population displaced in Damascus 53% of the whole. Further, In Yemen also happened the same. The country's humanitarian crisis, which remains the world's most acute, was aggravated further by devastating floods and storms during two intense rainy seasons between February and September 2020. Disasters triggered 223,000 new displacements during the year, the highest figure on record for Yemen. That was highlighted that how the impacts of disasters and conflict overlap in the country. This event is on top of more than 3.6 million displaced by the crisis and violence. As disasters prolonged compound the country's displacement, it is urgent to implement durable solutions for Yemen's IDPs.

From the top ten displacements, category seven places were acquired by Asian countries. This situation is not envisaged by any who dealt with the subject.

To the severity and density of the overwhelming extreme displacement in Asia, people from global kept eyes and ears to this region. While much of them are propose resettlement guidelines and policy framework instruments, this research paper presents 05 most credible elements to catalyst new and innovative solutions for overcoming upcoming issues in the decade. It is recorded 1.2 million new displaced populations in Asia due to climate events.

In Asia, around 95 percent of displacement records by natural hazards and violence. In the first half of 2020, four of the five countries which accounted for nearly 75 percent of the new internal displacements due to disasters were in the Asia region: India (2.7 million people), Bangladesh (2.5 million people), Philippines (811,000 people), and China (791,000 people). (Figure 1) Displaced populations have multiple vulnerabilities, and amidst COVID-19, these populations are additionally susceptible. Displacement is also likely to increase as a result of climate change, particularly in the Asian region.

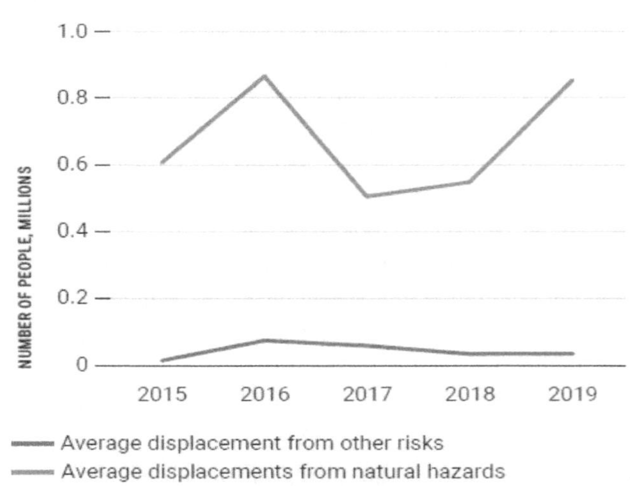

Fig. 38.1 Shows displacement population from 2015-2019 by natural hazards & other risks

Source: (IDMC-GRID, 2021)

1.2 Research Aims and Objectives

Against this background, and the extensive documentation and analysis that is already available on the displacement and resettlement this report has been commissioned by the Author **Specific Aims** to:

document and assess available solutions to prevent displacement and also give innovative solutions to tackle displacement in Asia extending conceptual experience to uplift policy frameworks.

The **objectives** of the study are

To provide conclusions and recommendations to support robust development and more representative displacement prevention policy engagement and regional advocacy strategy.

And

To showcase and harmonize available thousands of policy documents, guidelines, conventions, agreements, and Articles on all form of human mobility (IDPs, Refugees, forced displacements, Migrants Statelessness and Asylum seekers) in to 05 small pieces of product to reduce robustly future displacement risk in Asia.

2. Methodology

The methodology for the study comprised two processes.

2.1 A Desk Study Review

In this method, a wide range of kinds of literature (academic literature, Gray works of literature), secondary data, web portals studied Attended difference of zoom meetings events which has been organized by operational level and policy level parties. The government's legislation, policies, and reports were studied. Studied reports, policy documents revaluations by and individuals, organizations. Went through reports, policy documents, and evaluations of national regional, and international actors in displacement. Read general and country-specific academic and think-thank literature and reports related to the topic.

2.2 Questionnaires Filling

This method was with a sample of specialists, related actors in the region, and several country's individuals, coordinating networks and platforms, displaced popular especially. The questionnaire survey provided information and data complementary to the study. The sample was based on the snowball method, 98% more conducted by E-mails, Skype, and WhatsApps. The context of questioners was open-ended and closed-ended (70% and 30%). Selected sample responses via these instruments were succeeding than envisaged. Because most of the responders were at home, as the covid-19 travel restrictions or quarantine purposes. Though they had enough and convenient time to reply.

03 Synthesis of Findings and Contextual Discussion of Policy Making Environment

Following, this paper presents 05 credible elements to tackle displacement as research findings whilst synthesis contextual mapping for policy decisions.

3.1 Recurring—After Disaster Returning Home has to Transfer in Local Integration or Planned Relocation

In many Asian countries as their usual practice, pre-evacuation is doing or frequently affected by disasters and gathered to evacuation centers and transitional centers consequently adding to the previous location. This bad habit recurring twice or more yearly by adding numbers to the death toll and disaster victims. Therefore, need to implement a well-planned relocation/ permanent resettlement system and inclusive local integration approach to eliminate this bad and dangerous habit. In some cases, establishing and facilitating safer movement through regular mobility and humanitarian migration pathways may be sufficient (not everywhere) to support and protect those at risk and compelled to move due to rapid on- set disasters impact. States need to address their domestic challenges in supporting community via planned relocation by providing leadership on technical support, financial support, and standardization of human rights-based guiding principles for planned relocation otherwise affirmative interest will not come from the community who subject to the incident, States ought to have enabled every resource and ensure community envisage to use planned relocation as a resilience strategy.

"Planned Relocation" is defined as a planned process to prevent recurring displacement of any form.

Planned Relocation as a process

Planned Relocation is a process comprised of certain identifiable phases.

- The first step, and a prerequisite to undertaking Planned Relocation, is to determine whether it is necessary. This decision should be based on scientific evidence and none of the interest of political or other parties. Once the decision in favor of Planned Relocation is made,
- As the second step, analytical studies and assessments are needed, followed by an intensive participatory process of formulating the necessary plans.

- Next, as the third step, preparations must be made for the physical move, including, in some cases, acquisition of land and provision of infrastructure.
- In the fourth step, there must also be focused work with the affected persons, those who may take part in Planned Relocation, those in host communities, and others likely to be adversely affected by the Planned Relocation.
- As the fifth step, a transitional period is needed which time the livelihoods and standards of living of relocated persons are restored and any adverse impacts on other groups are mitigated.
- In the final stage of a successful Planned Relocation, relocated persons are incorporated into all aspects of life in the new setting and no longer have needs or vulnerabilities stemming from the Planned Relocation.

Different contexts of Planned Relocation

Without an understanding of different contexts of planned relocation not be sustainable. Therefore, there are several contexts to be highlighted in the Asian setup.

- Choose appropriate Shelter options until comes permanent relocation in situations of internal displacement Humanitarian interventions usually include shelter programs that support access to emergency shelter and/or transitional shelter, depending on the stage of the operation. An Enabling Environment like a political and social setting in which decisions on Planned Relocation are made is a factor that cannot be over-emphasized.
- Planned Relocation is a challenging undertaking for States, regardless of their level of economic development
- It is also important to note that successful implementation of Planned Relocation is expensive and international financial support will be needed in many cases.
- Planned Relocation is likely to be needed in urban as well as rural areas.
- Planned Relocation will be initiated by persons or groups of persons and will reflect their level of risk tolerance.
- In all types of Planned Relocation, distinctions between 'forced' versus 'voluntary' movement are somewhat artificial. Arguably, all those who participate in Planned Relocation are being compelled to move by forces beyond their control in Asia more examples like disasters and environmental change, including the effects of climate change.
- Planned Relocation is a crosscutting issue, therefore, needs to incorporate insights expertise in a number of areas such as disaster risk management, development, human rights, protection, humanitarian response, migration, and climate change adaptation.

Compulsory element is local/hot site integration need.

Otherwise in a certain period again going to happen the leaving from relocations, this was perceptional by 2004 tsunami in Asia, whilst leaving buffer zones origin places are acquired and now disaster recurring victims are reportedly recorded. Hence relocation and local/host site integration are synchronized. Inevitable several items are the majority of Asian countries would be adopted followings.

- Enhance Legal and Regulatory Frameworks to Integrate Displacement
- Data and Evidence for Policy on Displacement
- Partnership Building
- Localization, Contextualization, and Innovation
- Sustainable urban development for social inclusion and addressing poverty of locals.

Fulfill their social function, including the social and ecological function of land, and foster social cohesion, inclusion, and safety in peaceful and pluralistic societies, where the needs of all inhabitants are met, recognizing the specific financial needs of those in vulnerable situations. When local becomes global- address permanent relocation.

Local nature-based solutions have the potential to contribute to transformational change, even at the global level— for two reasons. First, many local and community decisions can add up to substantial global impact. Second, planetary and socioeconomic systems are interconnected, and local decisions can have impacts elsewhere and at multiple scales. It should not diverge in the situation of relocation also in Asia

The international legal framework and the responsibility of the State towards integrated relocation.

All the 51 states of the Asian Region need to recognize their responsibility and abiding by international legal systems in the perspective of displacement restitutions, which enabling countries to own a framework for sustainable relocation. None of the country nor right to escape by these issues and its enormous right of the victims as well as an inevitable obligation of the state. The right to adequate shelter is one aspect of the right of every person to enjoy an adequate standard of living.

The right to adequate housing was first recognized in the Universal Declaration of Human Rights and subsequently included in various international human rights instruments. While the right to adequate shelter applies to all persons, specific provisions exist at the international level emphasizing its applicability to IDPs. For instance, this is the case of the Guiding Principles on Internal Displacement. The UN Principles on Housing and Property Restitution for Refugees and Displaced Persons, known as the Pinheiro Principles, also contain specific reference to the right to adequate housing and request that States take positive measures to alleviate the difficulties faced by displaced persons living in inadequate shelters.

Finally, Asian region leaders ought to have considered their own relocation previews to prevent displacement recurrent. And the author explores its need for an urgent appeal to compliance of these concepts by the leaders and actors who involving in Bangladesh Cox's Bazar displacement, Syrian Damascus situation, and Yemen massive influxes to enable sustainable relocation measures before recurring the already existing. Further Myanmar Rohingya state issues also need to ensure the same appeal. Amidst Afghanistan has been suffering for more than 40 years – the current crisis adds to the already desperate humanitarian situation. more than half of the population lives on less than one dollar a day. Over 3.5 million people are displaced by conflict within Afghanistan, including more than 600,000 people this year alone. Many international actors involve with urgent relief items and other lifesaving aid since the start of 2021. But the needs continue to grow: many displaced people are living rough, in the open, and urgently need shelter without discriminating and heating to stay safe and warm during the coming winter.

3.2 Climate Change is Not the One & Only Factor that Drives Displacement Risk. A Range of Social and Economic Drivers must be Addressed

Internal displacement tends to severely disrupt the lives of those affected. Sometimes it presents them with new opportunities, but most often it undermines their welfare and wellbeing. As IDPs are uprooted from their homes and separated from their assets, livelihoods, and networks, their ability to earn a living may be compromised The average economic impact per IDP for one year of displacement globally is about $370, based on data from 18 countries. The figure ranges from $109 in Afghanistan to $830 in Syria. The variation arises from differences in the level of needs across affected populations and the estimated cost of meeting them. In countries where the national income is higher, the economic impact resulting from loss of livelihoods is also greater. The highest economic impacts stem from the loss of income and the cost of providing IDPs with housing and healthcare. In the case of large- scale, protracted displacement crises that take place in countries with smaller economies, these impacts can amount to a significant proportion of GDP, Somalia's case around 20 percent. The economic impacts of smaller-scale, shorter-term displacements triggered by disasters tend to go unrecorded. We have not yet been able to calculate these impacts, but aggregated at the global level they would run to billions of dollars. The best example is Myanmar. Myanmar continues to experience multiple shocks, both natural and human-induced. According to UN OCHA, Myanmar's requirements are characterized by a complex combination of vulnerabilities to natural disasters. Urged humanitarian assistance for 941000 protracted displaced people as armed conflict in Myanmar. About 244,000 individuals are internally displaced in Kachin, Shan, Rakhine, and Kayin States, of which about 77 percent are women or children.

Myanmar refugees have been displaced across multiple borders and seek, or are afforded, different levels of protection depending on their countries of refuge. There are estimated to be over 900,000 Myanmar refugees in Bangladesh. Thailand hosts around 97,000 Myanmar refugees in nine border camps, plus many more hidden and undocumented refugees in addition to economic migrants. Refugees from the Northeast often cross the border into China as they refuse refugee status, periodically returned across the border.

Differentiated impacts determine differentiated solutions

According to the understanding of internal displacement becomes more complicated every year. We must, however, continue to expand our monitoring of the scale and scope of this global phenomenon, and enough resources are being dedicated to addressing affected population needs. One considerable information gap remains in painting a more precise portrait of IDPs

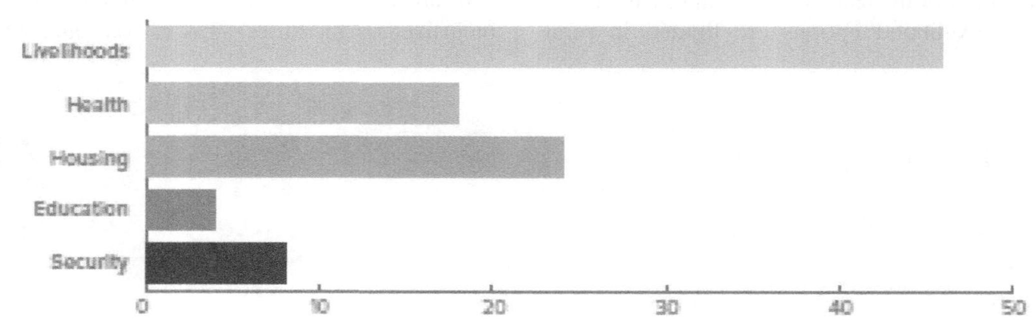

Fig. 38.2 Shows displacement population needs in Mn at 2020

Source: (IDMC-GRID, 2021)

Their sex, age, disability status, and other characteristics are often unknown. An increasing number of data collectors gather and publish local-level information disaggregated by sex and age, however, age groups are not standardized, where it impossible to compile national-level figures in most of the country. In the absence of consistent categories, national-level demographic data can be applied to the number of IDPs in each country to establish estimates by age group. Doing so suggests that 23.3 million people under 18 were living internal displacement worldwide as of the end of 2020. The figure includes 7.2 million children under five and 12.8 million of primary and early secondary school age. There were also 10.5 million IDPs aged between 15 and 24 and 2.6 million over 65. This type of information is vital to ensure targeted and inclusive responses. Each demographic group faces specific risks and has needs that interventions should take into account. Infants may need vaccination and nutritional supplements, for example, school-age children's educational support and young people's vocational training. Women and girls displaced by conflict face particulars risks, and those displaced by disasters tend to be disproportionally affected. They also have adaptive capacities that are often overlooked and cannot be fully harnessed without disaggregated data.

Less voice, less power, more suffering than climate change event impact.

Some groups suffer disproportionally from the continued pressures humans put on the planet, which exacerbates group-based inequalities, also known as horizontal inequalities. It happens through several main channels:

- Disadvantaged groups' increased exposure to less socially secure factors, their higher susceptibility to potential damages caused by the pressure humans put on the planet, and their lower ability to cope and recover from adverse events.
- Most affected populations depend on natural resources, such as coastal agricultural, pastoral, and forest communities, because of adverse effects on food, water, and infrastructure.
- This increases multidimensional horizontal inequalities. For instance, livelihood options are limited in rural communities of Burkina and mountainous areas in Nepal. And many people rely on weather-dependent agriculture for their food safety and generate resources to cover other living expenses.
- Adaptive capacity is typically low in these communities due to low levels of education and information.

Socioeconomic indicators

These should compare displaced people and their hosts with the population affected by disaster and need to be cover employment status and income, housing, social and capital assets, school enrolment, physical and mental health, discrimination, and prevalence of violence on this, the inevitable factors are,

- Consider disaster displacement and designate institutional leadership, authority, and budgetary allocations as required.
- Mandate collaboration across Government ministries with clear designations for provisions addressing displacement.
- Include specific references and measures to reduce disaster displacement risk, respond to disaster displacement and strengthen the resilience of people displaced by disasters. Strategies and plans should also include baseline data, targets, indicators, and timeframes to monitor progress in implementing measures.
- Be reviewed to ensure alignment with national, regional, or international legal instruments that address disaster displacements, such as the Guiding Principles on Internal Displacement and National laws and policies on internal displacement and other forms of related human mobility.
- Designate roles and responsibilities for addressing disaster displacement and other related forms of human mobility to facilitate coordination.

Keeping eyes on small-scale events and learning to tackle slow on-set events.

Many countries already prepared to tackle massive events and all the stakeholders are ongoing to response with inclusiveness measures. This is good and wish to keeping.

Fig. 38.3 Shows displacement population in different age category at 2020

Source: (IDMC-GRID,2021)

However inevitable responsibility of every citizens and residents is to prevent frequent and often occurrence's repercussions. Beyond monstrous and devastating disasters in Asian Region significant proportion of displacement happens through triggered by localized small scale events. 2019 analyzed data shows 54% of all event effected to displaced fewer than 100 people. Data repositories widely used for more than 100 people displacing like EM-DAT. But this fewer than 100 people displacing are not events of other parts, and these low visibilities explicitly caused to not only to lack assistance for recovery those victims but also less lead effort for prepare and to prevent such small scale disasters in future. (Fig. 38.4)

Other propound study is most countries in Asia remarkably from Northeast and south is pumping more two, three of fold displacements as disabling to tackle slow-onset events smartly.

Know on Climate change is already influencing patterns and trends related to human mobility.

In 2020, the Internal Many countries are already prepared to tackle massive events, and all the stakeholders are ongoing to respond with inclusiveness measures. 2019 analyzed data shows 54% of all events affected displaced fewer than 100 people. Data repositories are used widely for the events of more than 100 people displacement like EM-DAT. But these fewer than 100 people displacement are not events of other parts that low visibilities explicitly caused to not only to lack assistance for recovery those victims but also less lead effort for preparation and to prevent such small scale disasters in future. (Fig. 38.4)

Another propound study is most countries in Asia from the Northeast and south are pumping more than two, three of the fold displacements as disabling to tackle slow-onset events smartly. The world is amidst a climate crisis. Credible and evidenced-based predictions indicate that global temperatures will rise by between 2.7 to 3.1°C by the end of this century under current policies. This scale of change means that disruptive and deadly events such as extreme heatwaves and precipitation will be more frequent and intense in the future. It also means devastating and irreversible impacts on the Asian Region are being expected. This year's featured example was showcased from Bangladesh Cox Bazar's flood, which directly caused displaced people displacement further. and Indonesia massive flood in twice January and February in 2020 uprooted more than 190000 people. Both incidents occurred in the last two months of the rainy season (November-March), which was exacerbated by a strong La Niña that brought on between 20- 40% more rainfall than usual. And global warming at more modest levels, resulting from successful efforts at mitigation, will still result in an acceleration of natural hazards exacerbated by climate change. To tackle these unprecedented events, need to know,

Know on Climate change is already influencing patterns and trends related to human mobility.

In 2020, the Internal Displacement Monitoring Centre (IDMC) calculated that weather-related disasters displaced more than 30 million people—three times more than those displaced by conflict in the same year. In South Asia, 2050 will be expected 40 million internal climate migrants. (Groundswell, World Bank Group-2018) "Floods come every

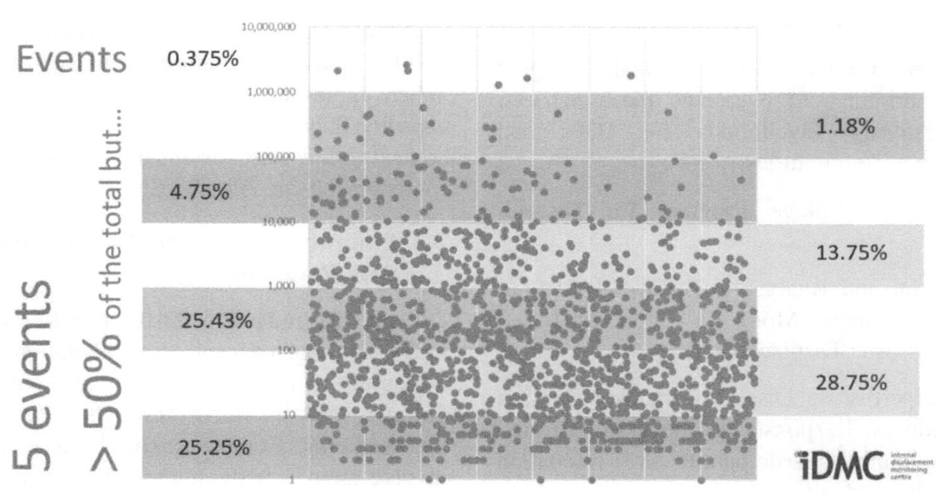

Fig. 38.4 Shows Data from IDMC's disaster displacement database shows that 54% of all events in 2019 displaced fewer than 100 people

Source: (IDMC-GRID, 2021)

year, but this year the situation is worse. Now all my family is living in one relative's house. I don't want to go back to my village, mainly because of the flood. In Dhaka, I can work and have a good secure life." —Monoara Khatun (23) Bangladesh.

South Asia could see an average of 35.7 million climate migrants by 2050 under the pessimistic reference scenario. By 2050, climate migrants make up about 23 percent of all internal migrants The Southern Indian highlands, especially between Bangalore and Chennai will be climate in-migration hotspots. Parts of Nepal, as well as northwestern India, also see climate in-migration. Climate out-migration hotspots include eastern and northern Bangladesh, the northern part of the Gangetic Plain, and the corridor from Delhi to Lahore. Climate out-migration will also occur from coastal metropolitan areas such as Mumbai, Dhaka, and Chennai as a result of sea-level rise and storm surge impacts. Irrigated areas and rice-growing areas are likely to see population dampening as a result of out-migration. In contrast, rain-fed cropping areas are likely to see population increases. (Groundswell, World Bank Group-2018)

Know on those already displaced or at risk of displacement in the context of climate change need nationally adopted practices or policy solutions today.

Refugee International(RI) has reported the Climate Displacement happened in dozens of countries where slow-onset disasters are becoming more frequent and intense over the last decade. Disasters are leading to protracted displacements, People who were displaced by the slow-onset disasters are receiving inadequate support from their governments and the international humanitarian community on a range of issues by providing durable shelter, planned relocation, and livelihoods alternatives.

For instance, some Asian countries use nationally adopted practices or policy briefings to overcome these barriers. Afghanistan - Restoring Livelihoods for IDPs and Displacement-Affected Communities.

Indonesia - The Sister Village Program for Disaster Preparedness.

Iraq - Data Collection and Analysis to Inform Efforts to End Protracted Displacement. Mongolia - Forecast-Based Financing to Avoid Disaster Displacement.

Nepal - Building Expertise on Disability-Inclusive Design in Disaster Preparedness, Response and Recovery Efforts. Philippines - Community Participation in Evacuation Planning in Metropolitan Manila.

Philippines - Disaster Displacement Data from Preparedness to Recovery. Philippines - Practical Solutions for Protecting IDPs' Right to Vote.

India – Digital Health Framework.

Know on targeted policies/Activities can help mitigate the impacts of future climate-related displacement and environmental mobility crises.

Projections of future climate-related migration trends are sobering. The World Bank predicts that unchecked carbon emissions and unequal development policies could lead to 143 million internally displaced "climate migrants" by 2050 in Latin America and the Caribbean, South Asia, and Sub-Saharan Africa. Changes in habitability could affect large parts of the planet.

Know on Slow-onset events can erode the capacity of ecosystems of providing critical services such as the availability of fresh water, food, shelter, and energy production.

This Scarcity of vital resources can lead to a serious disruption of livelihoods and acute recurring displacement. The regions with the highest vulnerability to undernutrition are also areas where yield losses are related to Slow-onset events. For example, mangrove destruction due to sea-level rise can result in higher exposure of coastal communities to storm surges as the capacity of mangroves to break waves and protect low-lying coastal areas from flooding diminishes. When become the population coping capacity is weakened by underlying slow-onset events it may turn into disasters. (Fig. 38.5)

Know on Slow-onset events may turn into disasters prompted by rapid-onset events to lead to massive displacement. Many impacts of slow-onset events are turned into disasters by contradicting with rapid-onset events. For instance, when sea-level rise turns into a flood when desertification turns into wildfire when temperature intensifies turn into heat waves. When rapid-onset disasters overwhelm the population's capacity, they may increase the risk of displacement. In these contexts, slow-onset events are one of the underlying factors for displacement. (Fig. 38.6)

3.4 Embrace Uncertainty Challenges Focusing on Risk Reduction

Disaster risk reduction (DRR) is a Key tool, encompassing the broad range of local, regional, national, and international measures to prevent or reduce the damage resulting from hazards such as floods, droughts, cyclones, earthquakes, and tsunamis. Comprehensive DRR policies and practices involve short-, medium-, and long-term actions, including early warning systems, public education, development of environmentally sensitive building codes, and other forms of structural mitigation, as well as land-use planning.

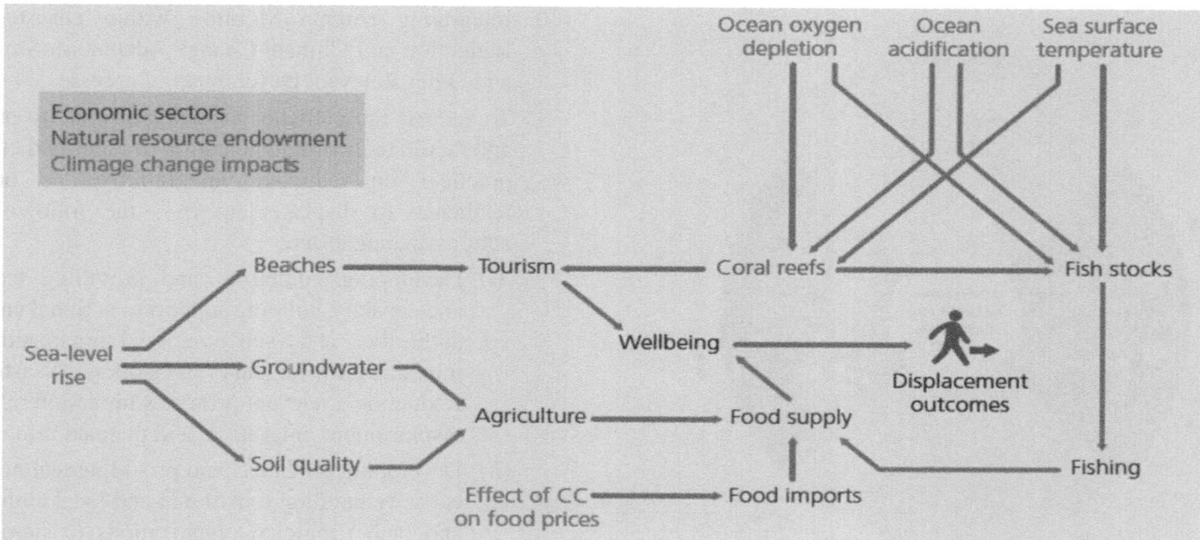

Fig. 38.5 Shows The impacts of slow-onset events to ecosystem and driven to displacement
Source: Groundswell, World Bank Group-2018

Fig. 38.6 Shows Slow-onset events which effected to create disasters
Source: Groundswell, World Bank Group-2018

Due to the overwhelming uncertainty in Asia many more Displacements occurred as a course of different scale risks cascading disasters by responding to the pandemic, Asia-Pacific countries have also had to contend with their regular sequence of other natural hazards, including cyclones, typhoons, storm surges, floods, droughts, heatwaves, glacial lake outbursts, locust swarms, earthquakes and volcanic eruptions, most of which have been hydro-meteorological. Tropical cyclones, such as Amphan, Nisarga, and Tauktae, have been hit countries in South and South-West Asia. Major flood events were reported in China, Japan, Papua New Guinea, Pakistan, the Islamic Republic of Iran, Kazakhstan, and Uzbekistan. After all these disasters, many established measures for prevention, response, and recovery were interrupted. And lockdowns, travel restrictions, and other containment measures were imposed as a response to COVID -19. At the same time, these natural hazards have hampered the response mechanism to the pandemic and facilitated the

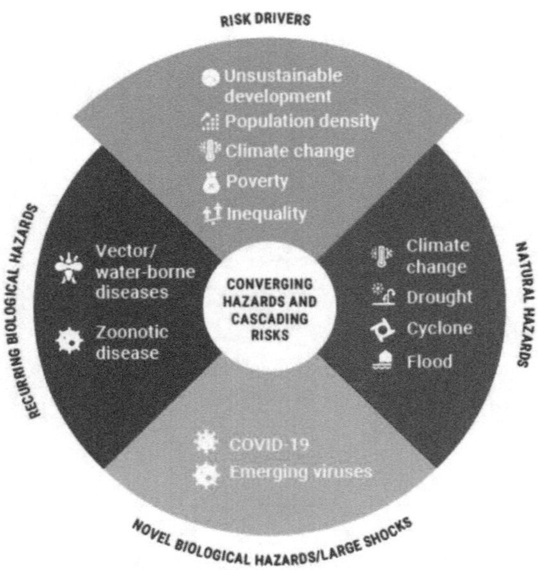

Source: United Nations, Economic and Social Commission of Asia and the Pacific, "Weaving a stronger fabric: Managing cascading risks for climate resilience", Policy Brief, 26 January 2021b. Available at https://www.unescap.org/kp/2021/weaving-stronger-fabric-managing-cascading-risks-climate-resilience

Fig. 38.7 Shows cascading risks including biological hazards and with major risk drivers

Source: UNESCAP, 2021 (www.unescap.org/kp/2021/weaving-stronger-fabric-managing-cascading-risks-climate-resilience)

spread of the virus, as people often had to crowd together in emergency shelters. Disasters have also caused economic damage. To embrace these foam of uncertainty challenges several DRR actions could be proposed.

1. Forecast-based Financing and Early Action on Displacement
2. Risk-aware development from national to sub-national levels
3. Continued Preparedness and Response Simulations at All Levels
4. Cross-sectoral coordination at the national level to address Complex Issues of uncertainty.
5. Enhance Legal and Regulatory Frameworks to Integrate Displacement uncertainty challenges.
6. Data and Evidence for Policy on Uncertainty drove Displacement prevention.
7. Reducing uncertainty, targeting groups
8. Minimizing actions related to climate change uncertainties in the underlying planetary processes and their implications.
9. Strengthening the Management of Disaster Displacement Risk in the Country of Origin.
10. Integrating Human Mobility within Disaster Risk Reduction and Climate Change Adaptation Strategies, and Other Relevant Development Processes

    To address these challenges and gaps, and to promote and facilitate the implementation of identified effective practices on reducing vulnerability and building resilience to displacement risk, the following key actions are suggested:

    (i) Developing guidelines and providing technical and capacity-building support to national and local authorities and (sub-) regional organizations to implement the Sendai Framework on Disaster Risk Reduction's relevant priorities for action related to displacement, migration, and planned relocation.

    (ii) Developing guidance, and providing technical and capacity building to national and local authorities, and (sub-) regional organizations to support the inclusion of human mobility considerations within climate change adaptation strategies and policies, including through collaboration, as appropriate, with UNFCCC bodies and processes.

    (iii) Continuing to consider the issues of displacement, migration, and planned relocation within the ongoing activities of the Conferences of Parties to the UNFCCC.

    (iv) Revising laws, policies, strategies, and plans on disaster risk reduction, climate change adaptation, and overall resilience building to integrate human mobility aspects.

    (v) Ensuring that access to funding is secured and facilitated to finance human mobility-related measures within local, national, and regional climate change adaptation, disaster risk reduction, and resilience-building plans and activities.
11. Facilitating Migration with Dignity as a Potentially Positive Way to Cope with the Effects of Natural Hazards and Climate Change
12. From theory to change practical norms

    We take several steps back to observe how and if pro-planet values have been formed and whether they have challenged and changed existing social norms throughout society and, if not, how it can be accomplished in a new setup and how to embrace uncertainty challenges to minimize displacement fears.
13. Alignment of domestic laws and policies with the guiding principles to embrace uncertainty challenges.

    In the period between 1992 and 2020, 43 states with situations of conflict-induced displacement adopted laws and policies on internal displacement. A significant majority of these laws and policies were introduced following the adoption of the Guiding Principles. There are, however, significant regional variations in the number of states introducing laws and policies. But something is greater than nothing, therefore, now most Asian countries have a trend to align national policies with guiding principles.

14. Challenge of Change and Uncertainty
15. Challenges and Gaps

- In the absence of improved resilience to future disasters and environmental stress, returnees may continue to be at a high risk of repeated crises and recurrent displacement.
- Slow or inefficient recovery and reconstruction efforts
- resolving land tenure issues
- financing construction with higher building standards to withstand future disasters,
-
- provide support for the development of diversified livelihood opportunities as quickly as possible to support self- reliance.
- Local governance models generally face challenges to ensure the inclusion of disaster displacements
- lasting solutions is often insufficient or ineffective due to significant conceptual, operational and institutional differences in the respective approaches between humanitarian and development actors.
- when movement is unavoidable, policy options include facilitating migration and planned relocation to move people away from hazardous areas to safer areas before a disaster occurs.

Need to put the community in the center, with joint force to force mobilize political will and identify the nexus between conflict and climate change.

Migration includes both short-term and long-term movements. The number of people who may be displaced by climate change-related phenomena by 2050, ranges from 50 million to 1 billion people. On the other hand, one argument is migration catalyzed by conflict. An example of non-climate related migration is the utmost of tensions towards conflict, particularly over international borders. Examples from mass migration from Bangladesh into Northern India over the last few decades have resulted in tension and conflict between ethnic groups. The tension persists between Pakistan and India in on Kashmir most crucial to the region. More, new incidents raised in the 2021 year molted with Myanmar military junta coup consequently reported 239000 from February to August 2021 displacement. More and more, Syria also triggered new displacement since the continuity of fuel while Yemen records displacement 143000 from crisis and with the flood, both are 223000 in 2020. Between Israel and Palestine were intensified tension and sought shelters, 113000 people in May 2021. This wave of violence forced to deteriorate the peace of the region. The latest incident after a long period of conflict of Afghanistan exacerbated and still happening internal migration as well as border crossing without aspiration. It was reported 338000 from crisis and 22500 January to June from flood displacements. Meanwhile, Indonesia declared a state of emergency as acute of a major flood in 11 regencies/cities. It was informed after 50 years the largest flood caused 557000 displacements during the first half of the year 2021. Joint forces and Assistances need to identify the trajectory of the political will and conflict and climate nexus as driving forces to the displacement of more than 95% of events.

Focusing on these circumstances recommendations have been given by UN Secretary General's high-level panel on IDPs in their final report of September 2021. This recommendation has been shown by the GP20 report also with more or low differences. (The plan of actions for celebrating of 20th anniversary of the guiding principle of IDPs).

The GP20 report- Plan of Action focused on four interconnected areas of work:

1. IDPs participation
2. national law and policy
3. data and analysis
4. protracted displacement and supporting durable solutions.

In this mentioned as priority 01 IDP participation means to put the community in the center.

In the recommendations explicitly explain the basic concept that are need to tackle displacement where this paper is being discussed rationale same.

(i) Achieving sustained political will amongst relevant government authorities at all levels to address internal displacement;
(ii) Establishing government leadership and designated roles and responsibilities across line ministries and at all levels of government;
(iii) Building effective partnerships and coordinated approaches between Governments, international actors, and civil society;
(iv) Ensuring adequate data to inform responses and monitor progress towards durable solutions;
(v) Effectively engaging IDPs, displacement-affected communities, and persons at risk of displacement;

(vi) Meeting IDPs' needs at scale, particularly concerning programs seeking to prevent and find durable solutions to displacement that include livelihoods, housing, land and property, and social cohesion elements;

(vii) Anticipating the future impacts of climate change on population movements;

(viii) Securing sufficient and flexible financial resources, particularly when faced with competing priorities.

This note is visible in the high-level panel on IDPs final report also with some areas difference more and less, but the main concept is parallel which this paper discusses as 05 credible elements. The UN-Secretary General's panel's report explores ten recommendations,

1: Make solutions a nationally owned, whole-of-government priority 2: Political will is key and should be catalyzed

3: Ensure the whole of society is invested

4: Make better use of the capacities of the private sector for solutions Durable solutions 5: Make the UN fit for purpose and accountable for solutions

6: Harness international financing for solutions

7: Strengthen the effective use of internal displacement data

8: Address the drivers of displacement and reduce displacement risks

9: Strengthen the quality of protection and assistance to IDPs and host communities

10: Sustain momentum and ensure robust follow-up

These recommendations reveal what we need in Asia to tackle internal displacement where is being discussed by this paper manage of political will is a key strategy to prevent and bring solutions to the subject. In the preceding time, even many Asian countries fight to overcome displacement issues that arise by politically fragile and conflicts. All citizens and residents of Asian countries are being urged by global actors giving their priority attention to operationalize and maintain these credible elements to plug long-lasting solutions for the occurrence and recurrence of displacements in Asia.

4. Conclusion

Profiling in most credible elements to tackle displacement in Asia in the nearest ten years is conceptually different from other profiling settings in terms of Acceptance and best practices, Nevertheless, when the phrase of operationalizing, the methodological challenges that are needed to be addressed in the common ground as Asian citizens. Because driving forces for IDPs are mostly general to the region itself. Therefore, to bring long-lasting solutions regional actors necessary play a critical and inevitable role. This study endeavored to comprehend various success factors for 05 credible elements to tackle displacement in the nearest ten years in Asia. The impact of these successful thoughts and words may bring into action by adhering to your national plans on IDPs as the Kampala convention and lake great papers done. The findings indicate that the successful implementation of policies and plans of actions nationally on IDPs to a great extent can be attributed to function durable solution's success in the dimensions such as institutional mechanism and legal and integral parts of the country.

The conceptual model developed in this study comprehends the IDPs management process and emphasizes a nationally built concept with the consultation of global guidelines, approach for a long-lasting solution in the county setup.

The Guiding Principles on IDPs come with no monitoring or enforcement machinery, but a global effort has begun to monitor and oversee compliance with their provisions. Governments, UN agencies, regional bodies, international and local NGOs, and the displaced themselves are all involved. Over time and with sustained usage, the Guiding Principles are bounded to make a difference in the lives of displaced people throughout the world. Therefore, not showcase much difference between The final report of the UN Secretary-General's high-level panels on IDPs recommendations and the GP20- Plan of actions report. Hence author denotes in this study's recommendations to follow both final reports when operationalizes these credible 05 elements in the IDPs management process.

No doubt uneasily caused by the process is more related to the fact that the principles reinforce the idea that governments have an obligation for protection and assistance to their displaced populations that if they lack capacity or are unwilling to do so the international community must get involved.

The author applauds the High-level panel's report and GP20's report conclusion on Migration. However, ambition and the vision of human mobility and statelessness people are not easy to understand. As the study has indicated, many IDPs are not sobering hence, predictions about this issue are critical and crucial, and most of them are beyond the capacity therefore urgently need thoughts and words into actions.

The study poses certain limitations. The study attempted to research and model success factors through stories of IDP's participation physically. Covid-19 pandemic has constrained on the pure objective of the research, though technology brought unique solutions fasted and cost less.

This document has attempted to put forward possible solutions. Some of the key lessons which emerged subject experts are regional and global practices in the human mobility context. By making available reliable and adaptable, credible 05 elements present synthesis practical endeavors to the subject. The findings may help prioritize and shape policies, frameworks, circulars and guidelines, any other form documents on internal displacements in the Asian region.

5. Acknowledgement

The author is grateful to Barbara Essig – The Policy Expert, and Bina Desai – Head of the program at Internal Displacement Monitoring Center (IDMC) for extending their technical support by establishing fundamental insights into the topic.

References

1. National Disaster Management Plan (2016). A publication of the National Disaster Management Authority, Government of India. May 2016, New Delhi.
2. Internal Displacement Monitoring Centre (IDMC), Global Report on Internal Displacement 2021.
3. Report of the Special Rapporteur on the human rights of internally displaced persons, Persons with disabilities in the context of internal displacement, A/HRC/44/41, May 2020, para 17.
4. Norwegian Refugee Council, NRC Submission to the High-Level Panel on Internal Displacement, May 2020, para 15.
5. UNDP/SOAS University of London, Towards Development Solutions to Internal Displacement: A Political Economy Approach, June 2021
6. World Bank, Groundswell: Preparing for Internal Climate Migration, 2018
7. Inter-Agency Standing Committee, IASC Framework on Durable Solutions for Internally Displaced Persons, April 2010
8. United Nations Office for Disaster Risk Reduction, Words into Action, Engaging for resilience in support of the Sendai Framework for Disaster Risk Reduction 2015-2030, 2019, p.08
9. Journal of South Asian Disaster Studies, Journal of the SAARC Disaster Management Centre, Vol 3 No 2 December 2010, p.2-5
10. https://www.unescap.org/kp/2021/asia-pacific-disaster-report-2021
11. Eva-Lotta Hedman, Refugees, IDPs, and regional security in the Asia-Pacific 2009 CSCAP, http://eprints.lse.ac.uk/26550/
12. United Nations, Economic and Social Commission for Asia and the Pacific (ESCAP) (2020). Asia-Pacific Migration Report 2020: Assessing Implementation of the Global Compact for Migration (ST/ESCAP/2801).
13. Report of the UN Secretary-General's High-Level Panel on Internal Displacement, Shining a Light on Internal Displacement a Vision for the Future, September 2021, Report available at https://internaldisplacement-panel.org
14. UN Special Rapporteur on the human rights of internally displaced persons, GP20 COMPILATION OF NATIONAL PRACTICES, 2020, Available at https://www.globalprotectioncluster.org/gp20/
15. United Nation –OCHA, Guiding Principles on Internally Displacement, September 2014, Available at https://www.OCHA.org.

CHAPTER 39

Climate Change and its Interconnectedness with Natural Disasters: A Global Perspective

Areeba Naaz
Student, Centre for Disaster Management, JMI (A Central University), New Delhi, India

Parvez Hayat IPS
Advisor/Faculty, Centre for Disaster Management, JMI (A Central University), New Delhi, India

Abstract

Natural disasters have been increasing in frequency and magnitude over decades. Economic losses also have been increasing; however, loss of life and injury have been fairly less due to huge investment in early warning systems. It is also being observed that besides the increase in hydro-metrological disasters, climatological disasters are on increase over a decade. It refers to an increase in climate change-triggered hazards such as an increase in sea level, fire incidents, melting of glaciers, flash floods, and so on. Such hazards are to be addressed throughout human history. Disasters, whether natural or human-induced are common phenomena around the globe. Disasters continue to occur and the same is constantly increasing in its magnitude and frequency impacting the economy globally for multivariate reasons. Globalization and industrialization have led to consequences of the destruction of flora and fauna. Scholars argue that geographical location is less responsible than increasing settings of technological and industrial development which causes threats to the environment. The impact of disaster is more in under-developed and developing countries as compared to developed countries where population density is relatively much higher. Losses in terms of economy, human and cultural are many times higher in developing countries which are more vulnerable and technologically and resource-wise at a disadvantageous position.

Keywords: Disaster Management, Police Response, Climate Change Vulnerability, Global Warming, Greenhouse gases

1. Introduction

Earth's climate system comprises the air, land surface, snow and ice, oceans and different waterways, and living things. Climate is frequently characterized as 'average weather'. Climate is normally depicted in terms of temperature mean and variability, precipitation, and wind throughout some period, ranging from months (usually 30 years). External forces include natural processes like solar variations and volcanic eruptions, and also human-induced changes in the climatic systems. The fundamental ways to change the radiation balance of the Earth are: 1) by changing the incoming solar radiation (e.g., by changes in the orbit of the Sun or the Earth); 2) by changing the fraction of solar radiation that is reflected from Earth back towards space by (e.g., by changing greenhouse gas concentrations). Such changes alter the climate directly and indirectly, through various mechanisms.

Global warming caused by human action is called Anthropogenic global warming. It is the result of increasing greenhouse gas emissions; it shows up in the form of hydrometeorological events but with increased intensity and frequency. It usually occurs in the form of floods, drought, heatwaves, and tropical cyclones, besides other environmental disasters such as degradation of Earth's ecosystem, food unavailability, absence of drinking water, and impacts on the livelihoods of the community (IPCC-ARC 5). The impacts are further observed through slow-onset events such as sea-level rise, increasing temperature, ocean acidification, glacier retreat, loss of biodiversity, and extreme weather events. Since 1860, an anomaly in global air temperature is observed to rise by 0.4 °C. Also detected is an increase in ocean temperature by 0.10 °C (0–700 m depth) during 1961–2003 (IPCC 2007). Urbanization in the coastal area is thus in danger due to sea-level rise, flood events, cyclonic storm, and storm surge. Global warming will lead to many expected and unexpected changes in the coming years. Warming will be larger over the land, and places situated on higher latitudes. The expected temperature increase will lead to many consequential changes like the changes we saw in the case of Central Europe- Melting of ice sheets, thermal ocean expansion, an increase of extreme weather events, increased intensity of the hydrological cycle. The adaptation to such changes will be costly for societies all over the world, and in the cases of undeveloped countries, impossible.

2. Green House Effects

About 30% of the sunlight that enters the atmosphere is reflected back to space. Two-third of this reflectivity is due to clouds and small particles in the atmosphere known as 'aerosols'. Light-shaded spaces of Earth's surface – basically snow, ice, and deserts reflect back the remaining one-third of the daylight and the discharge from volcanic eruptions travel high into the atmosphere. The air alone is capable of trapping thermal radiation. Because the Earth is much colder than the Sun, it radiates at much longer wavelengths, primarily in the infrared part of the spectrum.

The world's largest producer of CO2 is the US, followed by China at 2nd place and India as the third-largest producer of CO2. Besides, the planet's average global surface temperature is also increasing by 0.8°C over the past 100 years (GISS-NASA). It has been noted that 9 out of 10 warmest years on record were since 1995, 2005 was the warmest year on record following after 1998 (which was an El Nino year) (GISS-NASA). Over the last decade the weather scientists have observed a 0.5-degree increase in global warming with emissions too showing an increasing trend. Therefore, countries must address climate change and its effects that have been causing huge economic losses by contributing to natural disasters.

3. The Geographical Warming Pattern

The most vivid evidence for global warming is the retreating ice bodies from the poles to the tropics. Sea-level rise from the melting of ice is the largest potential threat of global warming. Global warming in polar regions may reach levels of 130,000 years ago by the year 2100, when sea levels were several meters above compared to what it is today. Arctic Polar Ice Cap has shrunk more than 20 percent in size since 1979. The size of the summer polar ice cap has shrunk more than 20 percent. On Sept. 21, 2005, sea ice extent dropped to 2.05 million sq. miles, the lowest extent yet recorded. This loss is twice the size of Texas.

4. Future Emissions

The Current global CO2 emissions are 34 billion metric tons per year. The rapid rise of this CO2 remains in the atmosphere for decades. Instabilities cause the climate system to inherit natural variabilities, which makes detecting relatively small climate change signals difficult. These changes only come under the radar when these signals get stronger over the time. In the case of climate change, the majority of scientists agree that our planet is getting warmer and we humans are the main cause. But we can't dismiss the fact that the past global warming would have occurred anyway because of natural factors.

The increase in temperature has caused the number of hot days to rise, and the number of cold/frost days have decreased. At mid and high latitudes, heavy precipitation events have been recorded. In some regions, such as parts of Africa and South Asia, droughts have become more frequent and intense, over the past few decades.

Prediction for the coming century suggests that there will be a continuous rise in the number of hot and very hot days, and the number of cold and very cold days will continue to decrease. Moreover, precipitation frequency and intensity will increase over many regions, however, the return period of extreme rainfall events will decline, which could suggest a rise in the number of floods and landslides. As changes in temperature extremes are quite certain and changes in precipitation or drought occurrence are certain, wildfire can also be predicted, but projections for small-scale atmospheric phenomena, such as storms are subject to greater uncertainty.

5. Heat Waves

The European summer of 2003 was unusually hot. The intense and prolonged heat-killed between 22,000 and 35,000 people, the majority of which were elders, and caused economic losses of over USD 13 billion (IFRC, 2004). The heat was repeatedly linked to climate change. Even though no single event can be directly responsible for climate change, the indirect, evidence that was gathered in various publications at the time is strong. In particular, Stott et al. (2004) showed that just a half a degree increases in the average summer temperature, as over the past century, for example, strongly influences the probability of the occurrence of a dramatic heatwave, such as in 2003. In fact, it was estimated that it is very likely that human-induced climate change has already at least doubled the risk of a heatwave like that in 2003, and that by the 2040s, every other summer could well be as hot as the one of 2003. According to some emission scenarios, at the end of the twenty-first century a summer like 2003 would be considered a cold one (Stott et al., 2004). However, just like and hazard, heatwaves only become disasters due to societies' inability to respond properly to the meteorological conditions. Heatwaves can be predicted at least three days in advance, and societies must be ready to read that signal and take action. Accordingly, the best way to deal with the increasing risks of heat waves is early warning and preparedness.

6. Floods

Several unusual flood events have made us debate about potential climate change impacts on natural disasters. In Europe, the record precipitation and flood events in the winter of 2000–01 in England and Wales (Marsh, 2001) and the larger-scale flood events of the Oder (1997), Elbe (2002) and Rhone (2002) are one of the many examples. The intensity of precipitation in the river basin is not the only cause of such flood events, but also the flow regime, which is shaped by land use in the watershed area. Moreover, heavy flood events are generally rare, so it is difficult to separate 'freak events' from real changes in the probability of a particular event. These two factors make it difficult to attribute changes in flood occurrence firmly to global warming.

7. Disaster and Climate Change Vulnerability

The SAARC region is a hotspot of almost all events related to climate change. It doesn't help the case that the Great Himalayas lie in the region which shapes the monsoon dynamics of the area. It is also a fact that the glacier runs the network of rivers in the region that sustain the livelihoods, economy and culture of the people situated in the area. The flood plains of major rivers e.g., Brahmaputra, Ganges and Indus draw their sustenance from the Himalayan ecosystem and support the life of over 1.5 billion people (IPCC, 2007a-c; 2012). The Himalayan region is under great stress due to the retreating glacier which increases threats of GLOFs (Glacier Lake Outburst Flood) in countries such as Bhutan and Nepal. The melting of the Himalayas poses a serious risk to the sustainability of the water resources as 21% of the world population lives in the South Asian region which comprises of only 8.3% of the global water resources.

The deadliest examples of climate change with unplanned urbanization are The Mumbai floods of 2005; the Leh Cloudburst in 2010; the Uttarakhand flash floods of 2013; the Kashmir floods of 2014; Cyclone Phailin of Odisha and Andhra Pradesh in 2013; and the Chennai Urban floods of 2015. The whole of South Asian regions faces the same problem with varying intensities.

Climate change effects can be seen mostly in hydro-meteorological events. Its impacts can be observed in the form of increases in the intensity and frequency of floods, drought, tropical cyclones, and heatwaves besides the environmental disasters such as of destruction of ecosystems, reduced availability of food and drinking water, and other impacts on the livelihoods of communities. South Asian region is the most disaster-prone region in the world, with 50% percent of all-natural calamities being hydro-meteorological ones, which arise due to the impact of the monsoon and cause maximum damage.

The global temperature since 1860 is observed to have risen by 0.4 centigrade. Besides, there have been increases in the ocean temperature by 0.10 centigrade (0-700 meter depth) during 1961-2003 (Intergovernmental Panel on Climate Change, 2007). Carbon emissions and urbanization are the reason for the above.

In South Asian countries such as India, rapid demand-driven urbanisation takes place and people tend to settle in the region without a proper sewage system, which is the major cause of insufficient sewerage. Inefficient waste management leads to greenhouse gas emissions. This creates an unhealthy environment for people to live in along with respiratory issues among the population.

Natural and hydro metrological disasters are the ideal setting for a variety of geo-hydrological disasters and even for the emergence of new disasters. Such changes can be seen around the world in the form of sea-level rise, coastal floods, coastal degradation, receding of glaciers, degradation of forest and the ecosystem etc. Therefore SAARC members need to work together on integrating the DRR plan to mitigate the climate change effects.

8. Climate Change, Disaster Risk Reduction and Resilience

The structure of Disaster Risk Reduction (DRR) has changed phenomenally over the period of time. The organization of Climate Change Adaptation (CCA), evolving from diplomacy to the government's domain-specific development agenda, has brought in legislation to govern the elements responsible for climate change, besides building several action plans and missions for its adaptation and mitigation. CCA, DRR and urbanisation, the measures that are suggested are mapping vulnerability and risk profiles. Climate change may be worsening the hazards, the solution to this problem lies in addressing known vulnerabilities to cyclones as they occur frequently. Moreover, risk reduction programs need to keep the sea-level rise in account, as it causes much higher damage if a cyclone occurs. Such methods can be factored into coastal zone management and infrastructure planning system.

The magnitude of the impact of any disaster depends on the choices we make for our lives and also for our environment. These choices are regarding where and how we build our homes and other infrastructure. When it comes to resilience, the political system we have mattered, how our financial system functions and even what we teach in our schools and colleges count. Each decision and choice make us more vulnerable - or more resilient to disasters. For understanding the impact of disasters one must have an understanding of society's vulnerability and resilience to make choices on how to intervene.

9. Climate Change Induced Vulnerability

Climate trends and variability in the SAARC region are characterized with increasing air temperatures that are more pronounced during winter than in the summer. In past two decades inter-seasonal, inter-annual and spatial variability in rainfall trends have been observed and there is an increase in the intensity of rainfall and decreasing the number of rainy days. The current rate of sea-level rise in coastal areas of Asia is reported to be 1–3 mm/year, which is marginally greater than the global average (Cruz et. al., 2007). The following Table 39.1 provides an observed characteristic in the surface air temperature and rainfall for the member countries of SAARC.

The climate models show an increase in the average warming during summer and there is general consistency among the models in their output for winter. Warming in this region is projected to increase by 2–4°C.

India is prone to extreme climatic events and related disasters due to its unique geo-climatic condition till the '90s of the last century Agriculture Ministry was looking after a disaster which subsequently was transferred to the Ministry of Home Affairs in 2002 following the Bhuj Earthquake of 2001. A legal framework was created through Disaster Management Act 2005, for addressing different activities at the national and sub-national level with a budget for a response, capacity development and for creating the response reserve, Act also provided for Authorities at the National, State and district level for formulating the plan and policy. It further created a framework for a response, preparedness, mitigation etc. at the various level through the Executive Committee (EC). It also provided for National Disaster Response Force (NDFR) for response purposes and the National Institute of Disaster Management (NIDM) for capacity development.

Table 39.1 Past and the present climate trends and variability among SAARC countries

SAARC Member States	Change in temperature	Change in precipitation
Afghanistan	Since 1960, the average temperature increased by 0.6°C and there were 25 more hot days and nights than usual.	A shortfall of 2% per decade, particularly during the months of March, April and May and declining by 6.6% per decade since 1960.
Bangladesh	The annual mean maximum temperature increased in 1961-1990 by0.05°C per year;	Decadal rain anomalies above long-term averages since the 1960s; Rainfall increased and became irregular causing recurring floods.
Bhutan	No long-term data available; increase in mean monthly temperature during 1990-2012	Increase in rainfall variability during last two decades causing drought and floods
India	0.68°C increase per century with increasing trends in annual mean temperature and warming	Increase in extreme rains in the northwest during the summer monsoon in recent decades and lower number of rainy days along the east coast
Maldives	Annual sea surface temperatures increasing 0.2 to 1.6°C at different places	No significant trends
Nepal	The temperature increased by 1.8 °C during 1975-2006; increased annual frequency of hot nights by 2.5%	Annual rainfall decreased by 10 mm per decade
Pakistan	0.57°C increase n mean annual temperature throughout 1901-2000; 0.24°C per decade during 1960-2007;	Rainfall increased by 25% during last century; Increase in Extreme events.
Sri Lanka	0.016°C increase per year between 1961 to 90 over the entire country	An increasing trend in February and decreasing in June respectively

Source: Cruz et al. 2007, UNDP, 2008; IPCC, 2007a-c; GoB, 2005

Focus on climate change in India came due to various international treaties and conventions, such as, Vienna Convention 1985; Montreal Protocol 1987; UNFCC 1992; and Kyoto Protocol 1997. The Legislation such as the Air (Prevention and Control of Pollution) Act 1986, Forest (Conservation) Act 1980, Energy Conservation Act 2001 and several policy measures have been put in place to regulate and mainstream the environment and climate related issues. The Ministry of Environment and Forest (MOEF) is the nodal Ministry for all the climate change related activities in the country. A Prime Minister's Council on Climate Change (PMCCCC) is constituted in 2008 to coordinate the national action for assessment, adaptation and investigation of climate change. A National Action on Plan on Climate Change (NAPCC) was launched in the same year (NAPCC, 2008).

10. Paris Climate Agreement

The Paris Agreement is a legally binding international treaty on climate change. It was adopted by 196 Parties at COP 21 in Paris, on 12 December 2015 and entered into force on 4 November 2016. Its goal is to limit global warming to well below 2, preferably to 1.5 degrees Celsius, compared to pre-industrial levels.

This should be done by reducing emissions as soon as possible, to achieve a balance between anthropogenic emissions by sources and removals by sinks of Greenhouse gases in the second half of the 21st century. It also aims to increase the ability of parties to adapt to the adverse impacts of climate change, and make finance flows consistent with a pathway towards low greenhouse gas emissions and climate-resilient development.

Under the Paris Agreement, there are no obligations for a country to set a specific emissions target or fulfill the target by a specific date, but each target should go beyond previously set targets. But a country must determine a plan and regularly report on its contribution regarding global warming mitigation. The Paris Agreement worked on a 5-year cycle of increasingly ambitious climate action carried out by countries. In 2020, countries submit their plans for climate action known as nationally determined contributions (NDCs).

11. Nationally Determined Contributions (NDCs)

In their NDCs, countries communicate actions they will take to reduce their Greenhouse Gas emissions to reach the goals of the Paris Agreement. Countries also communicate in the NDCs actions they will take to build resilience to adapt to the impacts of rising temperatures. The contributions should be set every five years and are to be registered by the UNFCCC Secretariat.

The five main targets that have been set in the Climate Change Action Plan are Strengthening Resilience and adaptive capacity to climate related disasters, integrating climate change measures into policies and Planning, building knowledge and capacity to meet climate change, promoting mechanisms to raise capacity for planning and management, reducing emissions/Reducing Vulnerabilities/integrating Resilience & Adaptation capacity to climate related hazards.

12. Glasgow Climate Change Conference (COP26)

The United Nations Climate Change Conference or COP26 was held in Glasgow, Scotland, the United Kingdom from 31 October to 13 November, 2021. Around 120 world leaders and over 40,000 participants came together to cover all facets of climate change.

In maybe the most contested decision in Glasgow, nations consented to a provision requiring a phase down of coal power and a phase out of inefficient fossil fuel subsidies. The Glasgow Pact settled after doubling of finance to help developing countries in adjusting to the effects of climate change and build resilience. Countries agreed on major issues of the Paris climate agreement.

103 nations, including 15 significant producers, signed up to the Global Methane Pledge, the primary aim of which is to restrict methane emissions by 30% by 2030, compared to 2020 levels. More than 30 nations, six significant vehicle manufacturers and others, set out their assurance for all new car and van deals to be zero-emission vehicles by 2040 around the world, accelerating the decarbonization of road transport. Pioneers from South Africa, the United Kingdom, the United States, France, Germany, and the European Union announced a partnership to help South Africa in making a transition away from coal, to a low-carbon economy.

India has pledged to cut the carbon emission and achieve a zero net carbon emission target by the year 2070. The claims to reach the target looks illusionary as 2070 lies in the distant future, whereas New Zealand plans on achieving zero net emission by the year 2050 and US by the year 2030. India has a point since the country holds an argument that the advanced countries that had agreed to the transfer of technology and fund at the Paris agreement didn't come as agreed.

Moreover, India also has a point saying since the advanced countries have already burnt a lot of carbon and injected the atmosphere to reach the present level development. These countries have spent millions of dollars for development of their decarbonization projects, to reach the same level of development India needs a Carbon budget which the advanced countries should allocate. At the same time, India stands firm on the prior pledge taken at the previous summit- cutting down greenhouse gas emissions gradually and increasing the forest coverage substantially and also increasing the energy production of renewable resources and migrating to non-fossil fuel-based energy production such as solar, wind and hydro energy.

Amidst all the chaos some countries are carbon negative, meaning they absorb more carbon than they emit. Bhutan, Suriname and Panama become the first carbon negative countries due to their rich forest cover and protected conservation area. Bhutan's main source of energy is hydropower, while Panama intends to phase out heavy fuels and coal by 2030. At COP26 the three countries signed a formal alliance for preferential trade and international support.

13. Conclusion

The global climate is already changing, and will continue to change over the coming decades and centuries. In many places, local trends in average temperature and precipitation due to climate change are already being observed, and fairly reliable projections for the future can improve planning decisions. However, climate change is not just about gradual changes. Key impacts of climate change will be due to changes in climate variability and weather extremes.

Many of these trends in variability and weather extremes are clear enough to warrant attention in regional and local risk reduction strategies. To benefit from good information on such local information on climate change, organizations working on disaster risk reduction and development will need to establish linkages with new partners, such as national meteorological offices or global centers of expertise on climate research.

Disaster risk reduction and more robust development planning are crucial in adapting to the increased risks associated with climate change. This is particularly important in the face of mounting vulnerability to natural hazards, as reflected, for instance, in rising numbers of people affected and escalating levels of economic damage. In almost all cases, Climate Change is just an additional factor to consider, which can be embedded in existing risk reduction strategies.

We have seen that the global climate is already changing and will continue to change for decades. Although different countries over the period of time have made claims about cutting CO2 emission, it hasn't been followed by them. At the Glasgow Climate change conference, we get to see the repetition of the same commitments. Although, France, the UK, and Germany came up with good prepositions regarding cutting down the exploitation of fossil fuel to attain net-zero carbon dioxide emissions by the years 2030 and 2050 and migrate to green energy. India confirms that it is taking measures to increase forest coverage and increase the share of green energy such as solar wind. Nations pledged to switch from Fossil fuels to Hydro, Wind, and Solar energy and use biofuel for energy production. India pledged to achieve net-zero carbon dioxide emission by the year 2070. Global warming, geophysical changes due to urbanization, and flawed development has led to global warming resulting in the melting of glaciers in the Himalayan region causing landslides and flash floods.

Advanced countries have been alluding to contributing desired funds as committed in Paris Agreement and also technology sharing seems a distant dream. The participants identified open access to easily available data as one of the main challenges of climate change adaptation.

Green Hydrogen holds an alternative and truly, clean fuel that can help the world's decarbonization efforts to migrate towards non-fossil fuels energy resources. The only challenge with green hydrogen is its liquefication and storage which is far below the temperature of 163 degrees Celsius (used to store liquified natural gas). But, at a time like this India needs to buckle up and switch to environmentally safe fuel for the future generation.

References

1. Climate change 2007: Impacts, adaptation and vulnerability by Rex oafallas Cruz https://www.researchgate.net/publication/284683271_Asia_Climate_change_2007_Impacts_adaptation_and_vulnerability
2. Climate Change in Indonesia by Michael Case http://awsassets.panda.org/downloads/inodesian_climate_change_impacts_report_14nov07.pdf
3. Impacts, vulnerabilities and adaptation in developing countries by UNFCCC United Nations Framework Convention on Climate Change https://unfccc.int/resource/docs/publications/impacts.pdf
4. Fifth Assessment Report by ipcc https://www.ipcc.ch/assessment-report/ar5/
5. GISS-NASA- News and Features https://www.giss.nasa.gov/
6. World Disaster Report WDR 2004 / International Federation of Red Cross and Red Crescent Societies Web: http://www.ifrc.org
7. The human contribution to the European heatwave of 2003 by Peter A. Stott https://www.nature.com/articles/nature03089

CHAPTER 40

Climate-Induced Displacement and Climate Disaster Law: Challenges and Opportunities

Sunil Kumar Chaudhary
Executive Engineer, Road Construction Department
Vaishali Road Division, Hazipur
sunil228@cimp.ac.in

Abstract

Climate disasters occur at the intersection of natural climate variability, influenced by climate change, exposure and vulnerability. There are many challenges when Climate Disaster Law attempts to address climate-induced migration and displacement. This Paper considers climate-induced migration and displacement following climate disasters. It focuses on disaster risk reduction and climate change adaptation; relocation and resettlement in the event of displacement; and financing, compensation and risk transfer. It also highlights three major new initiatives which are likely to better protect the rights of climate displaced persons including: the new synergies between the Paris Agreement, the Sendai Framework for Disaster Risk Reduction 2015-2030, and the 2015 Sustainable Development Goals; the 2016 Task Force on Displacement under Warsaw International Mechanism on Loss and Damage; and the proposed 2018 UN Global compact for safe and orderly and regular migration.

Keywords: Climate,Disaster, Challenges, Opportunities, Law

1. Introduction

To speak about climate change induced displacement as a form of forced migration and the challenges such displacement creates for international law is not an easy task. While we know that climate change causes glaciers to melt, sea-levels to raise, deserts to expand and wind-storms to grow stronger and more frequent, lawyers have difficulties to define what climate change is. Still, they may rely on the definition enshrined in Article 1 paragraph 2 of the 1992 United Nations Framework Convention on Climate Change (UNFCCC) stating that „Climate change means a change of climate which is attributed directly or indirectly to human activity that alters the composition of the global atmosphere and which is in addition to natural climate variability observed over comparable time periods."

Still the law does not provide an answer as to whether and to what extent the climate is changing, or whether the causes are indeed human-made. And while international law – in the form of the 1998 Kyoto Protocol United Nations Framework Convention on Climate Change and hopefully more efficiently on the basis of a future strong successor international agreement – may tell States what measures to adopt to fight climate change, it cannot determine whether these measures will have the desired effects. These questions are best left to scientists who have studied the complexities of weather and climate. However, issues of climate change are relevant for students of forced migration as we are increasingly becoming aware that while climate change as such does not cause displacement, certain of its effects already do and are likely to so do even more in the coming decades.

2. Climate Change: General Obligations of States Under International Law

In the context of climate change States face three types of challenges under international law, namely mitigation, i.e. the task of mitigating the degree of climate change, in particular by reducing greenhouse gas emissions, adaption, i.e. the challenge of how best to adapt to the threats caused by the effects of climate change, and protection, i.e. the obligation to secure the rights and addressing the humanitarian needs of people affected by negative effects of climate change.

2.1 Addressing the Cause: Mitigating Climate Change.

Parties to the UNFCCC and its Kyoto Protocol have committed themselves to reducing the emission of greenhouse gases. These mitigation measures aim at slowing down and eventually stopping the change of climate and its disastrous consequences. As such, they have an important preventive effect on displacement. This preventive character hopefully will be strengthened in the successor agreement to the Kyoto Protocol in particular by introducing a more effective climate regime based on comprehensive quantified emission limitations.

2.2 Addressing the Effects: Adapting to the Effects of Climate Change by Reducing Risks Created by Climate Change and Vulnerabilities Caused by it

Environmental and human impacts of climate change are already felt today. Additional global warming, even if it stays below two degree centigrade, will have substantial additional negative impacts. Faced with this reality, governments are expected to take measures to adapt to these new realities by reducing the adverse effects of climate change, e.g. by addressing the impact of natural hazards through reducing vulnerabilities or enhancing resilience capacities. According to this case law, the right to life and its corresponding state obligation to protect life require that, with regard to natural disasters, including those caused by climate change, competent authorities must: (i) enact and implement laws dealing with all relevant aspects of disaster risk mitigation and set up the necessary mechanisms and procedures; (ii) take the necessary administrative measures, including supervising potentially dangerous situations; (iii) inform the population about possible dangers and risks; (iv) evacuate potentially affected populations; (v) conduct criminal investigations and prosecute those responsible for having neglected their duties in case of deaths caused by a disaster; and (vi) compensate surviving relatives of victims killed as a consequence of neglecting these duties.

2.3 Addressing the Impact: Protecting Individuals Displaced by the Effects of Climate Change

Mitigation and ex-ante adaptation measures are often insufficient to protect individuals from suffering the impact of climate change, including from being displaced by effects such as suddenonset hydro-meteorological disasters. In a wider sense, adaptation measures must therefore also cover protection of and assistance for persons negatively affected by the effects of climate change. States as primary duty bearers are bound by human rights law to protect the rights of those affected. Inter alia, the 1998 United Nations Guiding Principles on Internal Displacement8 play an important role in addressing the protection needs of those displaced by the effects of climate change. Another relevant instrument for such settings, the Operational Guidelines on Human Rights and Natural Disasters, has been adopted by the Inter-Agency Standing Committee in 2006 and revised in 2010.9 They apply to all disasteraffected persons, including displaced persons.

3. Understanding Climate Change Induced Displacement: Causes and Magnitude

Before looking at the protection challenges in the context of climate-induced displacement and examining whether international law provides a sufficient normative framework to address them, we need to get a better understanding of how and in what contexts climate change and its effects may trigger displacement.

1. Climate change is likely to reduce water availability, particularly in parts of the tropics, the Mediterranean and Middle Eastern regions and the Southern tips of Africa and Latin America. In contrast, water availability may increase in parts of Eastern Africa, South Asia, China, and the Northern Latitudes. Hundreds of millions of people will experience water stress, whether due to too little or too much water, i.e. drought or flooding.
2. A decrease in crop yields is projected, increasing the likelihood that additional tens of millions of people will be at risk of hunger. The most affected region is likely to be Africa.
3. Due to rising sea-levels, the densely populated "mega-deltas," especially in Asia, including South Asia, and Africa, and small islands are at greatest risk from floods, storms and coastal flooding and eventual submerging, again with a potential impact on tens of millions of people.

Based on this, we can identify at least five scenarios indicating direct or indirect effects of climate change that may trigger displacement:

(i) Sudden-onset hydro-meteorological disasters, such as flooding, cyclones or mudslides can cause large-scale displacement and incur huge economic costs.

(ii) Slow onset environmental disasters such as reduced water availability and droughts, desertification, long-term effects of recurrent flooding, rising sea levels destroying costal zones, or increasing salination of ground-water and soil may also trigger population movements.

(iii) The case of "sinking" Small Island States caused by rising sea levels constitutes a particular challenge

(iv) Disasters will increase the need for governments to designate areas as high-risk zones too dangerous for human habitation.

(v) A decrease in essential resources such as water or fertile land due to climate change may trigger unrest seriously disturbing public order, violence and armed conflict:

Despite the incertitude regarding the magnitude and dynamics of what will happen, we can conclude that

1. Climate change per se does not trigger movement of persons, but some of its effects do, including sudden and slow on-set disasters and certain ways of governments or communities to react to such disasters.
2. To establish strict causalities between climate change and such events, however, is very difficult or even impossible in most cases; events described in the five scenarios should therefore be the point of reference for discussions about displacement in the context of climate related natural or human-made disasters, not climate change as such;
3. Such movement may be voluntary, or it may be forced; and
4. It may take place inside a country or across international borders.

4. Protecting People Displaced by Natural Disasters: Normative Frameworks and their Gaps

The five scenarios outlined above can help (1) to identify the character of population movements, i.e., whether they are forced or voluntary, (2) to qualify those who move: Are they migrants, internally displaced persons (IDPs), refugees, stateless persons, other something else altogether?), and (3) to assess whether and to what extent present international law is equipped and provides adequate normative frameworks to address the protection and assistance needs of such persons. Regarding the five scenarios, I would like to highlight the following:

1. *Sudden-onset hydro-meteorological disasters* can trigger forced displacement
2. *Situations of slow onset environmental disasters trigger several types of movements of persons*
3. *The "sinking" of Small Islands States will be gradual*
4. *The designation of high risk zones too dangerous for human habitation may trigger (forced) evacuations and displacement*
5. *"Climate change-induced" unrest, violence and armed conflict trigger forced displacement*

5. Addressing the Gap: People Displaced Across International Frontiers

The above analysis allows the following conclusions: First: Existing human rights norms and the Guiding Principles on Internal Displacement provide sufficient protection for those forcibly displaced inside their own country by sudden-onset disasters or because their place of origin has become uninhabitable as a consequence of a slow-onset disaster (scenario ii), or been declared too dangerous for human habitation. The challenge here is proper implementation, not the lack of appropriate norms.

First: Existing human rights norms and the Guiding Principles on Internal Displacement provide sufficient protection for those forcibly displaced inside their own country by sudden-onset disasters or because their place of origin has become uninhabitable as a consequence of a slow-onset disaster , or been declared too dangerous for human habitation . The challenge here is proper implementation, not the lack of appropriate norms.

Second: Existing international law is, if properly implemented by all concerned, sufficient to protect persons displaced by a breakdown of law and order, violence or armed conflict triggered by the effects of climate change, regardless of whether they cross an internationally recognized state border.

Third: Normative gaps in present international law exist for persons crossing borders due to the effects of climate change.

Fourth: The main challenge is to clarify or even develop the normative framework applicable to persons crossing internationally recognized state borders in the wake of sudden-onset disasters (scenario i), or in the wake of designation of their place of origin as high risk zone too dangerous for human habitation (scenario iv). In these cases, questions to be addressed include the following issues:

1. Should those moving voluntarily, on the one hand, and those being forcibly displaced across borders, on the other hand, be treated differently not only as regards assistance and protection while away from their homes but also as regards their possibility to be admitted to other countries and remain there at least temporarily? The answer seems obvious: Present international law, while recognizing that all human beings are entitled to the full enjoyment of human rights, does in fact differentiate between persons who move voluntarily and those forcibly displaced for whom special

normative regimes (refugee law; Guiding Principles on Internal Displacement) have been developed at least in some cases.

2. Therefore, what criteria would be appropriate to distinguish, in the context of climate change, between those who voluntarily leave their homes or places of habitual residence and those who are forced to leave or – even if they left voluntarily in the first place - can no longer return because of the effects of climate change and therefore need protection abroad? What would be the respective entitlements to assistance and protection of those leaving voluntarily and those forcibly displaced?

As regards the second question, there are different ways to develop criteria to determine when a movement across borders is no longer voluntary, but happens under compulsion. Let us look at a moment to refugee law to see whether it provides us with an answer. Except in certain cases described above, people displaced by effects of climate change are not refugees as defined by international law. While this suggests a different approach, some inspiration, nevertheless, can be taken from the three key elements of the refugee definition in article 1A(2) of the 1951 Refugee Convention. These are: (i) being outside the country of origin, (ii) because of persecution on account of specific reasons (race, religion, nationality, membership of a particular social group or political opinion), and (iii) being unable or unwilling to avail oneself of the protection of one's country.

People displaced across borders by the effects of climate change obviously fulfil the first criterion of having crossed a border. It is also obvious that, except in the case of scenario (v) and some other cases that will be rather exceptional, such people are not refugees because they do not fulfil the criterion of being persecuted on account of any of the relevant reasons. However, similar to persecution, climate change related disasters and the unavailability of adequate food, drinking water or health services in their aftermath may constitute serious threats to life, limb and health. In this broader sense, refugees and those displaced by the effects of climate change face similar dangers as refugees, albeit for different reasons. The third criterion may also help to conceptualize solutions for these people. Exactly as we do for refugees, we should ask: Under what circumstances should those displaced across borders by negative effects of climate related disasters not be expected to go back to their country of origin, and therefore remain in need of some form of surrogate international protection, whether temporary or permanent? In general, the answer will, as for refugees, depend on the elements of inability or unwillingness of the authorities in the country of origin or habitual residence to provide the necessary protection—and in the case of natural disasters, assistance to the people concerned.

The question whether people admitted in the aftermath of sudden- or slow-onset disasters can be obliged to return to their country of origin once the immediate danger is over is more complex. Here, the point of departure should not be the subjective motives of individuals or communities behind their decision to move, but rather whether, in light of the prevailing circumstances and the particular vulnerabilities of those concerned, they can be required to return to their country of origin. This 'returnability' test helps to better identify those in need of protection in another country. It covers not only those who actually flee to another country, but also those whose initial movement was voluntary but who now cannot be expected to return because the situation has deteriorated to such an extent that return is no longer an option.

The returnability of the person concerned should be analyzed on the basis of a three-pronged test that asks whether it is legally permissible, factually feasible and morally reasonable to oblige the person concerned to return to his or her country of origin or permanent residence.

1. **Legal Impediments: The Criterion of Permissibility:** There are certain cases where human rights law, by analogy to the refugee law principle of non-refoulement, prohibits return of certain persons. Such prohibition exists where there are substantial grounds to believe that an individual would face a real risk of torture or cruel, inhuman or degrading treatment or punishment, or arbitrary deprivation of life if sent back to a particular country. Arguably, this prohibition could be made fruitful for cases where rejection at the border or return would expose an individual to an imminent danger for life and limb related to the disaster causing their displacement.
2. **Factual Impediments**: The Criterion of Feasibility Return may be factually impossible due to temporary technical impediments, such as when roads are cut off by floods or airports in the country of origin closed. Return is also impossible for administrative reasons if the country of origin refuses readmission for technical or legal reasons: during an emergency, a country may lack the capacity to absorb large return flows, or it may prevent readmission of persons whose travel documents or proof of citizenship was destroyed, lost or left behind when they fled.
3. **Humanitarian Impediments: The Criterion of Reasonableness:** Even where return would be lawful and reasonable, people should not, on the basis of compassionate and humanitarian grounds, be expected to go back if the country of origin does not provide any assistance or protection, or if what is provided falls far below international standards of what would be considered adequate. The same is true where authorities do not provide any kind of durable solutions

to the displaced that are in line with international standards and would allow them to resume normal lives, especially where areas of land have become (or have been declared) uninhabitable and people have been unable to find an acceptable alternative themselves.

If the answer to one of these questions—is return permissible? is it feasible? can it reasonably be required? —is 'no', then individuals concerned should be regarded as victims of forced displacement in need of protection and assistance in another State. In this case, they should be granted at least a temporary stay in the country where they have found refuge until the conditions for their return are fulfilled. Permanent solutions on the territory of other States must be found particularly where vast parts of a country have become uninhabitable that it can no longer host its entire population.

6. The Way Forward

The Way Forward would consist of developing proposals for protection regimes applicable to those who were forced to seek refuge inside their own country or across an international border due to the effects of climate change. Their entitlements to assistance and protection, as well as their obligations, should also be elaborated. This can be done at different levels.

At the domestic level, we may look at existing provisions in domestic law addressing subsidiary or temporary protection that provide for or may be interpreted in a way to provide for such protection in the case of persons displaced by the effects of climate change and other environmental factors. The regional level would allow harmonizing differing domestic approaches. Furthermore, States must "enact national legislation to domesticate the Guiding Principles fully and to provide a legal framework for their implementation within national legal systems" and, in this context, to ensure that such legislation specifies the governmental organs responsible not only "for providing protection and assistance to internally displaced persons" but also for "disaster preparedness". At the international level, besides the Guiding Principles on Internal Displacement no normative framework exists yet that would address cross-border displacement. Nevertheless, a first step towards a discussion of such framework was made at the 2010 Cancún Conference on Climate Change.

This agreement is relevant in several regards. First, the international community recognizes for the first time the humanitarian consequences of climate change are as an adaptation challenge. Second, displacement can be expected to become part of national adaptation plans foreseen by the agreement, 34 thus providing an entry point for protection issues. Finally, the agreement recognizes that efforts addressing displacement need to be undertaken not only at the national but also regional and international levels, thus putting internal as well as cross-border displacement on the international agenda. This is an important, albeit still limited step that needs to be followed up by discussions on appropriate normative regimes closing current protection gaps. Besides cross-border displacement in the aftermath of sudden- or slow-onset disasters such discussions must, in particular, address the case of "sinking" Small Island States that cease to exist: The question of the responsibility of the international community, in particular regarding relocation, must be clarified. In other words, new law will be required if we are to avoid these populations becoming marginalized and disenfranchised inhabitants of their countries of refuge.

7. Conclusion

- Laws and policies addressing climate related displacement should not ask whether climate change has triggered the movement.
- It is impossible to determine whether a particular disaster would or would not have happened without climate change
- Moreover, an exclusive focus on climate change may incite us to neglect other causes of natural disasters and environmental changes such as volcano eruptions, tsunamis or earthquakes and thus amount to discrimination against persons having equally urgent protection needs.
- In determining whether and how to provide temporary or permanent international L 34 protection for persons fleeing their country of origin in the aftermath of a disaster, it is enough instead to consider the environmental factors combined with the temporary or permanent unwillingness or inability of the country of origin to protect affected persons.

References

1. Intergovernmental Panel on Climate Change, Fourth Assessment Report, Synthesis, Geneva 2007, p. 30 (available at: http://www.ipcc.ch/pdf/assessmentreport/ar4/syr/ar4_syr.pdf
2. Framework Convention on Climate Change – Secretariat, Press Release: UN Climate Change Conference in Cancún delivers balanced package of decisions, restores faith in multilateral process, Cancún 11 November 2010
3. 2010 UN Climate Change Conference in Cancún, Mexico, Decision*/CP.16, Outcome of the work of the Ad Hoc Working Group on long-term Cooperative Action under the

Convention, available at http://unfccc.int/files/meetings/cop_16/application/pdf/cop1 5_lca.pdf (last visited 27 December 2010).

4. European Court of Human Rights, Budayeva and others v. Russia, Applications nos. 15339/02, 21166/02, 20058/02, 11673/02 and 15343/02, judgment of 20 March 2008.
5. Walter Kälin/Claudine Haenni, Disaster risk mitigation – why human rights matter, in: Forced Migration Review, Issue 31, October 2008, pp. 38–39, at 39.
6. IASC Operational Guidelines on the Protection of Persons in Situations of Natural Disasters. December 2010, available at http://www.brookings.edu/projects/idp.aspx.
7. Intergovernmental Panel on Climate Change, Fourth Assessment Report, Synthesis, Geneva 2007, pp. 48–53.
8. Stern, N., The Economics of Climate Change – The Stern Review, Cambridge, Cambridge University Press 2006, p. 77.
9. 2005 World Summit Outcome, General Assembly resolutions 60/1, para. 132; 62/153, para. 10; and 64/162, para. 11; and Human Rights Council resolutions 6/32, para. 5; and 14/6, para. 9
10. Aliens Act of Finland 301/2004 with amendments up to 973/2007, Section 109.
11. Aliens Act of Finland 301/2004 with amendments up to 973/2007, Section 88
12. Aliens Act of Sweden (2005:716), of 29 September 2005, Chapter 4, Section 2.

CHAPTER 41

Climate Hazards and the Role of Local Authorities: Exploring the Indian Legal Framework

Chandrika Mehta
Assistant Professor, School of Law, Bennett University
chandrika.mehta@bennett.edu.in
+91 7585012267

Abstract

Climate change can severely impact disaster risks in multiple ways by altering the frequency and intensity of hazard events. Managing the risks of climate extremes would require a more thoughtful approach and fundamental variations in our socio-legal models. The vulnerability of communities in dealing with natural hazards owing to climate change is closely linked to adaptation and mitigation measures and capacity building.

Although numerous efforts to close this gap have been made at the global and national front, not much has been achieved at the local level. The Indian legal framework recognizes local government as the third tier of governance and establishes its role in a typical democratic set-up. In constitutional sense, democracy is the system of Government, in the administration of which, every adult citizen of the country enjoys some direct or indirect share. Keeping in view the real spirit and high ideas of democracy, Local Government forms an indispensable part of governance and administration in India. Before 1992, local level institutions did not have a constitutional status and had only a statutory status under state law. Thus, for instance, governance of urban areas was directly within the jurisdiction of the state government. This structure underwent a major change with the enactment of the 73rd and 74th Amendments of the Constitution of India in 1992, which marked a new era in the nature of local governance in rural and urban areas respectively.

It is against this backdrop that the paper seeks to delve into the conceptual understanding of the Indian legal framework with specific emphasis on the institutionalization of local self-governments and their contributions to disaster risk management.

Keywords: Climate Change, Disaster Risk, Local self-government, municipal governance, urban agglomeration

1. Introduction

Climate conditions have undoubtedly been driven and intrinsically tied by the ways in which we have built our cities. Cities are centres of planning and development; homes to knowledge externalities, which along with agglomeration externalities, act as powerful drivers of growth[1]. Amid profound economic, social and technological changes in the era of globalization, cities around the world are faced with the challenge of reconciling competitiveness with long-term sustainable development[2]. With rapid urbanization, cities soon came to be seen as a means to address environmental issues. The need for local governments to respond to the problem of climate change and its impacts has since long been a tenet of green political thought[3]. Policies on climate governance are required to be effective in their domestic capacities so as to ensure an efficacious working mechanism with special emphasis on emissions and energy use as well any future international emissions reduction commitments that would be of critical importance in the present as well as future.

Building resilience and adapting to climate change is increasingly a high priority for cities. A municipality is the first line of defence in preparing for weather disasters, given the relationship a local government bears to land use, infrastructure, and public health and safety[4]. The Hyogo Framework for Action 2005-2015 considers both communities and local authorities as key stakeholders in the management and reduction of disaster risk as they are concerned with the grassroots level management and have access to all the necessary information and resources. Disaster risks have rocketed in the recent past and led to extreme weather conditions, thus creating a hazardous environment.

2. Objective

In the light of these developments, this research was undertaken to study and analyse in detail, the municipal governance framework as a mechanism to address climate hazards. The research makes a detailed study of the impacts of climate hazards on urban agglomerations in India, and delves into the institutional mechanism that seeks to address the issue.

3. Methodology

The methodology adopted for this study is largely doctrinal. Use of analytical research in studying and analyzing various legislations namely municipal Acts, model laws and policy documents in the context of climate hazards has been greatly insightful. This methodology has helped develop a particular line of inquiry which is both conceptual and evaluative.

4. Urban Agglomerations and Local Self-governments in India

Cities in a wide-ranging spectrum reveal a three-fold link with issues concerning climate change. First, with a high level of energy consumption, concentration of economic activities and burning of fossil fuels, cities tend to be the cause of global climate change[5]. Second, with rise in the number of urban dwellers, and subsequent concentration of man-made infrastructure, cities are also vulnerable to various local manifestations of climate change impacts, e.g. sea-level rise, tropical cyclones, flooding and landslides, water crisis, and heat and cold waves[6]. Third, it is in cities where we find a higher concentration of resources and expertise[7].

The world is witnessing a surge in urban population and therefore, a rise in urban agglomerations. Cities emerged about 7,000 years ago owing to an increase in agricultural surplus following technological development[8]. As cities grew in size, they also grew in power and, as Smith notes, “Rulers used urban architecture to communicate messages about power, wealth, legitimacy, and other ideological themes”. The old system of rural living, in which human communities were dependent on a relationship with the land, changed with the rise of urban centres; now humans controlled the natural environment and bent the surrounding land to their will. While recognition of urban responses to climate change at the international policy level has been relatively recent, a burgeoning research community has studied the relationships between cities and climate change. Since the mid-1990s, research has focused on municipal strategies, policies and measures, and the challenges that municipal authorities face in terms of policy implementation and effectiveness[9]. The need for local governments to contribute to the problem of climate change and is impacts, has since long been a tenet of green political thought[10]. The Brundtland report of 1987 specifically contained a chapter discussing the environmental issues concerning cities. The report in vouching for cities to be a central part in the pursuit of sustainable development, premised its argument on the fact that majority of the world population would live in urban areas[11]. With rapid urbanization, cities soon came to be seen as a means to address environmental issues. The concern was later voiced through chapter 28 of Agenda 21 which called for local authorities to establish a Local Agenda 21 (LA21) through participation with their communities and encouraged the establishment of mechanisms to promote cooperation and coordination among local authorities internationally[12]. Building resilience and adapting to climate change is increasingly a high priority for cities.

India is emerging as one of the fastest urbanizing countries. Cities are being viewed as the engines of national development and economic growth. A planned city with effective governance in place makes good for a secured future.

Looking into the early rise of municipal institutions, one finds a close nexus between constitutional developments and emergence of municipalities. An exploration into the urban institutional system within the political arrangements of governance reveals a close linkage between constitutional development and the growth of municipal institutions in India. Madras received its municipal charter as early as 1687[13] and municipal institutions outside the Presidency cities were created from 1842 onward; it was only after the ‘Mutiny’ of 1857 that the British began seriously to think about the large scale municipalization of the country- partly to relieve the imperial exchequer- so that by 1870, there were around 200 municipalities in British India[14].

Over a period of time, the municipal institutions lost the credibility to deliver services in an efficient manner and began to shun their responsibility to service urban areas beyond their jurisdiction. This brought the emergence of Urban Development Authorities (UDAs) in India that spread rapidly during the 1980s, which inevitably led to confrontation between municipal authority and UDA[15]. The UDAs were created as Special Purpose Vehicle with a different governance structure and mandate that made it possible for them to perform the function efficiently. However, adequate emphasis was not laid down on creating it as a coordinating institution with a larger jurisdiction and to make its role without usurping the powers of municipal system and to take care of regional planning and development issues.

Before 1992, local level institutions did not have a constitutional status and had only a statutory status under state law. Thus, for instance, governance of urban areas was directly within the jurisdiction of the state government. This structure underwent a major change with the enactment of the 73rd and 74th Amendments of the Constitution of India in 1992, which marked a new era in the nature of local governance in rural and urban areas respectively. The 74th Amendment Act of the Constitution of India redefined the role, powers, functions and financial authority of urban local bodies (ULBs). At the beginning of the 8th Plan period, in 1992, the 74th Constitutional Amendment Act was promulgated. It was a landmark Act which sought to decentralize decision making in cities and towns through creation of elected urban local bodies (ULBs) as institutions of democratic self-governance and devolution of essential functions related to city planning and service provision to these bodies. The 74th CAA brought forth an introduction of the Twelfth Schedule which lists the functions of the ULBs, establishment of ward committees in areas having a population of over 3 lakhs, periodic and timely elections of ULBs, and devolution of finances to ULBs as per the suggestions of the State Finance Commissions (SFSs).

5. Climate Resilience and Indian Legal Framework

India released its Action Plan on Climate Change[16] as part of its Nationally Determined Contributions to the UNFCCC. NDCs embody efforts by each country to reduce national emissions and adapt to the impacts of climate change[17]. The Plan[18] emphasizes on implementation of the eight Mission Programmes under its umbrella through building of a proper and efficient institutional mechanism suited for effective delivery of each Mission's objectives and calls for greater public private partnership and civil society action.

NAPCC addresses the country's critical and urgent needs by directionally shifting the development path and enhancing the current and planned programmes and technologies. It identifies measures that promote our developmental goals and co-benefits by addressing climate change. Energy efficiency is one of the prime focus of the Plan. This is sought to be achieved specifically through two Missions- National mission for enhanced efficiency and National mission on sustainable habitat.

The National Mission for Enhanced Energy Efficiency builds on the 2001 Energy Conservation Act. This mission will look at ways to create a market based mechanism to enhance cost effectiveness of improvements on energy efficiency. This would also mean finding ways to shift to energy efficient products; further, the creation of mechanisms that promote demand side management will also be looked into under this mission. Switching to cleaner fuels, commercially viable technology transfers, capacity building needs etc will be the way forward for this mission. The National Mission on Sustainable Habitat has a manifold agenda. It looks at energy efficiency within buildings and also at waste disposal from these buildings. Another important aspect of this mission is to better the public transport system. In addition, it will promote energy efficiency in the residential and commercial sectors and also work towards management of solid waste through recycling and improvement of the urban transport system.

6. Institutional Mechanism

Speaking of local efforts, Municipal corporations deriving powers and authorities as the Urban Local Body, have immense energy saving potential. Enforcement of the institutional mechanism involves a lot of stakeholders at different levels of governance. Policy initiatives are meant to improve the quality of life of the people. Accordingly, their outreach and impacts must be at the very grassroots level. Building energy efficiency policies and programs in India are by large in an active design stage with limited implementation to date[19]. Decision-making authority at the national level is shared between several agencies and program design and implementation responsibilities are spread across a large number of state and municipal agencies and departments, resulting in a diversity of implementation regimes, thus creating jurisdiction conflicts and little coordination.

So essentially the mechanism is a top-down approach. The standards and codes are primarily developed at the central level. Later it trickles down to the state level and local level, depending upon the voluntary adoption of the codes and standards. So, as the hierarchical picture presents itself, the Ministry of Urban Development is the central authority that oversees the entire institutional mechanism. Although, the MoUD's role is largely related to policy formulations on matters such as urban development, urban water supply and sanitation, which are essentially state subjects, it acts as a coordinator for various programs and schemes funded by the central government and monitors such programs.

7. Findings and Conclusion

In a nutshell, this research draws a nexus between municipal powers and energy efficiency standards in an urban dwelling. The Municipal Commissioner, being the chief executive officer acts as the executive arm of the corporation. As such, the commissioner enjoys powers as the executive officer and it is through consultation with the council members and the Mayor that he executes policies and regulations so framed. It is thus understood that the corporation powers widely vest with the councillors and the standing committee, while the Mayor represents a titular head to the Corporation. The paper argues and questions the procedural system wherein although the structure of municipal corporations leave ample space for policy execution and decision making, there is evidently a lack of effective dialogue and discussion among the stakeholders.

REFERENCES

1. Prasanna Mohanty, *Cities and Public Policy: An urban Agenda for India* (Sage Publications India Pvt Ltd 2014)
2. Krassimira Paskaleva, 'E-Governance As An Enabler Of The Smart City' in Mark Deakin (ed), *Smart Cities: Governing, modeling and analyzing the transition* (Routledge 2014)
3. Michele Betsill and Harriet Bulkeley, 'Cities And Multilevel Governance Of Global Climate Change' (2006) 12(2) Global Governance http://sciencepolicy.colorado.edu/students/envs_4100/betsill_2006.pdf accessed 28 May 2015
4. Andrea McArdle, 'Lessons For New York: Comparative Urban Governance And The Challenge Of Climate Change' (2014) 42 Fordham Urban Law Journal http://heinonline.org accessed 22 May 2015
5. Mary Murphy(Ed), Up In Smoke? Asia And The Pacific: Thethreat From Climate Change To Human Development And The Environmet (IIED 2007)
6. Harriet Bulkeley, Cities And Climate Change (Routledge 2013)
7. Manoj Roy, 'Planning For Sustainable Urbanisation In Fast Growing Cities: Mitigation And Adaptation Issues Addressed In Dhaka, Bangladesh' (2009) 33 (3) Habitat International Http://Www.Elsevier.Com/Locate/Habitatint Accessed 22 May 2015
8. Paul Bairoch, 'International Industrialization Levels From 1750 To 1980' (1988) 11(2) Journal Of European Economic History Https://Search.Proquest.Com/Openview/B8f0ecee8c6f027c82a22281c0f46bcf/1?Pq-Origsite=Gscholar&Cbl=1819397 Accessed 12 December 2015
9. Vanesa Broto And Harriet Bulkeley, 'A Survey Of Urban Climate Change Experiments In 100 Cities' (2013) 23 (1) Global Environmental Change www.Sciencedirect.Com Accessed 22 May 2015
10. Michele Betsill And Harrietbulkeley, 'Cities And Multilevel Governance Of Global Climate Change' (2006) 12 (2) Global Governance Http://Sciencepolicy.Colorado.Edu/Students/Envs_4100/Betsill_2006.Pdf Accessed 28 May 2015
11. World Commission On Environment And Development, Our Common Future (OUP 1987)
12. Richard Gilbert And Others, Making Cities Work: The Role Of Local Authorities In The Urban Environment (Routledge 2009)
13. Abhijit Datta, 'Institutional Aspects Of Urban Governance' (1994) 40 (4) Indian Journal Of Public Administration Https://Doi.Org/10.1177%2F0019556119940404 Accessed 16 February 2016
14. Ibid
15. Ramakrishna Nallathiga, 'Metropolitan Urban Governance Approaches And Models: Some Implications For Indian Citie' (2011) 1 IPE Journal Of Governance And Public Policy Http://Www.Ipeindia.Org/Data/JGPP/JGPP_1_2_3.Pdf Accessed 3 April 2015
16. The Action Plan was released on 30th June 2008. It effectively pulls together a number of the government's existing national plans on water, renewable energy, energy efficiency, agriculture and others – bundled with additional ones – into a set of eight missions. The Prime Minister's Council on Climate Change is in charge of the overall implementation of the plan
17. 'Nationally Determined Contributions (NDCs)' (UNFCCC) https://unfccc.int/process/the-paris- agreement/nationally-determined-contributions accessed 24 June 2015
18. The eight National Missions on climate change include:
    (a) National Solar Mission
    (b) National Mission for Enhanced Energy Efficiency
    (c) National Mission on Sustainable Habitat
    (d) National Water Mission
    (e) National Mission for Sustaining the Himalayan Eco-system
    (f) National Mission for a Green India
    (g) National Mission for Sustainable Agriculture
    (h) National Mission on Strategic Knowledge for Climate Change

19 Christopher Williams and Mark Levine, 'Gauging improvements in urban building energy policy in India' (ACEEE Summer Study on Energy Efficiency in Buildings, Pacific Grove, California, August, 2012)

Part 6

TECHNOLOGICAL DISASTERS

CHAPTER 42

An Incident of Styrene Monomer Gas Poisoning at Visakhapatnam

Chandrasekhar Krishnamurti, M. D.
Associate Professor NRI Institute of Medical Sciences,
Sangivalasa, Visakhapatnam-531162, A.P., India
globeshaker@gmail.com
Mobil: 9701023746

Saurabh Dalal
Consultant, Medical Preparedness & Biological Disasters
NDMA, A1, NDMA Bhawan, Safdarjung Enclave, New Delhi 110029
Dr saurabhdalal@gmail.com
01126701800

Mounika Jonnavittula House Surgeon
GITAM Institute of Medical Sciences & Research Rushikonda,
Visakhapatnam-530045, A.P., India
Videha96@gmail.com

Abstract

Following a temporary and partial shutdown of the LG Polymers factory for nearly 44 days due to the COVID 19 lockdown, routine maintenance of the plant was affected. Compounded by a failure of the refrigeration system and the inhibitor tank attached to the styrene storage tank, auto polymerization reaction set in and half the 1800 tonnes styrene gas leaked into the atmosphere at around 3 am on 07th May 2020. Since the volatile organic compound detection system was defunct, no alert or alarm was raised and, driven by north easterly winds, a cloud of toxic styrene gas spread over a radius of nearly 3 km, affecting five villages in the vicinity.

The pungent smell was initially mistaken by the residents as emanating from a fire accident or a COVID sanitization measure in the vicinity. When many of the residents became symptomatic with nausea, dizziness, severe burning sensation in eyes and throat, suffocation and dizziness, panic ensued. There were 11 human and 22 animal fatalities in the immediate vicinity

Following initial triage, over 350 victims were hospitalized in various hospitals. 254 symptomatic adults and 64 children were shifted to the Government King George hospital. 165 others were admitted to 10 private and corporate hospitals. 25 victims were severely breathless and in a critical condition and 20 of them with pulmonary edema required endotracheal intubation and ventilatory support in the ICU. The main symptoms were burning sensation of eyes, skin and throat, nausea, vomiting, diplopia, muscle twitching, breathlessness and loss of consciousness. Nearly 10,000 residents from five villages were evacuated to relief camps set up in the perimeter and given first aid and food.

This mishap resulted in extremely high levels of styrene monomer exposure and constitute the first fatalities reported worldwide from its inhalation. The lack of facilities to monitor atmospheric, blood and urine levels of styrene and its metabolites has precluded a better assessment of its toxic effects. The rapid evacuation of residents within a 5 km radius was responsible for a lower mortality rate. Morbidity was, however, high and long term effects of the exposure remains to be seen in follow up monitoring.

Keywords: Styrene monomer gas, toxicity, chemical disaster

1. Introduction

The Hindustan Polymers factory was established in 1961 over 231 acres of land in RRV Puram, Gopalpatnam (15 km away from the Visakhapatnam city centre) to manufacture general purpose polystyrene. The unit was merged with McDowell & Co of the UB Group in 1978 and by LG Chem South Korea in 1997 and renamed LG Polymers. It was upgraded to manufacture high impact and expandable polysterene as well as engineering plastics. The site has refrigerated tanks capable of storing 2400 tonnes of styrene gas at temperatures below 17C.

Following a temporary and partial shut down for nearly 44 days due to the COVID 19 lockdown, routine maintenance of the plant was affected. Compounded by a failure of the refrigeration system, and failure of the inhibitor tank attached to the styrene storage tank, auto polymerization reaction set in and half the 1800 tonnes styrene gas leaked into the atmosphere at around 3 am on 07th May 2020. (Fig 1) Since the volatile organic compound detection system was defunct, no alert or alarm was raised and, driven by north easterly winds, a cloud of toxic styrene gas spread over a radius of nearly 3 km, affecting five villages in the vicinity. (Fig 2)

2. The Immediate Scenario

A cloud of styrene gas enveloped the villages of RR Venkatapuram I & II, SC/BC Colony, Padmanabhapuram and Nandamuri nagar. The pungent smell was initially mistaken by the residents as emanating from a fire accident or a COVID sanitization measure in the vicinity. When many of the residents became symptomatic with nausea, dizziness, severe burning sensation in eyes and throat, suffocation and dizziness, panic ensued. Scared for their lives, the villagers grabbed their valuables and headed out on two wheelers while others ran into the streets and hailed whatever available transport to distance themselves. Many of them locked themselves in their houses to avoid ingress of the gas. Blinded by the vapor, a 19 year old medical student lost his bearings and fell to his death from the second floor of his apartment while two others fell into a well and drowned. A six year old girl and seven adults developed severe breathlessness and succumbed.

The Gopalpatnam police station received an alert at 3:45 a.m., and by 4 a.m. the first team reached the spot to assess the situation. They radioed for help and 30 ambulances, police jeeps and patrol vehicles, as well as state transport buses were mobilized to evacuate victims and residents after initial triage. Loud hailers advised residents within a radius of 5 km to cover their face with a wet cloth or mask, abandon their residence and move to areas of safety. By 6 a.m., the State Disaster Relief teams supported by a Naval team donning protective suits and equipped with 50 breathing sets and portable air compressors conducted house to house search and rescue operations in the affected zones and rescued residents trapped in their residences.

3. Medical Management

Following initial triage, over 350 victims were hospitalized in various hospitals. 254 symptomatic adults and 64 children were shifted to the Government King George hospital. 165 others were admitted to 10 private and corporate hospitals. 25 victims were severely breathless and in a critical condition and 20 of them with pulmonary edema required endotracheal intubation and ventilatory support in the ICU. The main symptoms were burning sensation of eyes, skin and throat, nausea, vomiting, diplopia, muscle twitching, breathlessness and loss of consciousness. A pungent odor was appreciated in the breath and sweat for several days following admission. Those requiring ventilation exhibited elevated serum lactate and respiratory acidosis in the blood gas. Serum electrolytes and other blood parameters were normal and all of them could be weaned off supports within 48 h. Lungs were clear in the rest of those admitted and chest X rays normal. Urine was normal in color. Urine mandelic acid assay could not be done for any of the victims due to non availability of kit or reagents.

Treatment protocols observed were body wash with water, high flow oxygen inhalation and intravenous fluids. Those with obtunded levels of consciousness were kept nil oral while those with severe agitation were administered low dose midazolam or haloperidol. No specific antidote was advised or exhibited.

Nearly 10,000 residents from five villages were evacuated to relief camps set up in the perimeter and given first aid and food.

While 24 children were discharged on day 3, 38 others with persistent symptoms were under observation for delayed effects of styrene monomer poisoning.

4. The Aftermath

The SDRF and NDRF teams undertook water mist spraying to drive atmospheric styrene into the ground while the styrene storage tanks were cooled with water jet sprays. Despite these measures, the styrene gas continued to leak from the tanks when 500 kg para tertiary butyl catechol (PTBC) was airlifted from Vapi Daman at 1030 pm on 07 May Thursday to solidify the styrene gas remaining in the leaking tanks and prevent its escape. Other storage tanks in the plant containing

13,000 tonnes of styrene monomer gas were loaded into two vessels in two consignments of 3000 and 5000 tonnes to the parent company in South Korea by 15 May 2020.

About 22 domestic animals including cows and buffaloes and their calves were killed in the incident within a 1.5 kilometer radius of the factory. Additionally, six stray dogs and one cat also perished due to gas inhalation.

The gas leak disrupted the movement of 45 trains - nine Shramik Specials and the rest goods carriers - as rail workers suffering from irritation in their eyes and breathlessness halted work. Train services came to a halt at the Simhachalam North Railway Station (SCMN) in Visakhapatnam, near the chemical factory, between 8:35 hours and 12:00 hours, leading to several trains being stranded and prompting authorities to divert some others via Kondapalli and Ballarshah routes.

By 12 May, villagers began to return to their dwellings but complained of a pungent odour persisting in the water and foodstuff. The Government has established medical camps in the affected areas for long term health monitoring of the affected population.

5. Discussion

Styrene ($C_6H_5CH{=}CH_2$) is a high production chemical and a derivative of benzene (C_6H_6). Styrene, also known as vinylbenzene, ethylbenzene, cinnamene, or phenylethylene is a synthetic colourless, volatile, inflammable liquid with a sweet odour at very low concentrations, but sharp and disagreeable at higher concentrations. The boiling point of styrene monomer is 145.2°C, flash point 31°C and autoignition temp is 490°C and saturated vapor concentration is 6600 ppm at 20°C. It is insoluble in water.

Styrene is stored in factories at temperatures under 20°C.

Acute styrene poisoning occurs as a result of inhalation or absorption from the skin surface. "Styrene sickness" in industry describes a clinical picture after heavy exposure to styrene and other solvents and manifests as headache, nausea, vomiting, weakness, dizziness, fatigue and ataxia. (1). The inhaled level defined as immediately dangerous to life or health is 5000 ppm (2). The Lowest Lethal Concentration = 10000 ppm/30 min and Lowest Toxic Concentration = 600 ppm (3). Increasing concentrations >800ppm results in CNS depression associated with abnormal EEG patterns, asphyxia, muscular weakness, vcoma and death from respiratory paralysis. (4). Levels of Significant Exposure (LSE) is 376 ppm and Minimal Risk Levels (MRLs) is 60 ppb (1000ppb = 1ppm). Smell threshold is 320 ppb. There have been no reports of deaths in humans directly associated with exposure to styrene in the workplace (5,6).

6. Styrene Metabolism

The biological half-life of styrene, as measured by the appearance of metabolites like phenylglyoxylic acid and mandelic acid in urine is estimated between eight and nine hours in humans (7). The styrene concentration in blood is found to decay in a biexponential fashion typical of a two compartment kinetics model. Blood is proposed as the first component and tissues including adipose tissue as the second "slow" compartment. The half-life for styrene in blood in the initial phase is 0.58 hours and in the terminal phase 13 hours (8, 9, 10, 11). The half-life of concentration in adipose tissue is two to four days (12).

About 1 to 2% of the dose is exhaled unchanged. The majority of the absorbed material (about 90%) is metabolised in the liver by oxidation of the vinyl side group to styrene oxide. Styrene oxide is the active metabolite (13). The resulting major metabolites which are excreted are mandelic acid (60-80%) and phenylglyoxylic acid (about 30%). Only very small quantities of hippuric acid and 4-vinylphenol are produced and excreted in the urine (14). Urine samples of mandelic acid should not exceed 800 mg per gram of creatinine and phenylglyoxylic acid 240 mg per gram of creatinine. Accepted level of styrene in venous blood is 0.55 mg/L.

7. Management of Styrene Toxicity

Rapid, effective decontamination by staff wearing PPE is essential in the management of styrene exposure. Victims must be shifted to a safe, sheltered facility and a careful clinical evaluation done. All contaminated clothing and shoes are removed and eyes are irrigated with isotonic saline and skin surface washed thoroughly with water and non abrasive soap for at least 15 minutes.

Evaluation of vital signs and life-supporting measures are the main principles in poisoning management. Maintenance of a clear airway, administration of high flow oxygen 10-12 l/min by face mask, treatment of respiratory failure with intubation and IPPV, intravenous fluids to ensure good urine output > 2 ml/kg/h and vasoactive drugs if required.

Do not give any food containing fats and oils and gastric lavage may be considered in massive recent ingestion. Do not induce vomiting of consciousness is obtunded. There are no specific antidotes available.

Routine investigations like CBC, ABG, LFT, Creatinine and urine for styrene metabolites.

8. Conclusion

This mishap resulted in extremely high levels of styrene monomer exposure to those residing in the close vicinity and constitute the first fatalities reported worldwide from its inhalation. The lack of facilities to monitor atmospheric, blood and urine levels of styrene and its metabolites has precluded a better assessment of its toxic effects. The rapid evacuation of residents within a 5 km radius was responsible for a lower mortality rate. Morbidity was, however, high and long term effects of the exposure remains to be seen in follow up monitoring.

REFERENCES

1. O'Donoghue JL . Aromatic Hydrocarbons. Neurotoxicity of industrial and commercial chemicals 1985. Boca Raton, Florida. CRC, Press, Inc. vol II, p. 127–137.
2. Heiselman DE, Cannon LA. Benzene and the aromatic hydrocarbons. Haddad LM & Winchester JF eds. Clinical management of poisoning and drug overdose 1990. Philadelphia, W. Saunders Co. p. 1226–1227.
3. Sax NI, Lewis RJ. Dangerous properties of industrial materials, 7th ed 1989. New York, Van Nostrand Reinhold, vol III, p. 3127–3128.
4. ILO - International Labour Office. Encyclopedia of Occupational Health and Safety. 3 ed 1983 .Geneva, p. 2115.
5. Gosselin RE, Smith RP, Hodge HC. Clinical Toxicology of Commercial Products. 5th ed 1984. Baltimore, Williams & Wilkins, p.II-152.
6. The National Institute for Occupational Safety and Health (NIOSH) Workplace safety and Health Topics. Styrene CAS No 100-42-5.
7. Bond J: Review of the toxicology of styrene. Crit Rev Toxicol. 1989, 19: 227–49.
8. Guillemin MP, Bauer D (1979) Human exposure to styrene. Int Arch Occup Environ Health. 44: 249–63.
9. Harkonen H . Styrene, its experimental and clinical toxicology. Scand J Work Environ Health 1978; 4: 104–13.
10. Ramsey JC, Anderson ME. A physiologically based description of the inhalation pharmacokinetics of styrene in rats and humans. Toxicol Appl Pharmacol 1984; 73: 159–75
11. International Programme on Chemical Safety (IPCS) . Styrene. Environmental Health Criteria Document no 26. 1983 Geneva. World Health Organisation.
12. Engstrom J, Bjustrom R, Astrand I, Ovrum P. Uptake, Distribution and Elimination of Styrene in Man.Scand J Work Environ Health 1978; 4(4): 324–329
13. Baselt RC, Cravey RH. Disposition of toxic drugs and chemicals in man. 3d ed 1990. Chicago, Year Book Medical Publishers. p. 762–764.
14. Fielder RJ, Lowing R et al (1981) Styrene. Toxicity review 1. London, Her Majesty's Stationery Service.

CHAPTER 43

Chemical Disaster: Leakage of Styrene Gas at L. G. Polymers, Vishakhapatnam (Andhra Pradesh)

Anupam Srivastava
Commandant, NDRF, Pune, Maharashtra

Abstract

Occupational chemical accidents have increased in recent years in India. In the early hours of 7th May 2020, a chemical accident occurred in which a hazardous gas, styrene, was accidently released from a tank of LG Polymers, Vishakhapatnam (or Vizag), in the atmosphere causing several deaths and injuries to many and significant damage to the environment.

This article focussed on the accidental release of hazardous gas, styrene, from the tank of L. G. Polymers, Vizag, probable causes of accident, and existing relevant rules regulations on the subject in India. Data were obtained by making visits to accident place and nearby areas of the L.G. Polymers. Some weaknesses were seen in the safety management capacity of the Company. The main weaknesses were lack of an effective surveillance system for chemical accidents, low levels of safety in chemical plant, ineffective & inexperienced top- and low-level management to tackle incident of such type. Other weaknesses included the lack of awareness of residents of the nearby localities about chemical hazards. The Company was not prepared for such chemical accident both within industrialplant and in nearby residential area.

Action is needed to improve the areas of weakness so as to achieve the necessary capacities for chemical safety and preparedness and response to chemical incidents in line with the existing Rules & Regulations.

Keywords: Chemical Accident, industrial disasters, Styrene gas accident, Styrene gas Vizag, Accidental release, Chemical disaster

1. Introduction

Chemical Industry of India is one of the fastest growing industries. Chemical industry occupies a pivotal position in meeting basic needs and improving quality of life. The industry is a key enabler for industrial and agricultural development of the country and provides building blocks for several downstream industries, such as textiles, papers, paints, varnishes, soaps, detergents, and pharmaceuticals. It is also among the most diversified industrial sectors and covers over 80,000 commercial products.

The chemicals industry of India contributes 2.1% towards the nation's gross domestic product (GDP) and accounts for 15.95% of India's manufacturing sector.

On one hand, Chemical industries are the back bone of the development of the Nation and on the other hand they also pose severe risks of chemical accidents which may be a disaster at times. A chemical accident or emergency refers to an event which results in the release of a substance or substances that are hazardous to human, animal health and/or the environment in the short or long term.

These events can cause illness, injury, disability or death to living beings, often in large numbers, and can result in extensive damage to the environment with considerable human and economic costs. These disasters are threat to people, other living beings and environment. When these threats exceed human coping capabilities or the absorptive capacities of environmental systems, they give rise to industrial disasters.

India has a long history of Industrial Chemical accidents, including Bhopal gas tragedy, the worst industrial accident in the world so far, with the latest in the chain is Styrene gas leakage at LG Polymers in Vishakhapatnam (Vizag), which costs the lives of 12 persons in the immediate subsequent period and 585 people had to undergo treatment in hospitals, besides causing loss of livestock and vegetation.

At present Chemical industries have significant role, impact and contribution in the following sectors:

1. Agriculture
2. Environment
3. Hygiene & Health
4. Sustainable Energy Solutions
5. Food
6. Painting & Coating
7. Petro-Chemistry
8. Pharmaceuticals
9. Construction Industry
10. Automobile

Incontrovertibly, the Indian chemical industry is among the established traditional sectors of the country that play an integral role in the country's economic development. This sector forms a part of the basic goods industry and is a critical input for industrial and agricultural development.

It is also true that each development has its cost. With the established chemical industry sectors India has also seen some worst chemical accidents in the past. Few of them are following:

1. **Bombay Dock Explosion, 1944**

   The Bombay Explosion (or Bombay Dock Explosion) occurred on 14 April 1944, in the Victoria Dock of Bombay (now Mumbai) when the freighter SS Fort Stikine carrying a mixed cargo of cotton bales, gold, and ammunition including around 1,400 tons of explosives, caught fire and was destroyed in two giant blasts, scattering debris, sinking surrounding ships and setting fire to the area killing around 800 people.

2. **Chasnala Mining Disaster, 1975**

   The Chasnala Mine Disaster occurred on the evening of 27 December 1975, and killed 372 miners in Dhanbad, India. On 27 December 1975, an explosion rocked the Chasnala Colliery in Dhanbad, India. The explosion was most likely caused by sparks from equipment igniting a pocket of flammable methane gas. The Chasnala Colliery explosion was so severe that the mine collapsed, and millions of gallons of water from a nearby reservoir rushed into the pits at a rate of seven million gallons per minute. Those miners who weren't killed in the blast now found themselves trapped under debris, or drowned as the water quickly filled the mine. Rescue workers continued their efforts to dig out bodies and survivors until 19 January 1976. Sadly, there were no survivors, and most of the bodies were never recovered.

3. **Bhopal Gas Tragedy, 1984**

   The Bhopal disaster, also referred to as the Bhopal gas tragedy, was a gas leak incident in India, considered the world's worst industrial disaster. It occurred on the night of 2–3 December 1984 at the Union Carbide India Limited (UCIL) pesticide plant in Bhopal, Madhya Pradesh. Over 500,000 people were exposed to methyl isocyanate (MIC) gas and other chemicals. The toxic substance made its way in and around the shanty towns located near the plant. Estimates vary on the death toll. The official immediate death toll was 2,259. The government of Madhya Pradesh confirmed a total of 3,787 deaths related to the gas release. A government affidavit in 2006 stated the leak caused 558,125 injuries including 38,478 temporary partial injuries and approximately 3,900 severely and permanently disabling injuries.

4. **Jaipur Oil Depot Fire, 2009**

   The Jaipur oil depot fire broke out on 29 October 2009 at 7:30 PM (IST) at the Indian Oil Corporation (IOC) oil depot's giant tank holding 8,000 kilolitres (280,000 cu) of oil, in Sitapura Industrial Area on the outskirts of Jaipur, Rajasthan, killing 12 people and injuring over 200. The blaze continued to rage out of control for over a week after it started and during the period half a million people were evacuated from the area.

5. **Korba Chimney Collapse, 2009**

   The 2009 Korba chimney collapse occurred in the town of Korba in the Indian state of Chhattisgarh on 23 September 2009. It was under construction under contractfor the Bharat Aluminium Co Ltd. (BALCO). Construction had reached 240 m (790ft) when the chimney collapsed on top of more than 100 workers who had been taking shelter from a thunderstorm. At least 45 deaths were recorded.

6. **Mayapuri Radiological Incident, 2010**

   In April 2010, the locality of Mayapuri, Delhi was affected by a serious radiological accident.

   An AECL Gamma cell 220 research irradiator owned by Delhi University since 1968, but unused since 1985, was sold at auction to a scrap metal dealer in Mayapuri on February 26, 2010. The orphan source arrived at a scrap yard in Mayapuri during March, where it was dismantled by workers unaware of the hazardous nature of the device. The cobalt-60 source was cut into eleven pieces. These pieces of Co-60 radiologically affected several people.

7. **Vishakhapatnam Gas Leak, 2020**

   In the early hours of 7/5/20, a hazardous gas, Styrene leaked from one of the tankat LG Polymers in Vishakhapatnam, which coststhe lives of 12 persons in the immediate subsequent period and 585 people had to undergo treatment in hospitals, besides causing loss of livestock and vegetation.

Chemical accidents generally happened due to following major factors:

1. Ageing of the process plants and inadequate steps to pace with modern technologies in Indian chemical industry has increased vulnerability to chemical disasters.
2. A majority of the industrial accidents occurs due to human errors, as a result of non- compliance with standard operating procedure (SOP) that have been put into place by the company.
3. Chemical disasters have also occurred due to defects in design; absence of SOP to mitigate an early warning in the process, poor co-ordination between different departments within the chemical company.
4. Natural disasters such as flood and earthquakes have also caused major disasters in chemical industry.
5. Irregular improper maintenance of equipment in chemical industry, due to this equipment gets malfunction and ultimately fails, resulting in catastrophic disaster.
6. Not following the existing rules & regulations of the area.

Major Consequences of Chemical Accidents:

1. Loss of life/injuries
2. Impact on live stock
3. Damage of Flora/Fauna
4. Environmental Impact (Air, Soil, Water)
5. Financial losses

2. Incident of Styrene Gas Leakage, Vizag

In the wee hours of 7th May 2020, an incident of uncontrolled hazardous Styrene Gas Release occurred at LG Polymers, RR Venkatapuram, Visakhapatnam from one of the Styrene storage-tanks (M6Tank). The uncontrolled Styrene gas release from storage tank into the atmosphere occurred for the first time in India.

The accident took the life of 12 persons in the immediate subsequent period and 585 people had to undergo treatment in hospitals, besides causing loss of livestock and vegetation. This Styrene gas release, widely referred to as "Vizag Gas Leak", is one of the major Styrene gas release incidents from a bulk storage tank anywhere in the world. The Fig. 43.1 & Fig. 43.2 show the location of the factory along with the surrounding habitations.

Fig. 43.1 Google map of Factory Locations and surrounding habitation and deaths

Fig. 43.2 Google map of tanks in the premises of L. G. Polymers, Vizag

Fig. 43.3 Showing the accidental release of Styrene gas in L. G. Polymers, Vizag

2.1 Brief About Styrene Gas

Styrene is a chemical which is used to make latex, fibreglass, synthetic rubber and polystyrene resins. These resins are used to make plastic packaging, disposable cups and containers, insulation and other products. Styrene is found indoors in small quantities as a result of operating photocopiers and laser printers and from cigarette smoke.

Styrene gas vaporised from styrene liquid has following properties:

1. Colourless, mostly Odourless sometimes with mild sweet odour in low concentration.
2. Styrene (molecular weight of 104) vapours are 3.6 times heavier than air, but slight convection currents and common air movements easily dissipate the gas in larger areas diluting the concentration.
3. Styrene is quickly broken down in the air, usually within 24 hours.
4. Styrene readily evaporates from shallow soils and surface water.
5. Styrene is Class 'B' chemical having its flash point between 23°C and 65°C.
6. Styrene polymerizes above 45°C and gets into runaway reaction above 65°C.

Styrene has following health hazards:

1. **Acute Exposure:**
   (a) Styrene may be irritating to the eyes, skin, and mucous membranes. It can be ototoxic, nephrotoxic and hepatotoxic, and is a CNS depressant. Signs and symptoms of exposure may include nausea, fatigue, headache, loss of coordination, muscle weakness, a feeling of drunkenness, dizziness, and unconsciousness.
   (b) "Styrene sickness" with nausea, vomiting, and a sensation of drunkenness occurs with inhalation exposure.
   (c) Liver damage may occur with substantial chronic exposure (over 5 years).
   (d) Peripheral neuropathy and pulmonary edema may occur. Prolonged or repeated exposure may lead to defatting dermatitis. Fetotoxicity has been observed in experimental animals and genotoxicity has been observed in vitro.
   (e) Central nervous system depression can occur in serious acute exposures. Following chronic exposure, styrene can disrupt amino acid transport across the blood-brain barrier. Effects of styrene on the nervous system include CNS depression and peripheral neuropathy.
   (f) Irritation of the respiratory tract and occupational asthma may occur following acute exposure. Pulmonary edema has been reported in animals. Ongoing exposure to styrene can result in irritation or even obstructive lung disease.
   (g) Exposure in large concentrations causes lung complications, gastrointestinal effects and could lead to death due to damage to central nervous system.
2. **Reproductive Hazards**
   There are several studies suggestive of potential reproductive effects in humans. Styrene has also been extensively studied for possible reproductive effects in experimental animals.
3. **Carcinogenicity**
   Styrene is generally considered as metabolically toxic due to its metabolic conversion to styrene epoxide. There is inadequate evidence of Carcinogenicity in humans.

Fig. 43.4 Styrene liquid

3. L. G. Polymers Ltd., Vizag

Industrial activity at the current premises of M/s LG Polymers at R.R. Venkatapuram, Visakhapatnam started in 1961, under the name of M/s Sri Rama Mills to manufacture alcohol from molasses. Later, in 1962 M/s Sri Rama Mills was taken over by M/s Hindustan Polymers Ltd. In 1967, M/s Hindustan Polymers Ltd. replaced the manufacturing activity of alcohol with manufacturing activity of Polystyrene & Co-Polymer. Manufacturing of Styrene Monomer was reportedly initiated in 1973. In 1978, M/s Hindustan Polymers Ltd. merged with McDowell & Co. Ltd. of United Breweries Group. The Company terminated manufacturing of Styrene monomer and started importing it in the year 1993.LG Chemicals (South Korea) Ltd. set up the LG Chemicals India Pvt. Ltd., as a 100% subsidiary of LG Chemical, South Korea. The LG Chemicals India Pvt. Ltd. acquired M/s Hindustan Polymers Ltd. in 1997 and renamed the company as M/s LG Polymers India Private Limited, Visakhapatnam. After acquisition, the LG Polymers continued manufacturing of Polystyrene & Expandable Polystyrene and expanded the manufacturing capacities from time to time.

Fig. 43.5 Google map of Factory Locations and surrounding habitation

At present, LG Polymers is spread over a total area of 213 Acres and engaged in manufacturing of Polystyrene used for General Purpose Polystyrene (GPPS) & High Impact Polystyrene (HIPS) to the tune of 313 Tons per day, Expandable Polystyrene (EPS) to the tune of 102 Tons per day andEngineering Plastics 36.67 Tons per day. M/s LG Polymers is storing Styrene, Pentane and other raw materials within the plant premises.

4 Action Taken After the Accident

NDRF team, which was located 25 kms away from the incident site, reached the incident site immediately and found huge plume of gas emanating out of one of the tanks and it was spread all around in the premises of L. G. Polymers and in the vicinity. NDRF rescuers putting on SCBA suits rescued and shifted victims, who were found unconscious on the road or in the houses, to the nearby hospital. Thereafter, NDRF and other local agencies searched nearby houses and found many people trapped inside their houses and shifted them to the safe places.

The National Crisis Management Committee (NCMC) met under the Chairmanship of Cabinet Secretary to review the situation. It was decided that a team from CBRN (Chemical, Biological, Radiological and Nuclear) unit of NDRF from Pune, along with an expert team of National Environmental

Fig. 43.6 People of vicinity got unconscious due to inhalation of leaked hazardous Styrene gas

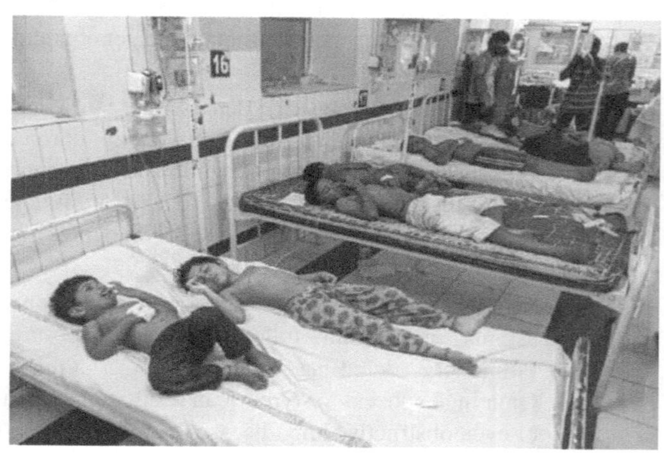

Fig. 43.7 Victims admitted to the hospital after getting affected by Styrene gas

Engineering Research Institute (NEERI), Nagpur would be rushed to Vishakhapatnam immediately to support the State Government in the management of the crisis on the ground.

5. Causes of Uncontrolled Styrene Gas Release

It was observed by the High-Power Committee that the uncontrolled release of Styrene vapour from Styrene storage Tank was due to the high increase in temperatures in the said Tank. The increase in temperatures led to polymerization and the heat generated due to polymerisation finally led to runaway reactions. Increase in temperature to the boiling point of Styrene monomer viz 145°C led to the boiling of the liquid Styrene, leading to uncontrolled vapour formation. Further increase in temperature led to increase in the pressure of the vapour which led to the uncontrolled release of vapour from the vents into the atmosphere.

6. Impact of Styrene Gas Release

Uncontrolled Styrene vapour release affected the surrounding habitations resulting in loss of lives of 12 persons while 585 were injured. Many were found unconscious, semi-conscious and faced nausea, breathing difficulties, skin rashes, sore eyes etc. About 600 residents of the neighbouring villages were rescued under the rescue and evacuation operation carried out by the NDRF, Police with assistance of local people. Those who were affected were immediately shifted to various hospitals by ambulances, NDRF, Police vehicles and other available vehicles. Nearly 20,000 people from 17,000 houses/ residences were evacuated.

All the trees near the M6 Tank have turned either fully or partially dry. The team observed that wherever the relatively heavy mixture of monomer and polymer Styrene flooded green plants, they got dried. All the trees were dried up to the level of Styrene vapour engulfing the affected area.

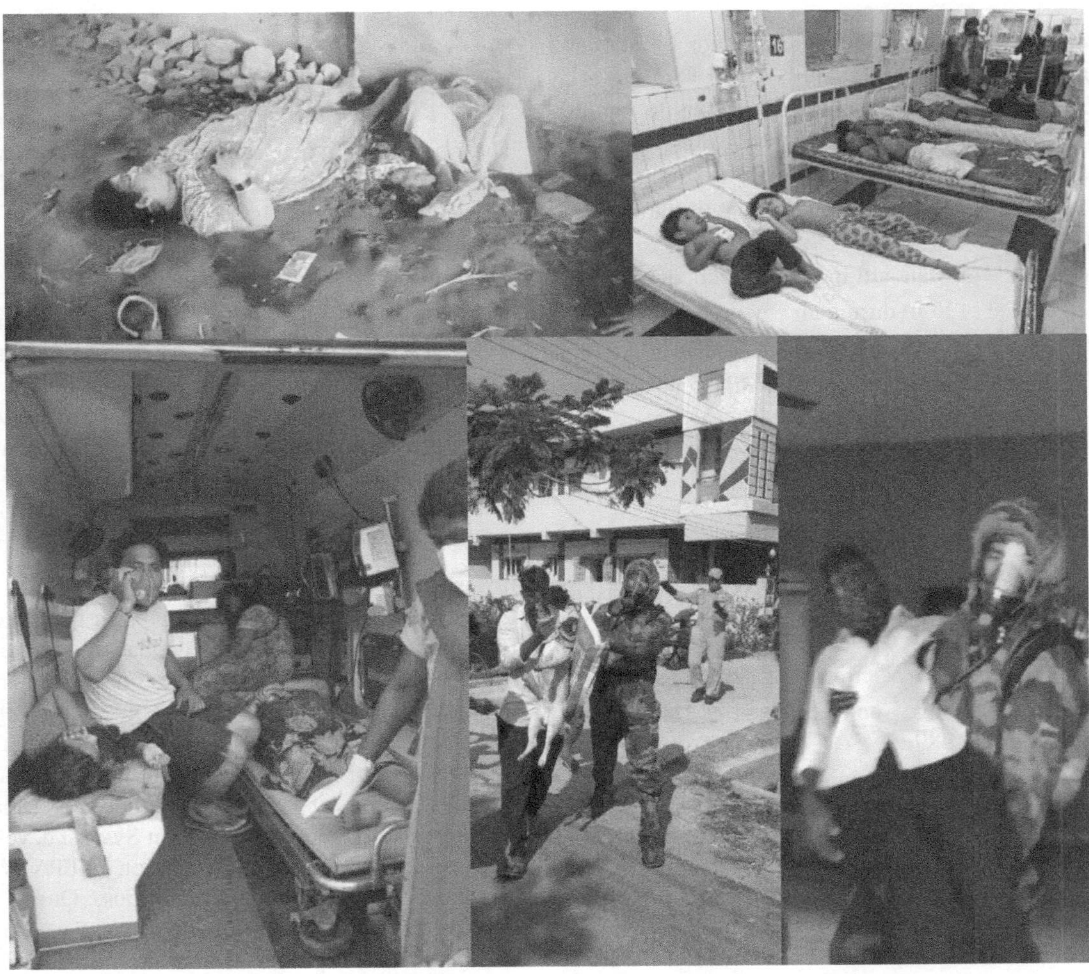

Fig. 43.8 Victims of the nearby villages rescued and admitted to the hospitals

Fig. 43.9 Styrene effect on trees: dried upto the height of moving of Styrene Plume

7. Negligence

1. No established protocol for the storage of styrene during extended lockdown periods.
2. Sufficient qualified and competent persons to handle such type of situation were not engaged by the company.
3. Poor design and poor maintenance of the storage tank.
4. Old technology for monitoring the safety parameters was used.
5. Non adherence of Standard Industry Practices.
6. Non adherence of SOP during an emergency situation.
7. Non compliance of Off-site emergency plan.
8. Lack of technical and emergency handling capabilities.
9. Human negligence and casualness.

8. Acts Related to Hazardous Chemical Management India

Following are few Rules & Acts which are in existence for the control of Plants using hazardous chemicals.

1. The Environmental (Protection) Act 1986
2. Manufacture, Storage and Import of Hazardous Chemical Rules, 1989
3. National Environment Tribunal Act, 1995
4. The Petroleum Act 1934
5. The Factories Act 1948
6. The Chemical Accidents (Emergency Planning, Preparedness and Response) Rules 1996
7. Air (Prevention and Control of Pollution) Act 1981
8. Water (Prevention and Control of Pollution) Act 1974
9. Andhra Pradesh Fire Service Act 1999
10. The Boilers Act 1923

9. Discussion

The above-mentioned Acts and Rules are comprehensive and cover most of the aspects needed for the scientific management of the hazardous Chemicals. The biggest problem with these Acts & Rules, however, are the lack of a deterrence mechanism or any penal provision.

The rules are silent on the provision to prosecute violators, levy fines or order closures. There is no mention of the handling of hazardous Chemicals if the license to store the chemical is not renewed.There is also a need to strengthen existing mechanisms of the disaster management plan, in case an accident occurs.There is no document in the public domain to point out whether provisions under the disaster management plan is implemented periodically.

It was observed that the Onsite Emergency Plan prepared by LG Polymers lacked the procedures to handle emergencies for toxic chemical release and runaway reactions in tanks/ reactors/ pipelines. Instead, the Onsite Emergency Plan was prepared only for fire occurrences and other accident scenarios. Thus, it is not surprising that though there was no fire, the LG Polymers management performed a firefighting action in place of onsite emergency for vapour release. According to Hazard Identification System developed by the National Fire Protection Association (NFPA) 704, Styrene Monomer is classified as hazardous. Onsite Emergency Plan for the LG Polymer Plant that handles Styrene in large storage tanks and processes Styrene should have been based on hazardous nature of Styrene rather than emphasising only on firefighting. It is observed that the LG Polymers did not follow any standard guidelines for Emergency Plan, which caused a lack of clarity in emergency response at time of accident.

After the accident no Emergency alarm was blown so as to make public aware about the emergency. The accident was further aggravated by the staff on shift duty who failed to alert the residents of the nearby colonies, through emergency siren, which led to the loss of human and animal lives and damaged flora and fauna in the nearby areas.

As per MSIHC Rules 1989 it is clearly mentioned that Plant will conduct Onsite Emergency mock drill once in six months and Offsite Emergency mock drill will be conducted once in a year, but these were not being conducted as per directions mentioned in MSIHC Rules 1989. Secondly, NDRF & SDRF are the stake holders in case of chemical emergency but they were never involved in such type of drills.

The management had neither conducted any awareness activities among the neighbourhoods, nor ensured the participation of the public in any mock drills. The local community were not informed about emergency measures to be taken in case of accidents like toxic releases, vapour release, fires and explosions. Company should have conducted local public awareness about Dos/Don'ts during an emergency, which was never done. If this was done, public would be more cautious of actions taken in such type of accidents.

It was also noted that L.G. Polymers which was using a hazardous chemical for its manufacturing of products was established in a residential area whereas Plants using such hazardous chemicals should be located far away from the habitation. If such Plant is located in residential area or habitations have already come up then a buffer zone around the Plant should be demarcated and these areas should be strictly marked as prohibited for residential area or any other type of habitations. If that is also not possible then such Plants should be relocated to the notified industrial zones, away from habitations. Another option is that storage facilities of hazardous substances can be located far away from habitations and transported through pipe lines or tankers.

This accident also severely affected the vegetation. Most of the trees/plants dried up completely or dried upto the level where styrene gas plume was present. Total of 34 animals died and 242 animals were affected due to the effect of released Styrene gas.

10. Conclusion

Disaster is a rarity in the chemical industry, but negligence or misfortune can so easily result in devastating consequences. The result of a chemical disaster has effects through generations of populations which are almost irreparable and the cost of that to the affected people is unimaginable. In addition to loss of life, the major consequences of chemical disasters include impact on livestock, flora/fauna, the environment (air, soil, water) and losses to industry.

Chemical industry is considered a societal problem solver; it synthesized crop- enhancing agricultural chemicals to ensure a constant and viable food supply, helped in the eradication of deadly diseases by developing life-saving pharmaceuticals and chemical pesticides, developed innovative plastics and synthetic fibres for use in a both industrial and consumer products. Over the years the chemical industry has been a vital sector of the modern industrialized economy. Despite of that no development should be based on the cost of lives of human or animals neither on the cost of degradation of the environment. Vizag incident is not just about the leak of a hazardous gas, it is also about contamination of the environment. This accident is also a reminder that though there arecomplex web of rules, regulations but there are various government departments which are not single tuned to control any industry as per the existing laws. In the above incident we found that different departments, though dealing with the same Company but not in understanding with each other, provided loop-holes for Company to avoid laws on the subject.

If we talk about the disaster management plan of the Company as per MSIHC Rules 1989, it is a complete disaster. They were supposed to have disaster management plans that local authorities—which include hospitals, fire stations, NDRF and the like—will know and were supposed to communicate to the local people what they were supposed to do in the event of a disaster. These plans were also supposed to be updated regularly. Many companies have disaster management plans for themselves, but what we learnt from Vizag gas leak was the importance of communication on disaster to the local population.

This has still not happened and people live in the vicinity of disaster and have no idea what a disaster is, what is likely to happen and what they are supposed to do in the event that it happens. People continue to live treating a hazardous industry as just another risk.

Chemical safety needs to be handled with extreme precautions. It is pertinent to have sound storage,protocols for chemical storage, including where it can be stored and where it cannot, handling and transport of chemicals in the country to avoid an accident similar to the Vizag gas leak.

What followed, however, was the realization that if industrial development was unregulated and reckless—without adequate safeguards—the consequences could be far-reaching. Alongside, the demand grew for accountability of industries that engage in potentially hazardous activities. Industrial disasters are not simply safety problems that need to be resolved, they also have wider significance because they offer important opportunities to learn about the "goodness of fit" between society, technology and environment and about how that fit can be strengthened or weakened by unexpected events. This is the kind of information that will be invaluable to humanity during an era of deep and far-reaching societal and environmental change. However, if we are to make optimal use of such opportunities it may be necessary to modify the way we think about industrial disasters.

References

1. Annual report 2015-16, Department of Chemicals and Petrochemicals, Ministry of Chemicals and Fertilizers, Government of India
2. Report of technofunc-chemical domain-importance of chemical industry
3. Newsletter of ENVIS-NIOH prepared by Vol.9, No.4, by Ms Prarthana Trivedi, Mr Deepak Purohit, Ms Annie Soju.
4. Ref: Report on Chemical Accidents by DMI, Bhopal
5. Report of High-Power Committee constituted by A.P.Govt.

CHAPTER 44

Managing a Blackout—The Consequences of an Oil Spill at Sea

Malini S. Shankar
Indian Maritime University, Chennai, India
malinishankar@imu.ac.in

K. M. Sivakholundu
Indian Maritime University, Chennai, India
kmsivakholundu@imu.ac.in

Abstract

The Ennore Oil Spill of 2017 triggered by a collision between two oil cargo vessels resulted in an environmental incident. This study is an attempt to cull the details from public domain and analyse the efficacy of the Government system to cope up with response. The information on events leading to the mishap indicates the suboptimal working conditions of seafarers prevailing on-board. The response activity drew negative publicity as the official communications were inconsistent. Besides declaring an under estimated quantum of oil spill, the perceived delay and inadequate resources seems to have aggravated the public sentiment. The episode of coastal spill clean-up came to close in two weeks with contributions from 30+ agencies of Government and industry along with 2000+ volunteers. The shoreline environment and fisheries activities returned to normalcy in three months' time. Compensation to affected public was settled in a years' time. However there is little information or unbiased evaluation of the episode available. An evaluation of such incidents will aid in addressing issues in future.

Keywords: Oil Spill; collision; Ennore; marine environment; contingency plan

1. Introduction

India with its growing economy ranked as the second largest importer of crude oil in the world in the year 2019 with an import of 205 million Metric Tonnes (MT). Combined with such volume of oil transport, the long coastline of the country remains vulnerable to oil spill incidents. For example, between the years 1982 and 2010, the Indian Coast Guard [1] has documented a total of 74 spill incidents – ie, more than 2 incidents per year. There are more than 200+ ports/terminals along our coast that are susceptible to such events unless the awareness among the stakeholders is enhanced. A combination of several factors – some preventable - are required for the incident of oil spill to take place. Once the spill occurs, the remedial actions can become complicated with serious environmental and legal issues. The oil spills can impact ecology and society in an adverse manner in both short and long terms in time depending on the size and response activities [7] [8] [9].

Considering the risks of maritime oil spill, India remains an active participant in international conventions (MARPOL, IOPC) and adopts policy instruments to prevent/respond to such events. The National Oil Spill Disaster Contingency Plan (NOS-DCP) is in operation since 1996 and updated periodically. NOS-DCP delineates the responsibilities of Government agencies at central, state and district level to coordinate the response in case of oil spill.

The effective implementation of the plan and the adequacy of response is tested in the present case involving an oil spill off Ennore Port near Chennai. The incident of Ennore Oil Spill (EOS here after) in the year 2017 is an opportunity to learn a few lessons to improve the system. The combined effect of command chain, the cooperation among the agencies, the size and nature of the spill, the climate conditions, the trapping of spill due to coastal structures (groins) seem to have diffused the situation in a reasonably short time. The spill effects were found to have spread over a coastal stretch of about 104 km in three coastal districts.

In the aftermath of the event, mopping up of oil on the coast proved to be laborious and not amenable to mechanical implements. Major portion of oil that settled on the shore on sand and boulders had to be removed manually by a large number of volunteers. The collected oil sludge and contaminated sand were treated using bio-remedial measures. The clean-up operations came to a close in June 2017 in the affected area. The process for the compensation claims was initiated with interim pay outs to the affected population and a settlement of Rs. 168 Crores has been realised from the insurers of the shipping company.

The present study considers the sequence of events, the environment, response, administrative glitches and claims for compensation. The reports submitted to National Green Tribunal by Government agencies [10] and details available in public domain form the basis for describing the event and the follow up.

The EOS episode – not comparable with other major oil spills on an international scale - lasted from January 2017 to April 2017 and highlighted few lacunae in the oil spill response and planning in our country.

2. Sequence of Events

2.1 The Accident

On the early hours of 28th January 2017, two ships – *MT Dawn Kanchipuram* and LPG/C *BW Maple* – collided in the navigation channel of Ennore Port. Both the ships are designed to carry petroleum products – LPG by *Maple* and Petrol & Diesel by *Dawn Kanchipuram*. While *Dawn Kanchipuram* was heading into the port with 32,750 MT of petroleum products, *Maple was* leaving the port after discharging its cargo of propane and butane.

The collision occurred due to improper manoeuvring of vessels in close proximity. This had resulted in *Maple's* front portion tearing into the *Dawn Kanchipuram's* port side (left side of ship), where fuel oil tanks were situated. Fortunately the main cargo tanks remained unaffected but the damage on the fuel tank initiated leakage of oil into the sea. Both the vessels were taken to anchorage area and assessments of damages were made. During the initial assessment of *Dawn Kanchipuram*, the extent of damage was thought to be localised and only 2 MT of oil was estimated to have leaked. However when the final estimates were made long after the accident, it emerged that a total of 251.46 MT of oil had been spilled. After the oil spill alert had been sounded to Kamarajar Port and Maritime Rescue Coordination Centre (MRCC), the response activities were initiated by Indian Coast Guard and a spill containment oil boom was rigged around vessel Dawn *Kanchipuram* within 6 hours of the accident. Reportedly, on the day of accident, in the initial assessment the damage on one of the fuel oil tanks was not noticed that lead to continuous leak. As a consequence Kamarajar Port and Indian Coast Guard did not deploy their oil response equipment to full capacity.

Dawn Kanchipuram was in anchorage for next two days to assess the damages and to decide on the next course of action. On 30th January 2017 the vessel was taken into the port after completion of procedural compliances and on the next day the main cargo of petrol and diesel was unloaded. Since the time of accident (03:35 28.Jan.17) to evacuation of cargo (01:30 hrs on 31. Jan.2017) almost three days had elapsed, risking a major spill.

2.2 The Spill Extents

Spill occurred at a distance of about 1.5 km from the coast. Due to the prevailing winds and South Westerly currents, the oil slick drifted towards south and started depositing partially on the beach. In the first 12 hours, a stretch of 6 km was affected [5]. In the next four days, the spill effects covered the sub-urban area of Chennai up to Palavakkam – about 32 km from the origin of spill. Final estimate by State Fisheries Department on the affected length of coastline is 104 km spread over three coastal districts.[11]

2.3 The Spill Effects and Clean-up

The spill resulted in widespread damage to local aquatic life and decline in fishing yield before returning to near-normalcy in about three months' time. Being a site for Olive Ridley Turtle breeding grounds, many turtle mortalities were reported by local fishermen. Fish catch and crabs in the area were impacted [3]. The weathering of oil spill includes processes like evaporation, emulsification, dispersion, dissolution, bio-degradation and forming tar balls. Among these processes, evaporation is the most effective means and depending on the nature of oil spill, it can take away as much as 80% of the volume during first two days [6].

The worst affected settlement called Eranavur lies immediate south of Ennore. The shoreline of Eranavur had a number of coastal protection structures called groins to mitigate sea erosion. These are shore-perpendicular bund like structures made up of boulders extending in to sea by 100-200m. Along the shore these groins are placed at 500m interval. This groin field has created a trap for the floating oil spill that otherwise could have affected much larger extent of shoreline towards south [4]. The oil sludge after getting trapped can only settle between adjacent groins, which made the clean-up efforts more concentrated. However the presence of boulders made the collection difficult for deployment of mechanical implements and manual clean up only was feasible. A large number of volunteers belonging to NGOs and general public helped to retrieve the oil sludge on the shore. In all, 37 agencies from Central and the State and from industry along with 2000+ volunteers were engaged to clean up the shore over a period of 14 days.[11]

The collected sludge and contaminated sand posed additional environmental hazard as it could not be disposed without treatment. Kamarajar Port Limited has provided a bio-remedial facility using a pit of 172m x 16.5m with poly-ethylene liner to treat the oil sludge and oil contaminated sand. The contents of the pit were treated using microbial degradation while monitoring the ground water quality. 16 piezo metric wells were installed around the treatment pit for monitoring purpose. The samples were extracted to test for the presence of Petroleum Hydro Carbon, Oil and Grease in the ground water in the area surrounding the treatment pit.

A long term monitoring plan for the marine environment costing Rs. 5.8 Crores was proposed by Integrated Coastal and Marine Area Management – Project Directorate of Ministry of Earth Sciences. This program plans to collect the samples and monitor the quality of marine life over a period of five years.

2.4 Compensation to Affected Entities

The environmental damage affected the marine species and was documented by studies [3] [4]. This resulted in significant reduction of fish catch in the region and loss of livelihood. Claimants for compensation included boat owners, fishing labourers, fisher women, fish vendors and allied activity labourers. For compensation purposes the number of persons affected was ascertained to be 112051 with a total claim of Rs. 203.225 Crores[10] Claims from the organisations involved in the spill like Chennai Port, Kamarajar Port, Indian Coast Guard, Ennore Tank Terminal Pvt Ltd, Bharat Petroleum Corporation, Indian Ouil Corporation amounted to US$ 13,80,864 [10]. Besides these compensations, a project for artificial reef made up of concrete blocks costing Rs. 10 Crores was also implemented to restore the environment and improve fish habitat.

After about a year of the incident, in August 2018, the vessel owner of *BW Maple* paid a compensation of Rs.168 Crores through their insurers.

Few operational and policy issues as indicated in the following sections are to be considered for evaluating the efficacy of the system as a whole.

3. Operational Issues

- Ship collisions are considered to be a major source of marine oil spills and has been estimated to account for more than 28% of all shipping accidents resulting in a major oil spills worldwide [2]. There were number of factors that contributed to the occurrence of collision. Most of the factors like loss of situational awareness, inadequate response of the Pilots, failure of watch keeping, the reported non-functional channel marker buoy, unilateral decisions by the Ship Masters disregarding the presence of another vessel in close proximity and more point to lack of training on the part of shipping personnel. It was reported that the pressure on the ship personnel from management and insufficient rest for the personnel could be contributing factors.
- The collision would have created on board panic and chaos. The inspection of damage by the Surveyor should have been more professional. The incorrect assessment of fuel leakage prompting the Master of vessel to make a misleading declaration of 2 MT spill, whereas the actual leakage continued from one of the fuel tanks (approximately 250 MT) for unknown duration.
- After the collision, *Dawn Kanchipuram* spent three days in the anchorage, while officials tried to assess the stability of the vessel and allowed it to berth in the port to empty the cargo of some 32000 MT petroleum product. Had the cargo tanks of the vessel been damaged (which was not verified at that time), the spill could have been much larger. The decision making process /procedure in such situation needs to be streamlined.
- The leakage from the fuel tank continued for days, initially in anchorage and then in the port. There seems to have been little effort to identify the source of leak even when the stability of vessel was assessed by surveyors. The operational procedure does not seem to have addressed this issue even after the event. It may be prudent to lay down a mandatory procedure to identify the source of leak and document the same as part of the damage assessment. This is vital for any decision during response activities.
- Inadequate work force for clean-up activities. The coordination among agencies needs to be strengthened. The work force allocated was reported to be far below the requirement [11]. The ICG vessels for the purpose were not helpful as the oil slick is in very shallow waters and on the shore. Smaller crafts were not available for deployment. Some of the equipments (such as suckers) were not effective. The availability of logistical support and suitability of deployed equipment for the intended clean-up seems to have not been planned. Periodic drills had not been conducted, leading to gaps in decision making.

4. Policy Issues

- NOS-DCP mandated rigorous contingency plans are not prepared/updated by ports and installation. Similarly the compliance reports are also not being followed up regularly [11].
- Scientific studies in the aftermath of spill needs to be effective. Apart from funding deficiency, there seems to be more issues when some studies threw up field data. For example, the ICMAM study on long term effects of Total Petroleum Hydrocarbons TPH was carried out. The analysis indicated that 21 species were contaminated with TPH of 4.51 μg/g. However there seems to be no information on definition of safe limit for TPH.
- The gap in operational level communication involving different organs of Government may require refinement. There seems to be a lack of coherence in response leading to difficulties at field operations – especially coastal clean-up that requires mobilisation and coordination among large number of agencies and personnel.
- Need for local contingency plan. Apart from District level plan, it may be desirable to have local level contingency plan. After the incident, the IIT-M has been requested to provide such plan. The interface between sea and land calls for a clearer demarcation of roles and responsibilities.
- Effective and verifiable procedures to monitor the compliance of plans and training the personnel for contingencies. Periodic drills and workshops should be conducted.
- The documentation on events and activities are spread over different agencies. A common point of accumulation and study of all available information will go a long way to understand the lacunae and address them more effectively.

5. Conclusion

The oils spill event of Ennore in 2017 resulted in widespread environmental damage affecting 104 km long coastal stretch. The damage to aquatic life was severe and showed up on many species like fish, turtle and crabs on the coastal region. The normal life of thousands of inhabitants linked with fisheries industry was disrupted. Compensations and insurance claims have been processed in reasonable time frame. The operational and policy issues that may need to be acted upon have been listed out. The long term effects on environment have not been followed up. There is little information on a systematic evaluation of the causes leading to the event and preparedness of the Government system as a whole by any agency that has access to all authentic information. It is suggested that a repository in public domain be initiated that can accumulate all information from official sources and contributions from stakeholders like government agencies, industry, scientific community and general public.

References

1. Indian Coast Guard https://www.indiancoastguard.gov.in/WriteReadData/Bulletin /201512290521575317057Spillssince1982.pdf
2. Sharma, D. R. (n.d.). Comparative study of implementation of existing measures for oil spill response in Northern Indian Ocean and investigate oil spill response in Northern Indian Ocean and investigate improvement mechanisms improvement mechanisms. https://commons.wmu.se /all_dissertations/122
3. Selvakumar, N., Dhanasekar, K., Whaiprib, P., & Munuswamy, N. (n.d.). Impact of January 2017 oil spill on the biota off Chennai, southeast coast of India with emphasis on histological impact on crab, Grapsus albolineatus. *J. Mar. Biol. Ass. India*, *59*(2). https://doi.org /10.6024/jmbai.2017.59.2.1989-08
4. Han, Y., Nambi, I. M., & Prabhakar Clement, T. (2018). Environmental impacts of the Chennai oil spill accident – A case study. *Science of the Total Environment*, *626*, 795–806. https://doi.org/10.1016/j.scitotenv.2018.01.128
5. Prasad, S. J., Balakrishnan Nair, T. M., Rahaman, H., Shenoi, S. S. C., & Vijayalakshmi, T. (2018). An assessment on oil spill trajectory prediction: Case study on oil spill off Ennore Port. *Journal of Earth System Science*, *127*(8). https://doi.org/10.1007/s12040-018-1015-3
6. Fingas, M. (2021). Introduction to Oil Spills and their Clean-up. In *Petrodiesel Fuels* (pp. 875–889). CRC Press. https://doi.org/10.1201/9780367456252-4
7. Liu, X. (2010). Integrated modelling of oil spill response strategies: a coastal management case study. *Environmental Science & Policy*, *13*(5), 415–422. https://doi.org/10.1016/J.ENVSCI .2010.04.006
8. Spaulding, M. L. (2017). State of the art review and future directions in oil spill modelling. *Marine Pollution Bulletin*, *115*(1–2), 7–19. https://doi.org/10.1016/J.MARPOLBUL.2017 .01.001
9. Chang, S. E., Stone, J., Demes, K., & Piscitelli, M. (2014). Consequences of oil spills: A review and framework for informing planning. *Ecology and Society*, *19*(2). https://[doi.org/10.5751/ES-06406-190226
10. Reports submitted to National Green Tribunal by Ministry of Environment, Forest and Climate Change, Tamilnadu Fisheries Department and Tamilnadu Pollution Control Board. https://greentribunal.gov.in/sites/default/files/news_updates/1417.pdf
11. Indian Coast Guard, Proceedings of the 22nd NOS-DCP and Preparedness Meeting 10.8. 2017. https://www.indiancoastguard.gov.in/WriteReadData/bookpdf/20171127051606811079822ndNOS-DCP.pdf

CHAPTER 45

Baghjan Fire: A Case Study of the '2020 Assam Gas and Oil Leak' at Baghjan Oil Field, Tinsukia, Assam

Swapnali Gogoi
Associate Professor, H.O.D. Department of Geography
North Lakhimpur College (Autonomous), Assam
swapnaligogoi77@gmail.com
+91 8011498024

Bubu Baruah
Ph. D., Department of Geography
North Lakhimpur College (Autonomous), Assam
bububoruahrgu@gmail.com
+91 7002022067

Annekha Chetia
M.Phil. Research Student Jamsetji Tata School of Disaster Studies
Tata Institution of Social Sciences, Mumbai
anweshachetia34@gmail.com
+91 7002172508

Abstract

The 2020 Assam gas and oil leak, also referred as the Baghjan gas leak, is a Natural gas blowout that happened in Oil India Limited's Baghjan Oilfield in the district of Tinsukia, Assam, India on 27 May 2020. The blowout occurring at Well No. 5 in the Baghjan Oil Field, resulting in a leak of natural gas, subsequently caught fire on 9 June 2020, and was finally doused on 15th of November after burning for 159 days, becoming India's longest Oil well fire. The disaster has resulted in three deaths (officially), large-scale local evacuations, and environmental damage and the locals losing their homes, belongings and livelihoods and separated families.

Thus, after documenting the disaster and emphasizing the fact that a disaster has its greatest impact at the local level; this paper aims at understanding the disaster at Baghjan from the perspective of the local disaster survivors. The study is guided by the experience of three women survivors with three different disaster response and recovery context of the same Disaster. While using qualitative methodologies, the study tries to document the unique personal accounts, in their own words and using their own personal time lines with respect to the disaster. This Phenomenological attempt draws out micro-histories of the three women survivors taken as individual subjects through unstructured interviews, and while doing so, it would shed light on how the interface of human, environment and disaster is of a complex and dynamic character which cannot be understood in a singular or linear aspect.

Keywords: Industrial Fire, phenomenology, women and disaster, disaster response, recovery, intersectionality

1. Introduction

A disaster is a sudden, calamitous event that seriously disrupts the functioning of a community or society and causes human, material, and economic or environmental losses that exceed the community's or society's ability to cope using its own resources. Though often caused by nature, disasters can have human origins. Disasters = (Vulnerability + Hazard)/ Capacity. Disasters are often used synonymously with the word Hazard. But presently, these two terms are understood in a much varied sense. Hazards are defined as phenomena that pose a threat to people, structures or economic assets and which may cause a disaster. Hazard is a threat, while disaster is an event. A disaster occurs when a hazard impacts on vulnerable people. The combination of hazards, vulnerability and inability to reduce the potential negative consequences of risk results in disaster.

Developing countries and the poorer sections of the society suffer the most when a disaster hits – more than 95% of all deaths caused by hazards occur in developing countries, and losses due to natural hazards are 20 times greater (as a percentage of GDP) in developing countries than in industrialized countries. No matter what society disasters occur in, they tend to induce change in government and social life. They may even alter the course of history by broadly affecting entire populations and exposing mismanagement or corruption regardless of how tightly information is controlled in a society. The word disaster is derived from Middle French désastre and that from Old Italian disastro, which in turn comes from the Ancient Greek pejorative prefix *dus,* meaning "bad" and *aster* meaning "star". The root of the word disaster ("bad star" in Greek) comes from an astrological sense of a calamity blamed on the position of planets.

Disasters are mainly divided into natural or man-made, although complex disasters, where there is no single root cause, are more common in developing countries. A specific disaster may spawn a secondary disaster that increases the impact. A classic example is an earthquake that causes a tsunami, resulting in coastal flooding. Some manufactured disasters have been ascribed to nature. Some researchers also differentiate between recurring events such as seasonal flooding, and those considered unpredictable. A natural disaster is a natural process or phenomenon that may cause loss of life, injury or other health impacts, property damage, loss of livelihoods and services, social and economic disruption, or environmental damage. Various phenomena like earthquakes, landslides, volcanic eruptions, floods, hurricanes, tornadoes, blizzards, tsunamis, cyclones and pandemics are all natural hazards that kill thousands of people and destroy billions of dollars of habitat and property each year. However, the rapid growth of the world's population and its increased concentration often in hazardous environments has escalated both the frequency and severity of disasters. With the tropical climate and unstable landforms, coupled with deforestation, unplanned growth proliferation, non-engineered constructions make the disaster-prone areas more vulnerable. Developing countries suffer more or less chronically from natural disasters due to ineffective communication combined with insufficient budgetary allocation for disaster prevention and management. Human-instigated disasters are the consequence of technological or human hazards. Examples include stampedes, fires, transport accidents, industrial accidents, oil spills, terrorist attacks, nuclear explosions/ nuclear radiation. War and deliberate attacks may also be put in this category. Other types of induced disasters include the more cosmic scenarios of catastrophic climate change, nuclear war, and bioterrorism. One opinion argues that all disasters can be seen as human-made, due to human failure to introduce appropriate emergency management measures.

Here in this study the focus is mainly on Industrial Disaster under the categories of man-made disasters, with special reference to the Industrial Fire. An industrial fire is a type of industrial disaster involving a conflagration which occurs in an industrial setting. Industrial fires often, but not always, occur together with explosions. They are most likely to occur in facilities where there is a lot of flammable material present. Such material can include petroleum, petroleum products such as petrochemicals, or natural gas. Processing flammable materials such as hydrocarbons in units at high temperature and/or high pressure makes the hazards more severe. Facilities with such combustible material include oil refineries, tank farms (oil depots), natural gas processing plants, and chemical plants, particularly petrochemical plants. Such facilities often have their own fire departments for firefighting. Sometimes dust or powder is vulnerable to combustion and their ignition can cause dust explosions. Severe industrial fires have involved multiple injuries, loss of life, costly financial loss, and/or damage to the surrounding community or environment.

The present work is a case study based on field surveys and unstructured interviews carried out in one of the most historic industrial fire disasters that took place in the Baghjan Oil Field situated in the Tinsukia district of Assam. An attempt has been made to understand the disaster through field survey, secondary literature and analysing first hand experiences narrated by the local people inhabiting the area. In this study, the experiences of the three women survivors are introduced and analyses from three perspectives. Firstly, how the understanding of the Disaster dynamics should start from the local level in a bottom-up approach; secondly, how the event itself is perceived, responded to and recovered from, differently from the standpoints of various intersectionalities and inequalities that exist in our society and lastly, how the same disaster is creating different aftermaths, which becomes more complex and diversified due to the intersectionality between Disaster Vulnerability, Risk, Risk Perception, Resource Distribution, Social categorizations such as Class, Caste and Gender, Disaster Response and Recovery. The study also aims to argue that a disaster event apart from being understood through a theoretical or technical lens should be understood from a more humanistic perspective, and how in doing so it transcends theoretical confinements as it spirals down to the life of the communities residing in a complex and dynamic natural as well a cultural environment.

2. Objectives and Methodology

The present study mainly revolves around two objectives. The first one is to document the disaster with its pre, post and during disaster stages. And the second objective is to understand the disaster and its effects in the local level in a more humanistic and qualitative perspective. The first objective is tried to be fulfilled by carrying out field visits, surveys and collecting information from primary and secondary sources. And for the second objective, a phenomenological methodology is taken into account. Phenomenology is an approach to qualitative research that focuses on the commonality of a lived experience of an individual or a group. This approach aims to study a phenomenon as it is experienced by the participant and to reveal what the phenomenon is rather than what causes it or why it is being experienced at all. This is achieved by interviewing three women survivors of the disaster through unstructured interviews. This also aims to bring out a gendered view of a disaster and the different intersectionalities that lie within our society that become a catalyst for creating different disaster aftermaths.

3. Documenting the Disaster at Baghjan

3.1 Study Area

The study area is the source area of the Industrial Fire disaster that occurred in the Oil No. 5in the Baghjan Oil Field situated in the Baghjan Gaon. According to Census 2011 information the location code or village code of Baghjan Gaon is 290252. Baghjan Gaon is located in Doom Dooma Tehsil of Tinsukia district in Assam, India. It is situated 20km away from sub-district headquarter Doom Dooma and 50km away from district headquarter Tinsukia. Baghjan is the gram panchayat of Baghjan Gaon village. The total geographical area of village is 533.84 hectares. Baghjan Gaon has a total population of 4,488 people, consisting of 2,244 male and 2,244 female population. There are about 872 houses in Baghjan Gaon. Average Sex Ratio of Baghjan village is 1000 which is higher than Assam state average of 958. Child Sex Ratio for the Baghjan as per census is 615, lower than Assam average of 962. Baghjan village has lower literacy rate compared to Assam. In 2011, literacy rate of Baghjan village was 57.60 % compared to 72.19 % of Assam. Male literacy stands at 58.33 % while female literacy rate was 56.92 %. The village is well connected with public and private bus services. The nearest railway station is available within 10+ km distance. As per constitution of India and Panchyati Raaj Act, Baghjan is administrated by Sarpanch (Head of Village) who is elected representative of village. Schedule Caste (SC) constitutes 5.61 % while Schedule Tribe (ST) was 1.09% of total population in Baghjan Gaon. In Baghjan village out of total population, 1915 were engaged in work activities. 72.43 % of workers describe their work as Main Work (Employment or Earning more than 6 Months) while 27.57 % were involved in Marginal activity providing livelihood for less than 6 months. Of 1915 workers engaged in Main Work, 339 were cultivators (owner or co-owner) while 54 were Agricultural labourers.

3.2 Baghjan Fire

The 2020 Assam gas and oil leak, also referred as the Baghjan gas leak, is a natural gas blowout that happened in Oil India Limited's Baghjan Oilfield in Tinsukia district, Assam, India on 27 May 2020. The blowout occurred at Well No. 5 in the Baghjan Oil Field, resulting in a leak of natural gas. The leaking well subsequently caught fire, and has resulted in three deaths (Officially), large-scale local evacuations, and environmental damage to the nearby Dibru-Saikhowa National Park. The region which is known for its resources like tea, natural gas, oil and coal was for burning for nearly six months with the night skies lit up with an orange hue and an ash cloud looming over the pristine landscape

The Background or Pre-Disaster Phase: The Baghjan Oil Field is located in Tinsukia district in the State of Assam, near Baghjan village, which has a resident population of 4,488 persons. The nearest towns are Doom Dooma and Tinsukia. Baghjan Oil Field is located near the Dibru-Saikhowa National Park in Assam, and is also in proximity to Maguri Motapung Beel. It is a natural wetland which is also the largest wetland in Upper Assam. Dibru-Saikhowa National Park is the only riverine island wildlife reserve, globally. It contains a variety of rare and endangered species, including the Gangetic dolphin, which is endangered, as well as herds of wild horses, tigers, capped langurs, hoolock gibbons, slow loris, and several rare butterflies and fish.[9] The Dibru-Saikhowa National Park is also connected to Namdapha National Park via the Dehing Patkai Wildlife Sanctuary. These regions are part of the Indo-Burma Biodiversity Hotspot. Baghjan Well No. 5, from which the leak occurred, is located at a distance of 900 meters from the park and adjoins a buffer forested region surrounding the park. It is also close to the Indo-Burma Biodiversity Hotspot.

On 18 January 2020, the Central Government exempted all oil and gas firms conducting exploratory drilling from legal requirements for environmental clearances. Prior to the leak, in 2020, the region witnessed protests after the Government of India announced that they had authorized Oil India Limited (OIL) to explore the area under Dibru-Saikhowa National Park for hydrocarbons. During May 2020, prior to and during the leak, Tinsukia district in Assam experienced heavy rains and flooding in two local rivers, Dibru and Lohit. The region was also placed in lockdown, following the COVID-19 pandemic in India.

Environmental Authorizations: A subsequent investigation by Assam's State Pollution Control Board found that Oil India Limited had begun drilling at the Baghjan Oil Field without obtaining prior environmental clearances in accordance with legal requirements under the Water (Prevention of Control of Pollution) Act, 1974, Air (Prevention of Control of Pollution) Act 1981 and Environment Protection Act, 1986. Oil India Limited had also failed to conduct mandatory public hearings before beginning drilling. The Eco-Sensitive Zone around Dibru-Saikhowa National Park was reduced in size at the request of Oil India Limited in order to allow the well to be established. In 2013, the Baghjan Oil Field was inspected by a team from the National Board for Wildlife, which criticised the Government of India for ratifying the breach of wildlife norms after Oil India Limited had already begun operating in an eco-sensitive zone.

Prior Leaks: In 2005, a blowout at an Oil India Limited well in Dikom in Assam resulted in the evacuation of 500 families. The leak subsequently caught fire, as well. On 3 February 2020 near Naharkatia in Assam, a stretch of the Burhi Dihing river, which is a tributary to the Brahmaputra river, caught fire after a punctured pipeline carrying crude oil from the Oil India Limited headquarters at Duliajan in Assam leaked, covering the water with a film of oil. The fire burned for a period of 48 hours before the leak was repaired and it could be extinguished. Oil India Limited officials stated that a technical error with their instruments had caused shut off valves in a storage tank to close, resulting in pressure building up in a pipeline carrying crude oil. This caused several leaks in the pipeline. Oil India Limited Officials blamed local residents for causing the fire, and stated that a number of leaks had previously occurred, as parts of attempts to steal quantities of crude oil from the delivery pipeline. The Central Pollution Control Board had directed Assam's State Pollution Control Board to investigate the damage to aquatic life and local environment caused by the leak.

Baghjan Oil Well No. 5: The Baghjan Oil Field has 21 active wells, of which 4 produce natural gas, while the remaining produces oil. Baghjan Oil Field's Well No. 5 was established in 2006 and produces natural gas. It drills down to a depth of 3,870 meters, and was producing between 80,000 to 100,000 standard cubic meters of gas per day before the leak. Drilling at the Baghjan 5 well, where the leak occurred, was being outsourced to an Ahmedabad-based firm, John Energy. In May 2020, a new reservoir was being tapped for additional production at a depth of 3,729 meters, with workover operations taking place. In addition, the infrastructure for the previous reservoir was being serviced, due to which production had been temporarily stopped. A device for pressure control, known as a blowout preventer, had been temporarily uninstalled while the extraction assembly was being serviced. In April 2020, following a lockdown caused by the COVID-19 pandemic in India, Oil India Limited suspended drilling at several of the wells in the Baghjan Oil Field, primarily because of the unavailability of employees. On 20 May 2020, before the leak, the Ministry of Environment, Forest and Climate Change announced that they would be authorizing exploration for hydrocarbons under the Dibru-Saikhowa National Park, at seven different locations. Oil India Limited stated that they would be using Extended Reach Drilling (ERD) to explore these wells without entering the National Park itself. Baghjan Oil Well 5, at which the leak occurred, will be one of the bases on which the ERD will extend under the National Park.

Blowout, Leak and Fire or During Disaster Phase:

Initial occurrence: On 27 May 2020, at 10:30 a.m., residents of Baghjan village in Assam reported hearing a loud sound from the nearby Baghjan Oil Field. Reports of the sound of the explosion were also confirmed from residents of Tinsukia town, which is located at a distance of 12 kilometres from the Oil Field. Local newspaper, the Sentinel, reported local accounts indicating that the gushing of natural gas from the well continued to produce an audible sound over the next few days. Although production had been temporarily stopped at Baghjan's Well No. 5 to explore new sand, and to service the well, workers noticed that gas was escaping from the capped well, and had begun to evacuate the Oil Field before the incident. At 10:30 a.m, the blowout occurred, resulted in the leak of natural gas from the well, and causing a complete suspension in operations. Oil India Limited's official statement indicated that the well had "suddenly became very active" while workover operations were ongoing. OIL officials confirmed that the leak occurred from the existing sand which had been capped to allow exploration of the new sand. This resulted in a "fountain of crude oil" from the sand. The leaked gas consisted of a naturally occurring mix of propane, methane, propylene and other gases.

Fires: On 9 June 2020, Well No. 5 caught fire after efforts to cap the leak were unsuccessful. An official from Oil India Limited stated, "We cannot say how and why it happened," in regard to the fire, also noting that fires, in the case of a blowout, were not unexpected. The fire occurred at the plinth of the well while cleaning operations were under way. Following the fire, people who had not yet evacuated from the local area left, as the fire spread to local grasslands. The fire quickly spread to a larger area and burned down nearby trees, crops and houses. 4 persons were injured, and 50 houses destroyed in the resulting fire. The Tinsukia District Collector stated that the condensate from the gas leak in surrounding areas had made the locale more vulnerable to fire and had allowed the fire to spread faster. On 10 June 2020, two firefighters, Durlov Gogoi and Tikheswar Gohain died fighting the fire. Their bodies were recovered from a water body near the site. Recently on 21 July 2020, an explosion occurred at the well number 5 and three foreign experts were injured in the incident.

Evacuation: Following the leak, 1610 local families, consisting of approximately 3000 persons were evacuated to relief camps. A safety zone consisting of a 1.5 km radius around the well was established. The National Disaster Relief Force was deployed to establish the relief camps. In addition to local residents, employees of Oil India Limited and their families were also evacuated from the area. The Indian Air Force and Indian Army also provided assistance.

Environmental damage: State Pollution Control Board officials stated that the leaked gas condensate had affected local agricultural crops and plants, including bamboo, tea, bananas and betel nuts. Wind conditions had carried the leaked gas towards the Dibru-Saikhowa National Park as well. On 29 May 2020, the carcass of a Gangetic dolphin covered in condensate oil was found in the Maguri Motapung Beel, a local wetland, and sent for a post-mortem by the Tinsukia Wildlife Division, to establish the cause of death. Wildlife Division officials noted that because of rain immediately following the leak, local waterbodies had been contaminated by condensate from the leak. On 31 May 2020, Assam Government officials confirmed that the State Pollution Control Board was investigating environmental damage as a result of the leak. Following the leak, the State Forest Department had also asked Oil India Limited to account for its actions after reports of dead fish in local waterbodies. On 5 June 2020, local residents protested near the Maguri Motapung Beel wetland, calling for the protection of the ecologically-sensitive Dibru-Saikhowa National Park, to mark World Environment Day, and called for compensation for their loss of livelihoods as a result of the leak.

Efforts to contain and stem the Leak or Post Disaster Phase:

Initial response: Operations at Baghjan Oil Field were suspended following the blowout and leakage of natural gas. OIL (Oil India Limited) requested assistance from the Crisis Management Team of the Oil and Natural Gas Corporation of India. IOC initially attempted to install a blowout preventer to bring the leak under control. OIL also sprayed the area with water. On 30 May 2020, four days after the initial leak, a spokesperson from Oil India Limited confirmed that "The well control operation is yet to start." Oil India Limited officials also stated that the process of controlling the well would take time, and that they were currently making arrangements to secure a water supply to help mitigate the leak.

Attempts to cap the well:

(i) **Placing a blowout preventer:** In June 2020, Oil India Limited officials stated that they had constructed a temporary reservoir of water in the vicinity of the well, by placing pipes from the Dangori River. Water from the reservoir would be used to spray on to the well area to protect workers while they attempted to place a blowout preventer, using hydraulic devices, followed by injecting mud to completely shut down production of gas, or "kill" the well. Singapore-based Alert Disaster Control experts were called in to assist with the process. Initial efforts to control the leak were delayed because of severe flooding in Assam, which caused widespread damage and resulted in the deaths of 16 persons. Oil India Limited said that because of the floods, working conditions at the Oil Field were temporarily unsafe, as the Dangori River which was being used to pump water to the field, was overflowing. The nearby Doom Dooma-Baghjan Bridge had collapsed due to the flooding, and consequently, access roads to the site were unavailable. There were two failed attempts to place the blowout preventer on Well No. 5. On 22 July 2020, during one such attempt, three Singaporean expert advisors were injured after a fire broke out. On 18 August 2020, a blowout preventer was successfully placed at the well head on a third attempt, but the process of 'killing the well' by injecting mud to completely stop production of natural gas from the well was not successful, and the well continued to leak. Oil India Limited officials stated that the reason for the failure was that a valve in the well casing collapsed.

(ii) **Snubbing and diversion:** On 3 September 2020, Assam's Commerce and Industry Minister, C. M.

Patowary informed the Assam Legislative Assembly that experts from Canada had been brought in to attempt to cap the well again. Oil India Limited's officials stated that they were also considering an attempt to divert the gas produced at Well No. 5 into two separate channels. If successful, this will result the diversion of part of the gas being produced to the Baghjan Early Production System facility, while the remaining gas will be flared. Once the diversion has occurred, an attempt will be made to cap the well at a greater depth, in a process called snubbing. On 13 September 2020, Oil India Limited succeeded in diverting the gas leaking from Well No. 5 to nearby flare pits on a second attempt. The first attempt at such a diversion had been unsuccessful. The well has not as yet been controlled, and on 30 September 2020, Sushil Chandra Mishra, the managing director of OIL stated that it would take two more months to kill the well and stop the leak. On 3 September 2020, Assam's Commerce and Industry Minister, C. M. Patowary stated that it may take another two months to control the leak. On 13 September 2020, Oil India succeeded in diverting some of the leaking gas to flare pits, but has yet to completely control the leak and 'kill' the well. On 5 November 2020, a second attempt to snub the well was initiated.

Fatalities and evacuations: Two firefighters employed by Oil India Limited died on 10 June 2020 while fighting the fire resulting from the gas leak. Four other fire fighters suffered injuries. Local reports indicate that there have been more deaths in the area; an inquiry by a local magistrate is currently investigating these claims. On 10 September 2020, an engineer working with Oil India Limited, Arnab Kishore Bordoloi, died while working at the well site. The cause of death is yet to be established. As of September 2020, 500 persons are currently still residing in relief camps, with the surrounding area still being affected by audible sounds of the gas leak as well as the ongoing fire. Local reports indicate that the sound of the leak can be heard from a distance of 5 kilometres from the site of the leak.

Litigation and compensation: On 25 June 2020, the National Green Tribunal found a prima facie case had been made out against Oil India Limited for damage to the environment and local livelihoods because of the leak. The Tribunal ordered Oil India Limited to deposit a sum of ₹250 million with the District Magistrate of Tinsukia towards compensation, and constituted an expert committee to investigate the disaster and recommend how much compensation was payable to affected persons. The expert committee constituted by the National Green Tribunal recommended that completely damaged houses would result in compensation of ₹2,500,000, while those that were severely and moderately damaged would receive ₹1,000,000 and ₹250,000 each. From 24 August 2020 onwards, local residents staged a peaceful protest, establishing a camp in front of the office of the local Deputy Commissioner, to ensure that all persons whose homes were damaged in the fire and leak received more compensation. The Assam Government confirmed to the Assam Legislative Assembly that some families had been issued compensation. Oil India Limited had filed objections to these orders for compensation in the National Green Tribunal, but the Tribunal dismissed their objections on 8 August 2020. OIL announced that they will give ₹30,000 as compensation to all those families who are affected. On 1 October 2020, the Gauhati High Court agreed to hear a public interest litigation petition filed by two local residents, and ordered OIL as well as a number of government authorities, including the Assam Government, the National Board for Wildlife, the State Board for Wildlife, and the State and Central Boards for Pollution Control, to respond to concerns raised about drilling inside the Dibru-Saikhowa National Park.

Internal investigation and criticism: On 30 May 2020, an Oil India Limited spokesperson stated to press that a five-member investigative committee had been established to ascertain the causes of the blowout and leak. On 10 June 2020, Oil India Limited confirmed that two employees had been placed on suspension in connection with the leak. The Indian Oil Workers Union criticized the use of private contractors at Well No.5, calling on Oil India Limited to use only trained corporation employees for such work.

3.3 Government Investigations and Environmental Damage

Assam Government: The initial leak at the Baghjan Oil Field resulted in the contamination of water bodies in the nearby Maguri Motapung Beel wetland, as well as in the Dibru-Saikhowa National Park. Days after the leak, local forest officials reported seeing a layer of gas condensate on water bodies in the Dibru-Saikhowa National Park. An Environmental Impact Assessment Report is currently being prepared to assess the damage caused by the leak and fire. On 12 June 2020, the Assam state government ordered an investigation into the environmental impact of the leak, and asked for a report within a month. An expert committee was formed by Assam's Forest Department to investigate the environmental impact. On 21 June 2020, Assam's State Pollution Control Board (SPCB) ordered Oil India Limited to shut down all production at the Baghjan Oil Field, following an investigation that revealed that the company had begun drilling operations there without obtaining prior environmental clearances. The SPCB also found that the company had failed to comply with continuing reporting and certification requirements under environmental laws. Oil

India Limited objected to this notice of closure, stating that it could potentially result in more blowouts at the wells. The closure notification was withdrawn three days after it was issued.

Union Government: The Indian Government's Ministry of Petroleum and Natural Gas has also ordered an investigation into the gas leak. On 2 August 2020, the Wildlife Institute of India, which operates under the Ministry of Environment, Forest and Climate Change, issued a report in which they stated that Oil India Limited had failed to address security concerns after two previous leaks in Dikom and Naharkatia in Assam. Their report indicated that between 60 to 70 hectares of land had been damaged by the leak, and noted particular concerns about the biodiversity in the Tinsukia district. The Wildlife Institute of India's report stated that the leak would have " prolonged ill effects on all life forms, including humans," in the area.

National Green Tribunal: On 26 June 2020, the National Green Tribunal constituted an expert committee headed by a former judge, A.P. Katakey to investigate the reasons for, and the impact, of the gas blowout. The expert committee was constituted after a preliminary report requested by the Tribunal revealed that there were errors in internal safety procedures that resulted in the incident. The preliminary report indicated that compliance with these safety procedures could have prevented the blowout. The Katakey Committee reported in November 2020 that Oil India Limited had repeatedly violated provisions of the Air (Prevention and Control of Pollution) Act, 1981, the Water (Prevention and Control of Pollution) Act, 1974, and the Environment Protection Act 1986. The Committee also found that Oil India Limited had failed to obtain the necessary clearances for the Baghjan Oil Field, and had not conducted Biodiversity Impact Assessment Study before beginning operations, which had been made mandatory by a Supreme Court order.

4. Baghjan: A Phenomenological Perspective

Although, the incident is now spoken of in the past, its effects and after effects are still deeply imprinted on the area and its inhabitants. The villagers, who were evacuated to relief camps promptly, now complain of anxiety, damage to their ears, and respiratory illnesses. Moreover, the rare Maguri-Motapung wetland and Dibru Saikhowa National Park, located next to the blowout site, have suffered long-lasting damage. According to OIL, seismological studies have not shown any surface deformity in the region, but researchers say the public sector unit is downplaying the disaster that affected hundreds of families. Last week, the Gauhati High Court stayed OIL plans to further explore oil wells in seven spots within the Dibru Saikhowa National Park. Earlier, the National Green Tribunal (NGT) had directed the state's Pollution Control Board to take legal action against OIL, besides revising compensation to affected families, on the recommendation of the Tinsukia deputy commissioner.

It had previously taken note of a report by the Wildlife Institute of India (WII), an autonomous institution under the Ministry of Environment Forest and Climate Change, which said no "restorative process" has been put in place. In accordance with the NGT direction, the state government has proposed relief of Rs 25 lakh to 161 families and Rs 20 lakh to 439 families to affected families in Baghjan, said Assam Chief Secretary Jishnu Barua. In the interim, the CM has directed compensation of Rs 15 lakh to 161 families and Rs 10 lakh to 349 families. OIL has deposited an amount of Rs 68.5 crore, which is being released to the affected people. The local people expressed their concerns by saying the Government of Assam is most sympathetic towards the citizens affected by the unfortunate incident of the well blowout at Baghjan and regains full commitment to mitigating the suffering caused by the same. But, the troubles faced by villagers in this region go beyond just damaged homes and lost crops. The WII report dated 15 July said the sound of the blowout travelled to 12 km and beyond. While sound up to 60 decibels (db) is safe for humans and animals, the blowout raised levels to between 72 and 112 db, thereby disrupting wildlife, birds and humans. Apart from all this, the real and concrete after-effects of the disaster can be understood when we see it from a different perspective, which is a phenomenological perspective. This draws the attention to the perception, response and recovery of the local survivors and how disasters are mostly affective in the immediate environment or in the local level. To bring into a gendered perspective, the lives of three women in the affected area are taken into account. Data and information is generated from the participants by conducting unstructured interviews and an observatory-conversational method. Conversations were audio-recorded with open-ended questions after getting consent from the participants and the names are changed for ethical reasons. Field notes of observations of the context and non-verbal clues (e.g. posters, charts, gestures, etc.) are also recorded and the conversations were either staged (interview type) or prompt (talk type). These three experiences gives us an understanding of the disaster in a more personalised manner and throw light on how a single disaster can create different aftermaths with respect to difference in gender, class, caste, vulnerabilities, risk, risk perception and economic, social, political and cultural backgrounds. And these different conditions give rise to difference in response and recovery measures taken by the survivors.

Story 1: *Kalyani* with her Sick husband

Though almost a year has passed, Kalyani still remembers the terror of the incident and said she had to take the responsibility of her family, as her husband was almost bed-ridden due to sickness and bad drinking habits. She was earning the little bit of money to sustain her family by selling tea and betelnut from the betelnut and tea plantations that her family owned. They had their rice-fields which was just enough for the whole year. But her homestead being just 45 metres away from the area of incident, the fire burned down her house, the entire plantations of tea, rice and betelnut, and her whole family had to evacuate to save just their lives. After the fire was put out and the area was safe for them to return, they came back to just ashes and land with grease and oil. "The air smelled of gas and burnt items", she said, "Only the frame of the house was there, everything else turned to ashes." After receiving the compensation, they slowly have built their houses but have no other livelihood options. She haven't been able to enrol her kids at school for this year's session and having a hard time trying to meet ends and look after her family. When asked about the present livelihood options, she said "A betelnut tree takes almost 7 to 8 years to grow and bear nuts. What am I supposed to sell now?" And she talked about how there were rumours going on about all the people who were now staying there and breathing the poisonous air would die soon of cancer and said "I hope that happens soon and we can be put out of this misery". She said she has nightmares of the incident and cannot sleep at night worrying about her husband and her kids' future.

Story 2: *Roopali* with her three daughters and a son

Roopali's house was just 50 metres away from the oil well. They have been living there for ages and she said she had never imagined even in her wildest dreams that a fire could change their lives to this extent. Even after the loud outburst, she was confident that they were not in danger and while remembering that day she says, "while my husband was yelling at me to pack our belongings and run for our lives, I scolded him and said that I won't leave the house without having lunch, as I had cooked fish that day for me and my kids after a long while." But they could not wait any longer as the fire spread very quickly and burned everything in its way. She had to send her son and daughters off to their relatives' house residing in the nearby villages while she and her husband stayed in the relief camps for almost a year. She said the separation from her kids worried her but she was at relief as they did not have to live in relief camps with not even the basic facilities. After they returned home, she said, "Everything was there, just the way we left them, clothes, utensils, books, but all were burnt to ashes. If you touch them they vanish." She had prepared jewellery, traditional utensils and saved up money for her elder daughter's marriage. Everything was burned down. The books of her children and other valuable items such as silk clothes were all lost to the fire. They had a huge betelnut plantation in and around their house which too was completely destroyed. Now, they are slowly re-building their house with the compensation money and are trying hard to meet ends. Her husband has taken to daily labour to feed the family. They are cultivating rice in their fields but are not sure if they will bear grain. "We had sown lentils in our fields. The plants looked good and had luxurious growth. But they did not bear any seeds. We planted cucumber trees, they too grew well and high, but they did not bear even a single fruit." They are now sceptical about their rice cultivation too and are worried for the near future.

Story 3: *Deepti* and her memory of Disaster

Deepti, a young girl in her early twenties, says her family was amongst the worst hit. "On the night of the incident, we couldn't understand what this noise was as we could barely hear ourselves in our own homes... OIL officials evacuated us to camps nearby, and kept us away from our homes for nearly two weeks," she said. Their family owned an orange plantation, a small tea garden and a fishery, and then all were burned down. While OIL has offered his family the highest compensation of Rs 25 lakh so far, she said her father worries about how they will get back their livelihoods a. She also worries about her mother's poor health. "My mother suffers from ache in the joints and severe headaches ever since... Medical care was basic and we haven't understood why she gets these terrible pains..." They haven't come back to their home ever since. When asked of the reason she said, "We can't think of going back there now. Without our home, the plantations, the bird songs, animals and life that flourished around our home, how we can live there?" They currently rent a home in Makum sub- division. She also talks about her not being able to go and re-join college as she has to look after her sick mother and her siblings. She wanted to study but she feels caring about her family in these tough times is her utmost priority. She also said how she and her friends were proud of their place as it had the rare Maguri-Motapung wetland and Dibru Saikhowa National Park — a rich biodiversity zone. But now, because, these were next to the blowout site, the only visuals that are left of here are burnt trees of betel nuts, burnt bamboo trees, and burnt tea gardens, natural ponds overrun by black muddy peat, no green vegetation, and just a few birds and the whole ground is covered in soot. She wishes for all this to disappear kike a bad dream but instead this has be carved in their minds and will be alive in their memories as a fire that never stopped burning.

5. Discussions and Conclusion

From the above observations we can understand that Disasters have their greatest impact at the local level which is now becoming more complex and diversified due to the intersectionality between Natural and Human-induced Disasters, Vulnerability, Risk, Risk Perception, Resource Distribution, Social categorizations such as Class, Caste and Gender and Government fronts. Therefore, the involvement and mobilising of the affected local communities has always and now more than ever, been an essential component of the whole Cycle of Disaster Management. Communities are a source of strength, experience and knowledge of living and adapting to the prevailing disasters in their own different ways. There are references to many incidences when local people predicted impending earthquakes or cyclones by observing natural sciences such as animal behaviours, changes in characteristics of ground water sources and abnormal weather phenomena[1]. But in case of Baghjan, the local communities were not prepared for such a disaster as they were never accustomed to or warned of such unforeseen incidents. They were never given any training, or awareness on how they should practice resilience and preparedness being communities that reside among active oil wells and natural gas outlets. And moreover, there is a growing realisation that community participation is the most effective element for achieving sustainable and long term results in disaster management and risk reduction. The government and the existing disaster management authorities have often failed to handle every end of the various disaster dynamics especially without the active participation of the people. There have been numerous instances where the top-down disaster management approach has not been affective according to their full potential and has created a gap between policy and practice. As a consequence, it was felt that it is high time to adopt a new strategy that will involve vulnerable people directly in planning and implementation of mitigation, preparedness, response, and recovery measures[1]. Community participation, in particular, is gaining significance as communities are becoming increasingly vocal and keen for policy makers to "hear" the voice of their experience.[2] When the communities are included in the entire process involving sequential stages of disaster management, a decentralized, bottom-up, intensive, micro and local level approach comes into being which decreases the gap between the affected community and the existing institutionalised disaster management frameworks. Incorporating indigenous knowledge and practices pertaining to Disaster Preparedness, Early Warning Systems, Resilience and Risk Reduction can be considered as one of the significant ways of community based frameworks.

6. Acknowledgement

We would like to thank the Department of Geography, North Lakhimpur College (Autonomous) for funding this field study and providing assistance wherever need be and would like to express our most sincere gratitude and appreciation to all those survivors of Baghjan and especially the three participants for so courageously sharing their story and letting us a glimpse of their life after the disaster.

References

1. Bhagat S. N. *Community Based Disaster Management Strategy in India: An experience sharing*. PDPU Journal of Energy & Management
2. Sankar M. (2005). *Bridging the Gap between Policy, Research and Practice: Experiences from a Community Economic Development Action Research Project in New Zealand*. Social Policy Journal of New Zealand Te Puna Whakaaro. Issue 26.
3. The India Times
4. The Assam Tribune

---

[1]Bhagat Sachinkumar N. *Community Based Disaster Management Strategy in India: An experience sharing.* PDPU Journal of Energy & Management

[2]Sankar Meenakshi. *Bridging the Gap between Policy, Research and Practice: Experiences from a Community Economic Development Action Research Project in New Zealand.* Social Policy Journal of New Zealand Te Puna Whakaaro. Issue 26. November 2005

Annexures

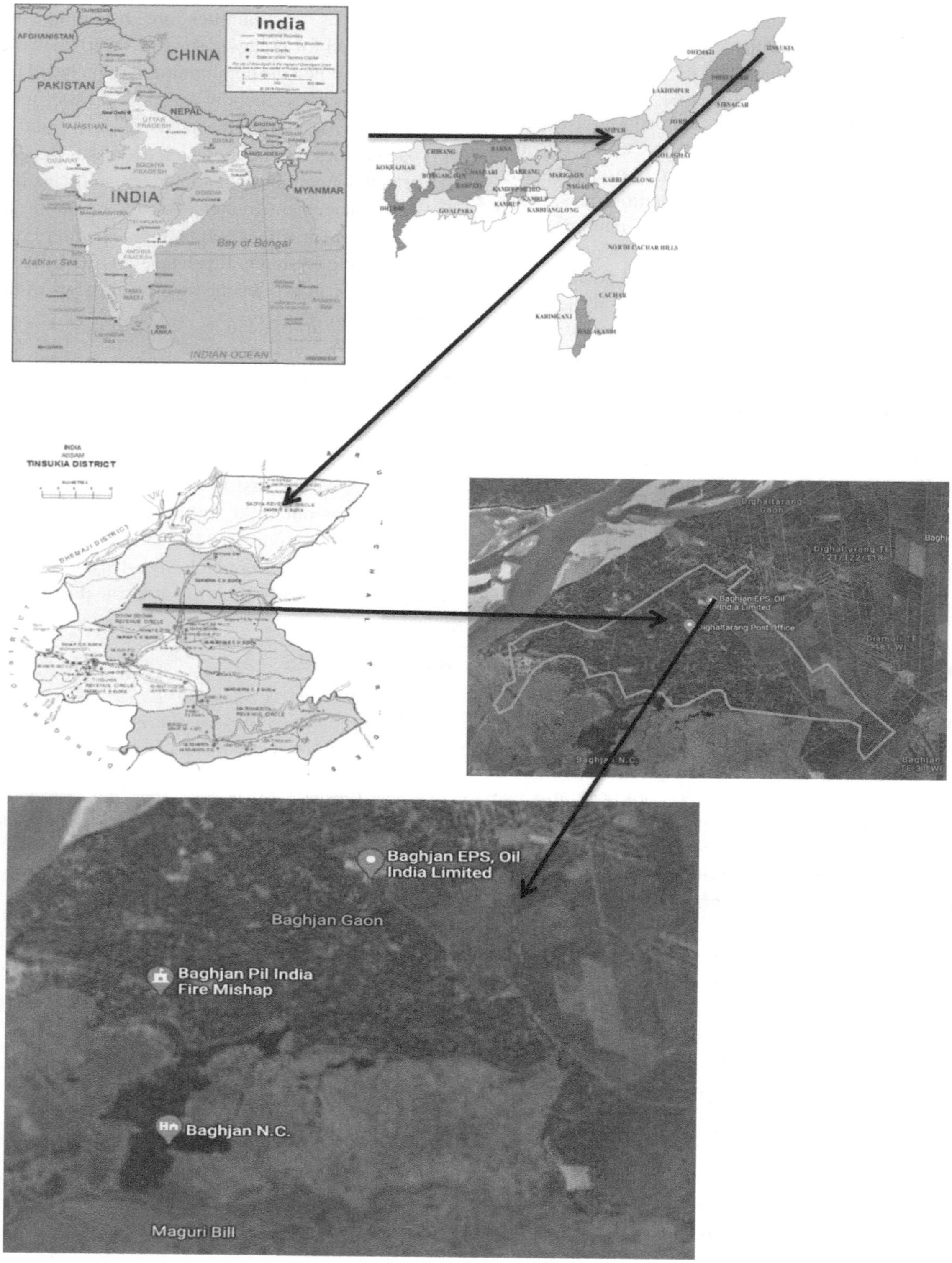

Fig. 45.1 Location Map

Source: Google, Google Map

Fig. 45.2 Baghjan while burning

Source: The Economic Times, The Telegraph and BusinessToday.in

Fig. 45.3 Baghjan Fire at night

Source: The Financial Express

Fig. 45.4 Baghjan after the fire: The main sourcepoint

Source: Primary Field Visit by the authors

Fig. 45.5 Burned trees, bamboo, betelnut and Tea plantations

Source: Primary Field Visit by the authors

Fig. 45.6 The ground in the aftermath

Source: Primary Field Visit by the authors

Fig. 45.7 Emergency compartments set up by OIL

Source: Primary Field Visit by the authors

Fig. 45.8 The damaged machineries

Source: Primary Field Visit by the authors

Fig. 45.9 Our team member with a participant

Source: Primary Field Visit by the authors

Fig. 45.10 Houses being rebuilt

Source: Primary Field Visit by the authors

CHAPTER 46

Assessing the Risk of Fire Hazard: A Case Study of Delhi

Kanika Bhatia
TERI School of Advanced Studies, Dept. of Natural and Applied Sciences
New Delhi

Aakash Upadhyay
Dept. of Geography, Delhi School of Economics,
University of Delhi

Abstract

Fire, a combustion process is an essential part of our day-to-day life. It can be a disaster when it notably increases the prospect of an already occurring fire, boosts any existing fire, or cause an increase in the potential harms by uncontrollable fire. Responders to fire hazards are not very much well organized in a developing world like India, with challenges mounting every passing day due to the frequency and intensity of the hazards India's capital city, New Delhi with its ever-expanding population, comprehensive crowded localities, and unplanned growth has always been at high risk to fire hazards. The vulnerability and risk have multiplied owing to growing anthropogenic activities Vis -a- Vis limited fire services and existing infrastructure. The present research will look to undertake and review the risk perception of fire hazards at the locality level and its presence in districts of Delhi, India. It will further look to outline the mitigation strategies and initiatives to combat fire hazards in the future. The primary data was collected using social media tools like Google forms, owing to the tough time of the pandemic. Based on the 250 respondents, the study has looked to analyse the Demographic, Economic, Physical, Social, and Psychological factors to understand the perception of risk and vulnerability. It was found that the majority of the population was unaware of the fire safety norms and risks involved. Though they perceived fire hazards and risk associated, unawareness made them highly vulnerable to fire hazards. Different Geospatial tools such as Quantum GIS, Google Earth Engine have been effectively used to co-relate the risk perception with an increasing number of fire accidents across Delhi. It was suggested that the city administration should look to increase the number of fire stations and staff and upgrade the existing infrastructure. Knowledge-based skill development initiatives, the idea of fire prevention inspectors at the locality level, and awareness campaigns at societies should be promoted and carried out to combat and prevent future fire events.

Keywords: Fire, Risk, Perception, Fire safety, urban growth, Built-up area, Social Media, disaster management.

1. Introduction

India's current population is close to 1.4 billion and unplanned urbanization and rapid population growth have increased the issue of urban fires in India. Moreover, the fire safety facilities are also not well organized. Setting up of industries at a rapid pace with wide-ranging treacherous materials and compact high-rise buildings has grown the existing snags of fire hazards. Fire services come under the 12th agenda of the constitution and have only ad-hoc progress in fire services in recent years. Presently, fire prevention is maintained and organized by concerned states and union territories, and urban local bodies (Hazard Analysis, GOI, 2011). Hazardous events pose danger and risks to life and have unfavourable impacts on Locality, hampering socio-economic development. The risk of hazardous events makes people and places susceptible and exposed to tragedy and therefore risk is often perceived as the product of Probability of Hazard and Degree of Vulnerability and exposure (Gupta, 2017).

Fire is a combustion process and its events are unexpected and unplanned in the building environment and can cause devastating effects. The present fire protection services do not justify all the possible hazards, due to which fire mitigation has become a growing concern (Kodur et al., 2019). The vulnerability of fire events consists of a wide range of physical, economic, social, and environmental factors besides, it is also closely linked to development. Its vulnerability depends upon the type of hazard, and its understanding with necessary safety measures can help lessen the cause to boost the resilience and inclusive assessment of Risk, Hazard, and Vulnerability that are crucial to evaluate the structural and non-structural capacity building. Risk perception navigates decisions about the accuracy of risk, it includes assessments of the probability along with outcome consequence. It is mostly people's perception that often has two main Dimensions- Cognitive and Emotional dimensions of risk perception (Paek and Hove, 2017). People often experience different aspects due to factors such as gender, age, profession, etc. In most cases, the fire hazards are not assessed correctly owing to unawareness of the severity, neither having any evacuation plan, nor adequate strategies and equipment's to control the fire. Furthermore, human actions sometimes worsen the situation through panic, intuitive reactions, or improper decision-making. Thus, hazard perception, one's capacity to anticipate and cope is a must for appropriate stability with physical installation (fire extinguisher, dry sand, water storage banks, and all other accessible equipment) with prior experience may decrease the threshold conditions of sensitivity to fire events (Kinateder et al., 2014). Its occurrence is often associated with commercial, and social losses, and the degree of loss depends on the type of material and its thermal property. Besides, in most cases, absolute safety in fire events is unachievable (Paulsen, 1984). Thus, the fire hazard is a wide- ranging category that has transformed over time. It is an action, material, or condition which significantly increases the probability of fire occurrence and causes potential danger its events in households, communities, and workplaces put people, property, and businesses at high risk, it can damage buildings, equipment and can injure or even kill the people around. Thus its prevention and management are vital.

Substantial events of fire hazard outbursts have been reported from different parts of the nation. Indian Risk survey 2018 places fire explosion as the third largest risk to businesses, and residential areas are the most predisposed to fire outbreaks (India-Risk-Survey-2019). The situations in towns are much worse than those in the countryside. According to the report of the National Crime Reports Bureau (NCRB) in the year, 2016- 16,695 cases of accidental fires and nearly 50% of deaths due to fire accidents from residential buildings have been reported (ADSI, NCRB, 2017).

India lacks insufficient fire stations and fire prevention services as equated with urbanization and population. Besides, as per the analysis by the Standing Fire and Advisory Council (SFAC), there is a shortage of 97.54% of fire stations, 80% of fire fighting's and rescue automobiles, and 96.28% skilled workers in the country respectively (User Training Manual – Fire Decision Support System). Moreover, as per the other reports, nearly 17,700 Indians have died due to the lack of preparedness towards fire threats which makes an average of 48 deaths a day (ADSI, NCRB, 2017).

There have been reports of series of fire incidents in Delhi in the past few years. January 2020 saw few fires incidents in Peeragarhi and Pratparganj taking more than 45 lives and injuring many. Also, fire accidents are common in festivals like Diwali in Delhi. Recently, fire incidents have been reported from All Indian Institute of Medical Sciences (AIIMS), Hotel in Dwarka taking 2 lives, Multi- storey showroom in Nehru place and many other accidents causing huge loss to life and property.

The Delhi fire Service reported about 8,700 fires have taken place in the first half of 2021, besides receiving over 8,700 fire-related calls in the first 6 months (NYOOOZ, 2021). The present study would look to analyze the spatial pattern and risk perception of fire incidents in Delhi using Geospatial technologies. Correspondingly, an online survey was conducted to acquire information on risk perception on fire across Delhi and it was found that most of the respondents were aware of fire as a hazard and have experienced it once in their lifetime. Increasing population, decreasing land area, less vegetation, aging infrastructure, and rising temperature along with carelessness, unawareness, and improper disastrous education and non-execution of policies were some of the prime reasons for fire hazards in Delhi.

2. Study Area

Geographically, Delhi is located between 28°.70' N, 77°.10' E, with a total geographical area of 1484 sq. km (Fig. 46.2). The population was 1.1 crores (census, 2011), which is estimated to cross 2.8 million in 2022 with a classified urban area of 1,113.65 sq km. More than 45% of Delhi's residents reside in the slum area, illegal colonies, and other unplanned settlements (Economic Survey of Delhi 2020-21). The city, Delhi is not stationary, its Land Use, Land Cover, Density, Population, Global Warming, Pollution, Urban Sprawl, Public Amenities, etc. keep on altering which shows a sharp temporary contrast. Its site has undertaken a major alteration from rural to urban during 1991 and 2011. Thus, the inhabitants of Delhi live in substantial danger. The Delhi fire service includes 64 fire stations including headquarters and training centre, the administrative department is situated in Connaught Lane, New Delhi and for the operations, the city is divided into three fire zones with two geographical divisions within each zone, namely New Delhi Fire Zone

Fig. 46.1 Location of Study Area (Delhi) in India map

Source: Survey of India prepared by Authors, 2021

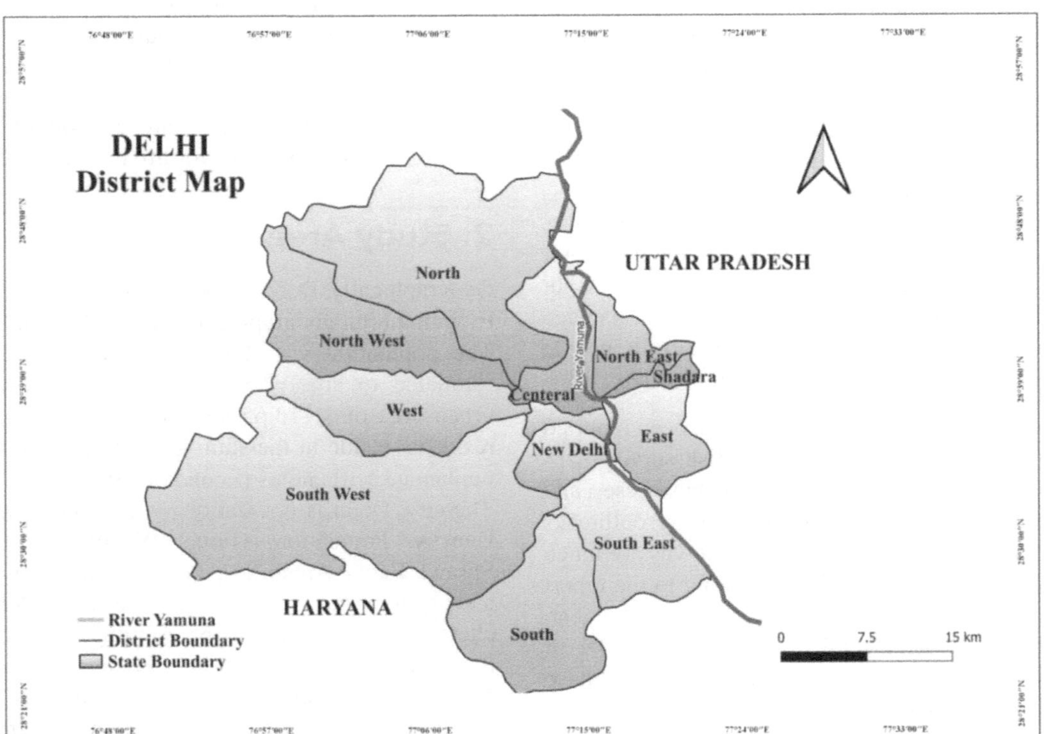

Fig. 46.2 Location of Study Area (Delhi) along with its Districts

Source: Open Street Map prepared by Authors, 2021

(East Division and Central Divisions), South Fire Zone (South Division and South West Division), and West Fire Zone (West Division and North West Divisions). They maintained a swift of 245 fire fighting automobiles and other supporting units (delhi.gov.in, 2021) Central 5 catering 578,671 population, the east has 6 catering 1707725 population, New Delhi has 6 catering 133,713, the north has 3 catering 883418, North-east has 3 catering 2240,749, the south has 6 catering 2,733,752, South-west has 5 catering 2292363, the west has 10 catering 2531583 (Fig. 46.9).

3. Database and Methodology

The database for the present study has been collected using both Primary and Secondary sources. The primary data for the fire risk perception was collected using online questionnaires like google forms, telephonic conversations, and personal observations using a simple random sampling method. The questionnaires were filled by using the social media tools like WhatsApp and Gmail, keeping in mind the pandemic and social distancing. About 250 responses were recorded from across different districts in Delhi. The survey was carried out in the month of May-July, 2021. Respondents from different age groups and professions were selected to have a holistic idea about the threat and risk perception. Out of total respondents, 64% were females and 36% were males, as generally during females encounter these problems at houses. Apart from the primary database, secondary data have been used in the study, which was collected from various secondary sources like, census.gov.in, Delhi government reports, economic survey reports, media reports, research articles, etc. They have been used to know and understand the level of urban growth, change in land use, vegetation cover in recent times. Other sources were used to determine population growth, density, the history of fire accidents in the city, fire stations availability, associated initiatives, and overall to determine the perception of fire risk at the locality level among the residents of the city Delhi.

Further, the Geo-spatial data was collected from different sources like data for land use and land cover, NDVI have been downloaded from USGS (www.usgs.gov), OpenStreetMap Data Extracts (Geofabrik Download Server) and processed further. Similarly, other thematic maps are based on the statistical data collected from various secondary sources.

The collected data were analysed using various statistical tools in MS Office like excel and word. Geo-informatics tools

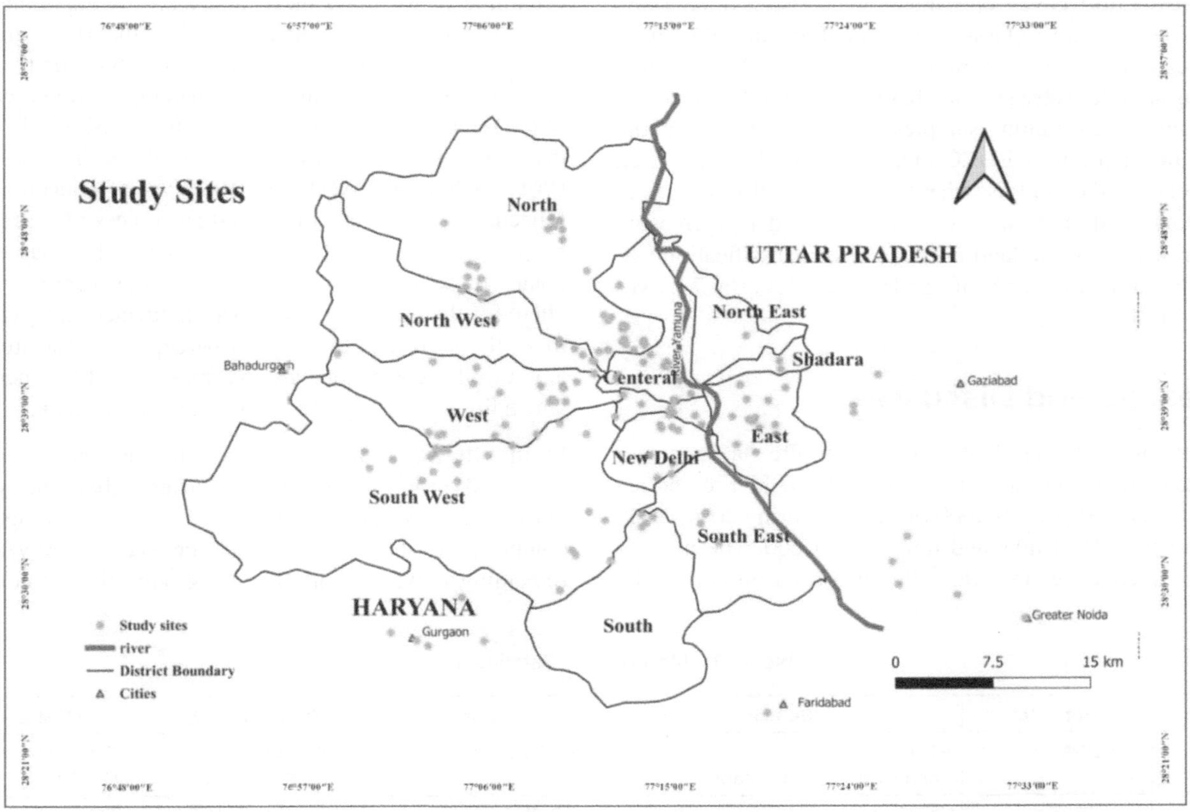

Fig. 46.3 Study Sites across Delhi

Source: Primary Survey, 2021

have been also used elaborately as it is considered as one of the most active tools to analyse, elucidate and visualize the spatial pattern and trends. Some of the Geospatial Software's that have been used are Q-GIS 3.15 and Google Earth Engine besides other Software's like Canva has also been sought as per the need. The obtained satellite data was pre-processed and processed to prepare the land-use/cover classes (LULC), and Normalized Differentiate Vegetation Index (NDVI). The same images were classified differently on a preference basis. For the present study, the land use and cover categories that have been taken are Vegetation cover, Water bodies, Built-up Area, Barren land, Agriculture land, Fallow land, etc. which could be associated with the fire accidents. Land use has been classified separately based on their spectral response.

Normalized Differential Vegetation Index was calculated using the two bands, visible infrared and near-infrared. NDVI value must range from -1 to +1

(NDVI) was calculated by using the two visible and near-infrared bands. (Solanki & Garg, n.d.) The equation of NDVI Includes: (NDVI) was calculated by using the two visible and near-infrared bands. The equation of NDVI includes:

$$NDVI = (NIR - RED)/(NIR + RED)$$

Land use Land cover classification has been done using the most common classification method in supervised classification, i.e. maximum likelihood classification. Training samples were selected based on spectral radiance. A minimum of 125 training samples was selected. The spatial distribution pattern of LULC obtained from the supervised classification, the land-use change with special reference to increasing built-up area would be co-related to increasing fire accidents. For the land use land cover classification two remotely sensed imageries of Landsat 8 and Landsat 7 ETM+ (Table 46.1).

4. Results and Discussion

The city of Delhi has had a history of fire hazards over the years, both domestic and industrial. Delhi fire service statistics say more than 95,000 fire events during 2007-2012, resulting in 1950 deaths and injury to 11,600. The number has been increasing since then, the major reason behind the demises is not burning but inhalation of deadly smokes which get intense in high density and less open spaces.

The fire hazards have been amplified more in manufacturing and built-up areas due to non-confirming productions, dangerous commercial activities, and congestions. (*Delhi Disaster Management Authority*, 2021).

The other reports from fire services suggest that the number of fire calls increased from 27089 in 2015- 16 to 31157 in 2019-20. Similarly, the number of deaths increased from 339 in 2015-16 to 346 in 2020- 21, which means the number of accidents is increasing along with the number of deaths taking place. The data suggests the increasing vulnerability along with high-risk factors for fire accidents across Delhi. When analysing the fire hazards and risk perception it is important to understand that it is a psychological component that differs from person to person and place to place. It is something qualitative that cannot be measured at all times, which is further influenced by age, gender, marital status, education, and level of awareness. In most cases such risk perception is directly related to human loss of life and livelihood, therefore, having a better understanding of risk perceived on fire hazards is important. During the study, it was found that demographic factors, physical, social, economic factors, and individual fear and experience were vital to understanding and analysing the risk perception. In the findings, it was found that about 68% of the female respondents were educated more than high school, with 30% of the female respondents being graduates. Similarly, 32% of the male respondents were educated more than class XIIth (Primary Survey, 2021). But, the level of education is not reflected in the preparedness and awareness of a fire hazard for female respondents, probably due to less say when it comes to decision making. However, the response may be altered if the direct interview was conducted in Covid's free time. It would lead to different perceptions of a fire hazard among different gender groups, making a difference in the way a fire hazard is perceived by men and women.

Further, it was found that people in the higher age group perceived more threat to fire hazards than the younger ones. Our respondents comprised 76% from the age group comprise less than 25 years, 21% between 25-50 years, and remaining above 50 years. The age group below 25 comprised

Table 46.1 Metadata of Satellite imageries

S. No.	Satellite	Sensors	Date	Row and Path	UTM Zone
1	LANDSAT 7	ETM+ Enhanced Thematic Mapper Plus	01-05-2000	WRS_PATH = 146 WRS_ROW = 040	UTM 43 N WGS 84
2	LANDSAT 8	OLI (Operational Land Imager) TIRS (Thermal Infrared Sensors)	03-05-2021	WRS_PATH = 146 WRS_ROW = 40	UTM 43 N WGS 84

of the youth and the working population, whose perception of a fire hazard was different than other age groups. They are more aware of fire hazards and do not perceive fire as a serious threat because they are aware of the dos and don'ts preparing them for emergencies. Response from the active working population was sought as it could help in identifying vulnerable groups and play an active role in the preparedness, mitigation, and especially responsiveness in case any disaster strikes. Generally, it is seen that children, females, and old age groups are more vulnerable at the household level, due to their dependency and tendency to stay more at home than the working population. The younger adults perceive fewer risks since they have to practice risk-related situations and decisions more frequently in their life.

The other factors that increase the risk and vulnerability to fire hazards were the economic factor. It was found that income level and occupation is an important determinant of how we perceive fire hazards. People from a lower income group would perceive fire as a great threat due to the lack of necessary resources/services (like fire alarms, fire extinguishers, medical expenditure, etc) as well as due to the economic loss that fire might incur. On the other hand, for someone who has a good level of income, fire hazard would not be a very serious threat as the available resource base, infrastructure and access to services is easier. Economically well-off people go for an economic investment that needs to be done for disaster preparedness.

The economic level of an individual is directly proportional to his/her vulnerability which affects the individual's risk perception for any hazard. A person's income determines his/her access to the resources influencing the risk perception of people. If a household has all the necessary equipment required to combat a fire situation and enough resources for relief and recovery post-disaster then the risk perception would be less serious and vice versa. During the survey, it was found that only about 27% of the respondents said to have at least 1 fire extinguisher on each floor of the building, only 7% claims for more than 1, and 64% had no extinguishers. The data indicates nearly 2/3 of the people have no access to fire extinguishers, which was directly proportional to their income, wellbeing, and awareness.

There are other factors as well which determine the vulnerability to any hazards and risk perceptions like **the physical or locational factors**. The biotic and abiotic factors in any given physical environment affect an individual's behaviour, perception, and way of living, cognition, fight and flight responses especially in urban settlements like Delhi. The place where we live and locality certainly affects how we perceive a fire hazard. A family living near industries, which are prone to fire hazards would have a different risk perception with high vulnerability. Likewise, congestion, proximity to chemical, paper, and cloth industries, housing without proper earthing systems make any area more prone to fire accidents. In the case of Delhi, we frequently hear about such fire hazards in industrial locations. Whereas people living in gated societies, RWAs, and less congested localities would perceive fire hazards differently and would be less vulnerable as building safety rules make it mandatory to follow fire safety rules and installation of fire extinguishers compulsorily. The availability of Metered Electric Connection, Mode of Gas Supply, Availability of amenities to combat fire accidents like a fire extinguisher, water storage, Installation of smoke alarm warning system in the community, Installation of earthing wiring system were enquired during the survey. The majority of respondents were aware of fire hazards (45%) and more than 2 fire extinguishers belonged to such locations. House types, single storey, multi-storey, kaccha houses also affect the risk and vulnerability. When asked about the type of housing structure, 68% lived in high-rise buildings/ flats (multi-storey), 30.5% in pucca but one storey and only 1.4% lived in the kutcha house as we are talking about Delhi. The risk perception for the high rise was less but it was found that multi storey was more vulnerable to hazards due to complex electrical wiring and the time needed for evacuation, as if the fire breaks out in one floor, it eventually increases the risk for others floors as well.

Apart from the locality and physical location household amenities also influence the risk perception and vulnerability. **Local Environments of Homes like** well-ventilated houses having open spaces, large escape routes and corridors, very well resourced in terms of emergency equipment like fire extinguishers, water storage tanks, etc. perceive fire hazards at a low-risk level. Contrary to this, people living in congested, poorly resourced, close-spaced houses with little space available for movement in staircases, corridors, etc., perceive fire to be a serious threat. Further, the types of cooking gas systems in place make the difference, with pipeline gas considered as the safest. It was found that around 66.36% of the respondents use LPG cylinders and the remaining 33.64% relied on pipeline supply which is 1/3 of the total respondents, thus exposing more people to fire risk. Further, it was found that about 81% had ventilation in the kitchen, 43.6% turned off their LPG regulator, and remaining attributing to most of the times, rarely and never respectively. However, it can be said that the majority of the people in Delhi live in high dense areas, with less or no open spaces and ventilation making them (mostly women and elderly) vulnerable to the kitchen and household-related accidents that can be interpreted by density map of 9 Districts of Delhi (Fig. 46.4). The North-east, east have the highest density followed by and west Delhi. The decreasing order of density could be directly related to decreasing order of vulnerability.

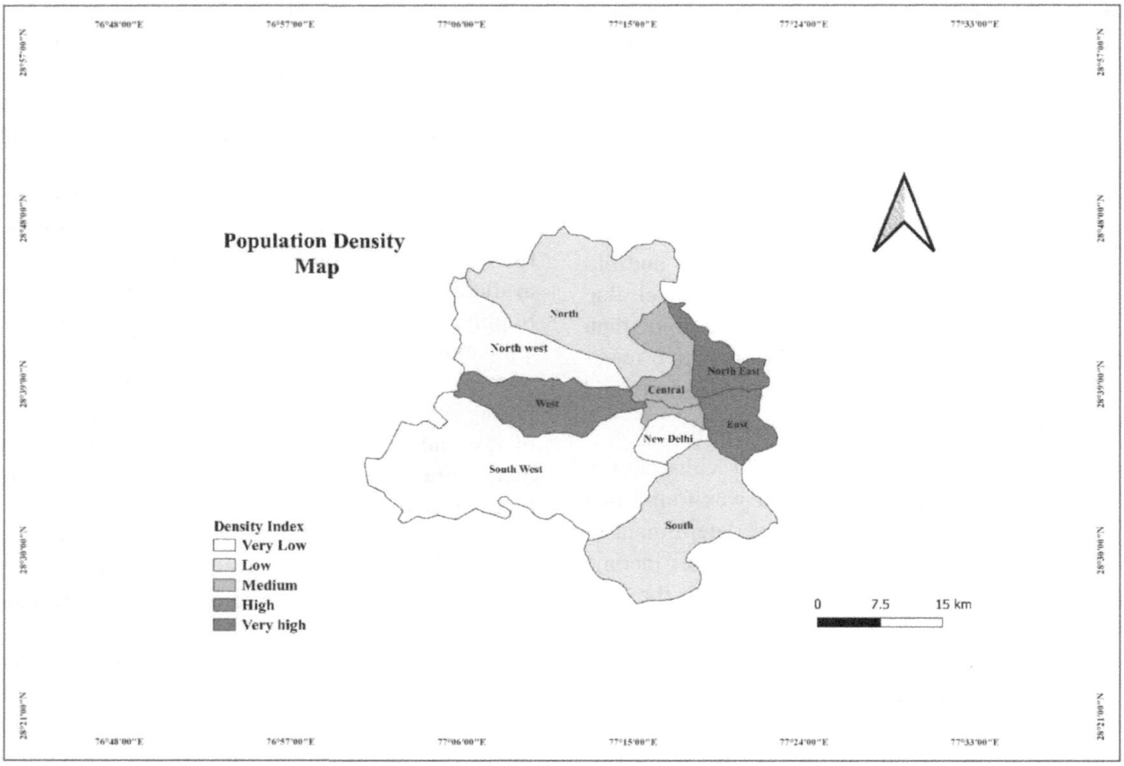

Fig. 46.4 District wise population density

Source:: Census of India, 2021

Household-level fire could be sparked from electrical wiring, short-circuiting, and other associated problems. So in this regard, it was found that about 98% of the respondents had metered connection in their household and out of which only half, i.e. 53% did inspection for meters and wiring of electric equipment's periodically and 47% did not inspection (Fig. 46.5).

Building safety says periodic maintenance and inspection are important, but it was a lack of awareness that nearly half of them did not do so, making them more vulnerable to such

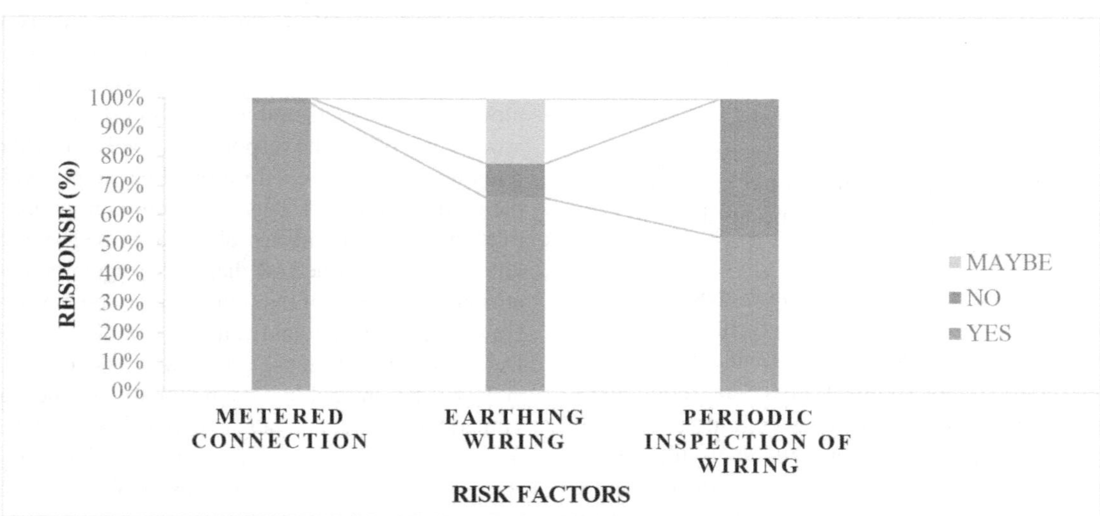

Fig. 46.5 Household Risk Perception

Source: Primary Survey, 2021

hazards. Also, earthing in the wiring system is important to prevent electric shock and further accidents, but it was surprising to know that only 66% of the respondents were certain about earthing in their house wiring, 22% had no idea about it and about 12% responded with no earthing in their system (Fig. 46.5). The availability of various resources influences the formation of one's risk perception.

The respondents lived in a different setting, locality, and environment depending on their social and economic freedom. They had a different level of risk perception in their houses, with 5% saying very high risk, 9.5% graded as high risk, 25.9% graded moderate risk, 35.5% graded low risk and 24.2% graded very low risk (Fig. 46.6). Here, it can be concluded that people who graded household risk levels from very high to moderate have a high-risk perception of fire hazard as compared to the people who graded household risk levels as low or very low.

Disaster risk in urban areas is impacted by access to various day services and the ability to bear its economic cost. Apart from risk perception it also plays a vital role in the relief and rescue for disaster management. Good accessibility facilitates easy access to emergency help for any rescue and relief, which in turn determines vulnerability. Good accessibility means less vulnerability to fire hazards or any kind of disaster. Services accessible to households and communities at large like the fire station, hospital and disaster management offices directly delineates the risk perception and level of vulnerability. During the survey it was found that 6.8% of respondents had fire stations within 1 km, 31.4% within 2-4 km, 27.7% has within 4 km and 34.1 % were not aware of the location of fire stations at all. Though the majority of respondents being literate were not aware of the location of fire stations, which means they may be more vulnerable to accidents, followed by people living within 4 km vicinity. Post-accident vulnerability increases for places far from fire

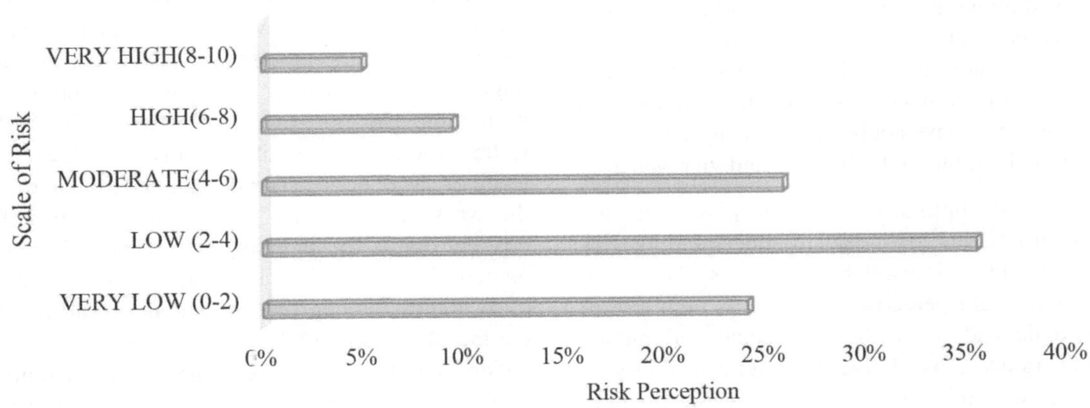

Fig. 46.6 Household Risk Perception

Source: Primary Survey, 2021

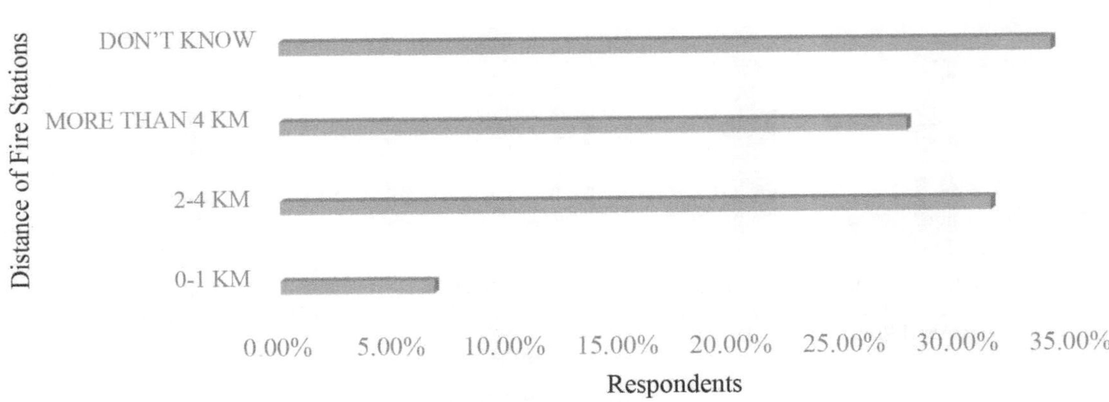

Fig. 46.7 Distance of Fire Stations

Source: Primary Survey, 2021

services or the people living at a greater distance from fire stations would perceive a higher risk of fire hazard, here about 27.7% are more vulnerable.

Psycho-Social factors are one of the most important factors in how we perceive and look at any kind of disaster. In the case of the corona, it has become more evident that to deal and cope up with any kind of disaster, it is important to have positive social behavior and strong psychology. Society is a social construct, which means where one lives, the group of people he/she identifies with and looks to be associated social behavior, interactions, lifestyle, religious groups, and race are some of its basic elements that determine how one participates in society and one cognitive behavior. A human being is a social animal, whose perception is largely influenced by the society he lives in and their engagement in the social process and event. Their perception changes with the level of awareness attained through various social events that are aimed at increasing and educating people about disaster management and general dos and don'ts of a disaster. People were asked about any such initiatives of action taken up in the recent past, about 52% responded saying there was part of some disaster mock drills and workshops on fire accidents, and the remaining 48% have not attended any such events. So, the ones who have not been part of any such social event were more vulnerable to the fire risk and vice-versa.

Further, individual psychological behaviour, psyche, mental perception, cognition, etc. is vital in understanding risk perception, as perception is itself something psychological. The way a person might perceive a disaster as high risk or low risk would depend on his/her emotional and mental states, past experiences, psychological traumas, ability to react in a stressful situation, daily exposure to a risk situation, resilience, etc. Based on the study it could be collected that the respondents were able to relate their response to the recent activity, fire events in the past, and the cognitive image that they have about the accident. Nearly 65.54% of the respondents have had experienced a fire hazard of different intensity in their past, whereas the remaining were not able to recall any such incident happening to them or their direct family. Microwave, AC Short Circuit, LPG Leakage, Short Circuit, Cylinder Blast, Crackers, Faulty Wiring, Transformer Malfunction, Geyser Blast, Candle Burn, Pipeline leakage, Faulty meter box, Room heater, Blowers, washing machine spark, etc. were some of the events the respondents could recall from the recent past. Also, it was found that the risk perception among those respondents was low who have had past experiences of fire hazards as they can more accurately perceive the threat, and his ability to cope with a similar situation in the future increases, making him more resilient than the others. The most important element in any hazard risk perception and vulnerability analysis is the level of awareness that a person has attained. The level of awareness depends on many factors like education, social, economic, and environmental factors. It is built by social conditioning, events in our social environment, and observation of the social functioning around an individual. Awareness is a huge factor for dealing with disasters, preparedness, and mitigation. The more aware an individual is, the less serious is the risk perception of an individual. Generally, a high level of awareness persuades an individual to prepare for the worst-case scenario and to educate himself about the dos and don'ts for an emergency. This makes an individual perceive less threat than a person who is not aware. The level of awareness among the respondents could be attributed to the fact that about 77.3% of the respondents were aware of the dos and don'ts of a fire hazard, 2.7% were not aware and 20% of respondents didn't know much about these things (Fig. 46.8). Further, when they were asked about the emergency helpline numbers, about 87.7% said they were

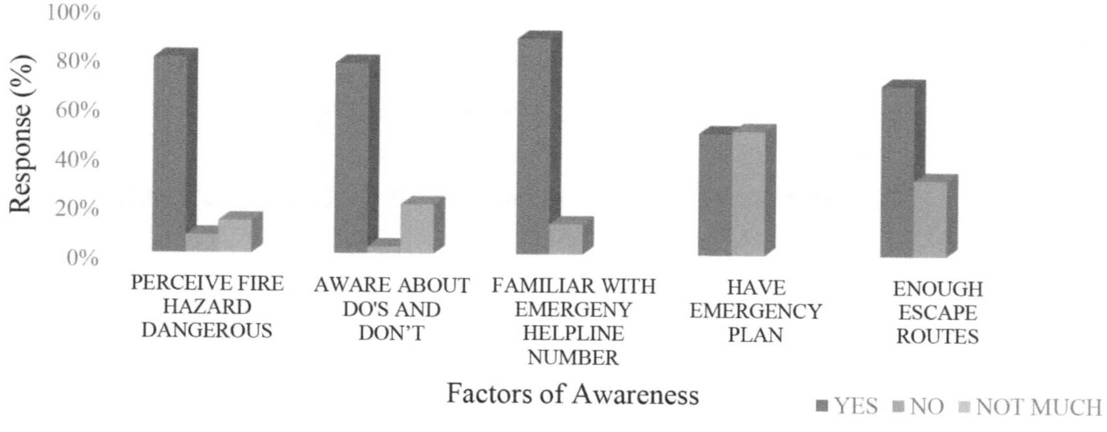

Fig. 46.8 Distance of Fire Stations

Source: Primary Survey, 2021

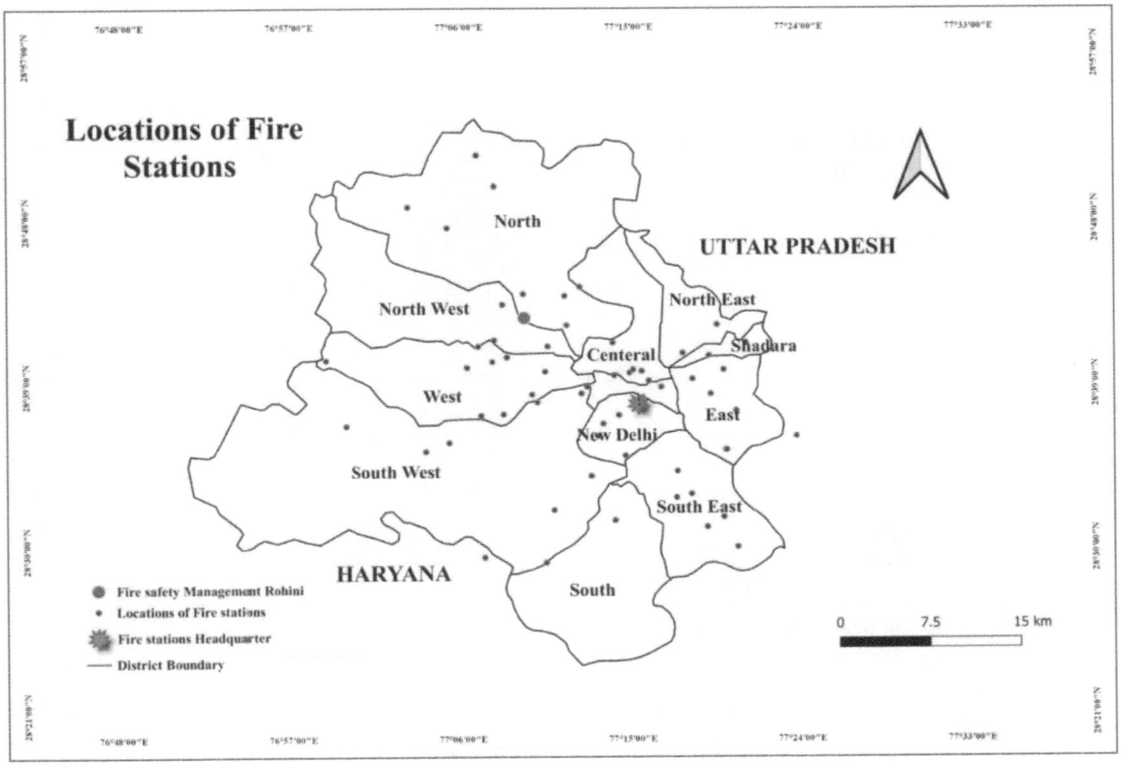

Fig. 46.9 Location of Fire-stations in Delhi

Source: Delhi Fire Services, 2020

aware of it but not sure about the number, while 12.3% were not even aware that such things do exist (Fig. 46.8). It can be said that the majority of the respondents being educated and staying in the metropolitan city were unaware of fire helpline services, making them vulnerable to such accidents. Besides, when they were asked about how to operate fire extinguishers, it was only 60.50% were aware whereas 39.5% had no idea how to use fire extinguishers during an emergency (Fig. 46.8). There is a lack of actions that need to be taken to prepare a locality or a society for a fire hazard. Therefore, people lacking basic awareness about disaster helpline numbers, fire extinguishers, hospitals, fire stations, evacuation plans, etc. would perceive fire hazards as a serious threat and vice versa.

The research, which was undertaken across all the districts in Delhi, it was analysed and concluded that the districts having more natural vegetation will have less built-up area and open spaces, and less vulnerable to household level fire accident. But, in the case of a forest fire or wildfire, the situation could be different whereby more vegetation means more prone to spread of any type of fire which may influence the level of risk perception. It is also observed that there is a lack of good policies, implications, preparedness, and overall management of aspects related to fire hazards which ultimately depicts a laid-back attitude and lack of awareness at all levels.

5. Geo-Spatial Analysis of Fire and Associated Factors

5.2 Land Use and Land Cover

Over the years the land use pattern in the city has been changing rapidly along with increasing urbanization. The change in the pattern is majorly attributed to anthropogenic activities, like the expansion of built-up areas, secondary, tertiary activities, and population growth. With time land use- land cover has seen a drastic change from 2000 to 2021. Here, a land-use map of two time periods May 2000 and May 2021 has been undertaken for the analysis of land-use types and how they could influence and increase the fire risk and hazard. It was found that in 2000, there was sparse vegetation along with low built-up areas and sufficient agricultural land (Fig. 46.10). It had more open spaces and less dense build-up areas. But, if we analyze the land use of 2021, large areas are covered by built-up land (Fig. 46.11). The agricultural land and wasteland have been converted to build-up areas. The settlement pattern besides the vegetative areas and agricultural fields has completely changed. Residential areas, buildings have eaten up large agricultural tracts, thereby indicating large-scale population growth along with increasing density.

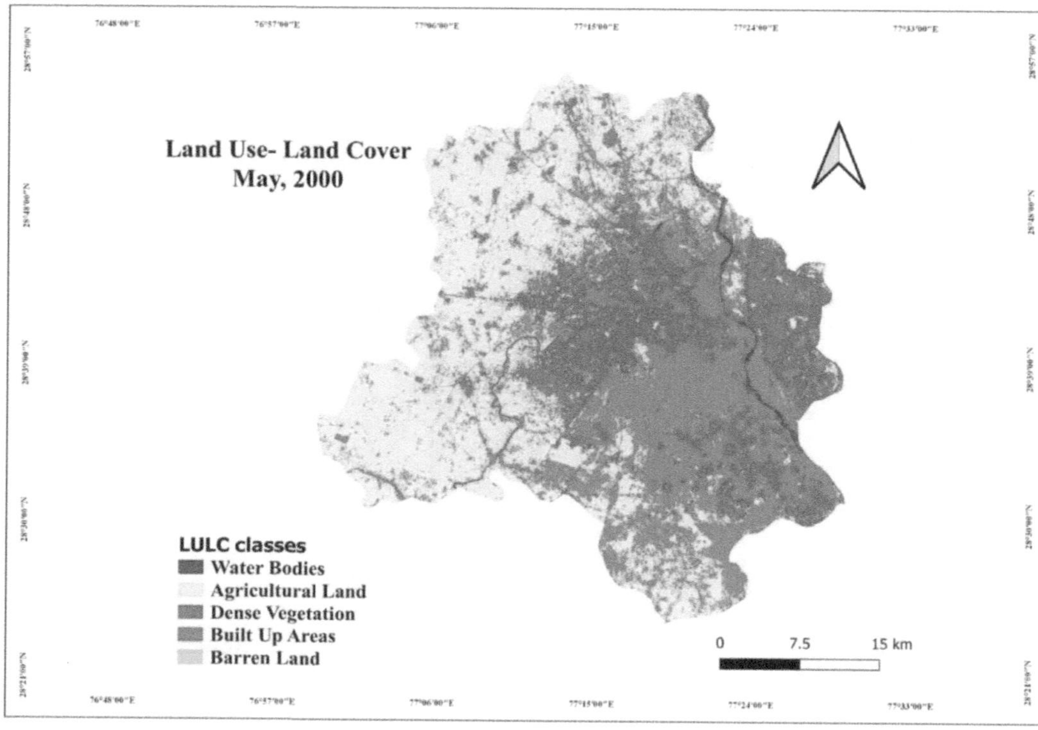

Fig. 46.10 Land Use – Land Cover, 2000

Source: Prepared by Authors, 2021

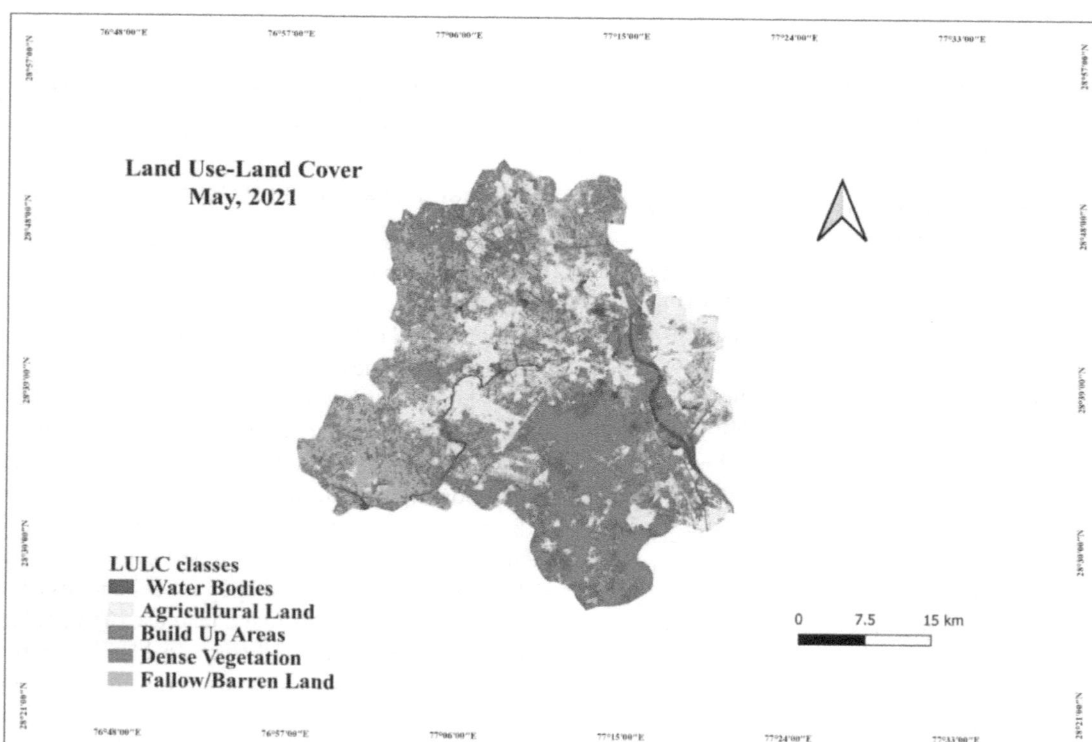

Fig. 46.11 Land Use – Land Cover, 2021

Source: Prepared by Authors, 2021

As a consequence of rapid population growth, urbanization, and the economy have led to a huge change in the land use and land cover scenario of Delhi in the past 20 years which in turn are one of the reasons for below-par infrastructure is making Delhi more and more vulnerable to fire accidents.

Table 46.2 Land use in Delhi

Categories of Land-use and distribution	Land (%)
Residential	45-55
Commercial	4-5
Industrial	4-5
Green/ recreational	15-20
Public and semi-public facilities	8-10
Circulation	10-12

Source: Delhi Development Authority Draft Plan (Master-Plan-for-Delhi-2021-(Updated 31.12.2020).

5.2 Normalized Difference Vegetation Index (NDVI)

Normalized difference Vegetation Index (NDVI) is the estimate of the degree of greenness over an area of land. NDVI is directly co-related to LST and they both, in turn, influence the fire hazard and vulnerability. The NDVI was computed for May 2000 and 2021 and it was found that overall, the NDVI has declined from 2000, as the area under high NDVI value has converted to NDVI of lower values. The ridge in Delhi is predominated with thorny type forest, the extent of which is around 111 km^2 of the total Delhi area besides this, the rest of the part in Delhi is categorized as plain Terrain (Mallick et al., 2008). The NDVI values estimated for the year 2000 are -0.1889 to 0.5168, having a mean value of 0.0695. The lower NDVI value that is less than 0 resembles the high built-up area on the eastern side of Yamuna River and at some parts of Northern Delhi, indicating increased vulnerability. A higher NDVI value of more than 0.5 is being observed at the central ridge, northwest and southwest part of the Delhi city, signifying less vulnerability in the context of household fires. The medium NDVI, i.e., between 0-0.5 is more common in the northern part, showing declining values over time with estimated values of -0.030018 to 0.445012, with a mean value of 0.192488. The lower NDVI generally indicates barren surfaces, water that is seen in the western part of Delhi. NDVI value between 0-0.5 represents shrubs and grasslands that have been seen in central and some parts of northern Delhi and around the river Yamuna. NDVI more than 0.5 represents relatively more dense vegetation which is nowhere in and around Delhi now due to high urbanization and industrialization, signifying overall increased vulnerability across the city.

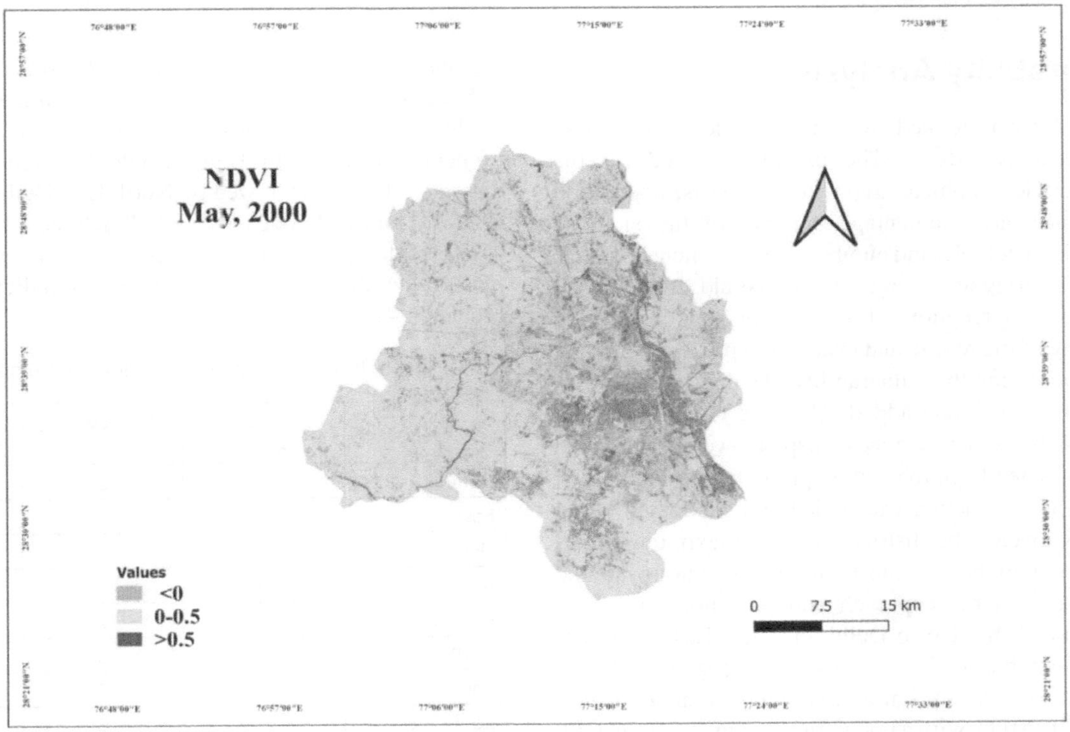

Fig. 46.12 NDVI, 2000

Source: Prepared by Authors, 2021

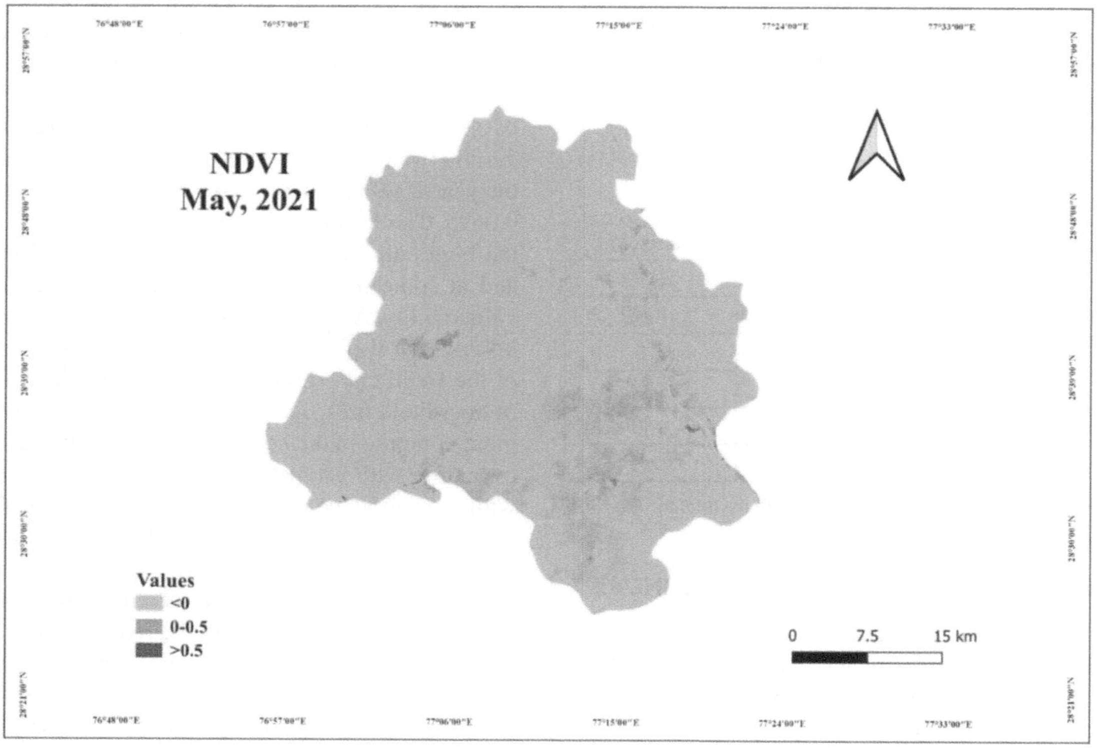

Fig. 46.13 NDVI, 2021

Source: Prepared by Authors, 2021

6. Vulnerability Analysis

The study further looked to analyze the district-wise vulnerability across Delhi. The parameters used for the analysis were Geographical area of the districts, population (2011), built-up area percentage, number of fire stations, rescue vehicles available, and number of government hospitals in the district during an emergency for first aid. It was based on the general perception of the respondents and basic understanding of the rescue and evacuation process, whereby the factors increasing the vulnerability like density and built-up areas were ranked and added. Then the factors reducing the vulnerability like fire services, hospitals were also ranked and then subtracted from the factors promoting vulnerability. The rank of all the factors was added and subtracted based on the ranks given. The district with the maximum values was considered to be the most vulnerable among all the districts. Based on the approach and indicators taken up, it was concluded that North Delhi and New East Delhi are the most vulnerable to fire hazards. The geographical area of North Delhi has 63 sq. km of Geographical area (Hazard Analysis, GOI, 2011), with a population density of 14,044.80, and 44.2% of the built-up area. It indicates that there are fewer open spaces and vegetation cover, along with only 3 fire stations within the district and 12 rescue vehicles, and 115 manpower. There are about 2 governments with basic facilities implies that during an emergency, the vulnerability of people from north Delhi would be higher than other districts. This was followed by North East Delhi with an area of 72 sq. km, with a density of 31208.20 people per sq. km., 41.4% built-up land and only 3 fire stations were the second most vulnerable to fire hazards based on indicators adopted (Table 46.4).

Table 46.3 District wise composite vulnerability

District	Composite Rank	Final Ranking	Degree of Vulnerability
Central	6	3	
East	–6	6	
New Delhi	0	4	
North	12	1	Most
North East	8	2	
North West	–11	8	Least
South	–7	7	
South West	-1	5	
West	-1	5	

Source: Prepared by Authors, 2021

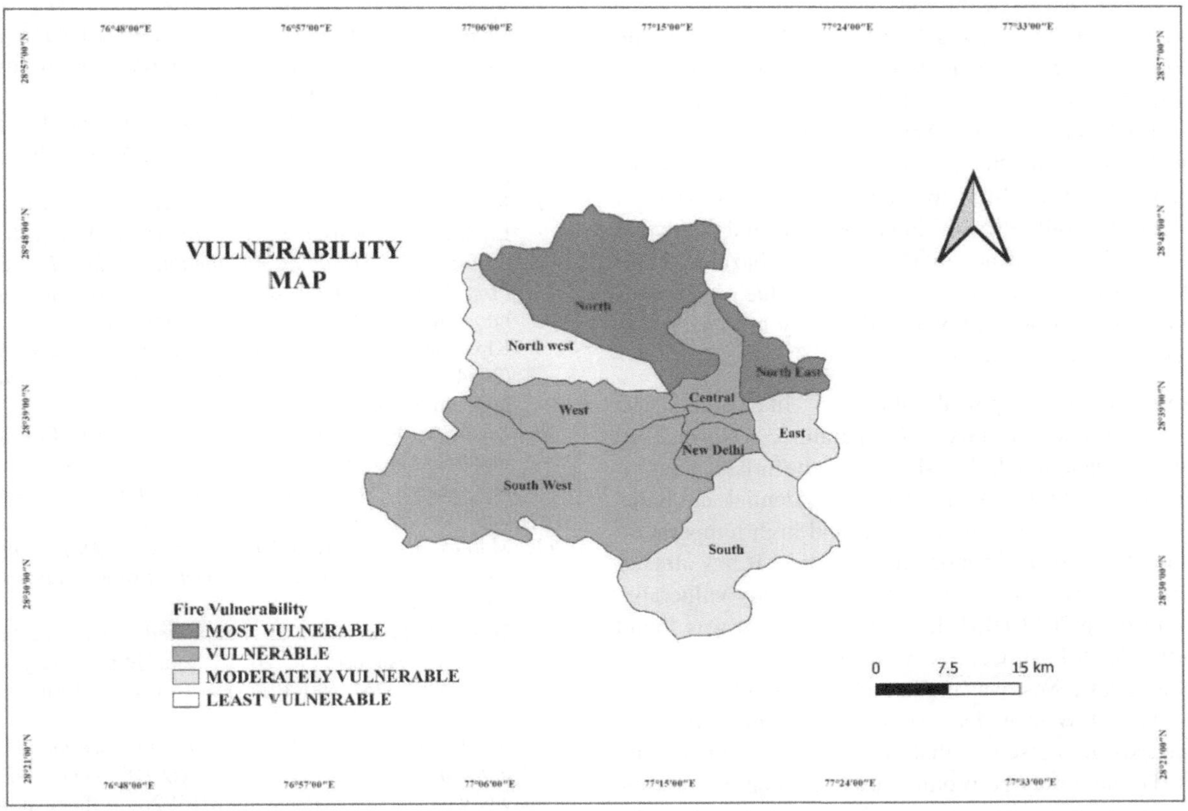

Fig. 46.14 Fire Vulnerability

Source: Based on Composite Index of various Factors. Prepared by Authors, 2021

The districts ranked based on the declining overall vulnerability are North, North East, Central, New Delhi, Southwest, West, East, South, and North West. North being the most vulnerable while North- West the least vulnerable (Table 4). Central Delhi has 5 fire stations for 578,671 population, with the highest area under built-up land. More built-up area and high density mean lesser open spaces for mobility during an emergency, which could trigger a bigger hazard within the district. This could be one of the reasons for the upsurge in the recent fire incidents in that area and hence, make central district vulnerable to fire hazards accidents. The districts which were moderately vulnerable based on the set of parameters were New Delhi and the west and southwest had better fire emergency infrastructure like fire stations and men powers. Men's land ratio in the form of geographical areas to a population made them relatively less vulnerable. East, south, and North West were found to be the least vulnerable as per the findings. Northwest district has more geographical area and low built-up percentage with 9 fire stations and 8 Governmental Hospitals, making evacuation, rescue, and first aid more accessible during any kind of emergency. Thus, North and north-east are the most vulnerable to fire hazards.

7. Conclusion

The incidents of fire and emergency services have been increasing tremendously, ill-equipped fire services like no or limited installations of fire extinguishers, sprinkler water systems, improper maintenance, no earthing systems, and various other defeat systems are some of the reasons for the rapid upsurge fire hazards causing substantial harm to life and property. Moreover, non-execution of the set fire norms, fire safety measures, and lack of awareness about the usage of fire fighting gear constitute a key issue and continue to be a concern. Any emergency involves attentive and swift response by the authorities, the system has been ignoring the fact that poor quality infrastructure, scarce planning, monitoring leads to an increase in the number of urban fires and other such disasters. There is a need for different categories and kinds of equipment depending upon the physical locations of different parts of the country to control such accidents. Besides, poor settlement, poor housing conditions in urban slums and agglomerations, closely built household structure, use of flammable materials in construction, liquid petroleum gas, faulty electric installation and wirings, inappropriate street lights, and connecting wires especially in the old Delhi areas, etc. have increased the risk of fires in the city at the manifold. The city has around 1,700 fire-fighters for around 25 million population whereas, New York City has eight times more numbers of firefighters and half the population of Delhi (www.ETRealty.com, 2020). Among the vulnerable groups children, females, and the elderly are the most vulnerable since they have very little awareness about any kind of hazards. It is generally the working population and youth that is the most active and aware about any kind of hazards. Their level of exposure to the outside world and education makes them much more aware and prepared for any fire hazards as compared to the other age groups.

Moreover, the geographical attributes like Land use Landcover, Land Surface Temperature, Normalised Differential Vegetation Index show that the urban areas of Delhi are under a high risk due to high residential and built-up cover, very little vegetation cover, and high temperature due to the effects of anthropogenic activities. It has already been attributed that 7 out of 9 districts are highly vulnerable to fire hazards in 2011 (RMSI, 2011). Further, it was found that North, North East, Central, were most vulnerable. New Delhi, Southwest, West was moderately vulnerable and East, South, and North West are least vulnerable among all districts to fire hazards. It is essential that the use of unmanned aerial vehicles (Drones) may be promoted, for estimating the fire-fighter's requirements and also can be used for finding the accurate locations of the victims, smoke sensors/ alarms to be linked with the nearest fire stations, there should be a phone application for providing accurate location detection, every locality should have enough amount of fire extinguishers and dry sands to control the increasing fires.

REFERENCES

1. *Accidental Deaths & Suicides in India (ADSI) | National Crime Records Bureau.* (n.d.). Retrieved July 28, 2021, from https://ncrb.gov.in/en/accidental-deaths-suicides-india-adsi
2. *Delhi Govt Portal.* (n.d.). Retrieved July 25, 2021, from http://www.web.delhi.gov.in/wps/wcm/connect/doit/Delhi+Govt/Delhi+Home/Departments/Citizen+ Services/Important+Telephone+Numbers/Fire+Services
3. *Department Of Delhi Disaster Management Authority.* (n.d.). Retrieved July 29, 2021, from http://ddma.delhigovt.nic.in/wps/wcm/connect/doit_dm/DM/Home/
4. *Economic Survey of Delhi 2020-21 (English) | Planning Department |.* (n.d.). Retrieved July 29, 2021, from http://delhiplanning.nic.in/content/economic-survey-delhi-2020-21-english
5. Gupta, S. (2017). Countrywide Fire Hazard and Risk Analysis for Revamping the Fire and Emergency Services in India. *FICCI Special Volume on Use of Geospatial Technology, Success Case Studies.*
6. *Hazard Analysis Directorate General Fire Services, Civil Defence & Home Guards Ministry Of Home Affairs Government of India.* (2011). REPORT. https://dgfscdhg.gov.in/fire-risk-and-hazard- analysis
7. *India-Risk-Survey-2019-ficci.* (n.d.). Retrieved July 28, 2021, from https://ficci.in/SEDocument/20487/India-Risk-Survey-2019-ficci.pdf
8. Käfer, P., Souza da Rocha, N., Veeck, G., Ribeiro Diaz, L., Costa, S., Kaiser, E., Robérti, D., & Rolim, S. (2021). *Assessing Uncertainties in Estimating Surface Energy Fluxes from Remote Sensing over Natural Grasslands in Brazil.* https://doi.org/10.21203/rs.3.rs-374033/v1
9. Kinateder, M., Kuligowski, E., Reneke, P., & Peacock, R. (2014). *Review of Risk Perception in Building Fire Evacuation.* https://doi.org/10.6028/NIST.TN.1840
10. Kodur, V., Kumar, P., & Rafi, M. M. (2019). Fire hazard in buildings: Review, assessment, and strategies for improving fire safety. *PSU Research Review*, *4*(1), 1–23. https://doi.org/10.1108/PRR- 12-2018-0033
11. Mallick, J., Kant, Y., & Bharath, B. D. (2008). *Estimation of land surface temperature over Delhi using Landsat-7 ETM+.* 10.
12. *Master-Plan-for-Delhi-2021-(updated 31.12.2020).* (n.d.). Retrieved August 27, 2021, from https://dda.gov.in/sites/default/files/Master-Plan-for-Delhi-2021-(updated%20 31.12.2020).pdf
13. NYOOOZ. (2021). *DFS responded to more than 8700 fires in the first half of 2021 | Delhi NYOOOZ.* NYOOOZ. https://www.nyoooz.com/news/delhi/1592892/dfs-responded-to-more-than-8700-fires-in- first-half-of-2021/
14. Paek, H.-J., & Hove, T. (2017, March 29). *Risk Perceptions and Risk Characteristics.* Oxford Research Encyclopedia of Communication. https://doi.org/10.1093/acrefore/9780190228613.013.283
15. Paulsen, R. L. (1984). Human behavior and fires: An introduction. *Fire Technology*, *20*(2), 15–27. https://doi.org/10.1007/BF02384147
16. Solanki, S., & Garg, J. K. (n.d.). *QUANTIFICATION OF SEASONAL CHARACTERISTICS OF LAND SURFACE TEMPERATURE OVER DELHI USING LANDSAT 8 SATELLITE IMAGERY.* 10.
17. Sumit Khandelwal, Rohit Goyal, Nivedita Kaul, Aneesh Mathew. (2017). *Assessment of land surface temperature variation due to change in elevation of the area surrounding Jaipur, India | Elsevier Enhanced Reader.* https://doi.org/10.1016/j.ejrs.2017.01.005
18. *User Training Manual – Fire Decision Support System.* (n.d.). Retrieved July 29, 2021, from https://dgfscdhg.gov.in/sites/default/files/User%20Training%20Manual%20 %E2%80%93%20Fire%2 0Decision%20Support%20System.pdf
19. www.ETRealty.com. (2020, February 14). *Apathetic approach towards fire incidents and measures to be initiated—RealtyCheck by Dr. Vinita Yadav | ET RealEstate* [Newspaper]. ETRealty.Com. https://realty.economictimes.indiatimes.com/realty-check/apathetic-approach-towards-fire-incidents- and-measures-to-be-initiated/4051

CHAPTER 47

CBRN Management in India

Pankaj Kumar
Commandant NDRF Academy, Nagpur

Abstract

CBRN is at best, an obscure acronym with knowledge limited to a few Scientists, Armed Forces and Disaster Managers in the entire country today. The spectre of gross ignorance of the subject among the masses in general, and the administrative authorities in particular, presents a dreadful scenario of too little too late.

It is not the case that technological disasters have not been witnessed. We have all been witness to such disasters in the past. They have been occurring with ever increasing frequency with greater ferocity; a fact, we can ignore at our own perils.

However, unfortunately the complacent world in general and the country in particular, subsist in a blissful ignorance and consequent apathy to a perpetual sword dangling over us.

The country has borne an industrial disaster of scale, nowhere seen in the world but still the inherent optimism bias lulls our anxieties on this count and results in the all-pervasive unpreparedness. It is true for the countries world over but the history has been harsher to our fellow countrymen than anywhere else.

The culpability can be shared by the administrative authorities and the scientific community in equal measures. But all said and done, there is no denying the fact that the morass we find ourselves in, is due to the rampant ignorance and the resultant apathy that has assumed a cyclical challenge. We are not prepared because we are ignorant and since weare not preparing, the ignorance refuses to go away.

This ignorance of the hazard and vulnerability has led us to a state where we are least prepared for a mishap or large-scale incident/ accident even as we are proceeding ahead with the chemical and nuclear technologies in our industries, leave alone the devious neighbourhood stockpiling the weapons of mass destruction.

CBRN threats, both accidental and intentional, looms large over global as well as at national levels. The non-state actors have been increasingly using innovative techniques to spread the divisive politics and creating a terrorized scenario. It is a well-known fact that in the past terrorist groups have tried to access and, in some cases, used the chemical, biological, radiological or nuclear (CBRN) means to spread terror. They evidently prefer more destructive and spectacular methods.

CBRN security in India is still in its nascent stages. There is a need to look at it from a broader perspective of internal and regional security challenges that manifest in t h e f o r m of CBRN terrorism, among others.

The paper assesses various aspects pertaining to Indian CBRN framework, vulnerabilities etc. The methodology used for completing the paper is, content analysis of existing literature, UNISDR, UNDRR, NDMA, NIDM reports and national/ international open-source material accessed from public domain. The paper is intended to contribute to CBRN Disaster Management as it focuses on role of training, technology, education and knowledge and attain the means of participatory approach for accomplishing CBRN Disaster Resilient Set-up..

Keywords: UNISDR, UNDRR, NDMA, NIDM, Disaster Management, Hazard & vulnerability, CBRN, EM-DAT, IAEA, ICRP,MoEF&CC, Disaster Resilience

1. Introduction

Grim realities and macabre politics in the current socio-political milieu present an invisible challenge, unimaginable to the uninitiated which unfortunately will impact majority of the humble humans. That technology is a double-edged sword, is known to us all, but the boundless miseries that it might heap upon the poor humans, is beyond the perceptible cognitive faculties of a common man.

Let there be someone who gives a clarion call to shake us out of our complacency, lest the powder keg we are sitting atop, gets lighted intentionally or otherwise.

The bombing of Hiroshima and Nagasaki had come as a major surprise to the international community primarily because the massive destructive capacity of the atom was fictional prior to that. Today, with technological development surpassing security preparedness; it is becoming increasingly difficult for a country to undertake an accurate threat assessment.

Chemical, Biological, Radiological and Nuclear (CBRN) security is a matter of grave concern for many nations' today. CBRN stands for Chemical, Biological, Radiological and Nuclear. It is the new acronym for NBC or WMD. It covers all toxic substances, the threats posed by them, their management and includes safety & security of the hazardous materials. CBRN threats, both accidental and intentional, looms large over global as well as at national level. The non-state actors have been increasingly using innovative techniques to spread the divisive politics and creating a terrorized scenario. It is a well- known fact, that in the past, the terrorist groups have tried to gain access to and, in some cases, used the chemical, biological, radiological or nuclear (CBRN) means to spread terror. They evidently prefer more destructive and spectacular methods.

At the global scenario, varied incidents have occurred in the past. The Tokyo subway Sarin attack was an act of domestic terrorism perpetrated on 20 March 1995, inTokyo, Japan, by members of the cult movement Aum Shinrikyo. In five coordinated attacks, the perpetrators released Sarin on three lines of the Tokyo Metro (then part of the Tokyo subway) during rush hour, killing 14 people, injuring 5,500 people causing injuries of varying degrees that included causing temporary vision problems for nearly 1,000 others.

The Ghouta chemical attack occurred in Ghouta, Syria, during the Syrian civil war, in the early hours of 21 August 2013. Two opposition- controlled areas in the suburbs around Damascus were struck by rockets containing the chemical agent Sarin. Estimates of the death toll range from 281 to 1,729.

Instances such as the March 2018 Novichok nerve agent poisoning of Sergei and Yulia Skripal in the UK and recurrent use of chemicals by ISIS in Syria raises fears of large scale CBRN threats to unassuming general public.

The current Covid-19 global pandemic has highlighted many issues. The paranoia created and panic generated shall have many repercussions and leave many lessons.

CBRN threats or weapons are low probability high impact ones. Each has its own peculiarities. From a purely single-instance impact point of view, Nuclear weapons carry the torch. However, Bio agents are over a period of time, the most frequent. We are witness to the havoc created by Covid-19. At the same time, probability wise, Chemicals (TICs) are also the likely threats as they are available in bulk in trade and industry. Wide use and knowledge of toxic chemicals make them easy and powerful threats.

CBRN Disasters-

Be it the accidents (Bhopal 1984 or Vizag 2020) or deliberate acts (terrorist, non-state or State sponsored) (Iraq 1988, Japan 1995, Syria 2012-19), Chemicals pose a very potent threat.

2. CBRN Disasters

2.1 Nuclear/Radiological Disasters

The risk of a nuclear incidents remains very high in South Asian subcontinent, where three nuclear weapon states, India-Pakistan-China, share common land boundaries with each other. India and China in recent years have emerged as bigger nuclear players by laying more emphasis on use of nuclear energy. India and Pakistan both are trying their best to acquire the membership of NuclearSuppliers Group (NSG) so that their participation in nuclear commerce may increase at international level.

The chances of a non-conventional war breaking out on the borders in the South Asian sub-continent with three nuclear powers remain very high.

There are 7 Nuclear Power plants across the country with 22 operational reactors in them. 8 more reactors are in the process of getting commissioned. An accident or sabotage at any of these facilities may not be ruled out.

Moreover, radiological threats have been increasing in the last decade. We had the Litvinenko poisoning in 2006, the Dirty Bomb scares (Chechen terror 1995 Moscow) and the Chernobyl (1986) and Fukushima (2011) disasters. The threats are real and monumental.

The radiological incidents, although more frequent, have largely been accidental with limited impact confined to smaller areas. However, the nuclear disasters, involving the fission or fusion chain reactions in the nuclear fuel, have been catastrophic with long lasting impact on the environment in a very vast area.

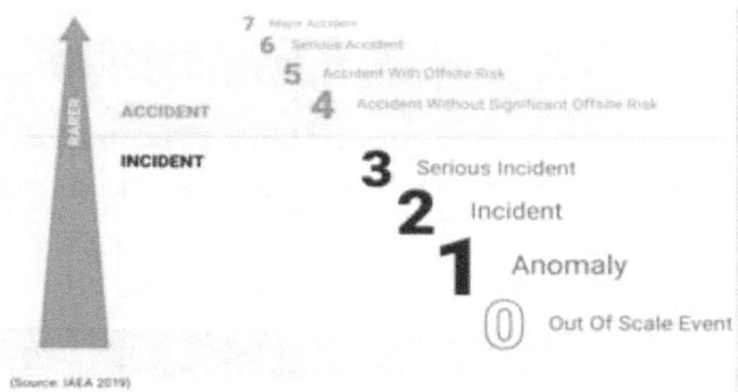

Fig. 47.1 Comparative probability chart for Nuclear accidents
Source: IAEA

Threats of Disasters due to accidents in Nuclear Power Plants seem most plausible. The Ghost town of Pripyat in erstwhile USSR, now in Ukraine, bears testimony to the cataclysmic impact of technological disaster seven today as the radioactivity in the area is not likely to be wished away for hundreds of years.

However, many scholars are of view that despite public concerns of onsite accidents, waste disposal and uncertainties over economics; switching to nuclear power currently remains the largest, proven, carbon- free generation option.

Nuclear disasters can also be caused from an outbreak of nuclear war between countries which may in turn be caused accidently through an error of judgment or misreading of the EN intent or plain technical malfunction or system failure. Experts are of view that threats of use of such weapons by the non-state actors in present era have increased considerably and they may resort to utilizing an IND (improvised nuclear device).

2.2 Radiological Disasters

Radiological emergencies involving radioactive material that can occur anywhere and include:

- Orphan Sources (Uncontrolled, abandoned, lost, stolen or found) dangerous sources;
- Misuse of dangerous industrial and medical sources (e.g. those used in radiotherapy);
- Public exposures and contamination from unknown origins;
- Serious overexposures;
- Malicious threats/acts; and
- Transport emergencies.

Experience shows that local emergency services (e.g., local medical, law enforcement, and fire brigades) will have the most important role in the early response to a radiological emergency. Within hours, national officials may also have an important role to play in supporting the response at the local level. Many aspects of the emergency scenarios expected to arise in the event of a radiological attack/accident may be similar to radiological accidents, but there may also be important differences. For instance, a radiological attack would probably be targeted at a public area, possibly in an urban environment, where the presence of radiation is not anticipated and the dispersion conditions commonly assumed for a nuclear or radiological emergency, such as at a nuclear installation, may not be applicable.

2.3 Industrial and Chemical Disasters

There are about 1900 MAH (major accidents hazard) units (declared) in the country which are largely concentrated in Maharashtra and Gujarat states and they pose a risk of industrial disasters if the stakeholders don't follow laid down guidelines and procedures pertaining to operation, storage and transportation of TICs (Toxic Industrial chemicals).

Susceptibility to industrial and chemical hazards pose a major threat to healthy and safe functioning of corporate sector. The industries employ production processes involving a wide range of chemical and hazardous raw materials, intermediates, waste and final products. These disasters, though normally caused by irresponsible handling of hazardous substances or due to inadequate attention to maintenance of manufacturing processes, have the potential to substantially undermine the very functioning of industries in the region in the aftermath of any untoward incident. Such incidents have widespread ramifications and long-term impact on the society and environment. Between 1970 and 1990, about 180 severe industrial accidents occurred worldwide, leading to the release of various chemical compounds into the environment which killed nearly eight thousand people, injured more than twenty thousand and led to hundreds of evacuations involving common masses.

The scenario becomes scary and dreadful if one takes into account the generational and genetic impact on the community for years to come. The Bhopal Gas Tragedy of 1984, involving a sudden release of about 30 tons of methyl Isocyanate (MIC) at the Union Carbide plant due to poor safety management practices, poor early warning systems and lack of community preparedness, led to death of nearly three thousand people, caused severe health and respiratory problems and birth of deformed and still-born children. The rapid increase in the transportation of hazardous chemicals has enhanced the accident potential during transit. In transit, the goods would be subjected to impact, vibration, compression and other adverse effects and exposure to harmful environment. The movement of hazardous substances by any mode of transport presents a greater risk of accidental release than in a static installation. It is also probable that appropriate information will not be readily available in a transport accident. The need for essential information to be clearly displayed in a transport emergency has always been accepted by both industry and the emergency services. The basis of many emergency information systems adopted in various parts of the world has been a combination of hazard classification and United Nations substance identification.

Chemical Warfare Agents

Unlike the Industrial chemicals, the warfare agents are the chemicals especially designed and optimized to kill/ maim. They are hundreds of times more toxic than the industrial chemicals. Although the required sophistication for designing and synthesizing the warfare agents are limited to the few state parties, the knowhow is not strictly limited as demonstrated by the chemical agent's use during the Syrian civil war. With every passing day there is an incremental threat of use of chemical weapons, not only by the state parties but also by the non-state entities which are not exactly obliged to own the scruples expected from a sovereign state.

Challenges:

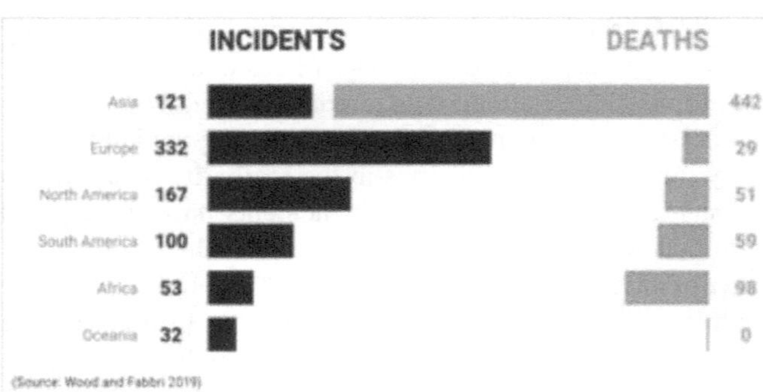

Fig. 47.2 Chemical Incidents in Media 2016- 17
Source: Wood and Fabbri 2019

1. According to the National Disaster Management Authority (NDMA), in the recent past, over 130 significant chemical accidents have been reported in the country, which has resulted in 259 deaths and caused major injuries to more than 560 people.
2. There are 1861 Major Accident Hazard (MAH) units spread across 301 districts and 25 states and three Union Territories across all zones of the country.
3. Furthermore, apart from the regulated and organized sectors, there are thousands of unregistered and hazardous factories in unorganized sectors dealing with range of hazardous materials posing serious and complex levels of disaster risks.

Fig. 47.3 Chart showing Occurrence Gas Global leak disasters from 1950 to 1975

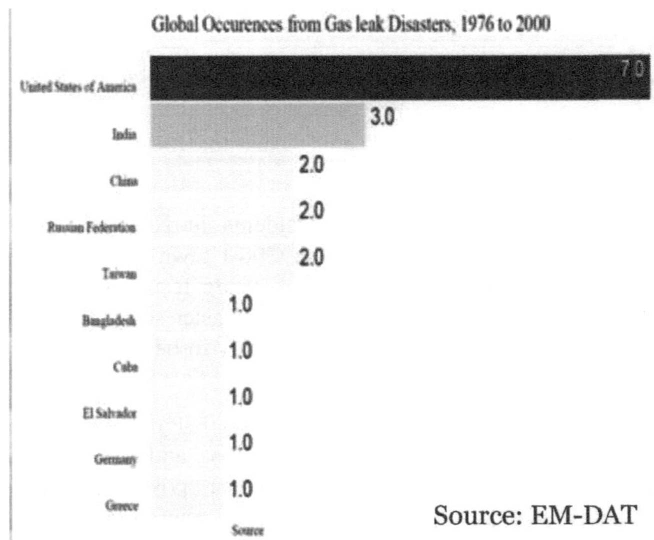

Fig. 47.4 Chart showing Occurrence Gas Global leak disasters from 1976 to 2000

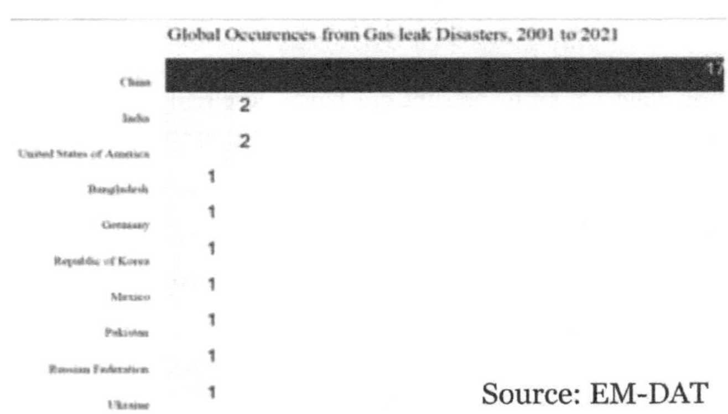

Fig. 47.5 Chart showing Global Occurrence Gas leak disasters from 2001 to 2021

As we can infer from the data sourced from the EM-DAT database, the developed countries although, do have their own share of incidents, there is little or no fatalities involved, Whereas, the developing countries' fatality is more. Thus, it can safely be assumed that the singular thing that makes the difference in casualties is the level of preparedness and advance systems to deal with the hazards.

Fig. 47.6 Chart showing Global deaths from Gas leak disasters from 1950 to 1975

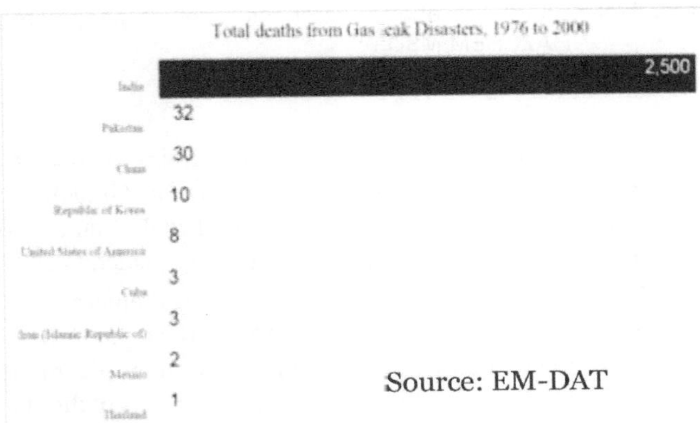

Fig. 47.7 Chart showing Global deaths from Gas leak disasters from 1976 to 2000

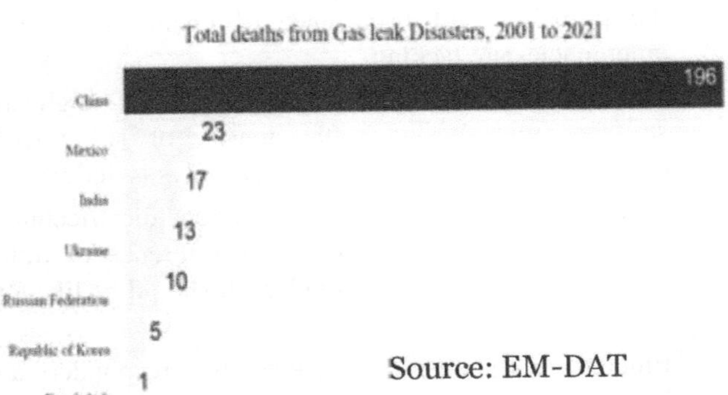

Fig. 47.8 Chart showing Global deaths from Gas leak disasters from2001 to 2021

Regulations:

At the time of the Bhopal gas tragedy,the Indian Penal Code (IPC) was the only relevant law specifying criminal liability for such incidents.

1. Section 304: Culpable homicide not amounting to murder.
2. Section 304A: Causing death due to negligence and imposes a maximum punishment of two years and a fine.

Soon after the tragedy, the government passed a series of laws regulating the environment, prescribing and specifying safeguards and penalties.

In order to regulate and prevent chemical disasters in India, a number oflegislations have been enacted:

The Explosives Act, 1884 (Amended in 1983);

The Poison Act, 1919;

The Civil Defence Act 1968;

The Factories Act, 1948 (Amended in 1987);

The Environment Protection Act, 1986:

Gives powers to the central government to undertake measures for improving the environment and set standards and inspect industrial units.

Manufacture, Storage and Import of Hazardous Chemicals Rules, 1989:

Importer must furnish complete product safety information to the competent authority and must transport imported chemicals in accordance with the amended rules.

The Public Liability Insurance Act, 1991:

Regulating Insurance meant to provide relief to persons affected by accidents that occur while handling hazardous substances.

Chemical Accidents (Emergency, Planning, Preparedness and Response) Rules, 1996:

Centre is required to constitute a central crisis group for management of chemical accidents; set up quick response mechanism termed as the crisis alert system. Each state is required to set up a crisis group and report on its work.

The National Environment Appellate Authority Act, 1997:

The National Environment Appellate Authority can hear appeals regarding the restriction of areas in which any industries, operations or processes or class of industries, operations or processes shall not be carried out or shall be carried out subject to certain safeguards under the Environment (Protection) Act, 1986.

The Environment Impact Assessment Notification, 2006:

Required construction of new projects or activities or the expansion or modernization of existing projects or listed activities as per the Schedule entailing capacity addition with change in process and or technology shall be undertaken in any part of India only after the prior environmental clearance from the Central Government or as the case may be, by the State Level Environment Impact Assessment Authority, duly constituted by the Central Government.

Amendments 2013: Guidelines for consideration of proposals for grant of environmental clearance Environmental Impact Assessment (EIA) Notification, 2006 and its amendments regarding categorization of Category 'B' projects/activities into Category 'B1' & 'B2'

Hazardous Waste (Management, Handling and Trans Boundary Movement) Rules, 2008:

Industry required to identify major accident hazards, take preventive measures and submit a report to the designated authorities.

Disaster Management Act 2005:

An act passed by the government of India for the efficient management of disasters and other matters connected to it.

National Green Tribunal Act, 2010:

Provides for the establishment of a National Green Tribunal for effective and expeditious disposal of cases related to environmental protection and conservation of forests.

New draft Indian Chemical (Management and Safety) Rules, 2020:

Framed and lays down those new Chemical substances have t o b e notified at least 60 days prior to the date they are placed in Indian Territory. Registration is required for identified substances that are listed in "priority substances required to be Registered" in Schedule II of the CMSR.

The Current Scenario:

As Explained earlier, the biggest challenge to any preparedness arises not due to the sophistication of technologies or the hazard causing potential of the agent involved, but it is the lack of awareness of the threat itself. Preparations can be made for any threat howsoever challenging, but what when we don't realize the looming threat.

Just to emphasize the point and to gauge the public awareness, a random sample survey of very small section of society representing the graduates/ under graduates/ employed was taken and the survey brought out the abysmal levels of awareness on the subject. Majority of the people were not even aware of the existence of HAZCHEM codes/ UN numbers or MSDS.

As per National Disaster Management Guidelines on strengthening of safety and security for transportation of pol tankers the analysis of existing regulatory framework revealed following gaps in the year 2010:

- Transport routes for HAZCHEM from the storage site to the delivery point with SOPs for transportation are not defined in detailed manner.

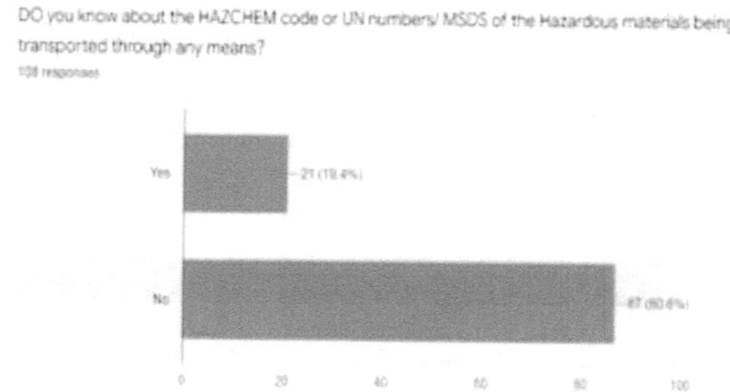

Fig. 47.9 Chart Indicating awareness level of people reg HAZCHEM/MSDS

Source: small size sample survey.

- The system of communication and training of persons involved National Disaster Management Guidelines: Strengthening of Safety and Securityof POL Tankers 20 in HAZCHEM transportation are grossly inadequate.
- Highways are prone to numerous chemical emergencies due to bulk transportation of HAZCHEM but still no appropriate highway DM Plan exists. It needs to be comprehensively addressed specially for road conditions, road signals for speed breakers, diversion, dividers, etc.
- Modification/ harmonization of legislations with respect to Disaster Management Act 2005 to reduce the probability of occurrence of chemical transport emergencies are lacking.
- The awareness material on the specific highways stretches with heavy traffic density of HAZCHEM carriers and other national/state highways is deficient.
- A national and state-wise directory of chemical/technical experts is not available for ready reference of traffic police and other service providers. It should be like Red Book published by MOEF.
- Emergency response guidance for first responders and highway DM Plans are not available.
- The guidance with respect to strict provisions on the part of state governments/district administration to identify specified parking places with full safety and security in their respective areas of control are not available. These places should be available after regular intervals during travel.

The vulnerability of the country as regards transport accidents has steadily increased with increase in the cross-country movement of goods, so has the number of accidents per quarter century. The EM-DAT database shows our country's unenviable position at the top of the list of countries

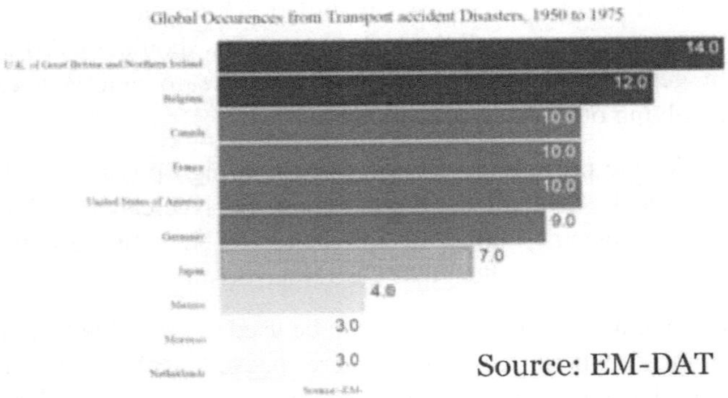

Fig. 47.10 Chart occurrence of showing Global Transport disasters from 1950 to 1975

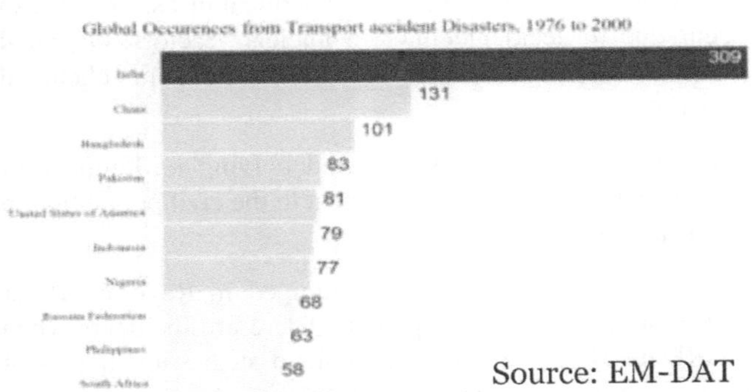

Fig. 47.11 Chart showing occurrence of Transport Global disasters from 1976 to 2000

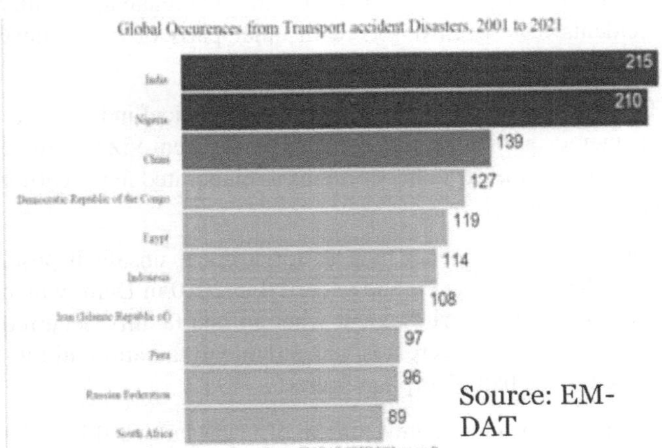

Fig. 47.12 Chart occurrence of showing Global Transport disasters from 2001 to 2021

holding highest number of transport accidents. Many of these accidents include the TICs being transported.

The threat arises not only from the accidents but also from the ignorance of the threat itself We are not sure whether we have the correct data. At present we have to rely on data being maintained by the foreign agencies and the UN organizations, but we don't maintain any national and regional database in public domain which impinges on any analytical attempts at zeroing on to the root causes and thereby precludes any possibility of quick remedial measures.

Even as the concerned agencies are seized of the matter (as can be seen in the reduction in the number of transport accidents), there is a palpable lack of coherent response to the malaise and the agencies are working in isolation, rather than making a comprehensive review bringing all stakeholders on the same platform.

State authorities have been entrusted with submitting the data on the following portals:

- Chemical Accident Information Reporting System (CAIRS)
- Accident Reporting Procedure (HSMD)

However, a cursory visit to the site brings out the glaring gaps in the planning and implementation of the measures to assess, analyze and initiate measures to improve the safety and security of the hazardous material handling, storage and transport.

Responders' Challenges

First responders to a CBRN incident and other rescuers need to be adequately trained and to have proper equipment for identifying radiation/radioactive contamination and other hazardous chemicals/materials. The specialists in radiological protection must be available to offer their advice. **It may beprudent for them to assume involvement of radiological, chemical, and/or biological agents in an attack until it is proven otherwise**. This calls for an all-hazard approach to the response. Responders involved in recovery, remediation and eventual restoration of radiological materials should be subject to the usual international standards for occupational radiological protection, which are based on ICRP (International Commission on Radiological Protection) WISER (Wireless Information System for Emergency Responders) recommendations, including the relevant requirements for occupational dose limitations established in such standards. Similar protocols must be followed in case of other hazardous materials as well.

The immediate countermeasures to protect the public in the rescue phase are to primarily care for people with traumatic injuries and controlling access. Subsequent actions include respiratory protection, personal decontamination, sheltering, iodine prophylaxis (if radioactive isotopes of Iodine are involved), and temporary evacuation.

A radiological attack could also be the cause of radioactive contamination of water, food and other widely consumed commodities. This possible outcome is considered unlikely (except to those in immediate vicinity)to lead to significant internal contamination of a large number of people owing to the large amounts of radioactive material that would be required to cause high levels of contamination of water, food, and other commodities. But all said and done, it would be enough to cause panic and the resultant paranoia among the masses would have severe economic and political fallouts.

An incident involving the chemicals would have a far serious impact onthose in immediate vicinity and there are much more chances for the water bodies and food contamination in case of an industrial accident involving hazardous chemicals.

The responses to radiological and chemical emergencies are very similar. In both cases, our senses (e.g., smell or sight) may not be able to detect hazardous levels of the material. Therefore, the initial response is often carried out based on secondary indications of the hazards such as labels, signs or placards indicating the presence of a hazardous material, the appearance of medical symptoms in exposed individuals or readings from specialized instruments.

In both radiological and chemical emergencies, the major goals of the response are:

- To protect the public; and
- To protect emergency personnel during response.

However, there are following differences:

- Responders generally have no experience with radiationemergencies as they are very rare; Even very low levels of radiation,that pose no significant risk, can bedetected rapidly with simple,commonly available instruments;
- Radioactive materials can cause radiation exposure even whenpersons are not in contact with them;
- The health effects resulting from radiation exposure may not appear for days, weeks or even after years (stochastic effects); and
- The public, media and responders often have an exaggerated fear of radiation.
- In case of presence of hazardous chemicals in the environment, the detection would be a challenge as the portable gas detectors have limitations on the number and type of gasses, they can detect based upon the detection principal being employed by that particular ü detector.

3. Enhancing CBRN Security

With both chemical and radiological emergencies, first responders at the initial stage of response are the same (local officials and Fire & emergency services personnel typically respond to such situations). The basic actions of first responders to radiological emergencies should not differ, in general, from those taken in response to emergencies involving other hazardous materials.

Under the prevailing scenario there is a high possibility of, Chemical, Biological, Radiological & Nuclear (CBRN) warfare agents falling in the hands of terrorist groups, imposing great challenge of CBRN terrorism.

Chemical and Nuclear agents can be used to assemble "Dirty Bombs" where these agents are packed with explosive material. Generally, the nuclear material used in the dirty bomb may have long half-life causing delayed radiation effect in the form of cancer, cataract, genetic damage, premature ageing and infertility. Nuclear or chemical disasters can also occur due to accident either in nuclear reactors, chemical plants or during transportation of nuclear or chemical material.

Chlorine gas leaked out from cylinders lying at Mumbai port in July 2010 posed significant threat to the civil population in the nearby areas.

In 2010 toxic chemicals were dumped in Rayoram River in Kerala affecting a few people. There are incidents when NDRF had been called to respond to such wanton acts of throwing toxic chemicals in open areas having no access control for the general public. One such incident was at Niwadi, Murad nagar where someone had thrown the toxic chemicals dangerously close to the Gangnahar. Similar incidents have been observed in other parts of the country as well.

There are possibilities of accidents in Chemical industries as the memories of Bhopal Gas tragedy and recent Vizag styrene gas leak are still fresh in our minds as elaborated in the earlier part of the paper.

Radiological emergency may also arise due to unsafe disposal of nuclear materials, as had occurred in 2010 in Delhi which resulted in Mayapuri incident wherein one fatality occurred and seven persons were hospitalized due to radiation injuries caused by Cobalt 60 orphan source.

Post great earthquake and tsunami of 11th March 2011 off the Japanese coast, there was a nuclear accident in the Fukushima Nuclear Power Plant of Japan which caused widespread radiation contamination in the surrounding areas.

The use of biological agents cannot be underestimated as they are cheap, easy to produce and only small amount is required to play havoc. Bacteria like anthrax, virus like small pox, toxins like botulinum toxin are most commonly used as biological agents. The current CoVID-19 pandemic has shown challenging and uncontrollable healthcare issues for general populaceand concerned Govt in equal measures.

Accidental release of biological agents from a Bio-safety lab will cause biological disasters of unimaginable proportions.

CBRN disasters will cause tremendous devastation to human and other living beings, flora, fauna and environment including air, water and soil.

All these challenges necessitate comprehensive concerted effort by bringing all the stakeholders on a single platform so as to work out all prevailing as well as emerging CBRN security challenges and devise a management plan to manage the CBRN hazards in all ways possible without jeopardizing the technological and industrial advancement in any way.

The lack of awareness on the subject was evident in a very small sample size survey wherein the subjects were all literate and highly successful people from all walks of life that included businessmen, Teachers, Professors, Judicial service officers and other persons in govt service **(95% Graduate and postgraduate; less than 5% School Students)**. Vast majority (More than 80%) of those surveyed did not think District authorities/ stakeholders other than NDRF have the necessary resources to deal with any Chemical/ Biological / Radiological /Nuclear emergencies.

More than 50% of the people did not think there are sufficient response mechanisms available in the country to tackle any Chemical incident and about 30% were not sure if it was there.

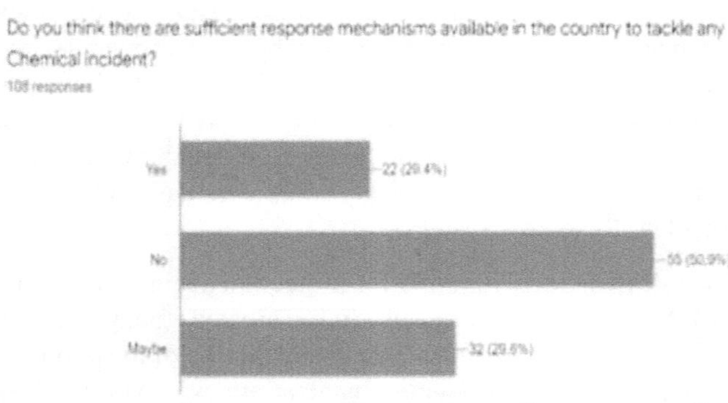

Fig. 47.13 Chart Indicating awareness level of people reg response mechanism to chemical incident

Source: Small size sample survey.

For any biological incident management, the opinion was even starker with more than 56% thinking, it was not sufficient and more than 28% not sure whether there are sufficient response mechanisms available in the country to tackle any Biological incident.

Similar results were obtained for question regarding response to any radiological incident as well. More than 56% thinking, it was not sufficient and more than 34%% not sure whether

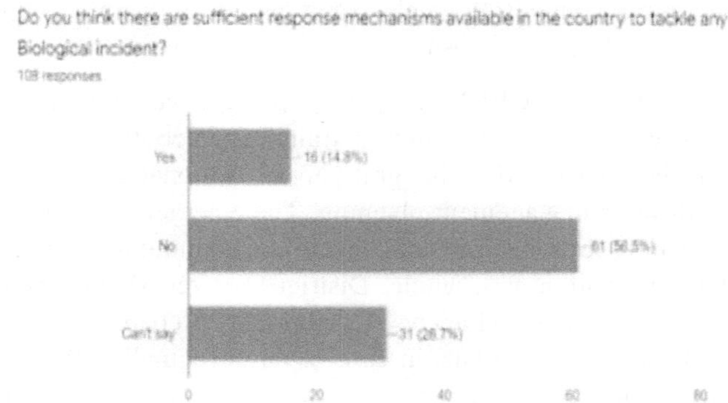

Fig. 47.14 Chart Indicating awareness level of people reg response mechanism to biological incident

Source: Small size sample survey

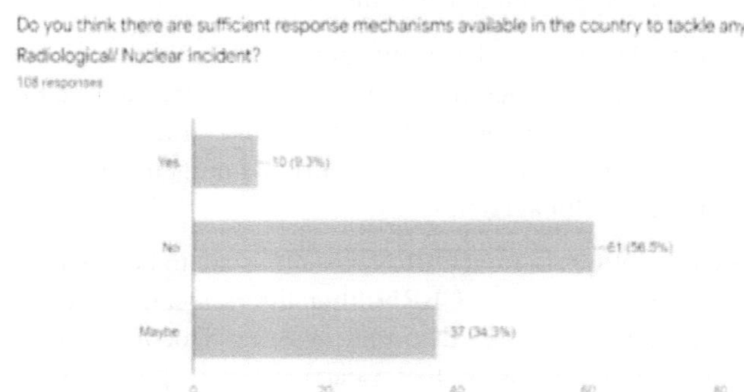

Fig. 47.15 Chart Indicating awareness level of people reg response mechanism to radiological incident

Source: Small size sample survey

there are sufficient response mechanisms available in the country to tackle any Radiological/Nuclear incident.

4. Management of CBRN Disasters

Management of disasters in Chemical, Biological, Radiological & Nuclear scenario poses the challenge to a great deal because of the contamination. In CBRN environment, search, rescue and quick reaction medical team has to put on protective clothing, respiratory masks, boots and gloves which reduces the work efficiency. Personal protection, detection, decontamination, vaccination and antidote administration are extra special requirement for management of CBRN casualties along with the usual care of heat and blast effects.

Nuclear agents mainly cause acute haemopoietic, gastrointestinal, cutaneous and CNS syndrome while chemical agents adversely affect eyes, skin, respiratory and nervous system. Biological agents can severely infect any part of body causing casualties of pandemic proportion.

For managing CBRN disasters, prevention, mitigation, preparedness, capacity building, training and communication is to be considered as integral part of National framework for disaster management planning. The emergency response plans have to be devised at all levels i.e., community, district, state and nation as a whole. District Disaster Management authority is the nodal agency for planning, preparedness and implementation of plans in case of any eventuality. Rescue and quick reaction medical team, doctors, paramedical and nursing staff are to be trained. Standard Operating Procedures (SOPs) need to be formalized at all levels. Mock drills and exercises need to be carried out regularly.

The greatest challenge may be the sudden presentation of large number of contaminated individuals.

The key elements of the healthcare facilities' response plan include prompt recognition of the incident, staff and facility protection, patient decontamination and triage, medical therapy and co-ordination with external emergency response and public health agencies.

Special facilities to treat CBRN casualties are to be created in the hospitals including CBRN filter fitted negative pressure wards and bio-waste disposal system. Documentations, research and analysis is to be carried out in post- disaster scenario for future lessons.

Keeping the numerous shortcomings and weaknesses in mind, there is a dire need of changing and revising the structure for CBRN Crisis and response mechanisms for CBRN incidents.

Following are the pre requisites for an effective CBRN Management structure:

- Presence of Legislative provisions in the form of relevant rule and regulations.
- SOPs to deal with the storage, handling and transport of the hazardous materials.
- Placement of these Rules and SOPs in public domain so that there is general awareness of the subject.
- Enforcement infrastructure. Clear-cut roles and mandates of each and every agency.
- Coordination among the agencies. Regular Onsite and Offsite mock exercises to be carried out with active and informed participation of all concerned including the vulnerable population.
- Creation of infrastructure for protection against hostile actions involving the CBRN agents.

Do we have the requisite SOPs and the systems to effectively deal with the CBRN incidents?

A cursory search on the internet would reveal a lot of SOPs/ Guidelines/Handbooks and tools for handling and responding to the hazardous material are freely available for use by the responders and we learn a lot from them.

A few examples are:

- ERG (Emergency Responder's Guide)
- WISER (Wireless Information System for Emergency Responders)
- CAMEO fm (Database and Information Management Tool)
- CAMEO Chemicals (Chemical Response Datasheets and Reactivity Prediction Tool)
- MARPLOT (Mapping Application for Response, Planning, and Local Operational Tasks)
- ALOHA (Areal Locations of Hazardous Atmospheres)
- NIOSH Guidelines on PPEs
- OSHA Guidelines on PPEs
- NFPA Guidelines and specifications on PPEs

However, the systems are yet to be indigenized and until that is done, the authenticity and applicability of the regulations/ guidelines/tools for CBRN management in the country would always be questionable.

We must strive for our own systems, adapted to our own peculiar circumstances with a countrywide database to be regularly updated by all concerned authorities so that they are readily available for reference by the responders and the CBRN management authorities. We already have the beautiful example of adaptation of an existing and tested system abroad (The ICS; Incident Command System) and its integration into the country's administrative setup as IRS (Incident Response System).

The infrastructures required for protection against the hostile CBRN attacks would need to be created today without creating undue panic and alarm in any section of the public or the world order. Just having a nuclear attack capability does not guarantee the safety and survivability of own population if the adversary decides to exercise the nuclear option in hostility. The toils of peace would reap dividends during the times of war.

According to Russia's Ministry of Emergency Situations, in the event of a nuclear attack on Moscow, there is space in underground facilities for all of the population. (Source: https://www.bbc.com/news/av/world-europe-38136169)

Unless the shelters are prepared in advance, there won't be any shelters for the civil population to save themselves from the CBRN attacks as building these structures require huge engineering efforts which is time taking process and cannot be built in the face of emergency.

Preparation of the community for such an eventuality is also imperative for the preparedness against the CBRN attacks. This aspect is also lacking somewhat asthe sample survey pointed to lack of sensitization about the matter. The following charts indicate the level of awareness of the population especially when we realize that the sample taken for survey was extremely erudite and active based on their current profile.

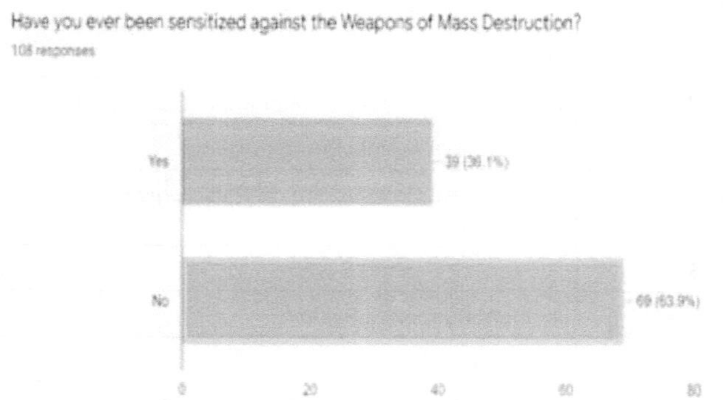

Fig. 47.16 Chart Indicating sens tization of people reg WMD

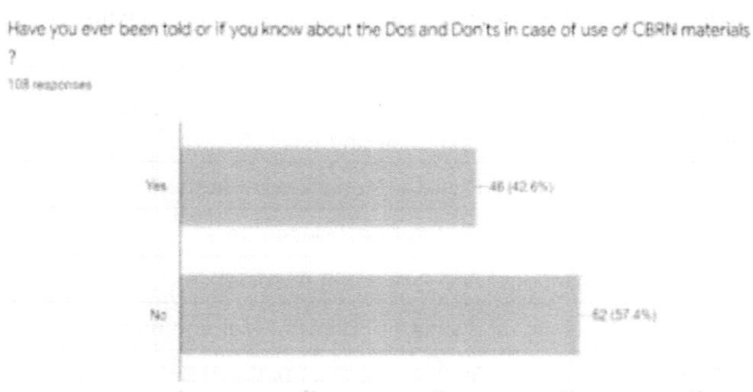

Fig. 47.17 Chart Indicating sensitization of people Dos and Don'ts in case of use of CBRN materials

Source: Small size sample survey

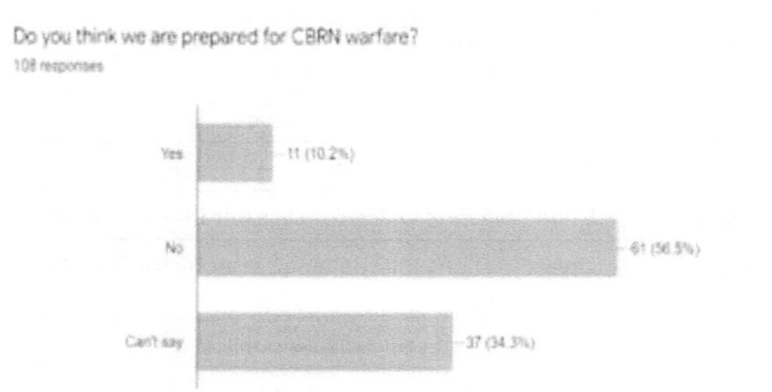

Fig. 47.18 Chart Indicating preparation of people against CBRN Warfare

Source: Small size sample survey

CBRN Incident Management Structure

The present structure is evolving but the general lack of awareness makes it grossly inadequate for effective CBRN preparedness and needs refinement and enhancement. We need to enact and constitute Nodal agencies for CBRN Crisis and Consequence Management not only at the highest levels but also at the state and District levels as well.

Proliferation Prevention and Border Control

We may need to emphasize the role of Customs, Excise and Border security agencies in CBRN nonproliferation in addition to imparting training and equipping them for the job.

CBRN Security for Critical Infrastructure

This is an area which requires a careful planning and further execution wherein we need to cater to the special security needs of the critical infrastructure which are vulnerable to such attacks giving a debilitating blow to their operation.

Industrial and Logistics Security

This is another area which needs proper emphasis. Industrial security, pilferage prevention and appropriate waste disposal need to be studied and re- examined. Strict surveillance and inspections at various levels need to be adequately enforced. More emphasis should be laid on mock drills in liaison with District/Municipal authorities and the planned exercises must b e religiously practiced with active participation of all concerned. Offsite and onsite emergency planning/training need to be standardized so as to achieve right coordination amongst various responders in a CBRN incident situation.

Enforcement of Laws and Protocols

While we may have the best of laws, enforcement is somewhat lax, weak and porous. The plethora of laws to deal with the regulatory mechanisms creates unwarranted confusion among the enforcement agencies.

Embracing Emerging/innovative CBRN Technologies

Emerging technologies are a force multiplier in preventing and effectively responding to a CBRN incident. This would help prevent casualties and save precious lives. Use of robotics, drones, unmanned ground vehicles and artificial intelligence need to be duly integrated in existing protocol/ procedures.

Various mobile and computer-based applications would need to be suitably adapted to the country's needs so that the people at large and the responders in particular find it easier to access the database and be confident to handle such incidents.

Societal Response and Community Participation

The victims in a CBRN incident would be the society. There is a dire need for the population at large to understand CBRN risks, threats response and the communication protocols associated with the response.

5. Conclusion

CBRN incidents cannot be prevented through our stereotypical reaction of posting trained manpower with equipment to deter threats. It's the efficient and effective management of the regulatory mechanisms which will save the day for us. God forbids, if something of the sorts does take place, it is the awareness of the public to the threats and responses that will prevent the losses.

CBRN induced disasters are bound to happen due to both accidental and intentional factors and the threat looms large over the nation. It is pertinent that we prepare for whole gamut of such threats in right earnest. We must strive to empower ourselves by inclusion of administration, law enforcement and the civil society to prevent such incidents and respond adequately. The sincerity of the efforts to deal with probable CBRN incidents will be underscored by not only putting in place the mechanisms to deal with the situations created by these hazardous chemicals and materials, but also by putting in place physical infrastructures like the CBRN shelters at vulnerable locations.

History is witness to the fact that even after our best efforts, the disasters may strike but the posterity shall not forgive the generation for our lapses in putting in place the requisite mechanisms and infrastructure to deal with such scenarios despite clear writings on the wall.

Source

1. DAE MoEF& CC
2. NDMA
3. NIDM
4. EM-DAT
5. NFPA
6. NIOSH
7. OSHA
8. OECD
9. Official Websites of The Ministries, Govt of India
10. A Very Small Size Sample Survey Through Google Form Conducted by the Author
11. Open-Source Internet

Abbreviations

Abbreviation	Full forms
AERB ALOHA	Atomic Energy Regulatory Board Areal Locations of Hazardous Atmospheres
BBC	British Broadcasting Corporation
CAMEO	Computer-Aided Management of Emergency Operations
CAIRS	Chemical Accident Information Reporting System
CBRN	Chemical Biological Radiological and Nuclear
DAE	Department of Atomic Energy
EIA	Environmental Impact Assessment
ERG	Emergency Response Guide
HAZCHEM	Hazardous Chemicals
HSMD	Hazardous Substances Management Division
IAEA	International Atomic Energy Agency
ICRP	International Commission on Radiological Protection
ICS	Incident Command System
IRS	Incident Response System
IND	Improvised Nuclear Device
MARPLOT	Mapping Application for Response, Planning, and Local Operational Tasks
MIC	Methyl Isocyanate
MOEF&CC	Ministry of Environment, Forests and Climate Change
MSDS	Material Safety Data Sheet
NBC	Nuclear Biological Chemical
NDMA	National Disaster Management Authority
NDRF	National Disaster Response Force
NFPA	National Fire Protection Association
NIDM	National Institute of Disaster Management
NIOSH	National Institute for Occupational Safety and Health certification
NSG	Nuclear Suppliers Group
OSHA	Occupational Safety and Health Administration
POL	Petroleum, Oil, & Lubricants
PPE	Personal Protective Equipment
SDRF	State Disaster Response Force
SOP	Standard Operating Procedures
TIC	Toxic Industrial Chemical
UNDRR	United Nations Office for Disaster Risk Reduction
UNISDR	United Nations International Strategy for Disaster Reduction
USSR	Union of Soviet Socialist Republics
WISER	Wireless Information System for Emergency Responders
WMD	Weapons of Mass Destruction

CHAPTER 48

Covid-19 Necessitates Review of Existing Framework for Management of CBRN Disasters/Emergencies in India

Brigadier Kamal Singh Chauhan
(Retd) Ph.D. Research Scholar, Department of Strategic and Regional Studies, University of Jammu
kamalchauhan410@gmail.com
9622222088

Abstract

Covid-19, despite being the biggest global disaster since last 100 years, has been visualised and handled across the world, more as a health and medical emergency rather than a disaster. However, keeping in view the peculiar challenges it posed because of prolonged physical, social and economic impact; dealing with the pandemic of such magnitude requires an integrated approach across all the sectors for planning, preparedness, response and recovery measures. Covid-19 or any other pandemic vastly differs from other disasters, especially in spatial and temporal domain. In fact, all the chemical, biological, radiological and nuclear (CBRN) disasters/emergencies have peculiar dimensions, and impact differently. Hence, CBRN disasters/emergencies are required to be managed differently in a holistic manner at the strategic and operational level during each cycle of disaster management. After enactment of Disaster Management (DM) Act 2005, an integrated institutional framework for disaster management for all types of disasters, including CBRN emergencies/disasters, has evolved in India. Consequently, disaster management authorities at national, state and district level have been instituted, and a comprehensive national disaster management policy; several guidelines for different disasters, including CBRN disasters/emergencies, and National Disaster Management plan have been issued by the National Disaster Management Authority (NDMA). A dedicated National Disaster Response Force (NDRF) to provide specialised response has also been created. For holistic capacity development, the National Institute for Disaster management (NIDM) has been established. The said institutional framework, which evolved in last 15 years, improved the preparedness and response to certain natural disasters; however, keeping in view the challenges encountered during Covid-19, it apparently falls short to handle CBRN disasters/emergencies. In this paper, besides analysing existing framework, distinct characteristics and likely impact of CBRN disasters/emergencies requiring different measures for preparedness, immediate response, mitigation and recovery have been deliberated. Accordingly, the adequacy of relevant aspects of the existing institutional framework for management of CBRN disaster/emergency in India have been analysed. Besides exploring various primary and secondary sources, in this paper the experience of Covid-19 has also been factored to arrive at findings and recommendations to improve upon the existing framework, which may enable holistic and effective management of CBRN disasters/emergencies in the future.

Keywords: Covid-19; Disaster Management; CBRN Disaster/ Emergency; Institutional Framework

1. Introduction

Covid-19 caught the world by surprise. Across the world, including India, the medical infrastructure and the government administration were found grossly inadequate and ill prepared to meet the challenges of Covid-19. Apparently, either the lessons from Spanish flu were not derived properly, or the same were forgotten; wherein the trends of spread including the re-emergence of waves after waves and methods for containment of the disease had a vast similarity to Covid-19. One reason for not realising the gravity of Covid-19 initially, could have also emanated from the fact that in the recent past, viruses such as Severe Acute Respiratory Syndrome (SARS), Middle East Respiratory Syndrome (MERS), and Ebola also had the potential to cause pandemic, but could be controlled because of the initiation of effective measures at national and international levels timely (Wong et al., 2015, p.398-400; Vincent et al., 2005; WHO, 2019). However, apparently, China did not share complete and accurate data with WHO, which delayed relevant notifications and declaration for raising alarms timely for realising the dimensions of the disaster, and for initiating actions for enhancing the preparedness to respond to Covid-19 (Chauhan 2020, 282).

Covid-19, as a global pandemic, posed severe health and medical emergency across the world. According to the Indian Disaster management (DM) Act 2005, it fits well in to the definition of disaster (Govt of India 2005, 4). Therefore, India is the only country that largely relied on the disaster management framework to deal with Covid-19. Even Japan, despite having a very advanced disaster management system, did not apply the provisions of disaster management and depended mostly on a high level of public awareness and public initiatives to achieve social distancing without having lockdown (NDMA, 2020). In absence of robust health and medical system, public health measures including lockdowns initiated in India by invoking the Disaster Management Act 2005, helped a great deal in containing the spread of Covid-19 during the initial phase, and strengthening the infrastructure and preparedness to deal with it on a long-term basis. Though the lockdown caused huge socioeconomic losses, but it did help in containing the spread of Covid-19 during the critical stage, and also optimised the economic growth in the long term. Notwithstanding adequate legal and financial provisions in the DM Act 2005, the application was diluted due to lack of adequate coordination at various levels during response to the first wave of Covid-19 to a certain extent (Chauhan 2020, 299). Despite the warnings of the second wave, the Covid protocols were widely ignored by the public. Also, the same were not strictly implemented by the Govt authorities. The worst part was absolute lack of adequate preparedness for the second wave, which proved very costly in terms of loss of valuable lives, besides other social and economic sufferings.

2. Objectives

This paper has following objectives:

1. Analysis of the peculiarities of the chemical, biological, radiological and nuclear (CBRN) emergencies/ disasters.
2. Analysis of existing framework and mechanism to deal with CBRN emergencies/disasters in India.
3. Analysing the inadequacies in existing DM framework and mechanism observed during Covid-19.
4. Arriving at findings and suitable recommendations to strengthen the existing DM framework for CBRN emergencies/disasters.

3. Review of Literature

Covid-19 has been seen and mostly studied as a pandemic only and not as a disaster. Largely, the attempts have been made to analyse the trends and impacts of the disease caused by Corona virus. For example, Yang and Wang (2020) analysed the characteristics of the virus causing Covid-19 and made recommendations for controlling and preventing the spread of the disease. Emanuel et al. (2020) in their study analysed the exponential rising trend of Covid-19, and expressed concerns that it would cause a scarcity of medical resources, which calls for a rational allocation of scarce medical resources. Liu et al. (2020) analysed the psychological impact of quarantine. Bevel et al. (2020) applied social and behavioural science while analysing the challenges and impact of Covid-19, and considered certain factors including threat perception, leadership, individual and collective interests of the people in a particular social context, and the mechanism for coping with the stress for adopting a meaningful approach to deal with Covid-19. Dasgupta et al. (2020) brought out that Covid-19 brought in an environment of uncertainty, and caused different impacts across different countries, economic sectors, sections of societies, and individuals. Ghosh (2020) used the concept of the pandemic index (PI) and economic policy uncertainty index (EPU) to underscore the asymmetric impact of uncertainty caused by Covid-19 on tourism. Yap et al. (2020) highlighted the lack of inclusiveness in the response measures to Covid-19, especially the absence of measures for disabled people across the world. Mohan and Alex (2020) while acknowledging the adequacy of legal provisions provided by the DM Act 2005 to deal with all types of disasters expressed concerns regarding challenges faced in following a collaborative and consultative approach, and coordinating the execution of necessary response measures at all levels of governance to address various social and economic problems during Covid-19. Chaturvedi (2020) also underlined certain drawbacks in implementing the legal and policy framework of DM, which resulted in skewed response and miseries to the economically weaker sections of the society, including migrant workers during Covid-19 initially. Covid-19 necessitates revisiting the DM framework for CBRN disaster/emergency.

4. Research Methodology

The primary data comprising information in the official documents duly authenticated by international, national, state, and district authorities, and the secondary data/information available in the books, journals, newspapers, and relevant websites have been used. The type of research is descriptive and analytical in nature. In view of the objectives of the study, the qualitative methodology has been followed.

5. Analysis

5.1 Peculiarities of CBRN Disasters/Emergencies

There are certain common social and economic impacts of almost all the disasters, but every type of disaster has some peculiarities which need to be factored for enhancing preparedness and planning for preventing, reducing risk, mitigating the impact, and responding effectively to the disaster. The enormous direct and indirect, wide ranging and long-lasting impacts of CBRN disasters on health and medical treatment stand out, besides other associated socio-economic and environmental impacts. The impact of CBRN disasters/emergencies on human health is fraught with the danger of long-term impact, which may be passed on to the next generations as well. Some of the studies carried out to study the impact of the nuclear attack during World War II substantiate the observation (Kwan 2017). More Studies would be required to find out if similar impacts are possible or not because of chemical and biological disasters.

For holistic management of CBRN disasters, it is also important to understand various hazards, which may cause these disasters. As a result of continuous industrial revolution and consequent enhanced energy and other needs, storage, processing and carriage of chemical and nuclear substances have been increasing at an accelerated pace. Similarly, medical research and analysis necessitates storage, use and carriage of hazardous biological substances in various laboratories across the world. These activities may cause onsite and offsite emergencies/disasters. CBRN disasters/emergencies may also be triggered because of natural disasters such as earthquake, floods, cyclone, tsunami, and extreme weather conditions etc. Melt down of Fukushima Nuclear Reactor and several oil refineries in Japan post tsunami in 2011, substantial chemical release from many industrial facilities in Turkey post-earthquake in 1999, and leakage of chemical substances because of floods in Czech Republic and Germany in 2002 are some of the examples of such disasters (WHO 2018, 3).

About 2,000 Major Accident Hazard (MAH) units and thousands of other registered hazardous units below the level of MAH are operational in India, and pose a serious chemical disaster risk. Despite legal framework, numerous safety regulations, liability, insurance and compensation, and guidelines issued by NDMA, the strict adherence remains a grey area (Govt of India, NDMA; Gupta et al. 2009, 11-12). Bhopal Gas tragedy in 1984, the worst industrial disaster in the world, which killed thousands of people and disabled many more for life, is a grim reminder of the enormity of the chemical disasters. (Cullinan 2004,1). Despite an international ban, threat of use of chemical weapon by some warring factions and terrorist organisation remains a reality. On January 24-25, 2019, the Maharashtra Anti-Terrorism Squad (ATS) arrested 8 people having links with ISIS handlers based at Syria with chemical substances and electronic devices, who had allegedly planned multiple chemical attacks to inflict mass causalities across India (PTI, January 23, 2019).

History is the witness to the fact that some viruses causing epidemics and pandemics inflicted irreparable damages including economic, social, political, and military impact by decimating population, thwarted military campaigns, and unsettled nations across the ages of human civilisation. Release of virulent microorganisms through biological agents such as anthrax, smallpox, etc., accidentally or through bioterrorism (BT) and biological warfare may also lead to biological disaster (Govt of India 2008, XXIII). There are several examples of use of biological weapon in the past. However, in the present context, use of biological attack by terrorists is more relevant. Biological weapon is very cheap, much lethal, and can be delivered without being discovered (Newman 2018). According to WHO studies, bio agents dispersed through a 50 kg bio bomb in an area inhabited by 500,000, people will kill about 95,000; and will incapacitate another over 125,000 people. (Govt of India 2008, 8). The repercussions of human-to-human transmission have been closely seen during Covid-19, the origin of which is still being debated. Advance biological warfare (ABW) agents have been developed using new technologies in biological science. These bio agents are engineered to target specific human biological systems at the molecular level, and require an innovative approach to develop counter measures (Kumar et al. 2011, 71).

From 1987 to 2010, four nuclear emergencies occurred in nuclear facilities: at Kalpakkam in 1987 and 2002, at Tarapur in 1992, and at Bulandshahr (Narora) in 1993, in India. Besides accidents in nuclear reactors and nuclear fuelled systems, other nuclear events caused because of terrorist activities such as an explosion of radiological dispersal device (RDD), improvised nuclear device (NID), and exchange of nuclear weapon will cause a sudden release of lethal heat, sound and radiological emission. Though, after World War II, there is no evidence of use of nuclear weapons, but the threat remained throughout the Cold War era. Both major neighbours, i.e., China and Pakistan, possess nuclear weapon and are belligerent towards India. Stated Nuclear Policy of Pakistan exhibit this capability as a major deterrence against

Indian major offensive operations. After Pakistan sponsored, Pulwama terror strike in February 2019, General Musharraf, ex-President of Pakistan said that India can destroy Pakistan by using 20 nuclear bombs, therefore, to ward off this threat Pakistan must attack India with 50 nuclear bombs (ANI February 26, 2019). Thus, the threat is real and cannot be wished away.

5.2 Institutional Mechanism for Management of CBRN Disasters/Emergencies in India

As mandated in Disaster management Act 2005, National Policy, National plan and guidelines for management of various disasters, including chemical, biological, and radiological and nuclear (CBRN) disasters/emergencies, have been issued by the NDMA. The national authority, i.e., NDMA, is responsible for all types of natural and manmade disasters except those which require intimate involvement of intelligence agencies and security forces. Disasters emanating from CBRN weapon system are required be handled by the Cabinet Committee on Security (CCS) and National Crisis Management Committee (NCMC) (Govt of India 2009, 9).

Management, especially the response to chemical disaster/emergency, differs vastly from that to the natural disasters. Detection of the presence and type of chemical agent, assessment of the extent and severity of impact, decontamination of affected place and people, protection of responders during the disaster, rescue and evacuation of victims and medical treatment, require a totally different equipment, training and approach. However, in India, there is a lack of requisite wherewithal, including adequate medical facility at the emergency sites, which prohibits availability of medical assistance to the victims in a reasonable time frame. Due to lack of adequate infrastructure, desirable first medical response, which is very critical in the event of any chemical disaster/emergency, is often not available to the victims. Also, Medical First responders (MFRs) do not have adequate knowledge about various chemical agents and their effects and antidotes for the same. Except few personnel of National Disaster Response Force, MFRs in civil are also are not trained to handle the chemical burns and other physical injuries. Most of the hospitals in civil, do not have adequate disaster management plans, including response plan. Hospitals also lack contingency plans for a scenario if the hospital itself is subjected to chemical attack. Hospitals lack standardised protocols for treating chemical burn injuries. There is an absolute lack of an institutional mechanism for long term follow up of the victims for treatment besides lack of protocols for crisis management and post disaster short and long-term rehabilitation. Often, the efforts are centred around immediate rescue and relief, and adequate attention is not paid for institutionalising the long-term rehabilitation and psychological fall outs of such disasters (Govt of India 2009, 31-32).

The impacts of biological disaster are amply visible during ongoing Covid-19. Despite comprehensive guidelines on management of biological disasters, which include management of pandemics, issued in 2008, and inclusion of the same in the National Disaster Management Plan (NDMP), absolute lack of preparedness was witnessed during the onset of covid. Subsequently, as an immediate response, several measures, including lockdown, were initiated and medical infrastructure was ramped up on a war footing. To a large extent, these measures did help in containing the spread and flattening the rising infection curve. But as the fresh cases of infection decreased, guards were loosened again, and Covid protocols and development and maintenance of medical infrastructure lost the focus of the State Govts as also of the Central Govt. Thus, during the second wave, India lost much more lives than in the previous one. Again, serious deficiency of basic requirements such as ICU Beds with ventilators and oxygen supply in the hospitals and enormous chaos could be seen across the country; and all the State Govts and Central Govt adopted firefighting kind of response. However, other socio-economic impacts were not as severe during the second wave as were in the first wave. The National Guidelines on Management of Biological Disasters mentions about the availability of Military Hospital across the country, which could be used for extending medical assistance to the victims during such emergencies (Govt of India 2008, 17). It was found that these facilities were overwhelmed by the serving and retired military personnel and their dependents, and there was hardly any spare capacity available for the use by the civilian population. Notwithstanding, Armed Forces did extend substantial help during Covid-19.

Application of radiological emissions in various medical and other industrial processes has been increasing exponentially. Accidental leakage of radiological substances may cause serious radiological emergency (Govt of India 2009, 7). Similarly, accidental release of radiological and nuclear substances in the nuclear plants may cause alarming emergency/disaster. All the nuclear plants in India have an emergency management plan to include personnel emergency, plant emergency, on-site emergency, and off-site emergency. The off-site emergency arises when the living beings and the environment outside the plant get impacted because of radiological and nuclear radiations. The Atomic Energy Regulatory Board is responsible for ensuring effective onsite and off-site emergency response plans in all the nuclear plants and radiation facilities. The Board also censures that the mock drills for all contingencies are conducted periodically by all the plants. 18 Emergency Response Centre (ERC) have been established by Bhabha Atomic Research Centre (BARC) across the country to handle radiological emergencies which may arise because of the accidents during transportation of radiological substances and at the orphan sources, and explosion of Radiological Dispersal Devices (RDD) etc. These centres, besides monitoring and detecting sources of radiation, also provide training to all the stakeholders, including responders; albeit adequacy and effectiveness of the training needs scrutiny. Adequate monitoring devices and protective gears are supposed to be available at these centres. They also provide technical advice to local authorities. However, in view of the vast expanses of India, 18 ERC are absolutely inadequate (Govt of India 2009, 19-23).

As of now, four NDRF Battalions are trained to respond to CBRN emergencies, and are located at Greater Noida, Kolkata, Pune and Arakkonam in Chennai for responding to CBRN emergencies across the country. Since the SOPs for responding to the nuclear disaster arising from any kind of weapon system are classified, the same are not available in the public domain (Govt of India 2009, 7-9). Notwithstanding the confidentiality, it is essential to make a realistic risk assessment of such events, and likely impact on health and other socio-economic sectors. The same should be shared with all the stakeholders on a need-to-know basis. Also, the efficacy of existing response capacity needs realistic assessment and proper gap analysis. The stores, equipment and requisite medical facilities required for response to nuclear and radiological emergency need to be located across the country in such a way that it facilitates quick response and avoids duplication of efforts by different response agencies. Existing medical facilities, including infrastructure and medical professionals, are absolutely inadequate to deal with the nuclear and radiological emergencies. There is a requirement of a proper network of hospitals for such emergencies, which doesn't exist. It is therefore essential to augment the existing medical facilities and ensure availability of proper health care system to deal with such emergency/ disaster at primary, secondary and tertiary levels. Medical First Responder (MFR) are also required to be trained accordingly. (Govt of India 2009, 60-72). Another important aspect of nuclear disaster is that it will have a severe impact on electronic devices and telecommunication (URSI, 1984, 425-428). Our dependence on networking and digitisation is rising exponentially. Therefore, sound strategy and preparatory measures are required to ensure that the digital space and networking domain remain protected from nuclear disasters.

National Guidelines for Management of Biological Disaster, Chemical Disasters, and Nuclear and Radiological Disaster issued in 2008 and 2009, respectively, have comprehensively covered the nuances of these disasters, existing gaps in the state of preparedness and response capacity; and have made several recommendations for initiating suitable actions to fill up the gaps. A closure scrutiny reveals that many vital recommendations in these guidelines have not been implemented for whatever reasons, which is a cause of concern. In a fast-changing scenario, scanning of the CBRN environment from emergency/disaster point of view is required to be carried out regularly. The revised risk assessment will bring out more glaring deficiencies in the preparedness and response mechanism of India, than mentioned in the said National Guidelines. The National Disaster Management Plan 2019 covers only the responsibility framework for various disasters, including CBRN. There is no specific response plan for CBRN disasters. In fact, it is the responsibility of the respective Nodal as well as other concerned Ministries/ Departments of Central and State Govts to prepare the specific preparedness and response plans, besides mainstreaming the disaster risk reduction measures in all the planned projects/ activities (Govt of India 2005,16-17). However, hardly any Ministry/Department has prepared such plans.

5.3 Inadequacies in Existing DM Framework Experienced During Covid-19

Consequences of ignoring the crucial deficiencies brought out in the national guidelines for management of CBRN disasters/emergencies have been seen during Covid-19. While it is good to learn from every disaster/setback, but it is not prudent to sit back and await another type of major CBRN disaster to learn more lessons for improving our resilience, preparedness and response capacity. Since, entire world, including India, was caught by surprise, the initial response to Covid-19 was absolutely knee-jerk type and reactive. However, the provisions of Disaster Management Act 2005 came in handy, while promulgating and enforcing various guidelines, instructions, and initiating certain public health measures, including lock down. While these measures helped in containing the spread of the disease, and buying time to improve health facilities and infrastructures across the country to some extent, but not without enormous chaos, confusion, and socio-economic sufferings (Chauhan 2020, 293-294).

Despite lockdown, thousands of migrant workers violated all norms of social distancing and gathered at various bus stations to go back to their native places, across many cities. Situation was very serious in Delhi, Mumbai and Surat. Since entire passenger trains in the country stopped operating, thousands of migrant workers along with their families, including children, walked hundreds of kilometres, some barefoot. On being stopped and quarantined along national and state highways, they started moving along the railways only to suffer more (Indian Express March 28, 2020). Exodus of migrant workers witnessed in the wake of Covid-19 was unprecedented. It reminded the mass migration and miseries of people during partition of India in 1947 (Bhagat et al. 2020, 1). Lack of sensitivity of some profiteers was also seen when they transported these helpless people like animals in trucks and tankers. Hundreds of these people lost their lives en route. Crushing of 18 sleeping people on the railway track by a goods train on May 8, 2020, and death of 25 migrant workers on May 16, 2020 because of a road accident at Auraiya in Uttar Pradesh speak volumes about national apathy (Ahmad et al. 2020, 2-3).

Despite comprehensive legal provisions and powers provided in the institutional framework for disaster management, there was a lack of forethought, adequate planning and coordination at every level. Since disaster management requires a multidisciplinary and multisectoral approach and measures, the coordinating agency/authority should be in overall control of various discipline/sectors of the Govt organs at all levels. Presently the National Executive committee (NEC) for DM is headed by the Home Secretary, who is required to coordinate all aspects with the nodal as well as other ministry/department. However, if the overall coordination is carried out by the cabinet secretary, it would be more effective. The same was also recommended by the Task Force set up in 2013 by the Govt of India to review the DM Act 2005 (Govt of India 2013, 52-63). During covid-19, powers of the chairperson of NEC regarding evaluation of preparedness for response, and laying down guidelines and issuing direction to all concerned to deal with it as provided in para 2 (i) and (l) of DM Act 2005 were delegated to the Secretary Ministry of Health and Family Welfare (MoH&FW) retrospectively as noticed in the Govt undertaking to Supreme Court on March 31, 2020(Govt of India, 2020). This apparently diluted the multisectoral approach and resulted in poor coordination and confusion initially. Later, when National Crisis Management Committee, headed by Cabinet Secretary, got involved, the coordination improved substantially.

During Covid-19, the need for revising the archaic Epidemic Diseases Act of 1897 (Govt of India, 1897) was felt. Several states enacted the law to deal with the pandemic. Central Govt also carried out amendments to the said Act through a Presidential Order and later it was processed in Parliament and Passed. However, it is pertinent to say that the said Act was made when there were no other legal provisions available to initiate public health measures, and impose certain restrictions required to deal with such disastrous pandemic. But in the present scenario, Section 75 of the DM Act 2005 provides authority to Central and State Govts to make rules to carry out the provisions of this Act, if necessary (Govt of India 2005, 28-29), which could have also been used to make necessary rules. COVID-19 caused huge socio-economic damage. Marginalised and poor people suffered the most. Several short- and long-term social and economic response measures helped a great deal in mitigating the impact to some extent and putting the national economy on track again (Chauhan 2020, 294-298). Notwithstanding, all the scientific warnings for the second wave were ignored, and the time between the two waves to strengthen medical infrastructure was lost, which proved very costly in terms of human life. Roll out and regulation of vaccination process also had initial hiccups, but got streamlined later (Bhatt et al. 2021). Covid-19 underlined the significance of health facilities and infrastructure to deal with CBRN disaster/emergency. National Health Policy and National Action Plan on Antimicrobial Resistance (NAP-AMR) formulated in 2017 helped to a large extent in augmenting the health facilities (Govt of India 2017; Rajalkar and Chandy 2019, 1828-1834).

6. Findings

Following findings flow out from the above analysis:

1. CBRN disasters/emergencies have certain peculiarities, which may cause different short- and long-term impact on health. Nuclear disasters/emergencies, besides causing health related crisis may severely impact electronic devices and telecommunication networks, which may paralyse all the socio-economic activities abruptly. This calls for proper risk analysis and preparedness.
2. Framework and mechanism for DM in India for usual natural disaster is comprehensive but implementation is quite challenging. However, the framework for CBRN disaster/emergency has deficiencies, besides challenges of implementation.
3. Legal and financial provisions of the Disaster management Act were found useful to contain the disease, as also to develop health facilities and infrastructure to some extent, but glaring coordination issues were noticed during Covid-19, and the reasons for the same have been identified. There is a requirement to revisit the institutional framework for DM, especially for the CBRN disaster/emergencies.
4. National Health Policy formulated in 2017 was helpful in augmenting the health facilities during Covid-19, in India to some extent, but its implementation in entirety is essential.

7. Recommendations

1. Proper risk analysis for CBRN emergencies/disaster must be carried out under the guidance of NDMA/

CCS. The risk must be shared with all the stakeholders on a need-to-know basis.

2. Respective nodal ministries/departments must prepare specific disaster management plan for CBRN disasters/emergencies to ensure preparedness for mitigation, response and risk reduction for all the contingencies minimising the generalities.
3. National guideline for management of CBRN disasters/emergencies must be revised and recommendations there in must be implemented.
4. National Health Policy 2017 must be implemented completely, and additional health facilities and infrastructure must be created to cater for specific health impact because of CBRN disasters/emergencies.
5. Strategy must be prepared and implemented to safeguard the digital space and networking, the lifeline of the day, from nuclear disaster/emergencies.

REFERENCES

1. Ahmad, Ritu Kumar; Mohammad Shakil Ahmad; Riyaz Ahamed Shaik. "India's Response to COVID-19 Pandemic and Its Impact on Migrant Workers – A Public Health Perspective". *International Journal of Current Research and Review,* 12, no. 21. 2020. https://www.researchgate.net/publication/341622321_India's_Response_to_COVID-19_Pandemic_and_Its_Impact_on_Migrant_Workers_-_A_Public_Health_Perspective
2. ANI. "If Pakistan attacks, India could finish us with 20 nuclear bombs: Musharraf". The economic Times Online, February 26, 2019. https://economictimes.indiatimes.com/news/international/world-news/it-pakistan-attacks-india-could-finish-us-with-20-nucl ear-bombs-pervez musharraf/articleshow/68149434.cms
3. Bhatt, M., Srivastava, S., Schmidt-Sane, M., & Mehta, L. "Key Considerations: India's Deadly Second COVID-19 Wave: Addressing Impacts and Building Preparedness against Future Waves" *Briefing, Brighton: Social Science in Humanitarian Action (SSHAP).* 2021. DOI: 10.19088/SSHAP.2021.031
4. Bavel, J. J.V., Baicker, K., Boggio, P.S., Capraro, V., Cichocka, A., Cikara, M., . . . Tucker, J. A. "Using social and behavioural science to support COVID-19 pandemic response". *Nature Human Behaviour* 4. 2020. https://doi.org/10.1038/s41562-020-0884-z
5. Chaturvedi, S. Pandemic Exposes Weaknesses in India's Disaster Management response. *The HINDU CENTRE for Politics and Public Policy.* 2020. https://www.thehinducentre.com/the-arena/currentissues/article32502100.ec
6. Chauhan, K.S. "COVID-19 As A Disaster and Its Management in India". *Administrative Development: A Journal of HIPA, Shimla,* VII no. 2. 2020. https://qtanalytics.in/journals/index.php/HIPA/article/view/950/496
7. Dasgupta, U., Jha, C.K., & Sarangi, S. "Procedural Rationality in the Time of Covid-19". *Economic & Political weekly,* 55, no. 25. 2020. https://www.epw.in/author/chandan-kumar-jha
8. Emanuel, E. J., Persad, G., Upshur, R., Thome, B., Parker, M., Glickman, A., Phillips, J. P. "Fair Allocation of Scarce Medical Resources in the Time of Covid-19". *The New England Journal of Medicines,* 382, no. 21, 2049- 2055. 2020. https://www.nejm.org/doi/pdf/10.1056/NEJMsb2005114?articleTools=true
9. Govt of India. *National Disaster Management Guidelines: Management of biological Disaster.* New Delhi: NDMA. 2008. https://ndma.gov.in/images/guidelines/biological_disasters.pdf
10. Govt of India. *National Health Policy.* Ministry of Health and Family Welfare: New Delhi. 2017. https://www.nhp.gov.in/nhpfiles/national_health_policy_2017.pdf
11. Govt of India. *Report of the Task Force, A Review of the Disaster Management Act 2005.* MHA. 2013. https://www.ndmindia.nic.in/images/TaskForce_report_DMact.pdf
12. Govt of India. Status Report. *In the Supreme Court of India, Diary No. 10789 of 2020.* https://www.mha.gov.in/sites/default/files/Status%20Report%20COVID19.pdf
13. Govt of India. *The Disaster Management Act 2005.* New Delhi: Ministry of Law and Justice Gazette of India, December 26, 2005. http://www.ndma.gov.in/en/disaster.html.
14. Govt of India. *The Epidemic disease Act.* 1897. https://legislative.gov.in/sites/default/files/A1897-03.pdfhttps://legislative.gov.in/sites/ default/files/A1897-03.pdf
15. Govt. of India. *National Disaster Management Guidelines: Management of Chemical (Terrorism) Disaster.* New Delhi: National Disaster Management Authority, June 2009. http://www.ndma.gov.in/en/ndma-guidelines.html.
16. Govt. of India. *National Disaster Management Guidelines: Management of Nuclear and Radiological Emergencies.* New Delhi: National Disaster Management Authority, February 2009. http://www.ndma.gov.in/en/ndma-guidelines.html
17. International Union of Radio Science (URSI). (1986) "Nuclear Electromagnetic Pulse (EMP) and Associated Effects". *IETE Technical Review,* 3, no.8. DOI: 10.1080/02564602.1986.11438008
18. Kumar, Alok; Archana Verma, Mukesh Yadav, Imran Sabri, and Ashish Asthana. "Biological Warfare, Bioterrorism and Biodefence". *J Indian Acad Forensic Med.* 33, no.1 (2011). http://medind.nic.in/jal/t11/i1/jalt11i1p69.pdf
19. Kwan, Kitty. "Long-term Effects of Nuclear Radiation". *Submitted as coursework for PH241.* Stanford University. 2017. http://large.stanford.edu/courses/2017/ph241/kwan2/
20. Liu, J.J., Bao Y., Huang, X., Shi, J., & Lu, L. "Mental Health Considerations for Children Quarantined Because of COVID-19". *Lancet Child Adolesc Health,* 4, no. 5, 347-349. 2020. https://doi:10.1016/S2352-4642(20)30096-1
21. Mohan, M.P. R., & Alex, J. P. "COVID-19 and the ambit of the Disaster Management Act". *THE WEEK.* 2020. https://www.theweek.in/news/india/2020/04/26/covid-19-and-the-ambit-of-the-disaste r-management-act.html
22. NDMA. "COVID-19 Impacts and Responses: The Indian Experience". *NDMA: New Delhi.* 2020. COVID-19-Indian-Experience.pdf (ndma.gov.in)
23. PTI. "I S suspects were planning mass casualty attacks using chemicals, claims ATS". *The Week Online. Jan 23, 2019.*

https://www.theweek.in/news/india/2019/01/23/is-suspects-planned-massive-attacks- was-kumbh-mela-in-target-list.html

24. Ranjalkar, J., & Chandy, S. J. "India's National Action Plan for antimicrobial resistance – An overview of the context, status, and way ahead". *Journal of family Care and Primary Medicines,* 8 no.6. 2019. https://dx.doi.org/10.4103/jfmpc.jfmpc_275_19
25. Vincent, M. J., Bergeron, E., Benjannet, S., Erickson, B. R., Rollin, P. E., Ksiazek, T., G., Seidah, N. G., & Nichol, S. T. "Chloroquine is a potent inhibitor of SARS coronavirus infection and spread". *Virology Journal*, 69, no. 2. 2005. https://doi.org/10.1186/1743-422X-2-69
26. WHO. "Chemical Releases Caused by Natural Hazard Events and Disasters – Information for Public Health Authorities. Geneva: World Health Organization". 2018. https://www.google.com/search?q=Chemical+disasters%2C+causes%2C+typical+charracteristics%2C+prevention%2C+mitigation+and+response+measures&oq=Che&aqs=chrome.0.69i59j69i57j46i131i433i512j0i131i433j46i199i433i465i512j0i131i433i512j69i60j69i61.10211j0j7&sourceid=chrome&ie=UTF-8#
27. WHO. "Worldwide reduction in MERS cases and deaths since 2016". 2019. https://www.who.int/emergencies/mers-cov/worldwide-reduction-in-mers-cases-and-d eaths-since-2016/en/
28. Wong, G., Liu, W., Liu, Y., Zhou, B., Bi, Y., Gao, G. F. "MERS, SARS and Ebola: The Role of Super Spreaders in Infectious Disease". *Cell Host & Microbe*, 18. 2015. https://www.researchgate.net/publication/282871294_MERS_SARS_and_Ebola_the_role_of_super-spreaders_in_infectious_disease
29. Yang, P., and Wang, X. "COVID-19: a new challenge for human beings". *Cellular and Molecular Immunology* 17, 555–557. 2020. https://doi.org/10.1038/s41423-020-0407-x
30. Yap, J., Chaudhry, V., Jha, C.K., Mani, S., &Mitra, S. (2020). "Are responses to the pandemic inclusive? A rapid virtual audit of COVID-19 press briefings in LMICs". *Elsevier Public Health Emergency*. doi: 10.1016/j.worlddev.2020.105122

Part 7

Other Disasters

CHAPTER 49

Are Strong Aftershocks Always Triggered by a Positive Coulomb Stress Change? : A Case Study of the 2018 Indonesia (Sulawesi) MW 7.5 Tsunamigenic Earthquake

Nazeel Sabah
Research Scholar, Department of Earthquake Engineering,
Indian Institute of Technology Roorkee
nsabah@eq.iitr.ac.in

Daya Shanker
Associate Professor, Department of Earthquake Engineering,
Indian Institute of Technology Roorkee
d.shanker@eq.iitr.ac.in

Abstract

September 28, 2018, Sulawesi Earthquake of MW 7.5 is known to be caused due to a strike- Slip along the NNW-SSE trending Palu-Koro fault. Around 150 aftershocks were associated with the mainshock. In this paper, we consider 6 aftershocks MW ≥ 5 earthquake and investigate whether these were caused as a result of positive Coulomb stress changes on receiver fault (the fault on which the aftershock acts). To generate a fault model of the Palu-Koro fault from focal mechanism solutions of the mainshock event as listed in global CMT catalogue are used. The Coulomb failure stress change on both the nodal planes of the focal mechanism are computed and the Coulomb failure stress change imparted by the MW 7.5 mainshock on subsequent aftershocks with the epicentre location of each aftershock are compared. In addition, shear stress and normal stress changes are also computed. Coulomb stress changes for the 6 aftershocks range from –0.454 to 0.036 bar. Only 2 computational results (around 17%) are positive, all others (10 in number) are negative. Here we find that a vast majority of MW ≥ 5 aftershocks occurred in regions of negative Coulomb stress areas triggered by the MW7.5 mainshock. Thus, our results suggest that Coulomb stress changes did not contribute to the development of MW 7.5 sequence of earthquakes.

Keywords: 2018 Sulawesi earthquake, Tsunami, Focal mechanism, Coulomb stress change, Palu- Koro Fault

1. Introduction

The physics of earthquake mechanism relates the generation of earthquake with time dependant stress accumulation along brittle crustal fault planes. Though the exact magnitude of existing stresses underneath cannot be quantified mathematically, the stress changes at tectonic level triggered by earthquakes can be calculated (Hainzl et al. 2010). The spatial and temporal characteristics of future earthquake events are found to greatly depend on coulomb stress changes beneath. Even a minor increase in Coulomb by as low as 0.1 bar on a potentially seismic fault plane could eventually trigger an earthquake (Stein 1999). Also, the changes in coulomb stresses can promote or alter seismic activity on neighbouring faults (Belardinelli et al. 2003).

The interdependence of stress changes (static) and spatial distribution of future aftershocks was initially suggested by Das and Scholz (1981). The popularity of this work was augmented when coulomb stress hypothesis was applied to Landers earthquake of 1992, which struck with a magnitude Mw 7.3. Stein et al. (1992) investigated the 1992 Landers earthquake and established the influence of several past earthquake events (over a period of seventeen years) on the Landers Earthquake of 1992. It was also evident that a vast majority of aftershocks, following the main event stroke along areas where there was a notable increase in Coulomb stresses. Since then, there has been numerous studies which reinforce the interdependence of the static stress changes and the coordinates of the subsequent mainshock and aftershock events (Ruth and Harris, 1998), (Steacy et al. 2005), (King, 2007). In this study, the coulomb stress changes accompanying the MW 7.6 mainshock event (Dziewonski et al. 1981), (Ekström et al. 2012) which hit the Island of Central Sulawesi on 28th September 2018 is examined and an analysis is carried out to check whether the positive Coulomb stresses generated by the mainshock triggered the series of subsequent aftershocks or not.

The Island of Sulawesi which was the worst affected Island during 2018 Indonesia tsunami slopes up from the sea shore to a high non volcanic mountainous interior. The northern Minahassa Peninsula houses active volcanoes as well. It is believed that the Island of Sulawesi came to existence when terranes (fragments of crustal materials) from the Australian plate (on the south-east) and the Asian plate (on west and southwest) collided with the island arcs previously in the Pacific (along the north and east Peninsulas). By virtue of complex and multiple tectonic origins, a large number of faults crisscross the region thereby making it significantly susceptible to earthquakes and hence tsunamis. The geological characteristics of Sulawesi are shown in Fig. 49.1. The Sulawesi province of Indonesia is located at a complex juncture between the Eurasian, the Philippine, the Pacific and the Indo-Australian plates. There are numerous microplates formed in this region because of the interaction between the major plates. Towards the northern Sulawesi, where the epicentre of the earthquake is located, the main active structure onshore is the Palu-Koro fault which acts as a line of demarcation between Makassar Strait and North Sula. It is estimated that along this fault the rate of slip per year is in the range of 32-45 mm (Bellier *et al.*, 2001). The Palu-Koro fault, being a major and active one was responsible for causing the 2018 Sulawesi earthquake and Tsunami. The fault trends along NNW-SSE direction and is categorised to be a strike-slip fault. The tectonic plates surrounding Indonesia can be observed from the tectonic map of the world in Fig. 49.2. The Palu-Koro fault extends from Dondowa (in north Luwu Regency) in the south to Matano fault which trends along WNW-ESE direction. Further the fault continues northwards, through the gulf of Palu and passes to the west of Minahassa Peninsula and eventually it links with the north Sulawesi subduction zone. The fault has witnessed several earthquakes in the past in the years 1905, 1907 1909, 1927, 1934, 1968, and 1993. Fig. 49.3 and Fig. 49.4 show the maps of significant historical earthquakes and tsunamis respectively in Sulawesi. Upon observing the historic earthquake database of Gusiakov (2005) it has been incurred that over the last 2000 years, three major earthquakes have been identified and this suggests the recurrence interval along this fault to be around 700 years. This is accountable with the long-term slip rate of 40-50 mm per year over the last five million years

2. Foreshocks, Main Event and Aftershocks

The sequence of foreshocks started around 3 hours prior to the strike of the main shock. The earliest foreshock took place at 15:00 (Indonesia Central Standard Time) with a magnitude of Mw 6.1 (as per USGS) and left Donggala Regency the worst affected. At least one person was killed and 10 others injured. The quake affected tens of structures which were either wrecked or damaged. Followed by this major foreshock, there were several others all of which were epicentred immediately south of the mainshock epicentre.

The mainshock occurred at 10:02:44 (UTC) and reportedly the magnitude was estimated as Mw 7.7 by USGS and BMKG. Later the estimated magnitude was revised to Mw 7.5 by USGS. The quake was so violent that the effects were felt in far off places like Samarinda (the capital of East Kalimantan) towards west, Makassar towards south and parts of Gorontalo towards north. Parts of Malaysia too experienced the effects of the quake. The earthquake resulted in an immediate blackout leaving telecommunication and transportation channels disrupted.

Within 24 hours of the mainshock, there were a series of aftershocks, 14 of them had magnitudes MW > 5. Thereafter at least 150 aftershocks of minor intensity and magnitude were recorded in the region. On October 2nd, just days after the mainshock, a strong aftershock with the magnitude of 6.0 (Mw) struck the east Sumba area in Indonesia which toppled a bridge in Wula Waijelu district. The locations of the mainshock and aftershocks are shown in Fig. 49.3.

3. Methodology

It is understood and widely accepted by the scientific community that earthquakes occur when the rocks are not able to withstand the accumulated stresses over time and they eventually release a lot of energy (Scholz, 2002). The concept of Coulomb failure stress (ΔS_f) is used to determine the proximity to failure. The Coulomb failure stress is found to depend on the slip of the fault, the fault geometry and effective coefficient of friction (Yadav *et al.*, 2012 ; Stein, 1999 ; King *et al.* 1994 ; Ishibe *et al.*, 2011). The Coulomb stress change is expressed mathematically as

$$\Delta S_f = \Delta r - \mu' \Delta \sigma_n$$

Where, ΔS_f represents the Coulomb stress change, Δr stands for the changes in shear stress on a particular fault

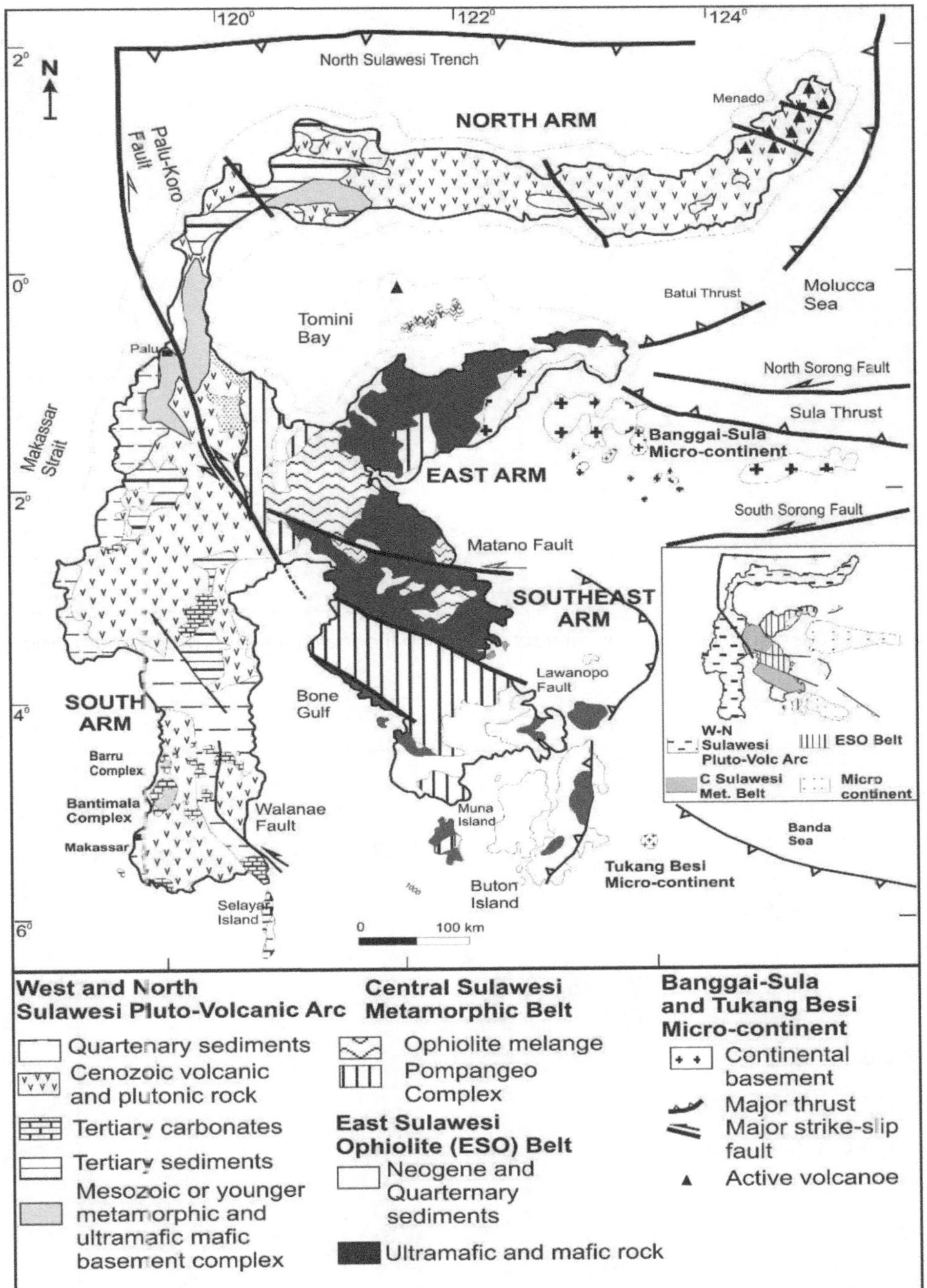

Fig. 49.1 Geology of Sulawesi Island (Maulana, 2018)

in consideration (positive along fault slip direction), $\Delta\sigma_n$ is the normal stress change (compression is taken as positive). μ' is the effective friction coefficient and is derived from $\mu' = \mu(1 - B)$, μ being the apparent friction coefficient and B ranges between 0-1 (Skempton's Coefficient).

For this study NNE-SSW Palu-Koro fault (Strike: 348, Dip: 57, Rake: –15) is taken as the triggering fault plane with a magnitude of MW 7.6 and the main event epicentred at 0.72°S latitude and 119.86°E longitude (Dziewonski *et al.* 1981; Ekström *et al.* 2012). For all our computations,

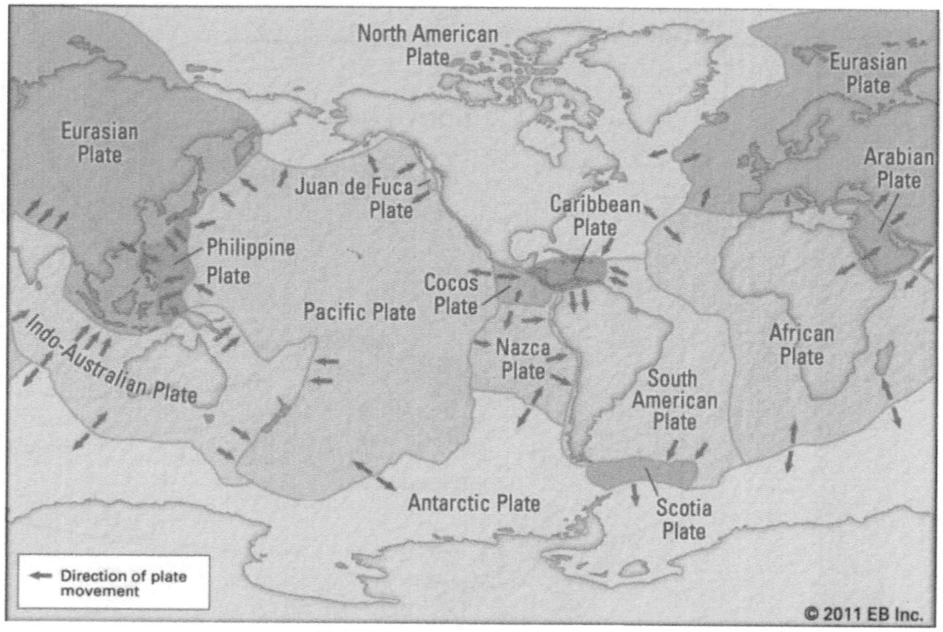

Fig. 49.2 Tectonic map of the world, the plates in interaction around Indonesia are clearly presented (Encyclopædia Britannica)

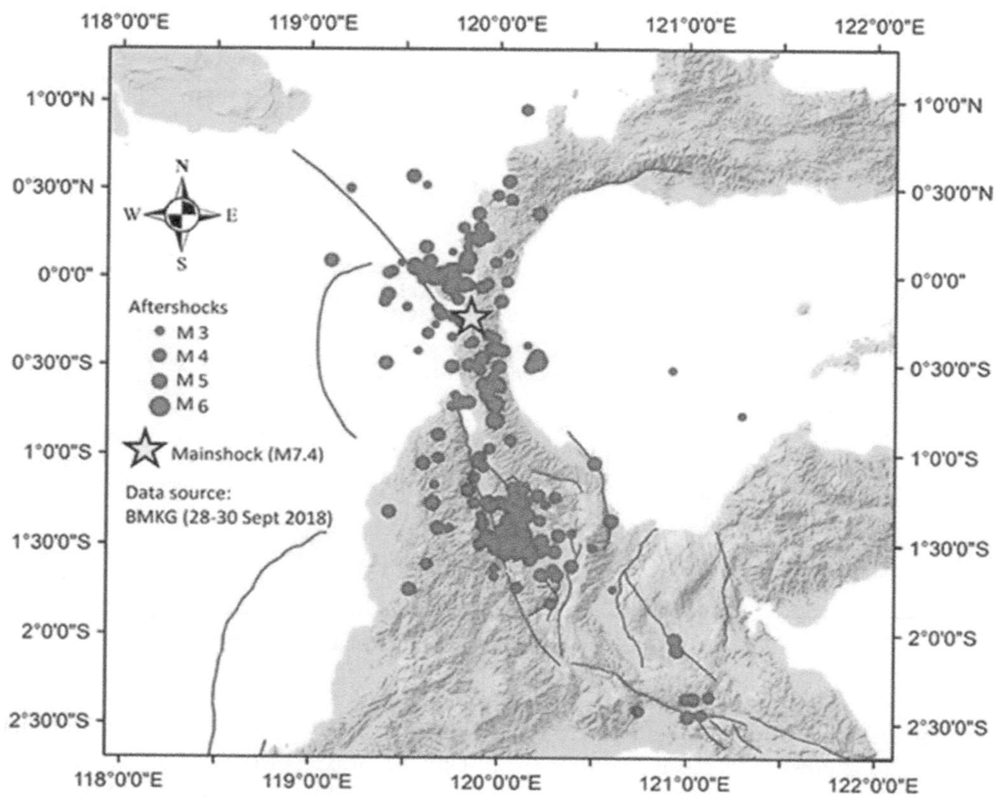

Fig. 49.3 Epicentre of Earthquakes (Main event and Aftershocks) and surface traces of surrounding main faults (Palu-Koro fault shown as disconnected red lines)

Source: BMKG LIPI: Indonesian Institute of Sciences, http://lipi.go.id/public/uploads/berita/berita_palu.jpg accessed on (27/08/2021) (Palu Mainshock and Aftershocks, 2018) we assume effective frictional coefficient as 0.4, Poisson's ratio as 0.25 and a Shear modulus of 3.32 X 10^{10} dyne-cm^2. Here a uniform slip model proposed by Gahalaut, 2006 of a rectangular fault plane is used to compute the stress changes in elastic half space. The faults are modelled using Coulomb 3.3 software developed by U.S.G.S using various parameters derived from the global CMT catalogue (Göran Ekström, 2006).

In our computation, we incorporate the Coulomb stress changes on both the nodal planes of each of the subsequent aftershocks (here 6 in number) in their rake directions (Fig. 49.4). The parameters which indicate a particular nodal plane are strike, dip and rake respectively. The strike, dip and rake of individual events are obtained from global CMT (Centroidal Moment Tensor) Catalogue (Dziewonski *et al.* 1981; Ekström *et al.* 2012). The red patch indicates the areas of Coulomb stress increase, whereas the blue patches indicate the areas of Coulomb stress release or dissipated. The Coulomb stresses are colour coded in units of bar, a scale of which is enclosed below the diagram that follows (Fig. 49.4).

4. Results and Discussion

The black pentagram in each figure represents the co-ordinate location of the aftershock event. In general, the Coulomb stress change across crustal faults range from 0.1 to 1 bar (for planes brought to near failure conditions). A stress change of merely 0.1 bar is considered enough as the causative reason for a series of aftershocks or future earthquakes. What matters is basically the temporal parameter of the stress change which keeps fluctuating and the trigger of any earthquake event is dependent on either the delay or advance of this triggering phenomena (Yadav *et al.*, 2012). It takes about 150-200 years for the rocks to accumulate a stress equivalent to 1 bar (King *et al.* 1994). Here in this study, we analyze the stress change on the receiver fault that is the fault which is most likely to receive the stress from the mainshock.

The table that follows (Table 49.1) represents the basic parameters of the MW ≥ 5 aftershocks chosen for this study. It also indicates the strike, dip and rake parameters as obtained from the focal plane solutions. Also, the result of stress change calculations in terms of shear, normal and coulomb stresses are clearly indicated.

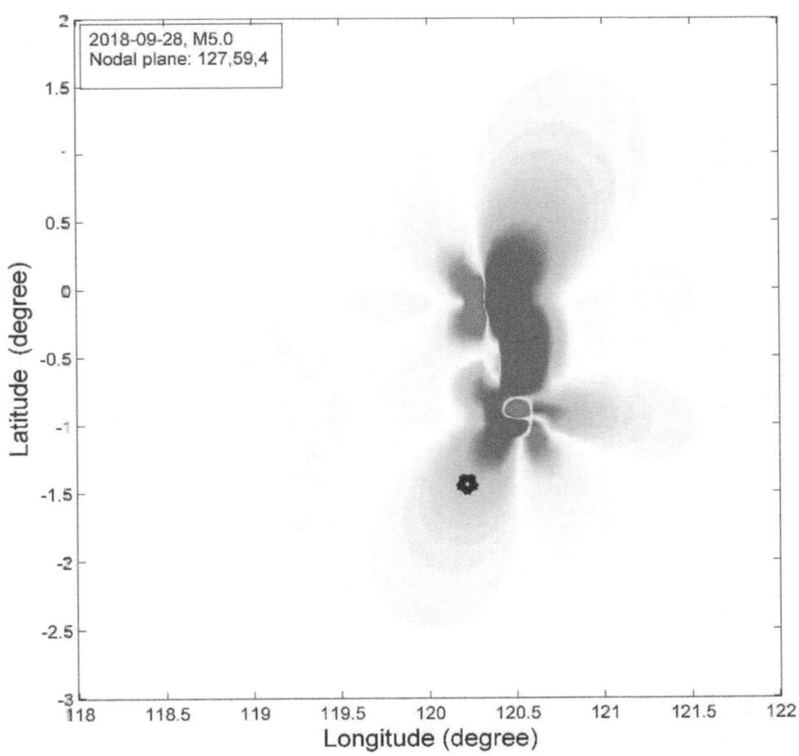

(a)

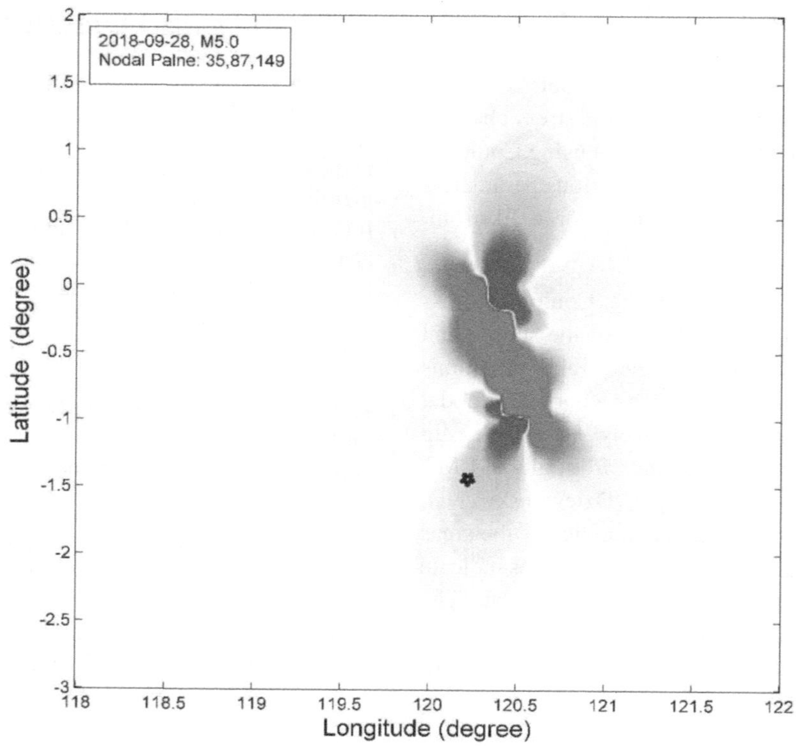
2018-09-28, M5.0
Nodal Palne: 35,87,149
Latitude (degree)
Longitude (degree)
2
1.5
1
0.5
0
-0.5
-1
-1.5
-2
-2.5
-3
118
118.5
119
119.5
120
120.5
121
121.5
122

(b)

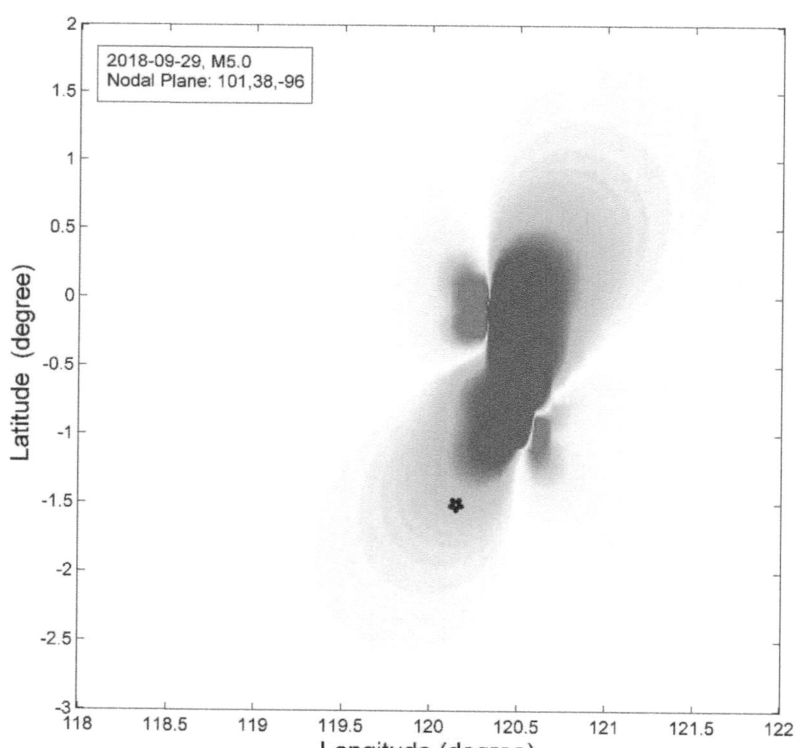
2018-09-29, M5.0
Nodal Plane: 101,38,-96
Latitude (degree)
Longitude (degree)
2
1.5
1
0.5
0
-0.5
-1
-1.5
-2
-2.5
-3
118
118.5
119
119.5
120
120.5
121
121.5
122

(c)

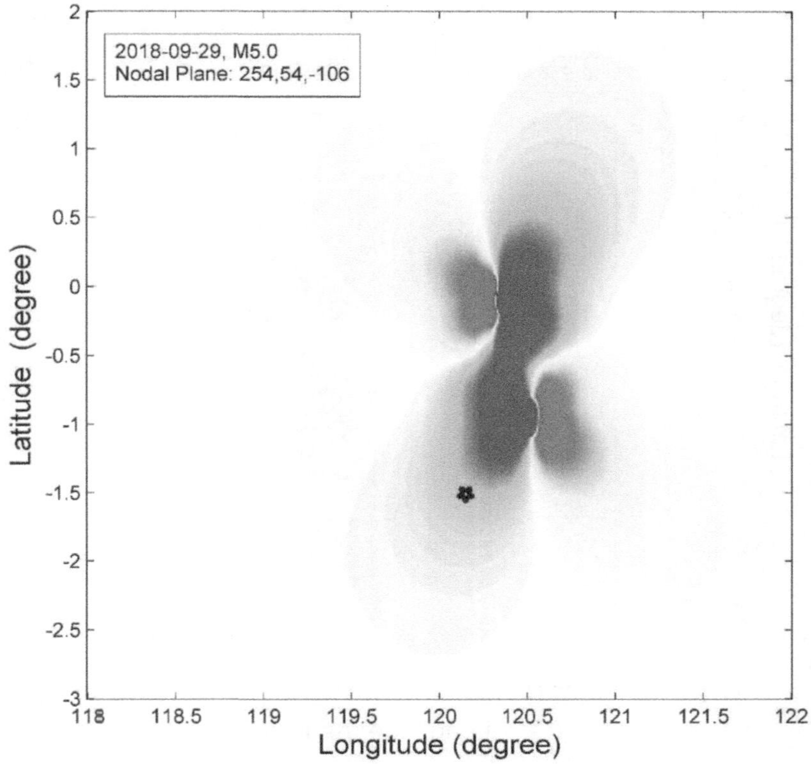
2018-09-29, M5.0
Nodal Plane: 254,54,-106
Latitude (degree)
Longitude (degree)
2
1.5
1
0.5
0
-0.5
-1
-1.5
-2
-2.5
-3
118
118.5
119
119.5
120
120.5
121
121.5
122

(d)

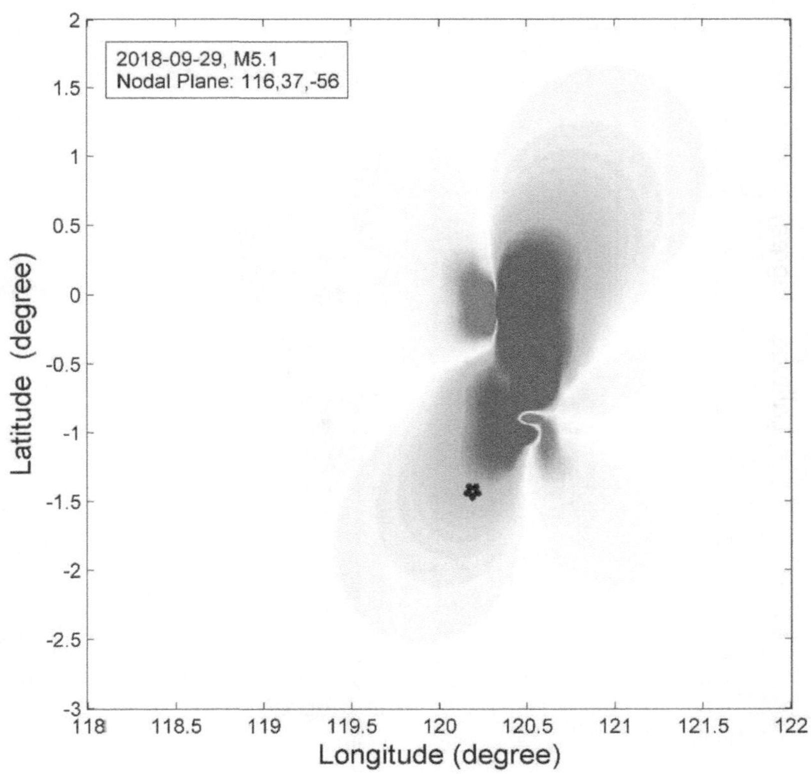
2018-09-29, M5.1
Nodal Plane: 116,37,-56
Latitude (degree)
Longitude (degree)
2
1.5
1
0.5
0
-0.5
-1
-1.5
-2
-2.5
-3
118
118.5
119
119.5
120
120.5
121
121.5
122

(e)

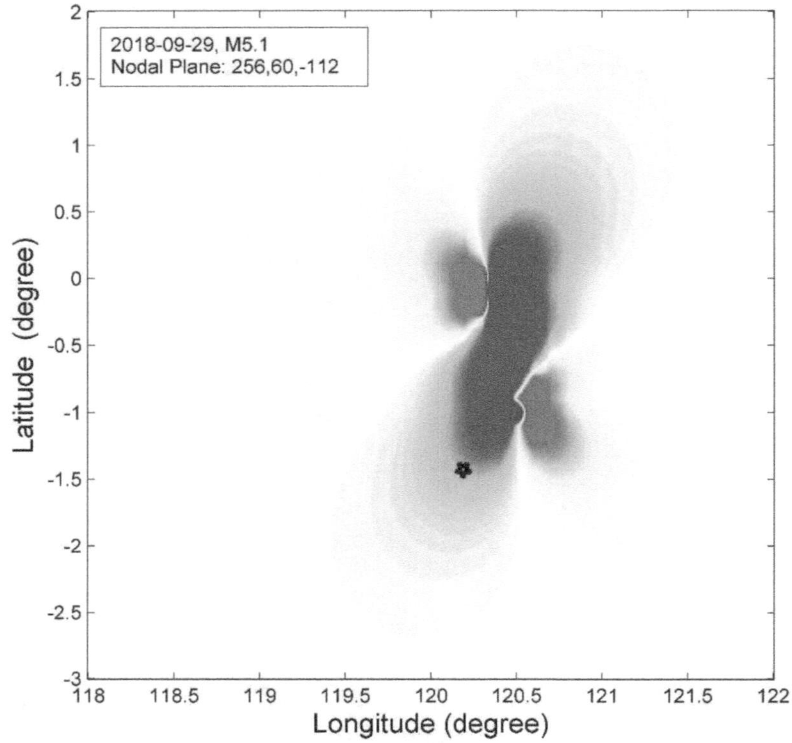
2018-09-29, M5.1
Nodal Plane: 256,60,-112
Latitude (degree)
Longitude (degree)
2
1.5
1
0.5
0
-0.5
-1
-1.5
-2
-2.5
-3
118
118.5
119
119.5
120
120.5
121
121.5
122

(f)

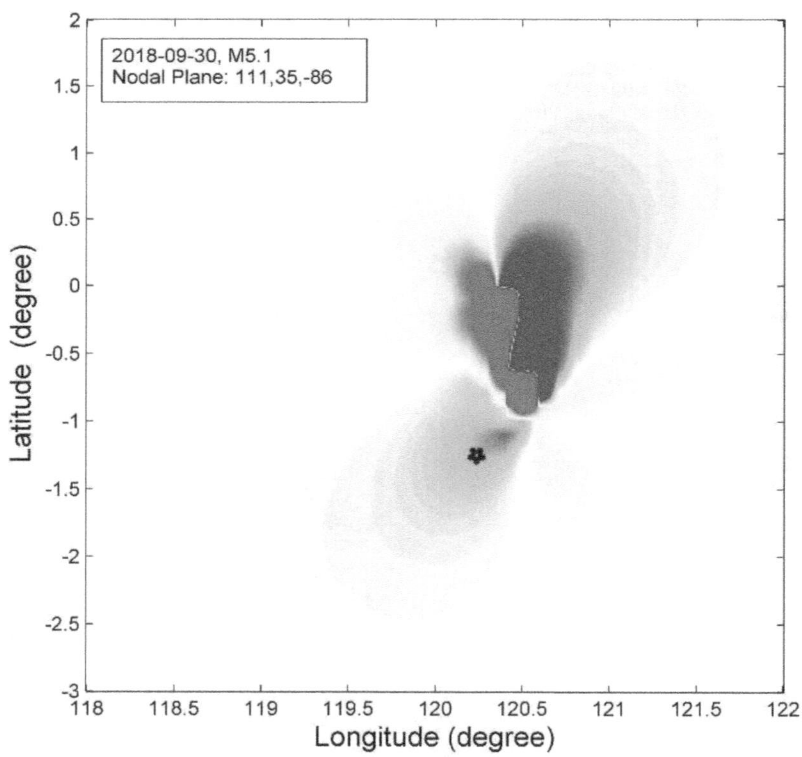
2018-09-30, M5.1
Nodal Plane: 111,35,-86
Latitude (degree)
Longitude (degree)
2
1.5
1
0.5
0
-0.5
-1
-1.5
-2
-2.5
-3
118
118.5
119
119.5
120
120.5
121
121.5
122

(g)

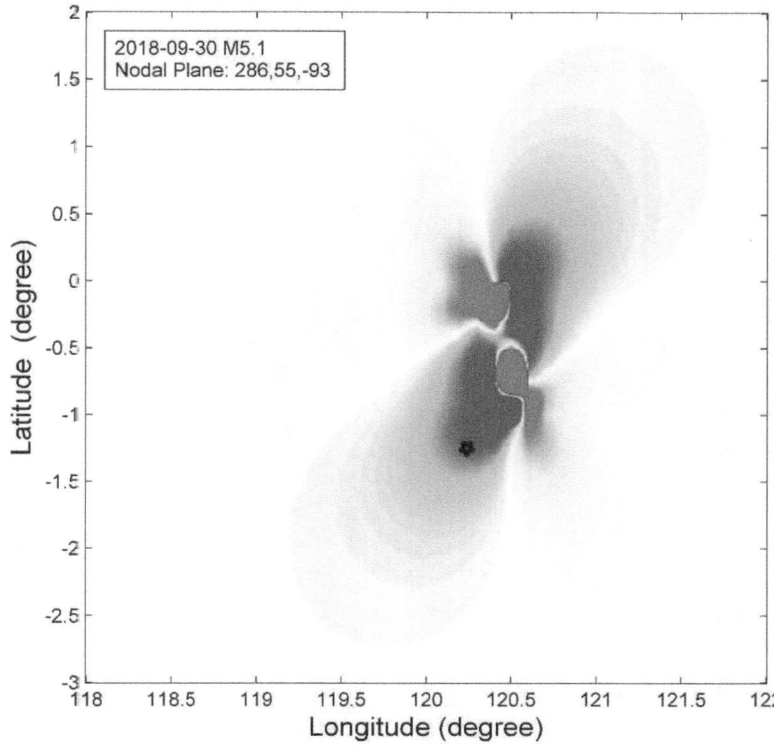
2018-09-30 M5.1
Nodal Plane: 286,55,-93
Latitude (degree)
Longitude (degree)
2
1.5
1
0.5
0
-0.5
-1
-1.5
-2
-2.5
-3
118
118.5
119
119.5
120
120.5
121
121.5
122

(h)

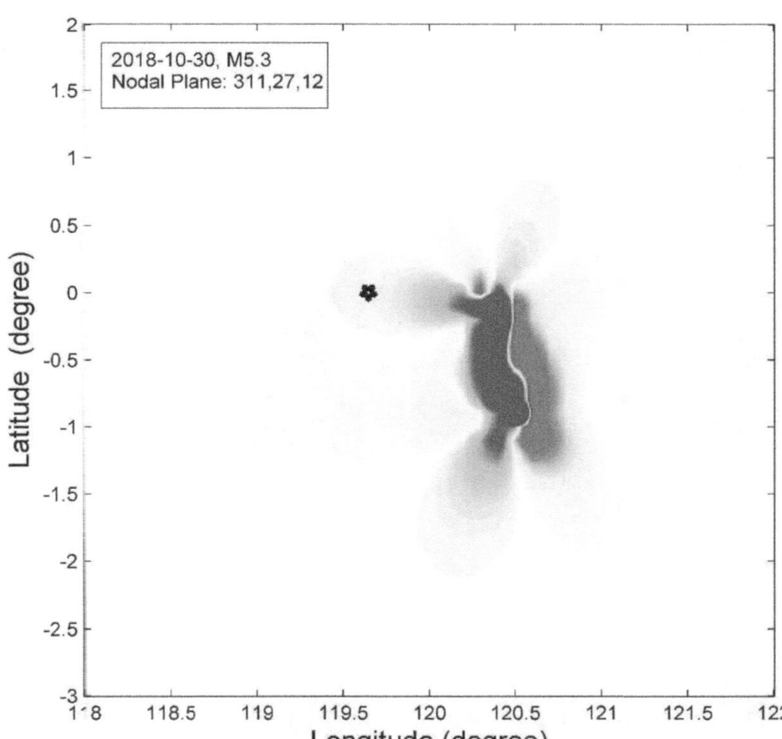
2018-10-30, M5.3
Nodal Plane: 311,27,12
Latitude (degree)
Longitude (degree)
2
1.5
1
0.5
0
-0.5
-1
-1.5
-2
-2.5
-3
118.5
119
119.5
120
120.5
121
121.5
122

(i)

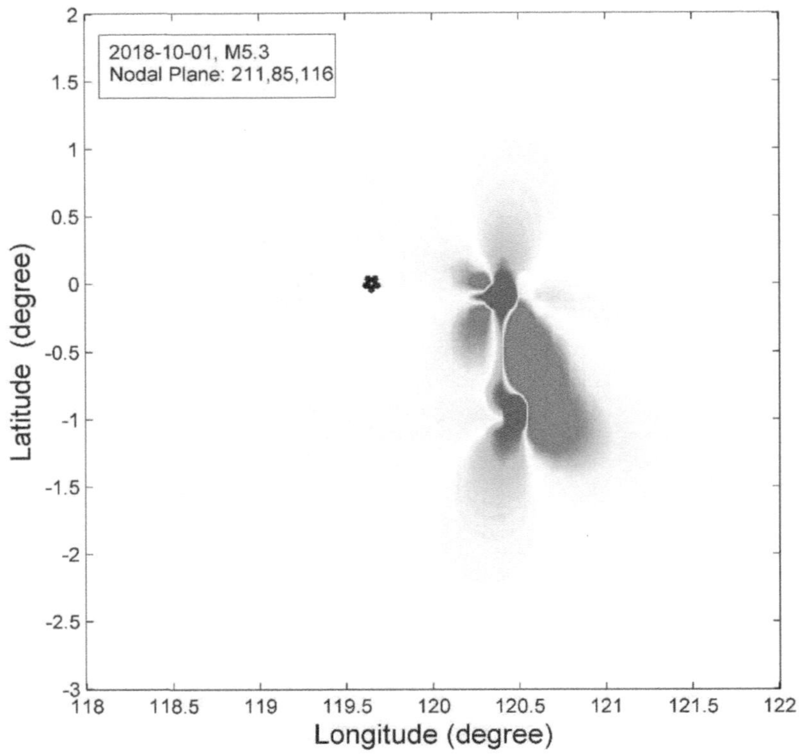
2018-10-01, M5.3
Nodal Plane: 211,85,116
Latitude (degree)
Longitude (degree)
2
1.5
1
0.5
0
-0.5
-1
-1.5
-2
-2.5
-3
118
118.5
119
119.5
120
120.5
121
121.5
122

(j)

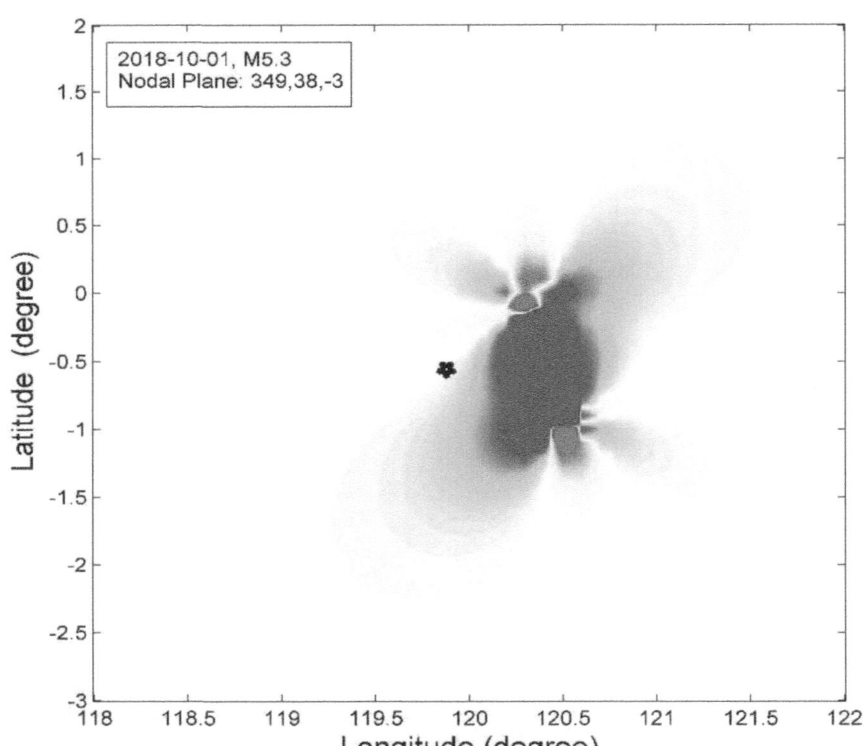
2018-10-01, M5.3
Nodal Plane: 349,38,-3
Latitude (degree)
Longitude (degree)
2
1.5
1
0.5
0
-0.5
-1
-1.5
-2
-2.5
-3
118
118.5
119
119.5
120
120.5
121
121.5
122

(k)

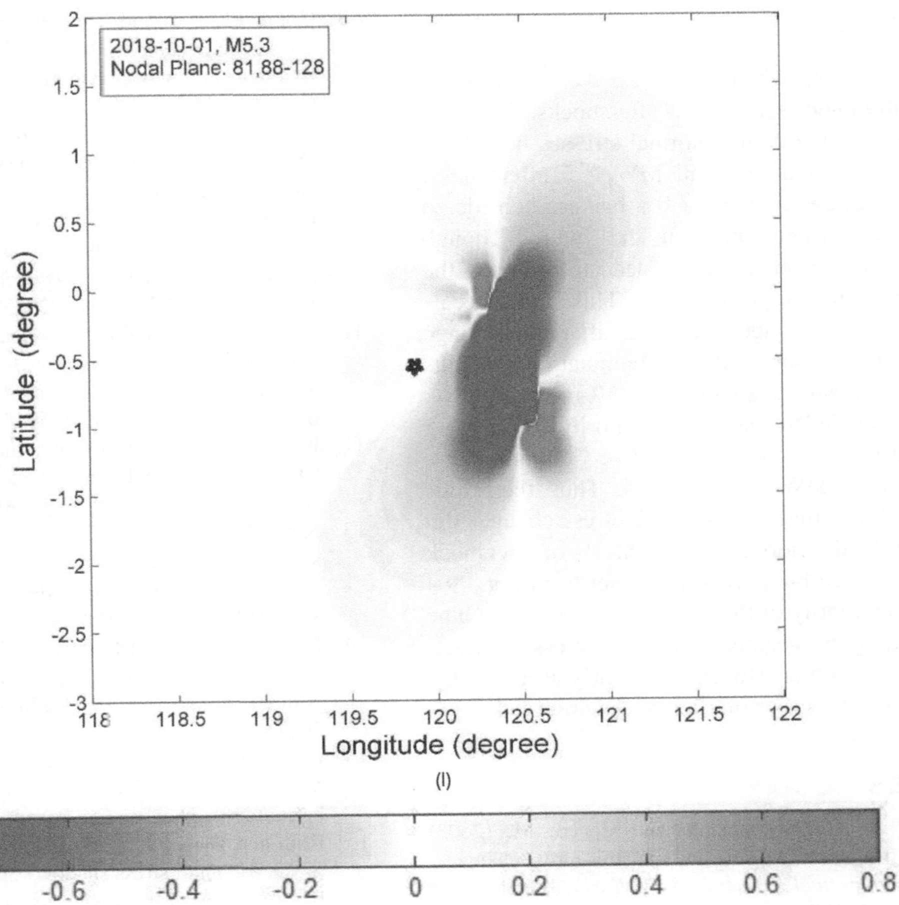

Fig. 49.4 (a-l) Coulomb Stress Change Illustration of MW ≥ 5 aftershocks on both nodal palnes as developed using Coulomb 3.3 software (Toda et al., 2011)

Table 49.1 Basic aftershock parameters obtained from Dziewonski et al. (1981) and Ekstrim et al. (2012); Stress change calculations computed using Coulomb 3.3 software (Toda et al., 2011)

Date	Aftershock	Latitude Longitude	Depth (Km)	Strike, Dip & Rake (°)	Stress Change (Bar)		
					Shear	Normal	Coulomb
28 Sept	M 5.0	–1.44	12	127, 59, 4	–0.129	–0.304	–0.251
		120.22		35, 87, 149	–0.13	–0.071	–0.159
29 Sept	M 5.0	–1.51	14.7	101, 38, –96	–0.185	–0.105	–0.227
		120.15		254, 54, –106	–0.161	–0.231	–0.254
29 Sept	M 5.1	–1.43	12	116, 37, –56	–0.22	–0.140	–0.278
		120.19		256, 60, –112	–0.221	–0.352	–0.361
30 Sept	M 5.1	–1.25	26.1	111, 35, –86	–0.307	–0.067	–0.334
		120.24		286, 55, –93	–0.189	–0.662	–0.695
1 Oct	M 5.3	0	15	311, 27, 12	–0.059	0.028	–0.048
		119.65		211, 85, 116	–0.058	0.193	0.019
1 Oct	M 5.3	–0.56	15	349, 38, –3	0.022	–0.076	–0.009
		119.88		81, 88, –128	0.022	0.034	0.036

5. Conclusion

It has been tested here whether the 7.6 MW mainshock triggered the subsequent series of 6 aftershocks. Also, the changes in Shear, Coulomb and Normal stresses have been recorded in the focal source of all MW ≥ 5 aftershocks. Coulomb stress changes as low as 0.1 bar are considered significant for triggering events in the future. Though few stress values are lesser, but could lead to failure if the particular fault is at the verge of failure. This study reveals that the Coulomb stress changes for the six aftershocks range from –0.695 to 0.036 bar. Only 2 computational results (around 17%) are positive, all others (83%) are negative. Hence it can be concluded that a vast majority of MW ≥ 5 aftershocks occurred in regions of negative Coulomb stress areas triggered by the MW 7.6 mainshock. Thus, the results suggest the possibility that the Coulomb stress changes did not contribute to the development of the series of aftershock events in the Palu-Koro fault. This is evident from Fig. 49.4 which indicates a majority of the epicentres lying in the blue areas which belong to negative coulomb stress changes. Further study needs to be performed to clearly ascertain the causative reason for the series of aftershocks observed

References

1. Belardinelli, M. E., Bizzarri, A. and Cocco, M. (2003) 'Earthquake triggering by static and dynamic stress changes', *Journal of Geophysical Research: Solid Earth*, 108(B3). doi: 10.1029/2002jb001779.
2. Bellier, O. *et al.* (2001) 'High slip rate for a low seismicity along the Palu-Koro active fault in Central Sulawesi (Indonesia)', *Terra Nova*, 13(6), pp. 463–470. doi: 10.1046/j.1365-3121.2001.00382.x.
3. Das, S. and Scholz, C. (1981) 'Off-fault aftershock clusters caused by shear stress increase?', *Bulletin of the Seismological Society of America*, 71(5), pp. 1669–1675.
4. Dziewonski, A. M., Chou, T. A. and Woodhouse, J. H. (1981) 'Determination of earthquake source parameters from waveform data for studies of global and regional seismicity.', *Journal of Geophysical Research*, 86(B4), pp. 2825–2852. doi: 10.1029/JB086iB04p02825.
5. Ekström, G., Nettles, M. and Dziewoński, A. M. (2012) 'The global CMT project 2004-2010: Centroid-moment tensors for 13,017 earthquakes', *Physics of the Earth and Planetary Interiors*, 200– 201, pp. 1–9. doi: 10.1016/j.pepi.2012.04.002.
6. Encyclopædia Britannica (no date) *Tectonic Map of The World, Britannica*. Available at: https://www.britannica.com/science/plate-tectonics/media/463912/172045 (Accessed: 11 October 2018).
7. Gahalaut, V. K. (2006) '2005 Kashmir earthquake: Not a Kashmir Himalaya seismic gap event', *Current Science*, 90(4), pp. 507–508.
8. Göran Ekström and Meredith Nettles (2006) 'Global CMT Catalogue', *Lamont-Doherty Earth Observatory (LDEO)*. Available at: www.globalcmt.org (Accessed: 2 January 2019).
9. Gusiakov, V. K. (2005) 'Historical tsunami database for the pacific, 47 B.C to present.', *Novosibirsk Tsunami Laboratory (NTL)*. Available at: http://tsun.sscc.ru/htdbpac/ (Accessed: 8 November 2018).
10. Hainzl, S., Steacy, S. and Marsan, D. (2010) 'Seismicity models based on Coulomb stress calculations', *Community Online Resource for Statistical Seismicity Analysis*, (November), pp. 1–25. doi: 10.5078/corssa-32035809.
11. Ishibe, T. *et al.* (2011) 'Correlation between Coulomb stress changes imparted by large historical strike-slip earthquakes and current seismicity in Japan', *Earth, Planets and Space*, 63(3), pp. 301–314. doi: 10.5047/eps.2011.01.008.
12. King, G. C. P. (2007) 'Fault Interaction, Earthquake Stress Changes, and the Evolution of Seismicity', in *Treatise on Geophysics*. doi: 10.1016/B978-044452748-6.00069-9.
13. King, G. C. P., Stein, R. S. and Jian Lin (1994) 'Static stress changes and the triggering of earthquakes', *Bulletin - Seismological Society of America*, 84(3), pp. 935–953. doi: 10.1016/0148- 9062(95)94484-2.
14. Maulana, A. (2018) *Geology of Sulawesi Island, WordPress.com*. Available at: https://adimaulana97.files.wordpress.com/2012/08/tectonic-setting2.jpg (Accessed: 22 August 2021).
15. *Palu Mainshock and Aftershocks* (2018) *Sciences, Indonesia Institute of*. Available at: http://lipi.go.id/public/uploads/berita/berita_palu.jpg (Accessed: 27 August 2021).
16. Ruth and Harris, A. (1998) 'Introduction to special section : Stress triggers , stress shadows , and implications for seismic hazard', *Journal of Geophysical Research*.
17. Scholz, C. H. (2002) *The Mechanics of Earthquakes and Faulting*. Cambridge University Press. doi: 10.1017/CBO9780511818516.
18. Steacy, S., Gomberg, J. and Cocco, M. (2005) 'Introduction to special section: Stress transfer, earthquake triggering, and time-dependent seismic hazard', *Journal of Geophysical Research: Solid Earth*. doi: 10.1029/2005JB003692.
19. Stein, R. S. (1999) 'The role of stress transfer in earthquake occurrence', *Nature*. doi: 10.1038/45144.
20. Stein, R. S., King, G. C. P. and Lin, J. (1992) 'Change in failure stress on the San Andreas and surrounding faults caused by the 1992 M=7.4 Landers earthquake', *Science*, 258(November), pp. 1328–1332.
21. Toda, Shinji, Stein, R. S., Sevilgen, Volkan, and Lin, Jian, 2011, Coulomb 3.3 Graphic-rich deformation and stress-change software for earthquake, tectonic, and volcano research and teaching—user guide: U.S. Geological Survey Open-File Report 2011–1060, 63 p., available at https://pubs.usgs.gov/of/2011/1060/.
22. Yadav, R. B. S. *et al.* (2012) 'Tectonic implications and seismicity triggering during the 2008 Baluchistan, Pakistan earthquake sequence', *Journal of Asian Earth Sciences*. Elsevier Ltd, 45, pp. 167–178. doi: 10.1016/j.jseaes.2011.10.003.

CHAPTER 50

Sedimentation Modeling of 2004 Indian Ocean Tsunami: Towards Early Warning Strategy

Abdullah Ansari
Research Scholar, Department of Civil Engineering,
Indian Institute of Technology Delhi, New Delhi – 110016
cez188391@iitd.ac.in

Javed N Malik
Professor, Department of Earth Sciences,
Indian Institute of Technology Kanpur, Kanpur – 208016
javed@iitk.ac.in

Shivam Tripathi
Professor, Department of Civil Engineering,
Indian Institute of Technology Kanpur, Kanpur – 208016
shiva@iitk.ac.in

Abstract

Natural hazard is a series of calamitous events that leads to quashing of lives in millions and mass destruction of properties in billions. These events are earthquakes, tsunamis, floods, landslides etc., which just takes only few minutes or hours to create such devastation that can affect countries at all fronts. It leads to socioeconomic breakdown and environmental degradation in that region. Thus, these catastrophic events require extensive research to figure out its nature, how it gets generated, its affect spatially-temporally and mitigation factors, which need to be considered. In the last few decades, Indian sub- continent experienced several catastrophic earthquakes. Sumatra Andaman earthquake (Mw 9.3) occurred on December 26th, 2004 resulted into the transoceanic tsunami event. This tsunami event directly affected the coastlines of Indian Mainland, Sri Lanka, Thailand, Indonesia, Maldives and, killing about 280,000 people. Transportation of sediments during tsunami flow cause either erosion or depositions of sediments. In this study, the tsunami sedimentation modeling technique was applied to 2004 tsunamigenic earthquake to estimate the tsunami flow height, flow speed and energy of tsunami waves. This study will help in formulation and design of mitigation measures with the early warning systems, which may eventually lead to less loss of life and damages to the property in seismically active coastal areas.

Keywords: Modeling, Indian Ocean Tsunami, tsunami flow speed, tsunami hazard, mitigation

1. Introduction

Natural hazard events are earthquakes, tsunamis, floods, landslides etc., which just takes only few minutes or hours to cause quashing of lives in millions and mass destruction of properties in billions resulting into the social-economic breakdown of any country. The 2001 Bhuj earthquake (Mw 7.6); 2004 Sumatra-Andaman earthquake (Mw 9.3); 2005 Muzaffarabad earthquake (Mw 7.6) and 2015 Gorkha earthquake (Mw 7.8) are the major earthquake events occurred in India. Among these all events, the 2004 event was most devastating occurred in Sumatra-Andaman Zone [1, 5, 6]. In Indian Ocean, there are chances of occurrence of many large magnitude earthquakes which are capable of generating tsunamis in the future [7].

It is very important to understand the ongoing tectonic activities and carryout the significant tasks in characterizing the tsunami waves. In present study, sedimentation modeling of tsunami waves carried out to estimate the tsunami flow speed and corresponding tsunami energy generated during 2004 Indian Ocean Tsunami. Furthermore tsunami hazard assessment is discussed to draw out the mitigation measures for future tsunamis in the Indian Ocean to protect coastal areas in terms of life and property.

2. Tsunami Sedimentation Modeling

Transportation of sediments during tsunami flow cause either erosion or depositions of sediments. As a result of spatial divergence in transport (more sediments leaving an area than coming) specifically due to lowering of flow speed and movement of sediments from bed into water column, erosion occurs. On the other hand, as a result of spatial convergence in transport (more sediments coming an area than leaving it) specifically due to falling of sediments in the water column to the bed, the process of deposition occurs [2]. Finally sediment settles out of suspension when flow speed drops to zero everywhere after tsunami flooding and narrow zone near the limit of inundation with neither erosion nor deposition is created by the tsunami flow speed dropping to a point where it is not able to transport the sediments. Width of these zones depend on the tsunami characterizations, sediments supply rate, the substrate over which the tsunami travels and topographical features of that area [3].

Tsunami flow speed calculated by matching the sediments suspended by the tsunami with the thickness and grain size distribution of tsunami sediment deposits at different locations of Andaman. The tsunami flow speed (Uz) can be calculated using following numerical integration [4].

$$U(z) = \int_{z_{o_{total}}}^{z} \frac{U_*^2}{K(Z)} dz \qquad (1)$$

In the above mentioned equation, $z_{o_{total}}$ and z are the bed roughness and elevation including the thickness of the tsunami sediments respectively. U^2 and K(Z) represent the shear velocity and eddy viscosity respectively.

After knowing about the tsunami flow speed, the next approach for hazard study is to estimate the tsunami energy at all defined locations. Tsunami energy is a parameter which directly gives an idea about the devastating nature of the tsunami waves at a particular location. The main reasons behind the variation of tsunami energy at different location could be due to local topography of the area, obstacles coming in the path of the high-speed tsunami waves. Most important point to be noted here is that the flow speed used to calculate the tsunami energy at desired locations was based on the depositional speed. Using the equations of hydrodynamics the corresponding tsunami energy can be estimated.

$$P = \frac{1}{2}\rho D\sqrt{gD}(U(z))^2 \qquad (2)$$

In the above Equation, P is tsunami energy along the shore line, D is the ocean depth, ρ is ocean water density and U(z) is the tsunami flow speed.

3. Result and Discussion

The tsunami sedimentation model is a mathematical model due to which lots of intermediate parameters like dimensionless particle size, eddy viscosity and sediment concentration etc. generated. Tsunami sediment samples collected from threes different sites of Andaman Island. To estimate the eddy viscosity, model ask for the kinematic viscosity of the ocean water. Kinematic viscosity based on dynamic viscosity and density of ocean water is calculated. This sedimentation model is run for all three locations to get

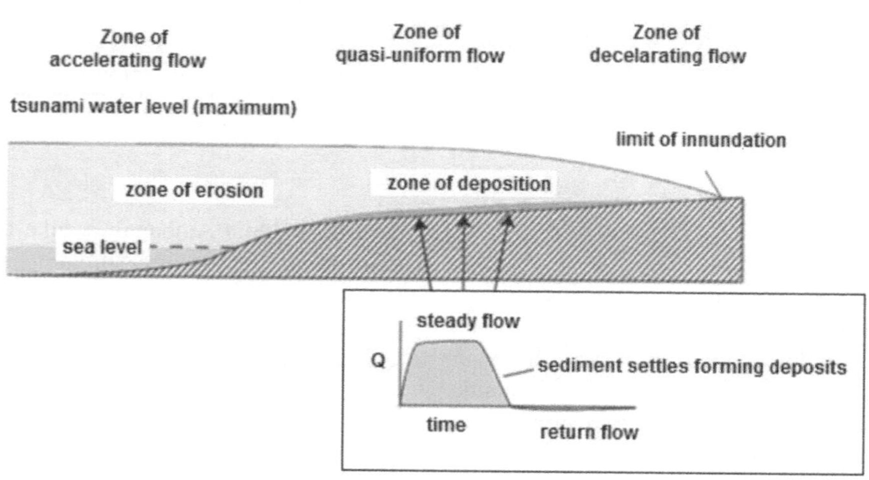

Fig. 50.1 Conceptual model of tsunami sedimentation modeling [2]

tsunami flow speed as final output. Samples collected from three different sites of Andaman labelled as Site-A, Site-B and Site-C. For each location particle size ranges between 63 μm to 2000 μm.

Tsunami sediments having particles sizes 63 μm, 125 μm, 150 μm, 180 μm, 250 μm, 355 μm, 500 μm, 710 μm and 2000 μm selected for modeling. For Site-A, the flow speed of tsunami for sediment size 63 μm,125 μm,150 μm, 180 μm, 250 μm, 355 μm, 500 μm, 710 μm and 2000 μm are 5.73 m/s, 5.91 m/s, 6.25 m/s, 6.42 m/s, 7.19 m/s, 7.28 m/s, 8.48 m/s, 8.96 m/s and 9.62 m/s respectively. Tsunami flow height for this location is 1.75 m. For Site-B, the flow speed of tsunami for particle size 63 μm, 125 μm, 150 μm, 180 μm, 250 μm, 355 μm, 500 μm, 710 μm and 2000 μm are 5.75 m/s, 5.98 m/s, 6.30 m/s, 6.38 m/s, 7.22 m/s, 7.25 m/s, 8.45 m/s, 8.92 m/s and 10.11 m/s respectively. For Site-C, the flow speed of tsunami for particle size 63 μm, 125 μm, 150 μm, 180 μm, 250 μm, 355 μm, 500 μm, 710 μm and 2000 μm are 5.71 m/s, 5.84 m/s, 6.28 m/s, 6.25 m/s, 7.20 m/s, 7.24 m/s, 8.15 m/s, 8.88 m/s and 9.12 m/s respectively. Tsunami flow height for Site-B and Site-C are 4.0 m and 3.0 m respectively. After estimating tsunami flow speed at three sites of Andaman, tsunami energy along shoreline calculated to plot the hazard contour line for coastal engineers to understand the level of damages due to upcoming tsunami. The average value of tsunami energy along the shoreline of Andaman is 3 MW/m.

Table 50.1 Tsunami flow data at different locations of Andaman [2]

Location	Grain Size (mm)	Tsunami Flow Height (m)	Average Flow Speed of Tsunami (m/s)	Tsunami Energy (MW/m)
Site A		1.75	7.32	3.155
Site B	63-2000	4.00	7.37	3.041
Site C		3.00	7.23	3.190

4. Conclusion

Population growth and industrialization in last few decades as well as lack of understanding of tsunami risk in seismically active coastal zones increased the invitation of catastrophic damages in terms of life and property. Now, it has become very important to learn and understand the tsunami hazard along the coastline of seismically active coastal cities. The sedimentation modeling based present study on tsunami deposits helped us to know about the tsunami wave height and tsunami flow speed at different locations of Andaman. Which further gives an idea about the energy of the tsunami waves on shoreline of Andaman. If a transoceanic tsunami occurs in Indian Ocean having epicenter near Andaman Island, it will cause major devastation of south eastern coast of Mainland India also. This study will help to formulate and design the mitigation measures with the early warning systems, which may eventually lead to less loss of life and damages to the property.

References

1. Ansari, A., Satake, K. and Malik, J. N.: Modelling the 2004 Indian Ocean Tsunami to estimate tsunami heights and its amplitude and to study its effects on coastal areas. ERI Earthquake Conference, University of Tokyo, Japan (2017).
2. Ameen, A. A. M. M.: Modeling the 2004 Andaman-Sumatra tsunami and historical tsunamis from Andaman and Nicobar Island: towards estimation of tsunami hazard along the adjoining areas of Indian Ocean. M.Tech. Thesis, Indian Institute of Technology Kanpur, India (2018).
3. Gelfenbaum, G. and Jaffe, B.: Erosion and sedimentation from the 17 July, 1998 Papua New Guinea tsunami. Pure and Applied Geophysics, 160 (10-11), 1969–1999 (2003).
4. Jaffe, B. E. and Gelfenbuam, G.: A simple model for calculating tsunami flow speed from tsunami deposits. Sedimentary Geology, 200 (3-4), 347–361 (2007).
5. Malik, J. N., Banerjee, C., Khan, A., Johnson, F. C., Shishikura, M., Satake, K. and Singhvi, A. K.: Stratigraphic evidence for earthquakes and tsunamis on the west coast of South Andaman Island, India during the past 1000 years. Tectonophysics, 661: 49–65 (2015).
6. Satake, K. and Atwater, B. F.: Long-term perspectives on giant earthquakes and tsunamis at subduction zones. Annu. Rev. Earth Planet. Sci., 35, 349–374 (2007).
7. Singh, A. P., Murty, T. S., Rastogi, B. K. and Yadav, R. B. S.: Earthquake generated tsunami in the Indian Ocean and probable vulnerability assessment for the east coast of India. Marine Geodesy, 35 (1), 49–65 (2012).

CHAPTER 51

Forest Fire Risk Zone Mapping in Parambikulam Tiger Reserve, Western Ghats, Kerala, India Using Geospatial Techniques and Fire Risk Index (FRI) Method

Shalu George
Department of Environmental Science, School of Earth Science Systems,
Central University of Kerala, Tejaswini Hills, Periya, Kasaragod – 671 316,
Kerala, India
shalu.george.963@gmail.com
9995561438

Kripa K.
Department of Environmental Science, School of Earth Science Systems,
Central University of Kerala, Tejaswini Hills, Periya, Kasaragod – 671 316,
Kerala, India

Anbazhagi S.
Department of Environmental Science, School of Earth Science Systems,
Central University of Kerala, Tejaswini Hills, Periya, Kasaragod – 671 316,
Kerala, India

Muthukumar Muthuchamy
Department of Environmental Science, School of Earth Science Systems,
Central University of Kerala, Tejaswini Hills, Periya, Kasaragod – 671 316,
Kerala, India
Corresponding Author: mmuthukumar@cukerala.ac.in

Abstract

Fires are a predominant cause of forest degradation and significantly affect resources, habitats, and biodiversity. The previous fire occurrence data show that the present study area Parambikulam Tiger Reserve, a part of Southern Western Ghats, is also prone to fires, where 48 fire incidences were recorded during the last decade. This present study aims to demarcate the fire risk zones in Parambikulam Tiger Reserve using geospatial techniques and the Fire Risk Index method. Factors such as land cover type, Normalized difference vegetation index, Land surface temperature, temperature, rainfall, wind speed, slope, aspect, drainage, elevation, distance from the road, distance from the settlement, and previous fire incidence points were selected for the mapping process. The final map was categorized into five zones: very low, low, moderate, high, and very high. The map is validated with previous fire incidence points and confirms that 66.66% of fire incidences fall in high and very high-risk zones, proving the methodology's reliability. The present methodology can be applied to similar areas, especially in Western Ghats regions. The generated maps will assist forest department officials in preventing or minimizing fire activities and taking appropriate measures when a fire occurs, resulting in improved forest fire management.

Keywords: Fire risk index; risk zones; forest fire; Geospatial techniques; Western Ghats

1. Introduction

The relatively undisturbed nature of the protected areas limits the broader utilization of the higher diversity of species that are difficult to thrive everywhere. Even though, under high anthropogenic pressure and climate change, these protected areas are susceptible to widespread depletion of forested areas. Forest fires are the most prevalent cause of forest destruction and have a significant environmental impact. Both natural and man-made reasons spark fires. Whether natural or man-made, fires frequently devastate larger areas of the forest. The severity of the fire, which substantially interrupts the equilibrium of vegetation and soil parameters in a post-fire environment, leads to land degradation (Escuin et al., 2008; Veraverbeke et al., 2010). Studies show that understory loss, vegetation loss, runoff, soil erosion, loss of organic matter, and soil moisture content are higher in regions with high fire severity than in areas that have not been burnt (Keeley, 2009; Cocke et al., 2005). Frequent fires in India's forests have deteriorated the forest; nevertheless, the extent of the overall impact is undefined (Ambat et al., 2006). Many of the protected areas in the Western Ghats are susceptible to fire and have burnt in the past. Researchers have performed numerous studies on forest fire risk zone mapping in the Western Ghats to identify forest fire risk zones by evaluating relevant factors using geospatial approaches (Sowmya and Somashekar, 2010; Ajin et al., 2015; Verma et al., 2015; Vinod et al., 2016). In risk evaluations, they efficiently used cost-effective and time-saving remote sensing and GIS techniques. Most calamities might have been prevented if proper measures had been taken in advance. The factual judgement and decision enable future awareness for the protection and post-fire management efforts that help the ecosystem and vegetation regeneration in impacted regions. Thus, the objectives of the present study are set to delineate the fire risk Zones in Parambikulam Tiger Reserve (PKTR) by considering the factors of land cover type, NDVI, LST, temperature, rainfall, wind speed, elevation, slope, aspect, distance from settlements, distance from roads, distance from previous fire incidence points, and distance from drainage using geospatial techniques and fire risk index (FRI) method.

2. Study Area

Parambikulam Tiger Reserve (PKTR) lies between 10° 15′ - 10° 35′ North latitudes and 76° 30′ - 76° 55′ East longitudes (Fig. 51.1). The Tiger Reserve is located in the Palakkad district of Kerala, India, and spans 643.662 km^2, with a core area of 390.89 km^2 and a buffer zone of 252.772 km^2. PKTR was declared a Tiger Reserve in 2010; earlier, it was a wildlife sanctuary. Four ranges of the Reserve are Orukomban, Parambikulam, Karimala, and Sungam. Kadar, Malasar, Malamalasar, and Muthuvan are the indigenous communities that live within the tiger reserve. Evergreen forest, semi-evergreen forest, moist deciduous forest, dry deciduous forest, grasslands, vayals, plantation, patches of bamboo and reed brakes are the vegetation types found in the tiger reserve.

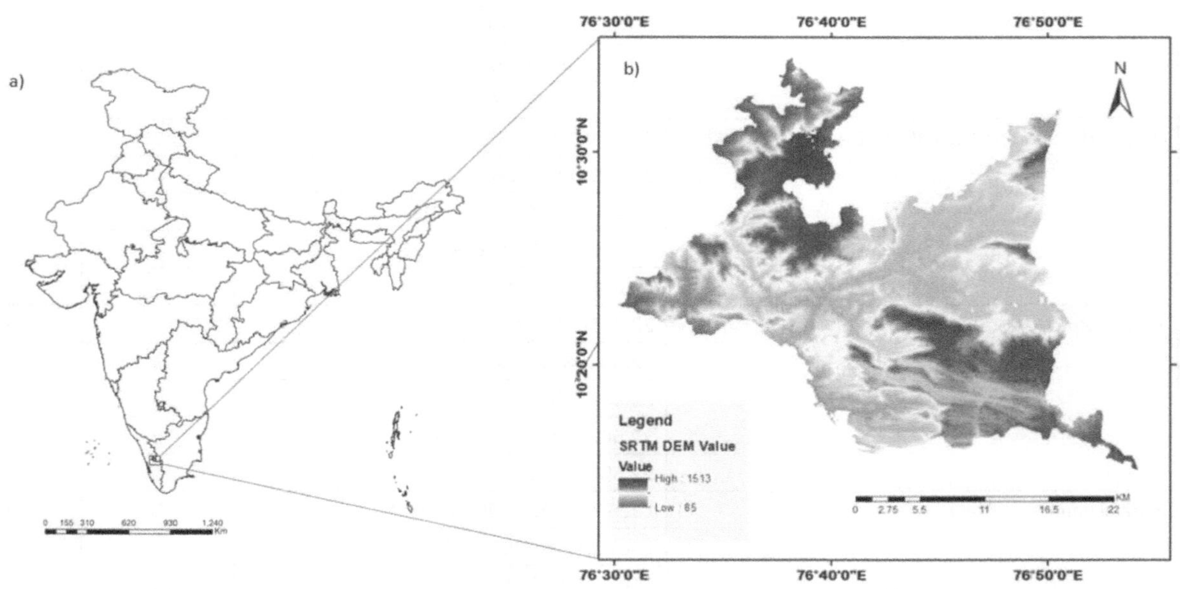

Fig. 51.1 (a) Map of India highlighting Parambikulam Tiger Reserve, (b) Digital Elevation Model of Parambikulam Tiger Reserve

2. Materials and Methods

For forest fire risk zone mapping, gathering and compiling forest fire causative factors are crucial and necessary stages. The entire study can be broken down into four sections: 1) obtaining satellite and auxiliary data, 2) gathering of historical fire incidents, 3) preparing the thematic map layers using remote sensing and GIS, generating the forest fire risk zone map, and 4) validating the map. India boundary shape file downloaded from DIVA GIS (www.divagis.org/gdata). The boundary shape file of PKTR was obtained from the concerned forest department.. Map layers such as DEM, slope, aspect, and drainage were generated from the SRTM DEM datasets using Arc GIS 10. 4. 1. The 10 m contour data were used as base data. The vegetation map and LST map were created using Landsat 8 OLI & TIRS data in the platform Arc GIS 10. 4. 1. The land cover type map was prepared using ERDAS Imagine 2013 software with Bhuvan and Google Earth as reference materials. Here combined classification was used to obtain the map. Natural Breaks (Jenks) method was used to reclassify the prepared thematic map layers. Each thematic map layer class was given a rank and weight depending on its sensitivity or ability to cause a fire. The auxiliary datasets collected for this study are the details of road networks, human settlements, drainage, rainfall, wind speed, and temperature. Road networks and human settlements were digitized from Google Earth. ArcGIS spatial analyst tools were used to prepare the distance from roads, settlements, and drainage. Rainfall, wind speed, and temperature data were obtained from the CHRS portal, Climatology Lab, and NASA Power data access, respectively. Previous fire information from 2008 to 2017 was obtained from the Fire Information for Resource Management System Moderate Resolution Imaging Spectroradiometer (FIRMS MODIS) data of NASA in shapefile as a point feature. The Fire Risk Index (FRI) method was used to prepare the study area's forest fire risk zone map. The fire index was calculated by multiplying the weight and rank, i.e., Fire Index = Weight x Rank. The details of weight, and rank are given in Table 51.1. The forest fire risk zone map was created by

Table 51.1 Rank and Weight assigned for different causative factors of fire

Sl. No.	Class	Factor	Rank	Weight
1	Land cover type	Deciduous forest	5	9
		Plantation, Grassland	4	
		Evergreen, semi-evergreen forest	3	
		Waterbody	0	
2	Normalized difference vegetation index (NDVI)	0.628707477 - 0.813438118	5	8
		0.498309377 - 0.628707477	4	
		0.302712228 - 0.498309377	3	
		-0.006983259 - 0.302712228	2	
		-0.57204169 - -0.006983259	1	
3	Land surface temperature (LST) (Degree Celsius)	26.54702229 - 33.80762863	5	7
		24.23991373 - 26.54702228	4	
		22.67922264 - 24.23991372	3	
		21.25424382 - 22.67922263	2	
		16.50431442 - 21.25424381	1	
4	Temperature (Degree Celsius)	22.5870172 - 22.58927345	5	6
		22.58038112 - 22.58701719	4	
		22.57109061 - 22.58038111	3	
		22.5629946 - 22.5710906	2	
		22.55542946 - 22.56299459	1	
5	Rainfall (mm)	1,326.756592 - 1,401.607547	5	6
		1,401.607548 - 1,488.277075	4	
		1,488.277076 - 1,588.734936	3	
		1,588.734937 - 1,685.253273	2	
		1,685.253274 - 1,829.045898	1	

Sl. No.	Class	Factor	Rank	Weight
6	Windspeed (m/s)	1.865865762 - 1.999970436	5	6
		1.776462645 - 1.865865761	4	
		1.644710683 - 1.776462644	3	
		1.512958721 - 1.644710682	2	
		1.400028467 - 1.51295872	1	
7	Slope (Degree)	33.4461465 - 77.53424072	5	5
		23.10824854 - 33.44614649	4	
		14.89874133 - 23.10824853	3	
		7.60140159 - 14.89874132	2	
		0.000006031 - 7.601401589	1	
8	Aspect	S, SW	5	5
		W	4	
		E, SE	3	
		NE	2	
		N, NW	1	
9	Distance from the settlement (m)	0-2202	5	4
		2202-4278	4	
		4278-6418	3	
		6418-9689	2	
		9689-16045	1	
10	Distance from road and trek paths (m)	0-1035	5	4
		1035-2341	4	
		2341-4233	3	
		4233-6890	2	
		6890-11483	1	
11	Distance from previous fire incidence points (m)	0-2220	5	3
		2220-4376	4	
		4376-6466	3	
		6466-10255	2	
		10255-16656	1	
12	Distance from drainage (m)	866-2026	5	2
		564-866	4	
		357-564	3	
		174-357	2	
		0-174	1	
13	Elevation (m)	1,013.000001 - 1,513	5	1
		817.0000001 - 1,013	4	
		641.0000001 - 817	3	
		414.0000001 - 641	2	
		85 - 414	1	

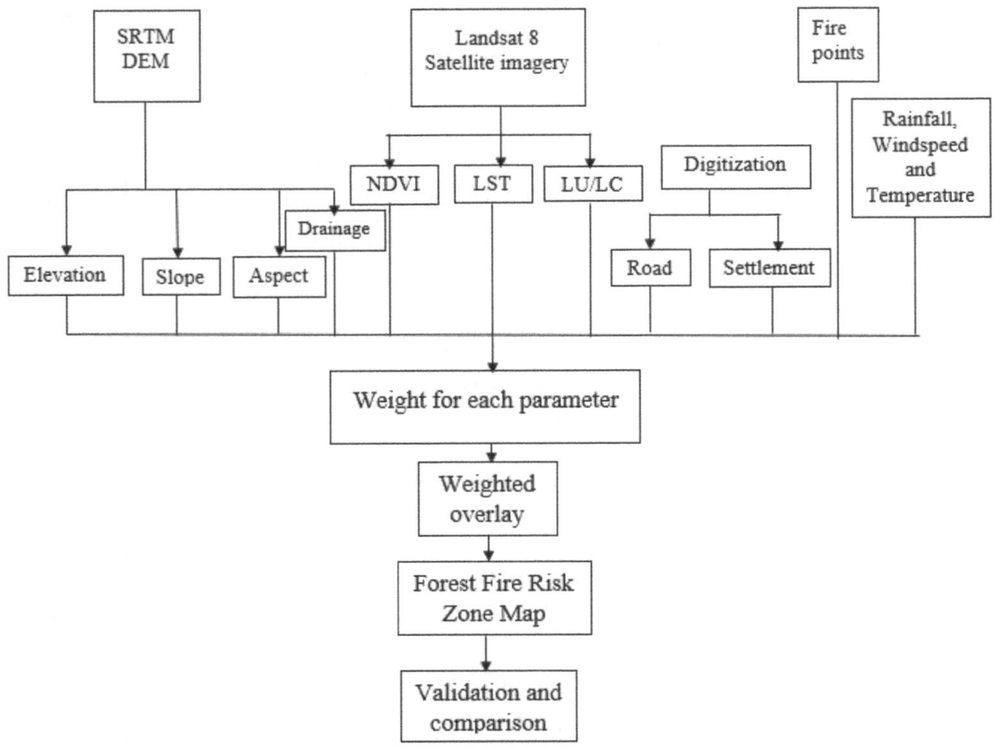

Fig. 51.2 Flow chart of the methodology adopted for the study

overlaying the map layers in Arc GIS. The final map was divided into five categories: very low, low, moderate, high, and very high. Finally, the generated fire risk zone map was validated with MODIS's previous fire incidence data for the past ten years. Fig. 51.2 shows the methodology adopted for this study.

3. Results and Discussion

A forest fire is initiated and spread by several factors, each playing a significant role. Land cover type, NDVI, LST, temperature, rainfall, wind speed, elevation, slope, aspect, distance from settlements, distance from road and trek paths, distance from previous fire incidence points, and distance from drainage were all taken into consideration in the current study on forest fire risk assessment. The following section briefly discusses their influences on fire.

3.1 Land Cover Type

The type of land cover is one of the most important elements influencing the occurrence and spread of forest fires. Forest fires occur in areas with dry vegetation than in areas with moist vegetation. Thus, the deciduous forest in PKTR receives a higher ranking because they are the areas where dry vegetation is predominant and are more susceptible to fire. In addition to this, the grassland and the plantation also receive a high ranking. Fig. 51.3 depicts the maps that reflect the various forms of land cover.

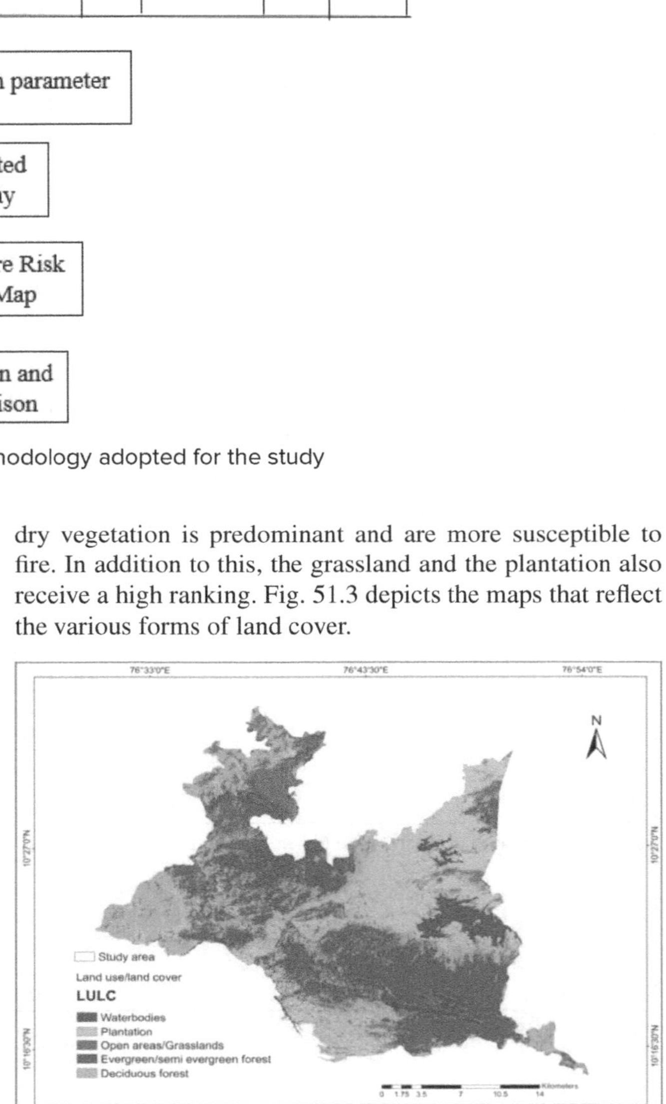

Fig. 51.3 Land cover type

3.2 Normalized Difference Vegetation Index (NDVI)

The biomass available for burning is represented by the NDVI, which gauges plant density (Yathish et al., 2019).

PKTR's NDVI value ranges from -0.57 to 0.81. The positive value denotes thick vegetation and, therefore, heavier fuel load. The negative value shows the water body. The NDVI of the study area was divided into five classes. The areas with the higher positive NDVI values received the highest ranking. Fig. 51.4 shows the NDVI map.

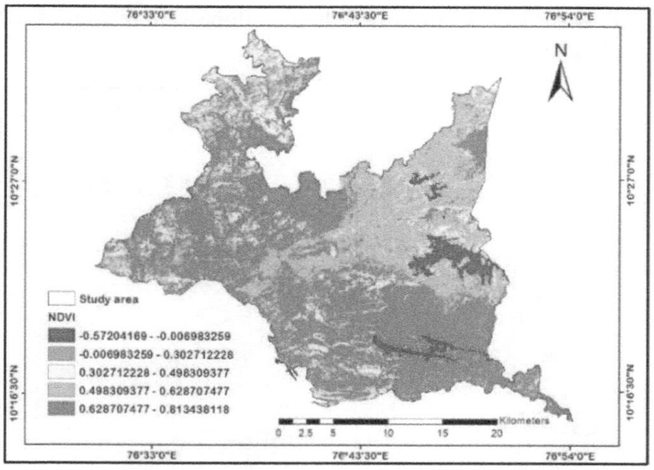

Fig. 51.4 Normalized Difference Vegetation Index

3.3 Land Surface Temperature (LST)

Higher temperatures decrease moisture content, which assists in plant combustion; therefore, LST was expected to be a significant determinant in fire hazards (Bisquert et al., 2012). The LST of this area was grouped into five classes, and the highest rank was given to areas with higher LST values. The LST map is provided in Fig. 51.5.

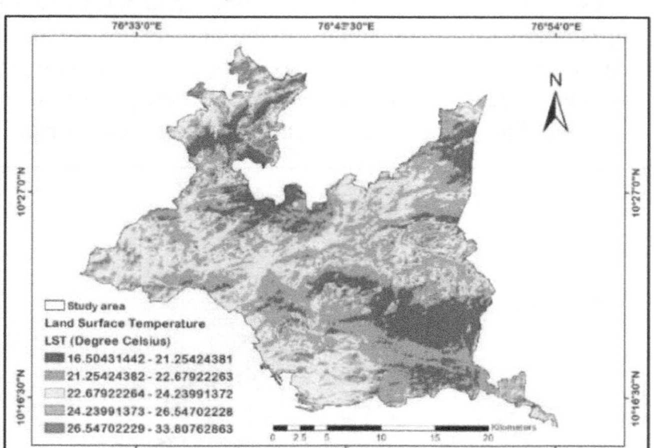

Fig. 51.5 Land Surface Temperature

3.4 Temperature

The temperature has a huge influence on how fire behaves. Temperature and moisture content are linked. High temperatures lower the moisture content of fuels and make them drier; this warmer temperature and fuel's drier condition assist in igniting and spreading fires (Flannigan et al., 2016). PKTR temperature data for ten years (2009-2018) were collected, interpolated and grouped into five classes. The highest rank was given to areas with higher temperature values. The temperature map is shown in Fig. 51.6.

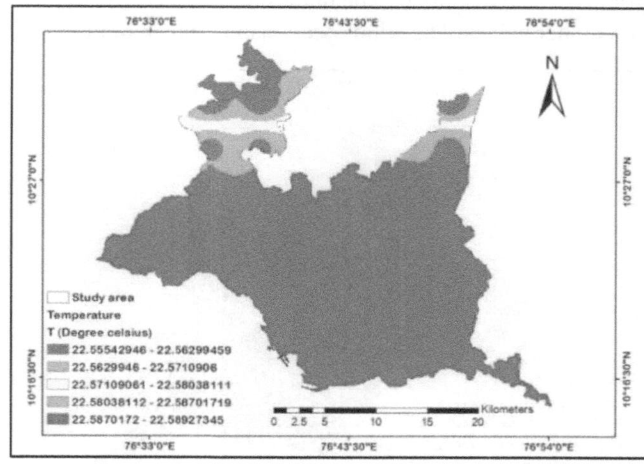

Fig. 51.6 Temperature

3.5 Rainfall

Rainfall and forest fires have an inverse connection. Lower rainfall lowers the moisture content of the fuels, causing plants to dry up, become more combustible, and raise the risk of forest fires (Mondal and Sukumar, 2016). Rainfall data of PKTR for ten years (2009-2018) were obtained, interpolated, and the data were grouped into five classes. The highest rank was given to areas with lower rainfall values. The rainfall map is shown in Fig. 51.7.

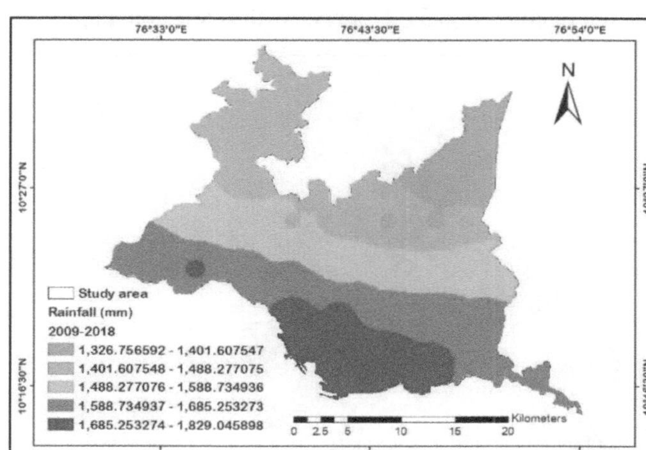

Fig. 51.7 Rainfall

3.6 Windspeed

The fire spread has a positive correlation with wind speed. The forest fire's broad damage and spread assisted by the combined effects of steeper topography, hot summer temperatures, and excellent wind velocity (Kanga et al., 2017). Windspeed data of PKTR for ten years (2009-2018) was obtained, interpolated, and the data of this area were grouped into five classes. The highest rank was given to areas with higher values. The wind speed map is shown in Fig. 51.8.

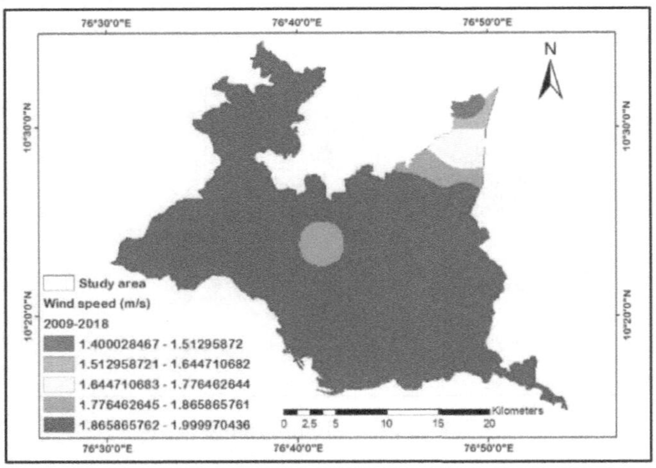

Fig. 51.8 Wind speed

3.7 Elevation

At higher elevations, fires are more likely to occur. Higher elevations have far more favourable conditions for lightning strikes than the lowlands (Saxena, 2012). Furthermore, humidity and temperature significantly influence higher elevations (Thompson et al., 2000). The study area was divided into five classes based on elevation. As per the observations, regions with greater elevation values received the top ranking. Fig. 51.9 represents the elevation map.

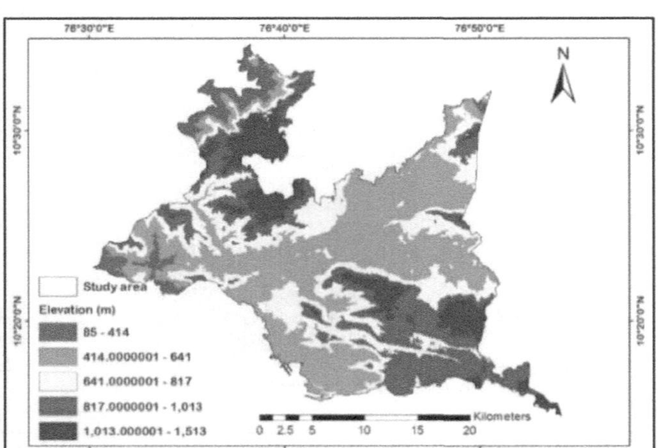

Fig. 51.9 Elevation

3.8 Slope

The degree of inclination or steepness of the land surface is termed the slope. It's another crucial component that influences the spread of fire. On upslope, fires spread much quicker than downslope (Rothermel et al., 1972). The slope of the study area was grouped into five classes, and the highest rank was given to areas with higher slope values. Fig. 51.10 depicts the slope map.

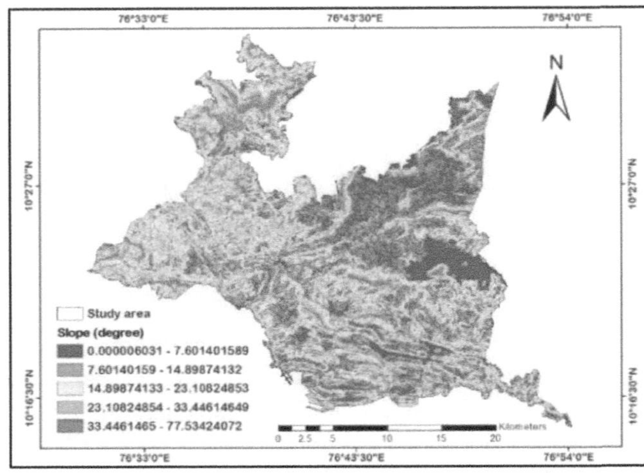

Fig. 51.10 Slope

3.9 Aspect

A terrain's aspect is defined as the direction of the slope toward the exposure to the sun's rays (Setiawan et al., 2004). In the northern hemisphere, the south and south-western slopes are more fire-prone than the northern and north-eastern slopes because they are inclined to the sun and receive more direct heat (Ambat et al., 2006). So, in this study south aspect has been assigned higher values and the north aspect with less value. Fig. 51.11 is a representation of the aspect map.

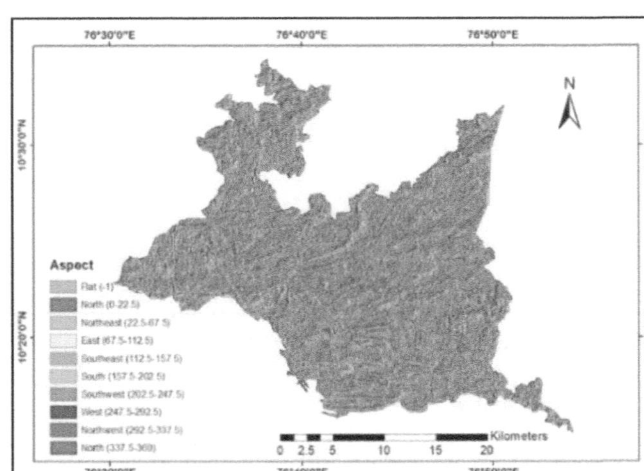

Fig. 51.11 Aspect

3.10 Distance from Settlements

Human settlements near the forest are the major element posing a threat to the forest. Forest fires can be started by individuals who are connected to the forest in two different ways: intentionally or unintentionally. The most common causes of forest fires include throwing lighted cigarette butts or match sticks, tourism, campfires, cooking activities, and the collection of non-timber forest products. Based on the distance from settlements, the study area was grouped into five classes, and the highest rank was assigned to areas closer to settlements. Fig. 51.12 depicts the distribution of these different classes.

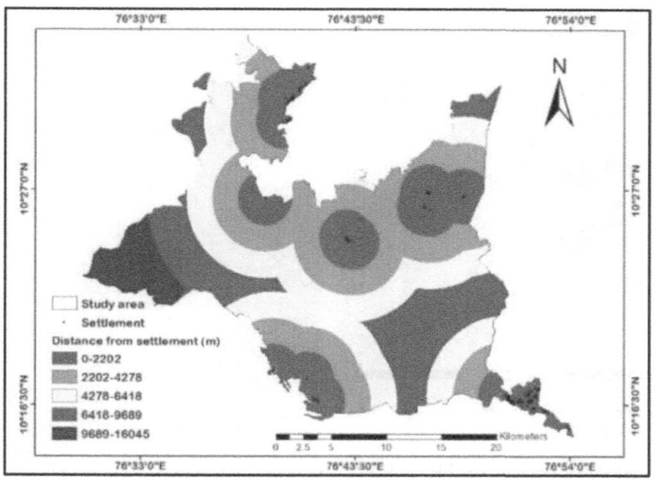

Fig. 51.12 Distance from settlement

3.11 Distance from Road

The proximity of roads and trek paths has a substantial impact on the occurrence of forest fires. Connecting roads and trek paths have been built in the PKTR to encourage trekking, tourism and departmental activities. Thereby, any tourist activities on these roadways, such as throwing unlit cigarette butts or match sticks into dry trash, dumping burning firewood from campfires, or roadside cooking, can initiate the fires. The entire study area was grouped into five different classes depending on the proximity to roads, and the highest rank was given to areas closer to road networks or trek paths. Fig. 51.13 is a map depicting the distribution of these various classifications.

3.12 Drainage

The humidity levels will be greater near water sources, so the rivers and streams will serve as a barrier, preventing forest fires (Chandra, 2005). The study area PKTR was divided into five classes based on its distance from drainage. Areas distant from drainage received the highest ranking. Fig. 51.14 depicts the distribution of these various classifications.

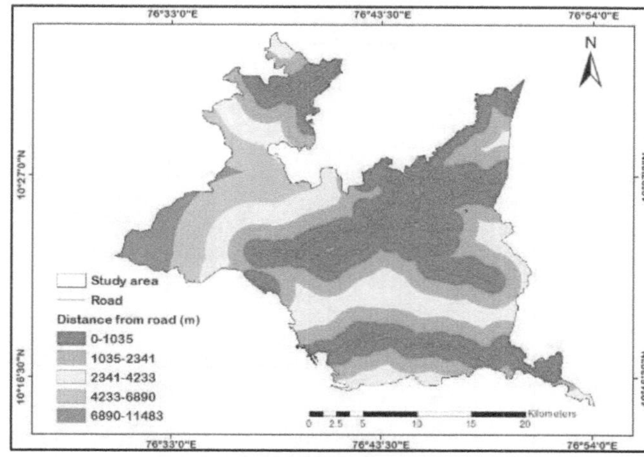

Fig. 51.13 Distance from road

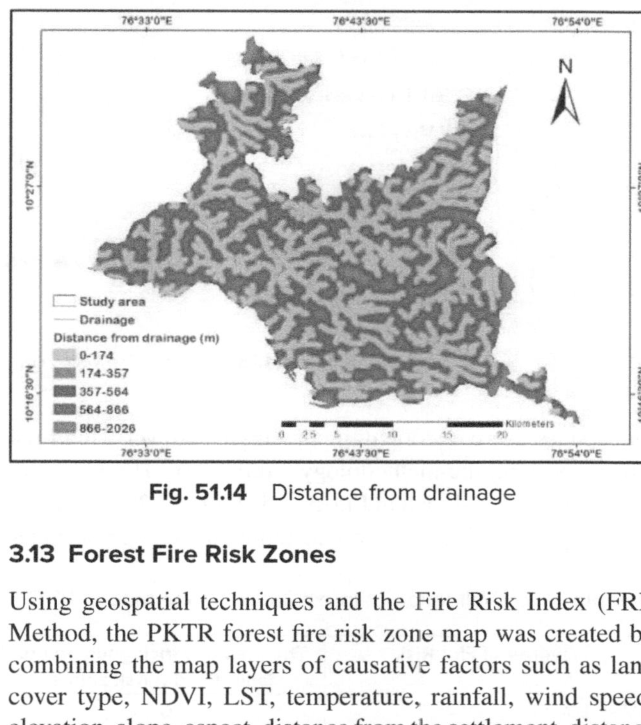

Fig. 51.14 Distance from drainage

3.13 Forest Fire Risk Zones

Using geospatial techniques and the Fire Risk Index (FRI) Method, the PKTR forest fire risk zone map was created by combining the map layers of causative factors such as land cover type, NDVI, LST, temperature, rainfall, wind speed, elevation, slope, aspect, distance from the settlement, distance from the road, distance from the drainage, and previous fire incidence points. Each class of these factors were assigned different weights based on their sensitivity to fire or their potential to cause a fire. The generated fire risk zone map of PKTR was divided into five zones: very low, low, moderate, high, and very high. The final map was validated with the previous fire incidence points for the past ten years, i.e., 2008-2017, obtained from MODIS. A total of 48 forest fires occurred in the concerned period. The result shows that, out of the 48 incidence points, 16.66% occurred in the low and moderate risk zones, respectively. High and very high-risk zones cover 66.66% points. It showed that most of the fire

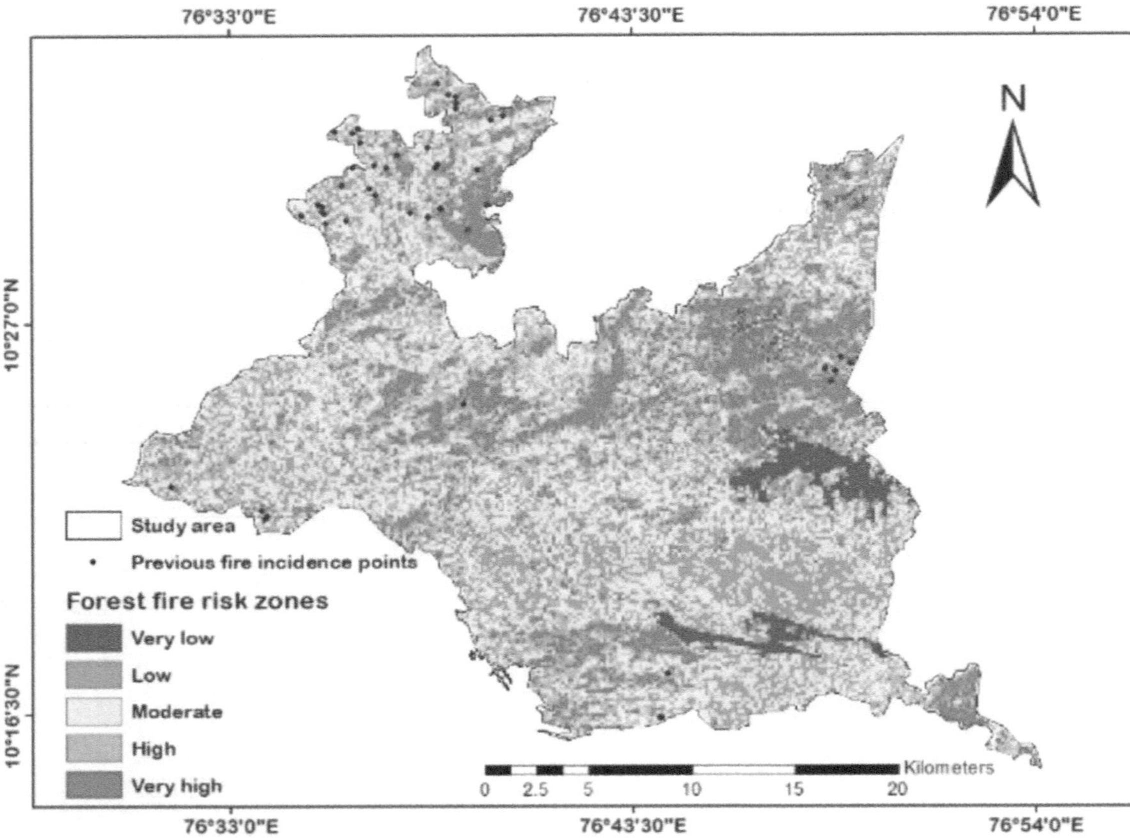

Fig. 51.15 Forest fire risk zone map of Parambikulam Tiger Reserve

incidences were recorded in the high and very high-risk zones, proving the methodology's reliability. The forest fire risk zone map is shown in Fig. 51.15. Table 51.2 provides the data of previous incidence points falling under the risk zones.

Table 51.2 Validation of forest fire risk zone map of PKTR

Degree of Fire Risk	The previous incidence points falling under risk zones
Very Low	0
Low	8
Moderate	8
High	8
Very High	24
Total	48

4. Conclusion

According to many studies, forest fires are a severe hazard to protected areas all over the world. Historical records confirm that the PKTR area has also been previously subjected to frequent fires. The present study's findings reveal that 66.66% of forest fires have occurred in the high and very high-risk zones. Therefore, regions with a high and very high fire risk must take prompt preventative steps to avoid future fires. The purpose of the forest fire risk zone map is to categorize the areas based on whether or not they pose a fire threat. So that officials can identify the risk zones and plan better to reduce fire risk and the spread of high-intensity fires.

5. Acknowledgement

We would like to thank the Kerala Forest Department for giving permission and assistance provided during the field visit. Also, we extend our thanks to SERB for providing financial aid.

REFERENCES

1. Ajin, R. S., Ciobotaru, A., Vinod, P. G., & Jacob, M. K. (2015). Forest and Wildland fire risk assessment using geospatial techniques: a case study of Nemmara forest division Kerala India. *Journal of Wetlands Biodiversity, 5*, 29–37.

2. Ambat, B., Nambudripad, K. D., Vinod, T. R., Sabu, T., Subin, K. J., & Samran, S. (2006). Development of a Forest Fire Forecasting System for Western Ghats, India using Web Based Geographical Information System (GIS) and Remote Sensing. *Technical Report*, Centre for Environment and Development (CED). 1–51.
3. Bisquert, M., Caselles, E., Sánchez, J. M., & Caselles, V. (2012). Application of artificial neural networks and logistic regression to the prediction of forest fire danger in Galicia using MODIS data. *International Journal of Wildland Fire, 21*, 1025–1029.
4. Chandra S. (2005). Application of Remote Sensing and GIS Technology in Forest Fire Risk Modeling and Management of Forest Fires: A Case Study in the Garhwal Himalayan Region. In: van Oosterom P., Zlatanova S., Fendel E.M. (eds) *Geo-information for Disaster Management*. Springer, Berlin, Heidelberg.
5. Cocke, A. E., Fulé, P. Z., & Crouse, J. E. (2005). Comparison of burn severity assessments using Differenced Normalized Burn Ratio and ground data. *International Journal of Wildland Fire*, *14*(2), 189–198.
6. Escuin, S., Navarro, R., & Fernandez, P. (2008). Fire severity assessment by using NBR (Normalized Burn Ratio) and NDVI (Normalized Difference Vegetation Index) derived from LANDSAT TM/ETM images. *International Journal of Remote Sensing*, *29*(4), 1053–1073.
7. Flannigan, M. D., Wotton, B. M., Marshall, G. A. *et al.* (2016). Fuel moisture sensitivity to temperature and precipitation: climate change implications. *Climatic Change, 134*, 59–71.
8. Kanga, S., Tripathi, G., Singh, S. K. (2017). Forest fire hazards vulnerability and risk assessment in Bhajji forest range of Himachal Pradesh (India): a geospatial approach. *J Remote Sens GIS*, *8*(1), 25–40
9. Keeley, J. E. (2009). Fire intensity, fire severity and burn severity: a brief review and suggested usage. *International Journal of Wildland Fire*, *18*(1), 116–126.
10. Mondal, N., Sukumar, R. (2016) Fires in seasonally dry tropical forest: testing the varying constraints hypothesis across a regional rainfall gradient. *PLoS ONE* 11:e0159691
11. Rothermel, R. C. (1972). A mathematical model for predicting fire spread in wildland fuels, USDA Forest Service Research Paper INT – 115, Ogden, Utah.
12. Saxena, A. (2012). Wildfire risk management in India and the community. In: Wildfire and community: facilitating preparedness and resilience. Paton D., Tedim F. (eds), Charles C Thomas Publisher, Ltd., Springfield, Illinois, USA. 129–149.
13. Setiawan, I., Mahmud, A. R., Mansor, S., Mohamed Shariff, A. R. & Nuruddin, A. A. (2004). GIS-grid-based and multi-criteria analysis for identifying and mapping peat swamp forest fire hazard in Pahang, Malaysia. *Disaster Prevention and Management*, *13*(5), 379–386.
14. Sowmya, S. V., Somashekar, R. K. (2010). Application of remote sensing and geographical information system in mapping forest fire risk zone at Bhadra wildlife sanctuary, India. *J Environ Biol, 31*(6), 969–974.
15. Thompson, W. A., Vertinsky, I., Schreier, H., & Blackwell, B. A. (2000). Using forest fire hazard modelling in multiple use forest management planning. *Forest Ecology and Management*, *134*(1-3), 163–176.
16. Veraverbeke, S., Lhermitte, S., Verstraeten, W. W., & Goossens, R. (2010). The temporal dimension of differenced Normalized Burn Ratio (dNBR) fire/burn severity studies: The case of the large 2007 Peloponnese wildfires in Greece. *Remote Sensing of Environment*, *114*(11), 2548–2563.
17. Verma, S., Vashum, K. T., Sathya, M., Jayakumar, S. (2015). Monitoring changes in Forest Fire Pattern in Mudumalai Tiger Reserve, Western Ghats India, using Remote Sensing and GIS. GJSFR-H.
18. Vinod P. G., Ajin R. S., Jacob M. K. (2016), RS and GIS based spatial mapping of forest fires in Wayanad Wildlife Sanctuary, Wayanad, North Kerala, India. *International Journal of Earth Sciences and Engineering, 9*(2), 498–502.
19. Yathish, H., Athira, K. V., Preethi, K. *et al.* (2019). A Comparative Analysis of Forest Fire Risk Zone Mapping Methods with Expert Knowledge. *J Indian Soc Remote Sens, 47*, 2047–2060.

Chapter 52

Forest Fire Early Detection System using Wireless Beacon Network and UAV based Object Detection

Amal Sujith
Department of Electronics and Communication
Amrita Vishwa Vidyapeetham
Amritapuri, India
amal.sujith7@gmail.com

Sagar Sajeev
Department of Electronics and Communication
Amrita Vishwa Vidyapeetham
Amritapuri, India
sagarsajeev453@gmail.com

Vishnu O. V.
Department of Electronics and Communication
Amrita Vishwa Vidyapeetham
Amritapuri, India
vishnuov1998@gmail.com

Abstract

Forest fire is one of the worst devastating events in the environment. It can harm the lives of plants and animals. This in turn results in climate change, the number of forest fires increased drastically in recent decades. Early-stage detection can reduce the impact of catastrophic fire. The Identification and detection of forest fire in the initial stage is an extremely difficult task even today. The satellite monitoring system is widely used in the detection of a forest fire, but it can only detect fire after a long time. The use of Unmanned Aerial vehicles for surveillance has increased tremendously in recent years. The delay in detection can be minimized by using Unmanned Aerial Vehicle and Wireless Sensor Network. In this paper, we wish to introduce a new approach for the detection of a forest fire. The aerial vehicle equipped with a camera will capture the image of fire, machine learning algorithms are used to predict the output. Wireless Sensor Network will increase the accuracy of prediction by combining the data from the sensor network. The key in this system is easy decision-making by analyzing the image from the Unmanned Aerial Vehicle within a short time.

Keywords: Communication Bridge, Coverage Path Planning, Detection Algorithm, Fixed-Wing, Ground Control Station, Image Processing, Mission Planner, Navigation, Unmanned Aerial Vehicle

1. Introduction

Fire detection and aversion are possible in areas with proper infrastructure and impeccable resources but, it remains a severe challenge in rural neighborhoods and large Forest covers. Forest Fires in India correspond to the largest number of fire accidents occurring every year. It becomes essential to highlight the devastating effect forest fires make on the lives of the Tribal community and other communities that solely depend upon the forest for their livelihood. The existing Detection system employed in the forest range involves Satellite imagery which takes landscape photos of the affected region and relays them to the Forest Officials after a gap of 2 days. This occurs after a foreign satellite orbiting at four instances of time in a day above India. The detailed spectroscopic images are forwarded to ISRO (Indian Space

Research Organization) where the data is configured for ease in the understanding of FSI (Forest Survey of India) Officials. The detection technique which we want to implement will cover the areas where fire management is difficult unless acted upon quickly hence, using our real-time detection technique we can alert the respective officials stationed in the vicinity and avert the danger as quickly as possible.

The system is distributed into two sections inclusive of Aerial detection and ground-based detection. The detection system presented in this paper leverages the accuracy of the ground detection system and large area coverage of UAV hence, making it an efficient and cost-effective detection technique. Unmanned Aerial Vehicle (Fixed Wing) involves a camera attached to its fuselage which also consists of a payload holding a Microprocessor board for detection and storing data during routine flight sessions. The UAV has a Flight Controller for optimum flight controls, precise navigation and GPS pinged for accurate data creation to aid in future analysis purposes such as Fire path recognition. The detection part consists of a microprocessor with a camera-enabled, which runs on a self-recognizing machine learning code internally. The detection of images will occur using a Machine Learning technique that can classify the forest fire with adequate precision. Once the UAV detects the fire by accessing the data collected by the microprocessor, this data will be sent to the Forest official's office, relayed through a ground station. The Ground system is a Wireless Sensor Network-based system with several sensors embedded in it. Wireless communication is done with the help of Seed Grove Lora Wan module 865MHz. And each wireless beacon setup comprises a temperature sensor LM-35, Humidity sensor DHT11, GPS, and Wind Sensor. Each of the beacons transmits the data. These data are collected by a main central module. The collected data is then uploaded to the cloud where the data is passed through a Machine Learning (ML) algorithm and predicts the result. ML Algorithm improvises throughout extensive usage. Additionally, each beacon comprises a GPS module to get the coordinates of the fire. And with the help of a wind sensor, we can predict the direction of the progressing fire. The whole beacon system is powered by individual solar panels and a rechargeable battery for power backup.

2. Motivation

The major motivation for us to design and implement such a model was to help the people or tribal community to be specific which utilises Forest resources for their daily livelihood. The Forest gets severely affected due to such a catastrophe which perishes not only the resources of the Forest as well as the animals surviving in such covers or sanctuaries. This causes severe ecological imbalance in the region. A real-time detection can prove to be handy for the forest authorities since, they provide the information at instant. The Satellite based detection heads forward to redundancy since, the relay time from Satellites is very poor. For detection of such hazardous events drones are going to be the front-runners. Also, the ideals which our University has put forth that inculcates the kind gesture of helping the needy and giving back to the society were pivotal for us in selection of such a project.

3. Related Literature Studies

This research paper [1] mainly uses the Sensors network, which uses dynamic variables and the main causes such as humidity, fuel type, the slope of the land, the direction and the speed of the wind, smoke, etc. This is mainly monitored using a GUI which provides alert messages on the developed application using some logical evaluation to generate the result. A combination of sensors is used for the data collection and the transmission of data between the different nodes is done with the help of the wireless MicaZ sensor mote. This data is fed into Artificial Neural Networks (ANN) which is already trained for the detection of fire and its pattern of spreading is proposed in the paper [2].

The author of the paper [3] discusses using the combination of sensors and Machine learning algorithms for fire detection. Here the whole set of the sensor is connected to Node MCU which is a low-cost WIFI-based IoT device and later logical regression algorithm of Machine learning is used for the detection.

This proposed system in paper [4] uses wireless sensors to measure physical as well as chemical parameters and they are connected using ZigBee based on IEEE 802.15.4 wireless technology. It is suitable for low power systems which use solar panels for battery charging, and suitable for low data rate. The system will predict the probability of fire by calculating an index using the given parameters from the sensors.

Forest fire detection using WSN integrated with Light and Temperature sensors and uses GSM for communication. The sensor network is connected to the cluster, the communication link is ISM and the central coordinator node (CCN) collects all the data from each header. Later it is fed into a model for analysis and detection based on previous data sets is proposed in the paper [5].

In this paper [6], an early fire detection model is proposed using Raspberry Pi 3B+ microcontroller and sensors like Temperature Sensor (LM35), Anemometer, Smoke Sensor (MQ2). A server is used for storing the data and monitoring that data. The integrated neural networking model is used for prediction purposes. Then, using GSM an alert message is sent to the officials and the people within the proximity.

The paper [7] proposes a system that detects the forest fire by using UAV, and incorporates other detection algorithms such as median filtering, Otsu threshold segmentation, and morphological operations.

The proposed system in the paper [8] has a Raspberry Pi 3 Model B which is a Single-board computer with remote LAN and Bluetooth connectivity. The method used in this paper converts the video record to frames and then passes it to the algorithm which will convert the image to an RGB format and the yellow region in the image will be converted to white then the image is grayscale and contours are found by contour detection. After fire detection, the message will be passed to APP or LAN.

The author of the paper [9] discusses the observation of the rate of error while detecting the fire through a combination of the traditional image feature detection techniques like Background Subtraction, RGB, HSV, YCbCr. Some of the unique techniques such as Morphology and Time frame selection are used which will remove the spots in the frame which may not be related to fire and also update the frame in regular time which helps to reduce the error.

A system that consists of an Unmanned Aerial vehicle, for collecting data and a fire source location module. The data receiving module will receive video and telemetry images. This data will be used by the fire detection module which enables it to detect the fire. The analysis module will analyze the data image and detect the fire source location. This geographical information is displayed to the user through the GIS display module, this is proposed in paper [10].

The system uses sensors such as flame sensors, smoke sensors, temperature sensor LM35, and wind direction sensors. The UAV is integrated with supervising software which includes a fire source detection module, GPS module, and fire range guess module. When the alert is received the UAV should spray fire extinguishers that put off the fire and the information regarding the affected area transmits to the control station and this is proposed in paper [11].

This proposed system in paper [12] uses computer vision-based autonomous Unmanned Aerial Vehicle systems. used to build a 3D representation of the surroundings. 3D representation of the surroundings and precise localization information to find the shortest path and reach the goal through a three-dimensional safe trajectory. It is built accumulating multiple observations coming from the 360 stereo vision sensing suite. Visual data is also used to extract information regarding the drone's altitude and position while exploring the environment.

4. Proposed Work Flow

This section provides the block diagram of the project and it portrays the operation to be performed by the Fixed-wing UAV during the mission. The Figure indicates the work cycle of the drone with logic block and detection condition. The ultimate goal is to detect with maximum accuracy during the mission and when faced any critical battery failure or draining, then to safely return back to the base with minimal damages. This workflow provides an idea that has been presented in paper [13], where the Swarm drones return back to the base in case of repairs and is predominantly used for Autonomous flights along with manual flight control capability for human intervention.

5. Methodology

Our proposed model is presented in Fig. 52.1. There are two major sections for our proposal: one is the section of Aerial Forest fire detection using UAV and the other is using WSN based forest fire detection to collect data and predict the temperature spikes for the future.

Aerial based forest fire detection system block diagram is presented in Fig. 52.2a. using UAV comprises a Remote controlled based Fixed-wing UAV shown in Fig. 52.2b, the vehicle is equipped with a high-resolution camera, Ardupilot based autopilot Pixhawk module Raspberry Pi microprocessor, and necessary electronic components required to power up the drone. A deep learning model is trained on an image set that was taken by creating an artificial scenario of a forest fire.

The generated data were annotated with fire as the label. And the training was done on Google Cloud. The obtained model was converted to the TensorFlow Lite model as it uses less computing power and it made the model run perfectly on the low computing power device Raspberry Pi. The real-time GPS-based waypoint navigation and total UAV guidance are controlled by the Pixhawk autopilot module. The drone is equipped with a Seed Grove Lora-WAN module 865MHz which transmits the location GPS coordinates and a message with Fire detected to the ground station.

In Fig. 52.2c, we've presented a screenshot from the Ardupilot software Mission Planner which is used for calibrating and pre-programming the flight controller. As part of the path planning, we've decided to use the pentagonal path as presented in the Screenshot. Google map was used as a reference for coverage planning because of its accurateness. The waypoints have been placed at the near implementation site in Banasura Sagar Forest Region. The waypoint radius is set to 2 meters hence when the UAV reaches 2 meters to

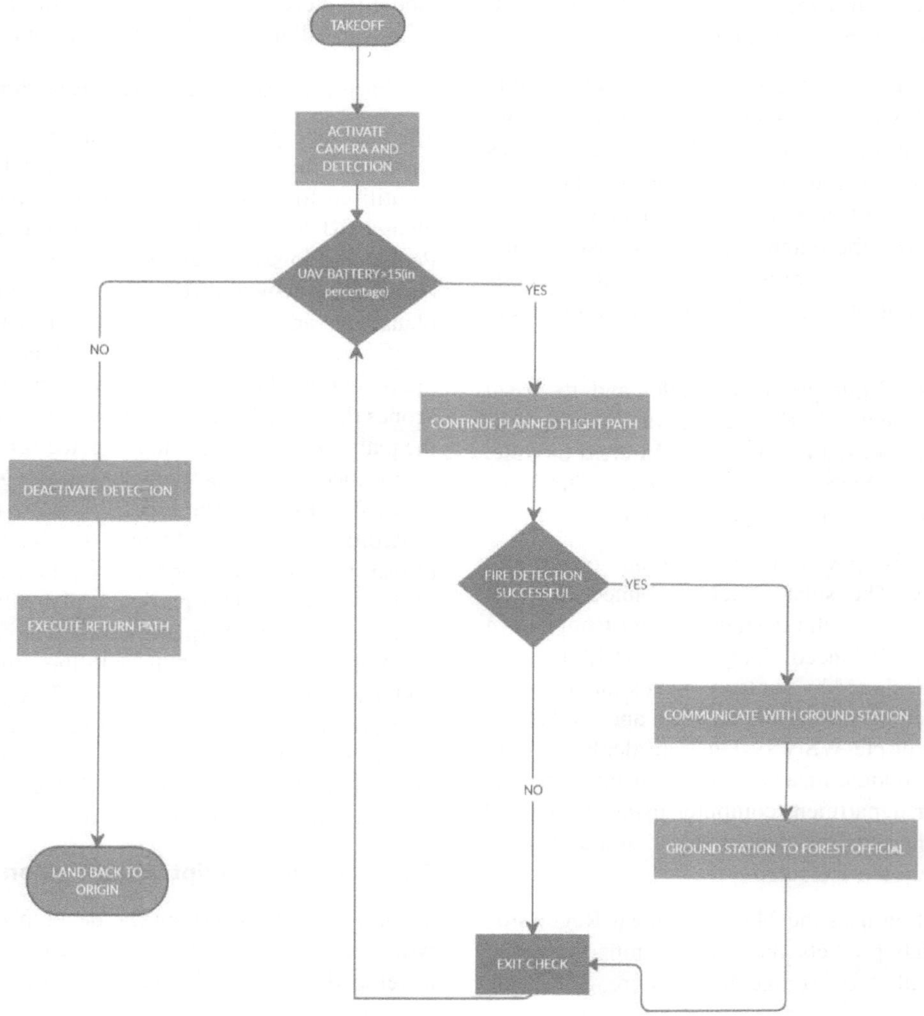

Fig. 52.1 Expected Flight Operation

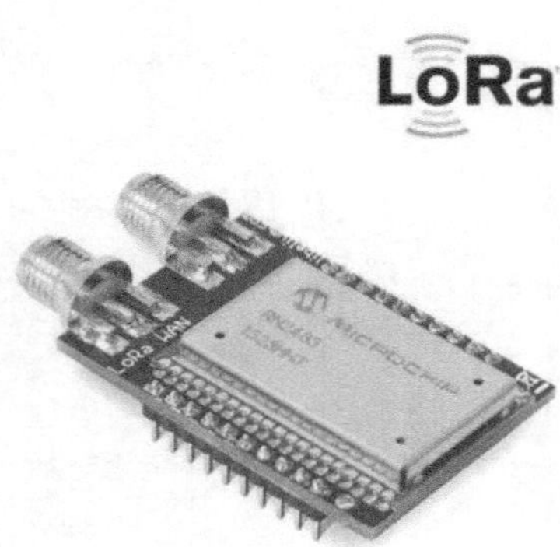

Fig. 52.2 LoRa Module

Fig. 52.3 LoRa Transmitter and Receiver

the set waypoint specified using Latitude and Longitude, it begins maneuver towards the next waypoint.

The flight controller receives the data from a GPS module embedded in the UAV which helps it in Navigation throughout the set path. Furthermore, real-time flight calibration takes place with demands to the user set altitude and airspeed. Rudders and Ailerons aid in-flight maneuver and maintain stable flight throughout the planned path. The take-off and landing waypoint can be pre-programmed. Such functions of the UAV's flight controller make the operation completely independent.

The second section of our proposed model and its basic working model is presented in Fig. 52.3a. which consists of multiple boxes which consist of an HC-12 Wireless module, Temperature Sensor (LM35), Humidity Sensor (DHT-11), Arduino Uno microcontroller board, and GPS module.

The WSN system is powered by a Solar panel and rechargeable battery. The solar panel is connected to an automatic solar charging circuit designed by our team shown in Fig 52.3b provides balanced charging of the lithium-ion battery. The circuit designed by our team makes sure the solar panel charges the battery during the daytime and the battery is used at night. Multiple WSN systems are deployed over the forest and each module measures and transmits the data to the Forest officer department computer using the HC-12 wireless module. The received data is stored and a custom-built python application is executed.

The python application uses the Multiple Linear Regression ML technique which predicts the future temperature and is cross-checked with the average temperature generating model to increase the accuracy. Both of system provides 24 hours surveillance with accuracy more than 75 percentage.

5.1 Flight Simulation - Coverage Plan

The coverage plan is decided based on the area required to cover and the Unmanned Aerial Vehicle frame which will be utilised for mission. According to the paper [13], several plans exist for varying frames such as Hexagonal mission, Pentagonal mission. The given plans are utilised for Rotary based drone which have freedom of motion in the air. These plans have area of detection present inside the area covered by it hence, forming the region of interest for user. Other plans such as Back-Forth, Spiral plans which are utilised by drones that have fixed camera and detection takes place along the path. The Coverage plan selected for this project is Back-Forth since, the Fixed-Wing detects along the path that will be scripted in Python. The Fig. 52.4 shows the Back-Forth plan in ArduPilot's Mission Planner 1.3.74. The Mission Planner utilises Google Maps for acquiring Latitudes and Longitudes of the waypoints. This particular plan allows the overall field coverage of 80,000 square metre. The swath width [15] of the drone is close to 700 m with the camera specification of 600 TVL and Field of View – 160 degrees. Hence, such a plan provides multiple overlapping of images which can be crucial in accurate detection and relaying precise warning to the Forest Official's office. Drone covers 200 metres in breadth and 400 metres in Length.

5.2 Indigenous Script Using Python DRONEKIT

The complete mission plan was developed using Python script with help of Drone-kit package. It is a package exclusively developed for Flight simulations and generating autonomous

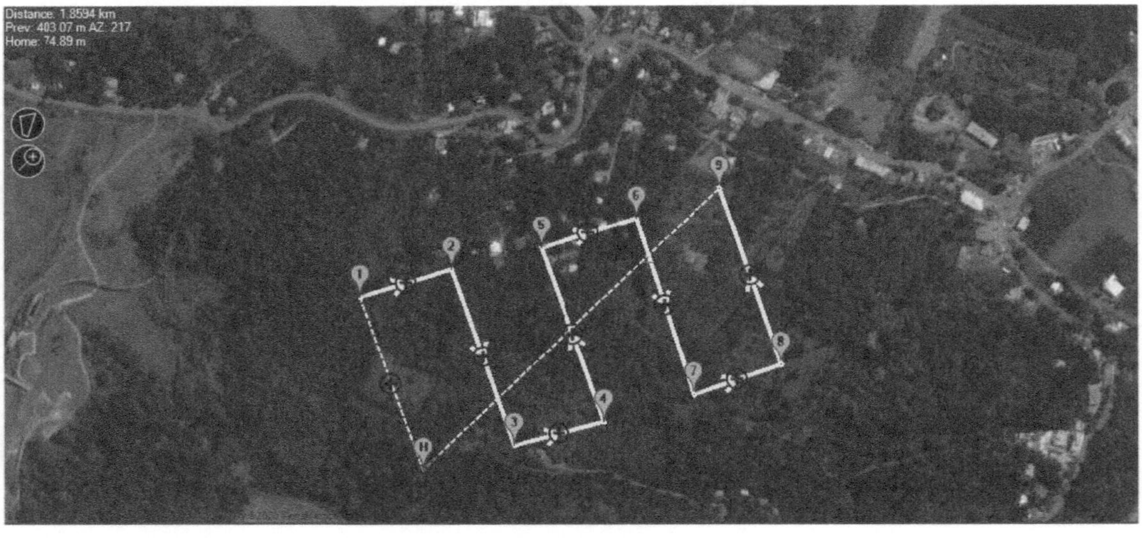

Fig. 52.4 Back-Forth Implemented near Banasura Sagar Dam

plans for respective drones. For real-time perspective, it can be implemented on a companion PC like Raspberry Pi [14] on board the drone. The Python script can send commands to the Flight controller which forms the heart of the drone as it manages flight stability, aerial control and flight propagation to planned waypoints with help of a GPS Module. For simulation perspective, the python script can connect to TCP Port 5760 and execute the program in Python shell indicating shell outputs as well as TCP(Transmission Control Protocol) Distribution port 5763 or 5762 can be connected with GCS(Ground Control Station). The GCS will provide the user with real-time behaviour of the drone and several parameters for analysis. Script contains several command functions which are compatible to the frame of the drone. The algorithm issues commands to the drone by making it follow waypoints depicted with help of Latitude and Longitude along with Elevation from ground. This designed path is followed by the drone until a critical battery failure or battery drains up to 15 percentage of total. Under such a situation, the Flight cuts off power to Microprocessor and Camera moreover, immediately returns to the launch site. This workflow is observed for all the mission despite change in frame of the drones. Figure 52.5 shows the connection across both the ports and shell outputs during mission phase of the drone. Figure 6 indicates the logic block of the program.

5.3 Flight Simulation Presets

The Script based simulation was performed on both Quad-copter as well as Fixed-Wing to showcase minor comparison and indicate the effectiveness of Fixed-Wing Drone. The Coverage plan for both the drones were same as mentioned above in the sub-sections. The frames were changed in the initialisation with help of Command Prompt. The changes and real-time reading were depicted in the GCS. Figure 52.7 indicates the Environment variables which were introduced such as Wind-speed and Orientation to bring an essence of real environment behaviour of the drones. HUD in the Mission Planner indicates the direction and elevation of the drone. It was configured include the battery readings as well as operating current. The wind speed of the region is close to 1.6667 m/s and direction of the wind is usually spread towards the East which is depicted by 90 degrees in simulation pane given below.

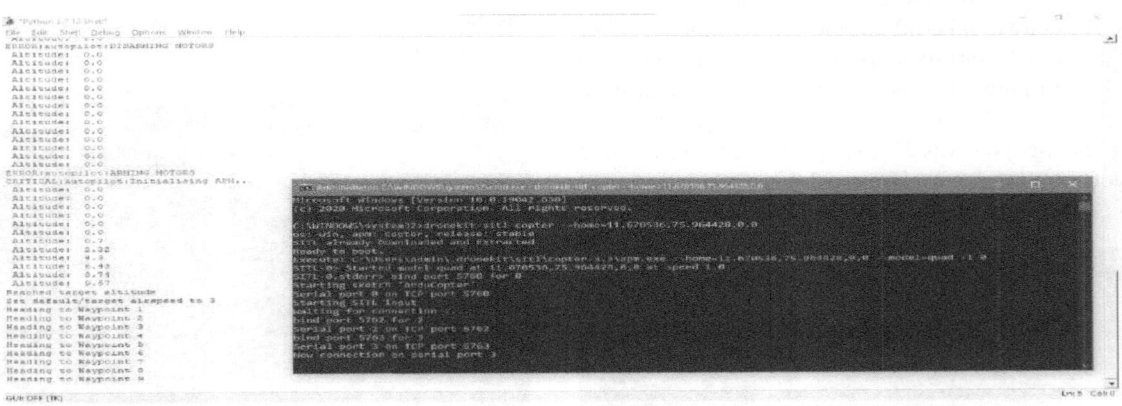

Fig. 52.5 Python Shell and Command Prompt window with base location

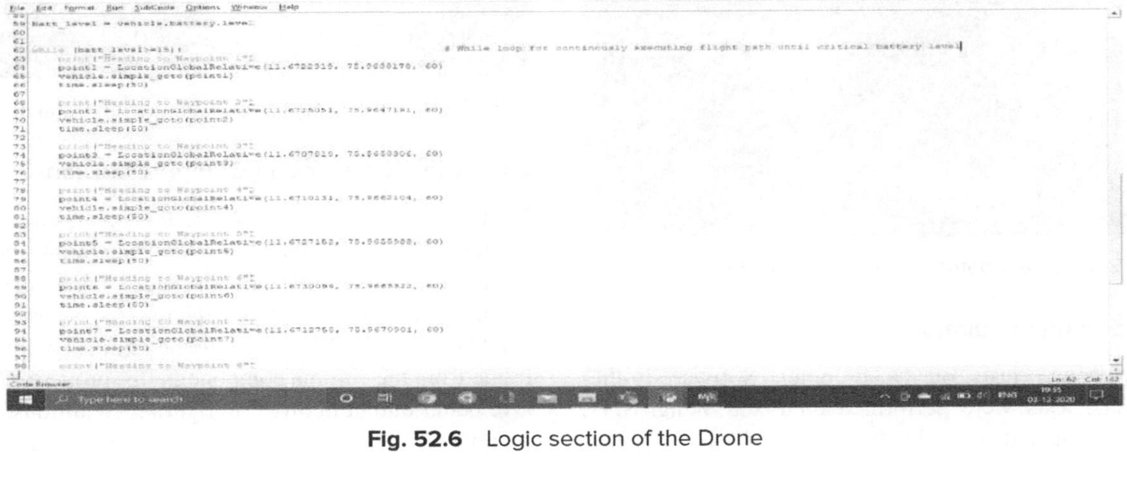

Fig. 52.6 Logic section of the Drone

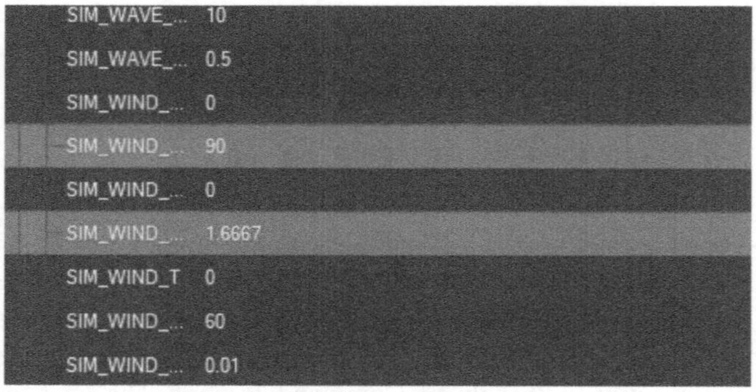

Fig. 52.7 Wind Direction and Speed

5.4 Quad-copter Simulation

Python script performed on Quad-copter showed optimal results, Quad-copter showed immense flight stability through the trail which was generated while following the path. Simulation time was considerably larger when compared to Fixed-Wing partly due to its' Average Air-speed of 4.37 m/s. Flight Altitude set was 60 metres since, the Tallest trees found in the region doesn't go beyond 45 metres. Ground Sampling Distance (GSD) [16] for the particular configuration was around 145.18 cm/px. Take-off Altitude was 10 metres, after which the drone will start its journey to the waypoints. Each Waypoint has a radius of 2 metres and the script consisted of 8 Waypoints excluding home location. Average Operating current for Quad-copter was 26.6 Amps. Distance covered in a single cycle is 1400 metres. Figure 52.8 gives user a snippet of a Quad-copter while performing the mission.

Fig. 52.8 Quadcopter Simulation behaviour on GCS

5.5 Fixed-Wing Simulation

After performing tests on Quad-copter, Software-in-the-Loop (SITL) tests were performed on Fixed Wing. SITL tests are conducted directly from the Mission Planner rather than connecting the Frame to Python script. Working on a constrained package support, the Python script was modified for Fixed-Wing along with few changes. Fixed-Wing performs horizontal take-off hence, those parameters had to be included with separate function declaration. Flight Stability along the coverage path varies a lot when the trail is compared to Quad-copter trail. Simulation time was very less when compared to Quad-copter simulation due to Average Airspeed throughout the simulation of 22.7 m/s. Flight Altitude set was 60 metres since, the Tallest trees found in the region doesn't go beyond 45 metres. Ground Sampling Distance (GSD) [8] for the particular configuration was around 145.18 cm/px. Take-off Altitude was 10 metres and the pitch was at 15 degrees, after which the drone will start its journey to the waypoints. Each Waypoint has a radius of 2 metres and the script consisted of 8 Waypoints excluding home location. Average Operating current for Quad-copter was 29.1 Amps. This indicated that the Fixed-Wing glider draws larger current when compared to Quad-copter. The reason would be partly due to the manoeuvring of Fixed-Wing displayed in the Fig. 52.9. Distance covered in a single cycle is 1400 metres. Since, the camera is fixed to Drone's fuselage, the Image detection is expected to perform along the coverage path.

Fig. 52.9 Fixed-Wing based Simulation behaviour on GCS

5.6 Detection Through Computer Vision

In the detection part, the sample fire images are captured at the rate of 30 fps, using a normal camera and the captured frames are converted from the RGB colour code to HSV format for the better detection. Here HSV is having the upper hand because HSV can separate the lumina component of the frame from the chroma component, this property of HSV will give better output in different lighting conditions. In order to set up a filter, masking is done in a particular range of colour spectrum including the combination of yellow and red. Later this mask and the real image is passed to a filter which will perform a bit-wise AND operation.

The output of AND operation in the black/white domain:

Fig. 52.10 Input Image

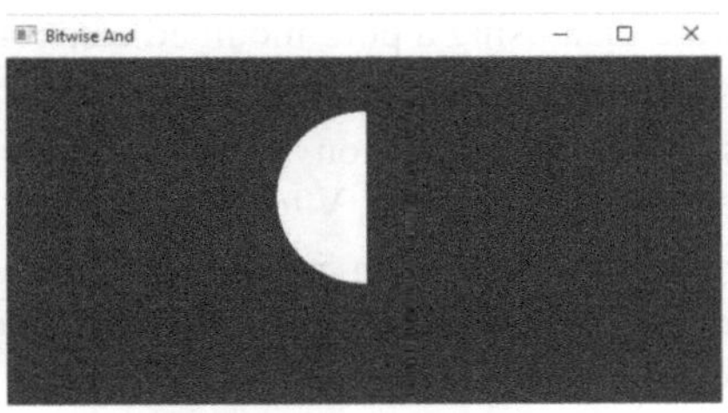

Fig. 52.11 Output Image

Filter response:

- "Colour in the range from frame" AND "colour in the range from mask" = same colour
- Same colour AND black = black

This operation will give the final filer output, later a non-zero-bit counter is used on the filter response to produce the final prediction of fire.

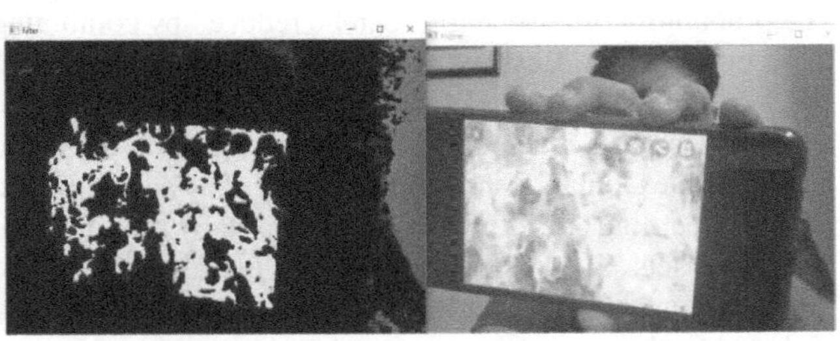

Fig. 52.12 Pre-processed Image and Original Image

5.7 Detection Through Wireless Sensor Network

The WSN system is powered by a Solar panel and rechargeable battery.The system uses multiple sensors such as cluster of LM35, MQ2 (Smoke Sensor) etc shown in Fig. 52.13.

Multiple WSN systems are deployed over the forest and each module measures and transmits the data to the Forest officer department computer using the HC-12 wireless module. The received data is stored and a custom-built python application is executed. The python application uses the Multiple Linear Regression ML technique which predicts the future temperature and is cross-checked with the average temperature generating model to increase the accuracy. The system provides 24 hours surveillance with accuracy more than 75 percentage shown in Fig. 52.14.

Fig. 52.13 WSN based temperature prediction system

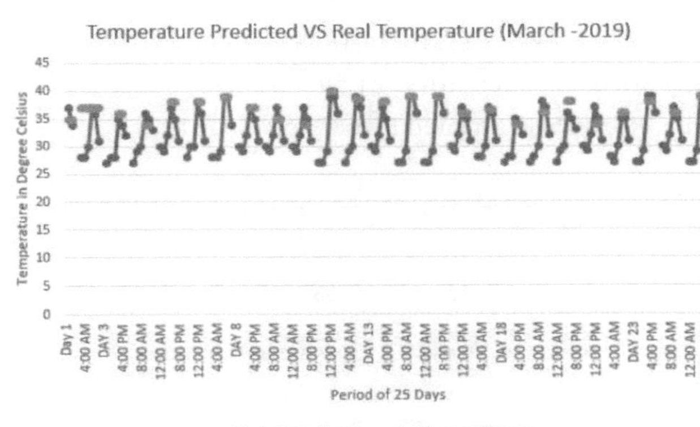

Fig. 52.14 Prediction result using Linear Regression model

6. Experiment and Results

The WSN system constitutes of six modules kept at a distance of 500m from the central pole which is common for UAV and WSN systems. Fig. 52.15 WSN system constitutes sensor cluster used for real-time temperature monitoring Seed Grove Lora Wan 865MHz module for communication between the routers. The collected data is sent to the central pole which stores the sensor values in Raspberry Pi 3 B+ which is then processed by the Linear Regression ML technique.

The Routers are implemented in the first phase for temperature data collection for six months shown in Fig. 52.15.

Fig. 52.15 Phase 1 implementation in Banasura Sagar, Wayanad, Kerala

Using the data collected from all sources, we generate a machine-specific model that maps an initial state of fire to a predicted one. Linear Regression model is used to predict daily temperature values based on the dataset built, by deploying the prototypic model kept on the site for 3 months and with the temperature data provided by Accuweather.

(i) Mean Absolute Error: 4.182859899
(ii) Mean Squared Error: 21.5987693072
(iii) Root Mean Squared Error: 4.6474476121

	Actual	Predicted
0	31	27.884145
1	29	35.732261
2	31	37.357018
3	30	26.794801
4	26	24.491033

Fig. 52.16 Prediction result table sample using Linear Regression model

Using the table shown in Fig. 52.16, Mean Absolute Error, Mean Squared Error, Root Mean Squared Error of the linear regression model is computed.

Our system which constitutes both Aerial based and ground-based prediction and detection system increase the accuracy of the system by 74.564 percentage. Figure 52.15 shows the prediction and the actual value plot. Hence the fire could be detected within minutes in the active region (where the sensors are deployed and the surveillance area covered by glider).

Table 52.1 Flight Simulation Results

Type	Airspeed	Operating Current	Flight Altitude
Quad-copter	4.37 m/s	26.1 Amps	60 m
Fixed-Wing	22.7 m/s	29.9 Amps	60 m

7. Advantages

1. The advantage of using a Fixed-wing RC Plane over other systems is it can cover 15km – 20km at a stretch which can be extended to 40km in the future.
2. The aerial view-based detection of the glider is more accurate than using a pole-mounted camera system for detection.
3. Less power consumption when compared to other gliders eg: Quadcopters, V-toll, Quad Plane, etc.
4. Alert the officials within 5min after the fire is detected which is better than the existing satellite-based detection system which alerts after 2 days.
5. The system enables 24 hours surveillance with an accuracy of more than 75 percentage.

8. Conclusion

The system provides fast and efficient results compared to existing solutions. Fixed-Wing based UAVs can cover a huge area of forest along with a faster speed and easy to detect fire in extreme terrain conditions. On the other hand, while the UAV system is offline WSN based compromises the system. The probability of false alarms can be reduced by combining these two systems. Besides, the current fire detection system can be intelligently tuned for the detection of both smoke and fire. This will enable the image processing unit to handle more complex situations in the real world. The accuracy of a machine learning algorithm can be increased by training with more data-sets. Even though our system improved the forest fire detection accuracy, yet the number of false alarms is still high and further research is required in this direction.

9. Acknowledgment

We would like to express our sincere gratitude to our Project Guide who supported us throughout the project work. We would also like to extend our thanks to all the faculties of Electronics and Communication Engineering department at Amrita Vishwa Vidyapeetham, Amritapuri Campus. Special Thanks to our College authorities for allowing us to perform tests during the Pandemic and giving us permission to implement our system during such adverse conditions. Thanks to our parents and friends who were supportive of this idea and to everyone who was involved directly or indirectly to bring the project to fruition.

References

1. Guruh Fajar Shidikl, Fajrian Nur Adnan, Catur Supriyanto, Ricardus Anggi Pramunendar, Pulung Nurtantio Andono, "Multi-ColorFeature,BackgroundSubtraction,andTimeFrame Selection for Fire Detection" 2013 International Conference on Robotics, Biomimetics, Intelligent Computational Systems. DOI: 10.1109/ROBIONETICS.2013.6743589
2. F. Valenti, D. Giaquinto, L. Musto, A. Zinelli, M. Bertozzi and Broggi, "Enabling Computer Vision-Based Autonomous Navigation for Unmanned Aerial Vehicles in Cluttered GPS-Denied Environments," 2018 21st International Conference on Intelligent Transportation Systems (ITSC), Maui, HI, 2018, pp. 3886-3891, DOI: 10.1109/ITSC.2018.8569695
3. B. C. Ko, S. J. Ham, and J. Y. Nam. "Modeling and formalization of fuzzy finite automata for detection of irregular fire flames," IEEE Trans. Circuits Syst. Video Technol., vol. 21, no. 12, pp. 1903–1912, Dec. 2011.
4. S. J. Pan and Q. Yang, "A survey on transfer learning," IEEE Trans. Knowl. Data Eng., vol. 22, no. 10, pp. 1345–1359, Oct. 2010.
5. K. Muhammad, M. Sajjad, M. Y. Lee, and S. W. Baik, "Efficient visual attention driven framework for key frames extraction from hysteroscopy videos," Biomed. Signal Process. Control, vol. 33, pp. 161–168, Mar. 2017.
6. A. Ullah, J. Ahmad, K. Muhammad, M. Sajjad, and S. W. Baik, "Action recognition in video sequences using deep Bi-directional LSTM with CNN features," IEEE Access, vol. 6, pp. 1155–1166, 2017.
7. A. Ullah et al., "Action recognition in movie scenes using deep features of keyframes," J. Korean Inst. Next Generation Comput., vol. 13, pp. 7–14, 2017.
8. A. Viseras, J. Marchal, M. Schaab, J. Pages and L. Estivill, "Wildfire Monitoring and Hotspots Detection with Aerial Robots: Measurement Campaign and First Results," 2019 IEEE International Symposium on Safety, Security, and Rescue Robotics (SSRR), Wu¨ rzburg, Germany, 2019, pp. 102-103, doi: 10.1109/SSRR.2019.8848961.
9. O. Surinta and S. Khruahong, "Tracking People and Objects with an Autonomous Unmanned Aerial Vehicle using Face and Color Detection," 2019 Joint International Conference on Digital Arts, Media and Technology with ECTI Northern Section Conference on Electrical, Electronics, Computer and Telecommunications Engineering (ECTI DAMT-NCON), Nan, Thailand, 2019, pp. 206–210, doi: 10.1109/ECTI-NCON.2019.8692269.
10. P. Foggia, A. Saggese, and M. Vento, "Real-time fire detection for videosurveillance applications using a combination of experts based on color, shape, and motion," IEEE Trans. Circuits Syst. Video Technol., vol. 25, no. 9, pp. 1545–1556, Sep. 2015.
11. V. Zope, T. Dadlani, A. Matai, P. Tembhurnikar and R. Kalani, "IoT Sensor and Deep Neural Network-based Wildfire Prediction System," 2020 4th International Conference on Intelligent Computing and Control Systems (ICICCS), Madurai, India, 2020, pp. 205-208, DOI: 10.1109/ICI- CCS48265.2020.9120949.
12. [1] C. Gomathi1, K. Vennila,2013/11/01, Wireless Sensor Network for Forest Fire Detection and Decision Making, Volume 2, International Journal of Advances in Engineering Science and Technology. 10.1109/ICIEA.2009.5138270.
13. Taua M. Cabreira, Lisanne B. Brisolara and Ferreira Paulo R. Jr., "Survey on Coverage Path Planning with Unmanned Aerial Vehicles" Drones Journal MDPI, January 2019. DOI: 10.3390/drones3010004
14. Hasan M. Qays, Dr. Bassam A. Juma, Dr. Ayman D. Salman, "Design and Implementation of Autonomous Quad-copter using SITL Simulator" Iraqi Journal of Computers, Communications and Control Systems Engineering (IJCCCE), January 2020. DOI: 10.33103/uot.ijccce.20.1.1
15. Erich Seifert, Stefan Seifert, Holger Vogt, David Drew, Jan van Aardt, Anton Kunneke and Thomas Seifert., "Multi-View Reconstruction of Forest Images" May 2019, Journal on Remote Sensing MDPI. DOI:10.3390/rs11101252
16. Kevin Sheng-Wei Lim, "Lidar remote sensing of forest canopy and stand structure" May 2006, Thesis Paper Queen's University, Ontario, Canada ISBN 2527-6638

CHAPTER 53

Patterns and Consequences: Australian Bushfires

Ritu Bir
Research Scholar, Galgotias University,
Greater Noida, UP, India

Meenakshi Singh
Professor in Chemistry, Galgotias University,
UP, India

Abstract

Bushfires are a regular occurrence in Australia and have played a significant role in shaping the continent. Australian flora in forests is dominated by eucalyptus which has adapted itself well to regular fires. Australia has witnessed havoc caused by these widespread fires of Black Saturday (2009), Ash Wednesday (1983) among many others. Gradual loss of moisture and increase of dryness has produced an ecosystem in Australia which is prone to fires and has brought adaptation in local flora and fauna but bushfires of (2019–20) are unprecedented in both their scale and intensity. Changing climate has been a reason of concern in Australia primarily caused by GHG emissions which are one of the highest per capita emissions, positive Indian Ocean Dipole which leads to warming of sea surface in the western part as compared to its eastern side and negative direction of Southern Annular mode had made Australia record its hottest and driest December. This extreme high temperature and prolonged drought conditions led to unprecedented bushfires in the states of Victoria, New South Wales, Western Australia, Queensland, Southern Australia and Australian Capital Territory. The first major bushfire started before the official arrival of spring in June 2019 and worsened at the beginning of September-2019, a series of lightning strikes further exacerbated these bushfires by the beginning of November 2019. Bushfires have burned more than eighteen million hectares of land in South Australia along with destruction of 2800 homes, killing 35 people, roughly 1.25 billion animals and evacuation of thousands of people from their homes. This paper highlights the changing climate conditions behind the Australian bushfires.

Keywords: Bushfire, changing climate, positive Indian ocean dipole, southern annular mode, greenhouse gases.

1. Introduction

Bushfires have long been part of the Australian landscape and have played an important role in shaping the distinct character of the continent. For millions of years, these regular widespread fires have created an environment which is prone to fires and has also brought adaptations in flora and fauna. Eucalyptus, a dominant flora in these forests, is able to open and release their seeds after being exposed to the heat of fire. These bushfires have brought significant amounts of damage to the continent, some exceeding much worse than others. Many of these fires have been named according to the day on which they peaked and caused much havoc-Black Saturday (2009) causing 173 fatalities and destroying 2000 homes; Ash Wednesday (1983) killing 75people and destroying 1900 homes; Black Friday (1939) killing 71 people and destroying 650 homes.

Bushfires are a regular feature in Australian calendar but bushfires of 2019-20 were unprecedented and alarming in both their scale and intensity but not unexpected. The first major bushfire started before the official arrival of spring in June 2019, worsened at the beginning of September-2019 and a series of lightning strikes further exacerbated these bushfires in November 2019. These fires were concentrated in the southeastern coast especially Victoria, North South Wales, Queensland, South Australia, and Western Australia but the effect of these was felt in every state and territory of Australia. These fires have burned an area of more than 18 million hectares releasing 430 tonnes of carbon dioxide in the atmosphere, causing 35 deaths, evacuating thousands of people from their homes and killing more than 1.25 billion animals.

2. Methodology

Extensive literature review was done to get information regarding Australia's changing climate pattern. Relevant information was also collected from Australian bureau of meteorological department to study how natural phenomenon triggerd with GHG gases emission and changing climate are causing havoc to Australia's ecosystem.

3. Results and Discussion

Changing climate is pummeling and has been a cause of concern in Australia (Beeson et al. 2013). The major reason for the spread of these fires is the combination of hot and dry weather which dries out the vegetation, makes it act as a fuel and presence of an ignition source (Deb et al. 2020). Since 1910 Australia's climate has seen an increase of more than 1°C annual average temperature with a warming rate twice as compared to the previous fifty years (McMichael et al. 2006). 2019 was Australia's hottest and driest ever season with the average daytime maximum temperature reaching 30.7°C (highest since 1910 and 2.1°C above average). Australia also recorded six of its hottest single days on record in 2019. All time high temperatures were witnessed during the month of December in Australia with an average maximum of 40.9C on 17th December which reached to 41.9C a day later, beating the 2013 record of 40.3C (Oldenborgh et al. 2020) and as temperature continues to rise due to climate change Australia is going to experience more intense heat waves and severe bushfires (Downie 2006).

At the same time, rainfall patterns in Australia are also changing. Since the 1990s there has been a decrease of approx 11% in April-October rainfall in Southeastern Australia with more frequent droughts (Head et al.2014) but an increase of rainfall in northern Australia. Overall rainfall is declining but whenever rain comes it falls as heavy downpours, increasing the risk of rising sea level thus making Australia swing between droughts and floods leading to weather whiplash.

4. Climate Drivers Hehind the Bushfires

Conditions of extreme heat and drought were evident in 2019 when Australia was hit by three of its main climate driving forces-El Nino generally associated with hot and dry conditions coincided with positive Indian Ocean Dipole(IOD) which caused a record strength temperature gradient highest in sixty years between warming western(African coast) Indian ocean faster as compared to eastern parts(Australian coast) of Indian ocean, leading to more than normal rainfall and floods in eastern African countries and pushing the moisture away from Australia in the month of spring. This positive

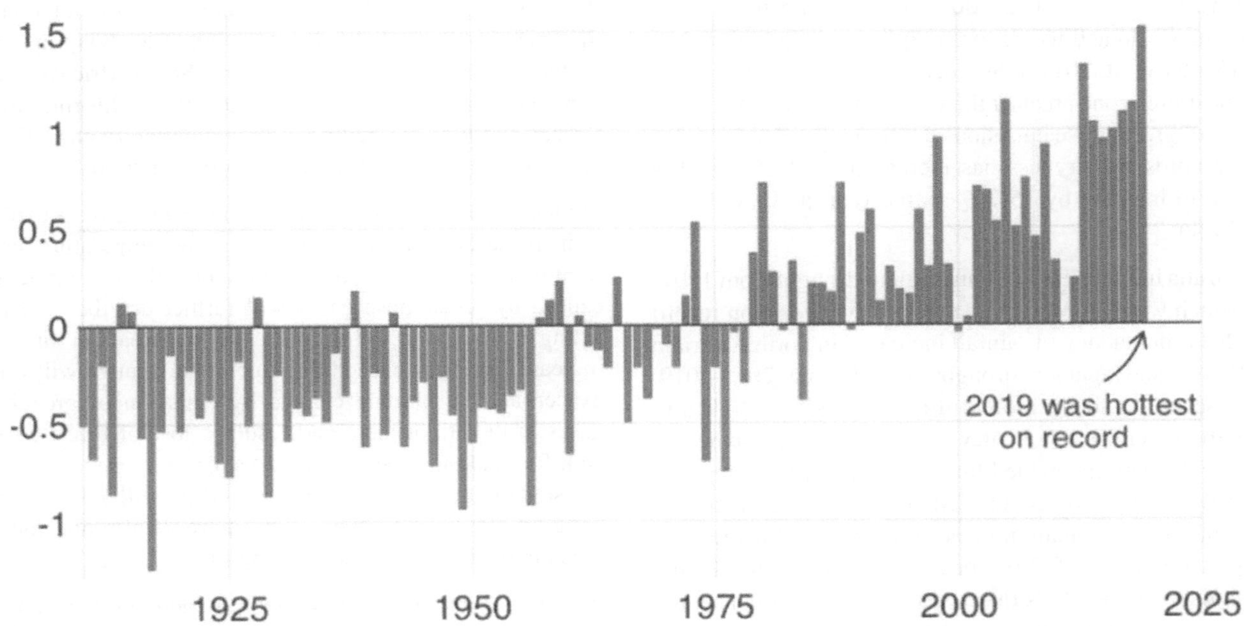

Fig. 53.1 Australian annual mean temperature above/below average (°C) (Australian Government Bureau of Meteorology)

dipole acted as a predecessor for the alarming bushfires and was responsible for less than average rainfall in southern Australia. Since the 1960s positive IOD have become more frequent and as climate will continue to warm this trend is expected to continue (Cai et al. 2009)

Another event, southern annular mode (SAM) was in the negative phase during the months of November-December when the bushfires were at their peak. This SAM was generated by warming of winds circling in the stratosphere over Antarctica in the month of September causing the stratosphere temperature to reach 40°C which added to the already hot and dry conditions present in Australia. Positive IOD and negative SAM combined with changing climate were responsible for Australia's record breaking heat and severe bushfires (Harris et al. 2019).

Strong El Nino along with positive IOD was seen in 2015, also 2013 and 2018 saw persistent negative SAM, these years witnessed earlier than usual and more fierce bushfires.

Changing climate has influenced two important parameters -rising temperature and drought (Jalaludin et al. 2020) both important for triggering bushfires. Global warming is an important part of climate change (Preston et al. 2006) which has made Australia hotter, more prone to droughts and longer duration of bushfire season (Bourne et al. 2018). Australia has one of the highest per capita emissions in the world (Parra et al. 2019). Australia is also among the leading exporter of coal and liquid natural gas in the world and is expanding its oil and gas exploration and production (Makhijani et al. 2015) Australia's carbon dioxide emissions are the highest among the world and its carbon footprint is even higher than USA (Miller et al. 2017)). Southeast Australia is among the three most fire prone areas of the world and even with a small increase of greenhouse emissions the forest fire danger index which records the dryness has increased by 4-25% and is expected to increase by 15-70% in the year 2050 (Williams et al. 2001).

SE Australia has witnessed a long spell of drought from 1997-2009 which was followed by the wettest two years on record 2010-2011, deficiency of rainfall increased in north Australia in 2013-14 and again a drought period from 2017-2019. Extent, severity and duration of droughts has been linked to the southern hemisphere hadley cell (Post et al.2014) which has expanded and intensified further by global warming (Hu et al.2018) leading to less rainfall. This decrease of rainfall further dried the vegetation creating more fuel for fire. Hot and dry conditions of 2019 had fallen for the third time consecutively, which tells there was hardly any water on the ground to absorb heat thus spreading fire in a shorter time.

5. Australian Bushfires-Generating their own weather

Large size and intensity of these 2019 bushfires trapped heat and generated pyrocumulonimbus clouds which generated lightning strikes and tornado strong winds which fuelled further fires and also generated large amounts of smoke. As this intense heat and smoke rose higher in the atmosphere, cooling of this plume resulted in scattering of ash over longer distances. Plumes of black carbon were seen as far as 7000 miles away from these deadly fires. Smoke coming from Australian bushfires was seen in Brazil, Chile, Argentina, Uruguay and as far as South America situated at a distance of 16,213 kms from Australia. These pyrocumulonimbus clouds had generated fire tornadoes in Canberra bushfires in 2003 (Dold et al.2005).

During Black Saturday (7 February-2009) a number of fire plumes had produced pyrocumulonimbus storms which had reached heights of 15 kms and generated massive lightning strikes near Melbourne in southeast Australia (Dowdy et al.2017). Intense heat and high temperature are leading to the rise of pyrocumulonimbus clouds. Global warming is increasing the frequency of bushfires which in turn is creating pyrocumulonimbus clouds which is making the fight against bushfires unpredictable and more dangerous to fight.

6. Conclusion

Australia is experiencing extreme weather conditions throughout the year but other countries must not relax as they may not be far behind. All the countries lying in similar latitude as Australia-Chile, Argentina, South Africa and areas experiencing mediterranean climate like California, Spain, Greece and Italy are getting drier and hotter. Even California had witnessed a record setting bushfires in 2020.

Frequency, intensity and area burnt under bushfires is expected to increase even further with increasing temperature, erratic rainfall pattern and frequent droughts. Rising temperature caused by greenhouse gases will further dry the vegetation acting as fuel and enhance the rapid spread of fire. This increased fire risk caused due to climate change will further exacerbate the climate change by emission of greenhouse gases in the atmosphere and result in loss of forest cover, a sink for soaking greenhouse gases, creating a vicious cycle. These climate changes and variability will have a direct bang-on effect on tourism, agriculture, health and increase the country's spend on devastating fires.

It is therefore the need of hour to address the changing climate to reduce the threat of fierce bushfires in Australia.

Fuel management to reduce the greenhouse gases emission is equally important. Without the strict implementation of laws, aiming to control greenhouse gases Australian continent is likely to suffer from frequent bushfires and these devastating bushfires would be the new norm rather than the exception!

7. Acknowledgements

Authors are thankful to WCDM for giving them a wonderful chance to write on a topic which needs public attention as well as to Galgotias University for giving them time and infrastructure to work for the conference.

References

1. Beeson, M., & McDonald, M. (2013). The politics of climate change in Australia. *Australian Journal of Politics & History*, *59*(3), 331–348.
2. Deb, P., Moradkhani, H., Abbaszadeh, P., Kiem, A. S., Engström, J., Keellings, D., & Sharma, A. (2020). Causes of the widespread 2019–2020 Australian bushfire season. *Earth's Future*, *8*(11), e2020EF001671.
3. McMichael, A. J., Woodruff, R. E., & Hales, S. (2006). Climate change and human health: present and future risks. *The Lancet*, *367*(9513), 859–869.
4. van Oldenborgh, G. J., Krikken, F., Lewis, S., Leach, N. J., Lehner, F., Saunders, K. R., ... & Otto, F. E. (2020). Attribution of the Australian bushfire risk to anthropogenic climate change. *Natural Hazards and Earth System Sciences Discussions*, 1–46.
5. Downie, C. (2006). *Heating up: bushfires and climate change*. Australia Institute. Australian Government Bureau of Meteorology, Fig: Australian annual mean temperature above/below average.
6. Head, L., Adams, M., McGregor, H. V., & Toole, S. (2014). Climate change and Australia. *Wiley Interdisciplinary Reviews: Climate Change*, *5*(2), 175-197.
7. Cai, W., Sullivan, A., & Cowan, T. (2009). Climate change contributes to more frequent consecutive positive Indian Ocean Dipole events. *Geophysical Research Letters*, *36*(23).
8. Harris, S., & Lucas, C. (2019). Understanding the variability of Australian fire weather between 1973 and 2017. *PloS one*, *14*(9), e0222328.
9. Jalaludin, B., Johnston, F., Vardoulakis, S., & Morgan, G. (2020). Reflections on the catastrophic 2019– 2020 Australian bushfires. *The Innovation*, *1*(1).
10. Preston, B. L., & Jones, R. N. (2006). *Climate change impacts on Australia and the benefits of early action to reduce global greenhouse gas emissions* (p. 41). Canberra: CSIRO.
11. Bourne, G., Stock, A., Steffen, W., Stock, P., & Brailsford, L. (2018). Australia's rising greenhouse gas emissions.
12. Parra, P. Y., Hare, B., Hutfilter, U. F., & Roming, N. (2019). Evaluating the significance of Australia's global fossil fuel carbon footprint.
13. Makhijani, S., & Doukas, A. (2015). G20 subsidies to oil, gas and coal production: Australia. *Background Paper for the Report Empty Promises: G20 subsidies to oil, gas and coal production.*
14. Miller, J., Du, L., & Kodjak, D. (2017). Impacts of world-class vehicle efficiency and emissions regulations in select G20 countries. *ICCT: Washington, DC, USA*, 24.
15. Williams, A. A., Karoly, D. J., & Tapper, N. (2001). The sensitivity of Australian fire danger to climate change. *Climatic change*, *49*(1), 171–191.
16. Post, D. A., Timbal, B., Chiew, F. H., Hendon, H. H., Nguyen, H., & Moran, R. (2014). Decrease in southeastern Australian water availability linked to ongoing Hadley cell expansion. *Earth's Future*, *2*(4), 231–238.
17. Hu, Y., Huang, H., & Zhou, C. (2018). Widening and weakening of the Hadley circulation under global warming. *Science Bulletin*, *63*(10), 640–644.
18. Dold, J. W., Weber, R. O., Gill, M., McRae, R., & Cooper, N. (2005). Unusual phenomena in an extreme bushfire.
19. Dowdy, A. J., Fromm, M. D., & McCarthy, N. (2017). Pyrocumulonimbus lightning and fire ignition on Black Saturday in southeast Australia. *Journal of Geophysical Research: Atmospheres*, *122*(14), 7342–7354.

CHAPTER 54

Global trend of Forest Fire and its Management

Rajesh Thakur
Commandant, 12 NDRF Itanagar,
Arunachal Pradesh

Abstract

Wildfire has been an important process affecting the Earth's surface and atmosphere for over 350 million years and human societies have coexisted with fire since their emergence.

Wildfire phenomenon plays a vital role in the atmospheric and terrestrial system. It poses threat not only to the bushes and forest but also to the entire regime of fauna and flora seriously disturbing the bio-diversity, ecology and environment of a region.

Many consider wildfire as an accelerating problem, with widely held perceptions both in the media and scientific papers of increasing fire occurrence, severity and resulting losses.

However, important exceptions aside, the quantitative evidence available does not support these perceived overall trends. Regarding fire severity, limited data are available. Direct fatalities from fire and economic losses also show no clear trends over the past three decades. Trends in indirect impacts, such as health problems from smoke or disruption to social functioning, remain insufficiently quantified to be examined.

In India as per the data from the National Forest Inventory program of Forest Survey of India (FSI), 9.89% of forest areas are heavily affected and 54.40% mildly affected due to forest fires. Therefore, almost two thirds of our forest areas are vulnerable to forest fires, with forests in the north-eastern region and central India being the most vulnerable. While the overall green cover has increased in the country, the forest cover in the north-east particularly in Mizoram, Arunachal Pradesh and Nagaland has decreased. Forest Fires could be one of the causes. As per data of FSI of last 15 years there is unclear trend of forest fire incident in India also.

Keywords: Forest fire; Jhum agriculture (slash and burn); Deforestation

1. Introduction

Every year fires occur on several hundred million hectares of forests and other vegetation types worldwide. Fires in forests and open vegetation (woodlands, bushland, grasslands, savannahs, steppes) are commonly designated as wildland fires or vegetation fires. Causes of uncontrolled wildfires and reasons for using fire as a management tool in agriculture, pastoralism, and forestry (land-use fires, prescribed burning) depend on the local environmental conditions like climate, vegetation type and the culture with its specific social and economic conditions. Lightning is the most important natural cause of wildfires, especially in the less populated regions of the forests in high northern latitudes and tropical savannahs. Human-caused wildfires, however, play the most significant role in most regions of the world. The major reason is negligence, including escaped land- use fires, and sometimes arson.

2. Worldwide Causes of Forest Fire Weather Anthropogenic or Natural?

Study and data says that more than 90 percent forest fires all over the world including India are initiated due to anthropogenic activities, in which majority is related to land use pattern. Conversion of forestland to agriculture, pastureland and plantation are the main eminent factors for such incidences. In tropics, two types of practices for conversion of forest land are common i.e. shifting cultivation/ Jhum and permanent conversion.

2.1 Jhum Cultivation

The Jhum cultivation is the farming practice where forest land is allowed to return to forest crop after a shorter period of use. In such practices, the forest vegetation is cut at the end of rainy season and allowed over to dry for some period and then burnt. The vegetation ash is left in the area, which provides sufficient nutrient for the agriculture crop. This cleared area is used for agriculture for few years until the crop production decline. After this the land is left for natural forest regeneration and new forest area is searched for agriculture purpose. The practice is commonly used by millions of people throughout the world causing immense loss to forest. In the recent past, due to increasing population pressure and lack of other alternative livelihood resources, this traditional land use techniques has not only increased but also lead to shortening of shifting cultivation cycle.

2.2 Permanent Conversion

With the increasing population pressure, forest areas in many countries are permanently converted to non forestry purposes, such as pastures, agriculture land, resettlement and orchards. Such conversion of forest also follows after clear felling and setting the permanent forest vegetation on fire. The resettlement in Tarai belt of U.P. and other states in India are the recent example of such conversion, where thousands of hectares of forest area are cut and burnt to resettle the migrant population and do agriculture practice.

The two practices discussed above are major causes of forest fire and degradation of forest cover dominant in tropical countries.

Though reliable data is not available, yet as per an estimate, almost one to two billion metric tons of forest vegetation is destroyed by these practices, resulting into enormous social, economical, ecological and environmental damage

2.3 Other causes

In addition to agricultural use, the forest fire issues are linked with other economical and social factors also. Collection of non timber forest produce (NTFP) like mahua, Sal-seeds, kendu leaves etc. are other major causes for initiating forest fire in many developing countries. Besides economical issues, in some tribal clans forest is ignited to celebrate some ritual customs. This ignited fire usually spreads in the adjoining forest and results into catastrophe.

The aggressive fire suppression has also been up to an extent responsible for extreme fire events by altering the composition of the tree species, which are more fire prone. As a result the forests, specially the dry forests are dominated by heavy ground fuels and high tree densities, which are much more likely to have severe fires. Conversion of fire resistant old forests to plantation is also up to an extent responsible for increased fire risk in the forests.

Among the developmental activities, road construction has been the most devastating factor for increase in forest-fire incidences. Roads are turned as ignition source that contribute to fire risk by providing access to forests, where people can ignite it. Increase in grazing by livestock is another common cause of forest fire throughout the world. Grazing removes the grassy fuels that carry frequent surface fires and often ushers in fire prone exotic plants and shrubs. Fires are initiated by picnickers, grazers, tourists etc. due to negligence. Throwing burning cigarette butts, camp fires, cooking food within or near the forest areas are examples when fire is caused merely by negligence.

3. Forest Fire: The Global Scenario

The problem of wild fire is a universal phenomenon which is a dominant disturbing factor in all types of vegetation throughout the world. Though in world-ecosystem fire is a natural phenomenon and helps the vegetation by organizing physical and biological attributes, influencing energy flows and biological cycles, yet its adverse impact in ecosystem is quite well known. Due to various natural and mostly human induced factors, the severity of forest fires in general is increasing day by day. The adverse impacts of this increased forest fire and its severity have brought this menace in the category of other natural disasters like floods, droughts, earthquakes etc.

Forest fires are common in almost all types of vegetation. In the temperate and northern boreal forests, it occurs regularly during the dry summers. On an average 5 to 20 million hectares of forest are burned annually in North America and Eurasia due to fire hazard. In the Mediterranean forests also the situation is similar and around 0.6 million hectare forest is burned here annually. Though equatorial rain forests are bit moist, however extreme droughts associated with other human induced activities make it too vulnerable to fire. In South Africa, cyclic climate variability caused by El-Nino-Southern Oscillation (ENSO) phenomenon is responsible for extreme fire events in the forests.

In tropical forests, fire is a regular phenomenon at the short interval of one to five years. The Tropical Sub-montane Conifers Forests due its specific vegetation (pine), rich in resin and susceptible to fire, are subjected to forest fire regularly. In addition to standing forest crop, fire is very common in tropical grass- land and Savannas. Tropical Savannas spread over an area of about 2.6 billion hectare (consisting of more or less continuous spread of grass with interrupted trees and shrubs are prone to fire and are burnt regularly. On an average hundred million hectares land in these grasslands are burnt annually, mostly in dry season. As per an estimate about 3 million of vegetation are burned here every year, resulting into three times more gas and particles emission than that by other burning. The intensity of fire in Savannas has increased in the recent past due to population pressure too. Though all forests bearing countries are hit by fire menace, one or other time, some are more prone.

3.1 Fire Statistics

The basic problem with fire research and development is lack of reliable statistics about the number of incidence, damage, area burnt, etc. This type of information is rarely available and if available, its reliability is question marked. The reliable data on occurrence of fires, area burned and damage are available only for limited countries. Economic Commission for Europe (ECE) publishes such data for Northern Hemisphere countries, including Eastern and Western European countries, U.S.A., Canada and Russia. For other countries data is collected from various sources and published regularly in different reports and journals. "International Forest Fire News" published from Germany, is a popular Journal publishing such reports regularly. "Global Fire Monitoring Centre" is another organisation, which is bringing out such reports on updated forest fire data worldwide.

4. Impact of Wildfires

Has the Fire Impacted All Types of Forest Regions of the Globe?

Wildfires can have both positive and negative effects on nature. In some ecosystems fires play an ecologically significant role in maintaining biogeochemical cycles. For instance, many savannahs in the tropics and subtropics burn annually or in intervals of several years (short fire cycles of 1 to 3 years). These fires are important to regenerate and stimulate the growth of grass and maintain the habitat structures of important wildlife species and domestic animals. Such "fire savannahs" are highly productive and are home to important plant and animal biodiversity. Similar adaptations are found in the natural northern coniferous forests which have a fire cycle ranging from several years or decades to several hundred years. Useful fires that improve the conditions for sustainable growing of crops are also set by experienced agriculturalists, herdsmen and hunters all over the world.

In other ecosystems fires are extremely destructive. The tropical rain forests are most vulnerable during extreme droughts such as those which repeatedly occur during El Nino years. During such extended dry spells the rain forest trees must protect themselves against the loss of water and shed their leaves. Consequently, the forest canopy becomes more open, sunlight penetrates down to the forest floor where the shed leaves become highly flammable. Fires escaping from agricultural sites or from pasture-burning are now able to spread into the forest and result in reduction of biodiversity and/or to long- term site degradation. During the strong El Nino of 1997- 1998 large areas of rain forest were destroyed by fire in Asia and the Americas. A major reason for these detrimental fires was the inappropriate use of fire in converting forest or secondary vegetation to plantations.

In the temperate and boreal forests major damages occur in places where people are living at the interface between metropolitan or other residential areas and the vegetated lands such as brushland or forests. Thus, in almost all regions of the world wildfires burning under extreme weather conditions affect economies, human health and safety, with consequences which are of a significance and severity comparable to other major natural hazards. The vast majority of destructive fires are caused by the carelessness of humans. Thus, unlike the majority of the geological and hydro-meteorological hazards, vegetation fires represent a hazard which can be predicted, controlled and, in many cases, prevented.

Have Fire Impacts Increased in Many Regions Around the Globe?

(i) Has the Fire intensity and severity increased?

While the trends in area burned explored above have implications for the effects of fire on global carbon emissions, ecosystems and society, the spatial extent of burning is not always closely linked to the impacts of a fire. From a perspective of fire ecology or risk to infrastructures, the intensity of a fire (i.e. its rate of energy output), its severity (its ecosystem impacts) and its spatial patterns (degree of patchiness) may be more important than the total area burned. For example, the degree of vegetation consumption, the depth of burning into the organic and mineral soil, and the proximity of areas less affected or not by fire are key in determining the length of time for a burned area to 'recover'. The notion that fire intensity and severity have increased in recent years pervades media reports and some of the literature. Whether or not this is the case is not easy to ascertain given that these parameters and associated trends are much more difficult to determine compared with area burned. All else being equal, fire intensity can indeed be expected to increase with air temperature, and it can be deduced that areas that are experiencing higher atmospheric temperatures in the fire season associated with global warming would experience more intense fires. For example, the catastrophic 2009 Black Saturday fires of Victoria (Australia) and 2019-2020

Australian bushfire seasons were reportedly associated, among other factors, with unprecedented high atmospheric temperatures and fire intensity. Whether or not these extreme events signifie a trend or may simply be the result of longer-term natural variability in fire behaviour remains an open question.

However, a few studies exist that have explicitly examined trends in fire severity. These have mainly focused on the western USA, an area where there are particular concerns about increased fire activity. Thus, while there is evidence of a recent increase in proportional fire severity for a specific forest type in California, these independent studies do not support the notion of an overall increase in fire severity over the past few decades in the fire-adapted forested landscapes in the western USA. (*Source:* US National Interagency Fire Centre)

In Australian bushfire season 2019–2020 over 18 million hectares have burned according to media reports, destroyed over 5,900 buildings including over 2,800 homes. In addition to human fatalities, many millions of animals are reported to have been killed. Thus if we analysis the data of Table 54.1, there are 3-4 fire incidents among 13 of severe intensity that also tends that there is no clear trend.

(ii) Impacts on society: weather forest fire has direct effects on people?

While the ecological impacts of fire or their interactions with climate are of concern to scientists, natural resource managers, policymakers and the public, policy and public perception regarding fire in the landscape is primarily shaped by the impacts of fire on people and society.

Lives lost, together with direct damage to homes and other infrastructures create wide media attention and are probably of greatest importance here. For example, the Black Saturday fires of 2009, in which 173 people lost their lives and loss in 2019-2020 Australian bushfire shook Australian society and led to major reconsideration of landscape fire related policy. These and other tragic losses to lives from fire may or may not have been preventable, but should be also seen in perspective to other risks to lives. When considering some of the extreme landscape fires as a form of natural disaster, the number of deaths is actually relatively low compared with other natural disaster types. For example, data by the Emergency Events Database (EM-DAT 2015) suggest that over the period 1901–2014, 3753 people have been killed by wildfire, compared with over 2.5 million from earthquakes and nearly 7 million from floods. These data indicate a comparatively low risk of death as a result of fire compared with that from other natural disaster types (Table 54.2), particularly considering that approximately 4% of the global vegetated land surface burns every year.

Table 54.1 Largest fires of the 21st-century

Rank	Name	Country	Area burned (km^2)	Deaths
1	2003 Russian wildfires	Russia	200,000	0
2	2019-2020 Australian bushfire season	Australia	180,000	34+
3	2019 Siberia wildfires	Russia	43,000	0
4	2014 Northwest Territories fires	Canada	34,000	0
5	2009 black Saturday bushfires	Australia	21,000	173
6	2020 California wildfires	United States	18,000	31
7	2010 Bolivia forest fires	Bolivia	15,000	0
8	2011–2012 Australian bushfire season	Australia	14,000	0
9	2006-2007 Australian bushfire season	Australia	13,000	5
10	2017 British Columbia wildfires	Canada	12,000	0
11	2015 Russian wildfires	Russia	11,000	33
12	2012–2013 Australian bushfire season	Australia	9,000	4
13	2019 Amazon rainforest wildfires	Brazil, Bolivia, Paraguay, Peru	9,000	2

Source: Media reports & UNISDR

Global comparison of human and economic losses derived from wildfire, earthquakes and flood disasters from 1901 to 2014. (*Source:* EM-DAT 2015)

(iii) Impacts on society: weather forest fire has direct impacts on economic?

Human losses aside, the direct financial costs, such as the damage to homes and other infrastructures, often dominate the perception of the fire impacts and an increase in these is often highlighted in the media or scientific papers and reports. These estimates of losses, however, only include damage to property, crops and livestock and do not reflect losses from fire events not classified as disasters. Other

Table 54.2

	wildfires	earthquakes	floods
No. of events	387	1291	4481
people killed	3753	2 574 627	6 947 908
people injured	6812	2 614 875	1 329 923
people affected (million)	6	190	3604
risk of death (%)a	0.06	1.4	0.02
total direct damage (million US$)	54 828	774 771	681 427
cost per event (million US$)	142	600	152
cost per person affected (US$)	9138	4078	189

Table 54.3

	No. events	people killed	total people affected	death rate/ event	economic costs (million US$)
Africa	25	272	21 672	11	440
America	118	234	1 229 175	9	25 229
Asia	50	748	3 188 257	30	11 892
Europe	89	462	1 295 562	18	12 619
Oceania	21	224	74 320	9	2121
total	303	1940	5 808 986	78	52 301

important economic parameters not included here are the costs arising from human losses, injuries and longer-term health implications. Furthermore, fire suppression costs are not considered in these figures. Data of EM-DAT of Table 54.2 & Table 54.3 shows that direct economic impact due to fire incidents is comparatively less to other disasters like earthquake and flood.

Human and economic losses from wildfire 'disasters' by global region from 1984 to 2013. Costs are based on the actual value of US$ in a given reporting year. (*Source:* EM-DAT 2013.)

Has forest fire increased in many regions around the globe?

Forest fire is often quoted as an increasing issue around the globe but some research and study has different opinion. One reason for this apparent contradiction may be that the global extent of fire is not necessarily correlated with impacts on human society. Another reason may be that our wider perception of fire is shaped by some widely publicized regional trends and a lack of discrimination between reported fire activity parameters. An important distinction regarding the latter is that between area burned (i.e. total ha or km^2) and fire occurrence (i.e. the number of fires for a given area and period).

Recent trends in area burned can now be derived from satellite observations and national records with reasonable accuracy at regional and global scales. If we analyse the data of FAO FRA-2010 Fire occurrence and associated area burned between 1990 to 2005 is in decreasing trend in some continents. However, the past two decades data indicates that there is total 13 major fire incidents reported (Table 54.1) all over the world but in 2003 Russian wildfires and 2019-2020 Australian bushfire reported unprecedented increase in

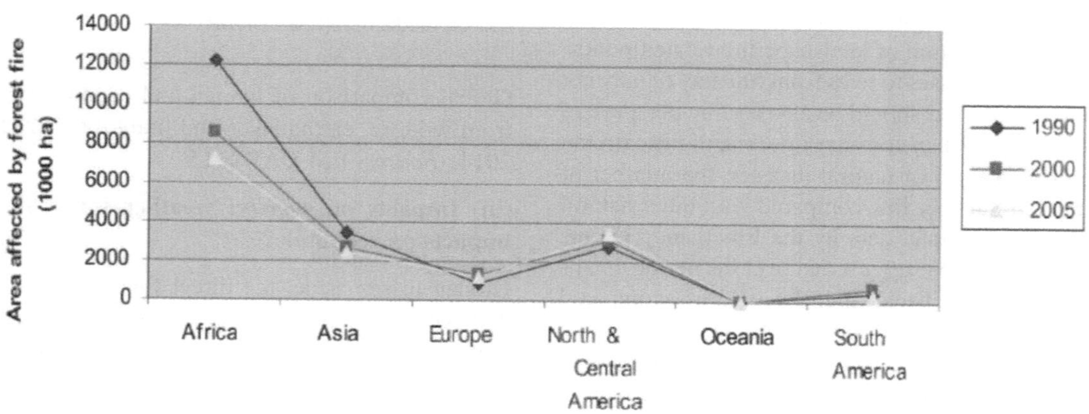

Fig. 54.1 Trends in Forest Annually Affected by Fire by Region, 1990-2005

Source: FAO FRA-2010

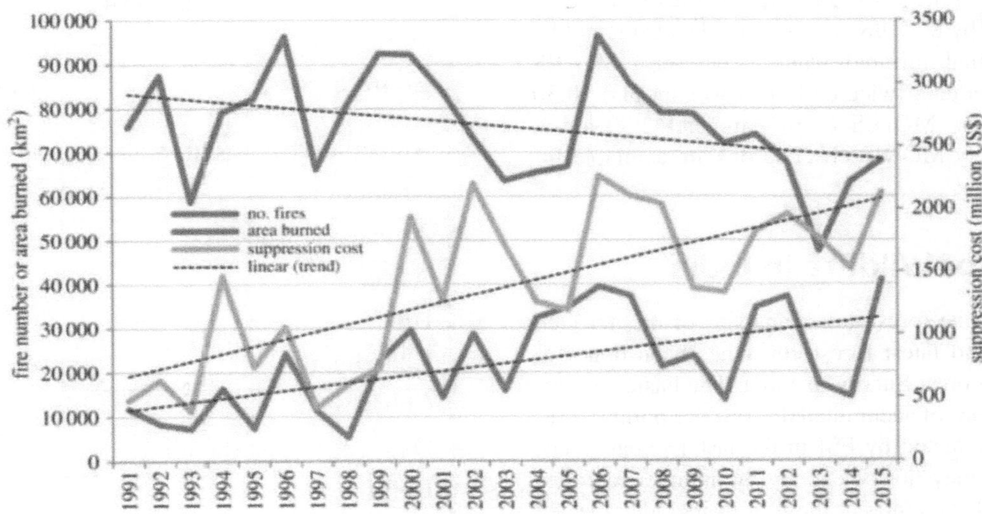

Fig. 54.2 Area burned, number of fires and suppression costs for the USA with linear trend lines (1991–2015)

Data: National Interagency Fire Centre.

area burn around 20 million hectares forest got burnt in both incidents. Again if we see the data of EM-DAT of last century (Table 54.2) then it indicates 3-4 forest fire incidents reported every year while in last two decades average 01 major forest fire reported every year.

As per data of US National Interagency Fire Centre in 20th century there is slight decrease in occurrence of fire but increasing trend in area burned in USA.

Thus, while there are clearly some noteworthy trends in area burned for specific recent periods and regions, the general perception of increasing fire occurrence around the world is not supported by the data available to date. However, in some major forest fire unprecedented increase in burnt area was reported.

5. Forest fire: The Indian Scenario

Forest fires are a recurrent annual phenomenon in India also. Almost all the fires in forest areas are manmade and usually the forest dependent communities are known to use fire for various purposes ranging from clearing community forests for shifting cultivation to clearing the forest floor to encourage grass growth. However, uncontrolled and unmanaged.

In recent years, there have been spurts in the number of forest fire incidents which is a cause of serious concern for all. Forest fire disaster of 2016 in Uttarakhand and Himachal Pradesh, Kurangani fire incident of Tamil Nadu, Mt. Abu and Vaishnodevi fires of 2018 are some of the recent examples from our country.

Forest fires cause large scale destruction of forest resources and biodiversity in the country every year. One forest fire incident may nullify all efforts of conservation and plantations done in the past several years. A recent analysis done by FSI, shows that nearly 36% of country's forests are prone to fires and of this, over 10% are highly prone. Forest fires in our country are largely of anthropogenic origin and therefore preventable to a large extent.

Apart from public awareness and their participation in controlling forest fires, use of all the technological options for prevention, detection, early warning, fire fighting and damage assessment should be employed under a scientifically prepared.

5.1 Recent Technological Initiative of Fire Mapping in India

FSI has been carrying out forest fire monitoring at national level since 2004 using Remote Sensing and GIS technology with the main objective of detecting active forest fires at nascent stage and to inform State Forest Departments about the forest fire incidences on near real time basis. FSI uses Moderate Resolution Imaging Spectro-radiometer (MODIS) data with 1km X 1km spatial resolution and Suomi-National Polar-orbiting Partnership (SNPP)-Visible Infrared Imaging Radiometer Suite (VIIRS) data with 375m X 375m spatial resolution.

MODIS is an instrument that is on-board two satellite platforms owned by NASA; Terra and Aqua. The MODIS instrument has 36 spectral bands available to view the Earth.

The image resolution used for detection of fires is 1 km. A hotspot is detected by MODIS sensor using data from middle infrared and thermal infrared bands. The Terra MODIS instrument acquires data twice daily (10:30 a.m. and 10:30 p.m.), as does Aqua MODIS (2:30 p.m. and 2:30 a.m.). Therefore, four daily MODIS observations are available for forest.

6. Statistics of Wildfire in India

Mapping of fire prone forest areas, analysis of the detected forest fire points and latest forest cover has been done for whole country by Forest Survey of India. The basic data for the study is the record of accumulated forest fire points (FFP) which have been detected by FSI in the last 13 years since 2004. These FFPs may not be exactly the number of forest fire occurrences in the country as these FFPs include repeated detections due to continuance of forest fires at many places.

The data of FFPs may not be uniform for all the years because threshold value of confidence for detecting forest fires and also forest fire season has not been kept same in all the past thirteen years. Never the less record of FFPs accumulated in the past thirteen years is an unbiased data which is strongly correlated with fire proneness of an area.

6.1 Forest Fire Points Detected by FSI

Total number of forest fire points in the last thirteen years in each State and UTsare shown in the following Table 54.4.

Has Forest Fire Increased in Many Regions of India?

Occurrence (frequency) of wildfire in India in last 13 years can be analyzed by following graph.

The year wise total number of detected forest fire points in the last thirteen years is show in the bar chart (Fig. 54.3).

Thus, if we analyze the data of FSI of last 13 years forest fire incidents in country there is no clear trend of increasing forest fire occurrence, there is increasing trend since 2004 to 2010 but it again decreases up to 2016 and in 2017 maximum no of forest fire occurrence reported in country.

Are many regions of country falls under extreme or highly fire prone zone?

Fire proneness has been categorized into different severity classes based on frequency of forest fire points of the last 13 years in each grid of the size 5 Km X 5 Km. Average annual frequency of FFPs in each grid has also been calculated by simple division of total frequency by 13 i.e. the number of years.

Table 54.4

State/UTs	Forest Fire Points
Andaman & Nicobar Islands	165
Andhra Pradesh	15,582
Arunachal Pradesh	6,422
Assam	20,862
Bihar	2,229
Chandigarh	1
Chhattisgarh	25,995
Dadra & Nagar Haveli	9
Delhi	8
Goa	70
Gujarat	2,393
Haryana	448
Himachal Pradesh	1,178
Jammu & Kashmir	977
Jharkhand	6,562
Karnataka	7,352
Kerala	1,719
Madhya Pradesh	24,422
Maharashtra	20,686
Manipur	17,907
Meghalaya	12,820
Mizoram	32,659
Nagaland	11,296
Odisha	26,719
Puducherry	9
Punjab	837
Rajasthan	1,159

Figure 54.4 map of Forest Survey of India depicts that North Eastern states and central India falls under extreme and highly fire prone zone of country.

However, the percentage wise forest cover in different fire prone classes is given in following Table. Nearly, 4 % of the country's forest cover is extremely prone to fire, where as 6 % of forest cover is highly prone. Thus about 10% of forest

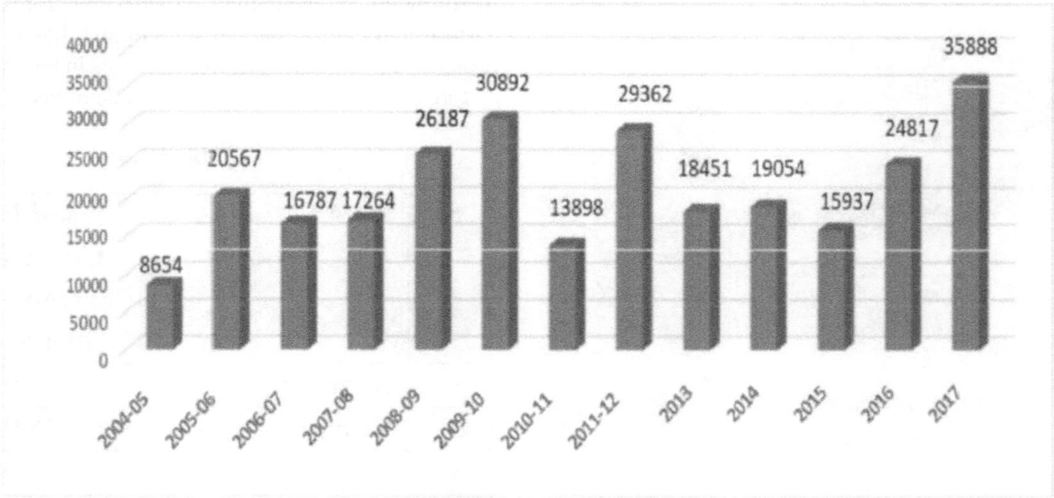

Fig. 54.3 Graphical Representation showing the total forest fires in 13 years

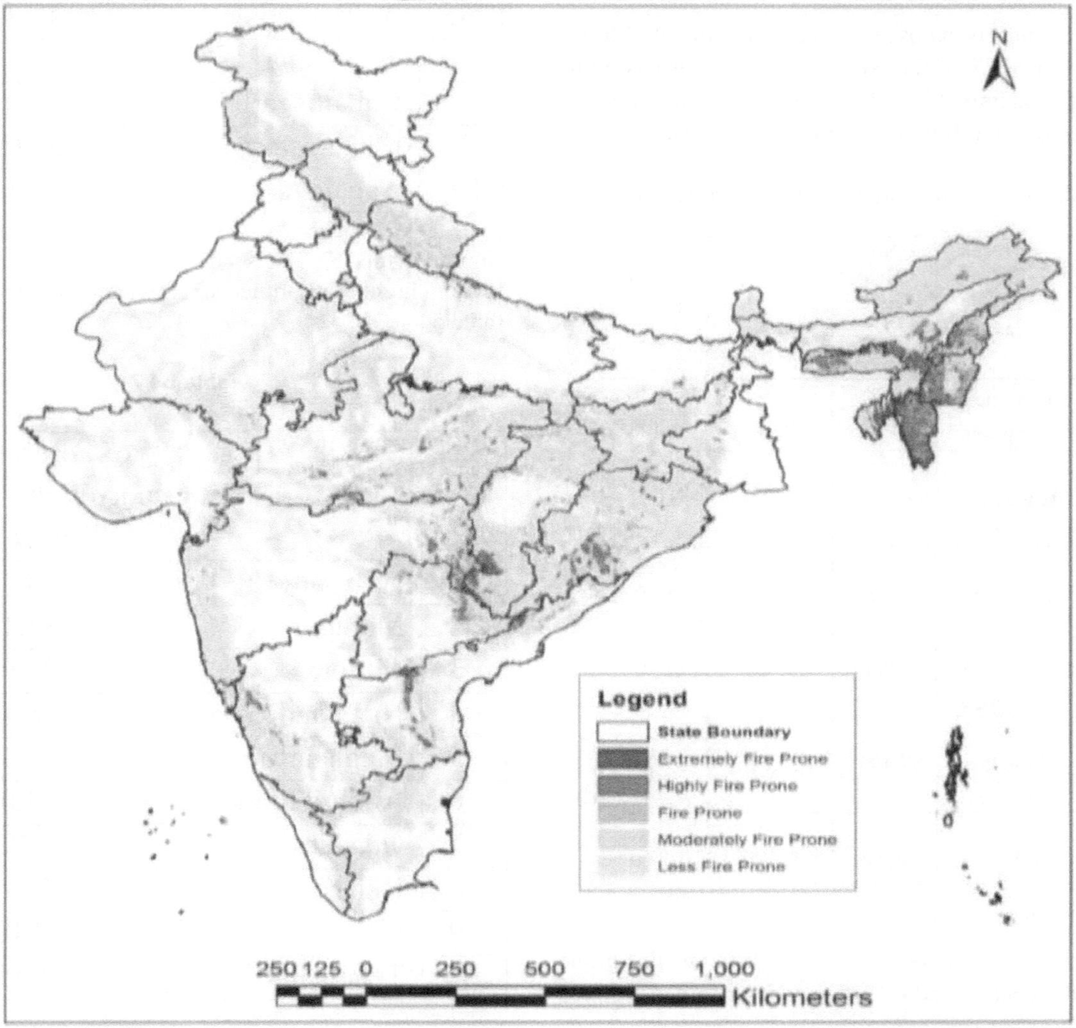

Fig. 54.4 Map Showing fire prone forest areas under different fire prone classes

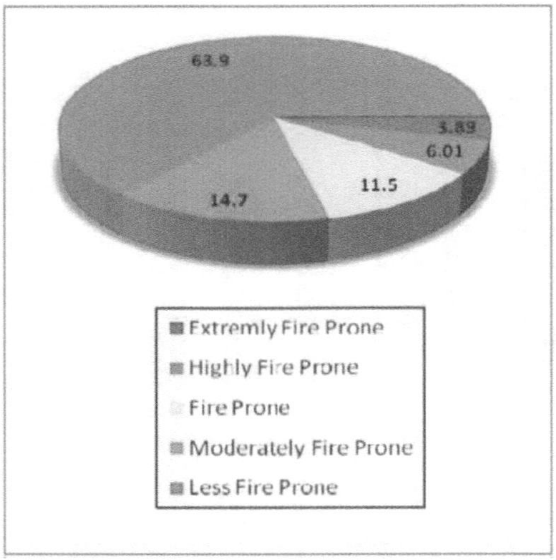

Fig. 54.5 Diagrammatic Representation showing the percentages of Fire Prone Areas

cover in the country is exposed to a very high threat of forest fire. In all, over 36 % of the country's forest cover is prone to frequent forest fires. Fig. 54.5 depicts distribution of fire prone forest areas in different categories in pie chart.

Table 54.5 Forest cover in different fire prone classes

S/N	Forest Fire Prone Classe	No. of Grids*	Forest Cover* *2 (in km)	% of Total forest cover
1	Extremely fire prone	665	25,617	3.89
2	Highly fire prone	2,259	39,500	6.01
3	Fire prone	3,708	75,952	11.50
4	Moderately fire prone	5,496	96,422	14.70
5	Less fire prone	57,489	4,20,625	63.90
	Total	69617	6,58,116	100.00

*Grid size 5km x 5km

Area and percentage of forest cover in different fire prone classes for all the States and UTs of the country are mentioned in following table. It may be seen that forest in the north eastern and central part of the country has most of the fire prone forest areas. List of top 10 States in terms of percentage of forest cover in extremely fire prone areas is given below.

Table 54.6

S/N	STATE / UTs	% of Forest Cover under extremely fire prone class
1	Mizoram	29.91
2	Tripura	26.95
3	Assam	21.98
4	Meghalaya	5.74
5	Manipur	4.48
6	Andhra Pradesh	4.27
7	Telangana	4.21
8	Chhattisgarh	3.90
9	Maharashtra	3.40
10	Nagaland	3.05

Similarly, top 10 States according to percentage of forest cover in highly fire prone category are given in the following Table 54.7.

Table 54.7

S/N	STATE / UTs	% of Forest Cover under highly prone class
1	Mizoram	38.46
2	Manipur	33.13
3	Tripura	21.90
4	Nagaland	18.48
5	Meghalaya	18.38
6	Andhra Pradesh	13.04
7	Odisha	7.73
8	Bihar	7.15
9	Uttar Pradesh	7.10
10	Telangana	6.89

Study of above data of FSI showing that around 40% region of country comes under moderate to extremely prone zone to fire and north eastern and central part of the country has most of the fire prone forest areas. This outcome is supported by latest study of forest fire in India i.e. 2019-2020 fire seasons.

Is forest fire frequency and intensity both high in same region of India?

Forest fire incidences detected by MODIS in the different physiographic zones of India during the last fire season (2019-20) are shown in Fig. 54.6.

The study has been carried out at the country level based on 8 physiographic zones of India namely Central Highlands, Deccan, Eastern Ghats, Eastern Himalayas, Eastern Plains, North-East Ranges, Western Ghats and Western Himalayas as shown in Fig. 54.7.

Apart from the assessment of burnt forest areas in the country, the study presents few interesting results.

- As seen from Fig. 54.6, the number of detected FFPs, like every year, are highest in the North-East region (12261) though the total burnt area in the region is not maximum (Fig. 54.8). This implies that the average burnt forest area per fire event is smaller in the region (17.59 ha) as seen from Fig. 54.9 This is possible due to the high content of moisture in the vegetation and soil in the forests of North-eastern region.
- On the other hand, average burnt area per fire event in the Deccan and Central Highlands is large i.e. 370.45 ha and 226.94 ha respectively (Fig. 54.4). Also, the maximum burnt area was seen in Deccan PZ followed by Central Highlands (Fig. 54.8). The reason for this could be preponderance of dry deciduous forest types, high temperature and dry conditions in the Deccan and Central Highlands of India which create conditions for faster spread of fires and thus, engulfing larger forest areas.

Figure 54.10 shows that fifty percent of the total fire affected area has been found in the Deccan physiographic zone followed by Central Highlands with 19.50%. North-eastern region with highest number of forest fire detections by MODIS (12261) has contributed 12.50% to the total fire affected forest areas. There are much lesser MODIS-based forest firs detections (5001) in the Deccan and Central Highlands PZs combined as compared to north-eastern ranges (12261), yet they have contributed to 70% of the total burnt forest area in the country.

The study and the data available to date do not support a general increase in forest fire occurrence worldwide and in India at least in last two decades. It is a mixed trend and area burned or fire severity for many regions of the world and India needs more specific data to come at any conclusion

Furthermore, the data evaluated here do not support the perception of increasing direct losses from fire. Over the past decades there is no clear trend of increasing direct losses such as losses of life or infrastructure. While any fire-related death can be seen as one too many, at least the risk of direct death from fire for the population as a whole is low compared with other natural hazards.

7. Can We Prevent Wildfires!

The prevention of destructive fires is an integral element of land-use policies and fire management strategies. Fire prevention must address a broad range of elements and sectors of society, natural resources and environmental management,

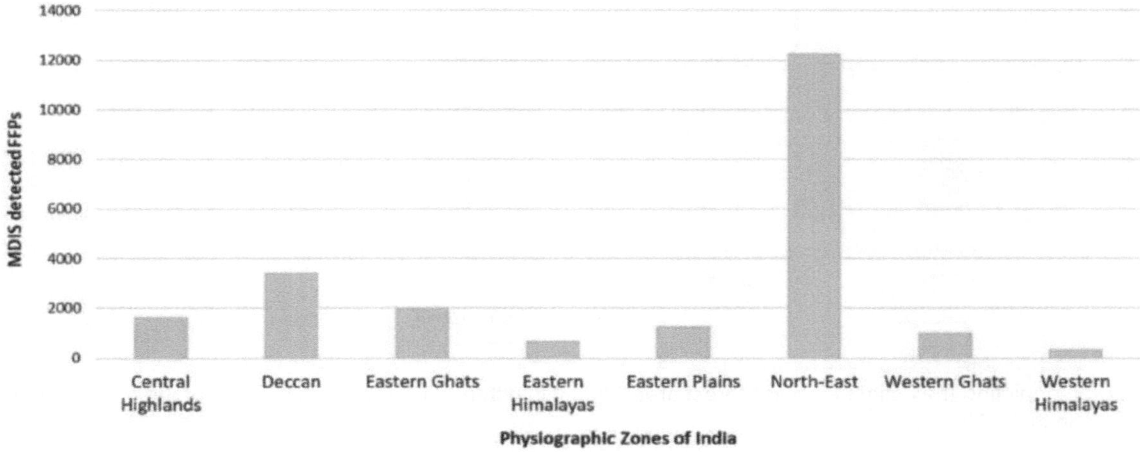

Fig. 54.6 MODIS-based forest fire detections in the country during the fire season 2019-2020

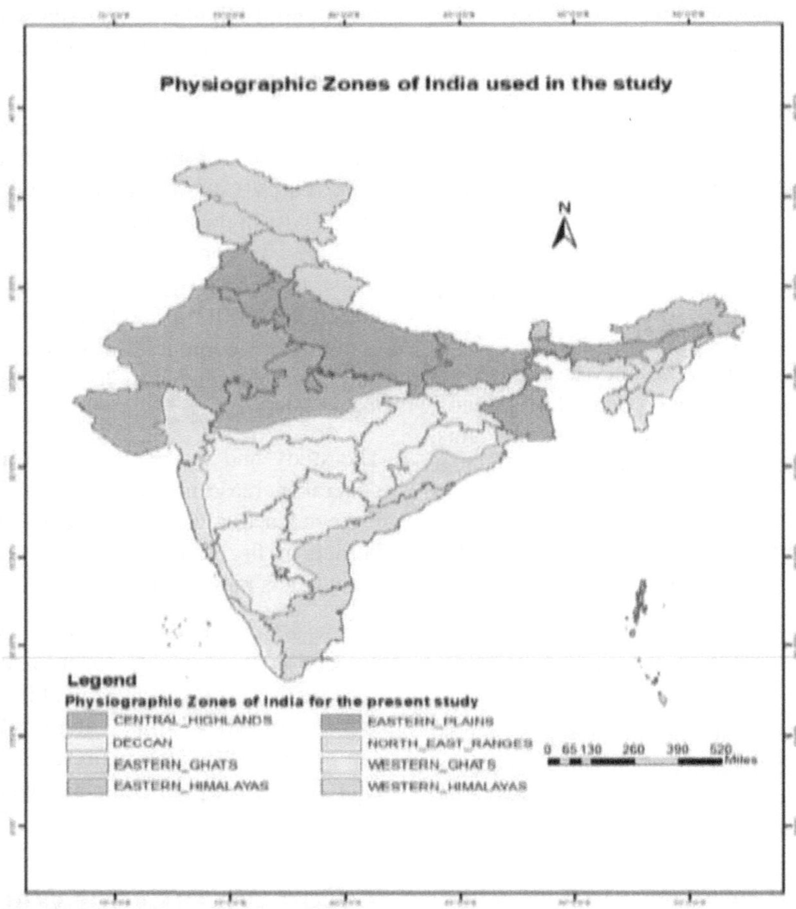

Fig. 54.7 Map showing the eight physiographic zones of the country

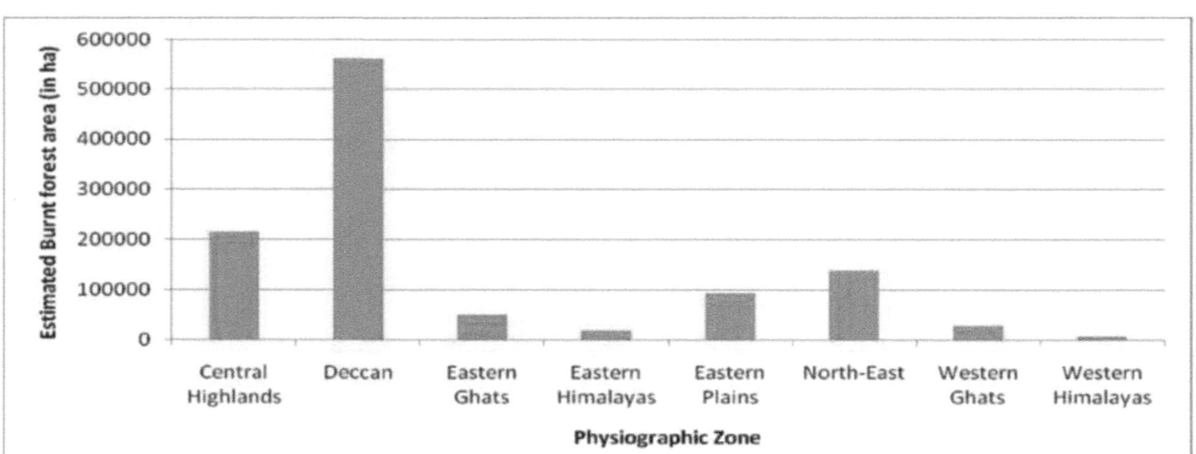

Fig. 54.8 Graph depicting the estimated burnt forest areas (in ha) in different physiographic zones of the country during the fire season 2019-20

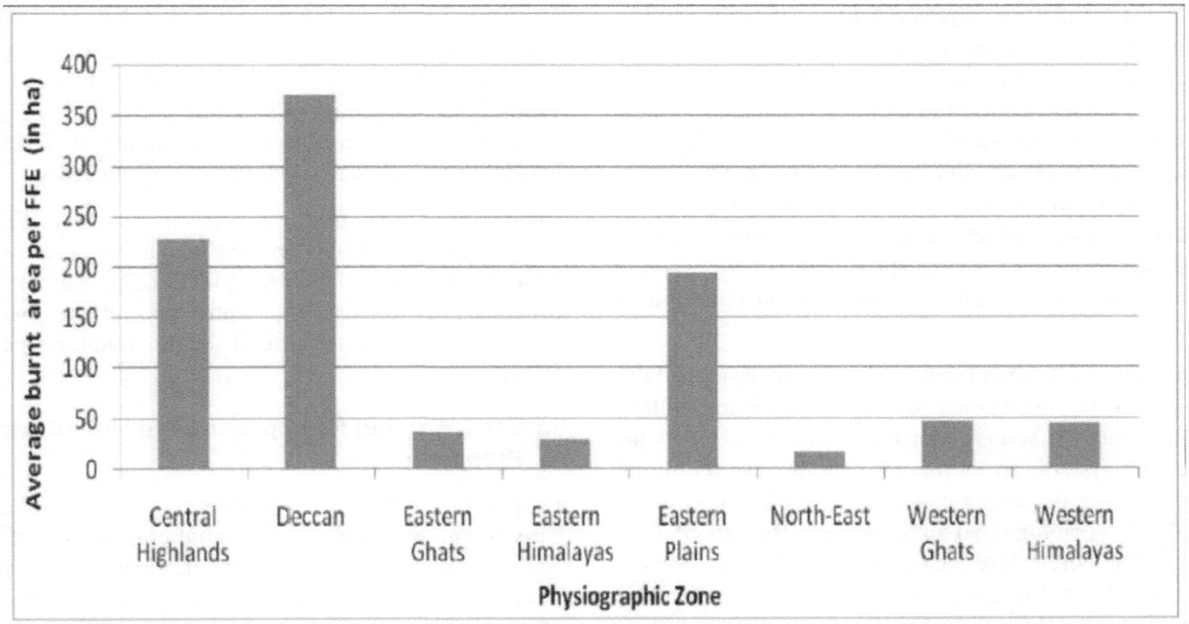

Fig. 54.9 Graph depicting the average burnt forest area per forest fire event (in ha) in different physiographic zones of the country during the fire season 2019-2020

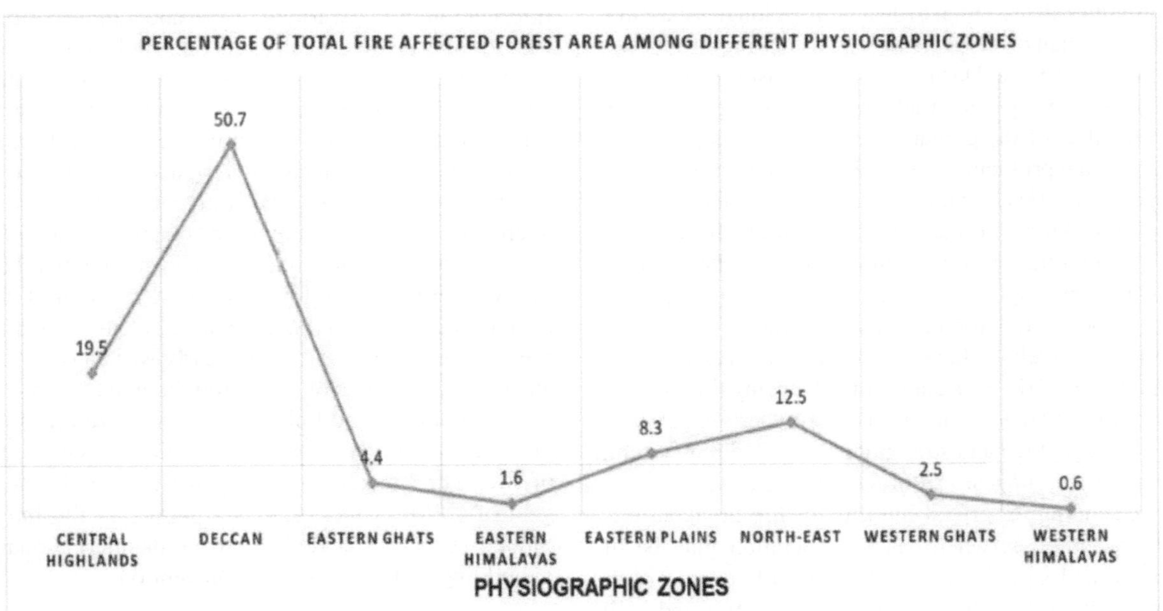

Fig. 54.10 Graph depicting percentage of total fire affected forest area (11094 sq. km) among different physiographic zones of the country

land-use planning, and technology development. Fire policies and strategies vary from country to country because of the different ecosystem characteristics and the cultural, social, and economic factors involved. Successful fire prevention must address the underlying causes of wrongful application of fire and other factors responsible for an increase in detrimental fires. For instance, in some regions of the tropics slash-and-burn agriculture is not sustainable because fire leads to impoverishment of soil fertility. Alternatively the combined use of agricultural and tree crop production (agroforestry), sometimes combined with grazing (agrosilvopastoral systems) and compost production of non-utilized plant biomass instead of wasteful burning will greatly improve the productivity and stability of soils. In the case of the need to convert forest and other wooded vegetation to agricultural and pastureland the woody biomass could be better used as an energy source (production of fuelwood or charcoal) instead of wasteful burning with uncontrolled air pollution.

Integrated fire management systems have been developed that are based on the involvement of communities and address fire at its roots. In some countries community- based fire management systems are in place which observe traditional rules of communities or are based on principles of advanced knowledge in sociology and anthropology. The reports from the Integrated Forest Fire Management (IFFN) projects in Indonesia and Namibia are good examples of experience in this regards.

What is Mechanism for International Cooperation in Forest Fire Management?

Global initiatives are essential to deal with regional issues, like global climate change, rise of sea levels, melting of glaciers and of course forest fire. During the recent past forest fire has emerged as a severe problem adversely affecting the health and sustainability of the global vegetation. Keeping in view the severity of the problem, a need has been felt worldwide to evolve an effective strategy for fire management planning and practices by seeking international cooperation. It has been felt that communities, being the first responder, be engaged in its management vigorously and experience and expertise of countries be shared for capacity building of individual countries to effectively tackle the problem. Joint research and development network, especially in advancing knowledge about forest fire and in learning from one another's experience carry a lot of advantage and strengthen the prevailing system. Early warning techniques, use of satellite data, suppression techniques, sending experts at the time of fire breakout etc. are the field activities, where such cooperation may be of immense use and significance. In the case of severe fire, it is natural for friendly countries to share expertise and rush helpful agencies that carry rich experiences and could be of great service in the hours of need.

The Global Fire Monitoring Centre (GFMC) is the foremost pioneer organization providing individual support and relevant information useful in strengthening an international cooperation. The Centre is hosted by the Fire Ecology and Biomass Burning Research Group (FEBBRG) of the Max Planck Institute of Chemistry, Bio geochemistry Department, Germany. The Centre was established initially by Government of Germany as a German contribution to the UN sponsored International Decade of Natural Disaster Reduction (IDNDR) 1990- 2000 programme. The GFMC, carrying out extensive research and demonstration in the subject of forest fire is providing consultancy services to many countries.

Cooperation with Deutsche Gesellschaft fur Technische Zusammenarbeit (GTZ) projects was implemented in Brazil 1980-82, Indonesia-1987, Algeria-1992, Argentina-1991-97, Srilanka-1991 etc. The Centre has close partnership and association with several international agencies and institutions.

7.1 Early Warning for Improving Prevention and Preparedness

Despite the importance of community-based fire management, early warning systems are essential components of fire and smoke prevention and management. They rely on evaluation of vegetation dryness and weather; detection and monitoring of active fires; integrating and processing of these data in fire information systems with other relevant information, e.g. vegetation cover and values at risk; modelling capabilities of fire occurrence and behaviour; and dissemination of information. Early warning of fire and atmospheric pollution hazards may involve locally generated indicators, such as local fire-weather forecasts and assessment of vegetation dryness. Advanced technologies, however, which rely on remotely sensed data, evaluation of synoptic weather information and international communication systems are now also available for remote locations. The Internet system and satellite telephone links are important channels for information flow. In addition, the international community has access to the international information system of the Global Fire Monitoring Center (GFMC), where scientists and operators prepare daily global to national early warnings on fire danger and monitor the state of fires throughout the world. Numerous other documents and links to sources in fire information can give the interested reader fast and reliable access to the world of fire. The prevention of destructive fires is the ultimate goal of GFMC which acts as a partner of the International Strategy for Disaster Reduction and its international partners in building a global culture of prevention.

8. Good practices of Wildfire Prevention in Australia

Wildfires, or bushfires as they are commonly known in Australia, are a naturally occurring phenomenon in the Australian environment. Before the arrival of humans on the Australian continent, bushfires were often started as a result of lightning strikes or volcanic eruption. While lightning still continues to provide the spark for bushfires, the actions of people (either deliberate or accidental) also contribute to the many bushfires that occur in Australia each year.

Every year, especially in summer, Australia experiences a large number of bushfires that occur as either 'grass fires' or 'forest fires'. Grass fires occur mainly on grazing, farming or remote scrub land country. Although these fires often destroy fences, livestock and some buildings, the nature of the country and its usage, generally mean that heavy losses (particularly human life) are rarely experienced.

Forest fires on the other hand often destroy a greater number of homes that border bushland areas, resulting in larger losses of human life and property. Australia's most devastating bushfires have happened where they have raged through the dense eucalypt forests in the southeast of the continent.

Most bushfires are not disasters, however, if not quickly detected and contained, even small fires can get out of control and result in a disaster. Thankfully, few bushfires earn the title of 'disaster', but repeated disastrous bushfires near the capital city in the states of Tasmania, South Australia, Victoria, New South Wales and Western Australia have occurred, in which many people have lost their lives, or their homes and property.

8.1 Bushfire Prevention Strategies

The raw materials for any potential bushfire are the presence of fuel such as grass, leaves and twigs, oxygen from the surrounding air, and heat or direct flame. Once a fire is started with these raw materials, the spread of the bushfire depends on a number of environmental factors including:

- The type and size of the fuel.
- The fuel's moisture content and its degree of compaction,
- The weather, and
- Topography

Australia's approach to bushfire prevention centres on lessening the possibility of a fire occurring and minimising the spread of bushfires. Fire prevention strategies fall into the following four main categories.

1. Land Management

Land management strategies are effective in:

- Lessening the presence of fuels in forests or grassland area;
- Slowing down and sometimes ceasing the spread of bushfires; and
- Providing easier access routes for fire fighters to reach and extinguish fires.

Fuel reduction is paramount to bushfire minimisation. By taking away the fuel, fires cannot start, nor can they continue to spread. One of Australia's strategies to reduce the build-up of fuels in forest and grassland areas involves the deliberate burning off of these fuels by various fire and land management agencies.

Not only do these 'fuel reduction burns' lessen the potential for future fires in these areas (especially during the dry hot summer months), if a fire does start or enter into such an area, the flame height and intensity will be reduced and the spread of the fire retarded.

The deliberate creation of fire lanes or firebreaks is another land management strategy. These firebreaks are generally areas of land that have been cleared of vegetation (and maintained) in order to provide 'strategic corridors' that act as fire suppression barriers and assist fire fighting operations.

Land management strategies for fire prevention also require a community-based approach. People living in rural areas or those who live in urban areas which are next to bushland have a responsibility both to themselves, their neighbours and the wider community in the prevention of bushfires. Fire agencies across Australia have developed programs, such as 'Community Fire Guard', which emphasise individual and shared responsibilities for the prevention of fires. Residents in these areas too are required to undertake similar land management strategies, which includes removing as much fuel as possible from around their house or property and the creation of suitable firebreaks particularly for properties that are situated next to bushland.

2. Building Management

The findings following the 'Ash Wednesday' fires showed that the sparks and embers emitted from the fires were what caused houses to catch fire. As such, building guidelines and standards are now being developed that are specifically aimed at making buildings more resistant to these fire emissions.

In some communities, local government authorities have regulations controlling home sitting, design and the use of building materials in bushfire prone areas. These controls serve not only to minimise the damage to home sites and reduce losses from fires, they also assist in the prevention and spread of bushfires.

3. Community Education

People and their actions (whether deliberate or not) are responsible for the majority of bushfires that occur in Australia. Some of the more common causes of bushfires in Australia have resulted from deliberate burning-off that gets out of control and fires escaping from burning rubbish heaps 1.

However it can also take only a single spark from machinery such as welding equipment, a campfire or outdoor cooking facility not properly extinguished or children playing with matches for a fire to begin.

With such a high bushfire incident rate resulting from the actions of people, community education in Australia is particularly important. Education takes on a number of forms and is generally designed to provide people with a better understanding of the risks they face from bushfires and the measures the community can take to minimise these risks. A range of information brochures are made available to the general public which cover such issues as what measures people (particularly those living in bushfire prevalent areas) can take to minimise the spread of a bushfire and the protection of their property and lives before and during a bushfire.

4. Preparation before the wildfire Season

- Where possible, prepare a firebreak around the home.
- Trim branches so that they are well clear of the house.
- Clear roof gutters of leaves and twigs.
- Store wood, fuel and paints etc well clear of the house.
- Remove rubbish, leaf litter and plants that are close to the house
- Keep grassed areas (especially those around the house) short/ green.
- Fit wire screens to doors, windows and vents (to prevent burning embers from entering the house).
- Enclose any gaps, roof eaves and the under area of the house.
- Keep a ladder nearby for roof access (both inside and out).
- Have water hoses available and ensure they will reach all parts of the house and garden. Where water is not connected, obtain a high pressure pump.
- Decide on a household plan to either leave early or stay to protect the properly prepared home during the bushfire. (*Source:* UNISDR report).

9. Conclusion

Forest fires are becoming an inherent part of the world's environment we cannot prevent them, but we can minimize the risks they pose to life, property and infrastructure, production systems, and the environment.

Geographical scale of our country, the large and expanding rural–urban interface, and the potential for rapid fire development and spread under adverse weather conditions mean that individual Indian cannot rely solely on fire agencies to protect their lives and property from bushfires.

Countries like Australia has a large and very capable force of volunteer and career firefighters, advanced firefighting technologies, and significant firefighting resources. Hence Indian rescuers of prominent rescue agencies may take training from California or Australian fire fighters.

Future roadmap for disaster management in India is inculcated in NDMP-2016 (Revised in 2019). Our national plan is important initiative in light of SFDRR 2015-2030 to manage the disaster through risk reduction, mitigation and preparedness measures for effective response and included "Build Back Better" concept in recovery, rehabilitation and reconstruction. The NDMP incorporates substantively the approach enunciated in the Sendai Framework and will help the country to meet the targets set in the Sendai framework. By 2030, the Sendai Framework aims to achieve substantial reduction of disaster risk and losses in lives, livelihoods, and health.

Glossary

Tarai. The Terai or Tarai is a lowland region in northern India and southern Nepal that lies south of the outer foothills of the Himalayas, the Sivalik Hills, and north of the Indo-Gangetic Plain.

Jhum. Jhum cultivation is a traditional farming method where first the land of trees and other vegetation is cleared, burned and then cultivated for a limited number of years.

Bush Fire. A fire in scrub or a forest, especially one that spreads rapidly.

Wild Fire. A large, destructive fire that spreads quickly over woodland or brush.

Metric Tons. A unit of weight, The metric ton used in most other countries is 1,000 kg, equivalent to 2,204.6 pounds avoirdupois.

Mahua. Mahua or Madhuca longifolia (Scientific name) is an Indian tropical tree found largely in the central and north Indian plains and forests.

Tree densities. The density, of a substance is its mass per unit volume in which it determine the number of trees in an area.

Hectares. unit of area in the metric system equal to 100 ares, or 10,000 square metres, and the equivalent of 2.471 acres in the British Imperial System and the United States Customary measure.

El Niño. A climate pattern in the Pacific Ocean that can affect weather worldwide.

MODIS. A Satellite sensor. The Moderate Resolution Imaging Spectroradiometer is a payload imaging sensor built by Santa Barbara Remote Sensing that was launched into Earth orbit by NASA in 1999 on board the Terra satellite, and in 2002 on board the Aqua satellite.

BIBLIOGRAPHY

1. Reports of National Disaster Management Authority (NDMA). Reports of National Institute Disaster Management (NIDM). Australian bushfire season 2019–2020; *https://en.wikipedia.org/wiki/2019%E2%80%9320_Australian_bushfire_season* Reports of US National Interagency Fire Centre (NIFC). Reports of Emergency Events Database (EM-DAT) UNISDR.
2. Reports of United Nations International Strategy for Disaster Reduction Secretariat (UNISDR). Reports of Media; Largest fires of the 21st-century-*https://en.wikipedia.org/wiki/List_of_largest_fires_of_the_21st-century* Reports of Forest Survey of India (FSI)
3. Reports of Food and Agriculture Organization Global Forest Resources assessment (FAO FRA-2010).

CHAPTER 55

Building Disaster Resilience to Cloud Burst Events in Uttrakhand, India

Anjali Saraswat
Malaviya National Institute of Technology Jaipur
anjalisaraswat7@gmail.com
8802707584

Satish Pipralia
Malaviya National Institute of Technology Jaipur
spipralia.arch@mnit.ac.in
9549658126

Abstract

The state of Uttarakhand in northern India has experienced innumerable events of cloud burst disaster, which being a violent downpour, can cause upto 100 mm of water fall in an area focussed upon in just a few square kilometres. Such large amounts of runoff from higher elevations causes large debris to flow leading to landslides and has the capacity to destroy large infrastructure, properties, agricultural lands and also kill people. Many regions of Uttarakhand such as Pithoragh, Chamoli, Almora, Rudraprayag, Tehri Garhwal and Haridwar have witnessed adversities of cloud burst in the past causing enormous loss of life and property. The paper aims at exploring the measures to build disaster resilience to cloudbursts events in Uttarakhand state India, for which firstly the four elements of disaster resilience viz robustness, redundancy, resourcefulness and rapidity are explored in the paper. Further cloudburst resilience is understood through a bifold process of tracking the city's current situation and predicting forecast. With thorough understanding of the same, framework for addressing cloudbursts in Uttarakhand is evolved, involving creating awareness, examining the topography and possessing real time data management. Exploring such a framework raises hope for building effective disaster resilience to cloudbursts in Uttarakhand.

Keywords: cloud burst, disaster, India, resilience, Uttarakhand

1. Introduction

Cloudbursts are instantaneous and substantially heavy rain showers that occur at a particular area over a brief span of time. The rain showers usually range from 200-1000 mm/day or could even be 100-250 mm/day lasting for a brief duration (Mueller et al. 2019; Sen 2013). The water usually hits the ground with a speed of 10m/s during a cloudburst with a droplet size of about 4-6 mm. They arrive erratically, lasting for just a limited period but leave behind a trail of destruction. Mutability along with rigorous environmental degradation by human beings has led to climate change, which has escalated the frequency and intensity of rainfall and consequently the recurrence of cloudbursts events has arisen. To add to this are the assorted anthropogenic activities especially on the hilly terrains that escalates further the immensity of any hazard (Satendra 2003; Kapur 2010). The data sources reveal that 70% of all the economic losses and more than half of the casualties are engendered by such disasters. The extent of cloudbursts disaster further soars up as after the downpour comes landslides and flash floods which move large debris on the slope and what follows after is large casualties and property damage. Indian Himalayas being the vital mountainous landscape of the country are largely affected by cloudburst disaster every year (Jena, Garg, and Azad 2020; Raveendranathan 2017; Gupta et al. 2014). Studies reveal that every square kilometre in the Himalayas shows up at least two landslide scars which are insinuated by the cloudbursts. In the aftermath of cloudburst, the bursting of glacial lakes in few situations worsens the scenarios as it further infuriates the flash floods intensity.

2. Cloudbursts in Uttrakhand

Uttarakhand is the state lying in the north-western part of India and is located between 28° 43"- 31° 27" N latitude and 77° 34"- 81° 02" E longitude. Lying at the centre, the state is an essential part of the Himalayas and is popularly known as the 'Indian Central Himalayan Region'. With a geographical cover of 53,483 sq. km, the state has 93% of its area as mountainous mainland and 64% covered by the forest. 16% area is snow-clad amongst the mountainous mainland of Uttarakhand and thus it incarnates the biggest river system of the country, the Ganga. With China in the North and Nepal in the East, the state acts as the geographical northern boundary for the nation (Arihant 2015; Kafaltia and Kafaltia 2019). Another mountain rich state Himachal surrounds the state on the west and on the South is the state of Uttar Pradesh, which earlier formally enclosed the state of Uttarakhand in its boundary. On 9 November 2000 the Uttaranchal (former name of Uttarakhand) state was carved out from Uttar Pradesh and it became the 27th State of the Republic of India. In January 2007, the name of the state was officially changed from Uttaranchal to Uttarakhand. The state of Uttarakhand is administratively divided into the Garhwal and Kumaon divisions and comprises total 13 districts. Uttarkashi, Rudraprayag, Chamoli, Tehri, Parui Garhwal, Haridwar and Dehradun districts join together to form Garhwal division and Pithoragarh, Bageshwar, Almora, Nainital, Champawat and Udham singh Nagar join to form Kumaun divisions. The city of Dehradun serves as the capital of the state Uttarakhand. The total population of the state is 1,01,16,752 with 51,54,178 male and 49,2,574 females. With a population density of 189 per square kilometres, the sex ratio in the state is 963 females for each 1000 male. The average literacy rate is 79.63 percent of which male literacy rate is 88.33 percent while female is 70.70 percent.

The Himalayas of Uttarakhand comprises delicate ecosystems, with undulating and fragile landscapes which makes them substantially vulnerable to natural hazards (Alam, n.d.; V. Pande 2018; P. C. Pande 1998). Also, the fact that they lie in IV and V seismic zones further adds to their disaster susceptibility. Unplanned construction on the mountain slopes such as the settlements, business avenues and institutions, especially along the rivers and streams escalates the catastrophe and damage that occurs after the disaster has happened. The heavy monsoon rains bring with them mass movements of soil on the slopes resulting in massive loss of lives and property. Being largely covered by high altitudes and mountains the occurrence of cloudburst disaster is quite usual in Uttarakhand generating enormous losses via flash floods, mass movements of debris flow and landslides (Mukherjee 2014; Lakshmi et al. 2016) . Undesirable developments, population pressure, deforestation and climate change are also contributing factors in squandering of the arable and forest land along with immense destruction of the various infrastructures. The usual events of cloudburst disaster

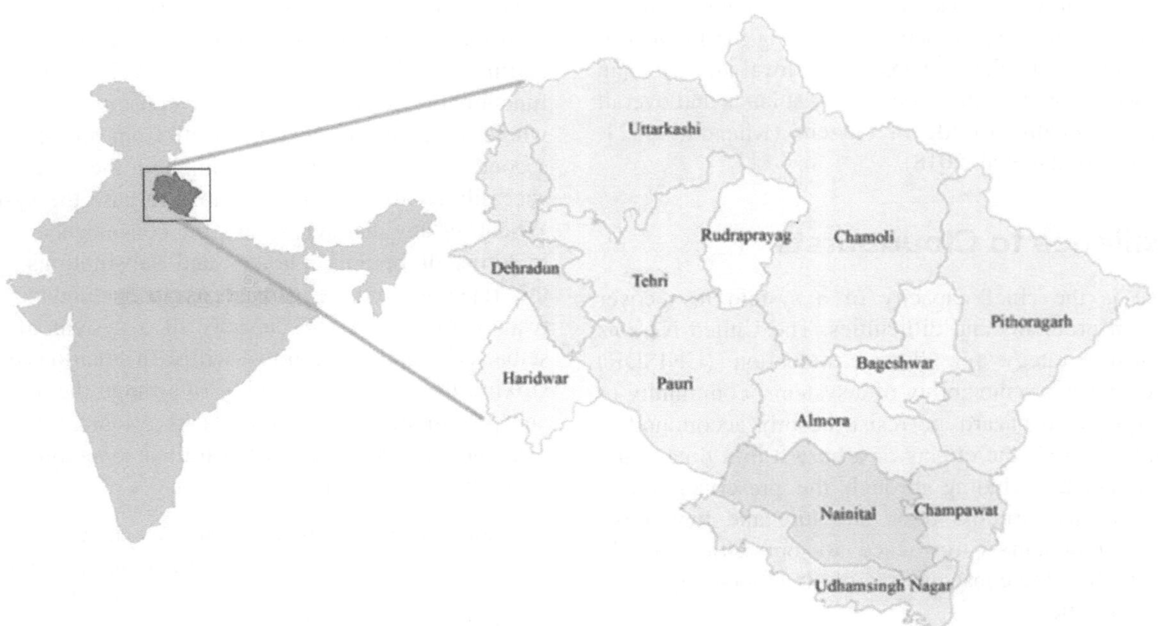

Fig. 55.1 Location of Uttarakhand state in India

Source: Google

Fig. 55.2 Destructions created by cloudburst events in the state of Uttarakhand

Source: Vishwambhar Prasad Sati (2018, January 26) 'Cloudburst-Triggered Natural Hazards in Uttarakhand Himalaya: Mechanism, Prevention, and Mitigation'. https://www.researchgate.net/publication/322714818

occurred only during monsoon season in Uttarakhand but the recent times have seen them even during the months of March to May in the elevated area of Garhwal and Kumaun region desecrating agriculture fields, forest lands and overall economy and quality of life of citizens (Khanduri 2021; Jayangondaperumal et al. 2018).

3. Resilience to Cloudbursts

Resilience is the chief capacity of a system to recover swiftly from problems and difficulties. The United Nations International Strategy for disaster reduction (UNISDR) defines resilience as 'the ability of a systems, community or society exposed to hazards to resist, absorb, accommodate to and recover from the effects of a hazard in a timely and efficient manner, including through the preservation and restoration of its essential basic structure and functions'. Developing cloudbursts resilience to our cities simply means regulating the consequences of cloudbursts by better preparing our cities.

Resilience of any system can be comprehended of its four elements viz robustness, redundancy, resourcefulness and rapidity (Haase, n.d.; Vahanvati 2017). The first element robustness is the inherent energy of a system to withstand extrinsic enforcements without dwindling or losing functionality. It is the essential capacity of society as a whole to resist and conquer itself from any disasters. The second element redundancy implied the availability and accessibility to tools and methods in case the system fails. Whenever any disaster occurs, the convenience of a society in terms of options, choices and substitutions to combat the repercussions is gauged as its redundancy. Thirdly resourcefulness, is the capacity of a system to overcome setbacks by opting solutions swiftly. In situations of disasters striking the cities, their ability to arrange the assets, ability and potentiality to overcome is its resourcefulness. Fourth element rapidity, signifies the trait of operating quickly to overcome a disruption or disaster.

Development of cloudburst resilience in cities is a bifold process and is based on, one- tracking the city's current situation and two- predicting forecast. Tracking the current situations of a city vulnerable to cloudburst disasters, helps to find out potential menaces and thrive better cognizance of what actually is ought to be. Furthermore, predicting the

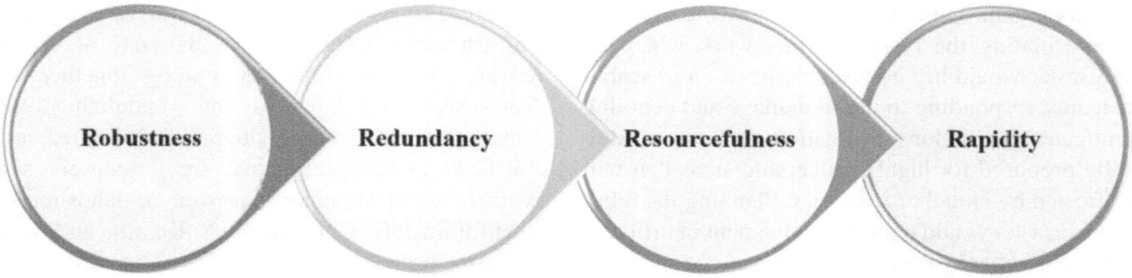

Fig. 55.3 Four elements of resilience

Source: Generated by author

forecast induces instances of how the real cloudburst situation and its related intensity is going to be, thus allowing the concerned authorities to plan mitigation measures effectively.

While tracing the city's current situation, the first step is to evaluate the threats that exist in the city which would amplify the ramifications of cloudburst. These threats could exist in terms of intensive anthropogenic activities, haphazard development, excessive impermeable cover and government negligence. Based on the evaluation of threats, priorities to be catered to the whole preparation of cloudbursts disaster need to be set. Additionally, based on the existing cloudbursts threat levels in the city, existing disaster mitigation standards need to be scrutinized and other updated with new ones enlisted. The distinctive organisations working in regards to several areas concerning the cloudburst disaster, need to communicate and convey their many discoveries and findings, so that each department is updated with the new information. Importantly stakeholder's participation in the whole process would be extremely beneficial, thus making the working institutions aware of the current on ground situations and drawbacks faced. Moreover, making the public aware through media campaigns and otherwise, would lead to effective disaster mitigation. Predicting the forecast about the aftermath of a cloudburst hitting the city involves anticipating patterns and trends based on tracking the city's current situation. This particularly involves accumulating

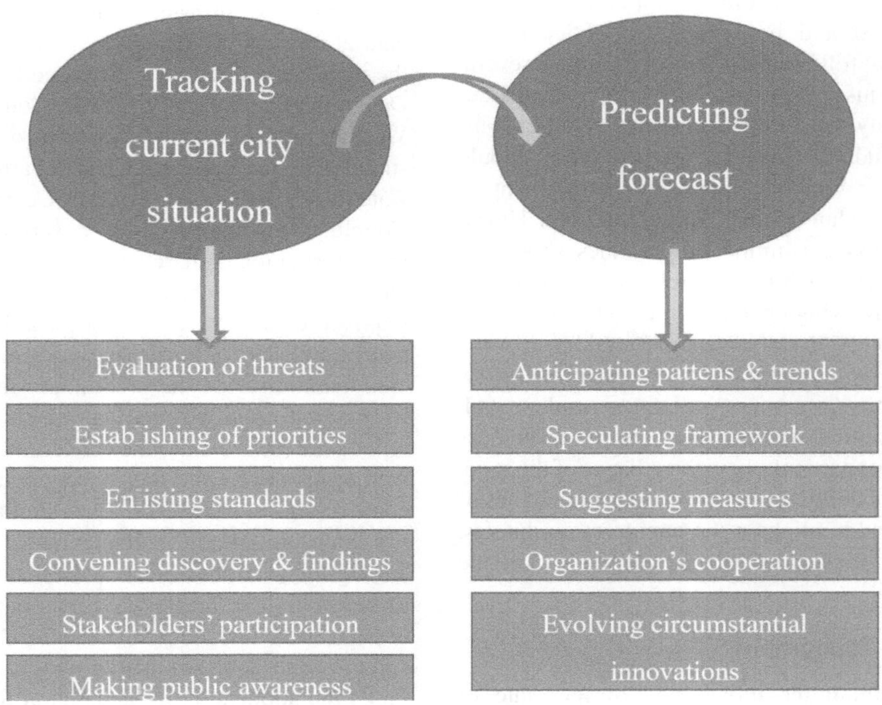

Fig. 55.4 Process for cloudburst resilience in cities

Source: Generated by author

various datasets together and envisioning how the situation would be. Speculating the response framework when the cloudburst disaster would hit, involves designation of search and rescue teams, responding to initial damage and denoting shelter for affected populations. Mitigation measures need to be essentially proposed for highly vulnerable areas that will be largely affected by cloudburst disaster. Planning the relief actions for the disaster would require involvement of different hierarchical government setups as well as cooperation among organisations. Keeping in mind the uncertainty of the outcomes of disaster, few circumstantial innovations need to be evolved so as to overcome the sudden shocks to come.

4. Framework for Cloudburst Resilience

Developing a framework in order to address the cloudburst disaster in the state of Uttarakhand requires collaborative efforts of various systems, administration and establishments. Several precautionary, preparatory and responsive efforts simultaneously would lead to better comprehension and tackling the disaster.

4.1 Creating Awareness

First and foremost, creating general consciousness about the cloudburst disaster, among the common masses and officials of the state is exceedingly important. Citizens and administrations should be brought to notice of the ponderous rainfall that follows the cloudbursts and its several repercussions in terms of threat of flooding, landslides as well as possibility of further tumbling ramifications. Significance of cloudburst disaster preparedness should be emphasized, along with the inevitable requirement of cloudburst emergency plan (Rofiah, Kawai, and Nur Hayati 2021). Mass awareness generation programmes should be commenced making communities entirely conscious of the cloudburst hazard. Involving chapters on cloudburst hazard, its risk and vulnerability in the state of Uttarakhand in the school curriculum, would prepare a well-informed new generation. Community awareness initiatives would entail, engendering mass rallies with slogans, many awareness exhibitions, workshops particularly in the villages of the state. Conducting regular meetings within the various hierarchies of state government would enlighten the engineers, architects, planners and policy makers of the developments on the disaster.

4.2 Examining the Topography

Knowing and examining the topography of the state of Uttarakhand, enables one to better prepare and respond to the cloudbursts disaster by becoming aware of the obstruction in the area, zones at escalated risks and placements of crucial infrastructure. Large parts of the state of Uttarakhand are covered by the Himalaya, putting together an exquisite landscape, accompanied by undulating and fragile environments. Parts of Uttarakhand located at the height of 1200 meters and above are immensely sensitive and vulnerable to cloudburst as well as landslides and floods that follow it (P. C. Pande 1998; Kafaltia and Kafaltia 2019). Disaster proneness of the whole state further elevates as it lies in IV & V seismic zones. Accumulating relevant details of the specific positioning of the community's sewage system is particularly important in regard to cloudburst disaster preparation as the large volumes of rain water flowing, needs to be directed through proper channels for disposal. Most parts of Uttarakhand Himalayas experience paramount rainfall and thus as a protection measure the valleys and inhabited drain channels should be avoided for construction activities and instead dense and solid grounds preferred. Due to mountainous terrain existing in large parts of the state, the drainage networks should be appropriately placed for effective disposal of rainwater. Both the surface and subsurface drainage channels shall be planned and executed in such a way that they are broad and steeped allowing large volumes of rainwater to flow

4.3 Real Time Data Management

Procuring and managing reliable data, immensely supports the cloudburst disaster management process. The major dataset required to assess cloudburst disasters in an area is the precipitation statistics, acquired from the Meteorological Department and Central Water Commission Uttarakhand. Weather forecasts for the coming days and rainfall reports are supplied by the Meteorological department, along with daily water levels and flood forecast by Central Water Commission. Coupling these datasets, early warning is created and then this information is shared amongst several technical agencies and committees, which work towards forecasting the other requisite details of cloudburst hazard (R. 2013; Li, Huang, and Emrich 2020). In case early warning fails to forecast the cloudburst disaster, the preliminary information is obtained by the area's police system, and with initial action taken the reports are passed on to the concerned officers. It is also very well necessary for the state to set up its emergency operation control centres, along with other similar hierarchical centres, which ensure an efficacious layout of status and knowledge amongst the various officers as well as share it with the media to reach to the masses. Geographic Information System is an efficacious data management tool that effortlessly captures and analyses spatial and geographic data and can find wide application in preparation for cloudbursts disaster management. Utilizing location, demographic, landscape related spatial and digital data, superior quality of results

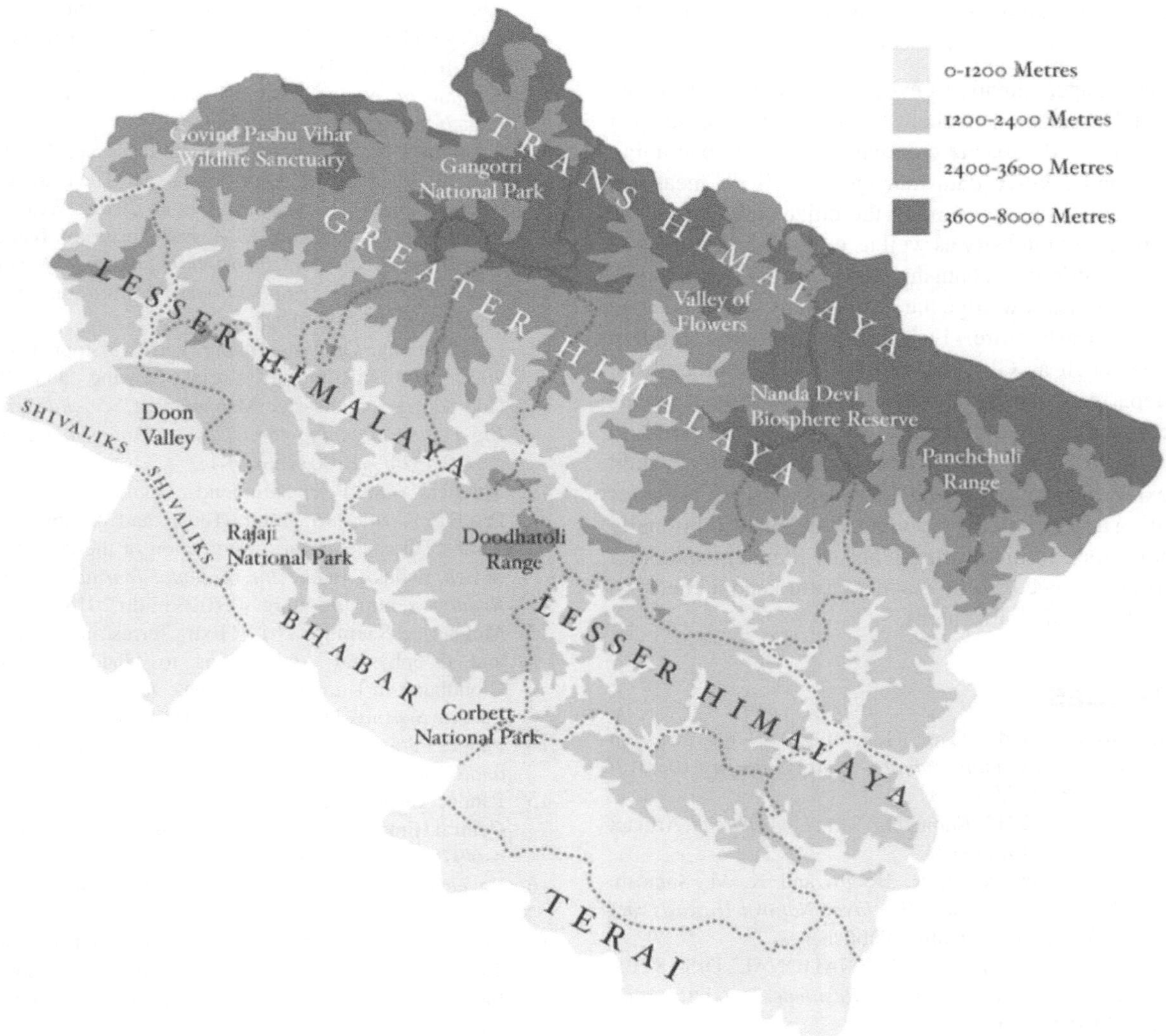

Fig. 55.5 Topography of Uttarakhand State

Source: uttrakhand.org

is calculated manually that facilitate decision making about cloudburst disaster. Arising at scientific inputs via GIS analysis enables functional cloudburst disaster planning, situation monitoring and strategies implementation. The digital maps originated from GIS analyses of the state enabling micro-zoning of cloudburst hazards and allowing the relevant authorities to take directive measures.

5. Conclusion

The impacts of cloudburst disaster on our cities are magnificent, in terms of the intensity of the after havoc it creates in our cities. Events of bursting of clouds is followed by the acts of massive flash floods and landslides which injures people, dismantles infrastructure and even subdues many to death. Tri contribution factors of, large parts being covered by the reign of Himalayas, frequent tectonic movements and seismic zones existence, makes the state of Uttarakhand largely vulnerable to the atrocities of the cloudbursts. Many districts of the state have suffered largely due to the reoccurring cloudbursts disaster events. Thus, it is imperative to build on the ways for making cities resilient to cloudburst disaster. Steps for building resilience to cloudburst disaster necessitates strengthening the robustness, redundancy, resourcefulness and rapidity of the several cities in the state of Uttarakhand at the time of vulnerabilities. Following the same would be tracking down the current situation scenario in the state and eventually building upon predicting the forecast for cloudburst disaster scenario.

The research paper intends to evolve a framework to make the state of Uttarakhand resilient to cloudburst and suggest threefold steps to do so, viz creating awareness, examining topography and relative data management. Such measures bring about consciousness among the citizens regarding the cloudbursts' event intensity as well as make them acquainted with various response mechanisms. Knowing the whereabouts of the city helps acknowledge the obstructions and also locate the pivotal infrastructure. Furthermore, utilizing modern technologies such as GIS and various data resources from several departments helps fabricate early warning systems and effective response mechanisms. Further research in building disaster resilience to cloudbursts disaster would require exploring the collaborative role of various agencies in preventing and mitigating cloudbursts events. Additionally investigation for city monitoring plans by planners and other directed officials, needs to be made so as to keep a check on urban growth in vulnerable areas.

References

1. Alam, Aniket. n.d. "The Geography of the Western Himalayas." *Becoming India.* https://doi.org/10.1017/upo9788175968387.004.
2. Arihant, Experts. 2015. *Know Your State Uttarakhand.* Arihant Publication India Limited.
3. Gupta, P., J. Sundaresan, R. Boojh, and K. M. Santosh. 2014. *Climate Change and Himalaya: Natural Hazards and Mountain Resources.* Scientific Publishers.
4. Haase, Thomas W. n.d. "INTERNATIONAL DISASTER RESILIENCE:" *Designing Resilience.* https://doi.org/10.2307/j.ctt5hjq0c.16.
5. Jayangondaperumal, R., V. C. Thakur, V. Joevivek, Priyanka Singh Rao, and Anil Kumar Gupta. 2018. *Active Tectonics of Kumaun and Garhwal Himalaya.* Springer.
6. Jena, Pravat, Sourabh Garg, and Sarita Azad. 2020. "Performance Analysis of IMD High- Resolution Gridded Rainfall (0.25° × 0.25°) and Satellite Estimates for Detecting Cloudburst Events over the Northwest Himalayas." *Journal of Hydrometeorology.* https://doi.org/10.1175/jhm-d-19-0287.1.
7. Kafaltia, Himanshu, and Gunjan Sharma Kafaltia. 2019. *A Comprehensive Study of UTTARAKHAND.* Notion Press.
8. Kapur, Anu. 2010. *Vulnerable India: A Geographical Study of Disasters.* SAGE Publications India.
9. Khanduri, Sushil. 2021. "Flash Flood Struck Dhauliganga Valley on February 7, 2021: A Case Study of Chamoli District of Uttarakhand Himalaya in India." *Academic Platform Journal of Natural Hazards and Disaster Management.* https://doi.org/10.52114/apjhad.903387.
10. Lakshmi, Deepa, Head &. Assistant Professor, Department of Computer Science & Applications, Sir Theagaraya College, Chennai, India, J. Karthikeyan, et al. 2016. "Weather Impact over Uttarakhand Using K-Means Clustering Technique for Cloudburst Prediction." *International Journal of Data Mining Techniques and Applications.* https://doi.org/10.20894/ijdmta.102.005.002.010.
11. Li, Zhenlong, Qunying Huang, and Christopher T. Emrich. 2020. "Introduction to Social Sensing and Big Data Computing for Disaster Management." *Social Sensing and Big Data Computing for Disaster Management.* https://doi.org/10.4324/9781003106494-1.
12. Mueller, Sonja, Peter Sammonds, Ghulam M. Bhat, Sundeep Pandita, Kavita Suri, Bindra Thusu, and Virginie Le Masson. 2019. "Disaster Scenario Simulation of the 2010 Cloudburst in Leh, Ladakh, India." *International Journal of Disaster Risk Reduction.* https://doi.org/10.1016/j.ijdrr.2018.09.004.
13. Mukherjee, Saumitra. 2014. "Extra Terrestrial Remote Sensing and Geophysical Applications to Understand Kedarnath Cloudburst in Uttarakhand, India." *Journal of Geophysics & Remote Sensing.* https://doi.org/10.4172/2169-0049.1000124.
14. Pande, Puran Ch. 1998. *The Himalayan Environment: Issues and Challenges.* Daya Books.
15. Pande, Vasudha. 2018. "Anthropogenic Landscapes of the Central Himalayas." *At Nature's Edge.* https://doi.org/10.1093/oso/9780199489077.003.0002.
16. Raveendranathan, D. 2017. *Himalayas Melting and Peninsular India Wilting.* Notion Press.
17. Rofiah, Nurul H., Norimune Kawai, and Elli Nur Hayati. 2021. "Key Elements of Disaster Mitigation Education in Inclusive School Setting in the Indonesian Context." *Jamba (Potchefstroom, South Africa)* 13 (1): 1159.
18. R., Paul. 2013. "Improved Disaster Management Using Data Assimilation." *Approaches to Disaster Management - Examining the Implications of Hazards, Emergencies and Disasters.* https://doi.org/10.5772/55840.
19. Satendra. 2003. *Disaster Management in the Hills.* Concept Publishing Company.
20. Sen, Debraj. 2013. "Coping in a Calamity: Radiology during the Cloudburst at Leh." *The Indian Journal of Radiology & Imaging* 23 (1): 106–9.
21. Vahanvati, Mittul. 2017. "Participation for Disaster Resilience." *Community Engagement in Post-Disaster Recovery.* https://doi.org/10.4324/9781315534213-6.

CHAPTER 56

Heat Wave and Outbreak of Encephalitis—A Case Study of Bihar

Sunil Kumar Chaudhary
Executive Engineer, Road Construction Department
Vaishali Road Division, Hazipur, Bihar
sunil228@cimp.ac.in

Abstract

Bihar is one of the most disaster prone states of the county. There are altogether 9 different types of disaster prevalent in the state of Bihar. Of these events so far recorded, Heat wave is the most crucial, common and a regular annual phenomenon in Bihar resulting in enormous loss of life. The heat wave is a possible aggravating factor in an encephalitis outbreak in Bihar which has killed hundreds of children in recent years. The purported causes behind such an outbreak are speculated to be multifold -- one of them being consumption of litchis, or lychees. The unhygienic conditions, Malnourished children and inadequate or no investment on mitigation/ Preparedness measures has further increased the vulnerability of the population to this hazards. In addition to extensive damage to life, this disaster over the years has also adversely affected economic development in the State. Hence there is an urgent need for a structured research plan and epidemiological surveys to understand the underlying cause of the outbreak that has killed hundreds of children in Bihar.

This Paper discusses the key reasons for the outbreak of encephalitis in Bihar, through stakeholder interviews and observations in the aftermath of the Heat wave. The institutional context comprises several challenges such as inefficient Hospital disaster management plan, inherent weaknesses of the health management plan, structural disease control strategy and the lack of connection between governmental decision-making bodies, implementation agencies and civil society.

Keywords: Heat Wave, encephalitis, outbreak, hazard, Disaster

1. Introduction

Heat wave is disastrous weather phenomenon which affects human comfort, agriculture, water management etc. particularly when it persists for longer period; thus, resulting in variety of problems to human beings/livestock, frequently even loss of human life/livestock. A large number of studies, primarily in developed country settings, show that heat wave is related to adverse health outcomes[1–5].A changing climate has heightened the risk of temperature related morbidity and mortality, globally. For instance, studies of recent heat waves across Greece in 2007[6], Australia in 2009[7], Russia in 2010[8] and US in 2012[9] were all associated with increased mortality. India has experienced a series of heat waves in the past[13] that reveal their significant mortality impacts. For instance, in the year 1998, the state of Orissa faced an unprecedented heat wave situation as a result of which 2042 people lost their lives. In another instance, 1421 people were killed in Andhra Pradesh from a heat wave in 2003[15]. More recently, 1344 excess deaths were recorded during the 2010 heat wave for the city of Ahmedabad 16. In India, AES was chiefly associated with Japanese encephalitis virus (JEV) before 1975. The JEV cases and outbreaks became more frequent and endemic regions developed between 1975 and 1999. After 1999, non-JEV cases and outbreaks of AES were increasingly reported caused by other viruses including Chandipura virus (CHPV), Nipah virus (NiV), and enteroviruses. After 2012, it is observed that the cause shifting to JEV.[12][13] Initial examination shows little sign of the JEV or other viral routes in the affected children.[5] A study by AIIMS Patna found the presence of enterovirus.[18]

An outbreak of acute encephalitis syndrome (AES), root cause behind which was heat wave in the northern regions of Bihar has been a matter of serious concern. High temperature, humidity, malnutrition, poor hygiene and lack of awareness are known aggravating factors of AES.[19][7][19][4] Cases of AES tend to occur during the Bihar's monsoon season.[14] In the month of June and July of year 2019, the temperature in Muzaffarpur remained above 40 °C (104 °F) and the rains were delayed which may have aggravated the situation.[7][18] Poverty and malnutrition is widespread among children in the region. Malnourished children lack a buffer stock of sugar as glycogen in the liver which puts them at higher risk of hypoglycemia.[18][20][16] Awareness campaigns were carried out in March–April 2019 but were not conducted later due to the 2019 Indian general elections. The local administration was not watchful due to the few cases of AES in recent years.[7]The purpose of this paper is to highlight the cause of Outbreak of Encephalitis, impact of heat wave on its severity, lapses and Way forward.

2. Outbreak of Encephalitis and Heat Wave

Acute encephalitis syndrome (AES), locally known as chamki bukhaar, is defined as the acute-onset of fever with change in mental status (including symptoms such as confusion, disorientation, coma, or inability to talk) and often with new onset of seizures. This syndrome is a major health problem in Asia. It includes illness caused by a wide variety of viruses, bacteria, protozoa, fungi. But, mostly considered as viral encephalitis. Cases of acute encephalitis syndrome are reported throughout the year, but there is an increase in these cases in June, which peak during July and August, and then decline in September-October. AES is classified into Japanese encephalitis (JE), known AES, and unknown AES. Whereas the vector of Japanese encephalitis is mosquito, in case of known acute encephalitis syndrome, it can be viral encephalitis or non-viral encephalitis (see chart)[7]. Between 2008 and 2013, acute encephalitis syndrome cases have recorded an increase in the country, from 3,855 to 7,485. As per news reports, last year 10,485 such cases were diagnosed and 632 deaths recorded across 17 states. India's fatality rate in acute encephalitis syndrome is 6 per cent, but in the case of children it is as high as 25 per cent.

This isn't the first time such an outbreak has hit Bihar. Since 1995, various teams have been studying the almost annual outbreak of the 'disease' in Bihar. The first case of AES in Muzaffarpur district was recorded in 1995.[4][5][6] There were 143 deaths in 2013, 355 in 2014, 11 in 2015, four in 2016, 11 in 2017 and 7 in 2018.[4][7] In June 2019, an outbreak of AES occurred in 222 blocks of Muzaffarpur and the adjoining districts in Bihar.[5][8]As a result of the outbreak,

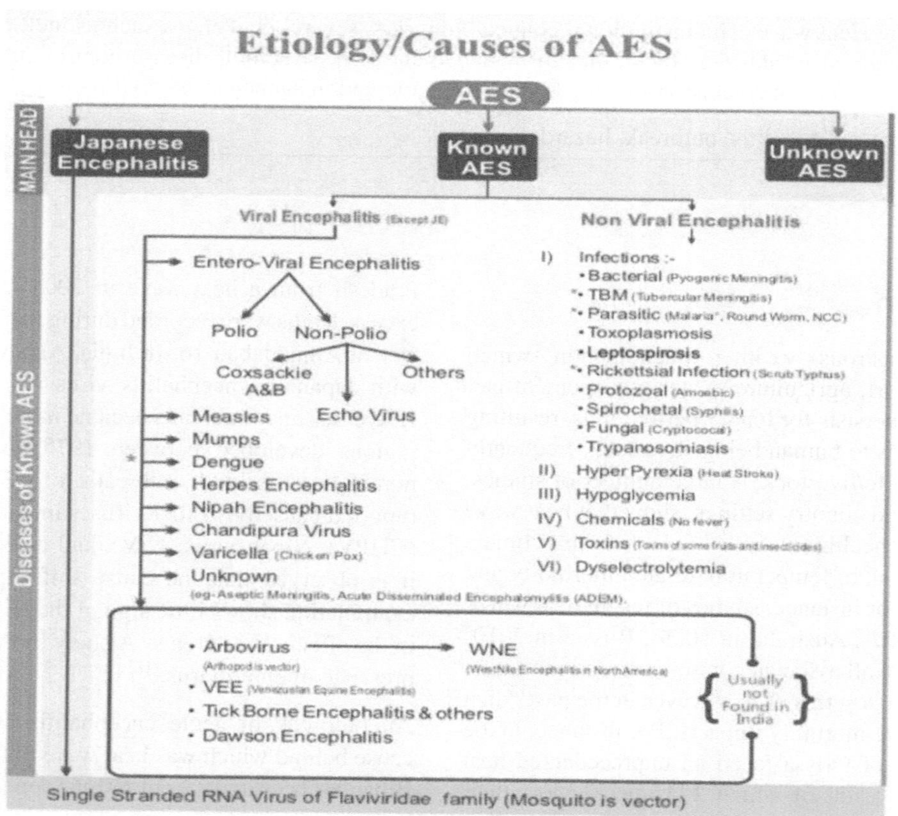

132 children died at the Sri Krishna Medical College and Hospital (SKMCH), the largest state-operated hospital in Bihar, while 18 children died at the Kejriwal Matrisadan, a trust-run hospital.[7][2] Most of them were aged between 1 and 10 years.[9][10][11] A total of 440 cases of AES were admitted to hospitals.[7] In 2016, of the total 424 acute encephalitis syndrome cases in the state, 52.8 per cent were unknown AES, 23.6 percent known AES, and 23.6 per cent Japanese encephalitis. The next year, in 2017, 50 per cent of the cases were classified as unknown AES, 32 per cent as Japanese encephalitis, and the rest known acute encephalitis syndrome cases. The etiology of 2017 cases showed 49 per cent were due to herpes encephalitis (encephalitis due to herpes simplex virus), 23 per cent due to hypoglycemia (a condition caused by a very low level of blood sugar), 13 per cent by dyselectrolytemia (acute muscle weakness), 10 per cent pyogenic meningitis (bacterial meningitis affecting brain or spinal cord), and others. Thus, there are hoards of viruses and conditions that can cause encephalitis.

A high-level team of specialists constituted by the Centre to review the surge in acute encephalitis syndrome (AES) cases in Bihar this year believes the heat wave was a factor behind more than 400 deaths, making it the first major AES outbreak in the state since 2014. Dr Arun Singh, who led the central team, says the affected districts had temperatures of over 40 degrees for several days at a stretch. "We know that the children played outdoors in high heat... The combination of extreme heat, exhaustion and micro-nutritional deficiency can be lethal." Muzaffarpur, Vaishali, East Champaran, Sheohar and Sitamarhi were the worst-hit districts this time, with Shri Krishna Medical College and Hospital (SKMCH), Muzaffarpur, alone recording more than 60 per cent of the casualties. Table 56.1 and 56.2 present the picture of no of days of heat wave and no of death in corresponding years[5][7].

Table 56.1 Number of heat wave (HW) over Bihar region

Year	Patna	Gaya	Bhagalpur	Purnia	Muzaffarpur
2012	17	15	13	3	16
2013	6	6	2	5	13
2014	22	11	10	1	17
2015	8	9	6	4	9
2016	10	9	9	3	8
2017	10	8	7	4	8
2018	11	10	9	7	12
2019	15	17	10	8	23

Table 56.2 Number of Death (HW) over Bihar region

Year	2012	2013	2014	2015	2016	2017	2018	2019
Number of Death	90	143	355	11	4	20	15	440

3. Methodology

In the present study both primary and secondary data sources has been explored. The process starts with review of available literature on impact of heat wave on outbreak of encephalitis and vulnerable people. A number of secondary sources have been used including research studies on cold wave as well as data from the website of govt of Bihar, Budget documents, speeches of the Finance Minister, block and district administration and other published sources. Data from Census and State election commission and other stakeholders has been also been used. Information collected through interviews and group discussions has been cross checked with other reliable sources. Local dialects have been used by the Field Investigators during the interview and FGD.

The study focus at qualitative & quantitative research methods, which aim at comparing and interpreting findings, rather than relying on direct, potentially conflictive questioning. Since impact of heat wave on outbreak of encephalitis and vulnerable people of Bihar is the focus of the present study, quantitative as well as qualitative techniques, focus group discussions, structured as well as unstructured interviewing and observations were used for collecting the data. Focus group discussions were conducted with members of NGO representatives, Government functionaries and village level leaders. The methodology also includes semi-structured in-depth interviews with different section of society.

The meteorological data used in the study is taken from National Data Centre (NDC) Pune and Meteorological Centre, Patna. The following criteria as per IMD guidelines for heat wave have been used.

The spells of abnormally hot weather occur every year during March to July in different parts of India. For declaration of heat wave condition the maximum temperature of a station must reach at least 40° C for plains and 30° C for Hilly regions. As per India Meteorological Department convention heat wave/severe heat wave is classified as follows:

(i) when the normal temperature of a station is <40 °C :
Heat wave: Departure from normal is +5° C to 6° C
Severe heat wave: Departure from normal is +7° C or more

(ii) when the normal temperature of a station is ≥ 40° C :
Heat wave: Departure from normal is +4° C to 5° C
Severe heat wave: Departure from normal is +6° C or more

The vulnerable population is defined in terms of age group such as children between1-1 5 years of age [17]. The three subsets of under-five children, such as neonates <1 months, infant <1 year, <5 years and <15 years, of which neonates is the most vulnerable [18].

The contribution of cold wave to the deaths of the vulnerable groups, and the mortality rate has been analysed by using multiple regression analysis. The multiple linear regression analysis has been used to develop a model for predicting mortality from the climatic variables at different time lags. This relationship is given by the equation [23]:

$$Y = \beta_o + \beta_1 X_1 + \beta_2 X_2 \ldots \beta_k X_k + \varepsilon$$

Where β_k is coefficients, X_i is the predictor, Y is mortality (predicted), β_o is a constant and ε is random error.

The fieldwork covered the 5 districts. Multi caste Panchayats were studied in each area. The perception about the death due to heat wave and status of vulnerable population groups were carried out by informal talking among 200 respondents selected randomly from each district: 40 each from different walks of life such as the local communities, government personnel, public health personnel, female community health volunteers and school teachers.

4. Findings

More than 200 children in Bihar's Muzaffarpur district have died of acute encephalitis syndrome (AES), with the state's medical authorities initially blaming the deaths on the heat wave, hypoglycemia (sudden drop in blood sugar levels) and lack of awareness. Now, belatedly, they have acknowledged the two most critical causes for the deaths — malnutrition and the inadequacy of primary health centre (PHCs).

The AES cases in Muzaffarpur, Bihar and adjoining litchi producing districts have been observed every year in summers(mostly from April to June), particularly in children who are undernourished with a history of visiting litchi orchards

What makes regions so vulnerable to AES

- It has been found that regions with high heat, humidity, unhygienic condition, malnutrition contributes to rise in AES
- AES has struck Muzaffarpur with regularity in Summers since 1995
- Incidence is higher in litchi field around which malnourished child live
- Malnutrition is high in Bihar,and malnourished children are prone to infection
- National family Health survey data show that 48% children less than 5 years were stunted-the highest in India.

Cause of Outbreak of Encephalitis

Expert says that the disease had spread because of excessive heat and humidity. Bihar is facing an intense heat wave. Again Litchi is being most commonly blamed for the Acute Encephalitis Syndrome (AES) outbreak in Bihar. However, experts are of the view that fruit is only a triggering factor for malnourished children as the toxin methylenecyclopropyl glycine (MCPG) can lead to hypoglycaemia (fall in sugar levels) in undernourished children that in turn results into unconsciousness, coma and even death.

- **Malnutrition cause hypoglycaemia:** Malnourished children have depleted glycogen store in the liver. So if there is no glycogen reserve, the glycogen breaks into glucose and when the shortage further increases, even fats start burning and this process produces by products like ketones and amino acids which are neurotoxic and are responsible for inflammation of the brain tissue.
- **Heat waves:** when temperature and humidity were recorded over consecutive days at more 38 degrees Celsius and 50% respectively, the epidemic had been at its worse. In other parts of northern India there is fluctuation in temperature and humidity, and the nights are cooler, but in Muzaffarpur humidity increases at night making it worse than the day, especially for children. Affected districts had temperatures of over 40 degrees for several days at a stretch.
- The main **causative agents** of acute encephalitis are the **viruses** like herpes viruses, enter viruses, West Nile, Japanese encephalitis, Eastern equine viruses, tick-borne viruses etc.
- Encephalitis is also caused due to **bacteria, fungi or parasites, chemicals, toxins and non-infectious agents.**

Mortality prediction

The prediction of number of deaths of people due to heat wave has been carried out by using multiple linear regressions, based on the data of three variables such as minimum temperature, maximum temperature and rainfall from 2012 to 2019. A significant regression equation was found at [$F(3, 38) = 4.258$, $p < 0.05$)], with an R2 of 0.252. Number of deaths is calculated at 32 people with a increase of 1°C of temperature and 0.56 deaths with each millimetre decrease in rainfall.

Lapses

The death of children in Muzaffarpur and other parts of Bihar due to AES, a preventable disease, shows the inaction of the state machinery, which has failed to take appropriate steps to prevent the seasonal outbreak of disease. Following lapses were significant:

- **Primary health centres**

  The first point of healthcare for most AES patients is **ill-equipped** to deal with the disease. Most of them do not have **virology lab** or adequate number of **paediatric beds** and **glucometers** to monitor blood sugar levels.

- **Standard operating procedure**

  Standard operating procedure to treat AES, have clearly not been implemented in Bihar, which mandates grassroots health workers, including auxiliary nurse-midwives, accredited social health activists, and anganwadi workers, to do daily **household-level surveys** to check children for JE/AES symptoms during the outbreak season from June to September.

- **Lack of preparedness**

  Lack of preparedness of the state government in terms **special nutrition programme** for AES-prone areas (as undernourished children are more prone to the disease).

- **Inadequate Machinery**

  Unavailability of **doctors, beds, intensive care units, medical professionals** to deal with an epidemic level situation which has resulted in more than 200 reported death of children in the state of Bihar particularly in Muzaffarpur and its adjoining areas.

Way Forward

- **Surveillance and Monitoring**

  Surveillance and Monitoring of AES vulnerable region (for identifying AES prone cases) is necessary for early **diagnosis and treatment** of disease. AES can be contained if the child is administered dextrose within 4 hours of the onset of symptoms.**10% Dextrose infusion** within the golden hours can speed up the pace of recovery.

- **Awareness campaigns and symptom management**

  Awareness campaigns and symptom management as pre-emptive measures must be put in place for dealing with the AES cases. State Government should learn from the ***dastak*** initiative (for treating Japanese Encephalitis) taken by Uttar Pradesh, that has brought together health, rural development and primary education officials to promote clean drinking water, sanitation, vaccination and early referral to hospitals for treatment through symptom management.

- **Infrastructural and Institutional set up**

  Primary Health Care centres must be equipped with all necessary equipment and amenities that are crucial for diagnosing and treating AES patients.

- **Medical Research and Studies:**

  Medical Research and Studies conducted in AES prone regions that are crucial for establishing link between causes and symptoms of the disease must be taken into account for formulating **treatment strategy.**

- **Special Nutrition programmes**

  Special Nutrition programmes and **mid-day meals** must be implemented in the AES prone regions as it has been identified that toxins in litchi is just a triggering factor in the already malnutritioned children.

REFERENCES

1. Anderson GB, Dominici F, Wang Y, McCormack MC, Bell ML, Peng RD. Heat-related Emergency hospitalizations for respiratory diseases in the medicare population American Journal of Respiratory and Critical Care Medicine1.8 72(01103):;1098-1103
2. "India's heatwave turns deadly". Al Jazeera. 18 June 2019. Retrieved 18 June 2019.
3. "Nitish Kumar Visits Encephalitis Ground Zero, "Go Back" Cry Protesters". NDTV.com. Archived from the original on 18 June 2019. Retrieved 18 June 2019.
4. "India encephalitis death toll rises to 103". BBC. 18 June 2019. Archived from the original on 17 June 2019. Retrieved 18 June 2019.
5. "Explained: What causes AES? What makes Bihar so vulnerable?". The Indian Express. Indian Express Group. 18 June 2019. OCLC 70274541. Archived from the original on 18 June 2019. Retrieved 18 June 2019.
6. Slater J (17 June 2019). "Could the humble litchi fruit be behind a mysterious sickness that has killed nearly 100 children in India?". The National Post. ISSN 1486-8008. Retrieved 17 June 2019.
7. Dutta, Prabhash K. (17 June 2019). "Bihar encephalitis deaths: What Nitish Kumar may learn from Yogi Adityanath govt". India Today. Aroon Purie. ISSN 0254-8399. Archived from the original on 17 June 2019. Retrieved 18 June 2019.
8. "Heat, lack of nutrition, awareness add to AES, Bihar kids toll over 100". The Indian Express. Indian Express Group. 18 June 2019. OCLC 70274541. Archivedfrom the original on 18 June 2019. Retrieved 18 June 2019.
9. "India - Acute Encephalitis Syndrome (AES) in Bihar (DG ECHO, media) (ECHO Daily Flash of 17 June 2019) - India". ReliefWeb. Archived from the original on 18 June 2019. Retrieved 18 June 2019.
10. "Encephalitis toll reaches 84 in Bihar's Muzaffarpur". India Today. Aroon Purie. 16 June 2019. ISSN 0254-8399. OCLC 321350054. Archived from the original on 17 June 2019. Retrieved 17 June 2019.
11. "Encephalitis: Death toll reaches 100 in Bihar". India Today. Aroon Purie. 17 June 2019. ISSN 0254-8399. OCLC 321350054. Archived from the original on 17 June 2019. Retrieved 17 June 2019.
12. "100 Children Die Of Encephalitis In 16 Days In Bihar's Muzaffarpur". NDTV.com. Archived from the original on 17 June 2019. Retrieved 17 June 2019.
13. Joshi R, Kalantri SP, Reingold A, Colford JM (July–August 2012). "Changing landscape of acute encephalitis syndrome in India: a systematic review". The National Medical Journal of India. 25 (4): 212–20. PMID 23278779.

14. "Encephalitis in India kills more than 90 children". BBC. 17 June 2019. Archivedfrom the original on 17 June 2019. Retrieved 17 June 2019.
15. Shrivastava A, Kumar A, Thomas JD, Laserson KF, Bhushan G, Carter MD, et al. (April 2017). "Association of acute toxic encephalopathy with litchi consumption in an outbreak in Muzaffarpur, India, 2014: a case-control study". The Lancet. Global Health. 5 (4): e458–e466. doi:10.1016/S2214-109X(17)30035-9. PMID 28153514.
16. "India encephalitis death toll rises to 103". 18 June 2019. Archived from the original on 17 June 2019. Retrieved 18 June 2019.
17. Chaudhuri A, Kennedy PG (October 2002). "Diagnosis and treatment of viral encephalitis". Postgraduate Medical Journal. 78 (924): 575–83. doi:10.1136/pmj.78.924.575. PMC 1742520. PMID 12415078. Archived from the original on 18 June 2019. Retrieved 18 June 2019.
18. "Acute encephalitis syndrome (AES) explained: Definition, cause, and its contrary theories". The Indian Express. 17 June 2019. Archived from the original on 17 June 2019. Retrieved 18 June 2019.
19. "Litchi Toxins, Malnutrition or Heat Wave? Doctors Explain What's Causing Deadly AES Epidemic in Bihar". News18. Archived from the original on 18 June 2019. Retrieved 18 June 2019.
20. "Is Litchi the Real Culprit Behind 1995's Mysterious Muzaffarpur Child Deaths?". The Wire. Archived from the original on 18 June 2019. Retrieved 18 June 2019.
21. "Encephalitis kills more than 100 children in India's Bihar state". www.aljazeera.com. Archived from the original on 18 June 2019. Retrieved 18 June 2019.
22. Ramashankar, Mr. (17 June 2019). "Encephalitis: 20 more kids die in 24 hours, Bihar encephalitis toll hits 93". The Times of India. Archived from the original on 18 June 2019. Retrieved 18 June 2019.
23. "Muzaffarpur Deaths: CM Calls for Hospital Expansion, Toll at 109". The Quint. 18 June 2019. Archived from the original on 18 June 2019. Retrieved 18 June 2019.
24. "Bihar encephalitis deaths: Harsh Vardhan constitutes high-level multi-disciplinary team to monitor patients, find cause of viral disease". Firstpost. Archived from the original on 18 June 2019. Retrieved 18 June 2019.
25. "Encephalitis deaths: SC directs Centre, Bihar govt to file response within 7 days".
26. Pachauri RK, Reisinger A. IPCC Fourth Assessment Report. Geneva: IPCC; 2007

For Product Safety Concerns and Information please contact our
EU representative GPSR@taylorandfrancis.com Taylor & Francis
Verlag GmbH, Kaufingerstraße 24, 80331 München, Germany